Some Important Organic Functional Groups

	Functional Group*	Example	IUPAC Name
Acid anhydride	$-C(=O)-O-C(=O)-$	$CH_3C(=O)OC(=O)CH_3$	Ethanoic anhydride (Acetic anhydride)
Acid chloride	$-C(=O)-Cl$	$CH_3C(=O)Cl$	Ethanoyl chloride (Acetyl chloride)
Alcohol	$-OH$	CH_3CH_2OH	Ethanol (Ethyl alcohol)
Aldehyde	$-C(=O)-H$	$CH_3C(=O)H$	Ethanal (Acetaldehyde)
Alkane	------	CH_3CH_3	Ethane
Alkene	$>C=C<$	$CH_2=CH_2$	Ethene (Ethylene)
Alkyne	$-C\equiv C-$	$HC\equiv CH$	Ethyne (Acetylene)
Amide	$-C(=O)-N<$	$CH_3C(=O)NH_2$	Ethanamide (Acetamide)
Amine, primary	$-NH_2$	$CH_3CH_2NH_2$	Ethylamine
Amine, secondary	$-NH-$	$(CH_3CH_2)_2NH$	Diethylamine
Amine, tertiary	$-N<$	$(CH_3CH_2)_3N$	Triethylamine

	Functional Group*	Example	IUPAC Name
Arene			Benzene
Carboxylic acid	$-C(=O)-O-H$	$CH_3C(=O)OH$	Ethanoic acid (Acetic acid)
Disulfide	$-S-S-$	CH_3SSCH_3	Dimethyl disulfide
Ester	$-C(=O)-O-C-$	$CH_3C(=O)OCH_3$	Methyl ethanoate (Methyl acetate)
Haloalkane	$-X$ X = F, Cl, Br, I	CH_3CH_2Cl	Chloroethane (Ethyl chloride)
Ketone	$-C(=O)-$	$CH_3C(=O)CH_3$	Propanone (Acetone)
Phenol	$-OH$	$-OH$	Phenol
Sulfide	$-S-$	CH_3SCH_3	Dimethyl sulfide
Thiol	$-S-H$	CH_3CH_2SH	Ethanethiol (Ethyl mercaptan)

* Where bonds to an atom are not specified, the atom is assumed to be bonded to one or more carbon or hydrogen atoms in the rest of the molecule.

What Reviewers Are Saying About *Introduction to Organic Chemistry*, Second Edition

"I consider the whole textbook excellent. We have been using the previous edition for the past three years. I have personally selected it for our organic course and the academic response of my students has been very convincing. The grades went up and the number of complaints about 'abstract' organic chemistry is nil. Thus, I am looking forward to introducing the second edition of the book to my students."

Miroslav Krumpolc,
University of Illinois at Chicago

"Chapter 12, on carboxylic acids, looks good. There is a good balance between each section without too much detail. The use of tables and graphics for each topic, nomenclature, physical properties, and acid strength make this chapter very readable and easy to understand at every level. The Chemical Connections, summary, and key reactions make all this information very accessible. Arrows showing mechanisms are very helpful. Also, the questions and problems fit in well with the material in the chapter."

Edward J. Parish,
Auburn University

"The style is somewhat informal, and both the students and I enjoyed this; it was not at all dry (especially chapters relating to biochemistry and materials chemistry)."

Michael J. Natan,
Pennsylvania State University

"In-chapter and end-of-chapter problems are excellent and they cover the topics evenly. I like problems through which students build upon what they read in the text (especially those involving real-world situations) and this book has lots of them. The Chemical Connections boxes are exceptionally informative and useful, as are the abundant photographs and other graphics."

David R. Benson,
University of Kansas

"I've found this book to be popular with the students, pedagogically sound for the level at which the course is handled, and a useful reference for those students who must complete the regular biochemistry course as the next step in their science courses. My favorite aspect of the book is its figures and pictures, which I think are most critical in this course. The prose is eloquent and enjoyable, and the frequent connections to society (and the personal interviews) is another very strong point about this book."

Jason R. Stenzel,
Southern Connecticut State University

"I like Brown the way it is. He is one of the few authors who can write a condensed book without having it sound like an outline. The level of the English is excellent. . . . I like the chapters on biological compounds, lipids, etc. Even if I don't cover the whole chapter, I use them for examples. They are also helpful as they relate to our labs."

Georgia Weinstein,
Boston University

"I am pleased with this text and plan to continue using it for the foreseeable future. Perhaps the best feature of the book is that the students find it readable. This fact actually becomes more noticable among the second-semester organic students, some of whom say they want a textbook more like Brown's. The students and I appreciate the color in the illustrations and the photographs. The photographs complement the demonstrations I use in the classroom in using nearly anything in the world as a fair example of organic chemistry."

Robert W. Chesnut,
Eastern Illinois University

"I think this book is fantastic. It clearly explains all essential points of the course using language that my students are able to grasp quite easily. My students commented that the length of chapters (in 1/e) was just right in that they were not too in-depth, yet covered material completely. . . . I especially appreciate the attention toward biological applications of organic chemistry, since many of my students are agriculture, health science, and biology majors."

Curtis J. Czerwinski,
University of Wyoming

"Brown's text seems more readable for students, and unsolicited comments from students on class evaluations have proved this to be true. The use of colorized tables, charts, and graphics really makes the text more enjoyable to read and use. I personally consider the Brown text the best of those I examined."

Robert Badger,
University of Wisconsin, Stevens Point

"The illustrations are very nice and useful in presenting information to the students. I use many transparencies of figures in the book. They are useful to me and I believe my students learn from them."

Marco Lopez,
California State University, Long Beach

Introduction to ORGANIC CHEMISTRY

SECOND EDITION

William H. Brown
Beloit College

SAUNDERS GOLDEN SUNBURST SERIES

Saunders College Publishing
A Division of Harcourt College Publishers

Fort Worth • Philadelphia • San Diego • New York • Orlando • Austin
San Antonio • Toronto • Montreal • London • Sydney • Tokyo

Publisher: John Vondeling
Marketing Strategist: Pauline Mula
Developmental Editor: Sandra Kiselica
Project Editor: Sarah Fitz-Hugh
Production Manager: Charlene Catlett Squibb
Art Director: Caroline McGowan
Illustrations: J/B Woolsey Associates

Front cover credits: © Andrew Henderson
© Frank Orel/Tony Stone Images

Back cover credit: Monarch butterfly, ©Frans Lanting/Minden Pictures

Frontispiece Credit: © Jerry Alexander/Tony Stone Images

Introduction to Organic Chemistry, second edition
ISBN: 0-03-025988-6
Library of Congress Catalog Card Number: 99-62547

Address for domestic orders:
Saunders College Publishing, 6277 Sea Harbor Drive, Orlando, Florida 32887-6777
1-800-782-4479

Address for international orders:
International Customer Service, Harcourt Brace & Company
6277 Sea Harbor Drive, Orlando, Florida 32887-6777
(407) 345-3800
Fax (407) 345-4060
e-mail hbintl@harcourtbrace.com

Address for editorial correspondence:
Saunders College Publishing, Public Ledger Building, Suite 1250, 150 S. Independence Mall West, Philadelphia, PA 19106-3412

Web Site Address
http://www.hbcollege.com

Printed in the United States of America

9012345678 032 10 987654321

To Carolyn,
with whom life is a joy.

About the Author

William H. Brown is Professor of Chemistry at Beloit College, where he has twice been named Teacher of the Year. He is also the author of the college textbook *Organic Chemistry,* 2nd edition, published in 1998. His regular teaching responsibilities include organic chemistry, advanced organic chemistry, and, more recently, special topics in pharmacology and drug synthesis. He received his Ph.D. from Columbia University under the direction of Gilbert Stork and did postdoctoral work at California Institute of Technology and the University of Arizona.

Bill and his wife Carolyn enjoy hiking in the Southwest and the study of petroglyphs and pictographs. Twice he has been the Director of Beloit College's World Outlook Seminar, a program coordinated with the University of Glasgow in Scotland.

Bill Brown in Capitol Reef National Park, Utah *(Carolyn S. Simonton)*

PREFACE

THE AUDIENCE

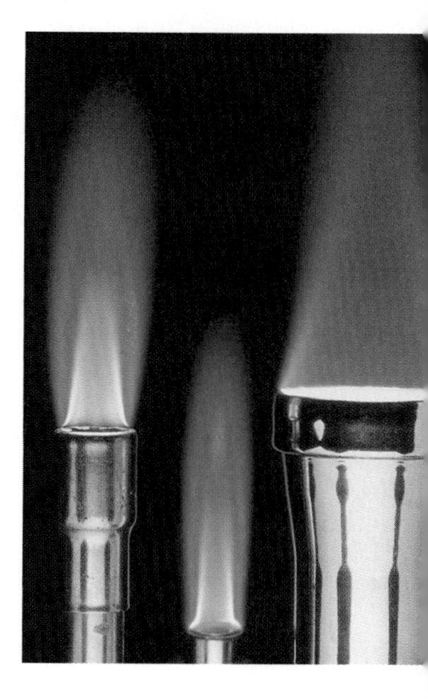

This book provides an introduction to organic chemistry for students who are aiming toward careers in the sciences and who require a grounding in organic chemistry. For this reason, special effort is made throughout to show the interrelation between organic chemistry and other areas, particularly the biological and health sciences. While studying with this book, students will see that organic chemistry is a tool for these many disciplines and that organic compounds, both natural and synthetic, are all around us—in pharmaceuticals, plastics, fibers, agrochemicals, surface coatings, toiletry preparations and cosmetics, food additives, adhesives, and elastomers. Furthermore, students will experience that organic chemistry in a dynamic and ever-expanding area of science waiting openly for those who are prepared, both by training and inquisitive nature, to ask questions and to explore.

ORGANIZATION: AN OVERVIEW

Chapters 1–14 lay a foundation for studying organic chemistry by first reviewing the fundamentals of covalent bonding, the shapes of molecules, and of acid/base chemistry. The structures and typical reactions of the important classes of organic compounds are then discussed; alkanes, alkenes, alcohols, benzene and its derivatives, amines, aldehydes, ketones, and finally carboxylic acids and their derivatives. Chapter 15 provides a brief introduction to organic polymer chemistry.

Chapters 16–20 present an introduction to the organic chemistry of carbohydrates, lipids, amino acids and proteins, and nucleic acids. Chapter 20, The Organic Chemistry of Metabolism, demonstrates how the chemistry developed to this point can be applied to an understanding of two major metabolic pathways—β-oxidation of fatty acids and glycolysis.

Chapters 21–22 introduce ^{1}H-NMR, ^{13}C-NMR, and IR spectroscopy. Discussions of spectroscopy require no more background than what students receive in general chemistry. These chapters are free-standing, and can be taken up in any order appropriate to a particular course. In the end-of-chapter problems of each functional group chapter, references are given to the appropriate end-of-chapter problems in the spectroscopy chapters.

Chapter-by-Chapter

Chapter 1 begins with a review of the electronic structure of atoms and molecules, and the use of the VSEPR model to predict shapes of molecules and polyatomic

ions. The theory of resonance is introduced midway through Chapter 1 and, with it, the use of curved arrows and electron pushing. A knowledge of resonance theory combined with a facility for pushing electrons gives students two powerful tools for writing reaction mechanisms and understanding chemical reactivity. The discussion of resonance is followed by an introduction to the valence bond description of covalent bonding. Chapter 1 concludes with an introduction to the hydroxyl, carbonyl, carboxyl, and amino groups, the functional groups encountered most frequently in Chapters 1–14.

Chapter 2 contains a general introduction to acid-base chemistry and concentrates on two major themes. The first is the qualitative determination of the position of equilibrium in acid-base reactions. The second theme is the relationship between structure and acidity.

Chapter 3 opens with a description of the structure, nomenclature, and conformations of alkanes and cycloalkanes. The IUPAC system is introduced in Section 3.3A through the naming of alkanes, and, in Section 3.5, the IUPAC system is presented as a general system of nomenclature. Beginning here and continuing throughout the text, a clear distinction is made between IUPAC and common names. Where names are introduced, the IUPAC name is given and the common name or names, where appropriate, follow in parentheses. The concept of stereoisomerism is introduced in this chapter with a discussion of cis-trans isomerism in cycloalkanes.

Chapter 4 introduces the concepts of chirality, enantiomerism, and diastereomerism. This material, coupled with the liberal use of molecular models, both in the text and on the accompanying interactive CD-ROM, encourages students to think of organic molecules as three-dimensional objects and to treat them as such in order to gain a deeper understanding of the chemistry of organic and biochemical reactions.

Chapter 5 presents the structure, nomenclature, and physical properties of alkenes and alkynes.

Chapter 6 opens with an introduction to chemical energetics and the concept of a reaction mechanism. The focus of this chapter is then on the reactions of alkenes, which are organized in the order: electrophilic additions, oxidation, and reduction. The twin concepts of regioselectivity and stereoselectivity are introduced in the context of electrophilic additions to alkenes.

Chapter 7 introduces alkyl halides and uses them as a vehicle for the discussion of nucleophilic substitution and β-elimination. The concepts of one-step and two-step nucleophilic substitutions along with S_N1 and S_N2 terminology are introduced first, and then the concepts of E1 and E2 reactions of alkyl halides.

Chapter 8 concentrates the structure and characteristic reactions of alcohols, including their conversion to alkyl halides, acid-catalyzed dehydration, and the oxidation of primary and secondary alcohols. There then follows a brief introduction to the structure, preparation, and acid-catalyzed ring opening of epoxides. This chapter concludes with a discussion of the acidity of thiols and their oxidation to disulfides.

Chapter 9 opens with the structure and nomenclature of aromatic compounds and several important heterocyclic aromatic compounds. Students are introduced to the unique properties of aromatic rings through the acid-base properties of phenols. The mechanism for electrophilic aromatic substitution, including the theory of directing effects, is then presented in detail.

Chapter 10 presents the structure and nomenclature of amines and concentrates on their most important chemical property, namely their basicity.

Chapters 11–14 develop the chemistry of carbonyl-containing compounds. First is the chemistry of aldehydes and ketones in Chapter 11, followed by carboxylic acids in Chapter 12 and their functional derivatives in Chapter 13. A major theme of these chapters is the addition of nucleophiles to the carbonyl group to form tetrahedral carbonyl addition products. Chapter 14 introduces the concept of an enolate anion and its involvement in aldol, Claisen, and Dieckmann reactions to form new carbon-carbon bonds.

Chapter 15, new to this edition, is a systematic introduction to organic polymer chemistry. Given the importance of organic polymers in the world around us, this material has been expanded in this edition.

Chapters 16–20 present the chemistry of carbohydrates in Chapter 16, lipids in Chapter 17, amino acids and proteins in Chapter 18, and nucleic acids in Chapter 19. Chapter 20 presents a discussion of two key metabolic pathways, namely β-oxidation of fatty acids and glycolysis. The purpose of this material is to show that the reactions of these pathways are the biochemical equivalents of organic functional group reactions already studied in previous chapters.

Chapters 21 and 22 present the fundamentals of the structure elucidation tools: ^{1}H-NMR and ^{13}C-NMR spectroscopy and then IR spectroscopy.

SPECIAL FEATURES

Full-Color Art Program

One of the most distinctive features of this text is its visual impact. The text's extensive full-color art program includes over 200 pieces of art by professional artists John and Bette Woolsey.

Photo Art

Photos, conceived and developed for this text, show organic chemistry as it occurs in the laboratory and in everyday life, and depict the natural sources of many organic compounds.

Molecular Models

Developed by William Brown is a set of more than 275 molecular models for incorporation into this text. Their purpose is to assist students to visualize organic molecules as three-dimensional objects. All models have been prepared using CambridgeSoft Corporation ChemDraw and Chem3D software, and are energy-minimized to 0.01RMS.

Interactive Organic Chemistry CD-ROM

Packaged with the text is a CD-ROM prepared by William Brown in conjunction with CambridgeSoft Corporation and containing over 300 molecular models rendered in Chem3D. With the plug-in supplied on the CD, students can rotate each

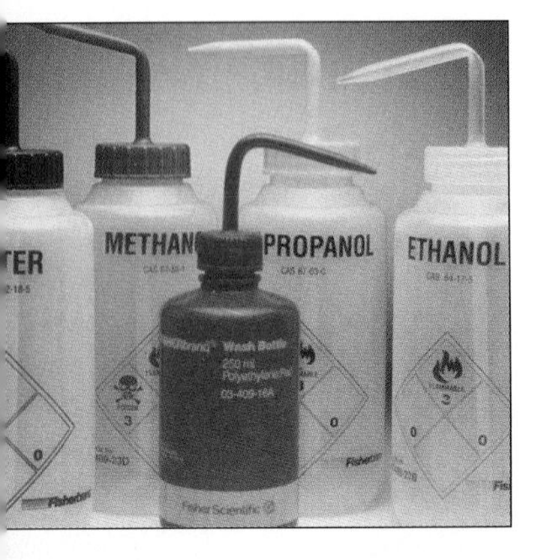

model, change model type from cylindrical bond to space-filling etc., measure bond angles, bond lengths, and the distance between nonbonded atoms. This icon [CD icon] signals students to go to the CD to view models seen in text.

Chemical Connections Boxes

Each of the 36 boxes illustrates an application of organic chemistry to an everyday setting. Among the Chemical Connections are Chiral Drugs in Chapter 4, Morphine as a Clue to Drug Design and Discovery in Chapter 10, From Willow Bark to Aspirin and Beyond in Chapter 12, and Recycling Plastics in Chapter 15.

Bioorganic Chemistry

Bioorganic chemistry is emphasized throughout the text, in the Chemical Connections Boxes, and in problems. For 116 compounds, there are references to *The Merck Index* (Susan Budavari, Editor, 12th Edition, Merck Research Laboratories, 1996), where students can read more about these compounds.

In-Chapter Examples

There is an abundance of in-chapter examples, each with a detailed solution. Following each in-chapter example is a comparable practice problem designed to give students the opportunity to solve a related problem.

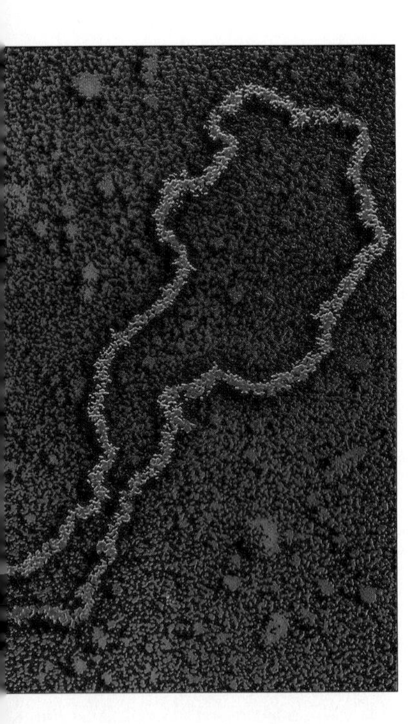

End-of-Chapter Problems

There are plentiful end-of-chapter problems. All problems are categorized by topic. A tetrahedral icon indicates an applied problem, and a number set in blue indicates a more challenging problem.

End-of-Chapter Summaries and Summaries of Key Reactions

End-of-chapter summaries highlight all important new terms and concepts found in the chapter. In addition each new reaction is annotated and keyed to the section where it is discussed.

Color

Color is used to highlight parts of molecules and to follow the course of reactions. The graphic here shows some of the colors used consistently in the artwork in this book.

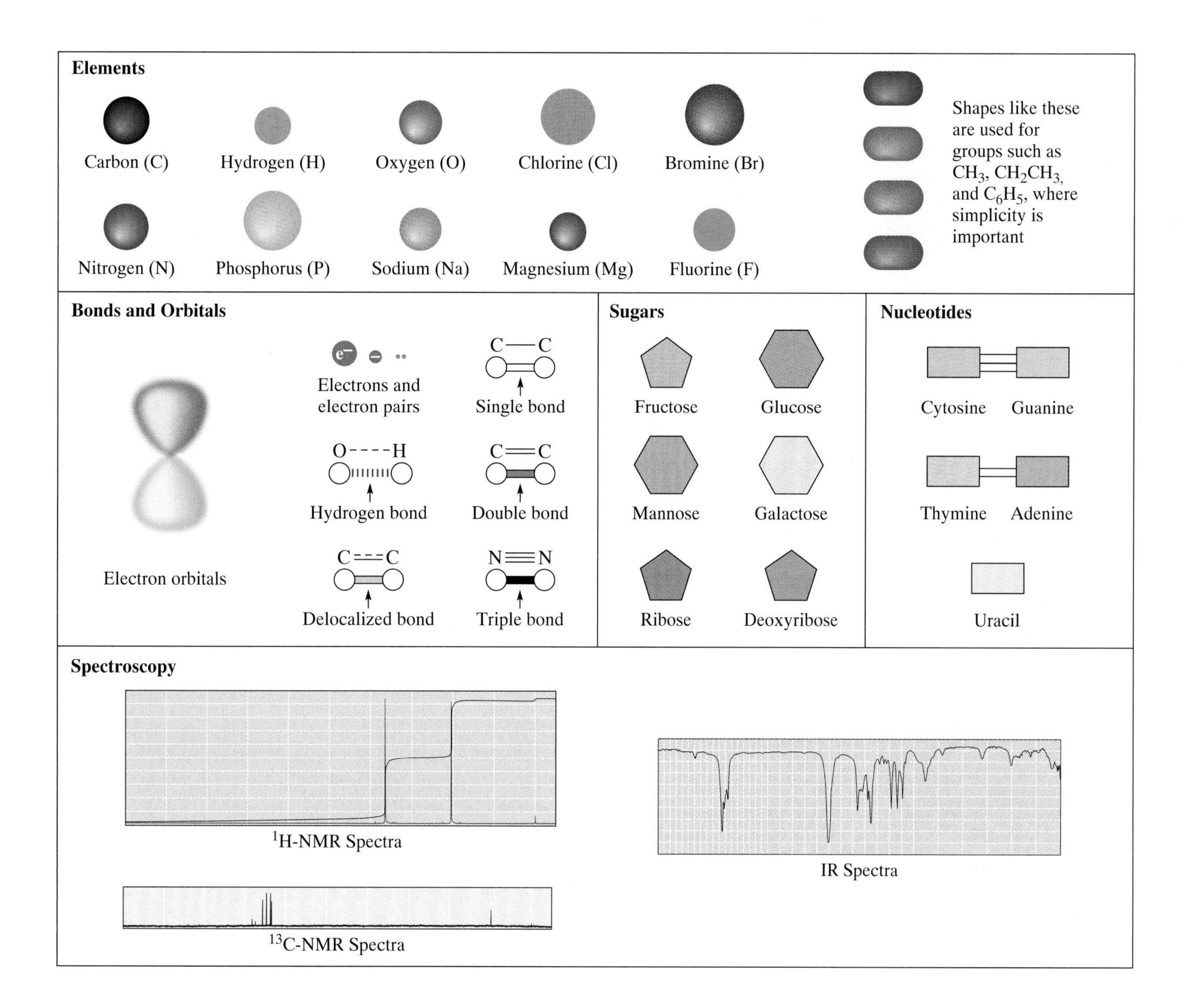

Interviews

Four interviews with prominent scientists describe how these people became interested in chemistry as a college major, then as an educator and/or research professional. Their enthusiasm for their work is evident, and they invite students to pursue similar interests in the sciences.

Margin Definitions and Glossary of Key Terms

Each new term has a free-standing definition placed in the margin adjacent to where the term is first introduced. All margin definitions are then collected in a

Glossary, with a notation added to show the section of the text in which the term is first introduced.

New to This Edition

- Extensive use of the evolving software for viewing molecular models in the form of over which almost 275 molecular models created in Chem3D especially for this text. The inclusion of these models is to further convince students that organic molecules are three-dimensional objects, and their interaction is in large part governed by their shape and their polarities.
- The Interactive Organic Chemistry CD-ROM packaged on the inside cover of the text contains over 300 models rendered in Chem3D.
- Acids and Bases has been moved forward to Chapter 2, immediately following the review of covalent bonding and shapes of molecules in Chapter 1.
- Stereochemistry has been moved forward to Chapter 3 and now is placed immediately before the chemistry of alkenes. With the earlier introduction of stereochemistry, the regioselectivity and stereoselectivity of alkene reactions can now be discussed in greater depth.
- New to this edition is a chapter on organic polymer chemistry. Material on chain-growth polymerization and step-growth polymerization has been collected from the chapters on alkenes and functional derivatives of carboxylic acids and grouped to form this new chapter. In addition, the introduction to organic polymer chemistry has been expanded.
- Mechanisms, key elements in the organization of information in the study of organic chemistry, are now set apart and highlighted by a special design feature. All steps in each mechanism are presented together in its box. A list of the mechanisms presented in the text can be found following the table of contents.

Support Package

- **PowerPoint™ Presentation.** Developed by William Brown is a pre-built set of approximately 700 PowerPoint™ lecture notes corresponding to every chapter in the text. These slides can be used in conjunction with the freely distributable PowerPoint™ Viewer, or edited and customized with the PowerPoint™ application program.
- **Student Study Guide and Problems** by David Benson, University of Kansas, and Brent and Sheila Iverson, University of Texas, contains detailed solutions to all problems.
- **Test Bank** by Jeffrey Elbert of South Dakota State University contains 25 multiple-choice questions per chapter for instructors to use as tests, quizzes, or homework assignments. Available also in computerized form for IBM-compatible and Macintosh computers.
- **Overhead Transparency Acetates.** A selection of 125 full-color figures from the text.

- **Instructor's Resource CD-ROM** contains all the images from this text as well as hundreds from other Saunders College Publishing chemistry texts.
- **CSC ChemOffice Limited, version 4.5.** Includes ChemDraw, Chem3D, and ChemFinder and is available at a very reasonable price from the publisher.
- **Pushing Electrons: A Guide for Students of Organic Chemistry,** third edition, by Daniel P. Weeks, Northwestern University. A paperback workbook designed to help students learn techniques of electron pushing. Its programmed approach emphasizes repetition and active participation.

Saunders College Publishing may provide complementary instructional aids and supplements or supplement packages to those adopters qualified under our adoption policy. Please contact your sales representative for more information. If as an adopter or potential user you receive supplements you do not need, please return them to your sales representative or send them to

Attn: Returns Department
Troy Warehouse
465 South Lincoln Drive
Troy, MO 63379

ACKNOWLEDGMENTS

Although one or a few persons are listed as "authors" of any textbook, the book is in fact the product of collaboration of many individuals, some obvious, others not so obvious. It is with gratitude that I acknowledge the contributions of the many. It is only fitting to begin with John Vondeling, Vice President and Publisher of Saunders College Publishing. John's keen sense of the marketplace has been primal in bringing this edition from rough manuscript to bound book form and to the development of the supplemental materials presented with this edition.

Sandi Kiselica has been a rock of support as Developmental Editor. I so appreciate her ability to set challenging but manageable schedules and her constant encouragement as I worked to meet those deadlines. She has also been an invaluable resource person with whom I could discuss everything from pedagogy to details of the art work.

Sarah Fitz-Hugh as Project Editor has been masterful in coordinating the transformation from manuscript to galleys and pages, along with incorporating the many components of the extensive art program. I have come to know Sarah well through this project and have great respect for her professionalism and judgment.

I also acknowledge the others at Saunders who contributed freely of their expertise to this project, in particular Caroline McGowan, Art Director; Pauline Mula, Marketing Strategist; and Charlene Squibb, Senior Production Manager.

Finally, I owe particular thanks to Professor Dana Chatellier, University of Delaware, for his involvement in this project, first as a reviewer of the entire manuscript, and then as an exceedingly careful and thorough proofer of both galleys and pages. Dana has provided a keen eye toward making sure that the book is as error-free as possible.

List of Reviewers

The following reviewers provided valuable critiques of this book in its many stages.

Robert Badger, *University of Wisconsin, Stevens Point*
David R. Benson, *University of Kansas*
Dana Chatellier, *University of Delaware*
Robert Chesnut, *Eastern Illinois University*
Curtis Czerwinski, *University of Wyoming*
Chip Frazier, *Virginia Polytechnic Institute and State University*
Bryant Gilbert, *Portland State University*
William D. Korte, *California State University, Chico*
Miroslav Krumpolc, *University of Illinois, Chicago*
Marco A. Lopez, *California State University, Long Beach*
Rita S. Majerle, *South Dakota State University*
Michael Natan, *Pennsylvania State University*
Edward J. Parish, *Auburn University*
Carmen Simone, *Casper College*
Ronald Starkey, *University of Wisconsin, Green Bay*
Jason R. Stenzel, *Southern Connecticut State University*

William H. Brown
Beloit College, Beloit, WI
July 1999

CONTENTS OVERVIEW

Contents

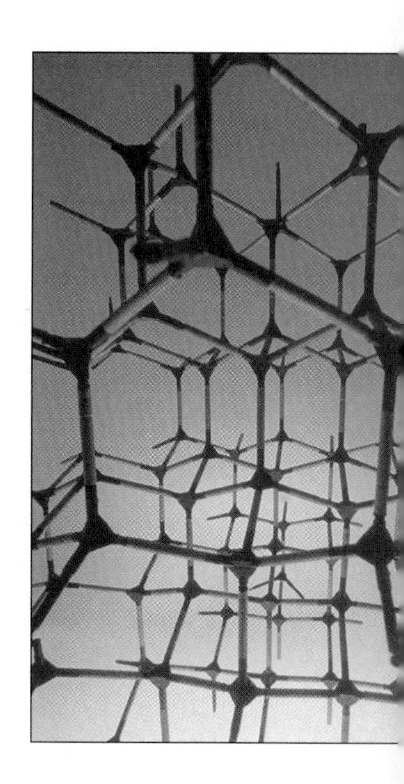

FILA
34

CHEMICAL CONNECTIONS

List of Mechanisms

Chapter 12 Carboxylic Acids

Chapter 13 Functional Derivatives of Carboxylic Acids

Chapter 14 Enolate Anions

Chapter 15 Organic Polymer Chemistry

Chapter 20 The Organic Chemistry of Metabolism

1

Covalent Bonding and Shapes of Molecules

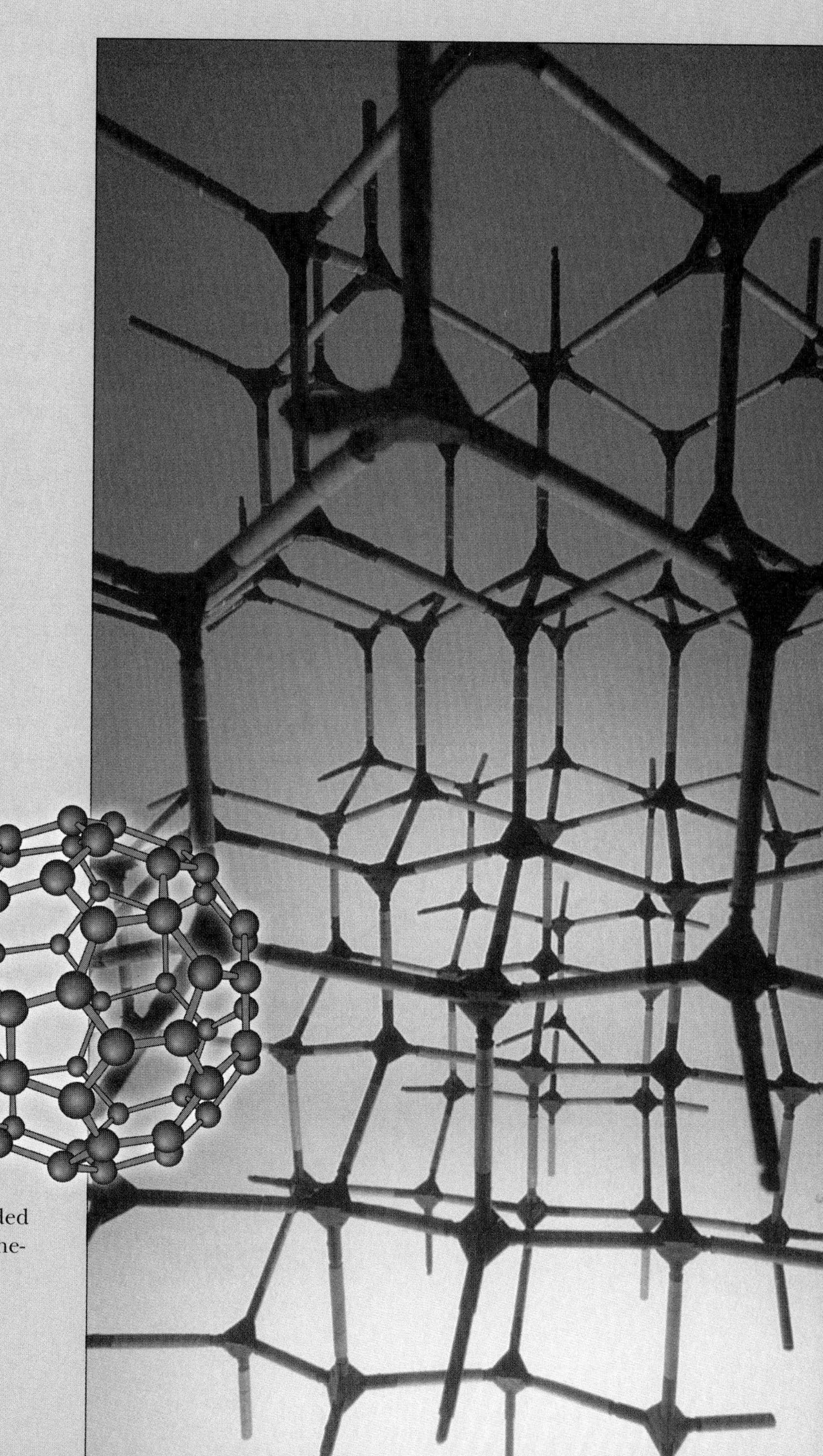

A model of the structure of diamond, one form of pure carbon. Each carbon atom in diamond is bonded to four other carbon atoms at the corners of a tetrahedron. Inset: A model of buckyball, a form of carbon with a molecular formula of C_{60}. *(Charles D. Winters)*

According to the simplest definition, **organic chemistry** is the study of the compounds of carbon. Perhaps the most remarkable feature of organic chemistry is that it is the chemistry of carbon and only a few other elements—chiefly, hydrogen, oxygen, and nitrogen. Chemists have discovered or made well over ten million compounds composed of carbon and these three other elements. Organic compounds are everywhere around us—in our foods, flavors, and fragrances; in our medicines, toiletries, and cosmetics; in our plastics, films, fibers, and resins; in our paints and varnishes; in our glues and adhesives; and, of course, in our bodies and those of all living things. Let us begin our study of organic chemistry with a review of how the elements of C, H, O, and N combine by sharing electron pairs to form molecules.

1.1 Electronic Structure of Atoms

You are already familiar with the fundamentals of the **electronic structure of atoms** from a previous study of chemistry. Briefly, an atom contains a small, dense nucleus made of neutrons and positively charged protons. Most of the mass of an atom is contained in its nucleus. The nucleus is surrounded by a much larger extranuclear space containing negatively charged electrons. The nucleus of an atom has a diameter of 10^{-14} to 10^{-15} meter (m). The extranuclear space where its electrons are found is a much larger area with a diameter of approximately 10^{-10} m (Figure 1.1).

Shell A region of space around a nucleus where electrons are found.

Electrons do not move freely in the space around a nucleus but are confined to regions of space called **principal energy levels** or, more simply, **shells.** Electron shells are identified by the numbers 1, 2, 3, and so on. Each shell can contain up to $2n^2$ **electrons,** where n is the number of the shell. Thus, the first shell can contain 2 electrons, the second 8 electrons, the third 18 electrons, the fourth 32 electrons, and so on (Table 1.1). Electrons in the first shell are nearest to the positively charged nucleus and are held most strongly by it; these electrons are said to be lowest in energy. Electrons in higher numbered shells are farther from the positively charged nucleus and are held less strongly to it; these electrons are said to be higher in energy.

Orbital A region of space where an electron or pair of electrons spends 90–95% of its time.

Shells are divided into subshells designated by the letters *s*, *p*, *d*, and *f* and, within these subshells, electrons are grouped in orbitals (Table 1.2). An **orbital** is a region of space that can hold two electrons. The first shell contains a single orbital called a 1*s* orbital. The second shell contains one *s* orbital and three *p* orbitals; these orbitals are designated 2*s*, $2p_x$, $2p_y$, and $2p_z$. The third shell contains one 3*s* orbital, three 3*p* orbitals, and five 3*d* orbitals. In this course, our focus is on com-

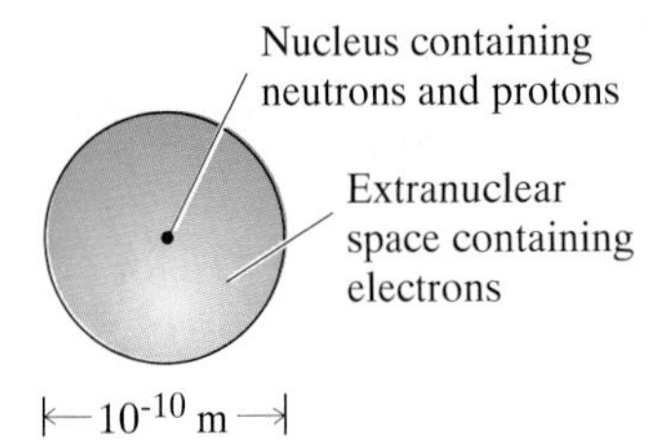

Figure 1.1
A schematic view of an atom. Most of the mass of an atom is concentrated in its small, dense nucleus.

TABLE 1.1 Distribution of Electrons in Shells

Shell	Number of Electrons Shell Can Hold	Relative Energies of Electrons in These Shells
4	32	higher
3	18	↑
2	8	↑
1	2	lower

pounds of carbon with hydrogen, oxygen, and nitrogen, all of which use only electrons in *s* and *p* orbitals for covalent bonding. Therefore, we are concerned only with *s* and *p* orbitals.

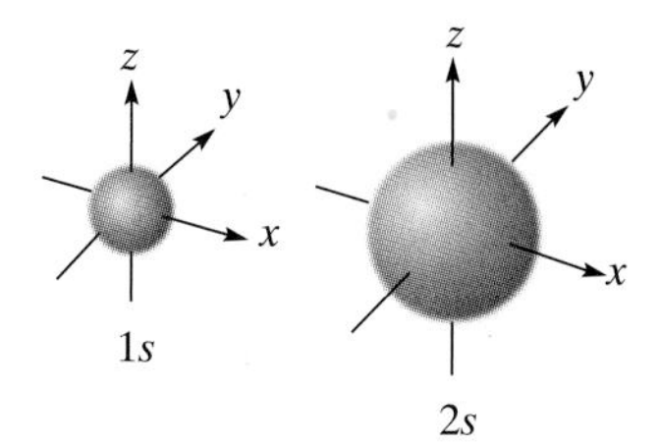

Figure 1.2
Shapes of 1*s* and 2*s* atomic orbitals.

A. Shapes of Atomic Orbitals

One way to visualize the electron density associated with a particular orbital is to draw a boundary surface around the region of space that encompasses some arbitrary percent of the negative charge associated with that orbital. Most commonly, we draw the boundary surface at 95%. When drawn in this manner, all *s* orbitals have the shape of a sphere, with the center of the sphere at the nucleus (Figure 1.2). Of the various *s* orbitals, the sphere representing the 1*s* orbital is the smallest. A 2*s* orbital is a larger sphere, and a 3*s* orbital is an even larger sphere.

Shown in Figure 1.3 are three-dimensional shapes of the three 2*p* orbitals, combined in one diagram to illustrate their relative orientations in space. Each 2*p* orbital consists of two lobes arranged in a straight line with the nucleus in the middle. The three 2*p* orbitals are mutually perpendicular and are designated $2p_x$, $2p_y$, and $2p_z$.

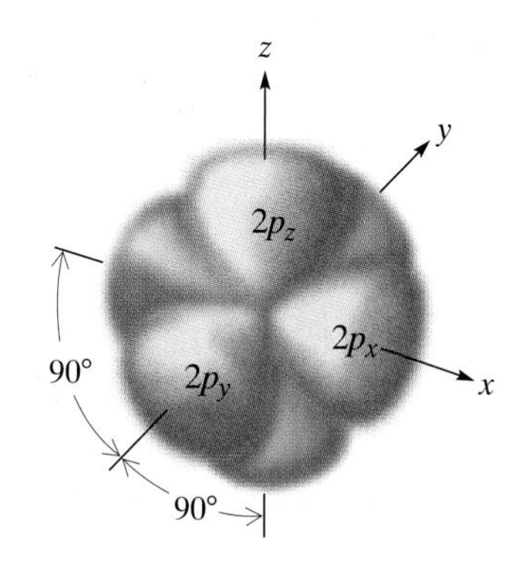

Figure 1.3
Shapes of $2p_x$, $2p_y$, and $2p_z$ atomic orbitals and their orientation in space relative to one another. The three 2*p* orbitals are mutually perpendicular.

B. Electron Configuration of Atoms

The electron configuration of an atom is a description of the orbitals its electrons occupy. Every atom has an infinite number of possible electron configurations. At this stage, we are concerned only with the **ground-state electron configuration**—the electron configuration of lowest energy. Table 1.3 shows ground-state electron

TABLE 1.2 Distribution of Orbitals Within Shells

Shell	Orbitals Contained in That Shell
3	$3s$, $3p_x$, $3p_y$, $3p_z$, plus five $3d$ orbitals
2	$2s$, $2p_x$, $2p_y$, $2p_z$
1	$1s$

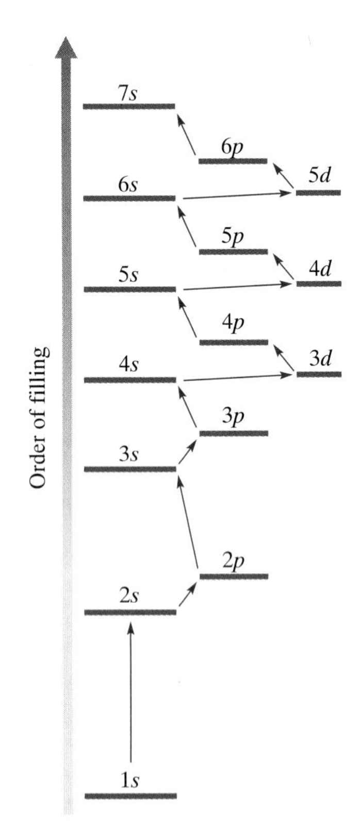

A representation of the order of filling of orbitals for many-electron atoms.

TABLE 1.3 Ground-State Electron Configurations for Elements 1–18

Element	Atomic Number	Orbital								
		$1s$	$2s$	$2p_x$	$2p_y$	$2p_z$	$3s$	$3p_x$	$3p_y$	$3p_z$
H	1	1								
He	2	2								
Li	3	2	1							
Be	4	2	2							
B	5	2	2	1						
C	6	2	2	1	1					
N	7	2	2	1	1	1				
O	8	2	2	2	1	1				
F	9	2	2	2	2	1				
Ne	10	2	2	2	2	2				
Na	11	2	2	2	2	2	1			
Mg	12	2	2	2	2	2	2			
Al	13	2	2	2	2	2	2	1		
Si	14	2	2	2	2	2	2	1	1	
P	15	2	2	2	2	2	2	1	1	1
S	16	2	2	2	2	2	2	2	1	1
Cl	17	2	2	2	2	2	2	2	2	1
Ar	18	2	2	2	2	2	2	2	2	2

Ground-state electron configuration The electron configuration of lowest energy for an atom, molecule, or ion.

configurations for the first 18 elements of the Periodic Table. We determine the ground-state electron configuration of an atom by using the following three rules.

Rule 1. Orbitals fill in order of increasing energy from lowest to highest.

Example: In this course, we are concerned primarily with elements of the first, second, and third periods of the Periodic Table. Orbitals in these elements fill in the order $1s$, $2s$, $2p$, $3s$, $3p$.

Rule 2. Each orbital can hold up to two electrons.

Example: With four electrons, the $1s$ and $2s$ orbitals are filled and are written $1s^2 2s^2$. With an additional six electrons, a set of three $2p$ orbitals is filled and is written $2p_x^2 2p_y^2 2p_z^2$. Alternatively, the set of three filled $2p$ orbitals may be written $2p^6$.

Rule 3. When orbitals of equivalent energy are available but there are not enough electrons to fill them completely, then one electron is added to each equivalent orbital before a second electron is added to any one of them.

Example: After the $1s$ and $2s$ orbitals are filled with four electrons, a fifth electron is added to the $2p_x$ orbital, a sixth to the $2p_y$ orbital, and a seventh to the $2p_z$ orbital. Only after each $2p$ orbital contains one electron is a second electron added to the $2p_x$ orbital. For ten electrons, the ground-state electron configuration is $1s^2 2s^2 2p_x^2 2p_y^2 2p_z^2$ or, alternatively, $1s^2 2s^2 2p^6$.

EXAMPLE 1.1

Write ground-state electron configurations for these elements.

(a) Lithium (b) Oxygen (c) Chlorine

SOLUTION

(a) Lithium (atomic number 3): $1s^22s^1$
(b) Oxygen (atomic number 8): $1s^22s^22p_x^22p_y^12p_z^1$. Alternatively, the four electrons of the $2p$ orbitals can be grouped, and the electron configuration is written $1s^22s^22p^4$.
(c) Chlorine (atomic number 17): $1s^22s^22p^63s^23p^5$.

Practice Problem 1.1

Write and compare the ground-state electron configurations for the elements in each set.

(a) Carbon and silicon (b) Oxygen and sulfur
(c) Nitrogen and phosphorus

C. Lewis Structures

When discussing the physical and chemical properties of an element, chemists often focus on the outermost shell of its atoms because electrons in this shell are the ones involved in the formation of chemical bonds and in chemical reactions. Carbon, for example, with a ground-state electron configuration of $1s^22s^22p^2$, has four outer shell electrons. Outer shell electrons are called **valence electrons,** and the energy level in which they are found is called the **valence shell.** To show the outermost electrons of an atom, we commonly use a representation called a **Lewis structure,** after the American chemist Gilbert N. Lewis who devised this notation. A Lewis structure shows the symbol of the element surrounded by a number of dots equal to the number of electrons in the outer shell of an atom of that element. In Lewis structures, the atomic symbol represents the nucleus and all filled inner shells. Table 1.4 shows Lewis structures for the first 18 elements of the Periodic Table.

Valence electrons Electrons in the valence (outermost) shell of an atom.

Valence shell The outermost electron shell of an atom.

Lewis structure of an atom The symbol of an element surrounded by a number of dots equal to the number of electrons in the valence shell of the atom.

The noble gases helium and neon have filled valence shells. The valence shell of helium is filled with 2 electrons; that of neon is filled with 8 electrons. Neon

TABLE 1.4 Lewis Structures for Elements 1–18 of the Periodic Table

IA	IIA	IIIA	IVA	VA	VIA	VIIA	VIIIA
H·							He:
Li·	Be:	·B:	·Ċ:	·Ṅ:	:Ȯ:	:F̣:	:Ṇe:
Na·	Mg:	·Al:	·Ṡi:	·Ṗ:	:Ṡ:	:C̣l:	:Ạr:

Gilbert N. Lewis (1875–1946) introduced the theory of the electron pair that extended our understanding of covalent bonding and of the concept of acids and bases. It is in his honor that we often refer to an "electron dot" structure as a Lewis structure. *(UPI/Bettmann)*

and argon have in common an electron configuration in which the *s* and *p* orbitals of their valence shells are filled with 8 electrons. The valence shells of all other elements shown in Table 1.4 contain fewer than 8 electrons.

Compare the Lewis structures given in Table 1.4 with the ground-state electron configurations given in Table 1.3. The Lewis structure of boron (B), for example, is shown in Table 1.4 with three valence electrons; these are the paired $2s$ electrons and the single $2p_x$ electron shown in Table 1.3. The Lewis structure of carbon (C) is shown in Table 1.4 with four valence electrons; these are the two paired $2s$ electrons and the single $2p_x$ and $2p_y$ electrons shown in Table 1.3.

Notice also from Table 1.4 that for C, N, O, and F in period 2 of the Periodic Table, the valence electrons belong to the second shell. With 8 electrons, this shell is completely filled. For Si, P, S, and Cl in period 3 of the Periodic Table, the valence electrons belong to the third shell. This shell is only partially filled with 8 electrons; the $3s$ and $3p$ orbitals are fully occupied, but the five $3d$ orbitals can accommodate an additional ten valence electrons. Because of the differences in number and kind of valence shell orbitals available to elements of the second and third periods, significant differences exist in the covalent bonding of oxygen and sulfur, and of nitrogen and phosphorus. For example, although oxygen and nitrogen can accommodate no more than 8 electrons in their valence shells, many phosphorus-containing compounds have 10 electrons in the valence shell of phosphorus, and many sulfur-containing compounds have 10 and even 12 electrons in the valence shell of sulfur.

1.2 Lewis Model of Bonding

A. Formation of Ions

In 1916, Lewis devised a beautifully simple model that unified many of the observations about chemical bonding and reactions of the elements. He pointed out that the chemical inertness of the noble gases indicates a high degree of stability of the electron configurations of these elements: helium with a valence shell of two electrons ($1s^2$), neon with a valence shell of eight electrons ($2s^22p^6$), and argon with a valence shell of eight electrons ($3s^23p^6$). The tendency of atoms to react in ways that achieve an outer shell of eight valence electrons is particularly common among elements of Groups IA–VIIA (the main-group elements) and is given the special name **octet rule.**

Octet rule The tendency among atoms of Group IA–VIIA elements to react in ways that achieve an outer shell of eight valence electrons.

Example 1.2

Show how sodium follows the octet rule in forming Na^+.

Solution

Following are ground-state electron configurations for Na and Na^+:

$$\text{Na (11 electrons): } 1s^22s^22p^63s^1$$

$$\text{Na}^+ \text{ (10 electrons): } 1s^22s^22p^6$$

Thus, Na^+ has a complete octet of electrons in its valence shell and has the same electron configuration as neon.

TABLE 1.5 Electronegativity Values for Some Atoms (Pauling Scale)

IA	IIA	IIIB	IVB	VB	VIB	VIIB	VIIIB	VIIIB	VIIIB	IB	IIB	IIIA	IVA	VA	VIA	VIIA
								H 2.1								
Li 1.0	Be 1.5											B 2.0	C 2.5	N 3.0	O 3.5	F 4.0
Na 0.9	Mg 1.2											Al 1.5	Si 1.8	P 2.1	S 2.5	Cl 3.0
K 0.8	Ca 1.0	Sc 1.3	Ti 1.5	V 1.6	Cr 1.6	Mn 1.5	Fe 1.8	Co 1.8	Ni 1.8	Cu 1.9	Zn 1.6	Ga 1.6	Ge 1.8	As 2.0	Se 2.4	Br 2.8
Rb 0.8	Sr 1.0	Y 1.2	Zr 1.4	Nb 1.6	Mo 1.8	Tc 1.9	Ru 2.2	Rh 2.2	Pd 2.2	Ag 1.9	Cd 1.7	In 1.7	Sn 1.8	Sb 1.9	Te 2.1	I 2.5
Cs 0.7	Ba 0.9	La 1.1	Hf 1.3	Ta 1.5	W 1.7	Re 1.9	Os 2.2	Ir 2.2	Pt 2.2	Au 2.4	Hg 1.9	Tl 1.8	Pb 1.8	Bi 1.9	Po 2.0	At 2.2

Legend: <1.0; 1.0 – 1.4; 1.5 – 1.9; 2.0 – 2.4; 2.5 – 2.9; 3.0 – 4.0

Practice Problem 1.2

Show that the following obey the octet rule:

(a) Sulfur forms S^{2-}. (b) Magnesium forms Mg^{2+}.

B. Formation of Chemical Bonds

According to Lewis' model, atoms bond together in such a way that each atom participating in a chemical bond acquires a valence shell electron configuration the same as that of the noble gas closest to it in atomic number. Atoms acquire completed valence shells in two ways.

1. An atom may lose or gain enough electrons to acquire a filled valence shell. An atom that gains electrons becomes an **anion** (a negatively charged ion), and an atom that loses electrons becomes a **cation** (a positively charged ion). A chemical bond between an anion and a cation is called an **ionic bond.**
2. An atom may share electrons with one or more other atoms to acquire a filled valence shell. A chemical bond formed by sharing electrons is called a **covalent bond.**

Anion An atom or group of atoms bearing a negative charge.

Cation An atom or group of atoms bearing a positive charge.

Ionic bond A chemical bond resulting from the electrostatic attraction of an anion and a cation.

Covalent bond A chemical bond formed by the sharing of one or more pairs of electrons.

C. Electronegativity and Chemical Bonds

How do we estimate the degree of ionic or covalent character to a chemical bond? One way is to compare the electronegativities of the atoms involved. **Electronegativity** is a measure of the force of an atom's attraction for electrons that it shares in a chemical bond with another atom. The most widely used scale of electronegativities (Table 1.5) was devised by Linus Pauling in the 1930s.

Electronegativity A measure of the force of an atom's attraction for electrons it shares in a chemical bond with another atom.

On the Pauling scale, fluorine, the most electronegative element, is assigned an electronegativity of 4.0, and all other elements are assigned values in relation to

Linus Pauling (1901–1994) was the first person ever to receive two unshared Nobel Prizes. He received the Nobel Prize for Chemistry in 1954 for his contributions to the nature of chemical bonding. He received the Nobel Prize for Peace in 1962 for his efforts on behalf of international control of nuclear weapons and against nuclear testing. *(UPI/Bettmann)*

fluorine. As you study the electronegativity values in this table, note that they increase from left to right within a period of the Periodic Table and decrease from top to bottom within a group. Values increase from left to right because of the increasing positive charge on the nucleus. They decrease from top to bottom because of the increasing distance of the valence electrons from the nucleus, which leads to weaker attraction.

You should be aware that the values given in Table 1.5 are only approximate. The electronegativity of a particular element depends not only on its position in the Periodic Table but also on its oxidation state. The electronegativity of Cu(I) in Cu_2O, for example, is 1.8, whereas the electronegativity of Cu(II) in CuO is 2.0. In spite of these variations, electronegativity is still a useful guide to the distribution of electrons in a chemical bond.

EXAMPLE 1.3

Judging from their relative positions in the Periodic Table, which element in each set has the larger electronegativity?

(a) Lithium or carbon (b) Nitrogen or oxygen (c) Carbon or oxygen

SOLUTION

The elements in these sets are all in the second period of the Periodic Table. Electronegativity in this period increases from left to right.

(a) C > Li (b) O > N (c) O > C

Practice Problem 1.3

Judging from their relative positions in the Periodic Table, which element in each set has the larger electronegativity?

(a) Lithium or potassium (b) Nitrogen or phosphorus
(c) Carbon or silicon

Ionic Bonds

An ionic bond is formed by the transfer of electrons from the valence shell of an atom of lower electronegativity to the valence shell of an atom of higher electronegativity. The more electronegative atom gains one or more valence electrons and becomes an anion; the less electronegative atom loses one or more valence electrons and becomes a cation. We say that a chemical bond is ionic if the difference in electronegativity between the bonded atoms is greater than 1.9. An example of an ionic bond is that formed between sodium (electronegativity 0.9) and fluorine (electronegativity 4.0). The difference in electronegativity between these two elements is 3.1. In the following equation, we use a single-headed arrow to show the transfer of one electron from sodium to fluorine.

$$Na\cdot + \cdot\ddot{\underset{..}{F}}: \longrightarrow Na^+ :\ddot{\underset{..}{F}}:^-$$

In forming Na^+F^-, the single $3s$ valence electron of sodium is transferred to the partially filled valence shell of fluorine.

$$Na(1s^22s^22p^6\mathbf{3s^1}) + F(1s^22s^2\mathbf{2p^5}) \longrightarrow Na^+(1s^22s^22p^6) + F^-(1s^22s^2\mathbf{2p^6})$$

As a result of this transfer of one electron, both sodium and fluorine form ions that have the same electron configuration as neon, the noble gas closest to each in atomic number.

Covalent Bonds

A **covalent bond** is a chemical bond formed by sharing electron pairs between two atoms whose difference in electronegativity is 1.9 or less. The simplest example of a covalent bond is that found in the hydrogen molecule. When two hydrogen atoms bond, the single electrons from each combine to form an electron pair. This shared pair then completes the valence shell of each hydrogen. According to the Lewis model, a pair of electrons in a covalent bond functions in two ways simultaneously; it is shared by two atoms and at the same time fills the valence shell of each. We use a line between the two hydrogens to symbolize the covalent bond formed by the sharing of a pair of electrons.

$$H\cdot + \cdot H \longrightarrow H{-}H \qquad \Delta H^0 = -104 \text{ kcal/mol } (-435 \text{ kJ/mol})$$

The Lewis model accounts for the stability of covalently bonded atoms in the following way. In forming a covalent bond, an electron pair occupies the region between two nuclei and serves to shield one positively charged nucleus from the repulsive force of the other positively charged nucleus. At the same time, an electron pair attracts both nuclei. In other words, an electron pair in the space between two nuclei bonds them together and fixes the internuclear distance to within very narrow limits. The distance between nuclei participating in a chemical bond is called a **bond length.** Every covalent bond has a definite bond length. In H—H, it is 0.74 Å, where 1 Å = 10^{-10} m.

Bond length The distance between atoms in a covalent bond.

Although all covalent bonds involve the sharing of electrons, they differ widely in the degree of sharing. We divide covalent bonds into two categories—polar and nonpolar—depending on the difference in electronegativity between the bonded atoms (Table 1.6).

A covalent bond between carbon and hydrogen, for example, is classified as nonpolar covalent because the difference in electronegativity between these two atoms is 2.5 − 2.1 = 0.4 unit. An example of a polar covalent bond is that of H—Cl. The difference in electronegativity between chlorine and hydrogen is 3.0 − 2.1 = 0.9 unit.

TABLE 1.6 Classification of Chemical Bonds

Difference in Electronegativity Between Bonded Atoms	Type of Bond
less than 0.5	nonpolar covalent
0.5 to 1.9	polar covalent
greater than 1.9	ionic

Nonpolar covalent bond A covalent bond between atoms whose difference in electronegativity is less than 0.5.

Polar covalent bond A covalent bond between atoms whose difference in electronegativity is between 0.5 and 1.9.

EXAMPLE 1.4

Classify each bond as nonpolar covalent, polar covalent, or ionic.

(a) O—H (b) N—H (c) Na—F (d) C—Mg

SOLUTION

Based on differences in electronegativity between the bonded atoms, three of these bonds are polar covalent and one is ionic.

Bond	Difference in Electronegativity	Type of Bond
(a) O—H	3.5 − 2.1 = 1.4	polar covalent
(b) N—H	3.0 − 2.1 = 0.9	polar covalent
(c) Na—F	4.0 − 0.9 = 3.1	ionic
(d) C—Mg	2.5 − 1.2 = 1.3	polar covalent

Practice Problem 1.4

Classify each bond as nonpolar covalent, polar covalent, or ionic.

(a) S—H (b) P—H (c) C—F (d) C—Cl

An important consequence of the unequal sharing of electrons in a polar covalent bond is that the more electronegative atom gains a greater fraction of the shared electrons and acquires a partial negative charge, indicated by the symbol $\delta-$. The less electronegative atom has a lesser fraction of the shared electrons and acquires a partial positive charge, indicated by the symbol $\delta+$.

EXAMPLE 1.5

Using the symbols $\delta-$ and $\delta+$, indicate the direction of polarity in these polar covalent bonds.

(a) C—O (b) N—H (c) C—Mg

SOLUTION

Electronegativity is given beneath each atom. The atom with the greater electronegativity has the partial negative charge; the atom with the lesser electronegativity has the partial positive charge.

(a) $\overset{\delta+}{\text{C}}—\overset{\delta-}{\text{O}}$ (b) $\overset{\delta-}{\text{N}}—\overset{\delta+}{\text{H}}$ (c) $\overset{\delta-}{\text{C}}—\overset{\delta+}{\text{Mg}}$

2.5 3.5 3.0 2.1 2.5 1.2

TABLE 1.7 Lewis Structures for Several Compounds

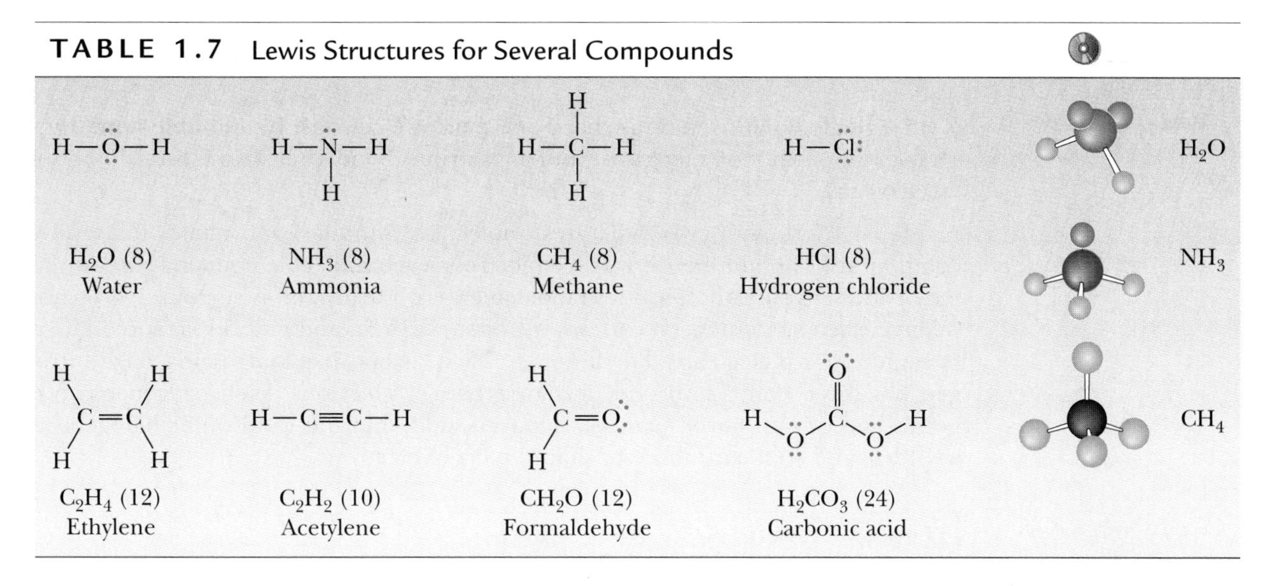

Practice Problem 1.5

Using the symbols δ− and δ+, indicate the direction of polarity in these polar covalent bonds.

(a) C—N (b) N—O (c) C—Cl

D. Guidelines for Drawing Lewis Structures of Molecules and Ions

The ability to write Lewis structures for molecules and ions is a fundamental skill for the study of organic chemistry. The following guidelines will help you to do this. As you study these guidelines, look at the examples in Table 1.7.

1. Determine the number of valence electrons in the molecule or ion. To do this, add the number of valence electrons contributed by each atom. For ions, add one electron for each negative charge on the ion, and subtract one electron for each positive charge on the ion. For example, the Lewis structure of the water molecule, H_2O, must show eight valence electrons: one from each hydrogen and six from oxygen. The Lewis structure for the hydroxide ion, OH^-, must also show eight valence electrons: one from hydrogen, six from oxygen, plus one for the negative charge on the ion.
2. Determine the arrangement of atoms in the molecule or ion. Except for the simplest molecules and ions, this arrangement must be determined experimentally. For some molecules and ions given as examples in the text, you are asked to propose an arrangement of atoms. For most, however, you are given the experimentally determined arrangement.
3. Connect the atoms with single bonds. Then arrange the remaining electrons in pairs so that each atom in the molecule or ion has a complete outer shell. Each hydrogen atom must be surrounded by two electrons. Each atom of carbon, oxygen, nitrogen, and halogen must be surrounded by eight electrons (per the octet rule).

Bonding electrons Valence electrons involved in forming a covalent bond, i.e., shared electrons.

Nonbonding electrons Valence electrons not involved in forming covalent bonds, i.e., unshared electrons.

4. A pair of electrons involved in a covalent bond **(bonding electrons)** is shown as a single bond; an unshared **(nonbonding)** pair of electrons is shown as a pair of dots.
5. In a **single bond** two atoms share one pair of electrons. In a **double bond** they share two pairs of electrons, and in a **triple bond** they share three pairs of electrons.

Table 1.7 shows Lewis structures, molecular formulas, and names for several compounds. The number of valence electrons each molecule contains is shown in parentheses. Notice that, in these molecules, each hydrogen is surrounded by two valence electrons, and each carbon, nitrogen, oxygen, and chlorine is surrounded by eight valence electrons. Furthermore, each carbon has four bonds, each nitrogen has three bonds and one unshared pair of electrons, each oxygen has two bonds and two unshared pairs of electrons, and chlorine (and other halogens as well) has one bond and three unshared pairs of electrons.

EXAMPLE 1.6

Draw Lewis structures, showing all valence electrons, for these molecules.

(a) CO_2 (b) CH_4O (c) CH_3Cl

SOLUTION

Under the Lewis structure for each molecule is the number of valence electrons it contains.

(a) O=C=O

Carbon dioxide
(16 valence electrons)

(b) H—C(H)(H)—O—H

Methanol
(14 valence electrons)

(c)

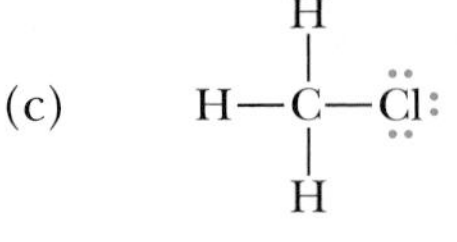

Chloromethane
(14 valence electrons)

Practice Problem 1.6

Draw Lewis structures, showing all valence electrons, for these molecules.

(a) C_2H_6 (b) CS_2 (c) HCN

E. Formal Charge

Throughout this course we deal not only with molecules but also with polyatomic cations and polyatomic anions. Examples of polyatomic cations are the hydronium ion, H_3O^+, and the ammonium ion, NH_4^+. An example of a polyatomic anion is the bicarbonate ion, HCO_3^-. It is important that you be able to determine which atom or atoms in a molecule or polyatomic ion bear the positive or negative charge. The charge on an atom in a molecule or polyatomic ion is called its **formal charge.** To derive a formal charge:

Formal charge The charge on an atom in a molecule or polyatomic ion.

1. Write a correct Lewis structure for the molecule or ion.
2. Assign to each atom all its unshared (nonbonding) electrons and one half its shared (bonding) electrons.

3. Compare this number with the number of valence electrons in the neutral, unbonded atom. If the number of electrons assigned a bonded atom is less than that assigned to the unbonded atom, then more positive charges are in the nucleus than counterbalancing negative charges, and the atom has a positive formal charge. Conversely, if the number of electrons assigned to a bonded atom is greater than that assigned to the unbonded atom, then the atom has a negative formal charge.

$$\text{Formal charge} = \begin{matrix}\text{Number of valence}\\ \text{electrons in neutral}\\ \text{unbonded atom}\end{matrix} - \left(\begin{matrix}\text{All unshared}\\ \text{electrons}\end{matrix} + \begin{matrix}\text{One half of all}\\ \text{shared electrons}\end{matrix}\right)$$

EXAMPLE 1.7

Draw Lewis structures for these ions, and show which atom in each bears the formal charge.

(a) H_3O^+ (b) HCO_3^-

SOLUTION

(a) The Lewis structure for the hydronium ion must show 8 valence electrons; 3 from the three hydrogens, 6 from oxygen, minus 1 for the single positive charge. An oxygen atom has 6 valence electrons. The oxygen atom in H_3O^+ is assigned two unshared electrons and one from each shared pair of electrons, giving it a formal charge of $6 - (2 + 3) = +1$.

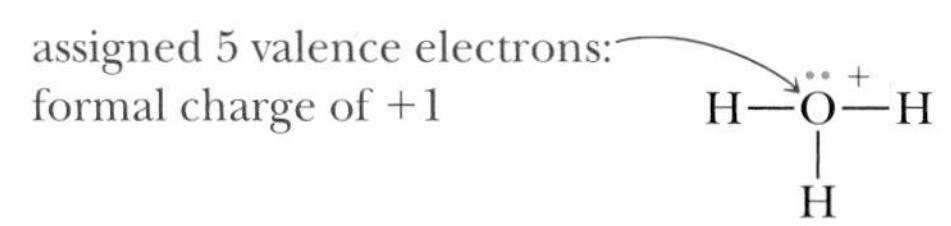

(b) The Lewis structure for bicarbonate ion must show 24 valence electrons; 4 from carbon, 18 from the three oxygens, 1 from hydrogen, plus 1 for the single negative charge. Loss of a hydrogen ion from carbonic acid (Table 1.7) gives the bicarbonate ion. Carbon is assigned 1 electron from each shared pair and has no formal charge $(4 - 4 = 0)$. Two oxygens are assigned 6 valence electrons each and have no formal charges $(6 - 6 = 0)$. The third oxygen is assigned 7 valence electrons and has a formal charge of $6 - (6 + 1) = -1$.

H—O—C(=O)—O—H   H—O—C(=O)—O: (assigned 7 valence electrons: formal charge of −1)

Carbonic acid, H_2CO_3 Bicarbonate ion, HCO_3^-

Practice Problem 1.7

Draw Lewis structures for these ions, and show which atom in each bears the formal charge(s).

(a) $CH_3NH_3^+$ (b) CO_3^{2-} (c) OH^-

When writing Lewis structures for molecules and ions, you must remember that elements of the second period, including carbon, nitrogen, and oxygen, can accommodate no more than eight electrons in the four orbitals ($2s$, $2p_x$, $2p_y$, and $2p_z$) of their valence shells. Following are two Lewis structures for nitric acid, HNO_3, each with the correct number of valence electrons, namely, 24:

$$\mathrm{H{-}\ddot{\underset{..}{O}}{-}\overset{\overset{\ddot{O}}{\|}}{\underset{+}{N}}{-}\ddot{\underset{..}{O}}\!:^{-}} \qquad \mathrm{H{-}\ddot{\underset{..}{O}}{-}\overset{\overset{\ddot{O}}{\|}}{N}{=}\ddot{O}\!:}$$

An acceptable Lewis structure — Not an acceptable Lewis structure (10 electrons in the valence shell of nitrogen)

The structure on the left is an acceptable Lewis structure. It shows the required 24 valence electrons, and each oxygen and nitrogen has a completed valence shell of 8 electrons. Further, it shows a positive formal charge on nitrogen and a negative formal charge on one of the oxygens. An acceptable Lewis structure must show these formal charges. The structure on the right is not an acceptable Lewis structure. Although it shows the correct number of valence electrons, it places 10 electrons in the valence shell of nitrogen. Yet the four orbitals of the second shell ($2s$, $2p_x$, $2p_y$, and $2p_z$) can hold no more than 8 valence electrons!

1.3 BOND ANGLES AND SHAPES OF MOLECULES

In Section 1.2 we used a shared pair of electrons as the fundamental unit of a covalent bond and drew Lewis structures for several small molecules and ions containing various combinations of single, double, and triple bonds (see, for example, Table 1.7). We can predict bond angles in these and other molecules in a very straightforward way using the **valence-shell electron-pair repulsion (VSEPR) model.** According to the VSEPR model, an atom is surrounded by an outer shell of valence electrons. These valence electrons may be involved in the formation of single, double, or triple bonds, or they may be unshared. Each combination creates a negatively charged region of space, and, because like charges repel each other, the various regions of electron density around an atom spread so that each is as far away from the others as possible.

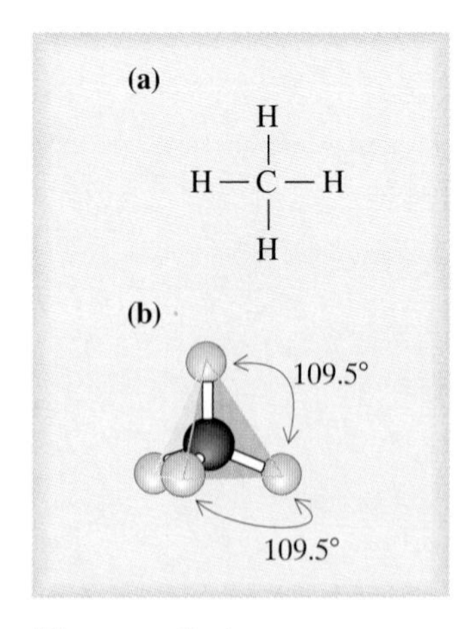

Figure 1.4
The shape of a methane molecule, CH_4. (a) Lewis structure and (b) shape. The hydrogens occupy the four corners of a regular tetrahedron and all H—C—H bond angles are 109.5°.

We use the VSEPR model in the following way to predict the shape of methane, CH_4. The Lewis structure for CH_4 shows a carbon atom surrounded by four regions of electron density, each of which contains a pair of electrons forming a bond to a hydrogen atom. According to the VSEPR model, the four regions radiate from carbon so that they are as far away from each other as possible. This occurs when the angle between any two pairs of electrons is 109.5°. Therefore, we predict all H—C—H bond angles to be 109.5°, and the shape of the molecule to be **tetrahedral** (Figure 1.4). The H—C—H bond angles in methane have been measured experimentally and found to be 109.5°. Thus, the bond angles and shape of methane predicted by the VSEPR model are identical to those observed.

We predict the shape of an ammonia molecule, NH_3, in exactly the same manner. The Lewis structure of NH_3 shows nitrogen surrounded by four regions of electron density. Three regions contain single pairs of electrons forming covalent bonds with hydrogen atoms. The fourth region contains an unshared pair of electrons [Figure 1.5(a)]. Using the VSEPR model, we predict that the four regions of electron density around nitrogen are arranged in a tetrahedral manner and that all H—N—H bond angles are 109.5°. The observed bond angles are 107.3°. This small difference between the predicted and observed angles can be explained by proposing that the unshared pair of electrons on nitrogen repels adjacent electron pairs more strongly than bonding pairs repel each other.

Figure 1.6 shows a Lewis structure and ball-and-stick model of a water molecule. In H_2O, oxygen is surrounded by four regions of electron density. Two of these regions contain pairs of electrons used to form covalent bonds to two hydrogens; the remaining two contain unshared electron pairs. Using the VSEPR model, we predict that the four regions of electron density around oxygen are arranged in a tetrahedral manner and that the H—O—H bond angle is 109.5°. Experimental measurements show the actual bond angle to be 104.5°, a value smaller than that predicted. This difference between the predicted and observed bond angle can be explained by proposing, as we did for NH_3, that unshared pairs of electrons repel adjacent pairs more strongly than do bonding pairs. Note that the distortion from 109.5° is greater in H_2O, which has two unshared pairs of electrons, than it is in NH_3, which has only one unshared pair.

A general prediction emerges from this discussion of the shapes of CH_4, NH_3, and H_2O molecules. If a Lewis structure shows four regions of electron density around an atom, the VSEPR model predicts a tetrahedral distribution of electron density and bond angles of approximately 109.5°.

In many of the molecules we encounter, an atom is surrounded by three regions of electron density. Shown in Figure 1.7 are Lewis structures for formaldehyde, CH_2O, and ethylene, C_2H_4.

According to the VSEPR model, a double bond is treated as a single region of electron density. In formaldehyde, carbon is surrounded by three regions of elec-

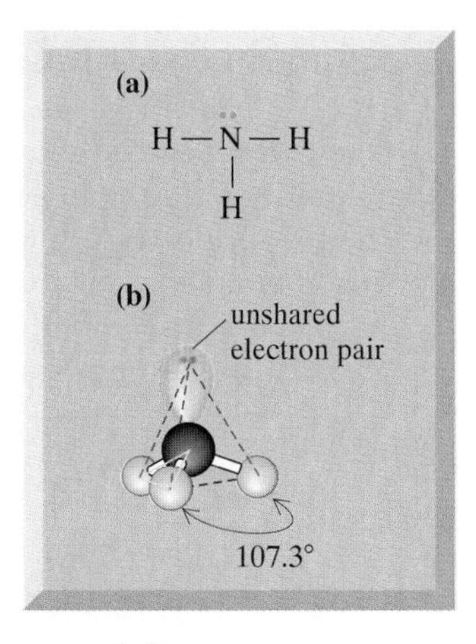

Figure 1.5
The shape of an ammonia molecule, NH_3. (a) Lewis structure and (b) ball-and-stick model. The H—N—H bond angle is 107.3°, slightly smaller than the H—C—H bond angle of methane.

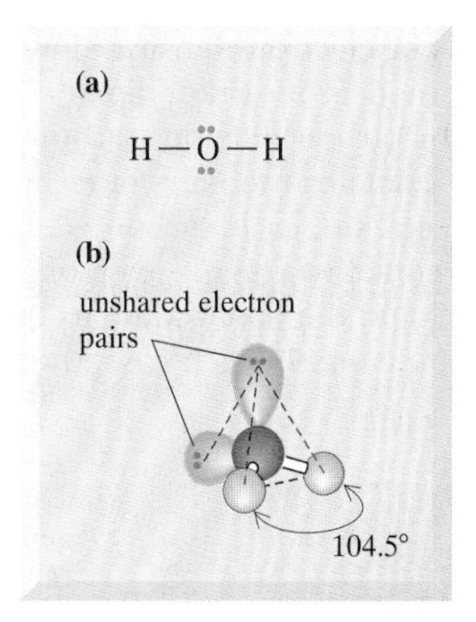

Figure 1.6
The shape of a water molecule, H_2O. (a) Lewis structure, (b) ball-and-stick model. The unshared pairs of electrons repel adjacent pairs of electrons giving the H—O—H bond angle of 104.5°.

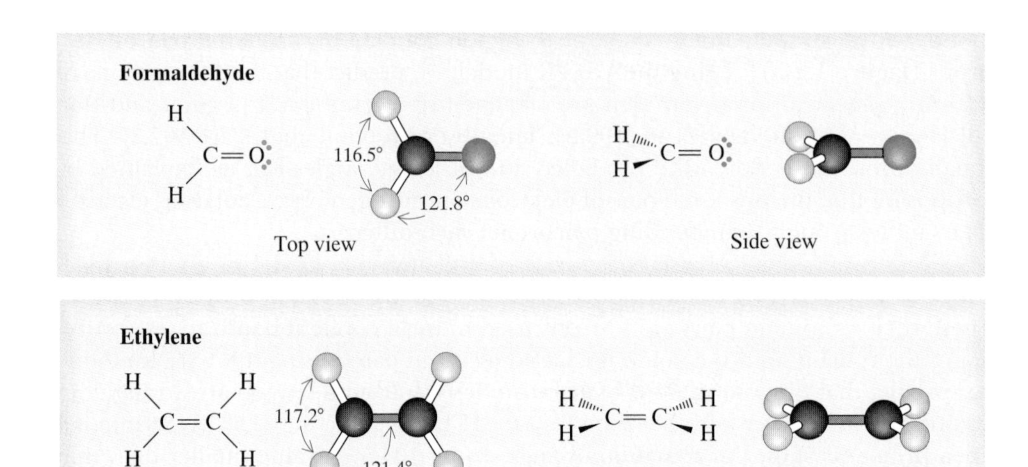

Figure 1.7
Shapes of formaldehyde, CH_2O, and ethylene, C_2H_4.

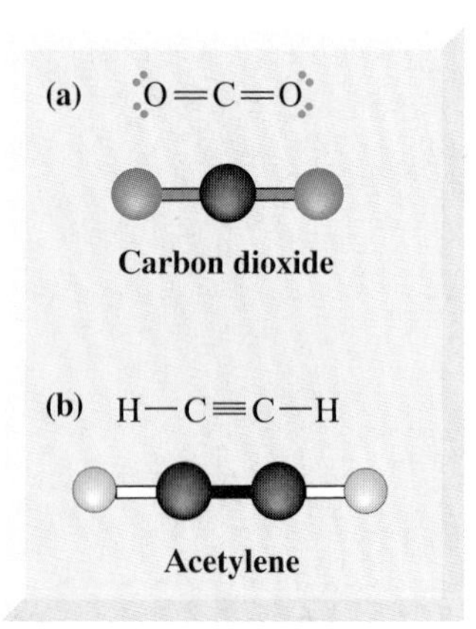

Figure 1.8
Shapes of (a) carbon dioxide, CO_2, and (b) acetylene, C_2H_2.

tron density: two regions contain single pairs of electrons forming single bonds to hydrogen atoms; the third region contains two pairs of electrons forming a double bond to oxygen. In ethylene, each carbon atom is also surrounded by three regions of electron density; two contain single pairs of electrons, and the third contains two pairs of electrons.

Three regions of electron density about an atom are farthest apart when they are coplanar and make angles of 120° with each other. Thus, the predicted H—C—H and H—C—O bond angles in formaldehyde and the H—C—H and H—C—C bond angles in ethylene are all 120°. Further, all bonds in each molecule lie in a plane. Both formaldehyde and ethylene are planar molecules.

In still other types of molecules, a central atom is surrounded by only two regions of electron density. Shown in Figure 1.8 are Lewis structures and ball-and-stick models of carbon dioxide, CO_2, and acetylene, C_2H_2.

In carbon dioxide, carbon is surrounded by two regions of electron density: each contains two pairs of electrons and forms a double bond to an oxygen atom. In acetylene, each carbon is also surrounded by two regions of electron density: one contains a single pair of electrons and forms a single bond to a hydrogen atom, and the other contains three pairs of electrons and forms a triple bond to a carbon atom. In each case, the two regions of electron density are farthest apart if

Three balloons tied together are trigonal. *(Charles D. Winters)*

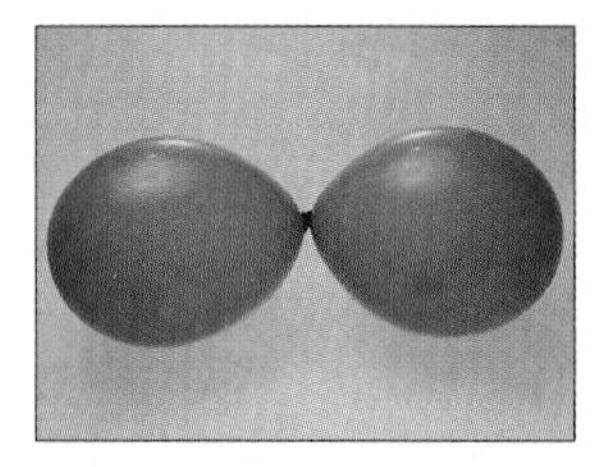

Two balloons tied together are linear. *(Charles D. Winters)*

TABLE 1.8 Predicted Molecular Shapes (VSEPR Model)

Regions of Electron Density Around Central Atom	Predicted Distribution of Electron Density	Predicted Bond Angles	Examples		
4	tetrahedral	109.5°	CH_4	NH_3	H_2O
3	trigonal planar	120°	$H_2C{=}CH_2$	$H_2C{=}O$	
2	linear	180°	$O{=}C{=}O$	$H{-}C{\equiv}C{-}H$	

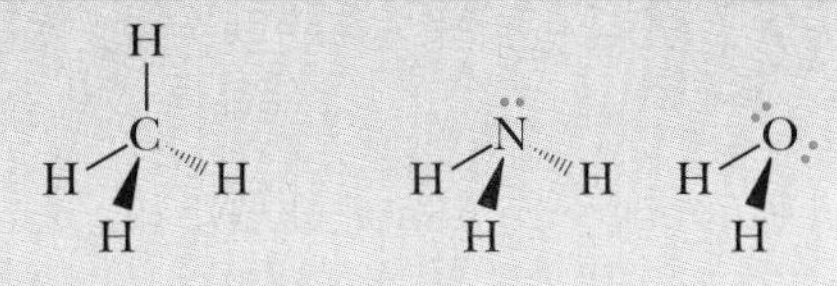

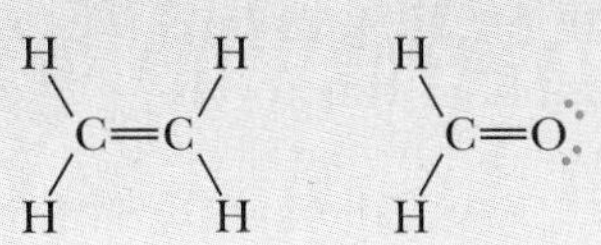

they form a straight line through the central atom and create an angle of 180°. Both carbon dioxide and acetylene are linear molecules. Predictions of the VSEPR model are summarized in Table 1.8.

EXAMPLE 1.8

Predict all bond angles in these molecules.

(a) CH_3Cl (b) $CH_2{=}CHCl$

SOLUTION

(a) The Lewis structure for CH_3Cl shows carbon surrounded by four regions of electron density. Therefore, predict the distribution of electron pairs about carbon to be tetrahedral, all bond angles to be 109.5°, and the shape of CH_3Cl to be tetrahedral.

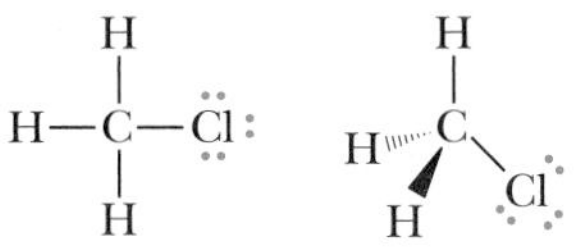

(b) The Lewis structure for $CH_2{=}CHCl$ shows each carbon surrounded by three regions of electron density. Therefore, predict all bond angles to be 120°.

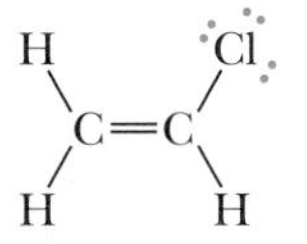

Practice Problem 1.8

Predict all bond angles for these molecules.

(a) CH_3OH (b) CH_2Cl_2 (c) H_2CO_3 (carbonic acid)

CHEMICAL CONNECTIONS

Buckyball—A New Form of Carbon

A favorite chemistry examination question is, What are the elemental forms of carbon? The usual answer is that pure carbon is found in two forms—graphite and diamond. These forms have been known for centuries, and it was generally believed that they are the only forms of carbon having extended networks of C atoms in well-defined structures.

But that is not so! The scientific world was startled in 1985 when Richard Smalley of Rice University and Harry W. Kroto of the University of Sussex, UK, and their coworkers announced that they had detected a new form of carbon with a molecular formula C_{60}. They suggested that the molecule has a structure that resembles a soccer ball; it has 12 five-membered rings and 20 six-membered rings arranged such that each five-membered ring is surrounded by six-membered rings. This structure reminded its discoverers of a geodesic dome, a structure invented by the innovative American engineer and philosopher R. Buckminster Fuller. Therefore, the official name of this allotrope of carbon has become buckminsterfullerene or, more simply, "fullerene," and many chemists refer to C_{60} simply as "buckyball." Kroto, Smalley, and Robert F. Curl were awarded the Nobel Prize for Chemistry in 1996 for this work. Many higher fullerenes, such as C_{70} and C_{84}, have also been isolated and studied.

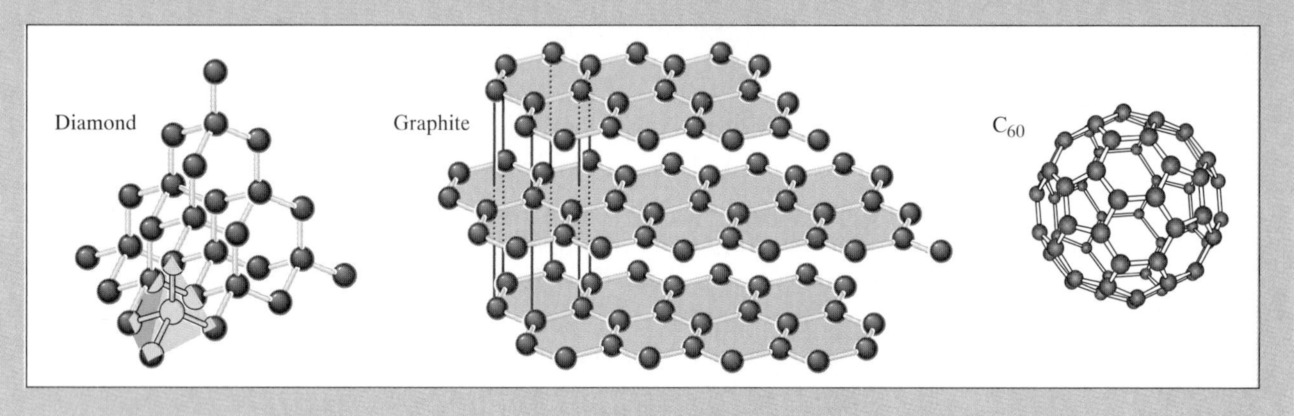

1.4 Polar and Nonpolar Molecules

In Section 1.2C, we used the term "polar" to describe a covalent bond in which one atom bears a partial positive charge and the other bears a partial negative charge. Further, we saw that we can use the difference in electronegativity between bonded atoms to determine the relative polarities of covalent bonds. We can now combine our understanding of bond polarity and molecular geometry (Section 1.3) to predict the polarity of polyatomic molecules.

A molecule will be polar if (1) it has polar bonds and (2) the center of partial positive charge lies at a different place within the molecule than the center of partial negative charge. Consider first carbon dioxide, CO_2. Each C—O bond is polar with oxygen, the more electronegative atom, bearing a partial negative charge and carbon bearing a partial positive charge. Because carbon dioxide is a linear molecule, the centers of negative and positive partial charge coincide, and, therefore, this molecule is nonpolar.

$$\overset{\delta-}{\ddot{O}}=\overset{\delta+}{C}=\overset{\delta-}{\ddot{O}}$$

Carbon dioxide
(a nonpolar molecule)

In a water molecule, each O—H bond is polar with oxygen, the more electronegative atom, bearing a partial negative charge and each hydrogen bearing a partial positive charge. Because water is a bent molecule, the center of its partial positive charge is between the two hydrogen atoms, and its center of partial negative charge is on oxygen. Thus, water has polar bonds and is a polar molecule. Ammonia has three polar N—H bonds, and, because of its geometry, the centers of partial positive and negative charges are at different places within the molecule. Ammonia is a polar molecule.

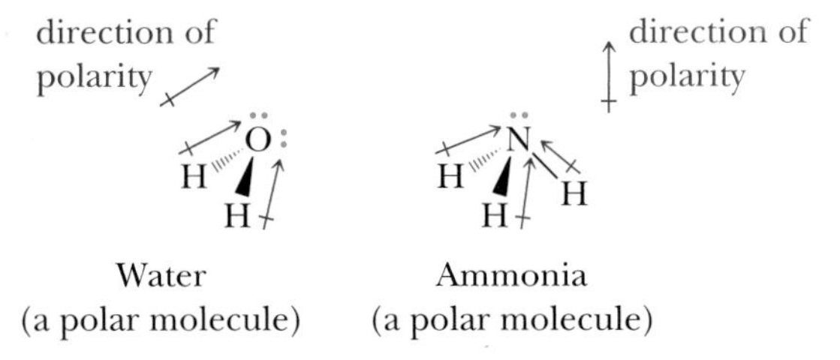

Water
(a polar molecule)

Ammonia
(a polar molecule)

EXAMPLE 1.9

Which of these molecules are polar? For each that is, specify the direction of its polarity.

(a) CH_3Cl (b) CH_2O (c) C_2H_2

SOLUTION

Both chloromethane, CH_3Cl, and formaldehyde, CH_2O, have polar bonds and, because of their geometry, are polar molecules. Because it contains no polar bonds, acetylene, C_2H_2, is a nonpolar molecule.

(a) Chloromethane (a polar molecule)

(b) Formaldehyde (a polar molecule)

(c) H—C≡C—H Acetylene (a nonpolar molecule)

Practice Problem 1.9

Both carbon dioxide, CO_2, and ozone, O_3, are triatomic molecules. Account for the fact that carbon dioxide is a nonpolar molecule whereas ozone is a polar molecule.

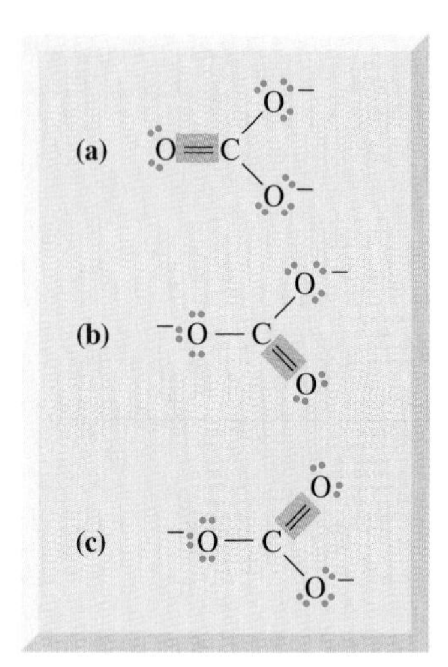

Figure 1.9
Three Lewis structures for the carbonate ion.

1.5 Resonance

As chemists developed more understanding of covalent bonding in organic compounds, it became obvious that for a great many molecules and ions, no single Lewis structure provides a truly accurate representation. For example, Figure 1.9 shows three Lewis structures for the carbonate ion, CO_3^{2-}, each of which shows carbon bonded to three oxygen atoms by a combination of one double bond and two single bonds. Each Lewis structure implies that one carbon-oxygen bond is different from the other two. This, however, is not the case. It has been shown that all three carbon-oxygen bonds are identical.

To describe the carbonate ion, as well as other molecules and ions for which no single Lewis structure is adequate, we turn to the theory of resonance.

A. The Theory of Resonance

The theory of resonance was developed by Linus Pauling in the 1930s. According to this theory, many molecules and ions are best described by writing two or more Lewis structures and considering the real molecule or ion to be a composite of these structures. Individual Lewis structures are called **contributing structures.** We show that the real molecule or ion is a **resonance hybrid** of the various contributing structures by interconnecting them with **double-headed arrows.**

Contributing structure Representations of a molecule or ion that differ only in the distribution of valence electrons.

Resonance hybrid A molecule or ion that is best described as a composite of a number of contributing structures.

Double-headed arrow A symbol used to connect contributing structures.

Three contributing structures for the carbonate ion are shown in Figure 1.10. These three contributing structures are said to be equivalent. **Equivalent contributing structures** have identical patterns of covalent bonding and are of equal energy.

The use of the term "resonance" for this theory of covalent bonding might suggest to you that bonds and electron pairs constantly change back and forth from one position to another over time. This notion is not at all correct. The carbonate ion, for example, has one and only one real structure. The problem is ours—how do we draw that one real structure? The resonance method is a way to describe the real structure and at the same time retain Lewis structures with electron-pair bonds. Thus, although we realize that the carbonate ion is not accurately represented by any one contributing structure shown in Figure 1.10, we continue to represent it as one of these for convenience. We understand, of course, that what is intended is the resonance hybrid.

B. Curved Arrows and Electron Pushing

Notice in Figure 1.10 that the only change from contributing structure (a) to (b) and then from (b) to (c) is a redistribution of valence electrons. To show how this redistribution of valence electrons occurs, chemists use a symbol called a curved

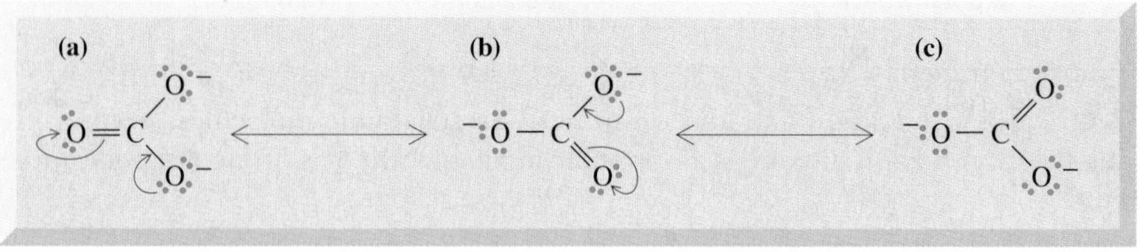

Figure 1.10
The carbonate ion represented as a resonance hybrid of three equivalent contributing structures. Curved arrows show the redistribution of valence electrons between one contributing structure and the next.

arrow. A **curved arrow** shows the repositioning of an electron pair from its origin (the tail of the arrow) to its destination (the head of the arrow). This repositioning may be from an atom to an adjacent bond, or from a bond to an adjacent atom.

Curved arrow A symbol used to show the redistribution of valence electrons.

A curved arrow is nothing more than a bookkeeping symbol for keeping track of electron pairs or, as some call it, **electron pushing.** Do not be misled by its simplicity. Electron pushing will help you see the relationship between contributing structures. Furthermore, it will help you follow bond-breaking and bond-forming steps in organic reactions. This type of electron pushing is a survival skill in organic chemistry.

C. Rules for Writing Acceptable Contributing Structures

These rules must be followed in writing acceptable contributing structures:

1. All contributing structures must have the same number of valence electrons.
2. All contributing structures must obey the rules of covalent bonding; no contributing structure may have more than 2 electrons in the valence shell of hydrogen or more than 8 electrons in the valence shell of a second-period element. Third-period elements, such as sulfur and phosphorus, may have up to 12 electrons in their valence shells.
3. The positions of all nuclei must remain the same; that is, contributing structures differ only in the distribution of valence electrons.

EXAMPLE 1.10

Which sets are pairs of contributing structures?

(a) $CH_3-C(=\ddot{O}:)-CH_3 \longleftrightarrow CH_3-\overset{+}{C}(-:\ddot{O}:^-)-CH_3$ (b) $CH_3-C(=\ddot{O}:)-CH_3 \longleftrightarrow CH_2=C(-\ddot{O}-H)-CH_3$

SOLUTION

(a) Contributing structures. They differ only in the distribution of valence electrons.

(b) Not a set of contributing structures. They differ in the arrangement of their atoms.

Practice Problem 1.10

Which sets are pairs of contributing structures?

(a) $CH_3-C(=\ddot{O}:)-\ddot{O}:^- \longleftrightarrow CH_3-C^+(-\ddot{O}:^-)-\ddot{O}:^-$ (b) $CH_3-C(=\ddot{O}:)-\ddot{O}:^- \longleftrightarrow CH_3-\bar{C}(=\ddot{O}:)=\ddot{O}:$

EXAMPLE 1.11

Draw the contributing structure indicated by the curved arrows. Be certain to show all valence electrons and all formal charges.

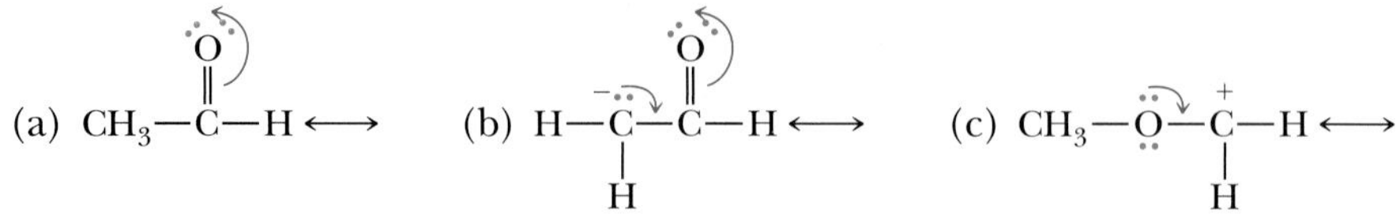

(a) $CH_3-C(=O)-H \longleftrightarrow$ (b) $H-\overset{-}{C}(H)-C(=O)-H \longleftrightarrow$ (c) $CH_3-\ddot{O}-\overset{+}{C}(H)-H \longleftrightarrow$

SOLUTION

(a) $CH_3-\overset{+}{C}(-\ddot{O}:^-)-H$ (b) $H-C(H)=C(-\ddot{O}:^-)-H$ (c) $CH_3-\overset{+}{O}=C(H)-H$

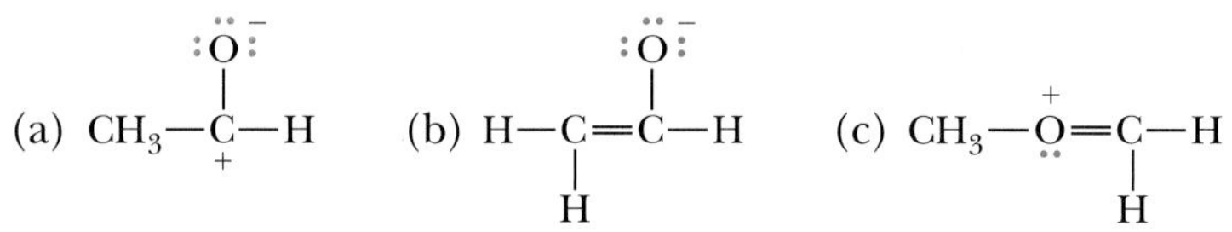

Practice Problem 1.11

Use curved arrows to show the redistribution of valence electrons in converting contributing structure (a) to (b), and then (b) to (c).

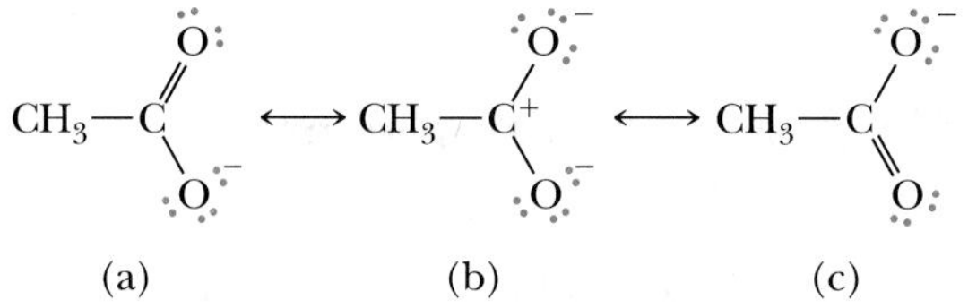

$CH_3-C(=O)-O^- \longleftrightarrow CH_3-C^+(-O^-)-O^- \longleftrightarrow CH_3-C(-O^-)=O$

(a) (b) (c)

1.6 VALENCE BOND MODEL OF COVALENT BONDING

As much as the Lewis and VSEPR models help us to understand covalent bonding and the geometry of molecules, they leave many important questions unanswered. The most important of these is the relation between molecular structure and chemical reactivity. For example, carbon-carbon double bonds are quite different in chemical reactivity from carbon-carbon single bonds. Most carbon-carbon single bonds are quite unreactive, but carbon-carbon double bonds react with a wide variety of reagents. The Lewis model gives us no way to account for these differences. Therefore, let us turn to a newer model of covalent bonding, namely formation of covalent bonds by the overlap of atomic orbitals.

A. Formation of a Covalent Bond by the Overlap of Atomic Orbitals

Sigma (σ) bond A covalent bond in which the overlap of atomic orbitals is concentrated along the bond axis.

According to the valence bond model, a covalent bond is formed when a portion of an atomic orbital of one atom overlaps a portion of an atomic orbital of another atom. In forming the covalent bond in H_2, for example, two hydrogens approach each other so that their 1*s* atomic orbitals overlap to form a sigma covalent bond (Figure 1.11). A **sigma (σ) bond** is a covalent bond in which orbital overlap of the bond occurs along the axis joining the two nuclei.

B. Hybridization of Atomic Orbitals

Formation of a covalent bond between two hydrogen atoms is straightforward. Formation of covalent bonds with second-period elements, however, presents the following problem. In forming covalent bonds, atoms of carbon, nitrogen, and oxygen (all second-period elements), use $2s$ and $2p$ atomic orbitals. The three $2p$ atomic orbitals are at angles of 90° to one another (Figure 1.3), and, if atoms of second-period elements used these orbitals to form covalent bonds, bond angles around each would be approximately 90°. Bond angles of 90°, however, are rarely observed in organic molecules. What we find, instead, are bond angles of approximately 109.5° in molecules with only single bonds, 120° in molecules with double bonds, and 180° in molecules with triple bonds.

Covalent bond formed by overlap of atomic orbitals

H + H → H H

1s 1s

Figure 1.11 Formation of the covalent bond in H_2 by overlap of the $1s$ atomic orbitals of each hydrogen.

109.5° 120° (approx.) 180°

H—C≡C—H

To account for these observed bond angles, Pauling proposed that atomic orbitals combine to form new orbitals, called **hybrid orbitals.** The number of hybrid orbitals formed is equal to the number of atomic orbitals combined. Elements of the second period form three types of hybrid orbitals, designated sp^3, sp^2, and sp, each of which can contain up to two electrons.

Hybridization The combination of two or more atomic orbitals to form a new set of atomic orbitals.

C. sp^3 Hybrid Orbitals—Bond Angles of Approximately 109.5°

Combination of one $2s$ atomic orbital and three $2p$ atomic orbitals forms four equivalent **sp^3 hybrid orbitals.** Because they are derived from four atomic orbitals, sp^3 hybrid orbitals always occur in sets of four. Each sp^3 hybrid orbital consists of a larger lobe pointing in one direction and a smaller lobe pointing in the opposite direction. The axes of the four sp^3 hybrid orbitals are directed toward the corners of a regular tetrahedron, and sp^3 hybridization results in bond angles of approximately 109.5° (Figure 1.12).

sp^3 Hybrid orbital A hybrid atomic orbital produced by the combination of one s atomic orbital and three p atomic orbitals.

Keep in mind that superscripts in the designation of hybrid orbitals tell you how many atomic orbitals have been combined to form the hybrid orbitals. The designation sp^3, for example, tells you *one s* atomic orbital and *three p* atomic orbitals are combined in forming the hybrid orbital. Do not confuse this use of superscripts with that used in writing a ground-state electron configuration, as for example $1s^22s^22p^5$ for fluorine. In the case of a ground-state electron configuration, superscripts tell you the number of electrons in each orbital or set of orbitals.

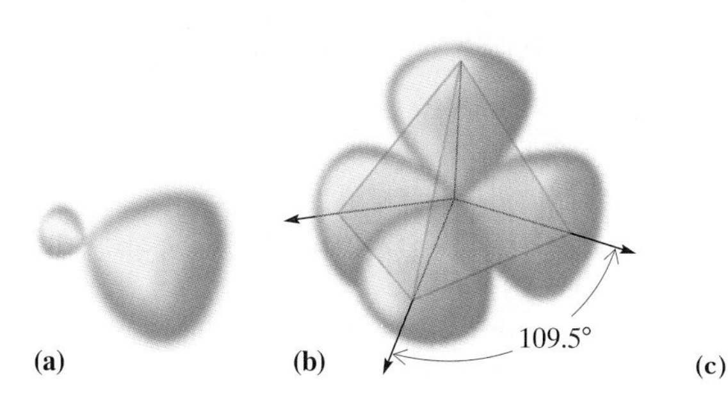

Figure 1.12 sp^3 Hybrid orbitals. (a) Representation of a single sp^3 hybrid orbital showing two lobes of unequal size. (b) Three-dimensional representation of four sp^3 hybrid orbitals directed toward the corners of a regular tetrahedron. The smaller lobes of each sp^3 hybrid orbital are hidden behind the larger lobes. (c) If four balloons of similar size and shape are tied together, they will naturally assume a tetrahedral geometry.

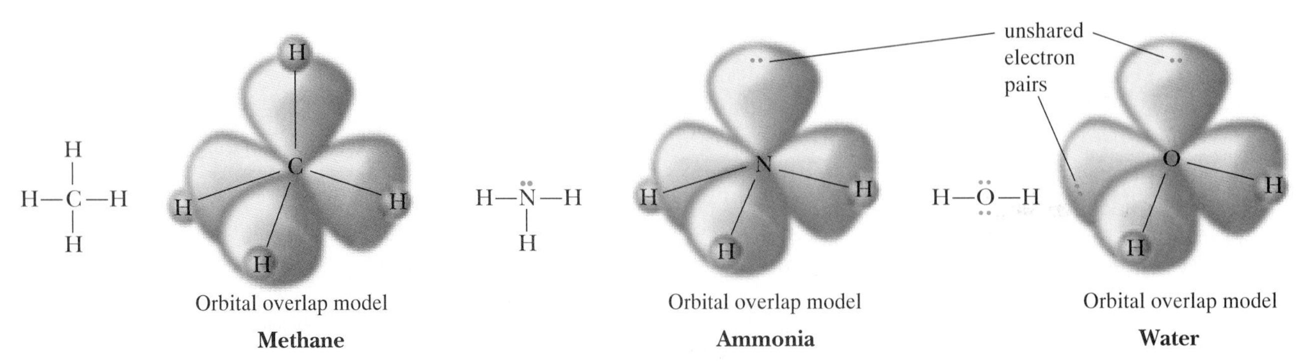

Figure 1.13
Orbital overlap pictures of methane, ammonia, and water.

In Section 1.2, we described the covalent bonding in CH_4, NH_3, and H_2O in terms of the Lewis model, and in Section 1.3, we used the VSEPR model to predict bond angles of approximately 109.5° in each molecule. Now let us consider the bonding in these molecules in terms of the overlap of atomic orbitals. To bond with four other atoms with bond angles of 109.5°, carbon uses sp^3 hybrid orbitals. Carbon has four valence electrons, and one electron is placed in each sp^3 hybrid orbital. Each partially filled sp^3 hybrid orbital then overlaps with a partially filled 1*s* atomic orbital of hydrogen to form a sigma (σ) bond, and hydrogen atoms occupy the corners of a regular tetrahedron (Figure 1.13).

In bonding with three other atoms, the five valence electrons of nitrogen are distributed so that one sp^3 hybrid orbital is filled with a pair of electrons and the remaining three sp^3 hybrid orbitals have one electron each. Overlapping of these partially filled sp^3 hybrid orbitals with 1*s* atomic orbitals of hydrogen atoms produces the NH_3 molecule (Figure 1.13).

In bonding with two other atoms, the six valence electrons of oxygen are distributed so that two sp^3 hybrid orbitals are filled and the remaining two have one electron each. Each partially filled sp^3 hybrid orbital overlaps with a 1*s* atomic orbital of hydrogen, and hydrogen atoms occupy two corners of a regular tetrahedron. The remaining two corners of the tetrahedron are occupied by unshared pairs of electrons (Figure 1.13).

sp^2 Hybrid orbital A hybrid atomic orbital produced by the combination of one *s* atomic orbital and two *p* atomic orbitals.

D. sp^2 Hybrid Orbitals—Bond Angles of Approximately 120°

Combination of one 2*s* atomic orbital and two 2*p* atomic orbitals forms three equivalent **sp^2 hybrid orbitals.** Because they are derived from three atomic orbitals,

Figure 1.14
sp^2 Hybrid orbitals. (a) A single sp^2 hybrid orbital showing two lobes of unequal size. (b) The three sp^2 hybrid orbitals with their axes in a plane at angles of 120°. (c) The unhybridized 2*p* atomic orbital perpendicular to the plane created by the three sp^2 hybrid orbitals.

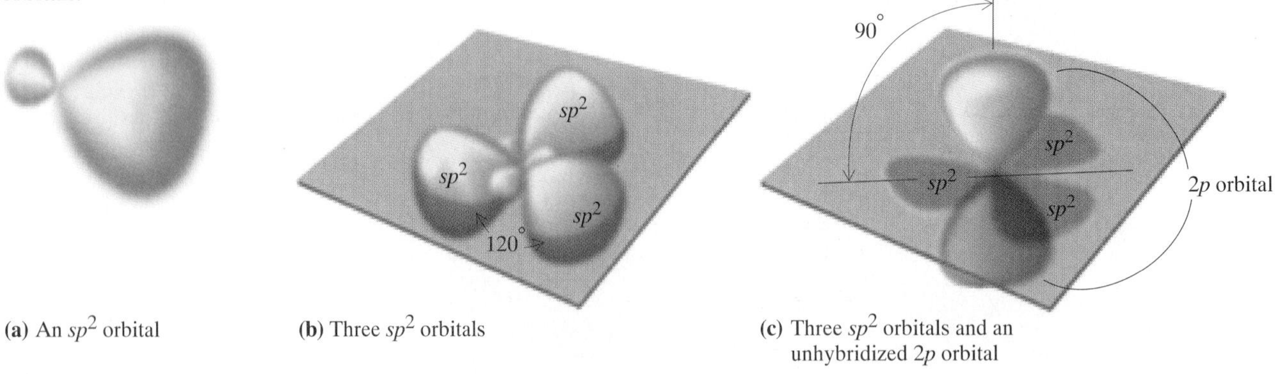

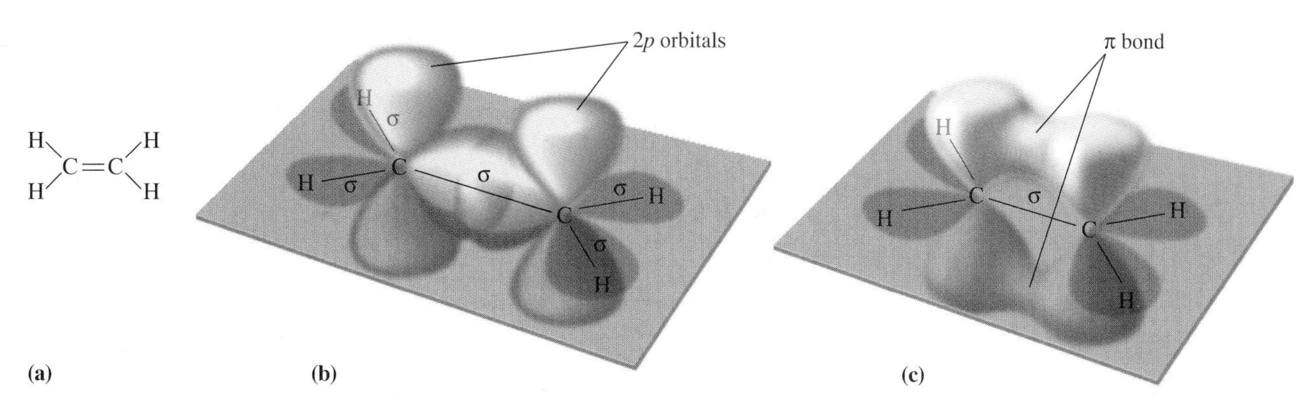

Figure 1.15
Covalent bond formation in ethylene. (a) Lewis structure, (b) overlap of sp^2 hybrid orbitals forms a sigma (σ) bond between the carbon atoms, and (c) overlap of parallel $2p$ orbitals forms a pi bond.

sp^2 hybrid orbitals always occur in sets of three. Each sp^2 hybrid orbital consists of two lobes, one larger than the other. The three sp^2 hybrid orbitals lie in a plane and are directed toward the corners of an equilateral triangle; the angle between sp^2 hybrid orbitals is 120°. The third $2p$ atomic orbital (remember $2p_x$, $2p_y$, $2p_z$) is not involved in hybridization and consists of two lobes lying perpendicular to the plane of the sp^2 hybrid orbitals. Figure 1.14 shows three equivalent sp^2 hybrid orbitals along with the remaining unhybridized $2p$ atomic orbital.

Second-period elements use sp^2 hybrid orbitals to form double bonds. Consider ethylene, C_2H_4, the Lewis structure for which is shown in Figure 1.15(a). A sigma bond between the carbons in ethylene is formed by overlap of sp^2 hybrid orbitals along a common axis as seen in Figure 1.15(b). Each carbon also forms sigma bonds to two hydrogens. The remaining $2p$ orbitals on adjacent carbon atoms lie parallel to each other and overlap to form a pi bond [Figure 1.15(c)]. A **pi (π) bond** is a covalent bond formed by overlap of parallel p orbitals. Because of the lesser degree of overlap of orbitals forming pi bonds compared with those forming sigma bonds, pi bonds are generally weaker than sigma bonds.

Pi (π) bond A covalent bond formed by the overlap of parallel p orbitals.

The valence bond approach describes all double bonds in the same manner as we have already used to describe carbon-carbon double bonds. In formaldehyde, $CH_2{=}O$, the simplest organic molecule containing a carbon-oxygen double bond, carbon forms sigma bonds to two hydrogens by overlap of sp^2 hybrid orbitals of carbon and $1s$ atomic orbitals of hydrogens. Carbon and oxygen are joined by a sigma bond formed by overlap of sp^2 hybrid orbitals and a pi bond formed by overlap of unhybridized $2p$ atomic orbitals (Figure 1.16).

Figure 1.16
A carbon-oxygen double bond. (a) Lewis structure of formaldehyde, $CH_2{=}O$, (b) the sigma (σ) bond framework and nonoverlapping parallel $2p$ atomic orbitals, and (c) overlap of parallel $2p$ atomic orbitals to form a pi (π) bond.

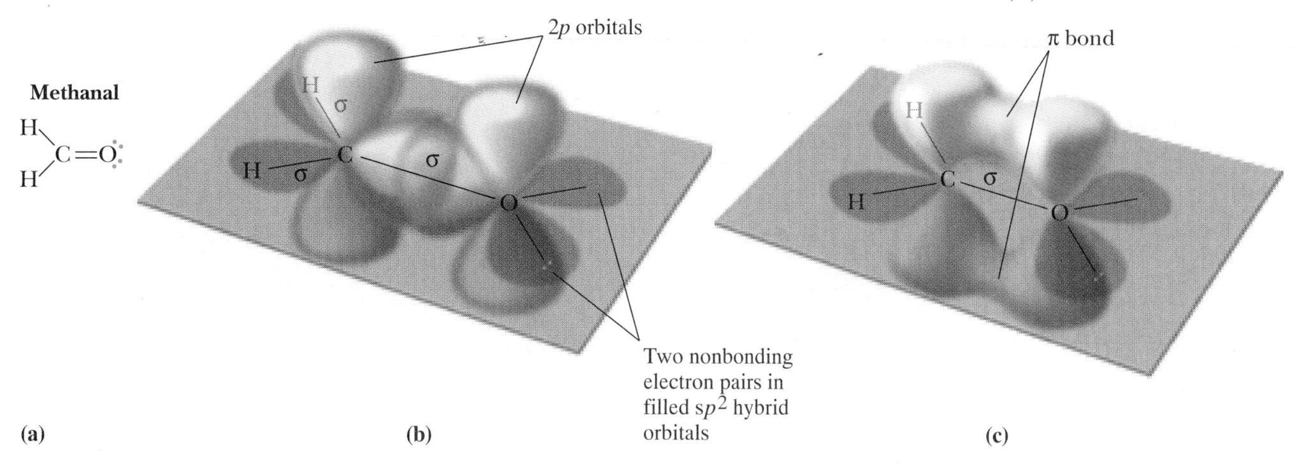

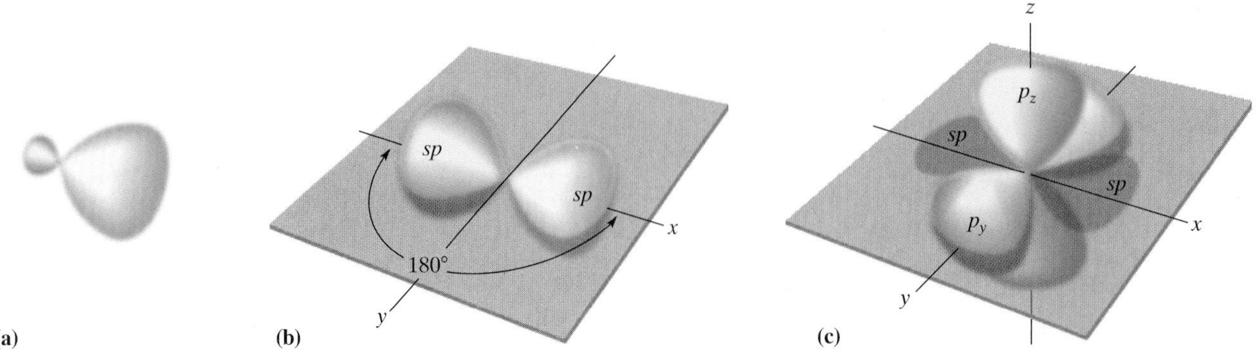

Figure 1.17
sp Hybrid orbitals. (a) A single *sp* hybrid orbital consisting of two lobes of unequal size. (b) Two *sp* hybrid orbitals in a linear arrangement. (c) Unhybridized $2p$ atomic orbitals are perpendicular to the line created by the axes of the two *sp* hybrid orbitals.

E. *sp* Hybrid Orbitals—Bond Angles of Approximately 180°

Combination of one $2s$ atomic orbital and one $2p$ atomic orbital forms two equivalent ***sp* hybrid orbitals.** Because they are derived from two atomic orbitals, *sp* hybrid orbitals always occur in sets of two. The two *sp* hybrid orbitals lie at an angle of 180°. The axes of the unhybridized $2p$ atomic orbitals are perpendicular to each other and to the axis of the two *sp* hybrid orbitals. In Figure 1.17, *sp* hybrid orbitals are shown on the *x* axis, and unhybridized $2p$ orbitals on the *y* axis and *z* axis.

sp Hybrid orbital A hybrid atomic orbital produced by the combination of one *s* atomic orbital and one *p* atomic orbital.

Figure 1.18 shows a Lewis structure and an orbital overlap diagram for acetylene, C_2H_2. A carbon-carbon triple bond consists of one sigma bond formed by overlap of *sp* hybrid orbitals and two pi bonds. One pi bond is formed by overlap of a pair of parallel $2p$ atomic orbitals. The second pi bond is formed by overlap of a second pair of parallel $2p$ atomic orbitals. The relationship among the number of groups bonded to carbon, orbital hybridization, and types of bonds involved is summarized in Table 1.9.

Example 1.12

Describe the bonding in acetic acid, CH_3CO_2H, in terms of the orbitals involved, and predict all bond angles.

Figure 1.18
Covalent bonding in acetylene. (a) The sigma bond framework shown along with nonoverlapping $2p$ atomic orbitals. (b) Formation of two pi bonds by overlap of two sets of parallel $2p$ atomic orbitals.

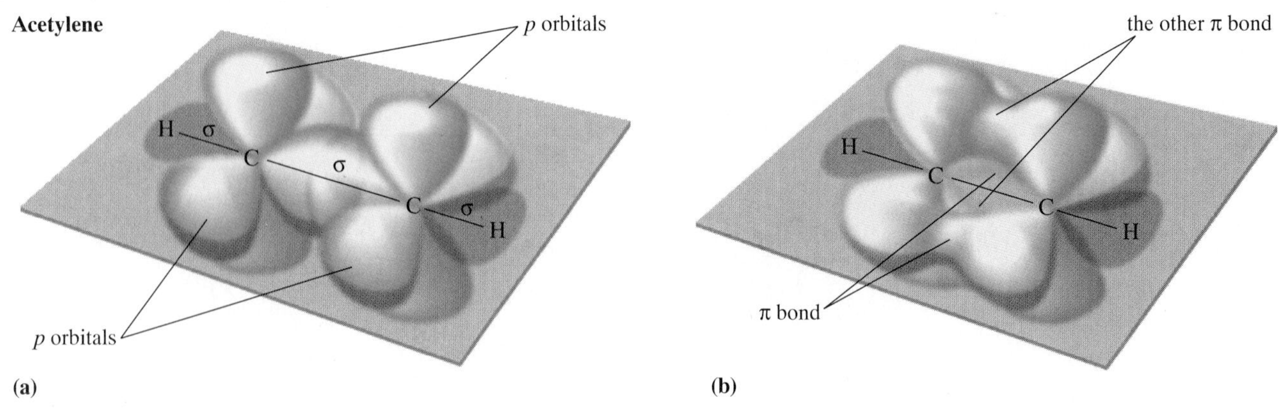

TABLE 1.9 Covalent Bonding of Carbon

Groups Bonded to Carbon	Orbital Hybridization	Predicted Bond Angles	Types of Bonds to Carbon	Example	Name
4	sp^3	109.5°	four sigma bonds	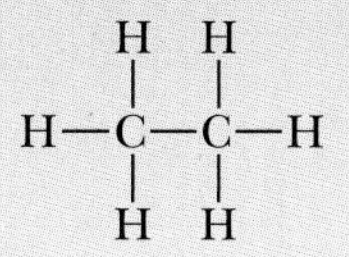 $H_3C{-}CH_3$	ethane
3	sp^2	120°	three sigma bonds and one pi bond	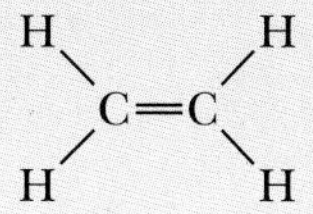 $H_2C{=}CH_2$	ethylene
2	sp	180°	two sigma bonds and two pi bonds	$H{-}C{\equiv}C{-}H$	acetylene

SOLUTION

Following are three identical Lewis structures. Labels on the first structure point to atoms and show hybridization. Labels on the second structure point to bonds and show the type of bond, either sigma or pi. Labels on the third structure point to atoms and show bond angles about each atom as predicted by the valence-shell electron-pair repulsion model.

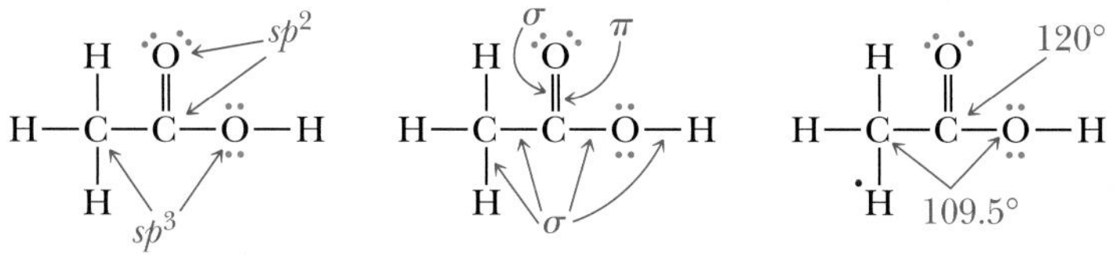

Practice Problem 1.12

Describe the bonding in these molecules in terms of the orbitals involved, and predict all bond angles.

(a) $CH_3CH{=}CH_2$ (b) CH_3NH_2

1.7 FUNCTIONAL GROUPS

Carbon combines with other atoms (e.g., C, H, N, O, S, halogens) to form structural units called **functional groups.** Functional groups are important for three reasons. First, they are the units by which we divide organic compounds into classes. Second, they are sites of chemical reaction; a particular functional group, in whatever compound it is found, undergoes the same types of chemical reactions. Third, functional groups serve as a basis for naming organic compounds.

Functional group An atom or group of atoms within a molecule that shows a characteristic set of physical and chemical properties.

Introduced here are several functional groups we encounter early in the course. At this point, our concern is nothing more than pattern recognition. We have more to say about the structure and the physical and chemical properties of these functional groups in following chapters. A complete list of the major functional groups we study in this text is presented at the front of the text.

A. Alcohols

Hydroxyl group An —OH group.

The functional group of an **alcohol** is an **—OH (hydroxyl)** group bonded to a tetrahedral (sp^3 hybridized) carbon atom.

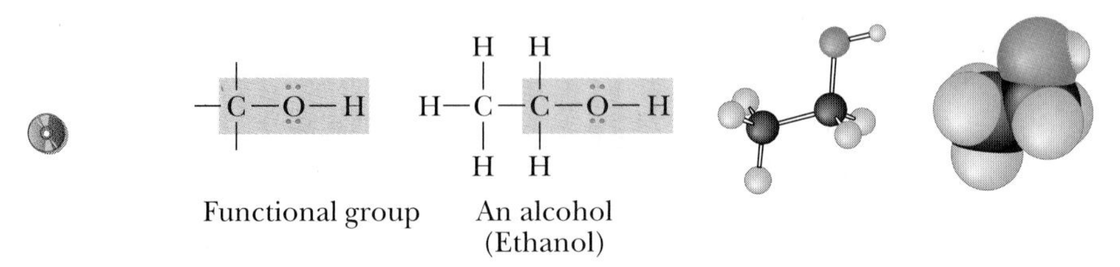

Functional group An alcohol (Ethanol)

We can write formulas for this alcohol in a more abbreviated form using what is called a **condensed structural formula.** In a condensed structural formula, CH_3 indicates a carbon with three attached hydrogens, CH_2 indicates a carbon with two attached hydrogens, and CH indicates a carbon with one attached hydrogen. Unshared pairs of electrons are generally not shown in a condensed structural formula. Following is a condensed structural formula for the alcohol of molecular formula C_2H_6O. It is also common to write these formulas in an even more condensed manner, by omitting all single bonds in the linear drawing.

$$CH_3—CH_2—OH \quad \text{or} \quad CH_3CH_2OH$$

Example 1.13

Write condensed structural formulas for the two alcohols of molecular formula C_3H_8O.

Solution

Bond the three carbon atoms in a chain with the —OH (hydroxyl) group attached to the end carbon of the chain or attached to the middle carbon of the chain. Then, to complete each structural formula, add seven hydrogens so that each carbon has four bonds to it.

H—C—C—C—O—H (each carbon bearing two H) or $CH_3CH_2CH_2OH$

1-Propanol

H—C—C—C—H (middle carbon bearing O—H; other positions H) or $CH_3CH(OH)CH_3$ (OH above the central CH)

2-Propanol

Practice Problem 1.13

Write condensed structural formulas for the four alcohols of molecular formula $C_4H_{10}O$.

B. Aldehydes and Ketones

Carbonyl group A C═O group.

Both aldehydes and ketones contain a **C═O (carbonyl)** group. The functional group of an **aldehyde** is a carbonyl group bonded through its carbon to two hydrogens in the case of formaldehyde, CH_2O, the simplest aldehyde, and to another carbon and a hydrogen in all other aldehydes. The functional group of a **ketone** is a carbonyl group bonded to two carbon atoms.

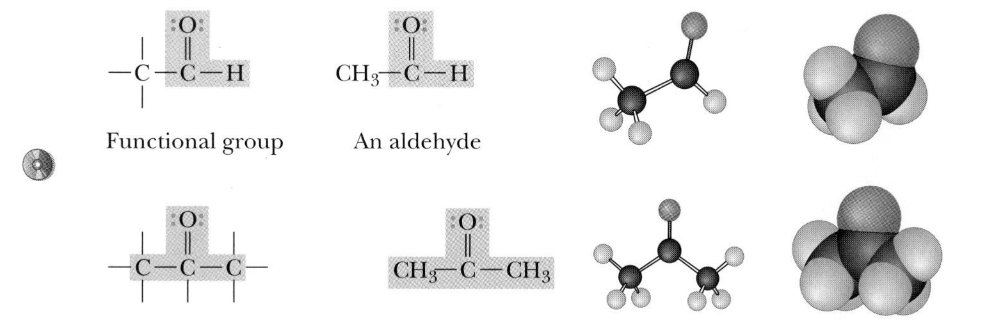

EXAMPLE 1.14

Write condensed structural formulas for the two aldehydes of molecular formula C_4H_8O.

SOLUTION

First draw the functional group of an aldehyde and then add the remaining carbons. These may be attached in two different ways. Then, add seven hydrogens to complete the four bonds of each carbon, and give the correct molecular formula. The aldehyde group may be written showing the carbon-oxygen double bond as C═O, or, alternatively, it may be written —CHO.

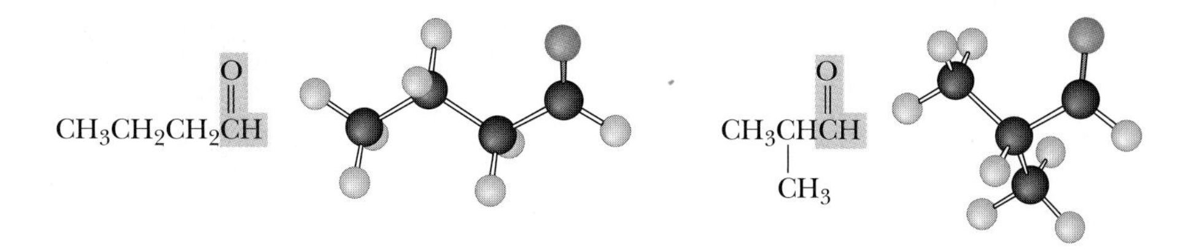

Practice Problem 1.14

Draw condensed structural formulas for the three ketones of molecular formula $C_5H_{10}O$.

C. Carboxylic Acids

Carboxyl group A —CO_2H group.

The functional group of a carboxylic acid is a **—CO_2H (carboxyl:** *carb*onyl + hydr*oxyl*) group.

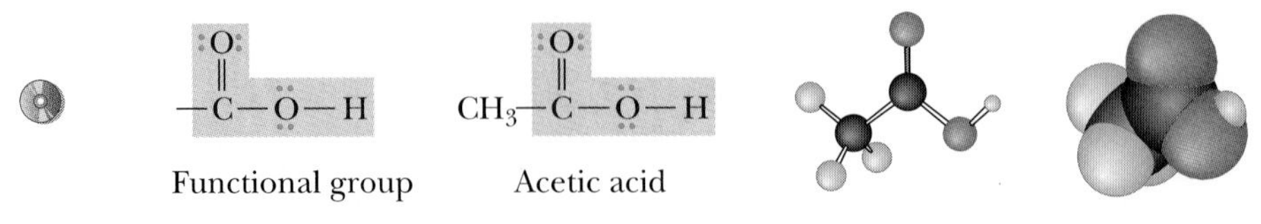

EXAMPLE 1.15

Draw a condensed structural formula for the single carboxylic acid of molecular formula $C_3H_6O_2$.

SOLUTION

$CH_3—CH_2—C(=O)—O—H$ or $CH_3CH_2CO_2H$

Practice Problem 1.15

Draw condensed structural formulas for the two carboxylic acids of molecular formula $C_4H_8O_2$.

D. Amines

Amino group An sp^3 hybridized nitrogen atom bonded to one, two, or three carbon groups.

The functional group of an amine is the **amino group,** an sp^3 hybridized nitrogen atom bonded to one, two, or three carbon groups. In a primary (1°) amine, one carbon group is bonded to nitrogen. In a secondary (2°) amine, two carbon groups are bonded to nitrogen; and in a tertiary (3°) amine, three carbon groups are bonded to nitrogen.

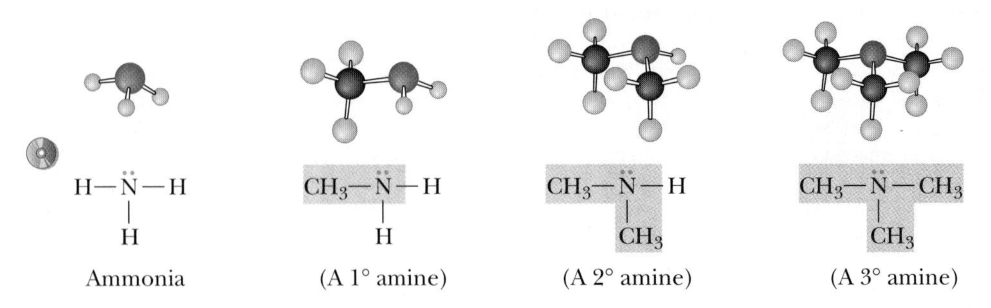

EXAMPLE 1.16

Draw structural formulas for the two primary (1°) amines of molecular formula C_3H_9N.

SOLUTION

For a primary amine, draw a nitrogen atom bonded to two hydrogens and one carbon. The carbons may be attached to each other in two different ways. Then add the seven hydrogens to give each carbon four bonds and give the correct molecular formula.

$$C{-}C{-}C{-}\ddot{N}{-}H \quad C{-}\underset{}{\overset{C}{\overset{|}{C}}}{-}\ddot{N}{-}H \longrightarrow CH_3CH_2CH_2{-}\ddot{N}{-}H \quad CH_3\overset{CH_3}{\overset{|}{C}}H{-}\ddot{N}{-}H$$

(each N also bears an H below)

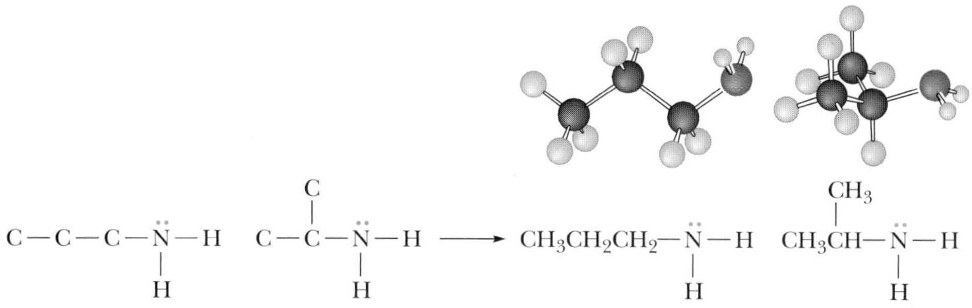

The three carbons may be attached to nitrogen in two ways

Add seven hydrogens to give each carbon four bonds and give the correct molecular formula

Practice Problem 1.16

Draw structural formulas for the three secondary amines of molecular formula $C_4H_{11}N$.

SUMMARY

Atoms consist of a small, dense nucleus and electrons concentrated about the nucleus in regions of space called **shells** (Section 1.1). Each shell can contain as many as $2n^2$ electrons, where n is the number of the shell. Each shell is subdivided into regions of space called **orbitals.** The first shell ($n = 1$) has a single s orbital and can hold $2 \times 1^2 = 2$ electrons. The second shell ($n = 2$) has one s orbital and three p orbitals and can hold $2 \times 2^2 = 8$ electrons. The **Lewis structure** (Section 1.1) of an element shows the symbol of the element surrounded by a number of dots equal to the number of electrons in its **valence shell.** According to the **Lewis model of bonding** (Section 1.2), atoms bond together in such a way that each atom participating in a chemical bond acquires a completed valence-shell electron configuration resembling that of the noble gas nearest it in atomic number. Atoms that lose sufficient electrons to acquire a completed valence shell become **cations;** those that gain sufficient electrons to acquire a completed valence shell become **anions.** An **ionic bond** is a chemical bond formed by the attractive force between an anion and a cation. A **covalent bond** is a chemical bond formed by the sharing of electron pairs between atoms. The tendency of main-group elements (those of Groups IA–VIIA) to achieve an outer shell of eight valence electrons is called the **octet rule.**

Electronegativity (Section 1.2C) is a measure of the force of attraction by an atom for electrons it shares in a chemical bond with another atom. Electronegativity increases from left to right and from bottom to top in the Periodic Table. A **nonpolar covalent bond** (Section 1.2C) is a covalent bond in which the difference in electronegativity between the bonded atoms is less than 0.5 units. A **polar covalent bond** is a covalent bond in which the difference in electronegativity between the bonded atoms is between 0.5 and 1.9 units. In a polar covalent bond, the more electronegative atom bears a partial negative charge ($\delta-$), and the less electronegative atom bears a partial positive charge ($\delta+$).

An acceptable **Lewis structure** (Section 1.2D) for a molecule or an ion must show: (1) the correct arrangement of atoms, (2) the correct number of valence electrons, (3) no more than two electrons in the outer shell of hydrogen, (4) no more than eight electrons in the outer shell of any second-period element, and (5) all formal charges. **Formal charge** is the charge on an atom on a molecule or polyatomic ion (Section 1.2E).

Bond angles of molecules and polyatomic ions can be predicted using Lewis structures and the **valence-shell electron-pair repulsion (VSEPR) model** (Section 1.3). For atoms surrounded by four regions of electron density, predict bond angles of 109.5°; by three regions of electron density, predict bond angles of 120°; and by two regions of electron density, predict bond angles of 180°.

A molecule is polar if it has one or more polar bonds and its centers of partial positive and negative charge are at different places within the molecule (Section 1.4).

According to the **theory of resonance** (Section 1.5A), molecules and ions for which no single Lewis structure is adequate are best described by writing two or more **contributing structures** and considering the real molecule or ion to be a resonance hybrid of the various contributing structures. Contributing structures to the resonance hybrid are interconnected by **double-headed arrows.** The manner in which valence electrons are redistributed from one contributing structure to the next is shown by **curved arrows** (Section 1.5B) that extend from where the electrons are initially shown (on an atom or in a covalent bond) to their new location (to an adjacent atom or an adjacent covalent bond). Use of curved arrows in this way is commonly referred to as **electron pushing.**

According to the **valence bond model,** formation of a covalent bond results from overlap of atomic orbitals (Section 1.6A). The greater the overlap, the stronger the resulting covalent bond. The combination of atomic orbitals is called **hybridization** (Section 1.6B), and the resulting orbitals are called **hybrid orbitals.** The combination of one $2s$ atomic orbital and three $2p$ atomic orbitals produces four equivalent **sp^3 hybrid orbitals,** each directed toward a corner of a regular tetrahedron at angles of 109.5°.

The combination of one $2s$ atomic orbital and two $2p$ atomic orbitals produces three equivalent **sp^2 hybrid orbitals,** the axes of which lie in a plane at angles of 120°. Most C=C and C=O double bonds are a combination of one **sigma (σ) bond** formed by the overlap of sp^2 hybrid orbitals and one **pi (π) bond** formed by overlap of parallel $2p$ atomic orbitals.

The combination of one $2s$ atomic orbital and one $2p$ atomic orbital produces two equivalent **sp hybrid orbitals,** the axes of which lie in a plane at an angle of 180°. All C≡C triple bonds are a combination of one sigma bond formed by the overlap of sp hybrid orbitals and two pi bonds formed by the overlap of two pairs of parallel $2p$ atomic orbitals.

Functional groups (Section 1.7) are characteristic structural units by which we divide organic compounds into classes and which serve as a basis for nomenclature. They are also sites of chemical reactivity; a particular functional group, in whatever compound it is found, undergoes the same types of reactions. Important functional groups for us at this stage in the course are the **hydroxyl group** of alcohols, the **carbonyl group** of aldehydes and ketones, the **carboxyl group** of carboxylic acids, and the **amino group** of 1°, 2°, and 3° amines.

ADDITIONAL PROBLEMS

Electronic Structure of Atoms

1.17 Write the ground-state electron configuration for each atom. After each is given its atomic number.

(a) Sodium (11) **(b)** Magnesium (12) **(c)** Oxygen (8) **(d)** Nitrogen (7)

1.18 Write the ground-state electron configuration for each atom.

(a) Potassium **(b)** Aluminum **(c)** Phosphorus **(d)** Argon

1.19 Which atom has the ground-state electron configuration of

(a) $1s^22s^22p^63s^23p^4$ **(b)** $1s^22s^22p^4$

1.20 Which element or ion does not have the ground-state electron configuration $1s^22s^22p^63s^23p^6$?

(a) S^{2-} **(b)** Cl^- **(c)** Ar **(d)** Ca^{2+} **(e)** K

1.21 Define valence shell and valence electron.

1.22 How many electrons are in the valence shell of each atom?

(a) Carbon **(b)** Nitrogen **(c)** Chlorine **(d)** Aluminum **(e)** Oxygen

Lewis Structures

1.23 Judging from their relative positions in the Periodic Table, which atom in each set is more electronegative?

(a) Carbon or nitrogen **(b)** Chlorine or bromine **(c)** Oxygen or sulfur

1.24 Which compounds have nonpolar covalent bonds, which have polar covalent bonds, and which have ionic bonds?

(a) LiF **(b)** CH_3F **(c)** $MgCl_2$ **(d)** HCl

1.25 Using the symbols $\delta-$ and $\delta+$, indicate the direction of polarity, if any, in each covalent bond.

(a) C—Cl **(b)** S—H **(c)** C—S **(d)** P—H

1.26 Write Lewis structures for these compounds. Show all valence electrons. None of them contains a ring of atoms.

(a) H_2O_2 Hydrogen peroxide	**(b)** N_2H_4 Hydrazine	**(c)** CH_3OH Methanol
(d) CH_3SH Methanethiol	**(e)** CH_3NH_2 Methylamine	**(f)** CH_3Cl Chloromethane
(g) CH_3OCH_3 Dimethyl ether	**(h)** C_2H_6 Ethane	**(i)** C_2H_4 Ethylene
(j) C_2H_2 Acetylene	**(k)** CO_2 Carbon dioxide	**(l)** CH_2O Formaldehyde
(m) CH_3COCH_3 Acetone	**(n)** H_2CO_3 Carbonic acid	**(o)** CH_3CO_2H Acetic acid

1.27 Write Lewis structures for these ions.

(a) HCO_3^- Bicarbonate ion	**(b)** CO_3^{2-} Carbonate ion	**(c)** $CH_3CO_2^-$ Acetate ion	**(d)** Cl^- Chloride ion

1.28 Why are the following molecular formulas impossible?

(a) CH_5 **(b)** C_2H_7

1.29 Following the rule that each atom of carbon, oxygen, and nitrogen reacts to achieve a complete outer shell of eight valence electrons, add unshared pairs of electrons as necessary to complete the valence shell of each atom in these ions. Then, assign formal charges as appropriate.

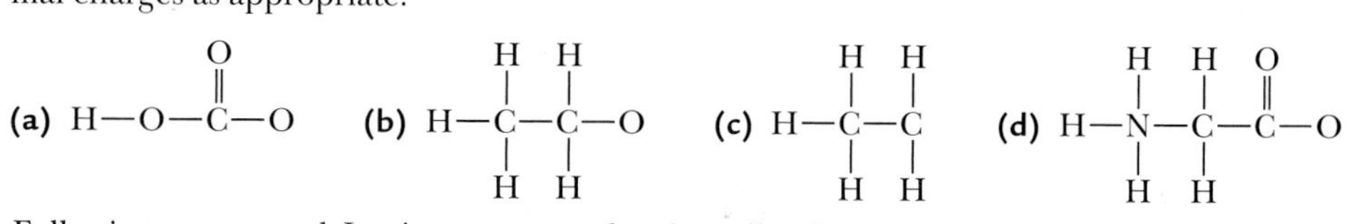
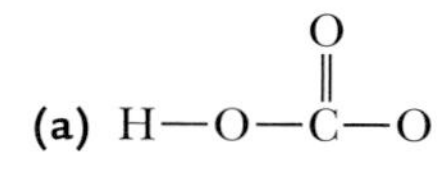
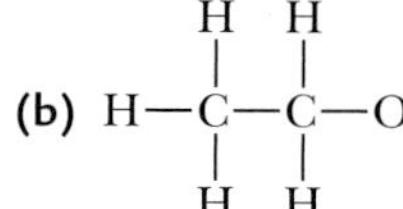
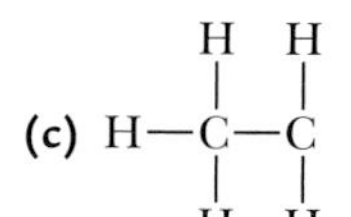

1.30 Following are several Lewis structures showing all valence electrons. Assign formal charges in each structure as appropriate.

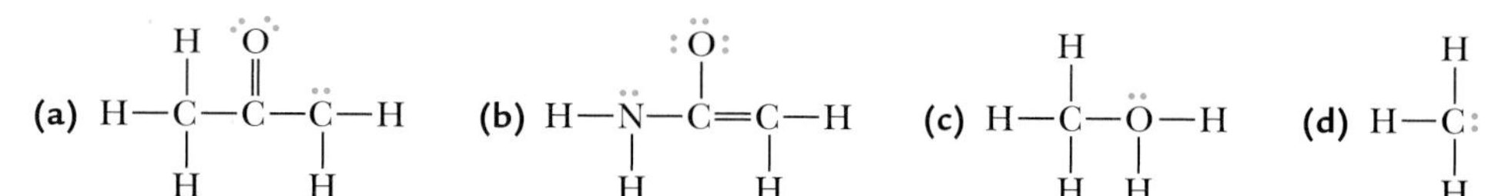
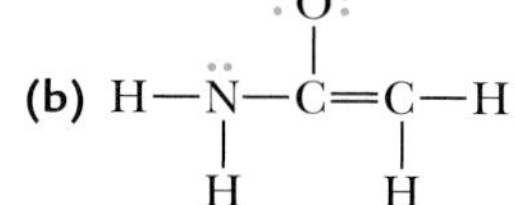
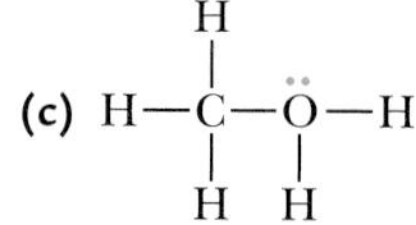
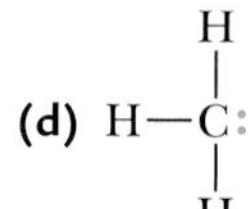

1.31 Each compound contains both ionic and covalent bonds. Draw a Lewis structure for each and show by charges which bonds are ionic and by dashes which bonds are covalent.

(a) NaOH **(b)** $NaHCO_3$ **(c)** NH_4Cl **(d)** CH_3CO_2Na **(e)** CH_3ONa

Polarity of Covalent Bonds

1.32 Which statement is true about electronegativity?

(a) Electronegativity increases from left to right in a period of the Periodic Table.

(b) Electronegativity increases from top to bottom in a column of the Periodic Table.

(c) Hydrogen, the element with the lowest atomic number, has the smallest electronegativity.

(d) The higher the atomic number of an element, the greater its electronegativity.

1.33 Why does fluorine, the element in the upper right corner of the Periodic Table, have the largest electronegativity of any element?

1.34 Arrange the single covalent bonds within each set in order of increasing polarity.

(a) C—H, O—H, N—H **(b)** C—H, C—Cl, C—I

(c) C—S, C—O, C—N **(d)** C—Li, C—Hg, C—Mg

1.35 Using the values of electronegativity given in Table 1.5, predict which indicated bond in each set is more polar and, using the symbols $\delta+$ and $\delta-$, show the direction of its polarity.

(a) CH_3—OH or CH_3O—H **(b)** H—NH_2 or CH_3—NH_2

(c) CH_3—SH or CH_3S—H **(d)** CH_3—F or H—F

1.36 Identify the most polar bond in each molecule.

(a) $HSCH_2CH_2OH$ **(b)** $CHCl_2F$ **(c)** $HOCH_2CH_2NH_2$

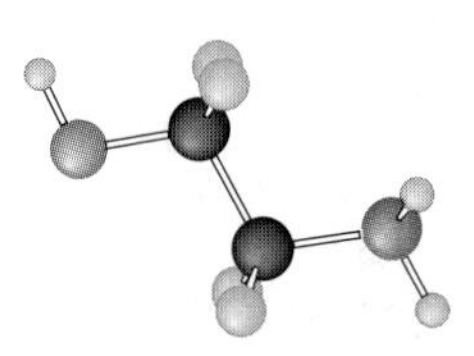

1.37 Predict whether the carbon-metal bond in these organometallic compounds is nonpolar covalent, polar covalent, or ionic. For each polar covalent bond, show its direction of polarity using the symbols $\delta+$ and $\delta-$.

(a) CH_3CH_2—Pb(CH_2CH_3)(CH_2CH_3)—CH_2CH_3

Tetraethyllead

(b) CH_3—Mg—Cl

Methylmagnesium chloride

(c) CH_3—Hg—CH_3

Dimethylmercury

Bond Angles and Shapes of Molecules

1.38 Use the VSEPR model to predict bond angles about each highlighted atom.

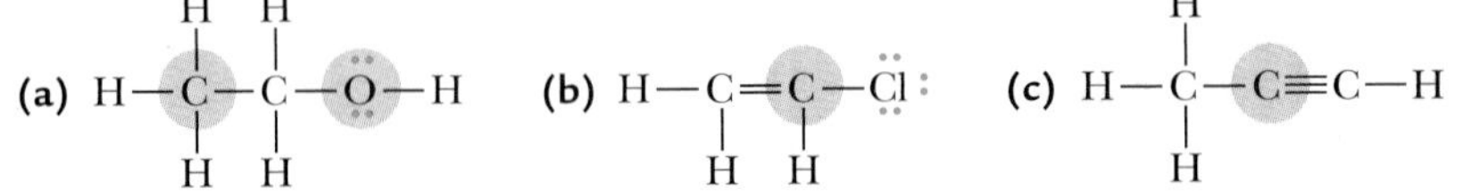

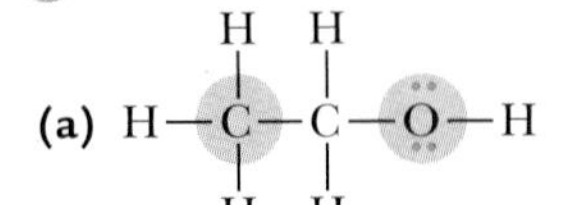

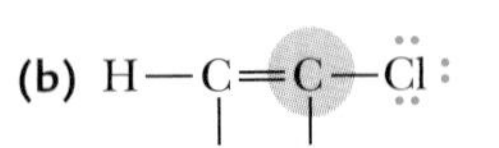

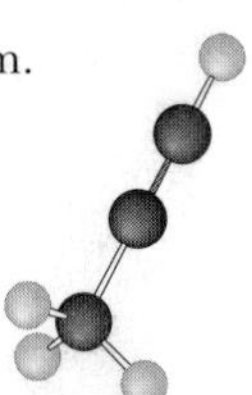

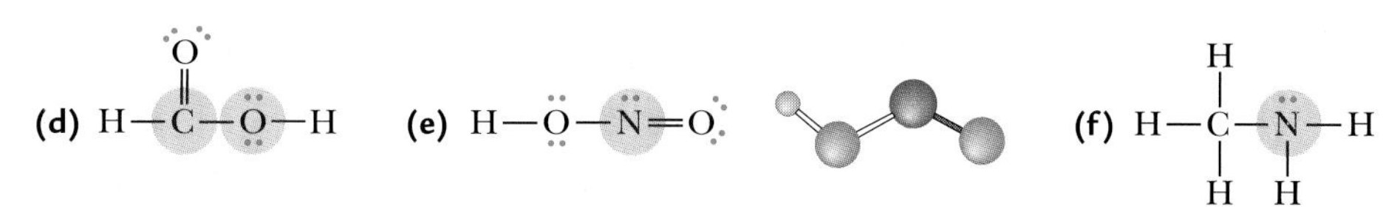

1.39 Use the VSEPR model to predict bond angles about each atom of carbon, nitrogen, and oxygen in these molecules. *Hint:* First add unshared pairs of electrons as necessary to complete the valence shell of each atom and then make your predictions of bond angles.

(a) $CH_3-CH_2-CH_2-OH$ **(b)** $CH_3-CH_2-C(=O)-H$ **(c)** $CH_3-CH=CH_2$

(d) $CH_3-C\equiv C-CH_3$ **(e)** $CH_3-C(=O)-O-CH_3$ **(f)** $CH_3-N(CH_3)-CH_3$

1.40 Silicon is immediately below carbon in the Periodic Table. Predict the C—Si—C bond angle in tetramethylsilane, $(CH_3)_4Si$.

Polar and Nonpolar Molecules

1.41 Draw a three-dimensional representation for each molecule. Indicate which molecules are polar and the direction of its polarity.

(a) CH_3F **(b)** CH_2Cl_2 **(c)** $CHCl_3$ **(d)** CCl_4

(e) $CH_2=CCl_2$ **(f)** $CH_2=CHCl$ **(g)** $CH_3C\equiv N$ **(h)** $(CH_3)_2C=O$

1.42 Tetrafluoroethylene, C_2F_4, is the starting material for the synthesis of the polymer poly(tetrafluoroethylene), commonly known as Teflon. Molecules of tetrafluoroethylene are nonpolar. Propose a structural formula for this compound.

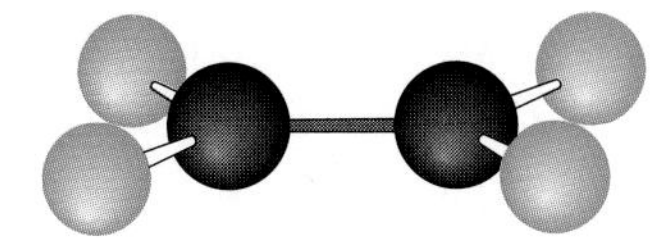

Tetrafluoroethylene

1.43 The two chlorofluorocarbons (CFCs) most widely used as heat transfer media for refrigeration systems were trichlorofluoromethane, CCl_3F, and dichlorodifluoromethane, CCl_2F_2. Draw a three-dimensional representation of each molecule, and indicate the direction of its polarity.

Resonance and Contributing Structures

1.44 Which of these statements are true about resonance contributing structures?

(a) All contributing structures must have the same number of valence electrons.

(b) All contributing structures must have the same arrangement of atoms.

(c) All atoms in a contributing structure must have complete valence shells.

(d) All bond angles in sets of contributing structures must be the same.

1.45 Draw the contributing structure indicated by the curved arrow(s) and assign formal charges as appropriate.

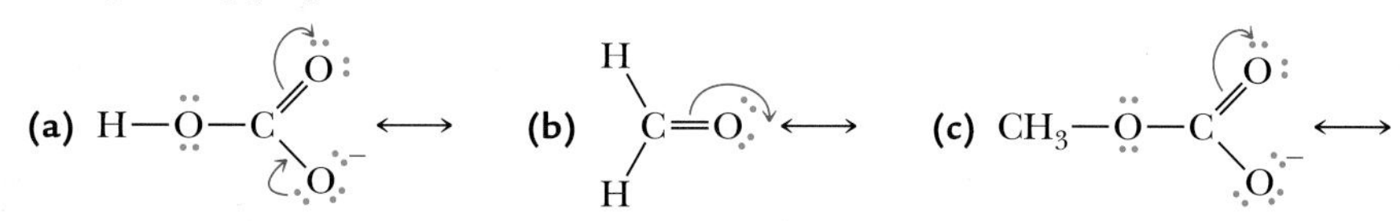

1.46 Using the VSEPR model, predict the bond angles about the carbon atom in each pair of contributing structures in Problem 1.45. In what way do the bond angles change from one contributing structure to the other?

Valence Bond Model

1.47 State the orbital hybridization of each highlighted atom.

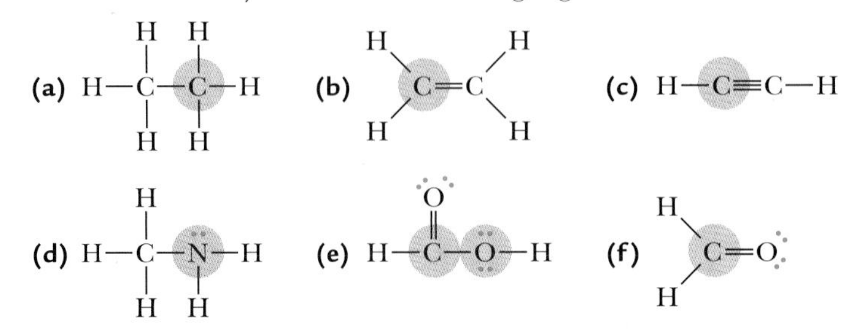

1.48 Describe each highlighted bond in terms of the overlap of atomic orbitals.

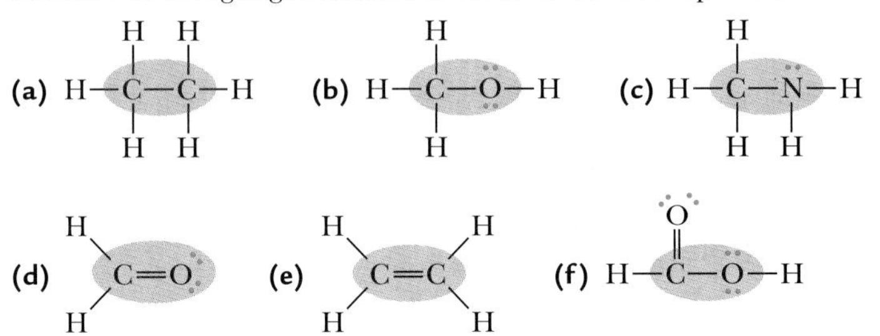

1.49 Following is a structural formula and a space filling model of benzene, C_6H_6.

(a) Predict each H—C—C and each C—C—C bond angle on benzene.

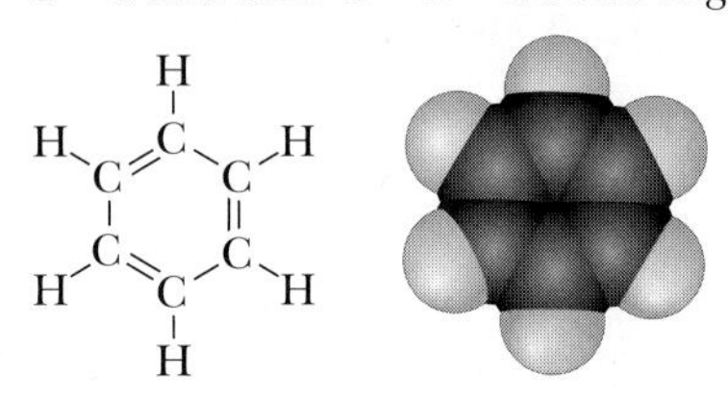

(b) State the hybridization of each carbon in benzene.

(c) Predict the shape of the benzene molecule.

1.50 In Chapter 6, we study a group of organic cations called carbocations. Following is the structure of one such carbocation, the *tert*-butyl cation.

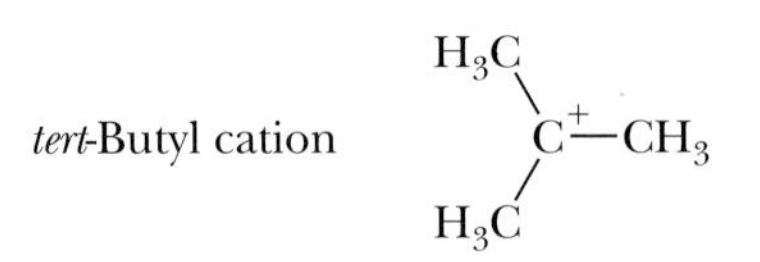

(a) How many electrons are in the valence shell of the carbon bearing the positive charge?

(b) Predict the bond angles about this carbon.

(c) Given the bond angles you predicted in (b), what hybridization do you predict for this carbon?

1.51 We also study the isopropyl cation, $(CH_3)_2CH^+$.

(a) Write a Lewis structure for this cation. Use a plus sign to show the location of the positive charge.

(b) How many electrons are in the valence shell of the carbon bearing the positive charge?

(c) Use the VSEPR model to predict all bond angles about the carbon bearing the positive charge.

(d) Describe the hybridization of each carbon in this cation.

Functional Groups

1.52 Draw Lewis structures for these functional groups. Be certain to show all valence electrons on each.

(a) Carbonyl group **(b)** Carboxyl group

(c) Hydroxyl group **(d)** Primary amino group

1.53 Draw the structure for a compound of molecular formula

(a) C_2H_6O that is an alcohol.

(b) C_3H_6O that is an aldehyde.

(c) C_3H_6O that is a ketone.

(d) $C_3H_6O_2$ that is a carboxylic acid.

(e) $C_4H_{11}N$ that is a tertiary amine.

1.54 Draw condensed structural formulas for all compounds of molecular formula C_4H_8O that contain

(a) A carbonyl group (there are two aldehydes and one ketone).

(b) A carbon-carbon double bond and a hydroxyl group (there are eight).

1.55 Draw structural formulas for

(a) The eight alcohols of molecular formula $C_5H_{12}O$.

(b) The eight aldehydes of molecular formula $C_6H_{12}O$.

(c) The six ketones of molecular formula $C_6H_{12}O$.

(d) The eight carboxylic acids of molecular formula $C_6H_{12}O_2$.

(e) The three tertiary amines of molecular formula $C_5H_{13}N$.

1.56 Identify the functional groups in each compound. We study each compound in more detail in the indicated section.

(a) $CH_3-CH(OH)-C(=O)-OH$

Lactic acid
(Section 20.7A)

(b) $HO-CH_2-CH_2-OH$

Ethylene glycol
(Section 8.2A)

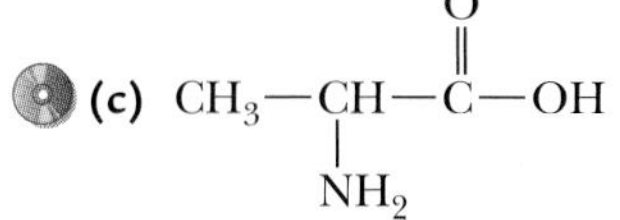

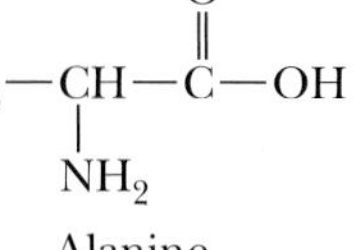

(c) $CH_3-CH(NH_2)-C(=O)-OH$

Alanine
(Section 18.1)

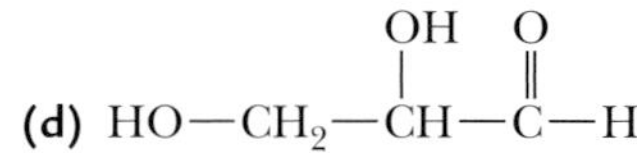

(d) $HO-CH_2-CH(OH)-C(=O)-H$

Glyceraldehyde
(Section 16.1)

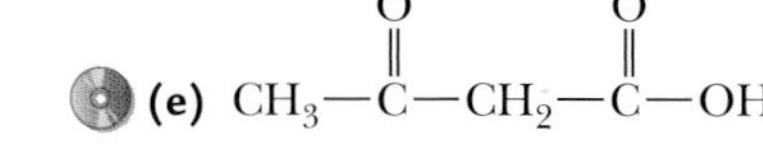

(e) $CH_3-C(=O)-CH_2-C(=O)-OH$

Acetoacetic acid
(Section 12.8A)

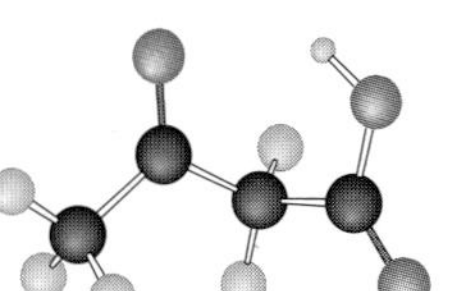

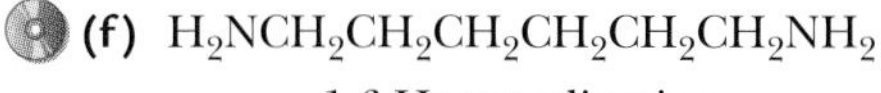

(f) $H_2NCH_2CH_2CH_2CH_2CH_2CH_2NH_2$

1,6-Hexanediamine
(Section 15.4A)

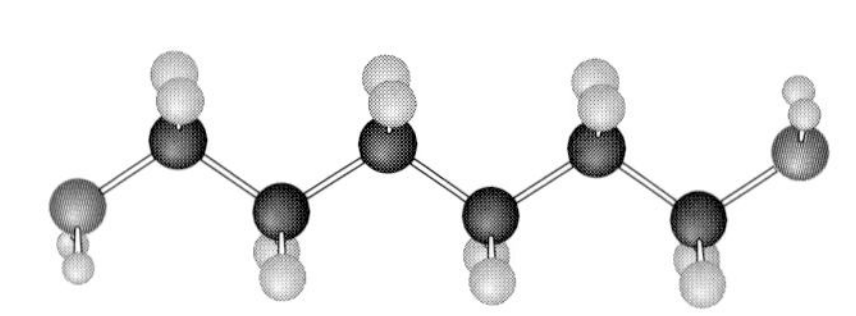

INTERACTIVE QUESTIONS

Refer to the CD accompanying the text.

1.57 Measure the carbon-carbon and carbon-oxygen bond lengths in these molecules and record your results in the appropriate table. What is the correlation between bond length and type of bond (single, double, triple)?

Compound	Name	Measured C—C Bond Length (Å)
Ethane	$CH_3{-}CH_3$	______
Ethylene	$CH_2{=}CH_2$	______
Acetylene	$HC{\equiv}CH$	______

Compound	Name	Measured C—O Bond Length (Å)
Methanol	CH_3OH	______
Formaldehyde	CH_2O	______

1.58 Measure all bond angles in each molecule and compare your results with the values predicted by the valence-shell electron-pair repulsion model.

(a) CH_3CH_3 **(b)** $CH_2{=}CH_2$ **(c)** CH_3OH **(d)** CH_3OCH_3
(e) $CH_2{=}O$ **(f)** $(CH_3)_2C{=}O$ **(g)** CH_3NH_2 **(h)** $CH_2{=}NH$
(i) H_2CO_3 **(j)** HCO_2H **(k)** CH_3CO_2H

1.59 Measure the carbon-oxygen bond lengths in acetic acid, CH_3CO_2H, and the acetate ion, $CH_3CO_2^-$. How do you account for any differences between them?

1.60 Identify the functional group(s) in each molecule. Do not be concerned with names; we discuss how to name organic compounds in subsequent chapters.

(a) 1-Butanol **(b)** 2-Butanamine **(c)** 2-Butanone
(d) 2-Hydroxybutanoic acid **(e)** Acetoacetic acid **(f)** Acrylic acid
(g) Alanine **(h)** Allyl alcohol **(i)** Dihydroxyacetone
(j) Ethanolamine **(k)** Propanal **(l)** Isoprene

2

Acids and Bases

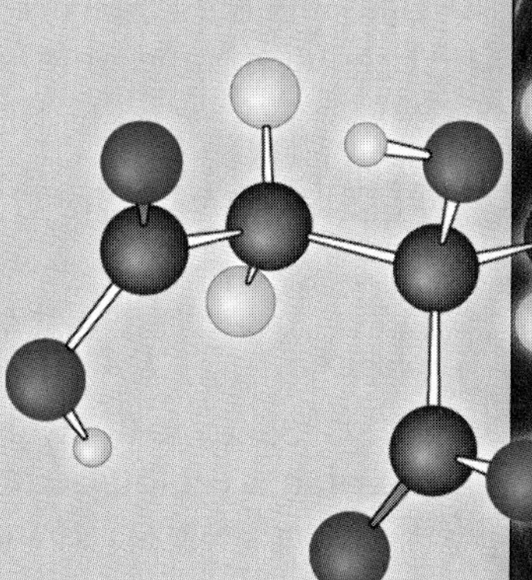

Citrus fruits are sources of citric acid. Lemon juice, for example, contains 5–8% citric acid. Inset: A model of citric acid. *(Charles D. Winters)*

A great many organic reactions are acid-base reactions. In this and later chapters, we study the acid-base properties of the major classes of organic compounds including alcohols, phenols, carboxylic acids, carbonyl compounds containing α-hydrogens, amines, amino acids and proteins, and finally nucleic acids. Furthermore, many organic reactions are catalyzed by proton-donating acids, such as H_3O^+. Others are catalyzed by Lewis acids, such as $AlCl_3$. It is essential, therefore, that you have a good grasp of the fundamentals of acid-base chemistry.

The first useful definition of acids and bases was put forth by Svante Arrhenius in 1884. According to the Arrhenius definition, an acid is a substance that dissolves in water to increase the concentration of H^+. A base is a substance that dissolves in water to increase the concentration of OH^-. The Arrhenius concept of acids and bases is so intimately tied to reactions that take place in water that it has no good way to deal with acid-base reactions in nonaqueous solutions. For this reason, we concentrate in this chapter on the Brønsted-Lowry definitions of acids and bases, which are more useful to us in our discussion of reactions of organic compounds.

2.1 Brønsted-Lowry Acids and Bases

Brønsted-Lowry acid A proton donor.

Brønsted-Lowry base A proton acceptor.

In 1923 the Danish chemist Johannes Brønsted and the English chemist Thomas Lowry independently proposed the following definitions: an acid is a **proton donor,** and a base is a **proton acceptor.** In the neutralization reaction, for example, between H_3O^+ and OH^-, a proton is transferred from H_3O^+, a Brønsted acid, to OH^-, a Brønsted base. We use curved arrows to show the flow of electrons in acid-base reactions. The curved arrow on the right in the following equation shows an unshared pair of electrons on oxygen forming a new bond with hydrogen; in donating electrons, this oxygen becomes neutral. The curved arrow on the left shows the electron pair from the O—H bond moving onto oxygen; this oxygen gains electrons and becomes neutral.

$$\underset{\text{Proton donor}}{H_2\ddot{O}^+{-}H} + \underset{\text{Proton acceptor}}{:\ddot{O}{-}H^-} \longrightarrow H_2\ddot{O}: + H{-}\ddot{O}{-}H$$

The acid-base reaction between H_3O^+ and NH_3 involves proton transfer from H_3O^+, a proton donor, to NH_3, a proton acceptor. In this reaction, the oxygen becomes neutral, and the nitrogen becomes positively charged.

$$\underset{\text{Proton donor}}{H_2\ddot{O}^+{-}H} + \underset{\text{Proton acceptor}}{:NH_3} \longrightarrow H_2\ddot{O}: + H{-}N^+H_3$$

Conjugate base The species formed when an acid donates a proton.

Conjugate acid The species formed when a base accepts a proton.

The reciprocal relationships in proton-transfer reactions are described by the following terminology. When an acid transfers a proton to a base, the acid is converted to its **conjugate base;** when a base accepts a proton, the base is converted to its **conjugate acid.** For example, in the reaction between hydronium ion and ammonia, H_3O^+ is converted into its conjugate base, H_2O, and NH_3 is converted into its conjugate acid, NH_4^+.

$$\underset{\text{Acid}}{H_3O^+} + \underset{\text{Base}}{NH_3} \longrightarrow \underset{\text{Conjugate base}}{H_2O} + \underset{\text{Conjugate acid}}{NH_4^+}$$

EXAMPLE 2.1

Write each acid-base reaction as a proton-transfer reaction. Label which reactant is the acid and which the base; which product is the conjugate base of the original acid and which is the conjugate acid of the original base. Use curved arrows to show the flow of electrons in each reaction.

(a) $H_2O + NH_4^+ \longrightarrow H_3O^+ + NH_3$

(b) $CH_3\overset{O}{\overset{\|}{C}}OH + NH_3 \longrightarrow CH_3\overset{O}{\overset{\|}{C}}O^- + NH_4^+$

SOLUTION

(a) First, complete the Lewis structure of each reactant by showing all valence electrons. Water is the base (proton acceptor), and ammonium ion is the acid (proton donor).

$$\underset{\text{Base}}{H{-}\ddot{O}{:}(H)} + \underset{\text{Acid}}{H{-}\overset{+}{N}H_3} \longrightarrow \underset{\text{Conjugate acid of } H_2O}{H{-}\overset{+}{\ddot{O}}(H){-}H} + \underset{\text{Conjugate base of } NH_4^+}{{:}NH_3}$$

(b) Acetic acid is the acid (proton donor), and ammonia is the base (proton acceptor).

$$\underset{\text{Acid}}{CH_3{-}\overset{O}{\overset{\|}{C}}{-}\ddot{O}{-}H} + \underset{\text{Base}}{{:}NH_3} \longrightarrow \underset{\text{Conjugate base of } CH_3CO_2H}{CH_3{-}\overset{O}{\overset{\|}{C}}{-}\ddot{O}{:}^-} + \underset{\text{Conjugate acid of } NH_3}{H{-}\overset{+}{N}H_3}$$

Practice Problem 2.1

Write each acid-base reaction as a proton-transfer reaction. Label which reactant is the acid and which is the base; which product is the conjugate base of the original acid and which is the conjugate acid of the original base. Use curved arrows to show the flow of electrons in each reaction.

(a) $CH_3SH + OH^- \longrightarrow CH_3S^- + H_2O$

(b) $CH_3OH + NH_2^- \longrightarrow CH_3O^- + NH_3$

2.2 Quantitative Measure of Acid and Base Strength

Strong acid An acid that is completely ionized in aqueous solution.

Strong base A base that is completely ionized in aqueous solution.

A **strong acid** or **strong base** is one that is completely ionized in aqueous solution. When HCl is dissolved in water, there is complete transfer of a proton from HCl to H_2O to form H_3O^+ and Cl^-. There is no tendency for the reverse reaction to occur, that is, for transfer of a proton from H_3O^+ to Cl^- to form HCl and H_2O. Therefore, when we compare the relative acidities of HCl and H_3O^+, we conclude that HCl is the stronger acid and H_3O^+ is the weaker acid. Similarly, H_2O is the stronger base, and Cl^- is the weaker base.

HCl	+	H_2O	$\longrightarrow$	Cl^-	+	H_3O^+
Acid (stronger acid)		Base (stronger base)		Conjugate base of HCl (weaker base)		Conjugate acid of H_2O (weaker acid)

Examples of strong acids in aqueous solution are HCl, HBr, HI, HNO_3, $HClO_4$, and H_2SO_4. Examples of strong bases in aqueous solution are LiOH, NaOH, KOH, $Ca(OH)_2$, and $Ba(OH)_2$.

A **weak acid** or **weak base** is one that is incompletely ionized in aqueous solution. Most organic acids and bases are weak. Among the most common organic acids we deal with are the carboxylic acids, which contain the carboxyl group, $—CO_2H$ (Section 1.7C).

$CH_3\overset{O}{\overset{\|}{C}}OH$	+	H_2O	$\rightleftharpoons$	$CH_3\overset{O}{\overset{\|}{C}}O^-$	+	H_3O^+
Acid (weaker acid)		Base (weaker base)		Conjugate base of CH_3CO_2H (stronger base)		Conjugate acid of H_2O (stronger acid)

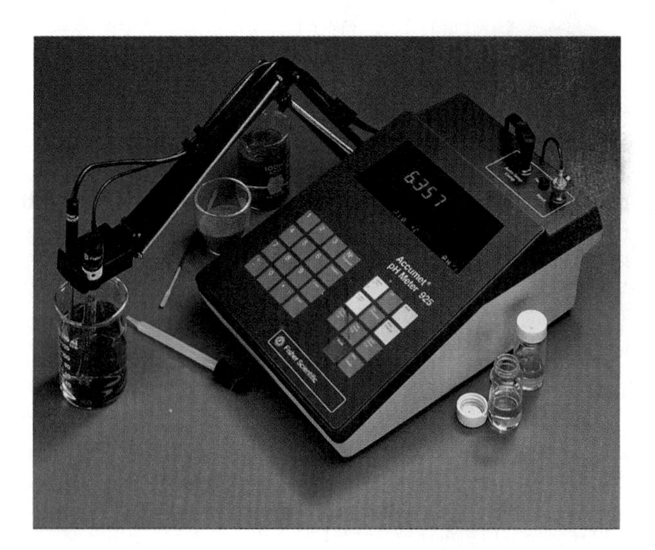

A pH meter is used to measure hydrogen ion concentration of aqueous solutions. *(Courtesy of Fischer Scientific)*

TABLE 2.1 pK_a Values for Some Organic and Inorganic Acids

	Acid	Formula	pK_a	Conjugate Base	
Weaker acid	ethane	CH_3CH_3	51	$CH_3CH_2^-$	Stronger base
	ammonia	NH_3	38	NH_2^-	
	ethanol	CH_3CH_2OH	15.9	$CH_3CH_2O^-$	
	water	H_2O	15.7	HO^-	
	methylammonium ion	$CH_3NH_3^+$	10.64	CH_3NH_2	
	bicarbonate ion	HCO_3^-	10.33	CO_3^{2-}	
	phenol	C_6H_5OH	9.95	$C_6H_5O^-$	
	ammonium ion	NH_4^+	9.24	NH_3	
	carbonic acid	H_2CO_3	6.36	HCO_3^-	
	acetic acid	CH_3CO_2H	4.76	$CH_3CO_2^-$	
	benzoic acid	$C_6H_5CO_2H$	4.19	$C_6H_5CO_2^-$	
	phosphoric acid	H_3PO_4	2.1	$H_2PO_4^-$	
	hydronium ion	H_3O^+	−1.74	H_2O	
	sulfuric acid	H_2SO_4	−5.2	HSO_4^-	
	hydrogen chloride	HCl	−7	Cl^-	
Stronger acid	hydrogen bromide	HBr	−8	Br^-	Weaker base
	hydrogen iodide	HI	−9	I^-	

The equation for the ionization of a weak acid, HA, in water and the acid ionization constant, K_a, for this equilibrium are

$$HA + H_2O \rightleftharpoons A^- + H_3O^+$$

$$K_a = K_{eq}[H_2O] = \frac{[H_3O^+][A^-]}{[HA]}$$

Because acid ionization constants for weak acids are numbers with negative exponents, acid strengths are often expressed as **pK_a** where p$K_a = -\log_{10} K_a$. Table 2.1 gives names, molecular formulas, and values of pK_a for some organic and inorganic acids. Note that the larger the value of pK_a, the weaker the acid. Note also the inverse relationship between the strengths of the conjugate acid-conjugate base pairs; the stronger the acid, the weaker its conjugate base.

EXAMPLE 2.2

For each value of pK_a, calculate the corresponding value of K_a. Which compound is the stronger acid?

(a) Ethanol, pK_a = 15.9 (b) Carbonic acid, pK_a = 6.36

SOLUTION

(a) For ethanol, $K_a = 1.3 \times 10^{-16}$ (b) For carbonic acid, $K_a = 4.4 \times 10^{-7}$

Because the value of pK_a for carbonic acid is smaller than that for ethanol, carbonic acid is the stronger acid, and ethanol is the weaker acid.

Practice Problem 2.2

For each value of K_a, calculate the corresponding value of pK_a. Which compound is the stronger acid?

(a) Acetic acid, $K_a = 1.74 \times 10^{-5}$ (b) Water, $K_a = 2.00 \times 10^{-16}$

Caution: In exercises such as Example 2.2 and Practice Problem 2.2, we ask you to select the stronger acid. You must remember that these and all other acids with acid ionization constants considerably less than 1.00 are weak acids. Although we may talk about one of them being stronger than another, nonetheless, they are all still weak acids, meaning that each is only slightly ionized in aqueous solution. Thus, even though acetic acid is a considerably stronger acid than water, it is only slightly ionized in water. The ionization of acetic acid in a 0.1*M* solution, for example, is only about 1.3%; the major form of this weak acid present in a 0.1*M* solution is the un-ionized acid!

$$CH_3\overset{\overset{O}{\|}}{C}OH + H_2O \rightleftharpoons CH_3\overset{\overset{O}{\|}}{C}O^- + H_3O^+$$

Forms present in 0.1*M* acetic acid: 98.7% (CH_3COOH) ... 1.3% (CH_3COO^-)

2.3 Molecular Structure and Acidity

Let us now examine briefly the relationship between the acidity of organic compounds and their molecular structure. These relationships can be understood by considering (1) the electronegativity of the atom bonded to H, (2) resonance, and (3) the inductive effect. We look at each of these factors briefly in this chapter, and then more fully in later chapters when we deal with particular functional groups.

A. Electronegativity—Acidity of HA Within a Period of the Periodic Table

The relative acidity of the hydrogen acids within a period (a horizontal row) of the Periodic Table is determined by the stability of A^-, the anion formed when a proton is transferred from HA to a base. The greater the electronegativity of A, the greater the stability of the anion, A^-, and the stronger the acid HA.

	$H_3C—H$	$H_2N—H$	$HO—H$	$F—H$
pK_a	51	38	15.7	3.5
Electronegativity of A in A—H	2.5	3.0	3.5	4.0
	Increasing acid strength →			

B. Resonance Effect—Delocalization of Charge in A^-

Carboxylic acids are weak acids. Values of pK_a for most unsubstituted carboxylic acids fall within the range 4 to 5. The value of pK_a for acetic acid, for example, is 4.76.

$$CH_3CO_2H + H_2O \rightleftharpoons CH_3CO_2^- + H_3O^+ \qquad pK_a = 4.76$$

Values of pK_a for most alcohols, compounds that also contain an —OH group, fall within the range 15 to 18; the value of pK_a for ethanol, for example, is 15.9. Thus, alcohols are slightly weaker acids than water (pK_a = 15.7) and much weaker acids than carboxylic acids.

$$\underset{\text{An alcohol}}{CH_3CH_2—\ddot{O}—H} + H_2O \rightleftharpoons \underset{\text{An alkoxide ion}}{CH_3CH_2—\ddot{O}:^-} + H_3O^+ \qquad pK_a = 15.9$$

We account for the greater acidity of carboxylic acids compared with alcohols in part using the resonance model and looking at the relative stabilities of the alkoxide ion and the carboxylate ion. Our guideline is this: The more stable the anion, the farther the position of equilibrium is shifted toward the right and the more acidic the compound.

There is no resonance stabilization in an alkoxide anion. Ionization of a carboxylic acid gives an anion for which we can write two equivalent contributing structures in which the negative charge of the anion is delocalized; that is, it is spread evenly over the two oxygen atoms. Because of the delocalization of its charge, a carboxylate anion is significantly more stable than an alkoxide anion. Therefore, the equilibrium for ionization of a carboxylic acid is shifted to the right relative to the ionization of an alcohol, and a carboxylic acid is a stronger acid than an alcohol.

$$CH_3—C(=\ddot{O}:)—\ddot{O}—H + H_2O \rightleftharpoons CH_3—C(=\ddot{O}:)—\ddot{O}:^- \longleftrightarrow CH_3—C(—\ddot{O}:^-)=\ddot{O}: + H_3O^+$$

These contributing structures are equivalent;
the carboxylate anion is stabilized
by delocalization of the negative charge

C. The Inductive Effect—Withdrawal of Electron Density from the HA Bond

The **inductive effect** is the polarization of electron density transmitted through covalent bonds by a nearby atom of higher electronegativity. We see the operation of the inductive effect in a carboxyl group in the following way. The carbonyl oxygen of a carboxyl group is more electronegative than the carbonyl carbon and polarizes the electrons of the C═O bond creating a partial negative charge on the carbonyl oxygen and a partial positive charge on the carbonyl carbon. The partial positive charge on the carbonyl carbon, in turn, withdraws electron density from the C—O and O—H bonds and promotes ionization of the carboxyl proton.

$$CH_3-\overset{\delta+}{C}(=\overset{\delta-}{O})-O-H$$

Thus, a carboxylic acid is more acidic than an alcohol because (1) the electron-withdrawing inductive effect of the adjacent carbonyl group weakens the O—H bond, promoting its ionization, and (2) the resonance effect stabilizes the carboxylate anion by delocalization of its negative charge.

2.4 Position of Equilibrium in Acid-Base Reactions

In acid-base reactions, the position of equilibrium favors reaction of the stronger acid with the stronger base to give the weaker acid and the weaker base. To determine the position of equilibrium for an acid-base reaction then, we need to know which is the stronger acid or, conversely, which is the stronger base.

Let us consider the equilibrium in which HA is the stronger acid and HB is the weaker acid. If HA is the stronger acid, then A^- must be the weaker base and, conversely, if HB is the weaker acid, then B^- must be the stronger base. We can now label the components in this acid-base equilibrium as follows:

$$\underset{\text{Stronger acid}}{HA} + \underset{\text{Stronger base}}{B^-} \rightleftharpoons \underset{\text{Weaker base}}{A^-} + \underset{\text{Weaker acid}}{HB}$$

The position of equilibrium in this acid-base reaction favors reaction of HA, the stronger acid, with B^-, the stronger base, to give A^-, the weaker base, and HB, the weaker acid. The position of equilibrium lies to the right, toward the weaker base and the weaker acid.

If, on the other hand, HB is the stronger acid and HA is the weaker acid, the position of equilibrium for this acid-base reaction lies to the left, in favor of HA, the weaker acid, and B^-, the weaker base.

$$\underset{\text{Weaker acid}}{HA} + \underset{\text{Weaker base}}{B^-} \rightleftharpoons \underset{\text{Stronger base}}{A^-} + \underset{\text{Stronger acid}}{HB}$$

We can predict the position of equilibrium in acid-base reactions of this type using the data in Table 2.1. For example, consider the reaction between aqueous solutions of acetic acid and ammonia. We must consider the following equilibrium:

$$CH_3CO_2H + NH_3 \rightleftharpoons CH_3CO_2^- + NH_4^+$$

CH_3CO_2H	NH_3	$CH_3CO_2^-$	NH_4^+
Acetic acid	Ammonia	Acetate ion	Ammonium ion
pK_a 4.76			pK_a 9.24
(stronger acid)			(weaker acid)

Acetic acid (pK_a 4.76) is a stronger acid than ammonium ion (pK_a 9.24); the position of equilibrium for this reaction, therefore, lies to the right.

Consider also the reaction between aqueous solutions of acetic acid and sodium bicarbonate to give sodium acetate and carbonic acid. In the equation for this equilibrium, we omit the sodium ion, Na^+, because it does not undergo a chemical change in this reaction. Instead, we write the equilibrium as a net ionic equation, which shows only the species undergoing chemical change.

$$CH_3CO_2H + HCO_3^- \rightleftharpoons CH_3CO_2^- + H_2CO_3$$

CH_3CO_2H	HCO_3^-	$CH_3CO_2^-$	H_2CO_3
Acetic acid	Bicarbonate ion	Acetate ion	Carbonic acid
pK_a 4.76			pK_a 6.36
(stronger acid)			(weaker acid)

Acetic acid is the stronger acid, and, therefore, the position of this equilibrium lies to the right. Carbonic acid is formed, which then decomposes to carbon dioxide and water.

Vinegar (which contains acetic acid) and baking soda (sodium bicarbonate) react to produce sodium acetate, carbon dioxide, and water. The carbon dioxide inflates the balloon. *(Charles D. Winters)*

EXAMPLE 2.3

Predict the position of equilibrium for each acid-base reaction. See Table 2.1 for pK_a values of each acid.

(a) $C_6H_5OH + HCO_3^- \rightleftharpoons C_6H_5O^- + H_2CO_3$

Phenol, Bicarbonate ion, Phenoxide ion, Carbonic acid

(b) $C_6H_5CO_2H + NH_2^- \rightleftharpoons C_6H_5CO_2^- + NH_3$

Benzoic acid, Amide ion, Benzoate ion, Ammonia

SOLUTION

(a) Carbonic acid is the stronger acid; the position of this equilibrium lies to the left. Phenol does not transfer a proton to bicarbonate ion to form carbonic acid.

$$C_6H_5OH + HCO_3^- \rightleftharpoons C_6H_5O^- + H_2CO_3$$

C_6H_5OH	H_2CO_3
pK_a 9.95	pK_a 6.36
(weaker acid)	(stronger acid)

(b) Benzoic acid is the stronger acid; the position of this equilibrium lies to the right.

$$C_6H_5CO_2H + NH_2^- \rightleftharpoons C_6H_5CO_2^- + NH_3$$

$C_6H_5CO_2H$	NH_3
pK_a 4.19	pK_a 38
(stronger acid)	(weaker acid)

Practice Problem 2.3

Predict the position of equilibrium for each acid-base reaction. See Table 2.1 for pK_a values of each acid.

(a) $CH_3NH_2 + CH_3CO_2H \rightleftharpoons CH_3NH_3^+ + CH_3CO_2^-$

Methylamine, Acetic acid, Methylammonium ion, Acetate ion

(b) $CH_3CH_2O^- + NH_3 \rightleftharpoons CH_3CH_2OH + NH_2^-$

Ethoxide ion, Ammonia, Ethanol, Amide ion

2.5 Lewis Acids and Bases

Lewis acid Any molecule or ion that can form a new covalent bond by accepting a pair of electrons.

Lewis base Any molecule or ion that can form a new covalent bond by donating a pair of electrons.

Gilbert Lewis, who proposed that covalent bonds are formed by sharing one or more pairs of electrons (Section 1.2C), further expanded the theory of acids and bases to include a group of substances not included in the Brønsted-Lowry concept. According to the Lewis definition, a **Lewis acid** is a species that can form a new covalent bond by accepting a pair of electrons; a **Lewis base** is a species that can form a new covalent bond by donating a pair of electrons. In the following general equation, the Lewis acid, A, accepts a pair of electrons in forming the new covalent bond and acquires a negative formal charge. The Lewis base, :B, donates the pair of electrons in forming the new covalent bond and acquires a positive formal charge.

$$\underset{\text{Lewis acid}}{A} + \underset{\text{Lewis base}}{:B} \rightleftharpoons \overset{-}{A}-\overset{+}{B}$$

new covalent bond formed in this Lewis acid-base reaction

Note that, although we speak of a Lewis base as "donating" a pair of electrons, the term is not fully accurate. Donating in this case does not mean that the electron pair under consideration is removed completely from the valence shell of the base. Rather, donating means that the electron pair becomes shared with another atom to form a covalent bond. Consider the reaction that occurs when this organic cation reacts with water.

$$CH_3-\overset{+}{C}H-CH_3 + H-\ddot{O}-H \longrightarrow CH_3-CH(\overset{+}{O}H_2)-CH_3$$

An organic cation (a Lewis acid); Water (a Lewis base); An oxonium ion

In this reaction, the organic cation is the electron pair acceptor (the Lewis acid), and a water molecule is the electron pair donor (the Lewis base).

EXAMPLE 2.4

Write an equation for the reaction between each Lewis acid-base pair, showing electron flow by means of curved arrows.

(a) $H^+ + NH_3 \longrightarrow$ (b) $CH_3-\overset{+}{C}H-CH_3 + Cl^- \longrightarrow$

SOLUTION

(a) H^+ is the Lewis acid. NH_3 has an unshared pair of electrons in the valence shell of nitrogen and is the Lewis base.

$$H^+ + :NH_3 \longrightarrow H-\overset{+}{N}H_3$$

Lewis acid Lewis base

(b) The trivalent carbon atom in the organic cation has an empty orbital in its valence shell and, therefore, is the Lewis acid. Chloride ion is the Lewis base.

$$CH_3-\overset{+}{C}H-CH_3 + :\ddot{Cl}:^- \longrightarrow CH_3-CH(Cl)-CH_3$$

Lewis acid Lewis base

Practice Problem 2.4

Write an equation for the reaction between each Lewis acid-base pair, showing electron flow by means of curved arrows. *Hint:* Aluminum is in Group IIIA of the Periodic Table, just under boron. Aluminum in $AlCl_3$ has only six electrons in its valence shell and has an incomplete octet.

(a) $Cl^- + AlCl_3 \longrightarrow$ (b) $CH_3Cl + AlCl_3 \longrightarrow$

SUMMARY

A **Brønsted-Lowry acid** is a proton donor, and a **Brønsted-Lowry base** is a proton acceptor (Section 2.1). Neutralization of an acid by a base is a **proton-transfer reaction** in which the acid is transformed into its **conjugate base,** and the base is transformed into its **conjugate acid.**

A **strong acid** or **strong base** is one that is completely ionized in water. A weak acid or weak base is one that is only partially ionized in water (Section 2.2). Among the most common weak organic acids are carboxylic acids, compounds that contain the $-CO_2H$ (carboxyl) group. The value of K_a (the acid ionization constant) for acetic acid, a representative carboxylic acid, is 1.74×10^{-5}; the value of **pK_a** (the negative logarithm of K_a) for acetic acid is 4.76.

The relative acidities of the organic acids, HA, are determined by (1) the electronegativity of A, (2) resonance stabilization of the conjugate base, A^- (Section 2.3), and (3) the electron-withdrawing inductive effect that weakens the H—A bond and promotes its ionization.

A **Lewis acid** (Section 2.5) is a species that forms a new covalent bond by accepting a pair of electrons; a **Lewis base** is a species that forms a new covalent bond by donating a pair of electrons.

KEY REACTIONS

1. Proton-Transfer Reaction (Section 2.1)

Involves transfer of a proton from a proton donor (a Brønsted-Lowry acid) to a proton acceptor (a Brønsted-Lowry base).

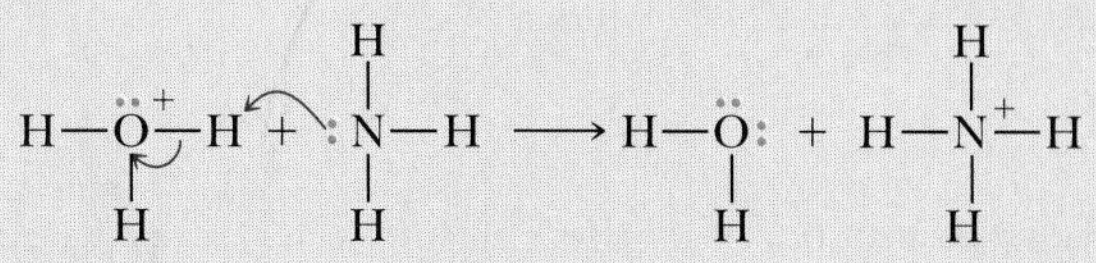

$$H_3O^+ + :NH_3 \longrightarrow H_2\ddot{O}: + NH_4^+$$

Proton donor — Proton acceptor

2. Position of Equilibrium in an Acid-Base Reaction (Section 2.4)

Equilibrium favors reaction of the stronger acid with the stronger base to give the weaker acid and the weaker base.

$$CH_3CO_2H + NH_3 \rightleftharpoons CH_3CO_2^- + NH_4^+$$

Acetic acid, pK_a 4.76 (stronger acid) — Ammonium ion, pK_a 9.24 (weaker acid)

3. Lewis Acid-Base Reaction (Section 2.5)

A Lewis acid-base reaction involves sharing an electron pair between an electron pair donor (a Lewis base) and an electron pair acceptor (a Lewis acid).

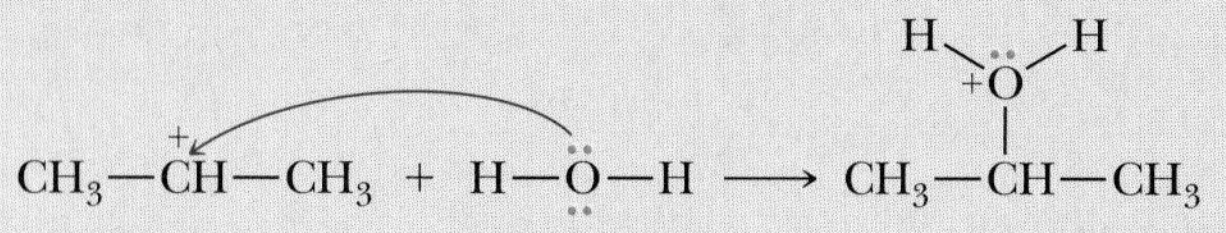

$$CH_3-\overset{+}{C}H-CH_3 + H-\ddot{O}-H \longrightarrow CH_3-CH(\overset{+}{O}H_2)-CH_3$$

A Lewis acid — A Lewis base

ADDITIONAL PROBLEMS

Brønsted-Lowry Acids and Bases

2.5 Complete a net ionic equation for each proton-transfer reaction using curved arrows to show the flow of electron pairs in each reaction. In addition, write Lewis structures for all starting materials and products. Label the original acid and its conjugate base; label the original base and its conjugate acid. If you are uncertain about which substance in each equation is the proton donor, refer to Table 2.1 for the pK_a values of proton acids.

(a) $NH_3 + HCl \longrightarrow$ **(b)** $CH_3CH_2O^- + HCl \longrightarrow$

(c) $HCO_3^- + OH^- \longrightarrow$ **(d)** $CH_3CO_2^- + NH_4^+ \longrightarrow$

(e) $NH_4^+ + OH^- \longrightarrow$ **(f)** $CH_3CO_2^- + CH_3NH_3^+ \longrightarrow$

(g) $CH_3CH_2O^- + NH_4^+ \longrightarrow$ **(h)** $CH_3NH_3^+ + OH^- \longrightarrow$

2.6 Each of these molecules and ions can function as a base. Complete the Lewis structure of each base, and write the structural formula of the conjugate acid formed by its reaction with HCl.

(a) CH_3CH_2OH **(b)**  HCH (with C=O) **(c)** $(CH_3)_2NH$ **(d)** HCO_3^-

2.7 Offer an explanation for the following observations.

(a) H_3O^+ is a stronger acid than NH_4^+.

(b) Nitric acid, HNO_3, is a stronger acid than nitrous acid, HNO_2 (pK_a 3.7).

(c) Ethanol and water have approximately the same acidity.

(d) Trichloroacetic acid, CCl_3CO_2H (pK_a 0.64), is a stronger acid than acetic acid, CH_3CO_2H (pK_a 4.74).

(e) Trifluoroacetic acid, CF_3CO_2H (pK_a 0.23), is a stronger acid than trichloroacetic acid, CCl_3CO_2H (pK_a 0.64).

2.8 As we shall see in Chapter 14, hydrogens on a carbon adjacent to a carbonyl group are far more acidic than those not adjacent to a carbonyl group. The highlighted H in propanone, for example, is more acidic than the highlighted H in ethane. Account for the greater acidity of propanone in terms of (a) the inductive effect and (b) the resonance effect.

$CH_3\overset{O}{\overset{\|}{C}}CH_2{-}H$	$CH_3CH_2{-}H$
Propanone	Ethane
pK_a = 22	pK_a = 51

Quantitative Measure of Acid Strength

2.9 Which has the larger numerical value

(a) The pK_a of a strong acid or the pK_a of a weak acid?

(b) The K_a of a strong acid or the K_a of a weak acid?

2.10 In each pair, select the stronger acid:

(a) Pyruvic acid (pK_a 2.49) or lactic acid (pK_a 3.85)

(b) Citric acid (pK_{a1} 3.08) or phosphoric acid (pK_{a1} 2.10)

(c) Nicotinic acid (niacin, K_a 1.4×10^{-5}) or acetylsalicylic acid (aspirin, K_a 3.3×10^{-4})

(d) Phenol (K_a 1.12×10^{-10}) or acetic acid (K_a 1.74×10^{-5})

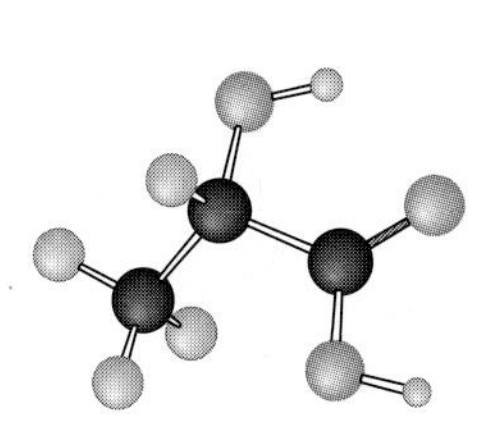

Lactic acid
(increases in muscle and blood after vigorous activity)

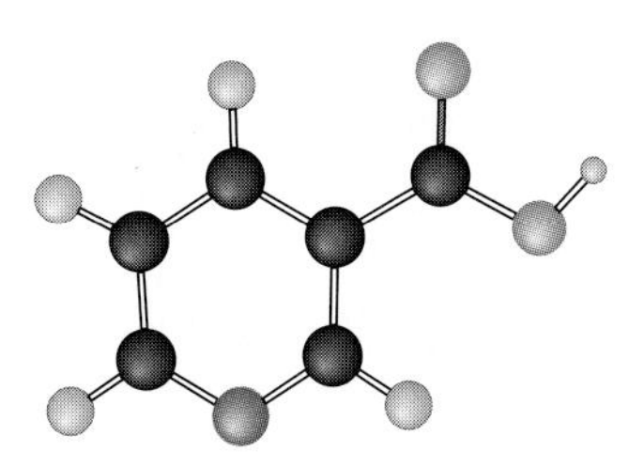

Nicotinic acid
(occurs in liver, fish, yeast, and cereal grains)

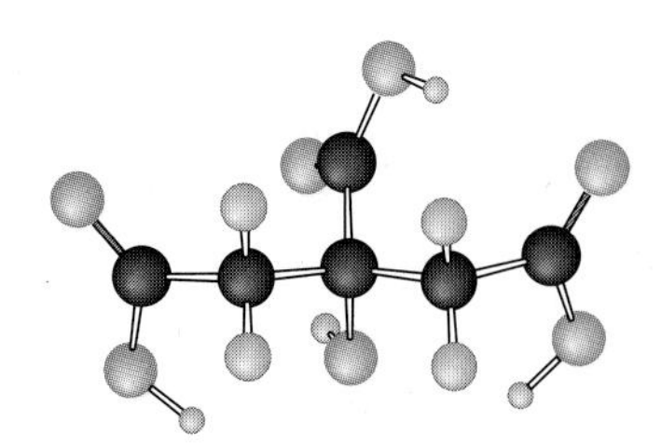

Citric acid
(in citrus fruits; lemons contain 5 to 8% by weight)

2.11 Arrange the compounds in each set in order of increasing acid strength. Consult Table 2.1 for pK_a values of each acid.

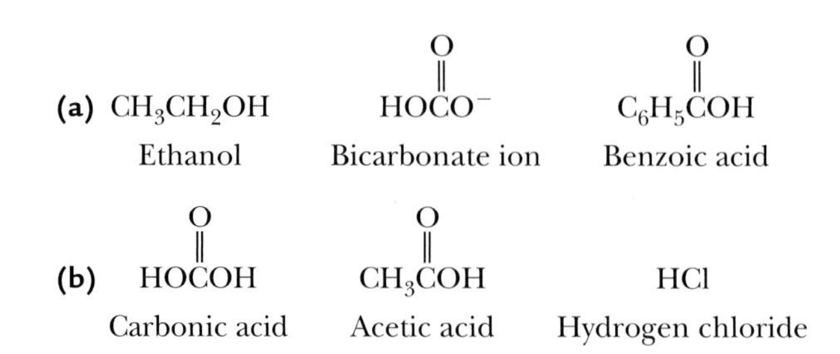

2.12 Arrange the compounds in each set in order of increasing base strength. Consult Table 2.1 for pK_a values of the conjugate acid of each base. (*Hint:* The stronger the acid, the weaker its conjugate base, and vice versa.)

(a) NH_3 $HOC(=O)O^-$ $CH_3CH_2O^-$ **(b)** OH^- $HOC(=O)O^-$ $CH_3C(=O)O^-$

(c) H_2O NH_3 $CH_3C(=O)O^-$ **(d)** NH_2^- $CH_3C(=O)O^-$ OH^-

Position of Equilibrium in Acid-Base Reactions

2.13 Unless under pressure, carbonic acid in aqueous solution breaks down into carbon dioxide and water, and carbon dioxide is evolved as bubbles of gas. Write an equation for the conversion of carbonic acid to carbon dioxide and water.

2.14 Will carbon dioxide be evolved when sodium bicarbonate is added to an aqueous solution of

(a) H_2SO_4? **(b)** CH_3CH_2OH? **(c)** NH_4Cl?

2.15 Acetic acid, CH_3CO_2H, is a weak organic acid, pK_a 4.76. Write equations for the equilibrium reactions of acetic acid with each base. Which equilibria lie considerably toward the left? Which lie considerably toward the right?

(a) $NaHCO_3$ **(b)** NH_3 **(c)** H_2O **(d)** $NaOH$

2.16 Alcohols are weak organic acids, pK_a 16–18. The pK_a of ethanol, CH_3CH_2OH, is 15.9. Write equations for the equilibrium reactions of ethanol with each base. Which equilibria lie considerably toward the right? Which lie considerably toward the left?

(a) $NaHCO_3$ **(b)** $NaOH$ **(c)** $NaNH_2$ **(d)** NH_3

2.17 Benzoic acid, $C_6H_5CO_2H$, is insoluble in water, but its sodium salt, $C_6H_5CO_2^-Na^+$, is quite soluble in water. Will benzoic acid dissolve in

(a) Aqueous sodium hydroxide?

(b) Aqueous sodium bicarbonate?

(c) Aqueous sodium carbonate?

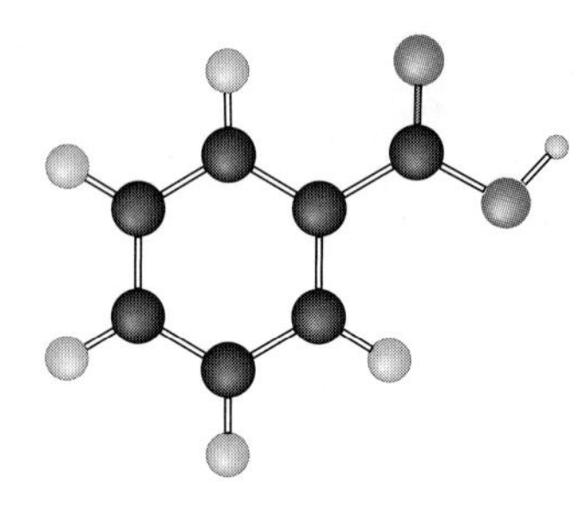

Benzoic acid (most berries contain about 0.05% by weight)

2.18 Phenol, C_6H_5OH (pK_a 9.95), is only slightly soluble in water, but its sodium salt, $C_6H_5O^-Na^+$, is quite soluble in water. Will phenol dissolve in

(a) Aqueous NaOH? **(b)** Aqueous $NaHCO_3$? **(c)** Aqueous Na_2CO_3?

2.19 For an acid-base reaction, one way to indicate the predominant species at equilibrium is to say that the reaction arrow points to the acid with the higher value of pK_a. For example,

$$\underset{\text{p}K_a\ 9.24}{NH_4^+} + H_2O \longleftarrow NH_3 + \underset{\text{p}K_a\ -1.74}{H_3O^+}$$

$$\underset{\text{p}K_a\ 9.24}{NH_4^+} + OH^- \longrightarrow NH_3 + \underset{\text{p}K_a\ 15.7}{H_2O}$$

Explain why this rule works.

Lewis Acids and Bases

2.20 Complete equations for these reactions between Lewis acid–Lewis base pairs. Label which starting material is the Lewis acid and which is the Lewis base, and use a curved arrow to show the flow of the electron pair in each reaction. In solving these problems, it is essential that you show all valence electrons for the atoms participating directly in each reaction.

(a) $CH_3-\underset{CH_3}{\overset{CH_3}{C}}-Cl + \underset{Cl}{\overset{Cl}{Al}}-Cl \longrightarrow$

(b) $CH_3-\underset{CH_3}{\overset{CH_3}{C^+}} + H-O-H \longrightarrow$

(c) $CH_3-\overset{+}{C}H-CH_3 + CH_3-O-H \longrightarrow$

(d) $CH_3-\overset{+}{C}H-CH_3 + Br^- \longrightarrow$

3

Alkanes and Cycloalkanes

Bunsen burners burn natural gas, which is primarily methane with small amounts of ethane, propane, and butane. Inset: A model of methane. *(Charles D. Winters)*

In this chapter we begin our study of organic compounds with the physical and chemical properties of alkanes and cycloalkanes, both saturated hydrocarbons and among the simplest types of organic compounds.

A **hydrocarbon** is a compound that is composed of only carbon and hydrogen. A **saturated hydrocarbon** contains only single bonds. Saturated in this context means that each carbon in the alkane has the maximum number of hydrogens bonded to it. An **alkane** is a saturated hydrocarbon whose carbon atoms are arranged in an open chain. Alkanes are commonly referred to as **aliphatic hydrocarbons** because the physical properties of the higher members of this class resemble those of the long carbon-chain molecules we find in animal fats and plant oils (Greek: *aleiphar,* fat or oil).

Hydrocarbon A compound that contains only carbon and hydrogen atoms.

Saturated hydrocarbon A hydrocarbon containing only carbon-carbon single bonds.

Alkane A saturated hydrocarbon whose carbon atoms are arranged in an open chain.

Aliphatic hydrocarbon An alternative word to describe an alkane.

3.1 STRUCTURE OF ALKANES

Methane, CH_4, and ethane, C_2H_6, are the first two members of the alkane family. Shown in Figure 3.1 are Lewis structures and ball-and-stick models for these molecules. The shape of methane is tetrahedral, and all H—C—H bond angles are 109.5°. Each carbon atom in ethane is also tetrahedral, and all bond angles are approximately 109.5°.

Although the three-dimensional shapes of larger alkanes are more complex than those of methane and ethane, the four bonds about each carbon atom are still arranged in a tetrahedral manner, and all bond angles are approximately 109.5°. The next members of the alkane family are propane, butane, and pentane.

Some camping stoves use butane as a fuel. *(Charles D. Winters)*

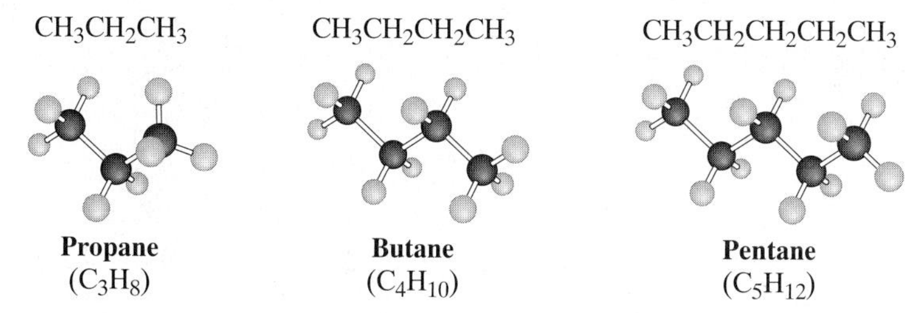

Condensed structural formulas for alkanes can also be written in an abbreviated form. For example, the structural formula of pentane contains three CH_2 **(methylene)** groups in the middle of the chain. They can be grouped together, and the structural formula written $CH_3(CH_2)_3CH_3$. Names and molecular formulas of the first 20 alkanes are given in Table 3.1. Note that the names of all these alkanes end in -ane. We will have more to say about naming alkanes in Section 3.3.

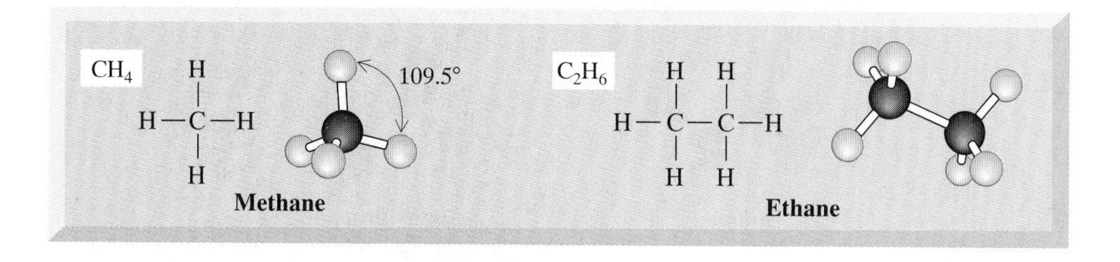

Figure 3.1
Methane and ethane.

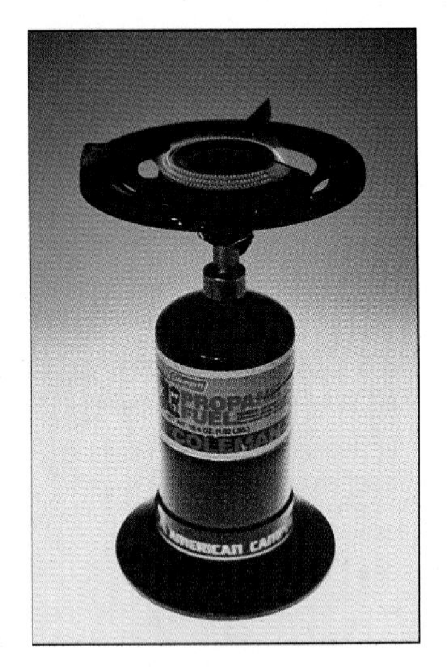

A tank of propane fuel.
(Charles D. Winters)

TABLE 3.1 Names, Molecular Formulas, and Condensed Structural Formulas for the First 20 Alkanes with Unbranched Chains

Name	Molecular Formula	Condensed Structural Formula	Name	Molecular Formula	Condensed Structural Formula
methane	CH_4	CH_4	undecane	$C_{11}H_{24}$	$CH_3(CH_2)_9CH_3$
ethane	C_2H_6	CH_3CH_3	dodecane	$C_{12}H_{26}$	$CH_3(CH_2)_{10}CH_3$
propane	C_3H_8	$CH_3CH_2CH_3$	tridecane	$C_{13}H_{28}$	$CH_3(CH_2)_{11}CH_3$
butane	C_4H_{10}	$CH_3(CH_2)_2CH_3$	tetradecane	$C_{14}H_{30}$	$CH_3(CH_2)_{12}CH_3$
pentane	C_5H_{12}	$CH_3(CH_2)_3CH_3$	pentadecane	$C_{15}H_{32}$	$CH_3(CH_2)_{13}CH_3$
hexane	C_6H_{14}	$CH_3(CH_2)_4CH_3$	hexadecane	$C_{16}H_{34}$	$CH_3(CH_2)_{14}CH_3$
heptane	C_7H_{16}	$CH_3(CH_2)_5CH_3$	heptadecane	$C_{17}H_{36}$	$CH_3(CH_2)_{15}CH_3$
octane	C_8H_{18}	$CH_3(CH_2)_6CH_3$	octadecane	$C_{18}H_{38}$	$CH_3(CH_2)_{16}CH_3$
nonane	C_9H_{20}	$CH_3(CH_2)_7CH_3$	nonadecane	$C_{19}H_{40}$	$CH_3(CH_2)_{17}CH_3$
decane	$C_{10}H_{22}$	$CH_3(CH_2)_8CH_3$	eicosane	$C_{20}H_{42}$	$CH_3(CH_2)_{18}CH_3$

Alkanes have the general molecular formula $\mathbf{C_nH_{2n+2}}$. Thus, given the number of carbon atoms in an alkane, it is easy to determine the number of hydrogens in the molecule and also its molecular formula. For example, decane with ten carbon atoms must have $(2 \times 10) + 2 = 22$ hydrogens and a molecular formula of $C_{10}H_{22}$.

3.2 Constitutional Isomerism in Alkanes

Constitutional isomers Compounds with the same molecular formula but a different order of attachment of their atoms.

For the molecular formulas CH_4, C_2H_6, and C_3H_8, only one order of attachment of atoms is possible. For the molecular formula C_4H_{10}, two orders of attachment of atoms are possible. In one of these, named butane, the four carbons are attached in a chain; in the other, named 2-methylpropane, three carbons are attached in a chain with the fourth carbon as a branch on the chain. Butane and 2-methylpropane are **constitutional isomers;** they are different compounds and have different physical and chemical properties.

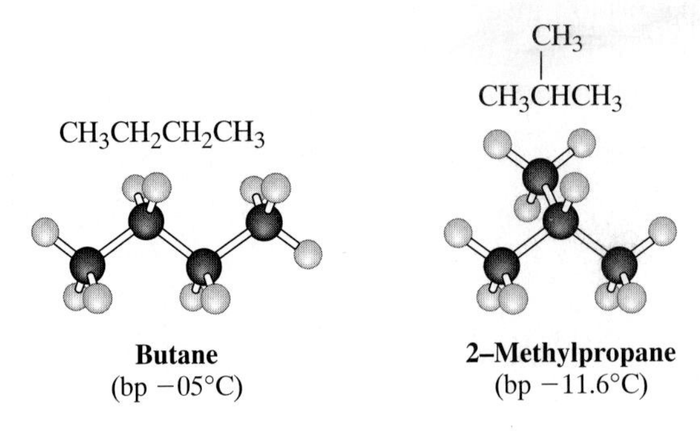

Butane
(bp −05°C)

2-Methylpropane
(bp −11.6°C)

In Section 1.7, we encountered several examples of constitutional isomers, although we did not call them that at the time. We saw that there are two alcohols of molecular formula C_3H_8O, two aldehydes of molecular formula C_4H_8O, and two carboxylic acids of molecular formula $C_4H_8O_2$.

To determine whether two or more structural formulas represent constitutional isomers, write the molecular formula of each and then compare them. All compounds that have the same molecular formula but different structural formulas are constitutional isomers.

EXAMPLE 3.1

Do the structural formulas in each set represent the same compound or constitutional isomers?

(a) $CH_3CH_2CH_2CH_2CH_2CH_3$ and $\begin{array}{l} CH_3CH_2CH_2 \\ \qquad\qquad | \\ \qquad\quad CH_2CH_2CH_3 \end{array}$ (each is C_6H_{14})

(b) $\begin{array}{l} \quad CH_3 \quad\; CH_3 \\ \quad\;\; | \qquad\quad | \\ CH_3CHCH_2CH \\ \qquad\qquad\;\; | \\ \qquad\qquad CH_3 \end{array}$ and $\begin{array}{l} \qquad\qquad CH_3 \\ \qquad\qquad\;\; | \\ CH_3CH_2CHCHCH_3 \\ \qquad\qquad\quad | \\ \qquad\qquad\; CH_3 \end{array}$ (each is C_7H_{16})

SOLUTION

To determine whether these structural formulas represent the same compound or constitutional isomers, first find the longest chain of carbon atoms. Note that it makes no difference if the chain is drawn straight or bent. As structural formulas are drawn in this problem, there is no attempt to show three-dimensional shapes. Second, number the longest chain from the end nearest the first branch. Third, compare the lengths of each chain and the size and locations of any branches. Structural formulas that have the same order of attachment of atoms represent the same compound; those that have different orders of attachment of atoms represent constitutional isomers.

(a) Each structural formula has an unbranched chain of six carbons; they are identical and represent the same compound.

$$\begin{array}{l} 1 \quad\; 2 \quad\; 3 \quad\; 4 \quad\; 5 \quad\; 6 \\ CH_3CH_2CH_2CH_2CH_2CH_3 \end{array} \quad \text{and} \quad \begin{array}{l} 1 \quad\; 2 \quad\; 3 \\ CH_3CH_2CH_2 \\ \qquad\quad |4 \quad 5 \quad 6 \\ \qquad\quad CH_2CH_2CH_3 \end{array}$$

(b) Each structural formula has a chain of five carbons with two CH_3 branches. Although the branches are identical, they are at different locations on the chains. Therefore, these structural formulas represent constitutional isomers.

$$\begin{array}{l} \qquad\qquad\quad 5 \\ \qquad\qquad CH_3 \\ 1 \quad 2 \quad 3 \quad 4| \\ CH_3CHCH_2CH \\ \qquad | \qquad\quad | \\ \quad CH_3 \quad CH_3 \end{array} \quad \text{and} \quad \begin{array}{l} \qquad\qquad CH_3 \\ 5 \quad 4 \quad 3| \quad 2 \quad 1 \\ CH_3CH_2CHCHCH_3 \\ \qquad\qquad\quad | \\ \qquad\qquad CH_3 \end{array}$$

Practice Problem 3.1

Do the structural formulas in each set represent the same compound or constitutional isomers?

(a)
$$\begin{array}{c}CH_2CH_3\\ |\\ CH_3CHCHCH_3\\ |\\ CH_2CH_3\end{array} \quad \text{and} \quad \begin{array}{c}CH_3CH_3\\ ||\\ CH_3CH_2CHCH_2CHCH_3\end{array}$$

(b)
$$\begin{array}{c}CH_3\\ |\\ CH_3CHCHCH_3\\ |\\ CH_2CH_3\end{array} \quad \text{and} \quad \begin{array}{c}CH_3\\ |\\ CH_3CHCHCH_2CH_3\\ |\\ CH_3\end{array}$$

EXAMPLE 3.2

Draw structural formulas for the five constitutional isomers of molecular formula C_6H_{14}.

SOLUTION

In solving problems of this type, you should devise a strategy and then follow it. Here is one such strategy. First, draw a structural formula for the constitutional isomer with all six carbons in an unbranched chain. Then, draw structural formulas for all constitutional isomers with five carbons in a chain and one carbon as a branch on the chain. Finally, draw structural formulas for all constitutional isomers with four carbons in a chain and two carbons as branches.

Six carbons in an unbranched chain:

$$\begin{array}{c}1\quad 2\quad 3\quad 4\quad 5\quad 6\\ CH_3CH_2CH_2CH_2CH_2CH_3\end{array}$$

Five carbons in a chain; one carbon as a branch:

$$\begin{array}{c}CH_3\\ 1\ \ 2|\ \ 3\ \ \ 4\ \ \ 5\\ CH_3CHCH_2CH_2CH_3\end{array} \qquad \begin{array}{c}CH_3\\ 1\ \ \ 2\ \ 3|\ \ 4\ \ \ 5\\ CH_3CH_2CHCH_2CH_3\end{array}$$

Four carbons in a chain; two carbons as branches:

$$\begin{array}{c}CH_3\\ 1\ \ 2|3\ \ \ 4\\ CH_3CCH_2CH_3\\ |\\ CH_3\end{array} \qquad \begin{array}{c}CH_3\\ 1\ \ 2|\ \ 3\ \ 4\\ CH_3CHCHCH_3\\ |\\ CH_3\end{array}$$

No constitutional isomers with only three carbons in the longest chain are possible for C_6H_{14}.

Practice Problem 3.2

Draw structural formulas for the three constitutional isomers of molecular formula C_5H_{12}.

The ability of carbon atoms to form strong, stable bonds with other carbon atoms results in a staggering number of constitutional isomers. As shown in the following table, there are 3 constitutional isomers of molecular formula C_5H_{12}.

CHEMICAL CONNECTIONS

Counting the Number of Constitutional Isomers

We have seen in this section that, for the hydrocarbon $C_{30}H_{62}$, there are over four billion constitutional isomers. The number of constitutional isomers becomes even larger for organic compounds with a functional group. For example, although there are 75 alkanes of molecular formula $C_{10}H_{22}$, there are 507 alcohols of molecular formula $C_{10}H_{22}O$.

Obviously, these numbers cannot be obtained by drawing and counting all the possible constitutional isomers. Instead, they are counted using a type of mathematics called **graph theory.** The English mathematician Arthur Cayley was the first person to use this concept for organic molecules, and his paper of 1874, titled "On the Mathematical Theory of Isomers," predates many fundamental organic concepts, including the idea that alkane carbons are tetrahedral. To read further about the concept of graph theory for counting the number of constitutional isomers, see P. J. Hansen and P. C. Jurs, *Journal of Chemical Education,* **65,** 661, 1988.

For molecular formula $C_{10}H_{22}$, there are 75 constitutional isomers, and for $C_{25}H_{52}$ there are almost 37 million.

Carbon Atoms	Constitutional Isomers
1	1
5	3
10	75
15	4,347
25	36,797,588
30	4,111,846,763

Thus, for even a small number of carbon and hydrogen atoms, a very large number of constitutional isomers is possible. In fact, the potential for structural and functional group individuality among organic molecules made from just the basic building blocks of carbon, hydrogen, nitrogen, and oxygen is practically limitless.

3.3 NOMENCLATURE OF ALKANES

A. The IUPAC System

Ideally every organic compound should have a name from which its structural formula can be drawn. For this purpose, chemists have adopted a set of rules established by an organization called the **International Union of Pure and Applied Chemistry (IUPAC).** The IUPAC name of an alkane with an unbranched chain of

TABLE 3.2 Prefixes Used in the IUPAC System To Show the Presence of 1 to 20 Carbons in an Unbranched Chain

Prefix	Number of Carbon Atoms	Prefix	Number of Carbon Atoms
meth-	1	undec-	11
eth-	2	dodec-	12
prop-	3	tridec-	13
but-	4	tetradec-	14
pent-	5	pentadec-	15
hex-	6	hexadec-	16
hept-	7	heptadec-	17
oct-	8	octadec-	18
non-	9	nonadec-	19
dec-	10	eicos-	20

carbon atoms consists of two parts: (1) a prefix that indicates the number of carbon atoms in the chain and (2) the suffix **-ane** to show that the compound is a saturated hydrocarbon. Prefixes used to show the presence of 1 to 20 carbon atoms are given in Table 3.2.

The first four prefixes listed in Table 3.2 were chosen by the IUPAC because they were well established in the language of organic chemistry. In fact, they were well established even before there were hints of the structural theory underlying the discipline. For example, the prefix *but-* appears in the name butyric acid, a compound of four carbon atoms formed by the air oxidation of butter fat (Latin: *butyrum*, butter). Prefixes to show five or more carbons are derived from Greek or Latin numbers. Names, molecular formulas, and condensed structural formulas for the first 20 alkanes with unbranched chains are given in Table 3.1.

IUPAC names of alkanes with branched chains consist of a parent name that indicates the longest chain of carbon atoms in the compound and substituent names that indicate the groups attached to the parent chain.

substituent

CH_3

1 2 3 4 5 6 7 8

$CH_3CH_2CH_2CHCH_2CH_2CH_2CH_3$

4-Methyloctane

Alkyl group A group derived by removing a hydrogen from an alkane. Given the symbol R-.

R- A symbol used to represent an alkyl group.

A substituent group derived from an alkane by removal of a hydrogen atom is called an **alkyl group.** The symbol **R** is commonly used to represent an alkyl group. Alkyl groups are named by dropping the -ane from the name of the parent alkane and adding the suffix *-yl.* Names and structural formulas for 8 of the most common alkyl groups are given in Table 3.3.

The rules of the IUPAC system for naming alkanes are as follows:

1. The general name of an open-chain saturated hydrocarbon is alkane.
2. For branched-chain hydrocarbons, the alkane corresponding to the longest chain of carbon atoms is taken as the parent alkane and its name as the root name.

TABLE 3.3 Names of the Most Common Alkyl Groups

Name	Condensed Structural Formula	Name	Condensed Structural Formula
methyl	$-CH_3$	isobutyl	$-CH_2CHCH_3$ | CH_3
ethyl	$-CH_2CH_3$	*sec*-butyl	$-CHCH_2CH_3$ | CH_3
propyl	$-CH_2CH_2CH_3$	*tert*-butyl	CH_3 | $-CCH_3$ | CH_3
isopropyl	$-CHCH_3$ | CH_3		
butyl	$-CH_2CH_2CH_2CH_3$		

3. Groups attached to the parent chain are called substituents. Each substituent is given a name and a number. The number shows the carbon atom of the parent chain to which the substituent is attached.

$$\begin{array}{l} \quad\;\; CH_3 \\ 1\;\; 2|\;\; 3 \\ CH_3CHCH_3 \end{array}$$

2-Methylpropane

4. If there is one substituent, number the parent chain from the end that gives it the lower number. The following alkane must be numbered as shown and named 2-methylpentane. Numbering from the other end of the chain gives the incorrect name 4-methylpentane.

$$\begin{array}{l} \qquad\qquad\quad CH_3 \\ 5\;\;\; 4\;\;\; 3\;\; 2|\;\; 1 \\ CH_3CH_2CH_2CHCH_3 \end{array}$$

2-Methylpentane

5. If the same substituent occurs more than once, the number of each carbon of the parent chain on which the substituent occurs is given. In addition, the number of times the substituent occurs is indicated by a prefix di-, tri-, tetra-, penta-, hexa-, and so on.

$$\begin{array}{l} \quad\;\; CH_3 \quad CH_3 \\ 1\;\; 2|\;\; 3\;\; 4|\;\; 5 \\ CH_3CHCH_2CHCH_3 \end{array}$$

2,4-Dimethylpentane

6. If there are two or more identical substituents, number the parent chain from the end that gives the lower number to the substituent encountered first.

$$\begin{array}{l} \qquad\quad CH_3 \quad CH_3 \\ 6\;\; 5\;\; 4|\;\; 3\;\; 2|\;\; 1 \\ CH_3CH_2CHCH_2CHCH_3 \end{array}$$

2,4-Dimethylhexane

7. If there are two or more different substituents, list them in alphabetical order and number the chain from the end that gives the lower number to the sub-

stituent encountered first. If there are different substituents in equivalent positions on the parent chain, the substituent of lower alphabetical order is given the lower number.

$$\begin{array}{c} CH_3 \\ 1\ 2\ \ 3\ \ 4\ \ 5|\ \ 6\ \ 7 \\ CH_3CH_2CHCH_2CHCH_2CH_3 \\ | \\ CH_2CH_3 \end{array}$$

3-Ethyl-5-methylheptane

8. The prefixes di-, tri-, tetra-, and so on are not included in alphabetizing. The names of substituents are alphabetized first, and then these multiplying prefixes are inserted. In this example, the alphabetizing parts are ethyl and methyl, not ethyl and dimethyl.

$$\begin{array}{c} CH_3\ \ CH_2CH_3 \\ 1\ \ 2|\ 3\ \ 4|\ \ 5\ \ 6 \\ CH_3CCH_2CHCH_2CH_3 \\ | \\ CH_3 \end{array}$$

4-Ethyl-2,2-dimethylhexane

9. Hyphenated prefixes, such as *sec-* and *tert-*, are not considered when alphabetizing. The prefix iso- is not a hyphenated prefix and is included when alphabetizing.

Example 3.3

Write IUPAC names for these alkanes.

(a) $CH_3CHCH_2CH_3$ with CH_3 on C-2 (b) $CH_3CHCH_2CHCH_2CH_2CH_3$ with CH_3 on C-2 and $CH(CH_3)_2$ on C-4

$$\text{(a)}\ \begin{array}{c} CH_3CHCH_2CH_3 \\ | \\ CH_3 \end{array} \qquad \text{(b)}\ \begin{array}{c} CH_3CHCH_2CHCH_2CH_2CH_3 \\ |\qquad\ \ | \\ CH_3\quad CH(CH_3)_2 \end{array}$$

Solution

The longest chain in each is numbered from the end of the chain toward the substituent encountered first (rule 4). The substituents in (b) are listed in alphabetical order (rule 7).

$$\text{(a)}\ \begin{array}{c} 1\ \ 2\ \ 3\ \ 4 \\ CH_3CHCH_2CH_3 \\ | \\ CH_3 \end{array} \qquad \text{(b)}\ \begin{array}{c} 1\ \ 2\ \ 3\ \ 4\ \ 5\ \ 6\ \ 7 \\ CH_3CHCH_2CHCH_2CH_2CH_3 \\ |\qquad\ \ | \\ CH_3\quad CH(CH_3)_2 \end{array}$$

2-Methylbutane   4-Isopropyl-2-methylheptane

Practice Problem 3.3

Write IUPAC names for these alkanes.

$$\text{(a)}\ \begin{array}{c} CH_3\qquad\qquad CH_3 \\ |\qquad\qquad\quad | \\ CH_3CHCH_2CH_2CHCHCH_3 \\ | \\ CH_2CH_2CH_3 \end{array} \qquad \text{(b)}\ \begin{array}{c} CH_2CH_2CH_3 \\ | \\ CH_3CH_2CH_2CCH_2CH_2CH_3 \\ | \\ CH_3CHCH_3 \end{array}$$

B. Common Names

In the older system of **common nomenclature,** the total number of carbon atoms in an alkane, regardless of their arrangement, determines the name. The first three alkanes are methane, ethane, and propane. All alkanes of formula C_4H_{10} are called butanes, all those of formula C_5H_{12} are called pentanes, and those of formula C_6H_{14} are called hexanes. For alkanes beyond propane, **iso** indicates that one end of an otherwise unbranched chain terminates in a $(CH_3)_2CH-$ group. Following are examples of common names.

$CH_3CH_2CH_2CH_3$	$CH_3CH(CH_3)CH_3$	$CH_3CH_2CH_2CH_2CH_3$	$CH_3CH_2CH(CH_3)CH_3$
Butane	Isobutane	Pentane	Isopentane

This system of common names has no good way of handling other branching patterns, and for more complex alkanes, it is necessary to use the more flexible IUPAC system of nomenclature.

In this text we concentrate on IUPAC names. We also use common names, however, especially when the common name is used almost exclusively in the everyday discussions of chemists and biochemists. When both IUPAC and common names are given in the text, we always give the IUPAC name first followed by the common name in parentheses. In this way, you should have no doubt about which name is which.

C. Classification of Carbon and Hydrogen Atoms

A carbon atom is classified as **primary (1°)**, **secondary (2°)**, **tertiary (3°)**, or **quaternary (4°)**, depending on the number of carbon atoms bonded to it. A carbon bonded to one carbon atom is a primary carbon; one bonded to two carbon atoms is a secondary carbon, and so forth. For example, propane contains two primary carbons and one secondary carbon. 2-Methylpropane contains three primary carbons and one tertiary carbon. 2,2,4-Trimethylpentane contains five primary carbons, one secondary carbon, one tertiary carbon, and one quaternary carbon.

Primary (1°) carbon A carbon bonded to one other carbon atom.

Secondary (2°) carbon A carbon bonded to two other carbon atoms.

Tertiary (3°) carbon A carbon bonded to three other carbon atoms.

Quaternary (4°) carbon A carbon bonded to four other carbon atoms.

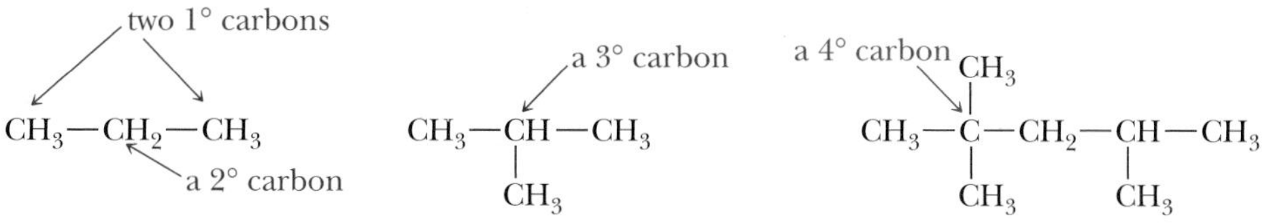

$CH_3-CH_2-CH_3$ (two 1° carbons; a 2° carbon)
Propane

$CH_3-CH(CH_3)-CH_3$ (a 3° carbon)
2-Methylpropane

$CH_3-C(CH_3)_2-CH_2-CH(CH_3)-CH_3$ (a 4° carbon)
2,2,4-Trimethylpentane

Similarly, hydrogens are also classified as primary, secondary, or tertiary depending on the type of carbon to which each is bonded. Those attached to a primary carbon are classified as primary hydrogens, those on a secondary carbon are secondary hydrogens, and those on a tertiary carbon are tertiary hydrogens.

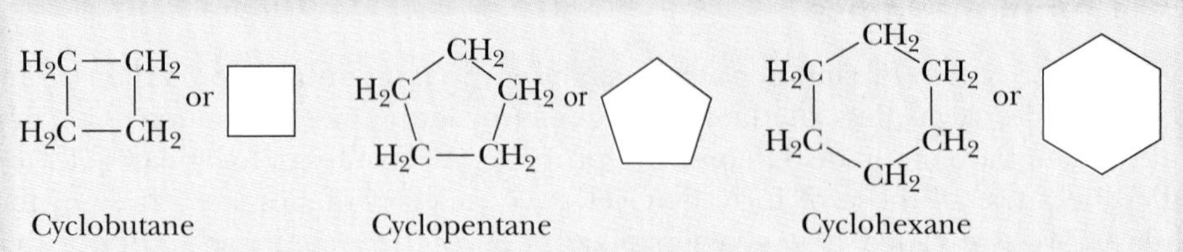

Figure 3.2
Examples of cycloalkanes.

Cyclobutane Cyclopentane Cyclohexane

3.4 CYCLOALKANES

Cycloalkane A saturated hydrocarbon that contains carbon atoms joined to form a ring.

A hydrocarbon that contains carbon atoms joined to form a ring is called a **cyclic hydrocarbon.** When all carbons of the ring are saturated, the hydrocarbon is called a **cycloalkane.** Cycloalkanes of ring sizes ranging from 3 to over 30 are found in nature, and, in principle, there is no limit to ring size. Five-membered (cyclopentane) and six-membered (cyclohexane) rings are especially abundant in nature and have received special attention.

Figure 3.2 shows structural formulas of cyclobutane, cyclopentane, and cyclohexane. As a matter of convenience, organic chemists often do not show all carbons and hydrogens when writing structural formulas for cycloalkanes. Rather, the rings are represented by regular polygons having the same number of sides. For example, cyclobutane is represented by a square, and cyclohexane is represented by a hexagon.

Line-angle drawing An abbreviated way to draw structural formulas in which each line ending represents a carbon atom and a line represents a bond.

The abbreviated structural formulas shown in Figure 3.2 are called line-angle drawings. In a **line-angle drawing** each angle and line terminus represents a carbon, each single line represents a C—C bond, each double line represents a C=C bond, and each triple line represents a C≡C bond. Thus, only the carbon framework of the molecule is shown and you are left to fill in hydrogen atoms as necessary to complete the tetravalence of each carbon.

Cycloalkanes contain two fewer hydrogen atoms than an alkane of the same number of carbon atoms. For example, compare the molecular formulas of cyclohexane (C_6H_{12}) and hexane (C_6H_{14}). The general formula of a cycloalkane is $\mathbf{C_nH_{2n}}$.

To name a cycloalkane, prefix the name of the corresponding open-chain hydrocarbon with cyclo-, and name each substituent on the ring. If there is only one substituent on the ring, there is no need to give it a number. If there are two substituents, number the ring beginning with the substituent of lower alphabetical order. If there are three or more substituents, number the ring to give them the lowest set of numbers, and then list them in alphabetical order.

EXAMPLE 3.4

Write the molecular formula and IUPAC name for each cycloalkane.

(a) 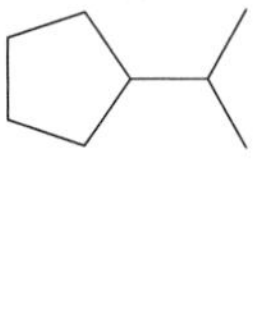 (b)

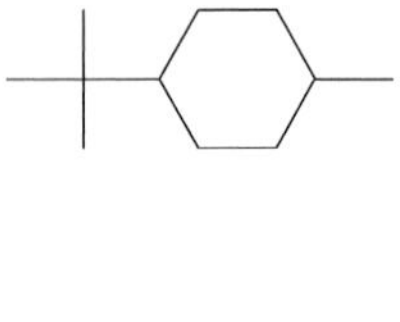

SOLUTION

(a) First replace each angle and line terminus by a carbon and then add hydrogens as necessary to give each carbon four bonds. The molecular formula of this compound is C_8H_{16}. Because there is only one substituent on the ring, there is no need to number the atoms of the ring. Its IUPAC name is isopropylcyclopentane.

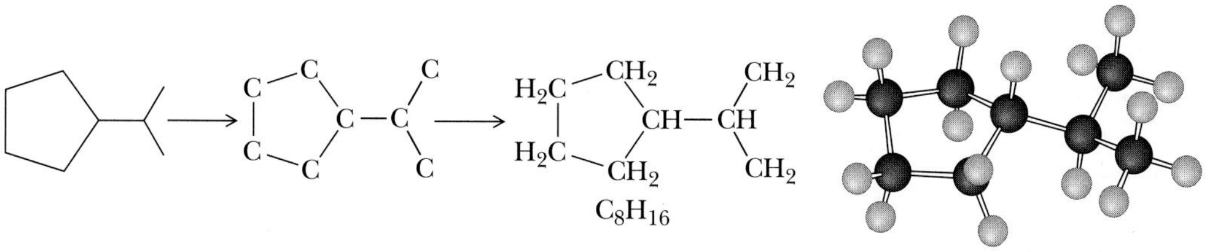

C_8H_{16}

(b) Number the atoms of the cyclohexane ring beginning with *tert*-butyl, the substituent of lower alphabetical order. Its name is 1-*tert*-butyl-4-methylcyclohexane, and its molecular formula is $C_{11}H_{22}$.

Practice Problem 3.4

Write the molecular formula and IUPAC name for each cycloalkane.

(a) 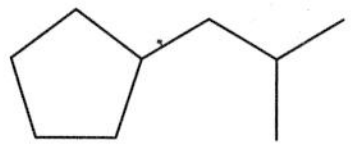 (b) 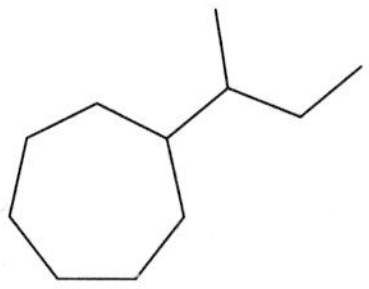 (c)

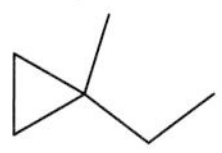

3.5 THE IUPAC SYSTEM—A GENERAL SYSTEM OF NOMENCLATURE

The naming of alkanes and cycloalkanes in Sections 3.3 and 3.4 illustrates the application of the IUPAC system of nomenclature to two specific classes of organic compounds. Now, let us describe the general approach of the IUPAC system. The name assigned to any compound with a chain of carbon atoms consists of three parts: a **prefix,** an **infix** (a modifying element inserted into a word), and a **suffix.** Each part provides specific information about the structural formula of the compound.

1. The prefix shows the number of carbon atoms in the parent chain. Table 3.2 gives prefixes to show the presence of 1 to 20 carbon atoms in a chain.
2. The infix shows the nature of the carbon-carbon bonds in the parent chain.

Infix	Nature of Carbon-Carbon Bonds in the Parent Chain
-an-	all single bonds
-en-	one or more double bonds
-yn-	one or more triple bonds

3. The suffix shows the class of compound to which the substance belongs.

Suffix	**Class of Compound**
-e	hydrocarbon
-ol	alcohol
-al	aldehyde
-one	ketone
-oic acid	carboxylic acid

EXAMPLE 3.5

Following are IUPAC names and structural formulas for four compounds. Divide each name into a prefix, an infix, and a suffix, and specify the information about the structural formula that is contained in each part of the name.

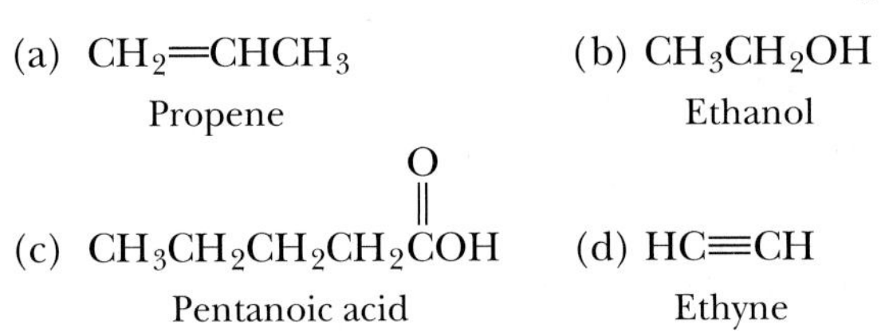

SOLUTION

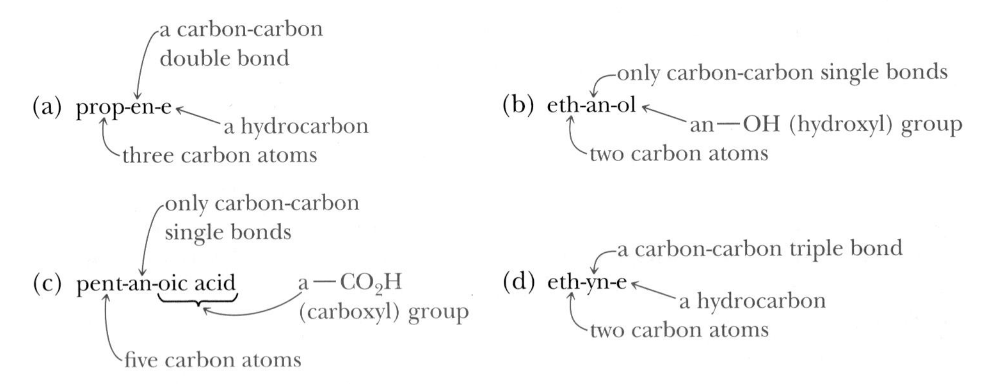

Practice Problem 3.5

Combine the proper prefix, infix, and suffix and write the IUPAC name for each compound.

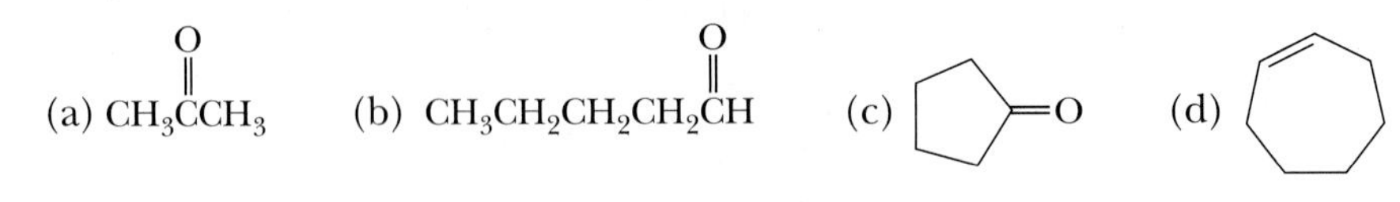

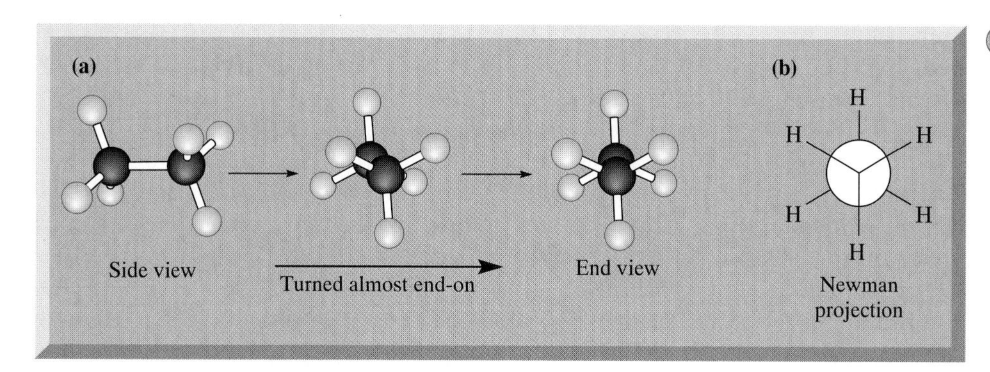

Figure 3.3
A staggered conformation of ethane. (a) Ball-and-stick model and (b) a Newman projection.

3.6 CONFORMATIONS OF ALKANES AND CYCLOALKANES

Conformation Any three-dimensional arrangement of atoms in a molecule that results by rotation about a single bond.

Even though structural formulas are useful to show the order of attachment of atoms, they usually do not show three-dimensional shapes. As chemists try to understand more and more about the relationships between structure and the chemical and physical properties of molecules, it becomes increasingly important to know more about the three-dimensional shapes of molecules.

In this section, we concentrate on ways to visualize molecules as three-dimensional objects and to visualize not only bond angles within molecules but also distances between various atoms and groups not bonded to each other. We also describe intramolecular strain, which we divide into two types—nonbonded interaction strain and angle strain. We urge you to physically build models and to study and manipulate the molecules prepared for you on the CD-ROM supplied with this text. Organic molecules are three-dimensional objects and it is essential that you become comfortable dealing with them as such.

A. Alkanes

Alkanes of two or more carbons can be twisted into a number of different three-dimensional arrangements of their atoms by rotation about a carbon-carbon bond or bonds. Any three-dimensional arrangement of atoms that results from rotation about a single bond is called a **conformation.** Figure 3.3(a) shows a ball-and-stick model of a **staggered conformation** of ethane. In this conformation, the three C—H bonds on one carbon are as far apart as possible from the three C—H bonds on the adjacent carbon. Figure 3.3(b) is a shorthand way to represent this conformation of ethane. It is called a **Newman projection.** In a Newman projection, a molecule is viewed along the axis of a C—C bond. The three atoms or groups of atoms nearer your eye are shown on lines extending from the center of the circle at angles of 120°. The three atoms or groups of atoms on the carbon farther from your eye are shown on lines extending from the circumference of the circle at angles of 120°. Remember that bond angles about each carbon in ethane are approximately 109.5° and not 120°, as this Newman projection might suggest.

Figure 3.4 shows a ball-and-stick model and a Newman projection for an **eclipsed conformation** of ethane. In this conformation, the three C—H bonds on

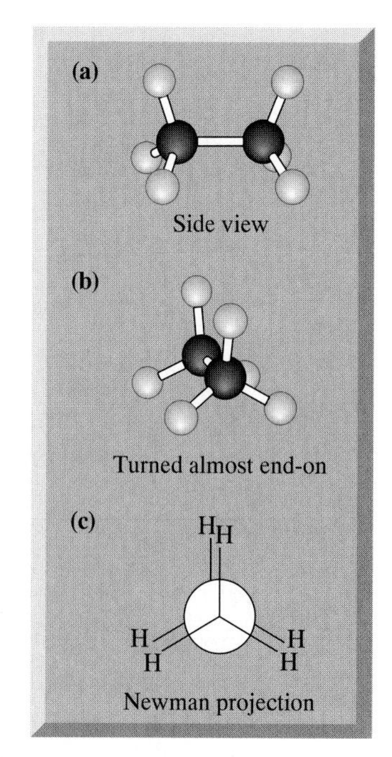

Figure 3.4
An eclipsed conformation of ethane. (a, b) Ball-and-stick models and (c) a Newman projection.

Staggered conformation A conformation about a carbon-carbon single bond, where the atoms on one carbon are as far apart as possible from atoms on the adjacent carbon.

Newman projection A way to view a molecule by looking along a carbon-carbon bond.

Eclipsed conformation A conformation about a carbon-carbon single bond where the atoms on one carbon are as close as possible to the atoms on the adjacent carbon.

Nonbonded interaction strain The strain that arises when atoms not bonded to each other are forced abnormally close to one another.

Angle strain The strain that arises when a bond angle is either compressed or expanded compared to its normal value.

one carbon are as close as possible to the three C—H bonds on the adjacent carbon. In other words, hydrogen atoms on the back carbon are eclipsed by the hydrogen atoms on the front carbon.

For a long time chemists believed that rotation about the C—C single bond in ethane was completely free. Studies of ethane and other molecules, however, have shown that a potential energy difference exists between staggered and eclipsed conformations and that rotation is not completely free. In ethane, the potential energy of the eclipsed conformation is a maximum and that of the staggered conformation is a minimum. The difference in potential energy between these two conformations is approximately 2.9 kcal/mol (12.1 kJ/mol), which means that, at room temperature, the ratio of ethane molecules in the staggered conformation to the eclipsed conformation is approximately 100 to 1.

The strain induced in the eclipsed conformation of ethane is an example of **intramolecular strain.** There are two types of intramolecular strain—nonbonded interaction strain and angle strain. **Nonbonded interaction strain** arises because atoms not bonded to each other are forced close to each other, as, for example, the eclipsed hydrogens of ethane. **Angle strain** arises from creation of abnormal bond angles. We will see examples of angle strain presently.

EXAMPLE 3.6

Draw Newman projections for one staggered conformation and one eclipsed conformation of propane.

SOLUTION

Following are Newman projections and ball-and-stick models for these conformations.

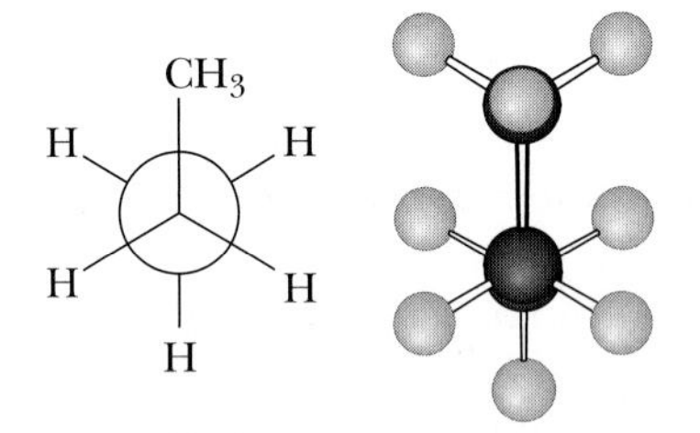

Staggered conformation

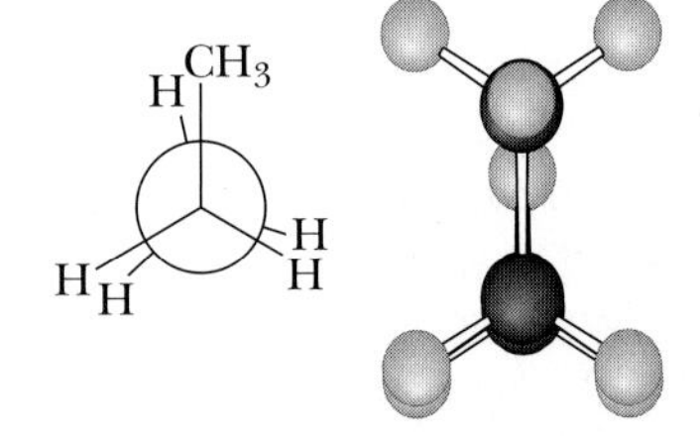

Eclipsed conformation

Practice Problem 3.6

Draw Newman projections for two staggered and two eclipsed conformations of 1,2-dichloroethane.

Figure 3.5 Cyclopentane. The most stable conformation is a puckered "envelope" conformation.

B. Cycloalkanes

We limit our discussion to the conformations of cyclopentanes and cyclohexanes because these are the most common carbon rings in the molecules of nature.

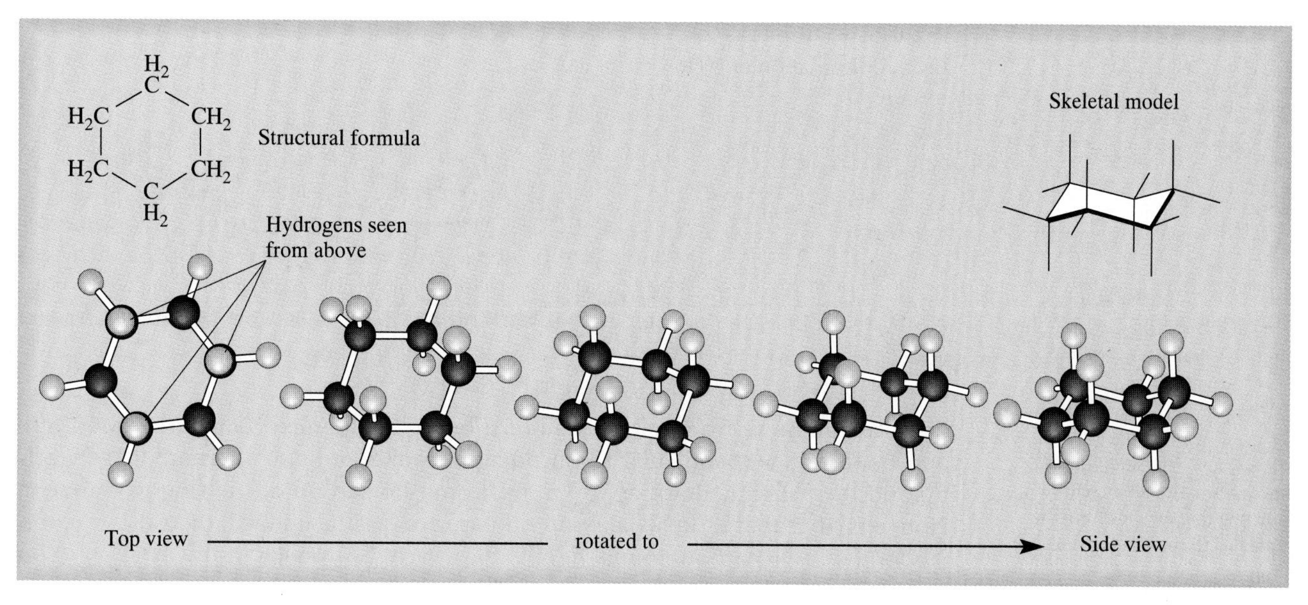

Figure 3.6
Cyclohexane. The most stable conformation is the puckered "chair" conformation.

Cyclopentane

Cyclopentane can be drawn as a planar conformation with all C—C—C bond angles equal to 108°. This angle differs only slightly from the tetrahedral angle of 109.5°, and, consequently, little angle strain occurs in the planar conformation of cyclopentane. There are, however, ten fully eclipsed C—H bonds creating nonbonded interaction strain of approximately 10 kcal/mol (42 kJ/mol). To relieve at least a part of this strain, the atoms of the ring twist into the **"envelope" conformation** shown in Figure 3.5. In this conformation, four carbon atoms are in a plane, and the fifth is bent out of the plane, rather like an envelope with its flap bent upward. In the envelope conformation, C—C—C bond angles are reduced (increasing angle strain), but the number of eclipsed hydrogen interactions is also reduced (decreasing nonbonded interaction strain). The observed C—C—C bond angles in cyclopentane are 105°, indicating that, in its conformation of lowest energy, cyclopentane is slightly puckered. The strain energy in cyclopentane is approximately 5.6 kcal/mol (23.4 kJ/mol).

Chair conformation The most stable puckered conformation of a cyclohexane ring; all bond angles are approximately 109.5°, and bonds to all adjacent carbons are staggered.

Cyclohexane

Cyclohexane adopts a number of puckered conformations, the most stable of which is the **chair conformation.** In a chair conformation (Figure 3.6), all C—C—C bond angles are 109.5° (minimizing angle strain), and hydrogens on adjacent carbons are staggered with respect to one another (minimizing nonbonded interaction strain). Thus, there is very little intramolecular strain in a chair conformation of cyclohexane.

In a chair conformation, the C—H bonds are arranged in two different orientations. Six C—H bonds are called **axial bonds,** and the other six are called **equatorial bonds.** One way to visualize the difference between these two types of bonds is to imagine an axis through the center of the chair, perpendicular to the floor (Figure 3.7). Axial bonds are parallel to this axis. Three axial bonds point up; the other three point down. Notice also that axial bonds alternate, first up and then

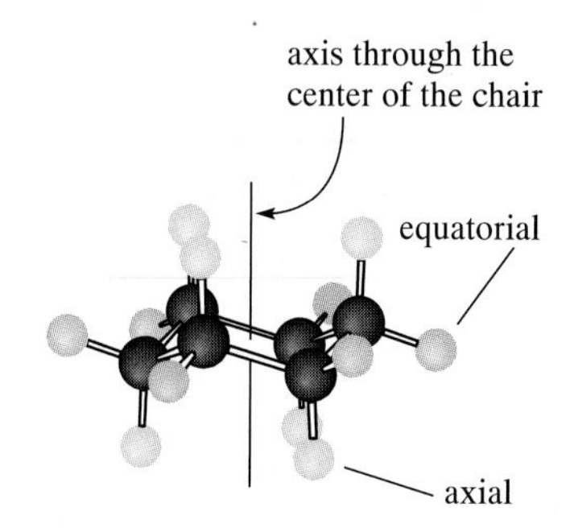

Figure 3.7
Chair conformation of cyclohexane, showing axial and equatorial C—H bonds.

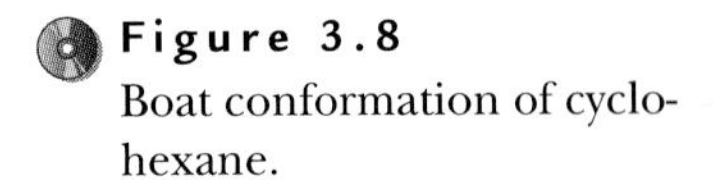

Figure 3.8
Boat conformation of cyclohexane.

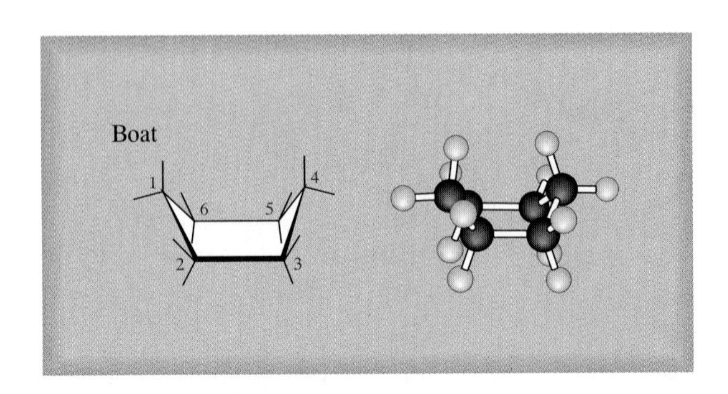

down as you move from one carbon of the ring to the next. Equatorial bonds are approximately perpendicular to our imaginary axis and also alternate first slightly up and then slightly down as you move from one carbon of the ring to the next. Notice further that if the axial bond on a carbon points upward, then the equatorial bond on that carbon points slightly downward. Conversely, if the axial bond on a particular carbon points downward, then the equatorial bond on that carbon points slightly upward.

Axial position A position on a chair conformation of a cyclohexane ring that extends from the ring parallel to the imaginary axis of the ring.

Equatorial position A position on a chair conformation of a cyclohexane ring that extends from the ring roughly perpendicular to the imaginary axis of the ring.

Boat conformation A puckered conformation of a cyclohexane ring in which carbons 1 and 4 of the ring are bent toward each other.

There are many other nonplanar conformations of cyclohexane, one of which is the **boat conformation** shown in Figure 3.8.

You can visualize interconversion of chair and boat conformations by twisting the ring as illustrated in Figure 3.9(a). A boat conformation is considerably less stable than a chair conformation because of five sets of nonbonded interactions: four sets of eclipsed hydrogen interactions and one set of **flagpole interactions.** The difference in potential energy between chair and boat conformations is approximately 6.5 kcal/mol (27 kJ/mol), which means that at room temperature approximately 99.99% of all cyclohexane molecules are in the chair conformation.

For cyclohexane, the two equivalent chair conformations can be interconverted by first twisting one chair to a boat and then to the other chair. When one chair is converted to the other, a change occurs in the relative orientations in space of the hydrogen atoms attached to each carbon. All hydrogen atoms axial in

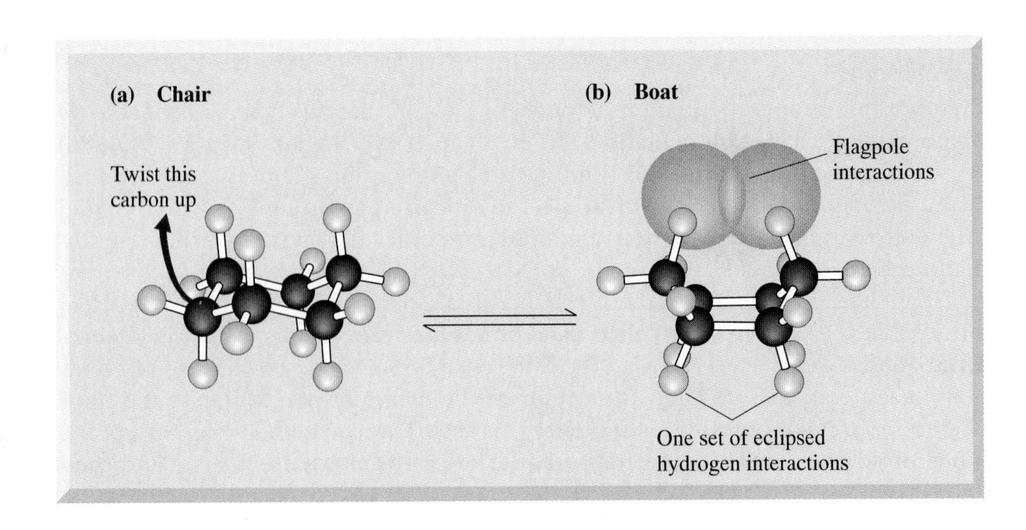

Figure 3.9
Conversion of (a) a chair conformation to (b) a boat conformation produces one set of nonbonded "flagpole" interactions and four sets of eclipsed C—H interactions.

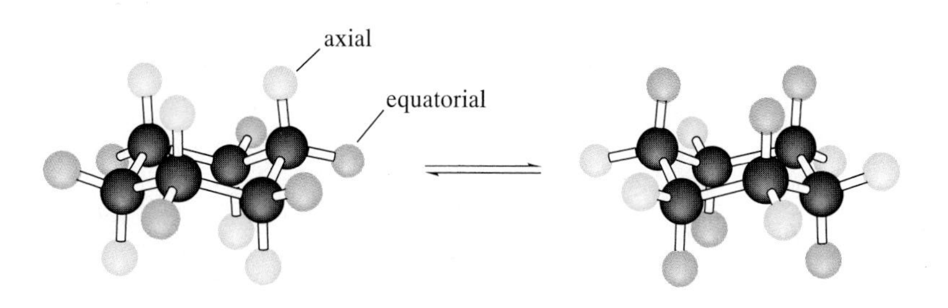

Figure 3.10
Interconversion of chair cyclohexanes. All C—H bonds equatorial in one chair are axial in the alternative chair, and vice versa.

one chair become equatorial in the other and vice versa (Figure 3.10). The conversion of one chair conformation of cyclohexane to the other occurs rapidly at room temperature.

EXAMPLE 3.7

Following is a chair conformation of cyclohexane showing a methyl group and one hydrogen.

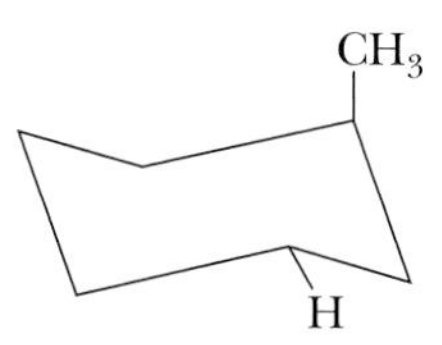

(a) Indicate by a label whether each group is equatorial or axial.
(b) Draw the other chair conformation and again label each group equatorial or axial.

SOLUTION

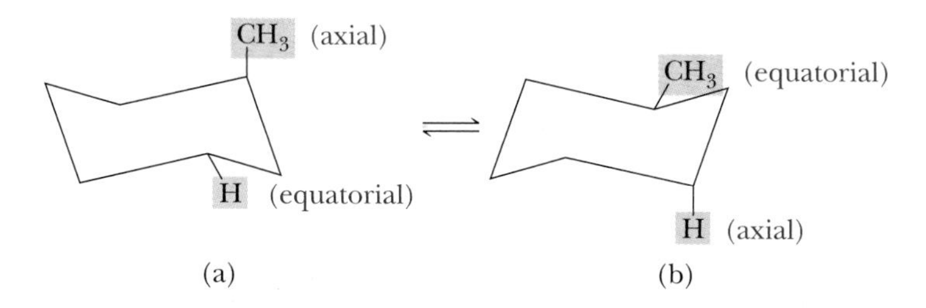

Practice Problem 3.7

Following is a chair conformation of cyclohexane with carbon atoms numbered 1 through 6.

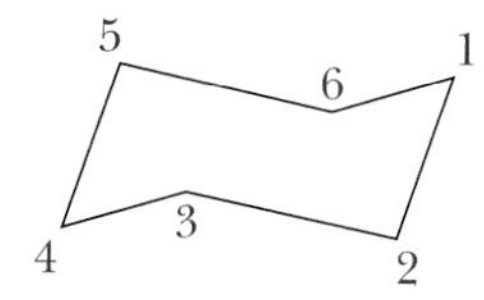

(a) Draw hydrogen atoms that are above the plane of the ring on carbons 1 and 2 and below the plane of the ring on carbon 4.
(b) Which of these hydrogens are equatorial? Which are axial?

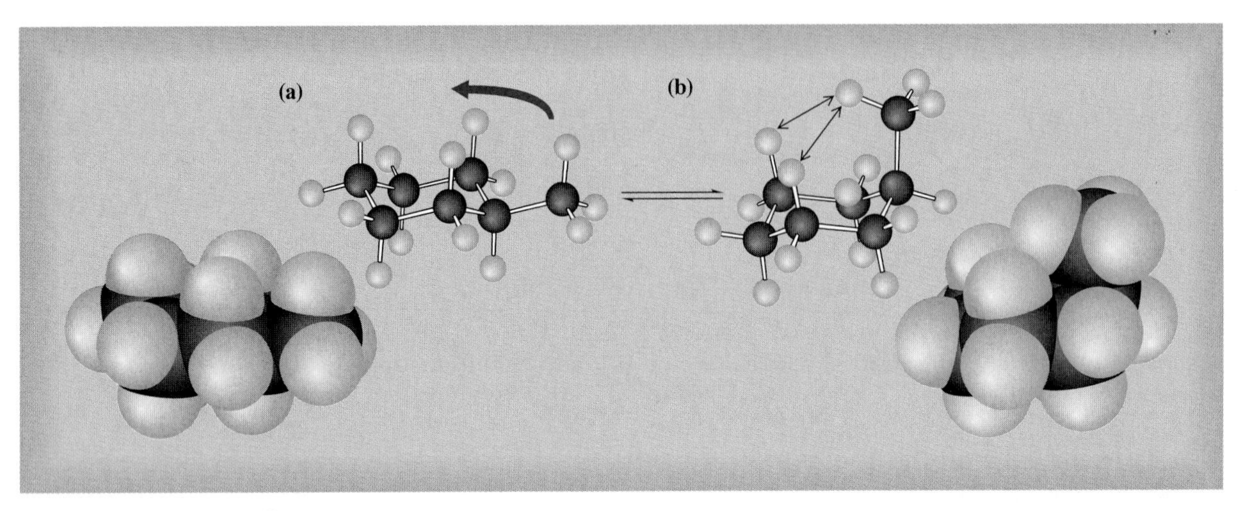

Figure 3.11
Two chair conformations of methylcyclohexane. The two axial-axial interactions make conformation (b) less stable than conformation (a) by approximately 1.74 kcal/mol (7.28 kJ/mol).

(c) Draw the other chair conformation. Now which hydrogens are equatorial? Which are axial? Which are above the plane of the ring, and which are below it?

If a hydrogen atom of cyclohexane is replaced by an alkyl group, the group occupies an equatorial position in one chair and an axial position in the other chair. This means that the two chairs are no longer equivalent and no longer of equal stability.

Axial-axial interactions Interactions between groups in parallel axial positions on the same side of a chair conformation of a cyclohexane ring.

A convenient way to describe the relative stabilities of chair conformations with equatorial or axial substituents is in terms of a type of nonbonded interaction strain called **axial-axial interaction.** Axial-axial interaction refers to the repulsion between an axial substituent and an axial hydrogen (or other group) on the same side of the ring. Consider methylcyclohexane (Figure 3.11). When the $—CH_3$ is equatorial, it is staggered with respect to all other groups on its adjacent carbon atoms. When the $—CH_3$ is axial, it is parallel to the axial C—H bonds on carbons 3 and 5. Thus, for axial methylcyclohexane, there are two unfavorable methyl-hydrogen axial-axial interactions. For methylcyclohexane, the equatorial methyl conformation is favored over the axial methyl conformation by approximately 1.74 kcal/mol (7.28 kJ/mol). At equilibrium at room temperature, approximately 95% of all methylcyclohexane molecules have their methyl group equatorial, and fewer than 5% have their methyl group axial.

As the size of the substituent increases, preference for conformations with the group equatorial increases. When the group is as large as *tert*-butyl, the equatorial conformation is approximately 4000 times more abundant at room temperature than the axial conformation, and, in effect, the ring is "locked" into a chair conformation with the *tert*-butyl group equatorial.

EXAMPLE 3.8

Label all axial-axial interactions in this chair conformation.

SOLUTION

There are four axial-axial interactions; each axial methyl group has two sets of axial-axial interactions with parallel hydrogen atoms on the same side of the ring. The equatorial methyl group has no axial-axial interactions.

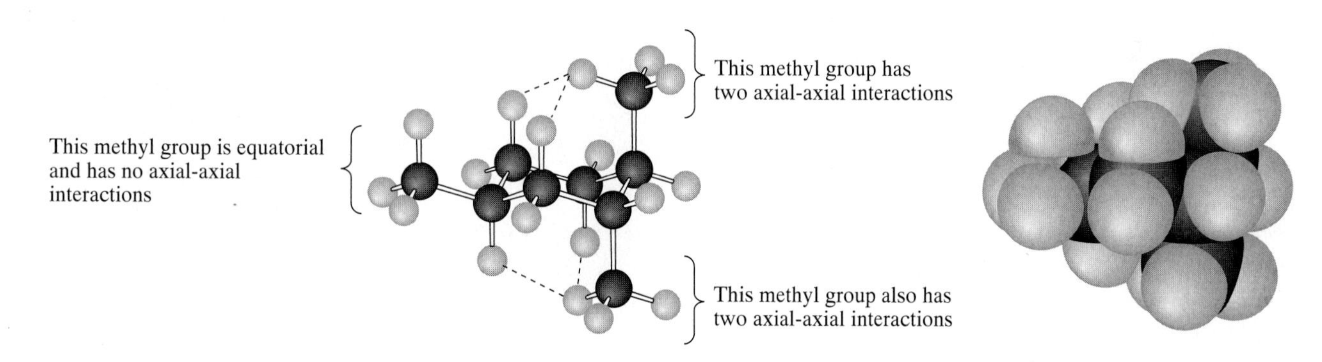
This methyl group has two axial-axial interactions
This methyl group is equatorial and has no axial-axial interactions
This methyl group also has two axial-axial interactions

Practice Problem 3.8

The conformational equilibria for methyl, ethyl, and isopropylcyclohexane are all about 95% in favor of the equatorial conformation, but the conformational equilibrium for *tert*-butylcyclohexane is almost completely on the equatorial side. Explain, by using the molecular models on the CD, why the conformational equilibria for the first three compounds are comparable but that for *tert*-butylcyclohexane lies considerably farther toward the equatorial conformation.

3.7 CIS-TRANS ISOMERISM IN CYCLOALKANES

Cycloalkanes with substituents on two or more carbons of the ring show a type of isomerism called **cis-trans isomerism.** Cis-trans isomers have (1) the same molecular formula, (2) the same order of attachment of atoms, and (3) an arrangement of atoms that cannot be interchanged by rotation about sigma bonds under ordinary conditions. By way of comparison, the potential energy difference between conformations is so small that they can be interconverted easily at or near room temperature by rotation about single bonds.

Cis-trans isomers Isomers that have the same order of attachment of their atoms but a different arrangement of their atoms in space due to the presence of either a ring or a carbon-carbon double bond.

CHEMICAL CONNECTIONS

The Poisonous Puffer Fish

Nature is by no means limited to carbon in six-membered rings. Tetrodotoxin, one of the most potent toxins known, is composed of a set of interconnected six-membered rings, each in a chair conformation. All but one of these rings have atoms other than carbon in them. Tetrodotoxin is produced in the liver and ovaries of many species of the family Tetraodontidae, especially the puffer fish, so called because it inflates itself to an almost spherical spiny ball when alarmed. It is evidently a species highly preoccupied with defense, but the Japanese are not put off. They regard the puffer, called *fugu* in Japanese, as a delicacy. To serve it in a public restaurant, a chef must be registered as sufficiently skilled in removing the toxic organs so as to make the flesh safe to eat.

Symptoms of tetrodotoxin poisoning begin with attacks of severe weakness, progressing to complete paralysis and eventual death. Tetrodotoxin exerts its severe poisoning effect by blocking Na^+ ion channels in excitable membranes. The $=NH_2^+$ end of tetrodotoxin lodges in the mouth of a Na^+ ion channel, thus blocking further transport of Na^+ ions through the channel.

Tetrodotoxin

Cis-trans isomerism in cyclic structures can be illustrated by models of 1,2-dimethylcyclopentane. In the following drawings, the cyclopentane ring is shown as a planar pentagon viewed through the plane of the ring. Carbon-carbon bonds of the ring projecting forward are shown as heavy lines. When viewed from this perspective, substituents attached to the ring project above and below the plane of the ring. In one isomer of 1,2-dimethylcyclopentane, the methyl groups are on the same side of the ring; in the other, they are on opposite sides of the ring. The prefix ***cis*** (Latin: on the same side) is used to indicate that the substituents are on the same side of the ring; the prefix ***trans*** (Latin: across) is used to indicate that they are on opposite sides of the ring. In each isomer, the configuration of the methyl groups is fixed, and, because of the restricted rotation about the ring C—C bonds, the cis isomer cannot be converted to the trans isomer or vice versa.

Cis A prefix meaning on the same side.

Trans A prefix meaning across from.

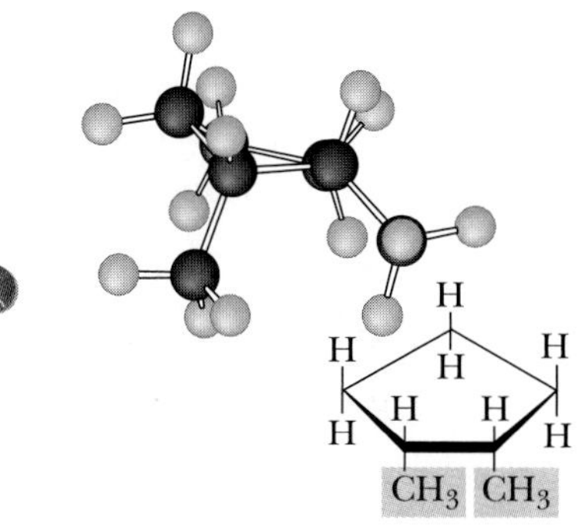

cis-1,2-Dimethylcyclopentane

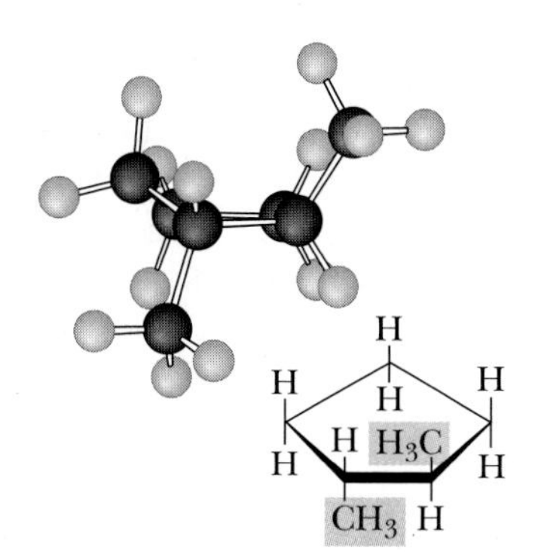

trans-1,2-Dimethylcyclopentane

Example 3.9

Which cycloalkanes show cis-trans isomerism? For each that does, draw both isomers.

(a) CH_3 (b) CH_3 CH_3 (c) H_3C CH_3

Solution

(a) Methylcyclopentane does not show cis-trans isomerism. It has only one substituent on the ring.

(b) 1,1-Dimethylcyclobutane does not show cis-trans isomerism. Only one arrangement is possible for the two methyl groups on the ring; they must be trans.

(c) 1,3-Dimethylcyclobutane shows cis-trans isomerism. Note that in these structural formulas, we show only the hydrogens on carbons bearing the methyl groups.

cis-1,3-Dimethylcyclobutane *trans*-1,3-Dimethylcyclobutane

Practice Problem 3.9

Which cycloalkanes show cis-trans isomerism? For each that does, draw both isomers.

(a) H_3C–cyclopentane–CH_3 (b) cyclopentane–CH_2CH_3 (c) cyclobutane with CH_2CH_3 and CH_3

For the purposes of determining the number of cis-trans isomers in substituted cycloalkanes, it is adequate to draw the cycloalkane ring as a planar polygon as is done in the following disubstituted cyclohexanes. Two cis-trans isomers exist for 1,4-dimethylcyclohexane.

H, CH_3, H_3C, H — *trans*-1,4-Dimethylcyclohexane

H, H, H_3C, CH_3 — *cis*-1,4-Dimethylcyclohexane

The cis and trans isomers of 1,4-dimethylcyclohexane can also be drawn as nonplanar chair conformations. In working with alternative chair conformations, it is helpful to remember that all groups axial on one chair are equatorial in the alternative chair, and vice versa. In one chair conformation of *trans*-1,4-dimethylcyclohexane, the two methyl groups are axial; in the alternative chair conformation, they are equatorial. Of these chair conformations, the one with both methyls equatorial is considerably more stable.

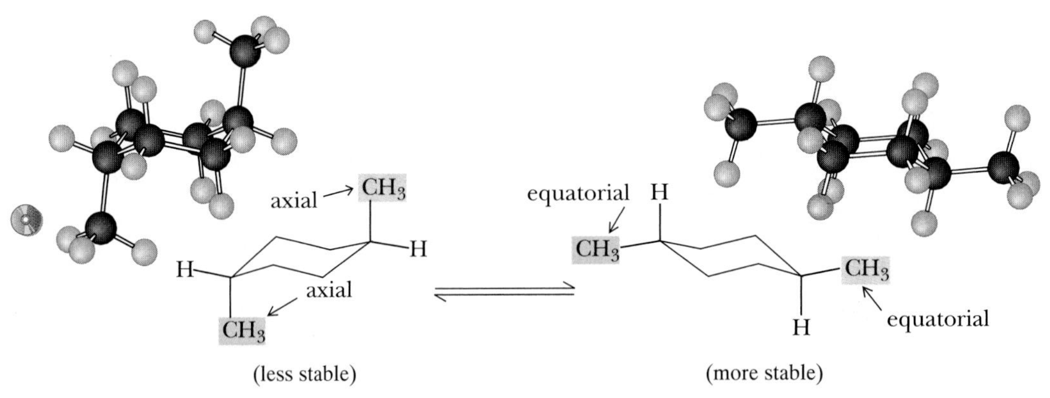

(less stable) (more stable)

trans-1,4-Dimethylcyclohexane

The alternative chair conformations of *cis*-1,4-dimethylcyclohexane are of equal energy. In one chair conformation of this molecule, one methyl group occupies an equatorial position, and the other occupies an axial position. In the other chair, the orientations in space of the —CH_3 groups are reversed.

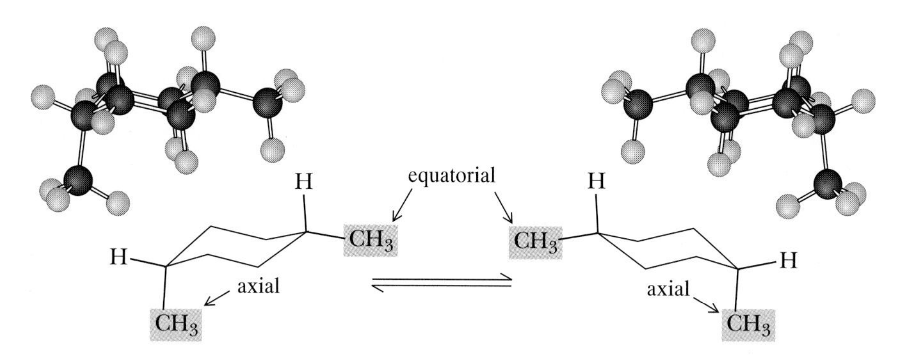

cis-1,4-Dimethylcyclohexane
(these conformations are of equal stability)

EXAMPLE 3.10

Following is a chair conformation of 1,3-dimethylcyclohexane.

H_3C CH_3 H H

(a) Is this a chair conformation of *cis*-1,3-dimethylcyclohexane or of *trans*-1,3-dimethylcyclohexane?
(b) Draw the alternative chair conformation. Of the two chair conformations, which is the more stable?
(c) Draw a planar hexagon representation for the isomer shown in this example.

SOLUTION

(a) The isomer shown is *cis*-1,3-dimethylcyclohexane; the two methyl groups are on the same side of the ring.

(b) H_3C CH_3 H H ⇌ CH_3 H CH_3 H

(more stable) (less stable)

(c) H_3C H H_3C H

Practice Problem 3.10

Following is a planar hexagon representation for one isomer of 1,2,4-trimethylcyclohexane. Draw alternative chair conformations of this compound, and state which chair conformation is the more stable.

H_3C CH_3 CH_3 H H H

3.8 Physical Properties of Alkanes and Cycloalkanes

You are already familiar with the physical properties of some alkanes and cycloalkanes from your everyday experiences. The low-molecular-weight alkanes, such as methane, ethane, propane, and butane, are gases at room temperature and atmospheric pressure. Higher-molecular-weight alkanes, such as those in gasoline and kerosene, are liquids. Very high-molecular-weight alkanes, such as those in paraffin wax, are solids. Melting points, boiling points, and densities of the first ten unbranched alkanes are listed in Table 3.4.

Methane is a gas at room temperature and atmospheric pressure. It can be converted to a liquid if cooled to −164°C and to a solid if further cooled to −182°C. The fact that methane (or any other compound, for that matter) can exist as a liquid or solid depends on the existence of **intermolecular forces of attraction** between particles of the pure compound. Although the forces of attraction between particles are all electrostatic in nature, they vary widely in their relative strengths. The strongest attractive forces are between ions, for example between Na^+ and Cl^- in NaCl (188 kcal/mol, 787 kJ/mol). Hydrogen bonding is a weaker

TABLE 3.4 Physical Properties of Some Unbranched Alkanes

Name	Condensed Structural Formula	mp (°C)	bp (°C)	Density of Liquid* (g/mL at 0°C)
methane	CH_4	−182	−164	(a gas)
ethane	CH_3CH_3	−183	−88	(a gas)
propane	$CH_3CH_2CH_3$	−190	−42	(a gas)
butane	$CH_3(CH_2)_2CH_3$	−138	0	(a gas)
pentane	$CH_3(CH_2)_3CH_3$	−130	36	0.626
hexane	$CH_3(CH_2)_4CH_3$	−95	69	0.659
heptane	$CH_3(CH_2)_5CH_3$	−90	98	0.684
octane	$CH_3(CH_2)_6CH_3$	−57	126	0.703
nonane	$CH_3(CH_2)_7CH_3$	−51	151	0.718
decane	$CH_3(CH_2)_8CH_3$	−30	174	0.730

* For comparison, the density of H_2O is 1 g/mL at 4°C.

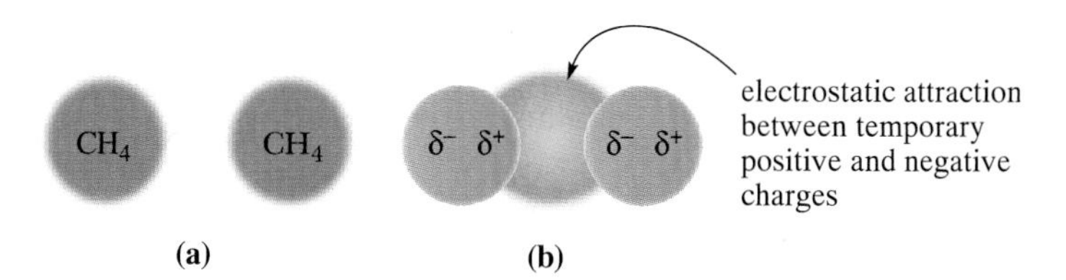

Figure 3.12 Dispersion forces. (a) The average distribution of electron density in a methane molecule is symmetrical, and there is no polarity. (b) Temporary polarization of one molecule induces temporary polarization in an adjacent molecule. Electrostatic attractions between temporary positive and negative charges are called dispersion forces.

attractive force (2–10 kcal/mol, 8–42 kJ/mol). We have more to say about hydrogen bonding in Chapter 8 where we discuss the physical properties of alcohols, compounds containing polar O—H groups.

Dispersion forces (0.02–2 kcal/mol, 0.08–8 kJ/mol) are the weakest intermolecular attractive forces. It is the existence of dispersion forces that accounts for the fact that low-molecular-weight, nonpolar substances such as methane can be liquefied. When we convert methane from a liquid to a gas at −164°C, for example, the process of separating its molecules requires only enough energy to overcome the very weak dispersion forces.

To visualize the origin of dispersion forces, it is necessary to think in terms of instantaneous distributions of electron density rather than average distributions. Over time, the distribution of electron density in a methane molecule is symmetrical [Figure 3.12(a)], and there is no separation of charge. However, at any instant, there is a probability that electron density is polarized (shifted) more toward one part of a methane molecule than toward another. This temporary polarization creates temporary partial positive and partial negative charges, which in turn induce temporary partial positive and negative charges in adjacent methane molecules [Figure 3.12(b)]. **Dispersion forces** are weak electrostatic attractive forces that occur between temporary partial positive and negative charges in adjacent atoms or molecules.

Dispersion forces Very weak intermolecular forces of attraction.

Because interactions between alkane molecules consist only of very weak dispersion forces, boiling points of alkanes are lower than those of almost any other type of compound of the same molecular weight. As the number of atoms and the molecular weight of an alkane increases, the strength of dispersion forces between a molecule and its neighbors increases, and, consequently, boiling point also increases.

Melting points of alkanes also increase with increasing molecular weight. The increase, however, is not as regular as that observed for boiling points because the ability of molecules to pack into ordered patterns of solids changes as molecular size and shape change.

The average density of the alkanes listed in Table 3.4 is about 0.7 g/mL; that of higher-molecular-weight alkanes is about 0.8 g/mL. All liquid and solid alkanes are less dense than water (1.0 g/mL), and, therefore, they float on water.

Alkanes that are constitutional isomers are different compounds and have different physical and chemical properties. Listed in Table 3.5 are boiling points, melting points, and densities of the five constitutional isomers of molecular formula C_6H_{14}. The boiling point of each of the branched-chain isomers of C_6H_{14} is lower than that of hexane itself, and the more branching there is, the lower the boiling point. These differences in boiling points are related to molecular shape in the following way. The only forces of attraction between alkane molecules are

TABLE 3.5 Physical Properties of the Isomeric Alkanes of Molecular Formula C_6H_{14}

Name	bp (°C)	mp (°C)	Density (g/mL)
hexane	68.7	−95	0.659
2-methylpentane	60.3	−154	0.653
3-methylpentane	63.3	−118	0.664
2,3-dimethylbutane	58.0	−129	0.661
2,2-dimethylbutane	49.7	−98	0.649

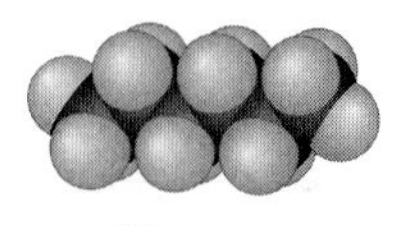
Hexane

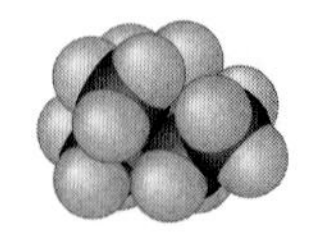
2,2-Dimethylbutane

dispersion forces. As branching increases, the shape of an alkane molecule becomes more compact, and its surface area decreases. As surface area decreases, the strength of dispersion forces decreases, and boiling points also decrease. Thus, for any group of alkane constitutional isomers, it is usually observed that the least branched isomer has the highest boiling point and the most branched isomer has the lowest boiling point.

EXAMPLE 3.11

Arrange the alkanes in each set in order of increasing boiling point.

(a) Butane, decane, and hexane
(b) 2-Methylheptane, octane, and 2,2,4-trimethylpentane

SOLUTION

(a) All compounds are unbranched alkanes. As the number of carbon atoms in the chain increases, dispersion forces between molecules increase, and boiling points increase. Decane has the highest boiling point, and butane has the lowest boiling point.

$CH_3CH_2CH_2CH_3$	$CH_3CH_2CH_2CH_2CH_2CH_3$	$CH_3(CH_2)_8CH_3$
Butane	Hexane	Decane
bp −0.5°C	bp 69°C	bp 174°C

(b) These three alkanes are constitutional isomers of molecular formula C_8H_{18}. Their relative boiling points depend on the degree of branching. 2,2,4-Trimethylpentane, the most highly branched isomer, has the smallest surface area and the lowest boiling point. Octane, the unbranched isomer, has the largest surface area and the highest boiling point.

$CH_3C(CH_3)_2CH_2CH(CH_3)CH_3$	$CH_3CH(CH_3)CH_2CH_2CH_2CH_2CH_3$	$CH_3(CH_2)_6CH_3$
2,2,4-Trimethylpentane	2-Methylheptane	Octane
bp 99°C	bp 118°C	bp 126°C

Practice Problem 3.11

Arrange the alkanes in each set in order of increasing boiling point.

(a) 2-Methylbutane, 2,2-dimethylpropane, and pentane
(b) 3,3-Dimethylheptane, 2,2,4-trimethylhexane, and nonane

3.9 REACTIONS OF ALKANES

Alkanes and cycloalkanes are quite unreactive toward most reagents, a behavior consistent with the fact that they are nonpolar compounds containing only strong sigma bonds. They do react, however, under certain conditions with oxygen, O_2. By far their most important reaction with oxygen is oxidation (combustion) to form carbon dioxide and water. Oxidation of saturated hydrocarbons is the basis for their use as energy sources for heat [natural gas, liquefied petroleum gas (LPG), and fuel oil] and power (gasoline, diesel fuel, and aviation fuel). Following are balanced equations for the complete combustion of methane, the major component of natural gas, and propane, the major component of LPG.

$$\underset{\text{Methane}}{CH_4} + 2O_2 \longrightarrow CO_2 + 2H_2O \qquad \Delta H^\circ = -212\ \text{kcal/mol}\ (-886\ \text{kJ/mol})$$

$$\underset{\text{Propane}}{CH_3CH_2CH_3} + 5O_2 \longrightarrow 3CO_2 + 4H_2O \qquad \Delta H^\circ = -530\ \text{kcal/mol}\ (-2220\ \text{kJ/mol})$$

3.10 SOURCES OF ALKANES

The three major sources of alkanes throughout the world are the fossil fuels—natural gas, petroleum, and coal. Fossil fuels account for approximately 90% of the total energy consumed in the United States. Nuclear electric power and hydroelectric power make up most of the remaining 10%. In addition, fossil fuels provide the bulk of the raw material for the organic chemicals consumed worldwide.

A. Natural Gas

Natural gas consists of approximately 90–95% methane, 5–10% ethane, and a mixture of other relatively low boiling alkanes—chiefly propane, butane, and 2-methylpropane. The current widespread use of ethylene as the organic chemical industry's most important building block is largely the result of the ease with which ethane can be separated from natural gas and cracked into ethylene. **Cracking** is a process whereby a saturated hydrocarbon is converted into an unsaturated hydrocarbon plus H_2. Ethane is cracked by heating it in a furnace at 800 to 900°C for a fraction of a second. Production of ethylene in the United States in 1997 was 51.1 billion pounds, making it, on a weight basis, the number one organic compound produced by the U.S. chemical industry. The bulk of it is used to create organic polymers, as described in Chapter 15.

$$\underset{\text{Ethane}}{CH_3CH_3} \xrightarrow[\text{(thermal cracking)}]{800\text{–}900^\circ C} \underset{\text{Ethylene}}{CH_2{=}CH_2} + H_2$$

A petroleum refinery. *(K. Straiton/Photo Researchers, Inc.)*

B. Petroleum

Petroleum is a thick, viscous liquid mixture of literally thousands of compounds, most of them hydrocarbons, formed from the decomposition of marine plants and animals. Petroleum and petroleum-derived products fuel automobiles, aircraft, and trains. They provide most of the greases and lubricants required for the machinery of our highly industrialized society. Furthermore, petroleum, along with natural gas, provides close to 90% of the organic raw materials for the synthesis and manufacture of synthetic fibers, plastics, detergents, drugs, dyes, and a multitude of other products.

It is the task of a petroleum refinery to produce usable products, with a minimum of waste, from the thousands of different hydrocarbons in this liquid mixture. The various physical and chemical processes for this purpose fall into two broad categories—**separation processes,** which separate the complex mixture into various fractions, and **reforming processes,** which alter the molecular structure of the hydrocarbon components themselves.

The fundamental separation process in refining petroleum is fractional distillation (Figure 3.13). Practically all crude oil that enters a refinery goes to distillation units where it is heated to temperatures as high as 370 to 425°C and separated into fractions. Each fraction contains a mixture of hydrocarbons that boils within a particular range.

1. Gases boiling below 20°C are taken off at the top of the distillation column. This fraction is a mixture of low-molecular-weight hydrocarbons, predominantly propane, butane, and 2-methylpropane, substances that can be liquefied under pressure at room temperature. The liquefied mixture, known as lique-

Figure 3.13
Fractional distillation of petroleum. The lighter, more volatile fractions are removed from higher up the column and the heavier, less volatile fractions from lower down.

Gases
Boiling point range below 20°C
Gasoline (naphthas)
20–200°C
Kerosene
175–275°C
Fuel oil
250–400°C
Lubricating oil
above 350°C
Crude oil and vapor are preheated
Residue (asphalt)

fied petroleum gas (LPG), can be stored and shipped in metal tanks and is a convenient source of gaseous fuel for home heating and cooking.
2. Naphthas, bp 20 to 200°C, are a mixture of C_5 to C_{12} alkanes and cycloalkanes. The naphthas also contain small amounts of benzene, toluene, xylene, and other aromatic hydrocarbons (Chapter 9). The light naphtha fraction, bp 20 to 150°C, is the source of straight-run gasoline and averages approximately 25% of crude petroleum. In a sense, naphthas are the most valuable distillation fractions because they are useful not only as fuel but also as sources of raw materials for the organic chemical industry.
3. Kerosene, bp 175 to 275°C, is a mixture of C_9 to C_{15} hydrocarbons.
4. Fuel oil, bp 250 to 400°C, is a mixture of C_{15} to C_{18} hydrocarbons. Diesel fuel is obtained from this fraction.
5. Lubricating oil and heavy fuel oil distill from the column at temperatures above 350°C.
6. Asphalt is the black, tarry residue remaining after removal of the other volatile fractions.

The two most common reforming processes are cracking, as illustrated by thermal conversion of ethane to ethylene (Section 3.10A), and **catalytic reforming,** as illustrated by the conversion of hexane first to cyclohexane and then to benzene.

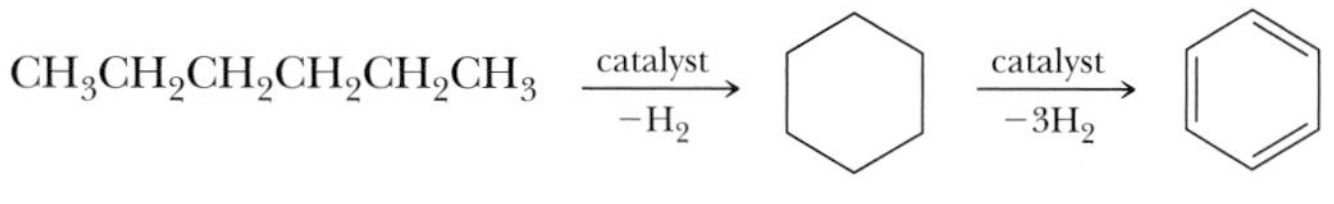

$$CH_3CH_2CH_2CH_2CH_2CH_3 \xrightarrow[-H_2]{\text{catalyst}} \text{Cyclohexane} \xrightarrow[-3H_2]{\text{catalyst}} \text{Benzene}$$

Hexane Cyclohexane Benzene

Gasoline is a complex mixture of C_6 to C_{12} hydrocarbons. The quality of gasoline as a fuel for internal combustion engines is expressed in terms of octane rating. Engine knocking occurs when a portion of the air-fuel mixture explodes prematurely (usually as a result of heat developed during compression) and independently of ignition by the spark plug. Two compounds were selected as reference fuels. One of these, 2,2,4-trimethylpentane (isooctane), has very good antiknock properties (the fuel-air mixture burns smoothly in the combustion chamber) and was assigned an octane rating of 100. (The name *isooctane* is a trivial name; its only relation to the name 2,2,4-trimethylpentane is that both show eight carbon atoms.) Heptane, the other reference compound, has poor antiknock properties and was assigned an octane rating of 0.

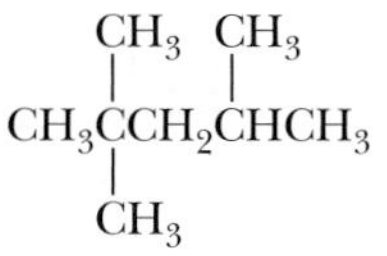

$$\begin{array}{c} CH_3\ \ CH_3 \\ |\ \ \ \ \ \ | \\ CH_3CCH_2CHCH_3 \\ | \\ CH_3 \end{array} \qquad CH_3(CH_2)_5CH_3$$

2,2,4-Trimethylpentane (octane rating 100) Heptane (octane rating 0)

The **octane rating** of a particular gasoline is that percent isooctane in a mixture of isooctane and heptane that has equivalent antiknock properties. For example, the antiknock properties of 2-methylhexane are the same as those of a mixture of 42% isooctane and 58% heptane; therefore, the octane rating of 2-methylhexane is 42. Octane itself has an octane rating of −20, which means that it produces even more engine knocking than heptane.

Octane rating The percentage of isooctane in a mixture of isooctane and heptane that has equivalent knock properties to a test gasoline.

C. Coal

To understand how coal can be used as a raw material for the production of organic compounds, it is necessary to discuss synthesis gas. **Synthesis gas** is a mixture of carbon monoxide and hydrogen in varying proportions depending on the means by which it is manufactured. Synthesis gas is prepared by passing steam over coal. It is also prepared by partial oxidation of methane with oxygen.

$$\underset{\text{Coal}}{C} + H_2O \xrightarrow{\text{heat}} CO + H_2$$

$$\underset{\text{Methane}}{CH_4} + \frac{1}{2}O_2 \xrightarrow{\text{catalyst}} CO + 2H_2$$

Two important organic compounds produced today almost exclusively from carbon monoxide and hydrogen are methanol and acetic acid. In the production of methanol, the carbon monoxide to hydrogen ratio is adjusted to 1:2, and the mixture is passed over a catalyst at elevated temperature and pressure.

$$CO + 2H_2 \xrightarrow{\text{catalyst}} \underset{\text{Methanol}}{CH_3OH}$$

Treatment of methanol, in turn, with carbon monoxide over a different catalyst gives acetic acid.

$$\underset{\text{Methanol}}{CH_3OH} + CO \xrightarrow{\text{catalyst}} \underset{\text{Acetic acid}}{CH_3\overset{\overset{\displaystyle O}{\|}}{C}OH}$$

Because the processes for making methanol and acetic acid directly from carbon monoxide are commercially proven, it is likely that the decades ahead will see the development of routes to other organic chemicals from coal via methanol.

SUMMARY

A **hydrocarbon** is a compound that contains only carbon and hydrogen. A **saturated hydrocarbon** contains only single bonds. Alkanes have the general formula C_nH_{2n+2}. **Constitutional isomers** (Section 3.2) have the same molecular formula but a different connectivity (a different order of attachment of their atoms). Alkanes are named according to a set of rules developed by the **International Union of Pure and Applied Chemistry (IUPAC)** (Section 3.3A). The IUPAC system is a general system of nomenclature (Section 3.5). The IUPAC name of a compound consists of three parts: (1) a **prefix** that tells the number of carbon atoms in the parent chain, (2) an **infix** that tells the nature of the carbon-carbon bonds in the parent chain, and (3) a **suffix** that tells the class to which the compound belongs. Substituents derived from alkanes by removal of a hydrogen atom are called **alkyl groups** and are given the symbol **R**. The name of an alkyl group is formed by dropping the suffix -ane from the name of the parent alkane and adding -yl in its place.

A carbon atom is classified as **primary (1°)**, **secondary (2°)**, **tertiary** (3°), or **quaternary** (4°) depending on the number of carbon atoms bonded to it (Section 3.3C). A hydrogen atom is classified as primary (1°), secondary (2°), or tertiary (3°) depending on the type of carbon to which it is bonded.

An alkane that contains carbon atoms bonded to form a ring is called a **cycloalkane** (Section 3.4). To name a cycloalkane, prefix the name of the open-chain hydrocarbon by cyclo-. Five-membered rings (cyclopentanes) and six-membered rings (cyclohexanes) are especially abundant in the biological world.

A **conformation** is any three-dimensional arrangement of the atoms of a molecule that results by rotation about a single bond (Section 3.6). One convention for showing conformations is the **Newman projection.** Staggered conformations are lower in energy than eclipsed conformations.

Intramolecular strain (Section 3.6A) is of two types: (1) **angle strain,** which arises from creation of either abnormally large or abnormally small bond angles, and (2) **nonbonded interaction strain,** which arises when atoms not bonded to each other are forced abnormally close to one another.

Cyclopentanes, cyclohexanes, and all larger cycloalkanes exist in dynamic equilibrium between a set of puckered conformations. The lowest energy conformation of cyclopentane is an envelope conformation. The lowest energy conformations of cyclohexane are two interconvertible **chair conformations** (Section 3.6B). In a chair conformation, six bonds are **axial** and six bonds are **equatorial.** Bonds axial in one chair are equatorial in the alternative chair, and vice versa. A **boat conformation** is higher in energy than chair conformations. The more stable conformation of a substituted cyclohexane is the one that minimizes **axial-axial interactions.**

Cis-trans isomers (Section 3.7) have the same molecular formula and the same order of attachment of atoms, but arrangements of atoms in space that cannot be interconverted by rotation about single bonds. **Cis** means that substituents are on the same side of the ring; **trans** means that they are on opposite sides of the ring. Most cycloalkanes with substituents on two or more carbons of the ring show cis-trans isomerism.

Low-molecular-weight alkanes, such as methane, ethane, and propane, are gases at room temperature and atmospheric pressure. Higher-molecular-weight alkanes, such as those in gasoline and kerosene, are liquids. Very high-molecular-weight alkanes, such as those in paraffin wax, are solids.

Alkanes are nonpolar compounds and the only forces of attraction between their molecules are **dispersion forces** (Section 3.8), which are weak electrostatic interactions between temporary partial positive and negative charges of atoms or molecules. Among a set of alkane constitutional isomers, the least branched isomer generally has the highest boiling point; the most branched isomer generally has the lowest boiling point.

Natural gas (Section 3.10A) consists of 90–95% methane with lesser amounts of ethane and other lower-molecular-weight hydrocarbons. **Petroleum** (Section 3.10B) is a liquid mixture of literally thousands of different hydrocarbons. **Synthesis gas,** a mixture of carbon monoxide and hydrogen, can be derived from natural gas and coal.

KEY REACTIONS

1. Oxidation of Alkanes (Section 3.9)

Oxidation of alkanes to carbon dioxide and water is the basis for their use as energy sources of heat and power.

$$CH_3CH_2CH_3 + 5O_2 \longrightarrow 3CO_2 + 4H_2O + \text{energy}$$

ADDITIONAL PROBLEMS

Constitutional Isomerism

3.12 Which statements are true about constitutional isomers?

(a) They have the same molecular formula.

(b) They have the same molecular weight.

(c) They have the same order of attachment of atoms.

(d) They have the same physical properties.

3.13 Which structural formulas represent identical compounds and which represent constitutional isomers?

(a) $CH_3CH_2CH(OH)CH_3$ (OH on the third carbon) **(b)** cyclobutyl–OH **(c)** $HOCH_2$–cyclopropyl **(d)** $CH_3CH(CH_2OH)CH_3$ (CH_2OH on the central carbon)

(e) $HOCH_2CHCH_3$ (with CH_3 on C-2) (f) $CH_3CH_2CH_2CH_2OH$ (g) CH_3CHOH (with CH_2CH_3 on C-2) (h) CH_3CCH_3 (with CH_3 and OH on C-2)

3.14 Name and draw structural formulas for the nine constitutional isomers of molecular formula C_7H_{16}.

3.15 Tell whether the compounds in each set are constitutional isomers.

(a) CH_3CH_2OH and CH_3OCH_3 **(b)** $CH_3\overset{O}{\overset{\|}{C}}CH_3$ and $CH_3CH_2\overset{O}{\overset{\|}{C}}H$

(c) $CH_3\overset{O}{\overset{\|}{C}}OCH_3$ and $CH_3CH_2\overset{O}{\overset{\|}{C}}OH$ **(d)** $CH_3\overset{OH}{\overset{|}{C}}HCH_2CH_3$ and $CH_3\overset{O}{\overset{\|}{C}}CH_2CH_3$

(e) 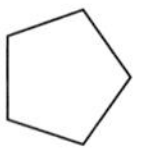 (cyclopentane) and $CH_3CH_2CH_2CH_2CH_3$ **(f)** 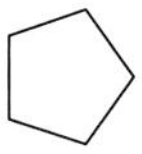 (cyclopentane) and $CH_2{=}CHCH_2CH_2CH_3$

3.16 Draw structural formulas for

(a) The four alcohols of molecular formula $C_4H_{10}O$.
(b) The two aldehydes of molecular formula C_4H_8O.
(c) The one ketone of molecular formula C_4H_8O.
(d) The three ketones of molecular formula $C_5H_{10}O$.
(e) The four carboxylic acids of molecular formula $C_5H_{10}O_2$.

Nomenclature of Alkanes and Cycloalkanes

3.17 Write IUPAC names for these alkanes and cycloalkanes.

(a) $CH_3\underset{CH_3}{\underset{|}{C}}HCH_2CH_2CH_3$ **(b)** $CH_3\underset{CH_3}{\underset{|}{C}}HCH_2CH_2\underset{CH_3}{\underset{|}{C}}HCH_3$ **(c)** $CH_3(CH_2)_4\underset{CH_2CH_3}{\underset{|}{C}}HCH_2CH_3$

(d) $(CH_3)_2CHC(CH_3)_3$ **(e)** 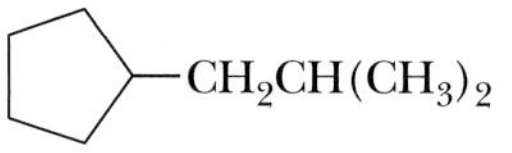 (cyclopentane)$-CH_2CH(CH_3)_2$ **(f)** 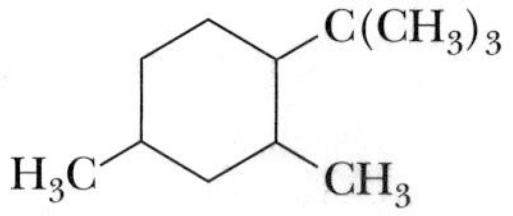 (cyclohexane with $C(CH_3)_3$, CH_3, and H_3C substituents)

3.18 Write structural formulas for these alkanes.

(a) 2,2,4-Trimethylhexane
(b) 2,2-Dimethylpropane
(c) 3-Ethyl-2,4,5-trimethyloctane
(d) 5-Butyl-2,2-dimethylnonane
(e) 4-Isopropyloctane
(f) 3,3-Dimethylpentane
(g) *trans*-1,3-Dimethylcyclopentane
(h) *cis*-1,2-Diethylcyclobutane

3.19 Explain why each is an incorrect IUPAC name. Write the correct IUPAC name for the intended compound.

(a) 1,3-Dimethylbutane
(b) 4-Methylpentane
(c) 2,2-Diethylbutane
(d) 2-Ethyl-3-methylpentane
(e) 2-Propylpentane
(f) 2,2-Diethylheptane
(g) 2,2-Dimethylcyclopropane
(h) 1-Ethyl-5-methylcyclohexane

The IUPAC System of Nomenclature

3.20 Draw a structural formula for each compound.

(a) Ethanol **(b)** Ethanal **(c)** Ethanoic acid
(d) Butanone **(e)** Butanal **(f)** Butanoic acid
(g) Propanal **(h)** Cyclopropanol **(i)** Cyclopentanol
(j) Cyclopentene **(k)** Cyclopentanone

3.21 Write the IUPAC names for each compound.

(a) $CH_3\overset{O}{\overset{\|}{C}}CH_3$ **(b)** $CH_3(CH_2)_3\overset{O}{\overset{\|}{C}}H$ **(c)** $CH_3(CH_2)_8\overset{O}{\overset{\|}{C}}OH$

(d) 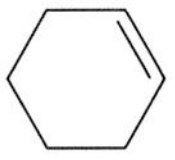 **(e)** 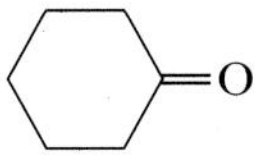 =O **(f)** 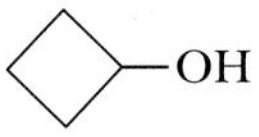 —OH

Conformations of Alkanes and Cycloalkanes

3.22 How many different staggered conformations are there for 2-methylpropane? How many different eclipsed conformations are there?

3.23 Looking along the bond between carbons 2 and 3 of butane, there are two different staggered conformations and two different eclipsed conformations. Draw Newman projections of each, and arrange them in order from most stable conformation to least stable conformation.

3.24 Demonstrate, using the molecular models on the CD, that, in cyclohexane, an equatorial substituent is almost equidistant from the axial and equatorial hydrogens on each adjacent carbon.

Cis-Trans Isomerism in Cycloalkanes

3.25 Name and draw structural formulas for the cis and trans isomers of 1,2-dimethylcyclopropane.

3.26 Name and draw structural formulas for all cycloalkanes of molecular formula C_5H_{10}. Be certain to include cis-trans isomers as well as constitutional isomers.

3.27 Using a planar pentagon representation for the cyclopentane ring, draw structural formulas for the cis and trans isomers of

(a) 1,2-Dimethylcyclopentane. **(b)** 1,3-Dimethylcyclopentane.

3.28 Draw the alternative chair conformations for the cis and trans isomers of 1,2-dimethylcyclohexane, 1,3-dimethylcyclohexane, and 1,4-dimethylcyclohexane.

(a) Indicate by a label whether each methyl group is axial or equatorial.

(b) For which isomer(s) are the alternative chair conformations of equal stability?

(c) For which isomer(s) is one chair conformation more stable than the other?

3.29 Use your answers from Problem 3.28 to complete the table showing correlations between cis, trans and axial, equatorial for disubstituted derivatives of cyclohexane.

Position of Substitution	cis	trans
1,4-	a,e or e,a	e,e or a,a
1,3-	____ or ____	____ or ____
1,2-	____ or ____	____ or ____

Peppermint plant *(Mentha piperita)*, a perennial herb with aromatic qualities used in candies, gums, hot and cold beverages, and garnish for punch and fruit. *(© John Kaprielian/Photo Researchers, Inc.)*

3.30 There are four cis-trans isomers of 2-isopropyl-5-methylcyclohexanol.

$CH(CH_3)_2$ 2 1 OH 5 CH_3

2-Isopropyl-5-methylcyclohexanol

(a) Using a planar hexagon representation for the cyclohexane ring, draw structural formulas for these four isomers.

(b) Draw the more stable chair conformation for each of your answers in part (a).

(c) Of the four cis-trans isomers, which is the most stable? (If you answered this part correctly, you picked the isomer found in nature and given the name menthol. *The Merck Index,* 12th ed., #5882.)

3.31 Draw alternative chair conformations for each substituted cyclohexane and state which chair is more stable.

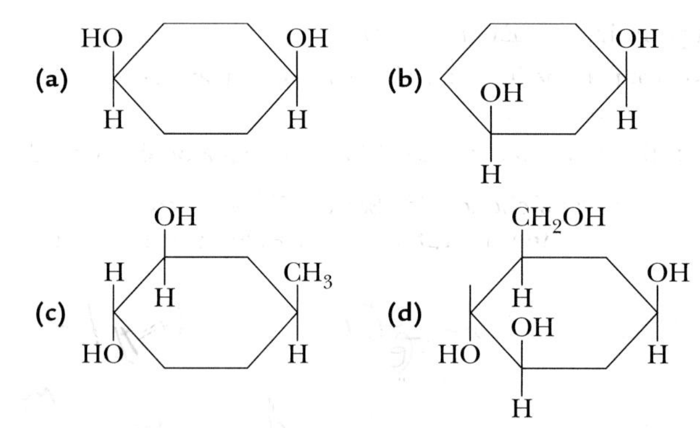

3.32 What kinds of conformations do the six-membered rings exhibit in adamantane?

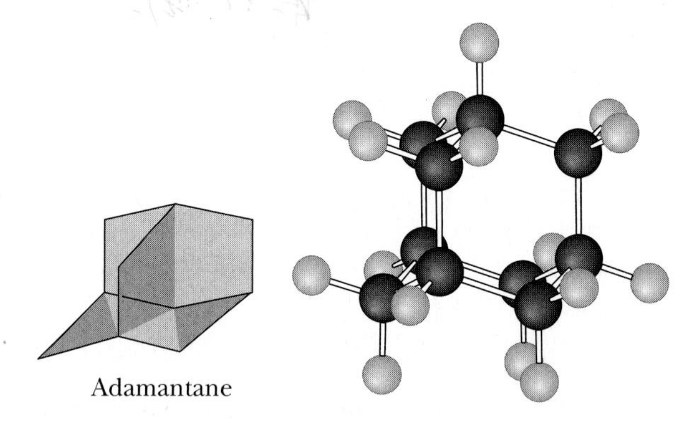

3.33 Glucose (Section 16.2B) contains a six-membered ring. In the more stable chair conformation of this molecule, all substituents on the ring are equatorial. Draw this more stable chair conformation using the ring skeleton on the right.

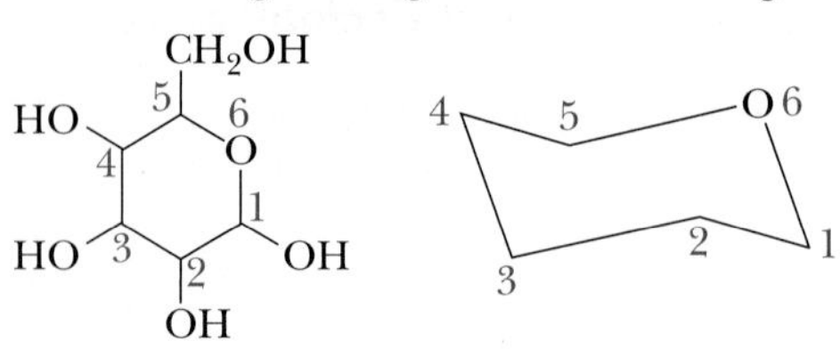

3.34 Following is the structural formula and ball-and-stick model of cholestanol (*The Merck Index,* 12th ed., #2255). The only difference between this compound and cholesterol (Section 17.4A) is that cholesterol has a carbon-carbon double bond in ring B.

Cholestanol

(a) Describe the conformation of rings A, B, C, and D in cholestanol.

(b) Is the hydroxyl group on ring A axial or equatorial?

(c) Consider the methyl group at the junction of rings A/B. Is it axial or equatorial to ring A? Is it axial or equatorial to ring B?

(d) Is the methyl group at the junction of rings C/D axial or equatorial to ring C?

3.35 Following is the structural formula of cholic acid (Section 17.4A), a component of human bile whose function is to aid in the absorption and digestion of dietary fats.

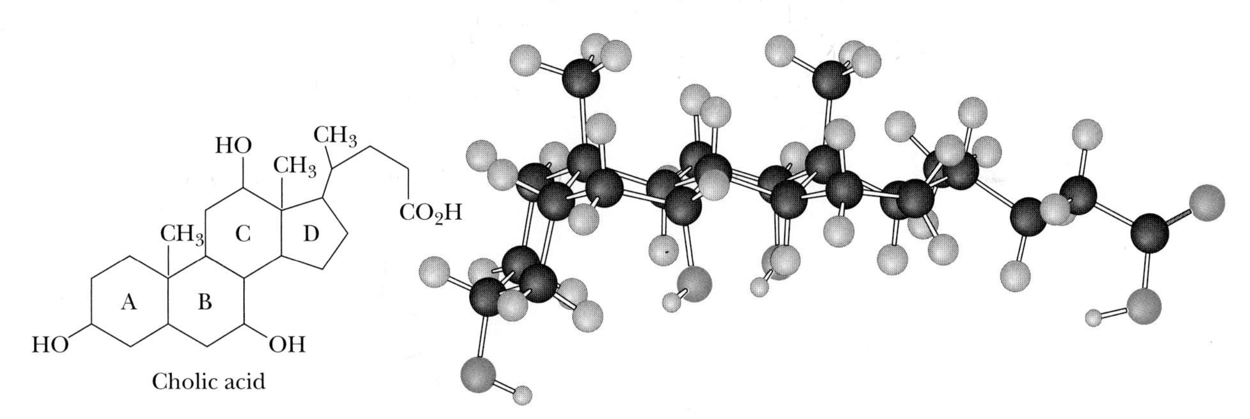

Cholic acid

(a) What is the conformation of ring A? of ring B? of ring C? of ring D?

(b) There are hydroxyl groups on rings A, B, and C. Tell whether each is axial or equatorial.

(c) Is the methyl group at the junction of rings A/B axial or equatorial to ring A? Is it axial or equatorial to ring B?

(d) Is the methyl group at the junction of rings C/D axial or equatorial to ring C?

Physical Properties of Alkanes and Cycloalkanes

3.36 In Problem 3.14 you drew structural formulas for all constitutional isomers of molecular formula C_7H_{16}. Predict which isomer has the lowest boiling point and which has the highest boiling point.

3.37 What generalizations can you make about the densities of alkanes relative to that of water?

3.38 What unbranched alkane has about the same boiling point as water (see Table 3.4)? Calculate the molecular weight of this alkane, and compare it with that of water.

Reactions of Alkanes

3.39 Write balanced equations for combustion of each hydrocarbon. Assume that each is converted completely to carbon dioxide and water.

(a) Hexane **(b)** Cyclohexane **(c)** 2-Methylpentane

3.40 Following are heats of combustion of methane and propane. On a gram-for-gram basis, which of these hydrocarbons is the best source of heat energy?

Hydrocarbon	Component of	$\Delta H°$ [kcal/mol (kJ/mol)]
CH_4	natural gas	−212 (−886 kJ/mol)
$CH_3CH_2CH_3$	LPG	−530 (−2220 kJ/mol)

4

Chirality

Tartaric acid is found in grapes and other fruits, both free and as its salts. Inset: Model of tartaric acid. *(Jerry Alexander/Tony Stone)*

Our goal in this chapter is to expand our awareness of molecules as three-dimensional objects. In particular, we explore the relationships between three-dimensional objects and their mirror images. When you look in a mirror, you see a reflection, or mirror image, of yourself. Now suppose your mirror image became a three-dimensional object. We could then ask what the relationship is between you and your mirror image. By relationship we mean, "Can your reflection be superposed on the original 'you' in such a way that every detail of the reflection corresponds exactly to the original?" The answer is that you and your mirror image are not superposable. If you have a ring on the little finger of your right hand, for example, your mirror image has the ring on the little finger of its left hand. If you part your hair on your right side, it will be parted on the left side in your reflection. Simply stated, you and your reflection are different objects. You cannot superpose one on the other.

An understanding of relationships of this type is fundamental to an understanding of organic chemistry and biochemistry. In fact, the ability to deal with molecules as three-dimensional objects is a survival skill in organic chemistry and biochemistry.

4.1 STEREOISOMERISM

Stereoisomers Isomers that have the same molecular formula and the same connectivity but different orientations of their atoms in space.

Stereoisomers have the same molecular formula and the same order of attachment of atoms in their molecules but different three-dimensional orientations of their atoms in space. The one example of stereoisomers we have seen thus far is that of cis-trans isomers in cycloalkanes (Section 3.7), which arise because of restricted rotation about the bonds in a ring.

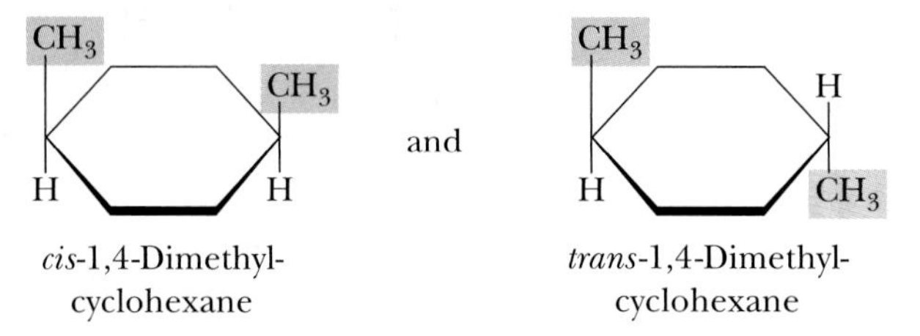

cis-1,4-Dimethylcyclohexane

trans-1,4-Dimethylcyclohexane

Mirror image The reflection of an object in a mirror.

Enantiomers Stereoisomers that are nonsuperposable mirror images; refers to a relationship between pairs of objects.

Diastereomers Stereoisomers that are not mirror images of each other; refers to relationships among objects.

Stereoisomers are divided into two groups—those that are mirror images of each other and those that are not. A **mirror image** is the reflection of an object in a mirror. Stereoisomers that are mirror images of each other are called **enantiomers** (Greek: *enantios* + *meros,* opposite + part). Stereoisomers that are not mirror images of each other are called **diastereomers.** Figure 4.1 shows these relationships.

4.2 CHIRALITY

Chiral From the Greek, *cheir,* meaning hand; objects that are not superposable on their mirror images.

Molecules that are not superposable on their mirror images are said to be **chiral** (pronounced ki-ral, to rhyme with spiral; from the Greek: *cheir,* hand); that is, they show handedness. Chirality is encountered in three-dimensional objects of all sorts. Your left hand is chiral and so is your right hand. A spiral binding on a notebook is chiral. A machine screw with a right-handed twist is chiral. A ship's propeller is chiral. As you examine the objects in the world around you, you will undoubtedly conclude that the vast majority of them are chiral as well.

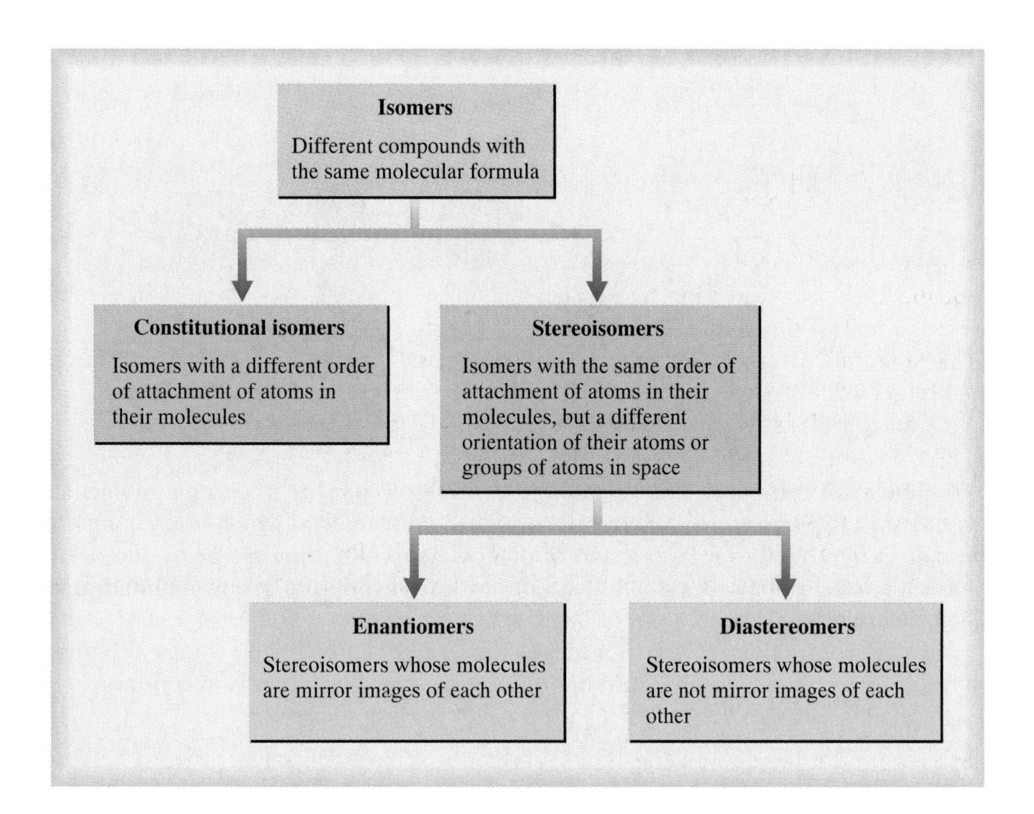

Figure 4.1
Relationships among isomers.

The contrasting situation to chirality occurs when an object and its mirror image are superposable. An object and its mirror image are superposable if one of them can be oriented in space so that all its features (corners, edges, points, designs, etc.) correspond exactly to those in the other member of the pair. If this can be done, the object and its mirror image are identical; the original object is achiral. An **achiral** object is one that lacks chirality. Examples of achiral objects are undecorated cups, a shell such as a sand dollar, a regular tetrahedron, a cube, and a perfect sphere.

Achiral An object that lacks chirality; an object that has no handedness.

An object is achiral if it possesses a plane of symmetry. A **plane of symmetry** (also called a **mirror plane**) is an imaginary plane passing through an object dividing it so that one half is the reflection of the other half. The beaker shown in Figure 4.2 has a single plane of symmetry, whereas the cube has several planes of symmetry. The compound bromochloromethane also has a single plane of symmetry.

Plane of symmetry An imaginary plane passing through an object dividing it such that one half is the mirror image of the other half.

We can illustrate the chirality of an organic molecule by considering 2-hydroxypropanoic acid, more commonly named lactic acid. Figure 4.3 shows three-dimensional representations for lactic acid and its mirror image. In these representations, all bond angles about the central carbon atom are approximately 109.5°, and the four bonds from this carbon are directed toward the corners of a regular tetrahedron.

A model of lactic acid can be turned and rotated in any direction in space, but as long as bonds on one model are not broken and rearranged, only two of the four groups attached to its central carbon can be made to coincide with those on its mirror image. Because lactic acid and its mirror image are nonsuperposable, they are classified as enantiomers.

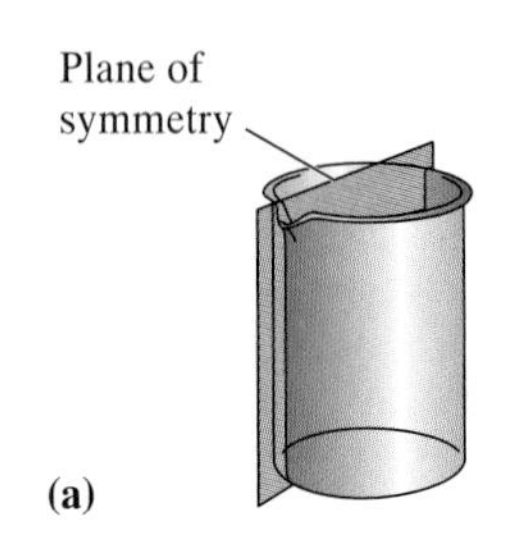

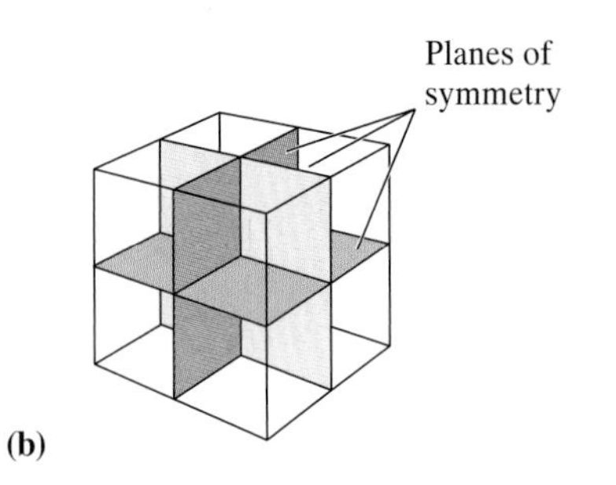

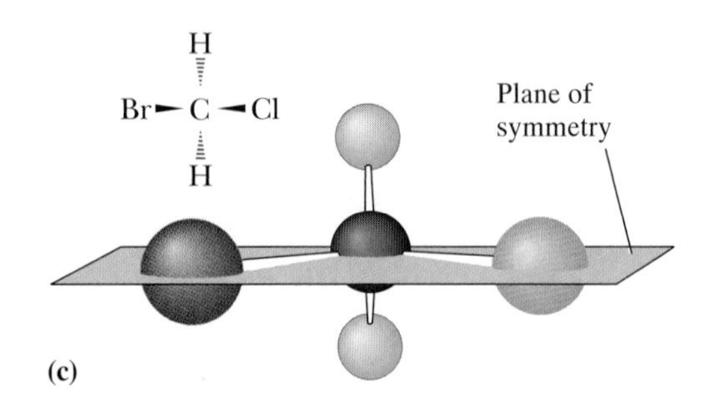

Figure 4.2
Planes of symmetry in (a) a beaker, (b) a cube, and (c) CH_2BrCl. The beaker and CH_2BrCl have one plane of symmetry; the cube has several planes of symmetry.

Stereocenter A tetrahedral atom that has four different groups attached to it.

The most common (but not the only) cause of chirality in organic molecules is the presence of a **stereocenter,** an atom in the molecule at which interchange of two groups bonded to it gives a stereoisomer. The most common type of stereocenter is a tetrahedral atom, usually a carbon, with four different groups bonded to it. For example, the carbon atom of lactic acid bearing the —OH, —H, $—CH_3$, and $—CO_2H$ groups (Figure 4.3) is a tetrahedral stereocenter. When we use the term *stereocenter,* we usually imply a tetrahedral stereocenter because it is so common.

EXAMPLE 4.1

Each molecule has one stereocenter. Draw stereorepresentations for the enantiomers of each.

(a) $CH_3CH(Cl)CH_2CH_3$ (b) 3-chlorocyclohexene (ring structure with Cl)

SOLUTION

You will find it helpful to study the CD-ROM models of each pair of enantiomers and to view them from different perspectives. As you work with these pairs of enan-

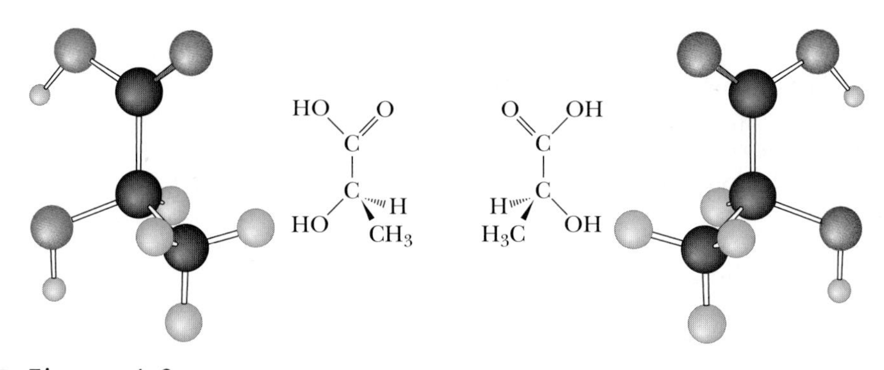

Figure 4.3
Stereorepresentations of lactic acid and its mirror image.

tiomers, notice that each has a carbon atom bonded to four different groups, which makes the molecule chiral.

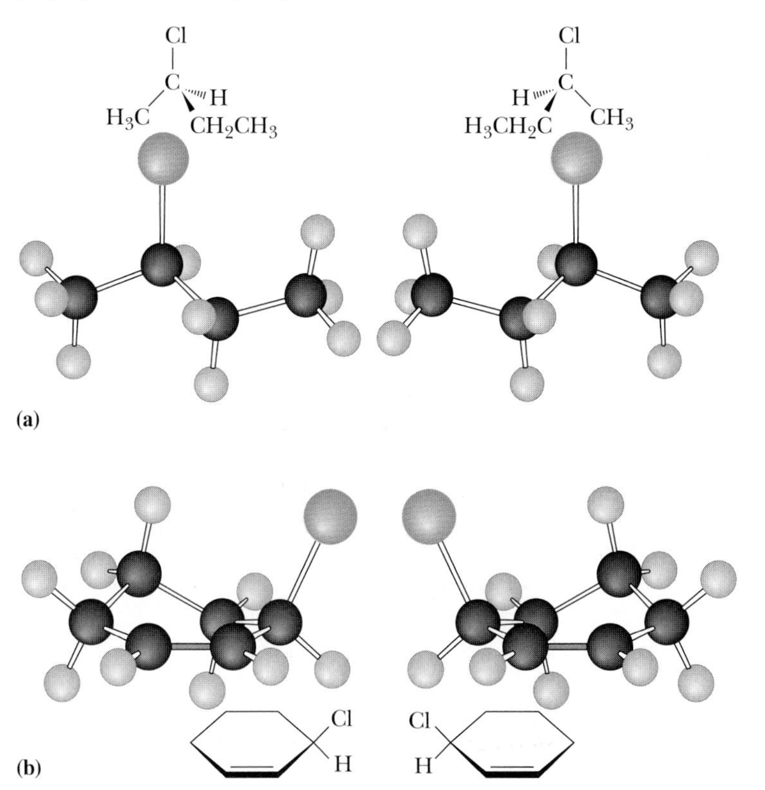

Practice Problem 4.1

Each molecule has one stereocenter. Draw stereorepresentations for the enantiomers of each.

(a) cyclopentyl–CH(OH)CH_3 (b) $CH_3CH(CH_3)CH(OH)CH_3$

4.3 NAMING STEREOCENTERS—THE R,S SYSTEM

The **R,S system** for designating the configuration of a stereocenter was devised in the late 1950s by R. S. Cahn and C. K. Ingold in England along with V. Prelog in Switzerland. The first step in assigning an R or S configuration to a stereocenter is to arrange the groups attached to it in order of priority.

R,S system A set of rules for specifying configuration about a stereocenter; also called the Cahn-Ingold-Prelog convention.

Priority Rules

1. Each atom bonded to the stereocenter is assigned a priority. Priority is based on atomic number—the higher the atomic number, the higher the priority. Fol-

lowing are several substituents arranged in order of increasing priority. The atomic number of the atom determining priority is shown in parentheses.

(1)	(6)	(7)	(8)	(16)	(17)	(35)	(53)
—H	$—CH_3$	$—NH_2$	—OH	—SH	—Cl	—Br	—I

Increasing priority →

2. If priority cannot be assigned on the basis of the atoms bonded directly to the stereocenter, look at the next set of atoms and continue until a priority can be assigned. Priority is assigned at the first point of difference. Following are a series of groups, arranged in order of increasing priority. Shown is the atomic number of the atom on which the assignment of priority is based.

(1)	(6)	(7)	(8)	(17)
$—CH_2—H$	$—CH_2—CH_3$	$—CH_2—NH_2$	$—CH_2—OH$	$—CH_2—Cl$

Increasing priority →

3. Atoms participating in a double or triple bond are considered to be bonded to an equivalent number of similar atoms by single bonds; that is, atoms of a double bond are replicated, and atoms of a triple bond are triplicated.

$—CH{=}CH_2$ is treated as $—CH(C)—CH_2(C)$

$—CH{=}O$ is treated as $—C(H)(O—C)—O$

$—C{\equiv}CH$ is treated as $—C(C)(C)—CH(C)(C)$

$—C{\equiv}N$ is treated as $—C(N)(N)—N(C)(C)$

EXAMPLE 4.2

Assign priorities to the groups in each set.

(a) $—\overset{O}{\overset{\|}{C}}OH$ and $—\overset{O}{\overset{\|}{C}}H$ (b) $—CH{=}CH_2$ and $—CH(CH_3)_2$

SOLUTION

(a) The first point of difference is O of the —OH in the carboxyl group compared to —H in the aldehyde group. The carboxyl group is higher in priority.

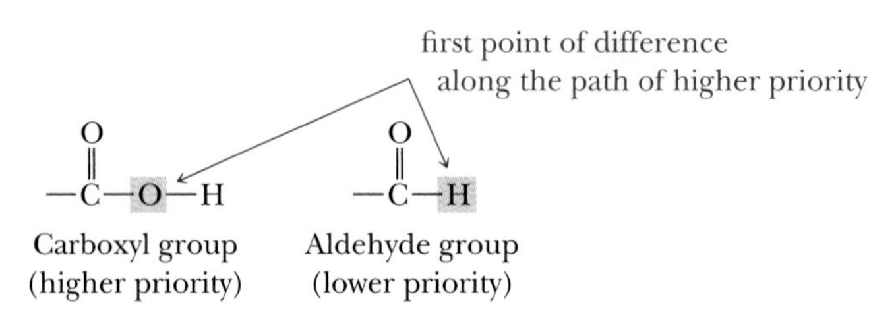

Carboxyl group (higher priority) Aldehyde group (lower priority)

(b) Carbon 1 in each group has the same pattern of atoms, namely C(C,C,H). Carbon 2 is the first point of difference. For the vinyl group, bonding at carbon 2 is C(C,H,H). For the isopropyl group, the bonding at carbon 2 is C(H,H,H). The vinyl group is higher in priority than is the isopropyl group.

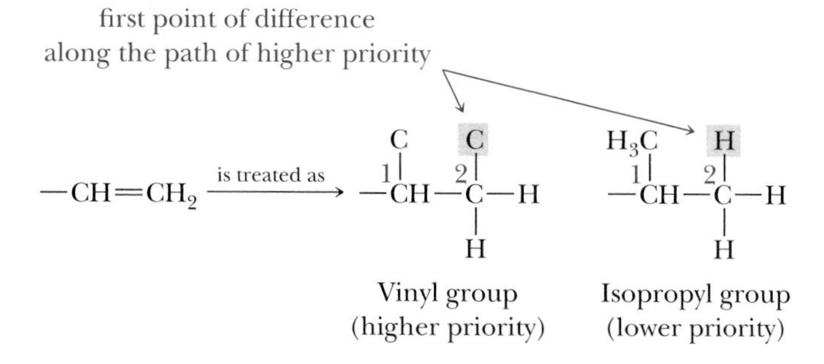

Practice Problem 4.2

Assign priorities to the groups in each set.

(a) $—CH_2OH$ and $—CH_2CH_2OH$
(b) $—CH_2OH$ and $—CH{=}CH_2$
(c) $—CH_2OH$ and $—C(CH_3)_3$

To assign an R or S configuration to a stereocenter:

1. Locate the stereocenter, identify its four substituents, and assign a priority from 1 (highest) to 4 (lowest) to each substituent.
2. Orient the molecule in space so that the group of lowest priority (4) is directed away from you as would be, for instance, the steering column of a car. The three groups of higher priority (1–3) then project toward you as would the spokes of a steering wheel.
3. Read the three groups projecting toward you in order from highest priority (1) to lowest priority (3).

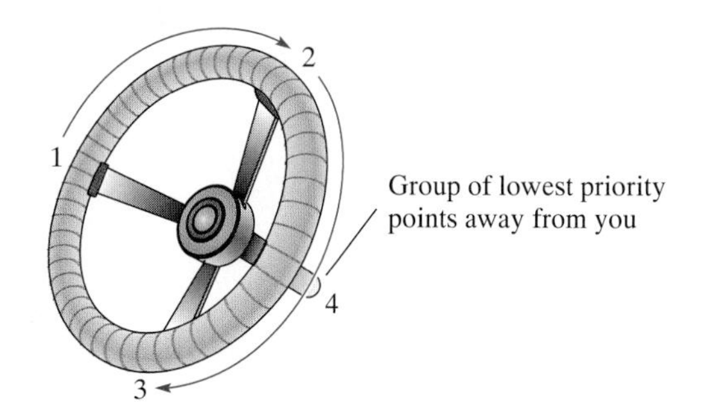

4. If reading the groups proceeds in a clockwise direction, the configuration is designated as **R** (Latin: *rectus,* right); if reading proceeds in a counterclockwise direction, the configuration is **S** (Latin: *sinister,* left). You can also visualize this as follows: turning the steering wheel to the right equals R, and turning it to the left equals S.

R From the Latin, *rectus,* meaning right; used in the R,S system to show that the order of priority of groups on a stereocenter is clockwise.

S From the Latin, *sinister,* meaning left; used in the R,S system to show that the order of priority of groups on a stereocenter is counterclockwise.

EXAMPLE 4.3

Assign an R or S configuration to each stereocenter.

(a) Cl, H, C, CH_3, CH_3CH_2 (b) Cl, H

SOLUTION

View each molecule through the stereocenter and along the bond from the stereocenter toward the group of lowest priority.

(a) The order of priority is $-Cl > -CH_2CH_3 > -CH_3 > -H$. The group of lowest priority, H, is pointed away from you. Reading the groups in the order 1, 2, 3 occurs in the counterclockwise direction, so the configuration is S.

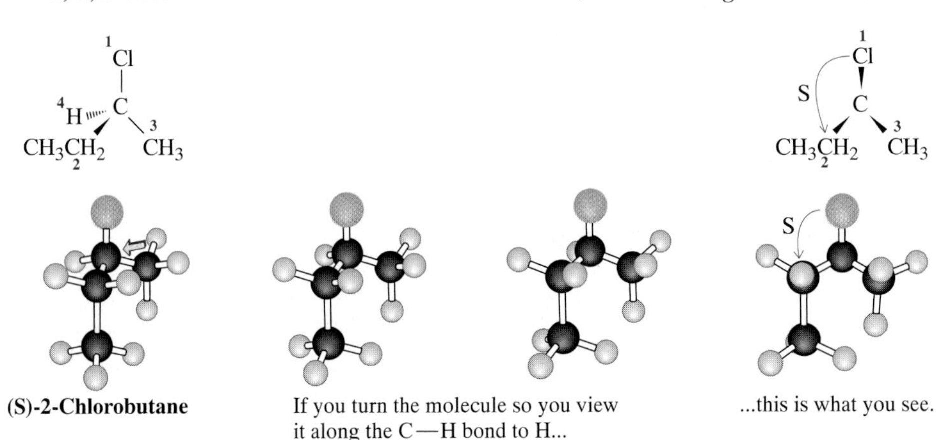

(S)-2-Chlorobutane If you turn the molecule so you view it along the C—H bond to H... ...this is what you see.

(b) The order of priority is $-Cl > -CH{=}CH > -CH_2{-}CH_2 > -H$. With hydrogen, the group of lowest priority pointing away from you, reading the groups in the order 1, 2, 3 occurs in the clockwise direction, so the configuration is R.

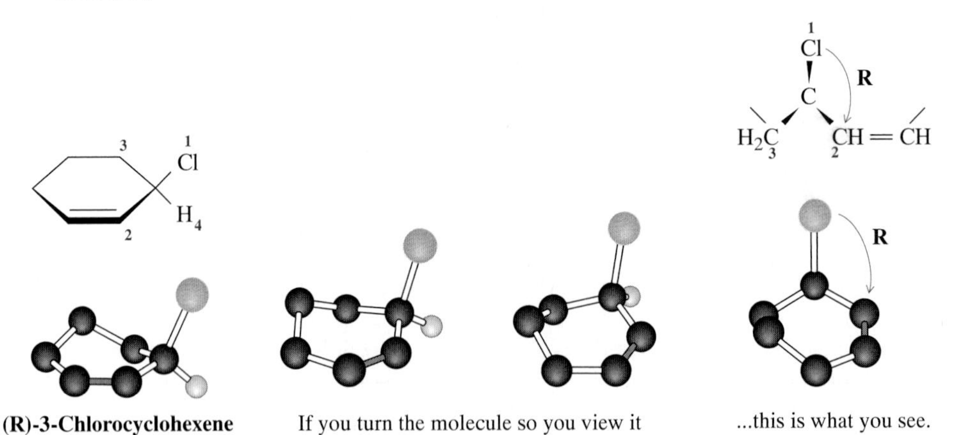

(R)-3-Chlorocyclohexene If you turn the molecule so you view it along the C—H bond to H... ...this is what you see.

Practice Problem 4.3

Assign an R or S configuration to each stereocenter.

(a) H OH, H_3C, H_3C (b) H, CH_3, C—OH, CH_3CH_2 (c) CH=O, H—C—OH, CH_2OH

4.4 ACYCLIC MOLECULES WITH TWO OR MORE STEREOCENTERS

We have now seen several examples of molecules with one stereocenter and verified that, for each, two stereoisomers (one pair of enantiomers) are possible. Now let us consider molecules with two or more stereocenters. To generalize, for a molecule with ***n*** stereocenters, the maximum number of stereoisomers possible is $\mathbf{2^n}$. We have already verified that, for a molecule with one stereocenter, $2^1 = 2$ stereoisomers are possible. For a molecule with two stereocenters, $2^2 = 4$ stereoisomers are possible; for a molecule with three stereocenters, $2^3 = 8$ stereoisomers are possible, and so forth.

A. Enantiomers and Diastereomers

We begin our study of molecules with two or more stereocenters by considering 2,3,4-trihydroxybutanal. Its two stereocenters are shown in color.

$HOCH_2—CH(OH)—CH(OH)—CH=O$

2,3,4-Trihydroxybutanal

The maximum number of stereoisomers possible for this molecule is $2^2 = 4$, each of which is drawn in Figure 4.4.

Stereoisomers (a) and (b) are nonsuperposable mirror images and are, therefore, a pair of enantiomers. Stereoisomers (c) and (d) are also nonsuperposable mirror images and are a second pair of enantiomers. One way to describe the four

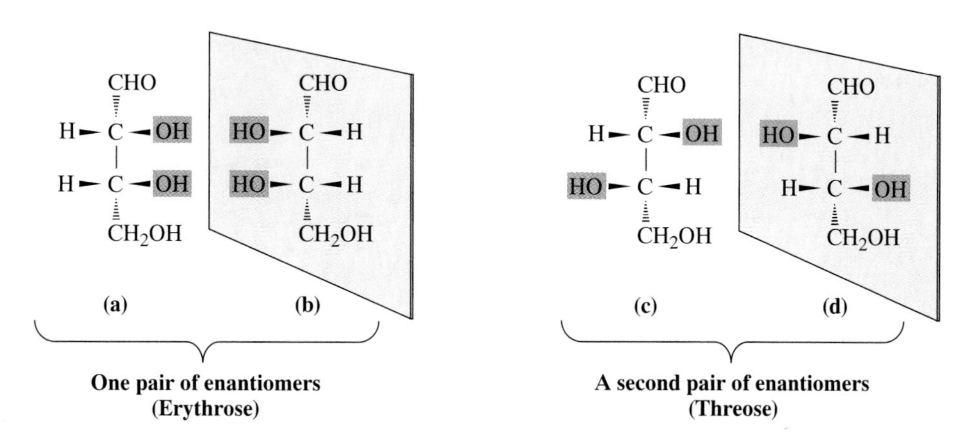

Figure 4.4 The four stereoisomers of 2,3,4-trihydroxybutanal, a compound with two stereocenters.

stereoisomers of 2,3,4-trihydroxybutanal is to say that they consist of two pairs of enantiomers. Enantiomer (a) is named (2R,3R)-erythrose, but it could be named (2R,3R)-2,3,4-trihydroxybutanal as well. Enantiomer (b) is named (2S,3S)-erythrose. Enantiomers (c) and (d) are named (2R,3S)-threose and (2S,3R)-threose. Erythrose and threose belong to the class of compounds called carbohydrates, which we discuss in Chapter 16. Erythrose is synthesized in erythrocytes (red blood cells), hence the derivation of its name.

We have specified the relationship between (a) and (b) and between (c) and (d). What is the relationship between (a) and (c), between (a) and (d), between (b) and (c), and between (b) and (d)? The answer is that they are diastereomers. **Diastereomers** are stereoisomers that are not enantiomers; that is, they are stereoisomers that are not mirror images of each other.

EXAMPLE 4.4

Following are stereorepresentations for the four stereoisomers of 1,2,3-butanetriol. R and S configurations are given for the stereocenters in (1) and (4).

(1): CH_2OH; H–C–OH (S); HO–C–H (S); CH_3
(2): CH_2OH; H–C–OH; H–C–OH; CH_3
(3): CH_2OH; HO–C–H; HO–C–H; CH_3
(4): CH_2OH; HO–C–H (R); H–C–OH (R); CH_3

(a) Which compounds are enantiomers?
(b) Which compounds are diastereomers?

SOLUTION

(a) Enantiomers are stereoisomers that are nonsuperposable mirror images. Compounds (1) and (4) are one pair of enantiomers, and compounds (2) and (3) are a second pair of enantiomers. Note that the configurations of the stereocenters in (1) are the opposite of those in (4), its enantiomer.
(b) Diastereomers are stereoisomers that are not mirror images. Compounds (1) and (2), (1) and (3), (2) and (4), and (3) and (4) are diastereomers.

Practice Problem 4.4

Following are stereorepresentations for the four stereoisomers of 3-chloro-2-butanol.

(1): CH_3; H–C–OH; Cl–C–H; CH_3
(2): CH_3; H–C–OH; H–C–Cl; CH_3
(3): CH_3; HO–C–H; H–C–Cl; CH_3
(4): CH_3; HO–C–H; Cl–C–H; CH_3

(a) Which compounds are enantiomers?
(b) Which compounds are diastereomers?

B. Meso Compounds

Certain molecules containing two or more stereocenters have special symmetry properties that reduce the number of stereoisomers to fewer than the maximum number predicted by the 2^n rule. One such molecule is 2,3-dihydroxybutanedioic acid, more commonly named tartaric acid.

$$\mathrm{HO\overset{\overset{\Large O}{\|}}{C}-\underset{\underset{\Large OH}{|}}{CH}-\underset{\underset{\Large OH}{|}}{CH}-\overset{\overset{\Large O}{\|}}{C}OH}$$

2,3-Dihydroxybutanedioic acid
(Tartaric acid)

Tartaric acid is a colorless, crystalline compound occurring largely in the vegetable kingdom, especially in grapes. During fermentation of grape juice, potassium bitartrate (one —CO_2H group is present as a potassium salt, —$CO_2^-K^+$) deposits as a crust on the sides of wine casks. When collected and purified, it is sold commercially as cream of tartar (*The Merck Index,* 12th ed., #7776).

In tartaric acid, carbons 2 and 3 are stereocenters, and, using the 2^n rule, the maximum number of stereoisomers possible is $2^2 = 4$, stereorepresentations for which are drawn in Figure 4.5. Structures (a) and (b) are nonsuperposable mirror images and are, therefore, a pair of enantiomers. Structures (c) and (d) are also mirror images, but they are superposable. To see this, imagine that you rotate (d) by 180° in the plane of the paper, lift it out of the plane of the paper, and place it on top of (c). If you do this mental manipulation correctly, you find that (d) is superposable on (c). Therefore, (c) and (d) are not different molecules; they are the same molecule, just oriented differently. Because (c) and its mirror image are superposable, (c) is achiral.

Another way to determine that (c) is achiral is to see that it has a plane of symmetry that bisects the molecule in such a way that the top half is the reflection of the bottom half. Thus, even though (c) has two stereocenters, it is achiral (Section 4.2).

The stereoisomer of tartaric acid represented by (c) or (d) is called a meso compound. A **meso compound** is an achiral compound that contains two or more stereocenters. We can now return to the original question: How many stereoisomers are there of tartaric acid? The answer is three—one meso compound and one pair of enantiomers. Note that the meso compound is a diastereomer of each of the other stereoisomers.

Meso compound An achiral compound possessing two or more stereocenters.

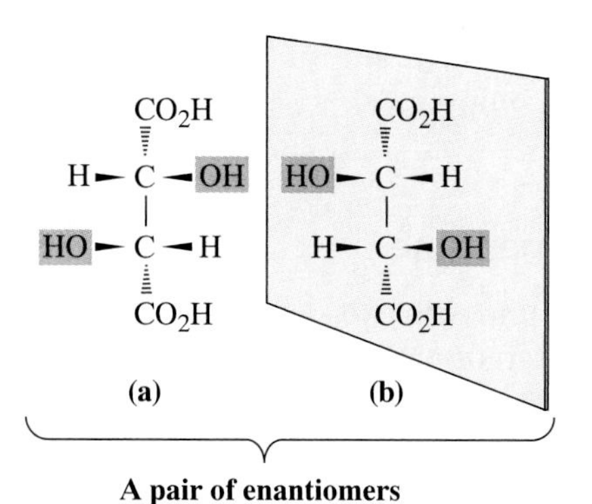

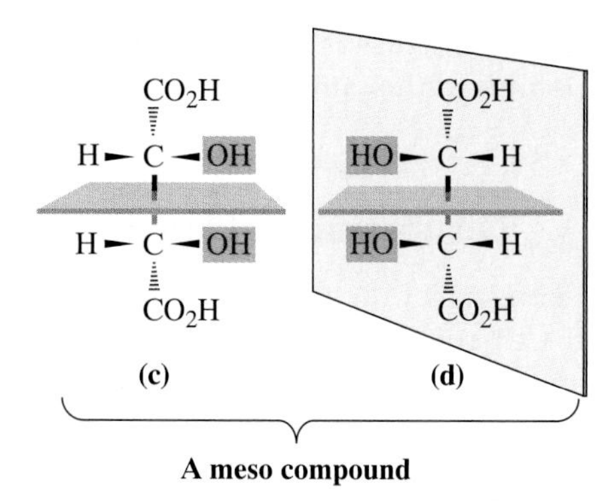

Figure 4.5
Stereoisomers of tartaric acid. One pair of enantiomers and one meso compound.

Example 4.5

Following are stereorepresentations for the three stereoisomers of 2,3-butanediol.

(1)	(2)	(3)
CH_3	CH_3	CH_3
H–C–OH	H–C–OH	HO–C–H
HO–C–H	H–C–OH	H–C–OH
CH_3	CH_3	CH_3

(a) Which are enantiomers?
(b) Which is the meso compound?

Solution

(a) Compounds (1) and (3) are enantiomers.
(b) Compound (2) is a meso compound.

Practice Problem 4.5

Following are four Newman projection formulas for tartaric acid.

	(1)	(2)	(3)	(4)
top	CO_2H	CO_2H	CO_2H	CO_2H
upper left	H	HO	H	HO
upper right	OH	H	CO_2H	CO_2H
lower left	H	H	H	H
lower right	OH	OH	OH	OH
bottom	CO_2H	CO_2H	OH	H

(a) Which represent the same compound?
(b) Which represent enantiomers?
(c) Which represent meso tartaric acid?

4.5 Cyclic Molecules with Two or More Stereocenters

In this section, we concentrate on derivatives of cyclopentane and cyclohexane containing two or more stereocenters. We can analyze chirality in these cyclic compounds in the same way we analyzed it in acyclic compounds.

A. Disubstituted Derivatives of Cyclopentane

Let us start with 2-methylcyclopentanol, a compound with two stereocenters. Using the 2^n rule, we predict a maximum of $2^2 = 4$ stereoisomers. Both the cis isomer and the trans isomer are chiral: the cis isomer exists as one pair of enantiomers, and the trans isomer exists as a second pair of enantiomers.

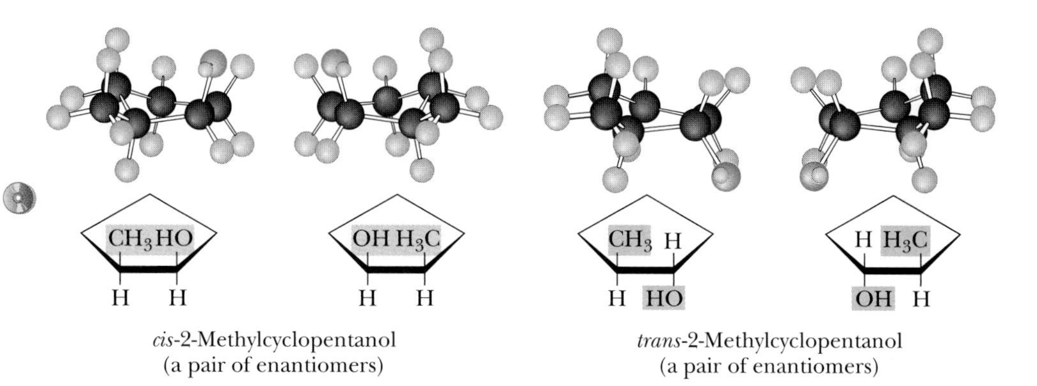

cis-2-Methylcyclopentanol
(a pair of enantiomers)

trans-2-Methylcyclopentanol
(a pair of enantiomers)

1,2-Cyclopentanediol also has two stereocenters; therefore, the 2^n rule predicts a maximum of $2^2 = 4$ stereoisomers. As seen in the following stereodrawings, only three stereoisomers exist for this compound. The cis isomer is achiral (meso) because it and its mirror image are superposable. Alternatively, the cis isomer is achiral because it possesses a plane of symmetry that bisects the molecule into two mirror halves. The trans isomer is chiral and exists as a pair of enantiomers.

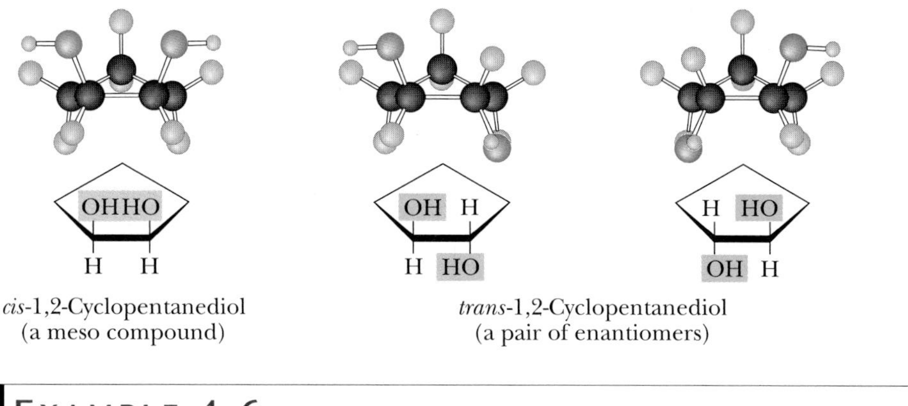

cis-1,2-Cyclopentanediol
(a meso compound)

trans-1,2-Cyclopentanediol
(a pair of enantiomers)

EXAMPLE 4.6

How many stereoisomers exist for 3-methylcyclopentanol?

SOLUTION

There are four stereoisomers of 3-methylcyclopentanol. The cis isomer exists as one pair of enantiomers; the trans isomer, as a second pair of enantiomers.

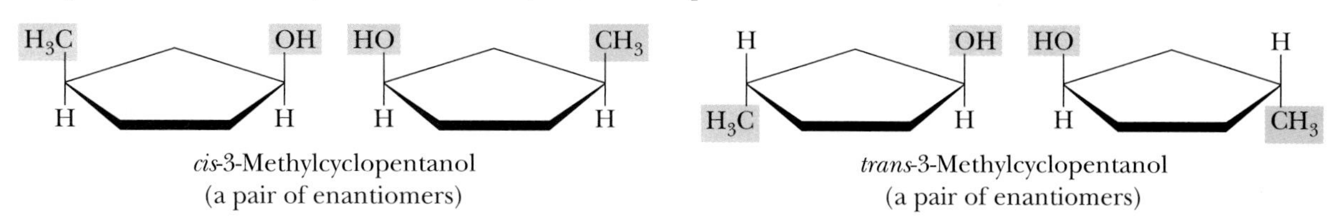

cis-3-Methylcyclopentanol
(a pair of enantiomers)

trans-3-Methylcyclopentanol
(a pair of enantiomers)

Practice Problem 4.6

How many stereoisomers exist for 1,3-cyclopentanediol?

B. Disubstituted Derivatives of Cyclohexane

As an example of a disubstituted cyclohexane, let us consider the methylcyclohexanols. 4-Methylcyclohexanol can exist as two stereoisomers—a pair of cis-trans isomers. Both the cis and the trans isomer are achiral. In each, a plane of symmetry runs through the CH_3 and OH groups and the two attached carbons.

3-Methylcyclohexanol has two stereocenters and exists as $2^2 = 4$ stereoisomers. The trans isomer exists as one pair of enantiomers, the cis isomer as a second pair of enantiomers. Similarly, 2-methylcyclohexanol has two stereocenters and exists as $2^2 = 4$ stereoisomers. The cis isomer exists as one pair of enantiomers, the trans isomer as a second pair of enantiomers.

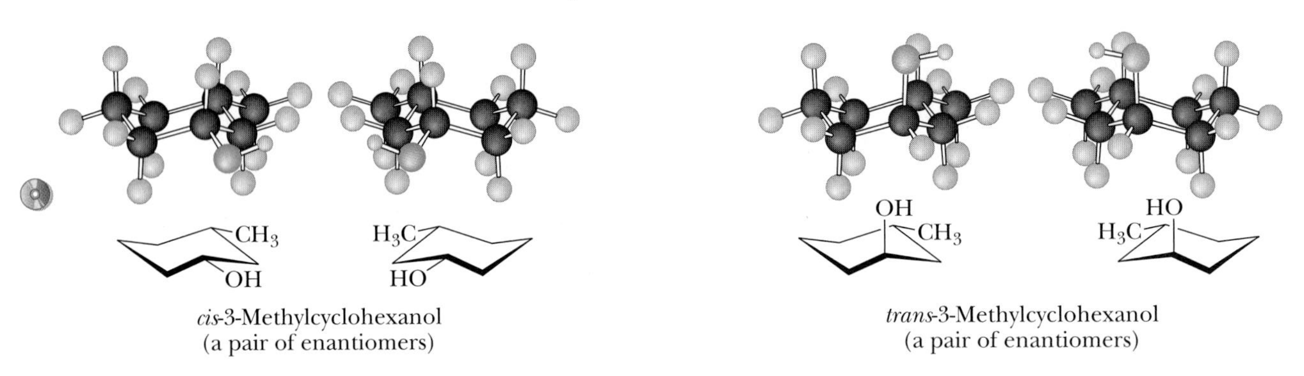

cis-3-Methylcyclohexanol (a pair of enantiomers)

trans-3-Methylcyclohexanol (a pair of enantiomers)

Example 4.7

How many stereoisomers exist for 1,3-cyclohexanediol?

Solution

1,3-Cyclohexanediol has two stereocenters and, according to the 2^n rule, has a maximum of $2^2 = 4$ stereoisomers. The trans isomer of this compound exists as a pair of enantiomers. The cis isomer has a plane of symmetry and is a meso compound. Therefore, although the 2^n rule predicts a maximum of four stereoisomers for 1,3-cyclohexanediol, only three exist—one meso compound and one pair of enantiomers.

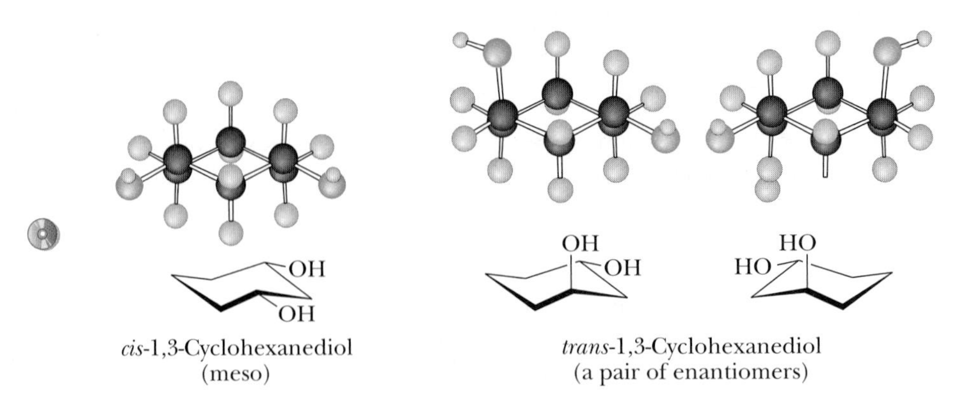

cis-1,3-Cyclohexanediol (meso)

trans-1,3-Cyclohexanediol (a pair of enantiomers)

TABLE 4.1 Some Physical Properties of the Stereoisomers of Tartaric Acid

	CO_2H H–C–OH HO–C–H CO_2H	CO_2H HO–C–H H–C–OH CO_2H	CO_2H H–C–OH H–C–OH CO_2H
	The pair of enantiomers		**Meso-tartaric acid**
Specific rotation*	+12.7	−12.7	0
Melting point (°C)	171–174	171–174	146–148
Density at 20°C (g/cm^3)	1.7598	1.7598	1.660
Solubility in water at 20°C (g/100 mL)	139	139	125
pK_1 (25°C)	2.98	2.98	3.23
pK_2 (25°C)	4.34	4.34	4.82

* Specific rotation is discussed in Section 4.7.

Practice Problem 4.7

How many stereoisomers exist for 1,4-cyclohexanediol?

4.6 PROPERTIES OF STEREOISOMERS

Enantiomers have identical physical and chemical properties in achiral environments. The enantiomers of tartaric acid (Table 4.1), for example, have the same melting point, the same boiling point, the same solubilities in water and other common solvents, and the same values of pK_a (the acid ionization constant), and they undergo the same acid-base reactions. The enantiomers of tartaric acid do, however, differ in optical activity (the ability to rotate the plane of polarized light), which is discussed in the following section.

Diastereomers have different physical and chemical properties, even in achiral environments. Meso-tartaric acid has different physical properties from those of the enantiomers.

4.7 OPTICAL ACTIVITY—HOW CHIRALITY IS DETECTED IN THE LABORATORY

As we have already established, enantiomers are different compounds, and we must expect, therefore, that they differ in some properties. One property that differs between enantiomers is their effect on the plane of polarized light. Each member of a pair of enantiomers rotates the plane of polarized light, and for this reason, enantiomers are said to be **optically active.**

Optically active Showing that a compound rotates the plane of polarized light.

The phenomenon of optical activity was discovered in 1815 by the French physicist Jean Baptiste Biot. To understand how it is detected in the laboratory, we must first understand something about plane-polarized light and a polarimeter, the device used to detect optical activity.

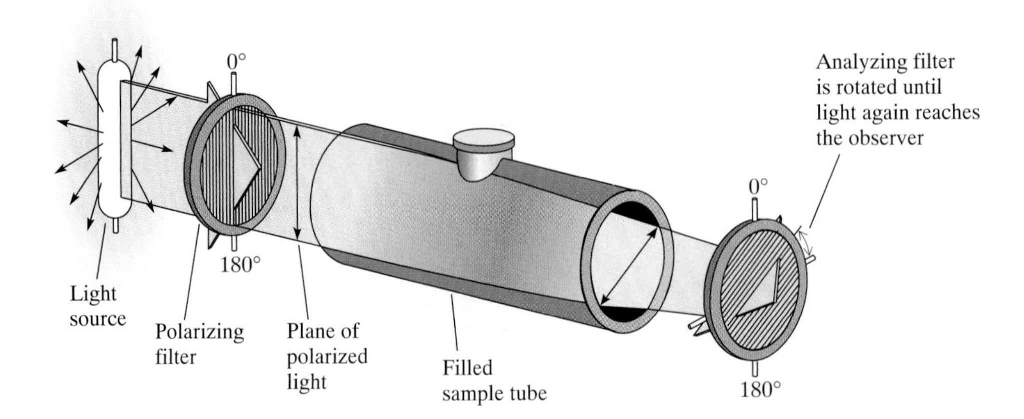

Figure 4.6
Schematic diagram of a polarimeter with its sample tube containing a solution of an optically active compound. The analyzing filter has been turned clockwise by α degrees to restore the light field.

A. Plane-Polarized Light

Ordinary light consists of waves vibrating in all planes perpendicular to its direction of propagation (Figure 4.6). Certain materials, such as calcite and Polaroid sheet (a plastic film containing properly oriented crystals of an organic substance embedded in it), selectively transmit light waves vibrating in parallel planes. Electromagnetic radiation vibrating in only parallel planes is said to be **plane polarized.**

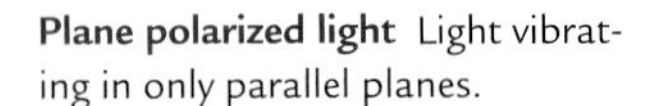

Plane polarized light Light vibrating in only parallel planes.

B. A Polarimeter

Polarimeter A device for measuring the ability of a compound to rotate the plane of polarized light.

A **polarimeter** consists of a light source, a polarizing filter and an analyzing filter (each made of calcite or Polaroid film), and a sample tube (Figure 4.6). If the sample tube is empty, the intensity of light reaching the detector (in this case, your eye) is at its maximum when the polarizing axes of the two filters are parallel. If the analyzing filter is turned either clockwise or counterclockwise, less light is transmitted. When the axis of the analyzing filter is at right angles to the axis of the polarizing filter, the field of view is dark. This position of the analyzing filter is taken as 0° on the optical scale.

The ability of molecules to **rotate the plane of polarized light** can be observed using a polarimeter in the following way. First, a sample tube filled with solvent is placed in the polarimeter. Then, the analyzing filter is adjusted so that no light passes through to the observer; that is, it is set to 0°. When a solution of an optically active compound is placed in the sample tube, a certain amount of light now passes through the analyzing filter; the optically active compound has rotated the plane of polarized light from the polarizing filter so that it is no longer at an angle of 90° to the analyzing filter. The analyzing filter is then rotated to restore darkness in the field of view (Figure 4.6). The number of degrees, α, through which the analyzing filter must be rotated to restore darkness to the field of view is called the **observed rotation.** If the analyzing filter must be turned to the right (clockwise) to restore the dark field, we say that the compound is **dextrorotatory** (Latin: *dexter,* on the right side). If the analyzing filter must be turned to the left (counterclockwise), we say that the compound is **levorotatory** (Latin: *laevus,* on the left side).

Observed rotation The number of degrees through which a compound rotates the plane of polarized light.

Dextrorotatory Rotation of the plane of polarized light in a polarimeter to the right.

Levorotatory Rotation of the plane of polarized light in a polarimeter to the left.

Specific rotation Observed rotation of the plane of polarized light when a sample is placed in a tube of 1.0 dm in length and at a concentration of 1 g/mL.

The magnitude of the observed rotation for a particular compound depends on its concentration, the length of the sample tube, the temperature, the solvent, and the wavelength of the light used. **Specific rotation, [α],** is defined as the observed rotation at a specific cell length and sample concentration expressed in grams per milliliter.

$$\text{Specific rotation} = [\alpha]_{\lambda}{}^{T} = \frac{\text{Observed rotation (degrees)}}{\text{Length (dm)} \times \text{Concentration (g/mL)}}$$

The standard cell length is 1 decimeter (1 dm = 0.1 m). For a pure liquid sample, the concentration is expressed in grams per milliliter (g/mL; density). The concentration of a sample dissolved in a solvent is also usually expressed as grams per milliliter of solution. The temperature (T, in degrees centigrade) and wavelength (λ, in nanometers) of light are designated, respectively, as a superscript and a subscript. The light source most commonly used in polarimetry is the sodium D line (λ = 589 nm), the same line responsible for the yellow color of sodium-vapor lamps.

A polarimeter is used to measure the rotation of plane-polarized light as it passes through a sample. *(Richard Megna, 1992, Fundamental Photographs)*

In reporting either observed or specific rotation, it is common to indicate a dextrorotatory compound with a plus sign in parentheses, (+), and a levorotatory compound with a minus sign in parentheses, (−). For any pair of enantiomers, one enantiomer is dextrorotatory, and the other is levorotatory. For each member, the value of the specific rotation is exactly the same, but the sign is opposite. Following are the specific rotations of the enantiomers of 2-butanol at 25°C using the D line of sodium.

(S)-(+)-2-Butanol	(R)-(−)-2-Butanol
OH; H, CH_3CH_2, CH_3 on C	HO; H, CH_3, CH_2CH_3 on C
$[\alpha]_D^{25}$ +13.52	$[\alpha]_D^{25}$ −13.52

Example 4.8

A solution is prepared by dissolving 400 mg of testosterone, a male sex hormone (Table 17.3), in 10.0 mL of ethanol and placing it in a sample tube 1.00 dm in length. The observed rotation of this sample at 25°C using the D line of sodium is + 4.36. Calculate the specific rotation of testosterone.

Solution

The concentration of testosterone is 400 mg/10.0 mL = 0.0400 g/mL. The length of the sample tube is 1.00 dm. Inserting these values in the equation for calculating specific rotation gives

$$\text{Specific rotation} = \frac{\text{Observed rotation (degrees)}}{\text{Length (dm)} \times \text{Concentration (g/mL)}} = \frac{+4.36}{1.00 \times 0.0400} = +109$$

Practice Problem 4.8

The specific rotation of progesterone, a female sex hormone (Table 17.3), is +172, measured at 20°C. Calculate the observed rotation for a solution prepared by dissolving 300 mg of progesterone in 15.0 mL of dioxane and placing it in a sample tube 1.00 dm long.

C. Racemic Mixtures

Racemic mixture A mixture of equal amounts of two enantiomers.

An equimolar mixture of two enantiomers is called a **racemic mixture,** a term derived from the name racemic acid (Latin: *racemus,* a cluster of grapes). Racemic acid is the name originally given to an equimolar mixture of the enantiomers of tartaric acid (Table 4.1). Because a racemic mixture contains equal numbers of the dextrorotatory and the levorotatory molecules, its specific rotation is zero. Alternatively, we say that a racemic mixture is optically inactive. A racemic mixture is indicated by adding the prefix (±) to the name of the compound.

4.8 Separation of Enantiomers—Resolution

Resolution Separation of a racemic mixture into its enantiomers.

The separation of a racemic mixture into its enantiomers is called **resolution.**

A. Resolution by Means of Diastereomeric Salts

One general scheme for separating enantiomers requires chemical conversion of the pair of enantiomers into two diastereomers with the aid of an enantiomerically pure chiral **resolving reagent.** This chemical resolution is successful because the diastereomers thus formed have different physical properties and can often be separated by physical means (most commonly fractional crystallization or column chromatography) and purified. The final step in this scheme for resolution is chemical conversion of the separated diastereomers back to individual enantiomers and recovery of the chiral resolving agent.

A reaction that lends itself to chemical resolution is salt formation because it is readily reversible.

$$\underset{\text{Carboxylic acid}}{RCO_2H} + \underset{\text{Base}}{:B} \longrightarrow \underset{\text{Salt}}{RCO_2^-\ HB^+}$$

Several enantiomerically pure bases available from plants have been used as chiral resolving agents for racemic acids. Examples are cinchonine and quinine.

Cinchona bark, the source of quinine, one of the first antimalarial drugs. *(© Walter H. Hodge/Peter Arnold, Inc.)*

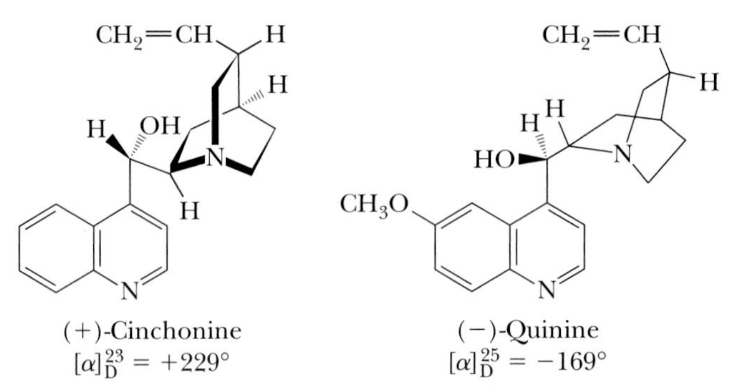

The base (+)-cinchonine (*The Merck Index,* 12th ed., #2346) is found in the bark of most species of *cinchona,* a genus of evergreen trees or shrubs growing in the tropical valleys of the Andes and now extensively cultivated for its bark in India,

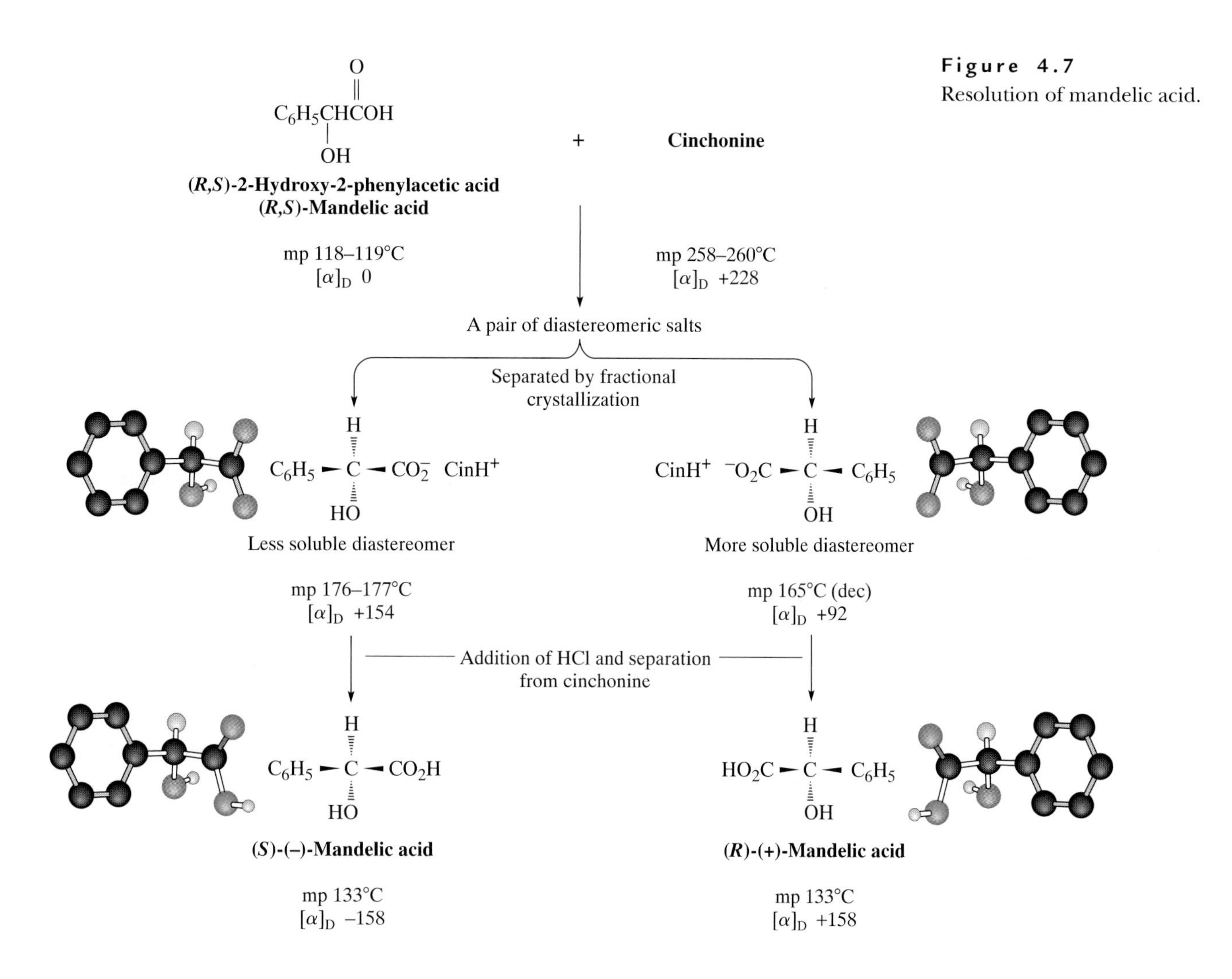

Figure 4.7
Resolution of mandelic acid.

Java, and parts of South America. Extracts of its bark have been used for centuries to cure the fevers associated with malaria. The genus was named after the Countess of Cinchon, wife of the viceroy of Peru, who was cured of fever by cinchona bark and later brought a supply of it back to Spain. Also found in cinchona bark is (−)-quinine (*The Merck Index,* 12th ed., #8245), an even more potent antimalarial drug than cinchonine.

The resolution of (±)-mandelic acid by way of its diastereomeric salts with cinchonine is illustrated in Figure 4.7. Racemic mandelic acid and optically pure (+)-cinchonine (Cin) are dissolved in boiling water, and the solution is allowed to cool, whereupon the less-soluble diastereomeric salt crystallizes. This salt is collected and purified by further recrystallization. The filtrates, richer in the more soluble diastereomeric salt, are concentrated to give this salt, which is also purified by further recrystallization. The purified diastereomeric salts are treated with aqueous HCl to precipitate the nearly pure enantiomers of mandelic acid. Cinchonine remains in the aqueous solution as the water-soluble salt.

Figure 4.8
Some carboxylic acids used as chiral resolving agents.

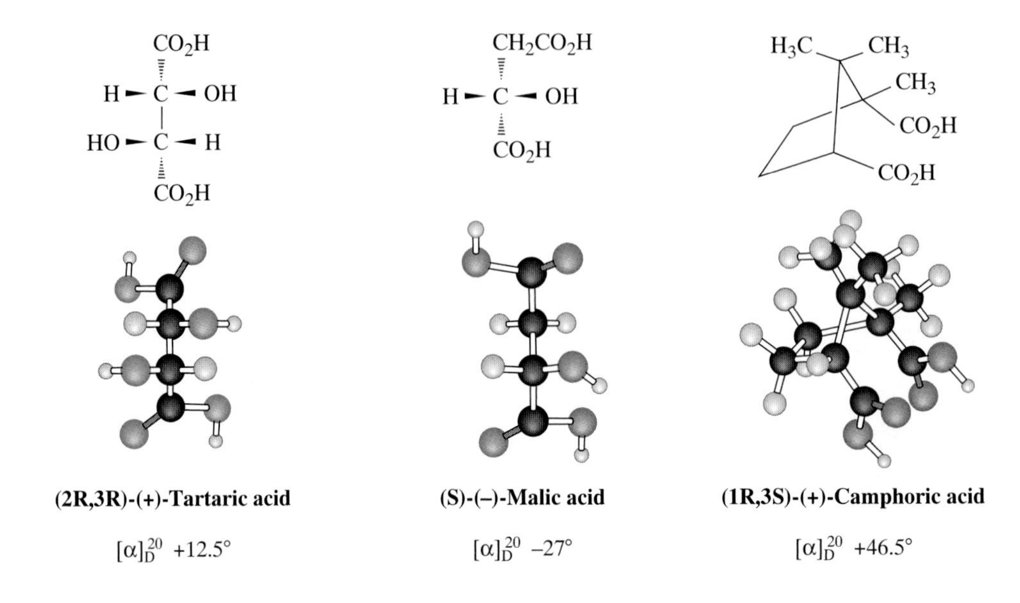

$[\alpha]_D^{20}$ +12.5° $[\alpha]_D^{20}$ –27° $[\alpha]_D^{20}$ +46.5°

Optical rotations and melting points of racemic mandelic acid, cinchonine, the purified diastereomeric salts, and the pure enantiomers of mandelic acid are given in Figure 4.7. Note the following three points: (1) The melting point of racemic mandelic acid is different from that of the pure enantiomers. (2) The diastereomeric salts have different specific rotations and different melting points. (3) The enantiomers of mandelic acid have identical melting points and have specific rotations that are identical in magnitude but opposite in sign.

Resolution of a racemic base with a chiral acid is carried out in a similar way. Acids that have been used as chiral resolving agents are (+)-tartaric acid, (−)-malic acid, and (+)-camphoric acid (Figure 4.8). These and other naturally occurring chiral resolving agents are produced in plant and animal systems as single enantiomers.

B. Enzymes as Resolving Agents

In their quest for enantiomerically pure compounds, organic chemists have developed several new techniques for chiral synthesis. One approach is to use enzymes as chiral catalysts for the large-scale synthesis of enantiomerically pure substances. A class of enzymes under study is the esterases, which catalyze the hydrolysis of esters to give an alcohol and a carboxylic acid.

The ethyl esters of naproxen crystallize in two forms—one containing the (R)-ester and the other containing the (S)-ester. Each form has a very low solubility in water. Chemists then use an esterase in alkaline solution to hydrolyze selectively the (S)-ester, which then goes into solution as the sodium salt of the (S)-carboxylic acid. The (R)-ester is unaffected by these conditions. Filtering the alkaline solution recovers the crystals of the (R)-ester. After crystals of the (R)-ester are removed, the alkaline solution is acidified to give enantiomerically pure (S)-naproxen (*The Merck Index,* 12th ed., #6540). The recovered (R)-ester is racemized (converted to an R,S mixture), and treated with the esterase. Thus, by recycling the (R)-ester, all the racemic ester is converted to (S)-naproxen. The sodium salt of (S)-naproxen is the active ingredient in Aleve and a score of other over-the-counter nonsteroidal anti-inflammatory preparations.

Ethyl ester of (*S*)-naproxen

Ethyl ester of (*R*)-naproxen (not affected by the esterase)

1. esterase 2. HCl, H_2O | NaOH, H_2O

(*S*)-Naproxen + CH_3CH_2OH

4.9 THE SIGNIFICANCE OF CHIRALITY IN THE BIOLOGICAL WORLD

Except for inorganic salts and a relatively few low-molecular-weight organic substances, the molecules in living systems, both plant and animal, are chiral. Although these molecules can exist as a number of stereoisomers, almost invariably only one stereoisomer is found in nature. Of course, instances do occur in which more than one stereoisomer is found, but these rarely exist together in the same biological system.

A. Chirality in Biomolecules

Perhaps the most conspicuous examples of chirality among biological molecules are the enzymes, all of which have many stereocenters. An illustration is chymotrypsin, an enzyme in the intestines of animals, which catalyzes the digestion of proteins (Section 18.4B). Chymotrypsin has 251 stereocenters. The maximum number of stereoisomers possible is 2^{251}, a staggeringly large number, almost beyond comprehension. Fortunately, nature does not squander its precious energy and resources unnecessarily; only one of these stereoisomers is produced and used by any given organism. Because enzymes are chiral substances, most either produce or react only with substances that match their stereochemical requirements.

B. How an Enzyme Distinguishes Between a Molecule and Its Enantiomer

An enzyme catalyzes a biological reaction of molecules by first positioning them at a **binding site** on its surface. An enzyme with specific binding sites for three of the

CHEMICAL CONNECTIONS

Chiral Drugs

Some of the common drugs used in human medicine, for example, aspirin (Section 14.4B), are achiral. Others are chiral and are sold as single enantiomers. The penicillin and erythromycin classes of antibiotics and the drug captopril are all chiral drugs. Captopril (*The Merck Index,* 12th ed., #1817), which is very effective for the treatment of high blood pressure and congestive heart failure, was developed in a research program designed to discover effective inhibitors for angiotensin-converting enzyme (ACE). It is manufactured and sold as the (S,S)-stereoisomer (*The Merck Index,* 12th ed., #1817). A large number of chiral drugs, however, are sold as racemic mixtures. The popular analgesic ibuprofen (the active ingredient in Motrin, Advil, and many other nonaspirin analgesics) is an example. Only the S enantiomer of the pain reliever ibuprofen is biologically active. The body, however, converts the inactive R enantiomer to the active S enantiomer.

Captopril

(S)-Ibuprofen

For racemic drugs, most often only one enantiomer exerts the beneficial effect, whereas the other enantiomer either has no effect or exerts a detrimental effect. Thus, enantiomerically pure drugs should, more often than not, be more effective than their racemic counterparts. A case in point is 3,4-dihydroxyphenylalanine, which is used in the treatment of Parkinson's disease. The active drug is dopamine (*The Merck Index,* 12th ed., #3479). Unfortunately, this compound does not cross the blood-brain barrier to the required site of action in the brain. What is administered, instead, is the prodrug, a compound that is not active by itself but is converted in the body to an active drug. 3,4-Dihydroxyphenylalanine (*The Merck Index,* 12th ed., #3478) is such a prodrug. It crosses the blood-brain barrier and then undergoes decarboxylation catalyzed by the enzyme dopamine decarboxylase to give dopamine. Dopamine decarboxylase is specific for the S enantiomer, which is commonly known as L-DOPA. It is essential, therefore, to administer the enantiomerically pure prodrug. Were the prodrug to be administered in a racemic form, there could be a dangerous buildup of the R enantiomer, which cannot be metabolized by the enzymes present in the brain.

enzyme-catalyzed decarboxylation

(S)-(−)-3,4-Dihydroxyphenylalanine
(L-DOPA)
$[\alpha]_D^{13}$ −13.1°

Dopamine $+ CO_2$

Recently, the U.S. Food and Drug Administration established new guidelines for the testing and marketing of chiral drugs. After reviewing these guidelines, many drug companies have decided to develop only single enantiomers of new chiral drugs. In addition to regulatory pressure, there are patent considerations. If a company has patents on a racemic drug, a new patent can often be taken out on one of its enantiomers.

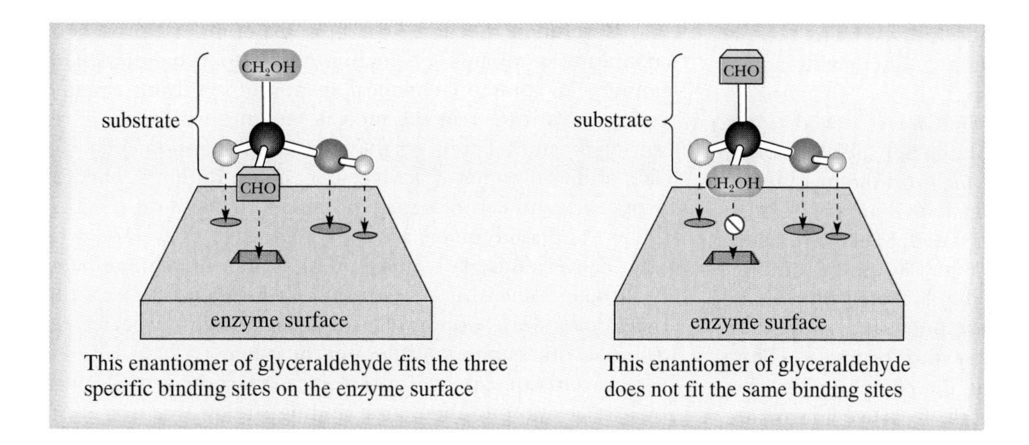

Figure 4.9
A schematic diagram of an enzyme surface capable of interacting with (R)-(+)-glyceraldehyde at three binding sites, but with (S)-(−)-glyceraldehyde at only two of these sites.

four groups on a stereocenter can distinguish between a molecule and its enantiomer or one of its diastereomers. Assume, for example, that an enzyme involved in catalyzing a reaction of glyceraldehyde has three binding sites—one specific for —H, a second specific for —OH, and a third specific for —CHO. Assume further that the three sites are arranged on the enzyme surface as shown in Figure 4.9. The enzyme can distinguish (R)-(+)-glyceraldehyde (the natural or biologically active form) from its enantiomer because the natural enantiomer can be absorbed with three groups interacting with their appropriate binding sites; for the S enantiomer, at best only two groups can interact with these binding sites.

Because interactions between molecules in living systems take place in a chiral environment, it should be no surprise that a molecule and its enantiomer or diastereomers elicit different physiological responses.

SUMMARY

Stereoisomers (Section 4.1) have the same order of attachment of atoms but a different three-dimensional orientation of their atoms in space. Stereoisomers are divided into enantiomers and diastereomers. **Enantiomers** are stereoisomers that are mirror images of each other. **Diastereomers** are stereoisomers that are not mirror images. A **mirror image** is the reflection of an object in a mirror.

A molecule that is not superposable on its mirror image is said to be **chiral** (Section 4.2). Chirality is a property of an object as a whole, not of a particular atom. An achiral object is one that lacks chirality; that is, it is an object that has a superposable mirror image. An object is achiral if it possesses a plane of symmetry. A **plane of symmetry** is an imaginary plane passing through an object dividing it such that one half is the reflection of the other half.

A **stereocenter** (Section 4.2) is an atom in a molecule at which the interchange of two atoms or groups of atoms bonded to it produces a different stereoisomer. The most common type of stereocenter is a tetrahedral stereocenter, an example of which is a carbon atom with four different groups bonded to it. The **configuration** at a stereocenter can be specified by the **Cahn-Ingold-Prelog convention,** known alternatively as the **R,S convention** (Section 4.3). To apply this convention, (1) each atom or group of atoms bonded to the stereocenter is assigned a priority and numbered from highest priority to lowest priority. (2) The molecule is oriented in space so that the group of lowest priority is directed away from the observer, and (3) the remaining three groups are read in order from highest priority to lowest priority. If the reading of groups is clockwise, the configuration is **R** (Latin: *rectus,* right). If the reading is counterclockwise, the configuration is **S** (Latin: *sinister,* left).

For a molecule with n stereocenters, the maximum number of stereoisomers possible is 2^n (Section 4.4). Certain molecules have special symmetry properties that reduce the number of stereoisomers to fewer than that predicted by the

2^n rule. A **meso** compound (Section 4.4B) contains two or more stereocenters assembled in such a way that its molecules are achiral.

Light that vibrates in only parallel planes is said to be **plane polarized** (Section 4.7A). A **polarimeter** (Section 4.7B) is an instrument used to detect and measure the magnitude of optical activity. **Observed rotation** is the number of degrees the plane of polarized light is rotated. **Specific rotation** is the observed rotation using a cell 1 dm long and a solution with a concentration of 1 g/mL. If the analyzing filter must be turned clockwise to restore the zero point, the compound is **dextrorotatory.** If the analyzing filter must be turned counterclockwise to restore the zero point, the compound is **levorotatory.** A compound is said to be **optically active** if it rotates the plane of polarized light. Each member of a pair of enantiomers rotates the plane of polarized light an equal number of degrees but opposite in direction (Section 4.7B). A **racemic mixture** (Section 4.7C) is a mixture of equal amounts of two enantiomers and has a specific rotation of zero.

Resolution (Section 4.8) is the experimental process of separating a mixture of enantiomers into two pure enantiomers. A common chemical means of resolving organic compounds is to treat the racemic mixture with a chiral resolving agent that converts the mixture of enantiomers into a pair of diastereomers. Diastereomers have different physical properties and can be separated based on these differences. Once the diastereomers are separated, each diastereomer is then converted back to a pure stereoisomer uncontaminated by its enantiomer. Another means of resolution is to treat the racemic mixture with an enzyme that catalyzes a specific reaction of one enantiomer but not the other.

An enzyme catalyzes a biological reaction of molecules by first positioning them at a binding site on its surface. An enzyme with a binding site specific for three of the four groups on a stereocenter can distinguish between a molecule and its enantiomer (Section 4.9B).

Additional Problems

Chirality

4.9 Which of these objects are chiral (assume there is no label or other identifying mark)?

(a) Pair of scissors **(b)** Tennis ball **(c)** Paper clip **(d)** Beaker

(e) The swirl created in water as it drains out of a sink or bathtub

4.10 Think about the helical coil of a telephone cord or the spiral binding on a notebook and suppose that you view the spiral from one end and find that it is a left-handed twist. If you view the same spiral from the other end, is it a right-handed twist or a left-handed twist from that end as well?

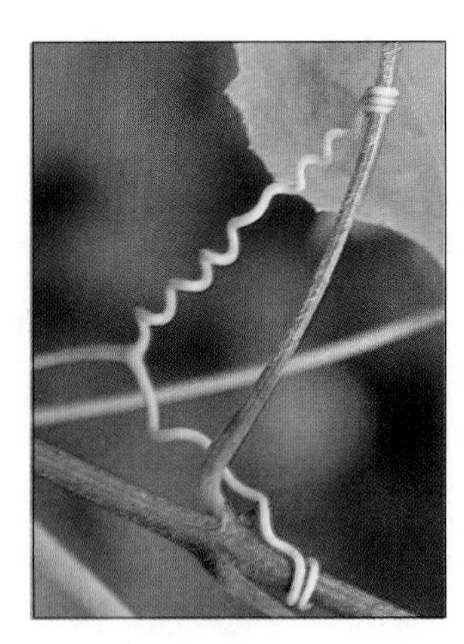

This plant climbs by sending out tendrils that twist helically. *(Charles D. Winters)*

4.11 Next time you have the opportunity to view a collection of augers or other seashells that have a helical twist, study the chirality of their twists. Do you find an equal number of left-handed and right-handed augers or, for example, do they all have the same handedness? What about the handedness of augers compared with other spiral shells?

4.12 One reason we can be sure that sp^3-hybridized carbon atoms are tetrahedral is the number of stereoisomers that can exist for different organic compounds.

(a) How many stereoisomers are possible for $CHCl_3$, CH_2Cl_2, and CHBrClF if the four bonds to carbon have a tetrahedral geometry?

(b) How many stereoisomers are possible for each of these compounds if the four bonds to the carbon have a square planar geometry?

Enantiomers

4.13 Which compounds contain stereocenters?

(a) 2-Chloropentane **(b)** 3-Chloropentane

(c) 3-Chloro-1-pentene **(d)** 1,2-Dichloropropane

This Atlantic auger shell has a right-handed helical twist. *(© Carolina Biological Supply/Phototake, NYC)*

4.14 Using only C, H, and O, write structural formulas for the lowest-molecular-weight chiral

(a) Alkane **(b)** Alcohol **(c)** Aldehyde
(d) Ketone **(e)** Carboxylic acid

4.15 Which alcohols of molecular formula $C_5H_{12}O$ are chiral?

4.16 Which carboxylic acids of molecular formula $C_6H_{12}O_2$ are chiral?

4.17 Write the structural formula of an alcohol of molecular formula $C_6H_{14}O$ that contains two stereocenters.

4.18 Draw mirror images for these molecules:

(a) OH, H_3C, H, CO_2H on C

(b) 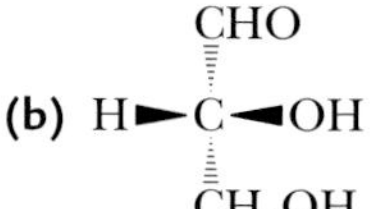

(c) 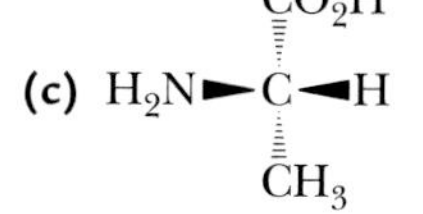

(d) 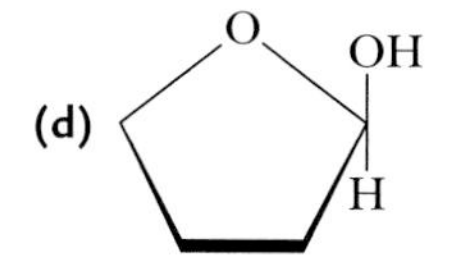

(e) 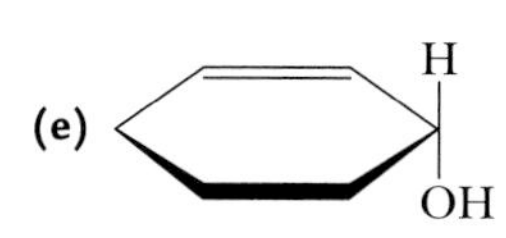

(f) OH, CH_3 on cyclohexane

(g) 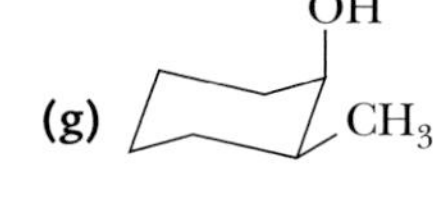

(h) CH_3, H, OH, H, OH, CH_3 (Newman projection)

4.19 Following are four stereorepresentations for lactic acid. Take (a) as a reference structure. Which stereorepresentations are identical with (a) and which are mirror images of (a)?

(a) CO_2H, H, CH_3, OH on C **(b)** CH_3, HO, CO_2H, H on C **(c)** CO_2H, HO, H, CH_3 on C **(d)** CH_3, H, HO, CO_2H on C

4.20 Mark each stereocenter in these molecules with an asterisk. Note that not all contain stereocenters.

(a) $CH_3C(CH_3)(OH)CH{=}CH_2$ **(b)** $HC(CO_2H)(CH_3)OH$ **(c)** $CH_3CH(CH_3)CH(NH_2)CO_2H$ **(d)** $CH_3C(=O)CH_2CH_3$

(e) $HC(CH_2OH)(CH_2OH)OH$ **(f)** $CH_3CH_2CH(OH)CH{=}CH_2$ **(g)** 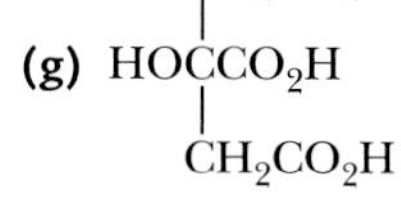 $HOC(CH_2CO_2H)(CH_2CO_2H)CO_2H$

Dill *(Anethum graveolens). (Barry Runk/Grant Heilman Photography, Inc.)*

Designation of Configuration—The R,S Convention

4.21 Assign priorities to the groups in each set.

(a) —H —CH_3 —OH —CH_2OH
(b) —$CH_2CH{=}CH_2$ —$CH{=}CH_2$ —CH_3 —CH_2CO_2H
(c) —CH_3 —H —CO_2^- —NH_3^+
(d) —CH_3 —CH_2SH —NH_3^+ —CO_2^-

Spearmint *(Mentha spicata)* is an invasive perennial aromatic herb used for jellies, sauces, and teas. *(© Betty Derig/Photo Researchers, Inc.)*

4.22 Which molecules have R configurations?

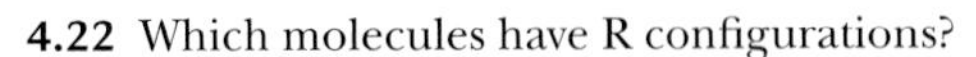

4.23 Following are structural formulas for the enantiomers of carvone. Each has a distinctive odor characteristic of the source from which it can be isolated. Assign an R or S configuration to each enantiomer (*The Merck Index,* 12th ed., #1925).

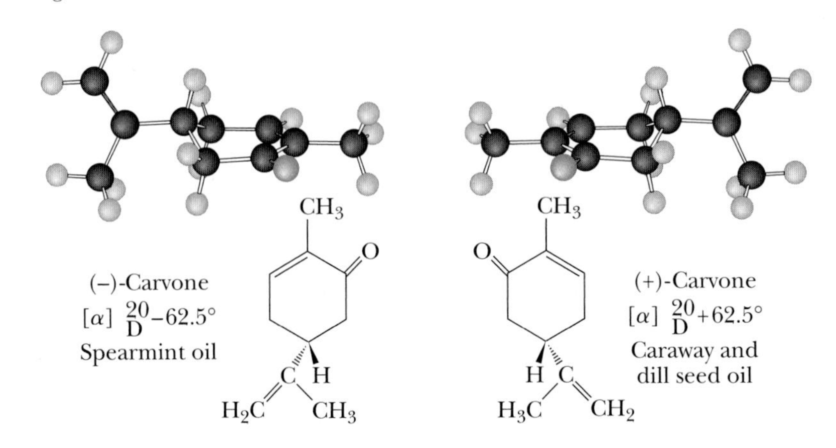

4.24 Following is a staggered conformation for one of the stereoisomers of 2-butanol.

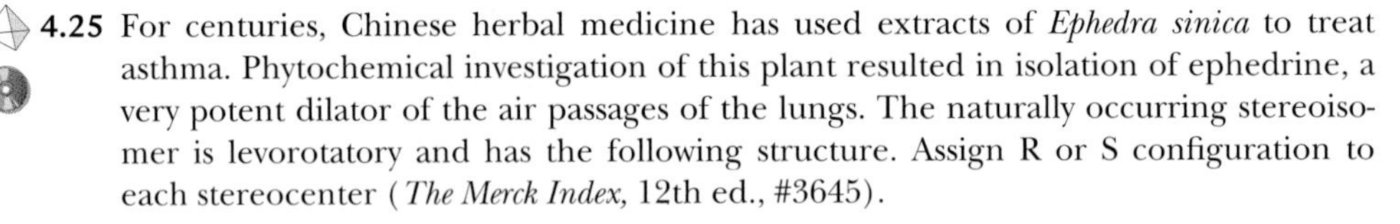

(a) Is this (R)-2-butanol or (S)-2-butanol?

(b) Draw a Newman projection for this staggered conformation, viewed along the bond between carbons 2 and 3.

(c) Draw a Newman projection for one more staggered conformation of this molecule. Which of your conformations is the more stable? Assume that —OH and $—CH_3$ are comparable in size.

Molecules with Two or More Stereocenters

Ephedra sinica, the source of ephedrine, a potent bronchodilator. *(© Paolo Koch/Photo Researchers, Inc.)*

4.25 For centuries, Chinese herbal medicine has used extracts of *Ephedra sinica* to treat asthma. Phytochemical investigation of this plant resulted in isolation of ephedrine, a very potent dilator of the air passages of the lungs. The naturally occurring stereoisomer is levorotatory and has the following structure. Assign R or S configuration to each stereocenter (*The Merck Index,* 12th ed., #3645).

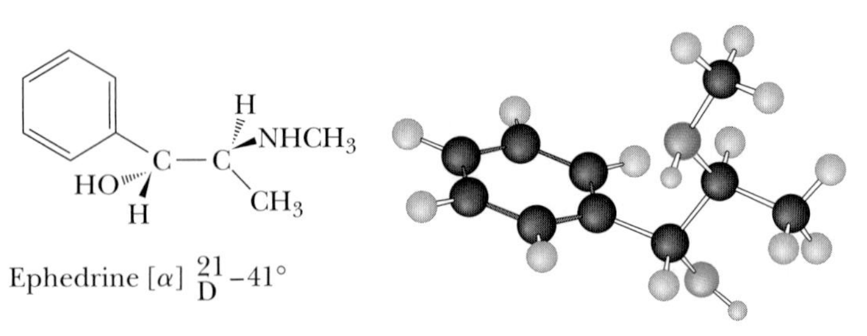

4.26 The specific rotation of naturally occurring ephedrine, shown in Problem 4.25, is $-41°$. What is the specific rotation of its enantiomer?

4.27 Label each stereocenter in these molecules with an asterisk. How many stereoisomers are possible for each molecule?

(a) 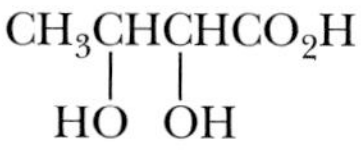 $CH_3CHCHCO_2H$ (HO, OH)

(b) 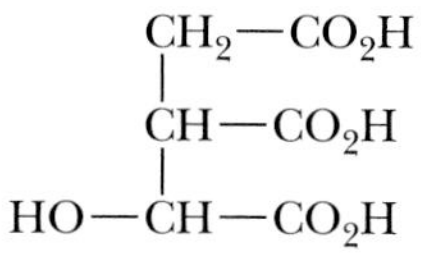 CH_2-CO_2H / $CH-CO_2H$ / $HO-CH-CO_2H$

(c) 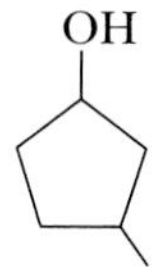 OH

(d)

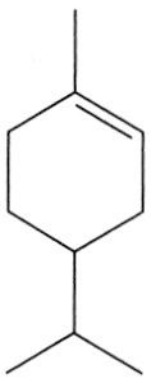

(e) 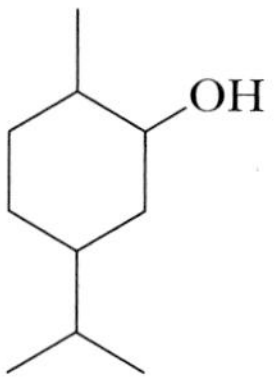 OH

(f) 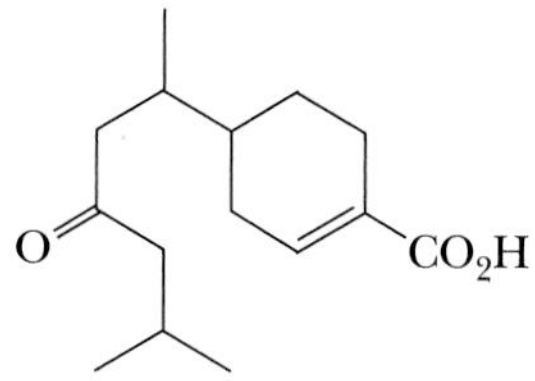 O, CO_2H

(g) 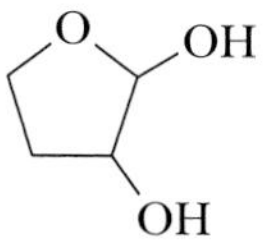 O, OH, OH

(h) 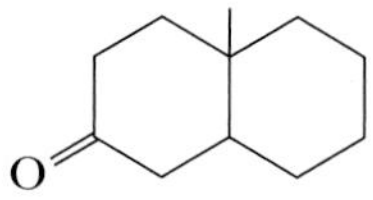 O

4.28 Label the eight stereocenters in cholesterol. How many stereoisomers are possible for a molecule of this structural formula?

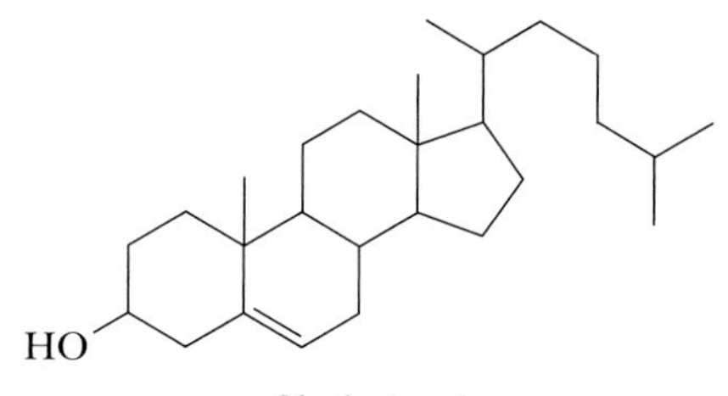

Cholesterol

4.29 Label the four stereocenters in amoxicillin, which belongs to the family of semisynthetic penicillins (*The Merck Index,* 12th ed., #617).

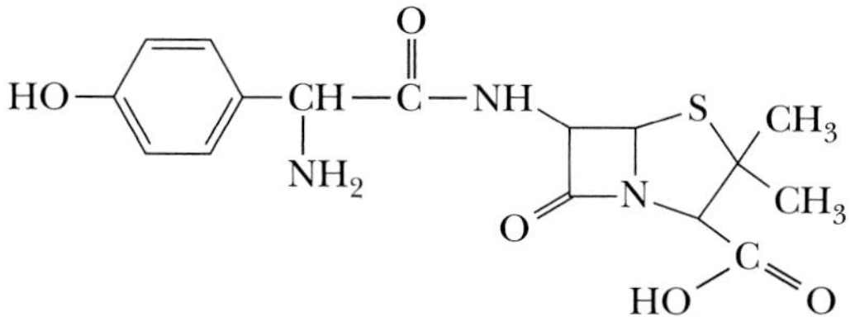

Amoxicillin

4.30 Label all stereocenters in the antihistamine terfenadine (*The Merck Index,* 12th ed., #9307). Terfenadine received FDA approval in May 1985 and by year's end had become the top-selling antihistamine in the United States. It provides relief from allergic disorders but, unlike so many of the earlier antihistamines, does not cause drowsiness.

Use of terfenadine in combination with certain other medications, however, has been linked to certain cardiac disorders, and, in 1997, terfenadine was withdrawn from the market.

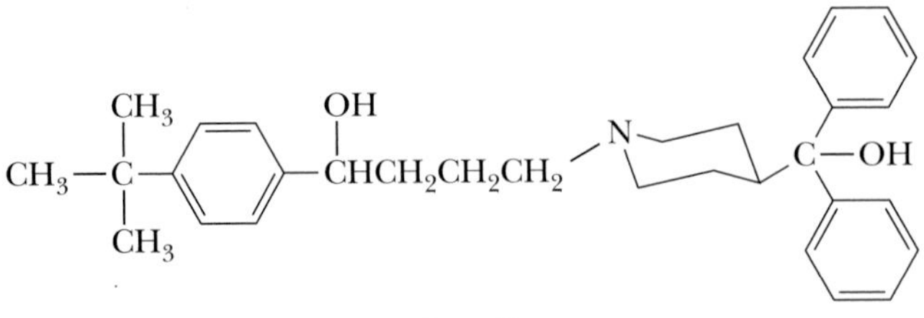

Terfenadine
(Seldane)

4.31 Label all stereocenters in loratadine (Claritin, *The Merck Index,* 12th ed., #5608), now the top-selling antihistamine in the United States. How many stereoisomers are possible for this compound?

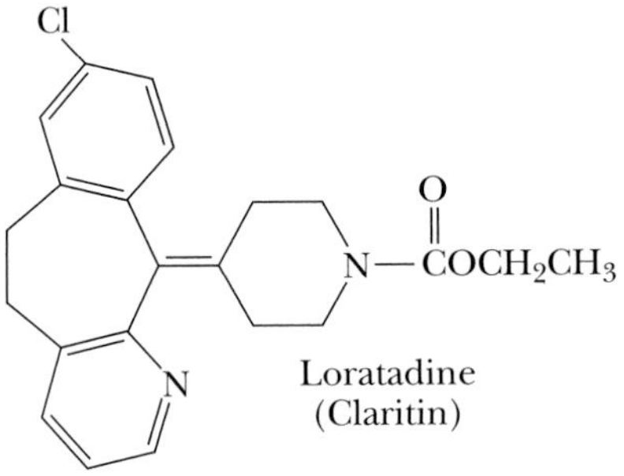

Loratadine
(Claritin)

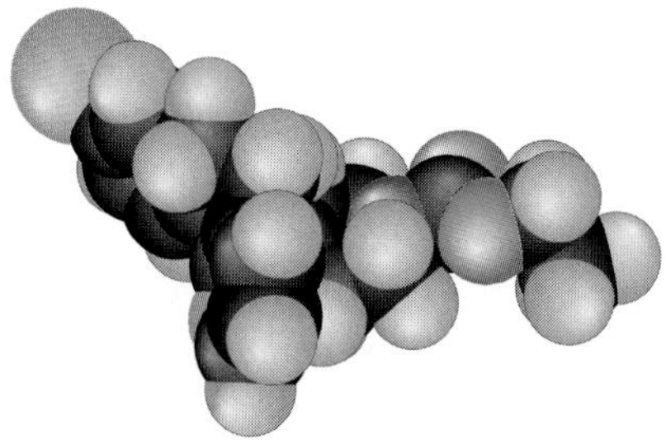

4.32 Which of these structural formulas represent meso compounds?

(a) 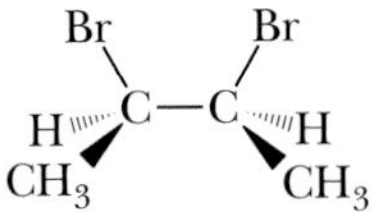 **(b)** 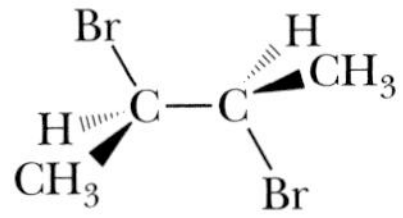 **(c)** 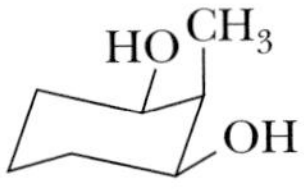 **(d)** 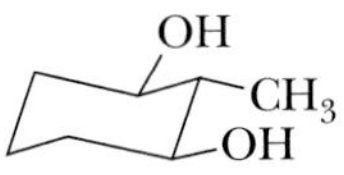 **(e)** 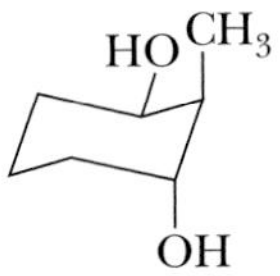 **(f)**

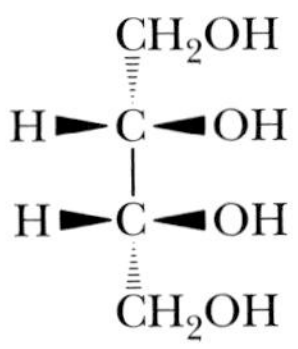

4.33 Draw a Newman projection, viewed along the bond between carbons 2 and 3, for both the most stable and the least stable conformations of meso-tartaric acid.

4.34 How many stereoisomers are possible for 1,3-dimethylcyclopentane? Which are pairs of enantiomers? Which are meso compounds?

4.35 In answer to Problem 3.33, you were asked to draw the more stable chair conformation of glucose, a molecule in which all groups on the six-membered ring are equatorial. Here is a drawing and molecular model of that conformation.

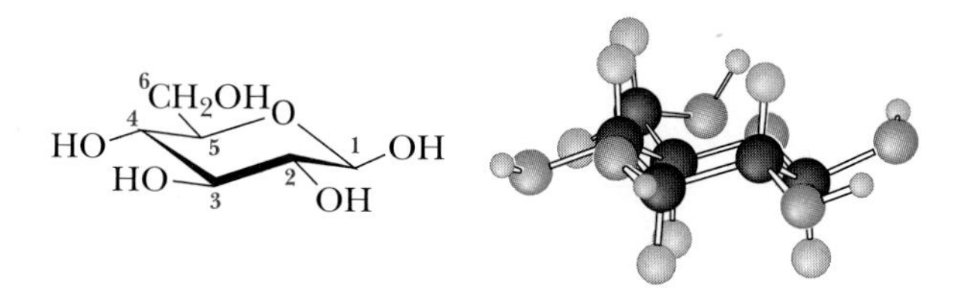

(a) Identify all stereocenters in this molecule.

(b) How many stereoisomers are possible?

(c) How many pairs of enantiomers are possible?

(d) What is the configuration (R or S) at carbons 1 and 5 in the stereoisomer shown?

5

Alkenes and Alkynes

5.1 Structure
5.2 Nomenclature
5.3 Physical Properties
5.4 Naturally Occurring Alkenes—The Terpenes

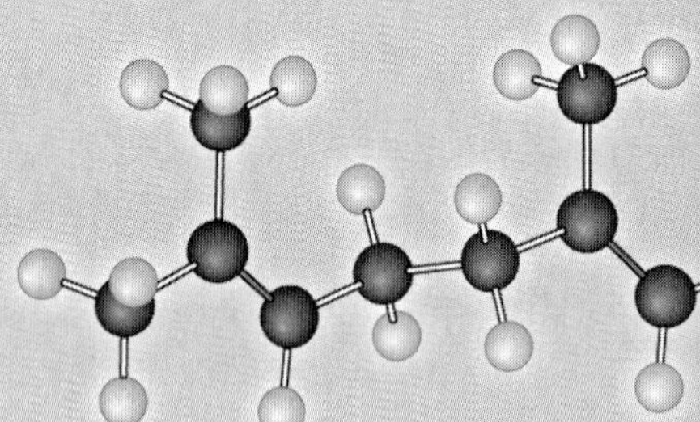

Lily-of-the-valley *(Convallaria majalis)*, from which farnesol, a terpene, is isolated. Inset: A model of farnesol. *(Barry L. Runk/Grant Heilman Photography, Inc.)*

An **unsaturated hydrocarbon** contains one or more carbon-carbon double or triple bonds. The term *unsaturation* shows that there are fewer hydrogens attached to carbon than in an alkane. There are three classes of unsaturated hydrocarbons: alkenes, alkynes, and arenes. **Alkenes** contain one or more carbon-carbon double bonds, and **alkynes** contain one or more carbon-carbon triple bonds. The simplest alkene is ethene, and the simplest alkyne is ethyne.

Alkene An unsaturated hydrocarbon that contains a carbon-carbon double bond.

Alkyne An unsaturated hydrocarbon that contains a carbon-carbon triple bond.

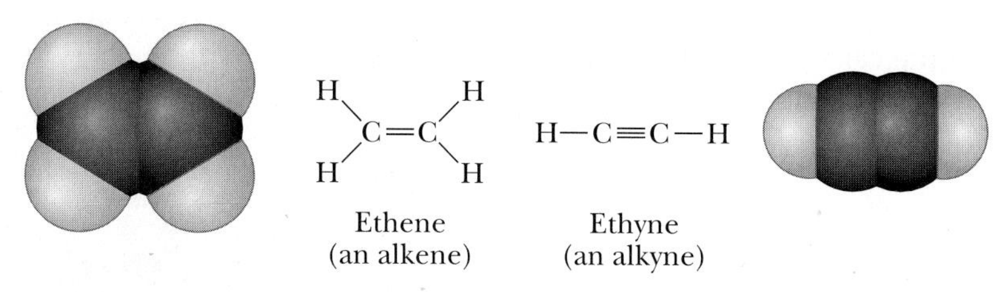

Ethene (an alkene)   Ethyne (an alkyne)

In this chapter we study the structure, nomenclature, and physical properties of alkenes and alkynes.

The third class of unsaturated hydrocarbons is the **arenes.** The Lewis structure of benzene, the simplest arene, is

Arene A compound containing one or more benzene rings.

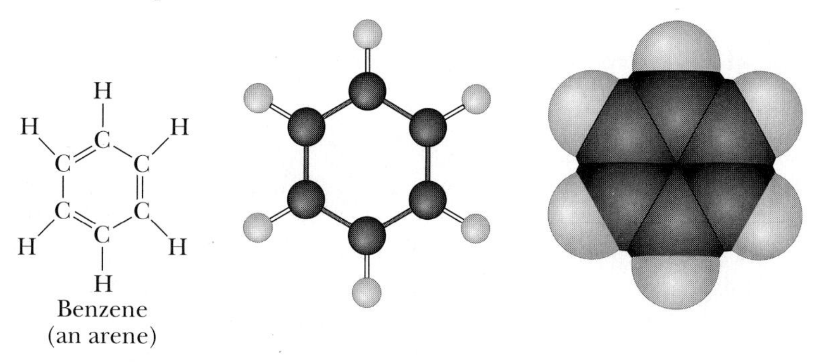

Benzene (an arene)

Just as a group derived by removal of an H from an alkane is called an alkyl group and given the symbol R— (Section 3.3A), a group derived by removal of an H from an arene is called an **aryl group** and is given the symbol **Ar**—.

When a benzene ring occurs as a substituent on a parent chain, it is named a **phenyl group.** You might think that, when present as a substituent, benzene would become benzyl, just as ethane becomes ethyl. This is not so! "Phene" is a now-obsolete name for benzene, and, although this name is no longer used, a derivative has persisted in the name **phenyl.** Following is a structural formula for the phenyl group and two alternative representations for it.

C_6H_5—   Ph—

Benzene   Alternative representations for the phenyl group

The chemistry of benzene and its derivatives is quite different from that of alkenes and alkynes, but, even though we do not study the chemistry of arenes until Chapter 9, we will show structural formulas of compounds containing aryl groups before that time. What you need to remember at this point is that an aryl group is not chemically reactive under any of the conditions we describe in Chapters 5–8.

5.1 STRUCTURE

A. Shapes of Alkenes

Using the valence-shell electron-pair repulsion model (Section 1.3), we predict a value of 120° for the bond angles about each carbon in a double bond. The observed H—C—C bond angle in ethylene is 121.7°, close to that predicted. In other alkenes, deviations from the predicted angle of 120° may be somewhat larger as a result of the nonbonded interaction strain between alkyl groups attached to the carbons of the double bond. The C—C—C bond angle in propene, for example, is 124.7°.

121.7° 124.7°

Ethene **Propene**

B. Orbital Overlap Model of a Carbon-Carbon Double Bond

In Section 1.6D, we describe the formation of a carbon-carbon double bond in terms of the overlap of atomic orbitals. A carbon-carbon double bond consists of one sigma bond and one pi bond. Each carbon of the double bond uses its three sp^2 hybrid orbitals to form sigma bonds to three atoms. The unhybridized $2p$ atomic orbitals, which lie perpendicular to the plane created by the axes of the three sp^2 hybrid orbitals, combine to form the pi bond of the carbon-carbon double bond.

It takes approximately 63 kcal/mol (264 kJ/mol) to break the pi bond in ethylene; that is, to rotate one carbon by 90° with respect to the other so that no overlap occurs between $2p$ orbitals on adjacent carbons (Figure 5.1). This energy is considerably greater than the thermal energy available at room temperature, and, as a consequence, rotation about a carbon-carbon double bond is severely re-

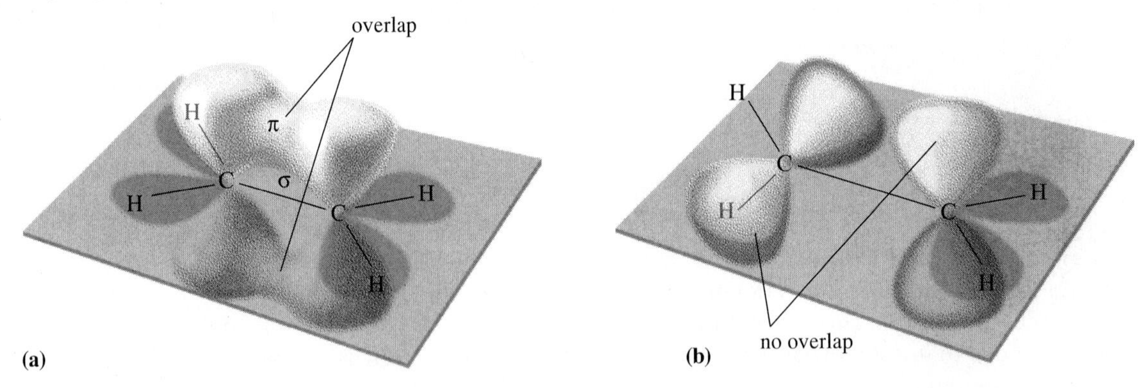

Figure 5.1 Restricted rotation about a carbon-carbon double bond in ethylene. (a) Orbital overlap model showing the pi bond. (b) The pi bond is broken by rotating the plane of one H—C—H group by 90° with respect to the plane of the other H—C—H group.

stricted. You might compare rotation about a carbon-carbon double bond, such as in ethylene, with that about a carbon-carbon single bond, such as in ethane (Section 3.6A). Whereas rotation about the carbon-carbon single bond in ethane is relatively free (energy barrier approximately 3 kcal/mol), rotation about the carbon-carbon double bond in ethylene is restricted (energy barrier approximately 63 kcal/mol).

C. Cis-Trans Isomerism in Alkenes

Cis-trans isomerism Isomers that have the same order of attachment of their atoms, but a different arrangement of their atoms in space due to the presence of either a ring (Chapter 3) or a carbon-carbon double bond (Chapter 5).

Because of restricted rotation about a carbon-carbon double bond, an alkene in which each carbon of the double bond has two different groups attached to it shows cis-trans isomerism. For example, 2-butene has two stereoisomers. In *cis*-2-butene, the two methyl groups are on one side of the double bond, and the two hydrogens are on the other side. In *trans*-2-butene, the two methyl groups are on opposite sides of the double bond. These two compounds cannot be converted into one another at room temperature because of the restricted rotation about the double bond; they are different compounds, with different physical and chemical properties.

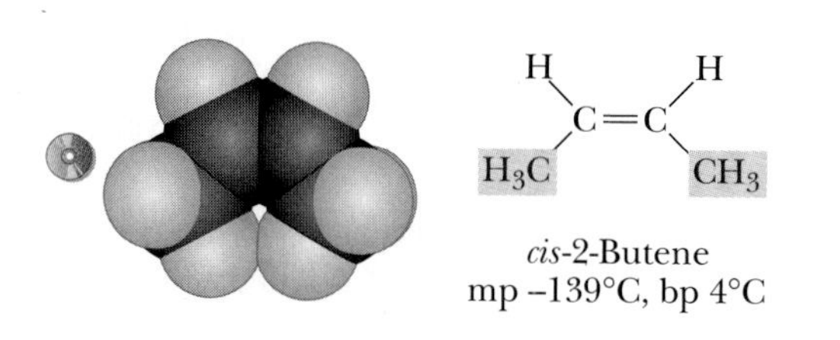

cis-2-Butene
mp −139°C, bp 4°C

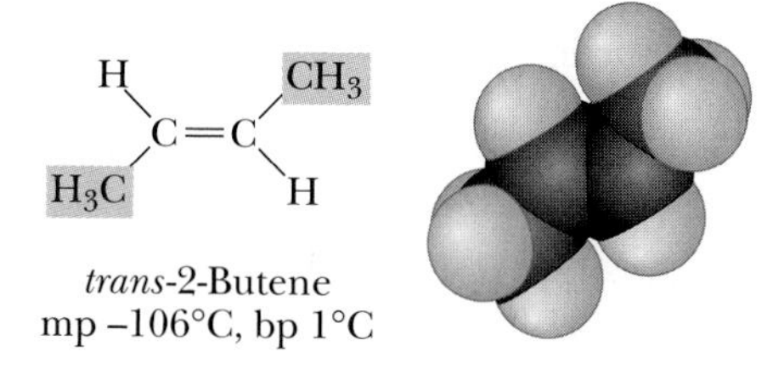

trans-2-Butene
mp −106°C, bp 1°C

Cis alkenes are less stable than their trans isomers because of nonbonded interaction strain between alkyl substituents on the same side of the double bond in the cis isomer, as can be seen in space-filling models of the cis and trans isomers of 2-butene. This is the same type of strain that results in the preference for equatorial methylcyclohexane over axial methylcyclohexane (Section 3.6B).

D. Shapes of Alkynes

The functional group of an alkyne is a **carbon-carbon triple bond.** The simplest alkyne is ethyne, C_2H_2. Ethyne is a linear molecule; all bond angles are 180°.

In valence bond theory, a triple bond is described in terms of the overlap of *sp* hybrid orbitals of adjacent carbons to form a sigma bond, overlap of parallel $2p_y$ orbitals to form one pi bond, and overlap of parallel $2p_z$ orbitals to form a second pi bond (Figure 1.18). In ethyne, each carbon-hydrogen bond is formed by the overlap of the 1*s* atomic orbital of hydrogen with an *sp* hybrid orbital of carbon.

Cutting with an oxyacetylene torch. *(Charles D. Winters)*

5.2 NOMENCLATURE

Alkenes are named using the IUPAC system, but, as we shall see, some are still referred to by their common names.

A. IUPAC Names

IUPAC names of alkenes are formed by changing the **-an-** infix of the parent alkane to **-en-** (Section 3.5). Hence, $CH_2{=}CH_2$ is named ethene, and $CH_3CH{=}CH_2$ is named propene. In higher alkenes, where isomers exist that differ in location of the double bond, a numbering system must be used. According to the IUPAC system, the longest carbon chain that contains the double bond is numbered in the direction to give the carbon atoms of the double bond the lower set of numbers. The location of the double bond is indicated by the number of its first carbon. Branched or substituted alkenes are named in a manner similar to alkanes. Carbon atoms are numbered, substituent groups are located and named, the double bond is located, and the main chain is named.

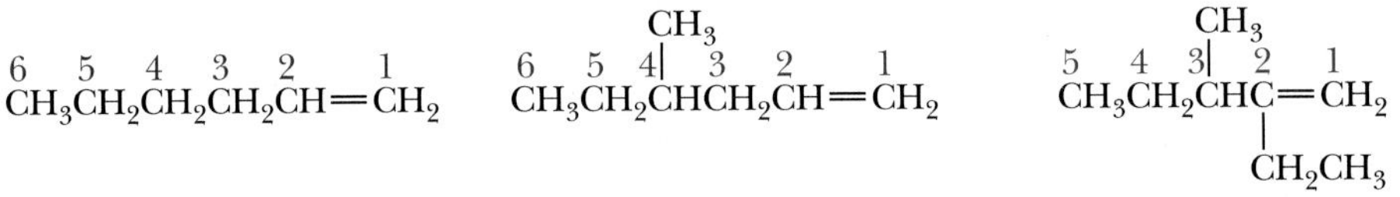

$$\overset{6}{C}H_3\overset{5}{C}H_2\overset{4}{C}H_2\overset{3}{C}H_2\overset{2}{C}H{=}\overset{1}{C}H_2$$

1-Hexene

$$\overset{6}{C}H_3\overset{5}{C}H_2\overset{4}{C}H(CH_3)\overset{3}{C}H_2\overset{2}{C}H{=}\overset{1}{C}H_2$$

4-Methyl-1-hexene

$$\overset{5}{C}H_3\overset{4}{C}H_2\overset{3}{C}H(CH_3)\overset{2}{C}(CH_2CH_3){=}\overset{1}{C}H_2$$

2-Ethyl-3-methyl-1-pentene

Note that there is a chain of six carbon atoms in 2-ethyl-3-methyl-1-pentene. However, because the longest chain that contains the double bond has only five carbons, the parent hydrocarbon is pentane, and the molecule is named as a disubstituted 1-pentene.

IUPAC names of alkynes are formed by changing the **-an-** infix of the parent alkane to **-yn-** (Section 3.5). Thus, $HC{\equiv}CH$ is named ethyne, and $CH_3C{\equiv}CH$ is named propyne. The IUPAC system retains the name acetylene; therefore, there are two acceptable names for $HC{\equiv}CH$: ethyne and acetylene. Of these two names, acetylene is used much more frequently. For larger molecules, the longest carbon chain that contains the triple bond is numbered from the end that gives the triply bonded carbons the lower set of numbers. The location of the triple bond is indicated by the number of the first carbon of the triple bond.

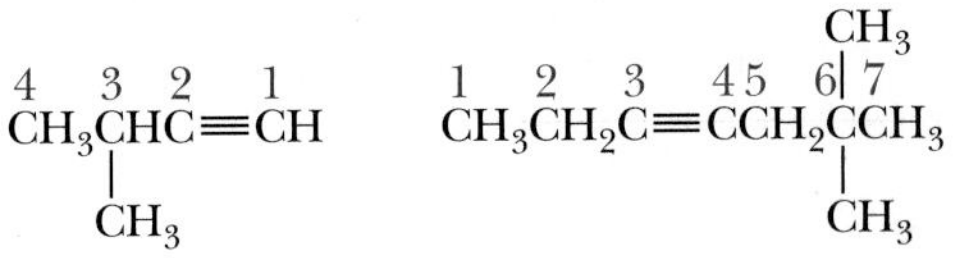

$$\overset{4}{C}H_3\overset{3}{C}H(CH_3)\overset{2}{C}{\equiv}\overset{1}{C}H$$

3-Methyl-1-butyne

$$\overset{1}{C}H_3\overset{2}{C}H_2\overset{3}{C}{\equiv}\overset{4}{C}\overset{5}{C}H_2\overset{6}{C}(CH_3)_2\overset{7}{C}H_3$$

6,6-Dimethyl-3-heptyne

EXAMPLE 5.1

Write the IUPAC name of each hydrocarbon.

(a) $CH_2{=}CH(CH_2)_5CH_3$

(b) 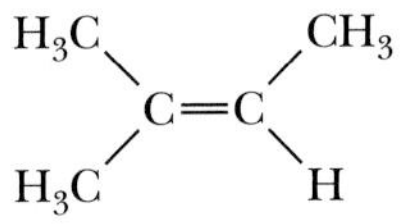 $(H_3C)_2C{=}C(CH_3)H$

(c) $CH_3(CH_2)_2C{\equiv}CCH_3$

SOLUTION

(a) 1-Octene (b) 2-Methyl-2-butene (c) 2-Hexyne

Practice Problem 5.1

Write the IUPAC name of each hydrocarbon.

$$\text{(a)}\ CH_3CH_2\overset{\overset{CH_3}{|}}{\underset{\underset{CH_3}{|}}{C}}CH{=}CH_2 \qquad \text{(b)}\ (CH_3)_2C{=}C(CH_3)_2 \qquad \text{(c)}\ CH_3\overset{\overset{CH_3}{|}}{\underset{\underset{CH_3}{|}}{C}}C{\equiv}CH$$

B. Common Names

Despite the precision and universal acceptance of IUPAC nomenclature, some alkenes and alkynes, particularly those of low molecular weight, are known almost exclusively by their common names, as illustrated by the common names of these alkenes.

	$CH_2{=}CH_2$	$CH_3CH{=}CH_2$	$CH_3\overset{\overset{CH_3}{\vert}}{C}{=}CH_2$
IUPAC name:	Ethene	Propene	2-Methylpropene
Common name:	Ethylene	Propylene	Isobutylene

Furthermore, the common names **vinyl** and **allyl** are often used to show the presence of these alkenyl groups:

Alkenyl Group	Common Name	Example
$CH_2{=}CH{-}$	vinyl	$CH_2{=}CHCl$ vinyl chloride
$CH_2{=}CHCH_2{-}$	allyl	$CH_2{=}CHCH_2Cl$ allyl chloride

Common names for alkynes are derived by prefixing the names of substituents on the carbon-carbon triple bond to the name acetylene as shown in these examples.

	$CH_3C{\equiv}CH$	$CH_3C{\equiv}CCH_3$	$CH_2{=}CHC{\equiv}CH$
IUPAC name:	Propyne	2-Butyne	1-Buten-3-yne
Common name:	Methylacetylene	Dimethylacetylene	Vinylacetylene

EXAMPLE 5.2

Write the common name of each compound.

$$\text{(a)}\ CH_3\overset{\overset{CH_3}{|}}{C}HCH_2C{\equiv}CCH_3 \qquad \text{(b)}\ CH_3CH_2\overset{\overset{CH_3}{|}}{C}HC{\equiv}CH \qquad \text{(c)}\ HC{\equiv}C\overset{\overset{CH_3}{|}}{\underset{\underset{CH_3}{|}}{C}}CH_3$$

SOLUTION

(a) Isobutylmethylacetylene (b) *sec*-Butylacetylene (c) *tert*-Butylacetylene

Practice Problem 5.2

Write the common name of each compound.

(a) $CH_3\underset{}{C}HC{\equiv}C\underset{}{C}HCH_3$ (with CH_3 on each CH) (b) cyclohexyl–$C{\equiv}CH$ (c) $HC{\equiv}CCH_2CH_2CH_2CH_3$

C. Systems for Designating Configuration in Alkenes

The Cis-Trans System

The most common method for specifying configuration in alkenes uses the prefixes cis and trans. There is no doubt whatsoever which isomer is intended by the name *trans*-3-hexene. For more complex alkenes, the orientation of the atoms of the parent chain determines whether the alkene is cis or trans. Following is a structural formula for the cis isomer of 3,4-dimethyl-2-pentene. In this example, carbon atoms of the main chain (carbons 1 and 4) are on the same side of the double bond, and, therefore, the configuration of this alkene is cis.

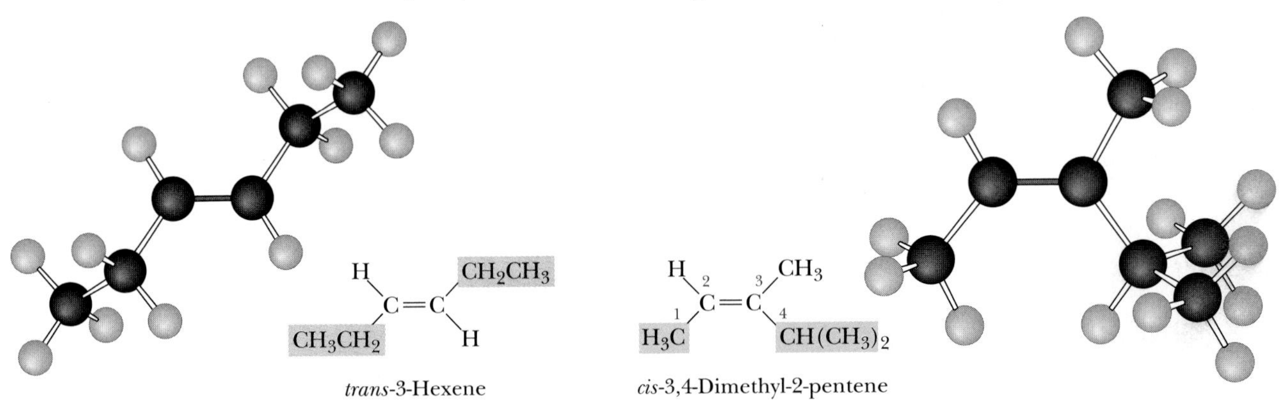

trans-3-Hexene

cis-3,4-Dimethyl-2-pentene

EXAMPLE 5.3

Name each compound and show the configuration about each double bond using the cis-trans system.

(a) $CH_3CH_2CH_2$ and H on one carbon, H and CH_2CH_3 on the other: $(CH_3CH_2CH_2)(H)C{=}C(H)(CH_2CH_3)$, with $CH_3CH_2CH_2$ top left, H top right, H bottom left, CH_2CH_3 bottom right

(b) $(CH_3CH_2CH_2)(H_3C)C{=}C(CH_2CH_3)(H)$, with $CH_3CH_2CH_2$ top left, CH_2CH_3 top right, H_3C bottom left, H bottom right

SOLUTION

(a) The chain contains seven carbon atoms and is numbered from the end that gives the lower number to the first carbon of the double bond. Its name is *trans*-3-heptene.

(b) The longest chain contains seven carbon atoms and is numbered from the right so that the first carbon of the double bond is carbon 3 of the chain. Its name is *cis*-4-methyl-3-heptene.

Practice Problem 5.3

Which alkenes show cis-trans isomerism? For each alkene that does, draw the trans isomer.

(a) 2-Pentene (b) 2-Methyl-2-pentene (c) 3-Methyl-2-pentene

Throughout this text, we use the cis-trans system for alkenes in which it is clear which is the main carbon chain. We use the E,Z system in all other cases.

The E,Z System

The **E,Z system** uses the priority rules of the R,S system (Section 4.3) to assign priorities to the substituents on each carbon of a double bond. If the groups of higher priority are on the same side of the double bond, the alkene is designated **Z** (German: *zusammen,* together). If the groups of higher priority are on opposite sides of the double bond, the alkene is designated **E** (German: *entgegen,* opposite).

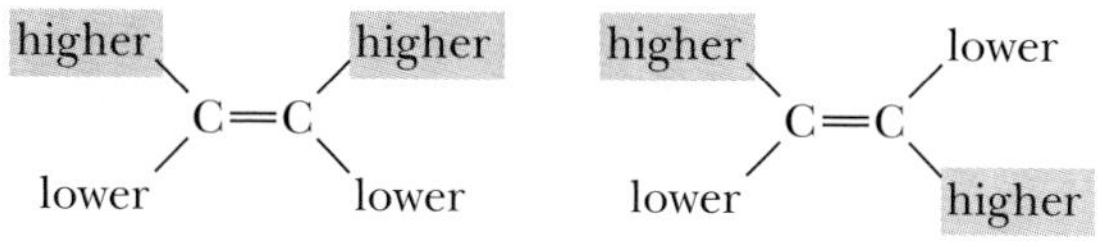

E,Z system A system to specify the configuration of groups about a carbon-carbon double bond.

Z From the German, *zusammen,* together. Specifies that groups of higher priority on the carbons of a double bond are on the same side.

E From the German, *entgegen,* opposite. Specifies that groups of higher priority on the carbons of a double bond are on opposite sides.

```
higher     higher        higher      lower
      \   /                    \    /
       C=C                      C=C
      /   \                    /    \
 lower     lower          lower      higher
  Z (zusammen)              E (entgegen)
```

EXAMPLE 5.4

Name each alkene and specify its configuration by the E,Z system.

```
     H       CH3                Cl      H
      \     /                     \    /
(a)    C=C              (b)        C=C
      /     \                     /    \
  H3C        CH(CH3)2          H3C      CH2CH3
```

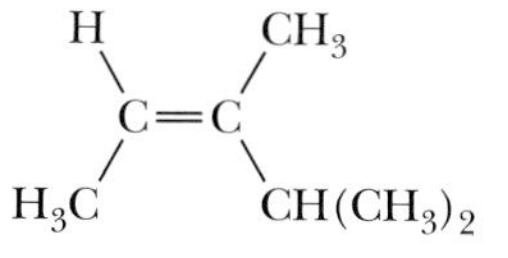

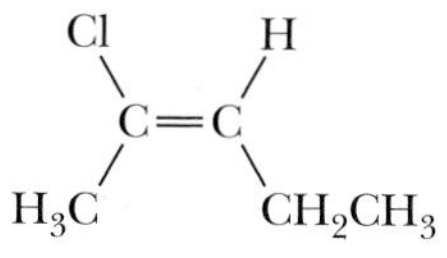

SOLUTION

(a) The group of higher priority on carbon 2 is methyl; that of higher priority on carbon 3 is isopropyl. Because the groups of higher priority are on the same side of the carbon-carbon double bond, the alkene has the Z configuration. Its name is (Z)-3,4-dimethyl-2-pentene. Using the cis-trans system, it is named *cis*-3,4-dimethyl-2-pentene.

(b) Groups of higher priority on carbons 2 and 3 are —Cl and $—CH_2CH_3$. Because these groups are on opposite sides of the double bond, the configuration of this alkene is E and its name is (E)-2-chloro-2-pentene. Using the cis-trans system, it is *cis*-2-chloro-2-pentene.

Practice Problem 5.4

Name each alkene and specify its configuration by the E,Z system.

(a) 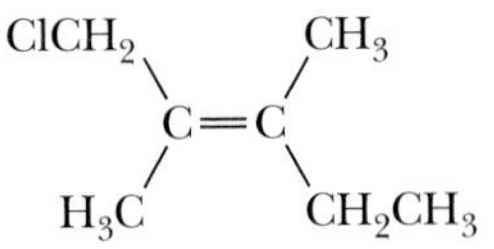 $ClCH_2(H_3C)C{=}C(CH_3)CH_2CH_3$ (b) $Cl(Br)C{=}C(H)CH_3$ (c) $CH_3CH_2CH_2(H_3C)C{=}C(CH_3)CH(CH_3)_2$

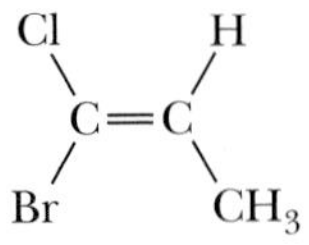

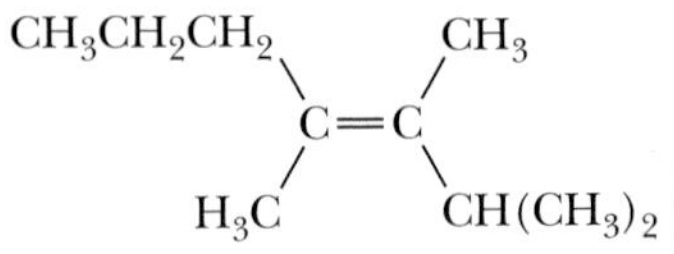

D. Cycloalkenes

In naming **cycloalkenes,** the carbon atoms of the ring double bond are numbered 1 and 2 in the direction that gives to the first-encountered substituent the smaller number. Substituents are listed in alphabetical order.

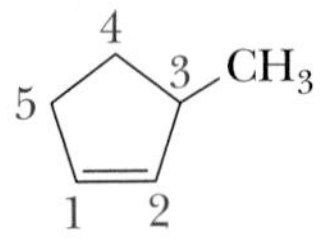

3-Methylcyclopentene 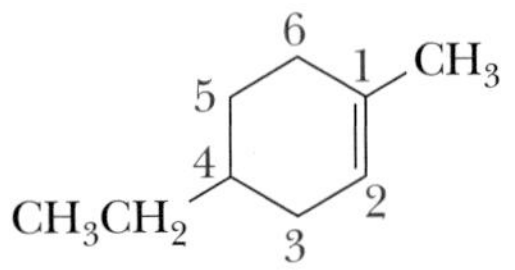 4-Ethyl-1-methylcyclohexene

EXAMPLE 5.5

Write the IUPAC name for each cycloalkene.

(a) 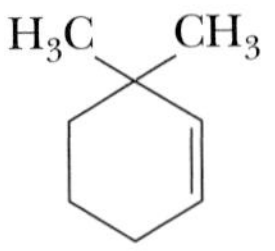 (b) 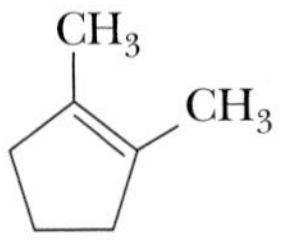 (c) 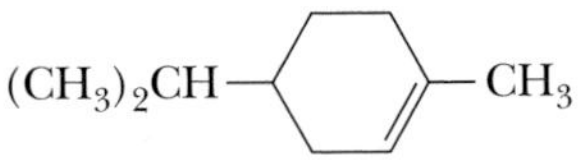 $(CH_3)_2CH$—⬡—CH_3

SOLUTION

(a) 3,3-Dimethylcyclohexene (b) 1,2-Dimethylcyclopentene
(c) 4-Isopropyl-1-methylcyclohexene

Practice Problem 5.5

Write the IUPAC name for each cycloalkene.

(a) 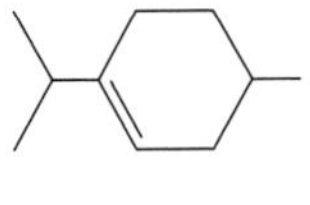 (b) 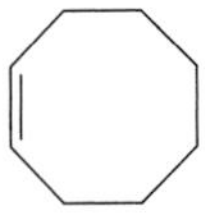 (c) 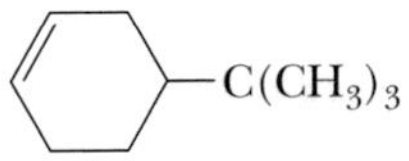 —$C(CH_3)_3$

E. Cis-Trans Isomerism in Cycloalkenes

Following are structural formulas for four cycloalkenes:

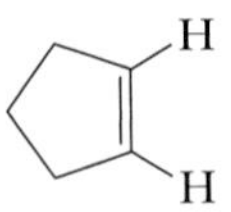

Cyclopentene 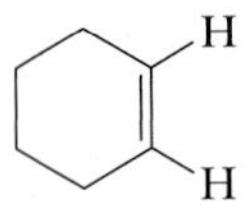 Cyclohexene 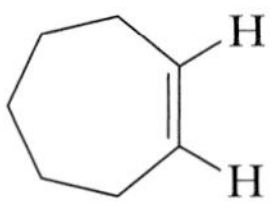 Cycloheptene 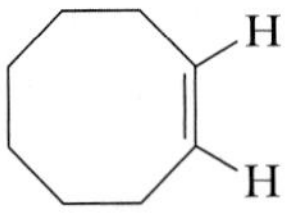 Cyclooctene

In these representations, the configuration about each double bond is cis. Is it possible to have a trans configuration in these and larger cycloalkenes? To date, *trans-*

cyclooctene is the smallest *trans*-cycloalkene that has been prepared in pure form and is stable at room temperature. Yet, even in this *trans*-cycloalkene, there is considerable intramolecular strain. *cis*-Cyclooctene is more stable than its trans isomer by 9.1 kcal/mol.

F. Dienes, Trienes, and Polyenes

Alkenes that contain more than one double bond are named as alkadienes, alkatrienes, and so on. Those that contain several double bonds are also referred to more generally as polyenes (Greek: *poly*, many). Following are examples of three dienes:

$CH_2{=}CHCH_2CH{=}CH_2$
1,4-Pentadiene

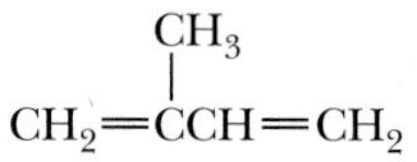

$CH_2{=}C(CH_3)CH{=}CH_2$
2-Methyl-1,3-butadiene
(Isoprene)

1,3-Cyclopentadiene

G. Cis-Trans Isomerism in Dienes, Trienes, and Polyenes

Thus far we have considered cis-trans isomerism in alkenes containing only one carbon-carbon double bond. For an alkene with one carbon-carbon double bond that can show cis-trans isomerism, two stereoisomers are possible. For an alkene with ***n*** carbon-carbon double bonds, each of which can show cis-trans isomerism, $\mathbf{2^n}$ stereoisomers are possible.

EXAMPLE 5.6

How many stereoisomers are possible for 2,4-heptadiene?

SOLUTION

This molecule has two carbon-carbon double bonds, each of which shows cis-trans isomerism. As shown in the table, there are $2^2 = 4$ stereoisomers. Two of these are drawn on the right.

Double Bond	
$C_2{-}C_3$	$C_4{-}C_5$
trans	trans
trans	cis
cis	trans
cis	cis

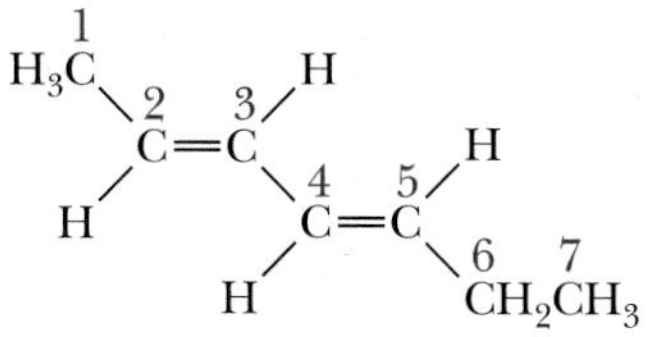

trans,trans-2,4-Heptadiene

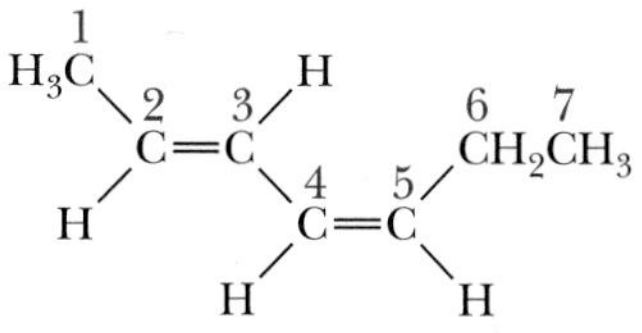

trans,cis-2,4-Heptadiene

Practice Problem 5.6

Draw structural formulas for the other two cis-trans isomers of 2,4-heptadiene.

CHEMICAL CONNECTIONS

The Case of the Iowa and New York Strains of the European Corn Borer

Although humans communicate largely by sight and sound, chemical signals are the primary means of communication for the vast majority of other species in the animal world. Often, communication within a species is specific for one of two configurational isomers. For example, a member of a given species may respond to a cis isomer of a chemical but not to the trans isomer. Or, alternatively, it might respond to a quite precise blend of cis and trans isomers but not to other blends of these same isomers.

Several groups of scientists have studied the components of the sex pheromones of both the Iowa and New York strains of the European corn borer. Females of these closely related species secrete the sex attractant 11-tetradecenyl acetate. Males of the Iowa strain show maximum response to a mixture containing 96% of the cis isomer and 4% of the trans isomer. When the pure cis isomer is used alone, males are only weakly attracted. Males of the New York strain show an entirely different response pattern. They respond maximally to a mixture containing 3% of the cis isomer and 97% of the trans isomer.

The European corn borer, *Pyrausta nubilalis*. *(Runk/Schoenberger/Grant Heilman Photography, Inc.)*

CH_3CH_2(H)C=C(H)$(CH_2)_{10}OCCH_3$ (C=O)

cis-11-Tetradecenyl acetate

(H)(CH_3CH_2)C=C(H)$(CH_2)_{10}OCCH_3$ (C=O)

trans-11-Tetradecenyl acetate

There is evidence that an optimal response to a narrow range of stereoisomers as we see here is widespread in nature and that most insects maintain species isolation for mating and reproduction by the stereochemistry of their pheromones. See "Sex Pheromones: Minor Amounts of Opposite Geometric Isomers Critical for Attraction" by J. A. Klun *et al.*, *Science* **181:** 661 (1973).

EXAMPLE 5.7

How many stereoisomers are possible for 10,12-hexadecadien-1-ol?

$$CH_3(CH_2)_2CH{=}CHCH{=}CH(CH_2)_8CH_2OH$$

10,12-Hexadecadien-1-ol

SOLUTION

Cis-trans isomerism is possible about both double bonds. Four isomers are possible.

Practice Problem 5.7

The sex pheromone of the silkworm is (10E,12Z)-10,12-hexadecadien-1-ol. Draw a structural formula for this compound.

Silk worms spinning cocoons on a loom, silk farm, Japan. *(Paul Chesley/Tony Stone Worldwide)*

An example of a biologically important compound for which a number of cis-trans isomers is possible is vitamin A. There are four carbon-carbon double bonds in the chain of carbon atoms attached to the substituted cyclohexene ring, and each has the potential for cis-trans isomerism. There are $2^4 = 16$ cis-trans isomers possible for this structural formula. Vitamin A is the all-trans isomer.

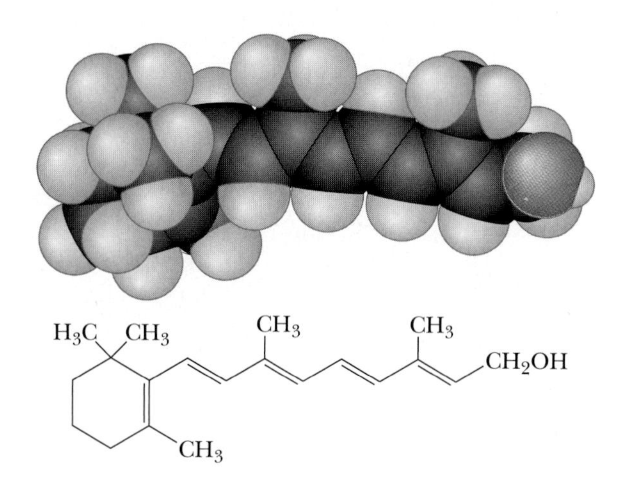

5.3 PHYSICAL PROPERTIES

Alkenes and alkynes are nonpolar compounds, and the only attractive forces between their molecules are dispersion forces. Therefore, their physical properties are similar to those of alkanes (Section 3.8) with the same carbon skeletons. Those that are liquid at room temperature have densities less than 1.0 g/mL (they are less dense than water). They are insoluble in water but soluble in one another, in other nonpolar organic liquids, and in ethanol.

5.4 NATURALLY OCCURRING ALKENES—THE TERPENES

A **terpene** is a compound whose carbon skeleton can be divided into two or more units that are identical with the carbon skeleton of isoprene. Carbon 1 of an **isoprene unit** is called the head, and carbon 4 is called the tail. Terpenes are formed by stringing together the tail of one isoprene unit to the head of another.

Terpene A compound whose carbon skeleton can be divided into two or more units identical with the carbon skeleton of isoprene.

CHEMICAL CONNECTIONS

Why Plants Emit Isoprene

Names like Virginia's Blue Ridge, Jamaica's Blue Mountain Peak, and Australia's Blue Mountains remind us of the bluish haze that hangs over wooded hills in the summertime. It was discovered in the 1950s that this haze is rich in isoprene, which means that isoprene is far more abundant in the atmosphere than anyone thought. The haze is caused by light-scattering from an aerosol produced by the photooxidation of isoprene and other hydrocarbons. Scientists now estimate that global emission of isoprene by plants is 3×10^{14} g/yr (3.3×10^{8} ton/yr), which represents approximately 2% of all carbon fixed by photosynthesis. A recent study of hydrocarbon emissions in the Atlanta area showed that plants were by far the largest emitters of hydrocarbons, with plant-derived isoprene accounting for almost 60% of the total.

Haze over the Blue Ridge Mountains. *(Derkel O'Hare/Tony Stone Images)*

Why do plants emit so much isoprene into the atmosphere rather than use it for the synthesis of terpenes and other natural products? Tom Sharkey, a University of Wisconsin plant physiologist, found that emission of isoprene is extremely sensitive to temperature. Plants grown at 20°C do not emit isoprene, but they begin to emit it when leaf temperature is increased to 30°C. In certain plants, isoprene emission can increase as much as tenfold for a 10°C increase in leaf temperature. Sharkey studied the relationship between temperature-induced leaf damage and isoprene concentration in leaves of the kudzu [*Pueraria lobata* (Wild.) Ohwi.] plant. He discovered that leaf damage, as measured by chlorophyll destruction, begins to occur at 37.5°C in the absence of isoprene, but not until 45°C in the presence of isoprene. Sharkey speculates that isoprene dissolves in leaf membranes and in some way increases their tolerance to heat stress. Because isoprene is made rapidly and also lost rapidly, its concentration correlates with temperature throughout the day. See "Why Plants Emit Isoprene" by T. D. Sharkey and E. L. Singsass, *Nature,* **374:** 27 (April 1995), p. 769.

$$CH_2{=}\overset{\displaystyle CH_3}{\overset{|}{C}}{-}CH{=}CH_2$$

2-Methyl-1,3-butadiene
(Isoprene)

$$\underset{\text{head}}{\rightarrow} \overset{1}{C}-\overset{\overset{\displaystyle C}{|}\,2}{C}-\overset{3}{C}-\overset{4}{C} \underset{\text{tail}}{\leftarrow}$$

Isoprene unit

Terpenes are among the most widely distributed compounds in the biological world, and a study of their structure provides a glimpse of the wondrous diversity

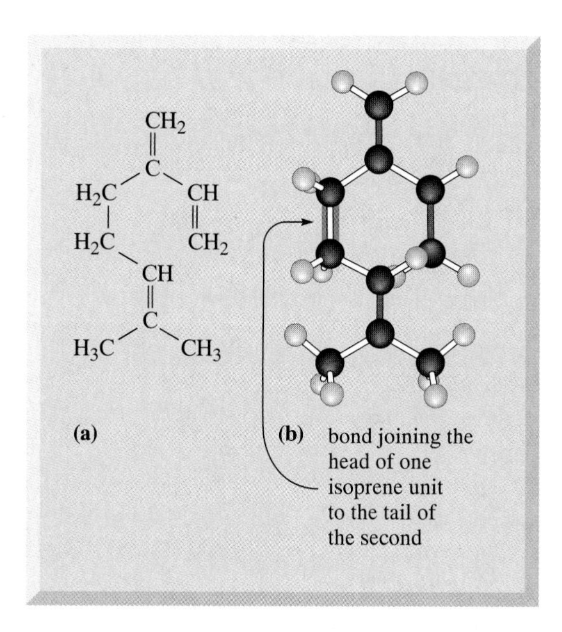

Figure 5.2
Myrcene. (a) The structural formula of myrcene and (b) a ball-and-stick model divided to show two isoprene units joined by a carbon-carbon bond between the head of one unit and the tail of the other.

that nature can generate from a simple carbon skeleton. Terpenes also illustrate an important principle of the molecular logic of living systems, namely that in building large molecules, small subunits are strung together enzymatically by an iterative process and then chemically modified by precise enzyme-catalyzed reactions. Chemists use the same principles in the laboratory, but our methods do not have the precision and selectivity of the enzyme-catalyzed reactions of cellular systems.

California laurel, *Umbellularia californica,* one source of myrcene. *(© Don Suzio)*

Probably the terpenes most familiar to you, at least by odor, are components of the so-called essential oils obtained by steam distillation or ether extraction of various parts of plants. Essential oils contain the relatively low-molecular-weight substances that are, in large part, responsible for characteristic plant fragrances. Many essential oils, particularly those from flowers, are used in perfumes.

One example of a terpene obtained from an essential oil is myrcene, $C_{10}H_{16}$, a component of bayberry wax and oils of bay and verbena. Myrcene is a triene with a parent chain of eight carbon atoms and two one-carbon branches (Figure 5.2).

Head-to-tail linkages of isoprene units are vastly more common in nature than are the alternative head-to-head or tail-to-tail patterns. Figure 5.3 shows structural formulas of six more terpenes, all derived from two isoprene units. Geraniol and the aggregating pheromone of the bark beetle have the same isoprene skeleton as myrcene. In addition, each has an —OH (hydroxyl) group. In the last four terpenes of Figure 5.3, the carbon atoms present in myrcene, geraniol, and the bark beetle pheromone are cross-linked to give cyclic structures. To help you identify the points of cross linkage and ring formation, the carbon atoms of the geraniol skeleton are numbered 1 through 8. This numbering pattern is used in the remaining terpenes to show points of cross-linking.

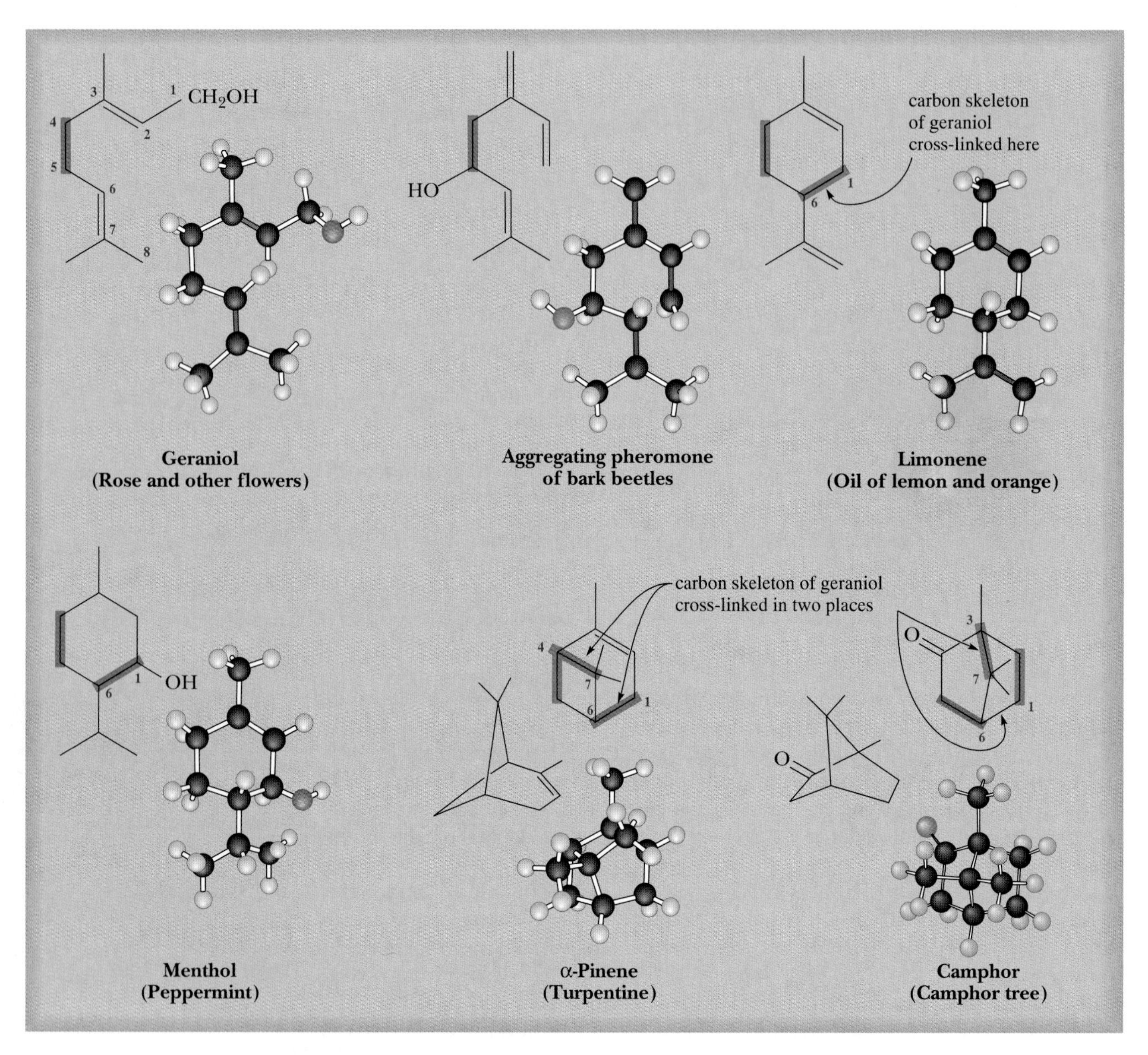

Figure 5.3
Six terpenes, each divisible into two isoprene units.

Shown in Figure 5.4 are structural formulas of three terpenes divisible into three isoprene units. For reference, the carbon atoms of the parent chain of farnesol are numbered 1 through 12. A bond between carbon atoms 1 and 6 of this skeleton gives the carbon skeleton of zingiberene. Try to discover for yourself what pattern of cross-linking gives the carbon skeleton of caryophyllene.

Vitamin A (Section 5.2G), a terpene of molecular formula $C_{20}H_{30}O$, consists of four isoprene units linked head-to-tail and cross-linked at one point to form a six-membered ring.

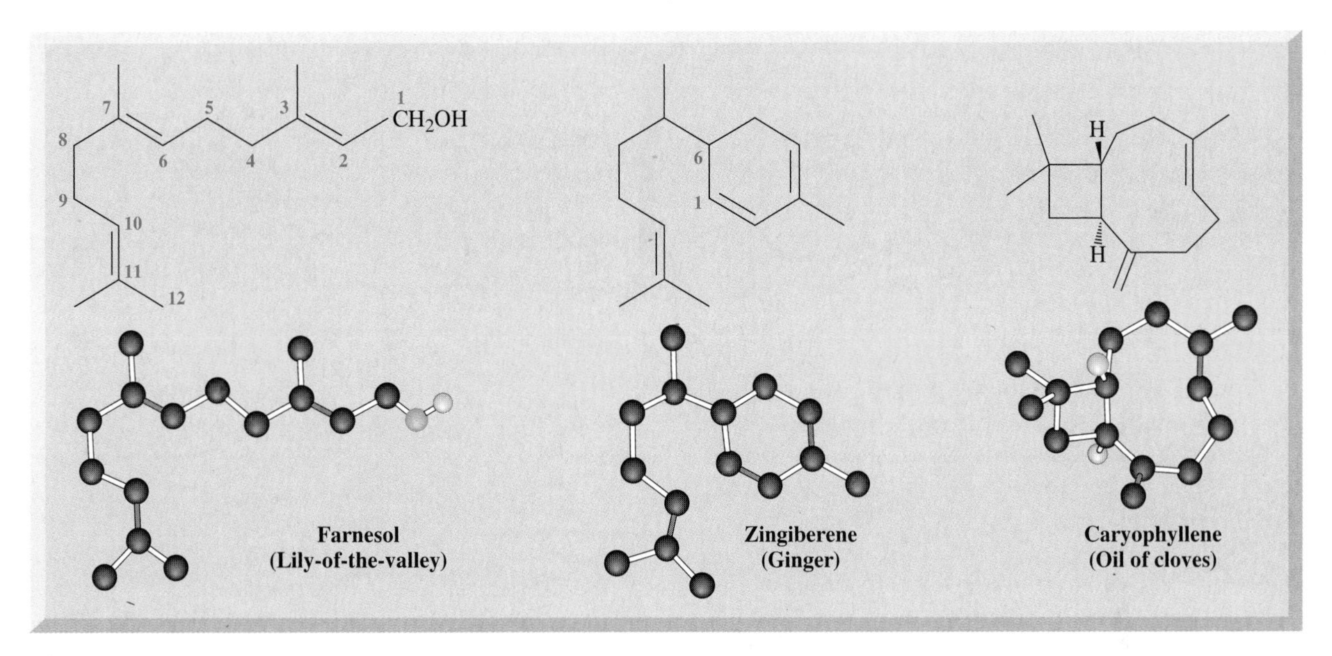

Figure 5.4
Three terpenes containing three isoprene units.

SUMMARY

An **alkene** is an unsaturated hydrocarbon that contains a carbon-carbon double bond. Alkenes have the general formula C_nH_{2n}. An **alkyne** is an unsaturated hydrocarbon that contains a carbon-carbon triple bond. Alkynes have the general formula C_nH_{2n-2}. According to the **orbital overlap model** (Section 5.1B), a carbon-carbon double bond consists of one sigma bond formed by the overlap of sp^2 hybrid orbitals and one pi bond formed by the overlap of parallel $2p$ atomic orbitals. It takes approximately 63 kcal/mol (264 kJ/mol) to break the pi bond in ethylene. A carbon-carbon triple bond consists of one sigma bond formed by the overlap of sp hybrid orbitals and two pi bonds formed by the overlap of pairs of parallel $2p$ orbitals.

The structural feature that makes **cis-trans isomerism** possible in alkenes is restricted rotation about the two carbons of the double bond (Section 5.1C). To date, *trans*-cyclooctene is the smallest trans cycloalkene that has been prepared in pure form and is stable at room temperature.

According to the IUPAC system (Section 5.2A), the presence of a **carbon-carbon double bond** is shown by changing the infix of the parent hydrocarbon from **-an-** to **-en-**. The names vinyl and allyl are commonly used to show the presence of $-CH{=}CH_2$ and $-CH_2CH{=}CH_2$ groups. The presence of a **carbon-carbon triple bond** is shown by changing the infix of the parent alkane from **-an-** to **-yn-**.

Whether the configuration of an alkene is cis or trans is determined by the orientation of the parent chain about the double bond (Section 5.2C). If atoms of the parent are on the same side of the double bond, the configuration of the alkene is cis; if they are on opposite sides, the configuration is trans. The configuration of a carbon-carbon double bond can also be specified by the **E,Z system** (Section 5.2C) using a set of priority rules. If the two groups of higher priority are on the same side of the double bond, the configuration of the alkene is **Z** (German: *zusammen,* together); if they are on opposite sides, it is **E** (German: **entgegen,** opposite).

For compounds containing two or more double bonds, the infix is changed to -adien-, -atrien-, and so on (Section 5.2F). Compounds containing several double bonds are called polyenes.

Because alkenes and alkynes are nonpolar compounds and the only interactions between their molecules are **dispersion forces,** their physical properties are similar to those of alkanes (Section 5.3).

The characteristic structural feature of a **terpene** (Section 5.4) is a carbon skeleton that can be divided into two or more **isoprene units.** Terpenes illustrate an important principle of the molecular logic of living systems, namely that in building large molecules, small subunits are strung together by an iterative process and then chemically modified by precise enzyme-catalyzed reactions.

CHEMICAL CONNECTIONS

Terpenes of the Cotton Plant

The floral fragrance of the cotton plant *(Gossypium)* is due in large part to a group of volatile, low-molecular-weight terpenes produced by leaf glands. At least a dozen scent components have been isolated and identified, including myrcene, limonene, and α-pinene. Among the terpenes divisible into three isoprene units are found spathulenol, gossonorol, and β-bisabolol.

What is notable about these three compounds is their diversity in structure. What is puzzling about them is why the cotton plant makes them. Are they, or were they at an earlier stage of evolution, important in the chemical defense of the cotton plant against insect predators or fungal infections? Are they important in these roles now? We do not know the answers to these questions. For the role of two other terpenes, hemigossypol and gossypol, in the cotton plant's defense against fungal infections, see the Chemical Connections box in Chapter 16, "Phytoalexins: A Natural Resistance of Plants to Pathogenic Fungi."

HO — Spathulenol

HO — Gossonorol

OH — β-Bisabolol

A field of cotton. *(© John Elk III)*

ADDITIONAL PROBLEMS

Structure of Alkenes and Alkynes

5.8 Predict all bond angles about each highlighted carbon atom. To make these predictions, use the valence-shell electron-pair repulsion (VSEPR) model (Section 1.3).

(a) (b) $-CH_2OH$ (c) $HC{\equiv}C{-}CH{=}CH_2$ (d)

5.9 For each highlighted carbon atom in Problem 5.8, identify which orbitals are used to form each sigma bond and which are used to form each pi bond.

5.10 Following is the structure of 1,2-propadiene (allene). In it, the plane created by H—C—H of carbon 1 is perpendicular to that created by H—C—H of carbon 3.

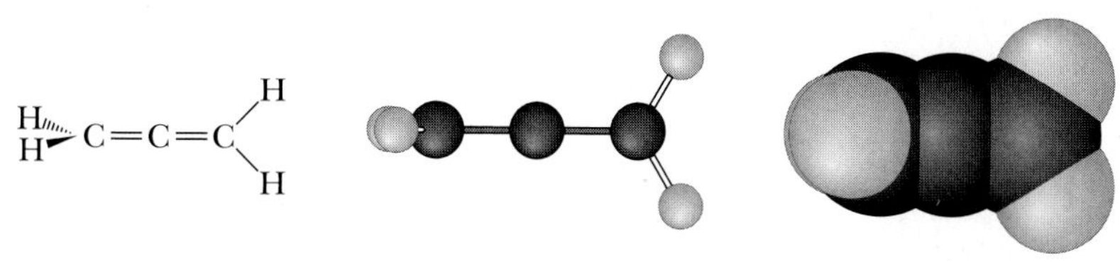

1,2-Propadiene (Allene) | Ball-and-stick model | Space-filling model

(a) State the orbital hybridization of each carbon of allene.

(b) Account for the molecular geometry of allene in terms of the orbital overlap model.

Nomenclature of Alkenes and Alkynes

5.11 Draw structural formulas for these compounds.

(a) *trans*-2-Methyl-3-hexene
(b) 2-Methyl-3-hexyne
(c) 2-Methyl-1-butene
(d) 3-Ethyl-3-methyl-1-pentyne
(e) 2,3-Dimethyl-2-butene
(f) *cis*-2-Pentene
(g) (Z)-1-Chloropropene
(h) 3-Methylcyclohexene
(i) 1-Isopropyl-4-methylcyclohexene
(j) (6E)-2,6-Dimethyl-2,6-octadiene
(k) Allylcyclopropane
(l) Diethylacetylene
(m) 2-Chloropropene
(n) Tetrachloroethylene

5.12 Name these compounds.

(a) $CH_2{=}C(CH_2)_4CH_3$ with $CH_2CH(CH_3)_2$ on the same carbon

(b) cyclopentene ring bearing H_3C and two CH_3 groups

(c) cyclopentadiene

(d) $(CH_3)_2CHCH{=}C(CH_3)_2$

(e) $CH_3(CH_2)_5C{\equiv}CH$

(f) $(CH_3)_2CHC{\equiv}CC(CH_3)_3$

5.13 Which alkenes exist as pairs of cis-trans isomers? For each alkene that does, draw the trans isomer.

(a) $CH_2{=}CHBr$
(b) $CH_3CH{=}CHBr$
(c) $BrCH{=}CHBr$
(d) $(CH_3)_2C{=}CHCH_3$
(e) $(CH_3)_2CHCH{=}CHCH_3$

5.14 Name and draw structural formulas for all alkenes of molecular formula C_5H_{10}. As you draw these alkenes, remember that cis and trans isomers are different compounds and must be counted separately.

5.15 Name and draw structural formulas for all alkenes of molecular formula C_6H_{12} that have these carbon skeletons. Remember cis and trans isomers!

(a) C—C(C)—C—C—C
(b) C—C(C)—C(C)—C
(c) C—C(C)(C)—C—C

5.16 Arrange the groups in each set in order of increasing priority. (Review Section 4.3).

(a) $-CH_3$ $-Br$ $-CH_2CH_3$

(b) $-OCH_3$ $-CH(CH_3)_2$ $-CH{=}CH_2$

(c) $-CH_2OH$ $-CO_2H$ $-OH$

5.17 Assign an E or Z configuration and a cis or trans configuration to these carboxylic acids, each of which is an intermediate in the tricarboxylic acid cycle. Under each is given its common name.

(a) H and CO_2H on the upper C; HO_2C and H on the lower C (C═C)

Fumaric acid

(b) HO_2C and CO_2H on top; H and CH_2CO_2H on the bottom (C═C)

Aconitic acid

5.18 Four stereoisomers exist for 3-penten-2-ol.

$$CH_3-CH{=}CH-\underset{}{CH}(OH)-CH_3$$

3-Penten-2-ol

(a) Explain how these four stereoisomers arise.

(b) Draw the stereoisomer having the E configuration about the carbon-carbon double bond and the R configuration at the stereocenter.

5.19 Draw the structural formula for at least one bromoalkene of molecular formula C_5H_9Br that shows

(a) Neither E,Z isomerism nor enantiomerism.

(b) E,Z isomerism but not enantiomerism.

(c) Enantiomerism but not E,Z isomerism.

(d) Enantiomerism and E,Z isomerism

5.20 For each molecule that shows cis-trans isomerism, draw the cis isomer.

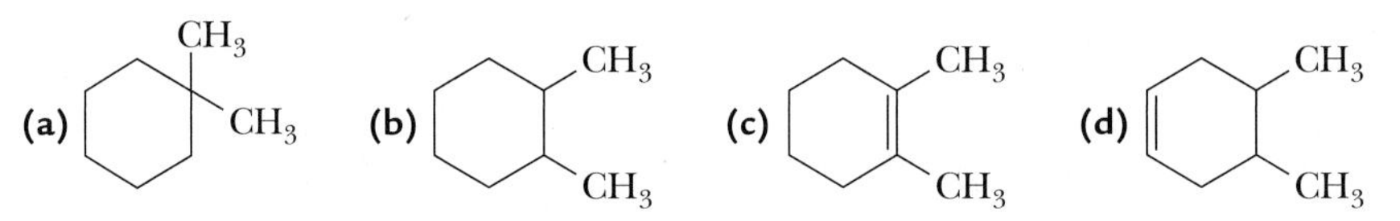

5.21 Draw structural formulas for all compounds of molecular formula C_5H_{10} that are

(a) Alkenes that do not show cis-trans isomerism.

(b) Alkenes that do show cis-trans isomerism.

(c) Cycloalkanes that do not show cis-trans isomerism.

(d) Cycloalkanes that do show cis-trans isomerism.

5.22 β-Ocimene (*The Merck Index,* 12th ed., #6837), a triene found in the fragrance of cotton blossoms and several essential oils, has the IUPAC name (3Z)-3,7-dimethyl-1,3,6-octatriene. Draw a structural formula for β-ocimene.

Terpenes

5.23 Show how the carbon skeleton of farnesol can be coiled and then cross-linked to give the carbon skeleton of caryophyllene (Figure 5.3).

5.24 Show that the structural formula of vitamin A (Section 5.3G) can be divided into four isoprene units joined by head-to-tail linkages and cross-linked at one point to form the six-membered ring.

5.25 Following is the structural formula of lycopene (*The Merck Index,* 12th ed., #5650), a deep-red compound that is partially responsible for the red color of ripe fruits, especially tomatoes. Approximately 20 mg of lycopene can be isolated from 1 kg of ripe tomatoes.

Lycopene

(a) Show that lycopene is a terpene, that its carbon skeleton can be divided into two sets of four isoprene units with the units in each set joined head-to-tail.

(b) How many of the carbon-carbon double bonds in lycopene have the possibility for cis-trans isomerism? Lycopene is the all trans isomer.

Carotene and carotene-like molecules are partnered with chlorophyll in nature in the role of assisting in the harvesting of sunlight. In autumn, green chlorophyll molecules are destroyed and the yellows and reds of carotene and related molecules are seen. The red color of tomatoes comes from lycopene, a molecule closely related to carotene. As a tomato ripens, its chlorophyll is destroyed and the green color is replaced by the red of lycopene. *(Charles D. Winters)*

5.26 The structural formula of β-carotene (*The Merck Index,* 12th ed., #1902), precursor to vitamin A, is given in Section 17.6A. As you might suspect, it was first isolated from carrots. Dilute solutions of β-carotene are yellow, hence its use as a food coloring. In plants, it is almost always present in combination with chlorophyll to assist in the harvesting of the energy of sunlight. As tree leaves die in the fall, the green of their chlorophyll molecules is replaced by the yellow and reds of carotene and carotene-related molecules. Compare the carbon skeletons of β-carotene and lycopene. What are the similarities? What are the differences?

5.27 α-Santonin (*The Merck Index,* 12th ed., #8509), isolated from the flower heads of certain species of artemisia, is an anthelmintic; that is, a drug used to rid the body of worms (helminths). It has been estimated that over one third of the world's population is infested with these parasites.

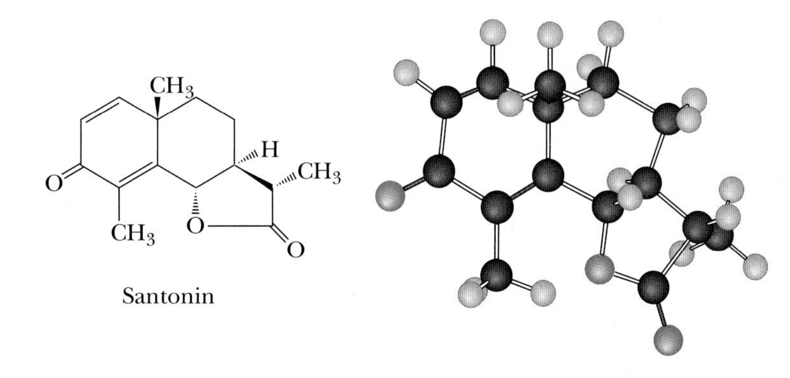

(a) Mark all stereocenters in santonin. How many stereoisomers are possible for it?

(b) Locate the three isoprene units in santonin and show how the carbon skeleton of farnesol might be coiled and then cross-linked to give santonin. Two different coiling patterns of the carbon skeleton of farnesol can lead to santonin. Try to find them both.

Santonin can be isolated from the flower heads of wormwood, *Artemisia absinthium.* Principal habitats are Chinese Turkistan and the Southern Urals. *(© Kenneth J. Stein, Phototake, NYC)*

5.28 In many parts of South America, extracts of the leaves and twigs of *Montanoa tomentosa* are used as a contraceptive, to stimulate menstruation, to facilitate labor, and as an abortifacient. Phytochemical investigations of this plant have resulted in isolation of a very potent fertility-regulating compound called zoapatanol (*The Merck Index,* 12th ed., #10318).

H_3C, O, HO, CH_2OH, C, H, O, CH_3, CH_3, H_3C

Zoapatanol

(a) Show that the carbon skeleton of zoapatanol can be divided into four isoprene units bonded head-to-tail and then cross-linked in one point along the chain.

(b) Specify the configuration about the carbon-carbon double bond to the seven-membered ring according to the E,Z system.

(c) How many cis-trans isomers are possible for zoapatanol? Consider the possibilities for cis-trans isomerism in cyclic compounds and about carbon-carbon double bonds.

5.29 Following is the structural formula of warburganal (*The Merck Index,* 12th ed., #10173), a crystalline solid isolated from the plant *Warburgia ugandensis* (Canellaceae). An important use of warburganal is its antifeeding activity against the African army worm. In addition, it acts as a plant growth regulator and has cytotoxic, antimicrobial, and molluscicidal properties.

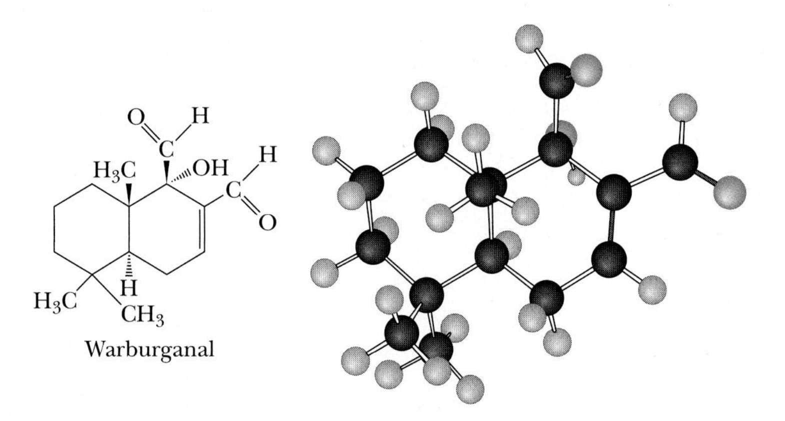

Warburganal

(a) Show that warburganal is a terpene.

(b) Label each stereocenter and specify the number of stereoisomers possible for a molecule of this structure.

5.30 Pyrethrin II and pyrethrosin are natural products isolated from plants of the chrysanthemum family. Pyrethrin II is a natural insecticide and is marketed as such (*The Merck Index,* 12 ed., #8148 and #8149).

(a) Label all tetrahedral stereocenters in each molecule and all carbon-carbon double bonds about which cis-trans isomerism is possible.

(b) State the number of stereoisomers possible for each molecule.

(c) Show that the ring system of pyrethrosin is composed of three isoprene units.

Pyrethrin II

Pyrethrosin

Pyrethrosin and pyrethrin II are natural products isolated from chrysanthemum blossoms. *(© Scott Camazine/Photo Researchers, Inc.)*

5.31 Show that the carbon skeletons of the three terpenes drawn in the Chemical Connections box "Terpenes of the Cotton Plant" can each be divided into three isoprene units bonded head-to-tail and then cross-linked at appropriate carbons.

A Conversation with . . .

CARL DJERASSI

Carl Djerassi, professor of chemistry at Stanford University, is one of a few modern-day Renaissance men. His prolific career includes many notable contributions as an industrial research chemist, a college professor, an author of poetry, drama, and fiction, and a patron and collector of fine art. Djerassi was born in Vienna in 1923.

His first job was with Ciba Pharmaceutical Corporation in New Jersey. He quickly established himself as a research chemist by synthesizing pyribenzamine, one of the first antihistamines used to control allergies.

At Syntex, a small pharmaceutical company then located in Mexico City, Djerassi and George Rosenkranz directed the research group that in 1951 first synthesized the important hormone cortisone, which is used as an anti-inflammatory drug. A short time later, he led the research team that synthesized the first oral contraceptive, norethindrone, one that is still widely used.

Djerassi has published more than 1000 research papers and seven books on steroids, alkaloids, antibiotics, lipids, and terpenoids. His many research awards include the National Medal of Science (1973) for the first synthesis of an oral contraceptive, the National Medal of Technology (1991) for promoting novel approaches to insect control, and the Priestley Medal (1992), the highest award given by the American Chemical Society. He is a member of the National Academy of Science and the National Inventors' Hall of Fame.

In addition to his career as a chemist, Djerassi has published a collection of short stories, five novels, a collection of poems, a collection of essays *(From the Lab into the World: A Pill for People, Pets, and Bugs),* and two autobiographies outlining his career: *The Pill, Pygmy Chimps, and Degas' Horse* and *Steroids Made It Possible* (http://www.djerassi.com).

Djerassi is in the process of expanding his literary endeavors to science-in-theater, where he plans a trilogy of plays that presents realistic science through an attention-gathering plot and the dialog of the actors. The first play, *An Immaculate Misconception,* premiered at the 1998 Edinburgh Fringe Festival.

As a collector and patron of the arts, Djerassi began to acquire in the 1960s the works of Swiss artist Paul Klee (1879–1940). The collection, one of the largest private holdings of Klee's works, is on permanent display in the San Francisco Museum of Modern Art. In 1979, Djerassi established a resident artists' colony near Woodside, California. This program, which has served 1000 artists thus far, provides residence and studio space for visual, literary, choreographic, and musical artists.

Early Scientific Endeavors

"I was not a kid who had any interest in chemistry. I didn't have a chemistry set at home. I didn't blow up any basements. When I lived in Vienna, my education was based much more in the humanities than in the sciences. But both my parents were physicians and it was always assumed that I would become a physician too. But then Hitler came and I left Vienna after the *Anschluss.*

"While I was waiting for my U.S. visa, I went to an American school for a year in Sofia, Bulgaria, where my father lived. There I learned English. Since I already had a very good education and could speak English very well, when in America, I went straight to Newark Junior College rather than completing the last two years of high school. I began a pre-med curriculum, which included chemistry. I had a superb chemistry teacher, Nathan Washton, and there is no question he got me very interested in chemistry."

Educational and Career Motivations

"After a year at Newark Junior College, I transferred for one semester to Tarkio College and then went on to Kenyon College. My father had stayed behind in Bulgaria and my mother, who didn't have an American license to practice medicine, took a job as a physician's assistant in upstate New York. So I was on my own.

"At Kenyon, there were only two chemistry professors, but I got a first-class education and began to do some research. I also applied to pharmaceutical companies for jobs, and I ended up as a junior chemist at Ciba. It was near the end of my days at Kenyon and certainly when I was at Ciba and did research in medicinal chemistry on antihistamines that I really got hooked on chemistry."

Rewards of Industrial Research

"Doing research for Ciba was really a seminal event. Consider that I had graduated from college in 1942 and wasn't quite 19 at the time. I was assigned to work with a senior chemist, Charles Huttrer, who himself was an immigrant from Vienna. He treated me much more like a colleague than an assistant. We started on a brand-new project to synthesize antihistamines because there were none in the United States at that time—the basic research had started in France. We were lucky. The research went very fast and within a year we had synthesized pyribenzamine, one of the first two antihistamines produced in this country.

"When the patent was filed I was one of the inventors and was one of the authors on a research paper in the *Journal of the American Chemical Society*. Thus in just one year I became a participant in the development of a drug that was very soon taken by hundreds of thousands of people to treat allergies. That was very heady stuff. It made me a real optimist. Success, of course, is the greatest incentive there is. So when I went to graduate school at Wisconsin I was already a person who had made an important discovery."

Interest in the Chemistry of Steroids

"I assumed all along that, after graduate work, I would go back to Ciba and didn't even consider doing anything else at that time—although in the back of my mind I knew I wanted to go into academia. Ciba's Swiss headquarters was a powerhouse in steroids at the time, and since steroid hormones were such a familiar concept in the organization—even though I was working on antihistamines—I started reading about steroids in Louis Fieser's famous text, *Natural Products Related to Phenanthrene.*

"I decided I wanted to do my Ph.D. thesis on steroids. There were two young assistant professors working in this area, William S. Johnson and Alfred L. Wilds, and I ended up working with the latter. I worked on the conversion of androgens to estrogens, a very tough chemical problem at that time. With the advantage of a full-time research fellowship and the year of experience at Ciba, I finished the program in slightly over two years."

Joining Syntex and Working in Mexico City

"I returned to Ciba as a senior chemist and continued working on antihistamines and antispasmodics. That was in the late 1940s when the therapeutic properties of cortisone were discovered and that became one of the hottest and most competitive topics in the field of organic chemistry. And I wanted to work on the synthesis of cortisone, but was not permitted to do so at Ciba since that was a project reserved primarily for the Swiss laboratories.

"Then one day I got a letter in the mail asking if I would consider a job at a Mexican firm called Syntex. They invited me for an interview and I was really charmed by Mexico City and by George Rosenkranz, the firm's technical director. He said that they wanted to develop a synthesis of cortisone from a Mexican yam and that I would be in charge of that project but could work on other steroid projects as well. The time I spent there from late 1949 until early 1952 was probably the most productive period of my scientific life."

I have always been very much an intellectual polygamist. In this age of tremendous specialization I think it is very important that people are interested in more than just one intellectual area.

Synthesizing an Oral Contraceptive

"Our initial goal was to develop an orally active progestational hormone—in other words, a compound that would mimic the biological properties of progesterone. At that time, progesterone was clinically used for menstrual disorders and infertility, but there were ideas about using it as a contraceptive because it is progestrone that naturally stops further ovulation after an ovum is fertilized. However, progesterone itself is not active by mouth, and daily injections would be needed. By combining ideas discovered by previous investigators in Europe and the States, we set out to synthesize a steroid that would not only be active by mouth but would also have enhanced progestational activity.

"This compound was 19-nor-17α-ethynyltestosterone (norethindrone), whose synthesis we completed on October 15, 1951. It was first tested for menstrual disorders and fertility problems and then as an oral contraceptive. Nearly 50 years after its synthesis, it is still the active ingredient in about a third of all the oral contraceptives used throughout the world."

From Contraceptives to Insect Control

"Conceptually, there was a relationship between our work on oral contraceptives and insect control. In a way, you could say that steroid oral contraceptives were true biorational methods of human birth control since progesterone—our conceptional lead compound—is really nature's contraceptive. That was a model on which insect control could be based on their counterpart, the insect juvenile hormones. At this time (the late 1960s), I was in charge of research at Syntex in addition to being a professor at Stanford University. Governments and the public realized that conventional methods of insect control—largely spraying with chlorinated hydrocarbons such as DDT—were damaging the environment.

"We formed a new company in the Stanford Industrial Park called Zoecon to synthesize insect-controlling hormones. In the 1960s, a juvenile hormone based on a sesquiterpene skeleton had been discovered, and we decided to focus on it. Insects pass through a juvenile stage controlled by the juvenile hormone, whose production is later shut off by another hormone so that the insect can then mature. Our biorational approach was to synthesize an artificial juvenile hormone that would continue to be applied to immature insects, so that the insect would never reach the stage at which it could reproduce. This new approach to controlling mosquitoes, fleas, cockroaches, and other insects that do their damage as adults, was approved by the Environmental Protection Agency for public use."

Exploring Scientific Issues with Fiction

"We as scientists pay very little attention to our behavior and practices. We do not teach such topics in our undergraduate or even in our graduate courses. We don't have courses on how to become a scientist or how to behave as a scientist. We learn how through osmosis from our mentors and colleagues. As chemists, we are very analytical about the work around ourselves, but we do very little introspective self-analysis.

"Writing a special type of fiction, which I call science-in-fiction, really did that for me. It was a very good idea for me to have started writing fiction later in life, because I think that fiction-writing is one area where it helps to have experience, a basis for comparison, and a historical perspective. In many respects writing fiction has enabled me to talk about aspects of my science that I really had not thought about that much. Then I became convinced of its importance pedagogically.

"Instead of telling readers, 'Let me tell you about my science,' because many nonscientists who pay no attention to it get scared the moment you say 'science,' I now say, 'Let me tell you a story.' And as I tell them a story, I teach them a lot about science and about scientists. And I am exceptionally accurate about this, which is why I call it science-in-fiction rather than science fiction. A recent graduate course I taught, 'Ethical Discourse Through Science-In-Fiction,' showed that it is much easier to demonstrate that point in an anonymous way [*Nature,* **393:** 511 (1998)]. By discussing these aspects of scientific behavior in the cloak of fiction we can illustrate ethical dilemmas that people would otherwise never discuss openly because of discretion, embarrassment, or fear of retribution."

Balancing a Three-Pronged Career

"I have always been very much an intellectual polygamist. In this age of tremendous specialization I think it is very important that people are interested in more than just one intellectual area. And I don't mean just in different areas of chemistry and not just in different areas of science. I have always been interested in doing different things at the same time, which is why I use the world 'polygamy,' because that refers to a man who has several wives at the same time. I'm not using it of course in a sexual context; I am using it in an intellectual one. In true polygamy, each wife is more or less equal. I would say that I don't have a favorite wife, even though I have several wives at this stage: academic, industrial, literary, artistic, and so on. They are all important components of my life."

6

Reactions of Alkenes

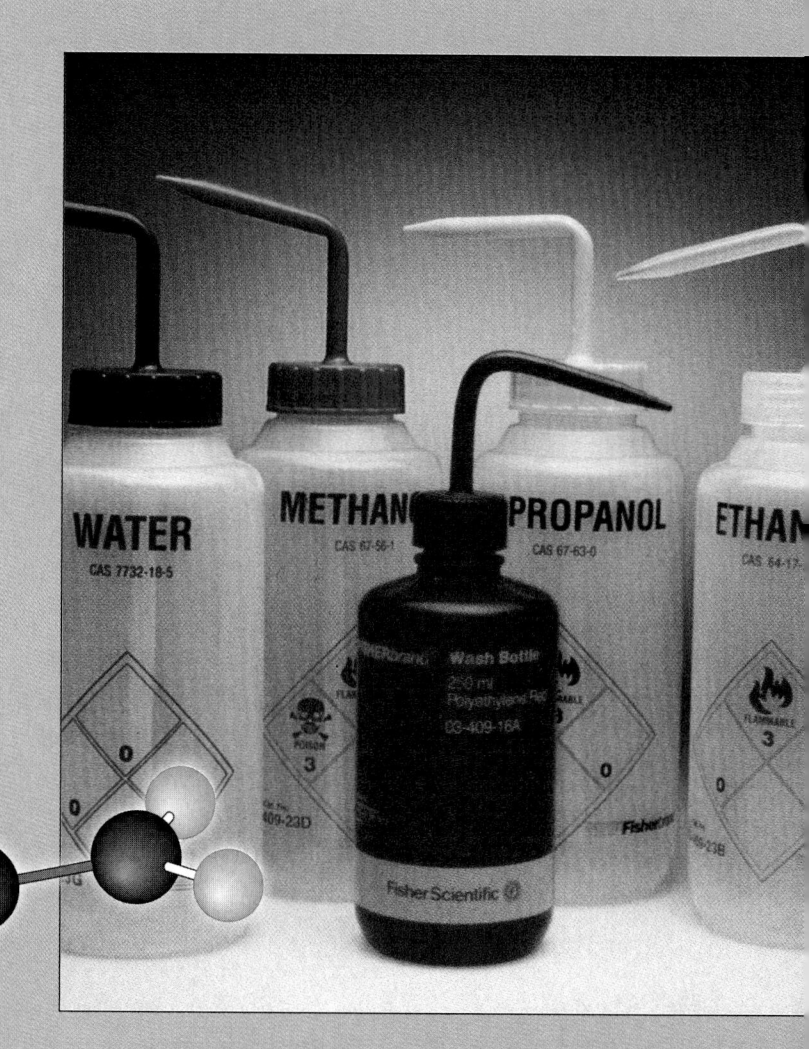

Polyethylene wash bottles. Inset: A model of ethylene.
(Charles D. Winters)

In this chapter we begin our systematic study of reaction mechanisms, one of the most important unifying concepts in organic chemistry. We use the reactions of alkenes as the vehicle to introduce this concept.

6.1 AN OVERVIEW

The most characteristic reaction of alkenes is **addition to the carbon-carbon double bond** in such a way that the pi bond is broken, and, in its place, sigma bonds are formed to two new atoms or groups of atoms. Several examples of reactions at the carbon-carbon double bond are shown in Table 6.1 along with the descriptive name(s) associated with each.

A second characteristic reaction of alkenes is the formation of **chain-growth polymers** (Greek words: *poly,* many, and *meros,* part). In the presence of certain catalysts called initiators, many alkenes form polymers made by the stepwise addition of **monomers** (Greek words: *mono,* one, and *meros,* part) to a growing polymer chain as illustrated by the formation of polyethylene from ethylene. In alkene polymers of industrial and commercial importance, n is a large number, typically several thousand.

$$n\mathrm{CH_2{=}CH_2} \xrightarrow{\text{initiator}} \mathrm{-\!\!(CH_2CH_2)\!\!-_{n}}$$

Ethylene (a monomer) → Polyethylene (a polymer)

We discuss this reaction in Chapter 15.

TABLE 6.1 Characteristic Addition Reactions of Alkenes

Reaction	Descriptive Name(s)
$\mathrm{>C{=}C<} + \mathrm{HCl}$ (HX) $\longrightarrow$ $\mathrm{-C(H)-C(Cl)-}$ (X)	hydrochlorination (hydrohalogenation)
$\mathrm{>C{=}C<} + \mathrm{H_2O} \longrightarrow \mathrm{-C(H)-C(OH)-}$	hydration
$\mathrm{>C{=}C<} + \mathrm{Br_2}$ (X_2) $\longrightarrow$ $\mathrm{-C(Br)-C(Br)-}$ (X) (X)	bromination (halogenation)
$\mathrm{>C{=}C<} + \mathrm{OsO_4} \longrightarrow \mathrm{-C(HO)-C(OH)-}$	hydroxylation (oxidation)
$\mathrm{>C{=}C<} + \mathrm{H_2} \longrightarrow \mathrm{-C(H)-C(H)-}$	hydrogenation (reduction)

6.2 REACTION MECHANISMS

A **reaction mechanism** describes how a reaction occurs. It describes which bonds are broken and which new ones are formed as well as the order and relative rates of the various bond-breaking and bond-forming steps. If the reaction involves a catalyst, the reaction mechanism also describes the role of the catalyst.

Reaction mechanism A step-by-step description of how a chemical reaction occurs.

A. Potential Energy Diagrams and Transition States

Let us consider a general reaction in which A and B react to give C and D. The **total energy,** E, of the reacting system is the sum of its **kinetic energy,** KE, and **potential energy,** PE.

$$A + B \longrightarrow C + D \qquad E = KE + PE$$

Reactants A and B have a certain kinetic energy. As they collide, a part of their kinetic energy is absorbed and converted to potential energy in the form of vibration motions of bonds. To understand the relationship between a chemical reaction and potential energy, think of a chemical bond as a spring. As a spring is stretched from its resting position, its potential energy increases. As it returns to its resting position, its potential energy decreases. Similarly, during a chemical reaction, bond breaking corresponds to an increase in potential energy, and bond forming corresponds to a decrease in potential energy. We use a **potential energy diagram** to show the changes in energy that occur in going from reactants to products. Potential energy is measured along the vertical axis, and the change in position of the atoms during a reaction is measured on the horizontal axis, the **reaction coordinate.**

Potential energy (PE) diagram A graph showing the changes in energy that occur during a chemical reaction; energy is plotted on the *y* axis, and reaction progress is plotted on the *x* axis.

Reaction coordinate A measure of the progress of a reaction, plotted on the *x* axis in a potential energy diagram.

Figure 6.1 shows a potential energy diagram for the proton-transfer reaction that occurs when hydrogen chloride is dissolved in water. This reaction occurs in one step, meaning that bond breaking in the reactants and bond forming to give the products occur simultaneously.

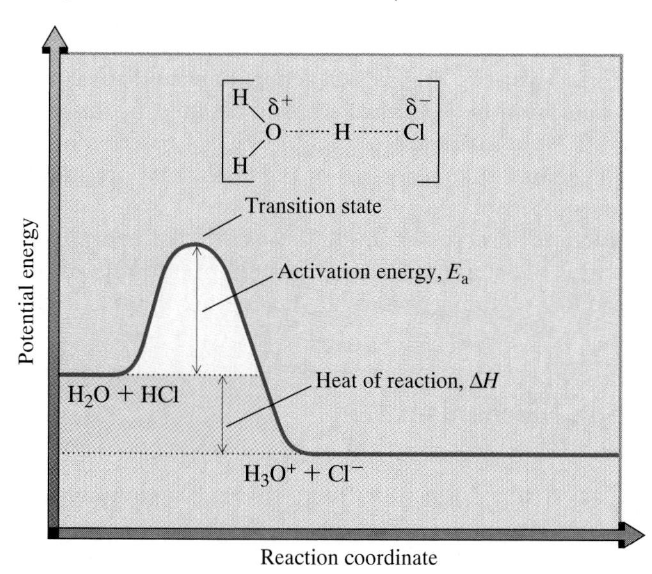

Figure 6.1
Potential energy diagram for a proton-transfer reaction between HCl and H_2O. The dashed lines indicate that, in the transition state, the new O—H bond is partially formed and the H—Cl bond is partially broken. The energy of the reactants is higher than that of products; the reaction is exothermic.

Heat of reaction The difference in potential energy between reactants and products.

Exothermic reaction A reaction in which the energy of the products is lower than the energy of the reactants; a reaction in which heat is liberated.

Endothermic reaction A reaction in which the energy of the products is higher than the energy of the reactants; a reaction in which heat is absorbed.

Transition state An unstable species of maximum energy formed during the course of a reaction; an energy maximum on a potential energy diagram.

The difference in potential energy between the reactants and products is called the **heat of reaction.** If the energy of products is lower than that of reactants, heat is released; the reaction is **exothermic.** If the energy of products is higher than that of reactants, heat is absorbed; the reaction is **endothermic.** The one-step proton-transfer reaction shown in Figure 6.1 is exothermic; the energy of H_3O^+ and Cl^- is lower than that of H_2O and HCl.

A **transition state** is the point on the reaction coordinate at which the potential energy is at a maximum. At the transition state, sufficient potential energy has become concentrated in the proper bonds so that bonds in reactants break and new bonds form, giving products. After the transition state is reached, the reaction proceeds to give products with the release of energy. A transition state has a definite geometry, a definite arrangement of bonding and nonbonding electrons, and a definite distribution of electron density and charge. Because a transition state is at an energy maximum on a potential energy diagram, it cannot be isolated, and its structure cannot be determined experimentally. Even though we cannot observe a transition state directly by any experimental means, we can often infer a great deal about its probable structure from other experimental observations.

For the proton-transfer reaction illustrated in Figure 6.1, we use dashed lines to show the partial bonding in the transition state. As an unshared pair of electrons on oxygen begins to form a new covalent bond with hydrogen (shown by the dashed line), oxygen develops a partial positive charge. Conversely, as the bonding pair of electrons in H—Cl becomes transferred to chlorine (shown by another dashed line), chlorine develops a partial negative charge.

Activation energy The difference in potential energy between reactants and the transition state.

The difference in potential energy between reactants and the transition state is called the **activation energy,** E_a. Activation energy is the minimum amount of potential energy that collisions between reacting particles must generate for a reaction to occur; it can be considered an energy barrier for reaction. E_a determines the rate of a reaction, that is, how fast the reaction occurs. If E_a is large, only a very few molecular collisions occur with sufficient energy to reach the transition state, and the reaction is slow. If E_a is small, many collisions generate sufficient energy to reach the transition state, and the reaction is fast.

Reaction intermediate An unstable species that lies in a potential energy minimum between two transition states.

In a reaction that occurs in two or more steps, each step has its own transition state and activation energy. Shown in Figure 6.2 is a potential energy diagram for conversion of reactants to products in two steps. A **reaction intermediate** corresponds to a potential energy minimum between two transition states, in this case between transition states 1 and 2. Note that because the energies of reaction intermediates we describe are higher than either reactants or products, they are highly reactive, and rarely, if ever, can one be isolated.

Rate-limiting step The step in a reaction sequence that crosses the highest potential energy barrier; the slowest step in a multistep reaction.

The step in a multistep reaction that crosses the highest potential energy barrier is the **rate-limiting step,** and is the step that determines the overall rate of the reaction. For the two-step reaction sequence shown in Figure 6.2, Step 1 is the rate-limiting step.

B. Developing a Reaction Mechanism

To develop a reaction mechanism, chemists begin by designing experiments that will reveal details of a particular chemical reaction. Next, through a combination of experience and intuition, they propose several sets of steps or mechanisms, each of which might account for the overall chemical transformation. Finally, each

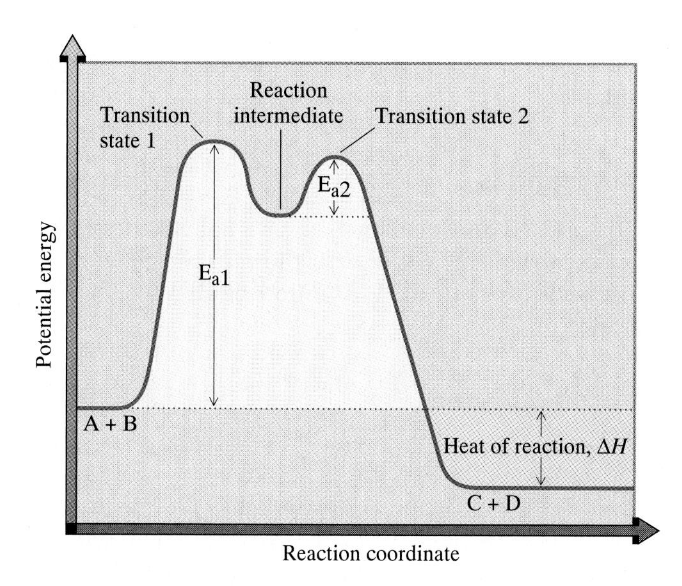

Figure 6.2
Potential energy diagram for a two-step reaction involving formation of an intermediate. The energy of the reactants is higher than that of the products, and heat is released in the conversion of A + B to C + D; the reaction is exothermic.

proposed mechanism is tested against the experimental observations to exclude those mechanisms that are not consistent with the facts.

A mechanism becomes generally established by excluding reasonable alternatives and by showing that it is consistent with every test that can be devised. This, of course, does not mean that a generally accepted mechanism is a completely accurate description of the chemical events but only that it is the best chemists have been able to devise. It is important to keep in mind that, as new experimental evidence is obtained, it may be necessary to modify a generally accepted mechanism or possibly even discard it and start all over again.

Before we go on to consider reactions and reaction mechanisms, we might ask why it is worth the trouble to establish them and your time to learn about them. One reason is very practical: mechanisms provide a theoretical framework within which to organize a great deal of descriptive chemistry. For example, with insight into how reagents add to particular alkenes, it is possible to make generalizations and then to predict how the same reagents might add to other alkenes. A second reason lies in the intellectual satisfaction derived from constructing models that accurately reflect the behavior of chemical systems. Finally, to a creative scientist, a mechanism is a tool to be used in the search for new knowledge and new understanding. A mechanism consistent with all that is known about a reaction can be used to make predictions about chemical interactions as yet unexplored, and experiments can be designed to test these predictions. Thus, reaction mechanisms provide a way not only to organize knowledge but also to extend it.

6.3 ELECTROPHILIC ADDITION REACTIONS

We begin our introduction to the chemistry of alkenes with an examination of three types of addition reactions: addition of hydrogen halides (HCl, HBr, and HI), water (H_2O), and halogens (Br_2, Cl_2). We first study some of the experimental observations about each addition reaction and then its mechanism. Through a study of these particular reactions, we develop a general understanding of how alkenes undergo addition reactions.

A. Addition of Hydrogen Halides

The hydrogen halides HCl, HBr, and HI add to alkenes to give haloalkanes (alkyl halides). These additions may be carried out either with the pure reagents or in the presence of a polar solvent such as acetic acid. Addition of HCl to ethylene gives chloroethane (ethyl chloride).

$$CH_2{=}CH_2 + HCl \longrightarrow \underset{|}{\overset{H}{}}CH_2-\overset{Cl}{CH_2}$$

Ethylene → Chloroethane ($CH_2(H)-CH_2(Cl)$)

Addition of HCl to propene gives 2-chloropropane (isopropyl chloride); hydrogen adds to carbon 1 of propene, and chlorine adds to carbon 2. If the orientation of addition were reversed, 1-chloropropane (propyl chloride) would be formed. The observed result is that 2-chloropropane is formed to the virtual exclusion of 1-chloropropane. We say that addition of HCl to propene is a highly **regioselective reaction.**

Regioselective reaction A reaction in which one direction of bond forming or bond breaking occurs in preference to all other directions.

$$CH_3CH{=}CH_2 + HCl \longrightarrow CH_3\overset{Cl}{CH}-\overset{H}{CH_2} + CH_3\overset{H}{CH}-\overset{Cl}{CH_2}$$

Propene; 2-Chloropropane; 1-Chloropropane (not observed)

This regioselectivity was noted by Vladimir Markovnikov, who made the generalization known as **Markovnikov's rule:** in the addition of H—X to an alkene, hydrogen adds to the doubly bonded carbon that has the greater number of hydrogens already bonded to it. Although Markovnikov's rule provides a way to predict the product of many alkene addition reactions, it does not explain why one product predominates over other possible products.

Markovnikov's rule In the addition of HX or H_2O to an alkene, hydrogen adds to the carbon of the double bond having the greater number of hydrogens.

EXAMPLE 6.1

Name and the draw a structural formula for the product of each alkene addition reaction.

(a) $CH_3\overset{CH_3}{C}{=}CH_2 + HI \longrightarrow$ (b) 1-methylcyclopentene (CH_3 on the double-bond carbon) $+ HCl \longrightarrow$

SOLUTION

Using Markovnikov's rule, predict that 2-iodo-2-methylpropane is the product in (a) and 1-chloro-1-methylcyclopentane is the product in (b).

(a) $CH_3\underset{I}{\overset{CH_3}{C}}CH_3$

2-Iodo-2–methylpropane

(b) cyclopentane bearing Cl and CH_3 on the same carbon

1-Chloro-1-methylcyclopentane

Practice Problem 6.1

Name and draw a structural formula for the product of each alkene addition reaction.

(a) $CH_3CH{=}CH_2 + HI \longrightarrow$ (b) (cyclohexane)$=CH_2 + HI \longrightarrow$

Chemists account for the addition of HX to an alkene by a two-step mechanism, which we illustrate by the reaction of 2-butene with hydrogen chloride to give 2-chlorobutane. Let us first look at this two-step mechanism in overview and then go back and study each step in detail. In overview, addition begins with the transfer of a proton from HCl to 2-butene, as shown by the two curved arrows on the left side of Step 1. The first curved arrow shows that the pi bond of the alkene is broken and that its electron pair is used to form a new covalent bond with the hydrogen atom of HCl. The second curved arrow shows that the polar covalent bond in HCl is broken and that its electron pair is given entirely to chlorine, forming chloride ion. Step 1 in this mechanism results in the formation of an organic cation and chloride ion. Step 2 is the reaction of the organic cation with chloride ion to form 2-chlorobutane.

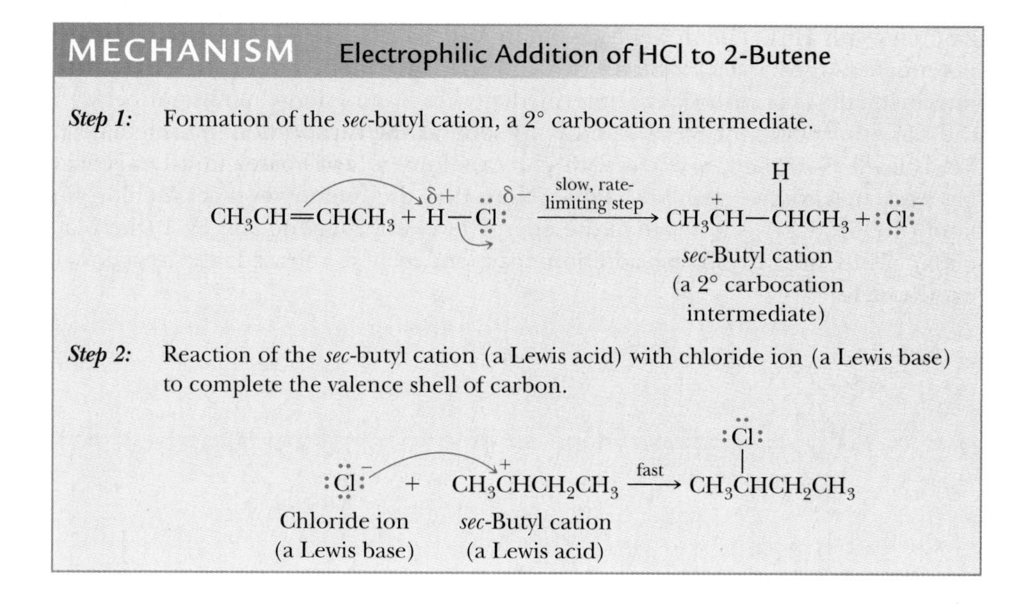

MECHANISM Electrophilic Addition of HCl to 2-Butene
Step 1: Formation of the sec-butyl cation, a 2° carbocation intermediate.
CH3CH=CHCH3 + H—Cl:
δ+
δ−
slow, rate-limiting step
H
CH3CH—CHCH3 + :Cl:
sec-Butyl cation
(a 2° carbocation intermediate)
Step 2: Reaction of the sec-butyl cation (a Lewis acid) with chloride ion (a Lewis base) to complete the valence shell of carbon.
:Cl:
+
CH3CHCH2CH3
fast
:Cl:
CH3CHCH2CH3
Chloride ion
(a Lewis base)
sec-Butyl cation
(a Lewis acid)

Now let us go back and look at the individual steps in more detail. There is a great deal of important organic chemistry embedded in these two steps, and it is important that you understand it now.

Step 1 results in the formation of an organic cation. One carbon atom in this cation has only 6 electrons in its valence shell and carries a charge of +1. A species containing a positively charged carbon atom is called a **carbocation** (*carbon* + *cation*). Carbocations are classified as **primary** (1°), **secondary** (2°), or **tertiary** (3°) depending on the number of carbon atoms bonded to the carbon bearing the positive charge. All carbocations are Lewis acids (Section 2.5). They are also electrophiles. The term **electrophile** quite literally means electron lover.

Carbocation A species containing a carbon atom with only three bonds to it and bearing a positive charge.

Electrophile Any molecule or ion that can accept a pair of electrons to form a new covalent bond; a Lewis acid.

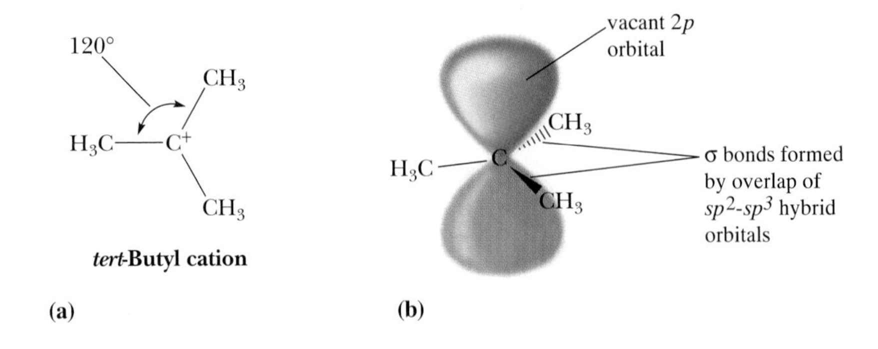

Figure 6.3
The structure of the *tert*-butyl cation. (a) Lewis structure and (b) an orbital picture.

In a carbocation, the carbon bearing the positive charge is bonded to three other atoms, and, as predicted by the valence-shell electron-pair repulsion (VSEPR) model, the three bonds about it are coplanar and form bond angles of approximately 120°. According to the valence bond model, the electron-deficient carbon of a carbocation uses its sp^2 hybrid orbitals to form sigma bonds to the three attached groups. Its unhybridized $2p$ orbital lies perpendicular to the sigma bond framework and contains no electrons. A Lewis structure and orbital overlap diagram for the *tert*-butyl cation are shown in Figure 6.3.

Figure 6.4 shows a potential energy diagram for the two-step reaction of 2-butene with HCl. The slower, rate-limiting step (the one that crosses the higher potential energy barrier) is Step 1, which leads to formation of the 2° carbocation intermediate. This carbocation intermediate lies in an energy minimum between the transition states for Steps 1 and 2. As soon as the carbocation intermediate (a Lewis acid) is formed, it reacts with chloride ion (a Lewis base) in a Lewis acid-base reaction to give 2-chlorobutane. Note that the energy level for 2-chlorobutane (the product) is lower than the energy level for 2-butene and HCl (the reactants). Thus, in this alkene addition reaction, heat is released; the reaction is exothermic.

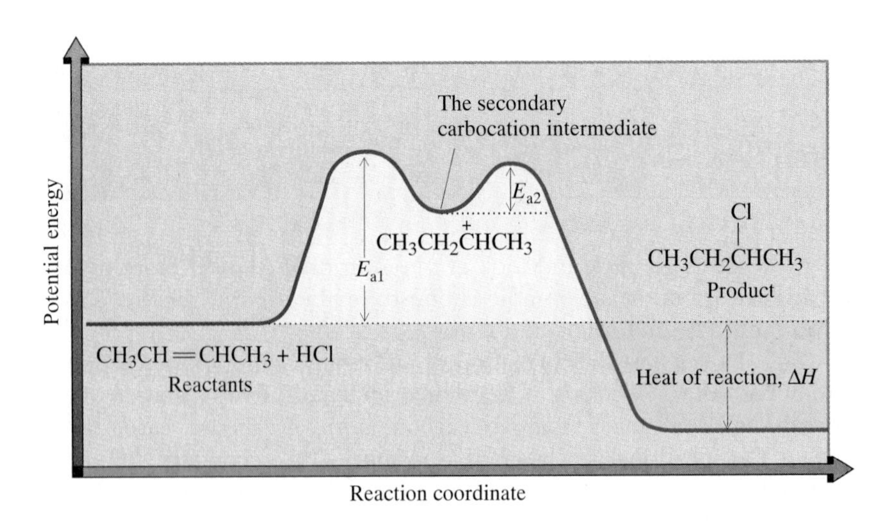

Figure 6.4
Potential energy diagram for the two-step addition of HCl to 2-butene. The reaction is exothermic.

Relative Stabilities of Carbocations — Regioselectivity and Markovnikov's Rule

Reaction of HX and an alkene can, at least in principle, give two different carbocation intermediates depending on which of the doubly bonded carbon atoms forms a bond with H^+, as illustrated by the reaction of HCl with propene.

$$CH_3CH{=}CH_2 + H{-}Cl: \longrightarrow CH_3CH_2\overset{+}{C}H_2 \xrightarrow{:Cl:^-} CH_3CH_2CH_2Cl:$$

Propene; Propyl cation (a 1° carbocation); 1-Chloropropane (not formed)

$$CH_3CH{=}CH_2 + H{-}Cl: \longrightarrow CH_3\overset{+}{C}HCH_3 \xrightarrow{:Cl:^-} CH_3CH(Cl)CH_3$$

Propene; Isopropyl cation (a 2° carbocation); 2-Chloropropane (product formed)

The propyl cation is a 1° carbocation, and the isopropyl cation is a 2° carbocation. The propyl cation reacts with chloride ion to give 1-chloropropane, and the isopropyl cation reacts with chloride ion to give 2-chloropropane. The observed product is 2-chloropropane, indicating that the 2° carbocation is formed in preference to the 1° carbocation.

Similarly, in the reaction of HCl with 2-methylpropene, addition of H^+ to the carbon-carbon double bond might form either the isobutyl cation (a 1° carbocation) or the *tert*-butyl cation (a 3° carbocation).

$$CH_3C(CH_3){=}CH_2 + H{-}Cl: \longrightarrow CH_3CH(CH_3)\overset{+}{C}H_2 \xrightarrow{:Cl:^-} CH_3CH(CH_3)CH_2Cl:$$

2-Methylpropene; Isobutyl cation (a 1° carbocation); 1-Chloro-2-methylpropane (not formed)

$$CH_3C(CH_3){=}CH_2 + H{-}Cl: \longrightarrow CH_3\overset{+}{C}(CH_3)CH_3 \xrightarrow{:Cl:^-} CH_3C(CH_3)(Cl)CH_3$$

2-Methylpropene; *tert*-Butyl cation (a 3° carbocation); 2-Chloro-2-methylpropane (product formed)

Reaction of the isobutyl cation with chloride ion gives 1-chloro-2-methylpropane (isobutyl chloride); reaction of the *tert*-butyl cation with chloride ion gives 2-chloro-2-methylpropane (*tert*-butyl chloride). The observed product is 2-chloro-2-methylpropane, indicating that the 3° carbocation is formed in preference to the 1° carbocation.

From such experiments and a great amount of other experimental evidence, we know that a 3° carbocation both is more stable and requires a lower activation energy for its formation than a 2° carbocation, which, in turn, is more stable and requires a lower activation energy for its formation than a 1° carbocation. It follows, then, that a more stable carbocation intermediate forms faster than a less stable carbocation intermediate. Following is the order of stability of four types of alkyl carbocations.

Figure 6.5
Methyl and *tert*-butyl cations. Delocalization of positive charge by the electron-withdrawing inductive effect of the trivalent, positively charged carbon according to molecular orbital calculations.

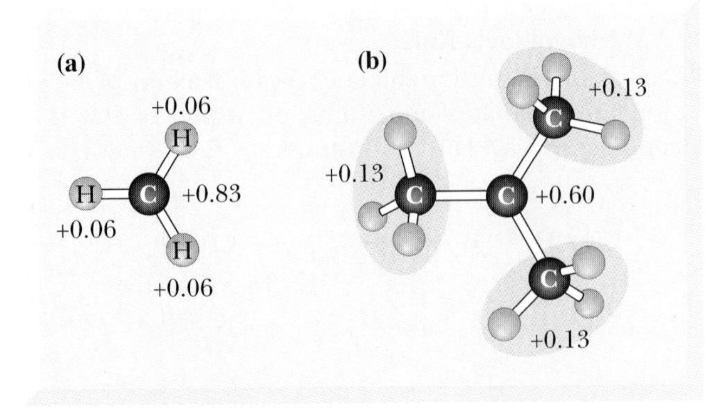

$$\underset{\substack{\text{Methyl cation} \\ \text{(methyl)}}}{H{-}\overset{+}{C}H_2} \qquad \underset{\substack{\text{Ethyl cation} \\ (1°)}}{H_3C{-}\overset{+}{C}H_2} \qquad \underset{\substack{\text{Isopropyl cation} \\ (2°)}}{H_3C{-}\overset{+}{C}(CH_3)H} \qquad \underset{\substack{\textit{tert}\text{-Butyl cation} \\ (3°)}}{H_3C{-}\overset{+}{C}(CH_3)_2}$$

Order of increasing carbocation stability →

Now that we know the order of stability of carbocations, how do we account for this order? The principles of physics teach us that a system bearing a charge (either positive or negative) is more stable if the charge is delocalized. Delocalization of the positive charge in carbocations is due to the **electron-withdrawing inductive effect** of the positively charged, trivalent carbon. This effect operates in the following way. The electron deficiency of the carbon atom bearing a positive charge draws electrons from adjacent sigma bonds toward it. Thus, the positive charge of the cation is not localized on the trivalent carbon but rather delocalized over nearby atoms. The larger the volume over which the positive charge is delocalized, the greater the stability of the cation. Thus, as the number of alkyl groups bonded to the cationic carbon increases, the stability of the cation increases. The electron-withdrawing inductive effect of the positively charged carbon and the resulting delocalization of charge are illustrated in Figure 6.5. According to quantum mechanical calculations, the charge on carbon in the methyl cation is approximately $+0.83$, and the charge on each of its hydrogen atoms is $+0.006$. Thus, even in the methyl cation, the positive charge is partially delocalized over the volume of space occupied by the entire ion. Polarization of electron density and delocalization of charge is even more extensive in the *tert*-butyl cation.

EXAMPLE 6.2

Arrange these carbocations in order of increasing stability.

(a) $CH_3\overset{+}{C}HC(CH_3)_2CH_3$ (b) $CH_3\overset{+}{C}(CH_3)CHCH_3$ with CH_3 on the CH (c) $CH_3C(CH_3)_2CH_2CH_2^{+}$

SOLUTION

Carbocation (a) is secondary, (b) is tertiary, and (c) is primary. In order of increasing stability they are c < a < b.

Practice Problem 6.2

Arrange these carbocations in order of increasing stability.

(a) cyclohexyl cation $-CH_3$ (b) cyclohexyl cation $-CH_3$ (c) cyclohexyl $-\overset{+}{C}H_2$

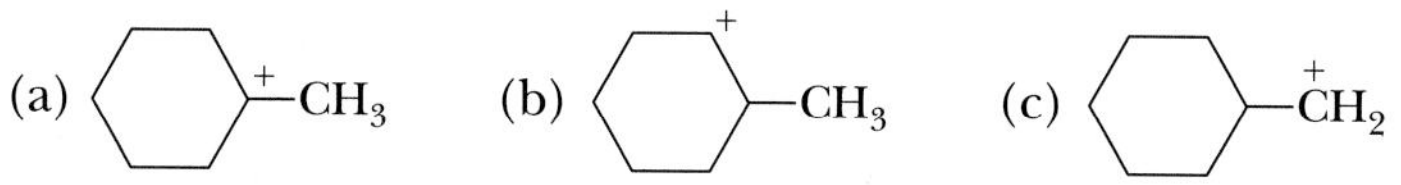

EXAMPLE 6.3

Propose a mechanism for the addition of HI to methylenecyclohexane to give 1-iodo-1-methylcyclohexane. Which step in your mechanism is rate limiting?

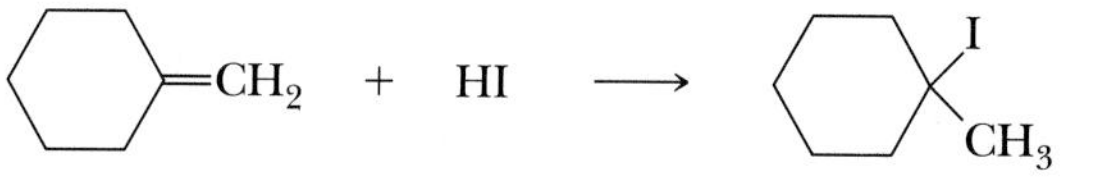

$$\text{Methylenecyclohexane } (=CH_2) + HI \longrightarrow \text{1-Iodo-1-methylcyclohexane } (I, CH_3)$$

SOLUTION

Propose a two-step mechanism similar to that proposed for the addition of HCl to propene.

Step 1: A rate-limiting proton transfer from HI to the carbon-carbon double bond gives a 3° carbocation intermediate.

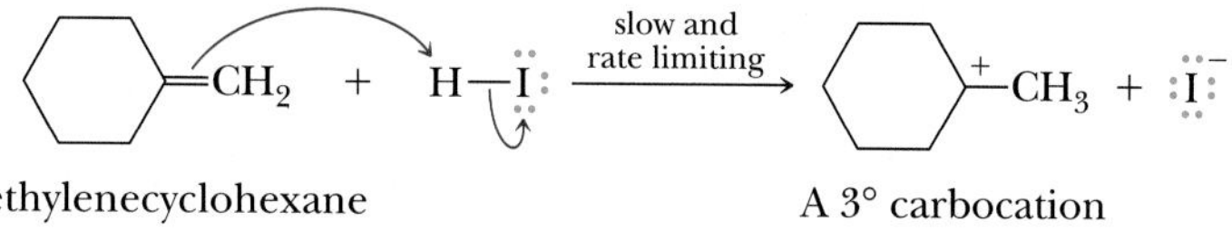

$$\text{Methylenecyclohexane } (=CH_2) + H{-}\ddot{I}: \xrightarrow{\text{slow and rate limiting}} \text{A 3° carbocation intermediate } (\overset{+}{}{-}CH_3) + :\ddot{I}:^-$$

Step 2: Reaction of the 3° carbocation intermediate (a Lewis acid) with iodide ion (a Lewis base) completes the valence shell of carbon and gives the product.

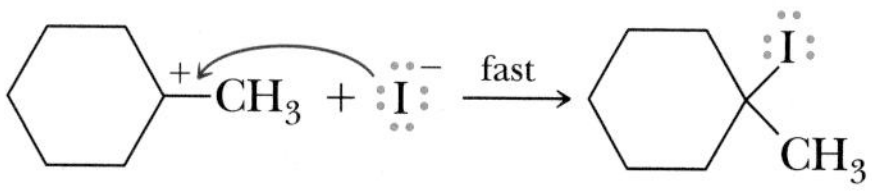

$$\overset{+}{}{-}CH_3 + :\ddot{I}:^- \xrightarrow{\text{fast}} (:\ddot{I}:, CH_3)$$

Practice Problem 6.3

Propose a mechanism for the addition of HI to 1-methylcyclohexene to give 1-iodo-1-methylcyclohexane. Which step in your mechanism is rate limiting?

B. Addition of Water—Acid-Catalyzed Hydration

In the presence of an acid catalyst, most commonly concentrated sulfuric acid, water adds to the carbon-carbon double bond of an alkene to give an alcohol. Addi-

Hydration Addition of water.

tion of water is called **hydration.** In the case of simple alkenes, —H adds to the carbon of the double bond with the greater number of hydrogens, and —OH adds to the carbon with the fewer hydrogens. Thus, H—OH adds to alkenes in accordance with Markovnikov's rule.

$$CH_3CH{=}CH_2 + H_2O \xrightarrow{H_2SO_4} CH_3\underset{OH}{CH}{-}\underset{H}{CH_2}$$

Propene → 2-Propanol

$$CH_3\overset{CH_3}{C}{=}CH_2 + H_2O \xrightarrow{H_2SO_4} CH_3\overset{CH_3}{\underset{HO}{C}}{-}\underset{H}{CH_2}$$

2-Methylpropene → 2-Methyl-2-propanol

EXAMPLE 6.4

Draw a structural formula for the product of acid-catalyzed hydration of 1-methylcyclohexene.

SOLUTION

1-Methylcyclohexene + H_2O $\xrightarrow{H_2SO_4}$ 1-Methylcyclohexanol

Practice Problem 6.4

Draw a structural formula for the product of each alkene hydration reaction.

(a) $CH_3\overset{CH_3}{C}{=}CHCH_3 + H_2O \xrightarrow{H_2SO_4}$ (b) $CH_2{=}\overset{CH_3}{C}CH_2CH_3 + H_2O \xrightarrow{H_2SO_4}$

The mechanism for acid-catalyzed hydration of alkenes is quite similar to what we already proposed for addition of HCl, HBr, and HI to alkenes and is illustrated by the hydration of propene. In Step 1, proton transfer from H_3O^+ to the alkene forms a 2° carbocation intermediate (a Lewis acid). This intermediate then completes its valence shell in Step 2 by forming a new covalent bond with an unshared pair of electrons of the oxygen atom of H_2O (a Lewis base) to give an **oxonium ion.** Finally, proton transfer to H_2O in Step 3 gives the alcohol and a new molecule of catalyst. Formation of the carbocation intermediate in Step 1 is the rate-limiting step. This mechanism is consistent with the fact that acid is a catalyst. An H_3O^+ is consumed in Step 1, but another is generated in Step 3.

Oxonium ion An ion in which oxygen is bonded to three other atoms and bears a positive charge.

MECHANISM Acid-Catalyzed Hydration of Propene

Step 1: Proton transfer from the acid catalyst to propene gives a 2° carbocation intermediate.

$$CH_3CH{=}CH_2 + H{-}\overset{+}{\ddot{O}}(H){-}H \xrightleftharpoons{\text{slow, rate-limiting step}} CH_3\overset{+}{C}HCH_3 + :\ddot{O}(H){-}H$$

A 2° carbocation intermediate

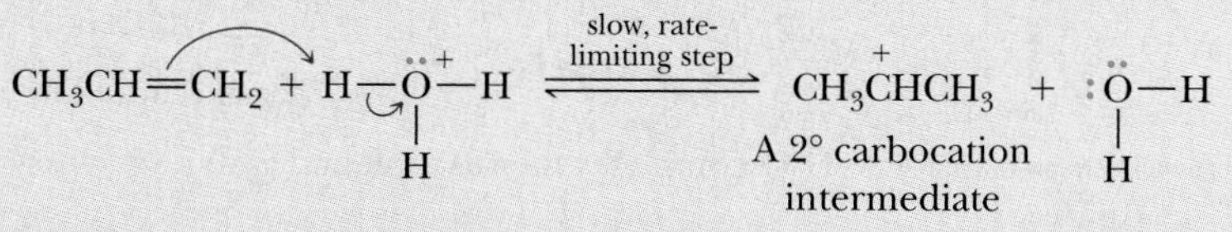

Step 2: Reaction of the carbocation intermediate (a Lewis acid) with water (a Lewis base) gives an oxonium ion.

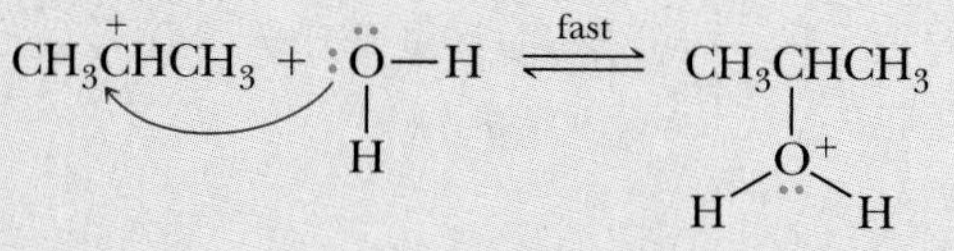

$$CH_3\overset{+}{C}HCH_3 + :\ddot{O}(H){-}H \xrightleftharpoons{\text{fast}} CH_3CH(\overset{+}{O}H_2)CH_3$$

An oxonium ion

Step 3: Proton transfer from the oxonium ion to water gives the alcohol and generates a new molecule of catalyst.

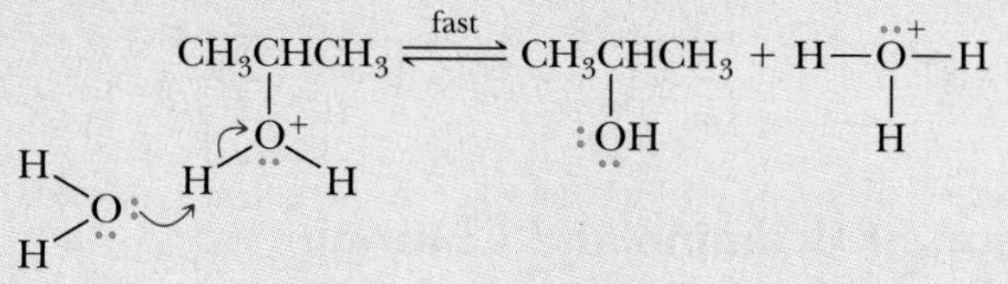

$$CH_3CH(\overset{+}{O}H_2)CH_3 + H_2\ddot{O}: \xrightleftharpoons{\text{fast}} CH_3CH(\ddot{O}H)CH_3 + H{-}\overset{+}{\ddot{O}}(H){-}H$$

EXAMPLE 6.5

Propose a mechanism for the acid-catalyzed hydration of methylenecyclohexane to give 1-methylcyclohexanol. Which step in your mechanism is the rate-limiting step?

SOLUTION

Propose a three-step mechanism similar to that for the acid-catalyzed hydration of propene. Formation of the 3° carbocation intermediate in Step 1 is rate limiting.

Step 1: Proton transfer from the acid catalyst to the alkene gives a 3° carbocation intermediate.

$$C_6H_{10}{=}CH_2 + H{-}\overset{+}{\ddot{O}}(H){-}H \xrightleftharpoons{\text{slow, rate-limiting step}} C_6H_{10}\overset{+}{C}{-}CH_3 + :\ddot{O}(H){-}H$$

A 3° carbocation intermediate

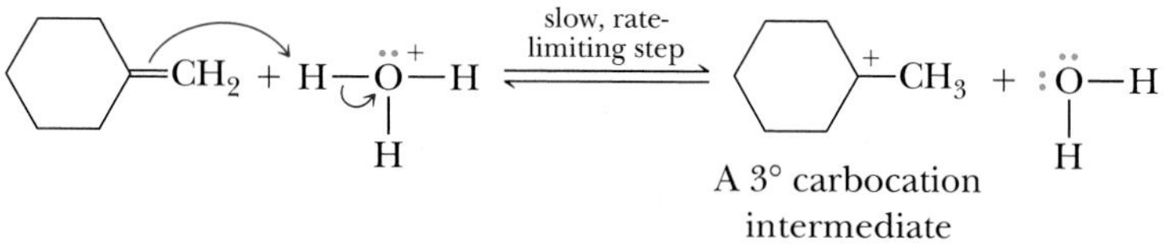

Step 2: Reaction of the carbocation intermediate (a Lewis acid) with water (a Lewis base) completes the valence shell of carbon and gives an oxonium ion.

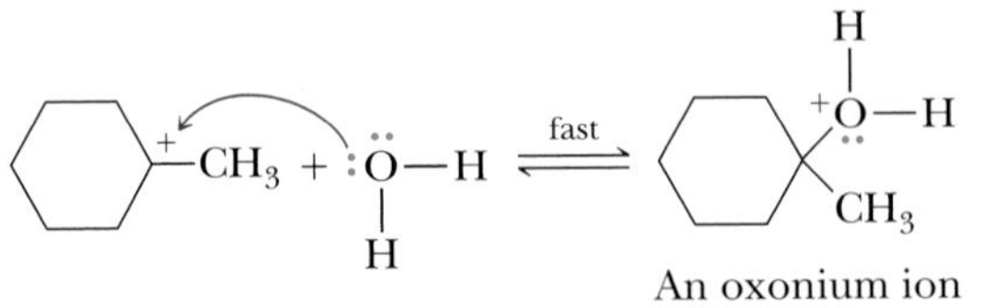

An oxonium ion

Step 3: Proton transfer to H_2O completes the reaction and generates a new molecule of catalyst.

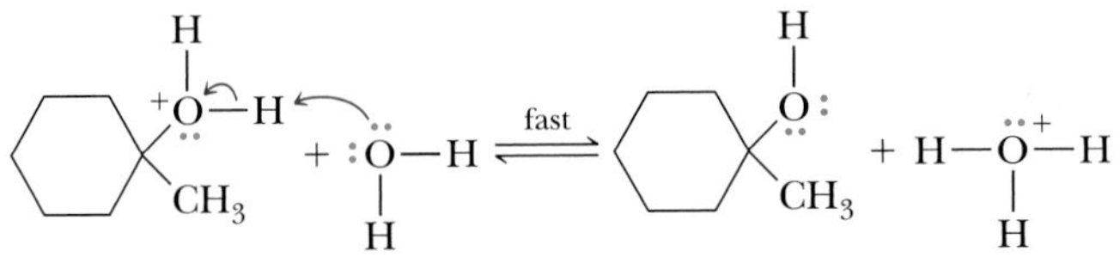

Practice Problem 6.5

Propose a mechanism for the acid-catalyzed hydration of 1-methylcyclohexene to give 1-methylcyclohexanol. Which step in your mechanism is the rate-limiting step?

C. Addition of Bromine and Chlorine

Chlorine, Cl_2, and bromine, Br_2, react with alkenes at room temperature by addition of halogen atoms to the two carbon atoms of the double bond with formation of two new carbon-halogen bonds. Reaction is generally carried out either with the pure reagents or by mixing them in an inert solvent, such as carbon tetrachloride, CCl_4, or dichloromethane, CH_2Cl_2.

$$CH_3CH{=}CHCH_3 + Br_2 \xrightarrow[CCl_4]{} CH_3CHBr{-}CHBrCH_3$$

2-Butene 2,3-Dibromobutane

Addition of bromine and chlorine to a cycloalkene gives a trans dihalocycloalkane. Addition of bromine to cyclohexene, for example, gives *trans*-1,2-dibromocyclohexane; the cis isomer is not formed. Thus, addition of a halogen to an alkene is **stereoselective**.

Stereoselective reaction A reaction in which one stereoisomer is formed or destroyed in preference to all others that might be formed or destroyed.

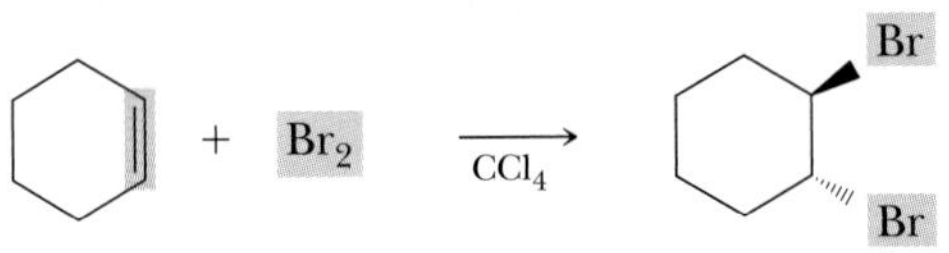

Cyclohexene + Br_2 $\xrightarrow[CCl_4]{}$ *trans*-1,2-Dibromocyclohexane

Addition of bromine to an alkene is a particularly useful qualitative test for the presence of a carbon-carbon double bond. If we dissolve bromine in carbon tetrachloride, the solution is red. Both alkenes and dibromoalkanes are colorless. If we now mix a few drops of the bromine solution with an alkene, a dibromoalkane is formed, and the solution becomes colorless.

A solution of bromine in carbon tetrachloride is red. Add a few drops of an alkene and the color disappears. *(Charles D. Winters)*

EXAMPLE 6.6

Complete these reactions, showing the stereochemistry of the product.

(a) cyclopentene + Br_2 $\xrightarrow[CH_2Cl_2]{}$ (b) 1-methylcyclohexene (CH_3) + Cl_2 $\xrightarrow[CH_2Cl_2]{}$

SOLUTION

Addition of both Br_2 and Cl_2 is stereoselective. The halogen atoms are trans to each other in each product.

(a) cyclopentene + Br_2 $\xrightarrow[CH_2Cl_2]{}$ trans-1,2-dibromocyclopentane (Br, Br)

(b) 1-methylcyclohexene (CH_3) + Cl_2 $\xrightarrow[CH_2Cl_2]{}$ trans-1,2-dichloro-1-methylcyclohexane (CH_3, Cl, Cl)

Practice Problem 6.6

Complete these reactions.

(a) $CH_3C(CH_3)_2CH{=}CH_2$ + Br_2 $\xrightarrow[CH_2Cl_2]{}$ (b) methylenecyclohexane ($=CH_2$) + Cl_2 $\xrightarrow[CH_2Cl_2]{}$

Stereoselectivity and Bridged Halonium Ion Intermediates

We explain the addition of bromine and chlorine to alkenes and their stereoselectivity by a two-step mechanism, illustrated by the reaction of bromine with an alkene. Reaction is initiated in Step 1 by interaction of the pi electrons of the alkene with bromine to form a reaction intermediate in which bromine bears a positive charge. A bromine atom bearing a positive charge is called a **bromonium ion,** and the cyclic structure of which it is a part is called a bridged bromonium ion. Comparable reaction with chlorine forms a bridged chloronium ion intermediate. This reaction intermediate is shown as a hybrid of three contributing structures, with the bridged bromonium ion being the most important. Then, in Step 2, a bromide ion reacts with this bridged intermediate from the side opposite that occupied by the bromine atom to give the dibromoalkane. Thus, bromine atoms are added from opposite faces of the carbon-carbon double bond. We say that this addition occurs with **anti stereoselectivity.**

Halonium ion An ion in which a halogen atom bears a positive charge.

Addition of chlorine or bromine to cyclohexene and its derivatives gives a trans diaxial product because only axial positions on adjacent atoms of a cyclohexane ring are anti and coplanar. The initial trans diaxial conformation of the product is in equilibrium with the trans diequatorial conformation, and, in simple derivatives of cyclohexane, the latter is more stable and predominates.

Anti addition Addition of atoms or groups of atoms from opposite sides or faces of a carbon-carbon double bond.

+ Br_2 ⟶ Br, Br ⇌ Br, Br

trans-Diaxial *trans*-Diequatorial (more stable)

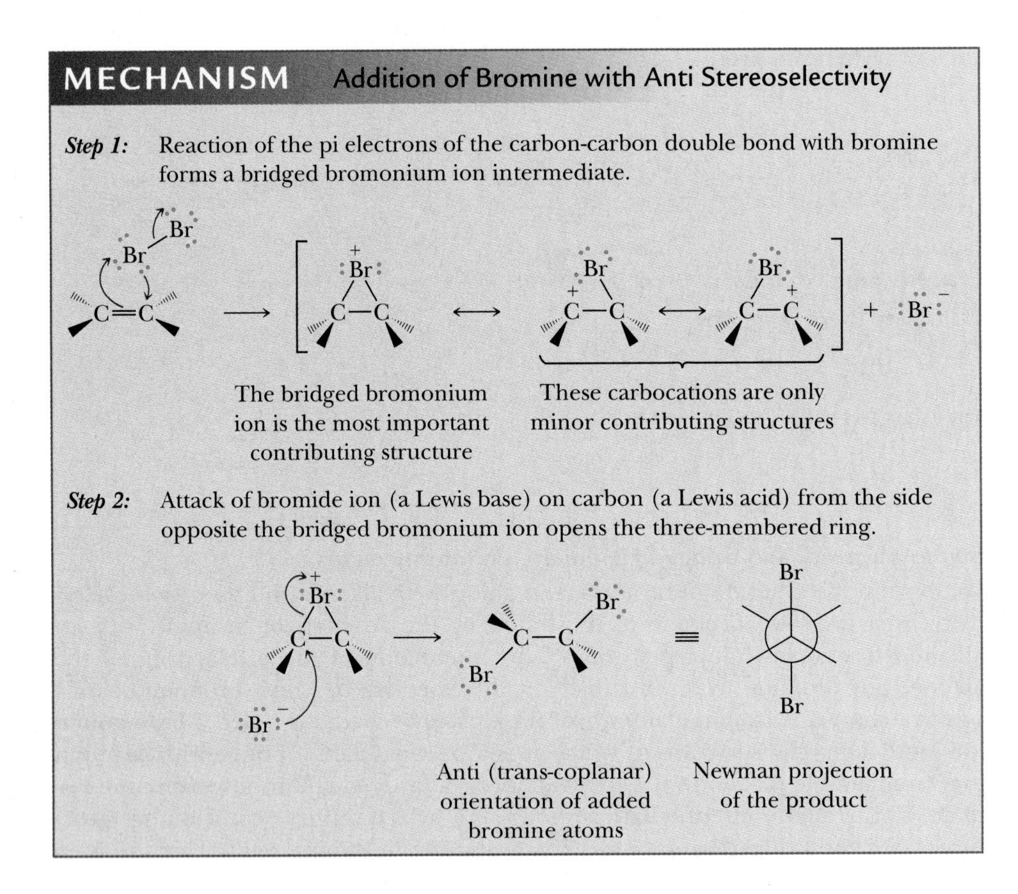

MECHANISM Addition of Bromine with Anti Stereoselectivity
Step 1: Reaction of the pi electrons of the carbon-carbon double bond with bromine forms a bridged bromonium ion intermediate.
Br
Br
C=C
Br
C—C
The bridged bromonium ion is the most important contributing structure
These carbocations are only minor contributing structures
Step 2: Attack of bromide ion (a Lewis base) on carbon (a Lewis acid) from the side opposite the bridged bromonium ion opens the three-membered ring.
Br
C—C
Anti (trans-coplanar) orientation of added bromine atoms
Newman projection of the product

6.4 Oxidation of Alkenes — Formation of Glycols

In this and the following section, we study oxidation and reduction of alkenes. We begin with a general method by which you can recognize oxidation-reduction reactions.

A. How to Recognize Oxidation-Reduction

In the following reactions, propene is transformed into three different compounds by reactions we study in this chapter. The first reaction involves reduction, the third involves oxidation, whereas the second involves neither oxidation nor reduction. These equations are not complete because they do not specify any reactant other than propene; that is, they do not specify what reagents are necessary to bring about the particular transformation. Each does specify, however, that the carbon atoms of the products are derived from those of propene.

$$CH_3CH{=}CH_2 \xrightarrow{\text{reduction}} CH_3\underset{|}{\overset{H}{C}}H{-}\overset{H}{\underset{|}{C}}H_2 \quad \text{Propane}$$

$$CH_3CH{=}CH_2 \xrightarrow{\text{neither}} CH_3\overset{OH}{C}H{-}\overset{H}{C}H_2 \quad \text{2-Propanol}$$

$$CH_3CH{=}CH_2 \xrightarrow{\text{oxidation}} CH_3\overset{OH}{C}H{-}\overset{OH}{C}H_2 \quad \text{1,2-Propanediol}$$

It is possible to decide if transformations such as these involve oxidation, reduction, or neither by using **balanced half-reactions.** To write a balanced half-reaction:

1. Write a half-reaction showing the organic reactant(s) and product(s).
2. Complete a material balance—that is, balance the number of atoms on each side of the half-reaction. To balance the number of oxygens and hydrogens for a reaction taking place in acid solution, use H_2O for oxygens and then H^+ for hydrogens. For a reaction taking place in basic solution, use OH^- and H_2O.
3. Complete a charge balance—that is, balance the charge on both sides of the half-reaction. To balance the charge, add electrons, e^-, to one side or the other. The equation completed in this step is a balanced half-reaction.

Oxidation The loss of electrons.

Reduction The gain of electrons.

Oxidation is the loss of electrons. If electrons appear on the right side of a balanced half-reaction, the reactant has given up electrons and has been oxidized. **Reduction** is the gain of electrons. If electrons appear on the left side of a balanced half-reaction, the reactant has gained electrons and has been reduced. If no electrons appear in the balanced half-reaction, then the transformation involves neither oxidation nor reduction. Let us apply these steps to the transformation of propene to propane.

Step 1: Half-reaction.

$$CH_3CH{=}CH_2 \longrightarrow CH_3CH_2CH_3$$

Step 2: Material balance.

$$CH_3CH{=}CH_2 + 2H^+ \longrightarrow CH_3CH_2CH_3$$

Step 3: Balanced half-reaction.

$$CH_3CH{=}CH_2 + 2H^+ + 2e^- \longrightarrow CH_3CH_2CH_3$$

Because two electrons appear on the left side of the balanced half-reaction (Step 3), conversion of propene to propane is a two-electron reduction. To bring it about requires use of a reducing agent.

Following is a balanced half-reaction for the transformation of propene to 2-propanol:

$$\text{Balanced half-reaction:} \quad \underset{\text{Propene}}{CH_3CH{=}CH_2} + H_2O \longrightarrow \underset{\text{2-Propanol}}{CH_3\overset{\overset{OH}{|}}{C}HCH_3}$$

Because no electrons are required to achieve an electrical balance, conversion of propene to 2-propanol is neither oxidation nor reduction.

A balanced half-reaction for the transformation of propene to 1,2-propanediol requires two electrons on the right side of the equation for a charge balance; this transformation is a two-electron oxidation.

$$\text{Balanced half-reaction:} \quad \underset{\text{Propene}}{CH_3CH{=}CH_2} + 2H_2O \longrightarrow \underset{\text{1,2-Propanediol}}{CH_3\overset{\overset{HO}{|}}{C}H\overset{\overset{OH}{|}}{C}H_2} + 2H^+ + 2e^-$$

It is important to realize that this strategy for recognizing oxidation and reduction is only that, a strategy. In no way does it indicate how a particular oxidation or reduction might be carried out in the laboratory. For example, the balanced half-reaction for the transformation of propene to propane requires $2H^+$ and $2e^-$. Yet by far the most common laboratory procedure for reducing propene to propane does not involve H^+ at all; rather it involves molecular hydrogen, H_2, and a transition metal catalyst (Section 6.5).

We use this method of balanced half-reactions throughout the text to recognize transformations that involve oxidation, those that involve reduction, and those that involve neither oxidation nor reduction.

EXAMPLE 6.7

Balance each half-reaction to show that each transformation involves an oxidation.

$$\text{(a)} \quad CH_3CH_2CH_2OH \longrightarrow CH_3CH_2\overset{\overset{O}{\|}}{C}H$$

$$\text{(b)} \quad CH_3CH{=}CH_2 \longrightarrow CH_3\overset{\overset{O}{\|}}{C}H + H\overset{\overset{O}{\|}}{C}H$$

SOLUTION

First complete a material balance and then a charge balance. Part (a) is a 2-electron oxidation; part (b) is a 4-electron oxidation. To bring each about requires an oxidizing agent.

$$\text{(a)} \quad CH_3CH_2CH_2OH \longrightarrow CH_3CH_2\overset{\overset{O}{\|}}{C}H + 2H^+ + 2e^-$$

$$\text{(b)} \quad CH_3CH{=}CH_2 + 2H_2O \longrightarrow CH_3\overset{\overset{O}{\|}}{C}H + H\overset{\overset{O}{\|}}{C}H + 4H^+ + 4e^-$$

Practice Problem 6.7

Balance each half-reaction to show that each transformation involves a reduction.

(a) 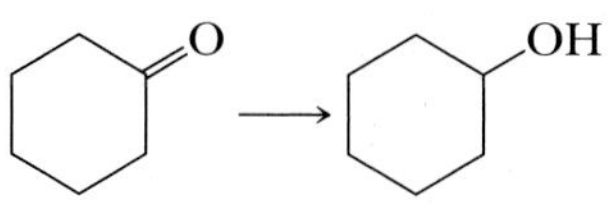 cyclohexanone ⟶ cyclohexanol (OH)

(b) $CH_3CH_2\overset{O}{\overset{\|}{C}}OH \longrightarrow CH_3CH_2CH_2OH$

B. OsO_4—Oxidation of an Alkene to a Glycol

Certain transition metal oxides, in particular osmium tetroxide, OsO_4, are effective oxidizing agents for the conversion of an alkene to a **glycol,** a compound with two hydroxyl groups on adjacent carbons. Oxidation of an alkene by osmium tetroxide is stereoselective in that it involves **syn addition** (addition from the same side) of —OH groups to the carbons of the double bond. For example, oxidation of cyclopentene by OsO_4 gives *cis*-1,2-cyclopentanediol, a **cis glycol.** Note that both cis and trans isomers are possible for this glycol.

Glycol A compound with two hydroxyl (—OH) groups on adjacent carbons.

Syn addition Addition of atoms or groups of atoms from the same side or face of a carbon-carbon double bond.

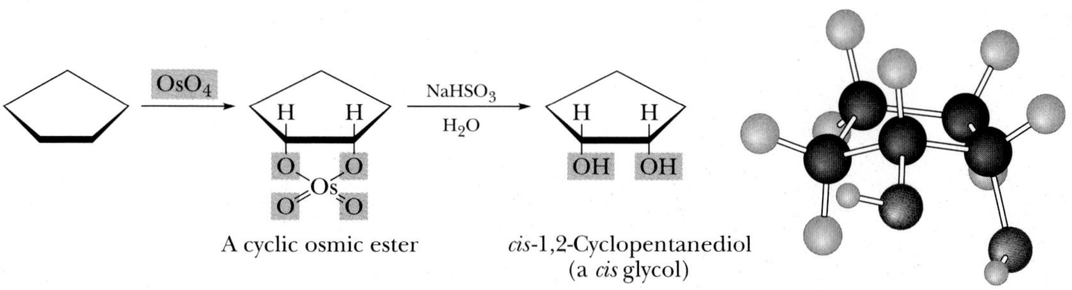

OsO_4 ⟶ A cyclic osmic ester ⟶ ($NaHSO_3$, H_2O) *cis*-1,2-Cyclopentanediol (a *cis* glycol)

The stereoselectivity of osmium tetroxide oxidation of alkenes is accounted for by the formation of a cyclic osmic ester in which oxygen atoms of OsO_4 form new covalent bonds with each carbon of the double bond in such a way that the five-membered osmium-containing ring is fused in a cis configuration to the original alkene. Osmic esters can be isolated and characterized. Usually, however, they are treated directly with a reducing agent, such as $NaHSO_3$, which cleaves osmium-oxygen bonds to give the cis glycol and reduced forms of osmium.

The drawbacks of OsO_4 are that it is both expensive and highly toxic. One strategy to circumvent the high cost of OsO_4 is to use it in catalytic amounts along with stoichiometric amounts of another oxidizing agent, which reoxidizes the reduced forms of osmium and thus recycles the osmium reagent. Oxidizing agents commonly used for this purpose are hydrogen peroxide and *tert*-butyl hydroperoxide. When this procedure is used, there is no need for a reducing step using $NaHSO_3$.

HOOH — Hydrogen peroxide

$(CH_3)_3COOH$ — *tert*-Butyl hydroperoxide (*t*-BuOOH)

6.5 REDUCTION OF ALKENES—FORMATION OF ALKANES

Virtually all alkenes react quantitatively with molecular hydrogen, H_2, in the presence of a transition metal catalyst to give alkanes. Commonly used transition metal catalysts include platinum, palladium, ruthenium, and nickel. Because conversion of an alkene to an alkane involves reduction by hydrogen in the presence of a catalyst, the process is called **catalytic reduction** or, alternatively, **catalytic hydrogenation.**

$$\text{Cyclohexene} + H_2 \xrightarrow[25^\circ C,\ 3\ \text{atm}]{Pd} \text{Cyclohexane}$$

The metal catalyst is used as a finely powdered solid, which may be supported on some inert material such as powdered charcoal or alumina. Reaction is carried out by dissolving the alkene in ethanol or another nonreacting organic solvent, adding the solid catalyst, and exposing the mixture to hydrogen gas at pressures from 1 to 100 atm. Alternatively, the metal may be complexed with certain organic molecules and used in the form of a soluble complex.

The most common pattern in catalytic reduction of an alkene is syn addition of hydrogens to the carbon-carbon double bond. Catalytic reduction of 1,2-dimethylcyclohexene, for example, yields predominantly *cis*-1,2-dimethylcyclohexane. Along with the cis isomer are formed lesser amounts of *trans*-1,2-dimethylcyclohexane.

$$\text{1,2-Dimethylcyclohexene} + H_2 \xrightarrow{Pt} \underset{70-85\%}{\textit{cis}\text{-1,2-Dimethylcyclohexane}} + \underset{30-15\%}{\textit{trans}\text{-1,2-Dimethylcyclohexane}}$$

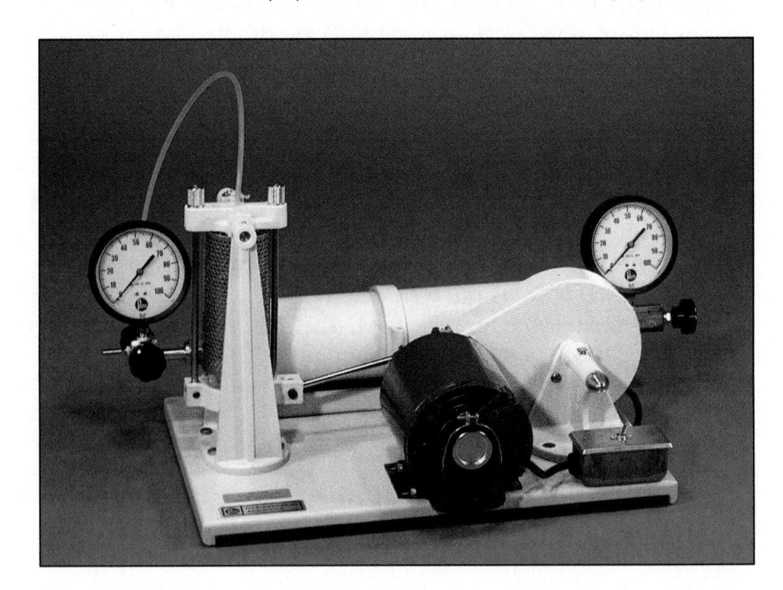

Paar shaker-type hydrogenation apparatus. *(Paar Instrument Co., Moline, IL)*

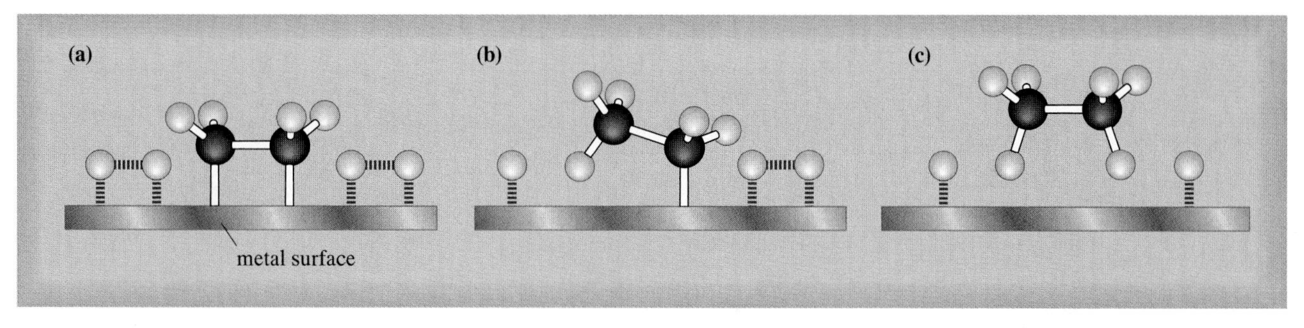

Figure 6.6
Addition of hydrogen to an alkene involving a transition metal catalyst. (a) Hydrogen and the alkene are adsorbed on the metal surface, and (b) one hydrogen atom is transferred to the alkene forming a new C—H bond. The other carbon remains adsorbed on the metal surface. (c) A second C—H bond is formed, and the alkane is desorbed.

A. Catalytic Reduction

The transition metals used in catalytic reduction are able to adsorb large quantities of hydrogen onto their surfaces, probably by forming metal-hydrogen sigma bonds. Similarly, alkenes are also adsorbed on metal surfaces with formation of carbon-metal bonds [Figure 6.6(a)]. Addition of hydrogen atoms to the alkene occurs in two steps. First, one new C—H bond is formed to give an intermediate in which the alkene remains partially adsorbed to the metal surface [Figure 6.6(b)]. The second hydrogen is then added from the same side as the first hydrogen.

B. Heats of Hydrogenation and the Relative Stabilities of Alkenes

The heat of hydrogenation of an alkene is defined as its heat of reaction, $\Delta H°$, with hydrogen to form an alkane. Table 6.2 lists heats of hydrogenation of several alkenes. Three important points are derived from the information given in Table 6.2.

1. Reduction of an alkene to an alkane is an exothermic process. This observation is consistent with the fact that, during hydrogenation, there is net conversion of a weaker pi bond to a stronger sigma bond; that is, one sigma bond (H—H) and one pi bond (C=C) are broken, and two new sigma bonds (C—H) are formed.
2. Heats of hydrogenation depend on the degree of substitution of the carbon-carbon double bond; the greater the substitution, the lower the heat of hydrogenation. Compare, for example, heats of hydrogenation of ethylene (no substituents), propene (one substituent), 1-butene (one substituent), and the cis and trans isomers of 2-butene (two substituents each).
3. The heat of hydrogenation of a trans alkene is lower than that of the isomeric cis alkene. Compare, for example, the heats of hydrogenation of *cis*-2-butene and *trans*-2-butene. Because reduction of each alkene gives butane, any difference in heats of hydrogenation must be due to a difference in relative energy between the two alkenes (Figure 6.7). The alkene with the lower (less negative) value of $\Delta H°$ is the more stable alkene.

We explain the greater stability of trans alkenes relative to cis alkenes in terms of nonbonded interaction strain. In *cis*-2-butene, the two $—CH_3$ groups are sufficiently close to each other that there is repulsion between their electron clouds.

TABLE 6.2 Heats of Hydrogenation of Several Alkenes

Name	Structural Formula	$\Delta H°$ [kcal/mol (kJ/mol)]	
ethylene	$CH_2{=}CH_2$	−32.8 (−137)	
propene	$CH_3CH{=}CH_2$	−30.1 (−126)	
1-butene	$CH_3CH_2CH{=}CH_2$	−30.3 (−127)	Ethylene
cis-2-butene	H_3C CH_3 C=C H H	−28.6 (−120)	
trans-2-butene	H_3C H C=C H CH_3	−27.6 (−115)	*trans*-2-Butene
2-methyl-2-butene	H_3C CH_3 C=C H_3C H	−26.9 (−113)	
2,3-dimethyl-2-butene	H_3C CH_3 C=C H_3C CH_3	−26.6 (−111)	2,3-Dimethyl-2-butene

This repulsion is reflected in the larger heat of hydrogenation (decreased stability) of *cis*-2-butene compared with *trans*-2-butene (approximately 1.0 kcal/mol).

6.6 Reactions That Produce Chiral Compounds

As the structure of an organic compound is altered in the course of a reaction, one or more stereocenters, usually at carbon, may be created, inverted, or destroyed. Let us consider two examples from this chapter to illustrate these possibilities.

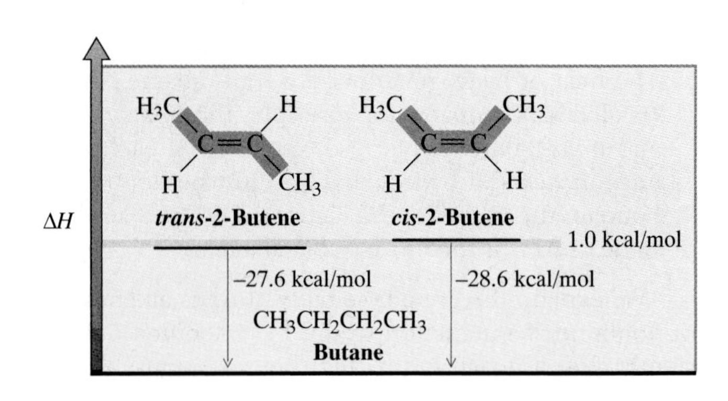

Figure 6.7
Heats of hydrogenation of *cis*-2-butene and *trans*-2-butene. *Trans*-2-butene is more stable than *cis*-2-butene by 1.0 kcal/mol (4.2 kJ/mol).

Addition of bromine to 2-butene (Section 6.3C) gives 2,3-dibromobutane, a molecule with two stereocenters. Three stereoisomers are possible for this compound: a meso compound and a pair of enantiomers (Section 4.4B). We now ask, "Is the product one enantiomer, a pair of enantiomers, the meso compound, or a mixture of all three stereoisomers?" A partial answer is that the product formed depends on the configuration of the alkene. Let us first examine addition of bromine to *cis*-2-butene.

Attack of bromine on *cis*-2-butene from either face of the planar part of the molecule gives the same bridged bromonium ion intermediate (Figure 6.8). Al-

Step 1: Formation of a meso bromonium ion intermediate

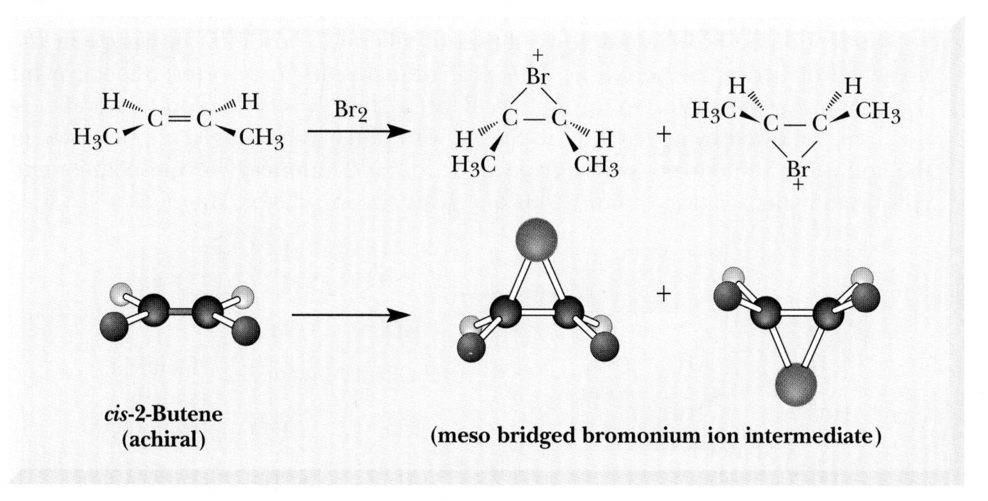

Step 2: Attack of bromide ion at carbon 2 gives one enantiomer; attack at carbon 3 gives the other enantiomer. Attack occurs with equal probability and rate at carbons 2 and 3 and gives a racemic mixture.

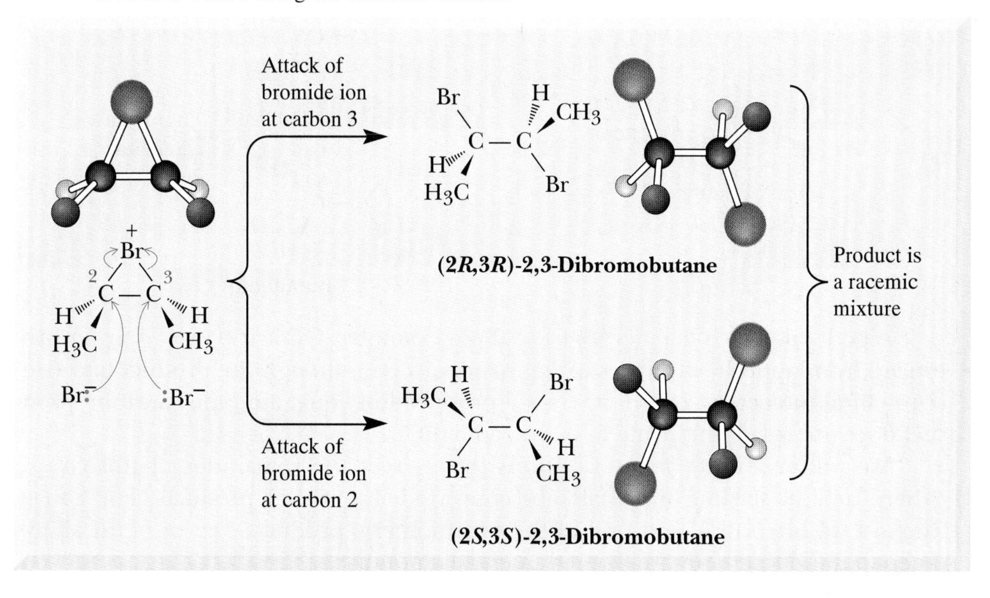

Figure 6.8
Anti addition of bromine to *cis*-2-butene gives 2,3-dibromobutane as a racemic mixture.

though this intermediate has two stereocenters, it has a plane of symmetry and is, therefore, meso. Attack of Br^- on this intermediate from the side opposite that of the bromonium ion bridge gives a pair of enantiomers. Attack of bromide ion on carbon 2 of this meso intermediate gives the (2S,3S) enantiomer. Attack of bromide ion on carbon 3 gives the (2R,3R) enantiomer. Attack of bromide ion occurs at equal rates at each carbon, and, therefore, the enantiomers are formed in equal amounts, and 2,3-dibromobutane is obtained as a racemic mixture (Figure 6.8). We have shown attack of Br_2 from one side of the carbon-carbon double bond. Attack of Br_2 from the opposite side followed by opening of the resulting bridged bromonium ion intermediate produces these same two stereoisomers.

Addition of Br_2 to *trans*-2-butene leads to two enantiomeric bridged bromonium ion intermediates. Attack by Br^- at either carbon atom of either bromonium ion intermediate gives the meso product, which is optically inactive (Figure 6.9).

In Section 6.4B, we studied oxidation of alkenes by osmium tetroxide in the presence of hydrogen peroxide. This oxidation results in syn stereoselective hydroxylation of the alkene to form a glycol. Step 1 involves formation of a cyclic osmic ester followed by its reaction in Step 2 with water to give a glycol. As shown in the following sequences, syn hydroxylation of *cis*-2-butene gives meso-2,3-butanediol. Syn hydroxylation of *trans*-2-butene gives racemic 2,3-butanediol.

cis-2-Butene (achiral) $\xrightarrow[ROOH]{OsO_4}$ (2S,3R)-2,3-Butanediol + (2R,3S)-2,3-Butanediol — Identical; a meso compound

trans-2-Butene (achiral) $\xrightarrow[ROOH]{OsO_4}$ (2S,3S)-2,3-Butanediol + (2R,3R)-2,3-Butanediol — A pair of enantiomers; a racemic mixture

A general principle emerges from these examples. Optical activity is never produced from optically inactive starting materials; even though the product may contain chiral molecules, enantiomers are produced in equal amounts, and the product does not rotate the plane of polarized light.

We will encounter many reactions throughout the remainder of this course where achiral starting materials are converted into chiral products. For convenience, we often draw just one of the enantiomeric products, but we must always keep in mind that both are formed in equal amounts.

Step 1: Reaction of *trans*-2-butene with bromine to form bridged bromonium ions which are enantiomers.

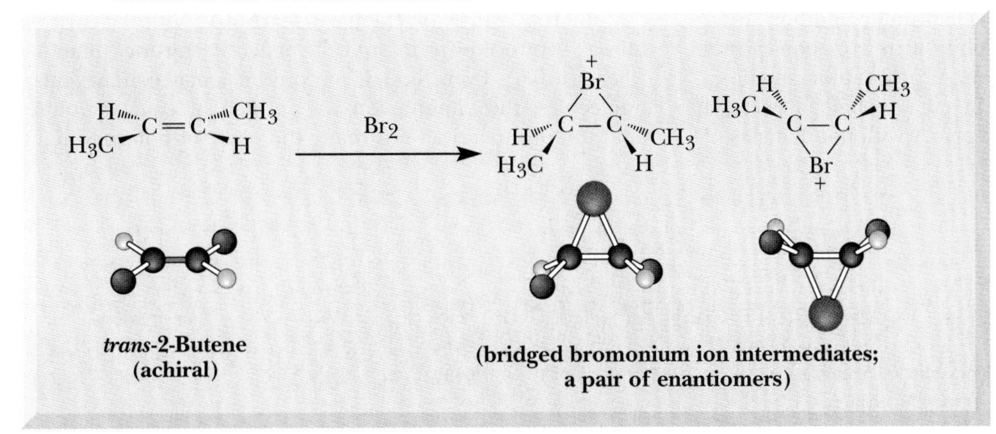

Step 2: Attack of bromine on either carbon of either enantiomer gives meso-2,3-dibromobutane.

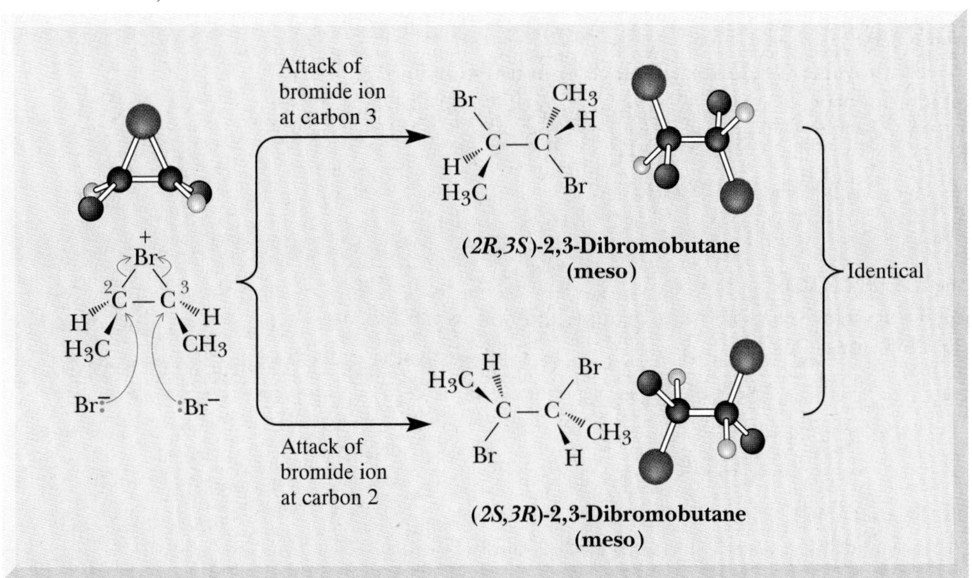

Figure 6.9
Anti addition of bromine to *trans*-2-butene gives meso-2,3-dibromobutane.

SUMMARY

A **reaction mechanism** (Section 6.2) is a description of how and why a chemical reaction occurs, which bonds are broken and which new ones are formed, the order in which the various bond-breaking and bond-forming steps take place and their relative rates, and the role of the catalyst if the reaction involves a catalyst. **Transition state theory** (Section 6.2A) provides a model for understanding the relationships among reaction rates, molecular structure, and energetics. A key postulate of transition state theory is formation of a **transition state.** The difference in potential energy between reactants and the transition state is called the **activation energy.** An **intermediate** is a potential energy minimum between two transition states. The slowest step in a multistep reaction, called the **rate-limiting step,** is the one that crosses the highest potential energy barrier.

A characteristic reaction of alkenes is **addition,** during which a pi bond is broken and sigma bonds to two new atoms or groups of atoms are formed.

An **electrophile** (Section 6.3A) is any molecule or ion that can accept a pair of electrons to form a new covalent bond. The rate-limiting step in **electrophilic addition** to an alkene is reaction of an electrophile with a carbon-carbon double bond to form a **carbocation.** A carbocation contains a carbon with only six electrons in its valence shell and bears a positive charge. Carbocations are planar with bond angles of 120° about the positive carbon. The order of stability of carbocations is 3° > 2° > 1° > methyl (Section 6.3A).

In the course of chemical reactions, optical activity is never produced from optically inactive starting materials (Section 6.6). Even though the products may contain chiral molecules, the enantiomers are formed in equal amounts, and the product does not rotate the plane of polarized light.

KEY REACTIONS

1. Addition of HX (Section 6.3A)

Addition is regioselective and follows Markovnikov's rule. Reaction occurs in two steps and involves formation of a carbocation intermediate.

1-methylcyclopentene CH_3 + HCl ⟶ 1-chloro-1-methylcyclopentane (Cl, CH_3)

2. Acid-Catalyzed Hydration (Section 6.3B)

Addition is regioselective and follows Markovnikov's rule. Reaction occurs in two steps and involves formation of a carbocation intermediate.

$$CH_3C(CH_3){=}CH_2 + H_2O \xrightarrow{H_2SO_4} CH_3C(CH_3)(OH)CH_3$$

3. Addition of Bromine and Chlorine (Section 6.3C)

Addition is stereoselective. Addition occurs in two steps and involves anti addition by way of a bridged bromonium or chloronium ion intermediate.

cyclohexene + Br_2 $\xrightarrow{CCl_4}$ trans-1,2-dibromocyclohexane (Br, Br)

4. Oxidation—Formation of Glycols (Section 6.4B)

Oxidation gives a glycol resulting from syn addition of —OH groups to the double bond via a cyclic osmic ester.

cyclopentene (H, H) $\xrightarrow[ROOH]{OsO_4}$ cis-cyclopentane-1,2-diol (H, H, HO, OH)

5. Reduction—Formation of Alkanes (Section 6.5)

Catalytic reduction involves predominantly syn addition of hydrogen.

cyclohexene + H_2 $\xrightarrow{\text{transition metal catalyst}}$ cyclohexane

ADDITIONAL PROBLEMS

Potential Energy Diagrams

6.8 Describe the differences between a transition state and a reaction intermediate.

6.9 Sketch a potential energy diagram for a one-step reaction that is very slow and only slightly exothermic. How many transition states are present in this reaction? How many intermediates?

6.10 Sketch a potential energy diagram for a two-step reaction that is endothermic in the first step, exothermic in the second step, and exothermic overall. How many transition states are present in this two-step reaction? How many intermediates?

Electrophilic Additions

6.11 From each pair, select the more stable carbocation.

(a) $CH_3CH_2CH_2^+$ or $CH_3\overset{+}{C}HCH_3$ **(b)** $CH_3CH(CH_3)\overset{+}{C}HCH_3$ or $CH_3\overset{+}{C}(CH_3)CH_2CH_3$

6.12 From each pair, select the more stable carbocation.

(a) 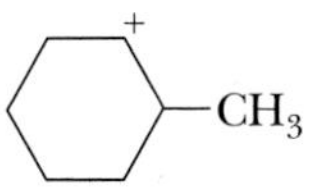 (cyclohexyl cation, + on ring carbon adjacent to C bearing CH_3)—CH_3 or (cyclohexyl cation, + on ring carbon bearing)—CH_3 **(b)** 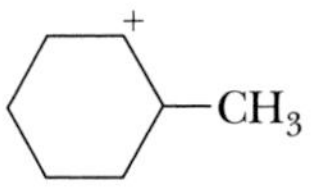 (cyclohexyl cation, + on ring carbon)—CH_3 or (cyclohexane)—CH_2^+

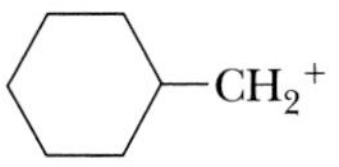

6.13 Draw structural formulas for the isomeric carbocation intermediates formed by reaction of each alkene with HCl. Label each carbocation primary, secondary, or tertiary, and state which of the isomeric carbocations is formed more readily, if either.

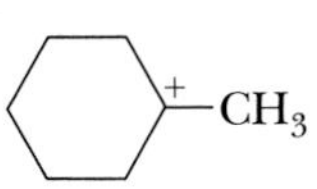

(a) $CH_3CH_2C(CH_3){=}CHCH_3$ **(b)** $CH_3CH_2CH{=}CHCH_3$ **(c)** 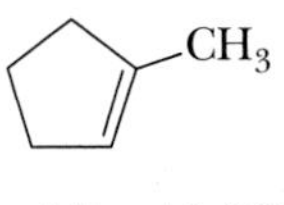 1-methylcyclopentene (CH_3) **(d)** 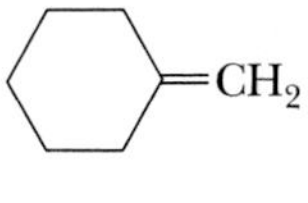 methylenecyclohexane ($=CH_2$)

6.14 From each pair of compounds, select the one that reacts more rapidly with HI. Draw the structural formula of the major product formed in each case, and explain the basis for your ranking.

(a) $CH_3CH{=}CHCH_3$ and $CH_3C(CH_3){=}CHCH_3$ **(b)** 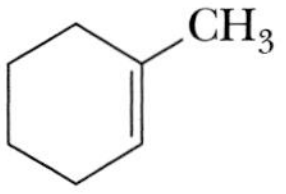 1-methylcyclohexene (CH_3) and cyclohexene

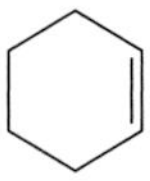

6.15 Complete these equations.

(a) 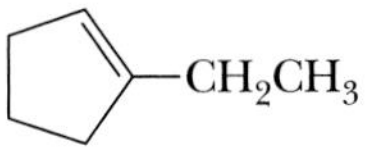 (1-ethylcyclopentene)—$CH_2CH_3 + HCl \longrightarrow$

(b) 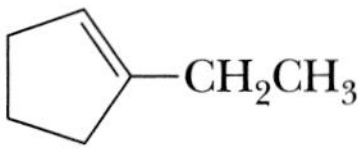 (1-ethylcyclopentene)—$CH_2CH_3 + H_2O \xrightarrow{H_2SO_4}$

(c) $CH_3(CH_2)_5CH{=}CH_2 + HI \longrightarrow$

(d) 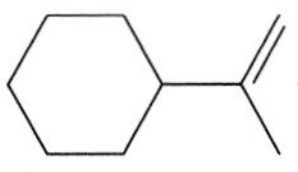 (cyclohexyl)—$C(CH_3){=}CH_2$ + HCl $\longrightarrow$

(e) $CH_3CH{=}CHCH_2CH_3 + H_2O \xrightarrow{H_2SO_4}$

(f) $CH_2{=}CHCH_2CH_2CH_3 + H_2O \xrightarrow{H_2SO_4}$

6.16 Reaction of 2-methyl-2-pentene with each reagent is regioselective. Draw a structural formula for the product of each of the reactions, and account for the observed regioselectivity.

(a) HI **(b)** H_2O in the presence of H_2SO_4

6.17 Addition of bromine and chlorine to cycloalkenes is stereoselective. Predict the stereochemistry of the product formed in each reaction, and account for your predicted stereoselectivity.

(a) 1-Methylcyclohexene + Br_2 **(b)** 1,2-Dimethylcyclopentene + Cl_2

6.18 Draw a structural formula for an alkene with the indicated molecular formula that gives the compound shown as the major product. Note that more than one alkene may give the same compound as the major product.

(a) $C_5H_{10} + H_2O \xrightarrow{H_2SO_4} CH_3C(CH_3)(OH)CH_2CH_3$ **(b)** $C_5H_{10} + Br_2 \longrightarrow CH_3CH(CH_3)CHBrCH_2Br$

(c) C_7H_{12} + HCl ⟶ 1-chloro-1-methylcyclohexane (ring with CH_3 and Cl on the same carbon)

6.19 Draw the structural formula for an alkene of molecular formula C_5H_{10} that reacts with Br_2 to give each product.

(a) $CH_3CBr(CH_3)-CHBrCH_3$ **(b)** $CH_2BrCBr(CH_3)CH_2CH_3$ **(c)** $CH_2BrCHBrCH_2CH_2CH_3$

6.20 Draw the structural formula for a cycloalkene of molecular formula C_6H_{10} that reacts with Cl_2 to give each compound.

(a) trans-1,2-dichlorocyclohexane (Cl wedged, Cl hashed) **(b)** 3,4-dichloro-1-methylcyclopentane (CH_3; Cl hashed, Cl wedged) **(c)** cyclopentane with H_3C and Cl (hashed) on one carbon and Cl (wedged) on the adjacent carbon **(d)** cyclopentane with Cl and CH_2Cl on the same carbon

6.21 Draw the structural formula for an alkene of molecular formula C_5H_{10} that reacts with HCl to give the indicated chloroalkane as the major product.

(a) $CH_3CCl(CH_3)CH_2CH_3$ **(b)** $CH_3CH(CH_3)CHClCH_3$ **(c)** $CH_3CHClCH_2CH_2CH_3$

6.22 Draw the structural formula of an alkene that undergoes acid-catalyzed hydration to give the indicated alcohol as the major product. More than one alkene may give each compound as the major product.

(a) 3-Hexanol **(b)** 1-Methylcyclobutanol

(c) 2-Methyl-2-butanol **(d)** 2-Propanol

6.23 Draw the structural formula of an alkene that undergoes acid-catalyzed hydration to give each alcohol as the major product. More than one alkene may give each compound as the major product.

(a) Cyclohexanol **(b)** 1,2-Dimethylcyclopentanol

(c) 1-Methylcyclohexanol **(d)** 1-Isopropyl-4-methylcyclohexanol

6.24 Terpin (*The Merck Index,* 12th ed., #9314) is prepared commercially by the acid-catalyzed hydration of limonene (Figure 5.3). Terpin is used medicinally as an expectorant for coughs.

$$\text{Limonene} + 2H_2O \xrightarrow{H_2SO_4} C_{10}H_{20}O_2 \ (\text{Terpin})$$

Limonene Terpin

(a) Propose a structural formula for terpin and a mechanism for its formation.

(b) How many cis-trans isomers are possible for the structural formula you propose?

6.25 Treatment of 2-methylpropene with methanol in the presence of a sulfuric acid catalyst gives *tert*-butyl methyl ether. Propose a mechanism for formation of this ether.

$$CH_3C(CH_3){=}CH_2 + CH_3OH \xrightarrow{H_2SO_4} CH_3C(CH_3)_2{-}OCH_3$$

6.26 Treatment of 1-methylcyclohexene with methanol in the presence of a sulfuric acid catalyst gives a compound of molecular formula $C_8H_{16}O$. Propose a structural formula for this compound and a mechanism for its formation.

$$\text{1-methylcyclohexene} + CH_3OH \xrightarrow{H_2SO_4} C_8H_{16}O$$

Oxidation-Reduction

6.27 Use balanced half-reactions to show which transformations involve oxidation, which involve reduction, and which involve neither oxidation nor reduction.

(a) $CH_3CH(OH)CH_3 \longrightarrow CH_3C(=O)CH_3$ **(b)** $CH_3CH(OH)CH_3 \longrightarrow CH_3CH{=}CH_2$

(c) $CH_3CH{=}CH_2 \longrightarrow CH_3CH_2CH_3$

6.28 Write a balanced equation for the combustion of 2-methylpropene in air to give carbon dioxide and water. The oxidizing agent is O_2, which makes up approximately 20% of air.

6.29 Draw the product formed by treatment of each alkene with aqueous $OsO_4/ROOH$.

(a) 1-Methylcyclopentene **(b)** Vinylcyclohexane **(c)** *cis*-2-Pentene

6.30 What alkene, when treated with $OsO_4/ROOH$, gives each glycol?

(a) 1-(hydroxymethyl)cyclohexanol (OH, CH_2OH) **(b)** cyclohexane with H, OH, OH, CH_3 **(c)** $(CH_3)_2CHCH_2CH(OH)CH_2OH$ (HO OH)

6.31 Draw the product formed by treatment of each alkene with H_2/Ni.

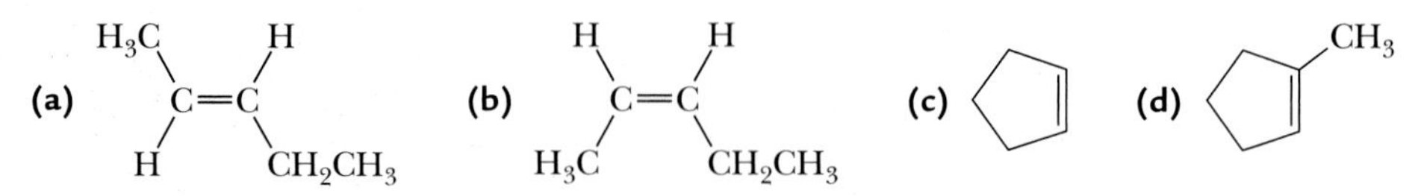

6.32 Hydrocarbon A, C_5H_8, reacts with 2 moles of Br_2 to give 1,2,3,4-tetrabromo-2-methylbutane. What is the structure of hydrocarbon A?

Synthesis

6.33 Show how to convert ethylene to these compounds.

(a) Ethane **(b)** Ethanol **(c)** Bromoethane

(d) 1,2-Dibromoethane **(e)** 1,2-Ethanediol **(f)** Chloroethane

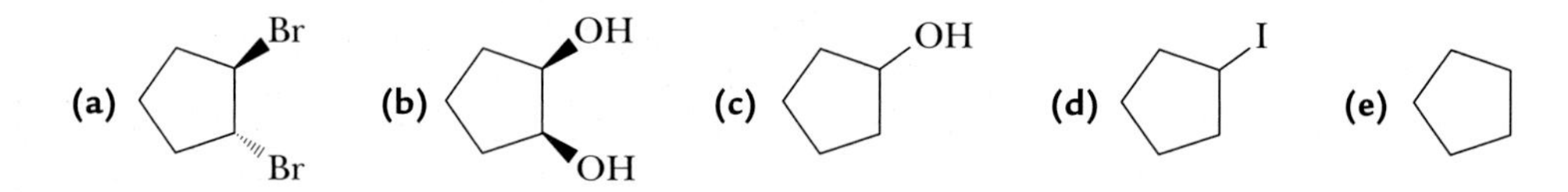

6.34 Show how to convert cyclopentene to these compounds.

Reactions That Produce Chiral Compounds

6.35 Show that acid-catalyzed hydration of 1-butene gives a racemic mixture of (R)-2-butanol and (S)-2-butanol and that the product is, therefore, optically inactive.

6.36 Consider the compound 1,2-cyclopentanediol.

(a) How many stereoisomers are possible for this compound?

(b) Which of the possible stereoisomers are formed when cyclopentene is treated with $OsO_4/ROOH$?

6.37 Consider the compound 1,2-dibromocyclopentane.

(a) How many stereoisomers are possible for this compound?

(b) Which of the possible stereoisomers are formed when cyclopentene is treated with bromine, Br_2?

6.38 At this point, the following spectroscopy problems may be assigned: for ^1H-NMR and ^{13}C-NMR, end-of-chapter problems 21.10–21.15, and for IR, end-of-chapter problems 22.3 and 22.4.

7

Haloalkanes

Compact disks are made of polyvinyl chloride. Inset: A model of vinyl chloride. *(Charles D. Winters)*

Compounds containing a halogen atom covalently bonded to an sp^3 hybridized carbon atom are named haloalkanes or, in the common system of nomenclature, alkyl halides. The general symbol for an **alkyl halide** is R—X, where —X may be —F, —Cl, —Br, or —I.

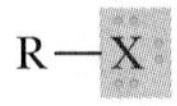

R—X

A haloalkane
(an alkyl halide)

In this chapter, we study two characteristic reactions of haloalkanes: nucleophilic substitution and β-elimination. By these reactions, alkyl halides can be converted to alcohols, ethers, thiols, amines, and alkenes. Thus, an understanding of nucleophilic substitution and β-elimination opens entirely new areas of organic chemistry for you.

7.1 NOMENCLATURE

A. IUPAC System

IUPAC names for haloalkanes are derived according to the rules given in Section 3.3A. The parent chain is located and numbered from the direction that gives the substituent encountered first the lower number. Halogen substituents are indicated by the prefixes fluoro-, chloro-, bromo-, and iodo- and are listed in alphabetical order along with other substituents. The location of each halogen atom on the parent chain is given by a number preceding the name of the halogen.

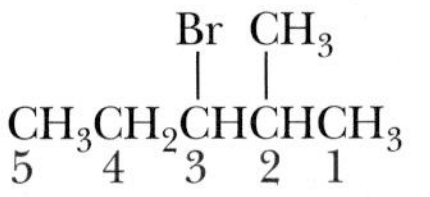

Br CH$_3$
CH$_3$CH$_2$CHCHCH$_3$
5 4 3 2 1

3-Bromo-2-methylpentane

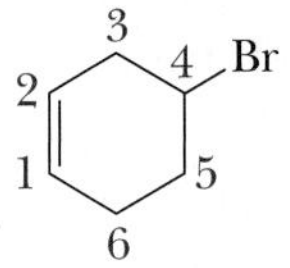

3, 4, Br, 2, 1, 5, 6

4-Bromocyclohexene

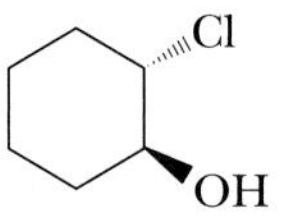

Cl, OH

trans-2-Chlorocyclohexanol

In haloalkenes, numbering the parent hydrocarbon is determined by the location of the carbon-carbon double bond. In molecules containing functional groups designated by a suffix (for example, -ol, -al, -one, -oic acid), numbering is determined by the location of the functional group indicated by the suffix.

B. Common Names

Alkyl halide A compound containing a halogen atom covalently bonded to an alkyl group; given the symbol RX.

Common names of haloalkanes consist of the common name of the alkyl group followed by the name of the halide as a separate word. Hence, the name **alkyl halide** is a common name for this class of compounds. In the following examples, the IUPAC name of the compound is given first and then, in parentheses, its common name.

Br
CH$_3$CHCH$_2$CH$_3$
2-Bromobutane
(*sec*-Butyl bromide)

CH$_2$=CHCl
Chloroethene
(Vinyl chloride)

CH$_2$=CHCH$_2$Cl
3-Chloropropene
(Allyl chloride)

CHEMICAL CONNECTIONS

Methyl Bromide, an Ozone-Depleting Pesticide

Although we may not think of methyl bromide as a useful pesticide, it is in fact just that. Worldwide production of this compound in 1996, for example, was 68,424 metric tons, over 70% of which was used as a pesticide for treatment of soils prior to planting. Another 20% of global production was as a pesticide to treat perishable products, such as flowers and fruits, and nonperishable products, such as nuts and timber. Methyl bromide's attractiveness for these purposes is due to its low cost relative to alternative pesticides, and its effectiveness against a large variety of pests.

The problem with its use is that it has been identified as an ozone-depleting substance. According to the provisions of the Montreal Protocol on Substances That Deplete the Ozone Layer, production and importation of methyl bromide by the United States and other developed countries must be reduced beginning in 1999, followed by a full ban on its production in 2005. Developing countries have until 2015 to phase out its production.

Finding non-ozone-depleting replacements for methyl bromide may not be easy. Alternatives currently under consideration are sulfuryl fluoride (SO_2F_2), phosphine (PH_3) dissolved in carbon dioxide, and carbonyl sulfide (COS). As a spokesperson for the U.S. methyl bromide producers put it, "Replacing methyl bromide is a tall order to fill—to find the same positive effects with no negative effects."

Several of the polyhalomethanes are common solvents and are generally referred to by their common, or trivial, names. Dichloromethane (methylene chloride) is the most widely used haloalkane solvent. Compounds of the type CHX_3 are called **haloforms.** The common name for $CHCl_3$, for example, is chloroform. It is from the name chloroform that the common name methyl chloroform is derived for the compound CH_3CCl_3. Methyl chloroform and trichloroethylene are common solvents for commercial dry cleaning.

CH_2Cl_2	$CHCl_3$	CH_3CCl_3	$CCl_2{=}CHCl$
Dichloromethane (Methylene chloride)	Trichloromethane (Chloroform)	1,1,1-Trichloroethane (Methyl chloroform)	Trichloroethylene (Trichlor)

EXAMPLE 7.1

Write the IUPAC name for each compound.

(a) $CH_3CH(CH_3)CH_2Br$

(b) $(H_3C)(H)C{=}C(CH_3)(CHBrCH_3)$ — H_3C and CH_3 on the same side; H and $CHCH_3$(Br) below

(c) $CH_3(CH_2)_6$—C(H)(Br)(CH_3), with H on a hashed bond and CH_3 on a wedged bond

SOLUTION

(a) 1-Bromo-2-methylpropane. Its common name is isobutyl bromide.
(b) (E)-4-Bromo-3-methyl-2-pentene or *trans*-4-bromo-3-methyl-2-pentene.
(c) (S)-2-Bromononane.

Practice Problem 7.1

Write the IUPAC name for each compound.

(a) $(CH_3)_2C{=}CHCH_2Cl$ (b) [cyclohexane with CH_3 and Br on the same carbon] (c) $CH_3CHClCH_2Cl$ (d) [skeletal structure with Cl]

Of all the fluoroalkanes, the **chlorofluorocarbons (CFCs)** manufactured under the tradename **Freons** are the most widely known. The CFCs are nontoxic, nonflammable, odorless, and noncorrosive and were ideal replacements for the hazardous compounds used as heat-transfer media in refrigeration systems. Among the CFCs most widely used for this purpose were trichlorofluoromethane (CCl_3F, Freon-11) and dichlorodifluoromethane (CCl_2F_2, Freon-12).

The CFCs found wide use as industrial cleaning solvents to prepare surfaces for coatings, to remove cutting oils and waxes from millings, and to remove protective coatings. The CFCs were also used as propellants for aerosol sprays.

Concern about the environmental impact of CFCs arose in the 1970s when it was shown that more than 4.5×10^5 kg/yr of CFCs were being emitted into the atmosphere. Then, in 1974 Sherwood Rowland of the University of California, Irvine, and Mario Molina of the Massachusetts Institute of Technology announced their theory of ozone destruction by these compounds. When released into the air, CFCs escape to the lower atmosphere, but, because of their inertness, they do not decompose there. Slowly, they find their way to the stratosphere where they absorb ultraviolet radiation from the sun and then decompose. As they do so, they set up a chemical reaction that leads to destruction of the stratospheric ozone layer, which acts as a shield for the earth against excess ultraviolet radiation from the sun. An increase in ultraviolet radiation reaching the earth, it is theorized, may lead to destruction of certain crops and agricultural species, and even increased incidence of skin cancer in sensitive individuals.

The results of this concern were two conventions—one in Vienna in 1985 and the other in Montreal in 1987—held by the United Nations Environmental Program. The 1987 meeting produced the "Montreal Protocol," which set limits on the production and use of ozone-depleting CFCs and urged complete phaseout of their production by the year 1996. The fact that an international agreement on the environment that set limits on the production of any substance could be reached is indeed amazing and bodes well for the health of our planet.

Rowland, Molina, and Paul Crutzen, a Dutch chemist at the Max Planck Institute for Chemistry, Germany, were awarded the Nobel Prize for Chemistry in 1995. As noted in the award citation by the Royal Swedish Academy of Sciences, "By explaining the chemical mechanisms that affect the thickness of the ozone layer, these three researchers have contributed to our salvation from a global environmental problem that could have catastrophic consequences."

CHEMICAL CONNECTIONS

Artificial Blood

Today, the best known use of fluorocarbons is in nonstick cookware. In the future, this same class of compounds may serve a life-saving role as a temporary substitute for blood.

Blood picks up oxygen using hemoglobin (an iron-containing protein) in red blood cells. Whole blood can transport approximately 20 mL of oxygen per 100 mL, or 20 vol % oxygen. Fluorinated hydrocarbons, such as perfluorodecalin and perfluorotripropylamine, are able to dissolve up to 50 vol % oxygen.

trans-Perfluorodecalin

$$CF_3CF_2CF_2-N(CF_2CF_2CF_3)-CF_2CF_2CF_3$$

Perfluorotripropylamine

Fluorocarbons cannot be used directly in the bloodstream because they are nonpolar and do not mix with water. They can, however, be used as a fluorocarbon-water emulsion. One such preparation is Fluosol DA, a 20% emulsion of perfluorodecalin and perfluorotripropylamine in water. It now appears that mammals can live with a majority of their hemoglobin replaced by fluorocarbons such as these. It has been found, for example, that dogs treated with a 25% perfluorotributylamine-water emulsion replacing 70% of their blood can still live a normal life span.

In 1989, a new drug application was approved in the United States for perflubron (LiquiVent), a water emulsion of 1-bromoperfluorooctane, as a blood substitute.

$$CF_3(CF_2)_6CF_2Br$$

1-Bromoperfluorooctane

In 1996, clinical trials on this blood substitute were undertaken for patients suffering from a variety of life-threatening lung illnesses and injuries, including infections, near drowning, and smoke inhalation. If results continue to be promising, we can expect that organic chemists will be able to synthesize new perfluorocarbon compounds optimized for use as blood substitutes (*The Merck Index,* 12th ed., #4221 and #7299).

The chemical industry is responding by developing non-ozone-depleting alternatives to CFCs, among which are the hydrofluorocarbons (HFCs) and hydrochlorofluorocarbons (HCFCs)

$$CH_2F-CF_3 \qquad CHCl_2-CF_3$$

HFC-134a HCFC-123

We must not assume, however, that haloalkanes are introduced into the environment only by human action. It is estimated, for example, that the annual production of bromomethane from natural sources is 2.7×10^8 kg, largely from marine algae, giant kelp, and volcanoes. Furthermore, global emission of chloromethane is estimated to be 4.5×10^9 kg/yr, most of it from terrestrial and marine biomass. These haloalkanes, however, have only short atmospheric lifetimes, and only a tiny fraction of them reach the stratosphere. The CFCs are the problem; they have longer atmospheric lifetimes, reach the stratosphere, and do their damage there.

7.2 Nucleophilic Aliphatic Substitution

Nucleophile An atom or group of atoms that can donate a pair of electrons to another atom or group of atoms to form a new covalent bond.

Nucleophilic substitution A reaction in which one nucleophile is substituted for another.

A **nucleophile** (nucleus-loving reagent) is any reagent that donates an unshared pair of electrons to form a new covalent bond. **Nucleophilic substitution** is any reaction in which one nucleophile is substituted for another. In the following general equations, $Nu{:}^-$ is the nucleophile, X is the leaving group, and substitution takes place on an sp^3 hybridized carbon atom.

$$\underset{\text{Nucleophile}}{Nu{:}^-} + -\overset{|}{\underset{|}{C}}-\overset{\text{leaving group}}{X} \xrightarrow{\text{nucleophilic substitution}} -\overset{|}{\underset{|}{C}}-Nu + {:}X^-$$

Nucleophilic substitution reactions of alkyl halides can lead to a wide variety of new functional groups, several of which are illustrated in Table 7.1. As you study the entries in this table, note these points.

1. If the nucleophile is negatively charged, as for example OH^- and Cl^-, then the atom donating the pair of electrons in the substitution reaction becomes neutral in the product.
2. If the nucleophile is uncharged, as for example NH_3 and CH_3OH, then the atom donating the pair of electrons in the substitution reaction becomes positively charged in the product.

Example 7.2

Complete these nucleophilic substitution reactions.

(a) $CH_3CH_2CH_2CH_2Br + Na^+OH^- \longrightarrow$
(b) $CH_3CH_2CH_2CH_2Cl + NH_3 \longrightarrow$

Solution

(a) Hydroxide ion is the nucleophile, and bromide is the leaving group.

$$\underset{\text{1-Bromobutane}}{CH_3CH_2CH_2CH_2Br} + \underset{\text{Sodium hydroxide}}{Na^+OH^-} \longrightarrow \underset{\text{1-Butanol}}{CH_3CH_2CH_2CH_2OH} + \underset{\text{Sodium bromide}}{Na^+Br^-}$$

TABLE 7.1 Some Nucleophilic Substitution Reactions

Reaction: $Nu:^- + CH_3Br \longrightarrow CH_3Nu + Br^-$			
Nucleophile		**Product**	**Class of Compound Formed**
$HO:^-$	⟶	CH_3OH	an alcohol
$RO:^-$	⟶	CH_3OR	an ether
$HS:^-$	⟶	CH_3SH	a thiol (a mercaptan)
$RS:^-$	⟶	CH_3SR	a sulfide (a thioether)
$:I:^-$	⟶	$CH_3I:$	an alkyl iodide
$:NH_3$	⟶	$CH_3NH_3^+$	an alkylammonium ion
HOH	⟶	$CH_3\overset{+}{O}(H)—H$	a protonated alcohol
CH_3OH	⟶	$CH_3\overset{+}{O}(H)—CH_3$	a protonated ether

(b) Ammonia is the nucleophile, and chloride is the leaving group.

$$CH_3CH_2CH_2CH_2Cl + NH_3 \longrightarrow CH_3CH_2CH_2CH_2NH_3^+Cl^-$$

1-Chlorobutane, Ammonia, Butylammonium chloride

Practice Problem 7.2

Complete these nucleophilic substitution reactions.

(a) cyclopentyl—Br $+ CH_3CH_2S^-Na^+ \longrightarrow$ (b) cyclopentyl—Br $+ CH_3CO_2^-Na^+ \longrightarrow$

7.3 MECHANISMS OF NUCLEOPHILIC ALIPHATIC SUBSTITUTION

There are two limiting mechanisms for nucleophilic substitutions. A fundamental difference between them is the timing of bond breaking between carbon and the leaving group and bond forming between carbon and the nucleophile. At one extreme, the two processes are concerted, meaning that bond breaking and bond forming occur simultaneously. This mechanism is designated $\mathbf{S_N2}$, where S stands for *S*ubstitution, N stands for *N*ucleophilic, and 2 stands for *bi*molecular. A **bimolecular reaction** is one in which two reactants (in this case, an alkyl halide and a nucleophile) are involved in the reaction leading to the transition state of the rate-limiting step.

Bimolecular reaction A reaction in which two species are involved in the reaction leading to the transition state of the rate-limiting step.

A. S_N2 Mechanism

Following is an S_N2 mechanism for the reaction of hydroxide ion and bromomethane to form methanol and bromide ion. The nucleophile is shown attacking the reactive center from the side of the molecule opposite from the leaving group; that is, reaction involves **backside attack** by the nucleophile. Backside attack at a stereocenter results in inversion of configuration.

MECHANISM An S_N2 Reaction

The nucleophile attacks the reactive center from the side opposite the leaving group, which results in inversion of configuration.

$HO^- + CH_3Br \longrightarrow [\,\overset{\delta-}{HO}\cdots C \cdots \overset{\delta-}{Br}\,] \longrightarrow HO{-}CH_3 + Br^-$

Transition state with simultaneous bond breaking and bond forming

Figure 7.1 shows a potential energy diagram for an S_N2 reaction. There is a single transition state and no reactive intermediate.

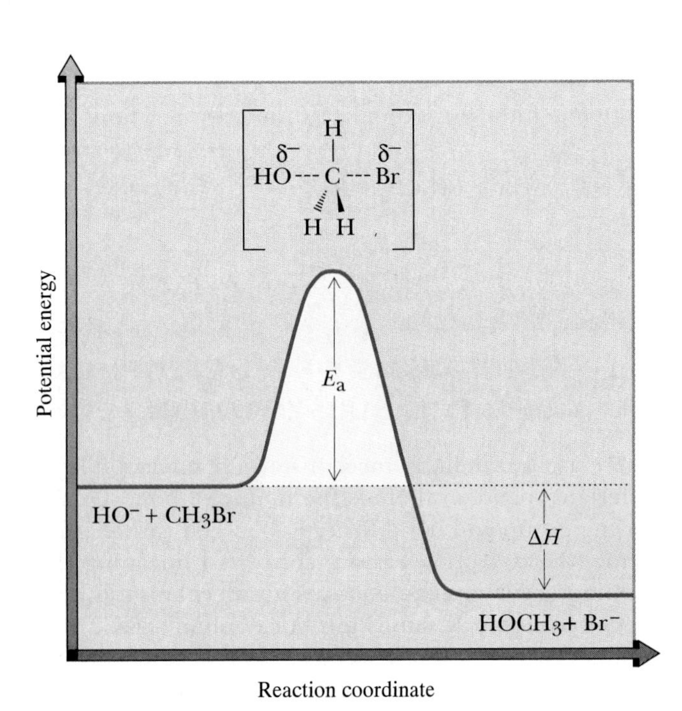

Figure 7.1
Potential energy diagram for an S_N2 reaction. There is one transition state and no reactive intermediate.

B. S_N1 Mechanism

In the S_N1 mechanism, bond breaking between carbon and the leaving group is entirely completed before bond forming with the nucleophile begins. In the designation **S_N1**, S stands for *S*ubstitution, N stands for *N*ucleophilic, and 1 stands for *uni*molecular. A **unimolecular reaction** is one in which only one reactant (in this case only the alkyl halide) is involved in the reaction leading to the transition state of the rate-limiting step. An S_N1 reaction is illustrated by the reaction between 2-bromo-2-methylpropane and methanol to form *tert*-butyl methyl ether and HBr. Ionization of the C—Br bond in Step 1 gives a 3° carbocation intermediate, which then reacts with methanol, the nucleophile, in Step 2 to give an oxonium ion. Attack of the nucleophile occurs with equal probability from either face of the planar carbon in the carbocation intermediate. Loss of H^+ by proton transfer from the oxonium ion in Step 3 gives the ether. This last step is an acid-base reaction after the S_N1 reaction is completed.

Unimolecular reaction A reaction in which only one species is involved in the reaction leading to the transition state of the rate-limiting step.

MECHANISM An S_N1 Reaction

Step 1: Ionization of C—X bond gives a carbocation intermediate.

A carbocation intermediate; carbon is trigonal planar

Step 2: Methanol, the nucleophile, reacts at either face of the planar carbon in the carbocation intermediate.

Step 3: Proton transfer to methanol (the solvent) gives *tert*-butyl methyl ether.

Figure 7.2 shows a potential energy diagram for the S_N1 reaction of 2-bromo-2-methylpropane and methanol. There is one transition state leading to formation of the carbocation intermediate in Step 1 and a second transition state for reaction of the carbocation intermediate with methanol to give the oxonium ion in Step 2. The reaction leading to formation of the carbocation intermediate crosses the higher potential energy barrier and, therefore, is the rate-limiting step.

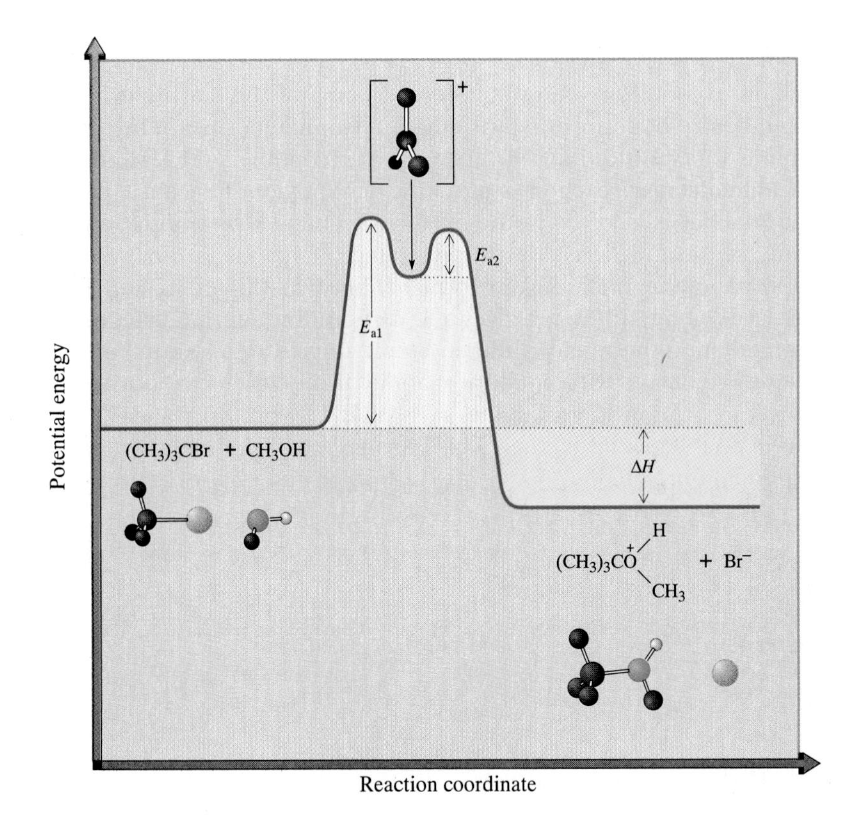

Figure 7.2
Potential energy diagram for the S_N1 reaction of 2-bromo-2-methylpropane and methanol. There are two transition states and one reactive intermediate. Step 1, which crosses the higher potential energy barrier, is the rate-limiting step.

If an S_N1 reaction is carried out at a tetrahedral stereocenter, the major product is a racemic mixture. We can illustrate this result with the following example. On ionization, the R enantiomer forms an achiral carbocation intermediate. Attack of the nucleophile from the left face of the carbocation intermediate gives the S enantiomer; attack from the right face gives the R enantiomer. Because attack of the nucleophile occurs with equal probability from either face of the planar carbocation intermediate, the R and S enantiomers are formed in equal amounts, and the product is a racemic mixture.

H—C—Cl $\xrightarrow{-Cl^-}$ [C^+ H Cl] $\xrightarrow[-H^+]{CH_3OH}$ CH_3O—C—H + H—C—OCH_3

Cl Cl Cl Cl

R enantiomer — Planar carbocation (achiral) — S enantiomer, R enantiomer: A racemic mixture

7.4 FACTORS THAT INFLUENCE THE RATE OF S_N1 AND S_N2 REACTIONS

Let us now examine some of the experimental evidence on which these two contrasting mechanisms are based. As we do, we consider the following questions:

A. What effect does the structure of the nucleophile have on the rate of reaction?
B. What effect does the structure of the alkyl halide have on the rate of reaction?
C. What effect does the structure of the leaving group have on the rate of reaction?
D. What effect does the solvent have on the reaction mechanism?

A. Structure of the Nucleophile

The effectiveness of a nucleophile in displacing a leaving group can be measured by the rate at which it attacks a reference compound under a standardized set of experimental conditions. For example, **relative nucleophilicities** for a series of nucleophiles can be established by measuring the rate at which each displaces bromide ion from ethyl bromide in ethanol at 25°C.

Relative Nucleophilicity The relative rate at which a nucleophile reacts in a reference nucleophilic substitution reaction.

$$CH_3CH_2Br + :Nu^- \longrightarrow CH_3CH_2Nu + Br^-$$

Table 7.2 gives several examples of good, moderate, and poor nucleophiles.

B. Structure of the Alkyl Halide

S_N1 reactions are governed mainly by **electronic factors,** namely the relative stabilities of carbocation intermediates. S_N2 reactions, on the other hand, are governed

TABLE 7.2 Examples of Common Nucleophiles and Their Relative Effectiveness

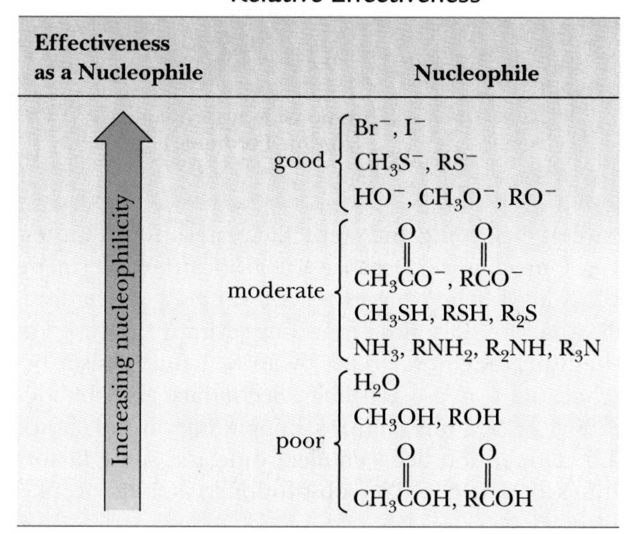

Effectiveness as a Nucleophile		Nucleophile
Increasing nucleophilicity (↑)	good	Br^-, I^-
		CH_3S^-, RS^-
		HO^-, CH_3O^-, RO^-
	moderate	$CH_3\overset{O}{\overset{\|}{C}}O^-$, $R\overset{O}{\overset{\|}{C}}O^-$
		CH_3SH, RSH, R_2S
		NH_3, RNH_2, R_2NH, R_3N
	poor	H_2O
		CH_3OH, ROH
		$CH_3\overset{O}{\overset{\|}{C}}OH$, $R\overset{O}{\overset{\|}{C}}OH$

Steric hindrance The ability of an atom or group of atoms, because of their size, to hinder access to a reaction site.

mainly by **steric factors,** and their transition states are particularly sensitive to crowding about the site of reaction.

1. *Relative stabilities of carbocations.* As we learned in Section 6.3A, among carbocations, tertiary carbocations are the most stable (lowest activation energy for their formation), whereas the methyl carbocation is the least stable (highest activation energy for its formation). Therefore, tertiary alkyl halides are most likely to react by carbocation formation, secondary alkyl halides are intermediate in reactivity, and primary and methyl halides never react by carbocation formation.
2. *Steric hindrance.* To complete a substitution reaction, the nucleophile must approach the substitution center and begin to form a new covalent bond to it. If we compare the ease of approach to the substitution center of a methyl halide with that to a tertiary alkyl halide, we see that approach is considerably easier in the case of the methyl halide. The backside of the substitution center of the methyl halide is screened by three hydrogen atoms, whereas the backside of the substitution center of a tertiary alkyl halide is screened by three alkyl groups. We use the term *steric hindrance* to indicate the ability of groups, because of their size, to hinder access to a reaction site within a molecule.

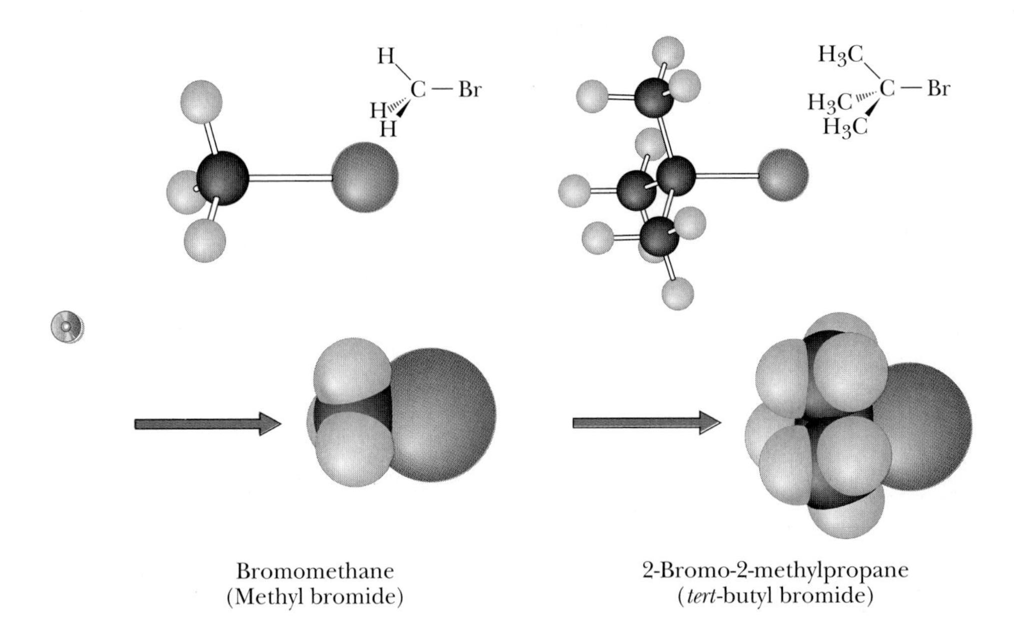

Bromomethane
(Methyl bromide)

2-Bromo-2-methylpropane
(*tert*-butyl bromide)

Given the competition between electronic and steric factors, we find that tertiary alkyl halides react by an S_N1 mechanism because 3° carbocation intermediates are particularly stable; tertiary alkyl halides never react by an S_N2 mechanism. Methyl halides and primary alkyl halides have little crowding around the reaction site and react by an S_N2 mechanism; they never react by an S_N1 mechanism because methyl and primary carbocations are so unstable. Secondary alkyl halides may be made to react by either S_N1 or S_N2 mechanisms, depending on the choice of nucleophile and solvent. The competition between electronic and steric factors and their effects on relative rates of nucleophilic substitution reactions of alkyl halides are summarized in Figure 7.3.

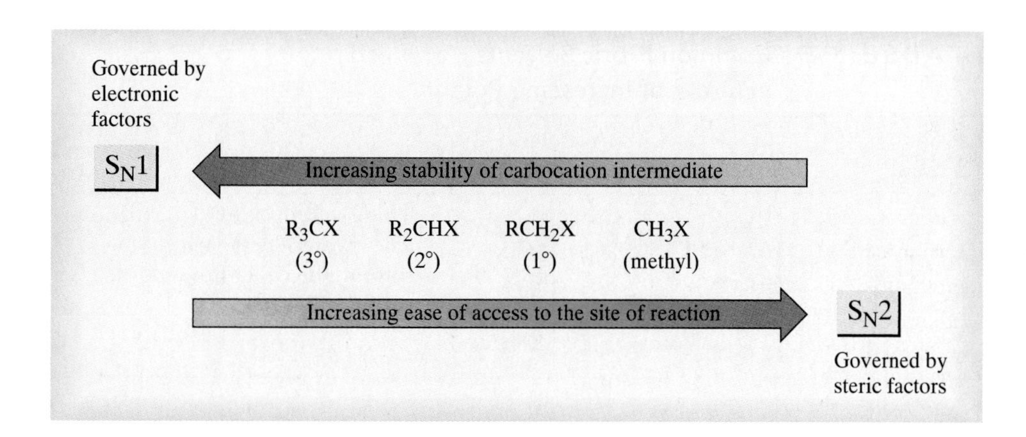

Figure 7.3 Effect of electronic and steric factors in competition between S_N1 and S_N2 reactions of alkyl halides.

C. Leaving Group

In the transition state for nucleophilic substitution on an alkyl halide, the leaving group develops a partial negative charge in both S_N1 and S_N2 reactions; therefore, the ability of a group to function as a leaving group is related to how stable it is as an anion. The most stable anions and the best leaving groups are the conjugate bases of strong acids. Thus, we can use the information on the relative strengths of organic and inorganic acids in Table 2.1 to determine which anions are the best leaving groups. This order is shown here.

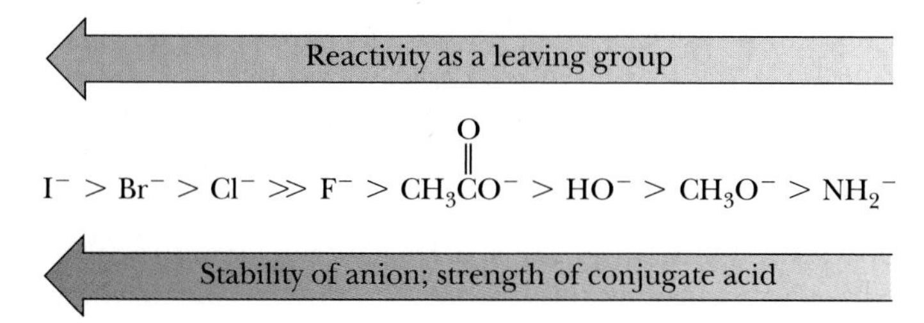

The best leaving groups in this series are the halogens, I^-, Br^-, and Cl^-. Hydroxide ion, OH^-, methoxide ion, CH_3O^-, and amide ion, NH_2^-, are such poor leaving groups that they rarely, if ever, are displaced in nucleophilic aliphatic substitution.

D. Solvent

Solvents provide the medium in which the reactants are dissolved and in which nucleophilic substitution reactions take place. Common solvents for these reactions are divided into two groups: protic solvents and aprotic solvents. **Protic solvents** contain —OH groups. The common protic solvents for nucleophilic substitution reactions are water, low-molecular-weight alcohols, and low-molecular-weight carboxylic acids (Table 7.3).

Protic solvent A hydrogen bond donor solvent, as for example water, ethanol, and acetic acid.

Protic solvents dissolve ionic compounds because of electrostatic interaction between their partially negatively charged oxygens and solute cations, and between their partially positively charged hydrogens and solute anions. These same properties aid in ionization of C—X bonds to give carbocations; thus, protic solvents are good solvents in which to carry out S_N1 reactions.

TABLE 7.3 Common Protic Solvents, Arranged in Order of Increasing Polarity

Protic Solvent	Structure	Polarity of Solvent	Notes
water	H_2O	↑ Increasing	These solvents favor S_N1 reactions. The greater the polarity of the solvent, the easier it is to form carbocations in it.
formic acid	HCO_2H		
methanol	CH_3OH		
ethanol	CH_3CH_2OH		
acetic acid	CH_3CO_2H		

Aprotic solvent A solvent that cannot serve as a hydrogen bond donor, as for example acetone, diethyl ether, and dichloromethane.

Aprotic solvents do not contain —OH groups. The aprotic solvents most commonly used for nucleophilic substitution reactions are given in Table 7.4. Dimethyl sulfoxide and acetone are classified as polar aprotic solvents. Dichloromethane and diethyl ether are classified as nonpolar aprotic solvents. The aprotic solvents listed in Table 7.4 are particularly good ones in which to carry out S_N2 reactions.

The factors favoring S_N1 or S_N2 reactions are summarized in Table 7.5. Also shown is the change in configuration when nucleophilic substitution takes place at a stereocenter.

7.5 An Analysis of Several Nucleophilic Substitution Reactions

Predictions about the mechanism for a particular nucleophilic substitution reaction must be based on considerations of the structure of the alkyl halide, the nucleophile, the leaving group, and the solvent. Following are three nucleophilic substitution reactions and an analysis of each.

TABLE 7.4 Common Aprotic Solvents

Aprotic Solvent	Structure	Polarity of Solvent	Notes
dimethyl sulfoxide (DMSO)	$CH_3S(=O)CH_3$	↑ Increasing	These solvents favor S_N2 reactions. Although solvents at the top of this list are polar, formation of carbocations in them is far more difficult than in protic solvents.
acetone	$CH_3C(=O)CH_3$		
dichloromethane	CH_2Cl_2		
diethyl ether	$(CH_3CH_2)_2O$		

TABLE 7.5 Summary of S_N1 Versus S_N2 Reactions of Alkyl Halides

Type of Alkyl Halide	S_N2	S_N1
methyl CH_3X	**S_N2 favored.**	S_N1 never occurs. The methyl cation is so unstable, it is never observed in solution.
primary RCH_2X	**S_N2 favored.**	S_N1 rarely occurs. Primary cations are so unstable, they are rarely observed in solution.
secondary R_2CHX	**S_N2 favored** in aprotic solvents with good nucleophiles.	**S_N1 favored** in protic solvents with poor nucleophiles.
tertiary R_3CX	S_N2 does not occur because of steric hindrance around the substitution center.	**S_N1 favored** because of the ease of formation of tertiary carbocations.
substitution at a stereocenter	**Inversion of configuration.** The nucleophile attacks the stereocenter from the side opposite the leaving group.	**Racemization.** The carbocation intermediate is planar, and attack of the nucleophile occurs with equal probability from either face.

Nucleophilic Substitution 1

$$CH_3\underset{\displaystyle Cl}{\underset{|}{C}}HCH_2CH_3 + CH_3OH \longrightarrow CH_3\underset{\displaystyle OCH_3}{\underset{|}{C}}HCH_2CH_3 + HCl$$

R configuration

Methanol is a polar protic solvent and a good one in which to form carbocations. 2-Chlorobutane ionizes in it to form a 2° carbocation intermediate. Methanol is a weak nucleophile. From this analysis, we predict that reaction is by an S_N1 mechanism. The 2° carbocation intermediate then reacts with methanol as the nucleophile to give the observed product. The product is formed as a 50:50 mixture of R and S configurations; that is, it is formed as a racemic mixture.

Nucleophilic Substitution 2

$$CH_3\overset{\displaystyle CH_3}{\overset{|}{C}}HCH_2Br + Na^+I^- \xrightarrow[\text{DMSO}]{} CH_3\overset{\displaystyle CH_3}{\overset{|}{C}}HCH_2I + Na^+Br^-$$

This is a primary alkyl bromide in the presence of iodide ion, a good nucleophile. Because primary carbocations are so unstable that they are never formed in solution, an S_N1 reaction is not possible. Dimethyl sulfoxide (DMSO), a polar aprotic solvent, is a good solvent in which to carry out S_N2 reactions. From this analysis, we predict that this reaction is by an S_N2 mechanism.

Nucleophilic Substitution 3

$$CH_3\overset{\displaystyle Br}{\overset{|}{C}}HCH_2CH_3 + CH_3S^-Na^+ \xrightarrow[\text{acetone}]{} CH_3\overset{\displaystyle SCH_3}{\overset{|}{C}}HCH_2CH_3 + Na^+Br^-$$

S configuration

Bromide ion is a good leaving group on a secondary carbon. The methylsulfide ion is a good nucleophile. Acetone, a polar aprotic solvent, is a good medium in which to carry out S_N2 reactions but a poor medium in which to carry out S_N1 reactions. We predict that reaction is by an S_N2 mechanism and that the product formed has the R configuration.

EXAMPLE 7.3

Write the product for each nucleophilic substitution reaction, and predict the mechanism by which it is formed.

(a) $\text{C}_5\text{H}_9\text{Cl (chlorocyclopentane)} + CH_3OH \xrightarrow{\text{methanol}}$

(b) $CH_3(CH_2)_5CH(I)CH_3 + CH_3C(=O)O^-Na^+ \xrightarrow{\text{DMSO}}$
R configuration

SOLUTION

(a) Methanol is a poor nucleophile and a polar protic solvent. Ionization of the alkyl chloride forms a 2° carbocation intermediate. Predict an S_N1 mechanism.

$$\text{C}_5\text{H}_9\text{Cl} + CH_3OH \xrightarrow[\text{methanol}]{S_N1} \text{C}_5\text{H}_9\text{OCH}_3 + HCl$$

(b) Iodide is a good leaving group on a secondary carbon. Acetate ion is a moderate nucleophile. DMSO is a particularly good solvent for S_N2 reactions. Predict substitution by an S_N2 mechanism with inversion of configuration at the stereocenter.

$$CH_3C(=O)O^- + CH_3(CH_2)_5CH(CH_3)\text{—I} \xrightarrow[\text{DMSO}]{S_N2} CH_3C(=O)O\text{—}CH(CH_3)(CH_2)_5CH_3 + I^-$$

R configuration → S configuration

Practice Problem 7.3

Write the product for each nucleophilic substitution reaction and predict the mechanism by which it is formed.

(a) $(CH_3)_3C\text{-cyclohexyl-Br} + Na^+SH^- \xrightarrow{\text{acetone}}$

(b) $CH_3CH(Cl)CH_2CH_3 + HC(=O)OH \xrightarrow{\text{formic acid}}$
R configuration

7.6 β-ELIMINATION

In the presence of a strong base, such as hydroxide ion or ethoxide ion, halogen can be removed from one carbon of an alkyl halide and hydrogen from an adjacent carbon to form a carbon-carbon double bond in a reaction called **dehydrohalogenation.** The carbon bearing the halogen is called the α-carbon, and the adjacent carbon is called the β-carbon.

Dehydrohalogenation Removal of —H and —X from adjacent carbons; a type of β-elimination.

$$\underset{\text{An alkyl halide}}{\overset{\beta\ \ \alpha}{-\underset{H}{C}-\underset{X}{C}-}} + \underset{\text{Base}}{CH_3CH_2O^-Na^+} \xrightarrow[CH_3CH_2OH]{} \underset{\text{An alkene}}{>C=C<} + CH_3CH_2OH + Na^+X^-$$

Removal of a small molecule, such as HCl, HBr, or HI, from a larger molecule is called an **elimination reaction,** and, because the elements of HX are removed from adjacent carbon atoms, this type of elimination reaction is called a **β-elimination.**

β-Elimination reaction Removal of atoms or groups of atoms from two adjacent carbon atoms, as for example removal of H and X from an alkyl halide or H and OH from an alcohol, to form a carbon-carbon double bond.

Following are three examples of base-promoted β-elimination reactions. In the second and third examples, base is a reactant but is shown over the reaction arrow.

$$\underset{\text{1-Bromodecane}}{CH_3(CH_2)_7\overset{\beta}{C}H_2\overset{\alpha}{C}H_2Br} + \underset{\text{Sodium ethoxide}}{CH_3CH_2O^-Na^+} \longrightarrow \underset{\text{1-Decene}}{CH_3(CH_2)_7CH=CH_2} + CH_3CH_2OH + Na^+Br^-$$

$$\underset{\text{2-Bromo-2-methylbutane}}{CH_3CH_2\underset{CH_3}{\overset{Br}{C}}CH_3} \xrightarrow[CH_3CH_2OH]{CH_3CH_2O^-Na^+} \underset{\substack{\text{2-Methyl-2-butene}\\\text{(major product)}}}{CH_3CH=\underset{CH_3}{C}CH_3} + \underset{\text{2-Methyl-1-butene}}{CH_3CH_2\underset{CH_3}{C}=CH_2}$$

1-Bromo-1-methylcyclopentane (ring with Br and CH_3) $\xrightarrow[CH_3OH]{CH_3O^-Na^+}$ 1-Methylcyclopentene (ring with CH_3) (major product) + Methylenecyclopentane (ring with $=CH_2$)

When isomeric alkenes are obtained in the dehydrohalogenation of an alkyl halide, the alkene having the greater number of substituents on the double bond generally predominates. This generalization is known as **Zaitsev's rule.**

Zaitsev's rule The major product from a β-elimination reaction contains the more highly substituted carbon-carbon double bond.

EXAMPLE 7.4

Predict the β-elimination product(s) formed when each bromoalkane is treated with sodium ethoxide in ethanol. If two products might be formed, predict which is the major product.

(a) $CH_3\overset{CH_3}{C}H\underset{Br}{C}HCH_3$ (b) $CH_3\underset{CH_3}{C}HCH_2CH_2Br$

SOLUTION

(a) There are two nonequivalent β-carbons in this bromoalkane, and two alkenes are possible. 2-Methyl-2-butene, the more substituted alkene, is the major product.

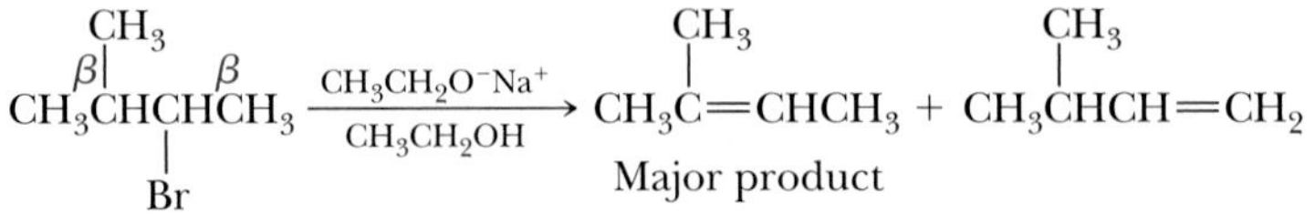

$$\underset{\text{Br}}{CH_3\overset{\beta}{C}H\overset{CH_3}{\underset{|}{C}}H\overset{\beta}{C}H_3}\xrightarrow[CH_3CH_2OH]{CH_3CH_2O^-Na^+} \underset{\text{Major product}}{CH_3\overset{CH_3}{C}{=}CHCH_3} + CH_3\overset{CH_3}{C}HCH{=}CH_2$$

(b) There is only one β-carbon in this bromoalkane, and only one alkene is possible.

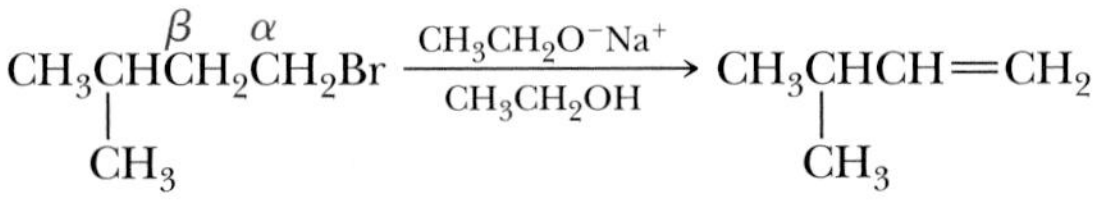

$$CH_3\overset{\beta}{\underset{CH_3}{C}}H\overset{\alpha}{C}H_2CH_2Br \xrightarrow[CH_3CH_2OH]{CH_3CH_2O^-Na^+} CH_3\underset{CH_3}{C}HCH{=}CH_2$$

Practice Problem 7.4

Predict the β-elimination product(s) formed when each chloroalkane is treated with sodium ethoxide in ethanol. If two products might be formed, predict which is the major product.

(a) 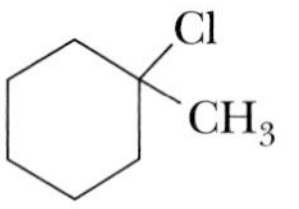 1-chloro-1-methylcyclohexane (ring carbon bearing Cl and CH_3) (b) 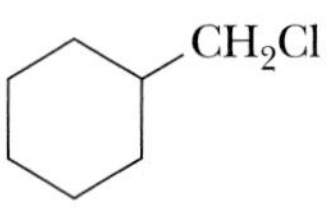 cyclohexyl–CH_2Cl (c) 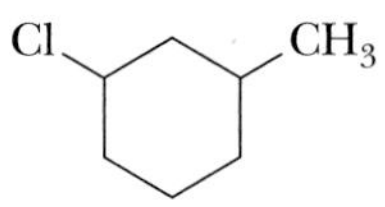 cyclohexane ring bearing Cl and CH_3

7.7 Mechanisms of β-Elimination

Chemists propose two limiting mechanisms for β-eliminations. A fundamental difference between them is the timing of the bond-breaking and bond-forming steps.

A. E1 Mechanism

At one extreme, breaking of the C—X bond is complete before any reaction occurs with base to lose a hydrogen and form the carbon-carbon double bond. This mechanism is designated **E1** where E stands for *E*limination and 1 stands for the fact that only *one* species (in this case the alkyl halide) is involved in the transition state of the rate-limiting step. The reaction of 2-bromo-2-methylpropane to form 2-methylpropene is an example of an E1 reaction.

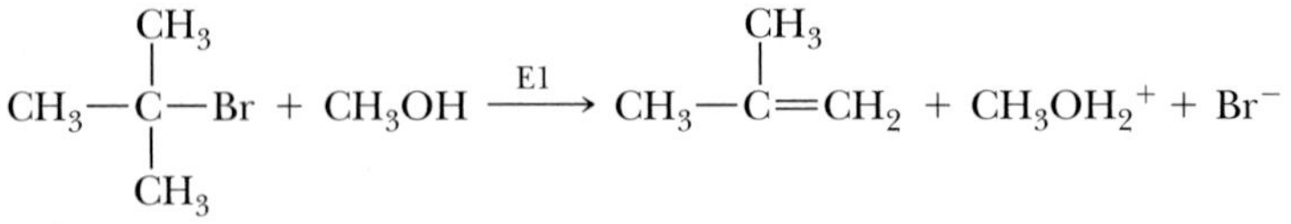

$$\underset{\text{2-Bromo-2-methylpropane}}{CH_3{-}\overset{CH_3}{\underset{CH_3}{C}}{-}Br} + CH_3OH \xrightarrow{E1} \underset{\text{2-Methylpropene}}{CH_3{-}\overset{CH_3}{C}{=}CH_2} + CH_3OH_2^+ + Br^-$$

An E1 mechanism involves two steps. Step 1 is a slow, rate-limiting ionization of the C—X bond to form a carbocation intermediate. Note that this is the same first step as in an S_N1 mechanism. In Step 2, the carbocation intermediate then reacts with a base to lose a hydrogen and form the alkene.

MECHANISM E1 Reaction of 2-Bromo-2-methylpropane

Step 1: Rate-limiting ionization of the C—Br bond forms a carbocation intermediate.

$$CH_3-\underset{Br}{\overset{CH_3}{C}}-CH_3 \xrightarrow{\text{slow, rate-limiting}} CH_3-\underset{+}{\overset{CH_3}{C}}-CH_3 + :\ddot{Br}:^-$$

A 3° carbocation intermediate

Step 2: Proton transfer from the carbocation intermediate to a molecule of solvent (in this case, methanol) gives the alkene.

$$(H)(H_3C)\ddot{O}: + H-CH_2-\underset{+}{\overset{CH_3}{C}}-CH_3 \xrightarrow{\text{fast}} (H)(H_3C)\overset{+}{O}-H + CH_2=\overset{CH_3}{C}-CH_3$$

B. E2 Mechanism

At the other extreme is a concerted process, designated **E2:** E because it is an *E*limination reaction, and 2 because *two* species, in this case the base and the alkyl halide, are involved in the transition state of the rate-limiting step. In the E2 mechanism, there is only one step. Base removes a hydrogen as a proton from the β-carbon at the same time as the double bond forms and the leaving group departs.

MECHANISM E2 Reaction of 1-Bromopropane

All bond-breaking and bond-forming steps occur simultaneously.

$$CH_3CH_2\ddot{O}:^- + H-\overset{CH_3}{CH}-CH_2-\ddot{Br}: \longrightarrow CH_3CH_2\ddot{O}-H + CH_3CH=CH_2 + :\ddot{Br}:^-$$

For both E1 and E2 reactions, the major product is that formed following Zaitsev's rule (Section 7.6), as illustrated by this E2 reaction.

$$\underset{\text{2-Bromohexane}}{CH_3(CH_2)_3\overset{Br}{C}HCH_3} + CH_3O^-Na^+ \xrightarrow[CH_3OH]{E2}$$

$$\underset{\substack{\text{2-Hexene}\\(74\%)}}{CH_3CH_2CH_2CH=CHCH_3} + \underset{\substack{\text{1-Hexene}\\(26\%)}}{CH_3CH_2CH_2CH_2CH=CH_2} + CH_3OH + Na^+Br^-$$

These generalizations about β-elimination reactions of alkyl halides are summarized in Table 7.6.

TABLE 7.6 Summary of E1 Versus E2 Reactions of Alkyl Halides

Halide	Reaction	Comments
primary	E2	Main reaction with strong bases such as HO^- and RO^-.
(RCH_2—X)	E1	Primary cations are rarely formed in solution; therefore, E1 reactions of primary halides are rarely observed.
secondary	E2	Main reaction with strong bases such as HO^- and RO^-.
(R_2CH—X)	E1	Common in reactions with weak bases such as $CH_3CO_2^-$.
tertiary	E2	Main reaction with strong bases such as HO^- and RO^-.
(R_3C—X)	E1	Main reaction with weak bases such as $CH_3CO_2^-$ and ROH.

Example 7.5

Predict whether each β-elimination reaction proceeds predominantly by an E1 or E2 mechanism. Write a structural formula for the major organic product.

(a) $CH_3C(CH_3)(Cl)CH_2CH_3 + NaOH \xrightarrow[H_2O]{80°C}$ (b) $CH_3C(CH_3)(Cl)CH_2CH_3 \xrightarrow[CH_3CO_2H]{}$

Solution

(a) A 3° alkyl halide is heated with a strong base. Elimination by an E2 reaction predominates to give 2-methyl-2-butene as the major product.

$$CH_3C(CH_3)(Cl)CH_2CH_3 + NaOH \xrightarrow[H_2O]{80°C} CH_3C(CH_3){=}CHCH_3 + NaCl + H_2O$$

(b) A 3° alkyl halide dissolved in acetic acid, a solvent that promotes formation of carbocations, forms a 3° carbocation that then loses a proton to give 2-methyl-2-butene as the major product. Reaction is by an E1 mechanism.

$$CH_3C(CH_3)(Cl)CH_2CH_3 \xrightarrow[CH_3CO_2H]{} CH_3C(CH_3){=}CHCH_3 + HCl$$

Practice Problem 7.5

Predict whether each elimination reaction proceeds predominantly by an E1 or E2 mechanism. Write a structural formula for the major organic product.

(a) $CH_3CH_2CH(I)CH_2CH_3 + CH_3O^-Na^+ \xrightarrow[methanol]{}$

(b) (cyclohexane with Cl and H_3C substituents) $+ CH_3CH_2O^-Na^+ \xrightarrow[CH_3CH_2OH]{}$

SUMMARY

Haloalkanes contain a halogen covalently bonded to an sp^3 hybridized carbon. In the IUPAC system, halogen atoms are named as fluoro-, chloro-, bromo-, or iodo- substituents and listed in alphabetical order with other substituents (Section 7.1A). In the common system, haloalkanes are named **alkyl halides.** Common names are derived by naming the alkyl group followed by the name of the halide as a separate word (Section 7.1B). Compounds of the type CHX_3 are named **haloforms.**

A **nucleophile** (Section 7.2) is any molecule or ion with an unshared pair of electrons that can be donated to another atom or ion to form a new covalent bond. An **S_N2 reaction** (Section 7.3A) occurs in one step. Departure of the leaving group is assisted by the incoming nucleophile, and both nucleophile and leaving group are involved in the transition state. An S_N2 reaction at a stereocenter proceeds with inversion of configuration.

An **S_N1 reaction** occurs in two steps (Section 7.3B). Step 1 is a slow, rate-limiting ionization of the C—X bond to form a carbocation intermediate, followed in Step 2 by rapid reaction of the carbocation intermediate with a nucleophile to complete the substitution. For S_N1 reactions taking place at a stereocenter, the major reaction occurs with racemization.

The **nucleophilicity** of a reagent is measured by the rate of its reaction in a reference nucleophilic substitution (Section 7.4A). S_N1 reactions are governed by **electronic factors,** namely the relative stabilities of carbocation intermediates. S_N2 reactions are governed by **steric factors,** namely the degree of crowding around the site of substitution.

The ability of a group to function as a leaving group is related to its stability as an anion (Section 7.4C). The most stable anions and the best leaving groups are the conjugate bases of strong acids.

Protic solvents contain —OH groups (Section 7.4D). Protic solvents interact strongly with polar molecules and ions and are good solvents in which to form carbocations; they favor S_N1 reactions. **Aprotic solvents** do not contain —OH groups. Common aprotic solvents are dimethyl sulfoxide, acetone, diethyl ether, and dichloromethane. Aprotic solvents do not interact as strongly with polar molecules and ions, and formation of carbocations in them is less likely; aprotic solvents favor S_N2 reactions.

Dehydrohalogenation, a type of **β-elimination reaction,** is the removal of H and X from adjacent carbon atoms (Section 7.6). β-Elimination to give the most highly substituted alkene is called **Zaitsev elimination.** An **E1 reaction** occurs in two steps: breaking the C—X bond to form a carbocation intermediate followed by loss of an H^+ to form the alkene. An **E2 reaction** occurs in one step: simultaneous reaction with base to remove an H^+, formation of the alkene, and departure of the leaving group.

KEY REACTIONS

1. Nucleophilic Aliphatic Substitution: S_N2 (Section 7.3A)

S_N2 reactions involve inversion of configuration at the substitution center. The nucleophile may be negatively charged or neutral.

$$:\ddot{I}:^- + \underset{CH_3}{\overset{CH_3CH_2}{H\text{—}C}}\text{—}\ddot{Cl}: \xrightarrow{S_N2} :\ddot{I}\text{—}CH_3\text{—}\overset{O}{\overset{\|}{C}}\text{—}O\text{—}\underset{CH_3}{\overset{CH_2CH_3}{C}}\text{—}H + Cl^-$$

$$CH_3\overset{O}{\overset{\|}{C}}\text{—}O^- +$$

2. Nucleophilic Aliphatic Substitution: S_N1 (Section 7.3B)

S_N1 reactions involve formation of a carbocation intermediate. An S_N1 reaction at a stereocenter gives a racemic product.

$$\text{1-chloro-1-methylcyclohexane } (CH_3, Cl) + CH_3CH_2OH \longrightarrow \text{1-ethoxy-1-methylcyclohexane } (CH_3, OCH_2CH_3) + HCl$$

3. β-Elimination: E1 (Section 7.7A)

E1 reactions involve elimination of atoms or groups of atoms from adjacent carbons. Reaction occurs in two steps and involves formation of a carbocation intermediate.

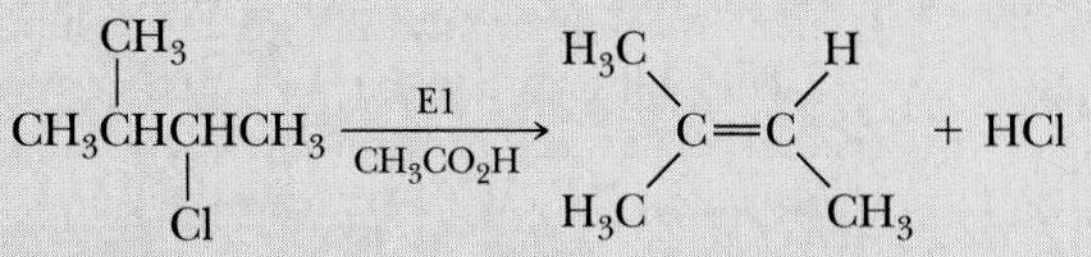

$$CH_3CH(CH_3)CH(Cl)CH_3 \xrightarrow[CH_3CO_2H]{E1} (H_3C)_2C{=}C(H)CH_3 + HCl$$

4. β-Elimination: E2 (Section 7.7B)

E2 reactions involve elimination of atoms or groups of atoms from adjacent carbon atoms. Reaction occurs in one step.

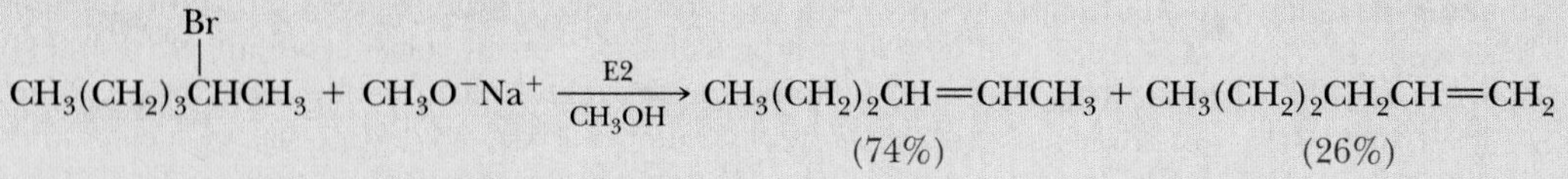

$$CH_3(CH_2)_3CH(Br)CH_3 + CH_3O^-Na^+ \xrightarrow[CH_3OH]{E2} \underset{(74\%)}{CH_3(CH_2)_2CH{=}CHCH_3} + \underset{(26\%)}{CH_3(CH_2)_2CH_2CH{=}CH_2}$$

Additional Problems

Nomenclature

7.6 Write the IUPAC name for each compound.

(a) $CH_2{=}CF_2$ **(b)** 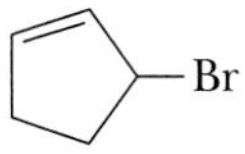 3-bromocyclopentene ring with Br **(c)**  $(CH_3)_2CHCH_2CH_2CH(Cl)CH_3$

(d) $Cl(CH_2)_6Cl$ **(e)** CF_2Cl_2 **(f)** 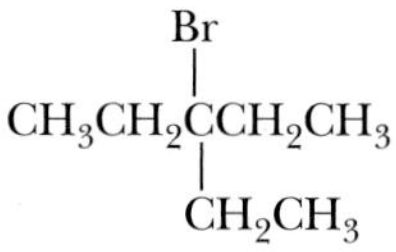 $CH_3CH_2C(Br)(CH_2CH_3)CH_2CH_3$

7.7 Write the IUPAC name for each compound. Be certain to include a designation of configuration in your answer.

(a) 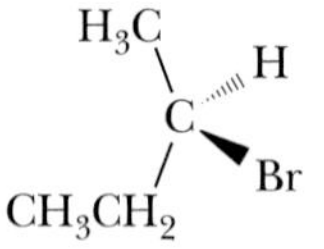 C with H_3C, H, Br, CH_3CH_2 **(b)** H_3C–cyclohexane–Br (chair) **(c)** Br, CH_3, H, C–C, H, H_3C, Br

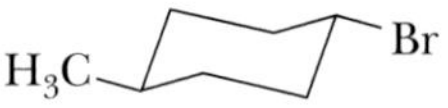

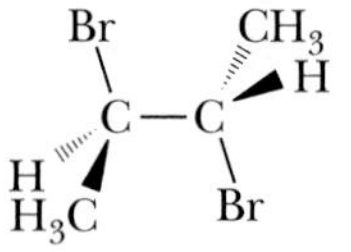

(d) 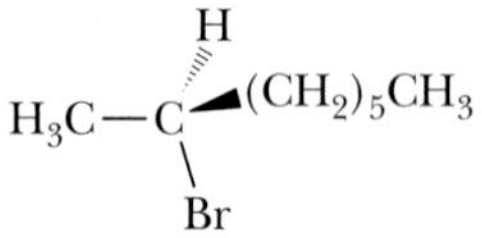 $H_3C{-}C(H)(Br){-}(CH_2)_5CH_3$ **(e)** 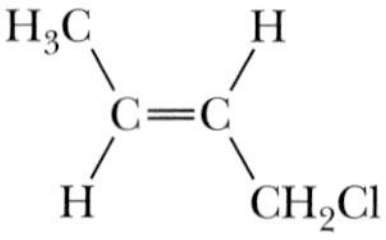 $H_3C(H)C{=}C(H)CH_2Cl$ **(f)** 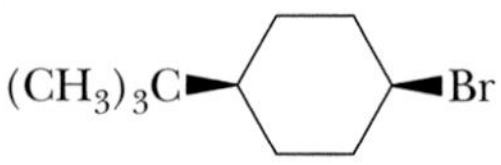 $(CH_3)_3C$–cyclohexane–Br

7.8 Draw a structural formula for each compound (given are IUPAC names).

(a) 3-Bromopropene **(b)** (R)-2-Chloropentane

(c) meso-3,4-Dibromohexane **(d)** *trans*-1-Bromo-3-isopropylcyclohexane

(e) 1,2-Dichloroethane **(f)** Bromocyclobutane

7.9 Draw a structural formula for each compound (given are common names).

(a) Isopropyl chloride **(b)** *sec*-Butyl bromide **(c)** Allyl iodide
(d) Methylene chloride **(e)** Chloroform **(f)** *tert*-Butyl chloride
(g) Isobutyl chloride

7.10 Which compounds are secondary (2°) alkyl halides?

(a) Isobutyl chloride **(b)** 2-Iodooctane
(c) *trans*-1-Chloro-4-methylcyclohexane

Synthesis of Alkyl Halides

7.11 What alkene or alkenes and reaction conditions give each alkyl halide in good yield?

(a) 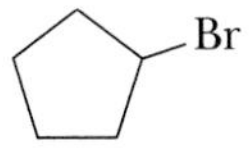 Br **(b)** $CH_3C(CH_3)(Br)CH_2CH_2CH_3$ **(c)** 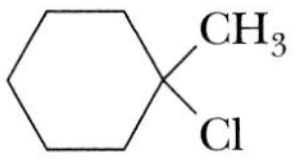 CH_3 Cl

7.12 Show reagents and conditions to bring about these conversions.

(a) 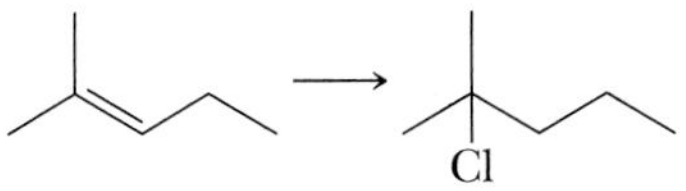 $\longrightarrow$ Cl **(b)** $CH_3CH_2CH{=}CH_2 \longrightarrow CH_3CH_2CH(I)CH_3$

(c) $CH_3CH{=}CHCH_3 \longrightarrow CH_3CH(Cl)CH_2CH_3$ **(d)** 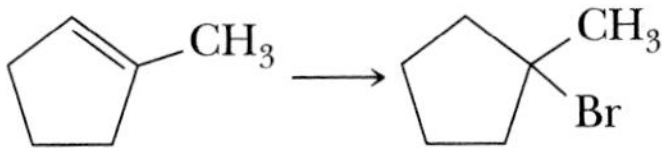 CH_3 $\longrightarrow$ CH_3 Br

Nucleophilic Aliphatic Substitution

7.13 Write structural formulas for these common organic solvents.

(a) Methylene chloride **(b)** Acetone **(c)** Ethanol
(d) Diethyl ether **(e)** Dimethyl sulfoxide

7.14 Arrange these protic solvents in order of increasing polarity.

(a) H_2O **(b)** CH_3CH_2OH **(c)** CH_3OH

7.15 Arrange these aprotic solvents in order of increasing polarity.

(a) Acetone **(b)** Pentane **(c)** Diethyl ether

7.16 From each pair, select the better nucleophile.

(a) H_2O or OH^- **(b)** $CH_3CO_2^-$ or OH^-
(c) CH_3SH or CH_3S^-

7.17 Which statements are true for S_N2 reactions of alkyl halides?

(a) Both the alkyl halide and the nucleophile are involved in the transition state.
(b) Reaction proceeds with inversion of configuration at the substitution center.
(c) Reaction proceeds with retention of optical activity.
(d) The order of reactivity is 3° > 2° > 1° > methyl.
(e) The nucleophile must have an unshared pair of electrons and bear a negative charge.
(f) The greater the nucleophilicity of the nucleophile, the greater the rate of reaction.

7.18 Complete these S_N2 reactions.

(a) $Na^+I^- + CH_3CH_2CH_2Cl \xrightarrow[\text{acetone}]{}$ **(b)** 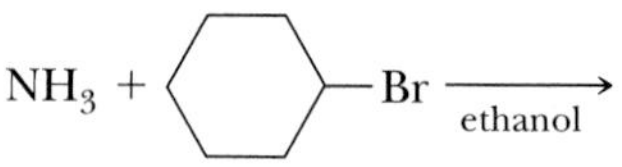 $NH_3 +$ (cyclohexyl)$-Br \xrightarrow[\text{ethanol}]{}$

(c) $CH_3CH_2O^-Na^+ + CH_2{=}CHCH_2Cl \xrightarrow[\text{ethanol}]{}$

7.19 Complete these S_N2 reactions.

(a) 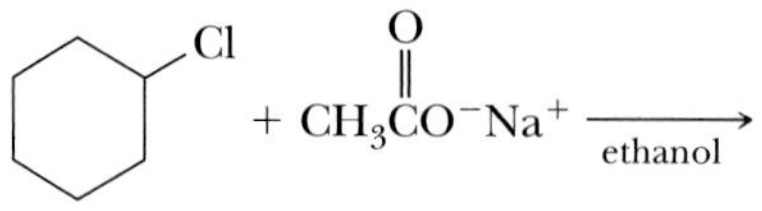 (cyclohexyl)$-Cl + CH_3\overset{\overset{\displaystyle O}{\|}}{C}O^-Na^+ \xrightarrow[\text{ethanol}]{}$

(b) $CH_3\overset{\overset{\displaystyle I}{|}}{C}HCH_2CH_3 + CH_3CH_2S^-Na^+ \xrightarrow[\text{acetone}]{}$

(c) $CH_3\overset{\overset{\displaystyle CH_3}{|}}{C}HCH_2CH_2Br + Na^+I^- \xrightarrow[\text{acetone}]{}$

(d) $(CH_3)_3N + CH_3I \xrightarrow[\text{acetone}]{}$

(e) 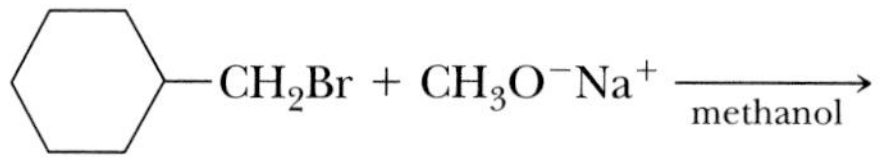 (cyclohexyl)$-CH_2Br + CH_3O^-Na^+ \xrightarrow[\text{methanol}]{}$

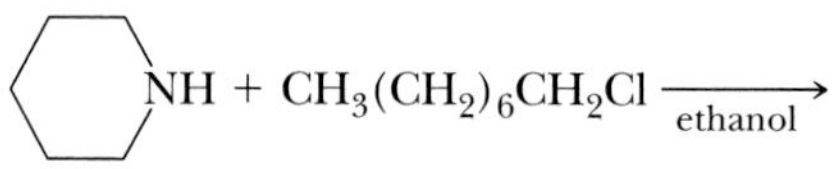

(f) 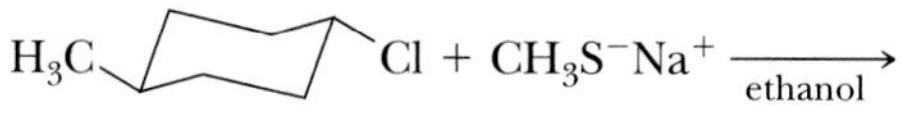 H_3C–(cyclohexane chair)–$Cl + CH_3S^-Na^+ \xrightarrow[\text{ethanol}]{}$

(g) (piperidine, ring NH) $NH + CH_3(CH_2)_6CH_2Cl \xrightarrow[\text{ethanol}]{}$

(h) (cyclopentyl)$-CH_2Cl + NH_3 \xrightarrow[\text{ethanol}]{}$

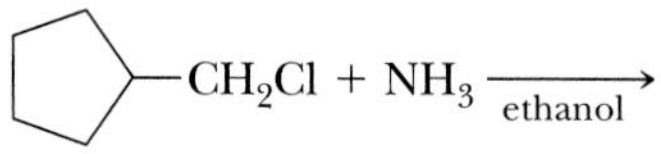

7.20 You were told that each reaction in Problem 7.19 proceeds by an S_N2 mechanism. Suppose that you were not told the mechanism. Describe how you could conclude from the structure of the alkyl halide, the nucleophile, and the solvent that each reaction is in fact an S_N2 reaction.

7.21 In these reactions, an alkyl halide is treated with a compound that has two nucleophilic sites. Select the more nucleophilic site in each part, and show the product of each S_N2 reaction.

(a) $HOCH_2CH_2NH_2 + CH_3I \xrightarrow[\text{ethanol}]{}$

(b) 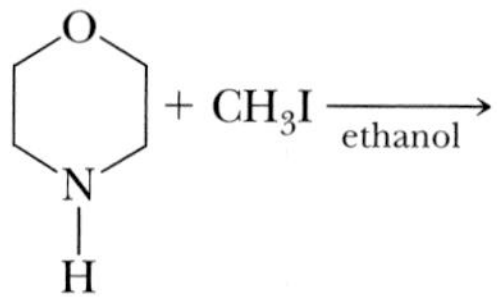 (morpholine: six-membered ring with O and N–H) $+ CH_3I \xrightarrow[\text{ethanol}]{}$ **(c)** $HOCH_2CH_2SH + CH_3I \xrightarrow[\text{ethanol}]{}$

7.22 Which statements are true for S_N1 reactions of alkyl halides?

(a) Both the alkyl halide and the nucleophile are involved in the transition state of the rate-limiting step.

(b) Reaction at a stereocenter proceeds with retention of configuration.

(c) Reaction at a stereocenter proceeds with loss of optical activity.

(d) The order of reactivity is 3° > 2° > 1° > methyl.

(e) The greater the steric crowding around the reactive center, the lower the rate of reaction.

(f) Rate of reaction is greater with good nucleophiles compared with poor nucleophiles.

7.23 Draw a structural formula for the product of each S_N1 reaction.

(a) 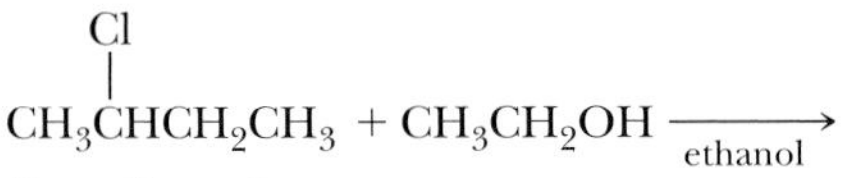 $CH_3CH(Cl)CH_2CH_3 + CH_3CH_2OH \xrightarrow{\text{ethanol}}$
S configuration

(b) 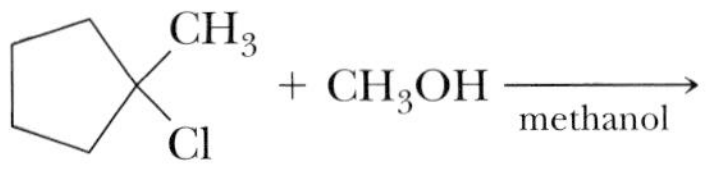 1-chloro-1-methylcyclopentane $+ CH_3OH \xrightarrow{\text{methanol}}$

(c) 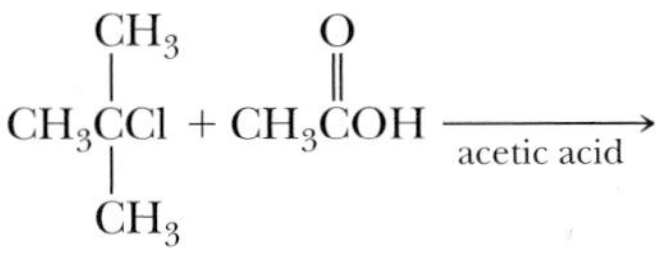 $(CH_3)_3CCl + CH_3COOH \xrightarrow{\text{acetic acid}}$

(d) 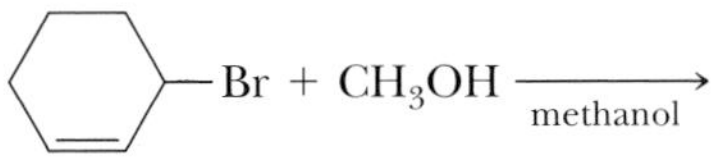 3-bromocyclohexene $+ CH_3OH \xrightarrow{\text{methanol}}$

7.24 You were told that each substitution reaction in Problem 7.23 proceeds by an S_N1 mechanism. Suppose that you were not told the mechanism. Describe how you could conclude from the structure of the alkyl halide, the nucleophile, and the solvent that each reaction is in fact an S_N1 reaction.

7.25 Select the member of each pair that undergoes nucleophilic substitution in aqueous ethanol more rapidly.

(a) 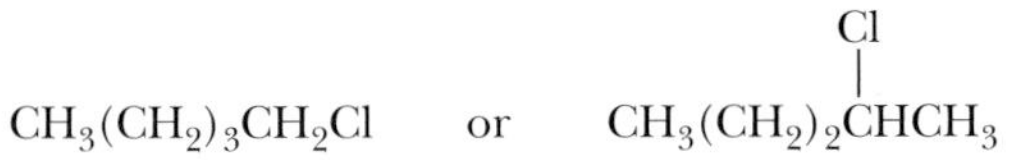 $CH_3(CH_2)_3CH_2Cl$ or $CH_3(CH_2)_2CH(Cl)CH_3$

(b) 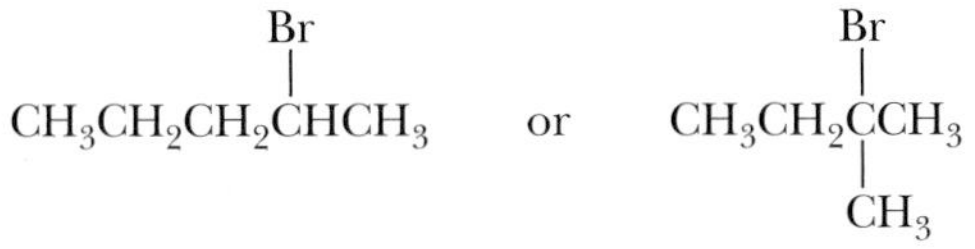 $CH_3CH_2CH_2CH(Br)CH_3$ or $CH_3CH_2C(Br)(CH_3)CH_3$

(c) 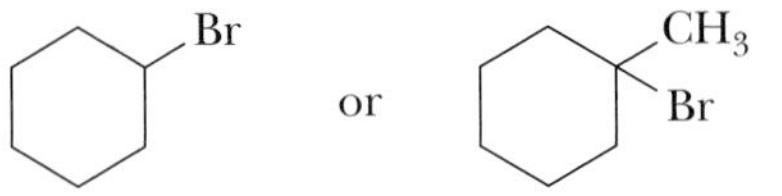 bromocyclohexane (Br) or 1-bromo-1-methylcyclohexane (CH_3, Br)

7.26 Propose a mechanism for the formation of the products (but not their relative percentages) in this reaction.

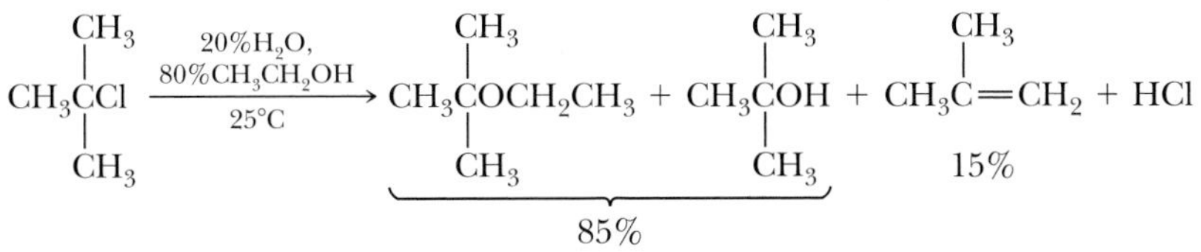

$$(CH_3)_3CCl \xrightarrow[25°C]{20\%H_2O,\ 80\%CH_3CH_2OH} \underbrace{(CH_3)_3COCH_2CH_3 + (CH_3)_3COH}_{85\%} + \underbrace{CH_3C(CH_3){=}CH_2}_{15\%} + HCl$$

7.27 The rate of reaction in Problem 7.26 increases by 140 times when carried out in 80% water : 20% ethanol compared with 40% water : 60% ethanol. Account for this rate difference.

7.28 Show how you might synthesize these compounds from an alkyl halide and a nucleophile:

(a) 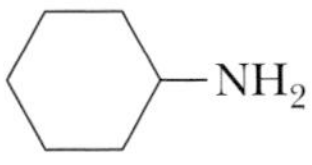 cyclohexyl$-NH_2$ **(b)** 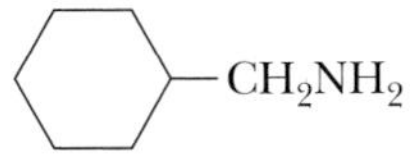 cyclohexyl$-CH_2NH_2$ **(c)** 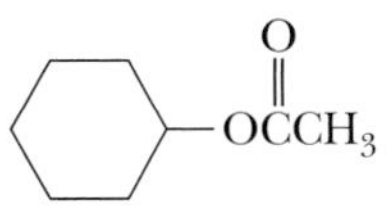 cyclohexyl$-OCCH_3$ (with C=O)

(d) $CH_3(CH_2)_3CH_2SH$ **(e)** 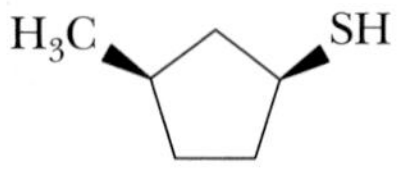 [cyclopentane with cis H_3C and SH substituents (wedge bonds)] **(f)** $CH_3CH_2OCH_2CH_3$

β-Eliminations

7.29 Draw structural formulas for the alkene(s) formed by treatment of each alkyl halide with sodium ethoxide in ethanol. Assume that elimination is by an E2 mechanism. Where two alkenes are possible, use Zaitsev's rule to predict which alkene is the major product.

(a) 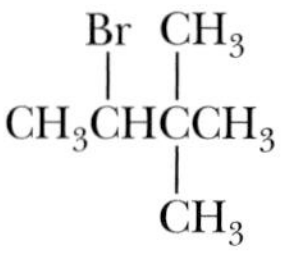 $CH_3CH(Br)C(CH_3)_2CH_3$ (Br on C-2; two CH_3 on C-3)

(b) 1-chloro-1-methylcyclohexane (cyclohexane ring bearing CH_3 and Cl on the same carbon)

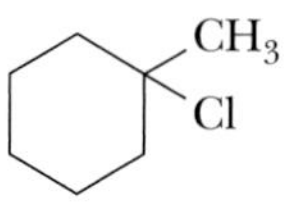

(c) cyclohexyl–$CH(Cl)CH_3$

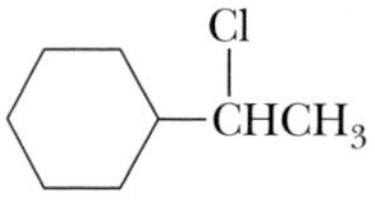

(d) 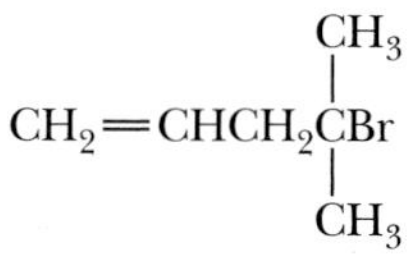 $CH_2{=}CHCH_2C(CH_3)_2Br$

7.30 Which alkyl halides undergo dehydrohalogenation to give alkenes that do not show cis-trans isomerism?

(a) 2-Chloropentane **(b)** 2-Chlorobutane **(c)** Chlorocyclohexane

(d) Isobutyl chloride

7.31 How many isomers, including cis-trans isomers, are possible for the major product of dehydrohalogenation of each haloalkane?

(a) 3-Chloro-3-methylhexane **(b)** 3-Bromohexane

7.32 What alkyl halide might you use as a starting material to produce each alkene in high yield and uncontaminated by isomeric alkenes?

(a) 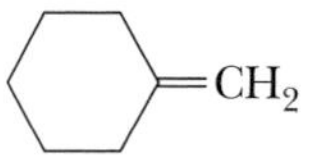 cyclohexane ring$=CH_2$ **(b)**  $CH_3CH(CH_3)CH_2CH{=}CH_2$

8

Alcohols, Ethers, and Thiols

Methanol is used as a fuel in cars of the type that race in the Indianapolis 500. Inset: A model of methanol. *(Stuart Westmorland/© Tony Stone Images)*

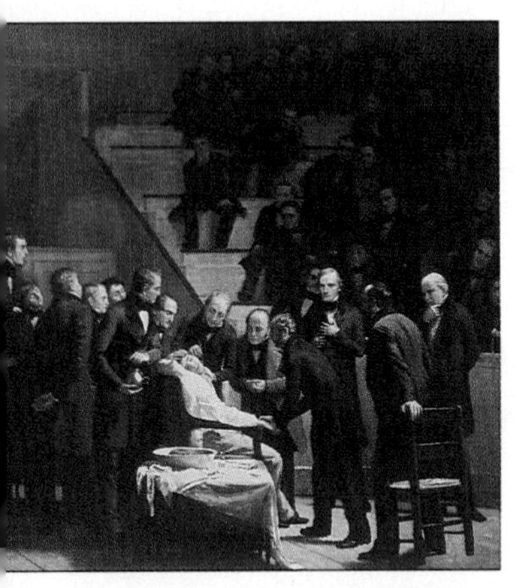

This painting by Robert Hinckley shows the first use of ether as an anesthetic in 1846. The patient, Gilbert Abbott, was having a tumor removed from his neck by Dr. Robert John Collins. The ether was administered by the dentist W. T. G. Morton, who discovered its anesthetic properties. *(Boston Medical Library in the Francis A. Countway Library of Medicine)*

In this chapter, we study the physical and chemical properties of alcohols and ethers, two classes of oxygen-containing compounds. We also study thiols, a class of sulfur-containing compounds. A thiol is like an alcohol in structure, except that it contains an —SH group rather than an —OH group.

CH_3CH_2OH	$CH_3CH_2OCH_2CH_3$	CH_3CH_2SH
Ethanol (an alcohol)	Diethyl ether (an ether)	Ethanethiol (a thiol)

Each of these three compounds is certainly familiar to you. Ethanol is the fuel additive in gasohol, the alcohol in alcoholic beverages, and an important industrial and laboratory solvent. Diethyl ether, or ether as it is also named, was the first inhalation anesthetic used in general surgery. It is also an important industrial and laboratory solvent. Ethanethiol, like other low-molecular-weight thiols, has a stench. Such smells as those from skunks, rotten eggs, and sewage are caused by thiols.

Alcohols are particularly important in laboratory and biochemical transformations of organic compounds. They can be converted into other types of compounds, such as alkenes, alkyl halides, aldehydes, ketones, carboxylic acids, and esters. Not only can alcohols be converted to these compounds, but they also can be prepared from them. Thus, alcohols play a central role in the interconversion of organic functional groups.

8.1 STRUCTURE

A. Alcohols

Alcohol A compound containing an —OH (hydroxyl) group bonded to an sp^3 hybridized carbon.

The functional group of an **alcohol** is an —OH (hydroxyl) group bonded to an sp^3 hybridized carbon atom (Section 1.7A). The oxygen atom of an alcohol is also sp^3 hybridized. Two sp^3 hybrid orbitals of oxygen form sigma bonds to atoms of carbon and hydrogen. The other two sp^3 hybrid orbitals of oxygen each contain an unshared pair of electrons. Figure 8.1 shows a Lewis structure and ball-and-stick model of methanol, CH_3OH, the simplest alcohol. The measured H—C—O bond angle in methanol is 108.9°, very close to the tetrahedral angle of 109.5°.

B. Ethers

Ether A compound containing an oxygen atom bonded to two carbon atoms.

The functional group of an **ether** is an atom of oxygen bonded to two carbon atoms. Figure 8.2 shows a Lewis structure and ball-and-stick model of dimethyl

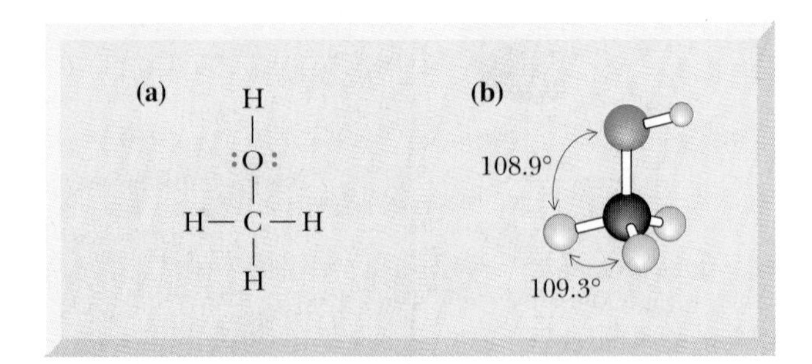

Figure 8.1
Methanol, CH_3OH. (a) Lewis structure and (b) ball-and-stick model.

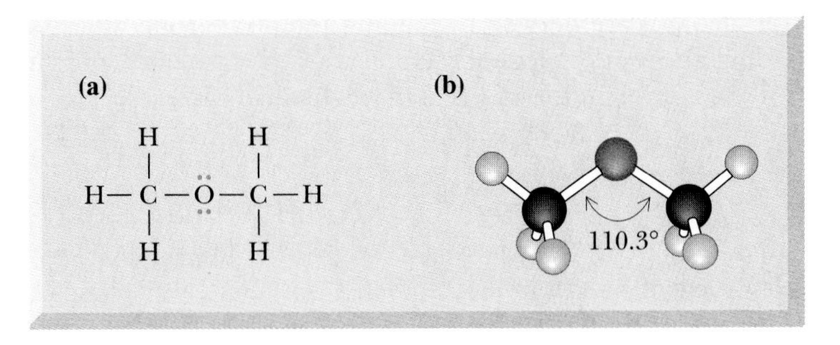

Figure 8.2
Dimethyl ether, CH_3OCH_3. (a) Lewis structure and (b) ball-and-stick model.

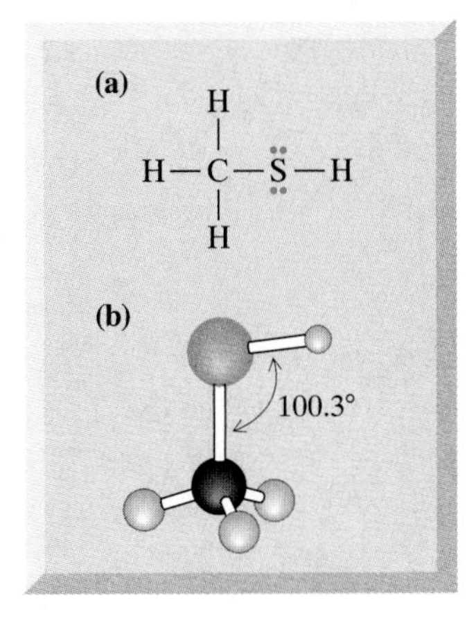

Figure 8.3
Methanethiol, CH_3SH. (a) Lewis structure and (b) ball-and-stick model.

ether, CH_3OCH_3, the simplest ether. In dimethyl ether, two sp^3 hybrid orbitals of oxygen form sigma bonds to carbon atoms. The other two sp^3 hybrid orbitals of oxygen each contain an unshared pair of electrons. The C—O—C bond angle in dimethyl ether is 110.3°, close to the predicted tetrahedral angle of 109.5°.

In ethyl vinyl ether, the ether oxygen is bonded to one sp^3 hybridized carbon and one sp^2 hybridized carbon.

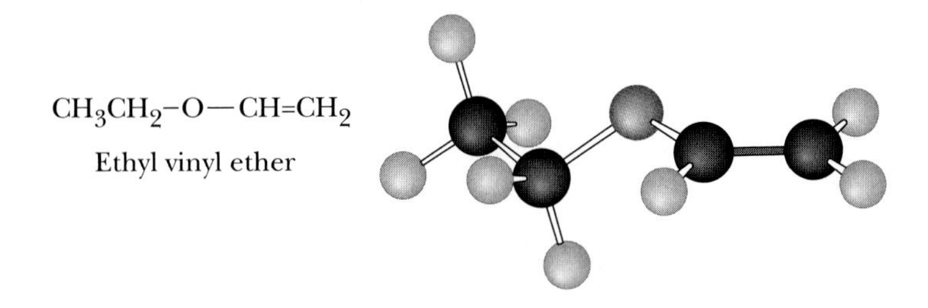

Thiol A compound containing an —SH (sulfhydryl) group.

C. Thiols

The functional group of a **thiol** is an —SH (sulfhydryl) group. Figure 8.3 shows a Lewis structure and a ball-and-stick model of methanethiol, CH_3SH, the simplest thiol. The C—S—H bond angle in methanethiol is 100.3°.

Mushrooms, onions, garlic, and coffee all contain sulfur compounds. One of these present in the aroma of coffee is

O CH_2SH

(Charles D. Winters)

8.2 NOMENCLATURE

A. Alcohols

In the IUPAC system, the longest chain of carbon atoms containing the —OH group is selected as the parent alkane and numbered from the end closer to the —OH group. To show that the compound is an alcohol, the suffix -e of the parent alkane is changed to -ol (Section 3.5), and a number is used to show the location of the —OH group. In the numbering of the parent chain, the location of the —OH group takes precedence over alkyl groups and halogen substituents. For cyclic alcohols, numbering begins at the carbon bearing the —OH group.

Common names for alcohols are derived by naming the alkyl group attached to —OH and then adding the word "alcohol." Here are IUPAC names and, in parentheses, common names for eight low-molecular-weight alcohols.

CH_3CH_2OH
Ethanol
(Ethyl alcohol)

$CH_3CH_2CH_2OH$
1-Propanol
(Propyl alcohol)

$CH_3CH(OH)CH_3$
2-Propanol
(Isopropyl alcohol)

$CH_3CH_2CH_2CH_2OH$
1-Butanol
(Butyl alcohol)

$CH_3CH_2CH(OH)CH_3$
2-Butanol
(*sec*-Butyl alcohol)

$CH_3CH(CH_3)CH_2OH$
2-Methyl-1-propanol
(Isobutyl alcohol)

$CH_3C(CH_3)_2OH$
2-Methyl-2-propanol
(*tert*-Butyl alcohol)

Cyclohexanol
(Cyclohexyl alcohol)

EXAMPLE 8.1

Write IUPAC names for these alcohols.

(a) $CH_3(CH_2)_6CH_2OH$ (b) $CH_3CH(CH_3)CH_2CH(OH)CH_3$ (c) [cyclohexane ring with OH (wedge) and CH_3 (hash)]

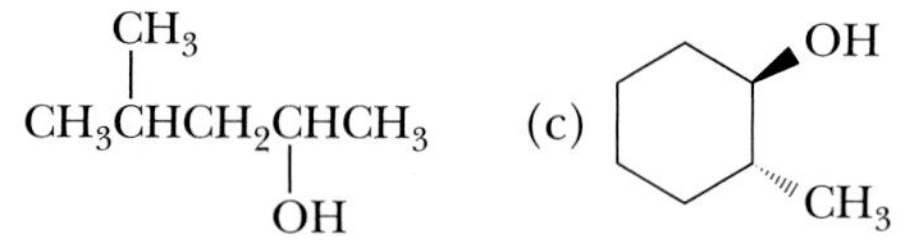

SOLUTION

(a) 1-Octanol (b) 4-Methyl-2-pentanol (c) *trans*-2-Methylcyclohexanol

Practice Problem 8.1

Write IUPAC names for these alcohols:

(a) [C bearing CH_2OH, H_3C, CH_2CH_3, H] (b) [cyclopentane with HO and CH_3 on same carbon] (c) $(CH_3)_3CCH_2OH$

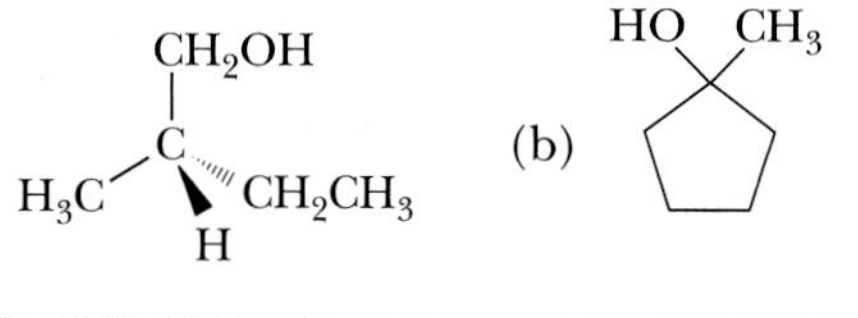

We classify alcohols as **primary (1°)**, **secondary (2°)**, or **tertiary (3°)** depending on whether the —OH group is on a primary, secondary, or tertiary carbon (Section 3.3C).

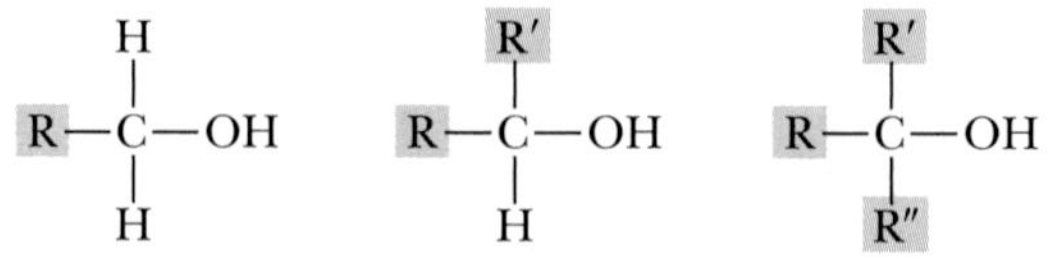

$R{-}CH_2{-}OH$
Primary (1°)

$R{-}CH(R'){-}OH$
Secondary (2°)

$R{-}C(R')(R''){-}OH$
Tertiary (3°)

EXAMPLE 8.2

Classify each alcohol as primary, secondary, or tertiary.

(a) cyclohexyl–C(OH)(H)CH_3 (b) $CH_3C(CH_3)_2OH$ (c) cyclopentyl–CH_2OH

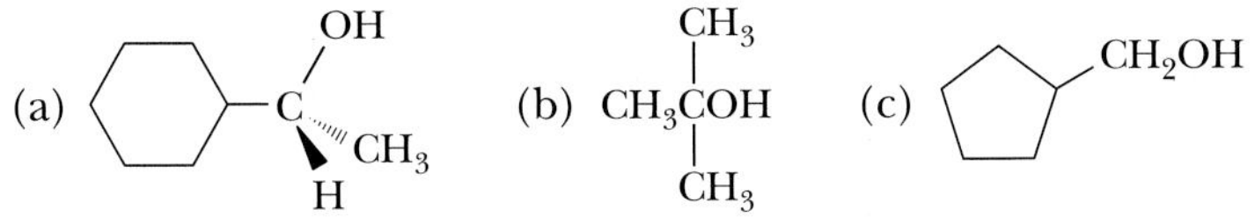

SOLUTION

(a) Secondary (2°) (b) Tertiary (3°) (c) Primary (1°)

Practice Problem 8.2

Classify each alcohol as primary, secondary, or tertiary.

(a) $CH_3C(CH_3)_2CH_2OH$ (b) cyclopropyl–OH (c) $CH_2{=}CHCH_2OH$ (d) 1-methylcyclopentanol (CH_3, OH)

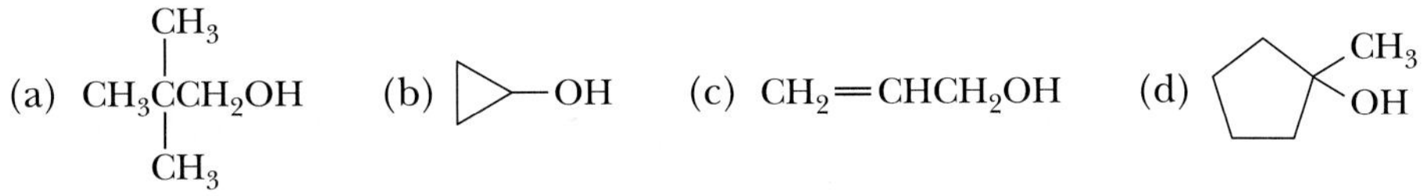

In the IUPAC system, a compound containing two hydroxyl groups is named as a **diol,** one containing three hydroxyl groups is named as a **triol,** and so on. In IUPAC names for diols, triols, and so on, the final -e (the suffix) of the parent alkane name is retained, as for example in the name 1,2-ethanediol. As with many organic compounds, common names for certain diols and triols have persisted. Compounds containing two hydroxyl groups on adjacent carbons are often referred to as **glycols** (Section 6.4B). Ethylene glycol and propylene glycol are synthesized from ethylene and propylene, respectively, hence their common names.

Glycol A compound with two hydroxyl (—OH) groups on adjacent carbons.

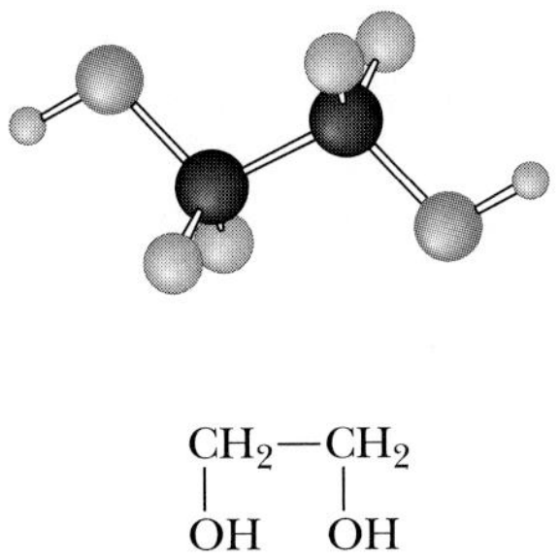

$CH_2(OH)—CH_2(OH)$
1,2-Ethanediol
(Ethylene glycol)

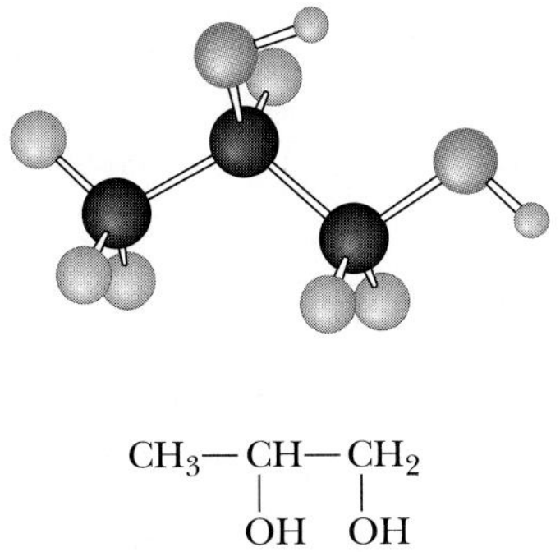

$CH_3—CH(OH)—CH_2(OH)$
1,2-Propanediol
(Propylene glycol)

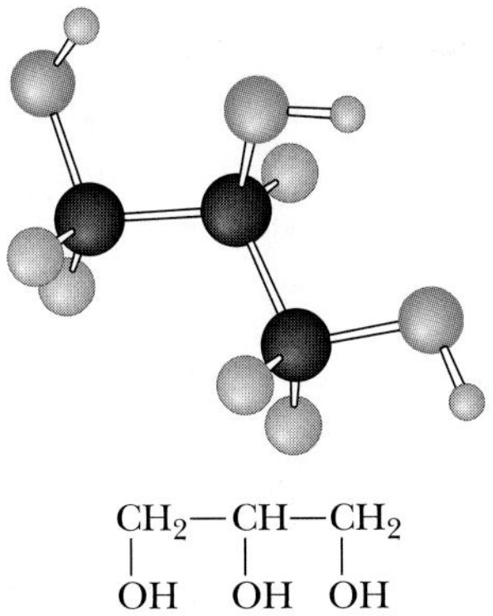

$CH_2(OH)—CH(OH)—CH_2(OH)$
1,2,3,-Propanetriol
(Glycerol, Glycerin)

Compounds containing —OH and C=C groups are often referred to as **unsaturated alcohols** because of the presence of the carbon-carbon double bond. In the IUPAC system, the parent alkane is numbered to give the —OH group the lowest possible number. The double bond is shown by changing the infix of the parent alkane from -an- to -en- (Section 3.5), and the alcohol is shown by changing the suffix of the parent alkane from -e to -ol. Numbers must be used to show the location of both the carbon-carbon double bond and the hydroxyl group. In numbering the parent chain of these difunctional molecules, priority is given to the functional group indicated by the suffix over the one indicated by an infix.

Example 8.3

Write the IUPAC name for each unsaturated alcohol.

(a) $CH_2{=}CHCH_2OH$ (b) 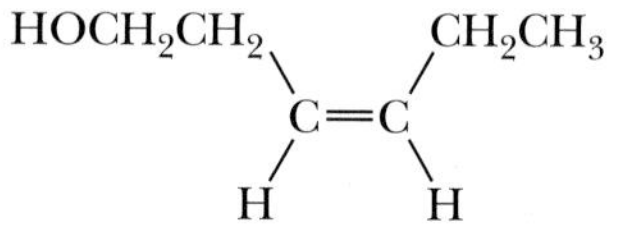 $HOCH_2CH_2$(H)C=C(H)CH_2CH_3

SOLUTION

(a) 2-Propen-1-ol. Its common name is allyl alcohol.
(b) *cis*-3-Hexen-1-ol. It is sometimes called leaf alcohol because of its occurrence in leaves of fragrant plants, including trees and shrubs (*The Merck Index,* 12th ed., #4737).

Practice Problem 8.3

Write the IUPAC name for each unsaturated alcohol.

(a) $CH_2{=}CHCH_2CH_2OH$ (b) 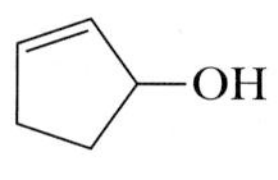 (cyclopentene ring)—OH

B. Ethers

Alkoxy group An —OR group, where R is an alkyl group.

In the IUPAC system, ethers are named by selecting the longest carbon chain as the parent alkane and naming the —OR group attached to it as an **alkoxy group.** Common names are derived by listing the alkyl groups attached to oxygen in alphabetical order and adding the word "ether." Following are the IUPAC names and, in parentheses, common names for three low-molecular-weight ethers.

$CH_3CH_2OCH_2CH_3$ — Ethoxyethane (Diethyl ether)

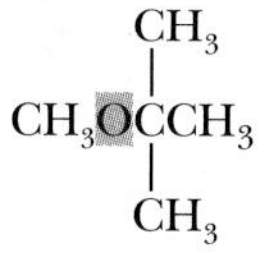

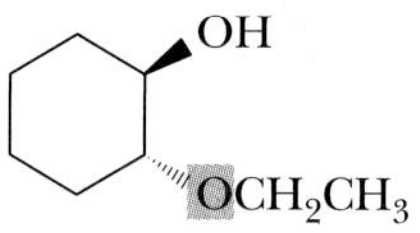

$CH_3OC(CH_3)_2CH_3$ — 2-Methoxy-2-methylpropane (methyl *tert*-butyl ether, MTBE)

(cyclohexane with OH and OCH_2CH_3) — *trans*-2-Ethoxycyclohexanol

Chemists almost invariably use common names for low-molecular-weight ethers. For example, although ethoxyethane is the IUPAC name for $CH_3CH_2OCH_2CH_3$, it is rarely called that but rather, diethyl ether, ethyl ether, or even more commonly, simply ether. The abbreviation for *tert*-butyl methyl ether, an ether that is becoming increasingly important as an octane-improving additive to gasolines, is MTBE after the common name of methyl *tert*-butyl ether.

Cyclic ether An ether in which the ether oxygen is one of the atoms of a ring.

Cyclic ethers are heterocyclic compounds in which the ether oxygen is one of the atoms in a ring. These ethers are generally known by their common names.

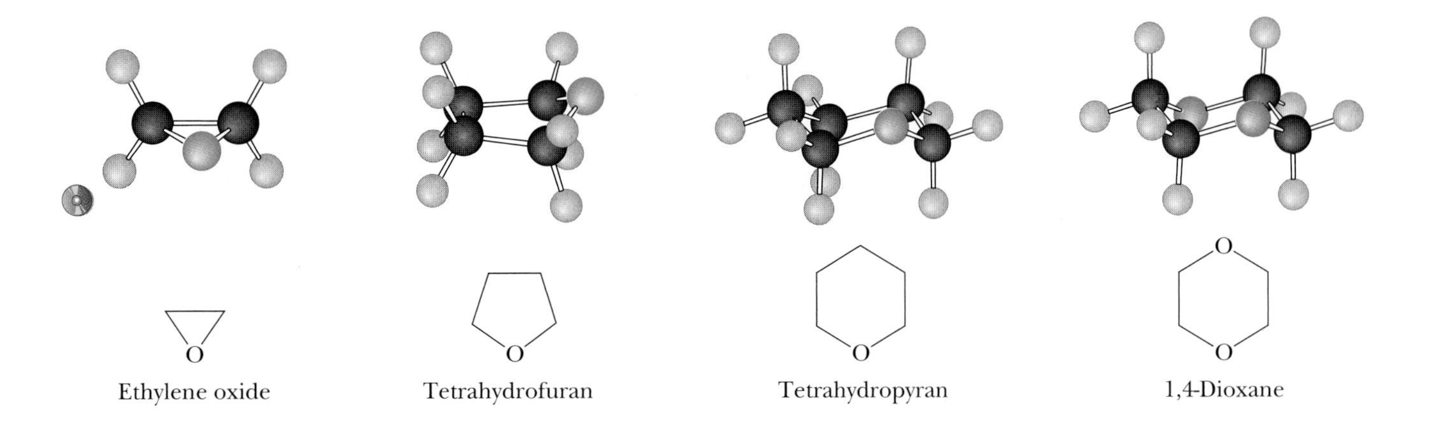

EXAMPLE 8.4

Write the IUPAC and common name for each ether.

(a) $CH_3C(CH_3)_2OCH_2CH_3$ (structure: CH_3 groups above and below the central C of $CH_3COCH_2CH_3$)

(b) cyclohexyl—O—cyclohexyl

SOLUTION

(a) 2-Ethoxy-2-methylpropane (*tert*-Butyl ethyl ether)
(b) Cyclohexoxycyclohexane (dicyclohexyl ether)

Practice Problem 8.4

Write the IUPAC and common name for each ether.

(a) $CH_3CH(CH_3)CH_2OCH_2CH_3$ (structure: CH_3 above the second C of $CH_3CHCH_2OCH_2CH_3$)

(b) cyclopentyl—OCH_3

C. Thiols

The sulfur analog of an alcohol is called a thiol (thi- from the Greek: *theion,* sulfur) or, in the older literature, a **mercaptan,** which literally means mercury capturing. Thiols react with Hg^{2+} in aqueous solution to give sulfide salts as insoluble precipitates. Thiophenol, C_6H_5SH, for example, gives $(C_6H_5S)_2Hg$.

Mercaptan A common name for any molecule containing an —SH group.

According to the IUPAC system, thiols are named by selecting as the parent alkane the longest chain of carbon atoms that contains the —SH group. To show that the compound is a thiol, the final -e in the name of the parent alkane is retained and the suffix -thiol is added. A number must be used to locate the —SH group on the parent chain.

Common names for simple thiols are derived by naming the alkyl group attached to —SH and adding the word "mercaptan." Here are IUPAC names and, in parentheses, common names for two low-molecular-weight thiols.

CH_3CH_2SH	$CH_3CH(CH_3)CH_2SH$	$HSCH_2CH_2OH$
Ethanethiol (Ethyl mercaptan)	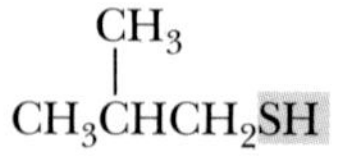 2-Methyl-1-propanethiol (Isobutyl mercaptan)	2-Mercaptoethanol

In compounds containing other functional groups, the presence of an —SH group is indicated by the prefix **mercapto-**. According to the IUPAC system, —OH takes precedence over —SH in both numbering and naming.

Sulfur analogs of ethers are named by using the word "sulfide" to show the presence of the —S— group. Following are common names of two sulfides:

CH_3SCH_3	$CH_3CH_2SCH(CH_3)CH_3$
Dimethyl sulfide	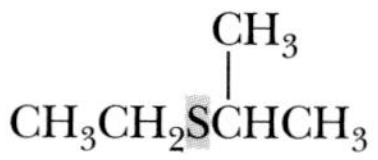 Ethyl isopropyl sulfide

EXAMPLE 8.5

Write names for these compounds.

(a) $CH_3CH_2CH_2CH_2CH_2SH$ (b) $CH_3CH_2SCH_2CH_3$ (c) 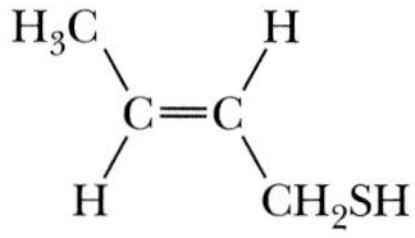 $H_3C(H)C{=}C(H)CH_2SH$ (trans: H_3C and CH_2SH on opposite sides)

SOLUTION

(a) 1-Pentanethiol (b) Diethyl sulfide (c) *trans*-2-Butene-1-thiol

Practice Problem 8.5

Write names for these compound.

(a) 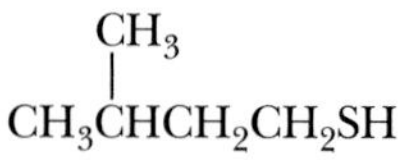 $CH_3CH(CH_3)CH_2CH_2SH$ (b) $CH_3SCH_2CH_3$ (c) 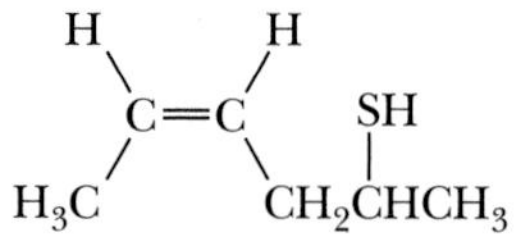 $H_3C(H)C{=}C(H)CH_2CH(SH)CH_3$ (cis: both H on the same side)

8.3 PHYSICAL PROPERTIES

A. Alcohols

Table 8.1 lists the boiling points and solubilities in water for five groups of alcohols and alkanes of similar molecular weight. Notice that, of the compounds compared in each group, the alcohol has the higher boiling point and is the more soluble in water. The higher boiling points of alcohols compared with alkanes of similar molecular weight is due to the fact that alcohols are polar molecules and are asso-

TABLE 8.1 Boiling Points and Solubilities in Water of Five Groups of Alcohols and Hydrocarbons of Similar Molecular Weight

Structural Formula	Name	Molecular Weight	bp (°C)	Solubility in Water
CH_3OH	methanol	32	65	infinite
CH_3CH_3	ethane	30	−89	insoluble
CH_3CH_2OH	ethanol	46	78	infinite
$CH_3CH_2CH_3$	propane	44	−42	insoluble
$CH_3CH_2CH_2OH$	1-propanol	60	97	infinite
$CH_3CH_2CH_2CH_3$	butane	58	0	insoluble
$CH_3CH_2CH_2CH_2OH$	1-butanol	74	117	8 g/100 g
$CH_3CH_2CH_2CH_2CH_3$	pentane	72	36	insoluble
$CH_3CH_2CH_2CH_2CH_2OH$	1-pentanol	88	138	2.3 g/100 g
$HOCH_2CH_2CH_2CH_2OH$	1,4-butanediol	90	230	infinite
$CH_3CH_2CH_2CH_2CH_2CH_3$	hexane	86	69	insoluble

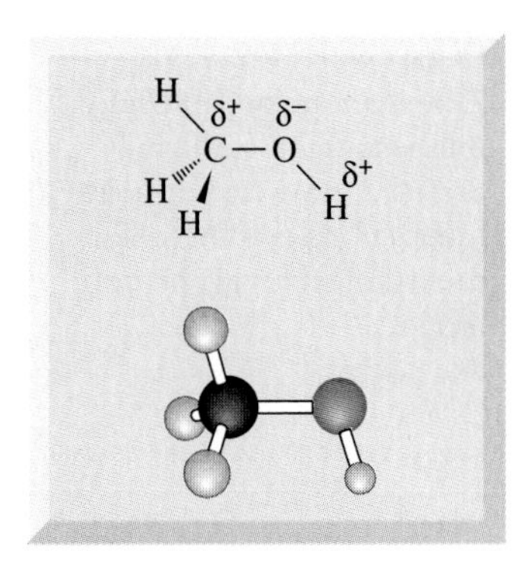

Figure 8.4
Polarity of the C—O—H bonds in alcohols.

ciated in the liquid state by a type of intermolecular attraction called **hydrogen bonding.** Figure 8.4 shows the polarity of the C—O—H bonds in an alcohol; there are partial positive charges on carbon and hydrogen and a partial negative charge on oxygen. The association of alcohols in the liquid state by hydrogen bonding is shown in Figure 8.5. The strength of hydrogen bonding between alcohol molecules is approximately 2 to 5 kcal/mol (8.4 to 21 kJ/mol). For comparison, the strength of the O—H covalent bond in an alcohol molecule is approximately 110 kcal/mol (460 kJ/mol). As can be seen by comparing these numbers, an O------H hydrogen bond is considerably weaker than an O—H covalent bond.

Hydrogen bonding The attractive force between a partial positive charge on hydrogen and partial negative charge of a nearby oxygen, nitrogen, or fluorine atom.

Because of hydrogen bonding between alcohol molecules in the liquid state, extra energy is required to separate each hydrogen-bonded alcohol molecule from

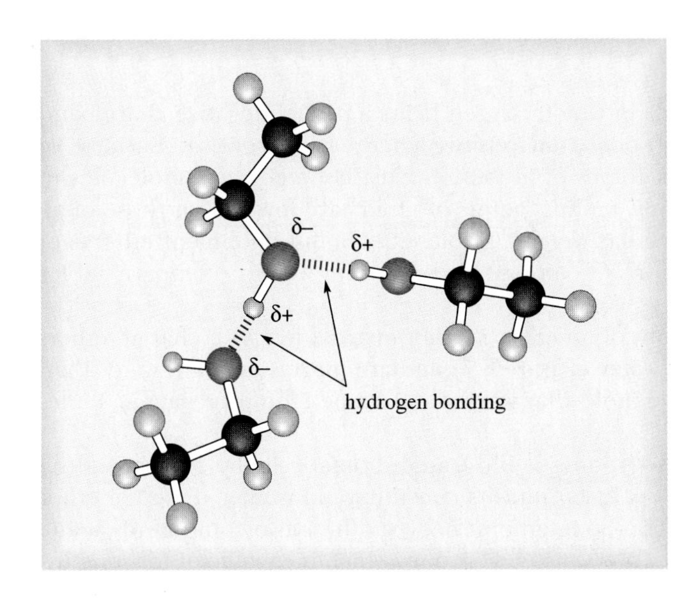

Figure 8.5
The association of ethanol molecules in the liquid state. Each O—H can participate in up to three hydrogen bonds (one through hydrogen and two through oxygen). Only two of these three possible hydrogen bonds per molecule are shown here. Hydrogen bonding gives alcohols an added attractive force between their molecules.

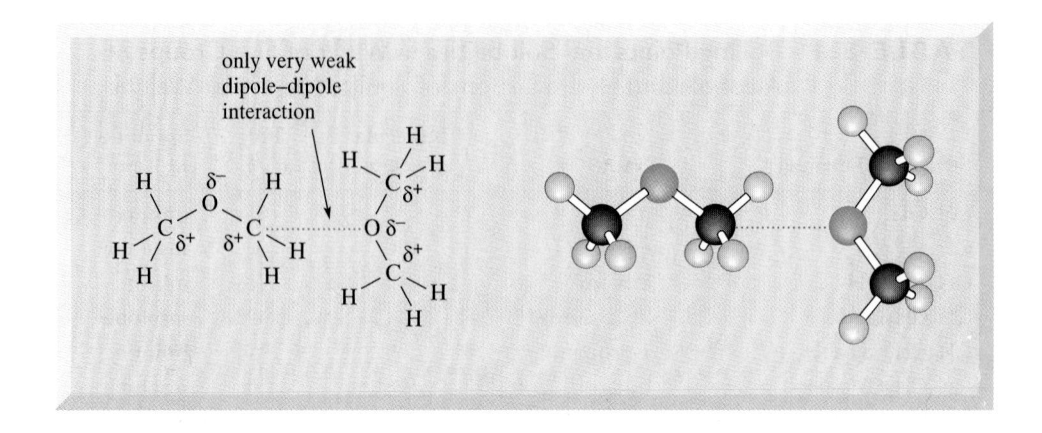

Figure 8.6
Ethers are polar molecules, but because of steric hindrance, only weak attractive interactions exist between their molecules in the pure liquid.

its neighbors, hence the relatively high boiling point of alcohols compared with alkanes. The presence of additional hydroxyl groups in a molecule further increases the extent of hydrogen bonding, as can be seen by comparing the boiling points of hexane (bp 69°C), 1-pentanol (bp 138°C), and 1,4-butanediol (bp 230°C), all of which have approximately the same molecular weight. Because of increased dispersion forces (Section 3.8) between larger molecules, boiling points of all types of compounds, including alcohols, increase with increasing molecular weight. Compare, for example, the boiling points of ethanol, 1-propanol, 1-butanol, and 1-pentanol.

Because alcohols can interact by hydrogen bonding with water, they are much more soluble in water than are alkanes, alkenes, and alkynes of comparable molecular weight. Methanol, ethanol, and 1-propanol are soluble in water in all proportions. As molecular weight increases, the physical properties of alcohols become more like those of hydrocarbons of comparable molecular weight. Alcohols of higher molecular weight are much less soluble in water because of the increase in size of the hydrocarbon portion of their molecules.

Commercial antifreeze contains ethylene glycol. *(Charles D. Winters)*

B. Ethers

Ethers are polar compounds in which oxygen bears a partial negative charge and each attached carbon bears a partial positive charge (Figure 8.6). Because of steric hindrance, only weak forces of attraction exist between their molecules in the pure liquid; consequently, boiling points of ethers are lower than those of alcohols of comparable molecular weight (Table 8.2). Boiling points of ethers are close to those of hydrocarbons of comparable molecular weight (compare Tables 3.4 and 8.2).

Because the oxygen atom of an ether carries a partial negative charge, ethers form hydrogen bonds with water (Figure 8.7) and are more soluble in water than hydrocarbons of comparable molecular weight and shape (compare data in Tables 3.4 and 8.2).

The effect of hydrogen bonding is illustrated dramatically by comparing the boiling points of ethanol (bp 78°C) and its constitutional isomer dimethyl ether (bp −24°C). The difference in boiling point between these two compounds is due to the presence in ethanol of a polar O—H group, which is capable of forming hy-

TABLE 8.2 Boiling Points and Solubilities in Water of Some Alcohols and Ethers of Comparable Molecular Weight

Structural Formula	Name	Molecular Weight	bp (°C)	Solubility in Water
CH_3CH_2OH	ethanol	46	78	infinite
CH_3OCH_3	dimethyl ether	46	−24	7 g/100 g
$CH_3CH_2CH_2CH_2OH$	1-butanol	74	117	8 g/100 g
$CH_3CH_2OCH_2CH_3$	diethyl ether	74	35	8 g/100 g
$CH_3CH_2CH_2CH_2CH_2OH$	1-pentanol	88	138	2.3 g/100 g
$HOCH_2CH_2CH_2CH_2OH$	1,4-butanediol	90	230	infinite
$CH_3CH_2CH_2CH_2OCH_3$	butyl methyl ether	88	71	slight
$CH_3OCH_2CH_2OCH_3$	ethylene glycol dimethyl ether	90	84	infinite

drogen bonds. This hydrogen bonding increases intermolecular associations and, thus, gives ethanol a higher boiling point than dimethyl ether.

CH_3CH_2OH	CH_3OCH_3
Ethanol	Dimethyl ether
bp 78°C	bp −24°C

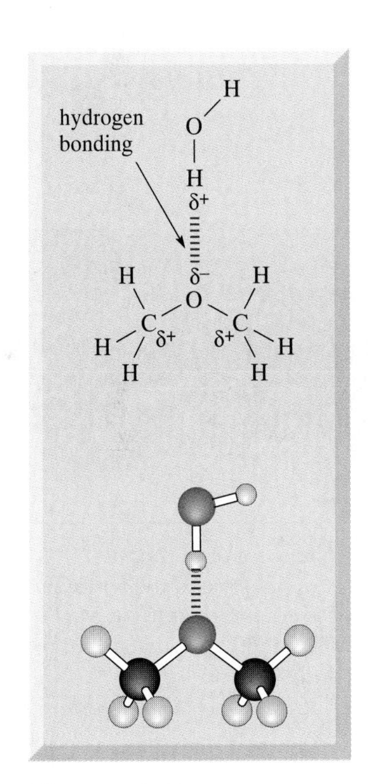

Figure 8.7
Ethers are hydrogen bond acceptors only. They are not hydrogen bond donors.

EXAMPLE 8.6

Arrange these compounds in order of increasing solubility in water:

$CH_3OCH_2CH_2OCH_3$	$CH_3CH_2OCH_2CH_3$	$CH_3CH_2CH_2CH_2CH_2CH_3$
Ethylene glycol dimethyl ether	Diethyl ether	Hexane

SOLUTION

Water is a polar solvent. Hexane, a nonpolar hydrocarbon, has the lowest solubility in water. Both diethyl ether and ethylene glycol dimethyl ether are polar compounds due to the presence of their polar C—O—C groups, and each interacts with water as a hydrogen bond acceptor. Because ethylene glycol dimethyl ether has more sites within its molecules for hydrogen bonding, it is more soluble in water than diethyl ether.

$CH_3CH_2CH_2CH_2CH_2CH_3$	$CH_3CH_2OCH_2CH_3$	$CH_3OCH_2CH_2OCH_3$
Insoluble	8 g/100 g water	Soluble in all proportions

Practice Problem 8.6

Arrange these compounds in order of increasing boiling point.

$CH_3OCH_2CH_2OCH_3$ $\quad$ $HOCH_2CH_2OH$ $\quad$ $CH_3OCH_2CH_2OH$

TABLE 8.3 Boiling Points of Three Thiols and Alcohols of the Same Number of Carbon Atoms

Thiol	bp (°C)	Alcohol	bp (°C)
methanethiol	6	methanol	65
ethanethiol	35	ethanol	78
1-butanethiol	98	1-butanol	117

C. Thiols

The most outstanding physical characteristic of low-molecular-weight thiols is their stench. Traces of low-molecular-weight thiols, most commonly ethanethiol (ethyl mercaptan), are added to natural gas so that gas leaks can be detected by the smell of the thiol. The scent of skunks is due primarily to these two thiols:

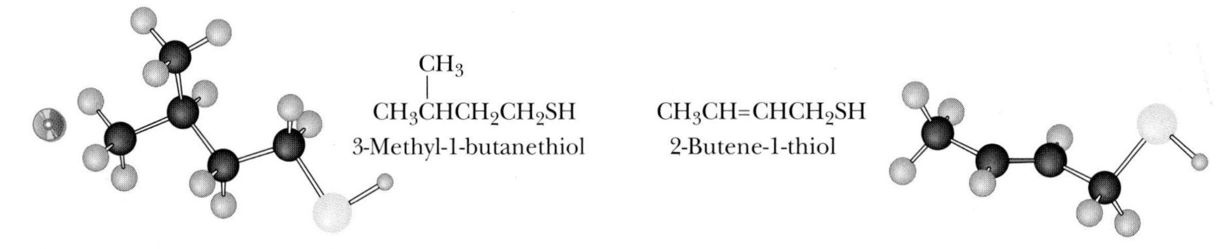

The scent of the spotted skunk, native to the Sonoran Desert, is a mixture of two thiols, 3-methyl-1-butanethiol and 2-butene-1-thiol. *(Stephan J. Krasemann/Photo Researchers, Inc.)*

Because of the very low polarity of the S—H bond, thiols show little association by hydrogen bonding. Consequently, they have lower boiling points and are less soluble in water and other polar solvents than alcohols of comparable molecular weights. Table 8.3 gives boiling points for three low-molecular-weight thiols. Shown for comparison are boiling points of alcohols of the same number of carbon atoms.

Earlier we illustrated the importance of hydrogen bonding in alcohols by comparing the boiling points of ethanol (bp 78°C) and its constitutional isomer dimethyl ether (bp −24°C). By comparison, the boiling point of ethanethiol is 35°C, and that of its constitutional isomer dimethyl sulfide is 37°C. The fact that the boiling points of these constitutional isomers are almost identical indicates that little or no association by hydrogen bonding occurs between thiol molecules.

CH_3CH_2SH	CH_3SCH_3
Ethanethiol	Dimethyl sulfide
bp 35°C	bp 37°C

8.4 Reactions of Alcohols

A. Acidity of Alcohols

In dilute aqueous solution, alcohols are very weakly acidic as illustrated by the ionization of methanol.

$$CH_3\ddot{O}{-}H + :\ddot{O}(H){-}H \rightleftharpoons CH_3\ddot{O}:^- + H{-}\ddot{O}^+(H){-}H \qquad K_a = \frac{[CH_3O^-][H_3O^+]}{[CH_3OH]} = 3.2 \times 10^{-16}$$

TABLE 8.4 pK_a Values for Selected Alcohols in Dilute Aqueous Solution*

Compound	Structural Formula	pK_a	
hydrogen chloride	HCl	−7	Stronger acid
acetic acid	CH_3CO_2H	4.8	↑
methanol	CH_3OH	15.5	
water	$\mathbf{H_2O}$	**15.7**	
ethanol	CH_3CH_2OH	15.9	
2-propanol	$(CH_3)_2CHOH$	17	Weaker
2-methyl-2-propanol	$(CH_3)_3COH$	18	acid

* Also given for comparison are pK_a values for water, acetic acid, and hydrogen chloride.

Shown in Table 8.4 are acid ionization constants for several low-molecular-weight alcohols. Methanol and ethanol are about as acidic as water. Higher-molecular-weight, water-soluble alcohols are slightly weaker acids than water. Thus, even though alcohols have some slight acidity, they are not strong enough acids to react with weak bases such as sodium bicarbonate or sodium carbonate. (At this point, it would be well to review Section 2.4 and the discussion of the position of equilibrium in acid-base reactions.) Note that, although acetic acid is a "weak acid" compared with acids such as HCl, it is still 10^{10} times stronger as an acid than alcohols.

B. Basicity of Alcohols

In the presence of strong acids, the oxygen atom of an alcohol is a weak base and reacts with an acid by proton transfer to form an oxonium ion.

$$\underset{\text{Ethanol}}{CH_3CH_2-\ddot{O}-H} + \underset{\substack{\text{Hydronium ion}\\(pK_a\ -1.7)}}{H-\overset{+}{O}(H)-H} \xrightleftharpoons{H_2SO_4} \underset{\substack{\text{Ethyloxonium ion}\\(pK_a\ -2.4)}}{CH_3CH_2-\overset{+}{O}(H)-H} + :\ddot{O}(H)-H$$

Thus, we see that alcohols can function as both weak acids and weak bases.

C. Reaction with Active Metals

Like water, alcohols react with Li, Na, K, Mg, and other active metals to liberate hydrogen and to form alkoxide salts.

$$2CH_3OH + 2Na \longrightarrow \underset{\text{Sodium methoxide}}{2CH_3O^-Na^+} + H_2$$

Sodium metal reacts with methanol with the evolution of hydrogen gas. *(Charles D. Winters)*

To name the salt of an alcohol, name the cation first followed by the name of the anion. The name of the alkoxide ion is derived from a prefix showing the number of carbon atoms and their arrangement (meth-, eth-, isoprop-, *tert*-but-, and so on) followed by the suffix -oxide.

Alkoxide ions are somewhat stronger bases than is the hydroxide ion. In addition to sodium methoxide, the following metal salts of alcohols are commonly used in organic reactions requiring a strong base in a nonaqueous solvent, as for example sodium ethoxide in ethanol.

$$CH_3CH_2O^-Na^+ \qquad CH_3\overset{\overset{CH_3}{|}}{\underset{\underset{CH_3}{|}}{C}}O^-K^+$$

Sodium ethoxide   Potassium *tert*-butoxide

EXAMPLE 8.7

Write a balanced equation for the reaction of cyclohexanol with sodium metal.

SOLUTION

$$2\,C_6H_{11}\text{—}OH + 2Na \longrightarrow 2\,C_6H_{11}\text{—}O^-Na^+ + H_2$$

Practice Problem 8.7

Predict the position of equilibrium for this acid-base reaction. (*Hint:* Review Section 2.4.)

$$CH_3CH_2O^-Na^+ + CH_3\overset{\overset{O}{\|}}{C}OH \rightleftharpoons CH_3CH_2OH + CH_3\overset{\overset{O}{\|}}{C}O^-Na^+$$

D. Conversion to Alkyl Halides

Conversion of an alcohol to an alkyl halide involves substitution of halogen for —OH at a saturated carbon. The most common reagents for this conversion are the halogen acids and $SOCl_2$. These reactions are examples of **nucleophilic aliphatic substitutions** (Section 7.2).

Reaction with HCl, HBr, and HI

Water-soluble tertiary alcohols react very rapidly with HCl, HBr, and HI. Mixing a tertiary alcohol with concentrated hydrochloric acid for a few minutes at room temperature results in conversion of the alcohol to an alkyl chloride. Reaction is evident by formation of a water-insoluble chloroalkane that separates from the aqueous layer. Low-molecular-weight, water-soluble primary and secondary alcohols are unreactive under these conditions.

$$CH_3\overset{\overset{CH_3}{|}}{\underset{\underset{CH_3}{|}}{C}}OH + HCl \xrightarrow{25°C} CH_3\overset{\overset{CH_3}{|}}{\underset{\underset{CH_3}{|}}{C}}Cl + H_2O$$

2-Methyl-2-propanol   2-Chloro-2-methylpropane

Water-insoluble tertiary alcohols are converted to tertiary halides by bubbling gaseous HX through a solution of the alcohol dissolved in diethyl ether or tetrahydrofuran (THF).

$$\text{1-Methylcyclohexanol} + HCl \xrightarrow[\text{ether}]{0°C} \text{1-Chloro-1-methylcyclohexane} + H_2O$$

Water-insoluble primary and secondary alcohols react only slowly under these conditions.

To account for the reaction between a tertiary alcohol and HX, chemists propose an S_N1 mechanism. Step 1 is a rapid, reversible proton transfer from HX (or H_3O^+ when aqueous acid is used) to the hydroxyl group to give an oxonium ion. This is followed in Step 2 by loss of a molecule of water (the leaving group) to give a 3° carbocation intermediate. The 3° carbocation intermediate then reacts with halide ion in Step 3 to give the product. Formation of the carbocation intermediate in Step 2 is the rate-limiting step.

MECHANISM Reaction of a Tertiary Alcohol with HCl: An S_N1 Reaction

Step 1: Rapid and reversible proton transfer to the OH group gives an oxonium ion.

$$(CH_3)_3C{-}\ddot{O}{-}H + H{-}\overset{+}{O}H_2 \xrightleftharpoons{\text{rapid and reversible}} (CH_3)_3C{-}\overset{+}{O}H_2 + :\ddot{O}H_2$$

2-Methyl-2-propanol (*tert*-Butyl alcohol) — An oxonium ion

Step 2: Loss of water from the oxonium ion gives a 3° carbocation intermediate.

$$(CH_3)_3C{-}\overset{+}{O}H_2 \xrightarrow[S_N1]{\text{rate-limiting step}} (CH_3)_3C^+ + :\ddot{O}H_2$$

A 3° carbocation intermediate

Step 3: Reaction of the 3° carbocation intermediate (a Lewis acid) with chloride ion (a Lewis base) gives the product.

$$(CH_3)_3C^+ + :\ddot{Cl}:^- \xrightarrow{\text{rapid}} (CH_3)_3C{-}\ddot{Cl}:$$

2-Chloro-2-methylpropane (*tert*-Butyl chloride)

Primary and secondary alcohols are converted to bromoalkanes and iodoalkanes by treatment with hydrobromic and hydroiodic acids. For example, when heated to reflux with concentrated HBr, 1-butanol is converted to 1-bromobutane.

$$CH_3CH_2CH_2CH_2OH + HBr \xrightarrow[\text{reflux}]{H_2O} CH_3CH_2CH_2CH_2Br + H_2O$$

1-Butanol — 1-Bromobutane

To account for the conversion of primary alcohols to alkyl bromides and iodides, chemists propose an S_N2 mechanism. Step 1 is a rapid, reversible reaction of the hydroxyl group with H_3O^+ to form an oxonium ion. In Step 2, halide ion reacts at the carbon bearing the oxonium ion to displace oxygen and to form a C—X bond. Displacement by halide ion is from the side of the primary carbon opposite that of the leaving oxonium ion. Step 2 is rate limiting.

MECHANISM Reaction of a Primary Alcohol with HBr: An S_N2 Mechanism

Step 1: Rapid and reversible proton transfer to the OH group gives an oxonium ion.

$$CH_3CH_2CH_2CH_2—\ddot{O}—H + H—\overset{+}{\ddot{O}}(H)—H \xrightleftharpoons{\text{rapid and reversible}} CH_3CH_2CH_2CH_2—\overset{+}{O}H_2 + :\ddot{O}(H)—H$$

An oxonium ion

Step 2: Nucleophilic displacement of H_2O by Br^- gives the alkyl bromide.

$$:\ddot{Br}:^- + CH_3CH_2CH_2CH_2—\overset{+}{O}H_2 \xrightarrow[S_N2]{\text{rate-limiting step}} CH_3CH_2CH_2CH_2—\ddot{Br}: + :\ddot{O}H_2$$

Why do tertiary alcohols react with HX by formation of a 3° carbocation intermediate, whereas primary alcohols react by direct displacement of —OH (more accurately, displacement of $—OH_2^+$)? The answer is a combination of the same two factors involved in nucleophilic substitution reactions of alkyl halides (Section 7.4B).

1. *Electronic factors* Tertiary carbocations are the most stable (lowest activation energy for their formation), whereas primary carbocations are the least stable (highest activation energy for their formation). Therefore, tertiary alcohols are most likely to react by carbocation formation; secondary alcohols are intermediate, and primary alcohols rarely, if ever, react by carbocation formation.
2. *Steric factors* To form a new carbon-halogen bond, halide ion must approach the substitution center and begin to form a new covalent bond to it. If we compare the ease of approach to the substitution center of a primary alcohol to that of a tertiary alcohol, we see that approach is considerably easier in the case of a primary alcohol. The backside of the substitution center of a primary alcohol is screened by two hydrogen atoms and one alkyl group, whereas the backside of the substitution center of a tertiary alcohol is screened by three alkyl groups.

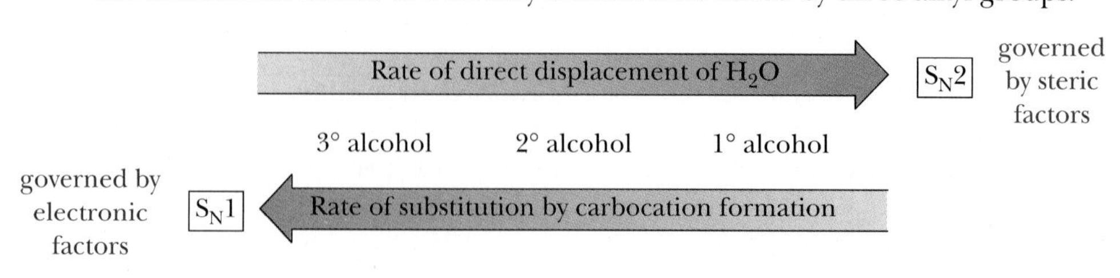

Reaction with Thionyl Chloride

The most widely used reagent for the conversion of primary and secondary alcohols to alkyl chlorides is thionyl chloride, $SOCl_2$. The by-products of this nucleophilic substitution reaction are HCl and SO_2, both given off as gases.

$$\underset{\text{1-Heptanol}}{CH_3(CH_2)_5CH_2OH} + \underset{\text{Thionyl chloride}}{SOCl_2} \longrightarrow \underset{\text{1-Chloroheptane}}{CH_3(CH_2)_5CH_2Cl} + \underset{\text{Sulfur dioxide}}{SO_2} + HCl$$

E. Acid-Catalyzed Dehydration of Alcohols to Alkenes

An alcohol can be converted to an alkene by elimination of a molecule of water from adjacent carbon atoms. Elimination of water is called **dehydration.** In the laboratory, dehydration of an alcohol is most often brought about by heating it with either 85% phosphoric acid or concentrated sulfuric acid. Primary alcohols are the most difficult to dehydrate and generally require heating in concentrated sulfuric acid at temperatures as high as 180°C. Secondary alcohols undergo acid-catalyzed dehydration at somewhat lower temperatures. Acid-catalyzed dehydration of tertiary alcohols often requires temperatures only slightly above room temperature.

Dehydration Elimination of a molecule of water from a compound.

$$CH_3CH_2OH \xrightarrow[180°C]{H_2SO_4} CH_2{=}CH_2 + H_2O$$

$$\underset{\text{Cyclohexanol}}{C_6H_{11}OH} \xrightarrow[140°C]{H_2SO_4} \underset{\text{Cyclohexene}}{C_6H_{10}} + H_2O$$

$$\underset{\substack{\text{2-Methyl-2-propanol} \\ (\textit{tert}\text{-Butyl alcohol})}}{CH_3C(CH_3)_2OH} \xrightarrow[50°C]{H_2SO_4} \underset{\substack{\text{2-Methylpropene} \\ \text{(Isobutylene)}}}{CH_3C(CH_3){=}CH_2} + H_2O$$

Thus, the ease of acid-catalyzed dehydration of alcohols is in this order:

1° alcohol < 2° alcohol < 3° alcohol

Ease of dehydration of alcohols →

When isomeric alkenes are obtained in the acid-catalyzed dehydration of an alcohol, the alkene having the greater number of substituents on the double bond generally predominates; that is, acid-catalyzed dehydration of alcohols follows **Zaitsev's rule** (Section 7.6).

$$\underset{\text{2-Butanol}}{CH_3CH_2CH(OH)CH_3} \xrightarrow[\text{heat}]{85\%\ H_3PO_4} \underset{\substack{\text{2-Butene} \\ (80\%)}}{CH_3CH{=}CHCH_3} + \underset{\substack{\text{1-Butene} \\ (20\%)}}{CH_3CH_2CH{=}CH_2}$$

EXAMPLE 8.8

Draw structural formulas for the alkenes formed on acid-catalyzed dehydration of each alcohol. For each, predict which alkene is the major product.

(a) 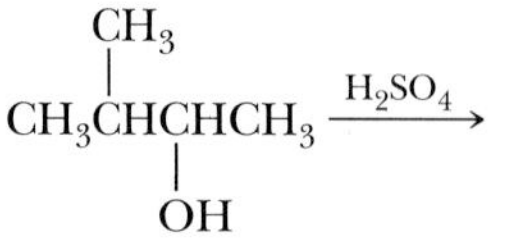 $CH_3CH(CH_3)CH(OH)CH_3 \xrightarrow{H_2SO_4}$

(b) 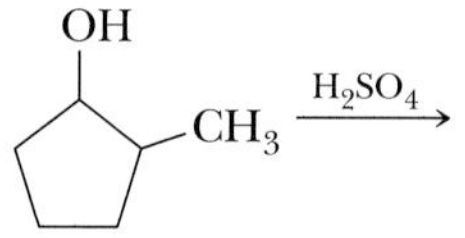 2-methylcyclopentanol (cyclopentane ring bearing OH and an adjacent CH_3) $\xrightarrow{H_2SO_4}$

SOLUTION

(a) Elimination of H_2O from carbons 2-3 gives 2-methyl-2-butene; elimination of H_2O from carbons 1-2 gives 3-methyl-1-butene. 2-Methyl-2-butene, with three alkyl groups (three methyl groups) on the double bond, is the major product. 3-Methyl-1-butene, with only one alkyl group (an isopropyl group) on the double bond, is the minor product.

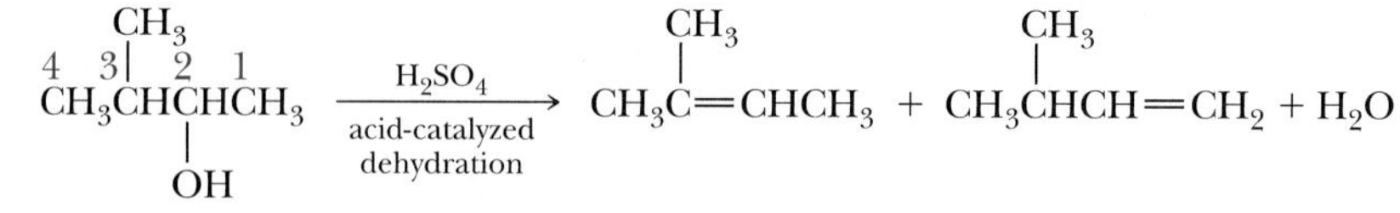

$$\underset{\text{3-Methyl-2-butanol}}{\overset{4}{C}H_3\overset{3}{C}H(CH_3)\overset{2}{C}H(OH)\overset{1}{C}H_3} \xrightarrow[\text{acid-catalyzed dehydration}]{H_2SO_4} \underset{\substack{\text{2-Methyl-2-butene}\\\text{(major product)}}}{CH_3C(CH_3){=}CHCH_3} + \underset{\text{3-Methyl-1-butene}}{CH_3CH(CH_3)CH{=}CH_2} + H_2O$$

(b) The major product, 1-methylcyclopentene, has three alkyl substituents on the double bond. The minor product, 3-methylcyclopentene, has only two alkyl substituents on the double bond.

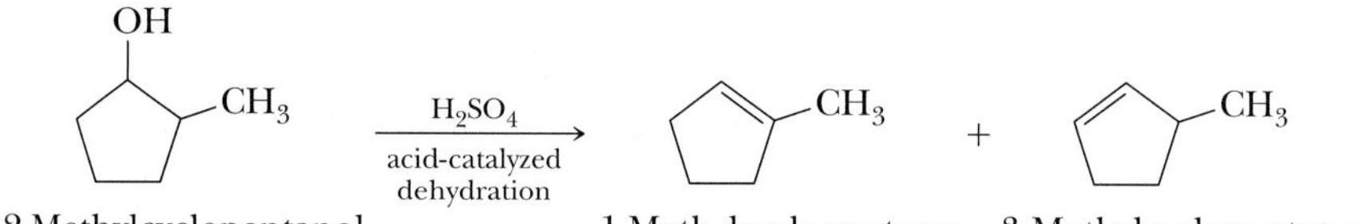

$$\text{2-Methylcyclopentanol} \xrightarrow[\text{acid-catalyzed dehydration}]{H_2SO_4} \underset{\text{(major product)}}{\text{1-Methylcyclopentene}} + \text{3-Methylcyclopentene}$$

Practice Problem 8.8

Draw structural formulas for the alkenes formed by acid-catalyzed dehydration of each alcohol. For each, predict which is the major product.

(a) 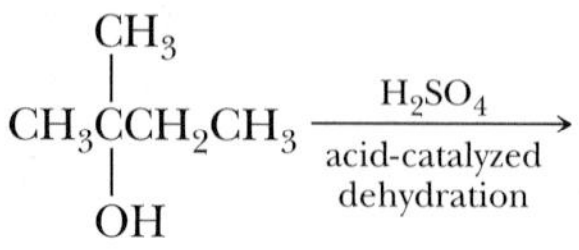 $CH_3C(CH_3)(OH)CH_2CH_3 \xrightarrow[\text{acid-catalyzed dehydration}]{H_2SO_4}$

(b) 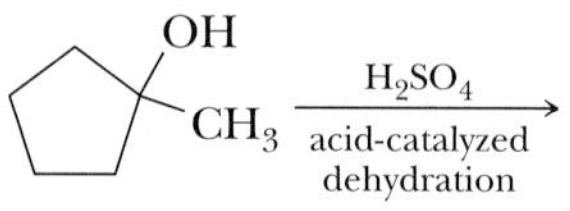 1-methylcyclopentanol (cyclopentane ring bearing OH and CH_3 on the same carbon) $\xrightarrow[\text{acid-catalyzed dehydration}]{H_2SO_4}$

Chemists propose the following three-step mechanism for the acid-catalyzed dehydration of secondary and tertiary alcohols. Step 2, the formation of the carbocation intermediate, is rate limiting.

MECHANISM Acid-Catalyzed Dehydration of 2-Butanol: An E1 Mechanism

Step 1: Proton transfer from H_3O^+ to the OH group of the alcohol gives an oxonium ion.

$$CH_3CH(OH)CH_2CH_3 + H{-}\overset{+}{O}(H){-}H \xrightarrow[\text{reversible}]{\text{rapid and}} CH_3CH(\overset{+}{O}H_2)CH_2CH_3 + {:}\ddot{O}(H){-}H$$

An oxonium ion

Step 2: Loss of H_2O from the oxonium ion gives a 2° carbocation intermediate.

$$CH_3CH(\overset{+}{O}H_2)CH_2CH_3 \xrightarrow[\text{rate limiting}]{\text{slow and}} CH_3\overset{+}{C}HCH_2CH_3 + H_2\ddot{O}{:}$$

A 2° carbocation intermediate

Step 3: Proton transfer from the carbocation intermediate to H_2O completes the elimination reaction. The sigma electrons of a C—H bond adjacent to the positively charged carbon become the pi electrons of the carbon-carbon double bond.

$$CH_3{-}\overset{+}{C}H{-}CH(H){-}CH_3 + {:}\ddot{O}(H){-}H \xrightarrow{\text{rapid}} CH_3{-}CH{=}CH{-}CH_3 + H{-}\overset{+}{O}(H){-}H$$

Because the rate-limiting step in acid-catalyzed dehydration of secondary and tertiary alcohols is formation of a carbocation intermediate, the relative ease of dehydration of these alcohols parallels the ease of formation of carbocations. Tertiary alcohols undergo acid-catalyzed dehydration more readily than secondary alcohols because tertiary carbocations are formed more readily than secondary carbocations.

Primary alcohols almost certainly do not undergo acid-catalyzed dehydration by a carbocation intermediate. Instead, they react by the following two-step mechanism in which Step 2 is the rate-limiting step.

MECHANISM Acid-Catalyzed Dehydration of a Primary Alcohol: E2 Mechanism

Step 1: Proton transfer from H_3O^+ to the OH group of the alcohol gives an oxonium ion.

$$CH_3CH_2{-}\ddot{O}{-}H + H{-}\overset{+}{O}(H){-}H \xrightarrow[\text{reversible}]{\text{rapid and}} CH_3CH_2{-}\overset{+}{O}H_2 + {:}\ddot{O}(H){-}H$$

Step 2: Simultaneous loss of H from the β-carbon and OH_2 from the α-carbon gives the alkene.

$$H-\ddot{O}: + H-\overset{H}{\underset{H}{C}}-CH_2-\overset{+}{O}(H)_2 \xrightarrow[\text{rate limiting}]{\text{slow and}} H-\overset{+}{\underset{H}{\ddot{O}}}-H + \; C{=}C \; + \; :\underset{H}{\ddot{O}}-H$$

In Section 6.3B we discussed the acid-catalyzed hydration of alkenes to give alcohols. In the present section, we discuss the acid-catalyzed dehydration of alcohols to give alkenes. In fact, hydration-dehydration reactions are reversible. Alkene hydration and alcohol dehydration are competing reactions, and the following equilibrium exists.

$$>C{=}C< + H_2O \underset{}{\overset{\text{acid catalyst}}{\rightleftharpoons}} -\underset{H}{\overset{|}{C}}-\underset{OH}{\overset{|}{C}}-$$

An alkene An alcohol

Large amounts of water (use of dilute aqueous acid) favor alcohol formation, whereas scarcity of water (use of concentrated acid) or experimental conditions where water is removed (heating the reaction mixture above 100°C) favor alkene formation. Thus, depending on experimental conditions, it is possible to use the hydration-dehydration equilibrium to prepare either alcohols or alkenes, each in high yields.

F. Oxidation of Primary and Secondary Alcohols

A primary alcohol can be oxidized to an aldehyde or a carboxylic acid, depending on the experimental conditions. Secondary alcohols are oxidized to ketones. Tertiary alcohols are not oxidized. Following is a series of transformations in which a primary alcohol is oxidized first to an aldehyde and then to a carboxylic acid. The fact that each transformation involves oxidation is indicated by the symbol O in brackets over the reaction arrow.

$$CH_3-\underset{H}{\overset{OH}{C}}-H \xrightarrow{[O]} CH_3-\overset{O}{\overset{\|}{C}}-H \xrightarrow{[O]} CH_3-\overset{O}{\overset{\|}{C}}-OH$$

A primary alcohol An aldehyde A carboxylic acid

Inspection of balanced half-reactions (Section 6.4A) shows that each transformation is a two-electron oxidation.

$$CH_3-\underset{H}{\overset{OH}{C}}-H \longrightarrow CH_3-\overset{O}{\overset{\|}{C}}-H + 2H^+ + 2e^-$$

$$CH_3-\overset{O}{\overset{\|}{C}}-H + H_2O \longrightarrow CH_3-\overset{O}{\overset{\|}{C}}-OH + 2H^+ + 2e^-$$

The reagent most commonly used in the laboratory for the oxidation of a primary alcohol to a carboxylic acid and a secondary alcohol to a ketone is chromic acid, H_2CrO_4. Chromic acid is prepared by dissolving either chromium(VI) oxide or potassium dichromate in aqueous sulfuric acid.

$$\underset{\text{Chromium(VI) oxide}}{CrO_3} + H_2O \xrightarrow{H_2SO_4} \underset{\text{Chromic acid}}{H_2CrO_4}$$

$$\underset{\text{Potassium dichromate}}{K_2Cr_2O_7} \xrightarrow{H_2SO_4} H_2Cr_2O_7 \xrightarrow{H_2O} \underset{\text{Chromic acid}}{2H_2CrO_4}$$

Oxidation of 1-octanol using chromic acid in aqueous sulfuric acid gives octanoic acid in high yield. These experimental conditions are more than sufficient to oxidize the intermediate aldehyde to a carboxylic acid.

$$\underset{\text{1-Octanol}}{CH_3(CH_2)_6CH_2OH} \xrightarrow[H_2SO_4,\ H_2O]{CrO_3} \left[\underset{\substack{\text{Octanal}\\\text{(not isolated)}}}{CH_3(CH_2)_6\overset{O}{\overset{\|}{C}}H}\right] \longrightarrow \underset{\text{Octanoic acid}}{CH_3(CH_2)_6\overset{O}{\overset{\|}{C}}OH}$$

The form of Cr(VI) commonly used for oxidation of a primary alcohol to an aldehyde is prepared by dissolving CrO_3 in aqueous HCl and adding pyridine to precipitate **pyridinium chlorochromate (PCC)** as a solid.

$CrO_3 + HCl +$ Pyridine $\longrightarrow$ Pyridinium chlorochromate (PCC) ($C_5H_5NH^+$ CrO_3Cl^-)

pyridinium ion

chlorochromate ion

Pyridine   Pyridinium chlorochromate (PCC)

Not only is this reagent very selective for the oxidation of primary alcohols to aldehydes, but it also has little effect on carbon-carbon double bonds or other easily oxidized functional groups. In the following example, geraniol is oxidized to geranial without affecting either carbon-carbon double bond.

Geraniol (CH_2OH) $\xrightarrow{PCC}$ Geranial ($\overset{O}{\overset{\|}{C}}H$)

Geraniol   Geranial

Secondary alcohols are oxidized to ketones by both chromic acid and PCC.

$$\text{Menthol} + H_2CrO_4 \xrightarrow{\text{acetone}} \text{Menthone} + Cr^{3+}$$

$CH(CH_3)_2$, OH, CH_3 → $CH(CH_3)_2$, O, CH_3

2-Isopropyl-5-methylcyclohexanol (Menthol)   2-Isopropyl-5-methylcyclohexanone (Menthone)

CHEMICAL CONNECTIONS

Blood Alcohol Screening

Potassium dichromate oxidation of ethanol to acetic acid is the basis for the original breath alcohol screening test used by law enforcement agencies to determine a person's blood alcohol content (BAC). The test is based on the difference in color between the dichromate ion (reddish orange) in the reagent and the chromium(III) ion (green) in the product. Thus, color change can be used as a measure of the quantity of ethanol present in a breath sample.

$$\underset{\text{Ethanol}}{CH_3CH_2OH} + \underset{\substack{\text{Dichromate ion} \\ \text{(reddish orange)}}}{Cr_2O_7^{2-}} \xrightarrow[H_2O]{H_2SO_4}$$

$$\underset{\text{Acetic acid}}{CH_3\overset{\overset{O}{\|}}{C}OH} + \underset{\substack{\text{Chromium(III) ion} \\ \text{(green)}}}{Cr_3^{+}}$$

In its simplest form, a breath alcohol screening test consists of a sealed glass tube containing a potassium dichromate-sulfuric acid reagent impregnated on silica gel. To administer the test, the ends of the tube are broken off, a mouthpiece is fitted to one end, and the other end is inserted into the neck of a plastic bag. The person being tested then blows into the mouthpiece until the plastic bag is inflated.

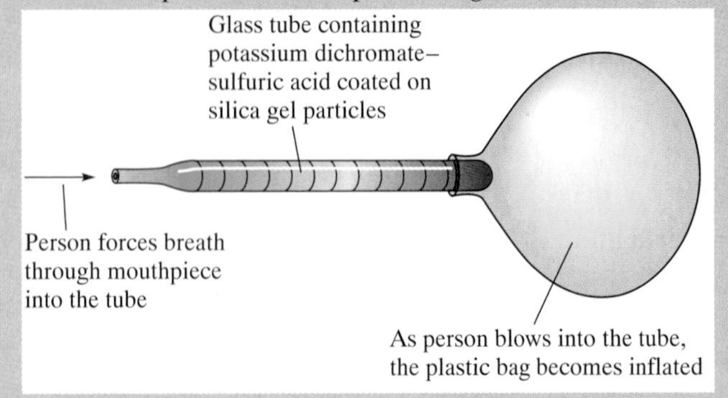

As breath containing ethanol vapor passes through the tube, reddish orange dichromate ion is reduced to green chromium(III) ion. The concentration of ethanol in the breath is then estimated by measuring how far the green color extends along the length of the tube. When it extends beyond the halfway point, the person is judged as having a sufficiently high blood alcohol content to warrant further, more precise testing.

The Breathalyzer, a more precise testing device, operates on the same principle as the simplified screening test. In a Breathalyzer test, a measured volume of breath is bubbled through a solution of potassium dichromate in aqueous sulfuric acid, and the color change is measured spectrophotometrically.

These tests measure alcohol in the breath. The legal definition of being under the influence of alcohol is based on blood alcohol content, not breath alcohol content. The chemical correlation between these two measurements is that air deep within the lungs is in equilibrium with blood passing through the pulmonary arteries, and an equilibrium is established between blood alcohol and breath alcohol. It has been determined by tests in persons drinking alcohol that 2100 mL of breath contains the same amount of ethanol as 1.00 mL of blood. See W. C. Timmer, *J. Chem. Ed.*, **63:** 897 (1986).

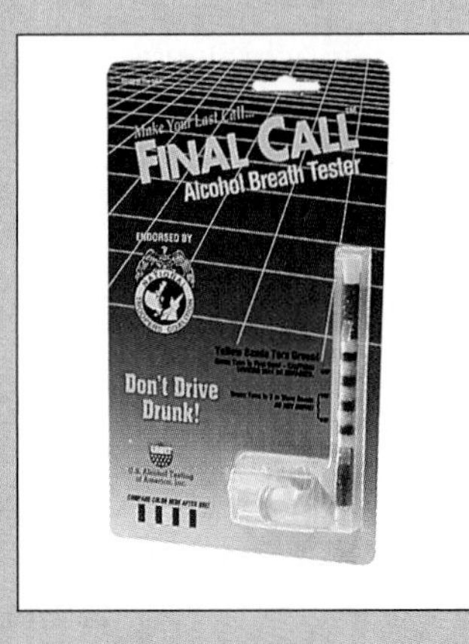

A device for testing the breath for the presence of ethanol. The test works because ethanol is oxidized by potassium dichromate. The yellow-orange color of dichromate ion turns to blue-green as it is reduced to chromium(III) ion. *(Charles D. Winters)*

Tertiary alcohols are resistant to oxidation because the carbon bearing the —OH is bonded to three carbon atoms and, therefore, cannot form a carbon-oxygen double bond.

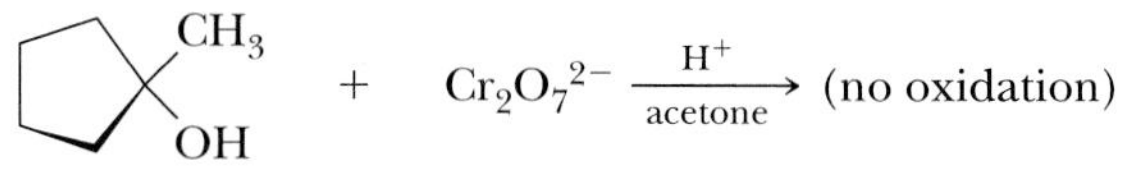

$$\text{1-Methylcyclopentanol} + Cr_2O_7^{2-} \xrightarrow[\text{acetone}]{H^+} \text{(no oxidation)}$$

1-Methylcyclopentanol

EXAMPLE 8.9

Draw the product of treatment of each alcohol with PCC.

(a) 1-Hexanol (b) 2-Hexanol (c) Cyclohexanol

SOLUTION

1-Hexanol, a primary alcohol, is oxidized to hexanal. 2-Hexanol, a secondary alcohol, is oxidized to 2-hexanone and cyclohexanol, a secondary alcohol, is oxidized to cyclohexanone.

(a) $CH_3(CH_2)_4CH{=}O$ (b) $CH_3(CH_2)_3C(=O)CH_3$ (c) 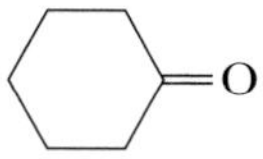 cyclohexane=O

Hexanal 2-Hexanone Cyclohexanone

Practice Problem 8.9

Draw the product of treatment of each alcohol in Example 8.9 with chromic acid.

8.5 REACTIONS OF ETHERS

Ethers, R—O—R, resemble hydrocarbons in their resistance to chemical reaction. They do not react with oxidizing agents, such as potassium dichromate or potassium permanganate. They are not affected by most acids or bases at moderate temperatures. Because of their good solvent properties and general inertness to chemical reaction, ethers are excellent solvents in which to carry out many organic reactions.

8.6 EPOXIDES

A. Structure and Nomenclature

An **epoxide** is a cyclic ether in which oxygen is one atom of a three-membered ring. Although epoxides are technically classed as ethers, we discuss them separately because of their exceptional chemical reactivity compared with other ethers.

Epoxide A cyclic ether in which oxygen is one atom of a three-membered ring.

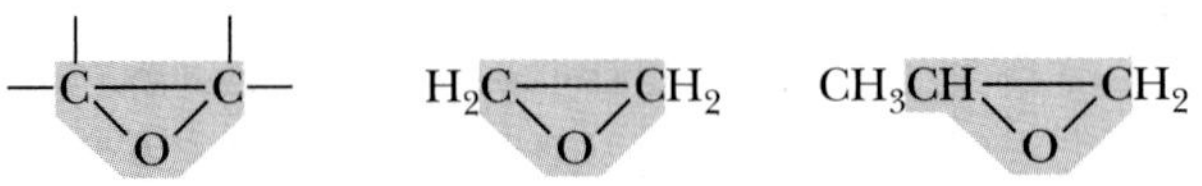

Functional group of an epoxide — Ethylene oxide — Propylene oxide

Common names for epoxides are derived by giving the common name of the alkene from which the epoxide might have been derived followed by the word "oxide"; an example is ethylene oxide.

B. Synthesis from Alkenes

Ethylene oxide, one of the few epoxides manufactured on an industrial scale, is prepared by passing a mixture of ethylene and air (or oxygen) over a silver catalyst.

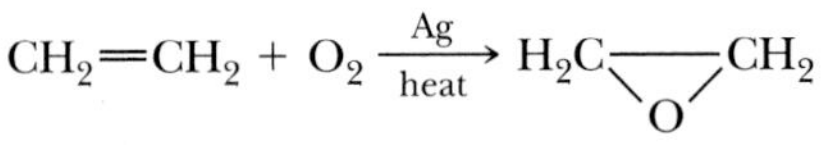

$$CH_2{=}CH_2 + O_2 \xrightarrow[\text{heat}]{\text{Ag}} H_2C\text{—}CH_2 \text{ (O bridge)}$$

Ethylene — Ethylene oxide

The most common laboratory method for the synthesis of epoxides from alkenes is oxidation with a peroxycarboxylic acid, RCO_3H, such as peroxyacetic acid.

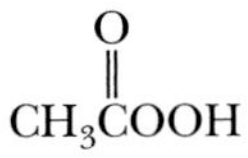

CH_3COOH (with C=O)

Peroxyacetic acid
(Peracetic acid)

Following is a balanced equation for the epoxidation of cyclohexene by a peroxycarboxylic acid. In the process, the peroxycarboxylic acid is reduced to a carboxylic acid.

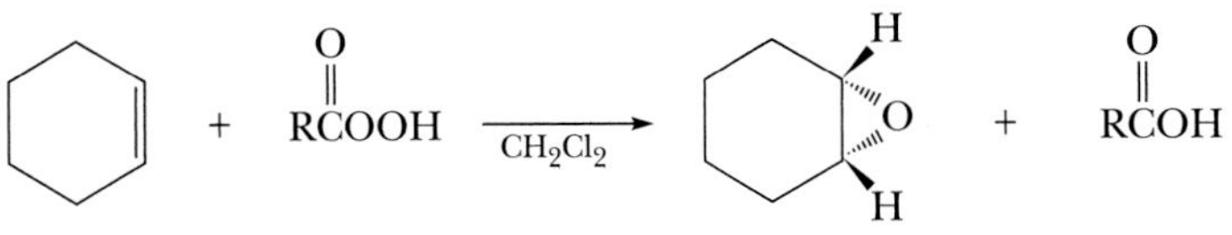

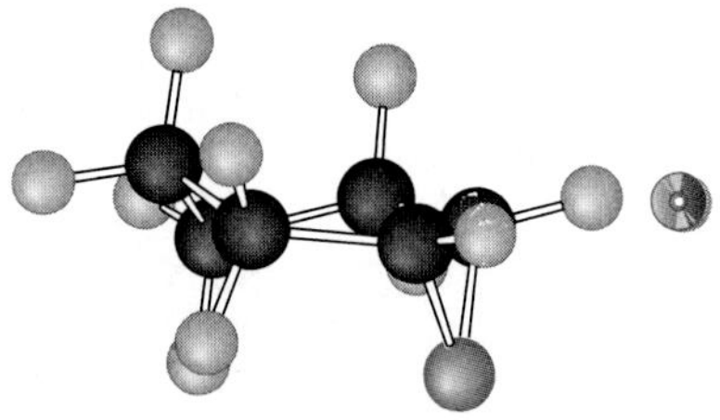

$$\text{Cyclohexene} + RCOOH \xrightarrow{CH_2Cl_2} \text{1,2-Epoxycyclohexane} + RCOH$$

Cyclohexene — A peroxycarboxylic acid — 1,2,-Epoxycyclohexane (Cyclohexene oxide) — A carboxylic acid — 1,2,-Epoxycyclohexane

Epoxidation of an alkene is stereoselective. Epoxidation of *cis*-2-butene, for example, yields only *cis*-2-butene oxide.

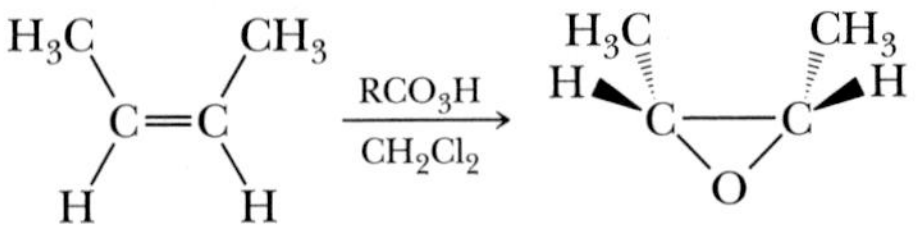

$$\xrightarrow[CH_2Cl_2]{RCO_3H}$$

cis-2-Butene — *cis*-2-Butene oxide

Example 8.10

Draw a structural formula of the epoxide formed by treating *trans*-2-butene with a peroxycarboxylic acid.

CHEMICAL CONNECTIONS

Ethylene Oxide—A Chemical Sterilant

Because ethylene oxide is such a highly strained molecule, it reacts with the types of nucleophilic groups present in biological materials. At sufficiently high concentrations, ethylene oxide reacts with enough molecules in cells to cause the death of microorganisms. This toxic property is the basis for using ethylene oxide as a chemical sterilant. In hospitals, surgical instruments and other items that cannot be made disposable are now sterilized by exposure to ethylene oxide.

SOLUTION

The oxygen of the epoxide ring is added by forming both carbon-oxygen bonds from the same side of the carbon-carbon double bond.

$$\underset{\textit{trans}\text{-2-Butene}}{(H_3C)(H)C{=}C(H)(CH_3)} \xrightarrow[CH_2Cl_2]{RCO_3H} \underset{\textit{trans}\text{-2-Butene oxide}}{(H_3C)(H)C\underset{O}{—}C(H)(CH_3)}$$

trans-2-Butene *trans*-2-Butene oxide

Practice Problem 8.10

Draw the structural formula of the epoxide formed by treating 1,2-dimethylcyclopentene with a peroxycarboxylic acid.

8.7 ACID-CATALYZED RING OPENING OF EPOXIDES

Ethers are not normally susceptible to reaction with aqueous acid. Epoxides, however, are especially reactive because of the angle strain in the three-membered ring. The normal bond angle about an sp^3 hybridized carbon or oxygen atom is 109.5°. Because of the strain associated with compression of bond angles in the three-membered epoxide ring from the normal 109.5° to 60°, epoxides undergo ring-opening reactions with a variety of reagents. In the presence of an acid catalyst, most commonly perchloric acid, epoxides are hydrolyzed to glycols. As an example, acid-catalyzed hydrolysis of ethylene oxide gives 1,2-ethanediol.

$$H_2C\underset{O}{—}CH_2 + H_2O \xrightarrow{H^+} HOCH_2CH_2OH$$

Ethylene oxide 1,2-Ethanediol (Ethylene glycol)

Under acidic conditions, the oxygen atom of the epoxide is protonated in Step 1 to form a bridged oxonium ion intermediate. The oxonium ion intermediate is then attacked by water in Step 2 from the side opposite the oxonium ion bridge, which results in the ring opening. Proton transfer in Step 3 completes the reaction.

MECHANISM Acid-Catalyzed Hydrolysis of an Epoxide

Step 1: Proton transfer from the acid catalyst to oxygen of the epoxide gives an oxonium ion.
Step 2: Backside attack of H_2O on the protonated epoxide opens the three-membered ring.
Step 3: Proton transfer to solvent completes formation of the glycol.

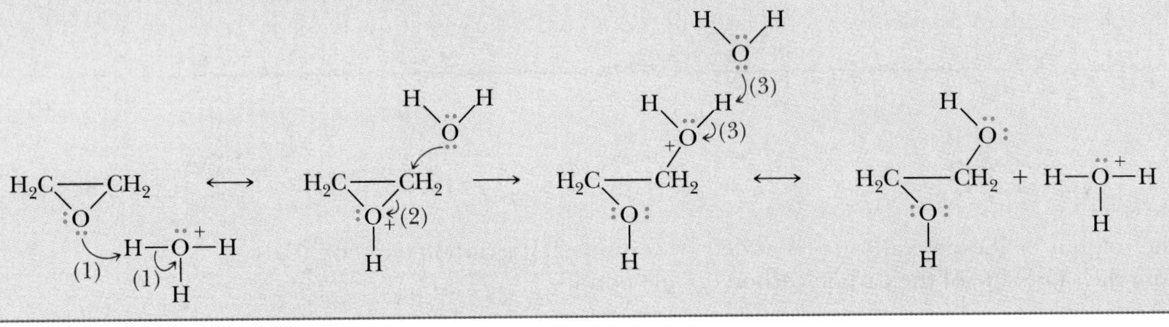

Attack of a nucleophile on a protonated epoxide shows a stereoselectivity typical of S_N2 reactions; the nucleophile attacks anti to the leaving hydroxyl group, and the —OH groups in the glycol thus formed are anti. As a result, hydrolysis of an epoxycycloalkane yields a *trans*-1,2-cycloalkanediol.

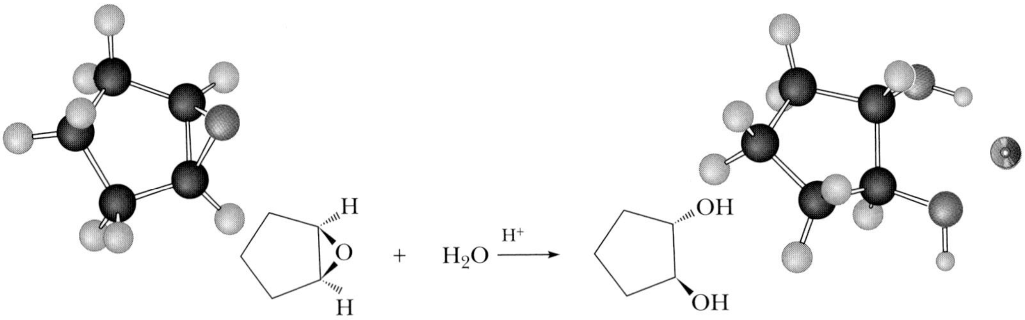

$$\text{1,2-Epoxycyclopentane} + H_2O \xrightarrow{H^+} \textit{trans}\text{-1,2-Cyclopentanediol}$$

1,2-Epoxycyclopentane (Cyclopentene oxide)

trans-1,2-Cyclopentanediol

At this point, let us compare the stereochemistry of the glycol formed by acid-catalyzed hydrolysis of an epoxide with that formed by oxidation of an alkene with osmium tetroxide (Section 6.4B). Each reaction sequence is stereoselective but gives a different stereoisomer. Acid-catalyzed hydrolysis of cyclopentene oxide gives *trans*-1,2-cyclopentanediol; osmium tetroxide oxidation of cyclopentene gives *cis*-1,2-cyclopentanediol. Thus, a cycloalkene can be converted to either a cis glycol or a trans glycol by the proper choice of reagents.

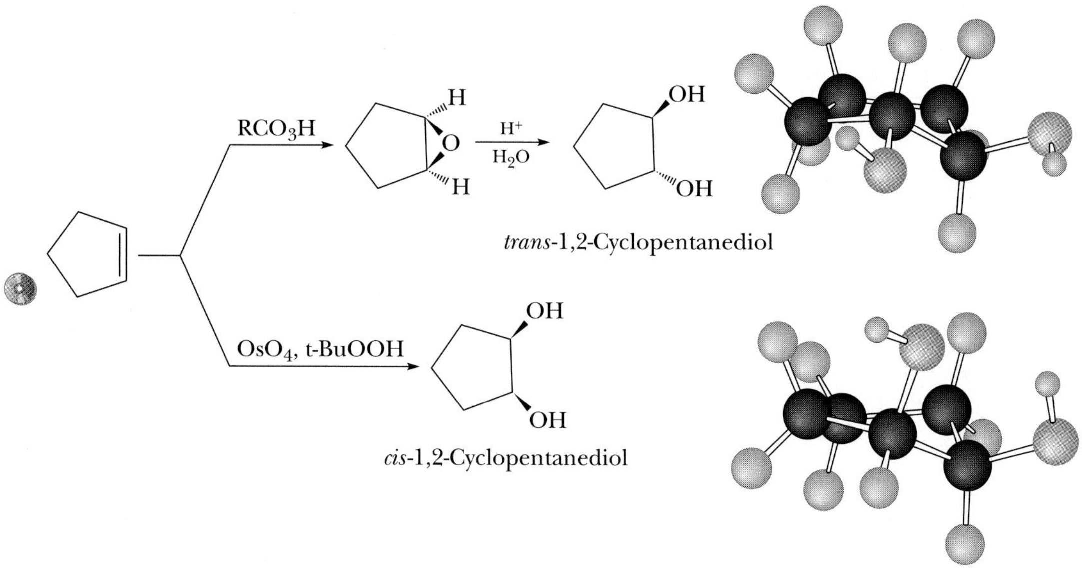

RCO$_3$H H$^+$ / H$_2$O OH OH

trans-1,2-Cyclopentanediol

OsO$_4$, t-BuOOH OH OH

cis-1,2-Cyclopentanediol

EXAMPLE 8.11

Draw the structural formula of the product formed by treatment of cyclohexene oxide with aqueous acid. Be certain to show the stereochemistry of the product.

SOLUTION

Acid-catalyzed hydrolysis of the three-membered epoxide ring gives a trans glycol.

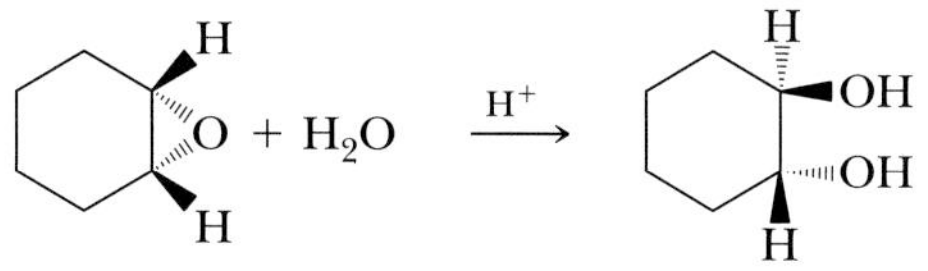

$$\text{(cyclohexene oxide)} + H_2O \xrightarrow{H^+} \text{(product)}$$

trans-1,2-Cyclohexanediol

Practice Problem 8.11

Show how to convert cyclohexene to *cis*-1,2-cyclohexanediol.

8.8 REACTIONS OF THIOLS

In this section, we discuss the acidity of thiols and their reaction with strong bases, such as sodium hydroxide, and with molecular oxygen.

A. Acidity

Hydrogen sulfide is a stronger acid than water.

$$H_2O + H_2O \rightleftharpoons HO^- + H_3O^+ \qquad pK_a = 15.7$$

$$H_2S + H_2O \rightleftharpoons HS^- + H_3O^+ \qquad pK_a = 7.0$$

Similarly, thiols are stronger acids than alcohols. Compare, for example, the pK_as of ethanol and ethanethiol in dilute aqueous solution.

$$CH_3CH_2OH + H_2O \rightleftharpoons CH_3CH_2O^- + H_3O^+ \qquad pK_a = 15.9$$

$$CH_3CH_2SH + H_2O \rightleftharpoons CH_3CH_2S^- + H_3O^+ \qquad pK_a = 8.5$$

Thiols are sufficiently strong acids that, when dissolved in aqueous sodium hydroxide, they are converted completely to alkylsulfide salts.

$$\underset{\substack{pK_a\ 8.5 \\ \text{Stronger acid}}}{CH_3CH_2SH} + \underset{\text{Stronger base}}{Na^+OH^-} \longrightarrow \underset{\text{Weaker base}}{CH_3CH_2S^-Na^+} + \underset{\substack{pK_a\ 15.7 \\ \text{Weaker acid}}}{H_2O}$$

To name salts of thiols, give the name of the cation first, followed by the name of the alkyl group to which is attached the suffix -sulfide. For example, the sodium salt derived from ethanethiol is named sodium ethylsulfide.

B. Oxidation of Thiols

Many of the chemical properties of thiols stem from the fact that the sulfur atom of a thiol is oxidized easily to several higher oxidation states. The most common oxidation-reduction reaction of sulfur compounds in biological systems is interconversion between a thiol and a disulfide. The functional group of a **disulfide** is an —S—S— group.

$$\underset{\text{A thiol}}{2R\text{—S—H}} \underset{\text{reduction}}{\overset{\text{oxidation}}{\rightleftharpoons}} \underset{\text{A disulfide}}{R\text{—S—S—}R}$$

Common names of disulfides are derived by listing the names of the groups attached to sulfur and adding the word "disulfide."

Thiols are readily oxidized to disulfides by molecular oxygen. In fact, thiols are so susceptible to oxidation that they must be protected from contact with air during storage.

$$\underset{\substack{\text{Methanethiol} \\ \text{(a thiol)}}}{2CH_3SH} + \tfrac{1}{2}O_2 \longrightarrow \underset{\substack{\text{Dimethyl disulfide} \\ \text{(a disulfide)}}}{CH_3SSCH_3} + H_2O$$

The disulfide bond is an important structural feature of many biomolecules, including proteins (Chapter 18).

SUMMARY

The functional group of an alcohol (Section 8.1) is an **—OH** (**hydroxyl**) group bonded to an sp^3 hybridized carbon. The functional group of an **ether** is an atom of oxygen bonded to two carbon atoms. A **thiol** (Section 8.1) is the sulfur analog of an alcohol; it contains an **—SH** (**sulfhydryl**) group in place of an —OH group. IUPAC names of alcohols (Section 8.2A) are derived by changing the suffix of the parent alkane from -e to -ol. The chain is numbered to give the carbon bearing —OH the lower number. Common names for alcohols are derived by naming the alkyl group bonded to —OH and adding the word "alcohol." Alcohols are classified as **1°**, **2°**, or **3°** (Section 8.2A) depending on whether the —OH group is bonded to a primary, secondary, or tertiary carbon.

In the IUPAC name of an ether (Section 8.2B), the parent alkane is named and then the —OR group is named as an alkoxy substituent. Common names are derived by naming the two groups attached to oxygen followed by the word "ether." For **thioethers,** name the two groups attached to sulfur followed by the word "sulfide."

Thiols (Section 8.2C) are named in the same manner as alcohols, but the suffix -e is retained, and -thiol is added. Common names for thiols are derived by naming the alkyl group bonded to —SH and adding the word "mercaptan." In compounds containing functional groups of higher precedence, the presence of —SH is indicated by the prefix mercapto-.

Alcohols are polar compounds (Section 8.3A) with oxygen bearing a partial negative charge and both the carbon and hydrogen bonded to it bearing partial positive charges. Because of intermolecular association by **hydrogen bonding,** the boiling points of alcohols are higher than those of hydrocarbons of comparable molecular weight. Because of increased dispersion forces, the boiling points of alcohols increase with increasing molecular weight. Alcohols interact with water by hydrogen bonding and, therefore, are more soluble in water than are hydrocarbons of comparable molecular weight. Ethers are weakly polar compounds (Section 8.3B). Their boiling points are close to those of hydrocarbons of comparable molecular weight. Because ethers are hydrogen bond acceptors, they are more soluble in water than are hydrocarbons of comparable molecular weight.

The S—H bond is nonpolar, and the physical properties of thiols are more like those of hydrocarbons of comparable molecular weight.

KEY REACTIONS

Alcohols

1. Acidity (Section 8.4A)

In dilute aqueous solution, methanol and ethanol are comparable in acidity to water. Secondary and tertiary alcohols are weaker acids.

$$CH_3OH + H_2O \rightleftharpoons CH_3O^- + H_3O^+ \qquad pK_a = 15.5$$

2. Reaction with Active Metals (Section 8.4C)

Alcohols react with Li, Na, K, and other active metals to form metal alkoxides, which are somewhat stronger bases than NaOH and KOH.

$$2CH_3CH_2OH + 2Na \longrightarrow 2CH_3CH_2O^-Na^+ + H_2$$

3. Reaction with HCl, HBr, and HI (Section 8.4D)

Primary alcohols react with HBr and HI by an S_N2 mechanism.

$$CH_3CH_2CH_2CH_2OH + HBr \longrightarrow CH_3CH_2CH_2CH_2Br + H_2O$$

Tertiary alcohols react with HCl, HBr, and HI by an S_N1 mechanism with formation of a carbocation intermediate.

$$CH_3\overset{\displaystyle CH_3}{\underset{\displaystyle CH_3}{C}}OH + HCl \xrightarrow{25°C} CH_3\overset{\displaystyle CH_3}{\underset{\displaystyle CH_3}{C}}Cl + H_2O$$

Secondary alcohols may react by an S_N2 or an S_N1 mechanism, depending on the alcohol and experimental conditions.

4. Reaction with $SOCl_2$ (Section 8.4D)

This is often the method of choice for converting an alcohol to an alkyl chloride.

$$CH_3(CH_2)_5OH + SOCl_2 \longrightarrow CH_3(CH_2)_5Cl + SO_2 + HCl$$

5. Acid-Catalyzed Dehydration (Section 8.4E)

When isomeric alkenes are possible, the major product is generally the more substituted alkene (Zaitsev's rule).

$$CH_3CH_2\overset{\displaystyle OH}{C}HCH_3 \xrightarrow[heat]{H_3PO_4} \underset{\text{Major product}}{CH_3CH{=}CHCH_3} + CH_3CH_2CH{=}CH_2 + H_2O$$

6. Oxidation of a Primary Alcohol to an Aldehyde (Section 8.4F)
This oxidation is most conveniently carried out using pyridinium chlorochromate (PCC).

$$\text{C}_5\text{H}_9\text{–CH}_2\text{OH} \xrightarrow{\text{PCC}} \text{C}_5\text{H}_9\text{–CH=O}$$

7. Oxidation of a Primary Alcohol to a Carboxylic Acid (Section 8.4F)
A primary alcohol is oxidized to a carboxylic acid by chromic acid.

$$CH_3(CH_2)_4CH_2OH + H_2CrO_4 \xrightarrow[\text{acetone}]{H_2O} CH_3(CH_2)_4COOH + Cr^{3+}$$

8. Oxidation of a Secondary Alcohol to a Ketone (Section 8.4F)
A secondary alcohol is oxidized to a ketone by chromic acid and by PCC.

$$CH_3(CH_2)_4CH(OH)CH_3 + H_2CrO_4 \longrightarrow CH_3(CH_2)_4C(=O)CH_3 + Cr^{3+}$$

Epoxides

9. Formation from Alkenes (Section 8.6B)
The most common method for the synthesis of an epoxide from an alkene is oxidation with a peroxycarboxylic acid, such as peroxyacetic acid.

$$\text{cyclohexene} + RC(=O)OOH \longrightarrow \text{cyclohexene oxide (cis H, H)} + RC(=O)OH$$

10. Acid-Catalyzed Hydrolysis (Section 8.7)
Hydrolysis of an epoxide derived from a cycloalkene gives a trans glycol.

$$\text{cyclopentene oxide} \xrightarrow[H_2O]{H^+} \text{trans-1,2-cyclopentanediol}$$

Thiols

11. Acidity (Section 8.8A)
Thiols are weak acids, pK_a 8–9, but are considerably stronger acids than alcohols, pK_a 16–18.

$$CH_3CH_2SH + H_2O \rightleftharpoons CH_3CH_2S^- + H_3O^+ \qquad pK_a = 8.5$$

12. Oxidation to Disulfides (Section 8.8B)
Oxidation of a thiol by O_2 gives a disulfide.

$$2RSH + \tfrac{1}{2}O_2 \longrightarrow RSSR + H_2O$$

ADDITIONAL PROBLEMS

Structure and Nomenclature

8.12 Which are secondary alcohols?

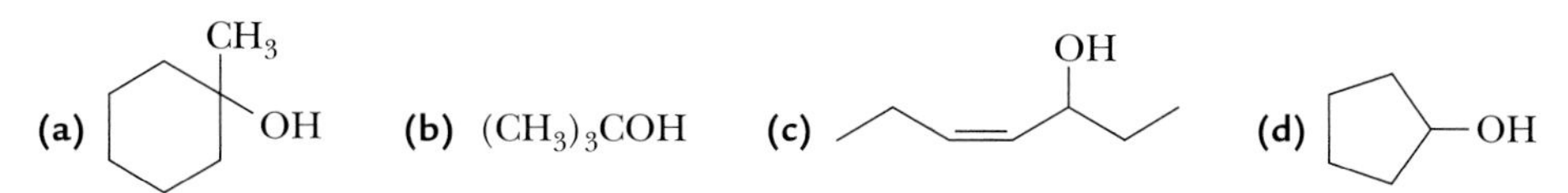

8.13 Name these compounds.

(a) $CH_3CH_2CH_2CH_2CH_2OH$ **(b)** $HOCH_2CH_2CH_2OH$ **(c)** $CH_2{=}CHCH_2CH_2OH$

(d) $HOCH_2CH_2\overset{\displaystyle CH_3}{\overset{|}{C}}HCH_3$ **(e)** OH OH **(f)** $CH_3CH_2CH_2CH_2SH$

8.14 Draw a structural formula for each alcohol.

(a) Isopropyl alcohol **(b)** Propylene glycol

(c) (R)-5-Methyl-2-hexanol **(d)** 2-Methyl-2-propyl-1,3-propanediol

(e) 2,2-Dimethyl-1-propanol **(f)** 2-Mercaptoethanol

(g) 1,4-Butanediol **(h)** (Z)-5-Methyl-2-hexen-1-ol

(i) *cis*-3-Pentene-1-ol **(j)** *trans*-1,4-Cyclohexanediol

8.15 Write names for these ethers.

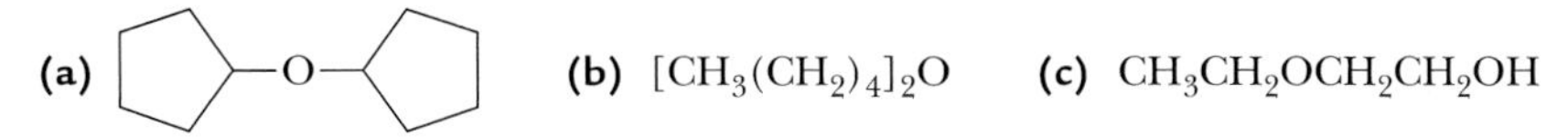

8.16 Name and draw structural formulas for the eight isomeric alcohols of molecular formula $C_5H_{12}O$. Which are chiral?

Physical Properties

8.17 Arrange these compounds in order of increasing boiling point. (Values in °C are − 42, 78, 117, and 198.)

(a) $CH_3CH_2CH_2CH_2OH$ **(b)** CH_3CH_2OH **(c)** $HOCH_2CH_2OH$

(d) $CH_3CH_2CH_3$

8.18 Arrange these compounds in order of increasing boiling point. (Values in °C are − 42, − 24, 78, and 118.)

(a) CH_3CH_2OH **(b)** CH_3OCH_3 **(C)** $CH_3CH_2CH_3$ **(d)** CH_3CO_2H

8.19 Propanoic acid and methyl acetate are constitutional isomers, and both are liquids at room temperature. One of these compounds has a boiling point of 141°C; the other has a boiling point of 57°C. Which compound has which boiling point?

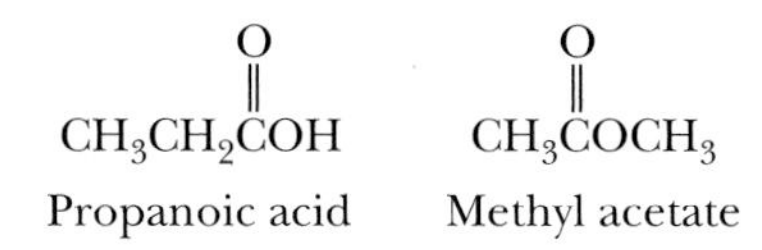

8.20 Compounds that contain an N—H group associate by hydrogen bonding.

(a) Do you expect this association to be stronger or weaker than that between compounds containing an O—H group?

(b) Based on your answer to part (a), which would you predict to have the higher boiling point, 1-butanol or 1-butanamine?

$CH_3CH_2CH_2CH_2OH$	$CH_3CH_2CH_2CH_2NH_2$
1-Butanol	1-Butanamine

8.21 Following are structural formulas for 1-butanol and 1-butanethiol. One of these compounds has a boiling point of 98.5°C, the other has a boiling point of 117°C. Which compound has which boiling point?

$CH_3CH_2CH_2CH_2OH$	$CH_3CH_2CH_2CH_2SH$
1-Butanol	1-Butanethiol

8.22 From each pair of compounds, select the one that is more soluble in water.

(a) CH_2Cl_2 or CH_3OH **(b)** $CH_3\overset{O}{\overset{\|}{C}}CH_3$ or $CH_3\overset{CH_2}{\overset{\|}{C}}CH_3$

(c) CH_3CH_2Cl or NaCl **(d)** $CH_3CH_2CH_2SH$ or $CH_3CH_2CH_2OH$

(e) $CH_3CH_2\overset{OH}{\overset{|}{C}}HCH_2CH_3$ or $CH_3CH_2\overset{O}{\overset{\|}{C}}CH_2CH_3$

8.23 Arrange the compounds in each set in order of decreasing solubility in water.

(a) Ethanol; butane; diethyl ether **(b)** 1-Hexanol; 1,2-hexanediol; hexane

8.24 Each compound given in this problem is a common organic solvent. From each pair of compounds, select the solvent with the greater solubility in water.

(a) CH_2Cl_2 or CH_3CH_2OH

(b) $CH_3CH_2OCH_2CH_3$ or CH_3CH_2OH

(c) $CH_3\overset{O}{\overset{\|}{C}}CH_3$ or $CH_3CH_2OCH_2CH_3$

(d) $CH_3CH_2OCH_2CH_3$ or $CH_3(CH_2)_3CH_3$

Synthesis of Alcohols

We have encountered three reactions for the synthesis of alcohols, including glycols.

1. Acid-catalyzed hydration of alkenes (Section 6.3B).
2. Oxidation of alkenes to glycols by OsO_4 (Section 6.4B).
3. Acid-catalyzed ring opening of epoxides to give glycols (Section 8.7).

8.25 Give the structural formula of an alkene or alkenes from which each alcohol or glycol can be prepared.

(a) 2-Butanol **(b)** 1-Methylcyclohexanol **(c)** 3-Hexanol

(d) 2-Methyl-2-pentanol **(e)** Cyclopentanol **(f)** 1,2-Propanediol

8.26 Addition of bromine to cyclopentene and acid-catalyzed hydrolysis of cyclopentene oxide are both stereoselective; each gives a trans product. Compare the mechanisms of these two reactions, and show how each accounts for the formation of the trans product.

Acidity of Alcohols and Thiols

8.27 From each pair, select the stronger acid. For each stronger acid, write a structural formula for its conjugate base.

(a) H_2O or H_2CO_3 **(b)** CH_3OH or CH_3CO_2H

(c) CH_3CH_2OH or CH_3CH_2SH

8.28 Arrange these compounds in order of increasing acidity (weakest to strongest).

$$CH_3CH_2CH_2OH \qquad CH_3CH_2\overset{O}{\overset{\|}{C}}OH \qquad CH_3CH_2CH_2SH$$

8.29 From each pair, select the stronger base. For each stronger base, write the structural formula of its conjugate acid.

(a) OH^- or CH_3O^- **(b)** $CH_3CH_2S^-$ or $CH_3CH_2O^-$

(c) $CH_3CH_2O^-$ or NH_2^-

8.30 Label the stronger acid, stronger base, weaker acid, and weaker base in each equilibrium. Also predict the position of each equilibrium. For pK_a values, see Table 2.1.

(a) $CH_3CH_2O^- + HCl \rightleftharpoons CH_3CH_2OH + Cl^-$

(b) $CH_3\overset{O}{\overset{\|}{C}}OH + CH_3CH_2O^- \rightleftharpoons CH_3\overset{O}{\overset{\|}{C}}O^- + CH_3CH_2OH$

8.31 Predict the position of equilibrium for each acid-base reaction; that is, does it lie considerably to the left or considerably to the right or are concentrations evenly balanced?

(a) $CH_3CH_2OH + Na^+OH^- \rightleftharpoons CH_3CH_2O^-Na^+ + H_2O$

(b) $CH_3CH_2SH + Na^+OH^- \rightleftharpoons CH_3CH_2S^-Na^+ + H_2O$

(c) $CH_3CH_2OH + CH_3CH_2S^-Na^+ \rightleftharpoons CH_3CH_2O^-Na^+ + CH_3CH_2SH$

(d) $CH_3CH_2S^-Na^+ + CH_3\overset{O}{\overset{\|}{C}}OH \rightleftharpoons CH_3CH_2SH + CH_3\overset{O}{\overset{\|}{C}}O^-Na^+$

Reactions of Alcohols

8.32 Show how to distinguish between cyclohexanol and cyclohexene by a simple chemical test. *Hint:* Treat each with Br_2 in CCl_4, and watch what happens.

8.33 Write equations for the reaction of 1-butanol, a primary alcohol, with these reagents.

(a) Na metal **(b)** HBr, heat **(c)** $K_2Cr_2O_7$, H_2SO_4, heat

(d) $SOCl_2$ **(e)** Pyridinium chlorochromate (PCC)

8.34 Write equations for the reaction of 2-butanol, a secondary alcohol, with these reagents.

(a) Na metal **(b)** H_2SO_4, heat **(c)** HBr, heat

(d) $K_2Cr_2O_7$, H_2SO_4, heat **(e)** $SOCl_2$

(f) Pyridinium chlorochromate (PCC)

8.35 When (R)-2-butanol is left standing in aqueous acid, it slowly loses its optical activity. When the organic material is recovered from the aqueous solution, only 2-butanol is found. Account for the observed loss of optical activity.

A *tert*-butyl methyl ether manufacturing plant. *(Ashland Petroleum)*

8.36 What is the most likely mechanism of this reaction? Draw a structural formula for the intermediate formed during the reaction.

$$CH_3CH_2C(CH_3)_2OH + HCl \longrightarrow CH_3CH_2C(CH_3)_2Cl + H_2O$$

8.37 Complete the equations for these reactions.

(a) $CH_3CH_2CH_2OH + H_2CrO_4 \longrightarrow$ **(b)** $CH_3CH(CH_3)CH_2CH_2OH + SOCl_2 \longrightarrow$

(c) 1-methylcyclohexanol (ring with CH_3 and OH on the same carbon) $+ HCl \longrightarrow$ **(d)** $HOCH_2CH_2CH_2CH_2OH + HBr \xrightarrow{heat}$

(e) cyclooctanol (ring with OH) $+ H_2CrO_4 \longrightarrow$ **(f)** cyclohexene $+ OsO_4, H_2O_2 \longrightarrow$

8.38 In the commercial synthesis of methyl *tert*-butyl ether (MTBE), an antiknock, octane-improving gasoline additive, 2-methylpropene and methanol are passed over an acid catalyst to give the ether. Propose a mechanism for this reaction.

$$CH_3C(CH_3){=}CH_2 + CH_3OH \xrightarrow{\text{acid catalyst}} CH_3C(CH_3)_2OCH_3$$

2-Methylpropene (Isobutylene) — Methanol — 2-Methoxy-2-methylpropane (Methyl *tert*-butyl ether, MTBE)

Syntheses

8.39 Show how to convert

(a) 1-Propanol to 2-propanol in two steps.

(b) Cyclohexene to cyclohexanone in two steps.

(c) Cyclohexanol to *cis*-1,2-cyclohexanediol in two steps.

(d) Propene to propanone (acetone) in two steps.

8.40 Show how to convert cyclohexanol to these compounds.

(a) Cyclohexene **(b)** Cyclohexane **(c)** Cyclohexanone

8.41 Show reagents and experimental conditions to synthesize these compounds from 1-propanol. Any derivative of 1-propanol prepared in an earlier part of this problem may then be used for a later synthesis.

(a) Propanal **(b)** Propanoic acid

(c) Propene **(d)** 2-Propanol

(e) 2-Bromopropane **(f)** 1-Chloropropane

(g) Propanone **(h)** 1,2-Propanediol

8.42 Show how to prepare each compound from 2-methyl-1-propanol (isobutyl alcohol). For any preparation involving more than one step, show each intermediate compound formed.

(a) $CH_3C(CH_3){=}CH_2$ **(b)** $CH_3C(CH_3)(OH)CH_3$ **(c)** $CH_3C(CH_3)(OH)CH_2OH$ **(d)** $CH_3CH(CH_3)CO_2H$

8.43 Show how to prepare each compound from 2-methylcyclohexanol. For any preparation involving more than one step, show each intermediate compound formed.

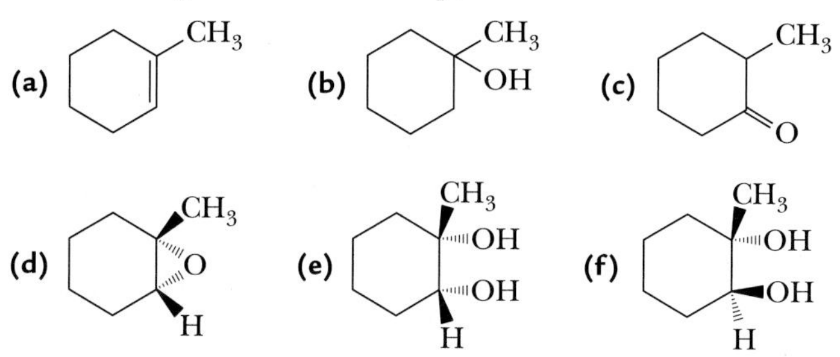

8.44 Show how to convert the alcohol on the left to compounds (a), (b), and (c).

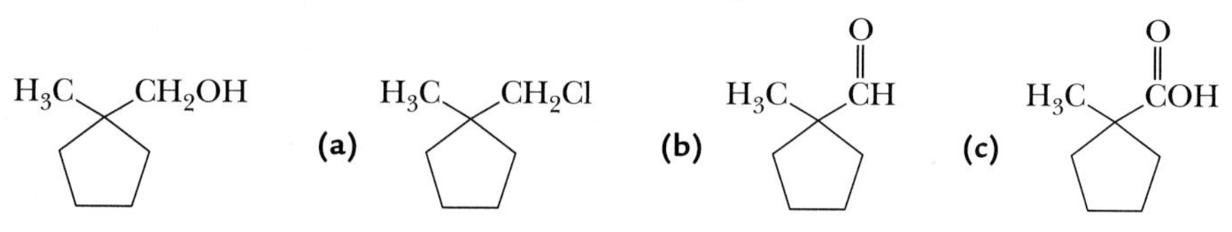

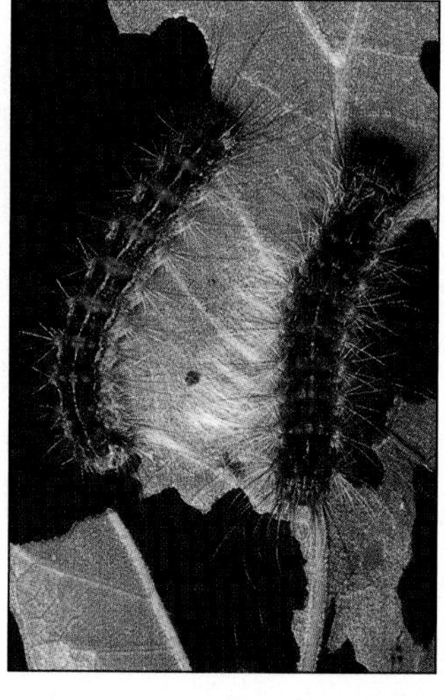

Disparlure is the sex attractant of the gypsy moth, *Porthetria dispar. (Animals, Animals/ © William D. Griffin)*

8.45 Disparlure, a sex attractant of the gypsy moth *(Porthetria dispar)*, has been synthesized in the laboratory from the following (Z)-alkene (*The Merck Index*, 12th ed., #3425).

(Z)-2-Methyl-7-octadecene → Disparlure

(a) How might the (Z)-alkene be converted to disparlure?

(b) How many stereoisomers are possible for disparlure? How many are formed in the sequence you chose?

8.46 The chemical name for bombykol, the sex pheromone secreted by the female silkworm moth to attract male silkworm moths, is *trans*-10-*cis*-12-hexadecadien-1-ol. (It has one hydroxyl group and two carbon-carbon double bonds in a 16-carbon chain.)

(a) Draw the structural formula for this bombykol, showing the correct configuration about each carbon-carbon double bond.

(b) How many cis-trans isomers are possible for this structural formula? All possible cis-trans isomers have been synthesized in the laboratory, but only the one named bombykol is produced by the female silkworm moth, and only it attracts male silkworm moths.

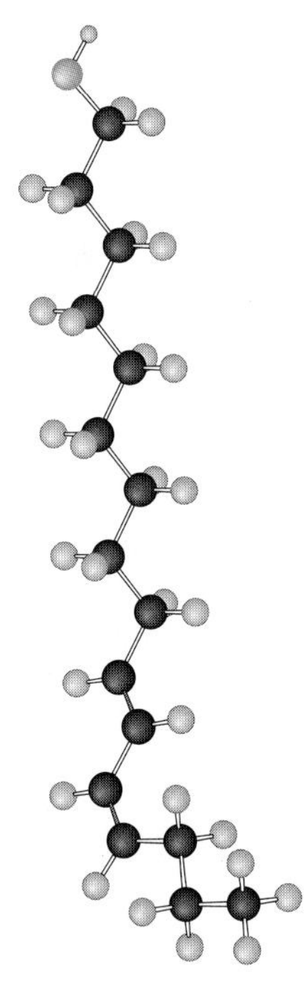

Bombykol

8.47 At this point, the following spectroscopy problems may be assigned:

^{1}H-NMR and ^{13}C-NMR: Additional Problems 21.15–21.18.

IR Problems: Additional Problems 22.5–22.6.

9

Benzene and Its Derivatives

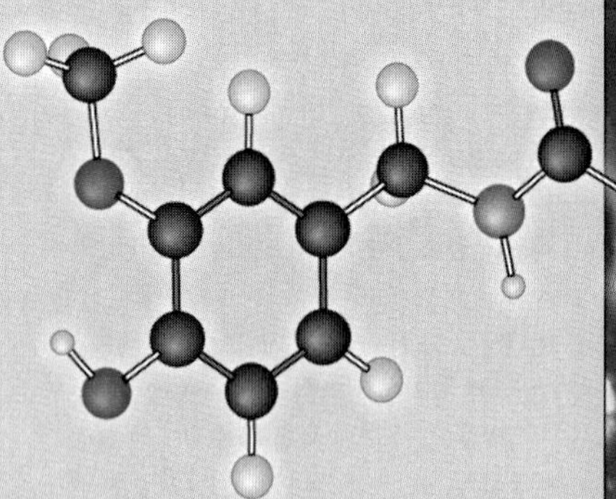

Peppers of the capsicum family. Inset: Model of capsaicin. *(Douglas Brown)*

Benzene is a colorless compound with a melting point of 6°C and a boiling point of 80°C. It was first isolated by Michael Faraday in 1825 from the oily residue that collected in the illuminating gas lines of London. Benzene's molecular formula, C_6H_6, suggests a high degree of unsaturation. For comparison, an alkane of six carbons has a molecular formula of C_6H_{14}, and a cycloalkane of six carbons has a molecular formula of C_6H_{12}. Considering benzene's high degree of unsaturation, it might be expected to show many of the reactions characteristic of alkenes. Yet, benzene is remarkably stable! It does not undergo the addition, oxidation, and reduction reactions characteristic of alkenes. For example, benzene does not react with bromine, hydrogen chloride, or other reagents that usually add to carbon-carbon double bonds. It is not oxidized by chromic acid or osmium tetroxide under conditions that readily oxidize alkenes. When benzene reacts, it does so by substitution in which a hydrogen atom is replaced by another atom or group of atoms.

The term **aromatic compound** was used to classify benzene and its derivatives because many of them have distinctive odors. It became clear, however, that a sounder classification for these compounds should be based on structure and chemical reactivity, not aroma. The term "aromatic" has been retained but now refers to the fact that these compounds are highly unsaturated and unexpectedly stable toward reagents that react with alkenes. The term **arene** is used to describe aromatic hydrocarbons, by analogy with alkane and alkene. Benzene is the parent arene. Just as a group derived by removal of an H from an alkane is called an alkyl group and given the symbol R—, a group derived by removal of an H from an arene is called an **aryl group** and given the symbol **Ar**—.

Aromatic compound A term used to classify benzene and its derivatives.

Arene An aromatic hydrocarbon.

Aryl group A group derived from an aromatic compound (an arene) by removal of an H; given the symbol Ar—.

Ar— The symbol used for an aryl group, by analogy with R— for an alkyl group.

9.1 THE STRUCTURE OF BENZENE

Let us put ourselves in the mid–19th century and examine the evidence on which chemists attempted to build a model for the structure of benzene. First, because the molecular formula of benzene is C_6H_6, it seemed clear that the molecule must be highly unsaturated. Yet benzene does not show the chemical properties of alkenes, the only unsaturated hydrocarbons known at that time. Benzene does undergo chemical reactions, but its characteristic reaction is substitution rather than addition. When benzene is treated with bromine in the presence of ferric chloride as a catalyst, for example, only one compound of molecular formula C_6H_5Br is formed. Chemists concluded from this observation that all six carbons and all six hydrogens of benzene must be equivalent.

$$\underset{\text{Benzene}}{C_6H_6} + Br_2 \xrightarrow{FeCl_3} \underset{\text{Bromobenzene}}{C_6H_5Br} + HBr$$

A. Kekulé's Model of Benzene

The first structure for benzene was proposed by August Kekulé in 1872 and consisted of a six-membered ring with alternating single and double bonds and with one hydrogen attached to each carbon. Because all carbons and hydrogens of Kekulé's structure are equivalent, substitution of bromine for any one of the hydrogens gives the same compound. Thus, Kekulé's proposed structure was consistent with the fact that treatment of benzene with bromine in the presence of ferric chloride gives only one compound of molecular formula C_6H_5Br.

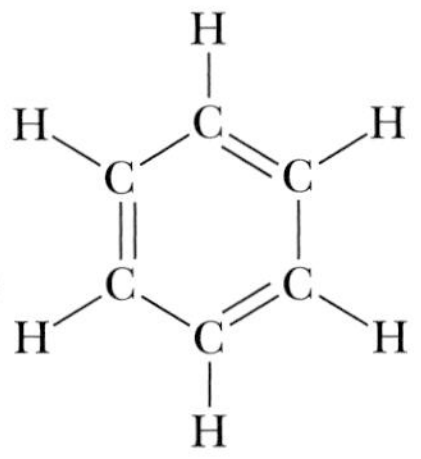

Kekulé structure, showing all atoms

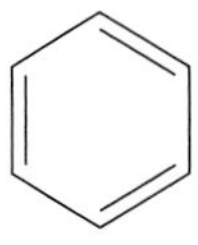

Kekulé structure, as line-angle drawing

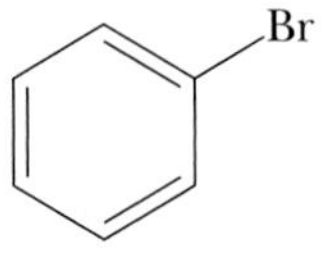

Kekulé structure for bromobenzene

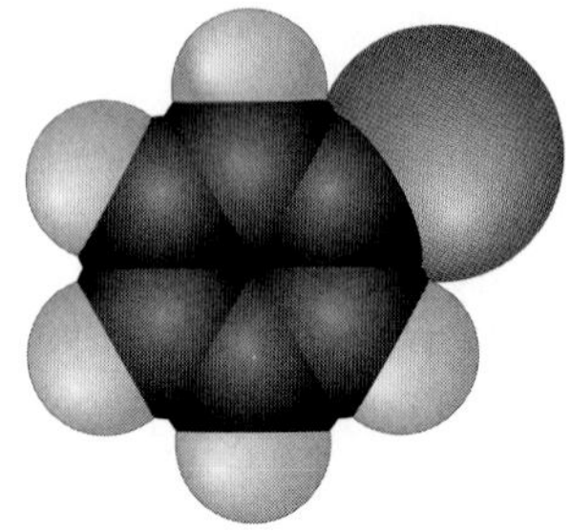

Although Kekulé's proposal was consistent with this and many other experimental observations, it was contested for years. The major objection was that it did not account for the unusual chemical behavior of benzene. If benzene contains three double bonds, Kekulé's critics asked, why does it not show reactions typical of alkenes? Why, for example, does benzene react with bromine by substitution rather than addition?

B. The Valence Bond Model of Benzene

The concepts of **hybridization of atomic orbitals** and the **theory of resonance,** developed in the 1930s, provided the first adequate description of the structure of benzene. The carbon skeleton of benzene forms a regular hexagon with C—C—C and H—C—C bond angles of 120°. For this type of bonding, carbon uses sp^2 hybrid orbitals (Section 1.6D). Each carbon forms sigma bonds to two adjacent carbons by overlap of sp^2-sp^2 hybrid orbitals, and one sigma bond to hydrogen by overlap of sp^2-$1s$ orbitals. As determined experimentally, all carbon-carbon bonds in benzene are the same length, 1.39 Å, a value midway between the length of a single bond between sp^3 hybridized carbons (1.54 Å) and a double bond between sp^2 hybridized carbons (1.33 Å).

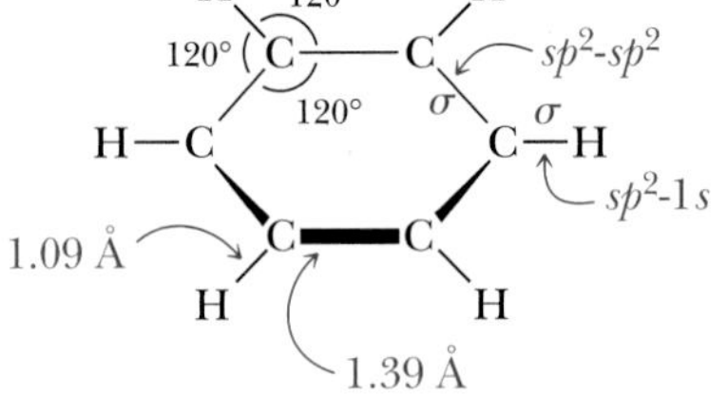

Each carbon also has a single unhybridized $2p$ orbital that contains one electron. These six $2p$ orbitals lie perpendicular to the plane of the ring and overlap to form a continuous pi cloud encompassing all six carbon atoms. The electron density of the pi system of a benzene ring lies in one torus (a doughnut-shaped region) above the plane of the ring and a second torus below the plane of the ring (Figure 9.1).

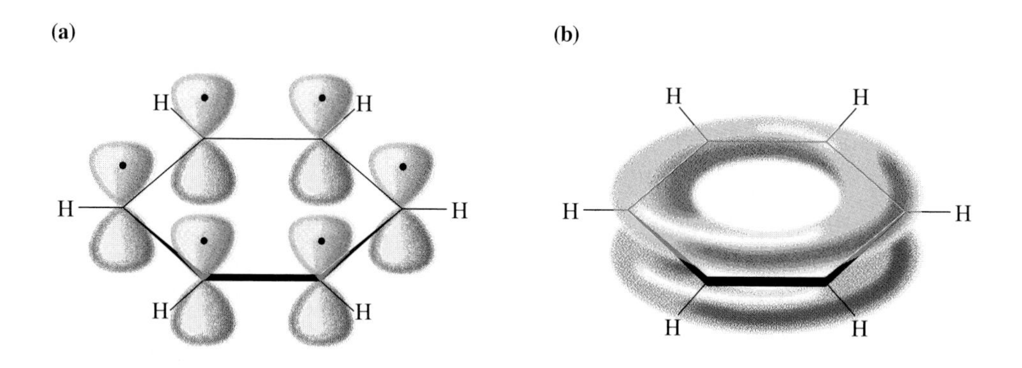

Figure 9.1 Valence bond model of bonding in benzene. (a) The carbon, hydrogen framework. The six $2p$ orbitals, each with one electron, are shown uncombined. (b) Overlap of parallel $2p$ orbitals forms a continuous pi cloud, shown by one torus above the plane of the ring and a second below the plane of the ring.

C. The Resonance Model of Benzene

One of the postulates of resonance theory is that, when a molecule or ion can be represented by two or more contributing structures, then it is not adequately represented by any single contributing structure. We represent benzene as a hybrid of two equivalent contributing structures, often referred to as Kekulé structures. Each Kekulé structure makes an equal contribution to the hybrid, and thus the C—C bonds are neither single nor double bonds but something intermediate.

Benzene as a hybrid of two equivalent contributing structures

We recognize that neither of these contributing structures exists; they are merely alternative ways to pair $2p$ orbitals with no reason to prefer one or the other. Nevertheless, chemists continue to use a single contributing structure to represent this molecule because it is as close as we can come to an accurate structure within the limitations of classical valence bond structures and the tetravalence of carbon.

D. The Resonance Energy of Benzene

Resonance energy is the difference in energy between a resonance hybrid and its most stable contributing structure. One way to estimate the resonance energy of benzene is to compare the heats of hydrogenation of cyclohexene and benzene. Cyclohexene is readily reduced to cyclohexane by hydrogen in the presence of a transition metal catalyst (Section 6.5).

Resonance energy The difference in energy between a resonance hybrid and the most stable of its hypothetical contributing structures.

$$\text{Cyclohexene} + H_2 \xrightarrow[1\text{–}2\text{ atm}]{\text{Ni}} \text{Cyclohexane} \qquad \Delta H^0 = -28.6\text{ kcal/mol } (-120\text{ kJ/mol})$$

Benzene is reduced only very slowly to cyclohexane under these conditions. It is reduced more rapidly when heated and under a pressure of several hundred atmospheres of hydrogen.

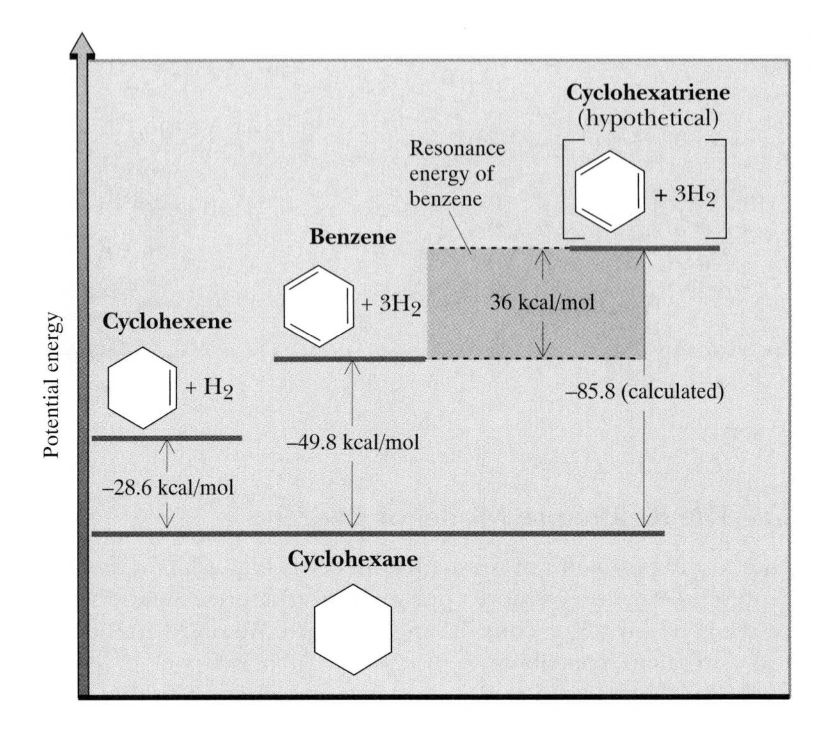

Figure 9.2
The resonance energy of benzene as determined by comparison of heats of hydrogenation of cyclohexene and benzene.

$$\text{Benzene} + 3H_2 \xrightarrow[\text{200–300 atm}]{\text{Ni}} \text{Cyclohexane} \qquad \Delta H^0 = -49.8 \text{ kcal/mol} \; (-208 \text{ kJ/mol})$$

Benzene Cyclohexane

Reduction of cyclohexene is exothermic by 28.6 kcal/mol (120 kJ/mol). Because benzene can be drawn with three carbon-carbon double bonds, we might expect its hydrogenation to be exothermic by 3×28.6 or 85.8 kcal/mol (359 kJ/mol). Instead, catalytic reduction of benzene to cyclohexane gives off only 49.8 kcal/mol (208 kJ/mol). The difference between the expected value and the experimentally observed value, 36.0 kcal/mol (151 kJ/mol), is the **resonance energy of benzene.** These experimental results are shown graphically in Figure 9.2.

For comparison, the strength of a carbon-carbon single bond is approximately 80–100 kcal/mol (333–418 kJ/mol), and that of hydrogen bonding in water and low-molecular-weight alcohols is approximately 2–5 kcal/mol (8.4–21 kJ/mol). Thus, although the resonance energy of benzene is less than the strength of a carbon-carbon single bond, it is considerably greater than the strength of hydrogen bonding in water and alcohols. We saw in Section 8.3 that hydrogen bonding has a dramatic effect on the physical properties of alcohols compared with alkanes. We see in this chapter that the resonance energy of benzene and other aromatic hydrocarbons has a dramatic effect on their chemical reactivity.

9.2 The Concept of Aromaticity

Many other types of molecules besides benzene and its derivatives show aromatic character; that is, they contain high degrees of unsaturation yet fail to undergo characteristic alkene addition and oxidation/reduction reactions. What chemists have sought to understand is, What are the principles underlying aromatic character? The answer to this problem was first recognized in the 1930s by the German chemical physicist, Eric Hückel.

Hückel's criteria are summarized as follows. To be aromatic, a five- or six-membered ring must

1. Be planar.
2. Have one $2p$ orbital on each atom of the ring.
3. Have six pi electrons (an aromatic sextet) in the cyclic arrangement of $2p$ orbitals.

Let us apply these criteria to the following heterocyclic compounds, all of which are aromatic.

A **heterocyclic compound** is one that contains more than one kind of atom in a ring. As the term "heterocyclic" is used in organic chemistry, it refers to a ring in which one or more of the atoms are different from carbon. Pyridine and pyrimidine are heterocyclic analogs of benzene. In pyridine, one CH group of benzene is replaced by a nitrogen atom, and in pyrimidine, two CH groups are replaced by nitrogen atoms. Each molecule meets the Hückel criteria for aromaticity. Each is cyclic and planar, has one $2p$ orbital on each atom of the ring, and has six electrons in the pi system. In pyridine, nitrogen is sp^2 hybridized, and its unshared pair of electrons occupies an sp^2 orbital perpendicular to the $2p$ orbitals of the pi system and, thus, is not a part of the pi system. In pyrimidine, neither unshared pair of electrons of nitrogen is part of the pi system. The resonance energy of pyridine is 32 kcal/mol (134 kJ/mol), slightly less than that of benzene. The resonance energy of pyrimidine is 26 kcal/mol (109 kJ/mol).

Heterocyclic compound An organic compound that contains one or more atoms other than carbon in its ring.

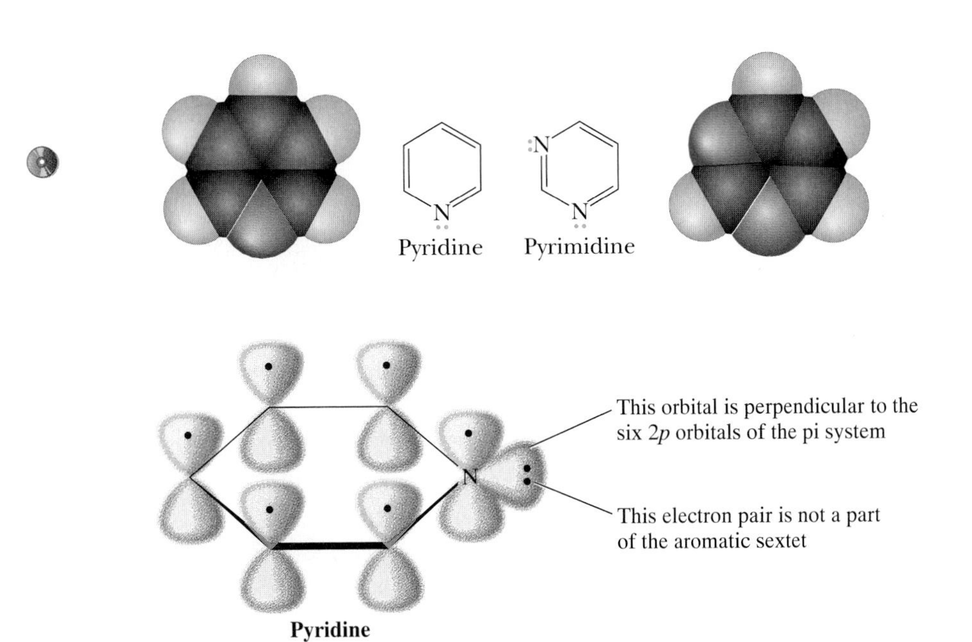

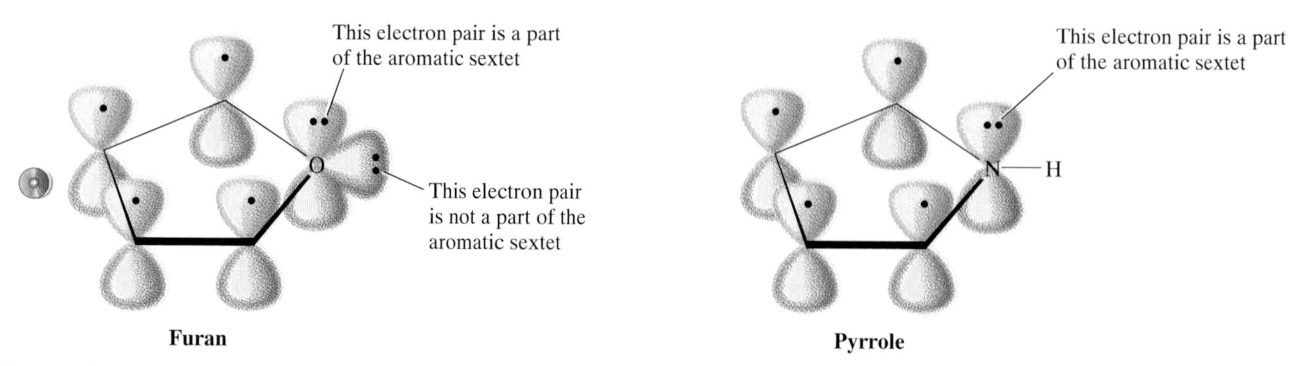

Figure 9.3
Origin of the 6 pi electrons (the aromatic sextet) in furan and pyrrole. The resonance energy of furan is 16 kcal/mol (67 kJ/mol); that of pyrrole is 21 kcal/mol (88 kJ/mol).

The five-membered-ring compounds furan, pyrrole, and imidazole are also aromatic.

Furan Pyrrole Imidazole

In these planar heterocyclic compounds, each heteroatom is sp^2 hybridized, and its unhybridized $2p$ orbital is part of a continuous cycle of five $2p$ orbitals. In furan, one unshared pair of electrons of the heteroatom lies in the unhybridized $2p$ orbital and is a part of the pi system (Figure 9.3). The other unshared pair of electrons lies in an sp^2 hybrid orbital, perpendicular to the $2p$ orbitals, and is not a part of the pi system. In pyrrole, the unshared pair of electrons on nitrogen is part of the aromatic sextet. In imidazole, the unshared pair of electrons on one nitrogen is part of the aromatic sextet; the unshared pair on the other nitrogen is not.

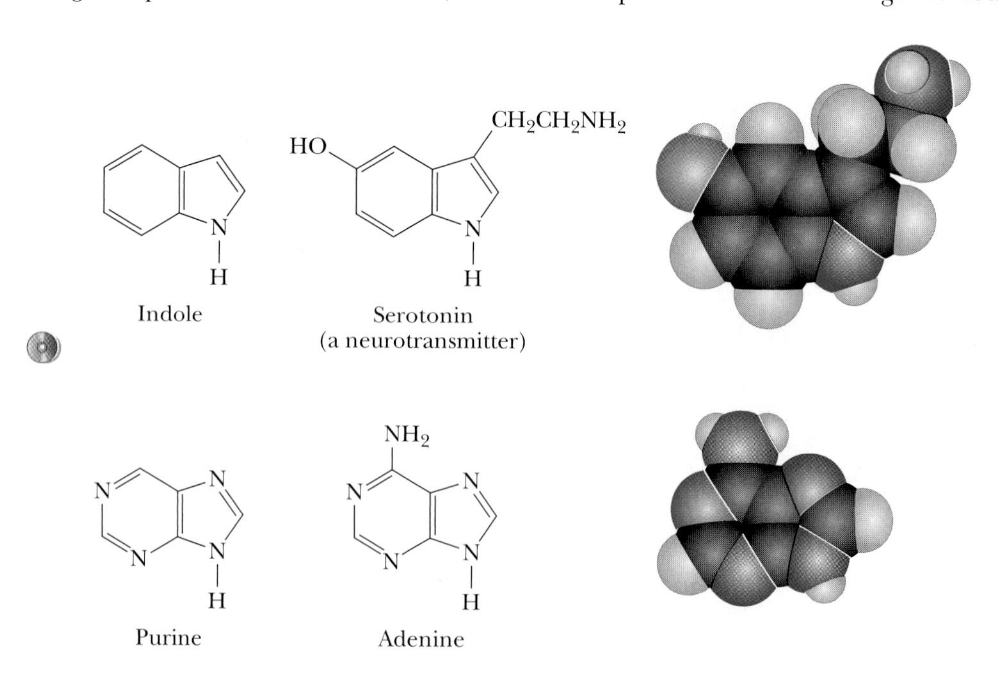

Nature abounds with compounds having a heterocyclic ring fused to one or more other rings. Two such compounds especially important in the biological world are indole and purine. Indole contains a pyrrole ring fused with a benzene ring. Compounds derived from indole include the amino acid L-tryptophan (Section 18.1A) and the neurotransmitter serotonin. Purine contains a six-membered pyrimidine ring fused with a five-membered imidazole ring. Adenine is one of the building blocks of deoxyribonucleic acids (DNA) and ribonucleic acids (RNA) as described in Chapter 19. It is also a component of the biological oxidizing agent nicotinamide adenine dinucleotide, abbeviated NAD^+ (Section 20.1B).

9.3 NOMENCLATURE

A. Monosubstituted Benzenes

Monosubstituted alkylbenzenes are named as derivatives of benzene, as for example ethylbenzene. The IUPAC system retains certain common names for several of the simpler monosubstituted alkylbenzenes. Examples are **toluene** (rather than methylbenzene) and **styrene** (rather than phenylethylene).

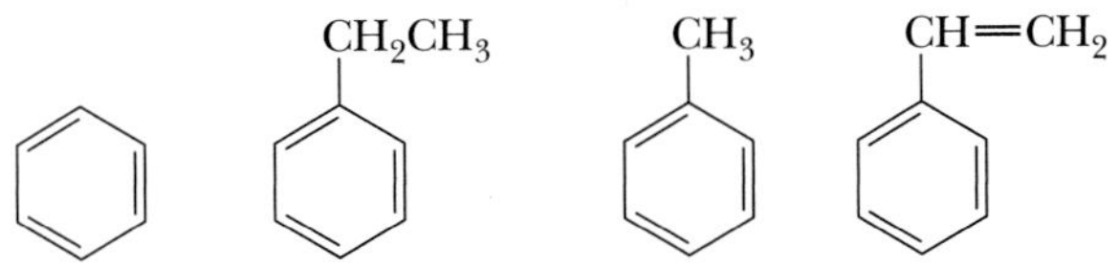

CH_2CH_3 CH_3 $CH{=}CH_2$

Benzene Ethylbenzene Toluene Styrene

The common names **phenol, aniline, benzaldehyde, benzoic acid,** and **anisole** are also retained by the IUPAC system.

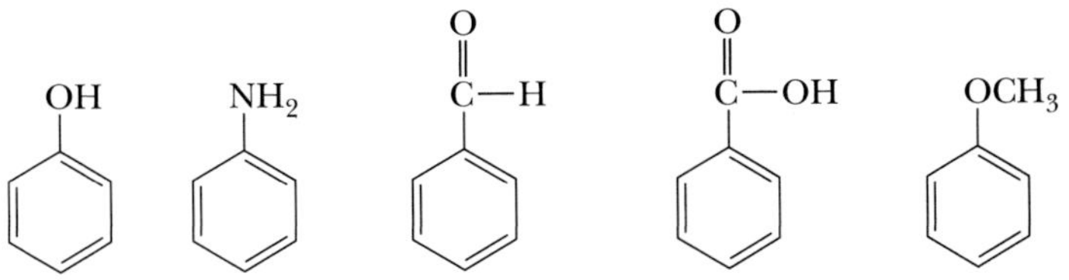

OH NH_2 O ‖ C—H O ‖ C—OH OCH_3

Phenol Aniline Benzaldehyde Benzoic acid Anisole

As noted in the introduction to Chapter 5, the substituent group derived by loss of an H from benzene is a **phenyl group;** that derived by loss of an H from the methyl group of toluene is a **benzyl group.**

Phenyl group C_6H_5—, the aryl group derived by removing a hydrogen from benzene.

Benzyl group $C_6H_5CH_2$—, the alkyl group derived by removing a hydrogen from the methyl group of toluene.

a benzylic carbon

—CH_3 —CH_2—

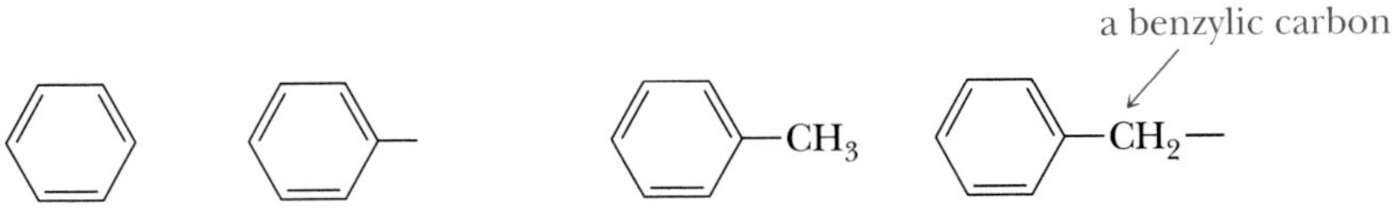

Benzene Phenyl group Toluene Benzyl group

In molecules containing other functional groups, phenyl groups and benzyl groups are often named as substituents.

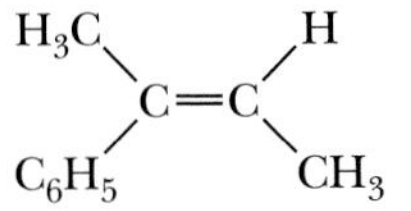

H_3C H C=C C_6H_5 CH_3 $C_6H_5CH_2CH_2OH$ $C_6H_5CH_2Cl$

(Z)-2-Phenyl-2-butene 2-Phenylethanol Benzyl chloride

B. Disubstituted Benzenes

Ortho (*o*) Refers to groups occupying 1,2 positions on a benzene ring.

Meta (*m*) Refers to groups occupying 1,3 positions on a benzene ring.

Para (*p*) Refers to groups occupying 1,4 positions on a benzene ring.

When two substituents occur on a benzene ring, three constitutional isomers are possible. The substituents may be located by numbering the atoms of the ring or by using the locators **ortho, meta,** and **para.** 1,2- is equivalent to ortho (Greek: straight); 1,3- is equivalent to meta (Greek: after); and 1,4- is equivalent to para (Greek: beyond).

When one of the two substituents on the ring imparts a special name to the compound, as for example toluene, phenol, and aniline, then the compound is named as a derivative of that parent molecule. The special substituent is assumed to occupy ring position number 1. The IUPAC system retains the common name **xylene** for the three isomeric dimethylbenzenes. Where neither group imparts a special name, the two substituents are located and listed in alphabetical order before the ending -benzene. The carbon of the benzene ring with the substituent of lower alphabetical ranking is numbered C-1.

CH_3 / Br — 4-Bromotoluene (*p*-Bromotoluene)

NH_2 / Cl — 3-Chloroaniline (*m*-Chloroaniline)

CH_3 / CH_3 — 1,3-Dimethylbenzene (*m*-Xylene)

CH_2CH_3 / Cl — 1-Chloro-4-ethylbenzene (*p*-Chloroethylbenzene)

C. Polysubstituted Benzenes

When three or more substituents are present on a ring, their locations are specified by numerals. If one of the substituents imparts a special name, then the molecule is named as a derivative of that parent molecule. If none of the substituents imparts a special name, then the substituents are located, numbered to give the smallest set of numbers, and listed in alphabetical order before the ending -benzene. In the following examples, the first compound is a derivative of toluene, and the second is a derivative of phenol. Because there is no special name for the third compound, its three substituents are listed in alphabetical order followed by the word "benzene."

CH_3 (1), NO_2 (2), Cl (4)

4-Chloro-2-nitrotoluene

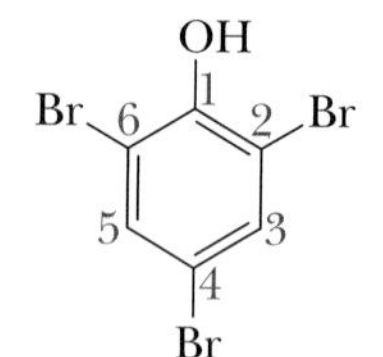

2,4,6-Tribromophenol

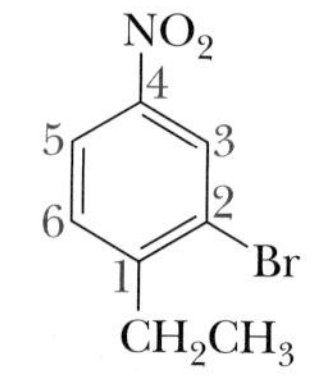

2-Bromo-1-ethyl-4-nitrobenzene

EXAMPLE 9.1

Write names for these compounds.

(a) H_3C, I (b) CO_2H, Br, Br (c) Cl, NO_2, NO_2 (d) $CH_2CH{=}CH_2$

SOLUTION

(a) 3-Iodotoluene or *m*-iodotoluene (b) 3,5-Dibromobenzoic acid
(c) 1-Chloro-2,4-dinitrobenzene (d) 3-Phenylpropene

Practice Problem 9.1

Write names for these compounds.

(a) OH, CCH_3, CH_3 (b) C_6H_5, CH_2CH_3, C=C, CH_3CH_2, C_6H_5 (c) CO_2H, CH_3

Polynuclear aromatic hydrocarbons (PAHs) contain two or more aromatic rings, each pair of which shares two ring carbon atoms. Naphthalene, anthracene, and phenanthrene, the most common PAHs, and substances derived from them are found in coal tar and high-boiling petroleum residues. At one time, naphthalene was used as a moth repellent and insecticide in preserving woolens and furs, but its use has decreased due to the introduction of chlorinated hydrocarbons such as *p*-dichlorobenzene.

Polynuclear aromatic hydrocarbon A hydrocarbon containing two or more fused aromatic rings.

Naphthalene Anthracene Phenanthrene

Also found in coal tar are lesser amounts of the following PAHs. These compounds can be found in the exhausts of gasoline-powered internal combustion engines (for example, automobile engines) and in cigarette smoke. Benzo[a]pyrene has attracted particular interest because it is a very potent carcinogen and mutagen.

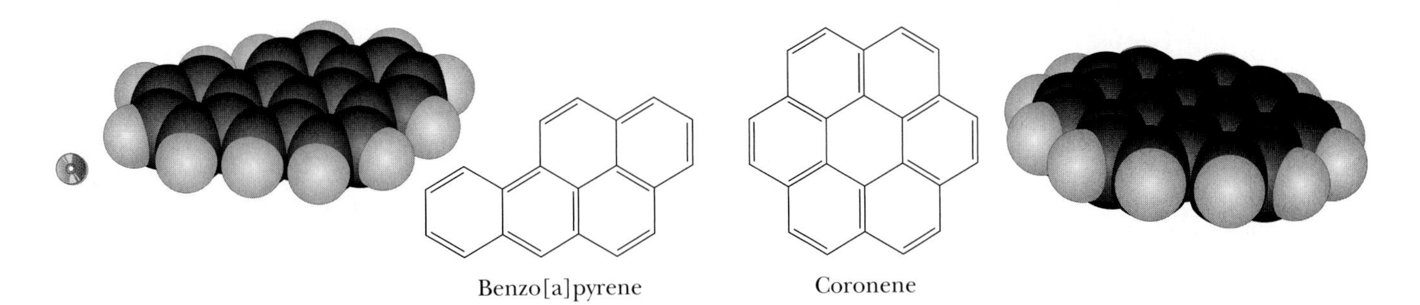

Benzo[a]pyrene Coronene

Thymol is a constituent of garden thyme, *Thymus vulgaris*. (*© Connie Toops*)

Phenol A compound that contains an —OH bonded to a benzene ring.

9.4 Phenols

A. Structure and Nomenclature

The functional group of a **phenol** is a hydroxyl group bonded to a benzene ring. Substituted phenols are named either as derivatives of phenol or by common names.

Phenol	3-Methylphenol (m-Cresol)	1,2-Benzenediol (Catechol)	1,3-Benzenediol (Resorcinol)	1,4-Benzenediol (Hydroquinone)

Phenols are widely distributed in nature. Phenol itself and the isomeric cresols (o-, m-, and p-cresol) are found in coal tar. Thymol and vanillin are important constituents of thyme and vanilla beans respectively.

2-Isopropyl-5-methylphenol (Thymol)

4-Hydroxy-3-methoxybenzaldehyde (Vanillin)

Phenol, or carbolic acid as it was once called, is a low-melting solid only slightly soluble in water. In sufficiently high concentrations, it is corrosive to all kinds of cells. In dilute solutions, it has some antiseptic properties and was introduced into the practice of surgery by Joseph Lister who demonstrated his technique of aseptic surgery in the surgical theater of the University of Glasgow School of Medicine in 1865. Phenol has been replaced by antiseptics that are both more powerful and have fewer undesirable side effects. Among these are hexylresorcinol (*The Merck Index,* 12th ed., #4750), which is widely used in nonprescription preparations as a mild antiseptic and disinfectant. Eugenol (*The Merck Index,* 12th ed., #3944), which can be isolated from the flower buds (cloves) of *Eugenia aromatica,* is used as a dental antiseptic and analgesic.

Eugenol, which is used in dentistry as an antiseptic agent, is isolated from the plant *Eugenia aromatica*. (*G. Büttner/Naturbild/OKAPIA/Photo Researchers, Inc.*)

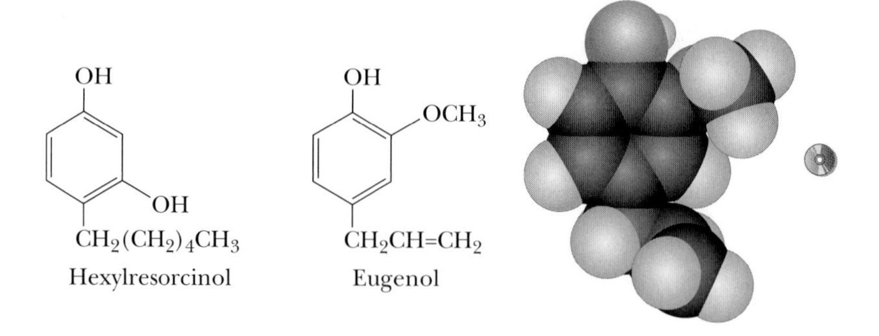

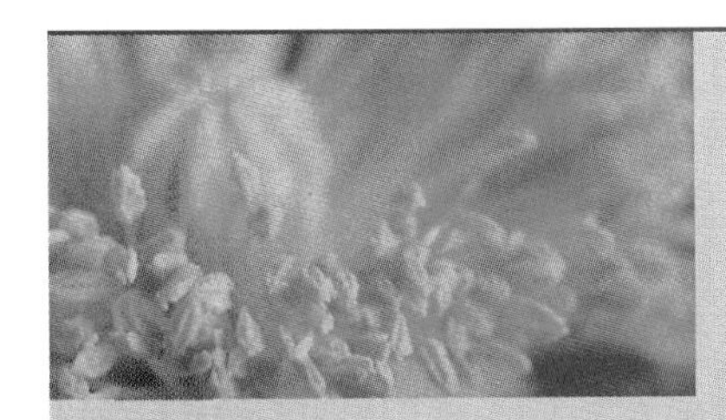

CHEMICAL CONNECTIONS

Capsaicin, for Those Who Like It Hot

Capsaicin, the pungent principle from the fruit of various species of peppers (Capsicum and Solanaceae), was isolated in 1876 and its structure was determined in 1919 (*The Merck Index,* 12th ed., #1811). The inflammatory properties of capsaicin are well known; as little as one drop in 5 L of water can be detected by the human tongue. We all know of the burning sensation in the mouth and sudden tearing in the eyes caused by a good dose of hot chili peppers. Capsaicin-containing extracts from these flaming foods are also used in sprays to ward off dogs or other animals that might nip at your heels while you are running or cycling.

Ironically, capsaicin is able to cause pain and relieve it as well. Currently, two capsaicin-containing creams, Mioton and Zostrix, are prescribed to treat the burning pain associated with postherpetic neuralgia, a complication of shingles. They are also prescribed for diabetics to relieve persistent foot and leg pain.

The mechanism by which capsaicin relieves these pains is not fully understood. It has been suggested, however, that, after application, the nerve endings in the area responsible for the transmission of pain remain temporarily numb. Capsaicin remains bound to specific receptor sites on these pain-transmitting neurons, blocking them from further action. Eventually, capsaicin is removed from these receptor sites, but in the meantime, its presence provides needed relief from pain.

CH_3O HO N H O

Capsaicin
(from various types of peppers)

B. Acidity of Phenols

Phenols and alcohols both contain a hydroxyl group, —OH. Phenols, however, are grouped as a separate class of compounds because their chemical properties are quite different from those of alcohols. One of the most important of these differences is that phenols are significantly more acidic than are alcohols. The acid ionization constant for phenol is 10^6 times larger than that of ethanol!

$$C_6H_5OH + H_2O \rightleftharpoons C_6H_5O^- + H_3O^+ \qquad K_a = 1.1 \times 10^{-10} \qquad pK_a = 9.95$$

Phenol — Phenoxide ion

$$CH_3CH_2OH + H_2O \rightleftharpoons CH_3CH_2O^- + H_3O^+ \qquad K_a = 1.3 \times 10^{-16} \qquad pK_a = 15.9$$

Ethanol — Ethoxide ion

TABLE 9.1 Relative Acidities of 0.1 *M* Solutions of Ethanol, Phenol, and HCl

Acid Ionization Equation	$[H^+]$	pH
$CH_3CH_2OH + H_2O \rightleftharpoons CH_3CH_2O^- + H_3O^+$	1×10^{-7}	7.0
$C_6H_5OH + H_2O \rightleftharpoons C_6H_5O^- + H_3O^+$	3.3×10^{-6}	5.4
$HCl + H_2O \rightleftharpoons Cl^- + H_3O^+$	0.1	1.0

Another way to compare the relative acid strengths of ethanol and phenol is to look at the hydrogen ion concentration and pH of a 0.1 *M* aqueous solution of each (Table 9.1). For comparison, the hydrogen ion concentration and pH of 0.1 *M* HCl are also included.

In aqueous solution, alcohols are neutral substances, and the hydrogen ion concentration of 0.1 *M* ethanol is the same as that of pure water. A 0.1 *M* solution of phenol is slightly acidic and has a pH of 5.4. By contrast, 0.1 *M* HCl, a strong acid (completely ionized in aqueous solution), has a pH of 1.0.

Phenols are more acidic than alcohols because the delocalization of the negative charge by resonance stabilizes the phenoxide ion relative to an alkoxide ion. Delocalization of the charge on the phenoxide ion can be seen in the following contributing structures. In the two contributing structures on the left, the negative charge is placed on oxygen. The three contributing structures on the right place the negative charge on the ortho and para positions of the ring. These contributing structures spread the negative charge of the phenoxide ion so that it is delocalized over four atoms. This delocalization of negative charge stabilizes the phenoxide ion relative to an alkoxide ion, for which no delocalization is possible.

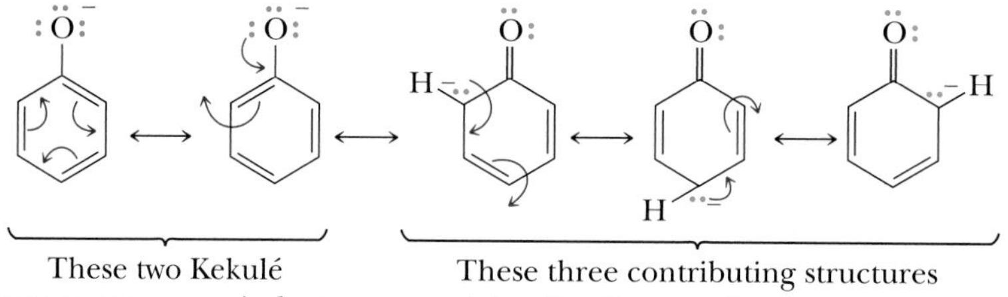

These two Kekulé structures are equivalent

These three contributing structures delocalize the negative charge onto carbon atoms of the ring

Note that although the resonance model gives us a way of understanding why phenol is a stronger acid than ethanol, it does not provide us with any quantitative means of predicting just how much stronger an acid it might be. To find out how much stronger one acid is compared with another, we must determine their pK_a values experimentally and compare them.

Ring substituents, particularly halogen and nitro groups, have marked effects on the acidities of phenols. Because the halogens are more electronegative than carbon, they withdraw electron density from the aromatic ring, weaken the O—H bond, and stabilize the phenoxide ion. Nitro groups also withdraw electron density from the ring, weaken the O—H bond, stabilize the phenoxide ion, and, thus, increase acidity.

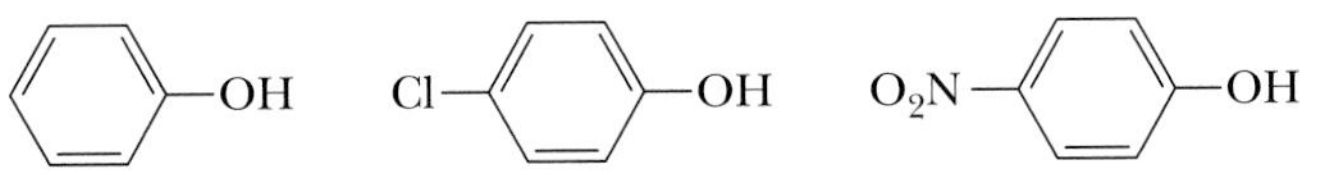

Phenol, pK_a 9.95 — 4-Chlorophenol, pK_a 9.18 — 4-Nitrophenol, pK_a 7.15

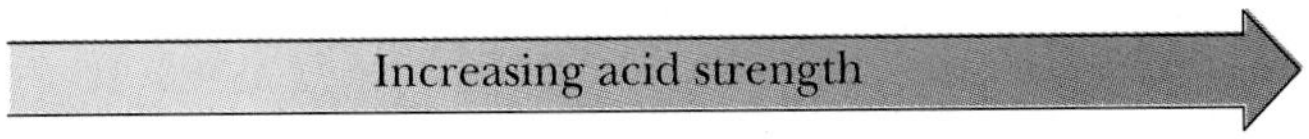

Increasing acid strength →

EXAMPLE 9.2

Arrange these compounds in order of increasing acidity: 2,4-dinitrophenol, phenol, and benzyl alcohol.

SOLUTION

Benzyl alcohol, a primary alcohol, has a pK_a of approximately 16–18 (Section 8.4A). The pK_a of phenol is 9.95. Nitro groups are electron-withdrawing and increase the acidity of the phenolic —OH group. In order of increasing acidity, these compounds are

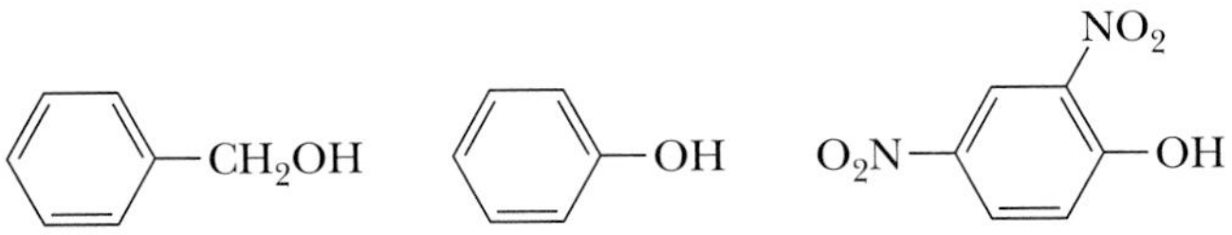

$C_6H_5CH_2OH$	C_6H_5OH	2,4-$(O_2N)_2C_6H_3OH$
Benzyl alcohol	Phenol	2,4-Dinitrophenol
pK_a 16–18	pK_a 9.95	pK_a 3.96

Practice Problem 9.2

Arrange these compounds in order of increasing acidity: 2,4-dichlorophenol, phenol, cyclohexanol.

C. Acid-Base Reactions of Phenols

Phenols are weak acids and react with strong bases such as NaOH to form water-soluble salts.

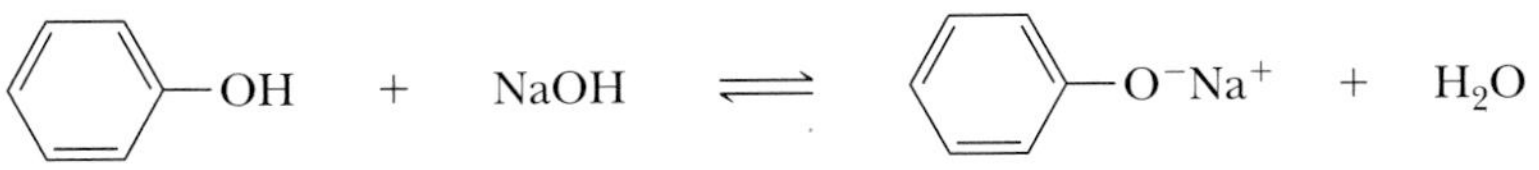

$$C_6H_5OH + NaOH \rightleftharpoons C_6H_5O^-Na^+ + H_2O$$

Phenol	Sodium hydroxide	Sodium phenoxide	Water
pK_a 9.95			pK_a 15.7
(stronger acid)	(stronger base)	(weaker base)	(weaker acid)

Most phenols do not react with weaker bases such as sodium bicarbonate; they do not dissolve in aqueous sodium bicarbonate. Carbonic acid is a stronger acid than most phenols and, consequently, the equilibrium for their reaction with bicarbonate ion lies far to the left. (Review Section 2.4.)

CHEMICAL CONNECTIONS

Phytoalexins—Natural Plant Antibiotics

Insect pests and fungal infections cause enormous losses for food and fiber crops throughout the world. Even though these losses attract attention, they often obscure the fact that resistance rather than susceptibility is the rule in nature. Although all plants are exposed to a very wide range of potentially pathogenic fungi, they are completely resistant to most of them.

Scientists have discovered that one mechanism of resistance is a buildup of protective chemicals at the site of attempted fungal infection. These natural plant antibiotics are given the name phytoalexins, a word derived from the Greek *phyto* (a plant) and *alexein* (to ward off, protect). Even though systematic investigations of these natural plant defense mechanisms are only now being carried out, it is clear that phytoalexin synthesis is a common mechanism of disease resistance in higher plants.

Among the several hundred phytoalexins characterized to date, there are wide variations in structure. Pisatin, for example, a fungitoxin induced in the sweet pea, has five phenol ether groups within its structure. Wyerone from the broad bean contains a carbon-carbon triple bond.

CH_3O ... OH

Pisatin
(from the sweet pea)

Wyerone
(from the broad bean)

The most thoroughly studied phytoalexins, those of cotton (Gossypium), include the terpene hemigossypol, its oxidation product gossypol (the yellow pigment of cottonseed), and a host of related terpene compounds. In addition to its role as a phytoalexin, gossypol is toxic to a variety of herbivores and various insects. There have also been reports from the Peoples Republic of China that it acts as a male infertility agent in humans.

$$C_6H_5OH + NaHCO_3 \rightleftharpoons C_6H_5O^-Na^+ + H_2CO_3$$

Phenol	Sodium bicarbonate	Sodium phenoxide	Carbonic acid
pK_a 9.95			pK_a 6.36
(weaker acid)	(weaker base)	(stronger base)	(stronger acid)

The fact that phenols are weakly acidic, whereas alcohols are neutral, provides a very convenient way to separate phenols from water-insoluble alcohols. Suppose that we want to separate phenol from cyclohexanol. Each is only slightly soluble in

Hemigossypol
(from cotton)

peroxidase →

Gossypol
(from cotton)

The failure of these plant antifungal agents to halt the progress of certain pathogenic fungi is not necessarily an indication of their ineffectiveness. What it means, instead, is that fungal parasites have coevolved with the plant and developed a biochemical machinery to detoxify the plant phytoalexins as soon as they are produced.

When we breed new resistant varieties of plants, for example, perhaps we are selecting an ability to produce more effective phytoalexins. If we can understand the biochemistry of their production, perhaps we can select and then transfer this ability by genetic engineering from resistant to susceptible varieties. Alternatively, an understanding of the chemical structure of phytoalexins and their modes of action may provide us with clues to the development of new and safer synthetic fungicides. See *Natural Resistance of Plants to Pests: Roles of Allelochemicals,* M. B. Green and P. A Hedin, editors, ACS Symposium Series No. 387, American Chemical Society, Washington, DC, 1986.

A sweet pea plant, *Lathyrus latifolius.* *(Runk/Schoenberger from Grant Heilman)*

water; therefore, they cannot be separated on the basis of their water solubility. They can be separated, however, on the basis of their differences in acidity. First, the mixture of the two is dissolved in diethyl ether or some other water-immiscible solvent. Next, the ether solution is placed in a separatory funnel and shaken with dilute aqueous NaOH. Under these conditions, phenol reacts with NaOH and is converted to sodium phenoxide, a water-soluble salt. The upper layer in the separatory funnel is now diethyl ether (density 0.74 g/cm^3) containing only dissolved cyclohexanol. The lower aqueous layer contains dissolved sodium phenoxide. The layers are separated, and distillation of the ether (bp 35°C) leaves pure cyclohexanol (bp 161°C). Acidification of the aqueous phase with 0.1 M HCl or other strong acid converts sodium phenoxide to phenol, which is water-insoluble and can be extracted with ether and recovered in pure form. These experimental steps are summarized in the flowchart.

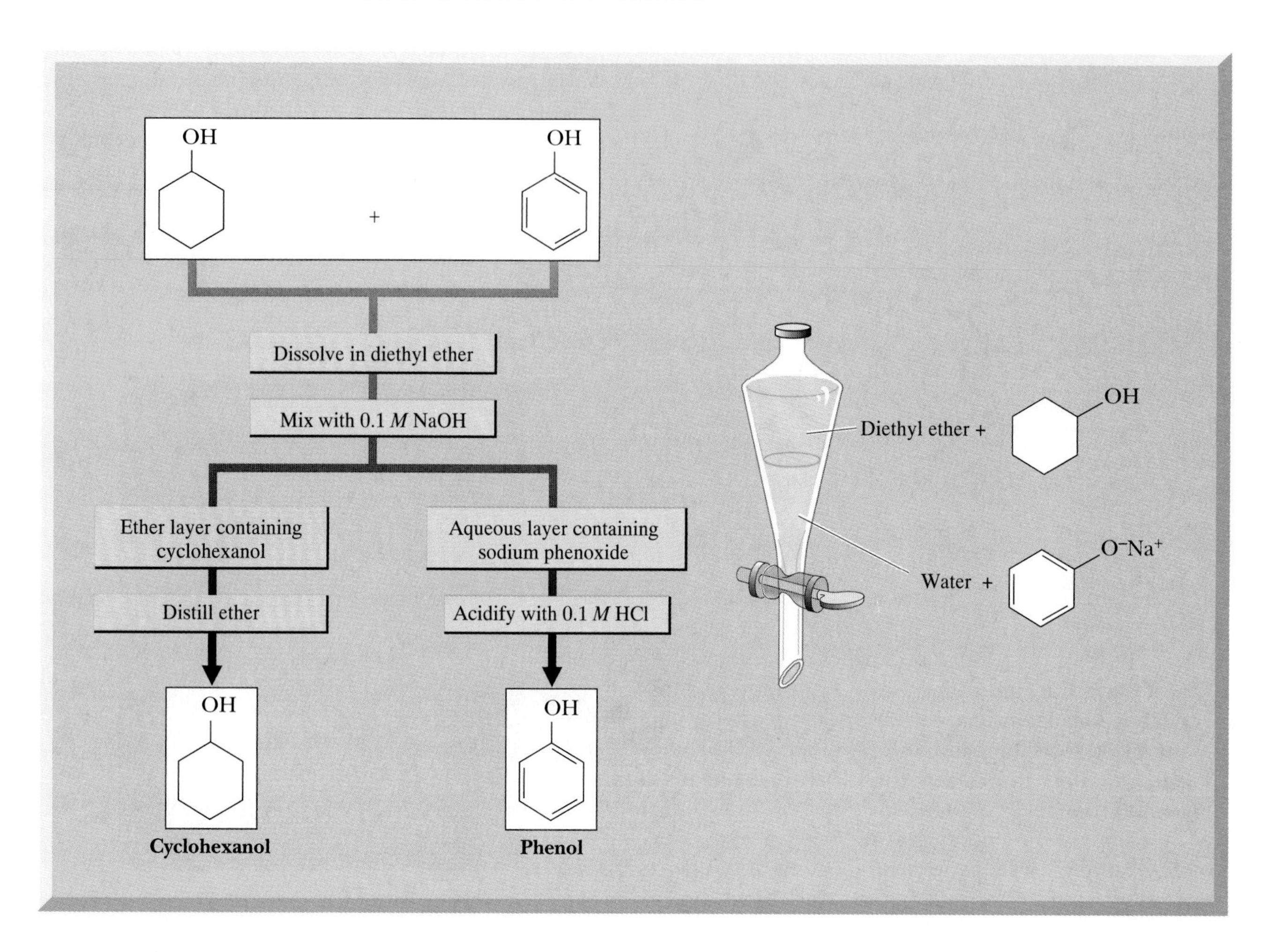

9.5 Oxidation at a Benzylic Position

Benzene is unaffected by strong oxidizing agents, such as H_2CrO_4 and $KMnO_4$. When toluene is treated with these oxidizing agents under vigorous conditions, the side-chain methyl group is oxidized to a carboxyl group to give benzoic acid. The fact that the side-chain methyl group is oxidized but the aromatic ring is unchanged illustrates the remarkable chemical stability of the aromatic ring.

$$C_6H_5CH_3 + H_2CrO_4 \longrightarrow C_6H_5CO_2H + Cr^{3+}$$

Toluene Benzoic acid

Halogen and nitro substituents on an aromatic ring are unaffected by these oxidations. 2-Chloro-4-nitrotoluene, for example, is oxidized to 2-chloro-4-nitrobenzoic acid.

$$\text{2-Chloro-4-nitrotoluene} \xrightarrow[H_2SO_4]{K_2Cr_2O_7} \text{2-Chloro-4-nitrobenzoic acid}$$

Ethylbenzene and isopropylbenzene are also oxidized under these conditions to benzoic acid. The side chain of *tert*-butylbenzene, which has no benzylic hydrogen, is not affected by these oxidizing conditions.

From these observations, we conclude that if a benzylic hydrogen exists, then the **benzylic carbon** (Section 9.3A) is oxidized to a carboxyl group, and all other carbons of the side chain are removed. If no benzylic hydrogen exists, as in the case of *tert*-butylbenzene, no oxidation of the side chain occurs.

Benzylic carbon An sp^3 hybridized carbon bonded to a benzene ring.

If more than one alkyl side chain exists, each is oxidized to $-CO_2H$. Oxidation of *m*-xylene gives 1,3-benzenedicarboxylic acid, more commonly named isophthalic acid.

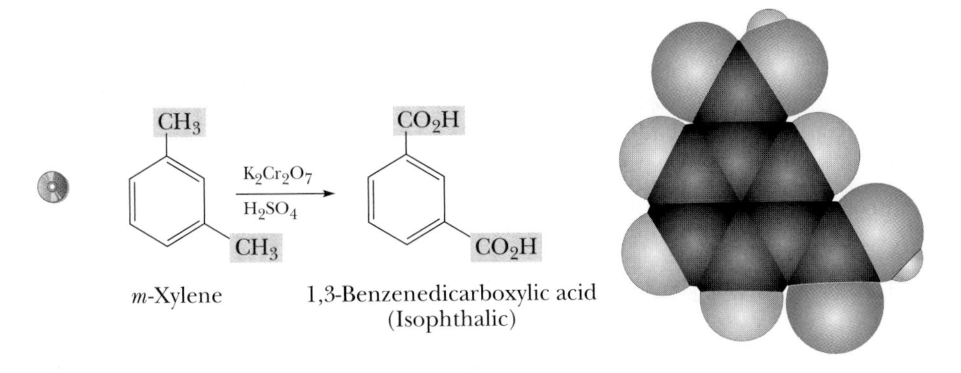

m-Xylene 1,3-Benzenedicarboxylic acid (Isophthalic)

EXAMPLE 9.3

Draw structural formulas for the product of vigorous oxidation of *p*-xylene by $K_2Cr_2O_7$ in aqueous H_2SO_4.

SOLUTION

Both alkyl groups are oxidized to $-CO_2H$ groups. The product is terephthalic acid, one of two monomers required for the synthesis of Dacron polyester and Mylar (Section 15.4B).

$$H_3C-C_6H_4-CH_3 \xrightarrow[H_2SO_4]{K_2Cr_2O_7} HOC(=O)-C_6H_4-C(=O)OH$$

1,4-Dimethylbenzene (*p*-Xylene) → 1,4-Benzenedicarboxylic acid (Terephthalic acid)

Practice Problem 9.3

Predict the products resulting from vigorous oxidation of each compound by $K_2Cr_2O_7$ in aqueous H_2SO_4.

(a) 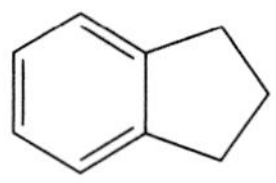 (b) 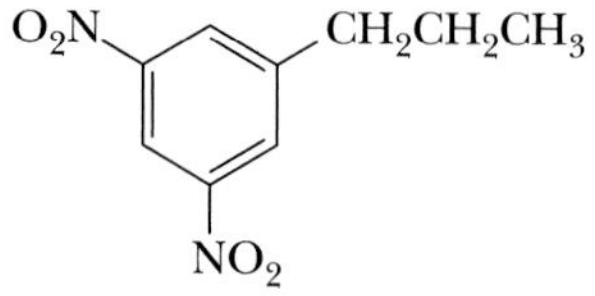 O_2N, $CH_2CH_2CH_3$, NO_2

9.6 Electrophilic Aromatic Substitution

By far the most characteristic reaction of aromatic compounds is substitution at a ring carbon. Some groups that can be introduced directly on the ring are the halogens, the nitro ($—NO_2$) group, the sulfonic acid ($—SO_3H$) group, alkyl ($—R$) groups, and acyl ($RCO—$) groups. Each of these substitution reactions is represented in the following equations.

Halogenation:

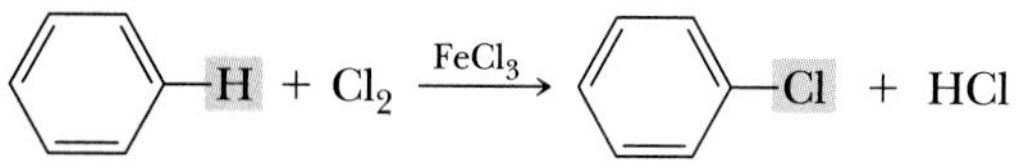

$$C_6H_5\text{—H} + Cl_2 \xrightarrow{FeCl_3} C_6H_5\text{—Cl} + HCl$$

Chlorobenzene

Nitration:

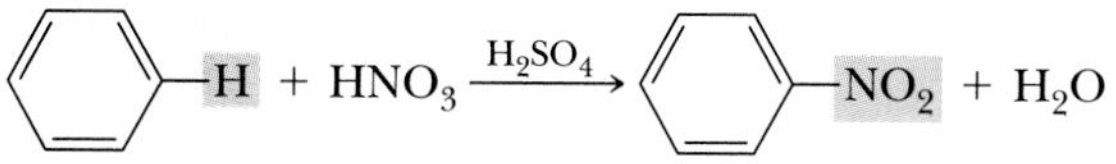

$$C_6H_5\text{—H} + HNO_3 \xrightarrow{H_2SO_4} C_6H_5\text{—}NO_2 + H_2O$$

Nitrobenzene

Sulfonation:

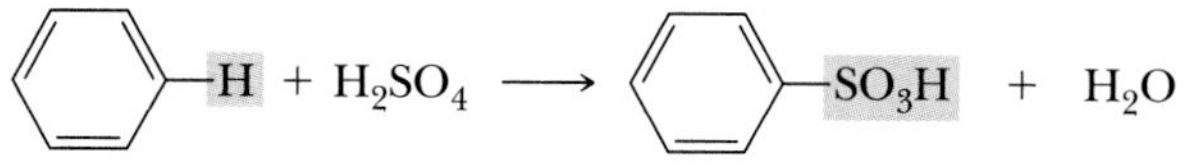

$$C_6H_5\text{—H} + H_2SO_4 \longrightarrow C_6H_5\text{—}SO_3H + H_2O$$

Benzenesulfonic acid

Alkylation:

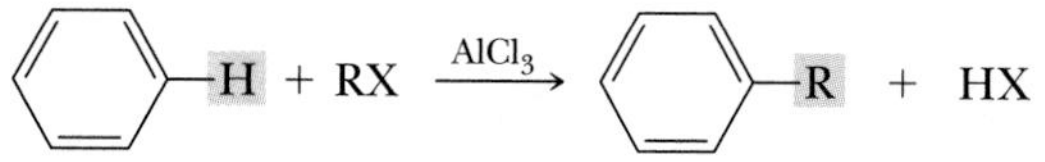

$$C_6H_5\text{—H} + RX \xrightarrow{AlCl_3} C_6H_5\text{—R} + HX$$

An alkylbenzene

Acylation:

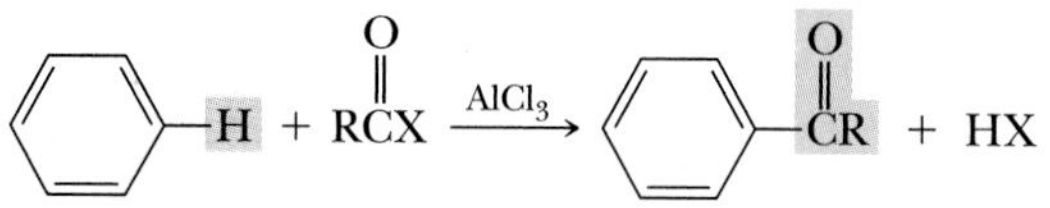

$$C_6H_5\text{—H} + R\overset{O}{\overset{\|}{C}}X \xrightarrow{AlCl_3} C_6H_5\text{—}\overset{O}{\overset{\|}{C}}R + HX$$

An acylbenzene

9.7 MECHANISM OF ELECTROPHILIC AROMATIC SUBSTITUTION

An **electrophile** is an electron-deficient species that can accept a pair of electrons from a nucleophile to form a new covalent bond. A reaction in which a hydrogen atom of an aromatic ring is replaced by an electrophile is called **electrophilic aromatic substitution.**

Electrophilic aromatic substitution A reaction in which there is substitution of an electrophile for a hydrogen on an aromatic ring.

$$C_6H_5{-}H + E^+ \longrightarrow C_6H_5{-}E + H^+$$

An electrophile

In this section, we study several common types of electrophiles including how each is generated and the mechanism by which it replaces hydrogen on an aromatic ring.

A. Chlorination and Bromination

Chlorine alone does not react with benzene, in contrast to its instantaneous addition to cyclohexene (Section 6.3C). In the presence of a Lewis acid catalyst, such as ferric chloride or aluminum chloride (Section 2.5), a reaction takes place to give chlorobenzene and HCl. The first step in this reaction involves interaction of chlorine and the Lewis acid catalyst to form a **chloronium ion, Cl^+**, as part of an ion pair. Reaction of the Cl_2—$FeCl_3$ ion pair with the pi electron cloud of the aromatic ring forms a resonance-stabilized cation intermediate, here represented as a hybrid of three contributing structures. Proton transfer from the cation intermediate to $FeCl_4^-$ forms HCl, regenerates the Lewis acid catalyst, and gives chlorobenzene.

MECHANISM Electrophilic Aromatic Substitution—Chlorination

Step 1: Reaction between chlorine (a Lewis base) and $FeCl_3$ (a Lewis acid) gives an ion pair containing a chloronium ion.

$$Cl{-}Cl + FeCl_3 \rightleftharpoons Cl^{+}{-}Cl{-}Fe^{-}Cl_3 \rightleftharpoons Cl^{+}\ \ Cl{-}Fe^{-}Cl_3$$

Chlorine (a Lewis base) | Ferric chloride (a Lewis acid) | A molecular complex with a positive charge on chlorine and a negative charge on iron | An ion pair containing a chloronium ion

Step 2: Reaction of the chloronium ion with the pi electrons of the aromatic ring gives a resonance-stabilized cation intermediate.

$$C_6H_6 + Cl^+ \xrightarrow{\text{rate-limiting step}} [C_6H_6Cl]^+ \longleftrightarrow [C_6H_6Cl]^+ \longleftrightarrow [C_6H_6Cl]^+$$

Resonance-stabilized cation intermediate

Step 3: Proton transfer from the cation intermediate regenerates the aromatic character of the ring.

Treatment of benzene with bromine in the presence of ferric chloride or aluminum chloride gives bromobenzene and HBr. The mechanism for this reaction is the same as we have drawn for chlorination of benzene.

We can write a general two-step mechanism for electrophilic aromatic substitution. The first and rate-limiting step is reaction of the electrophile, E^+, with the pi electrons of the aromatic ring to give a resonance-stabilized cation intermediate. In the second and faster step, loss of H^+ from the cation intermediate regenerates the aromatic ring and gives the product.

The major difference between addition of halogen to an alkene and substitution by halogen on an aromatic ring centers on the fate of the cation intermediate formed in the first step of each reaction. Recall from Section 6.3C that addition of chlorine to an alkene is a two-step process, the first and slower step of which is formation of a bridged chloronium ion intermediate. This intermediate then reacts with chloride ion to complete the addition. With aromatic compounds, the cation intermediate loses H^+ to regenerate the aromatic ring and regain its large resonance stabilization. There is no such resonance stabilization to be regained in the case of an alkene.

B. Nitration and Sulfonation

The sequence of steps for nitration and sulfonation of benzene is similar to that for chlorination and bromination. For nitration, the electrophile is the **nitronium ion,** NO_2^+, generated by reaction of nitric acid and sulfuric acid. In the following equation, nitric acid is written $HO—NO_2$ to show more clearly the origin of the nitronium ion.

MECHANISM Formation of the Nitronium Ion

Step 1: Proton transfer from sulfuric acid to the OH group of nitric acid gives the conjugate acid of nitric acid.

$$H-\ddot{O}-NO_2 + H-\ddot{O}-SO_3H \overset{(1)}{\rightleftharpoons} H-\overset{H}{\overset{+|}{\ddot{O}}}-NO_2 + HSO_4^-$$

Nitric acid — Conjugate acid of nitric acid

Step 2: Loss of water from this conjugate acid gives the nitronium ion.

$$H-\overset{H}{\overset{+|}{\ddot{O}}}-NO_2 \overset{(2)}{\rightleftharpoons} H-\overset{H}{\overset{|}{\ddot{O}}}: + NO_2^+$$

The nitronium ion

A particular value of nitration is that the resulting nitro group, $-NO_2$, can be reduced to a primary amino group, $-NH_2$, by hydrogenation in the presence of a transition metal catalyst such as nickel, palladium, or platinum. This method has the potential disadvantage that other susceptible groups, such as a carbon-carbon double bond, and the carbonyl group of an aldehyde or ketone may also be reduced. Note that neither the $-CO_2H$ nor the aromatic ring is reduced under these conditions.

$$O_2N-C_6H_4-CO_2H + 3H_2 \xrightarrow[3\ atm]{Ni} H_2N-C_6H_4-CO_2H + 2H_2O$$

4-Nitrobenzoic acid — 4-Aminobenzoic acid

Alternatively, a nitro group can be reduced to a primary amino group by a metal in acid. The most commonly used metal reducing agents are iron, zinc, and tin in dilute HCl. When reduced with a metal and hydrochloric acid, the amine is obtained as a salt, which is then treated with strong base to liberate the free amine.

$$CH_3C_6H_3(NO_2)_2 \xrightarrow[C_2H_5OH,\ H_2O]{Fe,\ HCl} CH_3C_6H_3(NH_3^+Cl^-)_2 \xrightarrow{NaOH,\ H_2O} CH_3C_6H_3(NH_2)_2$$

2,4-Dinitrotoluene — 4-Methyl-1,3-benzenediamine (2,4-Diaminotoluene)

Sulfonation of benzene is carried out using hot, concentrated sulfuric acid. The electrophile under these conditions is HSO_3^+, formed from sulfuric acid in the following way.

$$HO-\overset{O}{\overset{\|}{\underset{O}{\underset{\|}{S}}}}-\ddot{O}H + H^+ \rightleftharpoons HO-\overset{O}{\overset{\|}{\underset{O}{\underset{\|}{S}}}}-\overset{+}{O}H_2 \rightleftharpoons HO-\overset{O}{\overset{\|}{\underset{O}{\underset{\|}{S^+}}}} + H-\ddot{O}-H$$

Sulfuric acid — The electrophile

Example 9.4

Write a stepwise mechanism for nitration of benzene.

Solution

Step 1: Reaction of the nitronium ion with the pi electrons of the benzene ring gives a resonance-stabilized cation intermediate.

$$\text{C}_6\text{H}_6 + NO_2^+ \xrightarrow{(1)} \left[\text{H, NO}_2 \text{ cation} \longleftrightarrow \text{H, NO}_2 \text{ cation} \longleftrightarrow \text{H, NO}_2 \text{ cation} \right]$$

Resonance-stabilized cation intermediate

Step 2: Proton transfer from this intermediate to H_2O regenerates the aromatic ring and gives nitrobenzene.

$$H_2\ddot{O}: + \text{(H, NO}_2\text{ cation)} \xrightarrow{(2)} \text{C}_6\text{H}_5NO_2 + H_3O^+$$

Nitrobenzene

Practice Problem 9.4

Write a stepwise mechanism for the sulfonation of benzene.

C. Friedel-Crafts Alkylation and Acylation

Alkylation of aromatic hydrocarbons was discovered in 1877 by the French chemist Charles Friedel and a visiting American chemist, James Crafts. They discovered that mixing benzene, an alkyl halide, and $AlCl_3$ results in formation of an alkylbenzene and HX. **Friedel-Crafts alkylation** forms a new carbon-carbon bond between benzene and an alkyl group, as illustrated by reaction of benzene with 2-chloropropane in the presence of aluminum chloride.

$$\text{C}_6\text{H}_6 + CH_3CH(CH_3)Cl \xrightarrow{AlCl_3} \text{C}_6\text{H}_5\text{-}CH(CH_3)_2 + HCl$$

Benzene; 2-Chloropropane (Isopropyl chloride); Isopropylbenzene

Friedel-Crafts alkylation is among the most important methods for forming new carbon-carbon bonds to aromatic rings. It begins with formation of a complex between the alkyl halide and aluminum chloride in which aluminum has a negative formal charge and the halogen of the alkyl halide has a positive formal charge. Redistribution of electrons in this complex gives an alkyl carbocation as part of an ion pair. Reaction of the alkyl carbocation with an aromatic ring gives a resonance-stabilized carbocation intermediate, which then loses H^+ to give an alkylbenzene.

MECHANISM Friedel-Crafts Alkylation

Step 1: Reaction of an alkyl halide (a Lewis base) with aluminum chloride (a Lewis acid) gives an ion pair containing an alkyl carbocation.

$$R{-}\ddot{Cl}: + AlCl_3 \rightleftharpoons R{-}\overset{+}{Cl}{-}\overset{-}{Al}Cl_3 \rightleftharpoons R^+ \; :\ddot{Cl}{-}\overset{-}{Al}Cl_3$$

A molecular complex with a positive charge on chlorine and a negative charge on aluminum

An ion pair containing a carbocation

Step 2: Reaction of the alkyl carbocation with the pi electrons of the aromatic ring gives a resonance-stabilized cation intermediate.

The positive charge is delocalized onto three atoms of the ring

Step 3: Proton transfer regenerates the aromatic character of the ring and the Lewis acid catalyst.

$$:\ddot{Cl}{-}\overset{-}{Al}Cl_3 \longrightarrow C_6H_5{-}R + AlCl_3 + H{-}\ddot{Cl}:$$

Friedel and Crafts also discovered that treatment of an aromatic hydrocarbon with an acyl halide (Section 13.1A) in the presence of aluminum chloride gives a ketone. An **acyl halide** is a derivative of a carboxylic acid in which the —OH of the carboxyl group is replaced by a halogen, most commonly chlorine. Acyl halides are also referred to as acid halides. An RCO— group is known as an acyl group; hence, reaction of an acyl halide with an aromatic hydrocarbon is known as **Friedel-Crafts acylation,** as illustrated by reaction of benzene and acetyl chloride in the presence of aluminum chloride to give acetophenone.

Acyl halide A derivative of a carboxylic acid in which the —OH of the carboxyl group is replaced by a halogen, most commonly chlorine.

$$C_6H_6 + CH_3CCl(=O) \xrightarrow{AlCl_3} C_6H_5CCH_3(=O) + HCl$$

Benzene Acetyl chloride (an acyl halide) Acetophenone (a ketone)

Friedel-Crafts acylation begins with donation of a pair of electrons from the halogen of the acyl halide to aluminum chloride to form a molecular complex similar to what we drew for Friedel-Crafts alkylations. In this complex, halogen has a positive formal charge, and aluminum has a negative formal charge. Redistribution of electrons of the carbon-chlorine bond gives an ion pair containing an **acylium ion.**

MECHANISM Friedel-Crafts Acylation—Generation of an Acylium Ion

Step 1: Reaction between the halogen atom of the acid chloride (a Lewis base) and aluminum chloride (a Lewis acid) gives a molecular complex.

Step 2: Redistribution of valence electrons of this molecular complex gives an ion pair containing an acylium ion.

$$\text{R}-\overset{\overset{\displaystyle :\ddot{O}}{\|}}{C}-\ddot{\underset{\cdot\cdot}{Cl}}: + \overset{:\ddot{Cl}:}{\underset{:\underset{\cdot\cdot}{Cl}:}{Al}}-\ddot{\underset{\cdot\cdot}{Cl}}: \overset{(1)}{\rightleftharpoons} \text{R}-\overset{\overset{\displaystyle :\ddot{O}}{\|}}{C}-\overset{+}{\ddot{Cl}}-\overset{:\ddot{Cl}:}{\underset{:\underset{\cdot\cdot}{Cl}:}{Al^{-}}}-\ddot{\underset{\cdot\cdot}{Cl}}: \overset{(2)}{\rightleftharpoons} \text{R}-\overset{\overset{\displaystyle :\ddot{O}}{\|}}{C^{+}}\ :\ddot{\underset{\cdot\cdot}{Cl}}-\overset{:\ddot{Cl}:}{\underset{:\underset{\cdot\cdot}{Cl}:}{Al^{-}}}-\ddot{\underset{\cdot\cdot}{Cl}}:$$

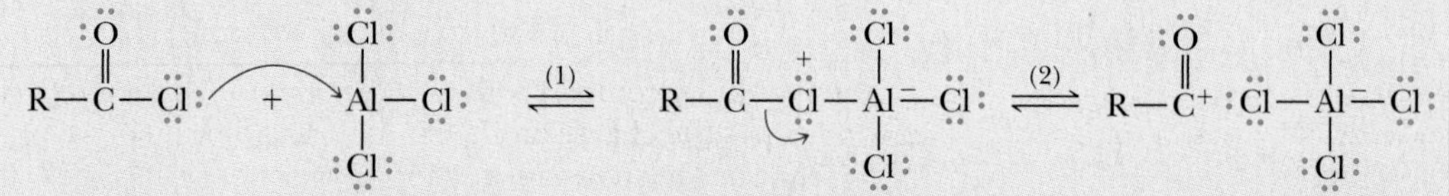

Acyl chloride (a Lewis base) | Aluminum chloride (a Lewis acid) | A molecular complex with a positive charge on chlorine and a negative charge on aluminum | An ion pair containing an acylium ion

EXAMPLE 9.5

Write a structural formula for the product formed by Friedel-Crafts alkylation or acylation of benzene with

(a) $C_6H_5CH_2Cl$ Benzyl chloride (b) $C_6H_5\overset{O}{\overset{\|}{C}}Cl$ Benzoyl chloride

SOLUTION

(a) Treatment of benzyl chloride with aluminum chloride gives the benzyl cation. Reaction of this cation with benzene followed by loss of H^+ gives diphenylmethane.

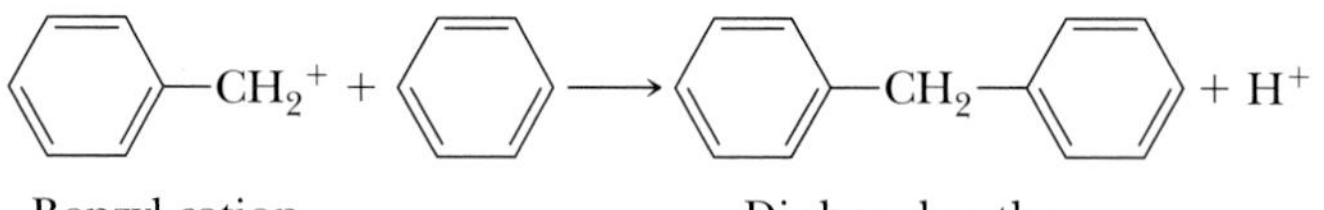

$$C_6H_5-CH_2^{+} + C_6H_6 \longrightarrow C_6H_5-CH_2-C_6H_5 + H^{+}$$

Benzyl cation Diphenylmethane

(b) Treatment of benzoyl chloride with aluminum chloride gives an acyl cation. Reaction of this cation with benzene followed by loss of H^+ gives benzophenone.

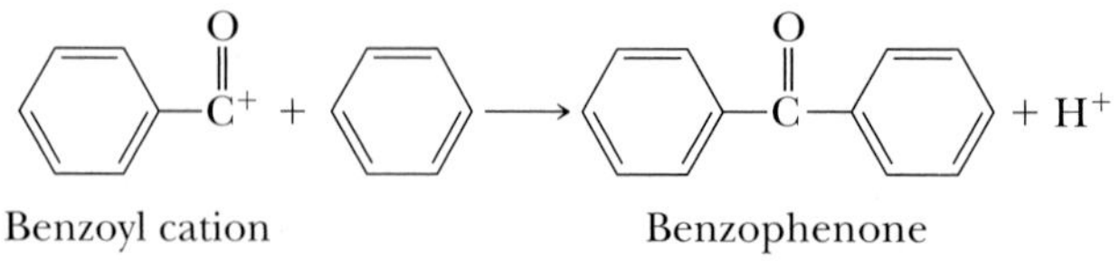

$$C_6H_5-\overset{O}{\overset{\|}{C^{+}}} + C_6H_6 \longrightarrow C_6H_5-\overset{O}{\overset{\|}{C}}-C_6H_5 + H^{+}$$

Benzoyl cation Benzophenone

Practice Problem 9.5

Write a structural formula for the product formed from Friedel-Crafts alkylation or acylation of benzene with

(a) $(CH_3)_3CCCl$ (with C=O) (b) $(CH_3)_2CHCl$ (c) C_6H_5–CHClCH_3

D. Other Electrophilic Aromatic Alkylations

Once it was discovered that Friedel-Crafts alkylations and acylations involve cation intermediates, it was realized that the same reactions can be accomplished by other combinations of reagents and catalysts. We study two of these in this section: generation of carbocations from alkenes and from alcohols.

As we saw in Section 6.3, treatment of an alkene with a strong acid, most commonly H_2SO_4 or H_3PO_4, generates a carbocation. Isopropylbenzene is synthesized industrially by reaction of benzene with propene in the presence of an acid catalyst.

$$C_6H_6 + CH_3CH{=}CH_2 \xrightarrow{H_3PO_4} C_6H_5{-}CH(CH_3)_2$$

Benzene Propene Isopropylbenzene

Carbocations are also generated by treatment of an alcohol with H_2SO_4 or H_3PO_4 (Section 8.4D).

$$C_6H_6 + (CH_3)_3COH \xrightarrow{H_3PO_4} C_6H_5{-}C(CH_3)_3 + H_2O$$

Benzene 2-Methyl-2-propanol (*tert*-Butyl alcohol) *tert*-Butylbenzene

The following example of electrophilic aromatic substitution involves intramolecular alkylation to form a six-membered ring. The product of this reaction is a derivative of naphthalene in which hydrogens have been added to carbons 1, 2, 3, and 4.

$$C_6H_5{-}CH_2CH_2CH_2CH_2OH \xrightarrow{H_3PO_4} C_{10}H_{12} + H_2O$$

4-Phenyl-1-butanol 1,2,3,4-Tetrahydronaphthalene (Tetralin)

EXAMPLE 9.6

Write a mechanism for the formation of isopropylbenzene (cumene) from benzene and propene in the presence of phosphoric acid.

SOLUTION

Step 1: Proton transfer from phosphoric acid to propene gives the isopropyl cation.

$$CH_3CH{=}CH_2 + H{-}O{-}\overset{\overset{O}{\|}}{\underset{\underset{OH}{|}}{P}}{-}O{-}H \xrightleftharpoons[]{\text{fast and reversible}} CH_3\overset{+}{C}HCH_3 + {}^{-}O{-}\overset{\overset{O}{\|}}{\underset{\underset{OH}{|}}{P}}{-}O{-}H$$

Step 2: Reaction of the isopropyl cation with benzene gives a resonance-stabilized carbocation intermediate.

$$C_6H_6 + {}^{+}CH(CH_3)_2 \xrightleftharpoons[]{\text{slow, rate limiting}} [C_6H_6{-}CH(CH_3)_2]^{+}$$

Step 3: Proton transfer from this intermediate to dihydrogen phosphate ion gives isopropylbenzene.

$$[C_6H_6{-}CH(CH_3)_2]^{+} + {}^{-}O{-}\overset{\overset{O}{\|}}{\underset{\underset{OH}{|}}{P}}{-}O{-}H \xrightarrow{\text{fast}} C_6H_5{-}CH(CH_3)_2 + H{-}O{-}\overset{\overset{O}{\|}}{\underset{\underset{OH}{|}}{P}}{-}O{-}H$$

Isopropylbenzene

Practice Problem 9.6

Write a mechanism for the formation of *tert*-butylbenzene from benzene and *tert*-butyl alcohol in the presence of phosphoric acid.

9.8 Disubstitution

A. Effects of a Substituent Group on Further Substitution

Ortho-para director Any substituent on a benzene ring that directs electrophilic aromatic substitution preferentially to ortho and para positions.

Meta director Any substituent on a benzene ring that directs electrophilic aromatic substitution preferentially to a meta position.

Activating group Any substituent on a benzene ring that causes the rate of electrophilic aromatic substitution to be greater than that for benzene.

Deactivating group Any substituent on a benzene ring that causes the rate of electrophilic aromatic substitution to be lower than that for benzene.

We can make the following generalizations about the manner in which an existing substituent influences further electrophilic aromatic substitution.

1. Substituents affect the orientation of new groups. Certain substituents direct a second substituent preferentially to the ortho and para positions; other substituents direct it preferentially to a meta position. In other words, substituents on a benzene ring can be classified as **ortho-para directing** or as **meta directing.**
2. Substituents affect the rate of further substitution. Certain substituents cause the rate of a second substitution to be greater than that for benzene itself, whereas other substituents cause the rate of a second substitution to be lower than that for benzene. In other words, groups on a benzene ring can be classified as **activating** or **deactivating** toward further substitution.

These directing and activating-deactivating effects can be seen by comparing the reactions of anisole and nitrobenzene. Bromination of anisole proceeds at a rate considerably greater than that for bromination of benzene (the methoxy group is activating), and the product is a mixture of *o*-bromoanisole and *p*-bromoanisole (the methoxy group is ortho-para directing).

OCH_3 + Br_2 $\xrightarrow{CH_3CO_2H}$ (OCH_3, Br) + (OCH_3, Br) + HBr

Anisole — *o*-Bromoanisole (4%) — *p*-Bromoanisole (96%)

Quite another situation is seen in the nitration of nitrobenzene. First, the reaction requires a nitrating mixture of nitric acid/sulfuric acid and a higher temperature. Because the nitration of nitrobenzene proceeds much more slowly than the nitration of benzene itself, we say that a nitro group is strongly deactivating. Second, the product consists of approximately 93% of the meta isomer and less than 7% of the ortho and para isomers combined; the nitro group is meta directing.

NO_2 + HNO_3 $\xrightarrow[100°C]{H_2SO_4}$ (NO_2, NO_2) + (NO_2, NO_2) + (NO_2, NO_2) + H_2O

Nitrobenzene — *m*-Dinitrobenzene (93%) — *o*-Dinitrobenzene, *p*-Dinitrobenzene (Less than 7% combined)

Listed in Table 9.2 are the directing and activating-deactivating effects for the major functional groups with which we are concerned in this text.

TABLE 9.2 Effects of Substituents on Further Electrophilic Aromatic Substitution

Ortho-para Directing	strongly activating	$-\ddot{N}H_2$, $-\ddot{N}HR$, $-\ddot{N}R_2$, $-\ddot{O}H$, $-\ddot{O}R$
	moderately activating	$-\ddot{N}HCR$ (C=O), $-\ddot{N}HCAr$ (C=O), $-\ddot{O}CR$ (C=O), $-\ddot{O}CAr$ (C=O)
	weakly activating	$-R$, $-C_6H_5$ (phenyl)
	weakly deactivating	$-\ddot{F}:$, $-\ddot{Cl}:$, $-\ddot{Br}:$, $-\ddot{I}:$
Meta Directing	moderately deactivating	$-CH$ (C=O), $-CR$ (C=O), $-COH$ (C=O), $-COR$ (C=O), $-CNH_2$ (C=O), $-SOH$ (O=S=O)
	strongly deactivating	$-NO_2$, $-NH_3^+$, $-CF_3$, $-CCl_3$

Relative importance in directing further substitution (arrow pointing upward)

If we compare these ortho-para and meta directors for structural similarities and differences, we can make the following generalizations:

1. Alkyl groups, phenyl groups, and substituents in which the atom bonded to the ring has an unshared pair of electrons are ortho-para directing. All other substituents are meta directing.
2. All ortho-para directing groups except the halogens are activating toward further substitution. The halogens are weakly deactivating.

We can illustrate the usefulness of these generalizations by considering the synthesis of two different disubstituted derivatives of benzene. Suppose we wish to prepare *m*-bromonitrobenzene from benzene. This conversion can be carried out in two steps: nitration and bromination. If the steps are carried out in just that order, the major product is indeed *m*-bromonitrobenzene. The nitro group is a meta director and, therefore, directs bromination to a meta position.

NO_2 NO_2

$\xrightarrow[H_2SO_4]{HNO_3}$ $\xrightarrow[FeCl_3]{Br_2}$

Br

Nitrobenzene *m*-Bromonitrobenzene

If, however, we reverse the order of the steps and first form bromobenzene, we now have an ortho-para directing group on the ring. Nitration of bromobenzene takes place preferentially at the ortho and para positions.

Br Br Br

NO_2

$\xrightarrow[FeCl_3]{Br_2}$ $\xrightarrow[H_2SO_4]{HNO_3}$ +

NO_2

Bromobenzene *o*-Bromonitrobenzene *p*-Bromonitrobenzene

As another example of the importance of order in electrophilic aromatic substitutions, consider the conversion of toluene to nitrobenzoic acid. The nitro group can be introduced with a nitrating mixture of nitric and sulfuric acids. The carboxyl group can be produced by oxidation of the methyl group (Section 9.5).

CH_3 CO_2H

$\xrightarrow[H_2SO_4]{HNO_3}$ $\xrightarrow[H_2SO_4]{K_2Cr_2O_7}$

CH_3

NO_2 NO_2

4-Nitrotoluene 4-Nitrobenzoic acid

CO_2H CO_2H

Toluene

$\xrightarrow[H_2SO_4]{K_2Cr_2O_7}$ $\xrightarrow[H_2SO_4]{HNO_3}$

NO_2

Benzoic acid 3-Nitrobenzoic acid

Nitration of toluene yields a product with the two substituents para to each other. Nitration of benzoic acid, on the other hand, yields a product with the substituents meta to each other. Again, we see that the order in which the reactions are performed is critical.

Note that, in this last example, we showed nitration of toluene producing only the para isomer. Because methyl is an ortho-para directing group, both ortho and para isomers are formed. In problems of this type in which you are asked to prepare one or the other of these isomers, we assume that both are formed but that there are physical methods by which they can be separated and the desired isomer obtained.

EXAMPLE 9.7

Complete the following electrophilic aromatic substitution reactions. Where you predict meta substitution, show only the meta product. Where you predict ortho-para substitution, show both products.

(a) 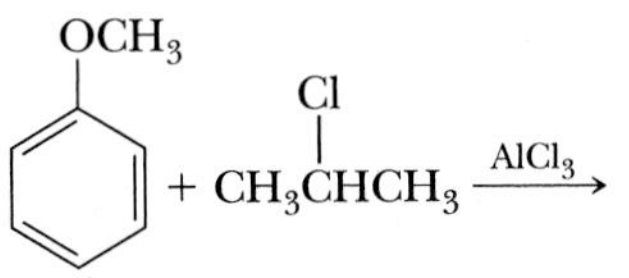 OCH_3 (benzene ring) + $CH_3CH(Cl)CH_3$ $\xrightarrow{AlCl_3}$ (b) 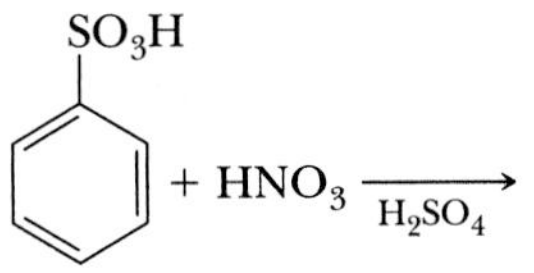 SO_3H (benzene ring) + HNO_3 $\xrightarrow[H_2SO_4]{}$

SOLUTION

The methoxyl group in (a) is ortho-para directing and strongly activating. The sulfonic acid group in (b) is meta directing and moderately deactivating.

(a) 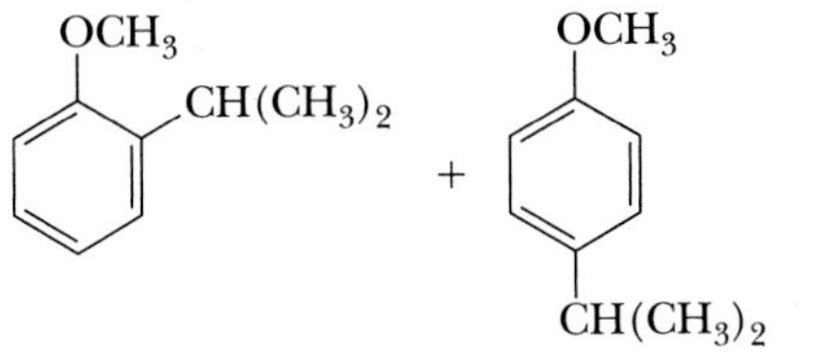 OCH_3 / $CH(CH_3)_2$ + OCH_3 / $CH(CH_3)_2$ (b) SO_3H / NO_2

2-Isopropylanisole 4-Isopropylanisole 3-Nitrobenzene-sulfonic acid

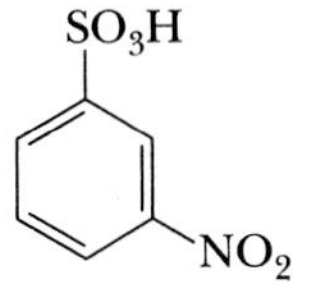

Practice Problem 9.7

Complete the following electrophilic aromatic substitution reactions. Where you predict meta substitution, show only the meta product. Where you predict ortho-para substitution, show both products.

(a) 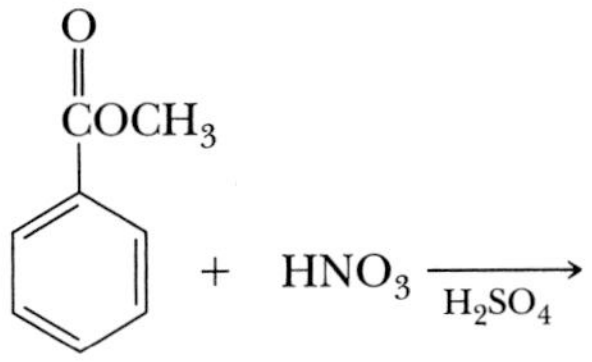 $C(=O)OCH_3$ (benzene ring) + HNO_3 $\xrightarrow[H_2SO_4]{}$ (b) 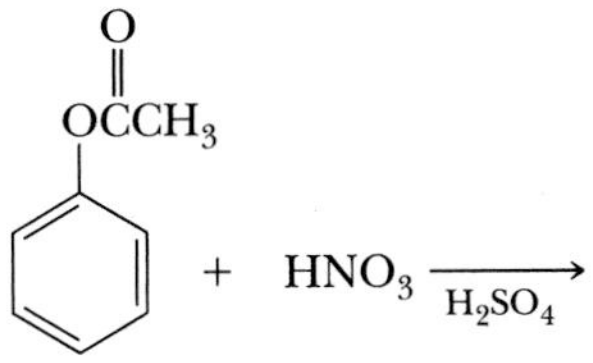 $OC(=O)CH_3$ (benzene ring) + HNO_3 $\xrightarrow[H_2SO_4]{}$

B. Theory of Directing Effects

As we have just seen, a group on an aromatic ring exerts a major effect on the patterns of further substitution. We account for these patterns by starting with the general mechanism first presented in Section 9.7 for electrophilic aromatic substitution. We now extend that mechanism to consider how a group already present on the ring might affect the relative stabilities of cation intermediates formed during a second substitution reaction.

We begin with the fact that the rate of electrophilic aromatic substitution is determined by the slowest step in the mechanism. For the substitutions we consider, the rate-limiting step is reaction of the electrophile with the aromatic ring to form a resonance-stabilized cation intermediate. We first draw contributing structures for the cation intermediate formed by reaction of the electrophile at a meta position, and then those for the cation formed by its reaction at an ortho or para position. We know that reaction proceeds by way of the cation intermediate more stabilized by resonance. Thus, what we have to do is determine whether the cation for ortho-para or meta substitution is the more stabilized by resonance.

Nitration of Anisole

The rate-limiting step is reaction of the nitronium ion with the aromatic ring to produce a resonance-stabilized cation intermediate. Shown in Figure 9.4 is the cation intermediate formed by reaction of the electrophile meta to the methoxy group. Also shown in Figure 9.4 is the cation intermediate formed by reaction para to the methoxy group. The cation intermediate formed by reaction at a meta position is a hybrid of three major contributing structures: (a), (b), and (c). The cation intermediate formed by reaction at the para position is a hybrid of four major contributing structures: (d), (e), (f), and (g). For each orientation, we can draw three contributing structures that place the positive charge on carbon atoms of the benzene ring. These three are the only important contributing structures

Figure 9.4
Nitration of anisole. Reaction of the electrophile meta and para to a methoxy group.

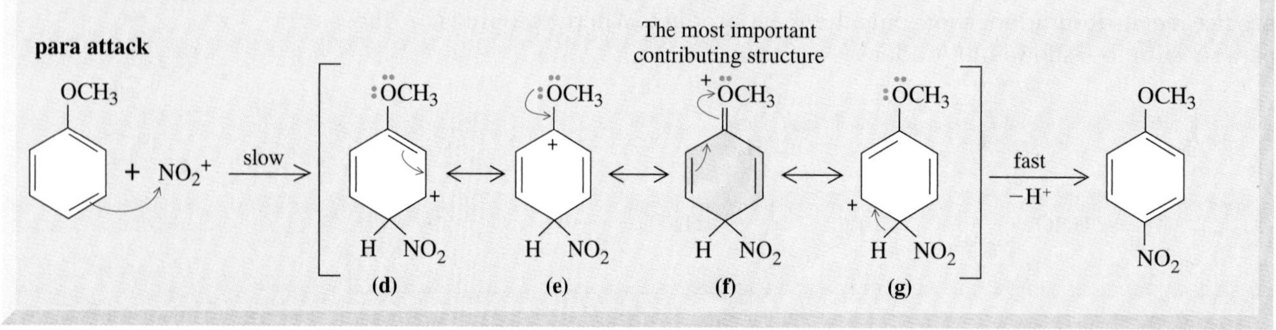

meta attack

OCH_3 + NO_2^+ —slow→ [(a) ↔ (b) ↔ (c)] —fast, $-H^+$→ m-nitroanisole (OCH_3, NO_2)

para attack

OCH_3 + NO_2^+ —slow→ [(d) ↔ (e) ↔ (f) ↔ (g)] —fast, $-H^+$→ p-nitroanisole (OCH_3, NO_2)

The most important contributing structure (f)

that can be drawn for reaction at a meta position. However, for reaction at the para position (and at an ortho position as well), a fourth contributing structure, (f), can be drawn. This structure involves an unshared pair of electrons on the oxygen atom of the methoxy group and places a positive charge on this oxygen. Structure (f) contributes more than structures (d), (e), or (g) because all atoms in it have complete octets. Because the cation formed by reaction at an ortho or para position on anisole has a greater degree of resonance stabilization and, hence, a lower activation energy for its formation, nitration of anisole occurs preferentially in the ortho and para positions.

Nitration of Nitrobenzene

Shown in Figure 9.5 are resonance-stabilized cation intermediates formed by reaction of the nitronium ion meta to the nitro group and also para to it. Each cation in Figure 9.5 is a hybrid of three contributing structures; no additional ones can be drawn. Now we need to compare the relative resonance stabilization of each hybrid. If we draw a Lewis structure for the nitro group showing the positive charge on nitrogen, we see that contributing structure (e) places positive charges on adjacent atoms. Because of the electrostatic repulsion thus generated, this structure makes only a negligible contribution to the hybrid.

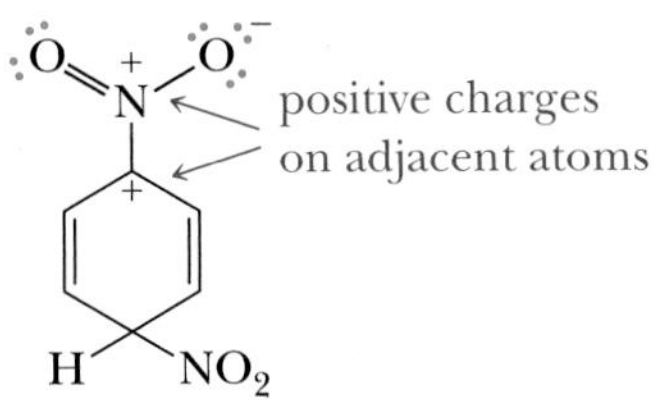

positive charges on adjacent atoms

H NO_2

None of the contributing structures for reaction at a meta position places positive charges on adjacent atoms. As a consequence, resonance stabilization of the cation formed by reaction at a meta position is greater than that formed by reac-

Figure 9.5
Nitration of nitrobenzene. Reaction of the electrophile meta and para to a nitro group.

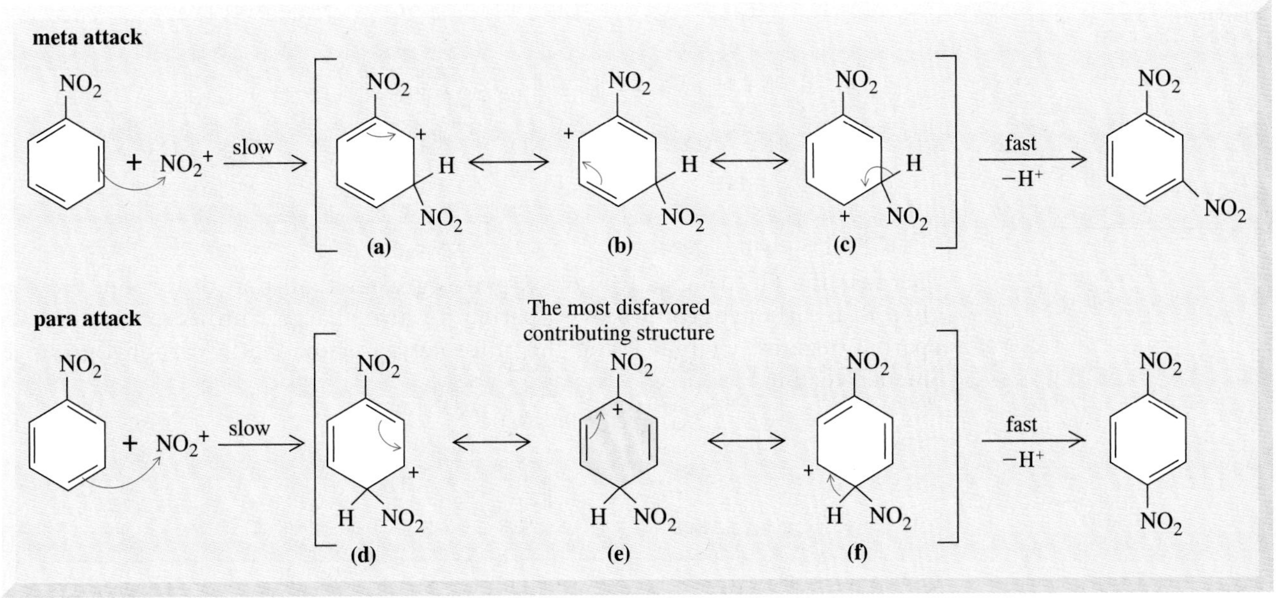

meta attack

NO_2 + NO_2^+ slow → [(a) ↔ (b) ↔ (c)] fast, $-H^+$ → NO_2, NO_2

para attack

NO_2 + NO_2^+ slow → [(d) ↔ (e) ↔ (f)] fast, $-H^+$ → NO_2, NO_2

The most disfavored contributing structure

tion at a para (or ortho) position. Stated alternatively, the activation energy for reaction at a meta position is less than that for reaction at a para position.

Comparison of the entries in Table 9.2 shows that almost all the ortho-para directing groups have an unshared pair of electrons on the atom bonded to the aromatic ring. Thus, the directing effect of most of these groups is due primarily to the ability of the atom bonded to the ring to delocalize further the positive charge on the aromatic ring of the cation intermediate.

The fact that alkyl groups are also ortho-para directing indicates that they too help to stabilize the cation intermediate. We saw in Section 6.3A that alkyl groups stabilize carbocation intermediates and that the order of stability of carbocations is $3° > 2° > 1° >$ methyl. Just as alkyl groups stabilize the carbocation intermediates formed in reactions of alkenes, they also stabilize the carbocation intermediates formed in electrophilic aromatic substitutions.

In summary, any substituent on an aromatic ring that further stabilizes the cation intermediate directs ortho-para. Conversely, any group that destabilizes the cation intermediate directs meta.

Example 9.8

The methoxy group is an ortho-para director because oxygen of the methoxyl group participates in stabilization of the cation intermediate. Draw contributing structures formed during nitration of chlorobenzene, and show how chlorine participates in a similar fashion to direct the incoming nitronium ion to ortho-para positions.

Solution

Contributing structures (a), (b), and (d) place the positive charge on atoms of the ring. Contributing structure (c) places it on chlorine and thus creates additional resonance stabilization for the cation intermediate.

:Cl: :Cl: +Cl: :Cl:
H NO$_2$ H NO$_2$ H NO$_2$ H NO$_2$
(a) (b) (c) (d)

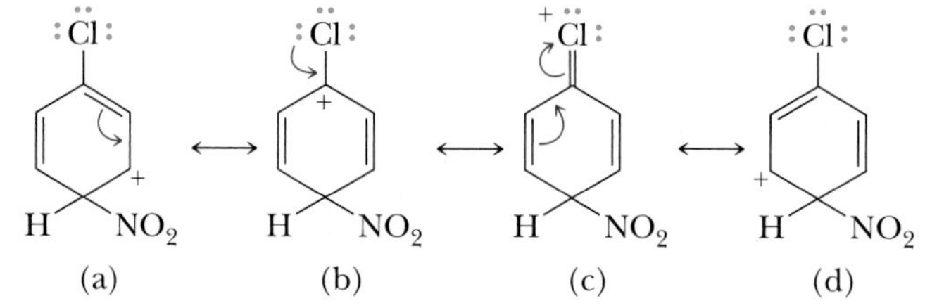

Practice Problem 9.8

Because the electronegativity of oxygen is greater than that of carbon, the carbon of a carbonyl group bears a partial positive charge, and its oxygen bears a partial negative charge. Using this information, show that a carbonyl group is meta directing.

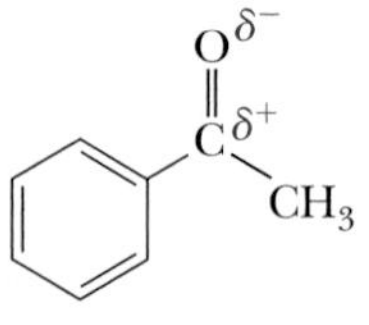

$O^{\delta-}$ $C^{\delta+}$ CH_3

SUMMARY

Benzene and its alkyl derivatives are classified as **aromatic hydrocarbons** or **arenes.** The concepts of **hybridization of atomic orbitals** and the **theory of resonance** (Section 9.1B), developed in the 1930s, provided the first adequate description for the structure of benzene. The **resonance energy** of benzene is approximately 36 kcal/mol (151 kJ/mol) (Section 9.1D).

According to the Hückel criteria for aromaticity, a five- or six-membered ring is aromatic if it: (1) has one *p* orbital on each atom of the ring, (2) is planar so that overlap of all *p* orbitals of the ring is continuous or nearly so, and (3) has 6 pi electrons in the overlapping system of *p* orbitals. A **heterocyclic aromatic compound** (Section 9.2) contains one or more atoms other than carbon in an aromatic ring.

Aromatic compounds are named by the IUPAC system. The common names toluene, xylene, styrene, phenol, aniline, benzaldehyde, and benzoic acid (Section 9.3) are retained. The C_6H_5— group is named **phenyl,** and the $C_6H_5CH_2$— group is named **benzyl.** Two substituents on a benzene ring may be located by numbering the atoms of the ring or by using the locators **ortho (*o*), meta (*m*),** and **para (*p*).**

Polynuclear aromatic hydrocarbons (Section 9.3) contain two or more fused benzene rings. Particularly abundant are naphthalene, anthracene, phenanthrene, and their derivatives.

The functional group of a **phenol is** an —OH group bonded to a benzene ring (Section 9.4A). Phenol and its derivatives are weak acids, pK_a approximately 10.0, but are considerably stronger acids than alcohols (pK_a 16–18).

A characteristic reaction of aromatic compounds is **electrophilic aromatic substitution** (Section 9.6). Substituents on an aromatic ring influence both the site of further substitution and its rate (Section 9.8A). Substituent groups that direct an incoming group preferentially to the ortho and para positions are known as **ortho-para directors.** Those that direct an incoming group preferentially to the meta positions are known as **meta directors.** Groups that cause the rate of further substitution to be faster than that for benzene are said to be **activating;** those that cause the rate of further substitution to be slower than that for benzene are said to be **deactivating.**

A mechanistic rationale for directing effects is based on the degree of resonance stabilization of the possible cation intermediates formed on reaction of the aromatic ring and the electrophile (Section 9.8B). Groups that stabilize the cation intermediate are ortho-para directors. Groups that destabilize it are meta directors.

KEY REACTIONS

1. Acidity of Phenols (Section 9.4B)

Phenols are weak acids. Substitution by electron-withdrawing groups, such as the halogens and the nitro group, increases the acidity of phenols.

$$C_6H_5\text{—OH} + H_2O \rightleftharpoons C_6H_5\text{—O}^- + H_3O^+ \qquad K_a = 1.1 \times 10^{-10} \quad pK_a = 9.95$$

Phenol — Phenoxide ion

2. Reaction of Phenols with Strong Bases (Section 9.4C)

Water-insoluble phenols react quantitatively with strong bases to form water-soluble salts.

$$C_6H_5\text{—OH} + NaOH \longrightarrow C_6H_5\text{—O}^-Na^+ + H_2O$$

Phenol	Sodium hydroxide	Sodium phenoxide	Water
pK_a 9.95			pK_a 15.7
(stronger acid)	(stronger base)	(weaker base)	(weaker acid)

3. Oxidation at a Benzylic Position (Section 9.5)

A benzylic carbon bonded to at least one hydrogen is oxidized to a carboxyl group.

$$CH_3-C_6H_4-CH(CH_3)_2 \xrightarrow[H_2SO_4]{K_2Cr_2O_7} HO_2C-C_6H_4-CO_2H$$

4. Chlorination and Bromination (Section 9.7A)

The electrophile is a halonium ion, Cl^+ or Br^+, formed by treatment of Cl_2 or Br_2 with $AlCl_3$ or $FeCl_3$.

$$C_6H_6 + Cl_2 \xrightarrow{AlCl_3} C_6H_5-Cl + HCl$$

5. Nitration (Section 9.7B)

The electrophile is the nitronium ion, NO_2^+, formed by treatment of nitric acid with sulfuric acid.

$$C_6H_5Br + HNO_3 \xrightarrow[H_2SO_4]{} o\text{-}BrC_6H_4NO_2 + p\text{-}BrC_6H_4NO_2 + H_2O$$

6. Reduction of a Nitro Group to a Primary Amino Group (Section 9.7B)

Reduction can be carried out using H_2/Ni or other transition metal catalyst. It can also be carried out using iron, zinc, or tin with HCl.

$$O_2N-C_6H_4-CO_2H + 3H_2 \xrightarrow[3\ atm]{Ni} H_2N-C_6H_4-CO_2H + 3H_2O$$

7. Sulfonation (Section 9.7B)

The electrophile is HSO_3^+.

$$C_6H_6 + H_2SO_4 \longrightarrow C_6H_5-SO_3H + H_2O$$

8. Friedel-Crafts Alkylation (Section 9.7C)

The electrophile is an alkyl carbocation formed by treatment of an alkyl halide with a Lewis acid.

$$C_6H_6 + (CH_3)_2CHCl \xrightarrow{AlCl_3} C_6H_5-CH(CH_3)_2 + HCl$$

9. Friedel-Crafts Acylation (Section 9.7C)

The electrophile is an acyl cation formed by treatment of an acyl halide with a Lewis acid.

$$C_6H_6 + CH_3\overset{O}{\overset{\|}{C}}Cl \xrightarrow{AlCl_3} C_6H_5-\overset{O}{\overset{\|}{C}}CH_3 + HCl$$

10. Alkylation Using an Alkene (Section 9.7D)

The electrophile is a carbocation formed by treatment of an alkene with H_2SO_4 or H_3PO_4.

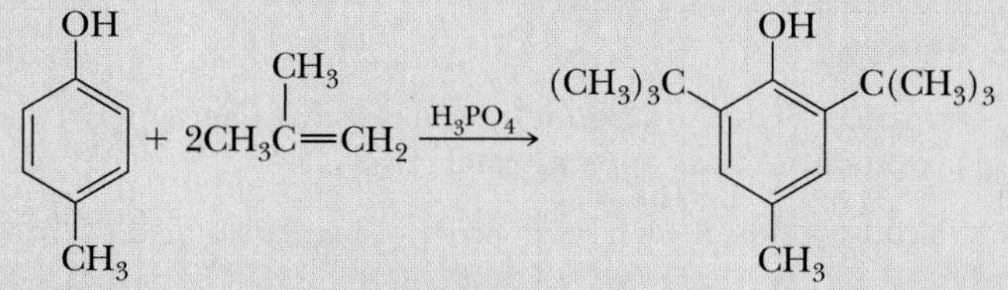

$$\text{4-methylphenol} + 2CH_3\overset{CH_3}{\overset{|}{C}}{=}CH_2 \xrightarrow{H_3PO_4} \text{2,6-di-}(CH_3)_3C\text{-4-methylphenol}$$

11. Alkylation Using an Alcohol (Section 9.7D)

The electrophile is a carbocation formed by treatment of an alcohol with H_2SO_4 or H_3PO_4.

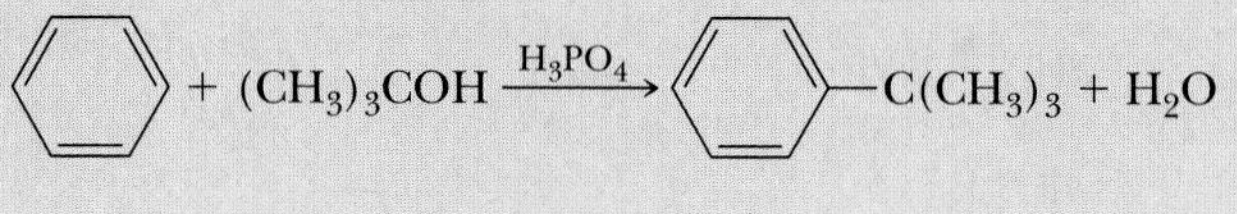

$$C_6H_6 + (CH_3)_3COH \xrightarrow{H_3PO_4} C_6H_5{-}C(CH_3)_3 + H_2O$$

ADDITIONAL PROBLEMS

Nomenclature and Structural Formulas

9.9 Name these compounds.

(a)  benzene ring with NO_2 and Cl para
(b) 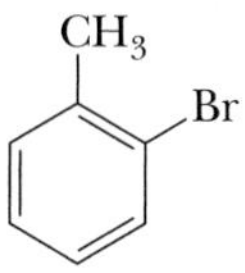 benzene ring with CH_3 and Br ortho
(c) $C_6H_5CH_2CH_2CH_2OH$
(d) 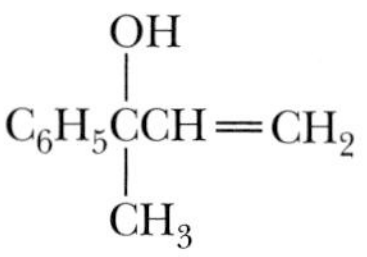 $C_6H_5\overset{OH}{\underset{CH_3}{C}}CH{=}CH_2$
(e) 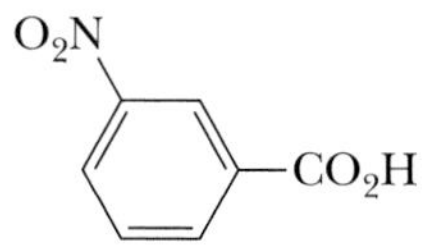 benzene ring with O_2N and CO_2H meta
(f) 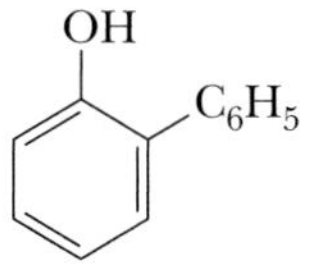 benzene ring with OH and C_6H_5 ortho
(g) 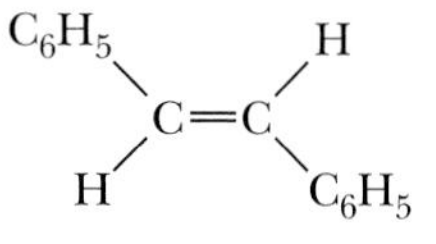 $C_6H_5(H)C{=}C(H)C_6H_5$ (trans)
(h) benzene ring with CH_3 (1), Cl (2), Cl (4)

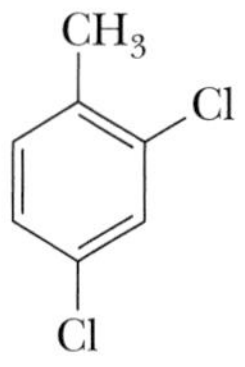

9.10 Draw structural formulas for these compounds.

(a) 1-Bromo-2-chloro-4-ethylbenzene
(b) 4-Iodo-1,2-dimethylbenzene
(c) 2,4,6-Trinitrotoluene
(d) 4-Phenyl-2-pentanol
(e) *p*-Cresol
(f) 2,4-Dichlorophenol
(g) 1-Phenylcyclopropanol
(h) Styrene (phenylethylene)
(i) *m*-Bromophenol
(j) 2,4-Dibromoaniline
(k) Isobutylbenzene
(l) *m*-Xylene

9.11 Show that pyridine can be represented as a hybrid of two equivalent contributing structures.

9.12 Show that naphthalene can be represented as a hybrid of three contributing structures, and show by the use of curved arrows how one contributing structure is converted to the next.

9.13 Draw four contributing structures for anthracene.

Acidity of Phenols

9.14 Use the resonance theory to account for the fact that phenol (pK_a 9.95) is a stronger acid than cyclohexanol (pK_a approximately 18).

9.15 Arrange the compounds in each set in order of increasing acidity (from least acidic to most acidic).

(a) —OH —OH CH_3CO_2H

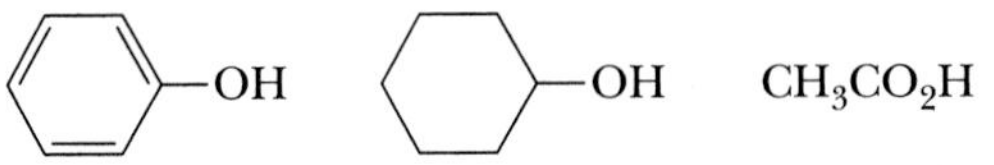

(b) —OH $NaHCO_3^-$ H_2O

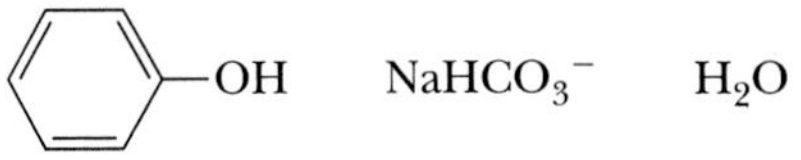

(c) O_2N— —OH —OH —CH_2OH

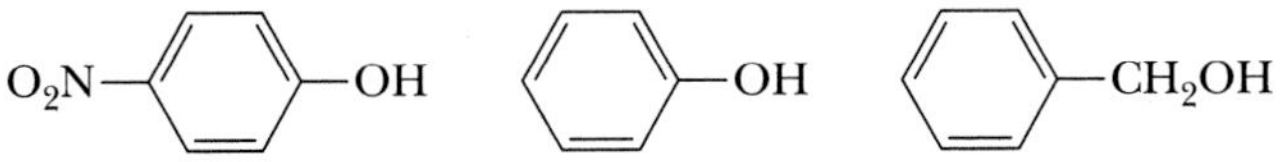

9.16 From each pair, select the stronger base.

(a) —O^- or OH^- **(b)** —O^- or —O^-

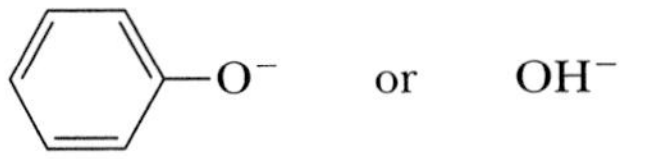

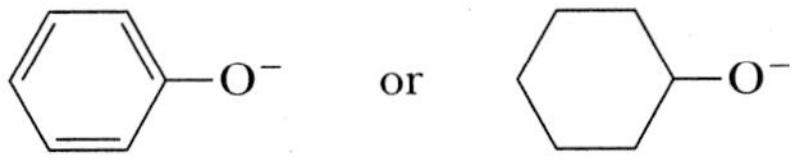

(c) —O^- or HCO_3^- **(d)** —O^- or $CH_3CO_2^-$

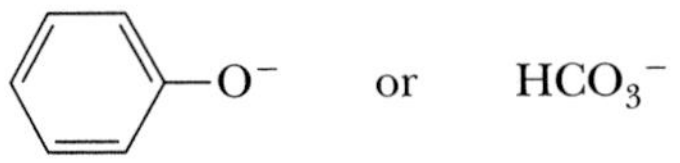

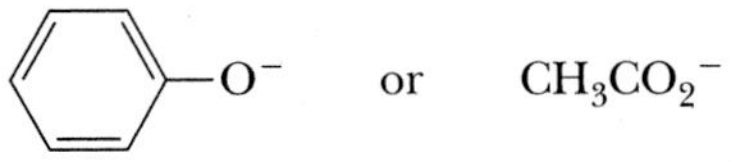

9.17 Account for the fact that water-insoluble carboxylic acids (pK_a 4–5) dissolve in 10% sodium bicarbonate with the evolution of a gas, but water-insoluble phenols (pK_a 9.5–10.5) do not dissolve in this solution.

9.18 Describe a procedure to separate a mixture of 1-hexanol and 2-methylphenol (*o*-cresol) and recover each in pure form. Each is insoluble in water but soluble in diethyl ether.

Electrophilic Aromatic Substitution: Monosubstitution

9.19 Draw a structural formula for the compound formed by treatment of benzene with each combination of reagents.

(a) $CH_3CH_2Cl/AlCl_3$ **(b)** $CH_2{=}CH_2/H_2SO_4$ **(c)** CH_3CH_2OH/H_2SO_4

9.20 Show three different combinations of reagents you might use to covert benzene to isopropylbenzene.

9.21 How many monochlorination products are possible when naphthalene is treated with $Cl_2/AlCl_3$?

9.22 Write a stepwise mechanism for this reaction. Use curved arrows to show the flow of electrons in each step.

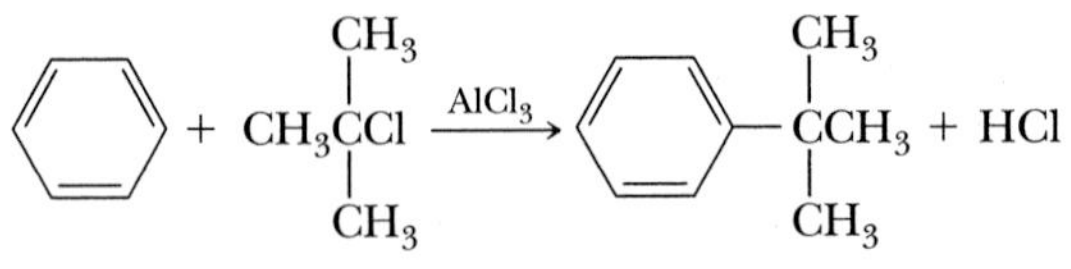

$$\text{C}_6\text{H}_6 + (CH_3)_3CCl \xrightarrow{AlCl_3} \text{C}_6\text{H}_5C(CH_3)_3 + HCl$$

9.23 Write a stepwise mechanism for the preparation of diphenylmethane by treating benzene with dichloromethane in the presence of an aluminum chloride catalyst.

Electrophilic Aromatic Substitution — Disubstitution

9.24 When treated with $Cl_2/AlCl_3$, *o*-xylene (1,2-dimethylbenzene) gives a mixture of two products. Draw structural formulas for these products.

9.25 How many monosubstitution products are possible when 1,4-dimethylbenzene (*p*-xylene) is treated with $Cl_2/AlCl_3$? When *m*-xylene (1,3-dimethylbenzene) is treated with $Cl_2/AlCl_3$?

9.26 Draw the structural formula for the major product formed on treatment of each compound with $Cl_2/AlCl_3$.

(a) Toluene **(b)** Nitrobenzene **(c)** Benzoic acid

(d) Chlorobenzene **(e)** *tert*-Butylbenzene

(f) $C_6H_5-\overset{O}{\overset{\|}{C}}CH_3$ **(g)** $C_6H_5-O\overset{O}{\overset{\|}{C}}CH_3$ **(h)** $C_6H_5-\overset{O}{\overset{\|}{C}}OCH_3$

9.27 Which compound undergoes electrophilic aromatic substitution more rapidly when treated with $Cl_2/AlCl_3$, chlorobenzene or toluene? Explain and draw structural formulas for the major product(s) from each reaction.

9.28 Arrange the compounds in each set in order of decreasing reactivity (fastest to slowest) toward electrophilic aromatic substitution.

(a) (A) C_6H_6 (B) $C_6H_5-O\overset{O}{\overset{\|}{C}}CH_3$ (C) $C_6H_5-\overset{O}{\overset{\|}{C}}OCH_3$

(b) (A) $C_6H_5-NO_2$ (B) $C_6H_5-CO_2H$ (C) C_6H_6

(c) (A) $C_6H_5-NH_2$ (B) $C_6H_5-NH\overset{O}{\overset{\|}{C}}CH_3$ (C) $C_6H_5-\overset{O}{\overset{\|}{C}}NHCH_3$

(d) (A) C_6H_6 (B) $C_6H_5-CH_3$ (C) $C_6H_5-OCH_3$

9.29 Account for the observation that the trifluoromethyl group is meta directing as shown in the following example.

$$C_6H_5CF_3 + HNO_3 \xrightarrow{H_2SO_4} m\text{-}O_2NC_6H_4CF_3 + H_2O$$

9.30 Show how to convert toluene to these carboxylic acids.

(a) 4-Chlorobenzoic acid

(b) 3-Chlorobenzoic acid

9.31 Show reagents and conditions to bring about these conversions.

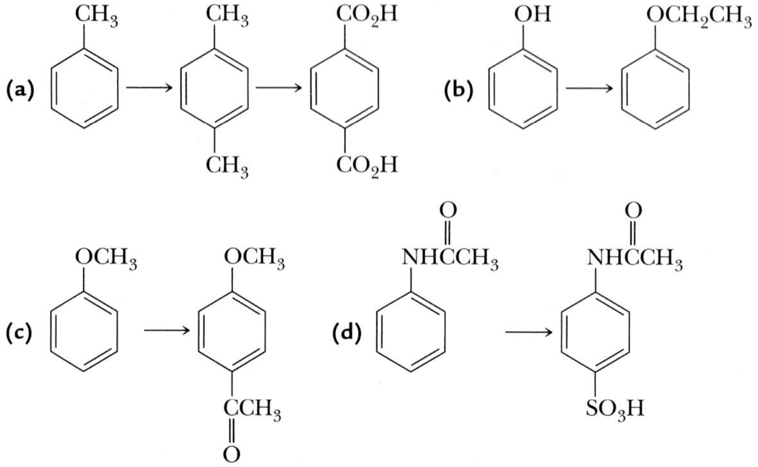

(a) toluene (CH_3) → p-xylene (CH_3, CH_3) → terephthalic acid (CO_2H, CO_2H)

(b) phenol (OH) → ethoxybenzene (OCH_2CH_3)

(c) anisole (OCH_3) → 4-methoxyacetophenone (OCH_3, CCH_3 with C=O)

(d) acetanilide ($NHCCH_3$ with C=O) → 4-acetamidobenzenesulfonic acid ($NHCCH_3$, SO_3H)

9.32 Propose a synthesis of triphenylmethane from benzene as the only source of aromatic rings. Use any other necessary reagents.

9.33 Reaction of phenol with acetone in the presence of an acid catalyst gives bisphenol A, a compound used in the production of polycarbonate and epoxy resins (Sections 15.4C and 15.4E). Propose a mechanism for the formation of bisphenol A.

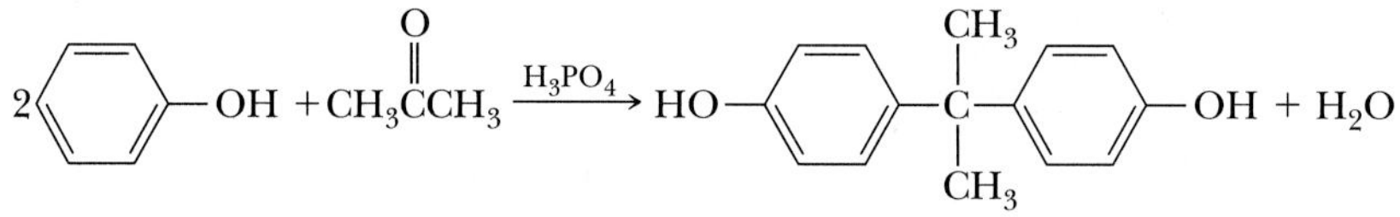

$$2\ C_6H_5OH + CH_3\overset{O}{\overset{\|}{C}}CH_3 \xrightarrow{H_3PO_4} HO{-}C_6H_4{-}C(CH_3)_2{-}C_6H_4{-}OH + H_2O$$

Bisphenol A

9.34 2,6-Di-*tert*-butyl-4-methylphenol, more commonly known as butylated hydroxytoluene or BHT, is used as an antioxidant in foods to "retard spoilage" (*The Merck Index,* 12th ed., #1583). BHT is synthesized industrially from 4-methylphenol (*p*-cresol) by reaction with 2-methylpropene in the presence of phosphoric acid. Propose a mechanism for this reaction.

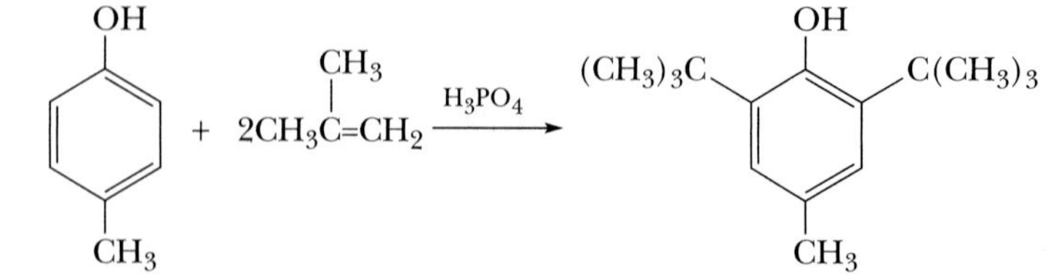

$$\text{4-Methylphenol (OH, } CH_3) + 2CH_3\overset{CH_3}{\overset{|}{C}}{=}CH_2 \xrightarrow{H_3PO_4} \text{2,6-Di-}tert\text{-butyl-4-methylphenol (OH, } (CH_3)_3C, C(CH_3)_3, CH_3)$$

4-Methylphenol
(p-Cresol)

2,6-Di-*tert*-butyl-4-methylphenol
"Butylated hydroxytoluene"
(BHT)

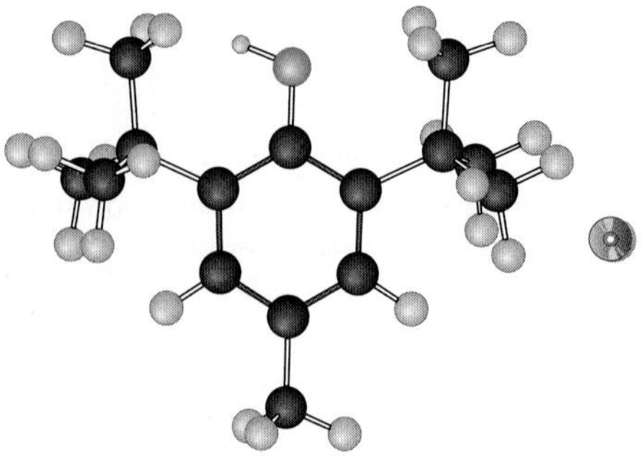

9.35 The first widely used herbicide for control of weeds was 2,4-dichlorophenoxyacetic acid (2,4-D). Show how this compound might be synthesized from 2,4-dichlorophenol and chloroacetic acid, $ClCH_2CO_2H$.

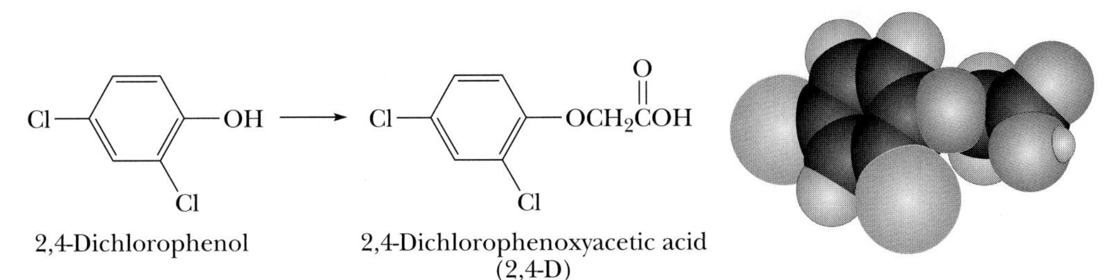

2,4-Dichlorophenol → 2,4-Dichlorophenoxyacetic acid (2,4-D)

Syntheses

9.36 Using styrene, $C_6H_5CH{=}CH_2$, as the only aromatic starting material, show how to synthesize these compounds. In addition to styrene, use any other necessary organic or inorganic chemicals. Any compound synthesized in one part of this problem may be used to make any other compound in the problem.

(a) $C_6H_5-\overset{\text{O}}{\overset{\|}{\text{C}}}\text{OH}$ **(b)** $C_6H_5-\overset{\text{Br}}{\overset{|}{\text{C}}}\text{HCH}_3$ **(c)** $C_6H_5-\overset{\text{OH}}{\overset{|}{\text{C}}}\text{HCH}_3$

(d) $C_6H_5-\overset{\text{O}}{\overset{\|}{\text{C}}}\text{CH}_3$ **(e)** $C_6H_5-CH_2CH_3$ **(f)** $C_6H_5-\overset{\text{OH}}{\overset{|}{\text{C}}}\text{HCH}_2\text{OH}$

9.37 Starting with benzene, toluene, or phenol as the only sources of aromatic rings, show how to synthesize these compounds. Assume in all syntheses that mixtures of ortho-para products can be separated to give the desired isomer in pure form.

(a) *m*-Bromonitrobenzene **(b)** 1-Bromo-4-nitrobenzene
(c) 2,4,6-Trinitrotoluene (TNT) **(d)** *m*-Bromobenzoic acid
(e) *p*-Bromobenzoic acid **(f)** *p*-Dichlorobenzene
(g) *m*-Nitrobenzenesulfonic acid **(h)** 1-Chloro-3-nitrobenzene

9.38 Starting with benzene or toluene as the only sources of aromatic rings, show how to synthesize these aromatic ketones. Assume in all syntheses that mixtures of ortho-para products can be separated to give the desired isomer in pure form.

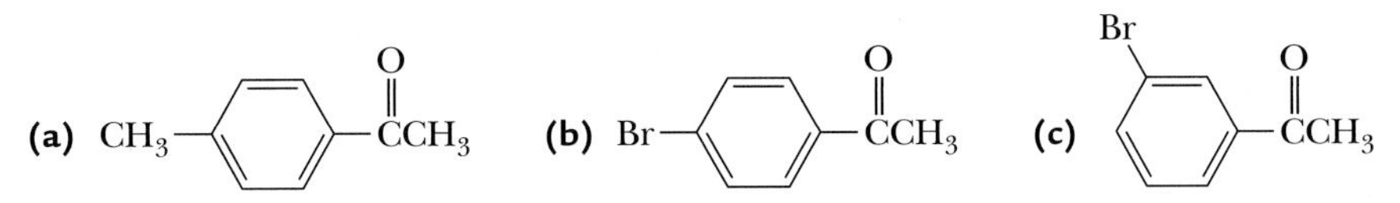

9.39 The following ketone, isolated from the roots of several members of the iris family, has an odor like that of violets and is used as a fragrance in perfumes. Describe the synthesis of this ketone from benzene.

$(CH_3)_2CH-C_6H_4-\overset{\text{O}}{\overset{\|}{\text{C}}}\text{CH}_3$

4-Isopropylacetophenone

9.40 The bombardier beetle generates *p*-quinone, an irritating chemical, by the enzyme-catalyzed oxidation of hydroquinone using hydrogen peroxide as the oxidizing agent. Heat generated in this oxidation produces superheated steam which is ejected, along with *p*-quinone, with explosive force.

$$\text{Hydroquinone} + H_2O_2 \xrightarrow{\text{enzyme catalyst}} p\text{-Quinone} + H_2O + \text{heat}$$

Hydroquinone (OH, OH)   *p*-Quinone (O, O)

(a) Balance the equation.

(b) Show by a balanced half-reaction that conversion of hydroquinone to *p*-quinone is a two-electron oxidation. (*Hint:* Review Section 6.4A on the use of half-reactions.)

A bombardier beetle ejects a mixture of superheated steam and *p*-quinone. *(Thomas Eisner and David Aneshansley, Cornell University)*

10

Amines

The opium poppy *(Papaver somniferum)*, from which morphine, codeine, and a score of related alkaloids are isolated. Inset: Model of morphine. *(Frank Orel/Tony Stone Images)*

Carbon, hydrogen, and oxygen are the three most common elements in organic compounds. Because of the wide distribution of amines in the biological world, nitrogen is the fourth most common component of organic compounds. The most important chemical property of amines is their basicity.

10.1 Structure and Classification

Amines are classified as **primary (1°), secondary (2°),** or **tertiary (3°)** depending on the number of hydrogen atoms of ammonia that are replaced by alkyl or aryl groups (Section 1.7D).

$:NH_3$	$CH_3-\ddot{N}H_2$	$CH_3-\ddot{N}H-CH_3$	$(CH_3)_3N:$
Ammonia	Methylamine (a 1° amine)	Dimethylamine (a 2° amine)	Trimethylamine (a 3° amine)

Aliphatic amine An amine in which nitrogen is bonded only to alkyl groups.

Aromatic amine An amine in which nitrogen is bonded to one or more aryl groups.

Amines are further divided into aliphatic amines and aromatic amines. In an **aliphatic amine,** all the carbons bonded directly to nitrogen are derived from alkyl groups; in an **aromatic amine,** one or more of the groups bonded directly to nitrogen are aryl groups.

$C_6H_5-\ddot{N}H_2$	$C_6H_5-\ddot{N}(CH_3)-H$	$C_6H_5-CH_2-\ddot{N}(CH_3)-CH_3$
Aniline (a 1° aromatic amine)	*N*-Methylaniline (a 2° aromatic amine)	Benzyldimethylamine (a 3° aliphatic amine)

Heterocyclic amine An amine in which nitrogen is one of the atoms of a ring.

Heterocyclic aromatic amine An amine in which nitrogen is one of the atoms of an aromatic ring.

An amine in which the nitrogen atom is part of a ring is classified as a **heterocyclic amine.** When the nitrogen is part of an aromatic ring (Section 9.2), the amine is classified as a **heterocyclic aromatic amine.** Following are structural formulas for two heterocyclic aliphatic amines and two heterocyclic aromatic amines.

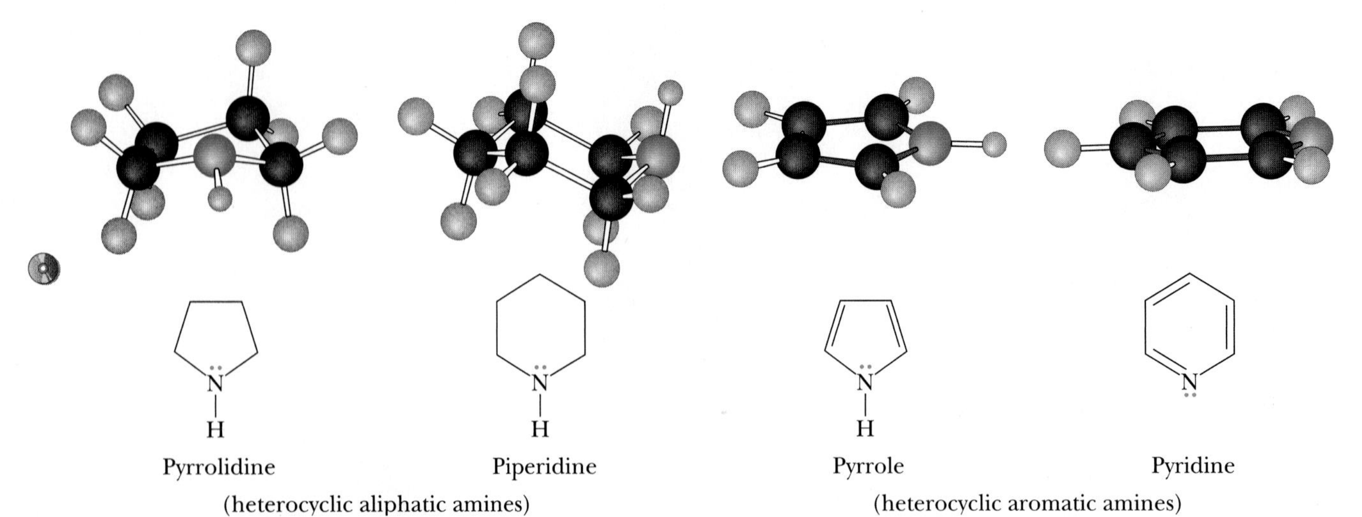

Pyrrolidine Piperidine (heterocyclic aliphatic amines)

Pyrrole Pyridine (heterocyclic aromatic amines)

EXAMPLE 10.1

Alkaloids are basic nitrogen-containing compounds of plant origin, many of which have physiological activity when administered to humans. Classify each amino group in these alkaloids according to type (that is, primary, secondary, tertiary, aliphatic, or aromatic). Coniine (*The Merck Index,* 12th ed., #2569), isolated from water hemlock, is highly toxic. Ingestion of it can cause weakness, labored respiration, paralysis, and eventually death. It is the toxic substance in "poison hemlock" used in the death of Socrates. Nicotine (*The Merck Index,* 12th ed., #6611) occurs in the tobacco plant. In small doses, it is an addictive stimulant. In larger doses, it causes depression, nausea, and vomiting. In still larger doses, it is a deadly poison. Solutions of nicotine in water are used as insecticides. Cocaine (*The Merck Index,* 12th ed., #2516) is a central nervous system stimulant obtained from the leaves of the coca plant.

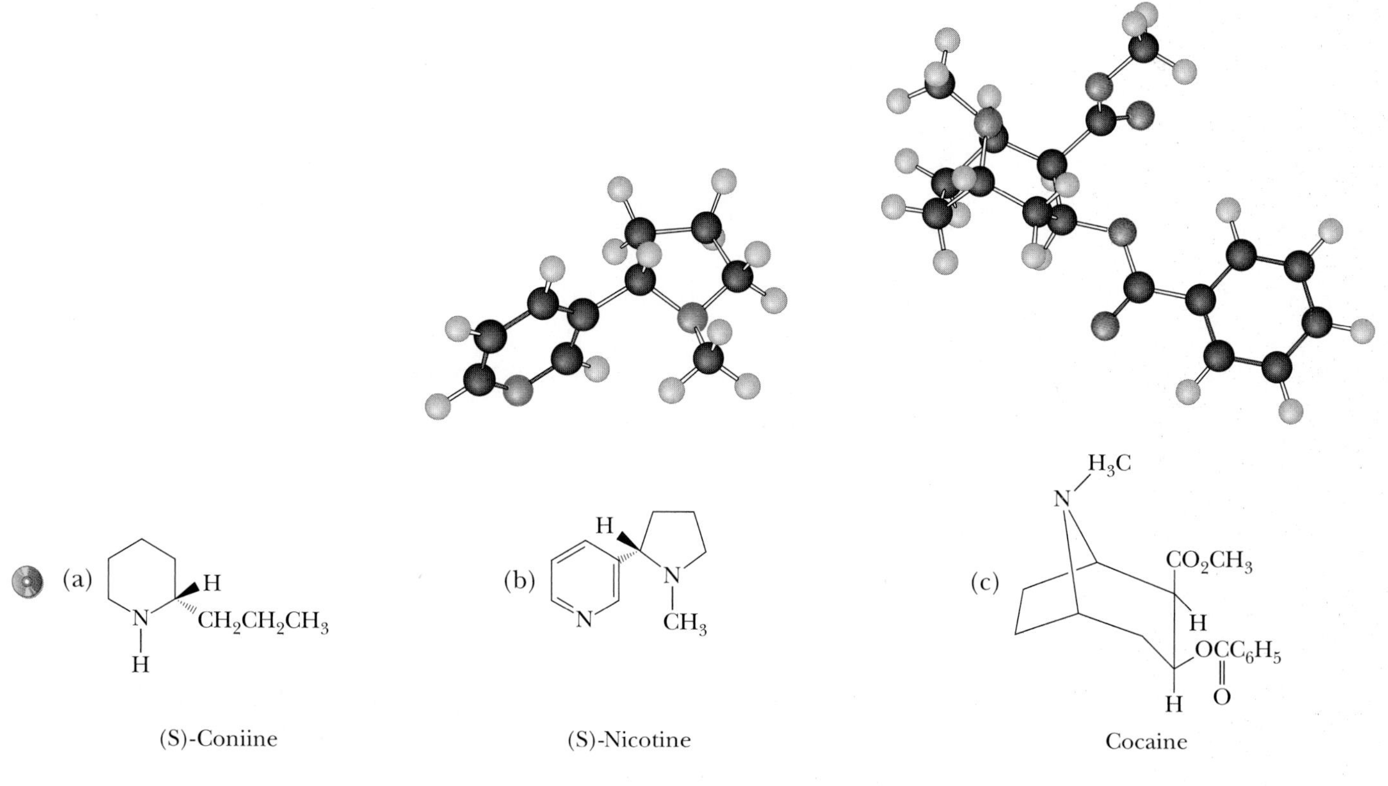

(S)-Coniine (S)-Nicotine Cocaine

SOLUTION

(a) A secondary heterocyclic aliphatic amine.
(b) Nicotine contains one tertiary heterocyclic aliphatic amine and one heterocyclic aromatic amine.
(c) A tertiary heterocyclic aliphatic amine.

Practice Problem 10.1

Identify all carbon stereocenters in coniine, nicotine, and cocaine.

CHEMICAL CONNECTIONS

Morphine and Related Analgesics

The analgesic, soporific, and euphoriant properties of the dried juice obtained from unripe seed pods of the opium poppy *Papaver somniferum* have been known for centuries. By the beginning of the 19th century, the active principal morphine (*The Merck Index,* 12th ed., #6359) had been isolated and its structure determined.

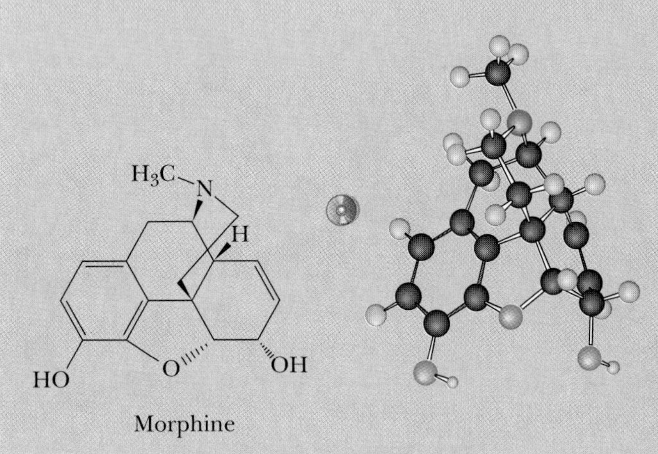

Morphine

Also occurring in the opium poppy is codeine (*The Merck Index,* 12th ed., #2525), a monomethyl ether of morphine. Heroin (*The Merck Index,* 12th ed., #3012) is the product of treatment of morphine with two moles of acetic anhydride.

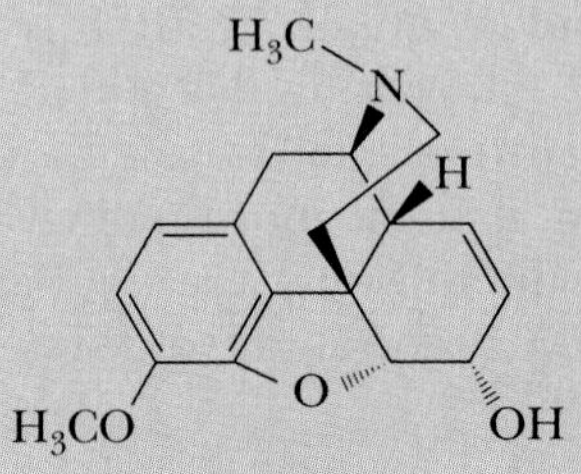

H_3C N H H_3CO O OH

Codeine

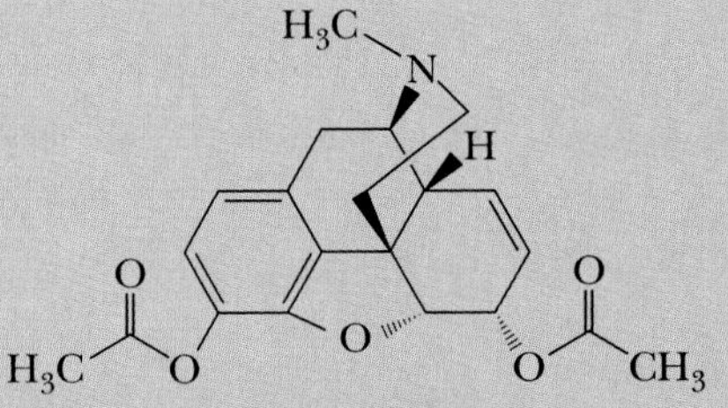

H_3C N H O O H_3C O O O CH_3

Heroin

Even though morphine is one of modern medicine's most effective pain killers, it has two serious disadvantages. First, it is addictive. Second, it depresses the respiratory control center of the central nervous system. Large doses of morphine (or heroin as well) can lead to death by respiratory failure. For these reasons, chemists have sought to produce pain killers related in structure to morphine, but without these serious disadvantages. One strategy in this on-going research has been to synthesize compounds related in structure to morphine in the hope that they would be equally effective analgesics but with reduced side effects. Following are struc-

10.2 NOMENCLATURE

A. Systematic Names

Systematic names for aliphatic amines are derived just as they are for alcohols. The suffix -e of the parent alkane is dropped and is replaced by -amine; that is, they are named as **alkanamines.**

tural formulas for two such compounds that have proven to be clinically useful. The levorotatory enantiomer of racemethorphan (*The Merck Index,* 12th ed., #8274) is a very potent analgesic. Interestingly, the dextrorotatory enantiomer, dextromethorphan, has no analgesic activity. It does, however, show approximately the same cough-suppressing activity as morphine and is, therefore, used extensively in cough remedies.

(+/−) mixture = Racemethorphan
(−)-enantiomer = Levomethorphan
(+)-enantiomer = Dextromethorphan

It has been discovered that there can be even further simplification in the structure of morphine-like analgesics. One such structural simplification is represented by meperidine, the hydrochloride salt of which is the widely used analgesic Demerol (*The Merck Index,* 12th ed., #5894).

Meperidine

It had been hoped that meperidine and related synthetic drugs would be free of many of the morphine-like undesirable side effects. It is now clear, however, that they are not. Meperidine, for example, is definitely addictive. In spite of much determined research, there are as yet no agents as effective as morphine for the relief of severe pain that are absolutely free of the risk of addiction.

How and in what regions of the brain does morphine act? In 1979, scientists discovered that there are specific receptor sites for morphine and other opiates, and that these sites are clustered in the brain's limbic system, the area involved in emotion and pain perception. Scientists then asked: Why does the human brain have receptor sites specific for morphine? Could it be that the brain produces its own opiates? In 1974 scientists discovered that opiate-like compounds are indeed present in the brain, and in 1975 isolated a brain opiate that was named enkephalin, meaning "in the brain." Scientists have yet to understand the role of these natural brain opiates. Perhaps when we do understand their biochemistry, we will discover clues that will lead to the design and synthesis of more potent but less addictive analgesics.

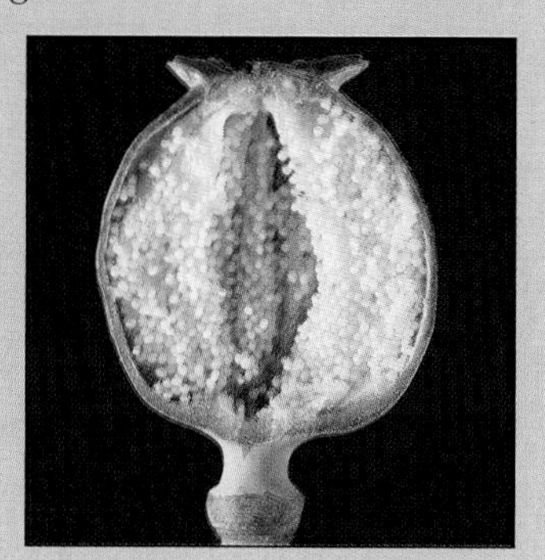

$CH_3CH(NH_2)CH_3$ — 2-Propanamine

(S)-1-Phenylethanamine

$H_2N(CH_2)_6NH_2$ — 1,6-Hexanediamine

EXAMPLE 10.2

Write the IUPAC names for these amine.

(a) $CH_3(CH_2)_5NH_2$ (b) $H_2N(CH_2)_4NH_2$

(c)

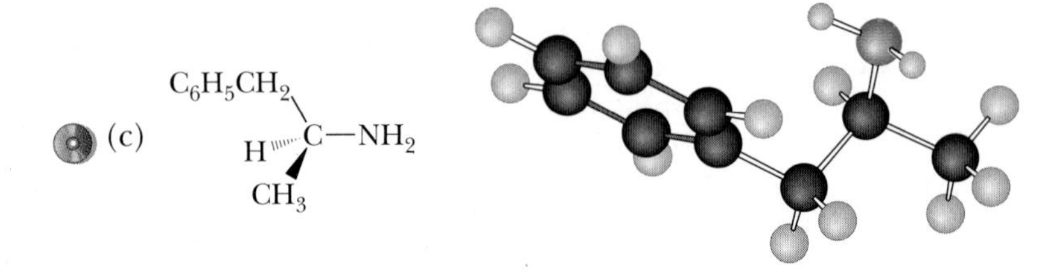

SOLUTION

(a) 1-Hexanamine (b) 1,4-Butanediamine

(c) Its systematic name is (S)-1-phenyl-2-propanamine. Its common name is amphetamine. The dextrorotatory isomer of amphetamine (shown here) is a central nervous system stimulant and is manufactured and sold under several trade names. The salt with sulfuric acid is marketed as Dexedrine sulfate (*The Merck Index,* 12th ed., #2996).

Practice Problem 10.2

Write a structural formula for each amine.

(a) 2-Methyl-1-propanamine (b) Cyclohexanamine

(c) (R)-2-Butanamine

IUPAC nomenclature retains the common name **aniline** for $C_6H_5NH_2$, the simplest aromatic amine. Its simple derivatives are named using the prefixes *o*-, *m*-, and *p*-, or numbers to locate substituents. Several derivatives of aniline have common names that are still widely used. Among these are **toluidine** for a methyl-substituted aniline and **anisidine** for a methoxy-substituted aniline.

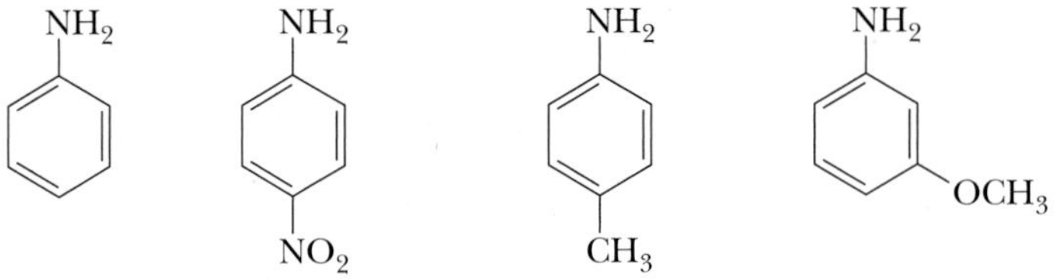

NH_2 NH_2 NH_2 NH_2

NO_2 CH_3 OCH_3

Aniline 4-Nitroaniline 4-Methylaniline (*p*-Toluidine) 3-Methoxyaniline (*m*-Anisidine)

Unsymmetrical secondary and tertiary amines are commonly named as *N*-substituted primary amines. The largest group is taken as the parent amine; the smaller group(s) attached to nitrogen are named, and their location is indicated by the prefix *N* (indicating that they are attached to nitrogen).

C_6H_5–$NHCH_3$ — *N*-Methylaniline

C_5H_9–$N(CH_3)_2$ — *N*,*N*-Dimethylcyclopentanamine

Following are names and structural formulas for four heterocyclic aromatic amines, the common names of which have been retained by the IUPAC.

Indole Purine Quinoline Isoquinoline

B. Common Names

Common names for most aliphatic amines are derived by listing the alkyl groups attached to nitrogen in alphabetical order in one word ending in the suffix **-amine;** that is, they are named as **alkylamines.**

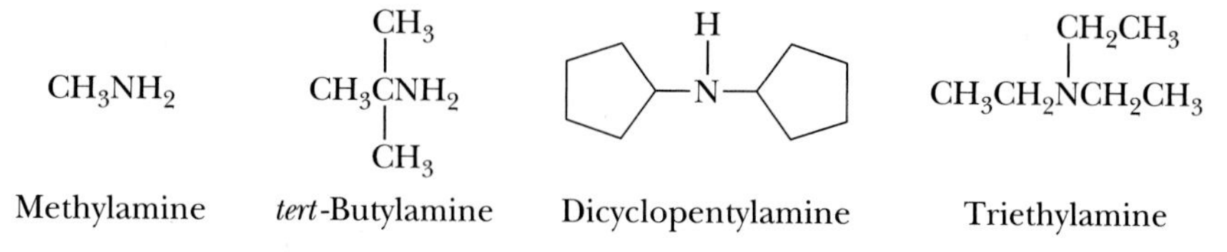

CH_3NH_2 — Methylamine

$(CH_3)_3CNH_2$ — *tert*-Butylamine

Dicyclopentylamine

$CH_3CH_2N(CH_2CH_3)CH_2CH_3$ — Triethylamine

EXAMPLE 10.3

Write a structural formula for each amine.

(a) Isopropylamine (b) Cyclohexylmethylamine (c) Benzylamine

SOLUTION

(a) $(CH_3)_2CHNH_2$ (b) C_6H_{11}–$NHCH_3$ (c) C_6H_5–CH_2NH_2

Practice Problem 10.3

Write a structural formula for each amine.

(a) Isobutylamine (b) Triphenylamine (c) Diisopropylamine

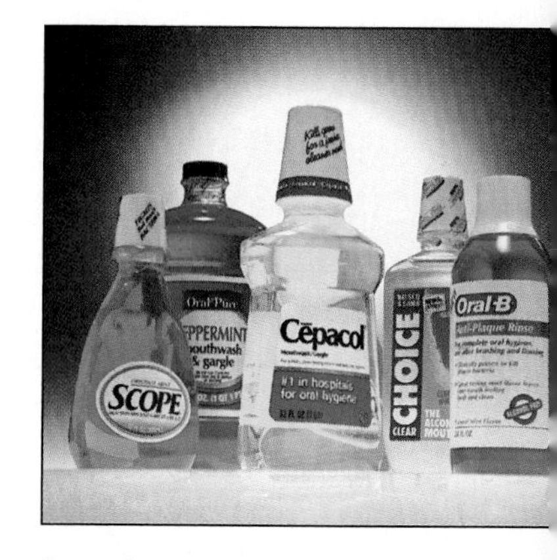

Several over-the-counter mouthwashes contain an *N*-alkylpyridinium chloride as an antibacterial agent. *(Charles D. Winters)*

When four atoms or groups of atoms are bonded to a nitrogen atom, the compound is named as a salt of the corresponding amine. The ending -amine (or aniline or pyridine or the like) is replaced by -ammonium (or anilinium or pyridinium or the like), and the name of the anion (chloride, acetate, and so on) is added. Compounds containing such ions have properties characteristic of salts.

$(CH_3)_4N^+ Cl^-$ — Tetramethylammonium chloride

$C_5H_5N^+CH_2(CH_2)_{14}CH_3\ Cl^-$ — Hexadecylpyridinium chloride (Cetylpyridinium chloride)

$C_6H_5CH_2N^+(CH_3)_3\ OH^-$ — Benzyltrimethylammonium hydroxide

Cetylpyridinium chloride is used as a topical antiseptic and disinfectant (*The Merck Index,* 12th ed., #2074).

10.3 Physical Properties

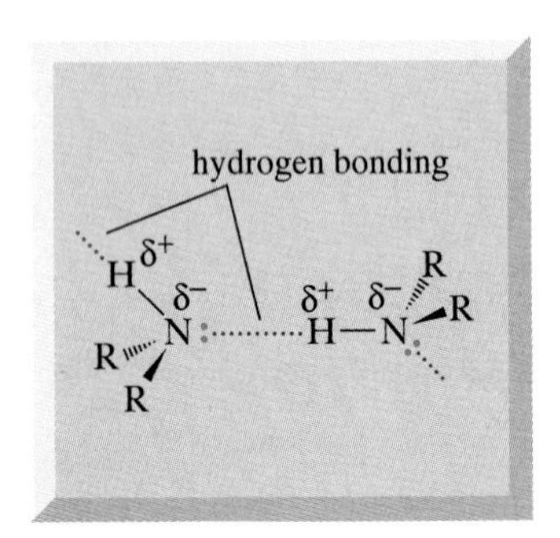

Figure 10.1
Intermolecular association by hydrogen bonding in primary and secondary amines. Nitrogen is approximately tetrahedral in shape with the axis of the hydrogen bond along the fourth position of the tetrahedron.

Amines are polar compounds, and both primary and secondary amines form intermolecular hydrogen bonds (Figure 10.1).

An N—H------N hydrogen bond is weaker than an O—H------O hydrogen bond because the difference in electronegativity between nitrogen and hydrogen $(3.0 - 2.1 = 0.9)$ is less than that between oxygen and hydrogen $(3.5 - 2.1 = 1.4)$. The effect of intermolecular hydrogen bonding can be illustrated by comparing the boiling points of methylamine and methanol. Both compounds have polar molecules and interact in the pure liquid by hydrogen bonding. Methanol has the higher boiling point because hydrogen bonding between its molecules is stronger than that between molecules of methylamine.

	CH_3NH_2	CH_3OH
MW (g/mol)	31.1	32.0
bp (°C)	−6.3	65.0

All classes of amines form hydrogen bonds with water and are more soluble in water than are hydrocarbons of comparable molecular weight. Most low-molecular-weight amines are completely soluble in water (Table 10.1). Higher molecular-weight amines are only moderately soluble or insoluble.

10.4 Basicity

Like ammonia, all amines are weak bases, and aqueous solutions of amines are basic. The following acid-base reaction between an amine and water is written using curved arrows to emphasize that, in this proton transfer reaction, the unshared pair of electrons on nitrogen forms a new covalent bond with hydrogen and displaces hydroxide ion.

$$CH_3NH_2 + H{-}\ddot{O}{-}H \rightleftharpoons CH_3NH_3^+ \ :\ddot{O}{-}H^-$$

Methylamine — Methylammonium hydroxide

TABLE 10.1 Physical Properties of Selected Amines

Name	Structural Formula	mp (°C)	bp (°C)	Solubility in Water
ammonia	NH_3	−78	−33	very soluble
Primary Amines				
methylamine	CH_3NH_2	−95	−6	very soluble
ethylamine	$CH_3CH_2NH_2$	−81	17	very soluble
propylamine	$CH_3CH_2CH_2NH_2$	−83	48	very soluble
isopropylamine	$(CH_3)_2CHNH_2$	−95	32	very soluble
butylamine	$CH_3(CH_2)_3NH_2$	−49	78	very soluble
benzylamine	$C_6H_5CH_2NH_2$	—	185	very soluble
cyclohexylamine	$C_6H_{11}NH_2$	−17	135	slightly soluble
Secondary Amines				
dimethylamine	$(CH_3)_2NH$	−93	7	very soluble
diethylamine	$(CH_3CH_2)_2NH$	−48	56	very soluble
Tertiary Amines				
trimethylamine	$(CH_3)_3N$	−117	3	very soluble
triethylamine	$(CH_3CH_2)_3N$	−114	89	slightly soluble
Aromatic Amines				
aniline	$C_6H_5NH_2$	−6	184	slightly soluble
Heterocyclic Aromatic Amines				
pyridine	C_5H_5N	−42	116	very soluble

The equilibrium constant for the reaction of an amine with water, K_{eq}, has the following form, illustrated for the reaction of methylamine with water to give methylammonium hydroxide:

$$K_{eq} = \frac{[CH_3NH_3^+][OH^-]}{[CH_3NH_2][H_2O]}$$

Because the concentration of water in dilute solutions of methylamine in water is essentially a constant ($[H_2O] = 55.5$ mol/L), it is combined with K_{eq} in a new constant called a base ionization constant, K_b. The value of K_b for methylamine is 4.37×10^{-4} ($pK_b = 3.36$).

$$K_b = K_{eq}[H_2O] = \frac{[CH_3NH_3^+][OH^-]}{[CH_3NH_2]} = 4.37 \times 10^{-4}$$

It is also common to discuss the basicity of amines by referring to the acid ionization constant of the corresponding conjugate acid as illustrated for the ionization of the methylammonium ion.

$$CH_3NH_3^+ + H_2O \rightleftharpoons CH_3NH_2 + H_3O^+$$

$$K_a = \frac{[CH_3NH_2][H_3O^+]}{[CH_3NH_3^+]} = 2.29 \times 10^{-11} \qquad pK_a = 10.64$$

CHEMICAL CONNECTIONS

The Poison Dart Frogs of South America—Lethal Amines

The Noanamá and Embrá peoples of the jungles of western Colombia have used poison blow darts for centuries, perhaps millennia. The poisons are obtained from the skin secretions of several highly colored frogs of the genus *Phyllobates* (*neará* and *kokoi* in the language of the native peoples). A single frog contains enough poison for up to 20 darts. For the most poisonous species (*Phyllobates terribilis*), just rubbing a dart over the frog's back suffices to charge the dart with poison.

Scientists at the National Institutes of Health (NIH) became interested in studying these poisons when it was discovered that they act on cellular ion channels, which would make them useful tools in basic research on mechanisms of ion transport. A field station was established in western Colombia to collect the relatively common poison dart frogs. From 5000 frogs, 11 mg of batrachotoxin and batrachotoxinin A was isolated. These names are derived from *batrachos*, the Greek word for frog.

Poison dart frog, *Phyllobates terribilis*. (*Animals, Animals/ ©Juan M. Renjifo*)

Batrachotoxin and batrachotoxinin A are among the most lethal poisons ever discovered. It is estimated that as little as 200 μg of batrachotoxin is sufficient to induce irreversible cardiac arrest in a human being. It has been determined that they act by causing voltage-gated Na^+ channels in nerve and muscle cells to be blocked in the open position, which leads to a huge influx of Na^+ ions into the affected cell.

Batrachotoxin

Batrachotoxinin A

The batrachotoxin story illustrates several common themes in drug discovery. First, information about the kinds of biologically active compounds and their sources is often obtained from the native peoples of a region. Second, tropical rain forests are a rich source of structurally complex, biologically active substances. Third, the entire ecosystem, not only the plants, is a potential source of fascinating organic molecules. See J. W. Daly *Progress in the Chemistry of Organic Natural Products,* Volume 41, edited by W. Herz, H. Grisebach, and G. W. Kirby, Springer-Verlag, Wien, p. 205, 1982.

TABLE 10.2 Base Strengths of Selected Amines and Acid Strengths of Their Conjugate Acids*

Amine	Structure	pK_b	pK_a
ammonia	NH_3	4.74	9.26
Primary Amines			
methylamine	CH_3NH_2	3.36	10.64
ethylamine	$CH_3CH_2NH_2$	3.19	10.81
cyclohexylamine	$C_6H_{11}NH_2$	3.34	10.66
Secondary Amines			
dimethylamine	$(CH_3)_2NH$	3.27	10.73
diethylamine	$(CH_3CH_2)_2NH$	3.02	10.98
Tertiary Amines			
trimethylamine	$(CH_3)_3N$	4.19	9.81
triethylamine	$(CH_3CH_2)_3N$	3.25	10.75
Aromatic Amines			
aniline	C_6H_5—NH_2	9.37	4.63
4-methylaniline	CH_3—C_6H_4—NH_2	8.92	5.08
4-chloroaniline	Cl—C_6H_4—NH_2	9.85	4.15
4-nitroaniline	O_2N—C_6H_4—NH_2	13.0	1.0
Heterocyclic Aromatic Amines			
pyridine	N (ring)	8.75	5.25
imidazole	N, N—H (ring)	7.05	6.95

* For each amine, $pK_a + pK_b = 14.00$.

Values of pK_a and pK_b for any acid-conjugate base pair are related by the following equation:

$$pK_a + pK_b = 14.00$$

Values of pK_a and pK_b for selected amines are given in Table 10.2.

EXAMPLE 10.4

Predict the position of equilibrium for this acid-base reaction.

$$CH_3NH_2 + CH_3CO_2H \rightleftharpoons CH_3NH_3^+ + CH_3CO_2^-$$

SOLUTION

Use the approach we developed in Chapter 2 to predict the position of equilibrium in acid-base reactions. Equilibrium favors formation of the weaker acid and the weaker base. Thus, in this reaction, equilibrium favors formation of methylammonium ion and acetate ion.

$$CH_3NH_2 + CH_3CO_2H \rightleftharpoons CH_3NH_3^+ + CH_3CO_2^-$$

CH_3NH_2	CH_3CO_2H	$CH_3NH_3^+$	$CH_3CO_2^-$
	pK_a 4.76	pK_a 10.64	
Stronger base	Stronger acid	Weaker acid	Weaker base

Practice Problem 10.4

Predict the position of equilibrium for this acid-base reaction.

$$CH_3NH_3^+ + H_2O \rightleftharpoons CH_3NH_2 + H_3O^+$$

Given information such as that in Table 10.2, we can make the following generalizations about the acid-base properties of the various classes of amines.

1. All aliphatic amines have about the same base strength, pK_b 3.0–4.0, and are slightly stronger bases than ammonia.
2. Aromatic amines and heterocyclic aromatic amines are considerably weaker bases than aliphatic amines. Compare, for example, values of pK_b for aniline and cyclohexylamine. The base ionization constant for aniline is smaller (the larger the value of pK_b, the weaker the base) than that for cyclohexylamine by a factor of 10^6.

$$C_6H_{11}NH_2 + H_2O \rightleftharpoons C_6H_{11}NH_3^+ \; OH^- \qquad pK_b = 3.34 \quad K_b = 4.5 \times 10^{-4}$$

Cyclohexylamine → Cyclohexylammonium hydroxide

$$C_6H_5NH_2 + H_2O \rightleftharpoons C_6H_5NH_3^+ \; OH^- \qquad pK_b = 9.37 \quad K_b = 4.3 \times 10^{-10}$$

Aniline → Anilinium hydroxide

Aromatic amines are weaker bases than aliphatic amines because of the resonance interaction of the unshared pair on nitrogen with the pi system of the aromatic ring. Because no such resonance interaction is possible for an alkylamine, the electron pair on its nitrogen is more available for reaction with an acid.

Interaction of the electron pair on nitrogen with the pi system of the aromatic ring

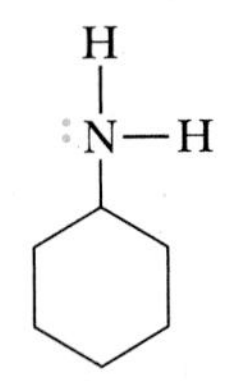

No resonance is possible with alkylamines

3. Electron-withdrawing groups such as halogen, nitro, and carbonyl decrease the basicity of substituted aromatic amines by decreasing the availability of the electron pair on nitrogen. Recall from Section 9.4B that these same substituents increase the acidity of phenols.

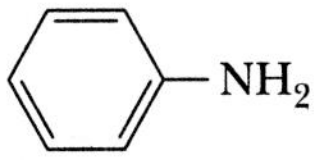

Aniline pK_b 9.37 — 4-Nitroaniline pK_b 13.0

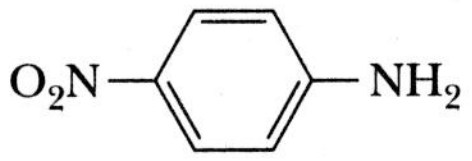

EXAMPLE 10.5

Select the stronger base in each pair of amines.

(a) 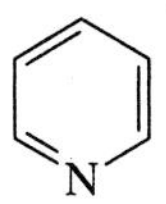 (A) or (B) — (b) 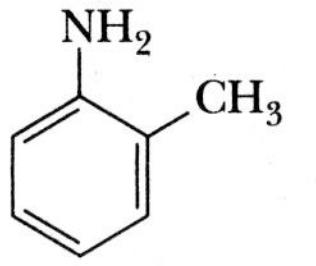 (C) or (D)

SOLUTION

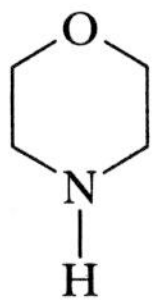

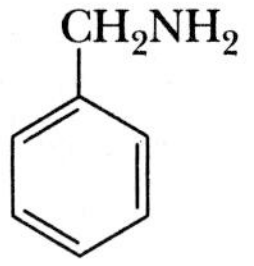

(a) Morpholine (B) is the stronger base (pK_b 5.79). It has a basicity comparable to that of secondary aliphatic amines. Pyridine (A), a heterocyclic aromatic amine (pK_b 8.75), is considerably less basic than aliphatic amines.

(b) Benzylamine (D), a primary aliphatic amine, is the stronger base (pK_b 3–4). *o*-Toluidine (C), an aromatic amine, is the weaker base (pK_b 9–10).

Practice Problem 10.5

Select the stronger acid from each pair of ions.

(a) 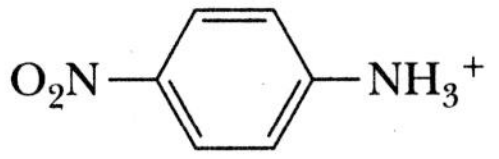 $O_2N-C_6H_4-NH_3^+$ (A) or $CH_3-C_6H_4-NH_3^+$ (B)

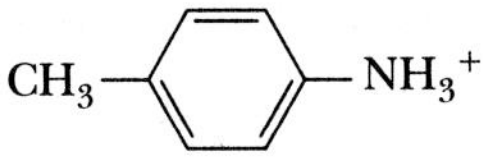

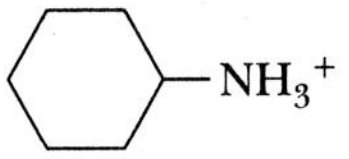

(b) 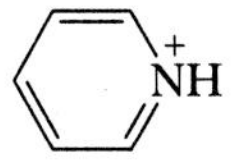 (C) or (D)

Guanidine (*The Merck Index,* 12th ed., #4591), pK_b 0.4, is the strongest base among neutral compounds.

$$H_2N-C(=NH)-NH_2 + H_2O \rightleftharpoons H_2N-C(={}^+NH_2)-NH_2 + OH^- \qquad pK_b = 0.4$$

Guanidine — Guanidinium ion

The remarkable basicity of guanidine is attributed to the fact that the positive charge on the guanidinium ion is delocalized equally over the three nitrogen atoms as shown by these three equivalent contributing structures.

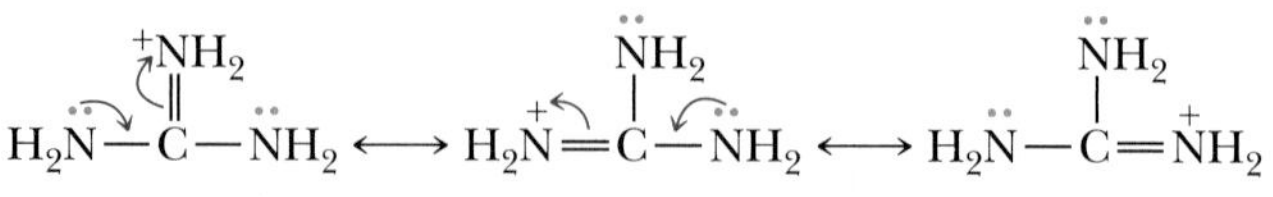

$$H_2N-C(=\overset{+}{N}H_2)-NH_2 \longleftrightarrow H_2\overset{+}{N}=C(NH_2)-NH_2 \longleftrightarrow H_2N-C(NH_2)=\overset{+}{N}H_2$$

Three equivalent contributing structures

10.5 Reaction with Acids

Amines, whether soluble or insoluble in water, react quantitatively with strong acids to form water-soluble salts as illustrated by the reaction of (R)-norepinephrine (noradrenaline; *The Merck Index,* 12th ed., #6788) with aqueous HCl to form a hydrochloride salt. Norepinephrine, secreted by the medulla of the adrenal gland, is a neurotransmitter. It has been suggested that it is a neurotransmitter in those areas of the brain that mediate emotional behavior.

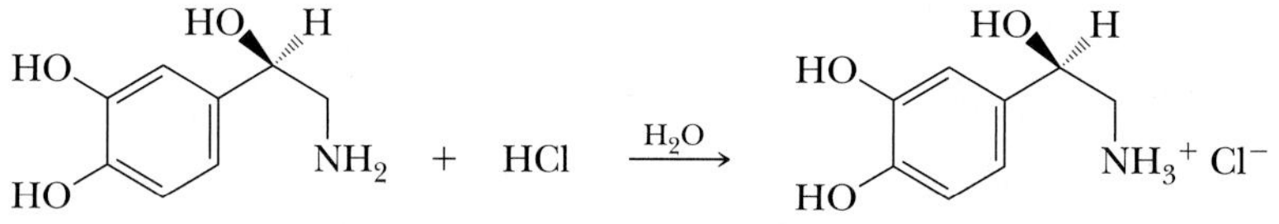

(R)-(−)-Norepinephrine (only slightly soluble in water) + HCl $\xrightarrow{H_2O}$ (R)-(−)-Norepinephrine hydrochloride (a water-soluble salt) ($NH_3^+\ Cl^-$)

Example 10.6

Complete each acid-base reaction, and name the salt formed.

(a) $(CH_3CH_2)_2NH + HCl \longrightarrow$ (b) 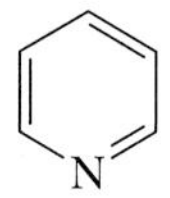 pyridine $+ CH_3CO_2H \longrightarrow$

Solution

(a) $(CH_3CH_2)_2NH_2^+\ Cl^-$
Diethylammonium chloride

(b) 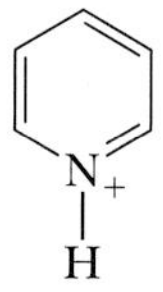 pyridinium (N⁺–H) $CH_3CO_2^-$
Pyridinium acetate

Practice Problem 10.6

Complete each acid-base reaction, and name the salt formed.

(a) $(CH_3CH_2)_3N + HCl \longrightarrow$ (b) 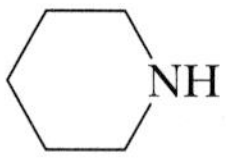 piperidine (NH) $+ CH_3CO_2H \longrightarrow$

The basicity of amines and the solubility of amine salts in water can be used to separate amines from water-insoluble, nonbasic compounds. The flowchart below is a flowchart for the separation of aniline from anisole. Note that aniline is recovered from its salt by treatment with NaOH.

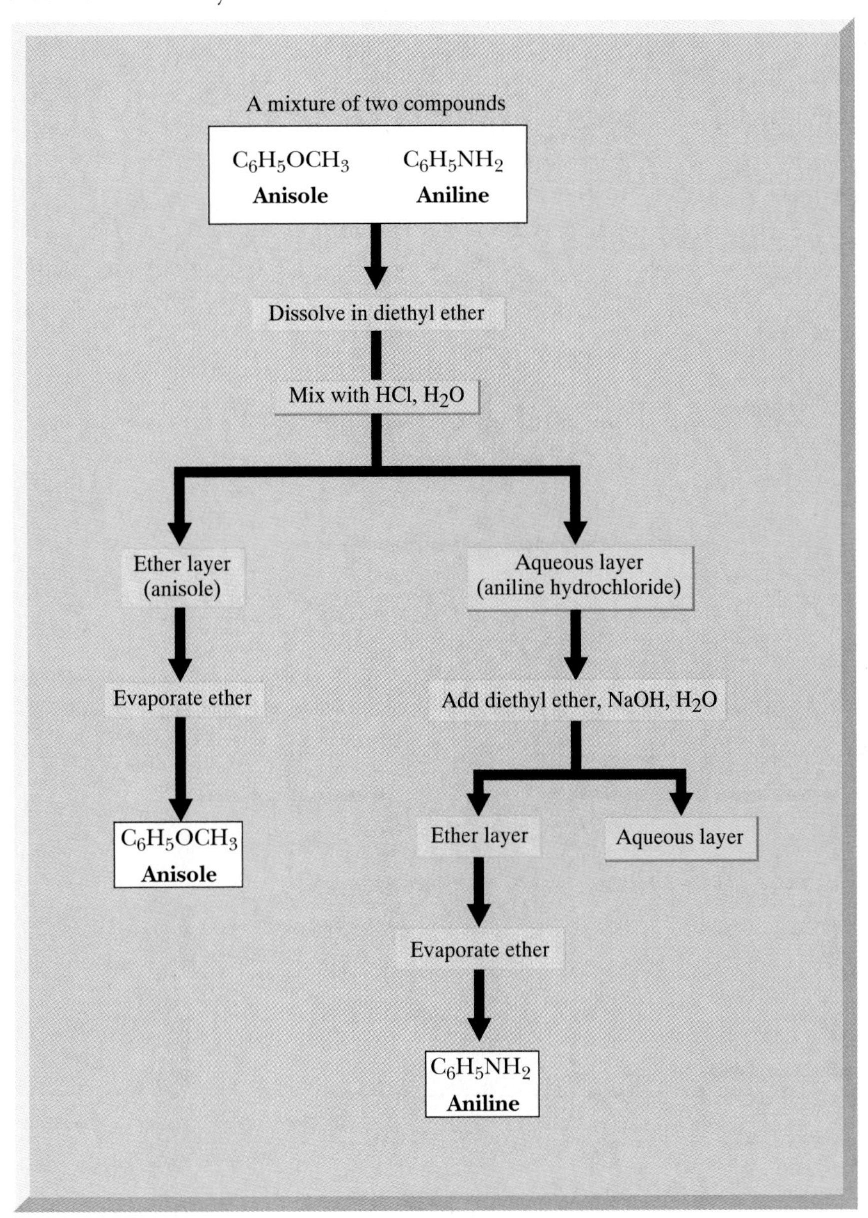

Separation and purification of an amine and a neutral compound.

EXAMPLE 10.7

Here is a flowchart for the separation of a mixture of a primary aliphatic amine (RNH_2, pK_a 10.8), a carboxylic acid (RCO_2H, pK_a 5), and a phenol (ArOH, pK_a 10). Assume that each is insoluble in water but soluble in diethyl ether. The mixture is separated into fractions A, B, and C. Which fraction contains the amine, which the carboxylic acid, and which the phenol? *Hint:* The pK_a of H_2CO_3 is 6.36 (Table 2.1).

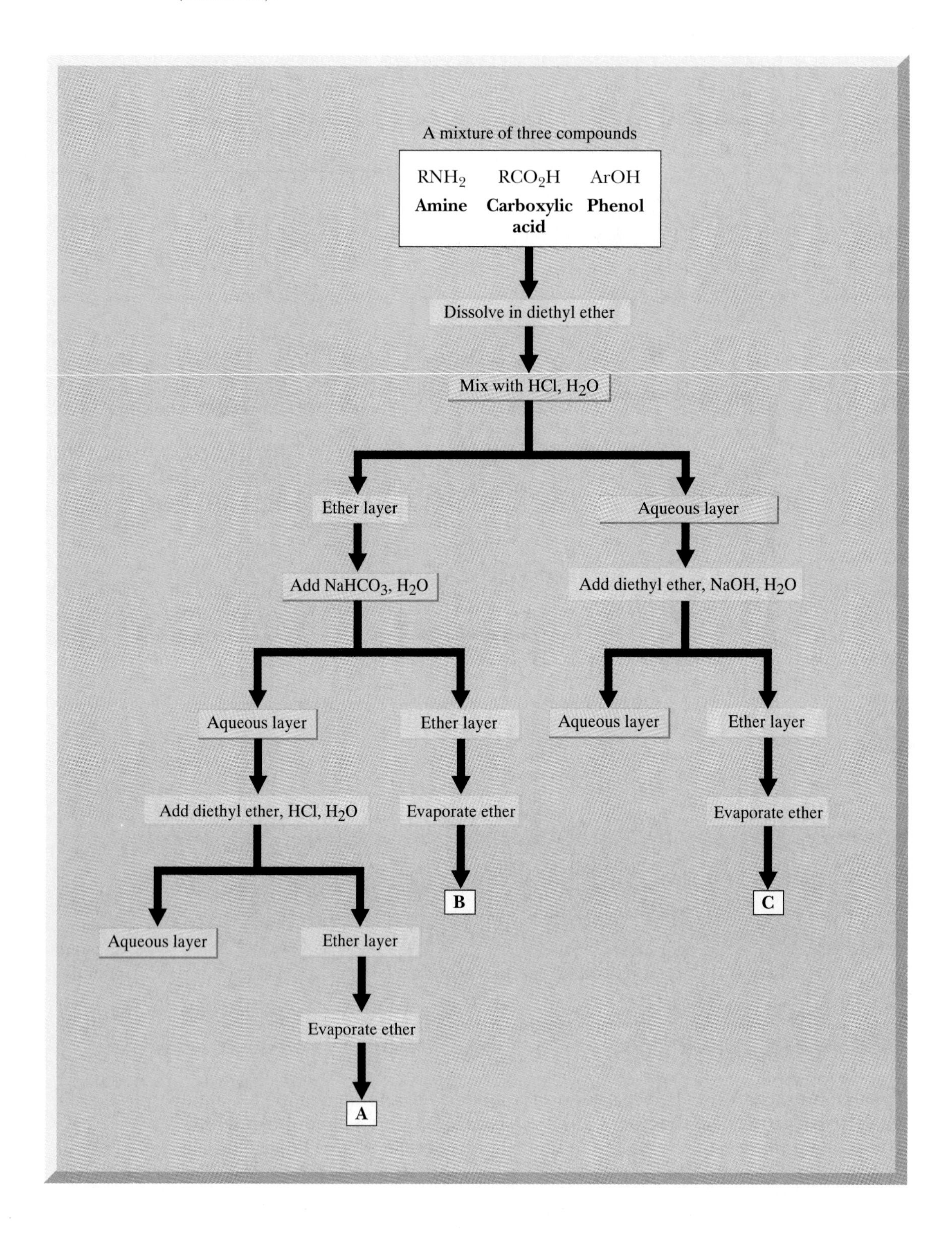

SOLUTION

Fraction C is RNH_2, fraction B is ArOH, and fraction A is RCO_2H.

Practice Problem 10.7

In what way(s) might the results of the separation and purification procedure outlined in Example 10.7 be different if

(a) Aqueous NaOH is used in place of aqueous $NaHCO_3$?
(b) The starting mixture contains an aromatic amine, $ArNH_2$, rather than an aliphatic amine, RNH_2?

SUMMARY

Amines are classified as **primary, secondary,** or **tertiary,** depending on the number of hydrogen atoms of ammonia replaced by alkyl or aryl groups (Section 10.1). In an **aliphatic amine,** all carbon atoms bonded to nitrogen are derived from alkyl groups. In an **aromatic amine,** one or more of the groups bonded to nitrogen are aryl groups. A **heterocyclic amine** is one in which the nitrogen atom is part of a ring. A **heterocyclic aromatic amine** is one in which the nitrogen atom is part of an aromatic ring.

In systematic nomenclature (Section 10.2), aliphatic amines are named **alkanamines.** In the common system of nomenclature, aliphatic amines are named **alkylamines;** the alkyl groups are listed in alphabetical order in one word ending in the suffix -amine. An ion containing nitrogen bonded to four alkyl or aryl groups is named as a **quaternary ammonium ion.**

Amines are polar compounds, and primary and secondary amines associate by intermolecular hydrogen bonding (Section 10.3). Because an N—H------N hydrogen bond is weaker than an O—H------O hydrogen bond, amines have lower boiling points than alcohols of comparable molecular weight and structure. All classes of amines form hydrogen bonds with water and are more soluble in water than hydrocarbons of comparable molecular weight.

Amines are weak bases, and aqueous solutions of amines are basic (Section 10.4). The base ionization constant for an amine in water is given the symbol K_b. It is also common to discuss the acid-base properties of amines by reference to the acid ionization constant, K_a, for its conjugate acid. Acid and base ionization constants for an amine in water are related by the equation $pK_a + pK_b = 14.0$.

KEY REACTIONS

1. Basicity of Aliphatic Amines (Section 10.4)

Most aliphatic amines have comparable basicity (pK_b 3.0–4.0) and are slightly stronger bases than ammonia.

$$CH_3NH_2 + H_2O \rightleftharpoons CH_3NH_3^+ + OH^- \qquad pK_b = 3.36$$

2. Basicity of Aromatic Amines (Section 10.4)

Aromatic amines (pK_b 9.0–10.0) are considerably weaker bases than aliphatic amines. Resonance stabilization from interaction of the unshared electron pair on nitrogen with the pi system of the aromatic ring decreases the availability of this electron pair for reaction with an acid. Substitution on the ring by electron-withdrawing groups decreases the basicity of the $—NH_2$ group.

$$C_6H_5\ddot{N}H_2 + H_2O \rightleftharpoons C_6H_5NH_3^+ + OH^- \qquad pK_b = 9.37$$

3. Reaction with Strong Acids (Section 10.5)
All amines react quantitatively with strong acids to form water-soluble salts.

$$C_6H_5-N(CH_3)_2 + HCl \longrightarrow C_6H_5-\overset{H}{\overset{|}{N^+}}(CH_3)_2\ Cl^-$$

Insoluble in water — A water-soluble salt

ADDITIONAL PROBLEMS

Structure and Nomenclature

10.8 Draw a structural formula for each amine.

(a) (R)-2-Butanamine **(b)** 1-Octanamine
(c) 2,2-Dimethyl-1-propanamine **(d)** 1,5-Pentanediamine
(e) 2-Bromoaniline **(f)** Tributylamine
(g) *N,N*-Dimethylaniline **(h)** Benzylamine
(i) *tert*-Butylamine **(j)** *N*-Ethylcyclohexanamine
(k) Diphenylamine **(l)** Isobutylamine

10.9 Draw a structural formula for each amine.

(a) 4-Aminobutanoic acid **(b)** 2-Aminoethanol (ethanolamine)
(c) 2-Aminobenzoic acid **(d)** (S)-2-Aminopropanoic acid (alanine)
(e) 4-Aminobutanal **(f)** 4-Amino-2-butanone

10.10 Classify each amino group as primary, secondary, or tertiary; as aliphatic or aromatic.

(a) HO, $CH_2CH_2NH_2$, N–H

Serotonin
(a neurotransmitter)

(b) $H_2N-C_6H_4-\overset{O}{\overset{\|}{C}}OCH_2CH_3$

Benzocaine
(a topical anesthetic)

(c) CH_3, $NHCH(CH_2)_3N(C_2H_5)_2$, Cl, N

Chloroquine
(a drug for the treatment of malaria)

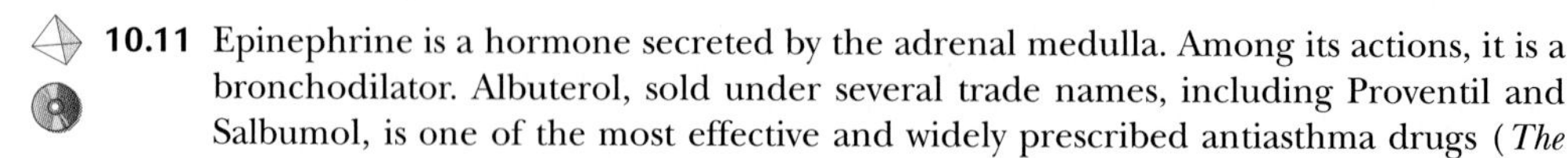

10.11 Epinephrine is a hormone secreted by the adrenal medulla. Among its actions, it is a bronchodilator. Albuterol, sold under several trade names, including Proventil and Salbumol, is one of the most effective and widely prescribed antiasthma drugs (*The*

Merck Index, 12th ed., #217). The R enantiomer of albuterol is 68 times more effective in the treatment of asthma than the S enantiomer.

(a) Classify each amino group as primary, secondary, or tertiary.

(b) List the similarities and differences between the structural formulas of these two compounds.

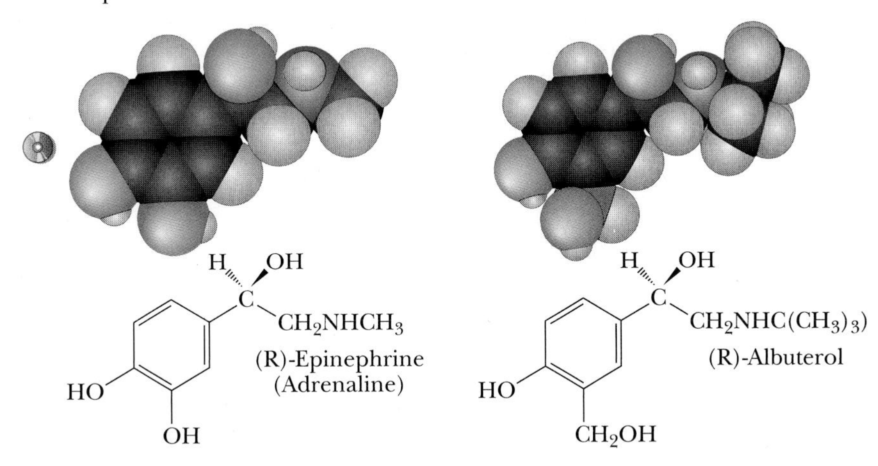

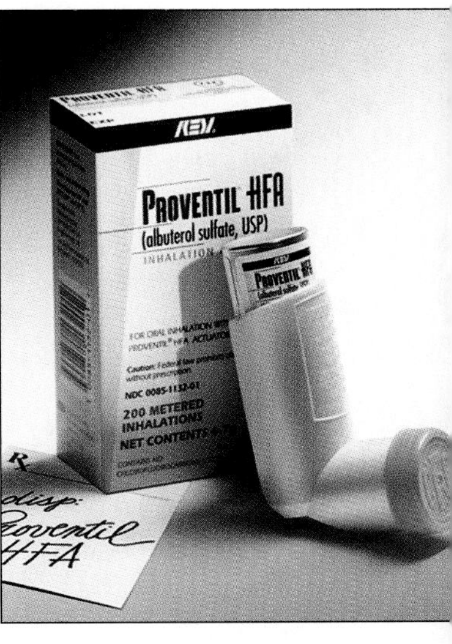

This inhaler delivers puffs of albuterol (Proventil), a potent bronchodilator whose structure is patterned after that of epinephrine (adrenaline). *(Charles D. Winters)*

10.12 There are eight constitutional isomers of molecular formula $C_4H_{11}N$. Name and draw structural formulas for each. Classify each amine as primary, secondary, or tertiary.

10.13 Draw a structural formula for each compound of the given molecular formula.

(a) A 2° arylamine, C_7H_9N **(b)** A 3° arylamine, $C_8H_{11}N$

(c) A 1° aliphatic amine, C_7H_9N **(d)** A chiral 1° amine, $C_4H_{11}N$

(e) A 3° heterocyclic amine, $C_5H_{11}N$ **(f)** A trisubstituted 1° arylamine, $C_9H_{13}N$

(g) A chiral quaternary ammonium salt, $C_9H_{22}NCl$

Physical Properties

10.14 Propylamine, ethylmethylamine, and trimethylamine are constitutional isomers of molecular formula C_3H_9N. Account for the fact that trimethylamine has the lowest boiling point of the three.

$CH_3CH_2CH_2NH_2$	$CH_3CH_2NHCH_3$	$(CH_3)_3N$
bp 48°C	bp 37°C	bp 3°C

10.15 Account for the fact that 1-butanamine has a lower boiling point than 1-butanol.

$CH_3CH_2CH_2CH_2OH$	$CH_3CH_2CH_2CH_2NH_2$
bp 117°C	bp 78°C

Basicity of Amines

10.16 Account for the fact that amines are more basic than alcohols.

10.17 From each pair of compounds, select the stronger base.

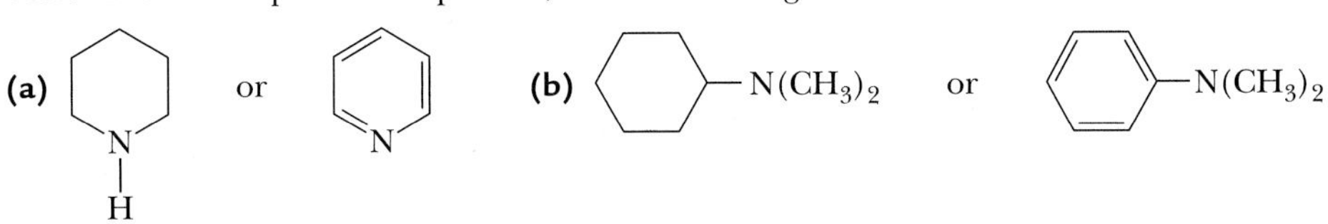

(c) 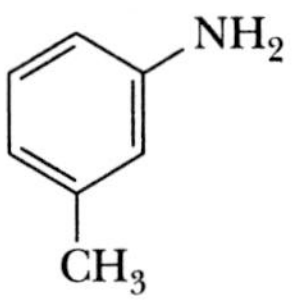 3-methylaniline (NH_2, CH_3) or benzylamine (CH_2NH_2) (d) 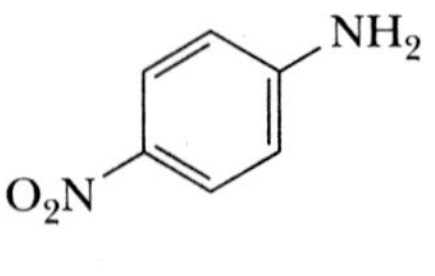 4-nitroaniline (NH_2, O_2N) or 4-methylaniline (NH_2, H_3C)

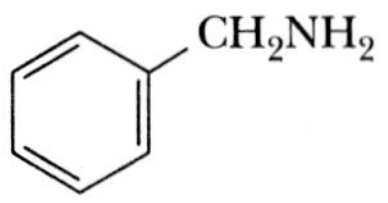

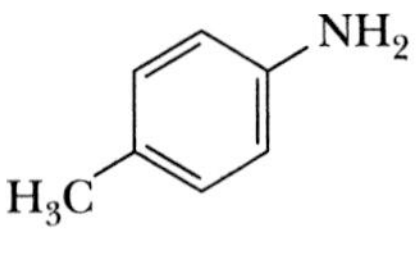

10.18 Account for the fact that substitution of a nitro group makes an aromatic amine a weaker base but makes a phenol a stronger acid. For example, 4-nitroaniline is a weaker base than aniline, but 4-nitrophenol is a stronger acid than phenol.

10.19 Select the stronger base in this pair of compounds.

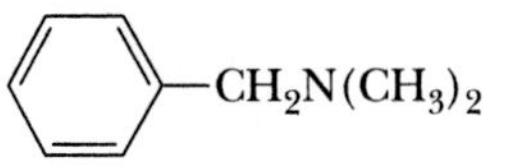

$C_6H_5-CH_2N(CH_3)_2$ or $C_6H_5-CH_2\overset{+}{N}(CH_3)_3\ OH^-$

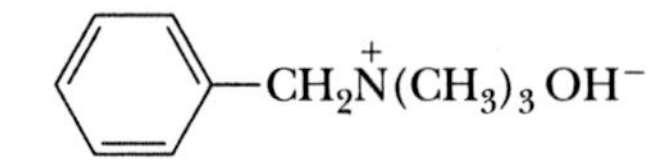

10.20 Following are two structural formulas for alanine (2-aminopropanoic acid), one of the building blocks of proteins (Chapter 18). Is alanine better represented by structural formula (A) or structural formula (B)? Explain.

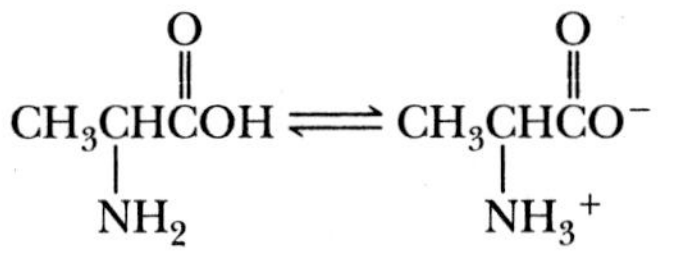

$$\underset{\text{(A)}}{CH_3CH(NH_2)COOH} \rightleftharpoons \underset{\text{(B)}}{CH_3CH(NH_3^+)COO^-}$$

10.21 Complete the following acid-base reactions and predict the position of equilibrium for each. Justify your prediction by citing values of pK_a for the stronger and weaker acid in each equilibrium. For values of acid ionization constants, consult Table 2.1 (pK_a values of some inorganic and organic acids), Table 8.4 (pK_a values of alcohols), Section 9.4B (acidity of phenols), and Table 10.2 (base strengths of amines). Where no ionization constants are given, make the best estimate from the information given in the reference tables and sections.

(a) 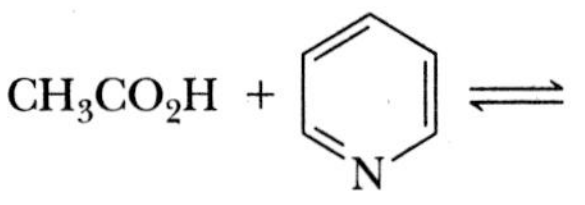 CH_3CO_2H + pyridine $\rightleftharpoons$
Acetic acid Pyridine

(b) 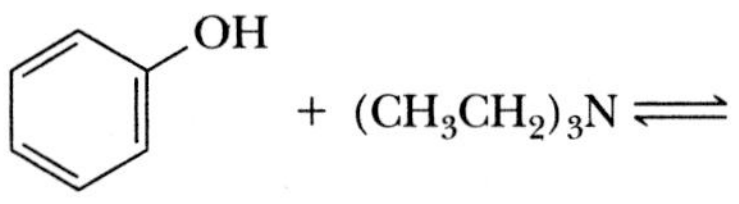 phenol + $(CH_3CH_2)_3N \rightleftharpoons$
Phenol Triethylamine

(c)  $PhCH_2CH(CH_3)NH_2$ + $CH_3CH(OH)COOH$ $\rightleftharpoons$
1-Phenyl-2-propanamine (Amphetamine) 2-Hydroxypropanoic acid (Lactic acid)

(d) 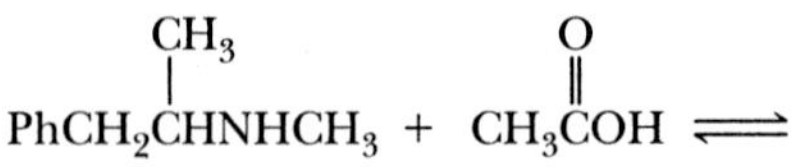 $PhCH_2CH(CH_3)NHCH_3$ + CH_3COOH $\rightleftharpoons$
Methamphetamine Acetic acid

10.22 The pK_a of morpholinium ion is 8.33.

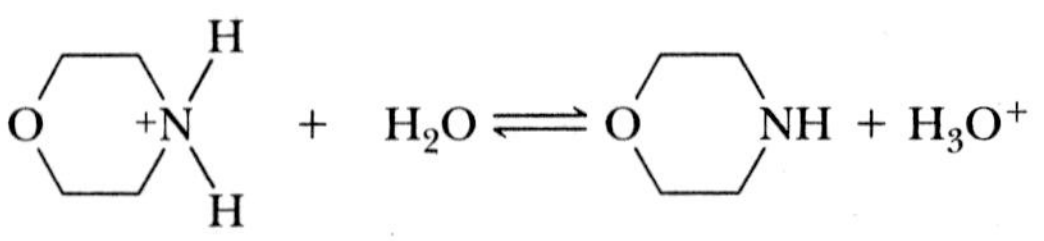

$$\text{Morpholinium ion} + H_2O \rightleftharpoons \text{Morpholine} + H_3O^+ \qquad pK_a = 8.33$$

(a) Calculate the ratio of morpholine to morpholinium ion in aqueous solution at pH 7.0.

(b) At what pH are the concentrations of morpholine and morpholinium ion equal?

10.23 The pK_b of amphetamine (Example 10.2) is approximately 3.2. Calculate the ratio of amphetamine to its conjugate acid at pH 7.4, the pH of blood plasma.

10.24 Calculate the ratio of amphetamine to its conjugate acid at pH 1.0, such as might be present in stomach acid.

10.25 Following is the structural formula of pyridoxamine, one form of vitamin B_6.

CH_2NH_2 HO CH_2OH CH_3 N

Pyridoxamine
(Vitamin B_6)

(a) Which nitrogen atom of pyridoxamine is the stronger base?

(b) Draw the structural formula of the hydrochloride salt formed when pyridoxamine is treated with 1 mole of HCl.

Poison arrow frog. *(Tom McHugh/Photo Researchers, Inc.)*

10.26 Epibatidine, a colorless oil isolated from the skin of the Ecuadorian poison frog *Epipedobates tricolor* has several times the analgesic potency of morphine. It is the first chlorine-containing, nonopioid (nonmorphine-like in structure) analgesic ever isolated from a natural source (*The Merck Index,* 12th ed., #3647).

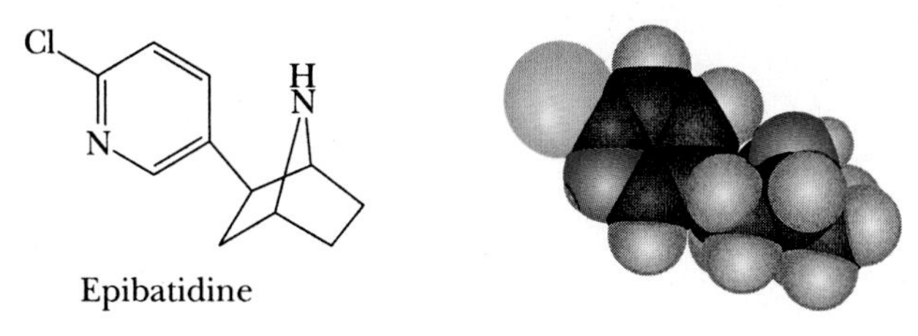

Epibatidine

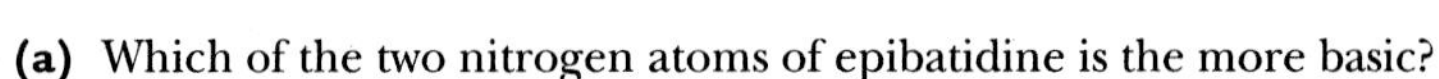

(a) Which of the two nitrogen atoms of epibatidine is the more basic?

(b) Mark all stereocenters in this molecule.

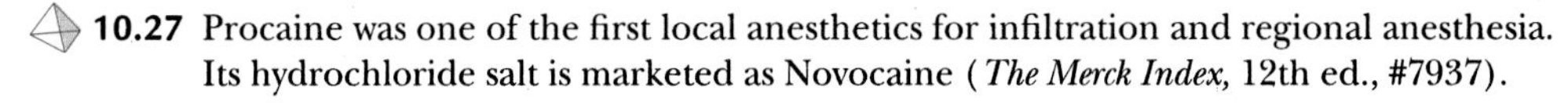

10.27 Procaine was one of the first local anesthetics for infiltration and regional anesthesia. Its hydrochloride salt is marketed as Novocaine (*The Merck Index,* 12th ed., #7937).

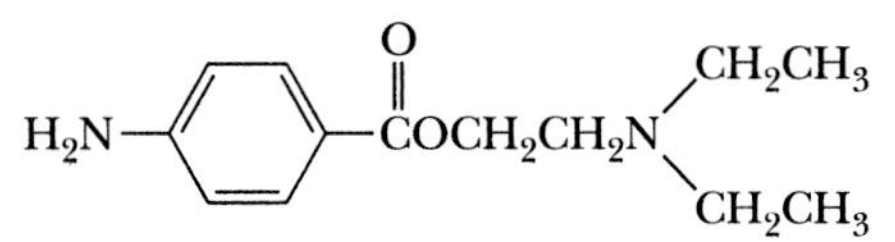

Procaine

(a) Which nitrogen atom of procaine is the stronger base?

(b) Draw the formula of the salt formed by treating procaine with 1 mole of HCl.

10.28 Treatment of trimethylamine with 2-chloroethyl acetate gives the neurotransmitter acetylcholine (*The Merck Index,* 12th ed., #88) as its chloride salt. Propose a structural formula for this quaternary ammonium salt and a mechanism for its formation.

$$(CH_3)_3N + CH_3\overset{O}{\overset{\|}{C}}OCH_2CH_2Cl \longrightarrow C_7H_{16}ClNO_2$$

Acetylcholine chloride

10.29 Aniline is prepared by catalytic reduction of nitrobenzene. Devise a chemical procedure based on the basicity of aniline to separate it from any unreacted nitrobenzene.

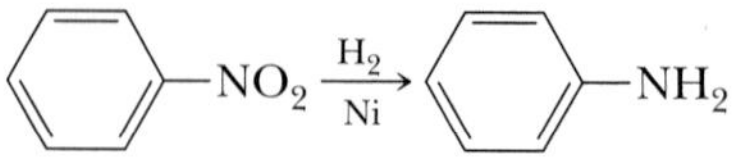

$$C_6H_5{-}NO_2 \xrightarrow[Ni]{H_2} C_6H_5{-}NH_2$$

10.30 Suppose that you have a mixture of these three compounds. Devise a chemical procedure based on their relative acidity or basicity to separate and isolate each in pure form.

$H_3C{-}C_6H_4{-}NO_2$ 4-Nitrotoluene (*p*-Nitrotoluene)

$H_3C{-}C_6H_4{-}NH_2$ 4-Methylaniline (*p*-Toluidine)

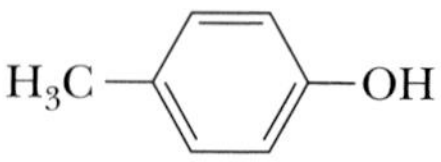

$H_3C{-}C_6H_4{-}OH$ 4-Methylphenol (*p*-Cresol)

A Conversation with . . .

JACQUELYN GERVAY

Jacquelyn Gervay (jur-vay) lives the life of a young faculty member, which is composed of a myriad of new responsibilities. Tasks include teaching several courses for the first time, establishing an independent research program, writing grant proposals, and working with graduate and undergraduate students in the lab.

Gervay is an organic chemist and an associate professor of chemistry at the University of Arizona in Tucson. She finds that teaching and research consume most of her days. However, Gervay still finds time for some of her other interests, such as running, inline skating, and even a game of basketball or tennis on occasion.

Gervay characterizes her research approach as one of "bridging" chemistry, biology, and immunology in an effort to understand biological processes at the molecular level. Her current projects include the synthesis of carbohydrates and novel helical materials; the development of targeted drug delivery systems for treatment of spinal meningitis, cancer, and HIV infections; and the development of new nuclear magnetic resonance techniques for understanding reaction mechanisms. For her research accomplishments, Gervay was named an Eli Lilly Academic awardee in 1997 and was appointed a Fellow of the Alfred P. Sloan Foundation in 1998. Gervay is also interested in chemical education. With a grant from the National Science Foundation, she is developing an approach that brings research into the classroom. In 1997, Gervay was awarded the University of Arizona, College of Science Innovation in Teaching Award in recognition of her teaching accomplishments in organic chemistry.

Gervay received B.S. and Ph.D. degrees in chemistry from the University of California at Los Angeles and held a postdoctoral position in the Yale University laboratory of synthetic organic chemist Samuel J. Danishefsky.

A Late Start

"My interest in chemistry got started late. I never took chemistry in high school; in fact, when I entered UCLA, I planned to major in psychobiology. The degree required chemistry, so I enrolled in the introductory courses my first year at UCLA. I did fine, although I considered it nothing more than work I had to complete to meet other goals. Then in my second year, I took organic chemistry. Even though I worked very hard, I just couldn't get it. So I understand students who have difficulty with organic chemistry in spite of their efforts. I waited until my senior year to finish the organic sequence. The instructor, Mike [Michael E.] Jung, was a terrific teacher. Suddenly organic chemistry made sense, and I became captivated by the subject matter. What turned me on was hearing about and discussing the research Mike was conducting in the laboratory. Research made chemistry a living science for me. It was during that course that I decided to become a chemistry professor. But since it was so late in my undergraduate career, I had to take another year and a half of chemistry courses to graduate as a chemistry major. That was a great time for me because I knew exactly what I wanted to do."

A Taste of Graduate School

"After I decided to become a chemistry major, I wanted to get some hands-on lab experience. I approached Professor Christopher Foote (of UCLA) and asked if I could work in his lab over the summer. He had an opening and let me join his group. There, I worked with singlet oxygen. I looked at the singlet oxygen 'ene' reaction and other kinds of photochemical experiments; I found the research fun and challenging. Following that experience, I joined Mike Jung's research group to get experience in the synthesis of organic compounds. When it came time to apply to graduate school, I applied only to UCLA. First of all, because I wanted to continue my work in Mike Jung's group and, second, because my husband had a business in Los Angeles, and it would have been difficult for me to go elsewhere for graduate school."

Organic Synthesis

"In graduate school, I spent a good two years trying to develop a route to the synthesis of reserpine [a natural product with tranquilizing properties] via an intramolecular Diels-Alder reaction.[1] Unfortunately, we could not overcome the unreactive nature of the Diels-Alder dienophile. Then Mike [Jung] came back from a conference with an interesting idea about doing an intramolecular Diels-Alder cyclization, and we decided to look at the effects of substituents on the rate of the reaction. I synthesized a number of test compounds with different alkyl groups near the reaction site and then studied their cyclization. We did all of the classic experiments to prove the reactions were essentially irreversible, and then we published a paper reporting this somewhat surprising result. Later I started doing the reactions in a different solvent and saw some amazing and unexpected differences in reaction rates. We were really intrigued, and in the end, we discovered that during the reaction some unfavorable dipole interactions took place, interactions accentuated in polar solvents that accelerated the reaction."

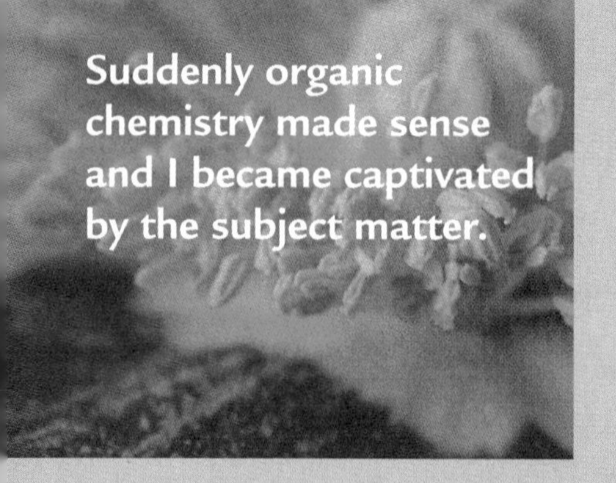
Suddenly organic chemistry made sense and I became captivated by the subject matter.

A New Interest Sends Gervay East

"For postdoctoral study, I went to Yale University to work with Samuel J. Danishefsky, a noted synthetic organic chemist. There I wanted to work on a molecule called sialic acid, which is a 3-deoxy sugar with a hydroxyl and a carboxyl group at the anomeric carbon.

Sialic acid

This compound is found in cell glycolipids, and I was captivated by sialic acid, both for its biological role and as a synthetic project.

"My years as a postdoctoral student were very formative. Almost everything I now study in my lab is based on the work I did at Yale. There I learned carbohydrate chemistry and helped to complete the total synthesis of Sialyl-Lewis X glycal [a tetrasaccharide that can be converted to compounds known to be important in tumor metastasis and in attracting white blood cells to damaged tissue]. Also exciting for me was interacting with graduate students and other postdocs, who had come there from all over the world."

On Being a Professor

"In the fall of 1992, I joined the faculty of the University of Arizona. The most rewarding part of being a professor is working with students. I always learn something; they always learn something. Using grant money from the National Science Foundation, one of my main goals in my undergraduate classes is to show the students that, given a basic understanding of chemistry, they can come up with original ideas for research, and they can learn to think critically about important areas of organic chemistry. I give them original research articles and ask them to propose some ideas for further research. Many of the ideas they come up with have been done and are in the literature, and that is exciting to them. They can't believe that they—just sophomores in organic chemistry—can think of ideas that scientists studied and wrote about in journal articles. Then I ask them to take their thinking a step further, and a step further, until they come up with questions that have not been explored. At that point, I ask them to design some experiments that might help answer their questions. Based on their ideas, the students are placed into research teams, and they present their combined ideas in a poster session. This endeavor constitutes the final exam of the first semester. In the second semester, the research teams conduct their proposed experiments and report their findings in an electronic conference on the Worldwide Web."

On Being a Woman Scientist

"I know that some women face barriers in science, but that hasn't been my personal experience. Some of my women students have told me it makes a difference to them that I'm a woman chemist. They say that they are much more comfortable talking to me than to a male professor, especially when they don't understand something. I have the sense that they don't feel intimidated when they approach me. If talking to women students helps them reach their potential, then I am happy to help."

Advice on Science for Students

"I don't want to make everyone a chemist. I don't think that's prudent. My goal is to make everybody in my classes appreciate chemistry. At the same time, I tell my sophomore students that chemistry, and certainly organic chemistry, is not strictly intuitive. If you have difficulty understanding the concepts, it's not because you're not smart; organic chemistry is challenging, much like a foreign language. So I think a large part of being successful is just sticking to it. For students who are contemplating a career in chemistry, one of the most important things to do is to get laboratory experience in a research setting, not just lab classes. That is how you'll know if a career in chemistry is for you."

---

[1] In a typical Diels-Alder reaction, a diene reacts with a dienophile to produce a six-membered ring compound. Here is an example.

$HC(=CH_2)-HC=CH_2$ + $HC(=CH_2)-C(=O)-CH_3$ ⟶ 4-Cyclohexenyl methyl ketone

1,3-Butadiene (a diene) | 3-Buten-2-one (a dienophile) | 4-Cyclohexenyl methyl ketone

11

Aldehydes and Ketones

Benzaldehyde is found in the kernels of bitter almonds. Cinnamaldehyde is found in Ceylon and Chinese cinnamon oils. Also shown are the molecular models of each. Inset: Molecular models of Cinnamaldehyde and Benzaldehyde. *(Charles D. Winters)*

In this and several following chapters, we study the physical and chemical properties of compounds containing the **carbonyl group, C=O**. Because the carbonyl group is the functional group of aldehydes, ketones, and carboxylic acids and their derivatives, it is one of the most important functional groups in organic chemistry. The chemical properties of this functional group are straightforward, and an understanding of its characteristic reaction themes leads very quickly to an understanding of a wide variety of organic reactions.

11.1 STRUCTURE AND BONDING

Aldehyde A compound containing a carbonyl group bonded to hydrogen (a CHO group).

Ketone A compound containing a carbonyl group bonded to two carbons.

The functional group of an **aldehyde** is a carbonyl group bonded to a hydrogen atom (Section 1.7B). In methanal, the simplest aldehyde, the carbonyl group is bonded to two hydrogen atoms. In other aldehydes, it is bonded to one hydrogen atom and one carbon atom. Following are Lewis structures for methanal and ethanal. Under each in parentheses is its common name. The functional group of a **ketone** is a carbonyl group bonded to two carbon atoms (Section 1.7B). Following is a Lewis structure for propanone, the simplest ketone.

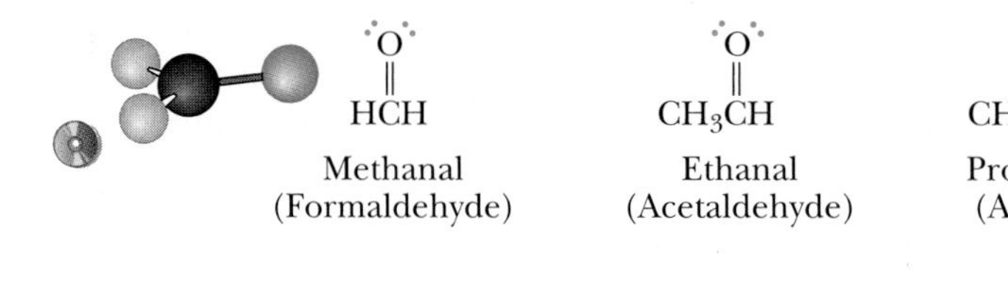

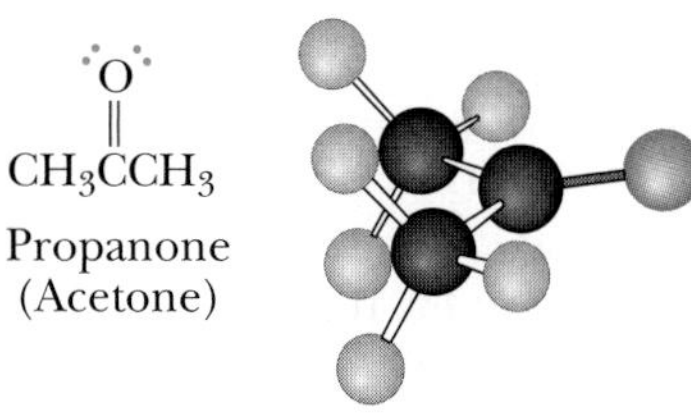

The carbon-oxygen double bond consists of one sigma bond formed by overlap of sp^2 hybrid orbitals of carbon and oxygen and one pi bond formed by the overlap of parallel $2p$ orbitals. The two nonbonding pairs of electrons on oxygen lie in the two remaining sp^2 hybrid orbitals (Figure 1.16).

11.2 NOMENCLATURE

A. IUPAC Nomenclature

The IUPAC system of nomenclature for aldehydes and ketones follows the familiar pattern of selecting as the parent alkane the longest chain of carbon atoms that contains the functional group. The aldehyde group is shown by changing the suffix -e of the parent alkane to -al , such as methanal (Section 3.5). Because the carbonyl group of an aldehyde can appear only at the end of a parent chain and numbering must start with it as carbon-1, its position is unambiguous, and there is no need to use a number to locate it.

For **unsaturated aldehydes,** the presence of a carbon-carbon double bond is indicated by the infix -en-. As with other molecules with both an infix and a suffix, the location of the suffix determines the numbering pattern.

$CH_3CH(CH_3)CH_2CHO$ (numbered 4, 3, 2, 1)
3-Methylbutanal

$CH_2{=}CHCHO$ (numbered 3, 2, 1)
2-Propenal
(Acrolein)

(2E)-3,7-Dimethyl-2,6-octadienal
(Geranial)

For cyclic molecules in which —CHO is attached directly to the ring, the molecule is named by adding the suffix -carbaldehyde to the name of the ring. The atom of the ring to which the aldehyde group is attached is numbered 1. Among the aldehydes for which the IUPAC system retains common names are benzaldehyde and cinnamaldehyde.

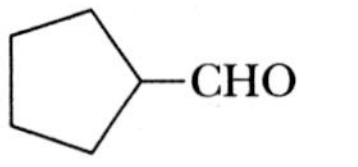
CHO

Cyclopentane-carbaldehyde

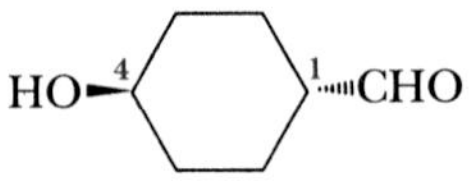
HO
4
1
CHO

trans-4-Hydroxycyclo-hexanecarbaldehyde

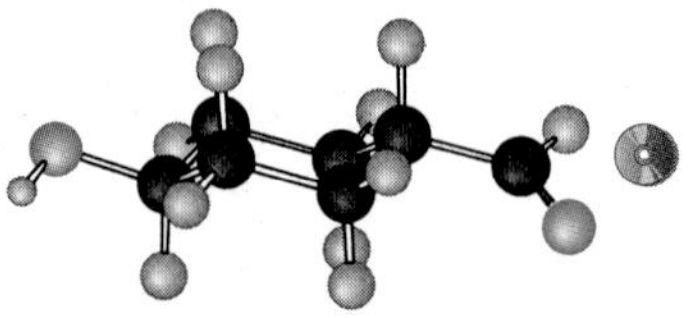

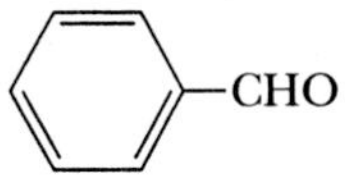
CHO

Benzaldehyde

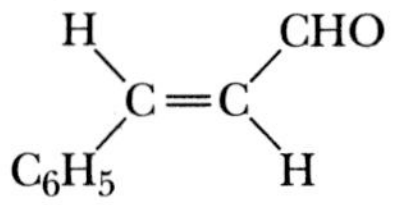
H
CHO
C=C
C6H5
H

trans-3-Phenyl-2-propenal (Cinnamaldehyde)

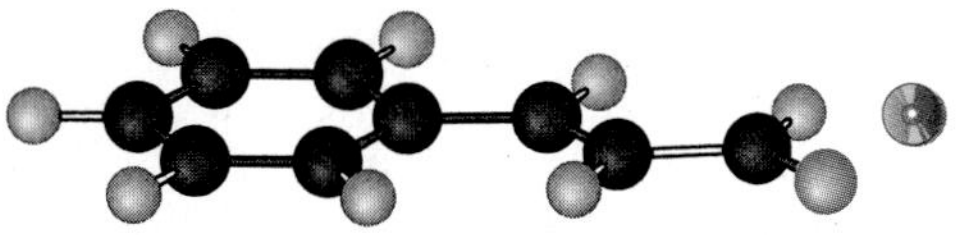

In the IUPAC system, ketones are named by selecting as the parent alkane the longest chain that contains the carbonyl group and then indicating its presence by changing the suffix from -e to -one (Section 3.5). The parent chain is numbered from the direction that gives the carbonyl carbon the smaller number. The IUPAC system retains the common names acetophenone and benzophenone.

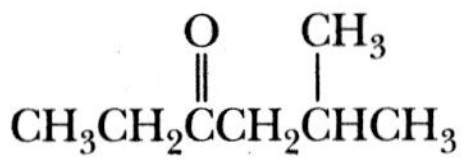
O
CH3
CH3CH2CCH2CHCH3

5-Methyl-3-hexanone

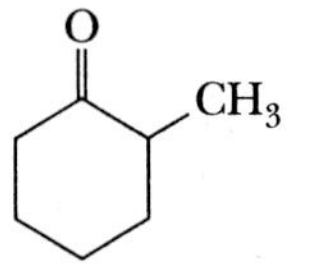
O
CH3

2-Methylcyclohexanone

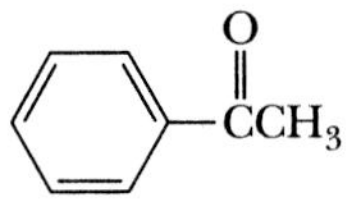
O
CCH3

Acetophenone

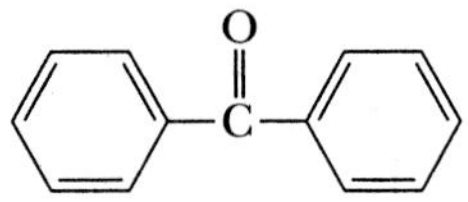
O
C

Benzophenone

EXAMPLE 11.1

Write the IUPAC name for each compound.

(a) 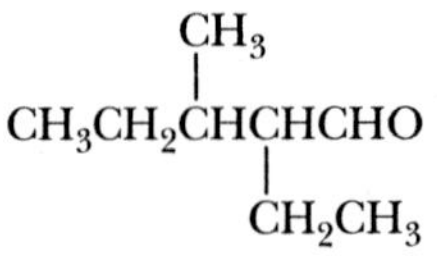
CH3
CH3CH2CHCHCHO
CH2CH3

(b) 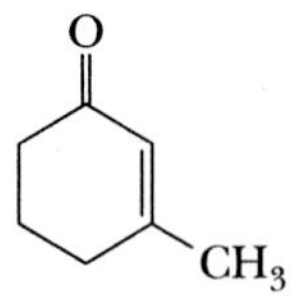
O
CH3

(c) 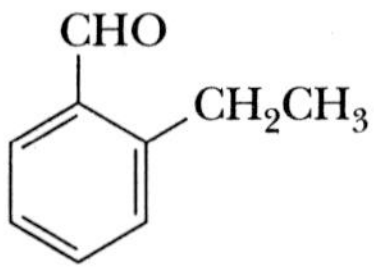
CHO
CH2CH3

SOLUTION

(a) The longest chain has six carbons, but the longest chain that contains the carbonyl group has five carbons. The name is 2-ethyl-3-methylpentanal.

(b) Number the six-membered ring beginning with the carbonyl carbon. The IUPAC name is 3-methyl-2-cyclohexenone.

(c) This molecule is derived from benzaldehyde. Its IUPAC name is 2-ethylbenzaldehyde.

Practice Problem 11.1

Write the IUPAC name for each compound. Specify the configuration of (c).

(a) 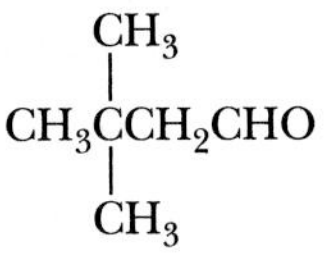 $CH_3C(CH_3)_2CH_2CHO$ (b) 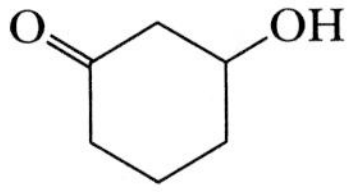 3-hydroxycyclohexanone structure (O=, OH) (c) 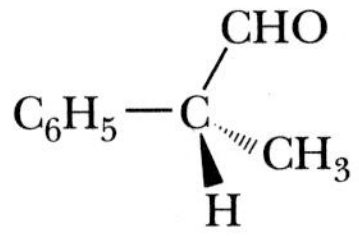 C_6H_5—C(CHO)(H)CH_3

EXAMPLE 11.2

Write structural formulas for all ketones of molecular formula $C_6H_{12}O$ and give each its IUPAC name. Which of these ketones are chiral?

SOLUTION

Following are line-angle drawings and IUPAC names for the six ketones of this molecular formula. Only 3-methyl-2-pentanone has a stereocenter and is chiral.

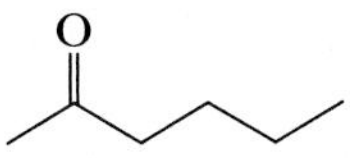

stereocenter

2-Hexanone 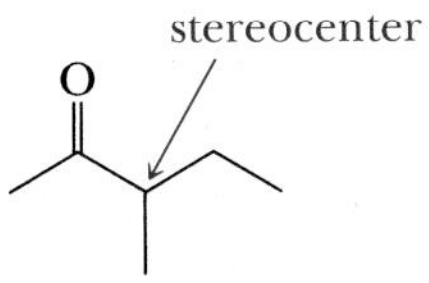 3-Methyl-2-pentanone 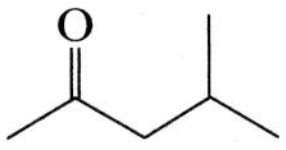 4-Methyl-2-pentanone

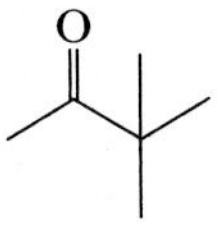

3,3-Dimethyl-2-butanone 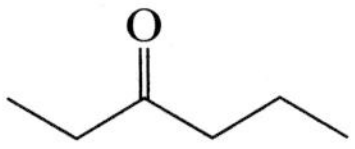 3-Hexanone 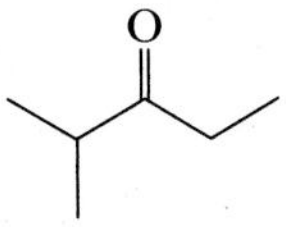 2-Methyl-3-pentanone

Practice Problem 11.2

Write structural formulas for all aldehydes of molecular formula $C_6H_{12}O$ and give each its IUPAC name. Which of these aldehydes are chiral?

B. IUPAC Names for More Complex Aldehydes and Ketones

In naming compounds that contain more than one functional group that might be indicated by a suffix, the IUPAC system has established an **order of precedence of functional groups.** The order of precedence for the functional groups we have studied so far is given in Table 11.1.

Order of precedence of functional groups A system for ranking functional groups in order of priority for the purposes of IUPAC nomenclature.

EXAMPLE 11.3

Write the IUPAC name for each compound.

(a) 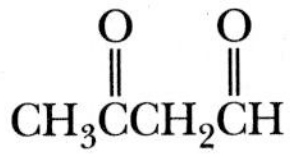 $CH_3\overset{O}{\overset{\|}{C}}CH_2\overset{O}{\overset{\|}{C}}H$ (b) H_2N—C_6H_4—CO_2H (c) H, H_3C, OH on C—$CH_2CH_2\overset{O}{\overset{\|}{C}}CH_3$

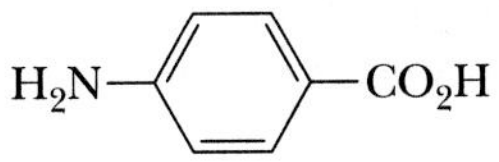

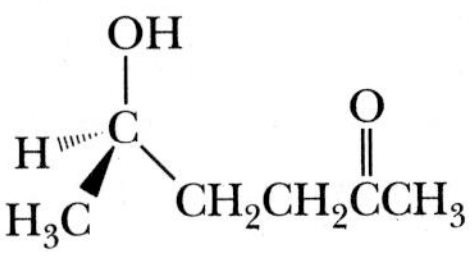

TABLE 11.1 Increasing Order of Precedence of Six Functional Groups

Functional Group	Suffix if Higher in Precedence	Prefix if Lower in Precedence
$-CO_2H$	-oic acid	—
$-CHO$	-al	oxo-
$>C=O$	-one	oxo-
$-OH$	-ol	hydroxy-
$-NH_2$	-amine	amino-
$-SH$	-thiol	mercapto-

(Arrow beside table: Increasing precedence, pointing upward)

SOLUTION

(a) An aldehyde is of higher precedence than a ketone. The presence of the carbonyl group of the ketone is indicated by the prefix oxo-. The IUPAC name of this compound is 3-oxobutanal.

(b) The carboxyl group is of higher precedence. The presence of the amino group is indicated by the prefix amino-. The IUPAC name is 4-aminobenzoic acid. Alternatively, it may be named *p*-aminobenzoic acid, abbreviated PABA. PABA, a growth factor of microorganisms, is required for the synthesis of folic acid (*The Merck Index,* 12th ed., #443).

(c) The C=O group has higher precedence than the —OH group. The —OH group is indicated by the prefix hydroxy-. The IUPAC name of this compound is (R)-5-hydroxy-2-hexanone.

Practice Problem 11.3

Write IUPAC names for these compounds, each of which is important in intermediary metabolism. Below each is the name by which it is more commonly known in the biological sciences.

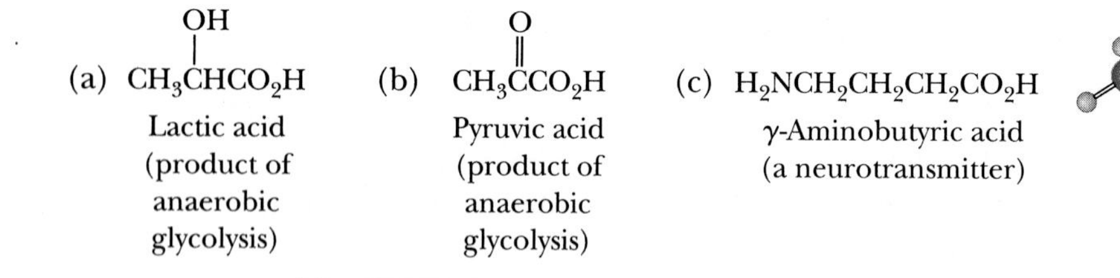

C. Common Names

The common name for an aldehyde is derived from the common name of the corresponding carboxylic acid by dropping the word "acid" and changing the suffix -ic or -oic to -aldehyde. Because we have not yet studied common names for carboxylic acids, we are not in a position to discuss common names for aldehydes. We can, however, illustrate how they are derived by reference to two common names of carboxylic acids with which you are familiar. The name formaldehyde is derived from formic acid; the name acetaldehyde is derived from acetic acid.

$$\underset{\text{Formaldehyde}}{\overset{\text{O}}{\overset{\|}{\text{HCH}}}} \qquad \underset{\text{Formic acid}}{\overset{\text{O}}{\overset{\|}{\text{HCOH}}}} \qquad \underset{\text{Acetaldehyde}}{\overset{\text{O}}{\overset{\|}{\text{CH}_3\text{CH}}}} \qquad \underset{\text{Acetic acid}}{\overset{\text{O}}{\overset{\|}{\text{CH}_3\text{COH}}}}$$

Common names for ketones are derived by naming each alkyl or aryl group attached to the carbonyl group as a separate word, followed by the word "ketone."

$$\underset{\text{Ethyl isopropyl ketone}}{\text{CH}_3\underset{\text{CH}_3}{\underset{|}{\text{CH}}}\overset{\text{O}}{\overset{\|}{\text{C}}}\text{CH}_2\text{CH}_3} \qquad \underset{\text{Diethyl ketone}}{\text{CH}_3\text{CH}_2\overset{\text{O}}{\overset{\|}{\text{C}}}\text{CH}_2\text{CH}_3} \qquad \underset{\text{Dicyclohexyl ketone}}{\text{C}_6\text{H}_{11}-\overset{\text{O}}{\overset{\|}{\text{C}}}-\text{C}_6\text{H}_{11}}$$

11.3 PHYSICAL PROPERTIES

Oxygen is more electronegative than carbon (3.5 compared with 2.5; Table 1.5); therefore, a carbon-oxygen double bond is polar, with oxygen bearing a partial negative charge and carbon bearing a partial positive charge.

$$\overset{\delta+}{\text{C}}=\overset{\delta-}{\ddot{\text{O}}}:$$

Polarity of a carbonyl group

Because of the polarity of the carbonyl group, aldehydes and ketones are polar compounds and have higher boiling points than do nonpolar compounds of comparable molecular weight. Table 11.2 lists boiling points of six compounds of comparable molecular weight.

Pentane and diethyl ether have the lowest boiling points of these six compounds. Both butanal and 2-butanone are polar compounds, and, because of the intermolecular attraction between carbonyl groups, their boiling points are higher than those of pentane and diethyl ether. Alcohols (Section 8.3A) and carboxylic

TABLE 11.2 Boiling Points of Six Compounds of Comparable Molecular Weight

Name	Structural Formula	Molecular Weight	bp (°C)
diethyl ether	$CH_3CH_2OCH_2CH_3$	74	34
pentane	$CH_3CH_2CH_2CH_2CH_3$	72	36
butanal	$CH_3CH_2CH_2CHO$	72	76
2-butanone	$CH_3CH_2COCH_3$	72	80
1-butanol	$CH_3CH_2CH_2CH_2OH$	74	117
propanoic acid	$CH_3CH_2CO_2H$	72	141

TABLE 11.3 Physical Properties of Selected Aldehydes and Ketones

IUPAC Name	Common Name	Structural Formula	bp (°C)	Solubility (g/100 g water)
methanal	formaldehyde	HCHO	−21	infinite
ethanal	acetaldehyde	CH_3CHO	20	infinite
propanal	propionaldehyde	CH_3CH_2CHO	49	16
butanal	butyraldehyde	$CH_3CH_2CH_2CHO$	76	7
hexanal	caproaldehyde	$CH_3(CH_2)_4CHO$	129	slight
propanone	acetone	CH_3COCH_3	56	infinite
2-butanone	ethyl methyl ketone	$CH_3COCH_2CH_3$	80	26
3-pentanone	diethyl ketone	$CH_3CH_2COCH_2CH_3$	101	5

acids (Section 12.3) are polar compounds, and their molecules associate by hydrogen bonding; their boiling points are higher than those of butanal and 2-butanone, compounds whose molecules cannot associate by hydrogen bonding.

Because the carbonyl groups of aldehydes and ketones interact with water molecules by hydrogen bonding, low-molecular-weight aldehydes and ketones are more soluble in water than are nonpolar compounds of comparable molecular weight. Listed in Table 11.3 are boiling points and solubilities in water for several low-molecular-weight aldehydes and ketones.

11.4 Reactions

The most common reaction theme of the carbonyl group is addition intermediate of a nucleophile to form a **tetrahedral carbonyl addition intermediate.** In the following general reaction, the nucleophilic reagent is written as $Nu:^-$ to emphasize the presence of its unshared pair of electrons.

$$Nu:^- + R_2C{=}O \longrightarrow Nu{-}C(O^-)(R)(R)$$

Tetrahedral carbonyl addition intermediate

11.5 Addition of Carbon Nucleophiles

From the perspective of the organic chemist, addition of carbon nucleophiles is the most important type of nucleophilic addition to a carbonyl group because these reactions form new carbon-carbon bonds. In this section, we describe the preparation of Grignard reagents, one common type of carbon nucleophile, and their reaction with aldehydes and ketones.

A. Formation and Structure of Organomagnesium Compounds

Alkyl and aryl halides react with Group I, Group II, and certain other metals to form **organometallic compounds.** Organomagnesium compounds are called **Grignard reagents** after Victor Grignard, who was awarded a Nobel Prize for Chemistry in 1912 for their discovery and their application to organic synthesis. Butylmagnesium chloride, for example, is prepared by treating 1-chlorobutane in diethyl ether with magnesium metal.

Organometallic compound A compound containing a carbon-metal bond.

Grignard reagent An organomagnesium compound of the type RMgX or ArMgX.

$$CH_3CH_2CH_2CH_2Cl + Mg \xrightarrow{\text{ether}} CH_3CH_2CH_2CH_2MgCl$$

1-Chlorobutane → Butylmagnesium chloride

The difference in electronegativity between carbon and magnesium is 1.3 units (2.5 − 1.2); therefore, the carbon-magnesium bond is polar covalent, as shown in the following structure on the left. In this organometallic bond, carbon bears a partial negative charge, and magnesium bears a partial positive charge. In the structure on the right, the carbon-magnesium bond is shown as an ionic bond. In all its reactions, a Grignard reagent behaves as a **carbanion,** an ion containing a carbon atom with an unshared pair of electrons and bearing a negative charge.

Carbanion An anion in which carbon has an unshared pair of electrons and bears a negative charge.

$$CH_3(CH_2)_2\overset{\delta-}{C}H_2-\overset{\delta+}{Mg}-Cl \qquad CH_3(CH_2)_2\overset{-}{C}H_2{:}\ \overset{+}{Mg}-Cl$$

A C—Mg bond shown as a polar covalent bond

A C—Mg bond shown as an ionic bond (a carbanion; a nucleophile and very strong base)

Grignard reagents are very strong bases and react with a wide variety of acids (proton donors) to form alkanes. Ethylmagnesium bromide, for example, reacts instantly with water to give ethane and magnesium salts. This reaction is an example of a stronger acid and a stronger base reacting to give a weaker acid and a weaker base (Section 2.4).

$$\overset{\delta-}{CH_3CH_2}-\overset{\delta+}{MgBr} + H-OH \longrightarrow CH_3CH_2-H + Mg^{2+} + OH^- + Cl^-$$

Stronger base	Stronger acid (pK_a 15.7)	Weaker acid (pK_a 51)	Weaker base

Following are several classes of proton donor acids that react readily with Grignard reagents.

R_2NH	ROH	HOH	ArOH	RSH	RCO_2H
pK_a 38–40	pK_a 16–18	pK_a 15.7	pK_a 9–10	pK_a 8–9	pK_a 4–5
Amines	Alcohols	Water	Phenols	Thiols	Carboxylic acids

EXAMPLE 11.4

Write an equation for the acid-base reaction between ethylmagnesium iodide and an alcohol. Use curved arrows to show the flow of electrons in this reaction. In addition, show that this reaction is an example of a stronger acid and stronger base reacting to form a weaker acid and weaker base.

SOLUTION

The alcohol is the stronger acid and ethyl carbanion is the stronger base.

$$CH_3CH_2{-}MgI + H{-}OR \longrightarrow CH_3CH_2{-}H + RO^-MgI^+$$

$CH_3CH_2{-}MgI$	$H{-}OR$	$CH_3CH_2{-}H$	RO^-MgI^+
Ethylmagnesium iodide (stronger base)	An alcohol pK_a 16–18 (stronger acid)	Ethane pK_a 51 (weaker acid)	A magnesium alkoxide (weaker base)

Practice Problem 11.4

Explain how these Grignard reagents react with molecules of their own kind to "self-destruct."

(a) $HO{-}C_6H_4{-}MgBr$ (b) $HOCCH_2CH_2CH_2MgBr$ (with C=O)

B. Addition of Grignard Reagents

The special value of Grignard reagents is that they provide excellent ways to form new carbon-carbon bonds. A carbanion is a good nucleophile and adds to the carbonyl group of an aldehyde or ketone to form a tetrahedral carbonyl addition compound. The driving force for these reactions is the attraction of the partial negative charge on the carbon of the organometallic compound to the partial positive charge of the carbonyl carbon. In the following examples, the magnesium oxygen bond is written $-O^-[MgBr]^+$ to emphasize its ionic character. The alkoxide ions formed in Grignard reactions are strong bases (Section 8.4C) and, when treated with an aqueous acid such as HCl, form alcohols.

Addition to Formaldehyde Gives a Primary Alcohol

Treatment of a Grignard reagent with formaldehyde followed by hydrolysis in aqueous acid gives a primary alcohol.

$$\overset{\delta-}{CH_3CH_2}{-}\overset{\delta+}{MgBr} + H{-}\overset{\delta+}{C}(=\overset{\delta-}{O}){-}H \xrightarrow{\text{ether}} CH_3CH_2{-}CH_2{-}O^-[MgBr]^+ \xrightarrow[H_2O]{HCl} CH_3CH_2{-}CH_2{-}OH + Mg^{2+}$$

Formaldehyde — A magnesium alkoxide — 1-Propanol (a primary alcohol)

Addition to an Aldehyde (Except Formaldehyde) Gives a Secondary Alcohol

Treatment of a Grignard reagent with any aldehyde other than formaldehyde followed by hydrolysis in aqueous acid gives a secondary alcohol.

$$C_6H_{11}\text{—}MgBr + CH_3\text{—}C(=O)\text{—}H \xrightarrow{\text{ether}} C_6H_{11}\text{—}CHCH_3(O^-[MgBr]^+) \xrightarrow[H_2O]{HCl} C_6H_{11}\text{—}CHCH_3(OH) + Mg^{2+}$$

Acetaldehyde — A magnesium alkoxide — 1-Cyclohexylethanol (a secondary alcohol)

Addition to a Ketone Gives a Tertiary Alcohol

Treatment of a Grignard reagent with a ketone followed by hydrolysis in aqueous acid gives a tertiary alcohol.

$$C_6H_5\text{—}MgBr + CH_3\text{—}C(=O)\text{—}CH_3 \xrightarrow{\text{ether}} C_6H_5\text{—}C(CH_3)_2(O^-[MgBr]^+) \xrightarrow[H_2O]{HCl} C_6H_5\text{—}C(CH_3)_2(OH) + Mg^{2+}$$

Acetone — A magnesium alkoxide — 2-Phenyl-2-propanol (a tertiary alcohol)

Addition to Carbon Dioxide Gives a Carboxylic Acid

Treatment of a Grignard reagent with carbon dioxide gives the magnesium salt of a carboxylic acid, which on hydrolysis in aqueous acid, gives a carboxylic acid. Thus, carbonation of a Grignard reagent is a convenient way to convert an alkyl or aryl halide to a carboxylic acid.

$$C_5H_9\text{—}MgBr + O\text{=}C\text{=}O \xrightarrow{\text{ether}} C_5H_9\text{—}C(=O)\text{—}O^-[MgBr]^+ \xrightarrow[H_2O]{HCl} C_5H_9\text{—}C(=O)\text{—}O\text{—}H + Mg^{2+}$$

Carbon dioxide — Cyclopentanecarboxylic acid

Addition to Ethylene Oxide Gives a Primary Alcohol

Grignard reagents add to epoxides by nucleophilic substitution. Treatment of a Grignard reagent with ethylene oxide is a very convenient way to lengthen a carbon chain by two atoms, as illustrated by the reaction of phenylmagnesium bromide with ethylene oxide.

$$C_6H_5\text{—}MgBr + CH_2\text{—}CH_2(\text{—}O\text{—}) \xrightarrow[\text{ether}]{S_N2} C_6H_5CH_2CH_2O^-[MgBr]^+ \xrightarrow[H_2O]{HCl} C_6H_5CH_2CH_2OH$$

Ethylene oxide — 2-Phenylethanol

Example 11.5

2-Phenyl-2-butanol can be synthesized by three different combinations of a Grignard reagent and a ketone. Show each combination.

SOLUTION

In each solution, curved arrows show formation of the new carbon-carbon bond.

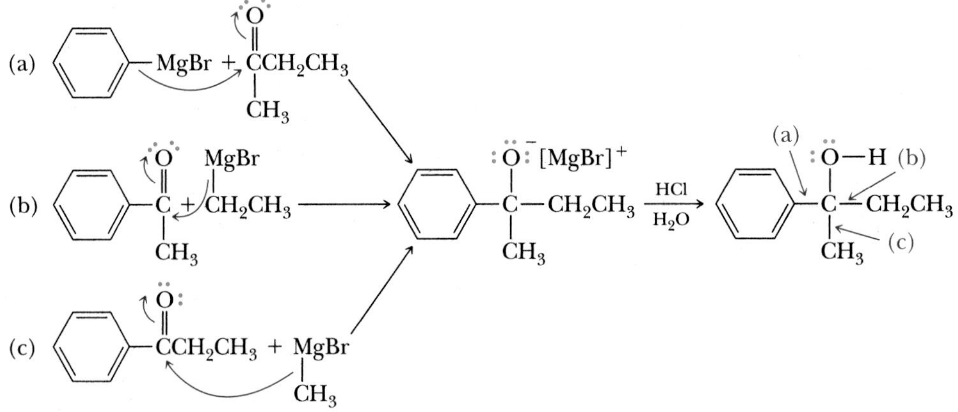

(a) C_6H_5—MgBr + $CH_3C(=O)CH_2CH_3$

(b) $C_6H_5C(=O)CH_3$ + CH_3CH_2MgBr ⟶ $C_6H_5C(O^-[MgBr]^+)(CH_3)CH_2CH_3$ $\xrightarrow[H_2O]{HCl}$ $C_6H_5C(OH)(CH_3)CH_2CH_3$

(c) $C_6H_5C(=O)CH_2CH_3$ + CH_3MgBr

(a), (b), (c) label the bonds formed in each route.

Practice Problem 11.5

Show how these four compounds can be synthesized from the same Grignard reagent.

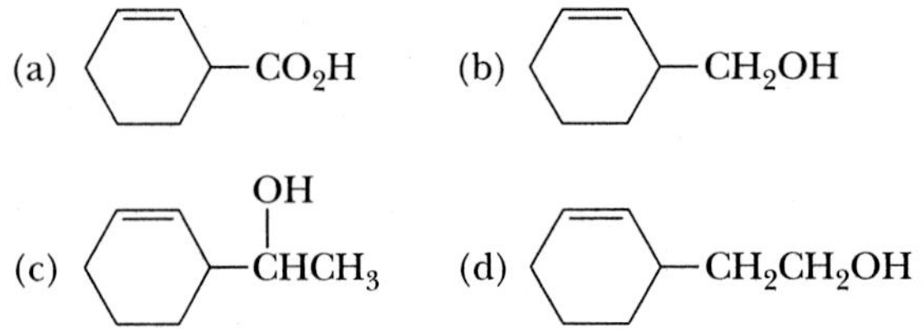

(a) cyclohexenyl—CO_2H (b) cyclohexenyl—CH_2OH

(c) cyclohexenyl—$CH(OH)CH_3$ (d) cyclohexenyl—CH_2CH_2OH

11.6 Addition of Oxygen Nucleophiles

Hemiacetal A molecule containing an —OH and an —OR or —OAr group bonded to the same carbon.

Addition of a molecule of alcohol to the carbonyl group of an aldehyde or ketone forms a **hemiacetal** (a half-acetal). Oxygen adds to the carbonyl carbon and hydrogen adds to the carbonyl oxygen.

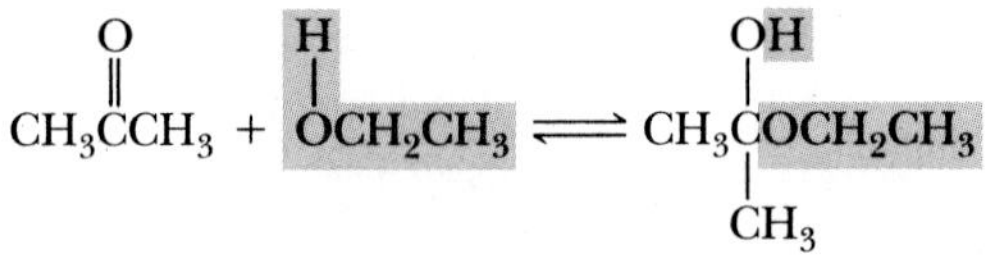

$$CH_3\overset{O}{\overset{\|}{C}}CH_3 + H{-}OCH_2CH_3 \rightleftharpoons CH_3\overset{OH}{\overset{|}{\underset{CH_3}{\underset{|}{C}}}}OCH_2CH_3$$

A hemiacetal

The functional group of a hemiacetal is a carbon bonded to an —OH group and an —OR or —OAr group.

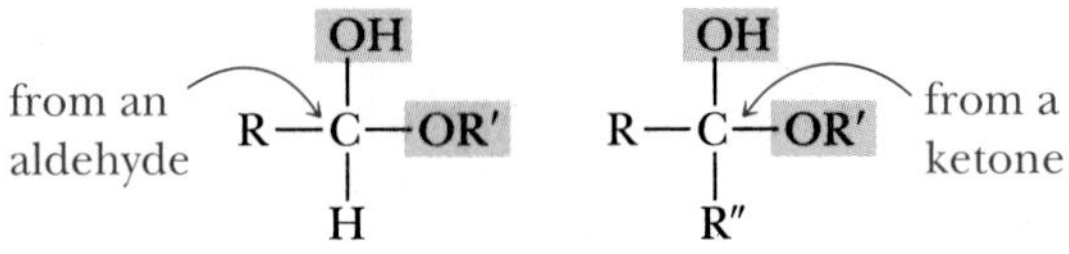

$$R{-}\overset{OH}{\overset{|}{\underset{H}{\underset{|}{C}}}}{-}OR' \qquad R{-}\overset{OH}{\overset{|}{\underset{R''}{\underset{|}{C}}}}{-}OR'$$

from an aldehyde — from a ketone

Hemiacetals

Hemiacetals are generally unstable and are only minor components of an equilibrium mixture except in one very important type of molecule. When a hydroxyl group is part of the same molecule that contains the carbonyl group, and a five- or six-membered ring can form, the compound exists almost entirely in a cyclic hemiacetal form. We have much more to say about cyclic hemiacetals when we consider the chemistry of carbohydrates in Chapter 16.

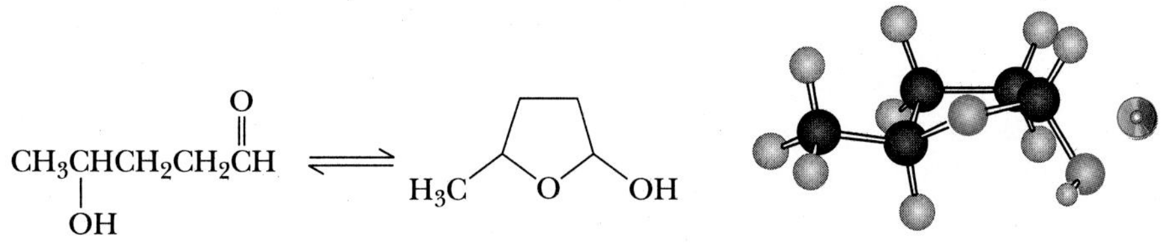

$$CH_3CH(OH)CH_2CH_2CHO \rightleftharpoons \text{(cyclic hemiacetal: } H_3C\text{–ring O–OH)}$$

4-Hydroxypentanal

A cyclic hemiacetal (major form present at equilibrium)

(the trans isomer)

Hemiacetals can react further with alcohols to form **acetals** plus a molecule of water. This reaction is acid-catalyzed.

Acetal A molecule containing two —OR or —OAr groups bonded to the same carbon.

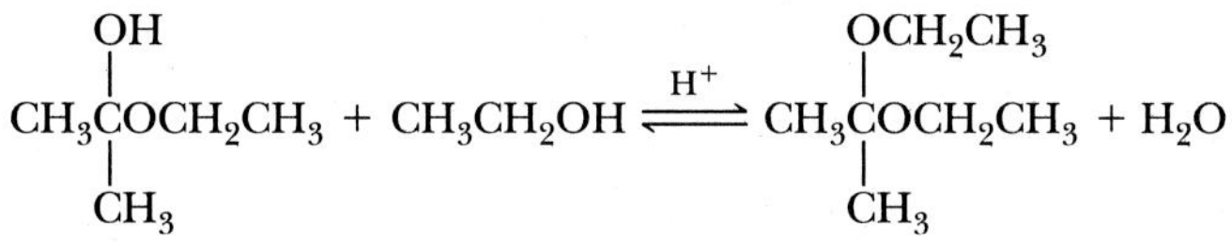

$$CH_3C(OH)(CH_3)OCH_2CH_3 + CH_3CH_2OH \overset{H^+}{\rightleftharpoons} CH_3C(OCH_2CH_3)(CH_3)OCH_2CH_3 + H_2O$$

A hemiacetal — A diethyl acetal

The functional group of an acetal is a carbon bonded to two —OR or —OAr groups.

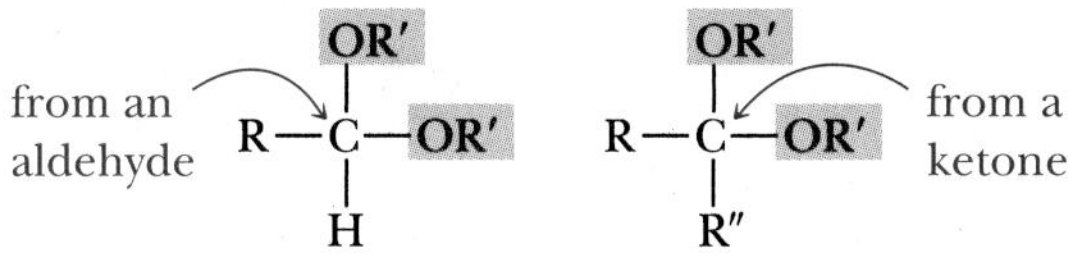

from an aldehyde: $R-C(OR')(OR')-H$ — from a ketone: $R-C(OR')(OR')-R''$

Acetals

The mechanism for the acid-catalyzed conversion of a hemiacetal to an acetal can be divided into four steps. Note that acid H—A is a true catalyst in this reaction. It is used in Step 1, but a replacement H—A is generated in Step 4.

MECHANISM Acid-Catalyzed Formation of an Acetal

Step 1: Proton transfer from the acid, H—A, to the hemiacetal OH group gives an oxonium ion.

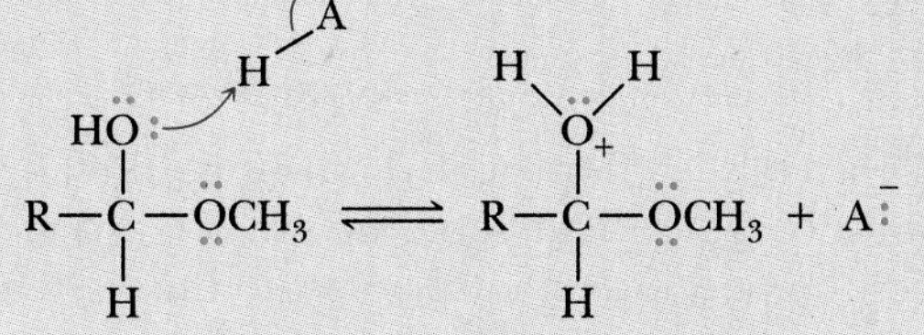

$$R-CH(OH)-OCH_3 + H-A \rightleftharpoons R-CH(OH_2^+)-OCH_3 + A^-$$

An oxonium ion

Step 2: Loss of water from the first oxonium ion gives a new oxonium ion.

$$R-\underset{H}{\overset{H_2O^+}{C}}-\ddot{O}CH_3 \rightleftharpoons R-\underset{H}{C}=\overset{+}{\ddot{O}}CH_3 \longleftrightarrow R-\underset{H}{\overset{+}{C}}-\ddot{O}CH_3 + H_2\ddot{O}:$$

A resonance-stabilized oxonium ion

Step 3: Reaction of the resonance-stabilized oxonium ion (a Lewis acid) with methanol (a Lewis base) gives the conjugate acid of the acetal.

$$CH_3-\underset{..}{\overset{H}{O}}: + R-\underset{H}{C}=\overset{+}{\ddot{O}}CH_3 \rightleftharpoons R-\underset{H}{\overset{H-\overset{+}{O}-CH_3}{C}}-\ddot{O}CH_3$$

A protonated acetal

Step 4: Proton transfer from the protonated acetal to A⁻ gives the acetal and generates a new molecule of H—A, the acid catalyst.

$$A^{-}: + R-\underset{H}{\overset{H-\overset{+}{O}-CH_3}{C}}-\ddot{O}CH_3 \rightleftharpoons R-\underset{H}{\overset{:\ddot{O}-CH_3}{C}}-\ddot{O}CH_3 + H-A$$

A protonated acetal An acetal

All steps in acetal formation are reversible. We drive this equilibrium reaction to the right, that is to formation of the acetal, either by using a large excess of alcohol or by removing water from the equilibrium mixture.

An excess of alcohol pushes the equilibrium toward formation of the acetal

Removal of water favors formation of the acetal

$$R-\overset{O}{\overset{\|}{C}}-R + 2CH_3CH_2OH \overset{H^+}{\rightleftharpoons} R-\underset{R}{\overset{OCH_2CH_3}{C}}-OCH_2CH_3 + H_2O$$

A diethyl acetal

Formation of acetals is often carried out using the alcohol as the solvent. Because the alcohol is both a reactant and the solvent, it is present in large molar excess, which forces the position of equilibrium to the right and favors acetal formation.

EXAMPLE 11.6

Show the reaction of the carbonyl group of each ketone with one molecule of alcohol to form a hemiacetal and then with a second molecule of alcohol to form an acetal. Note that in part (b), ethylene glycol is a diol and one molecule of it provides both —OH groups.

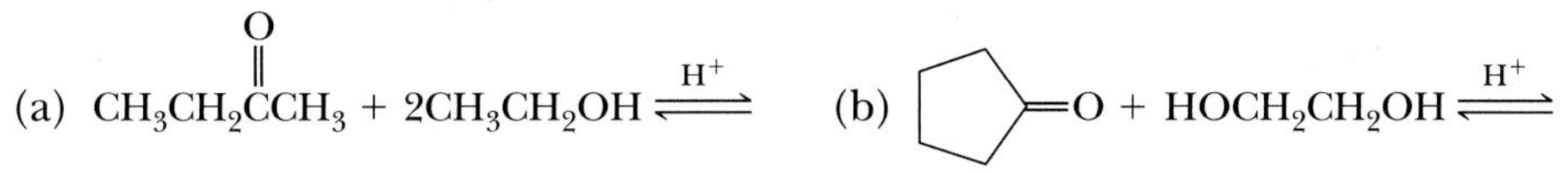

(a) $CH_3CH_2C(=O)CH_3 + 2CH_3CH_2OH \overset{H^+}{\rightleftharpoons}$ (b) cyclopentanone $+ HOCH_2CH_2OH \overset{H^+}{\rightleftharpoons}$

SOLUTION

Given are structural formulas of the hemiacetal and then the acetal.

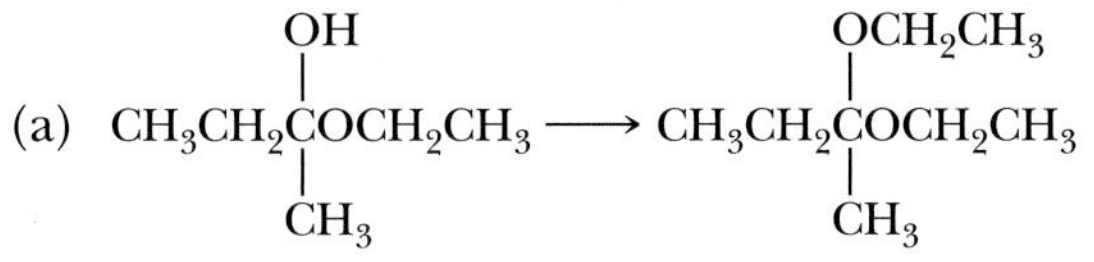

(a) $CH_3CH_2C(OH)(CH_3)OCH_2CH_3 \longrightarrow CH_3CH_2C(OCH_2CH_3)(CH_3)OCH_2CH_3$

A hemiacetal — An acetal

(b) 1-(2-hydroxyethoxy)cyclopentanol (OH, OCH_2CH_2OH) $\longrightarrow$ cyclopentane spiro-dioxolane (O–CH_2–CH_2–O)

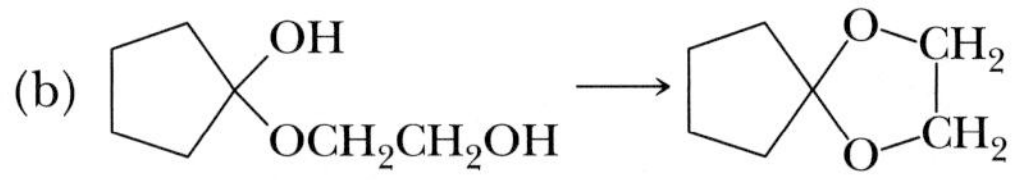

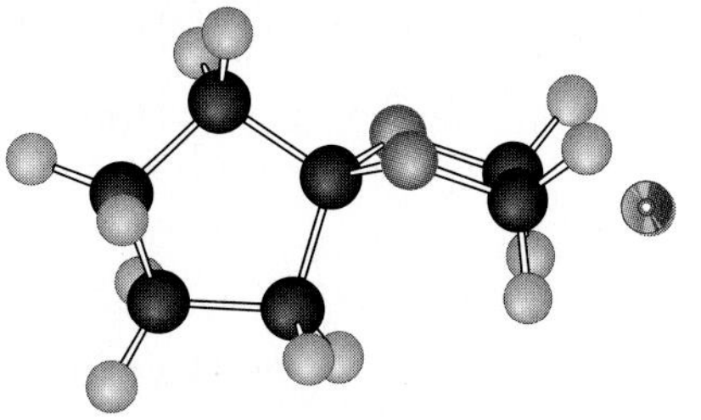

A hemiacetal — A cyclic acetal

Practice Problem 11.6

Hydrolysis of an acetal forms an aldehyde or ketone and two molecules of alcohol. Following are structural formulas for three acetals. Draw the structural formulas for the products of hydrolysis of each in aqueous acid.

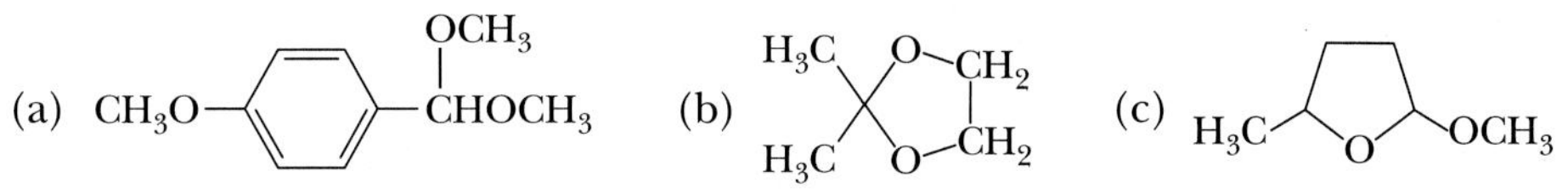

(a) $CH_3O-C_6H_4-CH(OCH_3)OCH_3$ (b) $(H_3C)_2C$ in a ring O–CH_2–CH_2–O (c) H_3C–(tetrahydrofuran ring, O)–OCH_3

11.7 ADDITION OF NITROGEN NUCLEOPHILES

Ammonia, primary aliphatic amines (RNH_2), and primary aromatic amines ($ArNH_2$) react with the carbonyl group of aldehydes and ketones in the presence of an acid catalyst to give a product that contains a carbon-nitrogen double bond. A molecule containing a carbon-nitrogen double bond is called an **imine** or, alternatively, a **Schiff base.**

Imine A compound containing a carbon-nitrogen double bond; also called a Schiff base.

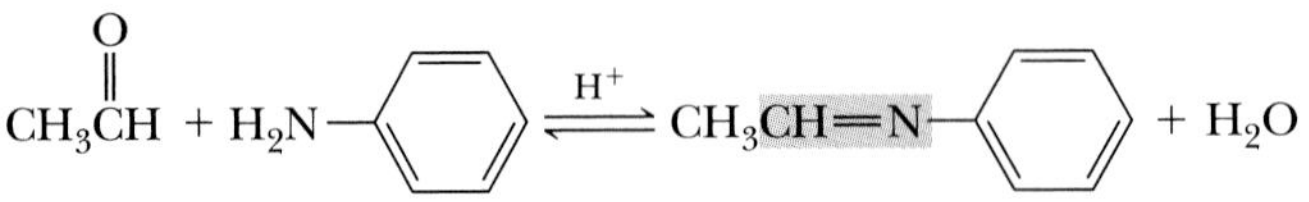

$$CH_3CHO + H_2N{-}C_6H_5 \overset{H^+}{\rightleftharpoons} CH_3CH{=}N{-}C_6H_5 + H_2O$$

Ethanal Aniline An imine (A Schiff base)

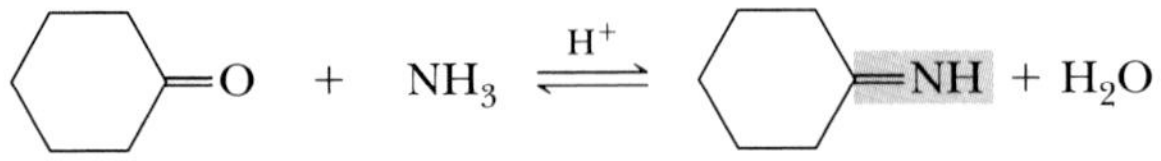

$$C_6H_{10}{=}O + NH_3 \overset{H^+}{\rightleftharpoons} C_6H_{10}{=}NH + H_2O$$

Cyclohexanone Ammonia An imine (A Schiff base)

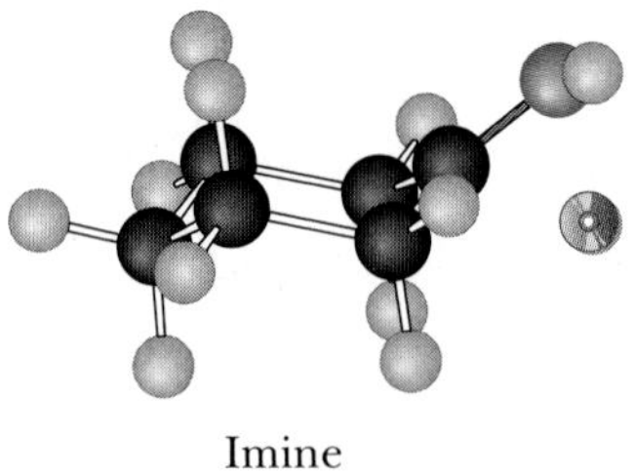

Imine

The mechanism of imine formation can be divided into two steps. In Step 1, the nitrogen atom of ammonia or a primary amine, both good nucleophiles, adds to the carbonyl carbon to form a tetrahedral carbonyl addition intermediate. Acid-catalyzed dehydration of this addition intermediate in Step 2 is the slow, rate-limiting step.

MECHANISM Formation of an Imine from an Aldehyde or Ketone

Step 1: Addition of the nucleophilic nitrogen to the carbonyl carbon followed by proton transfer gives a tetrahedral carbonyl addition intermediate.

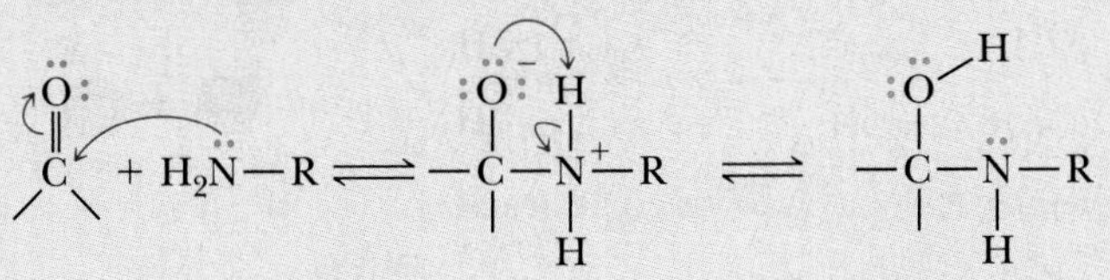

$$>C{=}O + H_2N{-}R \rightleftharpoons {-}C(O^-){-}N^+H_2{-}R \rightleftharpoons {-}C(OH){-}NH{-}R$$

A tetrahedral carbonyl addition intermediate

Step 2: Protonation of the OH group followed by loss of water and proton transfer to solvent gives the imine.

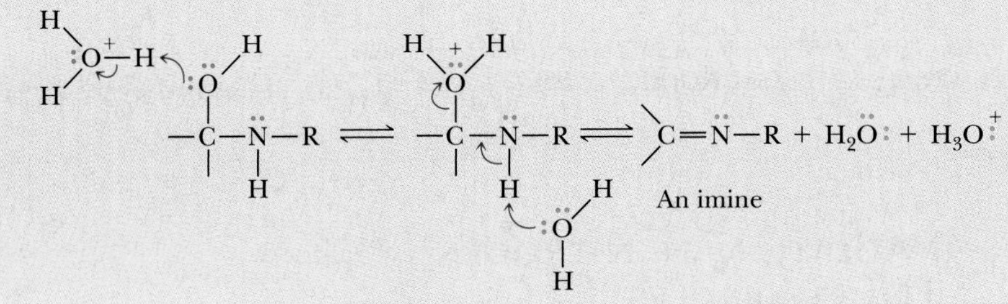

$$H_3O^+ + {-}C(OH){-}NH{-}R \rightleftharpoons {-}C(OH_2^+){-}NH{-}R \rightleftharpoons >C{=}N{-}R + H_2O + H_3O^+$$

An imine

To give but one example of the importance of imines in biological systems, the active form of **vitamin A aldehyde** (retinal) is bound to the protein **opsin** in the human retina in the form of an imine. The primary amino group for this reaction is provided by the side chain of the amino acid lysine (Table 18.1). The imine is called **rhodopsin** or **visual purple.**

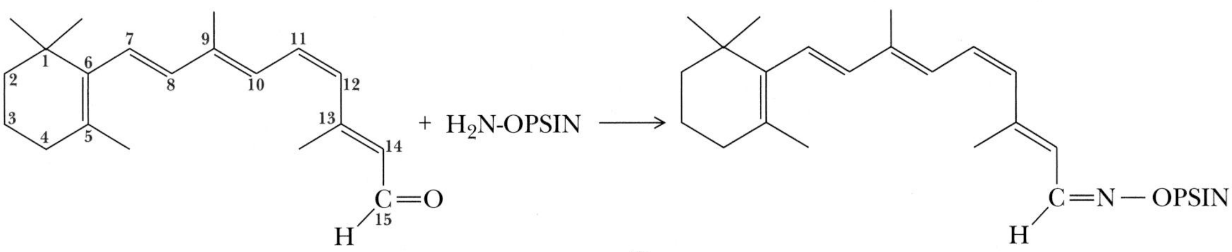

$$\text{11-}cis\text{-Retinal} + H_2N\text{-OPSIN} \longrightarrow \text{Rhodopsin (Visual purple)}$$

11-*cis*-Retinal

Rhodopsin
(Visual purple)

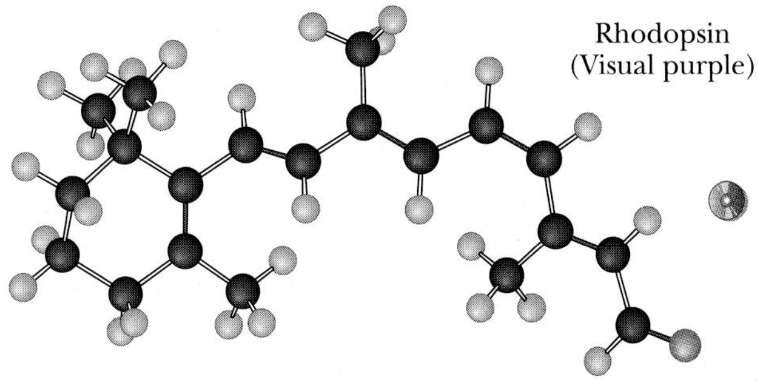

EXAMPLE 11.7

Write a structural formula for the imine formed in each reaction.

(a) 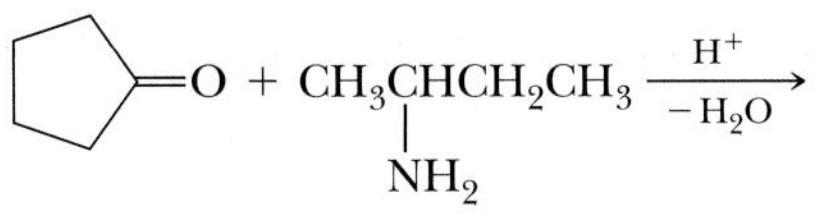 cyclopentanone $(=O) + CH_3CHCH_2CH_3$ with NH_2 on C-2 $\xrightarrow[-H_2O]{H^+}$

(b) 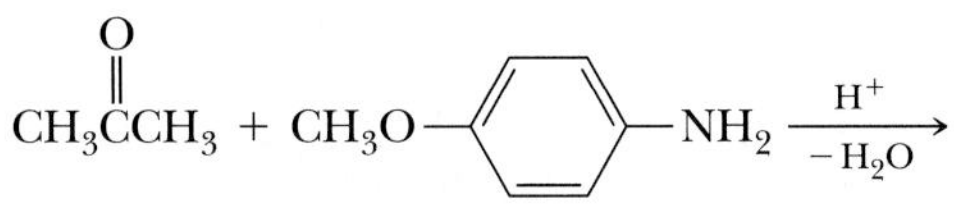 $CH_3\overset{O}{\overset{\|}{C}}CH_3 + CH_3O{-}C_6H_4{-}NH_2 \xrightarrow[-H_2O]{H^+}$

SOLUTION

Given is a structural formula for each imine.

(a) 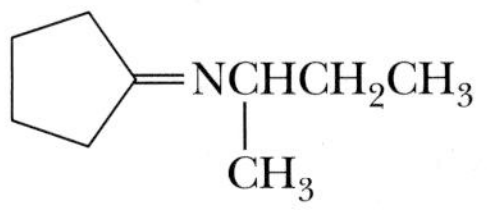 cyclopentylidene $=NCHCH_2CH_3$ with CH_3 on the N-bearing carbon (b) 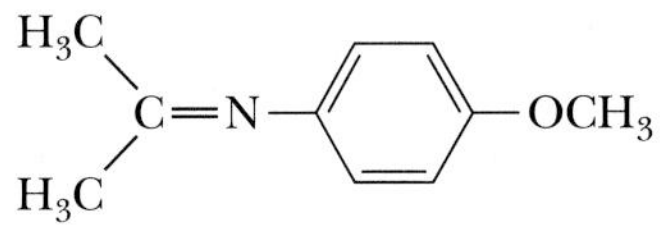 $(H_3C)(H_3C)C{=}N{-}C_6H_4{-}OCH_3$

Practice Problem 11.7

Acid-catalyzed hydrolysis of an imine gives an amine and an aldehyde or ketone. When one equivalent of acid is used, the amine is converted to its ammonium salt. Write structural formulas for the products of hydrolysis of each imine using one equivalent of HCl.

(a) 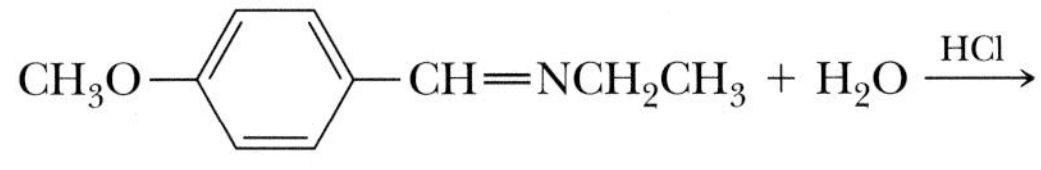 $CH_3O{-}C_6H_4{-}CH{=}NCH_2CH_3 + H_2O \xrightarrow{HCl}$

(b) 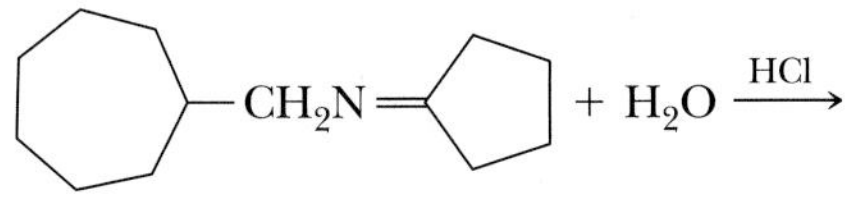 cycloheptyl$-CH_2N=$cyclopentylidene $+ H_2O \xrightarrow{HCl}$

11.8 KETO-ENOL TAUTOMERISM

A. Keto and Enol Forms

α-Carbon A carbon atom adjacent to a carbonyl group.

α-Hydrogen A hydrogen on an α-carbon.

A carbon atom adjacent to a carbonyl group is called an ***α*-carbon,** and a hydrogen atom bonded to it is called an ***α*-hydrogen.**

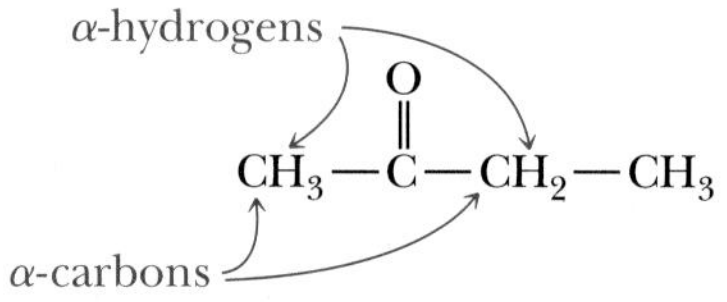

α-hydrogens

$CH_3-\overset{\overset{\displaystyle O}{\|}}{C}-CH_2-CH_3$

α-carbons

Enol A molecule containing an —OH group bonded to a carbon of a carbon-carbon double bond.

Tautomers Constitutional isomers that differ in the location of hydrogen and a double bond relative to O, N, or S.

A carbonyl compound that has a hydrogen on an α-carbon is in equilibrium with a constitutional isomer called an **enol.** The name enol is derived from the IUPAC designation of it as both an alkene (-en-) and an alcohol (-ol). Keto and enol forms are examples of **tautomers,** constitutional isomers in equilibrium with each other that differ in the location of a hydrogen atom and a double bond relative to a heteroatom, most often O, S, or N. This type of isomerism is called **tautomerism.**

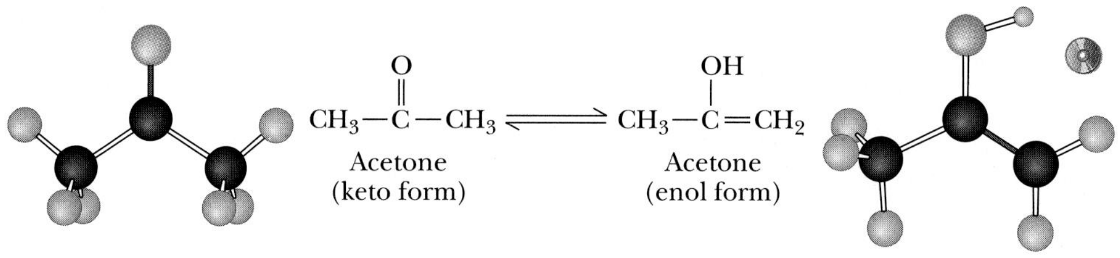

$$CH_3-\overset{\overset{\displaystyle O}{\|}}{C}-CH_3 \rightleftharpoons CH_3-\overset{\overset{\displaystyle OH}{|}}{C}=CH_2$$

Acetone (keto form) — Acetone (enol form)

For most simple aldehydes and ketones, the position of the equilibrium in keto-enol tautomerism lies far on the side of the keto form (Table 11.4) because a carbon-oxygen double bond is stronger than a carbon-carbon double bond.

TABLE 11.4 The Position of Keto-Enol Equilibrium for Four Aldehydes and Ketones*

	Keto Form		Enol Form	% Enol Present at Equilibrium
ethanal	$CH_3\overset{O}{\overset{\|}{C}}H$	$\rightleftharpoons$	$CH_2=\overset{OH}{\overset{\|}{C}}H$	6×10^{-5} 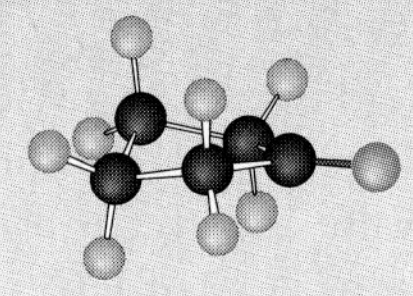
acetone	$CH_3\overset{O}{\overset{\|}{C}}CH_3$	$\rightleftharpoons$	$CH_3\overset{OH}{\overset{\|}{C}}=CH_2$	6×10^{-7}
cyclopentanone	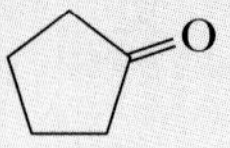 cyclopentanone (C=O)	$\rightleftharpoons$	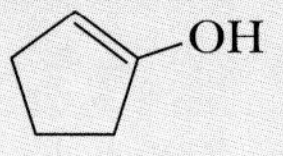 cyclopent-1-en-1-ol (OH)	1×10^{-6} 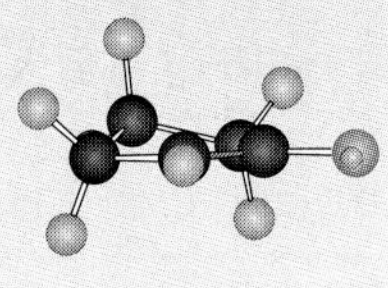
cyclohexanone	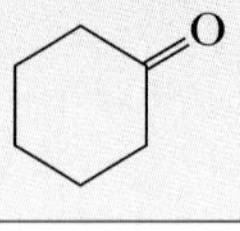 cyclohexanone (C=O)	$\rightleftharpoons$	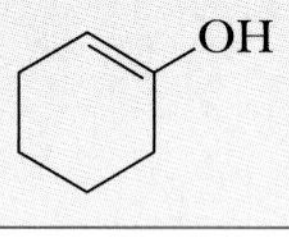 cyclohex-1-en-1-ol (OH)	4×10^{-5}

Cyclopentanone (keto form)

Cyclopentanone (enol form)

*Data from J. March, *Advanced Organic Chemistry,* 4th ed., (New York, Wiley Interscience), 1992, p. 70.

Equilibration of keto and enol forms is catalyzed by acid, as shown in the following two-step mechanism. Note that a molecule of H—A is consumed in Step 1 but another is generated in Step 2.

MECHANISM Acid-Catalyzed Equilibration of Keto and Enol Tautomers

Step 1: Proton transfer from the acid catalyst, H—A, to the carbonyl oxygen forms the conjugate acid of the aldehyde or ketone.

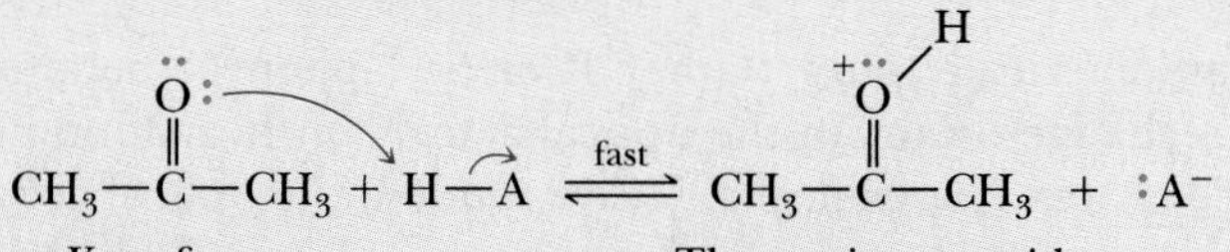

$$CH_3-\overset{\overset{\Large\ddot{O}:}{\|}}{C}-CH_3 + H-A \underset{}{\overset{\text{fast}}{\rightleftharpoons}} CH_3-\overset{\overset{\overset{+}{\ddot{O}}-H}{\|}}{C}-CH_3 + :A^-$$

Keto form — The conjugate acid of the ketone

Step 2: Proton transfer from the α-carbon to the base, A^-, gives the enol and generates a new molecule of the acid catalyst, H—A.

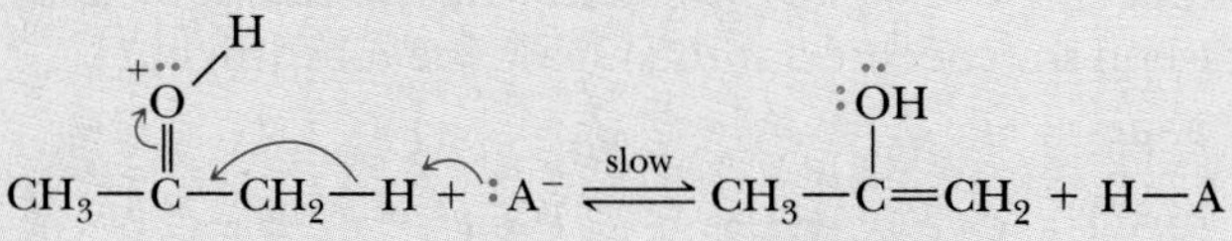

$$CH_3-\overset{\overset{\overset{+}{\ddot{O}}-H}{\|}}{C}-CH_2-H + :A^- \overset{\text{slow}}{\rightleftharpoons} CH_3-\overset{\overset{:\ddot{O}H}{|}}{C}=CH_2 + H-A$$

Enol form

EXAMPLE 11.8

Write two enol forms for each compound. Which enol of each predominates at equilibrium?

(a) 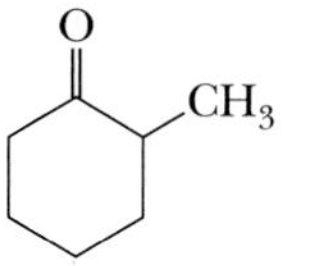 2-methylcyclohexanone (cyclohexanone ring with =O and CH_3 on adjacent carbon)

(b) $CH_3\overset{\overset{O}{\|}}{C}CH_2(CH_2)_2CH_3$

SOLUTION

(a) 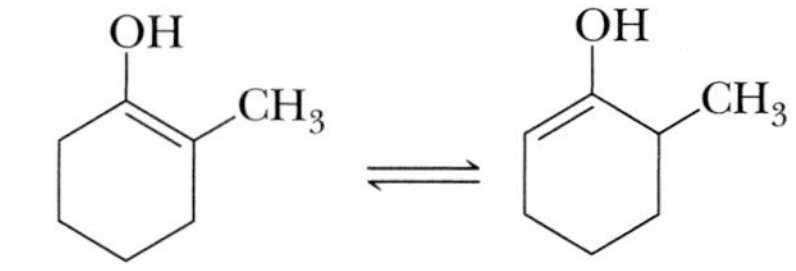 1-hydroxy-2-methylcyclohexene (OH, CH_3) $\rightleftharpoons$ 6-methylcyclohex-1-en-1-ol (OH, CH_3)

Major enol form (more substituted double bond)

(b) $CH_2=\overset{\overset{OH}{|}}{C}CH_2(CH_2)_2CH_3 \rightleftharpoons CH_3\overset{\overset{OH}{|}}{C}=CH(CH_2)_2CH_3$

Major enol form (more substituted double bond)

Practice Problem 11.8

Draw the structural formula for the keto form of each enol.

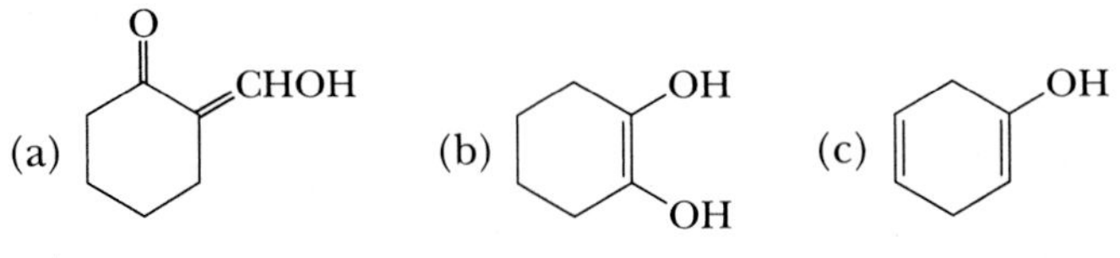

(a) (b) (c)

B. Racemization at an α-Carbon

When enantiomerically pure (either R or S) 3-phenyl-2-butanone is dissolved in ethanol, no change occurs in the optical activity of the solution over time. If, however, a trace of acid (for example, HCl) is added, the optical activity of the solution begins to decrease and gradually drops to zero. When 3-phenyl-2-butanone is isolated from this solution, it is found to be a racemic mixture (Section 4.7C). This observation can be explained by acid-catalyzed formation of an achiral enol intermediate. Tautomerism of the achiral enol to the chiral keto form generates the R and S enantiomers with equal probability. **Racemization** by this mechanism occurs only at α-carbon stereocenters with at least one α-hydrogen.

Racemization The conversion of a pure enantiomer into a racemic mixture.

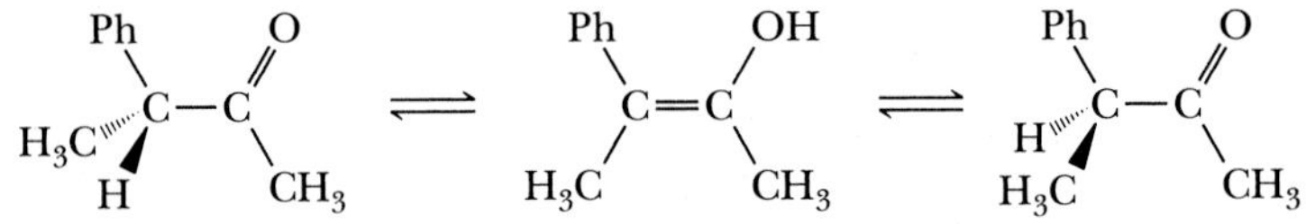

(R)-3-Phenyl-2-butanone $\rightleftharpoons$ An achiral enol $\rightleftharpoons$ (S)-3-Phenyl-2-butanol

11.9 OXIDATION

A. Oxidation of Aldehydes to Carboxylic Acids

Aldehydes are oxidized to carboxylic acids by a variety of common oxidizing agents, including chromic acid and molecular oxygen. In fact, aldehydes have one of the most easily oxidized of all functional groups. Oxidation by chromic acid is illustrated by conversion of hexanal to hexanoic acid.

$$\underset{\text{Hexanal}}{CH_3(CH_2)_4\overset{O}{\overset{\|}{C}}H} \xrightarrow{H_2CrO_4} \underset{\text{Hexanoic acid}}{CH_3(CH_2)_4\overset{O}{\overset{\|}{C}}OH}$$

Aldehydes are also oxidized to carboxylic acids by silver(I)) ion. One laboratory procedure is to shake a solution of the aldehyde dissolved in aqueous ethanol or tetrahydrofuran (THF) with a slurry of Ag_2O.

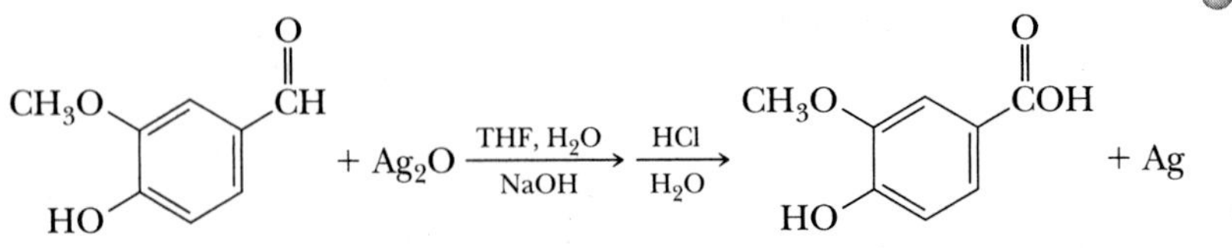

$$\text{Vanillin} + Ag_2O \xrightarrow[NaOH]{THF,\ H_2O} \xrightarrow[H_2O]{HCl} \text{Vanillic acid} + Ag$$

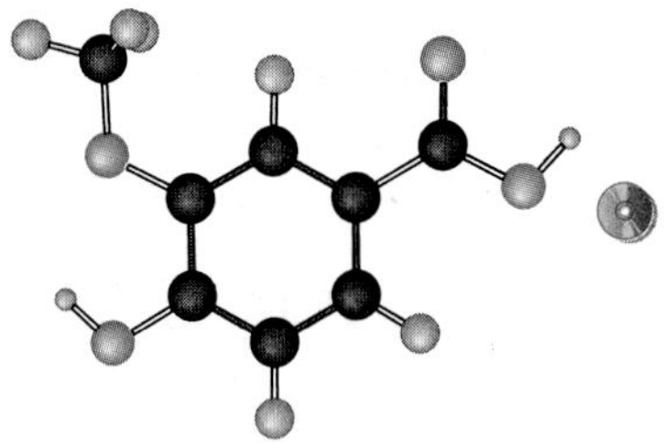

Tollens' reagent, another form of silver(I), is prepared by dissolving silver nitrate in water, adding sodium hydroxide to precipitate silver(I) as Ag_2O, and then adding aqueous ammonia to redissolve silver(I) as the silver-ammonia complex ion.

West Indian vanilla, *Vanilla pompona. (Jane Grushow from Grant Heilman)*

$$Ag^+NO_3^- + 2NH_3 \xrightleftharpoons{NH_3,\ H_2O} Ag(NH_3)_2^+NO_3^-$$

When Tollens' reagent is added to an aldehyde, the aldehyde is oxidized to a carboxylic anion, and silver(I) is reduced to metallic silver. If this reaction is carried out properly, silver precipitates as a smooth, mirror-like deposit, hence the name **silver-mirror test.** Silver(I) is rarely used at the present time for the oxidation of aldehydes because of the cost of silver and because other, more convenient methods exist for this oxidation. This reaction, however, is still used for silvering mirrors.

$$\underset{}{RCH}(=O) + 2Ag(NH_3)_2^+ \xrightarrow{NH_3,\ H_2O} RCO^-(=O) + \underset{\text{Precipitates as silver mirror}}{2Ag} + 4NH_3$$

Aldehydes are also oxidized to carboxylic acids by molecular oxygen and by hydrogen peroxide.

$$2\ C_6H_5CH(=O) + O_2 \longrightarrow 2\ C_6H_5COH(=O)$$

Benzaldehyde → Benzoic acid

Molecular oxygen is the least expensive and most readily available of all oxidizing agents, and, on an industrial scale, air oxidation of organic molecules, including aldehydes, is very common. Air oxidation of aldehydes can also be a problem. Aldehydes that are liquid at room temperature are so sensitive to oxidation by molecular oxygen that they must be protected from contact with air during storage. Often this is done by sealing the aldehyde in a container under an atmosphere of nitrogen.

EXAMPLE 11.9

Draw a structural formula for the product formed by treating each compound with Tollens' reagent followed by acidification with aqueous HCl.

(a) Pentanal (b) Cyclopentanecarbaldehyde

SOLUTION

The aldehyde group in each compound is oxidized to a carboxyl group.

(a) $CH_3(CH_2)_3COH$ (C=O) — Pentanoic acid

(b) C_5H_9COH (cyclopentyl–C(=O)OH) — Cyclopentanecarboxylic acid

A silver mirror has been deposited in the inside of this flask by the reaction of an aldehyde with Tollens' reagent. *(Charles D. Winters)*

Practice Problem 11.9

Complete these oxidations.

(a) Hexanedial + $H_2O_2 \longrightarrow$ (b) 3-Phenylpropanal + Tollens' reagent $\longrightarrow$

B. Oxidation of Ketones to Carboxylic Acids

Ketones are much more resistant to oxidation than are aldehydes. For example, ketones are not normally oxidized by chromic acid or potassium permanganate. In fact these reagents are used routinely to oxidize secondary alcohols to ketones in good yield (Section 8.4F).

Ketones undergo oxidative cleavage, via their enol form, by potassium dichromate and potassium permanganate at higher temperatures and higher concentrations of acid or base. The carbon-carbon double bond of the enol is cleaved to form two carboxyl or ketone groups, depending on the substitution pattern of the original ketone. An important industrial application of this reaction is the oxidation of cyclohexanone to hexanedioic acid (adipic acid), one of the two monomers required for the synthesis of the polymer nylon 66 (Section 15.4A).

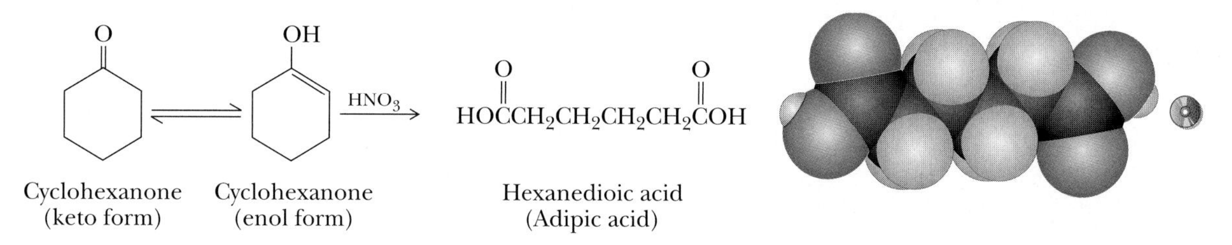

11.10 Reduction

Aldehydes are reduced to primary alcohols and ketones to secondary alcohols.

$$\underset{\text{An aldehyde}}{R\overset{O}{\overset{\|}{C}}H} \xrightarrow{\text{reduction}} \underset{\text{A primary alcohol}}{RCH_2OH} \qquad \underset{\text{A ketone}}{R\overset{O}{\overset{\|}{C}}R'} \xrightarrow{\text{reduction}} \underset{\text{A secondary alcohol}}{R\overset{OH}{\overset{|}{C}}HR'}$$

A. Catalytic Reduction

The carbonyl group of an aldehyde or ketone is reduced to a hydroxyl group by hydrogen in the presence of a transition metal catalyst, most commonly finely divided palladium, platinum, nickel, or rhodium. Reductions are generally carried out at temperatures from 25 to 100°C and at pressures of hydrogen from 1 to 5 atm. Under such conditions, cyclohexanone is reduced to cyclohexanol.

$$\text{Cyclohexanone} + H_2 \xrightarrow[25^\circ C,\ 2\ atm]{Pt} \text{Cyclohexanol}$$

CHEMICAL CONNECTIONS

A Green Synthesis of Adipic Acid

The current industrial production of adipic acid relies on the oxidation of a mixture of cyclohexanol and cyclohexanone by nitric acid.

$$4\ \text{Cyclohexanone} + 6HNO_3 \longrightarrow 4\ \text{Hexanedioic acid (Adipic acid)} + 3N_2O + 3H_2O$$

Cyclohexanone

Hexanedioic acid (Adipic acid) — Nitrous oxide

A by-product of this oxidation is nitrous oxide, a gas considered to play a role in global warming, ozone depletion, as well as contributing to acid rain and acid smog. Given the fact that worldwide production of adipic acid is approximately 2.2 billion metric tons per year, the production of nitrous oxide is enormous. In spite of technological advances that allow for its recovery and recycling, it is estimated that approximately 400,000 metric tons escape recovery and are released into the atmosphere each year.

Ryoji Noyori and coworkers at Nagoya University, Japan, have recently developed a "green" route to adipic acid, one that involves oxidation of cyclohexene by 30% hydrogen peroxide catalyzed by sodium tungstate, Na_2WO_4.

$$\text{Cyclohexene} + 4H_2O_2 \xrightarrow[{[CH_3(C_8H_{17})_3N]HSO_4}]{Na_2WO_4} \text{Hexanedioic acid} + 4H_2O$$

Cyclohexene

Hexanedioic acid

In this process, cyclohexene is mixed with aqueous 30% hydrogen peroxide. To the resulting two-phase system (cyclohexene is insoluble in water) are added sodium tungstate and methyltrioctylammonium hydrogen sulfate. Under these conditions, cyclohexene is oxidized to adipic acid in approximately 90% yield.

While this route to adipic acid is environmentally friendly, it is not yet competitive with the nitric acid oxidation route because of the high cost of 30% hydrogen peroxide. What will make it competitive is either a considerable reduction in the cost of hydrogen peroxide, or institution of more stringent limitations on the emission of nitrous oxide into the atmosphere, or a combination of these. See K. S. Aoki and R. Noyori, *Science,* **281,** 1647 (1998).

Catalytic reduction of aldehydes and ketones is simple to carry out, yields are generally very high, and isolation of the final product is very easy. A disadvantage is that some other functional groups are also reduced under these conditions, for example, carbon-carbon double bonds.

$$\underset{\textit{trans}\text{-2-Butenal (Crotonaldehyde)}}{H_3C(H)C{=}C(H)CHO} \xrightarrow[Ni]{2H_2} \underset{\text{1-Butanol}}{CH_3CH_2CH_2CH_2OH}$$

trans-2-Butenal (Crotonaldehyde) — 1-Butanol

B. Metal Hydride Reductions

Hydride ion A hydrogen atom with two electrons in its valence shell; $H:^-$.

By far the most common laboratory reagents for reduction of the carbonyl group of an aldehyde or ketone to a hydroxyl group are sodium borohydride and lithium aluminum hydride (LAH). These compounds behave as sources of **hydride ion,** a very strong nucleophile.

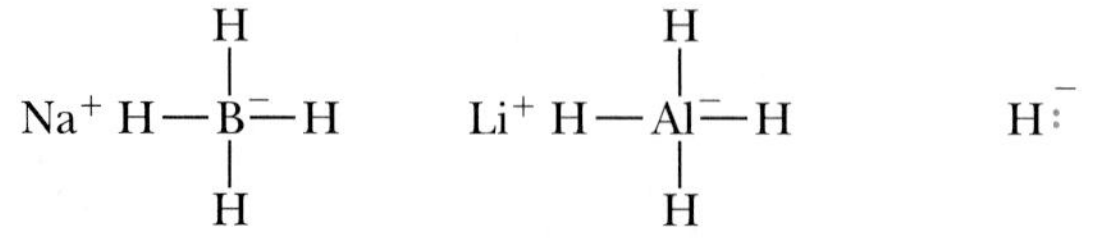

$$Na^+ \; H-\overset{H}{\underset{H}{B^-}}-H \qquad Li^+ \; H-\overset{H}{\underset{H}{Al^-}}-H \qquad H:^-$$

Sodium borohydride — Lithium aluminum hydride (LAH) — Hydride ion

Reductions using sodium borohydride are most commonly carried out in aqueous methanol, in pure methanol, or in ethanol. The initial product of reduction is a tetraalkyl borate, which, on treatment with water, is converted to an alcohol and sodium borate salts. One mole of sodium borohydride reduces 4 moles of aldehyde or ketone.

$$4R\overset{O}{\overset{\|}{C}}H + NaBH_4 \xrightarrow{\text{methanol}} \underset{\text{A tetraalkyl borate}}{(RCH_2O)_4B^- \; Na^+} \xrightarrow{H_2O} 4RCH_2OH + \text{borate salts}$$

The key step in the metal hydride reduction of an aldehyde or ketone is transfer of a hydride ion from the reducing agent to the carbonyl carbon to form a tetrahedral carbonyl addition intermediate. In the reduction of an aldehyde or ketone to an alcohol, only the hydrogen atom attached to carbon comes from the hydride reducing agent; the hydrogen atom attached to oxygen comes from water during hydrolysis of the metal alkoxide salt.

MECHANISM Sodium Borohydride Reduction of an Aldehyde or Ketone

Step 1: A hydride ion is transferred from the reducing agent to the carbonyl carbon.

$$Na^+ \; H-\overset{H}{\underset{H}{B^-}}-H + \overset{R'}{\underset{R}{}}C{=}\ddot{O}: \longrightarrow H-\overset{R'}{\underset{R}{C}}-\ddot{O}^--BH_3 \; Na^+ \xrightarrow[\text{times}]{\text{three more}} \left(H-\overset{R'}{\underset{R}{C}}-\ddot{O}-B^-\right)_4 Na^+$$

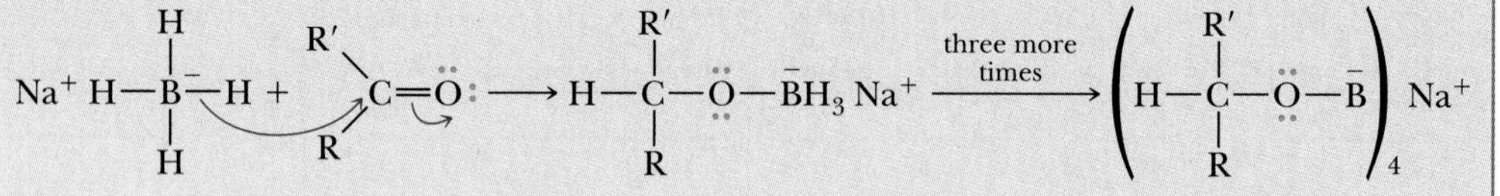

Step 2: Hydrolysis of the tetraalkyl borate gives the alcohol.

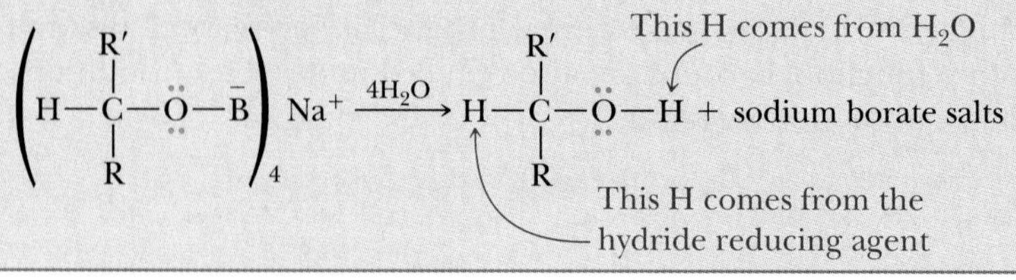

$$\left(H-\overset{R'}{\underset{R}{C}}-\ddot{O}-B^-\right)_4 Na^+ \xrightarrow{4H_2O} H-\overset{R'}{\underset{R}{C}}-\ddot{O}-H + \text{sodium borate salts}$$

This H comes from H_2O

This H comes from the hydride reducing agent

Lithium aluminum hydride is a very powerful reducing agent; it reduces not only the carbonyl groups of aldehydes and ketones rapidly but also those of carboxylic acids (Section 12.5) and their functional derivatives (Section 13.8).

Sodium borohydride is a much more selective reagent, reducing only aldehydes and ketones rapidly.

Lithium aluminum hydride reacts violently with water, methanol, and other protic solvents to liberate hydrogen gas and form metal hydroxides and alkoxides. Therefore, reductions of aldehydes and ketones using this reagent must be carried out in aprotic solvents, most commonly diethyl ether or tetrahydrofuran. The stoichiometry for LAH reductions is the same as that for sodium borohydride reductions: 1 mole of LAH per 4 moles of aldehyde or ketone.

$$4RC(=O)R + LiAlH_4 \xrightarrow{\text{ether}} \underset{\text{A tetraalkylaluminate}}{(R_2CHO)_4Al^- \ Li^+} \xrightarrow{H_2O} 4RCH(OH)R + \text{aluminum salts}$$

The following equations illustrate selective reduction of a carbonyl group in the presence of a carbon-carbon double bond and, alternatively, selective reduction of a carbon-carbon double bond in the presence of a carbonyl group.

Selective reduction of a carbonyl group

$$RCH{=}CHC(=O)R' \xrightarrow[2.\ H_2O]{1.\ NaBH_4} RCH{=}CHCH(OH)R'$$

Selective reduction of a carbon-carbon double bond

$$RCH{=}CHC(=O)R' + H_2 \xrightarrow{Rh} RCH_2CH_2C(=O)R'$$

EXAMPLE 11.10

Complete these reductions.

(a) $CH_3CH_2CH_2CH(=O) \xrightarrow[Pt]{H_2}$ (b) $CH_3O{-}C_6H_4{-}C(=O)CH_3 \xrightarrow[2.\ H_2O]{1.\ NaBH_4}$

SOLUTION

The carbonyl group of the aldehyde in (a) is reduced to a primary alcohol, and that of the ketone in (b) is reduced to a secondary alcohol.

(a) $CH_3CH_2CH_2CH_2OH$ (b) $CH_3O{-}C_6H_4{-}CH(OH)CH_3$

Practice Problem 11.10

What aldehyde or ketone gives these alcohols on reduction with $NaBH_4$?

(a) cyclohexyl–OH (b) $CH_3O{-}C_6H_4{-}CH_2CH_2OH$ (c) $CH_3CH(OH)(CH_2)_3CH(OH)CH_3$

C. Reductive Amination of Aldehydes and Ketones

Reductive amination The formation of an imine from an aldehyde or ketone followed by its reduction to an amine.

Aldehydes and ketones react with ammonia and primary amines (Section 11.7) to give **imines,** which, in turn, are reduced by hydrogen in the presence of a transition metal catalyst to amines. Formation of an imine followed by its reduction to an amine is called **reductive amination.** This two-step conversion of an aldehyde or ketone to an amine can be carried out in one laboratory operation by mixing together the carbonyl-containing compound, the amine or ammonia, hydrogen, and the transition metal catalyst, as illustrated by conversion of cyclohexanone to cyclohexanamine, a primary amine. The imine intermediate is not isolated.

$$C_6H_{10}{=}O \xrightarrow[(-H_2O)]{NH_3} [C_6H_{10}{=}NH] \xrightarrow{H_2,\ Ni} C_6H_{11}{-}NH_2$$

Cyclohexanone — An imine; not isolated — Cyclohexanamine (Cyclohexylamine)

Reductive amination can be used for the synthesis of secondary and tertiary amines as well. Secondary amines are prepared using an aldehyde or ketone and a primary amine in the presence of hydrogen and a hydrogenation catalyst. Tertiary amines are prepared in a similar manner using a secondary amine and an aldehyde or ketone.

$$C_6H_5CH{=}O + H_2NCH_2C_6H_5 \xrightarrow{H_2,\ Ni} C_6H_5CH_2NHCH_2C_6H_5 + H_2O$$

Benzaldehyde — Benzylamine — Dibenzylamine

EXAMPLE 11.11

Show how to synthesize each amine by a reductive amination.

(a) $C_6H_5CH(CH_3)NH_2$ (b) $[(CH_3)_2CH]_2NH$

SOLUTION

The appropriate compound is treated with ammonia or an amine in the presence of H_2/Ni.

(a) $C_6H_5C(=O)CH_3 + NH_3$ (b) $CH_3C(=O)CH_3 + CH_3CH(NH_2)CH_3$

Practice Problem 11.11

Show how you might convert piperidine to these compounds, using any other organic compounds and necessary reagents.

(a) N-ethylpiperidine ($C_5H_{10}NCH_2CH_3$) (b) N-cyclohexylpiperidine ($C_5H_{10}N{-}C_6H_{11}$)

SUMMARY

An **aldehyde** (Section 11.1) contains a carbonyl group bonded to a hydrogen atom and a carbon atom. A **ketone** contains a carbonyl group bonded to two carbons. An aldehyde is named by changing -e of the parent alkane to -al (Section 11.2). A CHO group bonded to a ring is indicated by the suffix -carbaldehyde. A ketone is named by changing -e of the parent alkane to -one and using a number to locate the carbonyl group. In naming compounds that contain more than one functional group, the IUPAC system has established an **order of precedence of functional groups** (Section 11.2B). If the carbonyl group of an aldehyde or ketone is lower in precedence than other functional groups in the molecule, it is indicated by the infix -oxo-.

Aldehydes and ketones are polar compounds (Section 11.3) and interact in the pure state by dipole-dipole interactions; they have higher boiling points and are more soluble in water than nonpolar compounds of comparable molecular weight.

The carbon-metal bond in **Grignard reagents** (Section 11.5A) has a high degree of partial ionic character. Grignard reagents behave as carbanions, and are both strong bases and good nucleophiles.

A carbon atom adjacent to a carbonyl group is called an ***α*-carbon** (Section 11.8A), and a hydrogen attached to it is called an ***α*-hydrogen.**

KEY REACTIONS

1. Reaction with Grignard Reagents (Section 11.5B)

Treatment of formaldehyde with a Grignard reagent followed by hydrolysis in aqueous acid gives a primary alcohol. Similar treatment of any other aldehyde gives a secondary alcohol.

$$CH_3\overset{O}{\overset{\|}{C}}H \xrightarrow[2.\ HCl,\ H_2O]{1.\ C_6H_5MgBr} C_6H_5\overset{OH}{\overset{|}{C}}HCH_3$$

Treatment of a ketone with a Grignard reagent gives a tertiary alcohol.

$$CH_3\overset{O}{\overset{\|}{C}}CH_3 \xrightarrow[2.\ HCl,\ H_2O]{1.\ C_6H_5MgBr} C_6H_5\overset{OH}{\overset{|}{C}}(CH_3)_2$$

2. Addition of Alcohols to Form Hemiacetals (Section 11.6)

Hemiacetals are only minor components of an equilibrium mixture of aldehyde or ketone and alcohol, except where the —OH and C=O groups are parts of the same molecule and a five- or six-membered ring can form.

$$CH_3\underset{OH}{\underset{|}{C}}HCH_2CH_2\overset{O}{\overset{\|}{C}}H \rightleftharpoons H_3C\text{-(cyclic, O)-}OH$$

4-Hydroxypentanal A cyclic hemiacetal

3. Addition of Alcohols to Form Acetals (Section 11.6)

Formation of acetals is catalyzed by acid. Acetals are stable in water and aqueous base but are hydrolyzed in aqueous acid.

$$\text{cyclopentanone}{=}O + HOCH_2CH_2OH \overset{H^+}{\rightleftharpoons} \text{(spiro acetal: } O{-}CH_2{-}CH_2{-}O) + H_2O$$

4. Addition of Ammonia and Its Derivatives: Formation of Imines (Section 11.7)

Addition of ammonia or a primary amine to the carbonyl group of an aldehyde or ketone forms a tetrahedral carbonyl addition intermediate. Loss of water gives an imine (a Schiff base).

$$\text{cyclopentanone}{=}O + H_2NCH_3 \xrightleftharpoons{H^+} \text{cyclopentylidene}{=}NCH_3 + H_2O$$

5. Keto-Enol Tautomerism (Section 11.8A)

The keto form generally predominates at equilibrium.

$$CH_3\overset{O}{\overset{\|}{C}}CH_3 \rightleftharpoons CH_3\overset{OH}{\overset{|}{C}}{=}CH_2$$

Keto form (Approx 99.9%) — Enol form

6. Oxidation of an Aldehyde to a Carboxylic Acid (Section 11.9A)

The aldehyde group is among the most easily oxidized functional groups. Oxidizing agents include $K_2Cr_2O_7$, Tollens' reagent, H_2O_2, and O_2.

$$\text{(2-hydroxybenzaldehyde, } C_6H_4(OH)\overset{O}{\overset{\|}{C}}H) + Ag_2O \xrightarrow[\text{2. } H_2O,\ HCl]{\text{1. THF, } H_2O,\ NaOH} \text{(2-hydroxybenzoic acid, } C_6H_4(OH)\overset{O}{\overset{\|}{C}}OH) + Ag$$

7. Catalytic Reduction (Section 11.10A)

Catalytic reduction of the carbonyl group of an aldehyde or ketone to a hydroxyl group is simple to carry out and yields of alcohols are high. A disadvantage of this method is that some other functional groups, including carbon-carbon double bonds, may also be reduced.

$$\text{cyclohexanone}{=}O + H_2 \xrightarrow[25°C,\ 2\ atm]{Pt} \text{cyclohexyl}{-}OH$$

8. Metal Hydride Reduction (Section 11.10B)

Both $LiAlH_4$ and $NaBH_4$ are selective in that neither reduces isolated carbon-carbon double bonds.

$$\text{cyclohex-3-enone}{=}O \xrightarrow[\text{2. } H_2O]{\text{1. } NaBH_4} \text{cyclohex-3-enyl}{-}OH$$

9. Reductive Amination to Amines (Section 11.10C)

The carbon-nitrogen double bond of an imine can be reduced by hydrogen in the presence of a transition metal catalyst to a carbon-nitrogen single bond.

$$\text{cyclohexanone}{=}O + H_2N{-}C_6H_{11} \xrightarrow[-H_2O]{} \left[C_6H_{10}{=}N{-}C_6H_{11}\right] \xrightarrow{H_2/Ni} C_6H_{11}{-}\overset{H}{\overset{|}{N}}{-}C_6H_{11}$$

ADDITIONAL PROBLEMS

Preparation of Aldehydes and Ketones

The methods covered to this point for preparation of aldehydes and ketones are oxidation of primary and secondary alcohols (Section 8.4F) and Friedel-Crafts acylation of arenes (Section 9.7C).

11.12 Complete these reactions.

(a) Cyclooctanol (ring–OH) $\xrightarrow[H_2SO_4]{K_2Cr_2O_7}$

(b) Cyclopentyl–CH_2OH $\xrightarrow[CH_2Cl_2]{PCC}$

(c) Cyclopentyl–CH_2OH $\xrightarrow[H_2SO_4]{K_2Cr_2O_7}$

(d) Benzene $\xrightarrow[AlCl_3]{(CH_3)_2CHCH_2CCl\ (C{=}O)}$

(e) C_6H_5–$CH_2CH(CH_3)_2$ $\xrightarrow[AlCl_3]{(CH_3)_2CHCCl\ (C{=}O)}$

11.13 Show how you would bring about these conversions.

(a) 1-Pentanol to pentanal
(b) 1-Pentanol to pentanoic acid
(c) 2-Pentanol to 2-pentanone
(d) 1-Pentene to 2-pentanone
(e) Benzene to acetophenone
(f) Styrene to acetophenone
(g) Cyclohexanol to cyclohexanone
(h) Cyclohexene to cyclohexanone

Structure and Nomenclature

11.14 Draw a structural formula for the one ketone of molecular formula C_4H_8O and for the two aldehydes of molecular formula C_4H_8O.

11.15 Draw structural formulas for the six ketones of molecular formula $C_6H_{12}O$. Which are chiral? Which contain two stereocenters?

11.16 Name these compounds.

(a) $(CH_3CH_2CH_2)_2C{=}O$

(b) Cyclopentanone with CH_3 (wedge) and H (hash) on the carbon adjacent to C=O

(c) H_3C(H)C=C(CH_3)CHO — H_3C and CHO on top, H and CH_3 on bottom

(d) CHO on C bearing H (hash), CH_3 (wedge), OH

(e) CH_3O–C_6H_4–$\overset{O}{\overset{\|}{C}}CH_3$

(f) $H\overset{O}{\overset{\|}{C}}(CH_2)_4\overset{O}{\overset{\|}{C}}H$

(g) Cyclopentane-1,3-dione with $CH_2CH_2CH_3$ at C-2

11.17 Draw structural formulas for these compounds.

(a) 1-Chloro-2-propanone **(b)** 3-Hydroxybutanal

(c) 4-Hydroxy-4-methyl-2-pentanone **(d)** 3-Methyl-3-phenylbutanal

(e) 1,3-Cyclohexanedione **(f)** 3-Methyl-3-buten-2-one

(g) 5-Oxohexanal **(h)** 2,2-Dimethylcyclohexanecarbaldehyde

(i) 3-Oxobutanoic acid

Addition of Carbon Nucleophiles

11.18 Write an equation for the acid-base reaction between phenylmagnesium iodide and a carboxylic acid. Use curved arrows to show the flow of electrons in this reaction. In addition, show that this reaction is an example of a stronger acid and stronger base reacting to form a weaker acid and weaker base.

11.19 Diethyl ether is prepared on an industrial scale by acid-catalyzed dehydration of ethanol. Explain why diethyl ether used in the preparation of Grignard reagents must be carefully purified to remove all traces of ethanol and water.

$$2CH_3CH_2OH \xrightarrow[180°C]{H_2SO_4} CH_3CH_2OCH_2CH_3 + H_2O$$

11.20 Draw structural formulas for the product formed by treatment of each compound with propylmagnesium bromide followed by hydrolysis in aqueous acid.

(a) CH_2O **(b)** 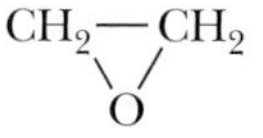 $CH_2{-}CH_2$ (epoxide, bridged by O) **(c)** $CH_3CH_2\overset{O}{\overset{\|}{C}}CH_2CH_3$ **(d)** 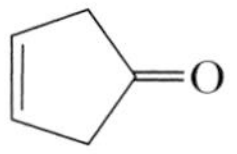 cyclopentanone (ring=O)

(e) CO_2 **(f)** 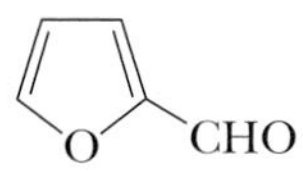 furan-2-CHO (ring O, CHO) **(g)** 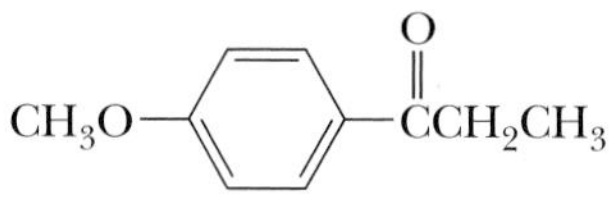 CH_3O–(benzene ring)–$\overset{O}{\overset{\|}{C}}CH_2CH_3$

11.21 Write structural formulas for all combinations of Grignard reagent and aldehyde or ketone that might be used to synthesize each alcohol.

(a) $CH_3CH_2CH_2CH_2\overset{OH}{\overset{|}{C}}HCH_3$ **(b)** 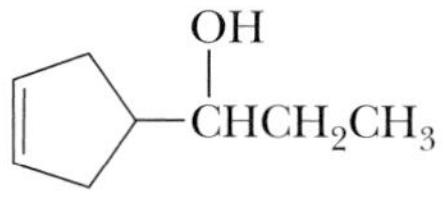 (cyclopentyl)–$\overset{OH}{\overset{|}{C}}HCH_2CH_3$

11.22 Show reagents to bring about this conversion.

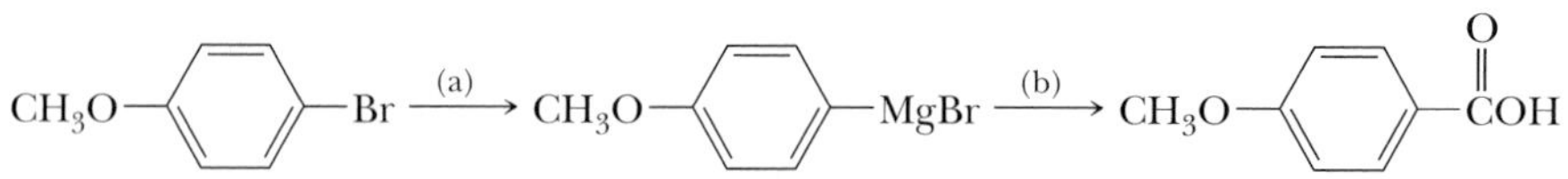

$$CH_3O\text{–(benzene ring)–}Br \xrightarrow{(a)} CH_3O\text{–(benzene ring)–}MgBr \xrightarrow{(b)} CH_3O\text{–(benzene ring)–}\overset{O}{\overset{\|}{C}}OH$$

11.23 Suggest a synthesis for these alcohols starting from an aldehyde or ketone and an appropriate Grignard reagent. In parentheses below each target molecule is shown the number of combinations of Grignard reagent and aldehyde or ketone that might be used.

(a) $CH_3\overset{OH}{\overset{|}{C}}CH_2CH_2CH_3$ with CH_2CH_3 on the central C

3 combinations

(b) $CH_3CH_2\overset{OH}{\overset{|}{C}}HCH{=}CHCH_3$

2 combinations

(c) 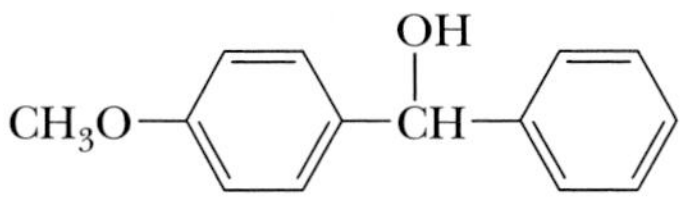 $CH_3O-C_6H_4-CH(OH)-C_6H_5$

2 combinations

11.24 Show how to synthesize 3-ethyl-1-hexanol using 1-bromopropane, propanal, and ethylene oxide as the only sources of carbon atoms. It can be done using each compound only once. (*Hint:* Carry out one Grignard reaction to form an alcohol, convert the alcohol to an alkyl halide, and then do a second Grignard reaction.)

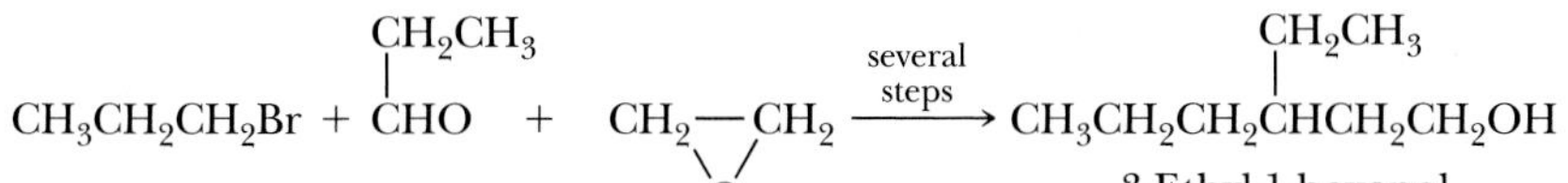

$$CH_3CH_2CH_2Br + CH_3CH_2CHO + \underset{O}{CH_2-CH_2} \xrightarrow[\text{steps}]{\text{several}} CH_3CH_2CH_2CH(CH_2CH_3)CH_2CH_2OH$$

3-Ethyl-1-hexanol

11.25 1-Phenyl-2-butanol is used in perfumery. Show how to synthesize this alcohol from bromobenzene, 1-butene, and any other needed reagents.

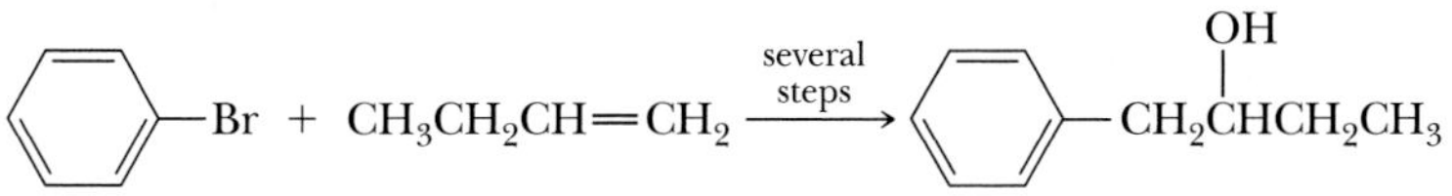

$$C_6H_5-Br + CH_3CH_2CH{=}CH_2 \xrightarrow[\text{steps}]{\text{several}} C_6H_5-CH_2CH(OH)CH_2CH_3$$

Bromobenzene 1-Butene 1-Phenyl-2-butanol

Addition of Oxygen Nucleophiles

11.26 5-Hydroxyhexanal forms a six-membered cyclic hemiacetal, which predominates at equilibrium in aqueous solution.

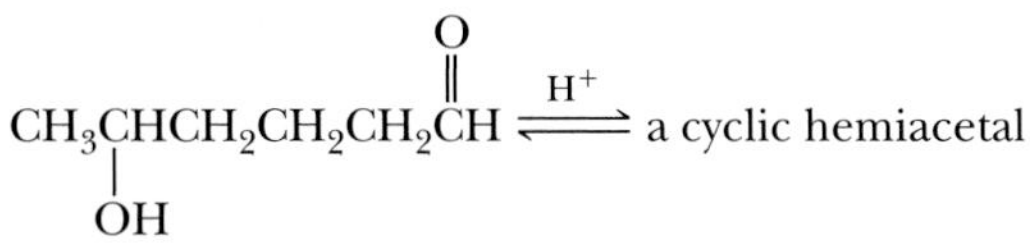

$$CH_3CH(OH)CH_2CH_2CH_2CH{=}O \underset{}{\overset{H^+}{\rightleftharpoons}} \text{a cyclic hemiacetal}$$

5-Hydroxyhexanal

(a) Draw a structural formula for this cyclic hemiacetal.

(b) How many stereoisomers are possible for 5-hydroxyhexanal?

(c) How many stereoisomers are possible for this cyclic hemiacetal?

(d) Draw alternative chair conformations for each stereoisomer.

(e) Which alternative chair conformation for each stereoisomer is the more stable?

11.27 Draw structural formulas for the hemiacetal and then the acetal formed from each pair of reactants in the presence of an acid catalyst.

(a) 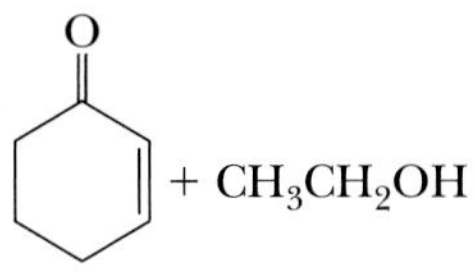 2-cyclohexenone + CH_3CH_2OH (b) 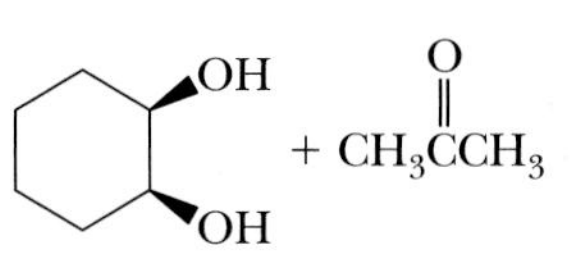 cis-1,2-cyclohexanediol (OH, OH) + $CH_3C(=O)CH_3$ (c) 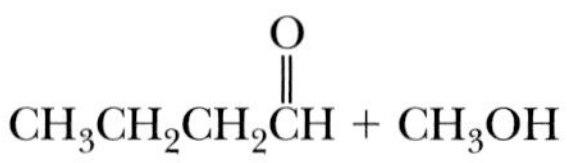 $CH_3CH_2CH_2CH{=}O + CH_3OH$

11.28 Draw structural formulas for the products of hydrolysis of each acetal in aqueous acid.

(a) 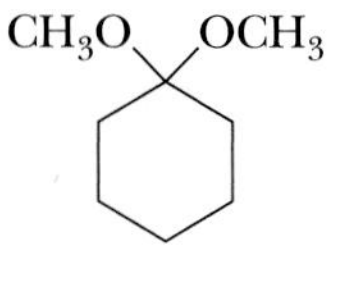 1,1-dimethoxycyclohexane (CH_3O, OCH_3) (b) 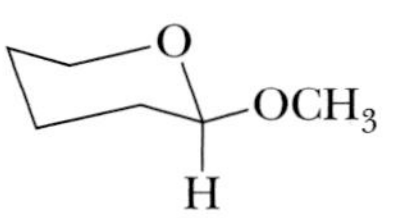 tetrahydropyran with OCH_3 and H on C-2 (c) CHO–HC–O, H_2C–O, C(CH_3)(CH_3)

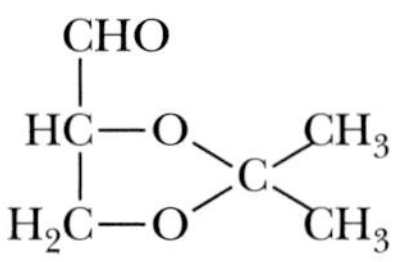

11.29 Propose a mechanism for formation of the cyclic acetal from treatment of acetone with ethylene glycol in the presence of an acid catalyst. Your mechanism must be consistent with the fact that the oxygen atom of the water molecule is derived from the carbonyl oxygen of acetone.

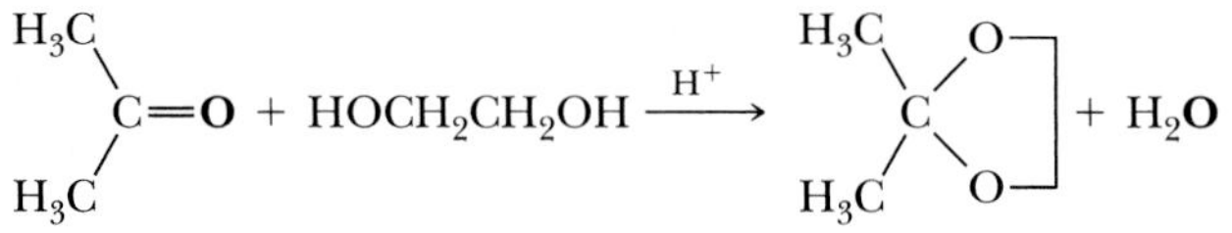

$$(CH_3)_2C{=}\mathbf{O} + HOCH_2CH_2OH \xrightarrow{H^+} (CH_3)_2C(OCH_2CH_2O) + H_2\mathbf{O}$$

11.30 Propose a mechanism for the formation of a cyclic acetal from 4-hydroxypentanal and one equivalent of methanol. If the carbonyl oxygen of 4-hydroxypentanal is enriched with oxygen-18, do you predict that the oxygen label appears in the cyclic acetal or in the water? Explain.

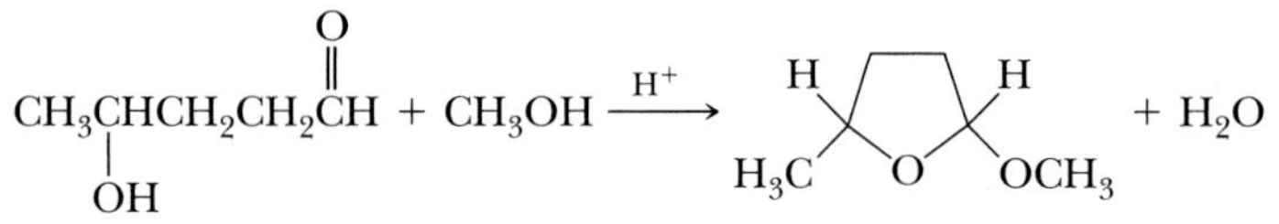

$$CH_3CH(OH)CH_2CH_2CH{=}O + CH_3OH \xrightarrow{H^+} \text{A cyclic acetal} + H_2O$$

4-Hydroxypentanal — A cyclic acetal

Addition of Nitrogen Nucleophiles

11.31 Show how this secondary amine can be prepared by two successive reductive aminations:

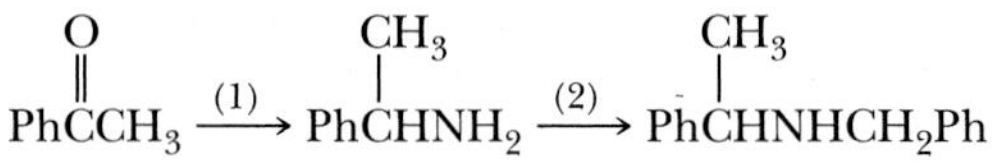

$$PhC(=O)CH_3 \xrightarrow{(1)} PhCH(CH_3)NH_2 \xrightarrow{(2)} PhCH(CH_3)NHCH_2Ph$$

11.32 Show how to convert cyclohexanone to these amines.

(a) 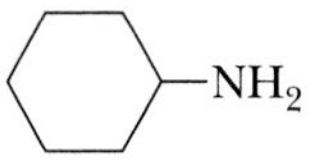 cyclohexyl–NH_2 (b) 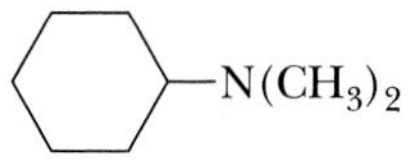 cyclohexyl–$N(CH_3)_2$ (c) 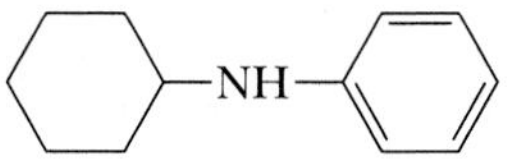 cyclohexyl–NH–phenyl

11.33 Following are structural formulas for amphetamine and methamphetamine (*The Merck Index,* 12th ed., #623 and #6015). The major central nervous system effects of amphetamine and amphetamine-like drugs are locomotor stimulation, euphoria and excitement, stereotyped behavior, and anorexia. Show how each drug can be synthesized by reductive amination of an appropriate aldehyde or ketone. Assign an R or S configuration to the stereocenter in the enantiomer of methamphetamine shown here.

(a) 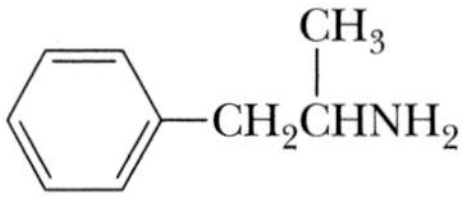 C_6H_5–$CH_2CH(CH_3)NH_2$

Amphetamine

(b) 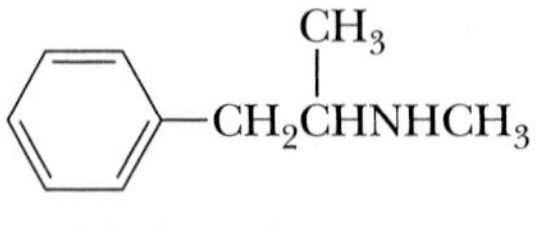 C_6H_5–$CH_2CH(CH_3)NHCH_3$

Methamphetamine

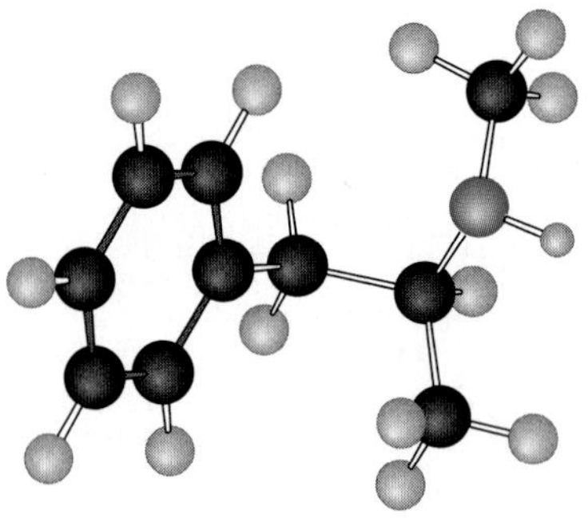

11.34 Rimantadine is effective in preventing infections caused by the influenza A virus and in treating established illness (*The Merck Index,* 12th ed., #8390). It is thought to exert its antiviral effect by blocking a late stage in the assembly of the virus. Following is the final step in the synthesis of this compound. Describe experimental conditions to bring about this conversion.

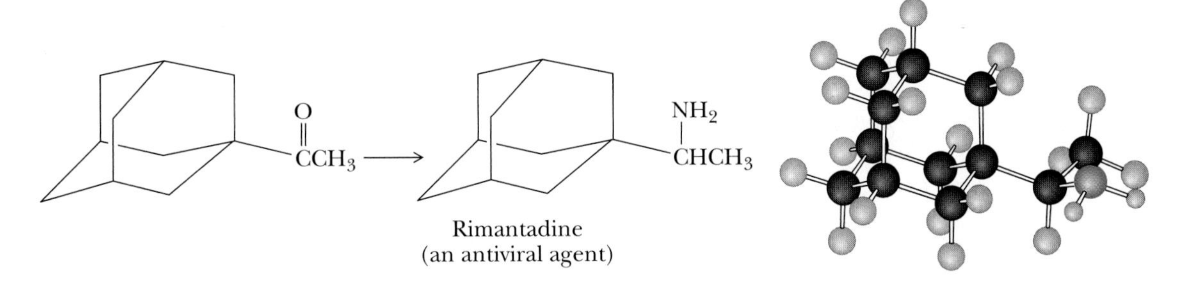

Rimantadine
(an antiviral agent)

11.35 Methenamine, a product of the reaction of formaldehyde and ammonia, is a prodrug, a compound that is inactive itself but is converted to an active drug in the body by a biochemical transformation. The strategy behind use of methenamine as a prodrug is that nearly all bacteria are sensitive to formaldehyde at concentrations of 20 mg/mL or higher. Formaldehyde cannot be used directly in medicine, however, because an effective concentration in plasma cannot be achieved with safe doses. Methenamine is stable at pH 7.4 (the pH of blood plasma) but undergoes acid-catalyzed hydrolysis to formaldehyde and ammonium ion under the acidic conditions of the kidneys and the urinary tract. Thus, methenamine can be used as a site-specific drug to treat urinary infections (*The Merck Index,* 12th ed., #6036).

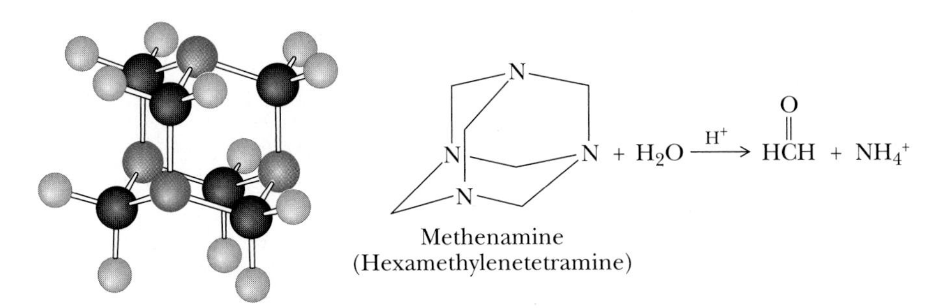

Methenamine
(Hexamethylenetetramine)

(a) Balance the equation for the hydrolysis of methenamine to formaldehyde and ammonium ion.

(b) Does the pH of an aqueous solution of methenamine increase, remain the same, or decrease as a result of its hydrolysis? Explain.

(c) Explain the meaning of the following statement: The functional group in methenamine is the nitrogen analog of an acetal.

(d) Account for the observation that methenamine is stable in blood plasma but undergoes hydrolysis in the urinary tract.

Keto-Enol Tautomerism

11.36 The following molecule belongs to a class of compounds called enediols; each carbon of the double bond carries an —OH group. Draw structural formulas for the α-hydroxyketone and the α-hydroxyaldehyde with which this enediol is in equilibrium.

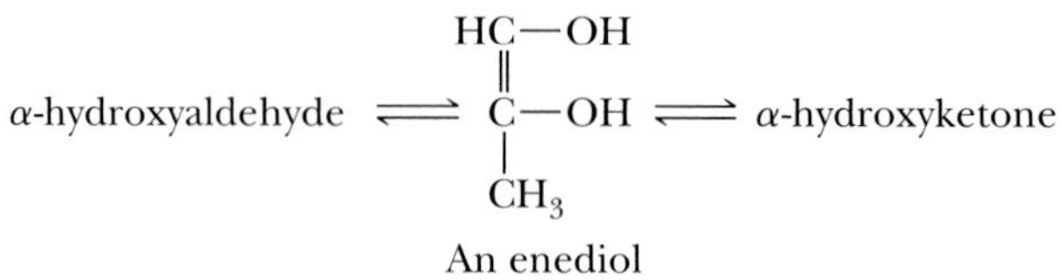

$$\alpha\text{-hydroxyaldehyde} \rightleftharpoons \begin{array}{c} HC{-}OH \\ \| \\ C{-}OH \\ | \\ CH_3 \end{array} \rightleftharpoons \alpha\text{-hydroxyketone}$$

An enediol

11.37 In dilute aqueous acid, (R)-glyceraldehyde is converted into an equilibrium mixture of (R,S)-glyceraldehyde and dihydroxyacetone. Propose a mechanism for this isomerization.

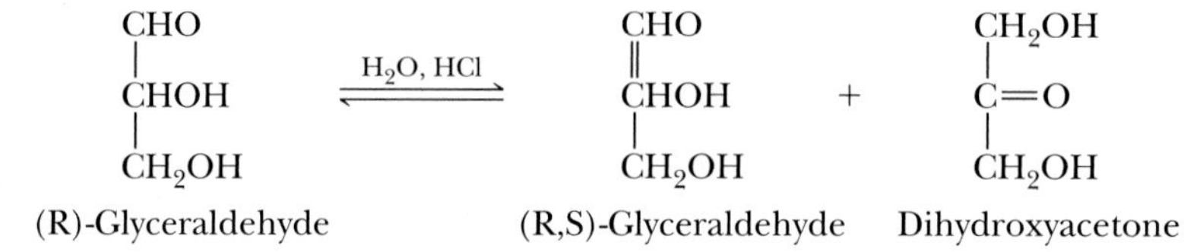

$$\begin{array}{c} CHO \\ | \\ CHOH \\ | \\ CH_2OH \end{array} \xrightleftharpoons{H_2O,\ HCl} \begin{array}{c} CHO \\ | \\ CHOH \\ | \\ CH_2OH \end{array} + \begin{array}{c} CH_2OH \\ | \\ C{=}O \\ | \\ CH_2OH \end{array}$$

(R)-Glyceraldehyde (R,S)-Glyceraldehyde Dihydroxyacetone

Oxidation/Reduction of Aldehydes and Ketones

11.38 Draw the structural formula for the product formed by treatment of butanal with each set of reagents.

(a) $LiAlH_4$ followed by H_2O
(b) $NaBH_4$ in CH_3OH/H_2O
(c) H_2/Pt
(d) $Ag(NH_3)_2^+$ in NH_3/H_2O, then HCl/H_2O
(e) $K_2Cr_2O_7/H_2SO_4$
(f) $C_6H_5NH_2$ in the presence of H_2/Ni

11.39 Draw the structural formula for the product for the reaction of *p*-bromoacetophenone with each set of reagents in Problem 11.38.

Synthesis

11.40 Show reagents and conditions to bring about the conversion of cyclohexanol to cyclohexanecarbaldehyde.

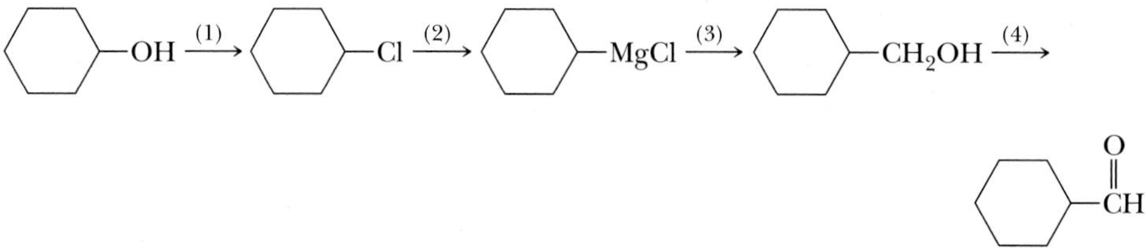

$$C_6H_{11}{-}OH \xrightarrow{(1)} C_6H_{11}{-}Cl \xrightarrow{(2)} C_6H_{11}{-}MgCl \xrightarrow{(3)} C_6H_{11}{-}CH_2OH \xrightarrow{(4)} C_6H_{11}{-}\overset{O}{\overset{\|}{C}}H$$

11.41 Starting with cyclohexanone, show how to prepare these compounds. In addition to the given starting material, use any other organic or inorganic reagents as necessary.

(a) Cyclohexanol
(b) Cyclohexene
(c) *cis*-1,2-Cyclohexanediol
(d) 1-Methylcyclohexanol
(e) 1-Methylcyclohexene
(f) 1-Phenylcyclohexanol
(g) 1-Phenylcyclohexene
(h) Cyclohexene oxide
(i) *trans*-1,2-Cyclohexanediol

11.42 Show how to bring about these conversions. In addition to the given starting material, use any other organic or inorganic reagents as necessary.

(a) $C_6H_5\overset{O}{\overset{\|}{C}}CH_2CH_3 \longrightarrow C_6H_5\overset{OH}{\overset{|}{C}}HCH_2CH_3 \longrightarrow C_6H_5CH{=}CHCH_3$

(b) 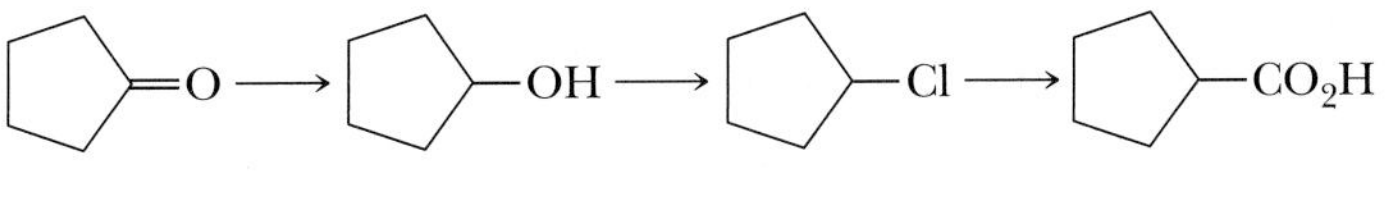 =O ⟶ —OH ⟶ —Cl ⟶ —CO_2H

(c) 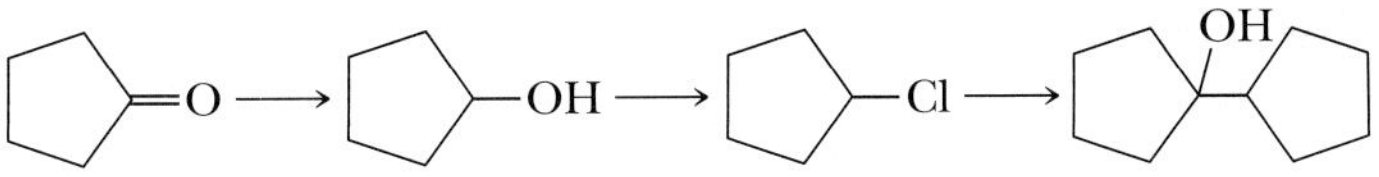 =O ⟶ —OH ⟶ —Cl ⟶ OH

(d) 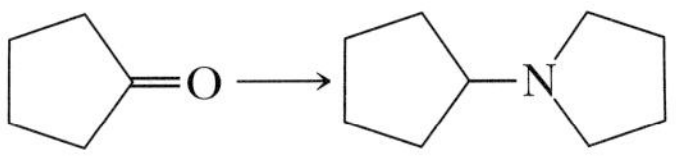 =O ⟶ —N

11.43 Many tumors of the breast are estrogen-dependent. Drugs that interfere with estrogen binding have antitumor activity and may even help prevent tumor occurrence. A widely used antiestrogen drug is tamoxifen (*The Merck Index,* 12th ed., #9216).

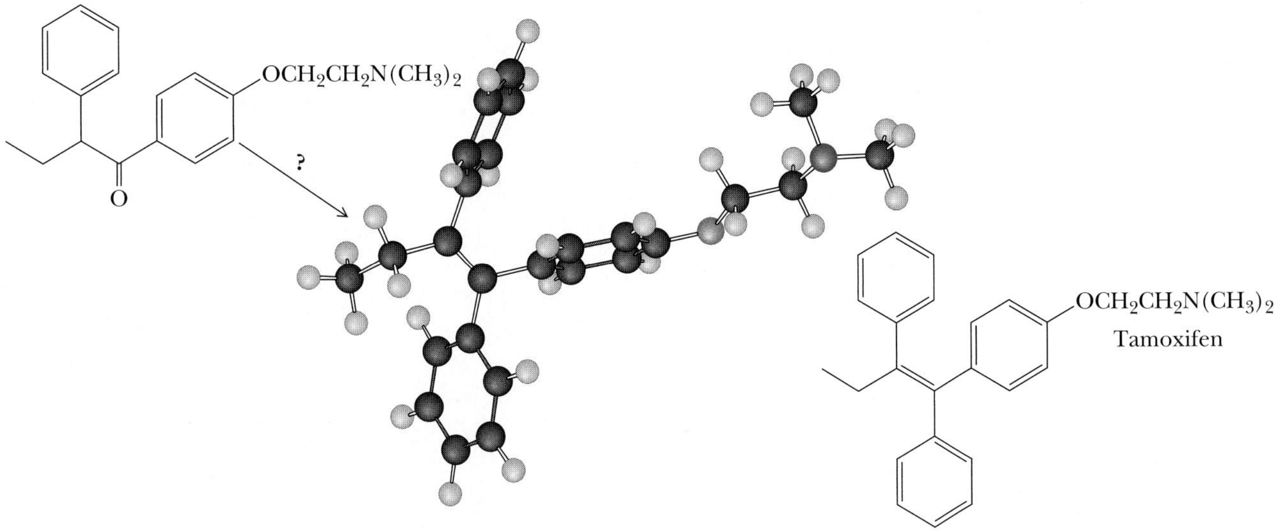

$OCH_2CH_2N(CH_3)_2$

O

?

$OCH_2CH_2N(CH_3)_2$

Tamoxifen

(a) How many stereoisomers are possible for tamoxifen?

(b) Specify the configuration of the stereoisomer shown here.

(c) Show how tamoxifen can be synthesized from the given ketone using a Grignard reaction followed by dehydration.

Spectroscopy

11.44 At this point, the following spectroscopy problems may be assigned:

^{1}H-NMR and ^{13}C-NMR: Additional Problems 21.24–21.27.

IR: Additional Problem 22.9.

12

Carboxylic Acids

Willow bark was the first source of salicylic acid, the compound from which aspirin is derived. Salicylic acid is now synthesized industrially from phenol. Inset: A model of salicylic acid. *(Charles D. Winters)*

The most important chemical property of carboxylic acids, another class of organic compounds containing the carbonyl group, is their acidity. Furthermore, carboxylic acids form numerous important derivatives, including esters, amides, anhydrides, and acid halides. In this chapter, we study carboxylic acids themselves; in Chapter 13, we study their derivatives.

12.1 STRUCTURE

The functional group of a carboxylic acid is a **carboxyl group,** so named because it is made up of a **carb**onyl group and a hydr**oxyl** group (Section 1.7C). Following is a Lewis structure for the carboxyl group as well as two alternative representations for it.

Carboxyl group A $—CO_2H$ group.

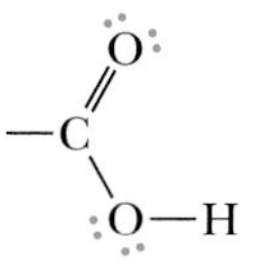

$$—C(=\ddot{O}:)—\ddot{O}—H \qquad —COOH \qquad —CO_2H$$

12.2 NOMENCLATURE

A. IUPAC System

The IUPAC name of a carboxylic acid is derived from that of the longest carbon chain that contains the carboxyl group by dropping the final -e from the name of the parent alkane and adding the suffix -oic followed by the word "acid" (Section 3.5). The chain is numbered beginning with the carbon of the carboxyl group. Because the carboxyl carbon is understood to be carbon 1, there is no need to give it a number. If the carboxylic acid contains a carbon-carbon double bond, change the infix from -an- to -en- to indicate the presence of the double bond and show the location of the double bond by a number. In the following examples, the common name is given in parentheses.

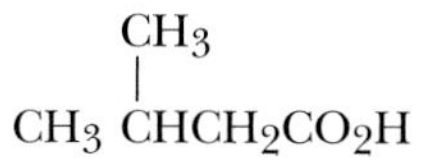

$CH_3CH(CH_3)CH_2CO_2H$

3-Methylbutanoic acid
(Isovaleric acid)

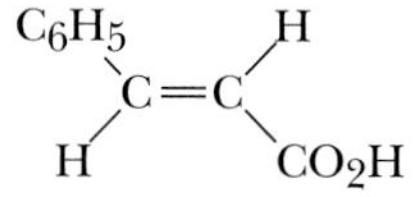

$C_6H_5(H)C{=}C(H)CO_2H$

trans-3- Phenylpropenoic acid
(Cinnamic acid)

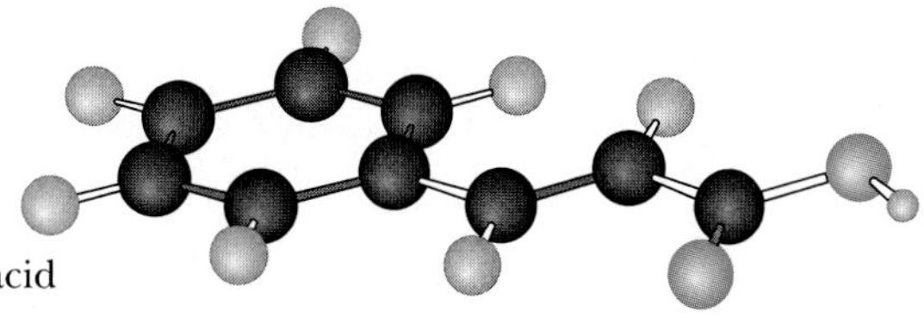

In the IUPAC system, a carboxyl group takes precedence over most other functional groups (Table 11.1), including hydroxyl and amino groups, as well as the carbonyl groups of aldehydes and ketones. As illustrated in the following examples, when a carboxyl group is present in a molecule, an —OH group of an alcohol is indicated by the prefix hydroxy-, an $—NH_2$ group of an amine is indicated by amino-, and an =O group of an aldehyde or ketone is indicated by oxo-.

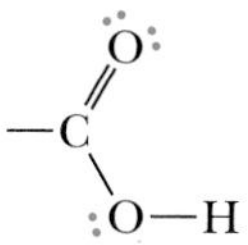

$$—C(=\ddot{O}:)—\ddot{O}—H \qquad —COOH \qquad —CO_2H$$

Leaves of the rhubarb plant contain the poison oxalic acid as its potassium or sodium salt. *(© Hans Reinhard/ OKAPIA/Photo Researchers, Inc.)*

Dicarboxylic acids are named by adding the suffix -dioic acid to the name of the carbon chain that contains both carboxyl groups. The numbers of the carboxyl carbons are not indicated because they can be only at the ends of the parent chain. Following are IUPAC names and common names for several important aliphatic dicarboxylic acids. The name oxalic acid is derived from one of its sources in the biological world, namely plants of the genus *Oxalis,* one of which is rhubarb. Adipic acid is one of the two monomers required for the synthesis of the polymer nylon 66. In 1995, the U.S. chemical industry produced 1.8 billion pounds of adipic acid, solely for the synthesis of nylon 66.

$HOC(=O)-C(=O)OH$ — Ethanedioic acid (Oxalic acid)

$HOC(=O)CH_2C(=O)OH$ — Propanedioic acid (Malonic acid)

$HOC(=O)CH_2CH_2C(=O)OH$ — Butanedioic acid (Succinic acid)

$HOC(=O)CH_2CH_2CH_2C(=O)OH$ — Pentanedioic acid (Glutaric acid)

$HOC(=O)CH_2CH_2CH_2CH_2C(=O)OH$ — Hexanedioic acid (Adipic acid)

A carboxylic acid containing a carboxyl group attached to a cycloalkane ring is named by giving the name of the ring and adding the suffix -carboxylic acid. The atoms of the ring are numbered beginning with the carbon bearing the $—CO_2H$ group.

2-Cyclohexenecarboxylic acid

trans-1,3-Cyclopentanedicarboxylic acid

The simplest aromatic carboxylic acid is benzoic acid. Derivatives are named by using numbers and prefixes to show the presence and location of substituents relative to the carboxyl group. Certain aromatic carboxylic acids have common names by which they are more usually known. For example, 2-hydroxybenzoic acid is more often called salicylic acid, a name derived from the fact that this aromatic carboxylic acid was first obtained from the bark of the willow, a tree of the genus *Salix.*

Benzoic acid

2-Hydroxybenzoic acid (Salicylic acid)

Aromatic dicarboxylic acids are named by adding the words "dicarboxylic acid" to "benzene." Following are structural formulas for 1,2-benzenedicarboxylic acid and 1,4-benzenedicarboxylic acid. Each is more usually known by its common name: phthalic acid and terephthalic acid, respectively. Terephthalic acid is one of

TABLE 12.1 Several Aliphatic Carboxylic Acids and Their Common Names

Structure	IUPAC Name	Common Name	Derivation
HCO_2H	methanoic acid	formic acid	Latin: *formica,* ant
CH_3CO_2H	ethanoic acid	acetic acid	Latin: *acetum,* vinegar
$CH_3CH_2CO_2H$	propanoic acid	propionic acid	Greek: *propion,* first fat
$CH_3(CH_2)_2CO_2H$	butanoic acid	butyric acid	Latin: *butyrum,* butter
$CH_3(CH_2)_3CO_2H$	pentanoic acid	valeric acid	Latin: *valeriana,* a flowering plant
$CH_3(CH_2)_4CO_2H$	hexanoic acid	caproic acid	Latin: *caper,* goat
$CH_3(CH_2)_6CO_2H$	octanoic acid	caprylic acid	Latin: *caper,* goat
$CH_3(CH_2)_8CO_2H$	decanoic acid	capric acid	Latin: *caper,* goat
$CH_3(CH_2)_{10}CO_2H$	dodecanoic acid	lauric acid	Latin: *laurus,* laurel
$CH_3(CH_2)_{12}CO_2H$	tetradecanoic acid	myristic acid	Greek: *myristikos,* fragrant
$CH_3(CH_2)_{14}CO_2H$	hexadecanoic acid	palmitic acid	Latin: *palma,* palm tree
$CH_3(CH_2)_{16}CO_2H$	octadecanoic acid	stearic acid	Greek: *stear,* solid fat
$CH_3(CH_2)_{18}CO_2H$	eicosanoic acid	arachidic acid	Greek: *arachis,* peanut

Formic acid was first obtained in 1670 from the destructive distillation of ants, whose Latin genus is *Formica.* It is one of the components of the venom injected by stinging ants. *(Ted Nelson/Dembinsky Photo Associates)*

the two organic components required for the synthesis of the textile fiber known as Dacron polyester (Section 15.4B).

CO_2H / CO_2H — 1,2-Benzenedicarboxylic acid (Phthalic acid)

CO_2H / CO_2H — 1,4-Benzenedicarboxylic acid (Terephthalic acid)

The peanut plant, from which arachidic acid was first isolated. *(Kenneth W. Fink/Photo Researchers, Inc.)*

B. Common Names

Aliphatic carboxylic acids, many of which were known long before the development of structural theory and IUPAC nomenclature, are named according to their source or for some characteristic property. Table 12.1 lists several of the unbranched aliphatic carboxylic acids found in the biological world along with the common name of each. Those of 16, 18, and 20 carbon atoms are particularly abundant in fats and oils (Section 17.1) and the phospholipid components of biological membranes (Section 17.5).

When common names are used, the Greek letters α, β, γ, and δ are often added as a prefix to locate substituents. The α-position in a carboxylic acid is the one next to the carboxyl group; an α-substituent in a common name is equivalent to a 2-substituent in an IUPAC name.

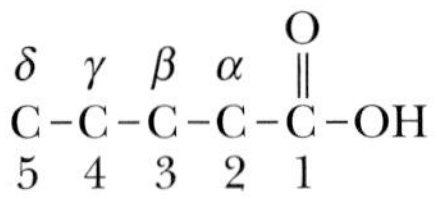

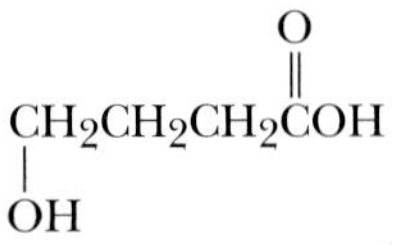

δ γ β α O ‖ C–C–C–C–C–OH 5 4 3 2 1

$CH_2CH_2CH_2COH$ with OH (4-Hydroxybutanoic acid (γ-Hydroxybutyric acid))

4-Hydroxybutanoic acid
(γ-Hydroxybutyric acid)

H_3C, H, NH_2, OH, O

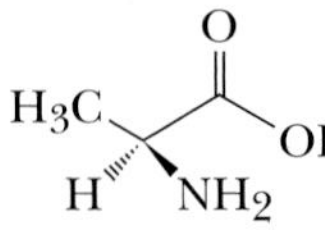

2-Aminopropanoic acid
(α-Aminopropionic acid,
Alanine)

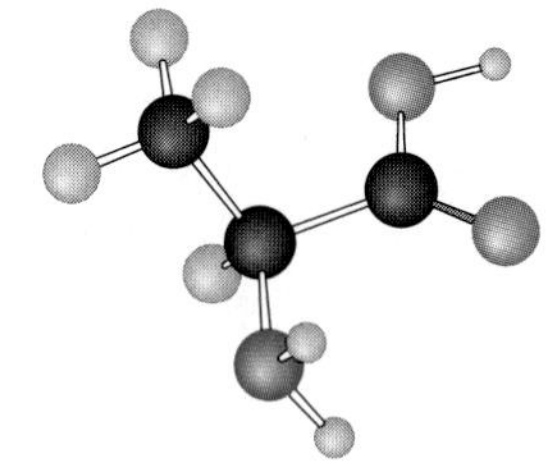

In common names, the presence of a ketone carbonyl in a substituted carboxylic acid is indicated by the prefix keto-, illustrated by the common name β-ketobutyric acid.

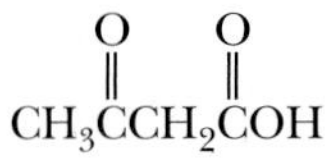

CH_3CCH_2COH

3-oxobutanoic acid
(β-Ketobutyric acid;
acetoacetic acid)

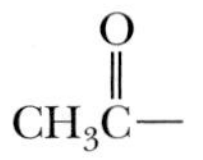

$CH_3C—$

Acetyl group
(Aceto group)

Aceto group A $CH_3CO—$ group.

This substituted carboxylic acid is also named acetoacetic acid. In deriving this name, 3-oxobutanoic acid is regarded as a substituted acetic acid, and the $CH_3CO—$ substituent is named an **aceto group.**

EXAMPLE 12.1

Write the IUPAC name for each carboxylic acid.

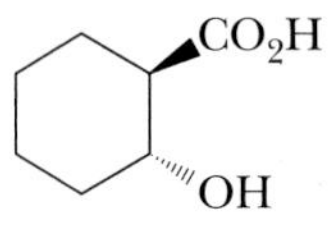

(a) 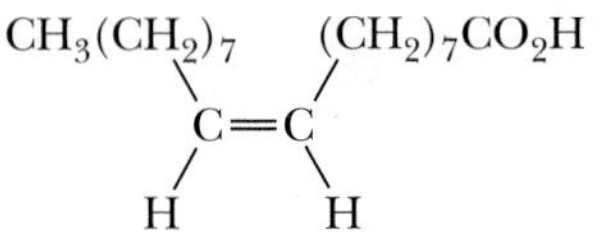 $CH_3(CH_2)_7$, $(CH_2)_7CO_2H$, C=C, H, H

(b) CO_2H, OH

(c) OH, H, C, CH_3, CO_2H

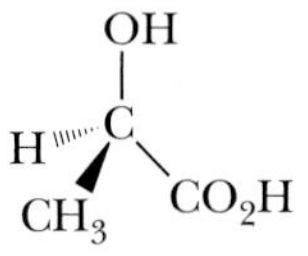

(d) $ClCH_2CO_2H$

SOLUTION

(a) *cis*-9-Octadecenoic acid (oleic acid)
(b) *trans*-2-Hydroxycyclohexanecarboxylic acid
(c) (R)-2-Hydroxypropanoic acid [(R)-lactic acid]
(d) Chloroacetic acid

Practice Problem 12.1

Each of these compounds has a well-recognized common name. A derivative of glyceric acid is an intermediate in glycolysis. Maleic acid is an intermediate in the tricarboxylic acid (TCA) cycle. Mevalonic acid is an intermediate in the biosynthesis of steroids (Section 17.4). Write the IUPAC name for each compound. Be certain to show configuration for each.

(a) 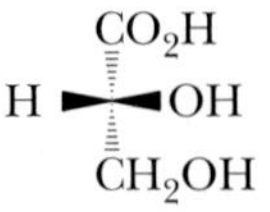 CO_2H, H, OH, CH_2OH

Glyceric acid

(b) 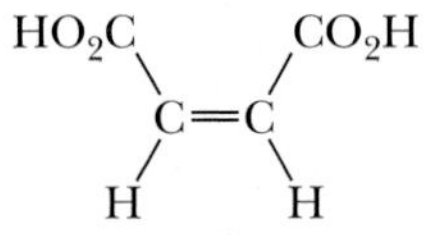 HO_2C, CO_2H, C=C, H, H

Maleic acid

(c) 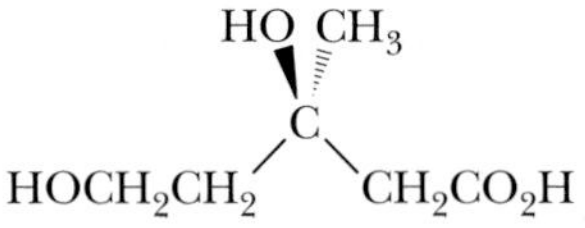 HO, CH_3, C, $HOCH_2CH_2$, CH_2CO_2H

Mevalonic acid

CHEMICAL CONNECTIONS

Abscisic Acid, a Potent Plant Growth Regulator

Throughout their growth, plants are exposed to environmental stresses such as drought, waterlogged soil, pathogenic fungi, and feeding insects and animals. It is not surprising, therefore, that they have developed a chemistry to protect themselves against these stresses. Abscisic acid (*The Merck Index,* 12th ed., #7), one of these chemicals, is probably universally distributed in higher plants and has a variety of actions. This substance promotes abscission (leaf fall), development of dormancy in buds, and formation of potato tubers. It has also been discovered that during times of lack of adequate water for growth, the concentration of abscisic acid in tomato plants increases by over 50-fold. Not only does the increased concentration of abscisic acid inhibit the hormones responsible for extension growth and cell division, but its presence in leaves also leads to closure to soma and thus retards loss of water. Thus, plants protect themselves against times of drought by producing a chemical that conserves energy, reduces the rate of growth, and reduces loss of water. (See S. T. C. Wright and R. W. P. Hiron, *Nature (London),* **224,** 719, 1969.)

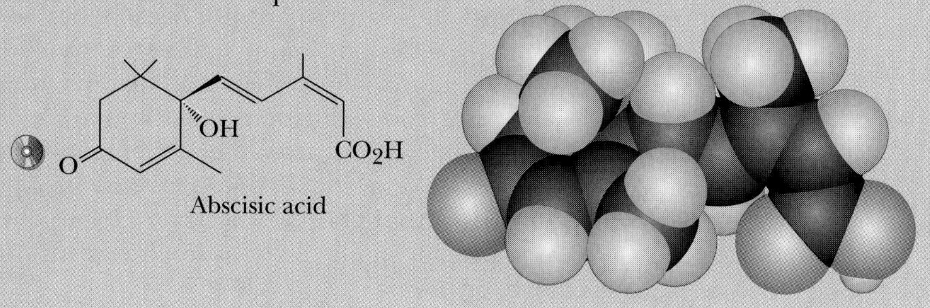

Abscisic acid

12.3 PHYSICAL PROPERTIES

In the liquid and solid states, carboxylic acids are associated by intermolecular hydrogen bonding into dimers, as shown for acetic acid. Carboxylic acids have significantly higher boiling points than other types of organic compounds of comparable molecular weight, such as alcohols, aldehydes, and ketones. For example, butanoic acid (Table 12.2) has a higher boiling point than either 1-pentanol or pentanal. The higher boiling points of carboxylic acids result from their polarity and from the fact that they form very strong intermolecular hydrogen bonds.

hydrogen bonding in the dimer

$H_3C-C(=O^{\delta-}\cdots H^{\delta+}-O)(-O-H^{\delta+}\cdots O^{\delta-}=)C-CH_3$

TABLE 12.2 Boiling Points and Solubilities in Water of Selected Carboxylic Acids, Alcohols, and Aldehydes of Comparable Molecular Weight

Structure	Name	Molecular Weight	Boiling Point (°C)	Solubility (g/100 mL H_2O)
CH_3CO_2H	acetic acid	60.5	118	infinite
$CH_3CH_2CH_2OH$	1-propanol	60.1	97	infinite
CH_3CH_2CHO	propanal	58.1	48	16
$CH_3(CH_2)_2CO_2H$	butanoic acid	88.1	163	infinite
$CH_3(CH_2)_3CH_2OH$	1-pentanol	88.1	137	2.3
$CH_3(CH_2)_3CHO$	pentanal	86.1	103	slight
$CH_3(CH_2)_4CO_2H$	hexanoic acid	116.2	205	1.0
$CH_3(CH_2)_5CH_2OH$	1-heptanol	116.2	176	0.2
$CH_3(CH_2)_5CHO$	heptanal	114.1	153	0.1

Carboxylic acids also interact with water molecules by hydrogen bonding through both the carbonyl and hydroxyl groups. Because of these hydrogen-bonding interactions, carboxylic acids are more soluble in water than are alcohols, ethers, aldehydes, and ketones of comparable molecular weight. The solubility of a carboxylic acid in water decreases as its molecular weight increases. We account for this trend in the following way. A carboxylic acid consists of two regions of distinctly different polarity: a polar hydrophilic carboxyl group and, except for formic acid, a nonpolar hydrophobic hydrocarbon chain. The **hydrophilic** carboxyl group increases water solubility; the **hydrophobic** hydrocarbon chain decreases water solubility.

Hydrophilic From the Greek meaning water-loving.

Hydrophobic From the Greek meaning water-hating.

The first four aliphatic carboxylic acids (methanoic, ethanoic, propanoic, and butanoic acids) are infinitely soluble in water because the hydrophobic character of the hydrocarbon chain is more than counterbalanced by the hydrophilic character of the carboxyl group. As the size of the hydrocarbon chain increases relative to the size of the carboxyl group, water solubility decreases. The solubility of hexanoic acid (six carbons) in water is 1.0 g/100 g water. That of decanoic acid (ten carbons) is only 0.2 g/100 g water.

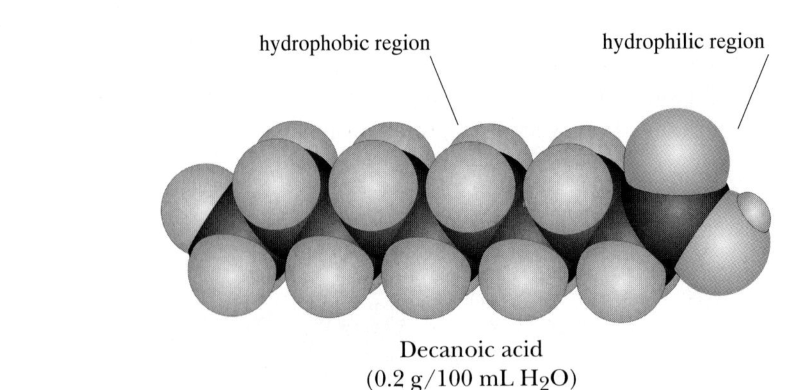

Decanoic acid
(0.2 g/100 mL H_2O)

12.4 ACIDITY

A. Acid Ionization Constants

Carboxylic acids are weak acids. Values of K_a for most unsubstituted aliphatic and aromatic carboxylic acids fall within the range 10^{-4} to 10^{-5}. The value of K_a for acetic acid, for example, is 1.74×10^{-5}. Its pK_a is 4.76.

$$CH_3CO_2H + H_2O \rightleftharpoons CH_3CO_2^- + H_3O^+$$

$$K_a = \frac{[CH_3CO_2^-][H_3O^+]}{[CH_3CO_2H]} = 1.74 \times 10^{-5}$$

$$pK_a = 4.76$$

As we discussed in Section 2.3, two factors contribute to the greater acidity of carboxylic acids (pK_a 4–5) compared with alcohols (pK_a 16–18). First, the electron-withdrawing inductive effect of the carbonyl group weakens the adjacent O—H bond and facilitates its ionization. Second, resonance stabilizes the carboxylate anion by delocalizing its negative charge. Neither of these effects exists in alcohols.

Electron-withdrawing substituents near the carboxyl group increase the acidity of carboxylic acids, often by several orders of magnitude. Compare, for example, the acidities of acetic acid and the halogen-substituted acetic acids. As the electronegativity of the halogen increases, its electron-withdrawing inductive effect increases, and the strength of the halogen-substituted acid increases. Fluoroacetic acid is the strongest of the monohalogenated acetic acids.

Formula:	CH_3CO_2H	ICH_2CO_2H	$BrCH_2CO_2H$	$ClCH_2CO_2H$	FCH_2CO_2H
Name:	Acetic acid	Iodoacetic acid	Bromoacetic acid	Chloroacetic acid	Fluoroacetic acid
pK_a:	4.76	3.18	2.90	2.86	2.59

Increasing acid strength →

To see the effects of multiple halogen substitution, compare the values of pK_a for the mono-, di-, and trichloroacetic acids. Both dichloroacetic acid and trichloroacetic acid are stronger acids than H_3PO_4 (pK_{a1} 2.1).

Formula:	CH_3CO_2H	$ClCH_2CO_2H$	Cl_2CHCO_2H	Cl_3CCO_2H
Name:	Acetic acid	Chloroacetic acid	Dichloroacetic acid	Trichloroacetic acid
pK_a:	4.76	2.86	1.48	0.70

Increasing acid strength →

The inductive effect of halogen substitution falls off rather rapidly when distance from the carboxyl group increases. Although the acid ionization constant for 2-chlorobutanoic acid ($K_a = 1.5 \times 10^{-3}$) is 100 times that for butanoic acid, the acid ionization constant for 4-chlorobutanoic acid ($K_a = 3.0 \times 10^{-5}$) is only about twice that for butanoic acid.

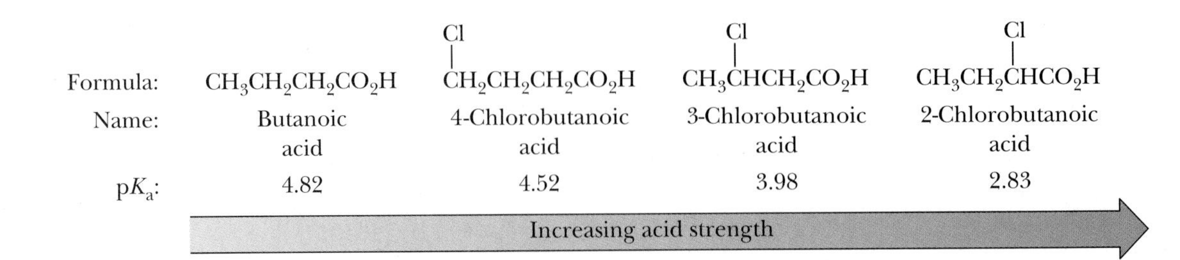

Formula: $CH_3CH_2CH_2CO_2H$ | $ClCH_2CH_2CH_2CO_2H$ | $CH_3CH(Cl)CH_2CO_2H$ | $CH_3CH_2CH(Cl)CO_2H$
Name: Butanoic acid | 4-Chlorobutanoic acid | 3-Chlorobutanoic acid | 2-Chlorobutanoic acid
pK_a: 4.82 | 4.52 | 3.98 | 2.83
Increasing acid strength

Example 12.2

Which acid in each set is the stronger?

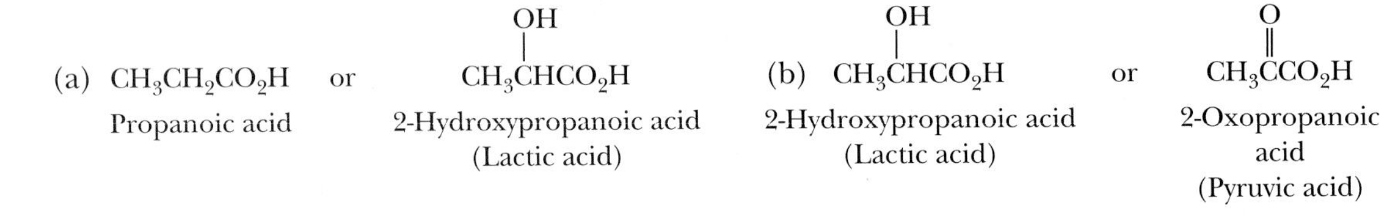

Solution

(a) 2-Hydroxypropanoic acid (pK_a 3.08) is a stronger acid than propanoic acid (pK_a 4.87) because of the electron-withdrawing inductive effect of the hydroxyl oxygen.

(b) 2-Oxopropanoic acid (pK_a 2.06) is a stronger acid than 2-hydroxypropanoic acid (pK_a 3.08) because of the greater electron-withdrawing inductive effect of the carbonyl oxygen compared with the hydroxyl oxygen.

Practice Problem 12.2

Match each compound with its appropriate pK_a value.

$CH_3C(CH_3)_2CO_2H$ 2,2-Dimethylpropanoic acid

CF_3CO_2H Trifluoroacetic acid

$CH_3CH(OH)CO_2H$ 2-Hydroxypropanoic acid (Lactic acid)

pK_a values = 5.03, 3.08, and 0.22

Sodium benzoate and calcium propanoate are used as preservatives in baked goods. *(Charles D. Winters)*

B. Reaction with Bases

All carboxylic acids, whether soluble or insoluble in water, react with NaOH, KOH, and other strong bases to form water-soluble salts. Sodium benzoate (*The Merck Index,* 12th ed., #8725), a fungal growth inhibitor, is often added to baked goods "to retard spoilage." Calcium propanoate (*The Merck Index,* 12th ed., #1745) is also used for the same purpose. Carboxylic acids also form water-soluble salts with ammonia and amines.

$$C_6H_5CO_2H + NaOH \xrightarrow[H_2O]{} C_6H_5CO_2^-Na^+ + H_2O$$

Benzoic acid (slightly soluble in water) — Sodium benzoate (60 g/100 mL water)

$$C_6H_5CO_2H + NH_3 \xrightarrow[H_2O]{} C_6H_5CO_2^-\ NH_4^+$$

Benzoic acid (slightly soluble in water) — Ammonium benzoate (20 g/100 mL water)

As described in Section 2.4, carboxylic acids react with sodium bicarbonate and sodium carbonate to form water-soluble sodium salts and carbonic acid (a weaker acid). Carbonic acid, in turn, decomposes to give water and carbon dioxide, which evolves as a gas.

$$CH_3CO_2H + Na^+HCO_3^- \longrightarrow CH_3CO_2^-Na^+ + CO_2 + H_2O$$

Salts of carboxylic acids are named in the same manner as are the salts of inorganic acids; the cation is named first and then the anion. The name of the anion is derived from the name of the carboxylic acid by dropping the suffix -ic acid and adding the suffix -ate. For example, $CH_3CH_2CO_2^-\ Na^+$ is named sodium propanoate and $CH_3(CH_2)_{14}CO_2^-Na^+$ is named sodium hexadecanoate (sodium palmitate).

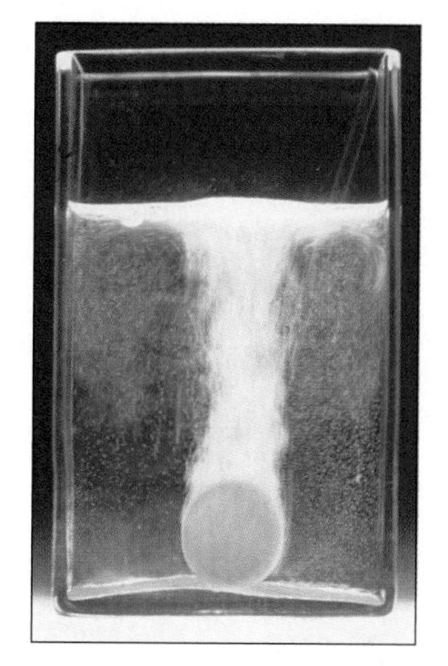

A commercial remedy for excess stomach acid. The bubbles are carbon dioxide, formed from the reaction between citric acid and sodium bicarbonate. *(Charles D. Winters)*

EXAMPLE 12.3

Complete each acid-base reaction and name the salt formed.

(a) $CH_3(CH_2)_2CO_2H + NaOH \longrightarrow$ (b) $CH_3CH(OH)CO_2H + NaHCO_3 \longrightarrow$

SOLUTION

Each carboxylic acid is converted to its sodium salt. In (b), carbonic acid is formed, which decomposes to carbon dioxide and water.

(a) $CH_3(CH_2)_2CO_2H + NaOH \longrightarrow CH_3(CH_2)_2CO_2^-Na^+ + H_2O$

Butanoic acid — Sodium butanoate

(b) $CH_3CH(OH)CO_2H + NaHCO_3 \longrightarrow CH_3CH(OH)CO_2^-Na^+ + H_2O + CO_2$

2-Hydroxypropanoic acid (Lactic acid) — Sodium 2-hydroxypropanoate (Sodium lactate)

Practice Problem 12.3

Write an equation for the reaction of each acid in Example 12.3 with ammonia and name the salt formed.

A consequence of the water solubility of carboxylic acid salts is that water-insoluble carboxylic acids can be converted to water-soluble ammonium or alkali metal salts and then extracted into aqueous solution. The salt, in turn, can be transformed into the free carboxylic acid by addition of HCl, H_2SO_4, or other strong acid. These reactions allow for easy separation of carboxylic acids from water-insoluble neutral compounds.

Shown in Figure 12.1 is a flow chart for separation of benzoic acid, a water-insoluble carboxylic acid, from benzyl alcohol, a nonacidic compound. First, the mixture of benzoic acid and benzyl alcohol is dissolved in diethyl ether. When the ether solution is shaken with aqueous NaOH, benzoic acid is converted to its water-soluble sodium salt. Then the ether and aqueous phases are separated. The ether solution is distilled, yielding first diethyl ether (bp 35°C) and then benzyl alcohol (bp 205°C). The aqueous solution is acidified with HCl, and benzoic acid precipitates as a water-insoluble solid (mp 122°C) and is recovered by filtration.

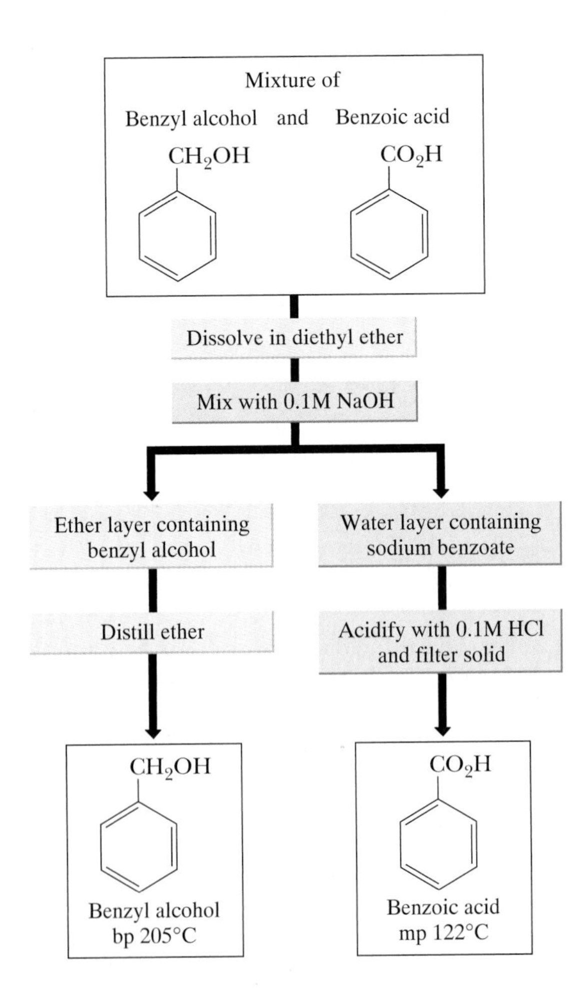

Figure 12.1
Flow chart for separation of benzoic acid from benzyl alcohol.

12.5 REDUCTION

The carboxyl group is one of the organic functional groups most resistant to reduction. It is not affected by catalytic reduction under conditions that easily reduce aldehydes and ketones to alcohols and that reduce alkenes to alkanes. The most common reagent for the reduction of a carboxylic acid to a primary alcohol is the very powerful reducing agent lithium aluminum hydride (Section 11.10B).

A. Reduction of a Carboxyl Group

Lithium aluminum hydride, $LiAlH_4$ (LAH), reduces a carboxyl group to a primary alcohol in excellent yield. Reduction is most commonly carried out in diethyl ether or tetrahydrofuran (THF). The initial product is an aluminum alkoxide, which is then treated with water to give the primary alcohol and lithium and aluminum hydroxides. These hydroxides are insoluble in diethyl ether or THF and are removed by filtration. Evaporation of the solvent yields the primary alcohol.

$$\text{3-Cyclopentenecarboxylic acid (C}_5\text{H}_7\text{-COOH)} \xrightarrow[\text{2. } H_2O]{\text{1. } LiAlH_4\text{, ether}} \text{4-Hydroxymethylcyclopentene (C}_5\text{H}_7\text{-CH}_2\text{OH)} + LiOH + Al(OH)_3$$

3-Cyclopentene-carboxylic acid → 4-Hydroxymethyl-cyclopentene

Alkenes are generally not affected by metal hydride-reducing reagents. These reagents function as hydride ion donors, that is, as nucleophiles, and alkenes are not attacked by nucleophiles.

B. Selective Reduction of Other Functional Groups

Because carboxyl groups are not affected by the conditions of catalytic reduction, which normally reduce aldehydes, ketones, and alkenes, it is possible to reduce these functional groups to alcohols or alkanes selectively in the presence of carboxyl groups.

$$CH_3\overset{O}{\overset{\|}{C}}CH_2CH_2CH_2\overset{O}{\overset{\|}{C}}OH + H_2 \xrightarrow[25°C,\ 2\ atm]{Pt} CH_3\overset{OH}{\overset{|}{C}}HCH_2CH_2CH_2\overset{O}{\overset{\|}{C}}OH$$

5-Oxohexanoic acid → 5-Hydroxyhexanoic acid

We saw in Section 11.10B that aldehydes and ketones are reduced to alcohols by both $LiAlH_4$ and $NaBH_4$. Only $LiAlH_4$, however, reduces carboxyl groups. Thus, it is possible to reduce an aldehyde or ketone carbonyl group selectively in the presence of a carboxyl group by using the less reactive $NaBH_4$ as the reducing agent.

$$C_6H_5\overset{O}{\overset{\|}{C}}CH_2CH_2CH_2\overset{O}{\overset{\|}{C}}OH \xrightarrow[2.\ H_2O]{1.\ NaBH_4} C_6H_5\overset{OH}{\overset{|}{C}}HCH_2CH_2CH_2\overset{O}{\overset{\|}{C}}OH$$

5-Oxo-5-phenylpentanoic acid → 5-Hydroxy-5-phenylpentanoic acid

CHEMICAL CONNECTIONS

From Willow Bark to Aspirin and Beyond

The first drug developed for widespread use was aspirin, today's most common pain reliever. Americans alone consume approximately 80 billion tablets of aspirin a year! The story of the development of this modern pain reliever goes back more than 2000 years. In 400 B.C., the Greek physician Hippocrates recommended chewing bark of the willow tree to alleviate the pain of childbirth and to treat eye infections. In 1763, the Reverend Edward Stone wrote, "There is a bark of an English tree, which I have found by experience to be a powerful astringent, and very efficacious in curing anguish and intermitting disorders."

The active component of willow bark was found to be salicin, a compound composed of salicyl alcohol joined to a unit of β-D-glucose (Section 16.4). It was discovered that salicin could be hydrolyzed in aqueous acid to give salicyl alcohol, which could then be oxidized to salicylic acid. Salicylic acid proved to be an even more effective reliever of pain, fever, and inflammation than salicin, and without its extremely bitter taste. Unfortunately, patients quickly recognized salicylic acid's major side effect; it causes severe irritation of the mucous membrane lining of the stomach.

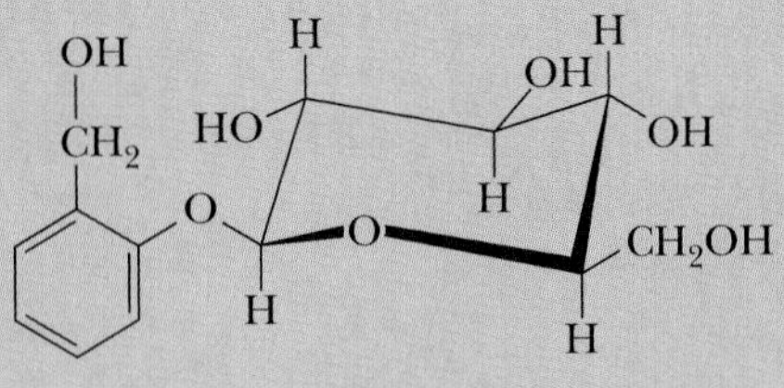

Salicin
(*The Merck Index*, 12th ed., #8476)

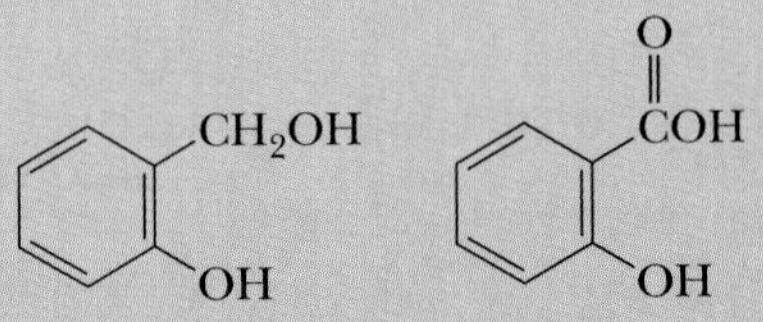

Salicyl alcohol   Salicylic acid

In the search for less irritating but still effective derivatives of salicylic acid, Felix Hofmann at the Bayer division of I. G. Farben in Germany in 1883 prepared acetylsalicylic acid. Heinrich Dreser, the director of research at Bayer, gave it the name aspirin, a word derived from the German spirsäure (salicylic acid) with the initial "a" for the acetyl group.

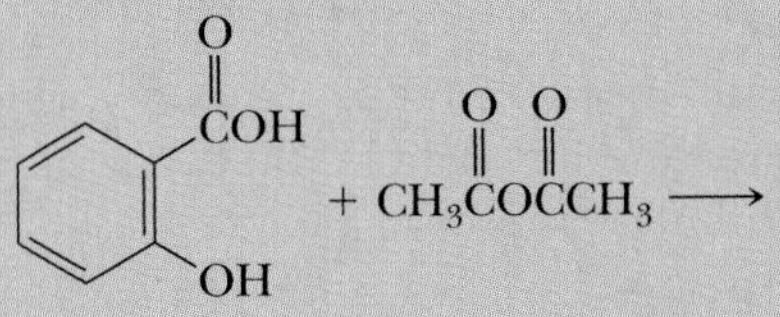

Salicylic acid   Acetic anhydride

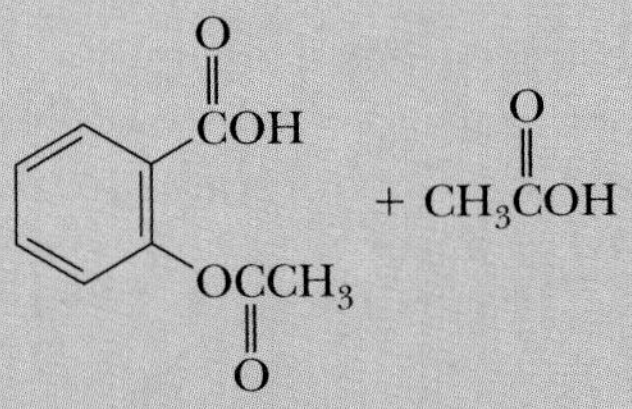

Acetyl salicylate
(Aspirin)

Aspirin proved to be less irritating to the stomach than salicylic acid and also more effective in relieving the pain and inflammation of rheumatoid arthritis. Large-scale production of aspirin began at Bayer in 1899.

In a search for even more effective and less irritating analgesics and anti-inflammatory drugs, in the 1960s the Boots Pure Drug Company in England studied compounds related in structure to salicylic acid. They discovered an even more potent compound, which they named ibuprofen. Soon thereafter, Syntex Corporation in the United States developed naproxen and Rhône-Poulenc in France developed ketoprofen. Notice that each compound has one stereocenter and can exist as a pair of enantiomers. For each drug, the physiologically active form is the S enantiomer. Even though the R enantiomer has none of the analgesic and anti-inflammatory activity, it is converted in the body to the active S enantiomer.

(S)-(+)-Ibuprofen
(The Merck Index,
12th ed., #4925)

(S)-Naproxen
(The Merck Index,
12th ed., #6504)

(S)-Ketoprofen
(The Merck Index,
12th ed., #5316)

In the 1960s, it was discovered that aspirin acts by inhibiting cyclooxygenase (COX), a key enzyme in the conversion of arachidonic acid to prostaglandins (Section 17.3). With this discovery, it became clear why only one enantiomer of ibuprofen, naproxen, and ketoprofen is active: only the S enantiomer of each has the correct handedness to bind to COX and inhibit its activity.

The discovery that these drugs owe their effectiveness to inhibition of COX opened an entirely new avenue for drug research. If we know more about the structure and function of this key enzyme, might it be possible to design and discover even more effective nonsteroidal anti-inflammatory drugs for the treatment of rheumatoid arthritis and other inflammatory diseases?

And so continues the story that began with the discovery of the beneficial effects of chewing willow bark.

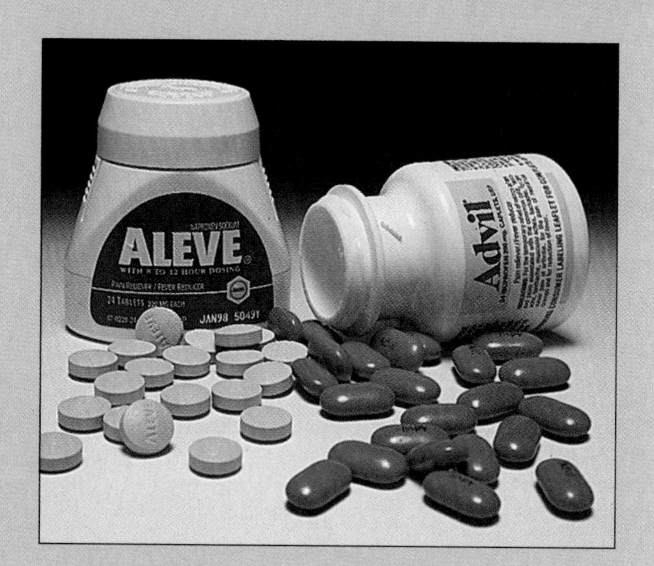

Two nonprescription pain relievers. *(Charles D. Winters)*

12.6 ESTERIFICATION

A. Fischer Esterification

Fischer esterification The process of forming an ester by refluxing a carboxylic acid and an alcohol in the presence of an acid catalyst, commonly sulfuric acid.

Esters can be prepared by treatment of a carboxylic acid with an alcohol in the presence of an acid catalyst, most commonly concentrated sulfuric acid. Conversion of a carboxylic acid and an alcohol to an ester is given the special name **Fischer esterification,** after the German chemist, Emil Fischer (1852–1919). As an example of Fischer esterification, treatment of acetic acid with ethanol in the presence of concentrated sulfuric acid gives ethyl acetate and water. We study the structure, nomenclature, and reactions of esters in detail in Chapter 13. In this chapter, we discuss only their preparation from carboxylic acids.

$$CH_3\overset{O}{\overset{\|}{C}}OH + CH_3CH_2OH \underset{}{\overset{H_2SO_4}{\rightleftharpoons}} CH_3\overset{O}{\overset{\|}{C}}OCH_2CH_3 + H_2O$$

Ethanoic acid (Acetic acid) — Ethanol (Ethyl alcohol) — Ethyl ethanoate (Ethyl acetate)

Acid-catalyzed esterification is reversible, and generally, at equilibrium, the quantities of remaining carboxylic acid and alcoholare appreciable. By control of reaction conditions, however, it is possible to use Fischer esterification to prepare esters in high yields. If the alcohol is inexpensive compared with the carboxylic acid, a large excess of the alcohol can be used to drive the equilibrium to the right and achieve a high conversion of carboxylic acid to its ester.

Household products containing ethyl acetate include nail polish remover and some glues. *(Charles D. Winters)*

EXAMPLE 12.4

Complete these Fischer esterification reactions.

(a) $C_6H_5\overset{O}{\overset{\|}{C}}OH + CH_3OH \overset{H^+}{\rightleftharpoons}$ (b) $HO\overset{O}{\overset{\|}{C}}CH_2CH_2\overset{O}{\overset{\|}{C}}OH + 2CH_3CH_2OH \overset{H^+}{\rightleftharpoons}$

SOLUTION

Here is a structural formula for the ester produced in each reaction.

(a) $C_6H_5\overset{O}{\overset{\|}{C}}OCH_3$ (b) $CH_3CH_2O\overset{O}{\overset{\|}{C}}CH_2CH_2\overset{O}{\overset{\|}{C}}OCH_2CH_3$

Practice Problem 12.4

Complete these Fischer esterification reactions.

(a) $HOCH_2CH_2CH_2\overset{O}{\overset{\|}{C}}OH \overset{H^+}{\rightleftharpoons}$ a cyclic ester (b) $CH_3\underset{CH_3}{\underset{|}{C}H}\overset{O}{\overset{\|}{C}}OH + HO\text{—}C_6H_{11} \overset{H^+}{\rightleftharpoons}$

CHEMICAL CONNECTIONS

Esters as Flavoring Agents

Flavoring agents are the largest class of food additives. At the present time, over a thousand synthetic and natural flavors are available. The majority of these are concentrates or extracts from the material whose flavor is desired and are often complex mixtures of from tens to hundreds of compounds. A number of ester flavoring agents are synthesized industrially. Many have flavors very close to the target flavor and adding only one or a few of them is sufficient to make ice cream, soft drinks, or candy taste natural.

Structure	Name	Flavor
$HCO_2CH_2CH_3$	ethyl formate	rum
$CH_3CO_2CH_2CH_2CH(CH_3)_2$	isopentyl acetate	banana
$CH_3CO_2CH_2(CH_2)_6CH_3$	octyl acetate	orange
$CH_3(CH_2)_2CO_2CH_3$	methyl butanoate	apple
$CH_3(CH_2)_2CO_2CH_2CH_3$	ethyl butanoate	pineapple
(benzene ring with CO_2CH_3 and NH_2)	methyl 2-aminobenzoate	grape

Synthetic flavoring agents have tastes very close to the target flavor. *(Charles D. Winters)*

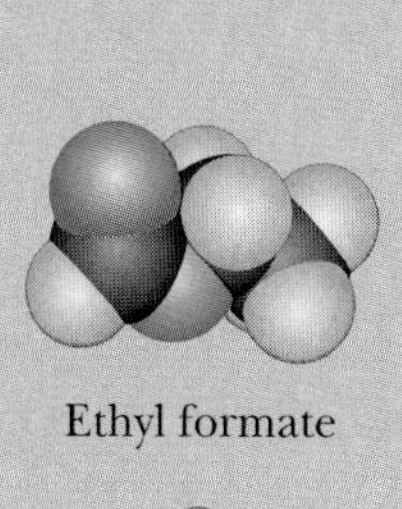

Ethyl formate

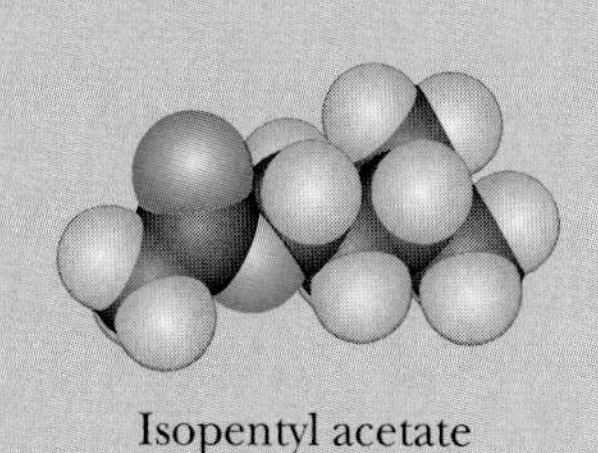

Isopentyl acetate

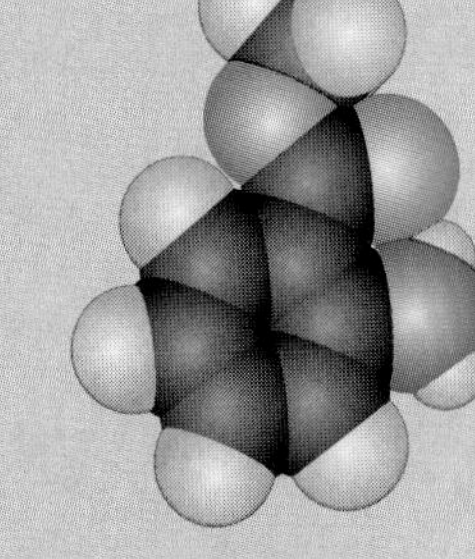

Methyl 2-aminobenzoate

B. The Mechanism of Fischer Esterification

It is important that you understand this mechanism thoroughly because it is a model for many of the reactions of the functional derivatives of carboxylic acids presented in Chapter 13.

MECHANISM Fischer Esterification

① Proton transfer from the acid catalyst to the carbonyl oxygen increases the electrophilicity of the carbonyl carbon . . .

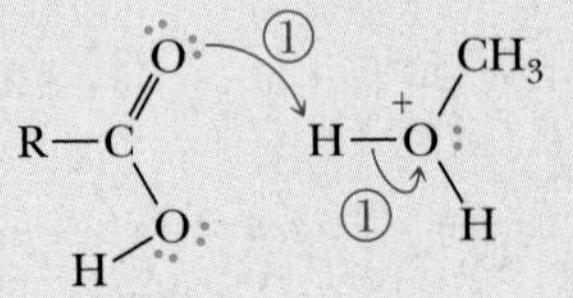

② which is then attacked by the nucleophilic oxygen atom of the alcohol . . .

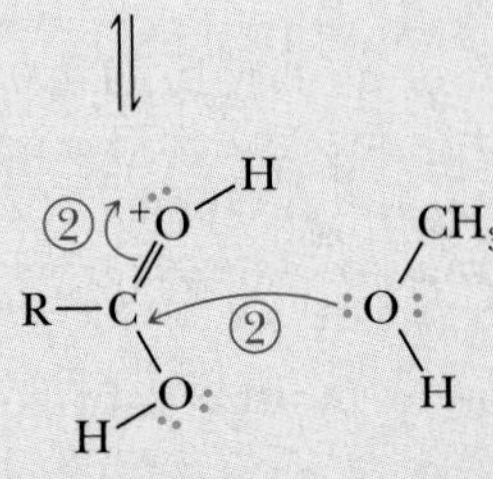

③ to form an oxonium ion.

④ Proton transfer from the oxonium ion to a second molecule of alcohol . . .

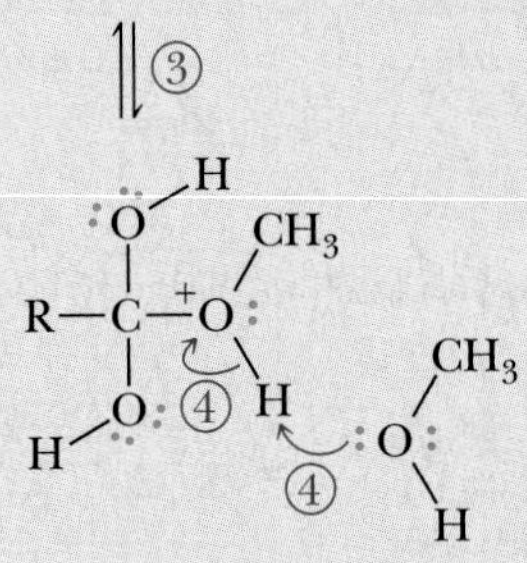

⑤ gives a tetrahedral carbonyl addition intermediate (TCAI).

⑥ Proton transfer to one of the —OH groups of the TCAI . . .

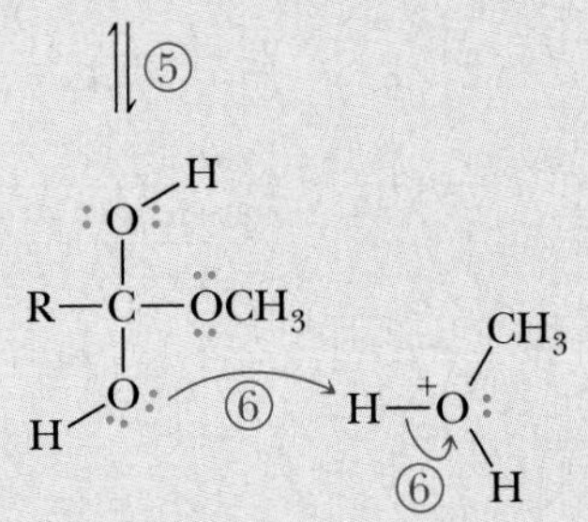

⑦ gives a new oxonium ion.

⑧ Loss of water from this oxonium ion . . .

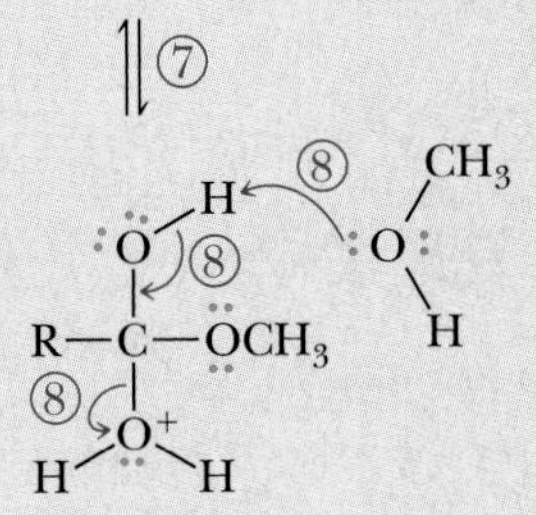

⑨ gives the ester and water.

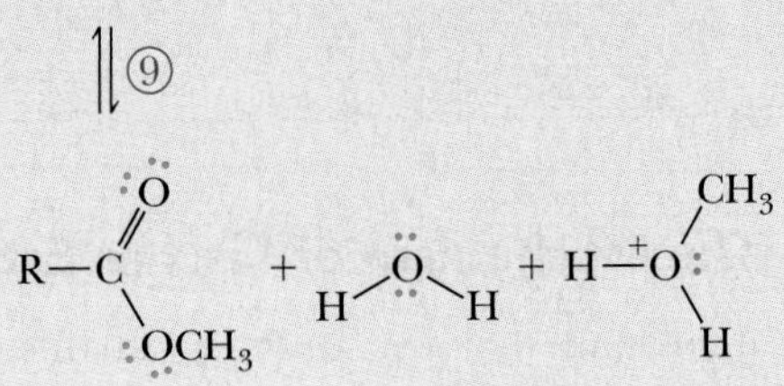

12.7 Conversion to Acid Halides

The functional group of an acid halide is a carbonyl group bonded to a halogen atom. Among the acid halides, acid chlorides are the most frequently used in the laboratory and in industrial organic chemistry.

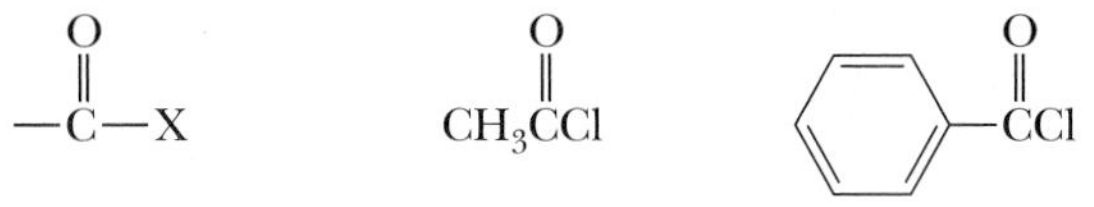

$$-\overset{\overset{\large O}{\|}}{C}-X \qquad CH_3\overset{\overset{\large O}{\|}}{C}Cl \qquad C_6H_5\overset{\overset{\large O}{\|}}{C}Cl$$

Functional group of an acid halide — Acetyl chloride — Benzoyl chloride

We study the nomenclature, structure, and characteristic reactions of acid halides in Chapter 13. In the present chapter, we are concerned only with their synthesis from carboxylic acids.

Acid chlorides are most often prepared by treatment of a carboxylic acid with thionyl chloride, the same reagent used to convert an alcohol to a chloroalkane (Section 8.4D).

$$\underset{\text{Butanoic acid}}{CH_3CH_2CH_2\overset{\overset{\large O}{\|}}{C}OH} + \underset{\text{Thionyl chloride}}{SOCl_2} \longrightarrow \underset{\text{Butanoyl chloride}}{CH_3CH_2CH_2\overset{\overset{\large O}{\|}}{C}Cl} + SO_2 + HCl$$

Example 12.5

Complete each equation.

(a) $CH_3(CH_2)_6CO_2H + SOCl_2 \longrightarrow$

(b) 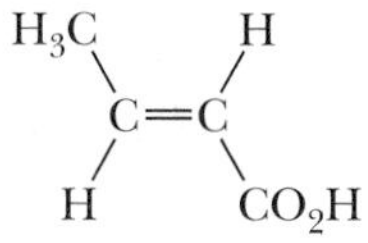 (H$_3$C)(H)C=C(H)(CO$_2$H) + $SOCl_2$ — trans: H_3C and H on one carbon, H and CO_2H on the other

Solution

Following are the products for each reaction.

(a) $CH_3(CH_2)_6\overset{\overset{\large O}{\|}}{C}Cl + SO_2 + HCl$

(b) 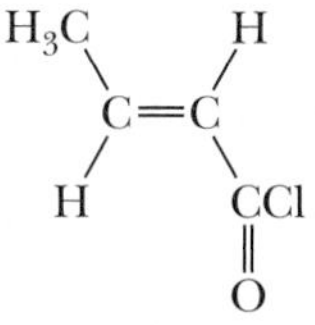 (H$_3$C)(H)C=C(H)(CCl=O) + SO_2 + HCl

Practice Problem 12.5

Complete each equation.

(a) 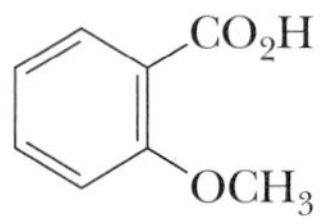 2-methoxybenzoic acid (benzene ring with CO_2H and OCH_3) $+ SOCl_2 \longrightarrow$

(b) 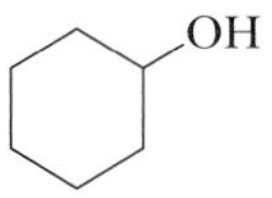 cyclohexanol (cyclohexane ring with OH) $+ SOCl_2 \longrightarrow$

12.8 DECARBOXYLATION

A. Decarboxylation of β-Ketoacids

Decarboxylation Loss of CO_2 from an organic molecule.

Decarboxylation is the loss of CO_2 from the carboxyl group. Almost any carboxylic acid, heated to a very high temperature, undergoes decarboxylation.

$$\mathrm{RCOH}\ (\text{C}=\text{O}) \xrightarrow{\text{decarboxylation}} \mathrm{RH} + \mathrm{CO_2}$$

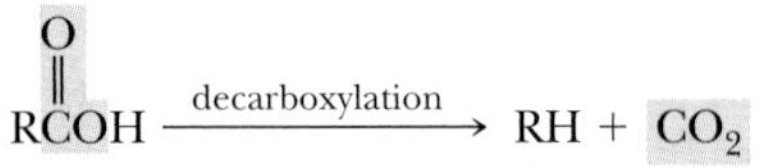

Most carboxylic acids, however, are quite resistant to moderate heat and melt or even boil without decarboxylation. Exceptions are carboxylic acids that have a carbonyl group β to the carboxyl group. This type of carboxylic acid undergoes decarboxylation quite readily on mild heating. For example, when 3-oxobutanoic acid (acetoacetic acid) is heated moderately, it undergoes decarboxylation to give acetone and carbon dioxide.

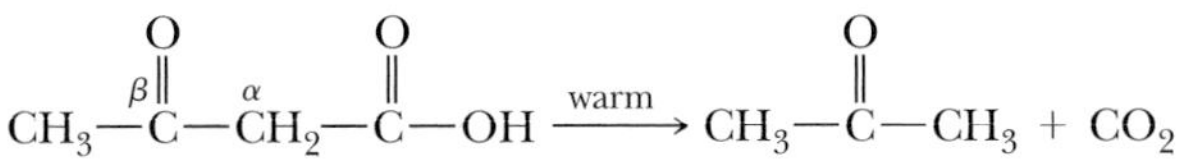

$$\underset{\substack{\text{3-Oxobutanoic acid}\\ \text{(Acetoacetic acid)}}}{\mathrm{CH_3-\overset{\overset{O}{\|}}{\underset{\beta}{C}}-\underset{\alpha}{CH_2}-\overset{\overset{O}{\|}}{C}-OH}} \xrightarrow{\text{warm}} \underset{\text{Acetone}}{\mathrm{CH_3-\overset{\overset{O}{\|}}{C}-CH_3}} + \mathrm{CO_2}$$

Decarboxylation on moderate heating is a unique property of 3-oxocarboxylic acids (β-ketoacids) and is not observed with other classes of ketoacids.

The mechanism for decarboxylation of a β-ketoacid involves the redistribution of six electrons in a cyclic six-membered transition state to give carbon dioxide and an enol. The enol then rearranges to the more stable keto form by keto-enol tautomerism (Section 11.8A).

MECHANISM Decarboxylation of a β-Ketocarboxylic Acid

Step 1: Rearrangement of six electrons in a cyclic six-membered transition state gives carbon dioxide and an enol.

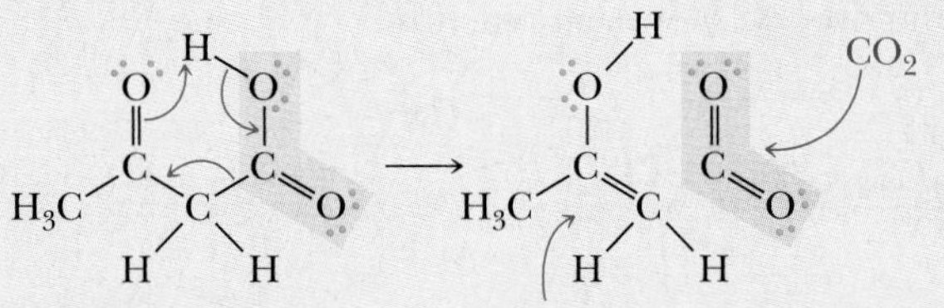

A cyclic six-membered transition state $\longrightarrow$ enol of a ketone + CO_2

Step 2: Keto-enol tautomerism of the enol gives the more stable keto form of the product.

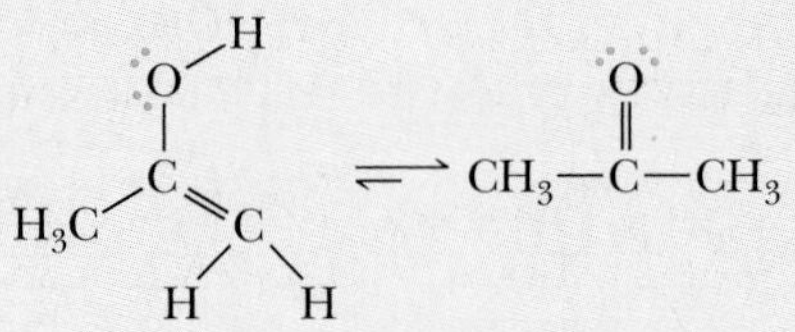

$$\mathrm{H_3C-C(OH)=CH_2} \rightleftharpoons \mathrm{CH_3-\overset{\overset{O}{\|}}{C}-CH_3}$$

CHEMICAL CONNECTIONS

Ketone Bodies and Diabetes

3-Oxobutanoic acid (acetoacetic acid) and its reduction product, 3-hydroxybutanoic acid, are synthesized in the liver from acetyl-CoA, a product of the metabolism of fatty acids (Section 20.3) and certain amino acids. 3-Hydroxybutanoic acid and 3-oxobutanoic acid are known collectively as ketone bodies. The concentration of ketone bodies in the blood of healthy, well-fed humans is approximately 0.01 mM/L. However, in persons suffering from starvation or diabetes mellitus, the concentration of ketone bodies may increase to as much as 500 times normal. Under these conditions, the concentration of acetoacetic acid increases to the point where it undergoes spontaneous decarboxylation to form acetone and carbon dioxide. Acetone is not metabolized by humans and is excreted through the kidneys and the lungs. The odor of acetone is responsible for the characteristic "sweet smell" on the breath of severely diabetic patients.

$CH_3C(=O)CH_2C(=O)OH$

3-Oxobutanoic acid
(Acetoacetic acid)

$CH_3CH(OH)CH_2C(=O)OH$

3-Hydroxybutanoic acid
(β-Hydroxybutyric acid)

An important example of decarboxylation of a β-ketoacid in the biological world occurs during the oxidation of foodstuffs in the tricarboxylic acid (TCA) cycle. One of the intermediates in this cycle is oxalosuccinic acid, which undergoes spontaneous decarboxylation to produce α-ketoglutaric acid. Only one of the three carboxyl groups of oxalosuccinic acid has a carbonyl group in the β-position to it, and it is this carboxyl group that is lost as CO_2.

$$HOC(=O)-CH(CH_2CO_2H)-C(=O)CO_2H \longrightarrow CH_2(CH_2CO_2H)-C(=O)CO_2H + CO_2$$

Only this carboxyl has a C=O beta to it

Oxalosuccinic acid $\qquad$ α-Ketoglutaric acid

B. Decarboxylation of Malonic Acid and Substituted Malonic Acids

The presence of a ketone or aldehyde carbonyl group on the carbon β to the carboxyl group is sufficient to facilitate decarboxylation. In the more general reaction, decarboxylation is facilitated by the presence of any carbonyl group on the

CHEMICAL CONNECTIONS

Sunscreens and Sunblocks

Ultraviolet (UV) radiation penetrating the earth's ozone layer can be divided into two regions: UVB (2900–3200 Å) and UVA (3200–4000 Å). UVB radiation interacts directly on biomolecules of the skin and eyes, causing skin cancer, skin aging, eye damage, and delayed sunburn that appears 12–24 hours after exposure. UVA radiation, by contrast, acts directly through reactive oxygen species and its role in promoting skin cancer is less well understood.

Commercial sunscreen products are rated according to their sun protection factor (SPF), which is defined as the minimum effective dose (MED) of UV radiation in units of joules per square meter (J/m^2) that produces a delayed sunburn on protected skin compared to unprotected skin. In the following equation, protected skin is defined as skin covered by 2 mg/cm^2 of sunscreen.

$$\text{SPF} = \frac{\text{MED on protected skin}}{\text{MED on unprotected skin}}$$

Two types of active ingredients are found in commercial sunblocks and sunscreens. The most common sunblock is titanium dioxide, TiO_2, which reflects and scatters UV radiation. (TiO_2 is also widely used in paints because it covers better than any other inorganic white pigment.) Chemical sunscreens, the second type of active ingredient, absorb UV radiation and then reradiate it as heat. These compounds are most effective in screening UVB radiation but less effective in screening UVA radiation. Given here are structural formulas for five chemical UVB-blocking sunscreens along with the name by which each is most commonly listed in the Active Ingredients label on commercial products.

O
H_3CO O CH_3
CH_3

Octyl p-methoxycinnamate
(*The Merck Index*, 12th ed. #6864)

β carbon, including that of a carboxyl group or ester. Malonic acid and substituted malonic acids, for example, undergo decarboxylation on heating, as illustrated by the decarboxylation of malonic acid when it is heated slightly above its melting point of 135–137°C.

$$\text{HO}\overset{\text{O}}{\overset{\|}{\text{C}}}\text{CH}_2\overset{\text{O}}{\overset{\|}{\text{C}}}\text{OH} \xrightarrow{140\text{–}150°\text{C}} \text{CH}_3\overset{\text{O}}{\overset{\|}{\text{C}}}\text{OH} + \text{CO}_2$$

Propanedioic acid
(Malonic acid)

The mechanism for decarboxylation of malonic acids is very similar to what we have just studied for the decarboxylation of β-ketoacids. Formation of a cyclic, six-membered transition state involving rearrangement of three electron pairs gives the enol form of a carboxylic acid, which, in turn, isomerizes to the carboxylic acid.

Sunscreens and sunblocks prevent some UV radiation from reaching the skin and damaging it. *(Bachmann/Photo Researchers, Inc.)*

H_3CO OH O

Oxybenzone
(*The Merk Index,* 12th ed. #7088)

H_3C CH_3 O O CH_3 OH

Homosalate
(*The Merck Index,* 12th ed. #4776)

O CH_3 O CH_3 H_3C N CH_3

Padimate A
(*The Merck Index,* 12th ed. #7088)

MECHANISM Decarboxylation of a β-Dicarboxylic Acid

Step 1: Rearrangement of six electrons in a cyclic six-membered transition state gives carbon dioxide and the enol form of a carboxyl group.

Step 2: Keto-enol tautomerism (Section 11.8) of the enol gives the more stable keto form of the carboxyl group.

$$\left[\text{HO}-\overset{\overset{\text{O}\cdots\text{H}\cdots\text{O}}{\|}}{\text{C}}-\underset{\text{H}\quad\text{H}}{\text{C}}-\text{C}=\text{O}\right] \xrightarrow{(1)} \text{HO}-\overset{\overset{\text{O}-\text{H}}{|}}{\text{C}}=\underset{\text{H}\quad\text{H}}{\text{C}} \;\; \text{O}=\text{C}=\text{O} \underset{}{\overset{(2)}{\rightleftharpoons}} CH_3-\overset{\overset{\text{O}}{\|}}{\text{C}}-\text{OH} + CO_2$$

A cyclic six-membered transition state

enol of a carboxyl group

EXAMPLE 12.6

Each of these carboxylic acids undergoes thermal decarboxylation. Draw a structural formula for the enol intermediate and final product formed in each reaction.

(a) 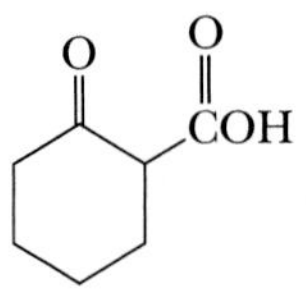 2-oxocyclohexanecarboxylic acid (O, O, COH)   (b) 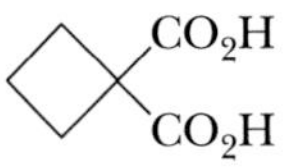 cyclobutane-1,1-dicarboxylic acid (CO_2H, CO_2H)

SOLUTION

(a) 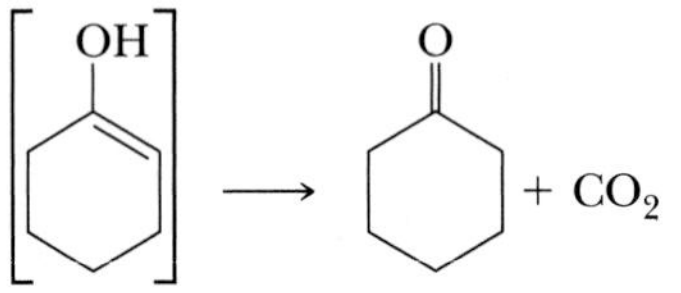 [OH enol] ⟶ cyclohexanone (O) + CO_2

Enol intermediate

(b) 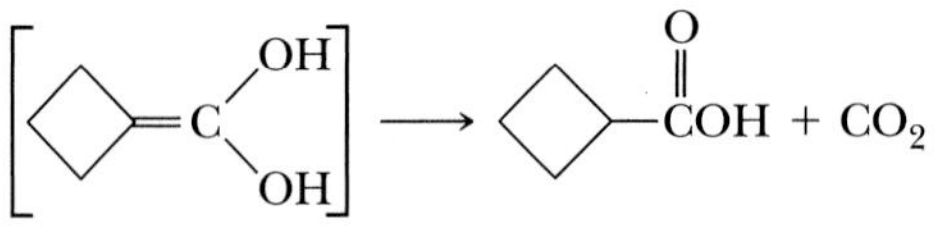 [cyclobutylidene=C(OH)(OH)] ⟶ cyclobutane–COH (O) + CO_2

Enol intermediate

Practice Problem 12.6

Draw the structural formula for the indicated β-ketoacid.

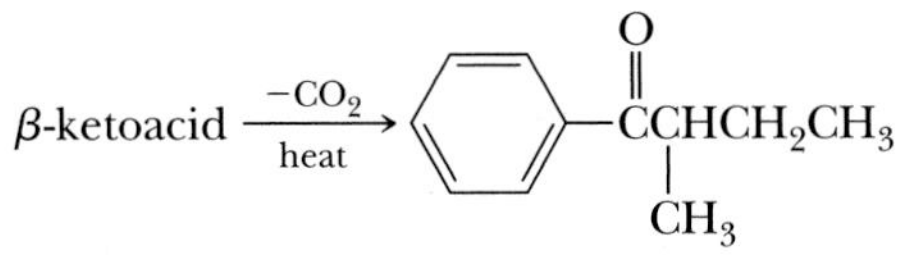

β-ketoacid $\xrightarrow[\text{heat}]{-CO_2}$ C_6H_5–C(=O)CHCH$_2$CH$_3$ (CH$_3$ on CH)

SUMMARY

The functional group of a **carboxylic acid** (Section 12.1) is the **carboxyl group, —CO_2H.** IUPAC names of carboxylic acids (Section 12.2) are derived from the parent alkane by dropping the suffix -e and adding -oic acid. Dicarboxylic acids are named as -dioic acids.

Carboxylic acids are polar compounds (Section 12.3) and, in the liquid and solid states, are associated by hydrogen bonding into dimers. Carboxylic acids have higher boiling points and are more soluble in water than alcohols, aldehydes, ketones, and ethers of comparable molecular weight.

A carboxylic acid consists of two regions of distinctly different polarity: a polar, **hydrophilic** carboxyl group, which increases solubility in water, and a nonpolar **hydrophobic** hydrocarbon chain, which decreases solubility in water. The first four aliphatic carboxylic acids are infinitely soluble in water because the hydrophilic carboxyl group more than counterbalances the hydrophobic hydrocarbon chain. As the size of the carbon chain increases, however, the hydrophobic group becomes dominant, and solubility in water decreases.

KEY REACTIONS

1. Acidity of Carboxylic Acids (Section 12.4A)

Values of pK_a for most unsubstituted aliphatic and aromatic carboxylic acids are within the range pK_a 4–5. Substitution by electron-withdrawing groups decreases pK_a (increases acidity).

$$CH_3\overset{O}{\overset{\|}{C}}OH + H_2O \rightleftharpoons CH_3\overset{O}{\overset{\|}{C}}O^- + H_3O^+ \qquad K_a = 1.74 \times 10^{-5}$$

2. Reaction of Carboxylic Acids with Bases (Section 12.4B)

Carboxylic acids form water-soluble salts with alkali metal hydroxides, carbonates, and bicarbonates, as well as with ammonia and amines.

$$C_6H_5{-}CO_2H + NaOH \xrightarrow[H_2O]{} C_6H_5{-}CO_2^-\ Na^+ + H_2O$$

3. Reduction by Lithium Aluminum Hydride (Section 12.5)

Lithium aluminum hydride reduces a carboxyl group to a primary alcohol.

$$\text{(cyclopentenyl)}{-}\overset{O}{\overset{\|}{C}}OH \xrightarrow[\text{2. } H_2O]{\text{1. } LiAlH_4} \text{(cyclopentenyl)}{-}CH_2OH$$

4. Fischer Esterification (Section 12.6)

Fischer esterification is reversible, and, in order to achieve high yields of ester, it is necessary to force the equilibrium to the right. One way to accomplish this is to use an excess of the alcohol.

$$CH_3\overset{O}{\overset{\|}{C}}OH + CH_3CH_2CH_2OH \xrightleftharpoons{H_2SO_4} CH_3\overset{O}{\overset{\|}{C}}OCH_2CH_2CH_3 + H_2O$$

5. Conversion to Acid Halides (Section 12.7)

Acid chlorides, the most common and widely used of the acid halides, are prepared by treatment of carboxylic acids with thionyl chloride.

$$CH_3CH_2CH_2\overset{O}{\overset{\|}{C}}OH + SOCl_2 \longrightarrow CH_3CH_2CH_2\overset{O}{\overset{\|}{C}}Cl + SO_2 + HCl$$

6. Decarboxylation of β-Ketoacids (Section 12.8A)

The mechanism of decarboxylation involves redistribution of bonding electrons in a cyclic, six-membered transition state.

$$\text{2-oxocyclohexane-}\overset{O}{\overset{\|}{C}}OH \xrightarrow{\text{warm}} \text{cyclohexanone} + CO_2$$

7. Decarboxylation of β-Dicarboxylic Acids (Section 12.8B)

The mechanism of decarboxylation of a β-dicarboxylic acid is similar to that for decarboxylation of a β-ketoacid.

$$HO\overset{O}{\overset{\|}{C}}CH_2\overset{O}{\overset{\|}{C}}OH \xrightarrow{\text{heat}} CH_3\overset{O}{\overset{\|}{C}}OH + CO_2$$

ADDITIONAL PROBLEMS

Structure and Nomenclature

12.7 Name and draw structural formulas for the four carboxylic acids of molecular formula $C_5H_{10}O_2$. Which of these carboxylic acids is chiral?

12.8 Write the IUPAC name for each compound.

(a) $-CO_2H$

(b) $CH_3CH(OH)CH_2CH_2CO_2H$

(c) $C(=O)OH$

(d) CH_3, CO_2H

(e) $CH_3(CH_2)_4CO_2^-NH_4^+$

(f) $HO_2CCH(OH)CH_2CO_2H$

12.9 Draw structural formulas for these carboxylic acids.

(a) 4-Nitrophenylacetic acid
(b) 4-Aminobutanoic acid
(c) 3-Chloro-4-phenylbutanoic acid
(d) *cis*-3-Hexenedioic acid
(e) 2,3-Dihydroxypropanoic acid
(f) 3-Oxohexanoic acid
(g) 2-Oxocyclohexanecarboxylic acid
(h) 2,2-Dimethylpropanoic acid

A black carpet beetle. *(Animals, Animals/© Oxford Scientific Films)*

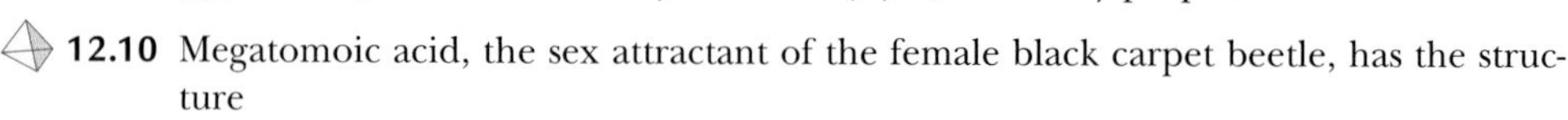

12.10 Megatomoic acid, the sex attractant of the female black carpet beetle, has the structure

$$CH_3(CH_2)_7CH{=}CHCH{=}CHCH_2CO_2H$$

Megatomoic acid

(a) What is its IUPAC name?

(b) State the number of stereoisomers possible for this compound.

12.11 The IUPAC name of ibuprofen is 2-(4-isobutylphenyl)propanoic acid. Draw the structural formula of ibuprofen.

12.12 Draw structural formulas for these salts.

(a) Sodium benzoate
(b) Lithium acetate
(c) Ammonium acetate
(d) Disodium adipate
(e) Sodium salicylate
(f) Calcium butanoate

12.13 The monopotassium salt of oxalic acid (*The Merck Index,* 12th ed., #7820) is present in certain leafy vegetables, including rhubarb. Both oxalic acid and its salts are poisonous in high concentrations. Draw the structural formula of monopotassium oxalate.

12.14 Potassium sorbate is added as a preservative to certain foods to prevent bacteria and molds from causing food spoilage and to extend the foods' shelf life. The IUPAC name of potassium sorbate is potassium 2,4-hexadienoate. Draw a structural formula of potassium sorbate (*The Merck Index,* 12th ed., #7841).

12.15 Zinc 10-undecenoate, the zinc salt of 10-undecenoic acid, is used to treat certain fungal infections, particularly *tineapedis* (athlete's foot). Draw a structural formula of this zinc salt (*The Merck Index,* 12th ed., #9983).

Physical Properties

12.16 Arrange the compounds in each set in order of increasing boiling point.

(a) $CH_3(CH_2)_5COOH$ $\quad CH_3(CH_2)_6CHO$ $\quad CH_3(CH_2)_6CH_2OH$

(b) CH_3CH_2COOH $\quad CH_3CH_2CH_2CH_2OH$ $\quad CH_3CH_2OCH_2CH_3$

Preparation of Carboxylic Acids

We have seen four general methods for the preparation of carboxylic acids.

1. Oxidation of primary alcohols (Section 8.4F)
2. Oxidation of arene side chains (Section 9.5)
3. Carbonation of Grignard reagents (Section 11.5B)
4. Oxidation of aldehydes (Section 11.9A)

12.17 Complete these oxidations.

(a) $CH_3(CH_2)_4CH_2OH + Cr_2O_7^{2-} \xrightarrow[\text{heat}]{H^+}$

(b) 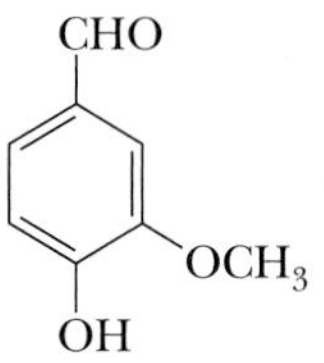 Vanillin (CHO, OCH_3, OH substituted benzene) $+ Ag(NH_3)_2^+ \xrightarrow{NH_3,\ H_2O}$

Vanillin

(c) 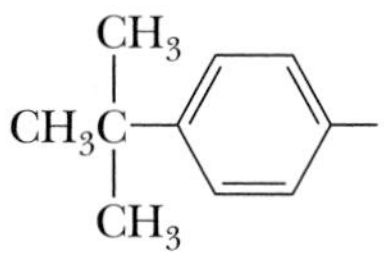 $(CH_3)_3C-C_6H_4-CH_3 + H_2CrO_4 \xrightarrow[\text{heat}]{}$

(d) 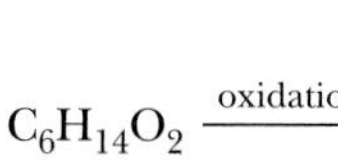 $HO-C_6H_{10}-CH_2OH + H_2CrO_4 \longrightarrow$

12.18 Draw the structural formula of a compound of the given molecular formula that, on oxidation by chromic acid, gives the carboxylic acid or dicarboxylic acid shown.

(a) $C_6H_{14}O \xrightarrow{\text{oxidation}} CH_3(CH_2)_4COOH$ $\qquad$ **(b)** $C_6H_{12}O \xrightarrow{\text{oxidation}} CH_3(CH_2)_4COOH$

(c) $C_6H_{14}O_2 \xrightarrow{\text{oxidation}} HOOC(CH_2)_4COOH$

12.19 Show reagents and experimental conditions to complete this synthesis.

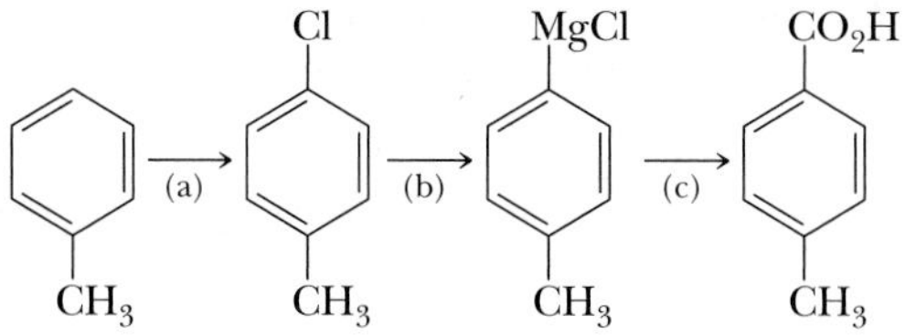

$C_6H_5CH_3 \xrightarrow{(a)} Cl-C_6H_4-CH_3 \xrightarrow{(b)} ClMg-C_6H_4-CH_3 \xrightarrow{(c)} HO_2C-C_6H_4-CH_3$

Acidity of Carboxylic Acids

12.20 Which is the stronger acid in each pair?

(a) Phenol (pK_a 9.95) or benzoic acid (pK_a 4.17)

(b) Lactic acid (K_a 8.4×10^{-4}) or ascorbic acid (K_a 7.9×10^{-5})

12.21 Assign the acid in each set its appropriate pK_a.

(a) $C_6H_5CO_2H$ (benzoic acid) and $p\text{-}O_2NC_6H_4CO_2H$ (pK_a 4.19 and 3.14)

(b) $p\text{-}O_2NC_6H_4CO_2H$ and $p\text{-}H_2NC_6H_4CO_2H$ (pK_a 4.92 and 3.14)

(c) $CH_3C(=O)CH_2CO_2H$ and $CH_3C(=O)CO_2H$ (pK_a 3.58 and 2.49)

(d) $CH_3CH(OH)CO_2H$ and $CH_3CH_2CO_2H$ (pK_a 4.78 and 3.08)

12.22 Complete these acid-base reactions:

(a) $C_6H_5CH_2CO_2H + NaOH \longrightarrow$

(b) $CH_3CH{=}CHCH_2CO_2H + NaHCO_3 \longrightarrow$

(c) $o\text{-}HOC_6H_4CO_2H + NaHCO_3 \longrightarrow$

(d) $CH_3CH(OH)CO_2H + H_2NCH_2CH_2OH \longrightarrow$

(e) $CH_3CH{=}CHCH_2CO_2^-Na^+ + HCl \longrightarrow$

12.23 The normal pH range for blood plasma is 7.35–7.45. Under these conditions, would you expect the carboxyl group of lactic acid (pK_a 4.07) to exist primarily as a carboxyl group or as a carboxylate anion? Explain.

12.24 The K_a of ascorbic acid (Section 12.6) is 7.94×10^{-5}. Would you expect ascorbic acid dissolved in blood plasma, pH 7.35–7.45, to exist primarily as ascorbic acid or as ascorbate anion? Explain.

12.25 Excess ascorbic acid is excreted in the urine, the pH of which is normally in the range 4.8–8.4. What form of ascorbic acid would you expect to be present in urine of pH 8.4, ascorbic acid or ascorbate anion?

12.26 The pH of human gastric juice is normally in the range 1.0–3.0. What form of lactic acid (pK_a 4.07) would you expect to be present in the stomach, lactic acid or its anion?

Reactions of Carboxylic Acids

12.27 Give the expected organic products formed when phenylacetic acid, $PhCH_2CO_2H$, is treated with each reagent.

(a) $SOCl_2$ **(b)** $NaHCO_3, H_2O$
(c) $NaOH, H_2O$ **(d)** NH_3, H_2O
(e) $LiAlH_4$ followed by H_2O **(f)** $CH_3OH + H_2SO_4$ (catalyst)

12.28 Show how to convert *trans*-3-phenyl-2-propenoic acid (cinnamic acid) to these compounds.

(a) (H)(C_6H_5)C=C(CH_2OH)(H) (trans) **(b)** $C_6H_5CH_2CH_2CO_2H$ **(c)** $C_6H_5CH_2CH_2CH_2OH$

12.29 Show how to convert 3-oxobutanoic acid (acetoacetic acid) to these compounds.

(a) $CH_3CH(OH)CH_2CO_2H$ **(b)** $CH_3CH(OH)CH_2CH_2OH$ **(c)** $CH_3CH{=}CHCO_2H$

12.30 Complete these examples of Fischer esterification. Assume an excess of the alcohol.

(a) $CH_3CO_2H + HOCH_2CH_2CH(CH_3)_2 \overset{H^+}{\rightleftharpoons}$ **(b)** (benzene-1,2-dicarboxylic acid, $C_6H_4(CO_2H)_2$) $+ CH_3OH \overset{H^+}{\rightleftharpoons}$

(c) $HO_2C(CH_2)_2CO_2H + CH_3CH_2OH \overset{H^+}{\rightleftharpoons}$

12.31 Methyl 2-hydroxybenzoate (methyl salicylate) has the odor of oil of wintergreen. This ester is prepared by Fischer esterification of 2-hydroxybenzoic acid (salicylic acid) with methanol. Draw the structural formula of methyl 2-hydroxybenzoate.

12.32 Benzocaine, a topical anesthetic, is prepared by treatment of 4-aminobenzoic acid with ethanol in the presence of an acid catalyst followed by neutralization. Draw the structural formula of benzocaine (*The Merck Index,* 12th ed., #1116).

12.33 From what carboxylic acid and alcohol is each ester derived?

(a) $CH_3C(=O)O$–(cyclohexane-1,4-diyl)–$OC(=O)CH_3$ **(b)** $CH_3OC(=O)CH_2CH_2C(=O)OCH_3$

(c) (cyclohexyl)–$C(=O)OCH_3$ **(d)** $CH_3CH_2CH{=}CHC(=O)OCH(CH_3)_2$

12.34 When 4-hydroxybutanoic acid is treated with an acid catalyst, it forms a lactone (a cyclic ester). Draw the structural formula of this lactone.

12.35 Draw the product formed on thermal decarboxylation of each compound.

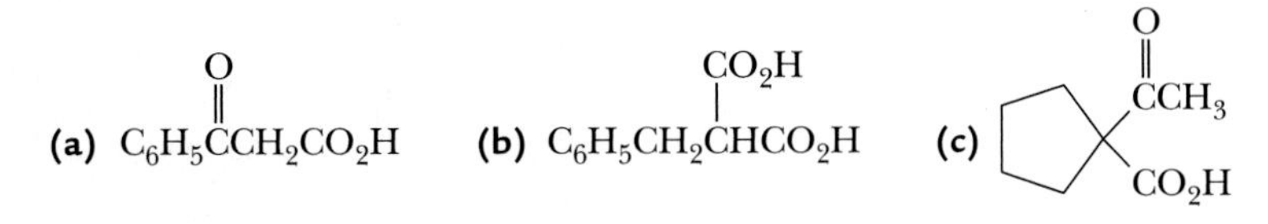

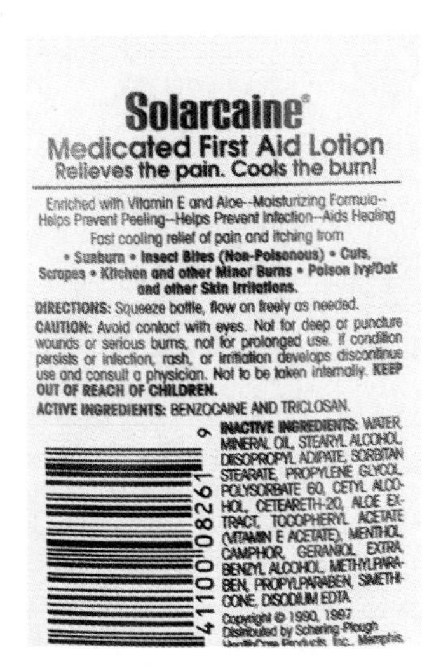

The active ingredient in Solarcaine is the topical anesthetic benzocaine, which is ethyl *p*-aminobenzoate. *(Charles D. Winters)*

13

Functional Derivatives of Carboxylic Acids

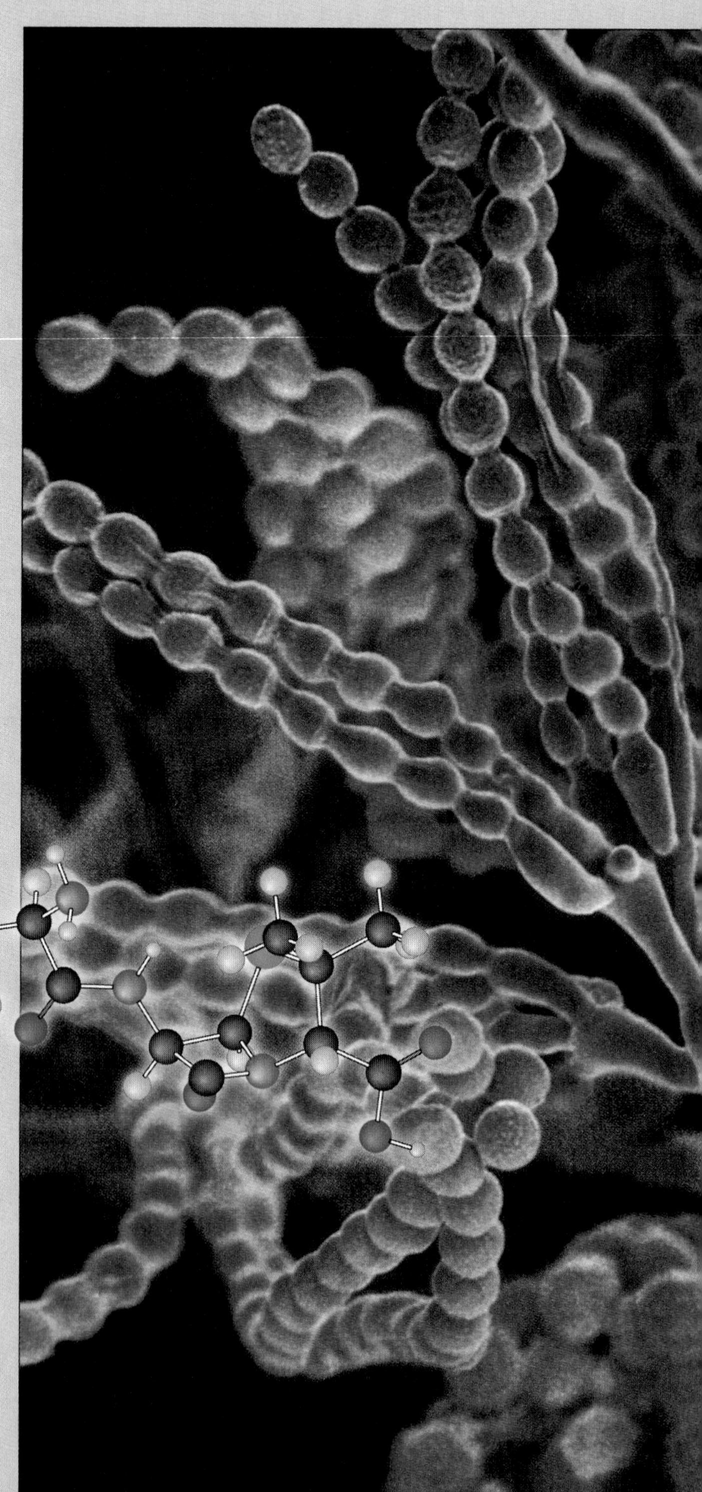

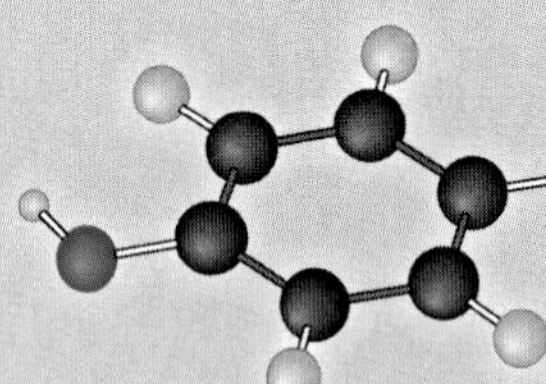

Colored scanning electron micrograph of Penicillium s. fungus. The stalk-like objects are condiophores to which are attached numerous round condia. The condia are the fruiting bodies of the fungus. Inset: A model of amoxicillin. *(© SCIMAT/Science Source/Photo Researchers, Inc.)*

In this chapter, we study four classes of organic compounds, all derived from the carboxyl group: acid halides, acid anhydrides, esters, and amides. Under the general formula of each functional group is a drawing to help you see how it is formally related to a carboxyl group. Loss of —OH from a carboxyl group and H— from H—Cl, for example, gives an acid chloride. Similarly, loss of —OH from a carboxyl group and H— from ammonia gives an amide.

$\overset{O}{\parallel}$ RCCl	$\overset{O}{\parallel}$ $\overset{O}{\parallel}$ RCOCR′	$\overset{O}{\parallel}$ RCOR′	$\overset{O}{\parallel}$ $RCNH_2$
An acid chloride	An acid anhydride	An ester	An amide
RC—OH H—Cl	RC—OH H—OCR′	RC—OH H—OR′	RC—OH H—NH_2

13.1 STRUCTURE AND NOMENCLATURE

A. Acid Halides

The functional group of an **acid halide** (acyl halide) is an **acyl group (RCO—)** bonded to a halogen atom (Section 12.7). Acid chlorides are the most common acid halides. Acid halides are named by changing the suffix -ic acid in the name of the parent carboxylic acid to -yl halide.

Acid halide A derivative of a carboxylic acid in which the —OH of the carboxyl groups is replaced by halogen, most commonly chlorine.

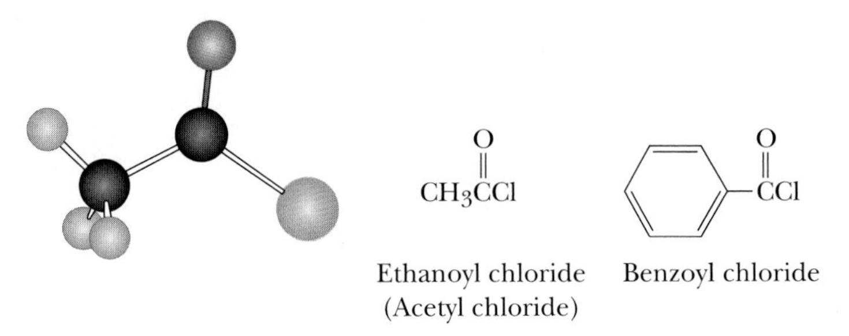

Ethanoyl chloride (Acetyl chloride) Benzoyl chloride

B. Acid Anhydrides

Carboxylic Anhydrides

The functional group of a **carboxylic anhydride** is two acyl groups bonded to an oxygen atom. The anhydride may be symmetrical (two identical acyl groups), or it may be mixed (two different acyl groups). Cyclic anhydrides are named from the dicarboxylic acids from which they are derived. Here are the cyclic anhydrides derived from succinic acid and phthalic acid.

Carboxylic anhydride A compound in which two acyl groups are bonded to an oxygen.

CH_3COCCH_3 (O O)

C₆H₅–COC–C₆H₅ (O O)

Acetic anhydride Benzoic anhydride

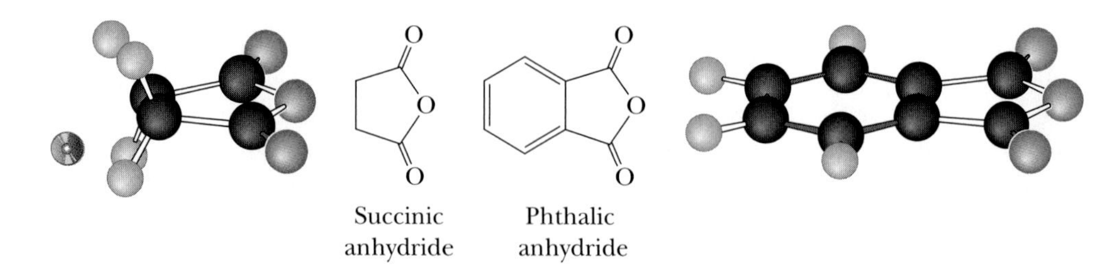

Succinic anhydride   Phthalic anhydride

Phosphoric Anhydrides

Because of the special importance of anhydrides of phosphoric acid in biochemical systems (Chapter 20), we include them here to show the similarity between them and the anhydrides of carboxylic acids. The functional group of a **phosphoric anhydride** is two phosphoryl groups bonded to an oxygen atom. Here are structural formulas for two anhydrides of phosphoric acid, H_3PO_4, and the ions derived by ionization of the acidic hydrogens.

$$\mathrm{HO{-}P(=O)(OH){-}O{-}P(=O)(OH){-}OH}$$

Diphosphoric acid
(Pyrophosphoric acid)

$$\mathrm{{}^{-}O{-}P(=O)(O^{-}){-}O{-}P(=O)(O^{-}){-}O^{-}}$$

Diphosphate ion
(Pyrophosphate ion)

$$\mathrm{HO{-}P(=O)(OH){-}O{-}P(=O)(OH){-}O{-}P(=O)(OH){-}OH}$$

Triphosphoric acid

$$\mathrm{{}^{-}O{-}P(=O)(O^{-}){-}O{-}P(=O)(O^{-}){-}O{-}P(=O)(O^{-}){-}O^{-}}$$

Triphosphate ion

C. Esters

Esters of Carboxylic Acids

The functional group of a **carboxylic ester** is an acyl group bonded to —OR or —OAr. Both IUPAC and common names of esters are derived from the names of the parent carboxylic acids. The alkyl or aryl group bonded to oxygen is named first, followed by the name of the acid in which the suffix -ic acid is replaced by the suffix -ate.

$$\mathrm{CH_3\overset{O}{\overset{\|}{C}}OCH_2CH_3}$$

Ethyl ethanoate
(Ethyl acetate)

$$\mathrm{CH_3CH_2O\overset{O}{\overset{\|}{C}}CH_2CH_2\overset{O}{\overset{\|}{C}}OCH_2CH_3}$$

Diethyl butanedioate
(Diethyl succinate)

Lactones—Cyclic Esters

Lactone A cyclic ester.

Cyclic esters are called **lactones.** The IUPAC system has developed a set of rules for naming these compounds. Nonetheless, the simplest lactones are still named by

From Moldy Clover to a Blood Thinner

In 1933, a disgruntled farmer delivered a bale of moldy clover, a pail of unclotted blood, and a dead cow to the laboratory of Dr. Carl Link at the University of Wisconsin. Six years and many bales of moldy clover later, Link and his collaborators isolated the anticoagulant dicoumarol (*The Merck Index,* 12th ed., #3140), a substance that delays or prevents blood clotting. When cows are fed moldy clover, they ingest dicoumarol, their blood clotting is inhibited, and they bleed to death from minor cuts and scratches.

The powerful anticoagulant dicoumarol was first isolated from moldy clover. *(Grant Heilman/Grant Heilman Photography, Inc.)*

Dicoumarol exerts its anticoagulation effect by interfering with vitamin K activity (Section 17.6D). Within a few years after its discovery, dicoumarol became widely used to treat victims of heart attack and others at risk for developing blood clots.

Dicoumarol is a derivative of coumarin, a lactone that gives sweet clover its pleasant smell. Coumarin, which does not interfere with blood clotting and has been used as a flavoring agent, is converted to dicoumarol as sweet clover becomes moldy.

O O

Coumarin
(from sweet clover)

as sweet clover becomes moldy →

OH OH

O O O O

Dicoumarol
(an anticoagulant)

In a search for even more potent anticoagulants, Link developed warfarin (named after the Wisconsin Alumni Research Foundation), now used primarily as a rat poison. When rats consume it, their blood fails to clot, and they bleed to death. Warfarin is also used as a blood thinner in humans. In 1989, physicians prescribed more than 2700 pounds of "rat poison" for human medical use (see *The Merck Index,* 12th ed., #10174). The S enantiomer, shown here, is more active than the R enantiomer. The commercial product is sold as a racemic mixture.

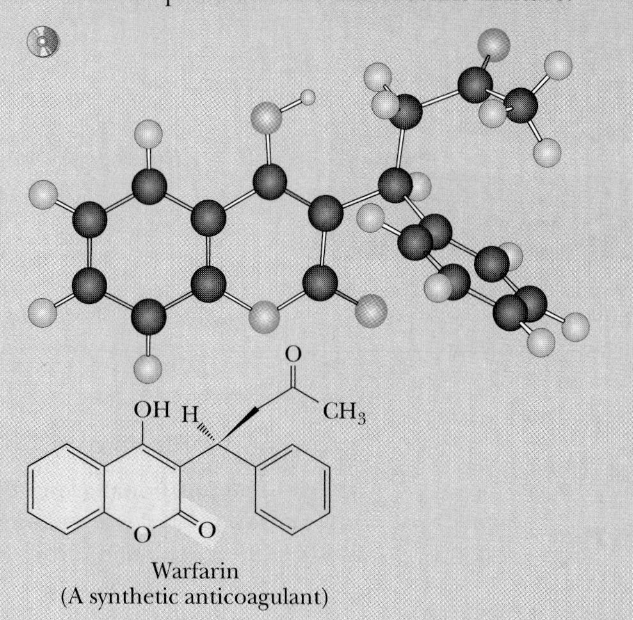

Warfarin
(A synthetic anticoagulant)

dropping the suffix -ic or -oic acid from the name of the parent carboxylic acid and adding the suffix -olactone. The location of the oxygen atom in the ring is indicated by a number if the IUPAC name of the acid is used, or by a Greek letter α, β, γ, δ, ε, and so forth if the common name of the acid is used.

4-Butanolactone
(γ-Butyrolactone)

6-Hexanolactone
(ε-Caprolactone)

Esters of Phosphoric Acid

Phosphoric acid has three —OH groups and forms mono-, di-, and triphosphoric esters, which are named by giving the name(s) of the alkyl or aryl group(s) bonded to oxygen followed by the word "phosphate," as for example dimethyl phosphate. In more complex phosphoric esters, it is common to name the organic molecule and then indicate the presence of the phosphoric ester using either the word "phosphate" or the prefix phospho-. Following on the right are two phosphoric esters, each of special importance in the biological world. The first reaction in the metabolism of glucose is formation of a phosphoric ester to give D-glucose-6-phosphate (Section 20.6). Pyridoxal phosphate is one of the metabolically active forms of vitamin B_6. Each of these esters is shown as it is ionized at pH 7.4, the pH of blood plasma; the two hydrogens of each phosphate group are ionized giving the phosphate group a charge of -2.

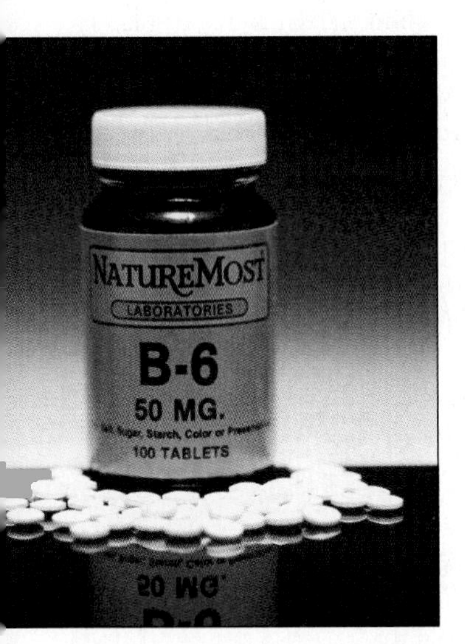

Vitamin B_6, pyridoxal. *(Charles D. Winters)*

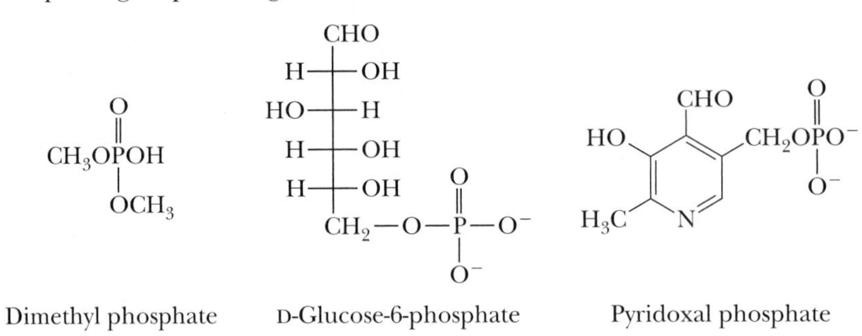

Dimethyl phosphate

D-Glucose-6-phosphate

Pyridoxal phosphate

D. Amides, Lactams, and Imides

The functional group of an **amide** is an acyl group bonded to a trivalent nitrogen atom. Amides are named by dropping the suffix -oic acid from the IUPAC name of the parent acid, or -ic acid from its common name, and adding -amide. If the nitrogen atom of an amide is bonded to an alkyl or aryl group, the group is named, and its location on nitrogen is indicated by *N*-. Two alkyl or aryl groups on nitrogen are indicated by *N,N*-di. Amide bonds are the key structural feature that joins amino acids together to form polypeptides and proteins (Chapter 18).

CHEMICAL CONNECTIONS

Insecticides of Plant Origin—Polyunsaturated *N*-Isobutylamides

Plant-derived substances, largely abandoned during the era of synthetic insecticides, such as DDT, are once again under study as safe and selective agents for insect control. It is now well recognized that almost every plant species has developed a unique set of chemicals that protects it from insect predation. In a search for natural means of small-scale control of mosquitoes in rural areas, scientists at the International Centre of Insect Physiology and Ecology (ICIPE) in Nairobi, Kenya, investigated *Spilanthes mauritiana* (Compositae), a medicinal plant used to treat mouth infections, stomachaches, diarrhea, and toothaches. A methanolic extract of its leaves yields the *N*-isobutylamide of a tetraunsaturated fatty acid, which, at a concentration of 10^{-5} mg/mL, causes 100% mortality in larvae of *Aedes aegypti*, the insect vector of yellow fever.

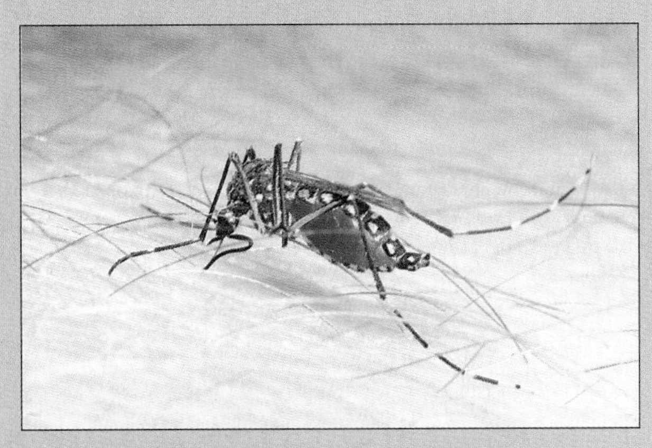

Aedes aegypti, the mosquito vector of yellow fever. *(D. R. Specker/Animals, Animals)*

N-isobutylamide portion

10 8 4 2 O N H

N-Isobutyl-(2E,4E,8E,10Z)-2,4,8,10,-dodecatetraenamide

It is hoped that such plant extracts may be useful in rural communities where mosquitoes breed in small collections of water, such as in temporary rain puddles, containers, and drums. Periodic treatment of these breeding pools with native plant-derived insecticides could considerably reduce the multiplication of disease-bearing mosquitoes. (See *Insecticides of Plant Origin,* J. T. Arnason, B. J. R. Philogène, and P. Morand, eds., ACS Symposium Series No. 387, American Chemical Society, Washington, D.C., 1989.)

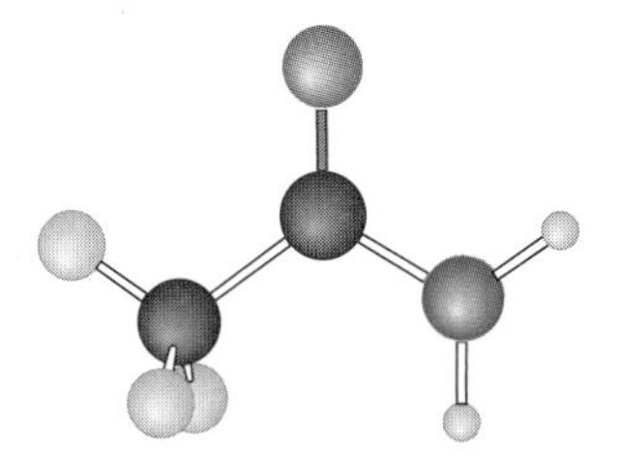

$CH_3\overset{O}{\overset{\|}{C}}NH_2$	$CH_3\overset{O}{\overset{\|}{C}}NHCH_3$	$H\overset{O}{\overset{\|}{C}}N(CH_3)_2$
Acetamide (a 1° amide)	*N*-Methylacetamide (a 2° amide)	*N,N*-Dimethylformamide (DMF) (a 3° amide)

Lactam A cyclic amide.

Cyclic amides are given the special name **lactam.** Their common names are derived in a manner similar to those of lactones, with the difference that the suffix -olactone is replaced by -olactam.

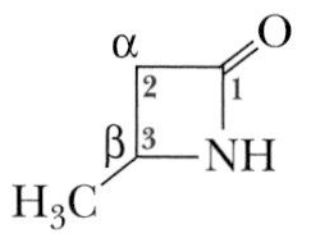

3-Butanolactam (β-Butyrolactam)

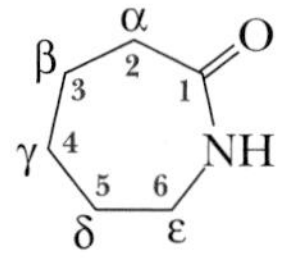

6-Hexanolactam (ε-Caprolactam)

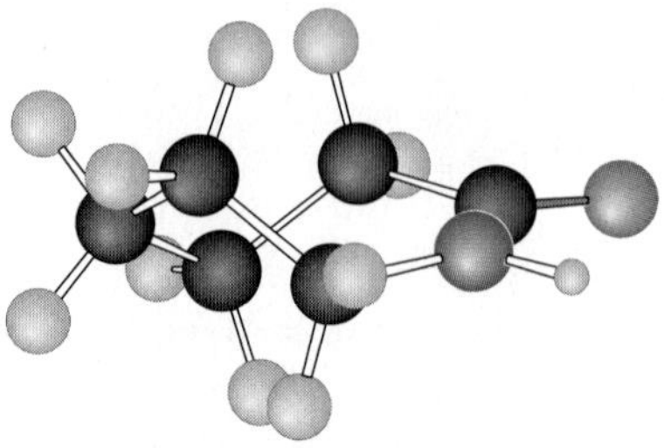

Imide A compound containing two acyl groups, RCO— or ArCO—, bonded to a nitrogen atom.

The functional group of an **imide** is two acyl groups bonded to nitrogen. Both succinimide and phthalimide are cyclic imides.

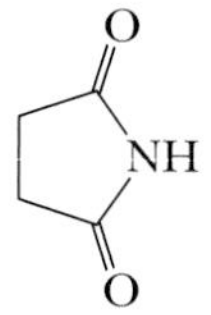

Succinimide 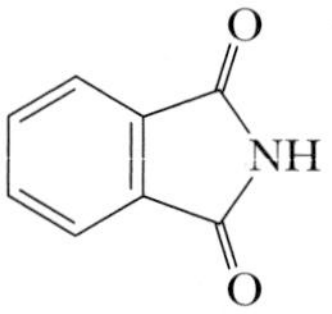 Phthalimide

EXAMPLE 13.1

Write the IUPAC name for each compound.

(a) $CH_3CH(CH_3)CH_2COOCH_3$ (b) $CH_3COCH_2COOCH_2CH_3$

(c) $H_2NCO(CH_2)_4CONH_2$ (d) $PhCH_2COOCOCH_2Ph$

SOLUTION

Given first are IUPAC names and then, in parentheses, common names.

(a) Methyl 3-methylbutanoate (methyl isovalerate, from isovaleric acid)
(b) Ethyl 3-oxobutanoate (ethyl β-ketobutyrate, from β-ketobutyric acid)
(c) Hexanediamide (adipamide, from adipic acid)
(d) Phenylethanoic anhydride (phenylacetic anhydride, from phenylacetic acid)

Practice Problem 13.1

Draw a structural formula for each compound.

(a) *N*-Cyclohexylacetamide (b) *sec*-Butyl acetate
(c) Cyclobutyl butanoate (d) *N*-(2-Octyl)succinimide
(e) Diethyl adipate (f) Propanoic anhydride

CHEMICAL CONNECTIONS

The Penicillins and Cephalosporins—β-Lactam Antibiotics

The **penicillins** were discovered in 1928 by the Scottish bacteriologist Sir Alexander Fleming. As a result of the brilliant experimental work of Sir Howard Florey, an Australian pathologist, and Ernst Chain, a German chemist who fled Nazi Germany, penicillin G was introduced into the practice of medicine in 1943. For their pioneering work in developing one of the most effective antibiotics of all time, Fleming, Florey, and Chain were awarded the Nobel Prize for Medicine and Physiology in 1945.

The mold from which Fleming discovered penicillin was *Penicillium notatum,* a strain that gives a relatively low yield of penicillin. It was replaced in commercial production of the antibiotic by *P. chrysogenum,* a strain cultured from a mold found growing on a grapefruit in a market in Peoria, Illinois. The structural feature common to all penicillins is a **β-lactam** ring fused to a five-membered thiazolidine ring. The penicillins owe their antibacterial activity to a common mechanism that inhibits the biosynthesis of a vital part of bacterial cell walls.

The penicillins differ in the group attached to the acyl carbon

β-lactam

CH_2-CNH, CH_3, CH_3, $CO_2^-K^+$

Penicillin G
(a β-lactam antibiotic)

Soon after the penicillins were introduced into medical practice, penicillin-resistant strains of bacteria began to appear and have since proliferated. One approach to combating resistant strains is to synthesize newer, more effective penicillins. Among those developed are ampicillin, methicillin, and amoxicillin. Another approach is to search for newer, more effective β-lactam antibiotics. So far the most effective of these discovered are the **cephalosporins,** the first of which was isolated from the fungus *Cephalosporium acremonium.* This class of β-lactam antibiotics has an even broader spectrum of antibacterial activity than the penicillins and is effective against many penicillin-resistant bacterial strains.

The cephalosporins differ in the group attached to the acyl carbon and the side chain attached to the thiazine ring

β-lactam

CH_2-CNH, CH_2OCCH_3, $CO_2^-K^+$

A cephalosporin, a newer generation β-lactam antibiotic

13.2 CHARACTERISTIC REACTIONS

The most common reaction theme of acid halides, anhydrides, esters, and amides is addition of a **nucleophile** to the carbonyl carbon to form a tetrahedral carbonyl addition intermediate. To this extent, the reaction of these functional groups is similar to nucleophilic addition to the carbonyl groups in aldehydes and ketones

Nucleophile A molecule or ion that donates a pair of electrons to another molecule or ion to form a new covalent bond; a Lewis base.

(Section 11.4). The tetrahedral carbonyl addition intermediate formed from an aldehyde or ketone then adds H^+ to give the product.

Nucleophilic acyl addition: $R_2C{=}O$ + $:Nu^-$ ⟶ [$R_2C(O^-)Nu$] $\xrightarrow{H^+}$ $R_2C(OH)Nu$

An aldehyde or ketone — Tetrahedral carbonyl addition intermediate — Addition product

For functional derivatives of carboxylic acids, the fate of the tetrahedral carbonyl addition intermediate is quite different. This intermediate collapses to regenerate the carbonyl group. The result of this addition-elimination sequence is **nucleophilic acyl substitution.**

Nucleophilic acyl substitution A reaction in which a nucleophile bonded to a carbonyl carbon is replaced by another nucleophile.

Nucleophilic acyl substitution: $RC({=}O)Y$ + $:Nu^-$ ⟶ [$RC(O^-)(Y)Nu$] ⟶ $RC({=}O)Nu$ + $:Y^-$

Tetrahedral carbonyl addition intermediate — Substitution product

The major difference between these two types of carbonyl addition reactions is that aldehydes and ketones do not have a group, Y, that can leave as a stable anion. They undergo only nucleophilic acyl addition. The four carboxylic acid derivatives we study in this chapter do have a group, Y, that can leave as a stable anion; they undergo nucleophilic acyl substitution.

In this general reaction, we show the nucleophile and the leaving group as anions. This need not be the case. Neutral molecules, such as water, alcohols, ammonia, and amines, may also serve as nucleophiles in the acid-catalyzed version of this reaction. We show the leaving group here as an anion to illustrate an important point about leaving groups: the weaker the base, the better the leaving group (Section 7.4C).

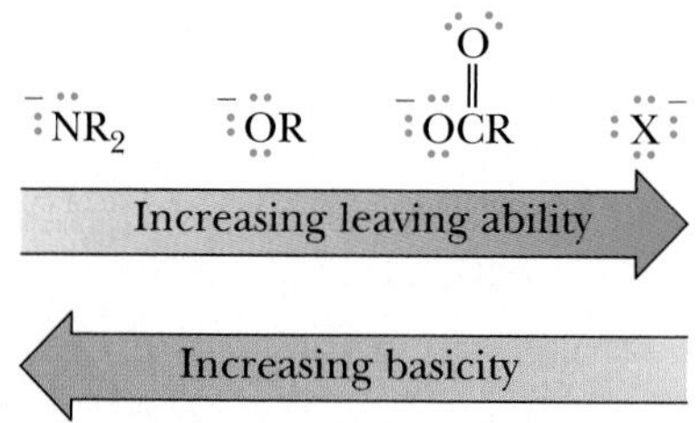

The weakest base in this series, and the best leaving group, is halide ion; acid halides are the most reactive toward nucleophilic acyl substitution. The strongest base, and the poorest leaving group, is amide ion; amides are the least reactive toward nucleophilic acyl substitution. Acid halides and acid anhydrides are so reactive that they are not found in nature. Esters and amides, however, are universally present.

$RC({=}O)NH_2$	$RC({=}O)OR'$	$RC({=}O)OC({=}O)R$	$RC({=}O)X$
Amide	Ester	Anhydride	Acid halide

Increasing reactivity toward nucleophilic acyl substitution ⟶

13.3 REACTION WITH WATER—HYDROLYSIS

A. Acid Chlorides

Low-molecular-weight acid chlorides react very rapidly with water to form carboxylic acids and HCl. Higher-molecular-weight acid chlorides are less soluble and, consequently, react less rapidly with water.

$$CH_3\overset{\overset{O}{\|}}{C}Cl + H_2O \longrightarrow CH_3\overset{\overset{O}{\|}}{C}OH + HCl$$

B. Acid Anhydrides

Acid anhydrides are generally less reactive than acid chlorides. The lower-molecular-weight anhydrides react readily with water to form two molecules of carboxylic acid.

$$CH_3\overset{\overset{O}{\|}}{C}O\overset{\overset{O}{\|}}{C}CH_3 + H_2O \longrightarrow CH_3\overset{\overset{O}{\|}}{C}OH + HO\overset{\overset{O}{\|}}{C}CH_3$$

C. Esters

Esters are hydrolyzed only very slowly, even in boiling water. Hydrolysis becomes considerably more rapid, however, when they are refluxed in aqueous acid or base. We already discussed acid-catalyzed (Fischer) esterification in Section 12.6B and pointed out that it is an equilibrium reaction. Hydrolysis of esters in aqueous acid is also an equilibrium reaction and proceeds by the same mechanism as esterification, except in reverse. The role of the acid catalyst is to protonate the carbonyl oxygen. In doing so, it increases the electrophilic character of the carbonyl carbon toward attack by water to form a **tetrahedral carbonyl addition intermediate.** Collapse of this intermediate gives the carboxylic acid and an alcohol. In this reaction, acid is a catalyst; it is consumed in the first step but another is generated at the end of the reaction.

$$R\overset{\overset{O}{\|}}{C}OCH_3 + H_2O \underset{}{\overset{H^+}{\rightleftharpoons}} \left[R-\overset{\overset{OH}{|}}{\underset{\underset{OCH_3}{|}}{C}}-OH \right] \overset{H^+}{\rightleftharpoons} R\overset{\overset{O}{\|}}{C}OH + CH_3OH$$

Tetrahedral carbonyl addition intermediate

Hydrolysis of esters may also be carried out using hot aqueous base, such as aqueous NaOH. Hydrolysis of esters in aqueous base is often called **saponification,** a reference to the use of this reaction in the manufacture of soaps (Section 17.2A). Although the carbonyl carbon of an ester is not strongly electrophilic, hydroxide ion is a good nucleophile and adds to the carbonyl carbon to form a tetrahedral carbonyl addition intermediate, which in turn collapses to give a carboxylic acid and an alkoxide ion. The carboxylic acid reacts with the alkoxide ion or other base present to form a carboxylic acid anion. Thus, each mole of ester hydrolyzed requires 1 mole of base as shown in the following balanced equation.

Saponification Hydrolysis of an ester in aqueous NaOH or KOH to an alcohol and the sodium or potassium salt of a carboxylic acid.

$$R\overset{\overset{O}{\|}}{C}OCH_3 + NaOH \xrightarrow{H_2O} R\overset{\overset{O}{\|}}{C}O^-Na^+ + CH_3OH$$

MECHANISM Hydrolysis of an Ester in Aqueous Base

Step 1: Addition of hydroxide ion to the carbonyl carbon of the ester gives a tetrahedral carbonyl addition intermediate.

Step 2: Collapse of this intermediate gives a carboxylic acid and an alkoxide ion.

Step 3: Proton transfer from the carboxyl group to the alkoxide ion gives the carboxylic anion.

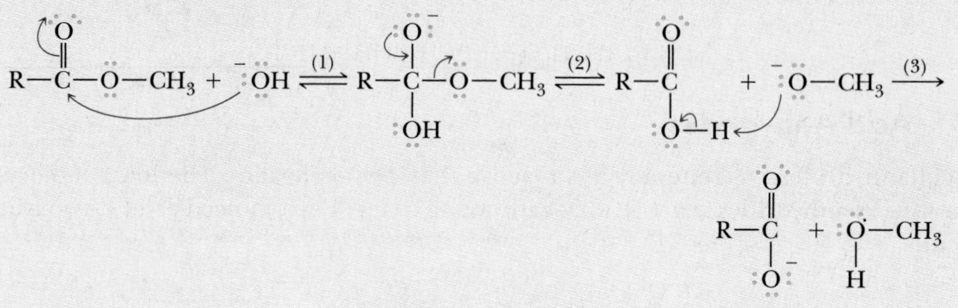

$$R{-}\overset{O}{\overset{\|}{C}}{-}O{-}CH_3 + {}^{-}OH \overset{(1)}{\rightleftharpoons} R{-}\underset{OH}{\underset{|}{\overset{O^-}{\overset{|}{C}}}}{-}O{-}CH_3 \overset{(2)}{\rightleftharpoons} R{-}\underset{O{-}H}{\underset{|}{\overset{O}{\overset{\|}{C}}}} + {}^{-}O{-}CH_3 \xrightarrow{(3)} R{-}\underset{O^-}{\underset{|}{\overset{O}{\overset{\|}{C}}}} + \underset{H}{\underset{|}{O}}{-}CH_3$$

There are two major differences between hydrolysis of esters in aqueous acid and aqueous base:

1. For hydrolysis in aqueous acid, acid is required in only catalytic amounts. For hydrolysis in aqueous base, base is required in equimolar amounts because it is a reactant, not just a catalyst.
2. Hydrolysis of an ester in aqueous acid is reversible; hydrolysis in aqueous base is irreversible. A carboxylic acid anion is not attacked by ROH.

EXAMPLE 13.2

Complete and balance equations for hydrolysis of each ester in aqueous sodium hydroxide. Show all products as they are ionized in aqueous NaOH.

(a) 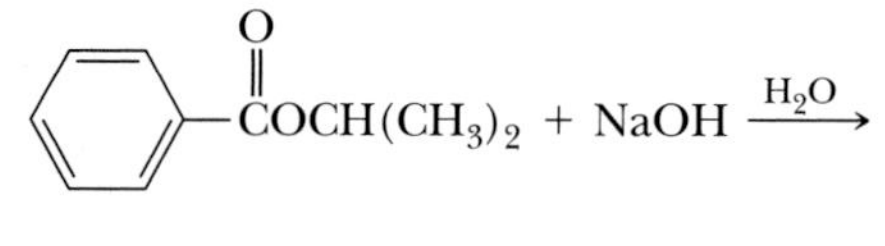 $C_6H_5{-}\overset{O}{\overset{\|}{C}}OCH(CH_3)_2 + NaOH \xrightarrow{H_2O}$ (b) $CH_3\overset{O}{\overset{\|}{C}}OCH_2CH_2O\overset{O}{\overset{\|}{C}}CH_3 + NaOH \xrightarrow{H_2O}$

SOLUTION

The products of hydrolysis of (a) are benzoic acid and 2-propanol. In aqueous NaOH, benzoic acid is converted to its sodium salt. Therefore, 1 mole of NaOH is required for hydrolysis of each mole of this ester. Compound (b) is a diester of ethylene glycol. Two moles of NaOH are required for its hydrolysis.

(a) 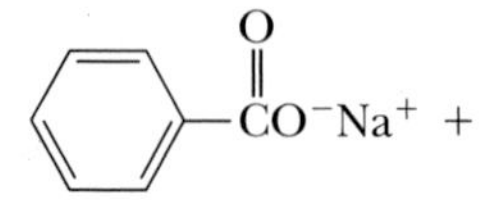 $C_6H_5{-}\overset{O}{\overset{\|}{C}}O^-Na^+ + HOCH(CH_3)_2$ (b) $2CH_3\overset{O}{\overset{\|}{C}}O^-Na^+ + HOCH_2CH_2OH$

Sodium benzoate; 2-Propanol (Isopropyl alcohol); Sodium acetate; 1,2-Ethanediol (Ethylene glycol)

Practice Problem 13.2

Complete and balance equations for hydrolysis of each ester in aqueous solution. Show each product as it is ionized under the given experimental conditions.

(a) $C_6H_4(CO_2CH_3)_2$ (benzene ring with two adjacent CO_2CH_3 groups) $+ \text{NaOH} \xrightarrow{H_2O}$

(b) $CH_3\overset{O}{\overset{\|}{C}}CH_2CH_2CH_2\overset{O}{\overset{\|}{C}}OCH_2CH_3 + H_2O \xrightarrow{HCl}$

D. Amides

Amides require considerably more vigorous conditions for hydrolysis in both acid and base than do esters. Amides undergo hydrolysis in hot aqueous acid to give a carboxylic acid and an ammonium ion. Hydrolysis is driven to completion by the acid-base reaction between ammonia or the amine and acid to form an ammonium ion. One mole of acid is required per mole of amide.

$$CH_3CH_2\underset{Ph}{\underset{|}{C}}H\overset{O}{\overset{\|}{C}}NH_2 + H_2O + HCl \xrightarrow[\text{heat}]{H_2O} CH_3CH_2\underset{Ph}{\underset{|}{C}}H\overset{O}{\overset{\|}{C}}OH + NH_4^+Cl^-$$

2-Phenylbutanamide 2-Phenylbutanoic acid

In aqueous base, the products of amide hydrolysis are a carboxylic acid salt, and ammonia or an amine. Base-catalyzed hydrolysis is driven to completion by the acid-base reaction between the carboxylic acid and base to form a salt. One mole of base is required per mole of amide.

$$CH_3\overset{O}{\overset{\|}{C}}NH-C_6H_5 + \text{NaOH} \xrightarrow[\text{heat}]{H_2O} CH_3\overset{O}{\overset{\|}{C}}O^-Na^+ + H_2N-C_6H_5$$

N-Phenylethanamide (*N*-Phenylacetamide, acetanilide) Sodium acetate Aniline

EXAMPLE 13.3

Write equations for hydrolysis of these amides in concentrated aqueous HCl. Show all products as they exist in aqueous HCl, and show the number of moles of HCl required for hydrolysis of each amide.

(a) $CH_3\overset{O}{\overset{\|}{C}}N(CH_3)_2$ (b) six-membered ring lactam (C=O adjacent to NH)

SOLUTION

(a) Hydrolysis of *N,N*-dimethylacetamide gives acetic acid and dimethylamine. Dimethylamine, a base, is protonated by HCl to form dimethylammonium ion and is shown in the balanced equation as dimethylammonium chloride.

$$CH_3\overset{O}{\overset{\|}{C}}N(CH_3)_2 + H_2O + HCl \xrightarrow[\text{heat}]{} CH_3\overset{O}{\overset{\|}{C}}OH + (CH_3)_2NH_2^+Cl^-$$

(b) Hydrolysis of this δ-lactam gives the protonated form of 5-aminopentanoic acid.

$$\text{(δ-lactam, six-membered ring with C=O and NH)} + H_2O + HCl \xrightarrow[\text{heat}]{} HO\overset{O}{\overset{\|}{C}}CH_2CH_2CH_2CH_2NH_3^+Cl^-$$

Practice Problem 13.3

Complete equations for the hydrolysis of the amides in Example 13.3 in concentrated aqueous NaOH. Show all products as they exist in aqueous NaOH, and show the number of moles of NaOH required for hydrolysis of each amide.

13.4 REACTION WITH ALCOHOLS

A. Acid Chlorides

Acid chlorides react with alcohols to give an ester and HCl. Because acid chlorides are so reactive toward even weak nucleophiles, such as alcohols, no catalyst is necessary for these reactions. Phenol and substituted phenols also react with acid chlorides to give esters.

$$CH_3CH_2CH_2\overset{O}{\overset{\|}{C}}Cl + HO\text{—}C_6H_{11} \longrightarrow CH_3CH_2CH_2\overset{O}{\overset{\|}{C}}O\text{—}C_6H_{11} + HCl$$

Butanoyl chloride — Cyclohexanol — Cyclohexyl butanoate

B. Acid Anhydrides

Acid anhydrides react with alcohols to give 1 mole of ester and 1 mole of a carboxylic acid. Thus, reaction of an alcohol with an anhydride is a useful method for the synthesis of esters.

$$CH_3\overset{O}{\overset{\|}{C}}O\overset{O}{\overset{\|}{C}}CH_3 + HOCH_2CH_3 \longrightarrow CH_3\overset{O}{\overset{\|}{C}}OCH_2CH_3 + CH_3\overset{O}{\overset{\|}{C}}OH$$

Acetic anhydride — Ethanol — Ethyl acetate — Acetic acid

Aspirin is synthesized on an industrial scale by reaction of acetic anhydride and salicylic acid.

$$\text{2-Hydroxybenzoic acid (Salicylic acid)} + CH_3\overset{O}{\overset{\|}{C}}O\overset{O}{\overset{\|}{C}}CH_3 \longrightarrow \text{Acetylsalicylic acid (Aspirin)} + CH_3\overset{O}{\overset{\|}{C}}OH$$

2-Hydroxybenzoic acid (Salicylic acid) — Acetic anhydride — Acetylsalicylic acid (Aspirin) — Acetic acid

13.5 REACTION WITH AMMONIA AND AMINES

A. Acid Chlorides

Acid chlorides react readily with ammonia and with 1° and 2° amines to form amides. For complete conversion of an acid chloride to an amide, two moles of ammonia or amine are used: one to form the amide and one to neutralize the hydrogen chloride formed.

$$CH_3(CH_2)_4\overset{O}{\overset{\|}{C}}Cl + 2NH_3 \longrightarrow CH_3(CH_2)_4\overset{O}{\overset{\|}{C}}NH_2 + NH_4^+Cl^-$$

Hexanoyl chloride — Ammonia — Hexanamide — Ammonium chloride

MECHANISM Reaction of Acetyl Chloride with Ammonia

Step 1: Nucleophilic addition of ammonia to the carbonyl carbon gives a tetrahedral carbonyl addition intermediate.

$$CH_3-\overset{O}{\overset{\|}{C}}-Cl + NH_3 \longrightarrow CH_3-\underset{NH_3^+}{\overset{O^-}{C}}-Cl$$

Tetrahedral carbonyl addition intermediate

Step 2: Collapse of this intermediate gives chloride ion and a protonated amide.

$$CH_3-\underset{NH_3^+}{\overset{O^-}{C}}-Cl \longrightarrow CH_3-\overset{O}{\overset{\|}{C}}-NH_3^+ + Cl^-$$

A protonated amide

Step 3: Proton transfer to ammonia gives the amide and ammonium ion.

$$CH_3-\overset{O}{\overset{\|}{C}}-\underset{H}{\overset{H}{N^+}}-H + NH_3 \longrightarrow CH_3-\overset{O}{\overset{\|}{C}}-\underset{H}{N}-H + NH_4^+$$

B. Acid Anhydrides

Acid anhydrides react with ammonia and with 1° and 2° amines to form amides. As with acid chlorides, 2 moles of ammonia or amine are required: one to form the amide and one to neutralize the carboxylic acid by-product.

$$\underset{\text{Acetic anhydride}}{CH_3\overset{O}{\overset{\|}{C}}O\overset{O}{\overset{\|}{C}}CH_3} + \underset{\text{Ammonia}}{2NH_3} \longrightarrow \underset{\substack{\text{Ethanamide}\\ \text{(Acetamide)}}}{CH_3\overset{O}{\overset{\|}{C}}NH_2} + \underset{\text{Ammonium acetate}}{CH_3\overset{O}{\overset{\|}{C}}O^-NH_4^+}$$

C. Esters

Esters react with ammonia and with 1° and 2° amines to form amides. Because an alkoxide anion is a poor leaving group compared with a halide or carboxylate ion, esters are less reactive toward ammonia, primary amines, and secondary amines than are acid chlorides or acid anhydrides.

$$\underset{\text{Ethyl phenylacetate}}{C_6H_5CH_2\overset{O}{\overset{\|}{C}}OCH_2CH_3} + NH_3 \longrightarrow \underset{\text{Phenylacetamide}}{C_6H_5CH_2\overset{O}{\overset{\|}{C}}NH_2} + CH_3CH_2OH$$

EXAMPLE 13.4

Complete these equations. The stoichiometry of each is given in the equation.

(a) $$\underset{\text{Ethyl 2-butenoate}}{CH_3CH{=}CH\overset{O}{\overset{\|}{C}}OCH_2CH_3} + NH_3 \longrightarrow$$

(b) $$\underset{\text{Diethyl carbonate}}{CH_3CH_2O\overset{O}{\overset{\|}{C}}OCH_2CH_3} + 2NH_3 \longrightarrow$$

SOLUTION

(a) $$\underset{\text{2-Butenamide}}{CH_3CH{=}CH\overset{O}{\overset{\|}{C}}NH_2} + CH_3CH_2OH$$

(b) $$\underset{\text{Urea}}{H_2N\overset{O}{\overset{\|}{C}}NH_2} + 2CH_3CH_2OH$$

Practice Problem 13.4

Complete these equations. The stoichiometry of each is given in the equation.

(a) $$CH_3\overset{O}{\overset{\|}{C}}O{-}C_6H_4{-}O\overset{O}{\overset{\|}{C}}CH_3 + 2NH_3 \longrightarrow$$

(b) (six-membered lactone ring, C=O adjacent to ring O) $+ NH_3 \longrightarrow$

CHEMICAL CONNECTIONS

Pets, Fleas, and Insect Juvenile Hormones

What pet owner hasn't lamented fleas, those almost invisible hopping black specks that raise welts on pets and cause rashes and miserable itching? One treatment is pyrethrin sprays (*The Merck Index,* 12th ed., #8148) that kill the adult insects but leave the thousands of eggs that each female lays unharmed. The most effective alternative treatments on the market today are so-called insect growth regulators, designed to interfere with a very specific phase of the flea's development. One of these growth regulators is (S)-methoprene (*The Merck Index,* 12th ed., #6063), manufactured by Sandoz Animal Health of Des Plaines, Illinois, and marketed under the trade name Precor. Precor has been available for over a decade in animal, house, and garden sprays and in flea and tick collars. (S)-Methoprene is a synthetic analog of insect juvenile hormone, and its role is to interfere with the normal stages of flea development. During the normal transition from larva to pupa stage, levels of juvenile hormone in the flea decrease dramatically. When the insect is exposed to (S)-methoprene, however, levels of juvenile hormone remain high, preventing pupa development. Additionally, (S)-methoprene is absorbed into the adult female flea's ovaries where it disrupts development of eggs before they are even laid. (See Elizabeth Wilson, "Flea Control Methods Capture Public's Fancy," *C&EN,* 31 July 1995.)

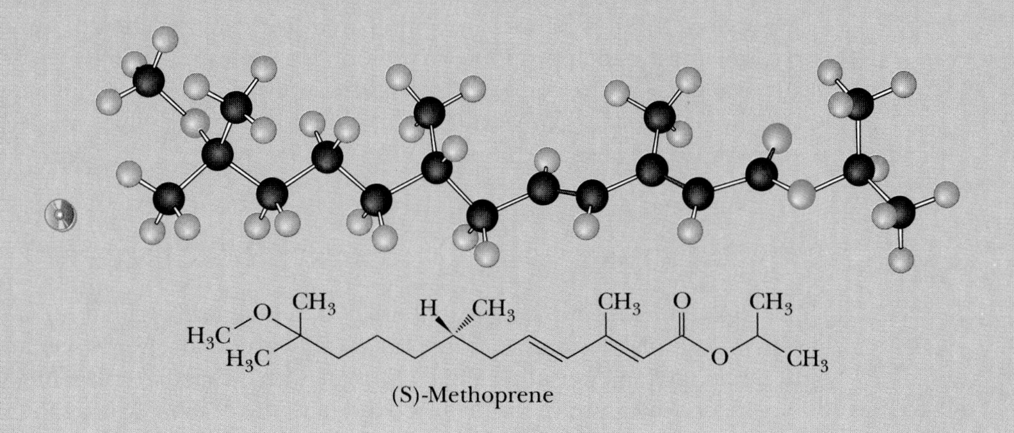

(S)-Methoprene

13.6 INTERCONVERSION OF FUNCTIONAL DERIVATIVES

Throughout these past several sections, we have seen that acid chlorides are the most reactive to nucleophilic acyl substitution and that amides are the least reactive.

Amide < Ester < Acid anhydride < Acid halide

Increasing reactivity toward nucleophilic acyl substitution

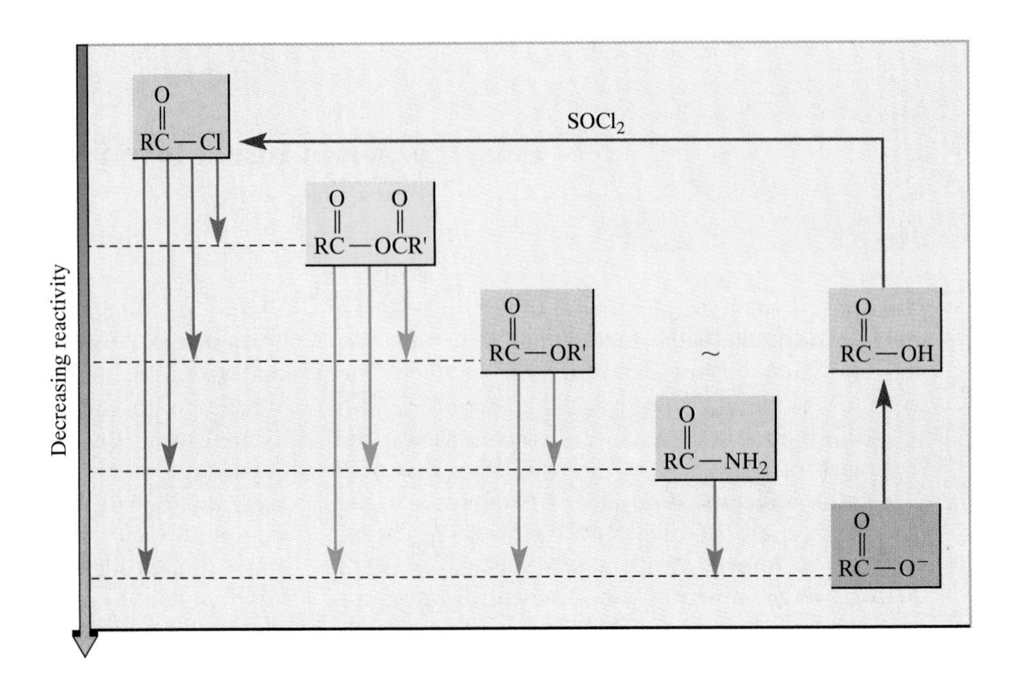

Figure 13.1 Relative reactivities of carboxylic acid derivatives toward nucleophilic acyl substitution. A more reactive derivative may be converted to a less reactive derivative by treatment with an appropriate reagent. Treatment of a carboxylic acid with thionyl chloride converts it to the more reactive acid chloride. Carboxylic acids are about as reactive as esters under acidic conditions but are converted to the unreactive carboxylate anions under basic conditions.

Another useful way to think about the relative reactivities of these four functional derivatives of carboxylic acids is summarized in Figure 13.1. Any functional group lower in this figure can be prepared from any functional group above it by treatment with an appropriate oxygen or nitrogen nucleophile. An acid chloride, for example, can be converted to an acid anhydride, an ester, an amide, or a carboxylic acid. An acid anhydride, ester, or amide, however, does not react with chloride ion to give an acid chloride.

13.7 Esters with Grignard Reagents

Treatment of a formic ester with 2 moles of a Grignard reagent followed by hydrolysis of the magnesium alkoxide salt in aqueous acid gives a secondary alcohol. Treatment of an ester other than a formate with a Grignard reagent gives a tertiary alcohol in which two of the groups bonded to the carbon bearing the —OH group are the same.

$$\underset{\text{An ester of formic acid}}{HC(=O)OCH_3} + 2RMgX \longrightarrow \text{magnesium alkoxide salt} \xrightarrow{H_2O,\ HCl} \underset{\text{A secondary alcohol}}{HC(OH)R-R} + CH_3OH$$

$$\underset{\text{An ester of any acid other than formic acid}}{CH_3C(=O)OCH_3} + 2RMgX \longrightarrow \text{magnesium alkoxide salt} \xrightarrow{H_2O,\ HCl} \underset{\text{A tertiary alcohol}}{CH_3C(OH)(R)-R} + CH_3OH$$

The reaction of an ester and a Grignard reagent begins with addition of 1 mole of Grignard reagent to the carbonyl carbon to form a tetrahedral carbonyl addition intermediate. This addition intermediate collapses to give a ketone (or aldehyde if the starting material was a formic ester) and a magnesium alkoxide salt. This new ketone then reacts with a second mole of Grignard reagent to form a second tetrahedral carbonyl addition intermediate, which, after hydrolysis in aqueous acid, gives a tertiary alcohol (or a secondary alcohol if the starting material was a formic ester). It is important to realize that it is not possible to use RMgX and an ester to prepare a ketone; the intermediate ketone is more reactive than the ester and reacts immediately with the Grignard reagent to give a tertiary alcohol.

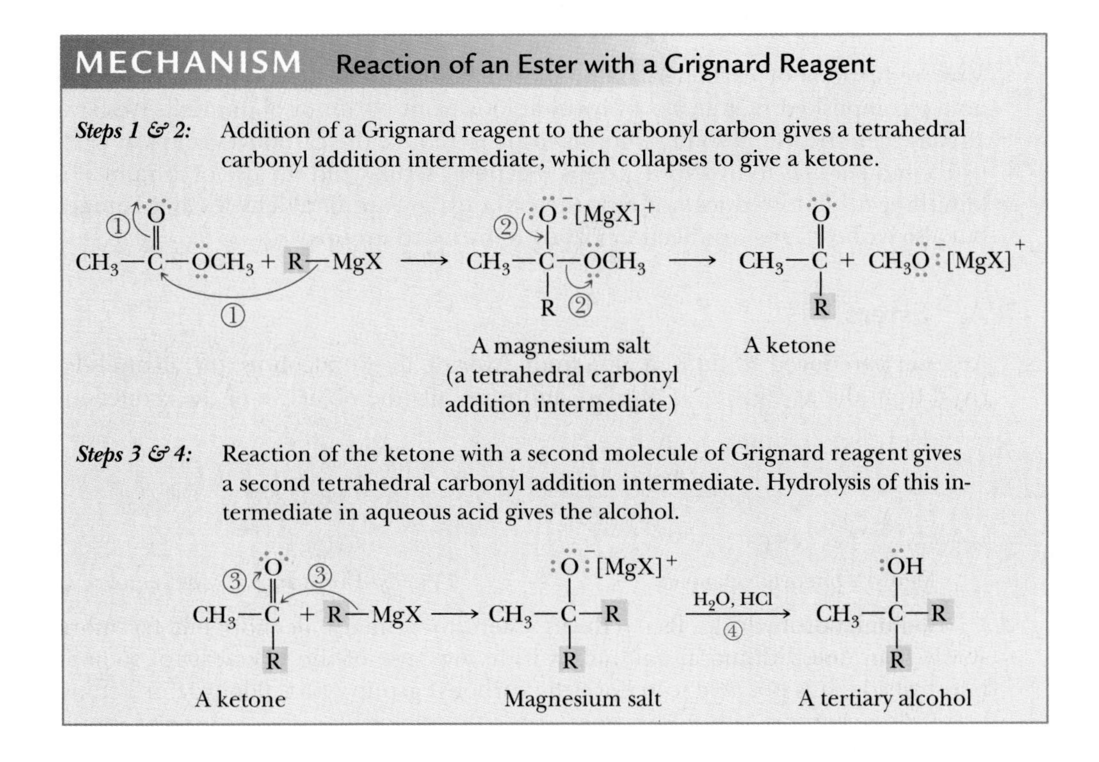

MECHANISM Reaction of an Ester with a Grignard Reagent
Steps 1 & 2: Addition of a Grignard reagent to the carbonyl carbon gives a tetrahedral carbonyl addition intermediate, which collapses to give a ketone.
CH3—C—OCH3 + R—MgX
CH3—C—OCH3
CH3—C + CH3O [MgX]+
A magnesium salt
(a tetrahedral carbonyl addition intermediate)
A ketone
Steps 3 & 4: Reaction of the ketone with a second molecule of Grignard reagent gives a second tetrahedral carbonyl addition intermediate. Hydrolysis of this intermediate in aqueous acid gives the alcohol.
CH3—C + R—MgX
CH3—C—R
H2O, HCl
CH3—C—R
A ketone
Magnesium salt
A tertiary alcohol

EXAMPLE 13.5

Complete each Grignard reaction.

(a) $\mathrm{HCOCH_3}$ (with C=O) $\xrightarrow[\text{2. } H_2O,\ HCl]{\text{1. } 2CH_3CH_2CH_2MgBr}$

(b) $\mathrm{CH_3CH_2CH_2COCH_3}$ (with C=O) $\xrightarrow[\text{2. } H_2O,\ HCl]{\text{1. } 2PhMgBr}$

SOLUTION

Sequence (a) gives a secondary alcohol, and sequence (b) gives a tertiary alcohol.

(a) $\mathrm{CH_3CH_2CH_2CH(OH)CH_2CH_2CH_3}$

(b) $\mathrm{CH_3CH_2CH_2C(OH)(Ph)Ph}$

Practice Problem 13.5

Show how to prepare each alcohol by treatment of an ester with a Grignard reagent.

(a) (cyclopentyl)$_2$CH(OH) — dicyclopentylmethanol, drawn as OH on CH between two cyclopentane rings

(b) $CH_2{=}CHCH_2C(OH)(Ph)CH_2CH{=}CH_2$

13.8 Reduction

Most reductions of carbonyl compounds, including aldehydes and ketones, are now accomplished by transfer of hydride ions from boron or aluminum. We have already seen the use of sodium borohydride to reduce the carbonyl groups of aldehydes and ketones to hydroxyl groups (Section 11.10B) and the use of lithium aluminum hydride to reduce not only the carbonyl groups of aldehydes and ketones but also carboxyl groups (Section 12.5A) to hydroxyl groups.

A. Esters

An ester is reduced by lithium aluminum hydride to two alcohols; the alcohol derived from the acyl group is primary and is usually the objective of the reduction.

$$C_6H_5CH(CH_3)\overset{O}{\overset{\|}{C}}OCH_3 \xrightarrow[\text{2. } H_2O,\ HCl]{\text{1. } LiAlH_4,\ \text{ether}} C_6H_5CH(CH_3)CH_2OH + CH_3OH$$

Methyl 2-phenylpropanoate → 2-Phenyl-1-propanol + Methanol

Sodium borohydride also reduces esters to primary alcohols but far more slowly than does lithium aluminum hydride. Because of the selectivity of sodium borohydride, it is possible to reduce the carbonyl group of an aldehyde or ketone to a hydroxyl group with this reagent without reducing an ester or carboxyl group in the same molecule.

$$CH_3\overset{O}{\overset{\|}{C}}CH_2\overset{O}{\overset{\|}{C}}OCH_2CH_3 \xrightarrow[\text{2. } H_2O]{\text{1. } NaBH_4,\ \text{ether}} CH_3\overset{OH}{\overset{|}{C}H}CH_2\overset{O}{\overset{\|}{C}}OCH_2CH_3$$

Ethyl 3-oxobutanoate (Ethyl acetoacetate) → Ethyl 3-hydroxybutanoate

B. Amides

Lithium aluminum hydride reduction of amides can be used to prepare primary, secondary, or tertiary amines, depending on the degree of substitution of the amide.

$$CH_3(CH_2)_6\overset{O}{\overset{\|}{C}}NH_2 \xrightarrow[\text{2. } H_2O]{\text{1. } LiAlH_4} CH_3(CH_2)_6CH_2NH_2$$

Octanamide → 1-Octanamine

$$\text{C}_6\text{H}_5\overset{\text{O}}{\overset{\|}{\text{C}}}\text{N(CH}_3)_2 \xrightarrow[\text{2. H}_2\text{O}]{\text{1. LiAlH}_4} \text{C}_6\text{H}_5\text{CH}_2\text{N(CH}_3)_2$$

N,N-Dimethylbenzamide　　*N,N*-Dimethylbenzylamine

Amides are also reduced by hydrogen in the presence of a transition metal catalyst. At one time, the major commercial preparation of 1,6-hexanediamine, one of the two monomers needed for the synthesis of nylon 66, was by catalytic reduction of hexanediamide.

$$\text{H}_2\text{N}\overset{\text{O}}{\overset{\|}{\text{C}}}(\text{CH}_2)_4\overset{\text{O}}{\overset{\|}{\text{C}}}\text{NH}_2 + 4\text{H}_2 \xrightarrow[\text{heat, pressure}]{\text{transition metal catalyst}} \text{H}_2\text{NCH}_2(\text{CH}_2)_4\text{CH}_2\text{NH}_2 + 2\text{H}_2\text{O}$$

Hexanediamide (Adipamide)　　1,6-Hexanediamine (Hexamethylenediamine)

EXAMPLE 13.6

Show how to bring about each conversion.

(a) $\text{C}_6\text{H}_5\overset{\text{O}}{\overset{\|}{\text{C}}}\text{OH} \longrightarrow \text{C}_6\text{H}_5\text{CH}_2\text{—N(pyrrolidine)}$　(b) cyclohexyl$\text{—}\overset{\text{O}}{\overset{\|}{\text{C}}}\text{OH} \longrightarrow$ cyclohexyl$\text{—CH}_2\text{NHCH}_3$

SOLUTION

The key in each part is to convert the carboxylic acid to an amide and then reduce the amide with $LiAlH_4$ (Section 13.8B). The amide can be prepared by treatment of the carboxylic acid with $SOCl_2$ to form the acid chloride (Section 12.7) and then treatment of the acid chloride with an amine (Section 13.5A). Alternatively, the carboxylic acid can be converted to an ethyl ester by Fischer esterification (Section 12.6), and the ester can be treated with an amine to give the amide. Solution (a) uses the acid chloride route, and solution (b) uses the ester route.

(a) $\text{C}_6\text{H}_5\overset{\text{O}}{\overset{\|}{\text{C}}}\text{OH} \xrightarrow{\text{SOCl}_2} \text{C}_6\text{H}_5\overset{\text{O}}{\overset{\|}{\text{C}}}\text{Cl} \xrightarrow{\text{HN(pyrrolidine)}} \text{C}_6\text{H}_5\overset{\text{O}}{\overset{\|}{\text{C}}}\text{—N(pyrrolidine)} \xrightarrow[\text{2. H}_2\text{O}]{\text{1. LiAlH}_4} \text{C}_6\text{H}_5\text{CH}_2\text{—N(pyrrolidine)}$

(b) cyclohexyl$\text{—}\overset{\text{O}}{\overset{\|}{\text{C}}}\text{OH} \xrightarrow[\text{2. CH}_3\text{NH}_2]{\text{1. CH}_3\text{CH}_2\text{OH, H}^+}$ cyclohexyl$\text{—}\overset{\text{O}}{\overset{\|}{\text{C}}}\text{NHCH}_3 \xrightarrow[\text{4. H}_2\text{O}]{\text{3. LiAlH}_4}$ cyclohexyl$\text{—CH}_2\text{NHCH}_3$

Practice Problem 13.6

Show how to convert hexanoic acid to each amine in good yield.

(a) $CH_3(CH_2)_5N(CH_3)_2$　　(b) $CH_3(CH_2)_5NHCH(CH_3)_2$

EXAMPLE 13.7

Show how to convert phenylacetic acid to these compounds.

(a) $\text{PhCH}_2\overset{\text{O}}{\overset{\|}{\text{C}}}\text{OCH}_3$　(b) $\text{PhCH}_2\overset{\text{O}}{\overset{\|}{\text{C}}}\text{NH}_2$　(c) $PhCH_2CH_2NH_2$　(d) $PhCH_2CH_2OH$

SOLUTION

Prepare methyl ester (a) by Fischer esterification (Section 12.6) of phenylacetic acid with methanol and then treat this ester with ammonia to prepare amide (b). Alternatively, treat phenylacetic acid with thionyl chloride (Section 12.7) to give an acid chloride, and then treat this acid chloride with two equivalents of ammonia to give amide (b). Reduction of amide (b) by $LiAlH_4$ gives the primary amine (c). Similar reduction of either phenylacetic acid or ester (a) gives alcohol (d).

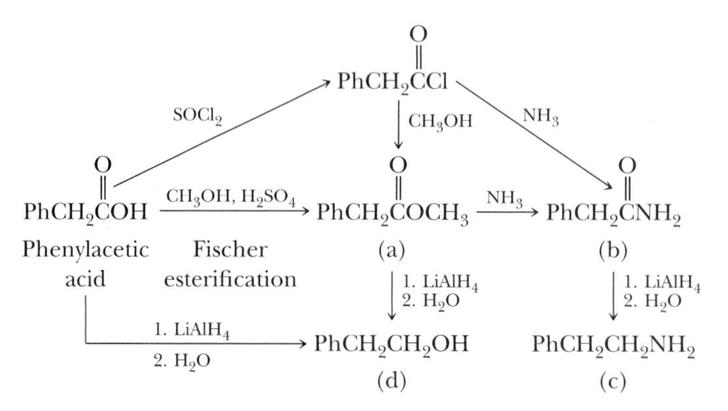

Practice Problem 13.7

Show how to convert (R)-2-phenylpropanoic acid to these compounds.

(a) (R)-PhCH(CH_3)CH_2OH — (R)-2-Phenyl-1-propanol

(b) (R)-PhCH(CH_3)CH_2NH_2 — (R)-2-Phenyl-1-propanamine

SUMMARY

The functional group of an **acid halide** (Section 13.1A) is an acyl group bonded to a halogen. The most common and widely used of these are the acid chlorides. The functional group of a **carboxylic anhydride** (Section 13.1B) is two acyl groups bonded to an oxygen. The functional group of an **ester** (Section 13.1C) is an acyl group bonded to —OR or —OAr. A cyclic ester is given the name **lactone.** Phosphoric acid has three —OH groups and can form mono-, di-, and triesters. The functional group of an **amide** (Section 13.1D) is an acyl group bonded to a trivalent nitrogen. A cyclic amide is given the name **lactam.** The functional group of an **imide** is a trivalent nitrogen bonded to two acyl groups.

A common reaction theme of functional derivatives of carboxylic acids is **nucleophilic acyl addition** to the carbonyl carbon to form a **tetrahedral carbonyl addition intermediate,** which then collapses to regenerate the carbonyl group. The result is **nucleophilic acyl substitution** (Section 13.2). Listed in order of increasing reactivity toward nucleophilic acyl substitution, these functional derivatives are

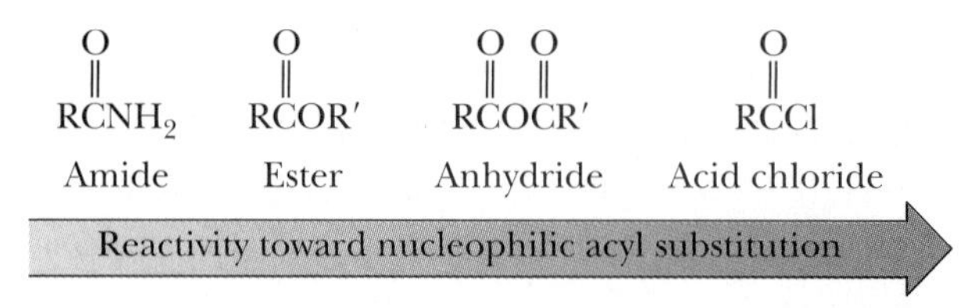

Any more reactive functional derivative can be converted to any less reactive functional derivative by reaction with an appropriate oxygen or nitrogen nucleophile (Section 13.6).

KEY REACTIONS

1. Hydrolysis of an Acid Chloride (Section 13.3A)

Low-molecular-weight acid chlorides react vigorously with water. Higher-molecular-weight acid chlorides react less rapidly.

$$CH_3\overset{O}{\overset{\|}{C}}Cl + H_2O \longrightarrow CH_3\overset{O}{\overset{\|}{C}}OH + HCl$$

2. Hydrolysis of an Acid Anhydride (Section 13.3B)

Low-molecular-weight acid anhydrides react readily with water. Higher-molecular-weight acid anhydrides react less rapidly.

$$CH_3\overset{O}{\overset{\|}{C}}O\overset{O}{\overset{\|}{C}}CH_3 + H_2O \longrightarrow CH_3\overset{O}{\overset{\|}{C}}OH + HO\overset{O}{\overset{\|}{C}}CH_3$$

3. Hydrolysis of an Ester (Section 13.3C)

Esters are hydrolyzed only in the presence of acid or base. Acid is a catalyst. Base is required in an equimolar amount.

$$CH_3\overset{O}{\overset{\|}{C}}O{-}C_6H_{11} + NaOH \xrightarrow{H_2O} CH_3\overset{O}{\overset{\|}{C}}O^-Na^+ + HO{-}C_6H_{11}$$

4. Hydrolysis of an Amide (Section 13.3D)

Either acid or base is required in an amount equivalent to that of the amide.

$$CH_3CH_2CH_2\overset{O}{\overset{\|}{C}}NH_2 + H_2O + HCl \xrightarrow[heat]{H_2O} CH_3CH_2CH_2\overset{O}{\overset{\|}{C}}OH + NH_4^+\ Cl^-$$

$$CH_3\overset{O}{\overset{\|}{C}}NH{-}C_6H_5 + NaOH \xrightarrow[heat]{H_2O} CH_3\overset{O}{\overset{\|}{C}}O^-Na^+ + H_2N{-}C_6H_5$$

5. Reaction of an Acid Chloride with an Alcohol (Section 13.4A)

Treatment of an acid chloride with an alcohol gives an ester and HCl.

$$CH_3CH_2CH_2\overset{O}{\overset{\|}{C}}Cl + HO{-}C_6H_{11} \longrightarrow CH_3CH_2CH_2\overset{O}{\overset{\|}{C}}O{-}C_6H_{11} + HCl$$

6. Reaction of an Acid Anhydride with an Alcohol (Section 13.4B)

Treatment of an acid anhydride with an alcohol gives an ester and a carboxylic acid.

$$CH_3\overset{O}{\overset{\|}{C}}O\overset{O}{\overset{\|}{C}}CH_3 + HOCH_2CH_3 \longrightarrow CH_3\overset{O}{\overset{\|}{C}}OCH_2CH_3 + CH_3\overset{O}{\overset{\|}{C}}OH$$

7. Reaction of an Acid Chloride with Ammonia or an Amine (Section 13.5A)

Reaction requires 2 moles of ammonia or amine: 1 mole to form the amide and 1 mole to neutralize the HCl by-product.

$$CH_3\overset{O}{\overset{\|}{C}}Cl + 2NH_3 \longrightarrow CH_3\overset{O}{\overset{\|}{C}}NH_2 + NH_4^+Cl^-$$

8. Reaction of an Acid Anhydride with Ammonia or an Amine (Section 13.5B)

Reaction requires 2 moles of ammonia or amine: 1 mole to form the amide and 1 mole to neutralize the carboxylic acid by-product.

$$CH_3\overset{O}{\overset{\|}{C}}O\overset{O}{\overset{\|}{C}}CH_3 + 2NH_3 \longrightarrow CH_3\overset{O}{\overset{\|}{C}}NH_2 + CH_3\overset{O}{\overset{\|}{C}}O^-NH_4^+$$

9. Reaction of an Ester with Ammonia or an Amine (Section 13.5C)

Treatment of an ester with ammonia, a primary amine, or a secondary amine gives an amide.

$$C_6H_5-CH_2\overset{O}{\overset{\|}{C}}OCH_2CH_3 + NH_3 \longrightarrow C_6H_5-CH_2\overset{O}{\overset{\|}{C}}NH_2 + CH_3CH_2OH$$

10. Reaction of an Ester with a Grignard Reagent (Section 13.7)

Treatment of a formic ester with a Grignard reagent followed by hydrolysis gives a secondary alcohol. Treatment of any other ester with a Grignard reagent gives a tertiary alcohol.

$$C_5H_9-\overset{O}{\overset{\|}{C}}OCH_3 \xrightarrow[2.\ H_2O,\ HCl]{1.\ 2CH_3CH_2MgBr} C_5H_9-\overset{OH}{\overset{|}{C}}(CH_2CH_3)_2 + CH_3OH$$

11. Reduction of an Ester (Section 13.8A)

Reduction by lithium aluminum hydride gives two alcohols.

$$C_6H_5-\underset{CH_3}{\underset{|}{C}}H\overset{O}{\overset{\|}{C}}OCH_3 \xrightarrow[2.\ H_2O]{1.\ LiAlH_4,\ ether} C_6H_5-\underset{CH_3}{\underset{|}{C}}HCH_2OH + CH_3OH$$

12. Reduction of an Amide (Section 13.8B)

Reduction by lithium aluminum hydride gives an amine.

$$C_6H_5-\overset{O}{\overset{\|}{C}}NH_2 \xrightarrow[2.\ H_2O]{1.\ LiAlH_4,\ THF} C_6H_5-CH_2NH_2$$

Additional Problems

Structure and Nomenclature

13.8 Draw a structural formula for each compound.

(a) Dimethyl carbonate **(b)** *p*-Nitrobenzamide

(c) Octanoyl chloride **(d)** Diethyl oxalate

(e) Ethyl *cis*-2-pentenoate **(f)** Butanoic anhydride

(g) Dodecanamide **(h)** Ethyl 3-hydroxybutanoate

13.9 Write the IUPAC name for each compound.

(a) $C_6H_5COOCC_6H_5$ (O=C–O–C=O between two benzene rings) **(b)** $CH_3(CH_2)_{14}COCH_3$ (C=O) **(c)** $CH_3(CH_2)_4CNHCH_3$ (C=O)

(d) H_2N–C_6H_4–CNH_2 (C=O) **(e)** $CH_2(CO_2CH_2CH_3)_2$ **(f)** $PhCH_2CCHCOCH_3$ (two C=O; CH_3 on CH)

13.10 When oil from the head of the sperm whale is cooled, spermaceti, a translucent wax with a white, pearly luster, crystallizes from the mixture. Spermaceti, which makes up 11% of whale oil, is composed mainly of hexadecyl hexadecanoate (cetyl palmitate). At one time, spermaceti was widely used in the making of cosmetics, fragrant soaps, and candles (*The Merck Index,* 12th ed., #8892). Draw a structural formula of cetyl palmitate.

Sperm whale, *Physterer macrocephalus,* diving, Kaikoura, NZ. *(Kim Westerskov/Tony Stone Images)*

Physical Properties

13.11 Acetic acid and methyl formate are constitutional isomers. Both are liquids at room temperature: one with a boiling point of 32°C, the other with a boiling point of 118°C. Which of the two has the higher boiling point?

13.12 Acetic acid has a boiling point of 118°C, whereas its methyl ester has a boiling point of 57°C. Account for the fact that the boiling point of acetic acid is higher than that of its methyl ester, even though acetic acid has a lower molecular weight.

Reactions

13.13 Arrange these compounds in order of increasing reactivity toward nucleophilic acyl substitution.

(1) cyclopentyl–C(=O)–OCH_2CH_3 (2) cyclopentyl–C(=O)–Cl (3) cyclopentyl–C(=O)–NH_2 (4) cyclopentyl–C(=O)–O–C(=O)–CH_3

13.14 A common method for preparing acid anhydrides is treatment of an acid chloride with the sodium salt of a carboxylic acid. For example, treatment of benzoyl chloride with sodium acetate gives acetic benzoic anhydride. Write a mechanism for this nucleophilic acyl substitution reaction.

$$CH_3CO^-Na^+ + ClCC_6H_5 \longrightarrow CH_3COCC_6H_5 + Na^+Cl^-$$

Sodium acetate Benzoyl chloride Acetic benzoic anhydride

13.15 Show how to prepare these mixed anhydrides using the method described in Problem 13.14. (*Hint:* HCOCl is too unstable to use.)

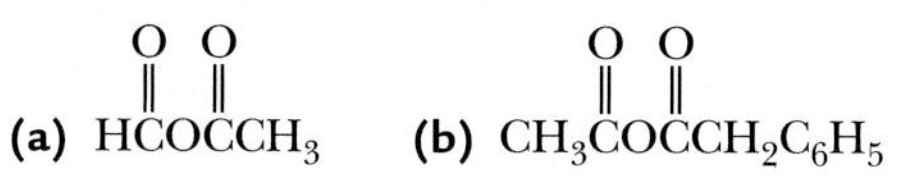

(a) $HCOCCH_3$ **(b)** $CH_3COCCH_2C_6H_5$

13.16 A carboxylic acid can be converted to an ester by Fischer esterification. Show how to synthesize each ester from a carboxylic acid and an alcohol by Fischer esterification.

(a) $C_6H_{11}OC(=O)(CH_2)_4CH_3$ **(b)** $(CH_3)_2CHC(=O)OCH_2CH_3$

13.17 A carboxylic acid can also be converted to an ester in two reactions by first converting the carboxylic acid to its acid chloride and then treating the acid chloride with an alcohol. Show how to prepare each ester in Problem 13.16 from a carboxylic acid and an alcohol by this two-step scheme.

13.18 Show how to prepare these amides by reaction of an acid chloride with ammonia or an amine.

(a) $C_6H_{11}NHC(=O)(CH_2)_4CH_3$ **(b)** $(CH_3)_2CHC(=O)N(CH_3)_2$ **(c)** $H_2NC(=O)(CH_2)_4C(=O)NH_2$

13.19 Write a mechanism for the reaction of butanoyl chloride and ammonia to give butanamide and ammonium chloride.

$$CH_3(CH_2)_2CCl(=O) + 2NH_3 \longrightarrow CH_3(CH_2)_2CNH_2(=O) + NH_4^+Cl^-$$

13.20 What product is formed when benzoyl chloride is treated with these reagents?

(a) C_6H_6, $AlCl_3$ **(b)** $CH_3CH_2CH_2CH_2OH$

(c) $CH_3CH_2CH_2CH_2SH$ **(d)** $CH_3CH_2CH_2CH_2NH_2$ (2 equivalents)

(e) H_2O **(f)** piperidine N—H (2 equivalents)

13.21 Write the product(s) of treatment of propanoic anhydride with each reagent.

(a) Ethanol (1 equivalent) **(b)** Ammonia (2 equivalents)

13.22 Write the product of treatment of succinic anhydride with each reagent.

(a) Ethanol (1 equivalent) **(b)** Ammonia (2 equivalents)

13.23 The analgesic phenacetin (*The Merck Index,* 12th ed., #7344) is synthesized by treating 4-ethoxyaniline with acetic anhydride. Write an equation for the formation of phenacetin.

13.24 The analgesic acetaminophen (*The Merck Index,* 12th ed., #45) is synthesized by treatment of 4-aminophenol with one equivalent of acetic anhydride. Write an equation for the formation of acetaminophen. (*Hint:* An —NH_2 group is a better nucleophile than an —OH group.)

13.25 Nicotinic acid (*The Merck Index,* 12th ed., #6612), more commonly named niacin, is one of the B vitamins. Show how nicotinic acid can be converted to ethyl nicotinate and then to nicotinamide.

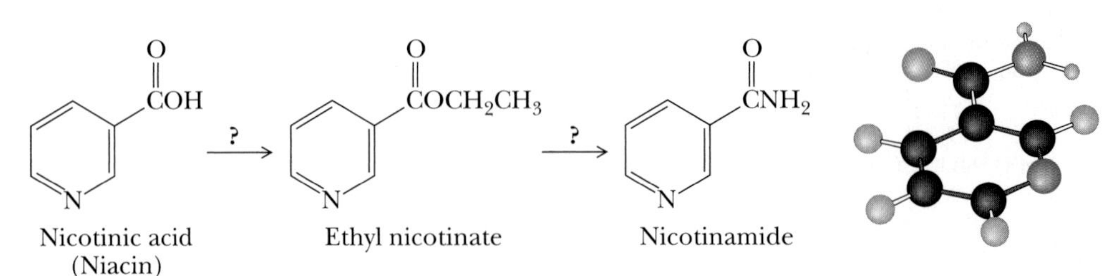

Nicotinic acid (Niacin) Ethyl nicotinate Nicotinamide

13.26 Complete these reactions.

(a) $CH_3O-C_6H_4-NH_2 + CH_3\overset{O}{\overset{\|}{C}}O\overset{O}{\overset{\|}{C}}CH_3 \longrightarrow$ **(b)** $CH_3\overset{O}{\overset{\|}{C}}Cl + 2HN(CH_2)_5 \longrightarrow$

(c) $CH_3\overset{O}{\overset{\|}{C}}OCH_3 + HN(CH_2)_5 \longrightarrow$ **(d)** $C_6H_5-NH_2 + CH_3(CH_2)_5\overset{O}{\overset{\|}{C}}H \longrightarrow$

13.27 What product is formed when ethyl benzoate is treated with these reagents?

(a) H_2O, NaOH, heat **(b)** $LiAlH_4$, then H_2O
(c) H_2O, H_2SO_4, heat **(d)** $CH_3CH_2CH_2CH_2NH_2$
(e) C_6H_5MgBr (2 moles), then H_2O/HCl

13.28 Show how to convert 2-hydroxybenzoic acid (salicylic acid) to these compounds.

(a) o-$HO-C_6H_4-\overset{O}{\overset{\|}{C}}OCH_3$ **(b)** o-$CH_3\overset{O}{\overset{\|}{C}}O-C_6H_4-\overset{O}{\overset{\|}{C}}OH$

Methyl salicylate (Oil of wintergreen) Acetylsalicylic acid (Aspirin)

13.29 What product is formed when benzamide is treated with these reagents?

(a) H_2O, HCl, heat **(b)** NaOH, H_2O, heat **(c)** $LiAlH_4$, then H_2O

13.30 Show the product of treatment of γ-butyrolactone with each reagent.

(a) NH_3 **(b)** $LiAlH_4$, then H_2O **(c)** NaOH, H_2O, heat

13.31 Show the product of treatment of *N*-methyl-γ-butyrolactam with each reagent.

(a) H_2O, HCl, heat **(b)** NaOH, H_2O, heat **(c)** $LiAlH_4$, then H_2O

13.32 Complete these reactions.

(a) $C_6H_5-\overset{O}{\overset{\|}{C}}OC_2H_5 \xrightarrow[\text{2. } H_2O/HCl]{\text{1. } 2CH_2{=}CHCH_2MgBr}$

(b) $C_6H_5-\overset{O}{\overset{\|}{C}}-C_6H_5 \xrightarrow[\text{2. } H_2O/HCl]{\text{1. } CH_3MgBr}$

(c) $CH_3CH_2CH_2CH_2\overset{O}{\overset{\|}{C}}OCH_2CH_3 \xrightarrow[\text{2. } H_2O/HCl]{\text{1. } 2CH_3MgBr}$

13.33 What combination of ester and Grignard reagent can be used to prepare each alcohol?

(a) 2-Methyl-2-butanol **(b)** 3-Phenyl-3-pentanol **(c)** 1,1-Diphenylethanol

13.34 Treatment of γ-butyrolactone with two equivalents of methylmagnesium bromide followed by hydrolysis in aqueous acid gives a compound of molecular formula $C_6H_{14}O_2$. Propose a structural formula for this compound.

$$\gamma\text{-butyrolactone} \xrightarrow[\text{2. } H_2O/HCl]{\text{1. } 2CH_3MgBr} C_6H_{14}O_2$$

13.35 Reaction of a primary or secondary amine with diethyl carbonate under controlled conditions gives a carbamic ester. Propose a mechanism for this reaction.

$$\underset{\text{Diethyl carbonate}}{\text{EtO}\overset{\text{O}}{\overset{\|}{\text{C}}}\text{OEt}} + \underset{\text{Butylamine}}{H_2NCH_2CH_2CH_2CH_3} \longrightarrow \underset{\text{A carbamic ester}}{\text{EtO}\overset{\text{O}}{\overset{\|}{\text{C}}}NHCH_2CH_2CH_2CH_3} + \text{EtOH}$$

13.36 Barbiturates are prepared by treatment of diethyl malonate or a derivative of diethyl malonate with urea in the presence of sodium ethoxide as a catalyst. Following is an equation for the preparation of barbital, a long-duration hypnotic and sedative, from diethyl diethylmalonate and urea. Barbital is prescribed under one of a dozen or more trade names (*The Merck Index,* 12th ed., #989).

$$\underset{\text{Diethyl diethylmalonate}}{(CH_3CH_2)_2C(COCH_2CH_3)_2} + \underset{\text{Urea}}{(H_2N)_2C{=}O} \xrightarrow[\text{2. } H_2O]{\text{1. } CH_3CH_2O^-Na^+} \underset{\substack{\text{5,5-Diethylbarbituric acid} \\ \text{(Barbital)}}}{\text{(ring structure, positions 1–6)}} + 2CH_3CH_2OH$$

(a) Propose a mechanism for this reaction.

(b) The pK_a of barbital is 7.4. Which is the most acidic hydrogen in this molecule and how do you account for its acidity?

Crystals of phenobarbital viewed under polarized light. *(© 1966 Mel Pollinger/Fran Heyl Associates)*

13.37 Draw structural formulas for the products of complete hydrolysis of meprobamate and phenobarbital in hot aqueous acid. Meprobamate is a tranquilizer prescribed under one or more of 58 different trade names. Phenobarbital is a long-acting sedative, hypnotic, and anticonvulsant (*The Merck Index,* 12th ed., #5908 and #7386) [*Hint:* Remember that when heated, β-dicarboxylic acids and β-ketoacids undergo decarboxylation (Section 12.8B).]

(a) $H_2N\overset{\text{O}}{\overset{\|}{\text{C}}}OCH_2\underset{CH_2CH_2CH_3}{\overset{CH_3}{C}}CH_2O\overset{\text{O}}{\overset{\|}{\text{C}}}NH_2$

Meprobamate

(b) Phenobarbital (5-ethyl-5-phenyl barbituric acid ring: C_6H_5, CH_3CH_2, NH, NH, O, O, =O)

Phenobarbital

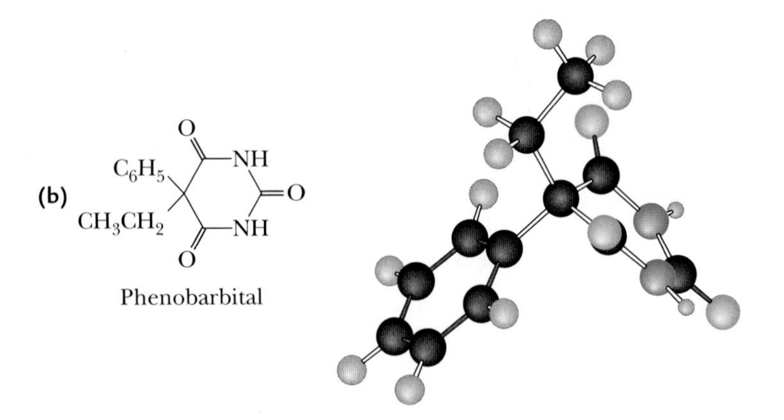

Synthesis

13.38 *N,N*-Diethyl-*m*-toluamide (Deet, *The Merck Index,* 12th ed., #2912), the active ingredient in several common insect repellents, is synthesized from 3-methylbenzoic acid (*m*-toluic acid) and diethylamine. Show how this synthesis can be accomplished.

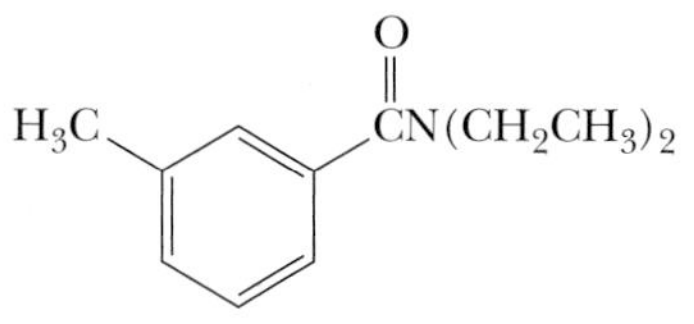

H_3C, $\overset{O}{\overset{\|}{C}}N(CH_2CH_3)_2$

N,N-Diethyl-*m*-toluamide (DEET)

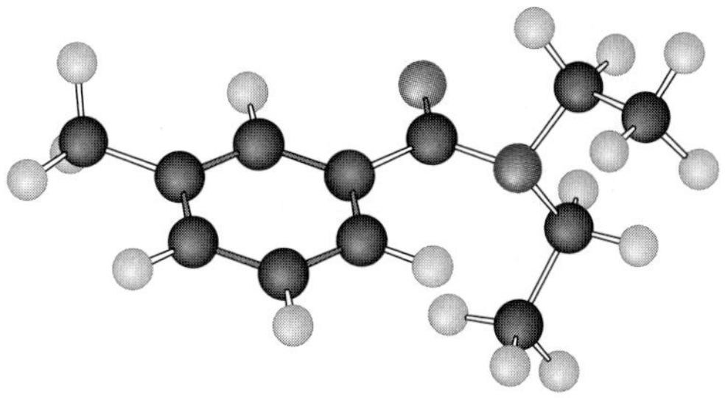

13.39 Show reagents for the synthesis of the following tertiary amine.

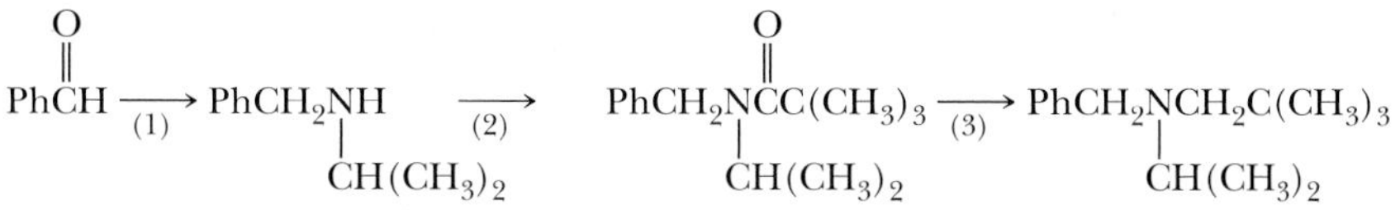

$$Ph\overset{O}{\overset{\|}{C}}H \xrightarrow{(1)} PhCH_2\underset{CH(CH_3)_2}{\underset{|}{N}}H \xrightarrow{(2)} PhCH_2\underset{CH(CH_3)_2}{\underset{|}{N}}\overset{O}{\overset{\|}{C}}C(CH_3)_3 \xrightarrow{(3)} PhCH_2\underset{CH(CH_3)_2}{\underset{|}{N}}CH_2C(CH_3)_3$$

13.40 Show how to convert ethyl 2-pentenoate to these compounds.

$$CH_3CH_2CH{=}CH\overset{O}{\overset{\|}{C}}OCH_2CH_3$$

Ethyl 2-pentenoate

(a) $CH_3CH_2CH_2CH_2\overset{O}{\overset{\|}{C}}OCH_2CH_3$ **(b)** $CH_3CH_2CH{=}CHCH_2OH$

(c) $CH_3CH_2\overset{HO}{\overset{|}{C}}H\overset{HO}{\overset{|}{C}}H\overset{O}{\overset{\|}{C}}OCH_2CH_3$

13.41 Procaine (its hydrochloride is marketed as Novocaine) was one of the first local anesthetics for infiltration and regional anesthesia. Show how to synthesize procaine using the given reagents as sources of carbon atoms.

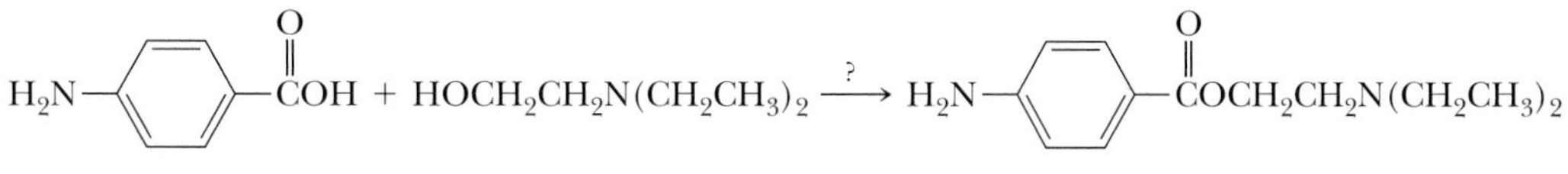

$$H_2N{-}C_6H_4{-}\overset{O}{\overset{\|}{C}}OH + HOCH_2CH_2N(CH_2CH_3)_2 \xrightarrow{?} H_2N{-}C_6H_4{-}\overset{O}{\overset{\|}{C}}OCH_2CH_2N(CH_2CH_3)_2$$

p-Aminobenzoic acid 2-Diethylaminoethanol Procaine (Novocaine)

13.42 Starting materials for the synthesis of the herbicide propranil, a weed killer used in rice paddies, are benzene and propanoic acid. Show reagents to bring about this synthesis.

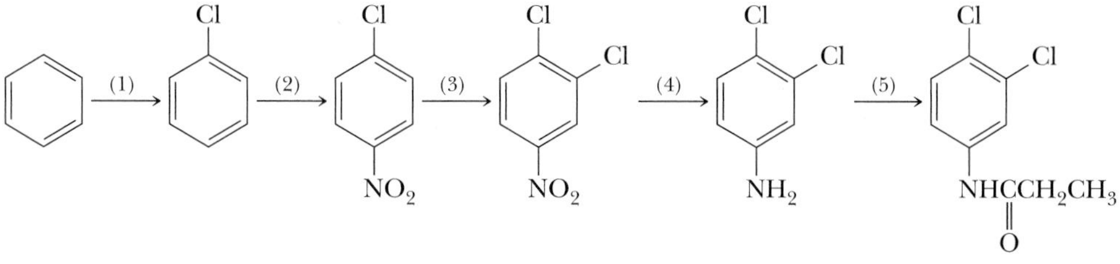

Benzene $\xrightarrow{(1)}$ Cl-benzene $\xrightarrow{(2)}$ 4-Cl-nitrobenzene (NO_2) $\xrightarrow{(3)}$ 3,4-dichloronitrobenzene (NO_2) $\xrightarrow{(4)}$ 3,4-dichloroaniline (NH_2) $\xrightarrow{(5)}$ $NH\overset{O}{\overset{\|}{C}}CH_2CH_3$

Propranil

13.43 Following are structural formulas for three local anesthetics. Lidocaine (*The Merck Index,* 12th ed., #5505) was introduced in 1948 and is now the most widely used local anesthetic for infiltration and regional anesthesia. Its hydrochloride is marketed under the name Xylocaine. Etidocaine (*The Merck Index,* 12th ed., #3907 hydrochloride marketed as Duranest) is comparable to lidocaine in onset, but its analgesic action lasts two to three times longer. Anesthetic action from mepivacaine (*The Merck Index,* 12th ed., #5905, hydrochloride marketed as Carbocaine) is faster and somewhat longer in duration than lidocaine.

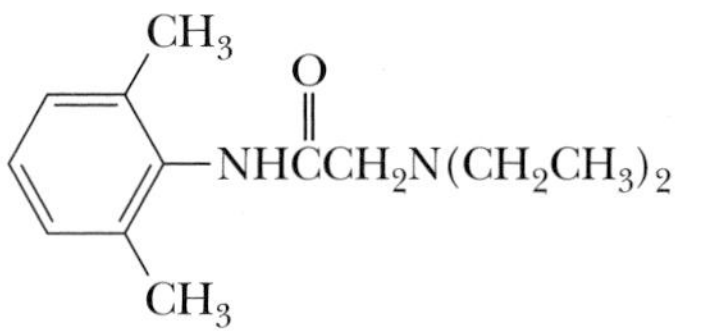

Lidocaine (Xylocaine) | Etidocaine (Duranest) | Mepivacaine (Carbocaine)

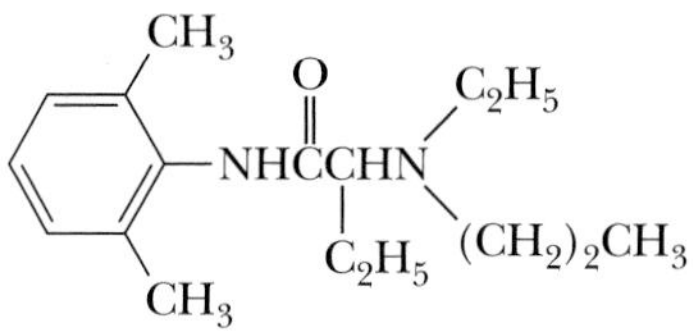

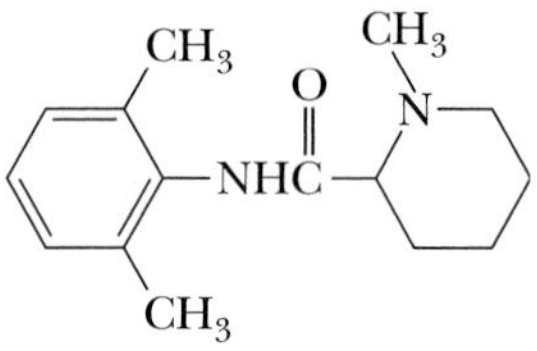

(a) Propose a synthesis of lidocaine from 2,6-dimethylaniline, chloroacetyl chloride ($ClCH_2COCl$), and diethylamine.

(b) Propose a synthesis of etidocaine from 2,6-dimethylaniline, 2-chlorobutanoyl chloride, and ethylpropylamine.

(c) What amine and acid chloride can be reacted to give mepivacaine?

13.44 At this point, all ^{1}H-NMR and ^{13}C-NMR spectroscopy Additional Problems (Problems 21.10–21.34) may be assigned.

13.45 At this point, all IR spectroscopy Additional Problems (Problems 22.1–22.13) may be assigned.

14

Enolate Anions

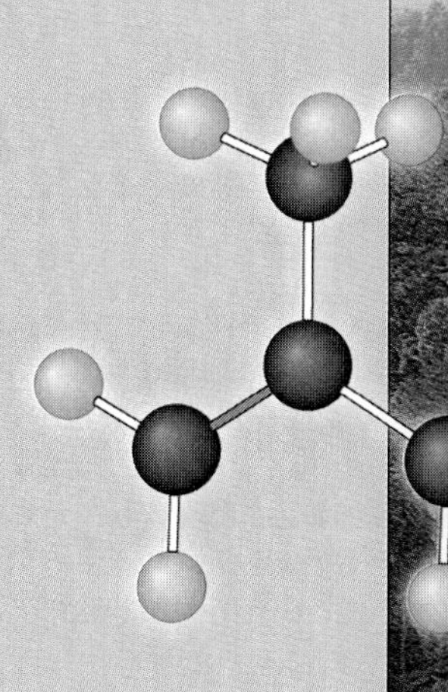

Haze over the Blue Ridge Mountains. Inset: A model of isoprene. *(Bob Thomason/Tony Stone Images)*

This chapter is a continuation of the chemistry of carbonyl compounds. In Chapters 11–13, we concentrated on the carbonyl group itself and on nucleophilic additions to it to form tetrahedral carbonyl addition compounds. In this chapter, we expand on the chemistry of carbonyl-containing compounds and consider the acidity of their α-hydrogens, and the formation and reactions of enolate anions.

14.1 Formation of Enolate Anions

A. Acidity of α-Hydrogens

A carbon atom adjacent to a carbonyl group is called an **α-carbon,** and a hydrogen atom bonded to it is called an **α-hydrogen.**

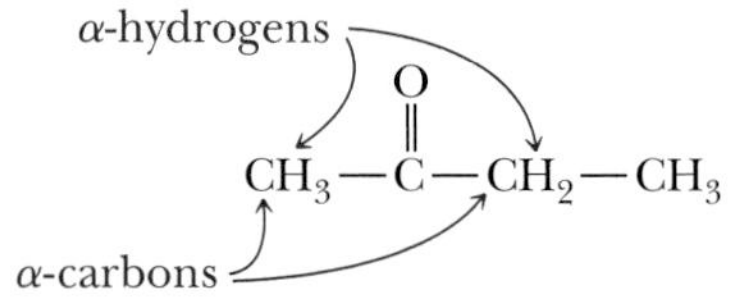

α-hydrogens

$$CH_3-\overset{\overset{\displaystyle O}{\|}}{C}-CH_2-CH_3$$

α-carbons

Because carbon and hydrogen have comparable electronegativities, a C—H bond normally has little polarity, and a hydrogen atom bonded to carbon shows very low acidity. The situation is different, however, for hydrogens alpha to a carbonyl group. As shown in the table, α-hydrogens are considerably more acidic than alkane hydrogens but less acidic than —OH hydrogens of alcohols.

Type of Bond	pK_a
CH_3CH_2O-H	16
$CH_3\overset{\overset{\displaystyle O}{\|}}{C}CH_2-H$	20
CH_3CH_2-H	51

Hydrogens alpha to the carbonyl group of an aldehyde or ketone are more acidic than hydrogens of alkanes because of two factors. First, the electron-withdrawing inductive effect of the adjacent carbonyl group weakens the bond to the alpha hydrogen and promotes its ionization. Second, the negative charge on the resulting **enolate anion** is delocalized by resonance, thus stabilizing it relative to the anion from an alkane. Recall that we used these same two factors in Section 2.3 to account for the greater acidity of carboxylic acids compared with alcohols.

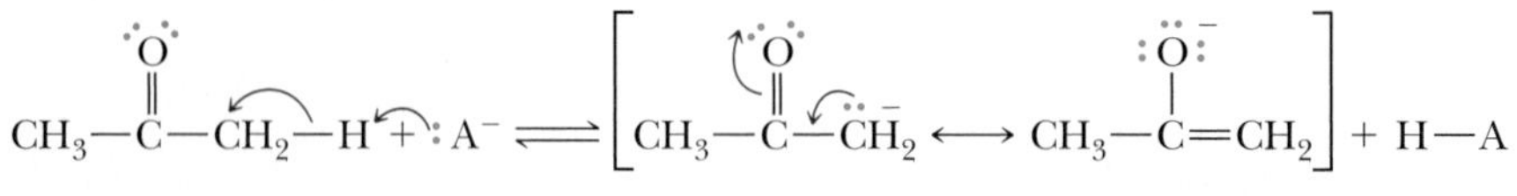

Enolate anion An anion formed by removal of an α-hydrogen from a carbonyl-containing compound.

$$CH_3-\overset{\overset{\displaystyle O}{\|}}{C}-CH_2-H + :A^- \rightleftharpoons \left[CH_3-\overset{\overset{\displaystyle O}{\|}}{C}-\ddot{C}H_2^- \longleftrightarrow CH_3-\overset{\overset{\displaystyle :\ddot{O}:^-}{|}}{C}=CH_2\right] + H-A$$

Resonance-stabilized enolate anion

EXAMPLE 14.1

Identify the acidic α-hydrogens in each compound.

(a) Butanal (b) 2-Butanone

SOLUTION

Hydrogen atoms on a carbon atom adjacent to a carbonyl group (α-hydrogens) show weak acidity compared with other alkane hydrogens. Butanal has one set of acidic α-hydrogens; 2-butanone has two sets of acidic α-hydrogens.

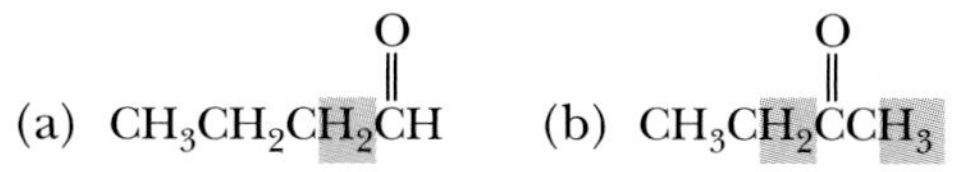

(a) $CH_3CH_2CH_2C(=O)H$ (b) $CH_3CH_2C(=O)CH_3$

Practice Problem 14.1

Identify the acidic hydrogens in each compound.

(a) Cyclohexanone (b) Acetophenone

B. Formation of Enolate Anions

Treatment of an aldehyde or ketone containing an acidic α-hydrogen with a strong base, such as sodium hydroxide or sodium ethoxide, gives an enolate anion, which is a hybrid of two major contributing structures.

$$CH_3C(=O)CH_3 + NaOH \rightleftharpoons \left[H-\overset{-}{C}H-C(=O)CH_3 \longleftrightarrow H-CH=C(O^-)CH_3 \right] Na^+ + H_2O$$

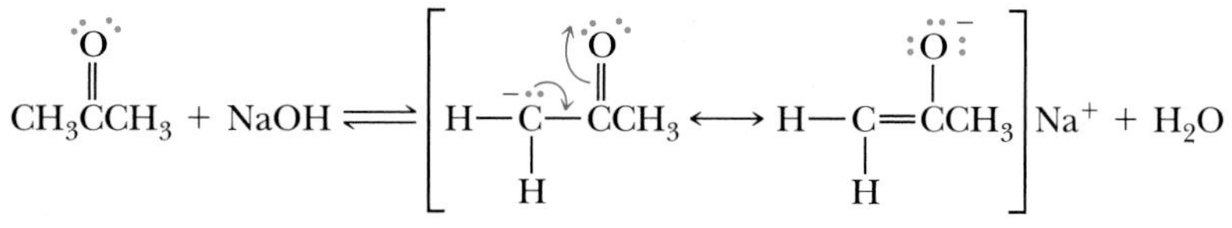

An enolate ion

EXAMPLE 14.2

Draw the enolate anion formed by treatment of each compound with base.

(a) Butanal (b) Cyclohexanone

SOLUTION

Each enolate anion is a hybrid of two contributing structures: one with the negative charge on the α-carbon, the other with the negative charge on oxygen.

(a) $CH_3CH_2\overset{-}{C}H-C(=O)H \longleftrightarrow CH_3CH_2CH=C(O^-)H$ (b) [cyclohexanone enolate with negative charge on α-carbon] $\longleftrightarrow$ [cyclohexenolate with negative charge on oxygen]

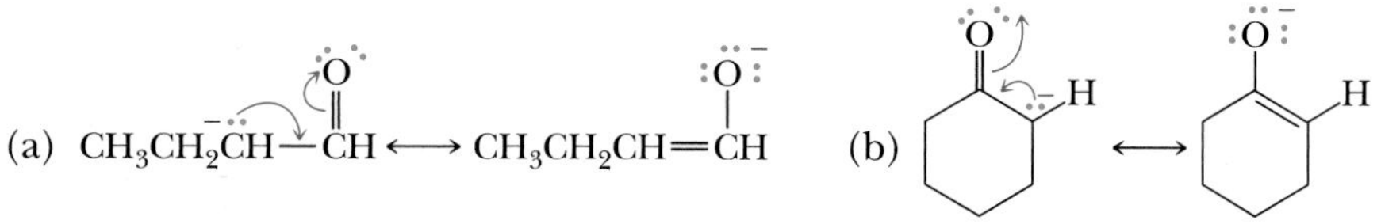

Practice Problem 14.2

Treatment of 2-butanone with base gives two enolate anions. Draw each enolate anion as a hybrid of two contributing structures.

EXAMPLE 14.3

Predict the position of equilibrium for this reaction. (*Hint:* The pK_a of an α-hydrogen of acetone is approximately 20.)

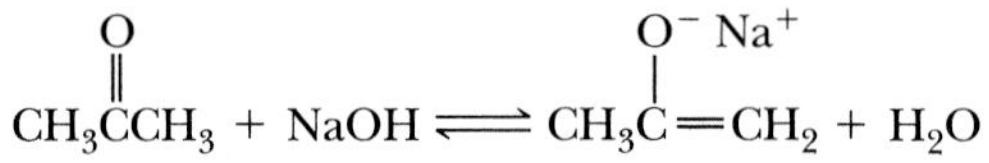

$$CH_3\overset{O}{\overset{\|}{C}}CH_3 + NaOH \rightleftharpoons CH_3\overset{O^- Na^+}{\overset{|}{C}}{=}CH_2 + H_2O$$

SOLUTION

Water is the stronger, acid and acetone is the weaker acid. The equilibrium lies to the left.

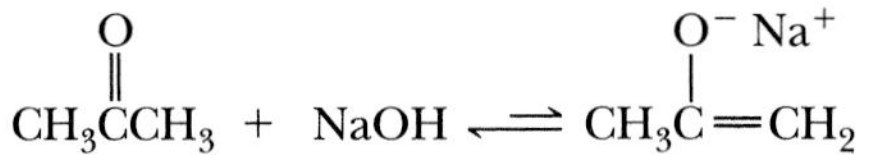

$$\underset{\substack{pK_a\ 20 \\ \text{(weaker acid)}}}{CH_3\overset{O}{\overset{\|}{C}}CH_3} + NaOH \rightleftharpoons CH_3\overset{O^- Na^+}{\overset{|}{C}}{=}CH_2 + \underset{\substack{pK_a\ 15.7 \\ \text{(stonger acid)}}}{H_2O}$$

Practice Problem 14.3

Predict the position of equilibrium for this reaction. Refer to Table 2.1 for the pK_a of NH_3.

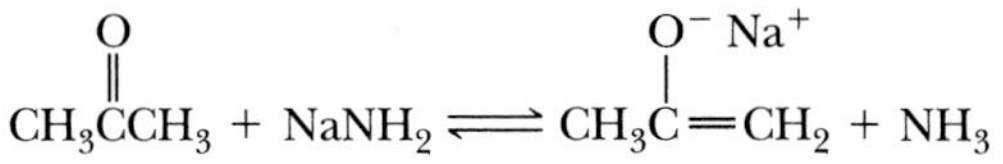

$$CH_3\overset{O}{\overset{\|}{C}}CH_3 + NaNH_2 \rightleftharpoons CH_3\overset{O^- Na^+}{\overset{|}{C}}{=}CH_2 + NH_3$$

The pK_a of α-hydrogens of an ester is approximately 22. The enolate anion of an ester is also a hybrid of two contributing structures in which the negative charge is shared by the α-carbon and the oxygen atom of the carbonyl group.

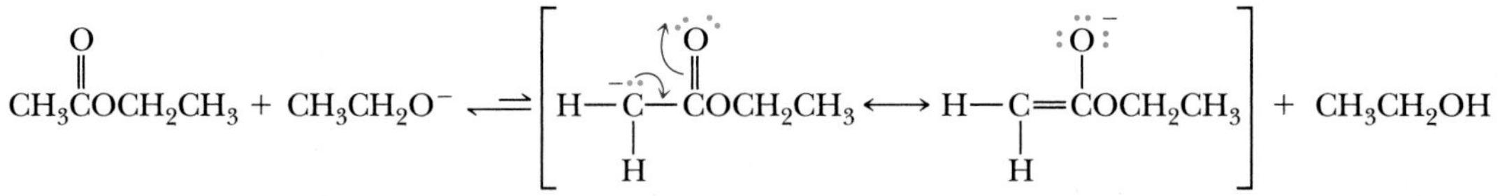

$$\underset{\substack{pK_a\ 22 \\ \text{(weaker acid)}}}{CH_3\overset{O}{\overset{\|}{C}}OCH_2CH_3} + CH_3CH_2O^- \rightleftharpoons \left[H{-}\underset{H}{\underset{|}{\overset{-}{C}}}{-}\overset{O}{\overset{\|}{C}}OCH_2CH_3 \longleftrightarrow H{-}\underset{H}{\underset{|}{C}}{=}\overset{:\ddot{O}:^-}{\overset{|}{C}}OCH_2CH_3 \right] + \underset{\substack{pK_a\ 15.9 \\ \text{(stronger acid)}}}{CH_3CH_2OH}$$

A resonance-stabilized enolate anion

EXAMPLE 14.4

Draw the enolate anion formed by treatment of ethyl butanoate with sodium ethoxide.

SOLUTION

In these contributing structures, the ethyl group of the ester is abbreviated Et.

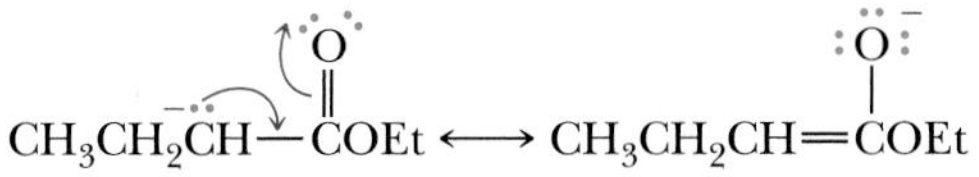

$$CH_3CH_2\overset{-}{C}H-\overset{O}{\overset{\|}{C}}OEt \longleftrightarrow CH_3CH_2CH=\overset{O^-}{\overset{|}{C}}OEt$$

Practice Problem 14.4

Draw the enolate anion formed by treatment of methyl phenylacetate with sodium methoxide.

Enolate anions undergo a very important type of reaction in which the α-carbon of the enolate anion adds to the carbonyl carbon of another carbonyl-containing compound to form a new carbon-carbon bond.

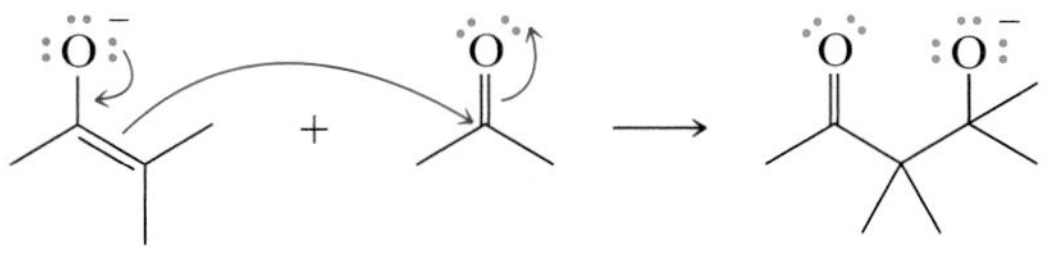

Enolate anion + Carbonyl group of another carbonyl-containing compound ⟶

In this chapter, we show how enolate anions derived from aldehydes, ketones, and esters can be used to form new carbon-carbon bonds.

14.2 THE ALDOL REACTION

The addition of the enolate anion derived from an aldehyde or ketone to the carbonyl group of another aldehyde or ketone is illustrated by these examples.

$$CH_3-\overset{O}{\overset{\|}{C}}-H + \overset{H}{\overset{|}{C}}H_2-\overset{O}{\overset{\|}{C}}-H \overset{NaOH}{\rightleftharpoons} CH_3-\overset{OH}{\overset{\beta|}{C}}H-\overset{\alpha}{C}H_2-\overset{O}{\overset{\|}{C}}-H$$

Ethanal (Acetaldehyde) + Ethanal (Acetaldehyde) ⟶ 3-Hydroxybutanal (a β-hydroxyaldehyde)

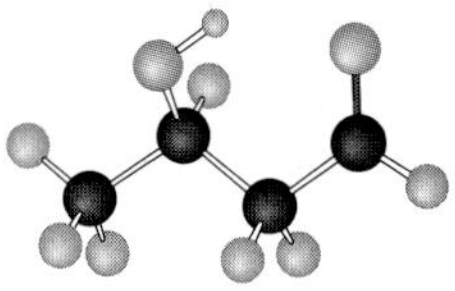

$$CH_3-\overset{O}{\overset{\|}{C}}-CH_3 + \overset{H}{\overset{|}{C}}H_2-\overset{O}{\overset{\|}{C}}-CH_3 \overset{Ba(OH)_2}{\rightleftharpoons} CH_3-\underset{CH_3}{\underset{|}{\overset{OH}{\overset{\beta|}{C}}}}-\overset{\alpha}{C}H_2-\overset{O}{\overset{\|}{C}}-CH_3$$

Propanone (Acetone) + Propanone (Acetone) ⟶ 4-Hydroxy-4-methyl-2-pentanone (a β-hydroxyketone)

The common name of the product derived from reaction of acetaldehyde in base is **aldol,** so named because it is both an **ald**ehyde and an alcoh**ol.** Aldol is also the generic name given to any product formed in this type of reaction. The functional group of the product of an **aldol reaction** is a β-hydroxyaldehyde or a β-hydroxyketone.

Aldol reaction A carbonyl condensation reaction between two aldehydes or ketones to give a β-hydroxyaldehyde or β-hydroxyketone.

The key step in a base-catalyzed aldol reaction is nucleophilic addition of the enolate anion from one carbonyl-containing molecule to the carbonyl group of another carbonyl-containing molecule to form a tetrahedral carbonyl addition intermediate. This mechanism is illustrated by the aldol reaction between two molecules of acetaldehyde. Notice in this three-step mechanism that OH^- is a catalyst: an OH^- is used in Step 1, but another OH^- is generated in Step 3.

MECHANISM Base-Catalyzed Aldol Reaction

Step 1: Removal of an α-hydrogen by base gives a resonance-stabilized enolate anion.

$$H{-}\ddot{O}{:}^- + H{-}CH_2{-}\overset{O}{\overset{\|}{C}}{-}H \rightleftharpoons H{-}\ddot{O}{-}H + \left[{:}\overset{-}{C}H_2{-}\overset{O}{\overset{\|}{C}}{-}H \longleftrightarrow CH_2{=}\overset{O^-}{\overset{|}{C}}{-}H \right]$$

Enolate anion

Step 2: Nucleophilic addition of the enolate anion to the carbonyl carbon of another aldehyde (or ketone) gives a tetrahedral carbonyl addition intermediate.

$$CH_3{-}\overset{O}{\overset{\|}{C}}{-}H + {:}\overset{-}{C}H_2{-}\overset{O}{\overset{\|}{C}}{-}H \rightleftharpoons CH_3{-}\overset{O^-}{\overset{|}{C}H}{-}CH_2{-}\overset{O}{\overset{\|}{C}}{-}H$$

A tetrahedral carbonyl addition intermediate

Step 3: Reaction of the tetrahedral carbonyl addition intermediate with a proton donor gives the aldol product and generates another hydroxide ion.

$$CH_3{-}\overset{O^-}{\overset{|}{C}H}{-}CH_2{-}\overset{O}{\overset{\|}{C}}{-}H + H{-}\ddot{O}H \rightleftharpoons CH_3{-}\overset{OH}{\overset{|}{C}H}{-}CH_2{-}\overset{O}{\overset{\|}{C}}{-}H + {:}\ddot{O}H^-$$

EXAMPLE 14.5

Draw the product of the base-catalyzed aldol reaction of each compound.

(a) Butanal (b) Cyclohexanone

SOLUTION

The aldol product is formed by nucleophilic addition of the α-carbon of one compound to the carbonyl carbon of another.

(a) $CH_3CH_2CH_2\overset{OH}{\overset{|}{C}H}{-}\underset{CH_2CH_3}{\underset{|}{C}H}\overset{O}{\overset{\|}{C}}H$

the new C—C bond

(b) [structure: 2-(1-hydroxycyclohexyl)cyclohexanone, OH, O]

the new C—C bond

Practice Problem 14.5

Draw the product of the base-catalyzed aldol reaction of each compound.

(a) Acetophenone (b) Cyclopentanone

β-Hydroxyaldehydes and β-hydroxyketones are very easily dehydrated, and often the conditions necessary to bring about an aldol reaction are sufficient to cause dehydration. Dehydration can also be brought about by warming the aldol product in dilute acid. The major product from dehydration of an aldol product is one in which the carbon-carbon double bond is conjugated with the carbonyl group, that is, the product is an α,β-unsaturated aldehyde or ketone.

$$CH_3CH(OH)CH_2CH{=}O \xrightarrow{\text{warm in either acid or base}} CH_3\overset{\beta}{C}H{=}\overset{\alpha}{C}HCH{=}O + H_2O$$

An α,β-unsaturated aldehyde

In base-catalyzed dehydration, an α-hydrogen is removed to form a new enolate anion, which then expels hydroxide ion. Following is a two-step mechanism for this reaction.

MECHANISM Base-Catalyzed Dehydration of an Aldol Product

Step 1: An acid-base reaction results in removal of an α-hydrogen to give an enolate anion.

Step 2: The enolate anion ejects hydroxide ion and gives the α,β-unsaturated carbonyl compound.

$$CH_3{-}CH(OH){-}CH_2{-}CH{=}O \xrightarrow[(1)]{HO^-} CH_3{-}CH(OH){-}CH{=}CH{-}O^- \xrightarrow[(2)]{-OH^-} CH_3{-}\overset{\beta}{C}H{=}\overset{\alpha}{C}H{-}CH{=}O$$

An enolate anion

An α,β-unsaturated aldehyde

We write the following three-step mechanism for acid-catalyzed dehydration of an aldol product.

MECHANISM Acid-Catalyzed Dehydration of an Aldol Product

Step 1: Proton transfer from the acid catalyst to the enol form of the aldol product gives an oxonium ion.

$$H_3O^+ + CH_3{-}C(OH)(CH_3){-}CH{=}C(OH){-}CH_3 \rightleftharpoons CH_3{-}C(OH_2^+)(CH_3){-}CH{=}C(OH){-}CH_3 + H_2O$$

The enol of the ketone

An oxonium ion

Step 2: The oxonium ion undergoes concerted loss of water to give the conjugate acid of the final product.

$$CH_3-\underset{H_3C}{\overset{H_2O^+}{C}}-CH=\overset{:OH}{C}-CH_3 \longrightarrow \underset{CH_3}{CH_3\overset{\beta}{C}}=\overset{\alpha}{C}H\overset{O^+H}{\overset{\|}{C}}CH_3 + H_2\ddot{O}:$$

Protonated α,β-unsaturated ketone

Step 3: Proton transfer from this conjugate acid to solvent completes the reaction.

$$\underset{CH_3}{CH_3\overset{\beta}{C}}=\overset{\alpha}{C}H\overset{O^+H}{\overset{\|}{C}}CH_3 + :\ddot{O}(H)-H \longrightarrow \underset{CH_3}{CH_3\overset{\beta}{C}}=\overset{\alpha}{C}H\overset{O}{\overset{\|}{C}}CH_3 + H_2\overset{+}{O}-H$$

EXAMPLE 14.6

Draw the product of base-catalyzed dehydration of each aldol product from Example 14.5.

SOLUTION

Loss of H_2O from aldol product (a) gives an α,β-unsaturated aldehyde; loss of H_2O from aldol product (b) gives an α,β-unsaturated ketone.

(a) $CH_3CH_2CH_2CH=\underset{CH_2CH_3}{C}\overset{O}{\overset{\|}{C}}H$ (b) [2-cyclohexylidenecyclohexanone structure with ball-and-stick model]

Practice Problem 14.6

Draw the product of acid-catalyzed dehydration of each aldol product from Practice Problem 14.5.

Base-catalyzed aldol reactions are readily reversible, and generally there is little aldol product present at equilibrium. Equilibrium constants for dehydration, however, are generally large so that, if reaction conditions are sufficiently vigorous to bring about dehydration, good yields of product can be obtained.

The reactants in the key step of an aldol reaction are an enolate anion and an enolate anion acceptor. In self-reactions, both roles are played by one kind of molecule. **Crossed aldol reactions** are also possible, as for example the crossed aldol reaction between acetone and formaldehyde. Formaldehyde cannot form an enolate anion because it has no α-hydrogen. It is, however, a particularly good enolate anion acceptor because its carbonyl group is unhindered. Acetone forms an enolate anion but its carbonyl group, which is bonded to two alkyl groups, is less reactive than that of formaldehyde. Consequently, the crossed aldol reaction between acetone and formaldehyde gives 4-hydroxy-2-butanone.

Crossed aldol reaction An aldol reaction between two different aldehydes and/or ketones.

$$CH_3\overset{O}{\overset{\|}{C}}CH_3 + H\overset{O}{\overset{\|}{C}}H \xrightleftharpoons{NaOH} CH_3\overset{O}{\overset{\|}{C}}CH_2CH_2OH$$

4-Hydroxy-2-butanone

As this example illustrates, for a crossed aldol reaction to be successful, one of the two reactants should have no α-hydrogen so that an enolate anion does not form. It also helps if the compound with no α-hydrogen has the more reactive carbonyl, for example an aldehyde. Following are examples of aldehydes that have no α-hydrogens and can be used in crossed aldol reactions. If these requirements are not met, a complex mixture of products results.

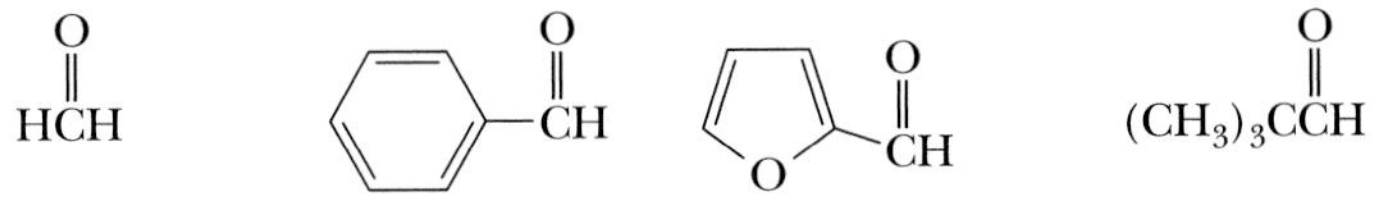

$H\overset{O}{\overset{\|}{C}}H$ — Methanal (Formaldehyde)

Benzaldehyde

Furfural

$(CH_3)_3C\overset{O}{\overset{\|}{C}}H$ — 2,2-Dimethylpropanal

EXAMPLE 14.7

Draw the product of the crossed aldol reaction between furfural and cyclohexanone and the product formed by its base-catalyzed dehydration.

SOLUTION

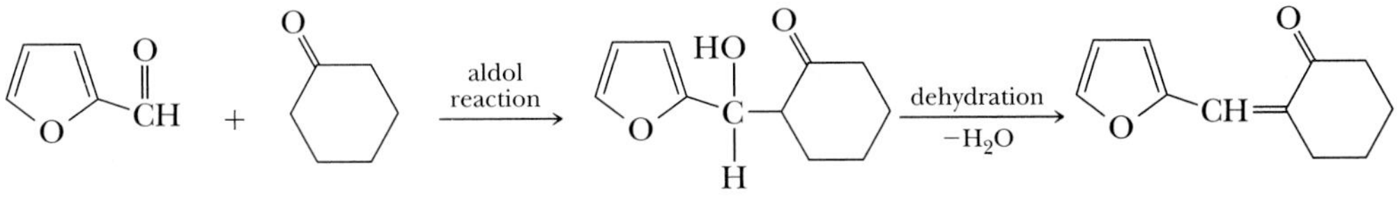

Furfural + Cyclohexanone —(aldol reaction)→ Aldol product —(dehydration, $-H_2O$)→ furyl-CH=cyclohexanone

Practice Problem 14.7

Draw the product of the crossed aldol reaction between benzaldehyde and 3-pentanone and the product formed by its base-catalyzed dehydration.

When both the enolate anion and the carbonyl group to which it adds are in the same molecule, aldol reaction results in formation of a ring. This type of **intramolecular aldol reaction** is particularly useful for formation of five- and six-membered rings. Intramolecular aldol reaction of 2,7-octanedione gives a five-membered ring.

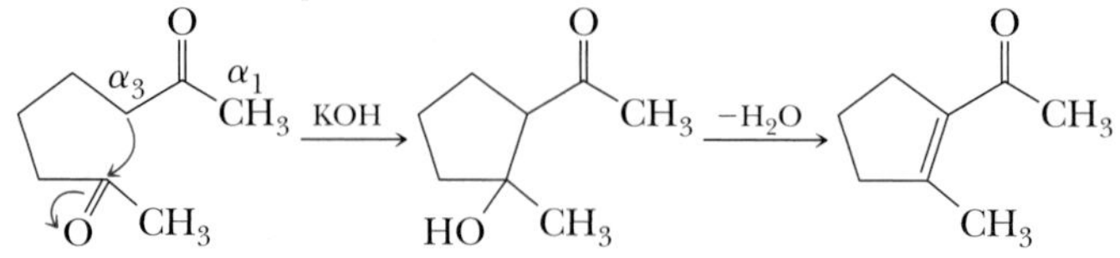

2,7-Octanedione

Following is another example in which either a four-membered ring (via an enolate anion at α_3) or a six-membered ring (via an enolate anion at α_1) could be formed. Because of the greater stability of six-membered rings compared to four-membered rings, the six-membered ring is formed in this intramolecular aldol reaction.

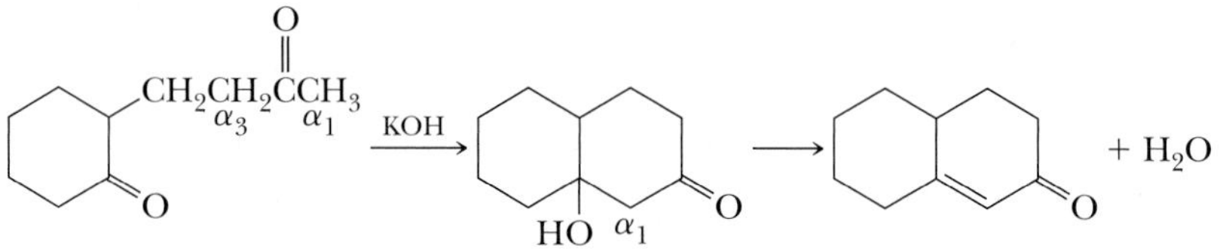

14.3 The Claisen and Dieckmann Condensations

A. Claisen Condensation

In this section we examine the formation of an enolate anion from one ester followed by its nucleophilic acyl substitution at the carbonyl carbon of another ester. One of the first of these reactions discovered was the **Claisen condensation,** named after its discoverer the German chemist Ludwig Claisen (1851–1930). A Claisen condensation is illustrated by the condensation of two molecules of ethyl acetate in the presence of sodium ethoxide, followed by acidification, to give ethyl acetoacetate.

Claisen condensation A carbonyl condensation reaction between two esters to give a β-ketoester.

$$2CH_3COEt \xrightarrow[2.\ H_2O,\ HCl]{1.\ EtO^-Na^+} CH_3CCH_2COEt + EtOH$$

Ethyl ethanoate (Ethyl acetate) → Ethyl 3-oxobutanoate (Ethyl acetoacetate) + Ethanol

Claisen condensations, like the aldol reaction, require a base. Aqueous bases, such as NaOH, however, cannot be used in Claisen condensations because their use would result in hydrolysis of the ester instead. Rather, the bases most commonly used in Claisen condensations are nonaqueous bases, such as sodium ethoxide in ethanol and sodium methoxide in methanol.

The functional group of the product of a Claisen condensation is a ***β*-ketoester** group.

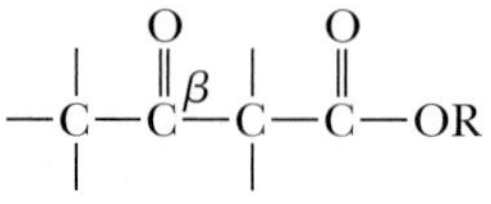

A β-ketoester

Claisen condensation of two molecules of ethyl propanoate gives the following β-ketoester.

$$CH_3CH_2\overset{O}{\overset{\|}{C}}OEt + \underset{CH_3}{CH}\overset{O}{\overset{\|}{C}}OEt \xrightarrow[2.\ H_2O,\ HCl]{1.\ EtO^-Na^+} CH_3CH_2\overset{O}{\overset{\|}{C}}\underset{CH_3}{CH}\overset{O}{\overset{\|}{C}}OEt + EtOH$$

Ethyl propanoate   Ethyl propanoate   Ethyl 2-methyl-3-oxopentanoate

The first steps of a Claisen condensation bear a close resemblance to the first steps of the aldol reaction (Section 14.2). In each, base removes a proton from the α-carbon in Step 1 to form a resonance-stabilized enolate anion. In Step 2 the enolate anion attacks the carbonyl carbon of another ester molecule to form a tetrahedral carbonyl addition intermediate, which in turn collapses in Step 3 to give a β-ketoester. The overall reaction is driven to completion because the β-ketoester formed is a stronger acid than ethanol. The β-ketoester (a stronger acid) reacts with ethoxide ion (a stronger base) in Step 4 to give ethanol (a weaker acid) and the anion of the β-ketoester (a weaker base).

MECHANISM Claisen Condensation

Step 1: Base removes an α-hydrogen from the ester to give an enolate anion.

$$C_2H_5O^- + H{-}CH_2{-}\overset{O}{\overset{\|}{C}}OEt \rightleftharpoons EtO{-}H + {}^-{:}CH_2{-}\overset{O}{\overset{\|}{C}}OEt \longleftrightarrow CH_2{=}\overset{O^-}{\overset{|}{C}}OEt$$

(weaker base); pK_a 22 (weaker acid); pK_a 15.9 (stronger acid); Resonance-stabilized enolate anion (stronger base)

Step 2: Attack of the enolate anion on the carbonyl carbon of another ester molecule gives a tetrahedral carbonyl addition intermediate.

$$CH_3{-}\overset{O}{\overset{\|}{C}}{-}OEt + {}^-{:}CH_2{-}\overset{O}{\overset{\|}{C}}OEt \rightleftharpoons \left[CH_3{-}\underset{OEt}{\overset{O^-}{\overset{|}{\underset{|}{C}}}}{-}CH_2{-}\overset{O}{\overset{\|}{C}}{-}OEt\right]$$

A tetrahedral carbonyl addition intermediate

Step 3: Collapse of the tetrahedral carbonyl addition intermediate gives a β-ketoester and generates a new ethoxide ion.

$$EtO^- + CH_3{-}\overset{O}{\overset{\|}{C}}{-}\overset{H}{\overset{|}{C}}H{-}\overset{O}{\overset{\|}{C}}{-}OEt \rightleftharpoons CH_3{-}\overset{O}{\overset{\|}{C}}{-}CH_2{-}\overset{O}{\overset{\|}{C}}{-}OEt + EtO^-$$

(stronger base); pK_a 10.7 (stronger acid)

Step 4: Formation of the enolate anion of the β-ketoester drives the Claisen condensation to the right.

$$CH_3-\overset{O}{\overset{\|}{C}}-\underset{}{\overset{H}{CH}}-\overset{O}{\overset{\|}{C}}-OEt + EtO^- \rightleftharpoons CH_3-\overset{O}{\overset{\|}{C}}-\overset{-}{CH}-\overset{O}{\overset{\|}{C}}-OEt + EtOH$$

pK_a 10.7 (stronger acid) — (stronger base) — (weaker base) — pK_a 15.9 (weaker acid)

Thus, the starting material required for a successful Claisen condensation is an ester with two α-hydrogens: one to form the initial enolate anion and the second to form the enolate anion of the resulting β-ketoester. The β-ketoester is formed and isolated upon acidification with aqueous acid during work-up.

$$CH_3-\overset{O}{\overset{\|}{C}}-\overset{-}{CH}-\overset{O}{\overset{\|}{C}}OC_2H_5 + H^+ \xrightarrow{HCl,\ H_2O} CH_3-\overset{O}{\overset{\|}{C}}-CH_2-\overset{O}{\overset{\|}{C}}OC_2H_5$$

EXAMPLE 14.8

Show the product of Claisen condensation of ethyl butanoate in the presence of sodium ethoxide followed by acidification with aqueous HCl.

SOLUTION

The new bond formed in a Claisen condensation is between the carbonyl group of one ester and the α-carbon of another.

$$CH_3CH_2CH_2\overset{O}{\overset{\|}{C}}-\underset{CH_2CH_3}{\underset{|}{CH}}\overset{O}{\overset{\|}{C}}OEt + EtOH$$

the new C—C bond

Ethyl 2-ethyl-3-oxohexanoate

Practice Problem 14.8

Show the product of Claisen condensation of ethyl 3-methylbutanoate in the presence of sodium ethoxide.

B. Dieckmann Condensation

Dieckmann condensation An intramolecular Claisen condensation of an ester of a dicarboxylic acid to give a five- or six-membered ring.

An intramolecular Claisen condensation of a dicarboxylic ester to give a five- or six-membered ring is given the special name of **Dieckmann condensation.** In the presence of one equivalent of sodium ethoxide, for example, diethyl hexanedioate (diethyl adipate) undergoes an intramolecular condensation to form a five-membered ring.

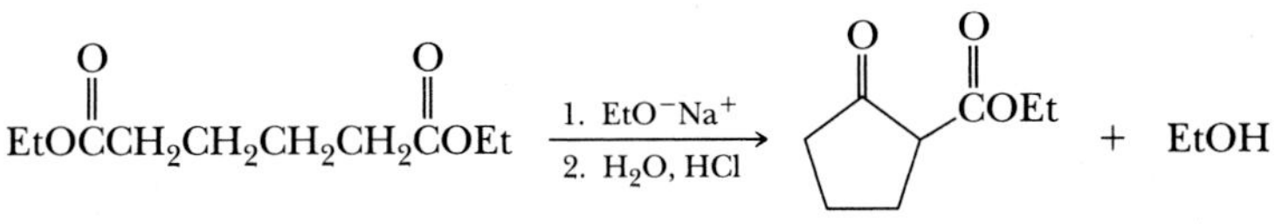

$$EtOC(=O)CH_2CH_2CH_2CH_2C(=O)OEt \xrightarrow[\text{2. } H_2O,\ HCl]{\text{1. } EtO^-Na^+} \text{(2-oxocyclopentyl)}C(=O)OEt + EtOH$$

Diethyl hexanedioate (Diethyl adipate) — Ethyl 2-oxocyclopentanecarboxylate

The mechanism of a Dieckmann condensation is identical to the mechanism we described for the Claisen condensation. An anion formed at the α-carbon of one ester group in Step 1 adds to the carbonyl of the other ester group in Step 2 to form a tetrahedral carbonyl addition intermediate. This intermediate ejects ethoxide ion in Step 3 to regenerate the carbonyl group. Cyclization is followed by formation of the conjugate base of the β-ketoester in Step 4, just as in the Claisen condensation. The β-ketoester is isolated after acidification with aqueous acid.

MECHANISM Dieckmann Condensation

Step 1: Proton transfer to ethoxide ion gives an enolate anion.

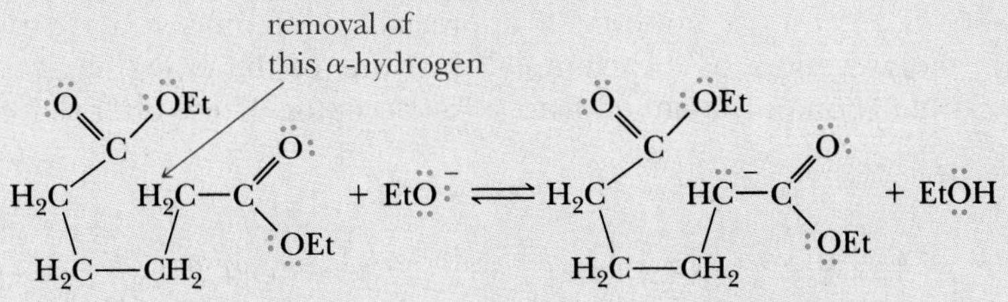

removal of this α-hydrogen

$$\text{diester} + EtO^- \rightleftharpoons \text{enolate ion} + EtOH$$

An enolate ion

Step 2: Attack of the enolate anion on the other carbonyl carbon of the diester gives a tetrahedral carbonyl addition intermediate.

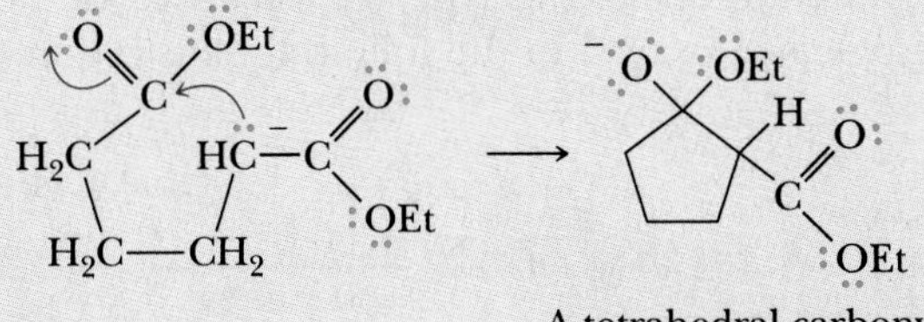

A tetrahedral carbonyl addition intermediate

Step 3: Collapse of the tetrahedral carbonyl addition intermediate gives a β-ketoester and generates a new ethoxide ion.

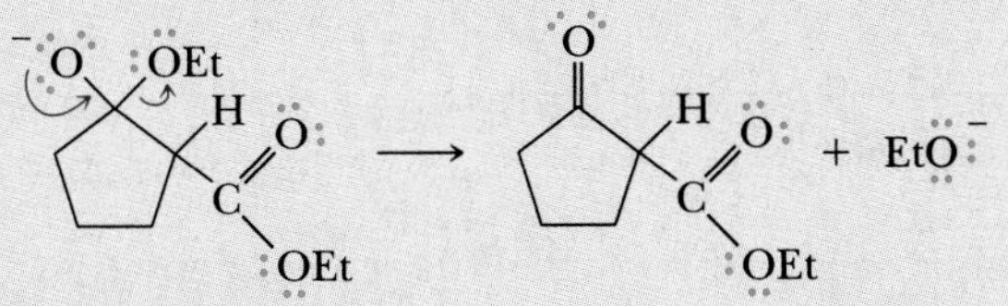

$$\longrightarrow \; + EtO^-$$

A tetrahedral carbonyl addition intermediate

Step 4: Proton transfer from the β-ketoester to ethoxide ion forms the anion of the β-ketoester, which drives the Dieckmann condensation to completion. Final work-up with aqueous acid gives the β-ketoester.

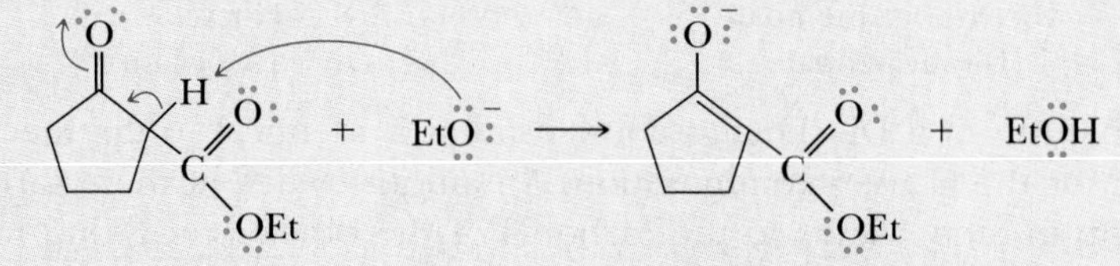

$$\text{(cyclopentanone-2-C(=O)OEt with H)} + EtO^- \longrightarrow \text{(enolate)} + EtOH$$

Resonance-stabilized enolate anion of the β-ketoester

C. Crossed Claisen Condensations

Crossed Claisen condensation A Claisen condensation between two different esters.

In a **crossed Claisen condensation** between two different esters, each with two α-hydrogens, a mixture of four β-ketoesters is possible, and, therefore, crossed Claisen condensations of this type are generally not synthetically useful. Such condensations are useful, however, if appreciable differences in reactivity exist between the two esters, as for example when one of the esters has no α-hydrogens and can function only as an enolate anion acceptor. These esters have no α-hydrogens:

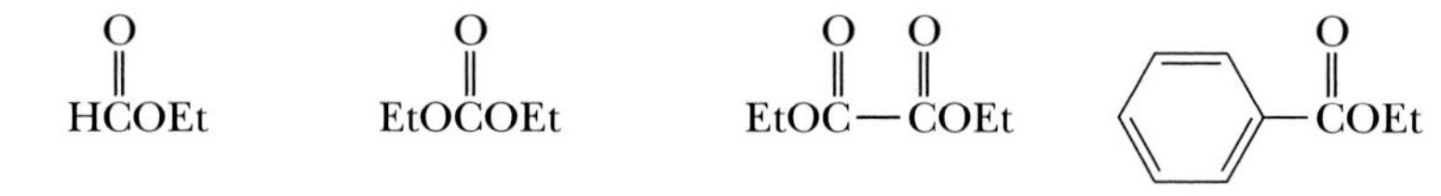

$$\underset{\text{Ethyl formate}}{HC(=O)OEt} \qquad \underset{\text{Diethyl carbonate}}{EtOC(=O)OEt} \qquad \underset{\text{Diethyl ethanedioate (Diethyl oxalate)}}{EtOC(=O){-}C(=O)OEt} \qquad \underset{\text{Ethyl benzoate}}{C_6H_5C(=O)OEt}$$

Crossed Claisen condensations of this type are usually carried out by using the ester with no α-hydrogens in excess. In the following illustration, methyl benzoate is used in excess.

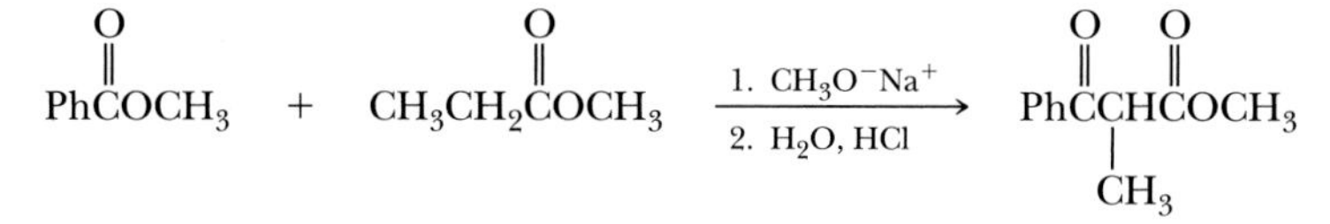

$$\underset{\text{Methyl benzoate}}{PhC(=O)OCH_3} + \underset{\text{Methyl propanoate}}{CH_3CH_2C(=O)OCH_3} \xrightarrow[2.\ H_2O,\ HCl]{1.\ CH_3O^-Na^+} \underset{\text{Methyl 2-methyl-3-oxo-3-phenylpropanoate}}{PhC(=O)CH(CH_3)C(=O)OCH_3}$$

Example 14.9

Complete the equation for this crossed Claisen condensation.

$$CH_3CH_2C(=O)OEt + HC(=O)OEt \xrightarrow[2.\ H_2O,\ HCl]{1.\ EtO^-Na^+}$$

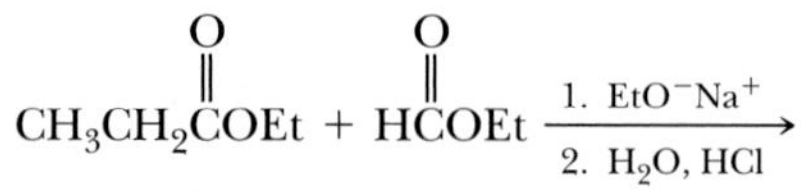

SOLUTION

$$\mathrm{HC(=O)CH(CH_3)C(=O)OEt + EtOH}$$

Practice Problem 14.9

Complete the equation for this crossed Claisen condensation.

$$\mathrm{C_6H_5COEt\ (C{=}O)} + \mathrm{C_6H_5CH_2COEt\ (C{=}O)} \xrightarrow[\text{2. } H_2O,\ HCl]{\text{1. } EtO^-Na^+}$$

D. Hydrolysis and Decarboxylation of β-Ketoesters

Recall from Section 13.3C that hydrolysis of an ester in aqueous sodium hydroxide **(saponification)** followed by acidification of the reaction mixture with HCl or other mineral acid converts an ester to a carboxylic acid. Recall also from Section 12.8 that β-ketoacids and β-dicarboxylic acids readily undergo **decarboxylation** (lose CO_2) when heated. The following equations illustrate the results of a Claisen condensation followed by saponification, acidification, and decarboxylation.

Saponification Hydrolysis of an ester in aqueous NaOH or KOH to give an alcohol and the sodium or potassium salt of a carboxylic acid.

Decarboxylation Loss of CO_2 from an organic molecule.

Claisen condensation:

$$\mathrm{2CH_3CH_2\overset{O}{\overset{\|}{C}}OEt \xrightarrow[\text{2. } H_2O,\ HCl]{\text{1. } EtO^-Na^+} CH_3CH_2\overset{O}{\overset{\|}{C}}\underset{CH_3}{\underset{|}{C}}H\overset{O}{\overset{\|}{C}}OEt + EtOH}$$

Saponification followed by acidification:

$$\mathrm{CH_3CH_2\overset{O}{\overset{\|}{C}}\underset{CH_3}{\underset{|}{C}}H\overset{O}{\overset{\|}{C}}OEt \xrightarrow[\text{4. } H_2O,\ HCl]{\text{3. } NaOH,\ H_2O,\ heat} CH_3CH_2\overset{O}{\overset{\|}{C}}\underset{CH_3}{\underset{|}{C}}H\overset{O}{\overset{\|}{C}}OH + EtOH}$$

Decarboxylation:

$$\mathrm{CH_3CH_2\overset{O}{\overset{\|}{C}}\underset{CH_3}{\underset{|}{C}}H\overset{O}{\overset{\|}{C}}OH \xrightarrow{\text{5. Heat}} CH_3CH_2\overset{O}{\overset{\|}{C}}CH_2CH_3 + CO_2}$$

The result of these five steps is reaction between two molecules of ester, one furnishing a carboxyl group and the other furnishing an enolate anion, to give a ketone and carbon dioxide. In the general reaction, both ester molecules are the same, and the product is a symmetrical ketone.

$$\mathrm{R{-}CH_2{-}\overset{O}{\overset{\|}{C}}{-}OR' + \underset{R}{\underset{|}{CH_2}}{-}\overset{O}{\overset{\|}{C}}{-}OR' \xrightarrow{\text{several steps}} R{-}CH_2{-}\overset{O}{\overset{\|}{C}}{-}CH_2{-}R + 2HOR' + CO_2}$$

from the ester furnishing the carboxyl group

from the ester furnishing the enolate anion

EXAMPLE 14.10

Each set of compounds undergoes (1, 2) Claisen condensation, (3) saponification followed by (4) acidification, and (5) thermal decarboxylation. Draw a structural formula of the product after completion of this reaction sequence.

(a) $PhC(=O)OEt + CH_3C(=O)OEt$ (b) $EtOC(=O)(CH_2)_4C(=O)OEt$

SOLUTION

Steps 1 and 2 bring about a crossed Claisen condensation in (a) and a Dieckmann condensation in (b) to form a β-ketoester. Steps 3 and 4 bring about hydrolysis of the β-ketoester to give a β-ketoacid, and Step 5 brings about decarboxylation to give a ketone.

(a) $\xrightarrow{1,2} PhC(=O)CH_2C(=O)OEt \xrightarrow{3,4} PhC(=O)CH_2C(=O)OH \xrightarrow{5} PhC(=O)CH_3$

(b) $\xrightarrow{1,2}$ 2-(CO_2Et)cyclopentanone $\xrightarrow{3,4}$ 2-(CO_2H)cyclopentanone $\xrightarrow{5}$ cyclopentanone (=O)

Practice Problem 14.10

Show how to convert ethyl benzoate to 3-methyl-1-phenyl-1-butanone (isobutyl phenyl ketone) using a Claisen condensation at some stage in the synthesis.

$PhC(=O)OEt \xrightarrow{?} PhC(=O)CH_2CH(CH_3)CH_3$

Ethyl benzoate 3-Methyl-1-phenyl-1-butanone

14.4 CLAISEN AND ALDOL CONDENSATIONS IN THE BIOLOGICAL WORLD

Carbonyl condensations are among the most widely used reactions in the biological world for the assembly of new carbon-carbon bonds in such important biomolecules as fatty acids, cholesterol, steroid hormones, and terpenes. One source of carbon atoms for the synthesis of these biomolecules is **acetyl-CoA,** a thioester of acetic acid and the thiol group of coenzyme A (Additional Problem 16.27). Note that in the discussions that follow, we will not be concerned with the mechanism by which each of these enzyme-catalyzed reactions occurs. Rather, our concern is on recognizing the type of reaction that takes place.

In the Claisen condensation catalyzed by the enzyme thiolase, acetyl-CoA is converted to its enolate anion, which then attacks the carbonyl group of a second molecule of acetyl CoA to form a tetrahedral carbonyl addition intermediate. Collapse of this intermediate by loss of CoA-SH gives acetoacetyl-CoA. The mechanism for this condensation reaction is exactly the same as that of the Claisen condensation (Section 14.3A).

$$\underset{\text{Acetyl-CoA}}{CH_3\overset{O}{\overset{\|}{C}}SCoA} + \underset{\text{Acetyl-CoA}}{CH_3\overset{O}{\overset{\|}{C}}SCoA} \xrightarrow[\text{Claisen condensation}]{\text{thiolase}} \underset{\text{Acetoacetyl-CoA}}{CH_3\overset{O}{\overset{\|}{C}}CH_2\overset{O}{\overset{\|}{C}}SCoA} + \underset{\text{Coenzyme A}}{CoASH}$$

Enzyme-catalyzed aldol reaction with a third molecule of acetyl-CoA on the ketone carbonyl of acetoacetyl-CoA gives (S)-3-hydroxy-3-methylglutaryl-CoA. Note three features of this reaction. First, creation of the new stereocenter is stereoselective; only the S enantiomer is formed. Although the acetyl group of each reactant is achiral, their condensation takes place in a chiral environment created by the enzyme, 3-hydroxy-3-methylglutaryl-CoA synthetase. Second, hydrolysis of the thioester group of acetyl-CoA is coupled with the aldol reaction. Third, the carboxyl group is shown as it is ionized at pH 7.4, the approximate pH of blood plasma and many cellular fluids.

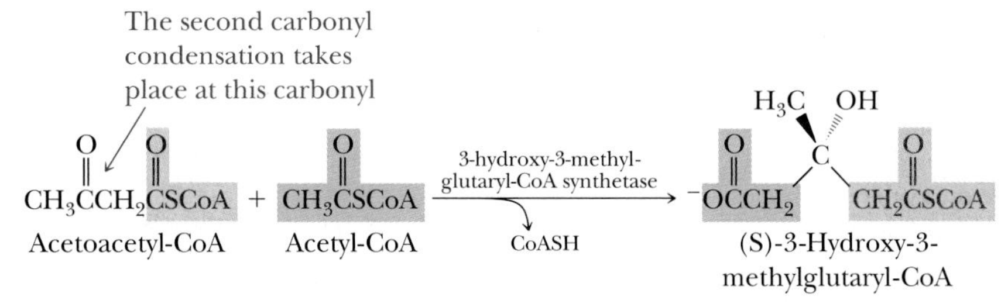

Enzyme-catalyzed reduction by NADH (Section 20.1B) of the thioester group of 3-hydroxy-3-methylglutaryl-CoA to a primary alcohol gives mevalonic acid, here shown as its anion. Note that, in this reduction, a change occurs in the designation of configuration from S to R, not because of any change in configuration at the stereocenter, but rather because of a change in priority among the four groups bonded to the stereocenter.

$$\underset{\text{(S)-3-Hydroxy-3-methylglutaryl-CoA}}{{}^{-}O\overset{O}{\overset{\|}{C}}CH_2C(CH_3)(OH)CH_2\overset{O}{\overset{\|}{C}}SCoA} \xrightarrow[\text{2NADH} \rightarrow \text{2NAD}^+]{\text{3-hydroxy-3-methyl-glutaryl-CoA reductase}} \underset{\text{(R)-Mevalonate}}{{}^{-}O\overset{O}{\overset{\|}{C}}CH_2C(CH_3)(OH)CH_2CH_2OH}$$

Enzyme-catalyzed transfer of a phosphate group from adenosine triphosphate (ATP, Section 19.1) to the 3-hydroxyl group of mevalonate gives a phosphoric ester at carbon 3. Enzyme-catalyzed transfer of a pyrophosphate group from a second molecule of ATP gives a pyrophosphoric ester at carbon 5. Enzyme-catalyzed β-elimination from this molecule results in loss of CO_2 and PO_4^{3-}, both good leaving groups.

a phosphoric ester; a pyrophosphoric ester

$$\underset{\text{3-Phospho-5-pyrophospho-(R)-mevalonate}}{{}^{-}O\overset{O}{\overset{\|}{C}}CH_2C(CH_3)(OPO_3^{2-})CH_2CH_2OP_2O_6^{3-}} \xrightarrow[CO_2 + PO_4^{3-}]{\beta\text{-elimination}} \underset{\text{Isopentenyl pyrophosphate}}{CH_2{=}\overset{CH_3}{\overset{|}{C}}CH_2CH_2OP_2O_6^{3-}}$$

Isopentenyl pyrophosphate has the carbon skeleton of isoprene, the unit into which terpenes can be divided (Section 5.4). This molecule is a key intermediate in the biosynthesis of isoprene, terpenes, cholesterol, steroid hormones, and bile acids.

$$CH_2{=}C(CH_3)CH_2CH_2OP_2O_6^{3-} \longrightarrow \text{cholesterol} \longrightarrow \text{steroid hormones, bile acids}; \quad \longrightarrow \text{terpenes}$$

Isopentenyl pyrophosphate

SUMMARY

An **enolate anion** is an anion formed by removal of an α-hydrogen from a carbonyl-containing compound. Aldehydes, ketones, and esters can be converted to their enolate anions by treatment with a metal alkoxide or other strong base. An **aldol reaction** (Section 14.2) is the addition of an enolate anion from one aldehyde or ketone to the carbonyl carbon of another aldehyde or ketone to form a ***β*-hydroxyaldehyde** or ***β*-hydroxyketone.** Dehydration of the product of an aldol reaction gives an ***α,β*-unsaturated aldehyde** or **ketone. Crossed aldol reactions** are useful only when appreciable differences in reactivity occur between the two carbonyl-containing compounds, as for example when one of them has no α-hydrogens and can function only as an enolate anion acceptor. When both carbonyl groups are in the same molecule, aldol reaction results in the formation of a ring. Intramolecular aldol reaction is particularly useful for the formation of five- and six-membered rings.

A key step in the **Claisen** and **Dieckmann condensations** (Section 14.3) is addition of an ester enolate anion to a carbonyl group of another ester to form a tetrahedral carbonyl addition intermediate followed by collapse of the intermediate to give a β-ketoester. Each condensation is an example of nucleophilic acyl substitution.

Acetyl-CoA (Section 14.4) is the source of the carbon atoms for the synthesis of terpenes, cholesterol, and fatty acids. Key intermediates in the synthesis of these biomolecules are mevalonic acid and isopentenyl pyrophosphate.

KEY REACTIONS

1. The Aldol Reaction (Section 14.2)

The aldol reaction involves nucleophilic addition of an enolate anion from one aldehyde or ketone to the carbonyl carbon of another aldehyde or ketone to give a β-hydroxyaldehyde or ketone.

$$2CH_3CH_2CH_2\overset{O}{\overset{\|}{C}}H \xrightleftharpoons{NaOH} CH_3CH_2CH_2\overset{OH}{\overset{|}{C}}H\underset{CH_2CH_3}{\underset{|}{C}}H\overset{O}{\overset{\|}{C}}H$$

2. Dehydration of the Product of an Aldol Reaction (Section 14.2)

Dehydration of the β-hydroxyaldehyde or ketone from an aldol reaction occurs very readily under acidic or basic conditions and gives an α,β-unsaturated aldehyde or ketone.

$$CH_3\overset{OH}{\overset{|}{C}}HCH_2\overset{O}{\overset{\|}{C}}H \xrightarrow{H^+} CH_3CH{=}CH\overset{O}{\overset{\|}{C}}H + H_2O$$

3. The Claisen Condensation (Section 14.3A)

The product of a Claisen condensation is a β-ketoester. Condensation occurs by nucleophilic acyl substitution in which the attacking nucleophile is the enolate anion of an ester.

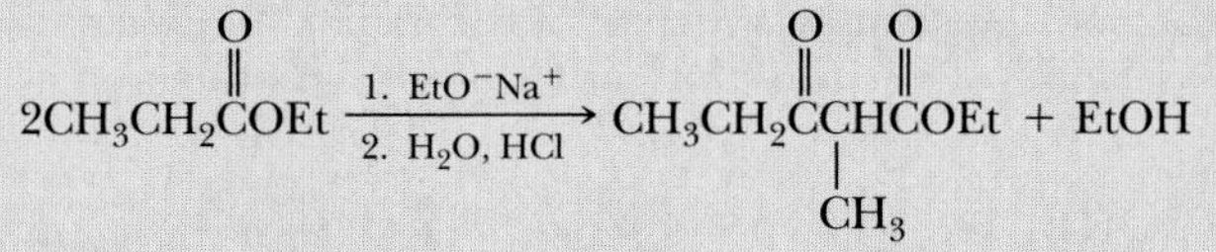

$$2CH_3CH_2\overset{O}{\overset{\|}{C}}OEt \xrightarrow[2.\ H_2O,\ HCl]{1.\ EtO^-Na^+} CH_3CH_2\overset{O}{\overset{\|}{C}}\underset{CH_3}{\underset{|}{C}H}\overset{O}{\overset{\|}{C}}OEt + EtOH$$

4. The Dieckmann Condensation (Section 14.3B)

An intramolecular Claisen condensation is called a Dieckmann condensation.

$$EtO\overset{O}{\overset{\|}{C}}(CH_2)_4\overset{O}{\overset{\|}{C}}OEt \xrightarrow[2.\ H_2O,\ HCl]{1.\ EtO^-Na^+} \text{2-(ethoxycarbonyl)cyclopentanone } (\text{ring C=O, } \overset{O}{\overset{\|}{C}}OEt) + EtOH$$

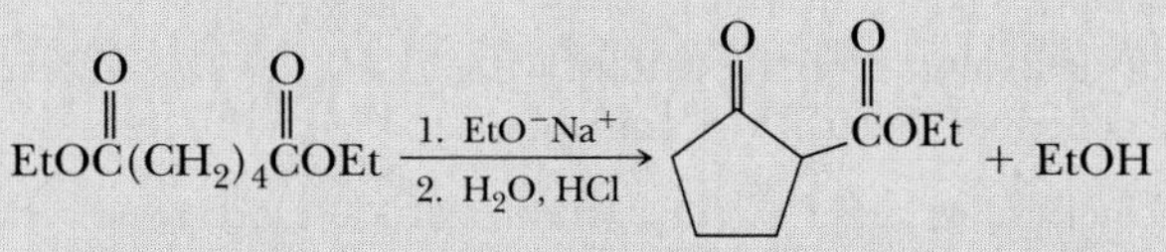

5. Crossed Claisen Condensations (Section 14.3C)

Crossed Claisen condensations are useful only where an appreciable difference exists in the reactivity between the two esters. Such is the case when an ester that has no α-hydrogens can function only as an enolate anion acceptor.

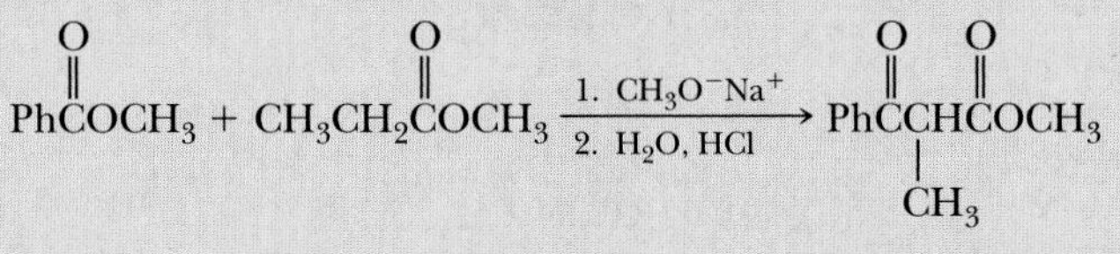

$$Ph\overset{O}{\overset{\|}{C}}OCH_3 + CH_3CH_2\overset{O}{\overset{\|}{C}}OCH_3 \xrightarrow[2.\ H_2O,\ HCl]{1.\ CH_3O^-Na^+} Ph\overset{O}{\overset{\|}{C}}\underset{CH_3}{\underset{|}{C}H}\overset{O}{\overset{\|}{C}}OCH_3$$

6. Hydrolysis and Decarboxylation of β-Ketoesters (Section 14.3D)

Hydrolysis of the ester followed by decarboxylation of the resulting β-ketoacid gives a ketone and carbon dioxide.

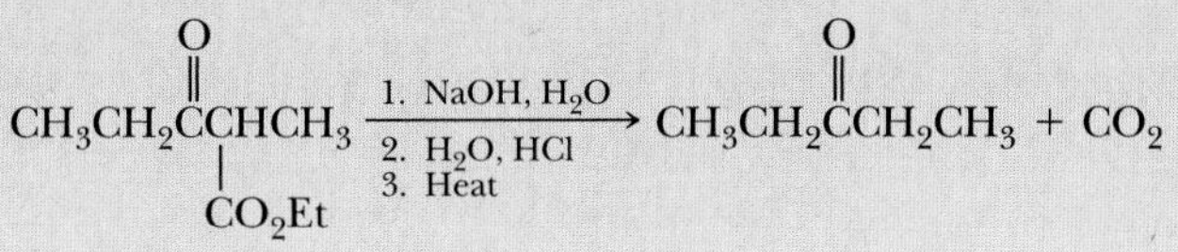

$$CH_3CH_2\overset{O}{\overset{\|}{C}}\underset{CO_2Et}{\underset{|}{C}H}CH_3 \xrightarrow[2.\ H_2O,\ HCl \quad 3.\ Heat]{1.\ NaOH,\ H_2O} CH_3CH_2\overset{O}{\overset{\|}{C}}CH_2CH_3 + CO_2$$

ADDITIONAL PROBLEMS

The Aldol Reaction

14.11 Estimate the pK_a of each compound and then arrange them in order of increasing acidity.

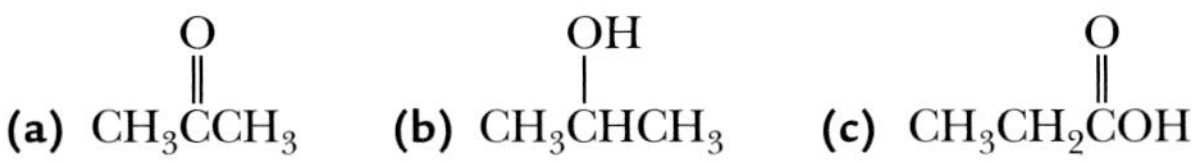

(a) $CH_3\overset{O}{\overset{\|}{C}}CH_3$ **(b)** $CH_3\overset{OH}{\overset{|}{C}}HCH_3$ **(c)** $CH_3CH_2\overset{O}{\overset{\|}{C}}OH$

14.12 Identify the most acidic hydrogen(s) in each compound.

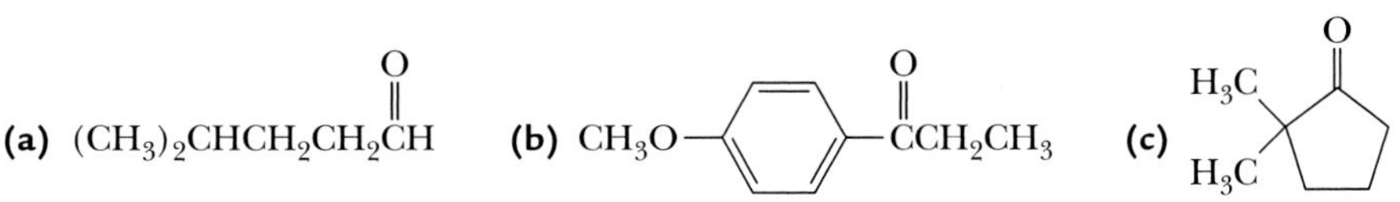

(a) $(CH_3)_2CHCH_2CH_2\overset{O}{\overset{\|}{C}}H$ **(b)** $CH_3O{-}C_6H_4{-}\overset{O}{\overset{\|}{C}}CH_2CH_3$ **(c)** 2,2-dimethylcyclopentanone (H_3C, H_3C on the carbon adjacent to C=O)

(d) 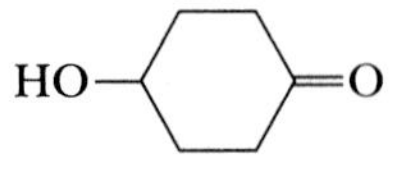 HO–(cyclohexane)=O **(e)** 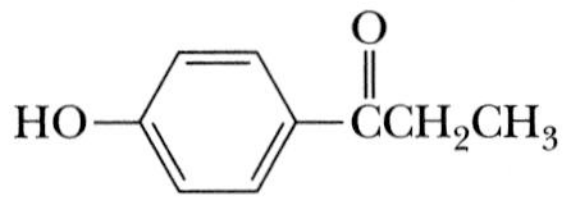 HO–(benzene ring)–$\overset{O}{\overset{\|}{C}}CH_2CH_3$

14.13 Write a second contributing structure of each anion and use curved arrows to show the redistribution of electrons that gives your second structure.

(a) 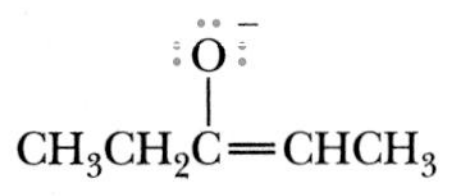 $CH_3CH_2\overset{:\ddot{O}:^-}{\overset{|}{C}}=CHCH_3$ **(b)** (cyclohexene with $:\ddot{O}:^-$ and CH_3) **(c)** (benzene ring)–$\overset{\ddot{O}}{\overset{\|}{C}}-\ddot{C}H_2^-$

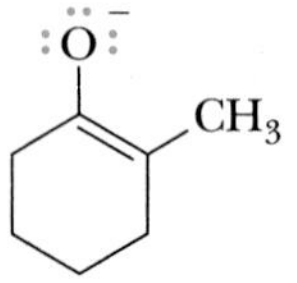

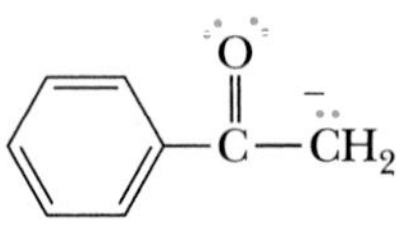

14.14 Treatment of 2-methylcyclohexanone with base gives two different enolate anions. Draw the contributing structure for each that places the negative charge on carbon.

14.15 Draw a structural formula for the product of the aldol reaction of each compound and for the α,β-unsaturated aldehyde or ketone formed by dehydration of each aldol product.

(a) 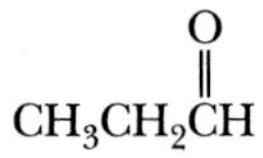 $CH_3CH_2\overset{O}{\overset{\|}{C}}H$ **(b)** (benzene ring)–$\overset{O}{\overset{\|}{C}}CH_3$

(c) (cyclopentanone) **(d)** 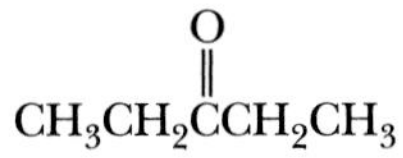 $CH_3CH_2\overset{O}{\overset{\|}{C}}CH_2CH_3$

14.16 Draw a structural formula for the product of each crossed aldol reaction and for the compound formed by dehydration of each aldol product.

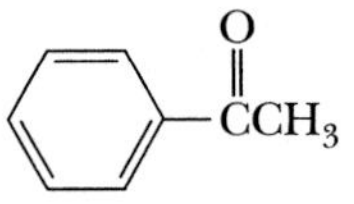

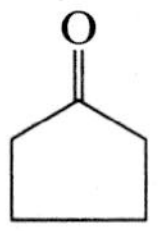

(a) 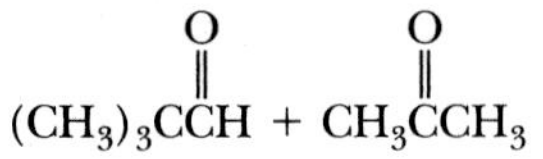 $(CH_3)_3C\overset{O}{\overset{\|}{C}}H + CH_3\overset{O}{\overset{\|}{C}}CH_3$ **(b)** (benzene ring)–$\overset{O}{\overset{\|}{C}}CH_3$ + (benzene ring)–$\overset{O}{\overset{\|}{C}}H$

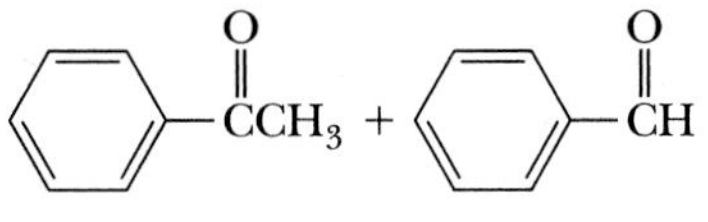

(c) (cyclohexanone) + $H\overset{O}{\overset{\|}{C}}H$ **(d)** 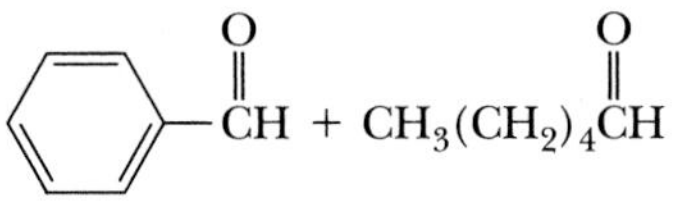 (benzene ring)–$\overset{O}{\overset{\|}{C}}H + CH_3(CH_2)_4\overset{O}{\overset{\|}{C}}H$

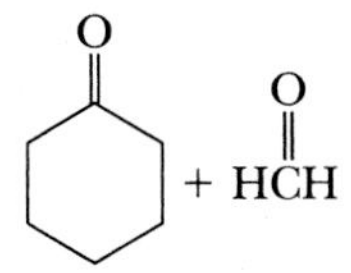

14.17 When a 1 : 1 mixture of acetone and 2-butanone is treated with base, six aldol products are possible. Draw a structural formula for each aldol product.

14.18 Show how to prepare each α,β-unsaturated ketone by an aldol reaction followed by dehydration of the aldol product.

(a) 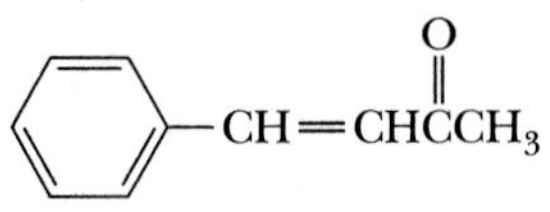 (benzene ring)–$CH=CH\overset{O}{\overset{\|}{C}}CH_3$ **(b)** 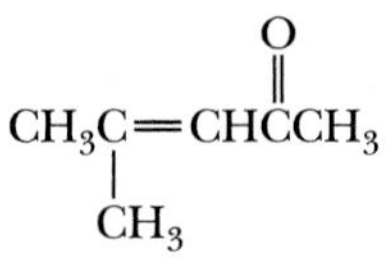 $CH_3\underset{CH_3}{\underset{|}{C}}=CH\overset{O}{\overset{\|}{C}}CH_3$

14.19 Show how to prepare each α,β-unsaturated aldehyde by an aldol reaction followed by dehydration of the aldol product.

(a) 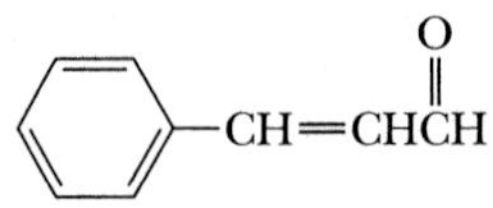 (benzene ring)–$CH=CH\overset{O}{\overset{\|}{C}}H$ **(b)** 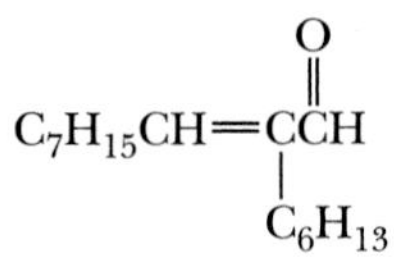 $C_7H_{15}CH=\underset{C_6H_{13}}{\underset{|}{\overset{O}{\overset{\|}{C}}}}H$

14.20 When treated with base, the following compound undergoes an intramolecular aldol reaction followed by dehydration to give a product containing a ring (yield 78%). Propose a structural formula for this product.

$$CH_3CH_2CH{=}CHCH_2CH_2\overset{O}{\overset{\|}{C}}CH_2CH_2\overset{O}{\overset{\|}{C}}H \xrightarrow[\text{aldol reaction}]{\text{base}} C_{10}H_{14}O + H_2O$$

14.21 Propose a structural formula for the compound of molecular formula $C_6H_{10}O_2$ that undergoes an aldol reaction followed by dehydration to give this α,β-unsaturated aldehyde.

$$C_6H_{10}O_2 \xrightarrow{\text{base}} \text{(cyclopentene–CHO)} + H_2O$$

1-Cyclopentenecarbaldehyde

14.22 Show how to bring about this conversion.

(cyclodecane-1,6-dione) → (bicyclic hydroxy ketone, OH) → (bicyclic enone)

14.23 Oxanamide (*The Merck Index,* 12th ed., #7053), a mild sedative, is synthesized from butanal in these five steps.

$$CH_3CH_2CH_2\overset{O}{\overset{\|}{C}}H \xrightarrow{(1)} CH_3CH_2CH_2CH{=}\underset{CH_2CH_3}{\underset{|}{C}}\overset{O}{\overset{\|}{C}}H \xrightarrow{(2)} CH_3CH_2CH_2CH{=}\underset{CH_2CH_3}{\underset{|}{C}}\overset{O}{\overset{\|}{C}}OH \xrightarrow{(3)}$$

Butanal — 2-Ethyl-2-hexenal — 2-Ethyl-2-hexenoic acid

$$CH_3CH_2CH_2CH{=}\underset{CH_2CH_3}{\underset{|}{C}}\overset{O}{\overset{\|}{C}}Cl \xrightarrow{(4)} CH_3CH_2CH_2CH{=}\underset{CH_2CH_3}{\underset{|}{C}}\overset{O}{\overset{\|}{C}}NH_2 \xrightarrow{(5)}$$

2-Ethyl-2-hexenoyl chloride — 2-Ethyl-2-hexenamide

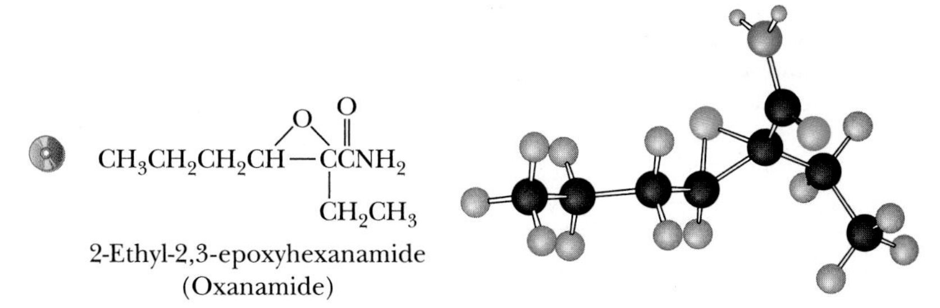

2-Ethyl-2,3-epoxyhexanamide (Oxanamide)

(a) Show reagents and experimental conditions to bring about each step in this synthesis.

(b) How many stereocenters are in oxanamide? How many stereoisomers are possible for this compound?

14.24 This reaction is one of the ten steps in glycolysis (Section 20.6), a series of enzyme-catalyzed reactions by which glucose is oxidized to two molecules of pyruvate. Show that this step is the reverse of an aldol reaction.

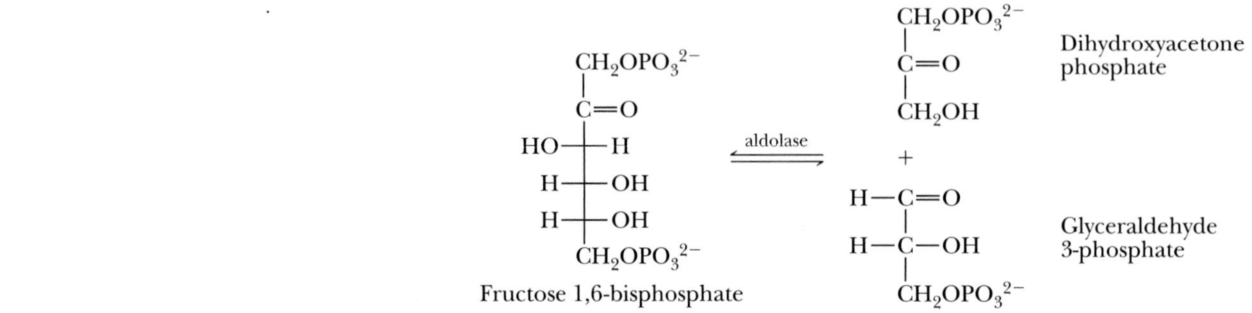

$$\begin{array}{c} CH_2OPO_3^{2-} \\ | \\ C{=}O \\ HO{-}C{-}H \\ H{-}C{-}OH \\ H{-}C{-}OH \\ CH_2OPO_3^{2-} \end{array} \underset{}{\overset{\text{aldolase}}{\rightleftharpoons}} \begin{array}{c} CH_2OPO_3^{2-} \\ | \\ C{=}O \\ | \\ CH_2OH \end{array} + \begin{array}{c} H{-}C{=}O \\ | \\ H{-}C{-}OH \\ | \\ CH_2OPO_3^{2-} \end{array}$$

Fructose 1,6-bisphosphate; Dihydroxyacetone phosphate; Glyceraldehyde 3-phosphate

The Claisen and Dieckmann Condensations

14.25 Show the product of Claisen condensation of each ester.

(a) Ethyl phenylacetate in the presence of sodium ethoxide.

(b) Methyl hexanoate in the presence of sodium methoxide.

14.26 When a 1:1 mixture of ethyl propanoate and ethyl butanoate is treated with sodium ethoxide, four Claisen condensation products are possible. Draw a structural formula for each product.

14.27 Draw a structural formula for the β-ketoester formed in the crossed Claisen condensation of ethyl propanoate with each ester.

(a) $EtOC(=O){-}C(=O)OEt$ **(b)** $PhC(=O)OEt$ **(c)** $HC(=O)OEt$

14.28 Draw a structural formula for the product of saponification, acidification, and decarboxylation of each β-ketoester formed in Additional Problem 14.27.

14.29 Complete the equation for this crossed Claisen condensation.

$$\text{3-pyridyl}{-}C(=O)OCH_2CH_3 + CH_3C(=O)OCH_2CH_3 \xrightarrow[\text{2. } H_2O,\ HCl]{\text{1. } CH_3CH_2O^-Na^+}$$

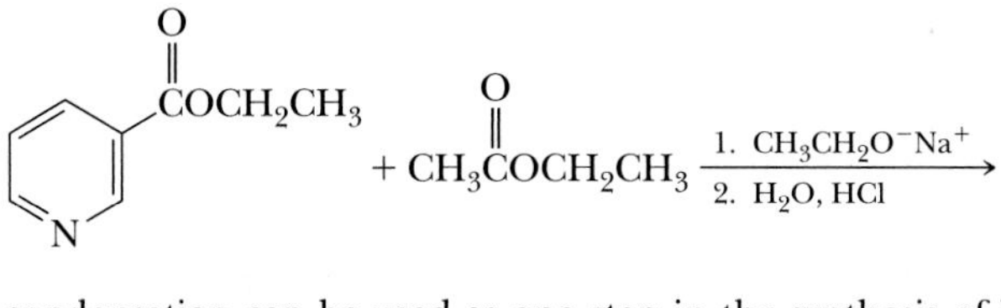

14.30 The Claisen condensation can be used as one step in the synthesis of ketones, as illustrated by this reaction sequence. Propose structural formulas for compounds A, B, and the ketone formed in this sequence.

$$2CH_3CH_2CH_2CH_2C(=O)OEt \xrightarrow[\text{2. } HCl,\ H_2O]{\text{1. } EtO^-Na^+} A \xrightarrow[\text{heat}]{NaOH,\ H_2O} B \xrightarrow[\text{heat}]{HCl,\ H_2O} C_9H_{18}O$$

14.31 Draw a structural formula for the ketone formed by treating each diester with sodium ethoxide followed by acidification with HCl. (*Hint:* These are Dieckmann condensations.)

(a) 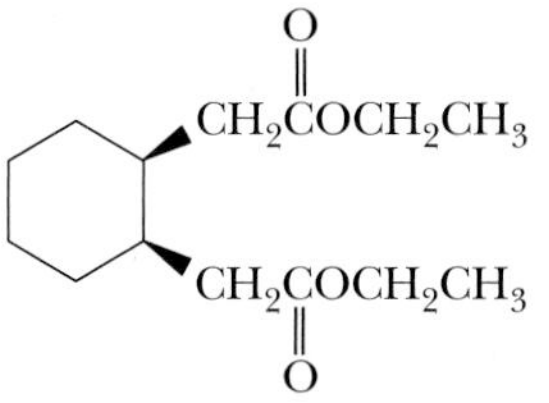 cyclohexane ring bearing two cis substituents: $CH_2COCH_2CH_3$ (C=O) and $CH_2COCH_2CH_3$ (C=O)

(b) 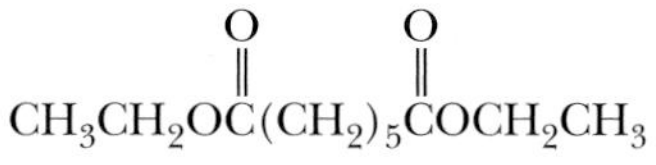 $CH_3CH_2O\overset{O}{\overset{\|}{C}}(CH_2)_5\overset{O}{\overset{\|}{C}}OCH_2CH_3$

14.32 Claisen condensation between diethyl phthalate and ethyl acetate followed by saponification, acidification, and decarboxylation forms a diketone, $C_9H_6O_2$. Propose structural formulas for compounds A, B, and the diketone.

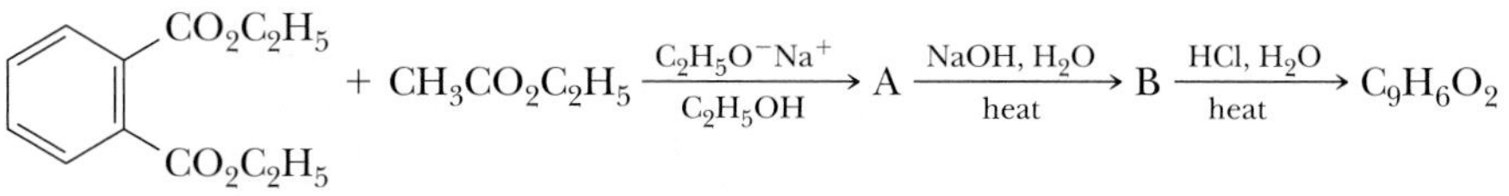

$$\text{Diethyl phthalate } (C_6H_4(CO_2C_2H_5)_2) + \underset{\text{Ethyl acetate}}{CH_3CO_2C_2H_5} \xrightarrow[C_2H_5OH]{C_2H_5O^-Na^+} A \xrightarrow[\text{heat}]{NaOH,\ H_2O} B \xrightarrow[\text{heat}]{HCl,\ H_2O} C_9H_6O_2$$

Diethyl phthalate Ethyl acetate

14.33 The rodenticide and insecticide pindone (*The Merck Index,* 12th ed., #7598) is synthesized by the following sequence of reactions. Propose a structural formula for pindone. (*Hint:* Pindone is a triketone.)

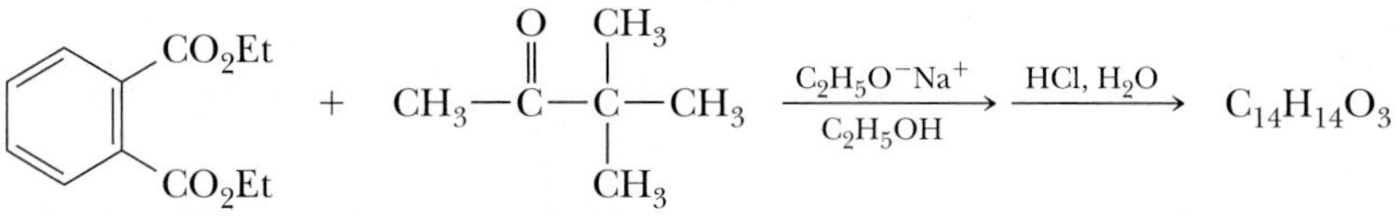

$$\text{Diethyl phthalate } (C_6H_4(CO_2Et)_2) + \underset{\text{3,3-Dimethyl-2-butanone}}{CH_3-\overset{O}{\overset{\|}{C}}-C(CH_3)_2-CH_3} \xrightarrow[C_2H_5OH]{C_2H_5O^-Na^+} \xrightarrow{HCl,\ H_2O} \underset{\text{Pindone}}{C_{14}H_{14}O_3}$$

Diethyl phthalate 3,3-Dimethyl-2-butanone Pindone

14.34 This reaction is the fourth in the set of four enzyme-catalyzed steps by which the hydrocarbon chain of a fatty acid (Section 20.3) is oxidized, two carbons at a time, to acetyl-coenzyme A. Show that this reaction is the reverse of a Claisen condensation.

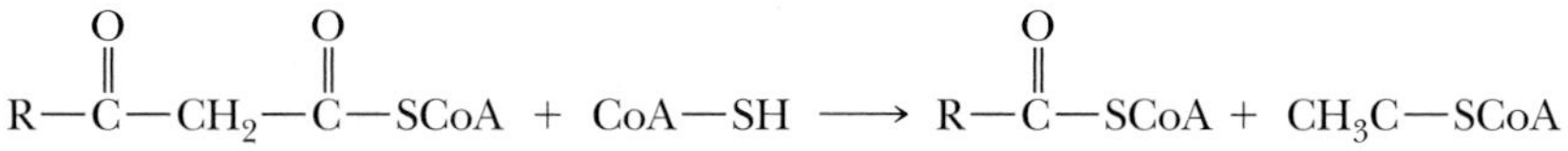

$$\underset{\beta\text{-ketoacyl-CoA}}{R-\overset{O}{\overset{\|}{C}}-CH_2-\overset{O}{\overset{\|}{C}}-SCoA} + \underset{\text{Coenzyme A}}{CoA-SH} \longrightarrow \underset{\text{An acyl-CoA}}{R-\overset{O}{\overset{\|}{C}}-SCoA} + \underset{\text{Acetyl-CoA}}{CH_3\overset{O}{\overset{\|}{C}}-SCoA}$$

15

Organic Polymer Chemistry

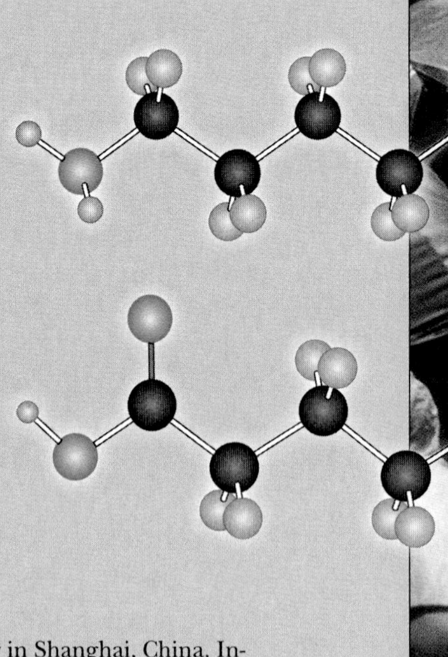

Sea of umbrellas on a rainy day in Shanghai, China. Inset: The two monomer units of nylon 66. *(Gavin Hellier/Tony Stone Images)*

The years since the 1930s have seen extensive research and development in organic polymer chemistry, and an almost explosive growth in plastics, coatings, and rubber technology has created a worldwide multibillion-dollar industry. A few basic characteristics account for this phenomenal growth. First, the raw materials for synthetic polymers are derived mainly from petroleum. With the development of petroleum-refining processes, raw materials for the synthesis of polymers became generally cheap and plentiful. Second, within broad limits, scientists have learned how to tailor polymers to the requirements of the end use. Third, many consumer products can be fabricated more cheaply from synthetic polymers than from such competing materials as wood, ceramics, and metals. For example, polymer technology created the water-based (latex) paints that have revolutionized the coatings industry; plastic films and foams have done the same for the packaging industry. The list could go on and on as we think of the manufactured items that are everywhere around us in our daily lives.

15.1 The Architecture of Polymers

Polymers (Greek: *poly* + *meros,* many parts) are long-chain molecules synthesized by linking **monomers** (Greek: *mono* + *meros,* single part) through chemical reactions. The molecular weights of polymers are generally high compared with those of common organic compounds and typically range from 10,000 g/mol to more than 1,000,000 g/mol. The architectures of these macromolecules can also be quite diverse. Types of polymer architecture include linear and branched chains as well as those with comb, ladder, and star structures (Figure 15.1). Additional structural variations can be achieved by introducing covalent crosslinks between individual polymer chains.

Polymer From the Greek, *poly* + *meros,* many parts. Any long-chain molecule synthesized by linking together many single parts called monomers.

Monomer From the Greek, *mono* + *meros,* single part. The simplest nonredundant unit from which a polymer is synthesized.

The term **plastic** refers to any polymer that can be molded when hot and retains its shape when cooled. **Thermoplastics** are polymers that can be melted and become sufficiently fluid that they can be molded into shapes that are retained when they are cooled. **Thermosetting plastics** can be molded when they are first prepared, but once they are cooled, they harden irreversibly and cannot be remelted. Because of their very different physical characteristics, thermoplastics and thermosets must be processed differently and are used in very different applications.

Plastic A polymer that can be molded when hot and retains its shape when cooled.

Thermoplastic A polymer that can be melted and molded into a shape that is retained when it is cooled.

Thermosetting plastic A polymer that can be molded when it is first prepared, but once cooled, hardens irreversibly and cannot be remelted.

The single most important property of polymers at the molecular level is the size and shape of their chains. A good example of the importance of size is a comparison of paraffin wax, a natural polymer, and polyethylene, a synthetic polymer. These two distinct materials have identical repeat units, namely $—CH_2—$, but differ greatly in chain size. Paraffin wax has between 25 and 50 carbon atoms per chain, whereas polyethylene has between 1000 and 3000 carbons per chain. Paraf-

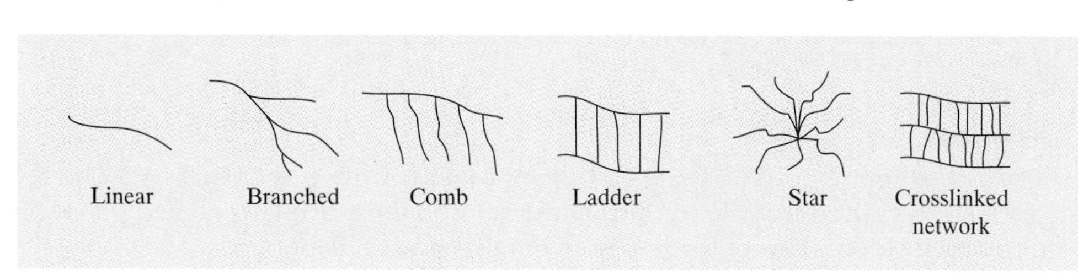

Figure 15.1 Various polymer architectures.

fin wax, such as in birthday candles, is soft and brittle, but polyethylene, such as in plastic beverage bottles, is strong, flexible, and tough. These vastly different properties arise directly from the difference in size and molecular architecture of the individual polymer chains.

15.2 Polymer Notation and Nomenclature

Average degree of polymerization, *n* A subscript placed outside the parentheses of the simplest nonredundant unit of a polymer to indicate that the unit repeats *n* times in the polymer.

The structure of a polymer is shown by placing parentheses around the **repeat unit,** which is the smallest molecular fragment that contains all the nonrepeating structural features of the chain. Thus, the structure of an entire polymer chain can be reproduced by repeating the enclosed structure in both directions. A subscript *n*, called the **average degree of polymerization,** is placed outside the parentheses to indicate that this unit is repeated *n* times.

$$\sim O-CH_2CH_2-O-CH_2CH_2-O-CH_2-CH_2\sim \qquad -(CH_2CH_2-O)_n-$$

Part of an extended polymer chain — The repeat unit

Exceptions to this notation are the polymers formed from symmetric monomers, such as polyethylene, $-(CH_2CH_2)_n-$, and polytetrafluoroethylene, $-(CF_2CF_2)_n-$. Although the simplest repeat units are $-CH_2-$ and $-CF_2-$, respectively, we show two methylene groups and two difluoromethylene groups because they originate from ethylene ($CH_2{=}CH_2$) and tetrafluoroethylene ($CF_2{=}CF_2$), the monomer units from which these polymers are derived.

The most common method of naming a polymer is to attach the prefix poly- to the name of the monomer from which the polymer is derived, as for example polyethylene and polystyrene. In the case of a more complex monomer or where the name of the monomer is more than one word, as for example the monomer vinyl chloride, parentheses are used to enclose the name of the monomer.

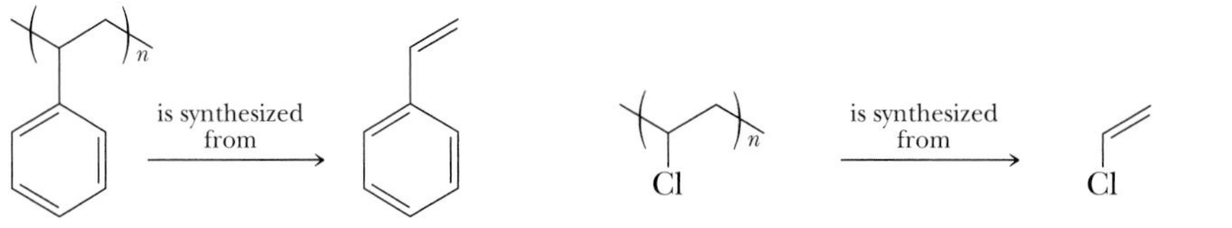

Polystyrene —is synthesized from→ Styrene

Poly(vinyl chloride) (PVC) —is synthesized from→ Vinyl chloride

Example 15.1

Given the following structure, determine the polymer's repeat unit, redraw the structure using the simplified parenthetical notation, and name the polymer.

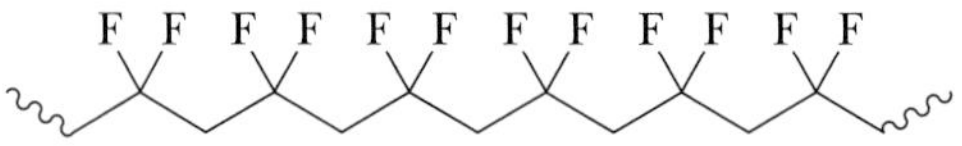

F F F F F F F F F F F F

Solution

The repeat unit is $-CH_2CF_2-$, and the polymer is written $-(CH_2CF_2)_n-$. The repeat unit is derived from 1,1-difluoroethylene, and the polymer is named poly(1,1-difluoroethylene). This polymer is used in microphone diaphragms.

Practice Problem 15.1

Given the following structure, determine the polymer's repeat unit, redraw the structure using the simplified parenthetical notation, and name the polymer.

15.3 POLYMER MORPHOLOGY—CRYSTALLINE VERSUS AMORPHOUS MATERIALS

Polymers, like small organic molecules, tend to crystallize upon precipitation or as they are cooled from a melt. Acting to inhibit this tendency are their very large molecules and sometimes complicated or irregular structures, which prevent efficient packing of the chains. The result is that polymers in the solid state tend to be composed of both ordered **crystalline domains** (crystallites) and disordered **amorphous domains.** The relative amounts of crystalline and amorphous domains differ from polymer to polymer and often depend upon the manner in which the material is processed.

Crystalline domains Ordered crystalline regions in the solid state of a polymer. Also called crystallites.

Amorphous domains Disordered, noncrystalline regions in the solid state of a polymer.

High degrees of crystallinity are most often found in polymers with regular, compact structures and strong intermolecular forces, such as hydrogen bonding. The temperature at which crystallites melt corresponds to the **melt transition temperature (T_m)** of the polymer. As the degree of crystallinity of a polymer increases, its T_m increases, and it becomes more opaque because of the scattering of light by its crystalline domains. With an increase in crystallinity, there is also a corresponding increase in strength and stiffness. For example, poly(6-aminohexanoic acid), known more commonly as nylon 6, has a $T_m = 223°C$. At and well above room temperature, this polymer is a hard, durable material that does not undergo any appreciable change in properties even on a very hot summer afternoon. Its uses range from textile fibers to shoe heels.

Melt transition temperature, T_m The temperature at which crystalline regions of a polymer melt.

Amorphous domains have little or no long-range order. Highly amorphous polymers are sometimes referred to as **glassy** polymers. Because they lack crystalline domains that scatter light, amorphous polymers are transparent. In addition, they are typically weak polymers, both in terms of their high flexibility and low mechanical strength. On being heated, amorphous polymers are transformed from a hard glass state to a soft, flexible, rubbery state. The temperature at which this transition occurs is called the **glass transition temperature (T_g)**. Amorphous polystyrene, for example, has a $T_g = 100°C$. At room temperature, it is a rigid solid used for drinking cups, foamed packaging materials, disposable medical wares, tape reels, and so forth. If it is placed in boiling water, it becomes soft and rubbery.

Glass transition temperature, T_g The temperature at which a polymer undergoes the transition from a hard glass to a rubbery state.

This relationship between mechanical properties and the degree of crystallinity can be illustrated by poly(ethylene terephthalate) (PET).

Poly(ethylene terephthalate)
(PET)

PET can be made with a percent of crystalline domains ranging from 0% to about 55%. Completely amorphous PET is formed by cooling from the melt quickly. By prolonging the cooling time, more molecular diffusion occurs, and crystallites form as the chains become more ordered. The differences in mechanical properties between these forms of PET are great. PET with a low degree of crystallinity is used for plastic beverage bottles, whereas fibers drawn from highly crystalline PET are used for textile fibers and tire cords.

Elastomer A material that, when stretched or otherwise distorted, returns to its original shape when the distorting force is released.

Rubber materials must have low T_g values in order to behave as **elastomers.** If the temperature drops below its T_g value, then the material is converted to a rigid glassy solid, and all elastomeric properties are lost. A poor understanding of this behavior of elastomers contributed to the *Challenger* spacecraft disaster in 1985. The elastomeric O-rings used to seal the solid booster rockets had a T_g value around 0°C. When the temperature dropped to an unanticipated low on the morning of the *Challenger* launch, the O-ring seals dropped below their T_g value and obediently changed from elastomers to rigid glasses, losing any sealing capabilities. The rest is tragic history. The physicist Richard Feynman sorted this out publicly in a famous televised hearing in which he put a *Challenger*-type O-ring in ice water and showed that its elasticity was lost!

15.4 STEP-GROWTH POLYMERIZATIONS

Step-growth polymerization A polymerization in which chain growth occurs in a stepwise manner between difunctional monomers, as for example between adipic acid and hexamethylenediamine to form nylon 66.

Condensation polymerization A polymerization in which chain growth occurs in a stepwise manner between difunctional monomers. Also called step-growth polymerization.

Polymerizations in which chain growth occurs in a stepwise manner are called **step-growth** or **condensation polymerizations.** Step-growth polymers are formed by reaction between difunctional molecules, with each new bond created in a separate step. During polymerization, monomers react to form dimers, dimers react with monomers to form trimers, dimers react with dimers to form tetramers, and so on.

There are two common types of step-growth processes: (1) reaction between A—A and B—B type monomers to give $\left(\mathrm{A{-}A{-}B{-}B}\right)_n$ polymers and (2) the self-condensation of A—B monomers to give $\left(\mathrm{A{-}B}\right)_n$ polymers. In each case, an A functional group reacts exclusively with a B functional group, and a B functional group reacts exclusively with an A functional group. New covalent bonds in step-growth polymerizations are generally formed by polar reactions, as for example nucleophilic acyl substitution. In this section, we discuss five types of step-growth polymers: polyamides, polyesters, polycarbonates, polyurethanes, and epoxy resins.

A. Polyamides

Polyamide A polymer in which each monomer unit is joined to the next by an amide bond, as for example nylon 66.

In the early 1930s, Wallace Carothers and his associates at E. I. DuPont de Nemours & Company began fundamental research into the reactions of aliphatic dicarboxylic acids and diols. From adipic acid and ethylene glycol, they obtained a polyester of high molecular weight that could be drawn into fibers.

$$n\,\mathrm{HO\overset{O}{\overset{\|}{C}}(CH_2)_4\overset{O}{\overset{\|}{C}}OH} + n\,\mathrm{HOCH_2CH_2OH} \longrightarrow \left(\mathrm{\overset{O}{\overset{\|}{C}}(CH_2)_4\overset{O}{\overset{\|}{C}}OCH_2CH_2O}\right)_n + 2n\mathrm{H_2O}$$

Hexanedioic acid (Adipic acid) — 1,2-Ethanediol (Ethylene glycol) — Poly(ethylene adipate)

These first polyester fibers had melt transition temperatures (T_m) too low for use as textile fibers, and they were not investigated further. Carothers then turned his attention to the reactions of dicarboxylic acids and diamines to form polyamides and, in 1934, synthesized nylon 66, the first purely synthetic fiber. Nylon 66 is so named because it is synthesized from two different monomers, each containing six carbon atoms.

In the synthesis of nylon 66, hexanedioic acid (adipic acid) and 1,6-hexanediamine (hexamethylenediamine) are dissolved in aqueous ethanol where they react to form a one-to-one salt called nylon salt. Nylon salt is then heated in an autoclave to 250°C where the internal pressure rises to about 15 atm. Under these conditions, $—CO_2^-$ groups from adipic acid and $—NH_3^+$ groups from hexamethylenediamine react by loss of H_2O to form a polyamide. Nylon 66 formed under these conditions melts at 250 to 260°C and has a molecular weight range of 10,000 to 20,000 g/mol.

$$n\text{HO}\overset{\text{O}}{\overset{\|}{\text{C}}}(\text{CH}_2)_4\overset{\text{O}}{\overset{\|}{\text{C}}}\text{OH} + n\text{H}_2\text{N}(\text{CH}_2)_6\text{NH}_2 \longrightarrow n\left[{}^-\text{O}\overset{\text{O}}{\overset{\|}{\text{C}}}(\text{CH}_2)_4\overset{\text{O}}{\overset{\|}{\text{C}}}\text{O}^-\ \overset{+}{\text{H}_3\text{N}}(\text{CH}_2)_6\text{NH}_3^+\right]$$

Hexanedioic acid (Adipic acid) | 1,6-Hexanediamine (Hexamethylenediamine) | Nylon salt

$$\xrightarrow{\text{heat}} \left(\overset{\text{O}}{\overset{\|}{\text{C}}}(\text{CH}_2)_4\overset{\text{O}}{\overset{\|}{\text{C}}}\text{NH}(\text{CH}_2)_6\text{NH}\right)_n + 2n\text{H}_2\text{O}$$

Nylon 66
(A polyamide)

In the first stage of fiber production, crude nylon 66 is melted, spun into fibers, and cooled. Next, the melt-spun fibers are **cold-drawn** (drawn at room temperature) to about four times their original length to increase their degree of crystallinity. As the fibers are drawn, individual polymer molecules become oriented in the direction of the fiber axis, and hydrogen bonds form between carbonyl oxygens of one chain and amide hydrogens of another chain (Figure 15.2). The effects of orientation of polyamide molecules on the physical properties of the fiber are dramatic; both tensile strength and stiffness are increased markedly. Cold-drawing is an important step in the production of most synthetic fibers.

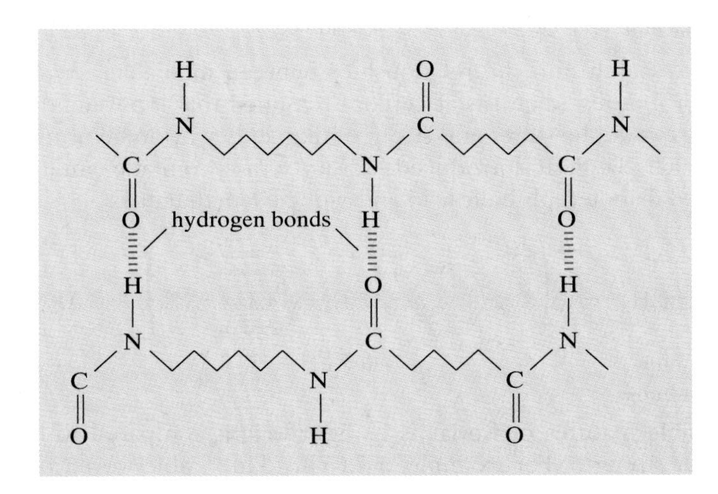

Figure 15.2
The structure of cold-drawn nylon 66. Hydrogen bonds between adjacent polymer chains provide additional tensile strength and stiffness to the fibers.

The current raw material base for the production of adipic acid is benzene, which is derived almost entirely from catalytic cracking and reforming of petroleum. Catalytic reduction of benzene to cyclohexane followed by catalyzed air oxidation gives a mixture of cyclohexanol and cyclohexanone. Oxidation of this mixture by nitric acid gives adipic acid.

$$\text{Benzene} \xrightarrow[\text{catalyst}]{3H_2} \text{Cyclohexane} \xrightarrow[\text{catalyst}]{\text{air}} [\text{Cyclohexanol} + \text{Cyclohexanone}] \xrightarrow{HNO_3} HO\overset{O}{\overset{\|}{C}}(CH_2)_4\overset{O}{\overset{\|}{C}}OH$$

Benzene, Cyclohexane, Cyclohexanol, Cyclohexanone, Hexanedioic acid (Adipic acid)

Adipic acid, in turn, is a starting material for the synthesis of hexamethylenediamine. Treatment of adipic acid with ammonia gives an ammonium salt, which, when heated, gives adipamide. Catalytic reduction of adipamide gives hexamethylenediamine. Thus, carbon sources for the production of nylon 66 are derived entirely from petroleum, which, unfortunately, is not a renewable resource.

$$NH_4^{+}\,{}^{-}O\overset{O}{\overset{\|}{C}}(CH_2)_4\overset{O}{\overset{\|}{C}}O^{-}\,NH_4^{+} \xrightarrow{\text{heat}} H_2N\overset{O}{\overset{\|}{C}}(CH_2)_4\overset{O}{\overset{\|}{C}}NH_2 \xrightarrow[\text{catalyst}]{4H_2} H_2N(CH_2)_6NH_2$$

Ammonium hexanedioate (Ammonium adipate), Hexanediamide (Adipamide), 1,6-Hexanediamine (Hexamethylenediamine)

Bulletproof vests have a thick layer of Kevlar. *(Charles D. Winters)*

Aramid A polyaromatic amide; a polymer in which the monomer units are an aromatic diamine and an aromatic dicarboxylic acid.

The nylons are a family of polymers, the members of which have subtly different properties that suit them to one use or another. The two most widely used members of this family are nylon 66 and nylon 6. Nylon 6 is so named because it is synthesized from caprolactam, a six-carbon monomer. In this synthesis, caprolactam is partially hydrolyzed to 6-aminohexanoic acid and then heated to 250°C to bring about polymerization. Nylon 6 is fabricated into fibers, brush bristles, rope, high-impact moldings, and tire cords.

$$n\ \text{Caprolactam} \xrightarrow[\text{2. heat}]{\text{1. partial hydrolysis}} \left(NH(CH_2)_5\overset{O}{\overset{\|}{C}} \right)_n$$

Caprolactam, Nylon 6

Based on extensive research into the relationships between molecular structure and bulk physical properties, scientists at DuPont reasoned that a polyamide containing aromatic rings would be stiffer and stronger than either nylon 66 or nylon 6, and, in early 1960, DuPont introduced Kevlar, a polyaromatic amide **(aramid)** fiber synthesized from terephthalic acid and *p*-phenylenediamine.

$$nHO\overset{O}{\overset{\|}{C}}-C_6H_4-\overset{O}{\overset{\|}{C}}OH + nH_2N-C_6H_4-NH_2 \longrightarrow \left(\overset{O}{\overset{\|}{C}}-C_6H_4-\overset{O}{\overset{\|}{C}}NH-C_6H_4-NH \right)_n + 2nH_2O$$

1,4-Benzenedicarboxylic acid (Terephthalic acid), 1,4-Benzenediamine (*p*-Phenylenediamine), Kevlar

One of the remarkable features of Kevlar is its light weight compared with other materials of similar strength. For example, a 7.6-cm (3-in.) cable woven of Kevlar has a strength equal to that of a similarly woven 7.6-cm (3-in.) steel cable.

Whereas the steel cable weighs about 30 kg/m (20 lb/ft), the Kevlar cable weighs only 6 kg/m (4 lb/ft). Kevlar now finds use in such articles as anchor cables for offshore drilling rigs and reinforcement fibers for automobile tires. Kevlar is also woven into a fabric that is so tough that it can be used for bulletproof vests, jackets, and raincoats.

B. Polyesters

Polyester A polymer in which each monomer unit is joined to the next by an ester bond, as for example poly(ethylene terephthalate).

Recall that, in the early 1930s, Carothers and his associates had concluded that polyester fibers from aliphatic dicarboxylic acids and ethylene glycol were not suitable for textile use because their melting points are too low. Winfield and Dickson at the Calico Printers Association in England further investigated polyesters in the 1940s and reasoned that a greater resistance to rotation in the polymer backbone would stiffen the polymer, raise its melting point, and thereby lead to a more acceptable polyester fiber. To create stiffness in the polymer chain, they used terephthalic acid. Polymerization of this aromatic dicarboxylic acid with ethylene glycol gives poly(ethylene terephthalate), abbreviated PET. Virtually all PET is now made by the following transesterification (exchange of esters) reaction.

$$n\mathrm{CH_3O\overset{O}{\overset{\|}{C}}\!-\!C_6H_4\!-\!\overset{O}{\overset{\|}{C}}OCH_3} + n\mathrm{HOCH_2CH_2OH} \xrightarrow{\text{heat}} \left(\!\mathrm{\overset{O}{\overset{\|}{C}}\!-\!C_6H_4\!-\!\overset{O}{\overset{\|}{C}}OCH_2CH_2O}\!\right)_n + 2n\mathrm{CH_3OH}$$

Dimethyl terephthalate; 1,2-Ethanediol (Ethylene glycol); Poly(ethylene terephthalate) (Dacron, Mylar)

The crude polyester can be melted, extruded, and then cold-drawn to form the textile fiber Dacron polyester, outstanding features of which are its stiffness (about four times that of nylon 66), very high strength, and remarkable resistance to creasing and wrinkling. Because the early Dacron polyester fibers were harsh to the touch due to their stiffness, they were usually blended with cotton or wool to make acceptable textile fibers. Newly developed fabrication techniques now produce less harsh Dacron polyester textile fibers. PET is also fabricated into Mylar films and recyclable plastic beverage containers.

Ethylene glycol for the synthesis of PET is obtained by air oxidation of ethylene to ethylene oxide (Section 8.6B) followed by hydrolysis to the glycol (Section 8.7). Ethylene is, in turn, derived entirely from cracking either petroleum or ethane derived from natural gas (Section 3.10A). Terephthalic acid is obtained by oxidation of *p*-xylene (Section 9.5), an aromatic hydrocarbon obtained along with benzene and toluene from catalytic cracking and reforming of naphtha and other petroleum fractions (Section 3.10B).

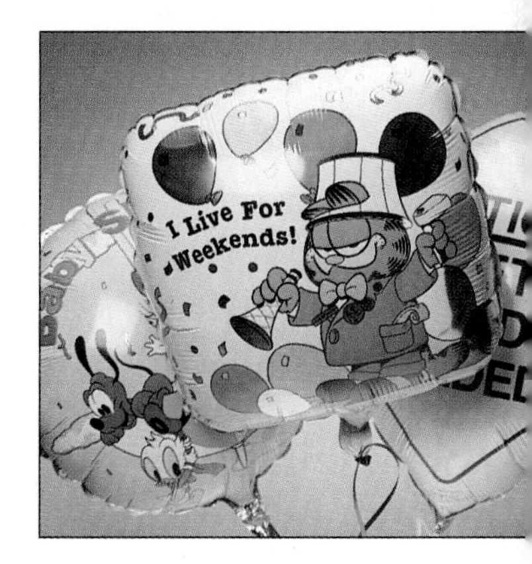

Mylar can be made into extremely strong films. Because the film has very tiny pores, it is used for balloons that can be inflated with helium; the helium atoms diffuse only slowly through the pores of the film. *(Charles D. Winters)*

$$\mathrm{CH_2{=}CH_2} \xrightarrow[\text{catalyst}]{O_2} \mathrm{\overset{O}{CH_2{-}CH_2}} \xrightarrow{H^+,\ H_2O} \mathrm{HOCH_2CH_2OH}$$

Ethylene; Ethylene oxide; 1,2-Ethanediol (Ethylene glycol)

$$\mathrm{H_3C\!-\!C_6H_4\!-\!CH_3} \xrightarrow[\text{catalyst}]{O_2} \mathrm{HO\overset{O}{\overset{\|}{C}}\!-\!C_6H_4\!-\!\overset{O}{\overset{\|}{C}}OH}$$

p-Xylene; Terephthalic acid

C. Polycarbonates

Polycarbonate A polyester in which the carboxyl groups are derived from carbonic acid.

Polycarbonates, the most familiar of which is Lexan, are a class of commercially important engineering polyesters. Lexan is formed by reaction between the disodium salt of bisphenol A (Problem 9.33) and phosgene.

$$Na^+O^- - C_6H_4 - C(CH_3)_2 - C_6H_4 - O^-Na^+ + Cl-C(=O)-Cl \longrightarrow \left(C_6H_4 - C(CH_3)_2 - C_6H_4 - O - C(=O) - O \right)_n + 2nNaCl$$

Disodium salt of bisphenol A — Phosgene — Lexan (a polycarbonate)

A polycarbonate hockey mask. *(Charles D. Winters)*

Lexan is a tough, transparent polymer with high impact and tensile strengths and retains its properties over a wide temperature range. It has found significant use in sporting equipment, such as bicycle, football, motorcycle, and snowmobile helmets as well as hockey and baseball catchers' face masks. In addition, it is used to make light, impact-resistant housings for household appliances and automobile and aircraft equipment and in the manufacture of safety glass and unbreakable windows.

D. Polyurethanes

A urethane, or carbamate, is an ester of carbamic acid, H_2NCO_2H. Carbamates are most commonly prepared by treatment of an isocyanate with an alcohol.

$$RN{=}C{=}O + R'OH \longrightarrow RNHC(=O)OR'$$

An isocyanate — A carbamate

Polyurethane A polymer containing the $-NHCO_2-$ group as a repeating unit.

Polyurethanes consist of flexible polyester or polyether units (blocks) alternating with rigid urethane units (blocks). The rigid urethane blocks are derived from a diisocyanate, commonly a mixture of 2,4- and 2,6-toluene diisocyanate. The more flexible blocks are derived from low-molecular-weight (MW = 1000–4000) polyesters or polyethers with —OH groups at each end of their polymer chains. Polyurethane fibers are fairly soft and elastic and have found use as Spandex and Lycra, the "stretch" fabrics used in bathing suits, leotards, and undergarments.

$$O{=}C{=}N - C_6H_3(CH_3) - N{=}C{=}O + nHO-\text{polymer}-OH \longrightarrow \left(C(=O)NH - C_6H_3(CH_3) - NHC(=O)O-\text{polymer}-O \right)_n$$

2,6-Toluene diisocyanate — Low-molecular-weight polyester or polyether with OH groups at each end of the chain — A polyurethane

Polyurethane foams for upholstery and insulating materials are made by adding small amounts of water during polymerization. Water reacts with isocyanate groups to produce gaseous carbon dioxide, which then acts as the foaming agent.

$$RN{=}C{=}O + H_2O \longrightarrow \left[RNH{-}\overset{\overset{\displaystyle O}{\|}}{C}{-}OH \right] \longrightarrow RNH_2 + CO_2$$

An isocyanate — A carbamic acid (unstable)

E. Epoxy Resins

Epoxy resins are materials prepared by a polymerization in which one monomer contains at least two epoxy groups. Within this range, there are a large number of polymeric materials possible, and epoxy resins are produced in forms ranging from low-viscosity liquids to high-melting solids. The most widely used epoxide monomer is the diepoxide prepared by treatment of 1 mole of bisphenol A (Problem 9.33) with 2 moles of epichlorohydrin.

Epoxy resin A material prepared by a polymerization in which one monomer contains at least two epoxy groups.

Epichlorohydrin + Disodium salt of bisphenol A (Na^+O^-–C₆H₄–C(CH_3)₂–C₆H₄–O^-Na^+) + Epichlorohydrin

↓

A diepoxide + 2NaCl

To prepare the following epoxy resin, the diepoxide monomer is treated with 1,2-ethanediamine (ethylene diamine). This component is usually called the catalyst in the two-component formulations that can be bought in any hardware store; it is also the component with the acrid smell.

A diepoxide + H_2N–CH₂CH₂–NH_2 ⟶

An epoxy resin (repeating unit with OH, O, CH_3, CH_3, OH, N–H, N–H groups)$_n$

Epoxy resins are widely used as adhesives and insulating surface coatings. They have good electrical insulating properties, which lead to their use for encapsulating electrical components ranging from integrated circuit boards to switch coils and insulators for power transmission systems. They are also used as composites with other materials, such as glass fiber, paper, metal foils, and other synthetic fibers to create structural components for jet aircraft, rocket motor casings, and so on.

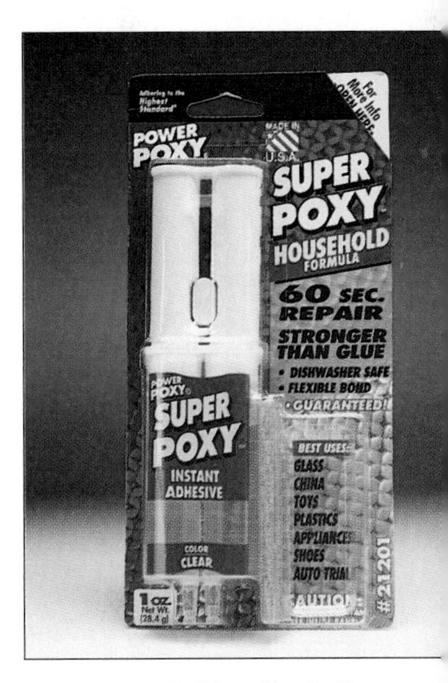

An epoxy resin kit. *(Charles D. Winters)*

CHEMICAL CONNECTIONS

Stitches That Dissolve

As the technological capabilities of medicine have grown, the demand for synthetic materials that can be used inside the body has increased as well. Polymers have many of the characteristics of an ideal biomaterial: they are lightweight and strong, are inert or biodegradable depending on their chemical structure, and have physical properties (softness, rigidity, elasticity) that are easily tailored to match those of natural tissues. Carbon-carbon backbone polymers are degradation-resistant and are used widely in permanent organ and tissue replacements.

Even though most medical uses of polymeric materials require biostability, applications have been developed that use the biodegradable nature of some macromolecules. An example is the use of poly(glycolic acid) and glycolic acid/lactic acid copolymers as absorbable sutures.

$$\underset{\text{Glycolic acid}}{HOCH_2\overset{O}{\overset{\|}{C}}OH} + \underset{\text{Lactic acid}}{HO\underset{CH_3}{\underset{|}{C}}H\overset{O}{\overset{\|}{C}}OH} \xrightarrow[-nH_2O]{\text{copolymerization}} \underset{\substack{\text{A polymer of}\\\text{poly(glycolic acid)-}\\\text{poly(lactic acid)}}}{\left(CH_2\overset{O}{\overset{\|}{C}}O\underset{CH_3}{\underset{|}{C}}H\overset{O}{\overset{\|}{C}}O \right)_n}$$

Traditional suture materials such as catgut must be removed by a health care specialist after they have served their purpose. Stitches of these hydroxyester polymers, however, are hydrolyzed slowly over a period of approximately 2 weeks, and by the time the torn tissues have fully healed, the stitches have fully degraded, and no suture removal is necessary. Glycolic and lactic acids formed during hydrolysis of the stitches are metabolized and excreted by existing biochemical pathways.

EXAMPLE 15.2

(a) By what type of mechanism does the reaction between the disodium salt of bisphenol A and epichlorohydrin take place?

SOLUTION

An S_N2 mechanism. The phenoxide ion of bisphenol A is a good nucleophile, and chlorine on the primary carbon of epichlorohydrin is the leaving group.

Practice Problem 15.2

Write the repeating unit of the epoxy resin formed from the following reaction.

[A diepoxide: 1,3-bis(glycidyloxy)benzene] + H_2N–C_6H_4–NH_2 ⟶

A diepoxide A diamine

15.5 CHAIN-GROWTH POLYMERIZATIONS

Chain-growth polymerization A polymerization that involves sequential addition reactions, either to unsaturated monomers or to monomers possessing other reactive functional groups.

Radical Any molecule that contains one or more unpaired electrons.

Fishhook arrow A single-barbed curved arrow used to show the change in position of a single electron.

From the perspective of the chemical industry, the single most important reaction of alkenes is **chain-growth polymerization,** a type of polymerization in which monomer units are joined together without loss of atoms. An example is the formation of polyethylene from ethylene.

$$\underset{\text{Ethylene}}{n\mathrm{CH_2{=}CH_2}} \xrightarrow{\text{catalyst}} \underset{\text{Polyethylene}}{\mathrm{-\!\!\left(CH_2CH_2\right)\!\!-}_n}$$

Table 15.1 lists several important polymers derived from ethylene and substituted ethylenes along with their common names and most important uses.

A. Radical Chain-Growth Polymerizations

The first commercial polymerizations of ethylene were initiated by radicals formed by thermal decomposition of organic peroxides, such as benzoyl peroxide. A **radical** is any molecule that contains one or more unpaired electrons. Radicals can be formed by cleavage of a bond in such a way that each atom or fragment participating in the bond retains one electron. In the following equation, **fishhook arrows** are used to show the change in position of single electrons.

TABLE 15.1 Polymers Derived from Substituted Ethylenes

Monomer Formula	Common Name	Polymer Name(s) and Common Uses
$CH_2{=}CH_2$	ethylene	polyethylene, Polythene; break-resistant containers and packaging materials
$CH_2{=}CHCH_3$	propylene	polypropylene, Herculon; textile and carpet fibers
$CH_2{=}CHCl$	vinyl chloride	poly(vinyl chloride), PVC; construction tubing
$CH_2{=}CCl_2$	1,1-dichloroethylene	poly(1,1-dichloroethylene); Saran Wrap is a copolymer with vinyl chloride
$CH_2{=}CHCN$	acrylonitrile	polyacrylonitrile, Orlon; acrylics and acrylates
$CF_2{=}CF_2$	tetrafluoroethylene	poly(tetrafluoroethylene), Teflon; nonstick coatings
$CH_2{=}CHC_6H_5$	styrene	polystyrene, Styrofoam; insulating materials
$CH_2{=}CHCO_2CH_2CH_3$	ethyl acrylate	poly(ethyl acrylate); latex paints
$CH_2{=}C(CH_3)CO_2CH_3$	methyl methacrylate	poly(methyl methacrylate), Lucite, Plexiglas; glass substitutes

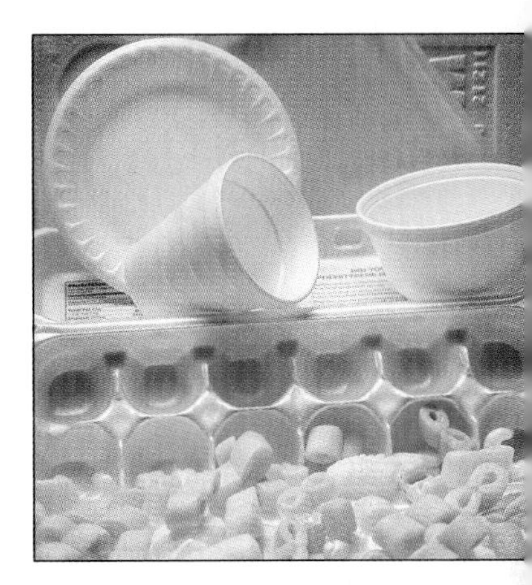

The low thermal conductivity of polystyrene makes it a good insulating material. *(Charles D. Winters)*

Petri dishes made of Plexiglas. *(© Dan McCoy, Rainbow)*

Some products made of polyvinyl chloride (PVC). *(Charles D. Winters)*

$$C_6H_5-C(=O)-O-O-C(=O)-C_6H_5 \xrightarrow{\text{heat}} C_6H_5-C(=O)-O\cdot + \cdot O-C(=O)-C_6H_5$$

Benzoyl peroxide — Benzoyloxy radicals

Radical polymerization of ethylene and substituted ethylenes involves three steps: (1) chain initiation, (2) chain propagation, and (3) chain termination. We show these three steps here and then discuss each separately in turn.

MECHANISM Radical Polymerization of Ethylene

Step 1: Chain initiation: formation of radicals from nonradical compounds.

$$In-In \longrightarrow In\cdot + \cdot In$$

$$In\cdot + CH_2{=}CH_2 \longrightarrow In-CH_2CH_2\cdot$$

An alkyl radical

Step 2: Chain propagation: reaction of a radical and a molecule to form a new radical.

The first chain-propagation step
$$In-CH_2CH_2\cdot + CH_2{=}CH_2 \longrightarrow In-CH_2CH_2CH_2CH_2\cdot$$

The product of $(n-1)$ chain-propagation steps
$$In-CH_2CH_2\cdot + (n-1)CH_2{=}CH_2 \longrightarrow In-(CH_2CH_2)_n\cdot$$

Step 3: Chain termination: destruction of radicals.

$$\sim CH_2CH_2\cdot + \cdot CH_2CH_2\sim \longrightarrow \sim CH_2CH_2-CH_2CH_2\sim$$

Chain-initiation step In radical polymerization, the formation of radicals from molecules containing only paired electrons.

The characteristic feature of a **chain-initiation step** is the formation of radicals from molecules with only paired electrons. In the case of peroxide-initiated polymerizations of alkenes, chain initiation is by thermally induced cleavage of the O—O bond of a peroxide to give two alkoxy radicals and reaction of an alkoxy radical with a molecule of alkene to give an alkyl radical. In the general mechanism shown in the mechanism box, the initiating catalyst is given the symbol In—In, and its radical is given the symbol In·.

The structure and geometry of carbon radicals are very similar to those of alkyl carbocations. They are planar or nearly so, with bond angles of approximately 120° about the carbon with the unpaired electron. The relative stabilities of alkyl radicals are similar to those of alkyl carbocations.

$$\text{methyl} < 1° < 2° < 3°$$

Increasing stability of alkyl radicals →

Chain-propagation step In radical polymerization, a reaction of a radical and a molecule to give a new radical.

The characteristic feature of a **chain-propagation step** is the reaction of a radical and a molecule to give a new radical. The product of each chain-propagation step in the radical polymerization of ethylene is a new alkyl radical increased in chain length by one CH_2CH_2 unit.

Propagation steps repeat over and over (propagate) with the radical formed in one step reacting with a monomer to produce a new radical, and so on. The number of times a cycle of chain propagation steps repeats is called **chain length** and is given the symbol *n*. In polymerization of ethylene, chain-lengthening reactions occur at a very high rate, often as fast as thousands of additions per second, depending on experimental conditions.

A low-density polyethylene (LDPE) bag. *(Charles D. Winters)*

In principle, chain-propagation steps can continue until all starting materials are consumed. In practice, they continue only until two radicals react with each other to terminate the process. The characteristic feature of a **chain-termination step** is the destruction of radicals. In the mechanism for radical polymerization of ethylene shown in the box, chain termination occurs by the coupling to two radicals to form a new carbon-carbon single bond.

Chain-termination step In radical polymerization, a reaction in which two radicals combine to form a covalent bond.

Propene and other substituted ethylene monomers can also be polymerized under a variety of experimental conditions, including radical chain polymerization. Radical polymerization of vinyl chloride gives poly(vinyl chloride), with chlorine atoms repeating regularly on every other carbon atom of the polymer chain.

$$n\text{ClCH}{=}\text{CH}_2 \xrightarrow{\text{initiator}} \left(\begin{matrix}\text{Cl}\\ |\\ \text{CHCH}_2\end{matrix}\right)_n$$

Vinyl chloride → Poly(vinyl chloride)

The first commercial process for ethylene polymerization used peroxide catalysts at temperatures of 500°C and pressures of 1000 atm and produced a soft, tough polymer known as low-density polyethylene (LDPE). LDPE has a density of between 0.91 and 0.94 g/cm^3 and a melt transition temperature (T_m) of about 115°C. Because its melting point is only slightly above 100°C, it cannot be used for products that will be exposed to boiling water. At the molecular level, chains of LDPE are highly branched.

Chain-transfer reaction In radical polymerization, the transfer of reactivity of an endgroup from one chain to another during a polymerization.

The branching on chains of low-density polyethylene results from a "back-biting" reaction in which the radical endgroup abstracts a hydrogen from the fourth carbon back (the fifth carbon in the chain). Abstraction of this hydrogen is particularly facile because the transition state associated with the process can adopt a conformation like that of a chair cyclohexane. In addition, the less stable 1° radical is converted to a more stable 2° radical. This side reaction is called a **chain-transfer reaction** because the activity of the endgroup is "transferred" from one chain to another. Continued polymerization of monomer from this new radical center leads to a branch four carbons long.

$\xrightarrow{n\text{CH}_2{=}\text{CH}_2}$

A six-membered transition state leading to 1,5-hydrogen abstraction

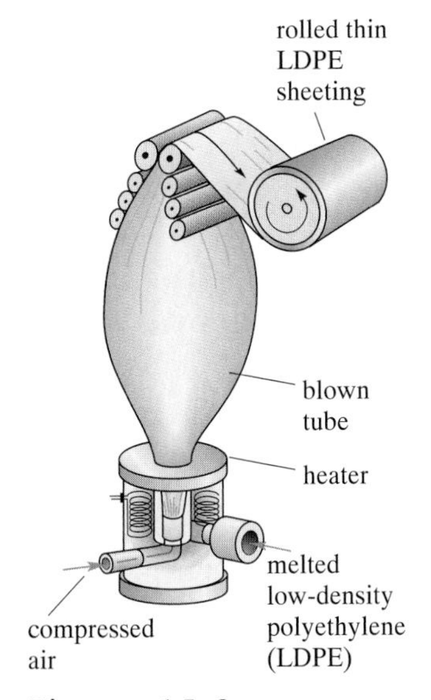

Figure 15.3 Fabrication of an LDPE film.

Approximately 65% of all low-density polyethylene is used for the manufacture of films by a blow-molding technique illustrated in Figure 15.3. A tube of melted LDPE along with a jet of compressed air is forced through an opening and blown into a giant, thin-walled bubble. The film is then cooled and taken up onto a roller. This double-walled film can be slit down the side to give LDPE film, or it can be sealed at points along its length to make LDPE bags. LDPE film is inexpen-

CHEMICAL CONNECTIONS

Recycling of Plastics

Polymers, in the form of plastics, are materials upon which our society is incredibly dependent. Durable and lightweight, plastics are probably the most versatile synthetic materials in existence; in fact, their current production in the United States exceeds that of steel. Plastics have come under criticism, however, for their role in the trash crisis. They comprise 21% of the volume and 8% of the weight of solid wastes, most of which is derived from disposable packaging and wrapping. Of the 1.1×10^8 kg of thermoplastic materials produced in 1993 in America, less than 2% was recycled.

Why aren't more plastics being recycled? The durability and chemical inertness of most plastics make them ideally suited for reuse. The answer to this question has more to do with economics and consumer habits than with technological obstacles. Because curbside pickup and centralized drop-off stations for recyclables are just now becoming common, the amount of used material available for reprocessing has traditionally been small. This limitation, combined with the need for an additional sorting and separation step, rendered the use of recycled plastics in manufacturing expensive compared with virgin materials. The increase in environmental awareness over the last decade, however, has resulted in a greater demand for recycled products. As manufacturers adapt to satisfy this new market, plastic recycling will eventually catch up with the recycling of other materials, such as glass and aluminum.

Six types of plastics are commonly used for packaging applications. In 1988, manufacturers adopted recycling code numbers developed by the Society of the Plastics Industry. Because the plastics recycling industry still is not fully developed, only polyethylene terephthalate (PET) and high-density polyethylene (HDPE) are currently being recycled in large quantities. Low-density polyethylene, which accounts for about 40% of plastic trash, has been slow in finding acceptance with recyclers. Facilities for the reprocessing of polyvinyl chloride (PVC), polypropylene, and polystyrene exist but are still rare.

The process for the recycling of most plastics is simple, with separation of the desired plastics from other contaminants the most labor-intensive step. For example, PET soft drink bottles usually have a paper label and adhesive that must be removed before the PET can be reused. The recycling process begins with hand or machine sorting, after which the bottles are shredded into small chips. An air cyclone then removes paper and other lightweight materials. Any remaining labels and adhesives are eliminated with a detergent wash, and the PET

These students are wearing jackets made from recycled PET soda bottles. *(Charles D. Winters)*

Recycling Code	Polymer	Common Uses	Uses of Recycled Polymer
1 PET	poly(ethylene terephthalate)	soft drink bottles, household chemical bottles, films, textile fibers	soft drink bottles, household chemical bottles, films, textile fibers
2 HDPE	high-density polyethylene	milk and water jugs, grocery bags, bottles	bottles, molded containers
3 V	poly(vinyl chloride), PVC	shampoo bottles, pipes, shower curtains, vinyl siding, wire insulation, floor tiles, credit cards	plastic floor mats
4 LDPE	low-density polyethylene	shrink wrap, trash and grocery bags, sandwich bags, squeeze bottles	trash bags and grocery bags
5 PP	polypropylene	plastic lids, clothing fibers, bottle caps, toys, diaper linings	mixed plastic components
6 PS	polystyrene	styrofoam cups, egg cartons, disposable utensils, packaging materials, appliances	molded items such as cafeteria trays, rulers, frisbees, trash cans, videocasettes
7	all other plastics and mixed plastics	various	plastic lumber, playground equipment, road reflectors

chips are then dried. The PET produced by this method is 99.9% free of contaminants and sells for about half the price of the virgin material. Unfortunately, plastics with similar densities cannot be separated with this technology, nor can plastics composed of several polymers be broken down into pure components. However, recycled mixed plastics can be molded into plastic lumber that is strong, durable, and graffiti-resistant.

An alternative to this process, which uses only physical methods of purification, is chemical recycling. Eastman Kodak salvages large amounts of its PET film scrap by a transesterification reaction. The scrap is treated with methanol in the presence of an acid catalyst to give ethylene glycol and dimethyl terephthalate. These monomers are purified by distillation or recrystallization and used as feedstocks for the production of more PET film.

$$\text{Poly(ethylene terephthalate) (PET)} \xrightarrow[H^+]{CH_3OH} HOCH_2CH_2OH + H_3CO_2C{-}C_6H_4{-}CO_2CH_3$$

Poly(ethylene terephthalate) (PET) — Ethylene glycol — Dimethyl terephthalate

Polyethylene films are produced by extruding the molten plastic through a ring-like gap and inflating the film into a balloon. *(The Stock Market)*

sive, which makes it ideal for packaging such consumer items as baked goods, vegetables, and other produce and for trash bags.

B. Ziegler-Natta Chain-Growth Polymerizations

An alternative method for polymerization of alkenes, which does not involve radicals, was developed by Karl Ziegler of Germany and Giulio Natta of Italy in the 1950s. The early Ziegler-Natta catalysts were highly active, heterogeneous materials composed of a $MgCl_2$ support, a Group IVB transition metal halide such as $TiCl_4$, and an alkylaluminum compound such as $Al(CH_2CH_3)_2Cl$. These catalysts bring about polymerization of ethylene and propylene at 1–4 atm and at temperatures as low as 60°C.

The active catalyst in a Ziegler-Natta polymerization is thought to be an alkyltitanium compound formed by alkylation of the titanium halide by $Al(CH_2CH_3)_2Cl$ on the surface of a $MgCl_2/TiCl_4$ particle. Once formed, this species repeatedly inserts ethylene units into the titanium-carbon bond to yield polyethylene.

MECHANISM Ziegler-Natta Catalysis of Ethylene Polymerization

Step 1: Formation of a titanium-ethyl bond.

$$\text{Ti—Cl} + Al(CH_2CH_3)_2Cl \longrightarrow \text{Ti—}CH_2CH_3 + Al(CH_2CH_3)Cl_2$$

Step 2: Insertion of ethylene into the titanium-carbon bond.

$$\text{Ti—}CH_2CH_3 + CH_2{=}CH_2 \longrightarrow \text{Ti—}CH_2CH_2CH_2CH_3$$

Over 60 billion pounds of polyethylene are produced worldwide every year using optimized Ziegler-Natta catalysts, and large-scale reactors can yield up to 1.3×10^5 kg of polyethylene per hour. Production of polymer at this scale is partly due to the mild conditions required for a Ziegler-Natta polymerization and the fact that the polymer obtained has substantially different physical and mechanical properties from that obtained by radical polymerization. Polyethylene from Ziegler-Natta systems, termed **high-density polyethylene (HDPE),** has a higher density (0.96 g/cm^3) and melt transition temperature (133°C) than low-density polyethylene, is three to ten times stronger, and is opaque rather than transparent. This added strength and opacity is due to a much lower degree of chain branching and the resulting higher degree of crystallinity of HDPE compared with LDPE.

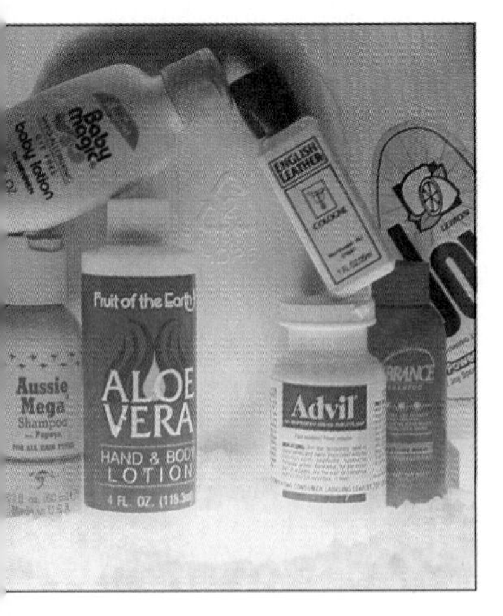

Some common products packed in high-density polyethylene (HDPE). *(Charles D. Winters)*

Approximately 45% of all HDPE used in the United States is blow molded. In blow molding, a short length of HDPE tubing is placed in an open die [Figure 15.4(a)], and the die is closed, sealing the bottom of the tube. Compressed air is then forced into the hot polyethylene/die assembly, and the tubing is literally blown up to take the shape of the mold [Figure 15.4(b)]. After cooling, the die is opened [Figure 15.4(c)], and there is the container!

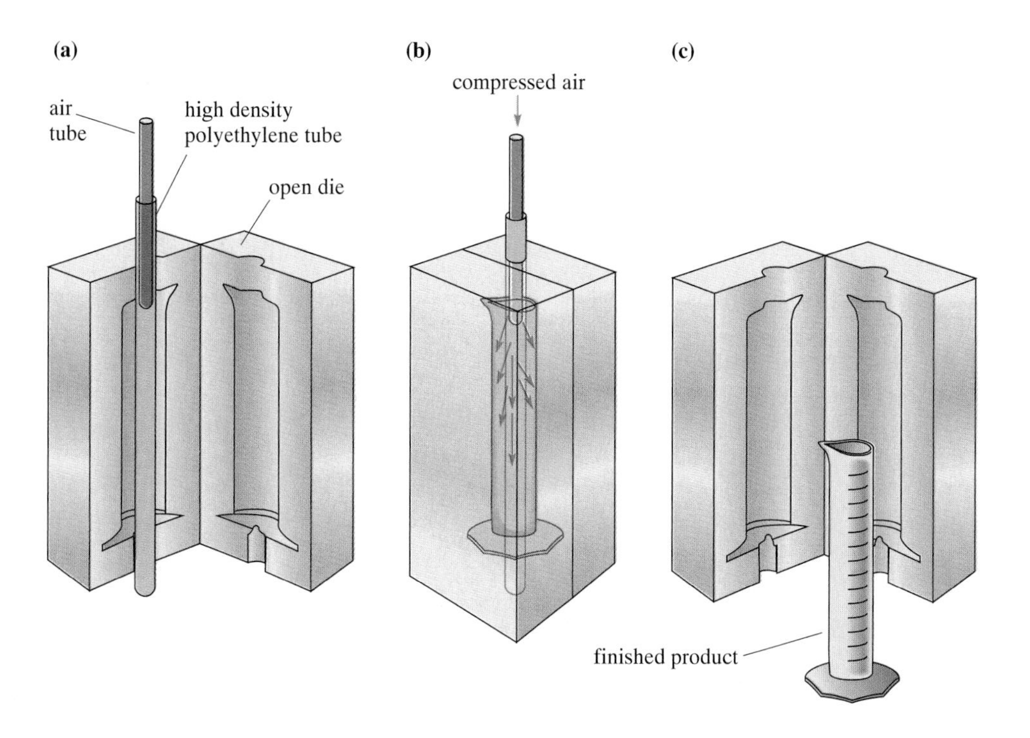

Figure 15.4 Blow molding of an HDPE container.

Even greater improvements in properties of HDPE can be realized through special processing techniques. In the melt state, HDPE chains adopt random coiled conformations similar to those of cooked spaghetti. Engineers have developed special extrusion techniques that force the individual polymer chains of HDPE to uncoil and adopt an extended linear conformation. These extended chains then align with one another to form highly crystalline materials. HDPE processed in this fashion is stiffer than steel and has approximately four times the tensile strength of steel! Because the density of polyethylene ($\approx 1.0\ g/cm^3$) is considerably less than that of steel ($8.0\ g/cm^3$), these comparisons of strength and stiffness are even more favorable if they are made on a weight basis.

SUMMARY

Polymerization is the process of joining together many small **monomers** into large, high-molecular-weight **polymers** (Section 15.1). The properties of polymeric materials depend on the structure of the repeat unit, chain architecture, and morphology (Section 15.2).

Step-growth polymerizations involve the stepwise reaction of difunctional monomers (Section 15.4). Important commercial polymers synthesized through step-growth processes include polyamides, polyesters, polycarbonates, polyurethanes, and epoxy resins.

Chain-growth polymerization proceeds by the sequential addition of monomer units to an active chain endgroup (Section 15.5). **Radical chain-growth polymerization** (Section 15.5A) consists of three stages: chain initiation, chain propagation, and chain termination. In **chain initiation,** radicals are formed from nonradical molecules. In **chain propagation,** a radical and a monomer react to give a new radical. **Chain length** is the number of times a cycle of chain propagation steps repeats. In **chain termination,** radicals are destroyed. Alkyl radicals are planar or almost so with bond angles of 120° about the carbon with the unpaired electron. **Ziegler-Natta chain-growth polymerizations** involve formation of an alkyl-transition metal compound and then repeated insertion of alkene monomers into the transition metal-to-carbon bond to yield a saturated polymer chain (Section 15.5B).

KEY REACTIONS

1. Step-growth polymerization of a dicarboxylic acid and a diamine gives a polyamide (Section 15.4A)

In this equation, M and M′ indicate the remainder of each monomer unit.

$$\mathrm{HOC(=O)-M-C(=O)OH + H_2N-M'-NH_2 \xrightarrow{heat} \left(C(=O)-M-C(=O)-NH-M'-NH \right)_n + 2nH_2O}$$

2. Step-growth polymerization of a dicarboxylic acid and a diol gives a polyester (Section 15.4B)

$$\mathrm{HOC(=O)-M-C(=O)OH + HO-M'-OH \xrightarrow{acid\ catalyst} \left(C(=O)-M-C(=O)-O-M'-O \right)_n + 2nH_2O}$$

3. Step-growth polymerization of phosgene and a diol gives a polycarbonate (Section 15.4C)

$$\mathrm{Cl-C(=O)-Cl + HO-M-OH \longrightarrow \left(O-C(=O)-O-M \right)_n + 2nHCl}$$

4. Step-growth polymerization of a diisocyanate and a diol gives a polyurethane (Section 15.4D)

$$\mathrm{OCN-M-NCO + HO-M'-OH \longrightarrow \left(C(=O)-NH-M-NH-C(=O)-O-M'-O \right)_n}$$

5. Step-growth polymerization of a diepoxide and a diamine gives an epoxy resin (Section 15.4E)

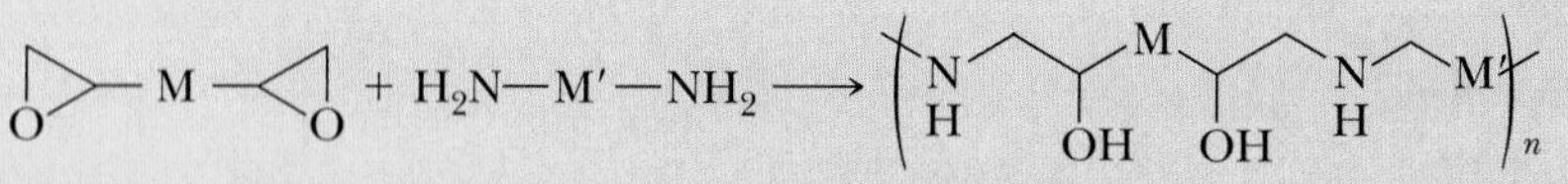

$$\mathrm{(epoxide)-M-(epoxide) + H_2N-M'-NH_2 \longrightarrow \left(NH-CH_2-CH(OH)-M-CH(OH)-CH_2-NH-M' \right)_n}$$

6. Radical chain-growth polymerization of ethylene and substituted ethylenes (Section 15.5A)

$$\mathrm{nCH_2{=}CHCO_2CH_3 \xrightarrow[heat]{peroxide} \left(CH_2CH(CO_2CH_3) \right)_n}$$

7. Ziegler-Natta chain-growth polymerization of ethylene and substituted ethylenes (Section 15.5B)

$$\mathrm{nCH_2{=}CHCH_3 \xrightarrow[MgCl_2]{TiCl_4/Al(C_2H_5)_2Cl} \left(CH_2CH(CH_3) \right)_n}$$

ADDITIONAL PROBLEMS

Step-Growth Polymers

15.3 Identify the monomers required for the synthesis of each step-growth polymer.

(a)

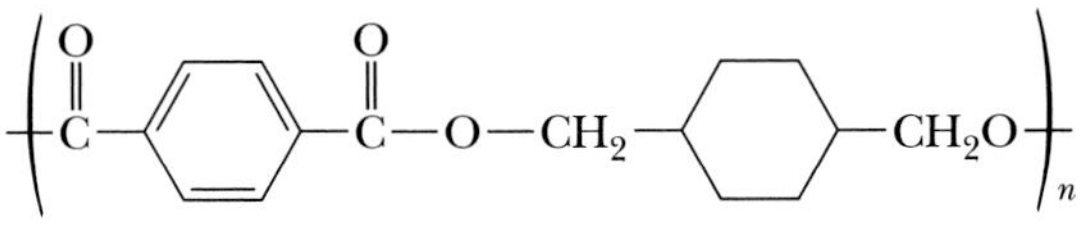

$$\left[\overset{O}{\overset{\|}{C}} - C_6H_4 - \overset{O}{\overset{\|}{C}} - O - CH_2 - C_6H_{10} - CH_2O \right]_n$$

Kodel
(a polyester)

(b)

$$\left[\overset{O}{\overset{\|}{C}}(CH_2)_6\overset{O}{\overset{\|}{C}} - NH - C_6H_{10} - CH_2 - C_6H_{10} - NH \right]_n$$

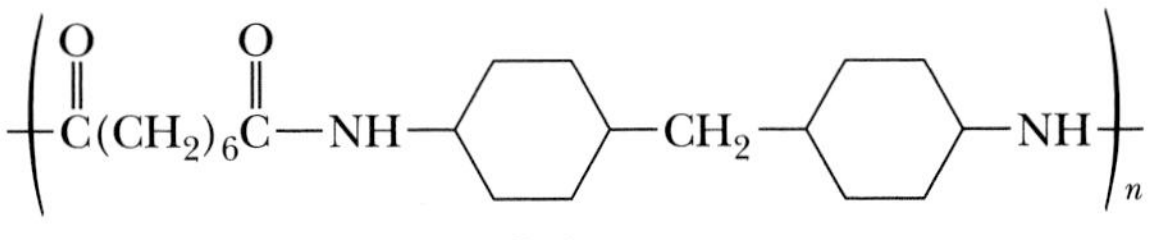

Quiana
(a polyamide)

(c)

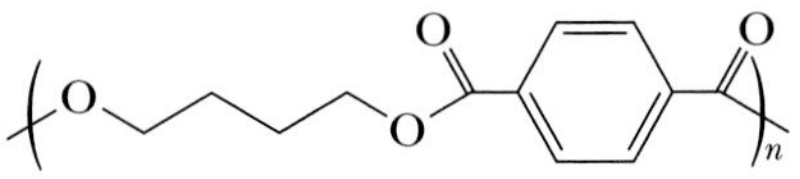

(a polyester)

(d)

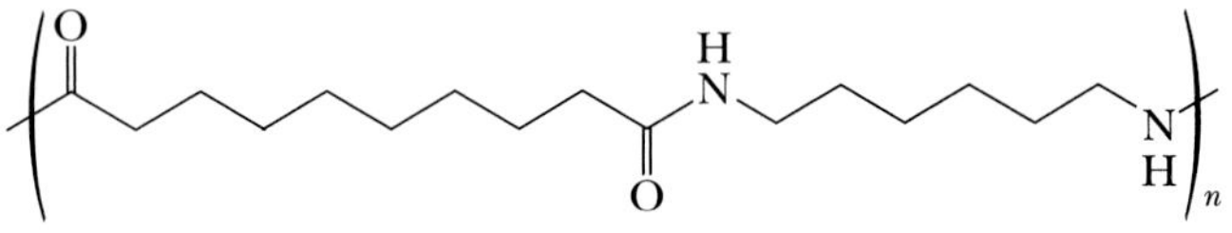

Nylon 6, 10
(a polymide)

15.4 Poly(ethylene terephthalate) (PET) can be prepared by this reaction. Propose a mechanism for the step-growth reaction in this polymerization.

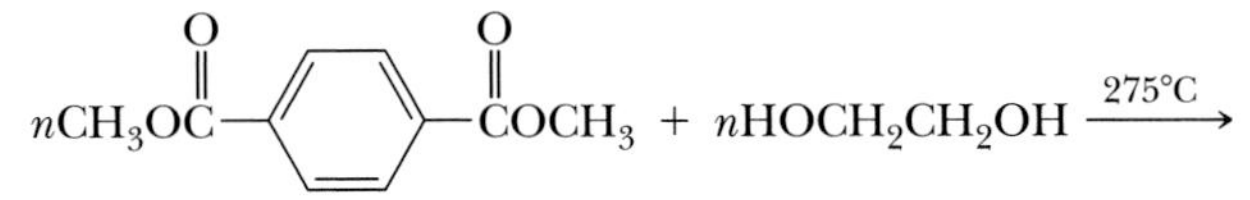

$$nCH_3O\overset{O}{\overset{\|}{C}} - C_6H_4 - \overset{O}{\overset{\|}{C}}OCH_3 + nHOCH_2CH_2OH \xrightarrow{275°C}$$

Dimethyl terephthalate   Ethylene glycol

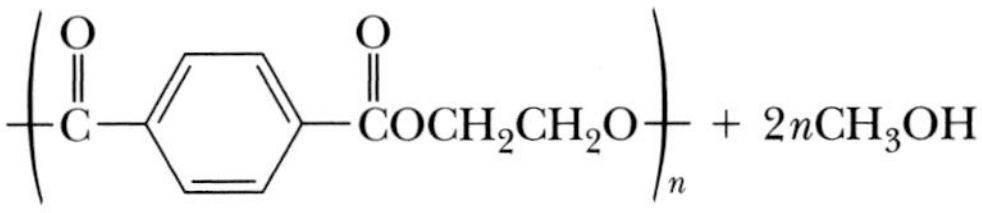

$$\left[\overset{O}{\overset{\|}{C}} - C_6H_4 - \overset{O}{\overset{\|}{C}}OCH_2CH_2O \right]_n + 2nCH_3OH$$

Poly(ethylene terephthalate)   Methanol

15.5 Currently about 30% of PET soft drink bottles are being recycled. In one recycling process, scrap PET is heated with methanol in the presence of an acid catalyst. The methanol reacts with the polymer, liberating ethylene glycol and dimethyl terephthalate. These monomers are then used as feedstock for the production of new PET products. Write an equation for the reaction of PET with methanol to give ethylene glycol and dimethyl terephthalate.

15.6 Nomex is an aromatic polyamide (aramid) prepared from polymerization of 1,3-benzenediamine and the acid chloride of 1,3-benzenedicarboxylic acid. The physical properties of the polymer make it suitable for high-strength, high-temperature applications such as parachute cords and jet aircraft tires. Draw a structural formula for the repeating unit of Nomex.

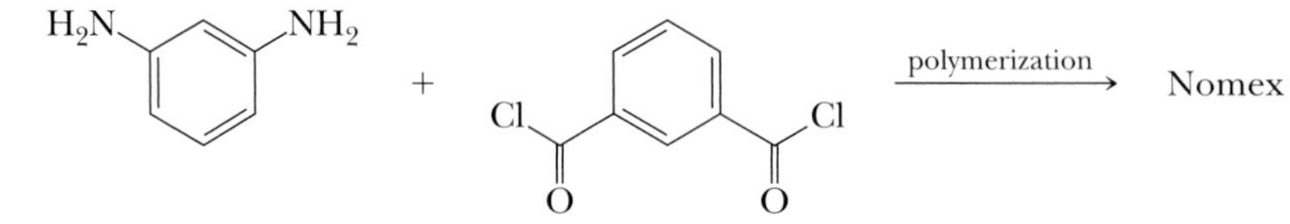

H_2N–(benzene ring)–NH_2 + Cl–C(=O)–(benzene ring)–C(=O)–Cl $\xrightarrow{\text{polymerization}}$ Nomex

1,3-Benzenediamine 1,3-Benzenedicarbonyl chloride

15.7 Nylon 6, 10 [Additional Problem 15.3(d)] can be prepared by reaction of a diamine and a diacid chloride. Draw the structural formula of each reactant.

Chain-Growth Polymerization

15.8 Following is the structural formula of a section of polypropylene derived from three units of propylene monomer.

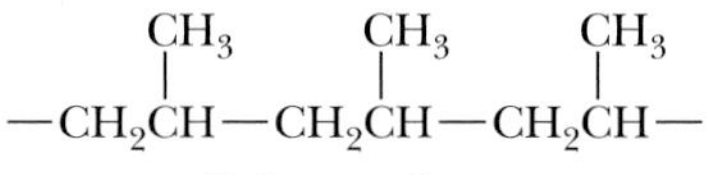

$$-CH_2\overset{\overset{\displaystyle CH_3}{|}}{C}H-CH_2\overset{\overset{\displaystyle CH_3}{|}}{C}H-CH_2\overset{\overset{\displaystyle CH_3}{|}}{C}H-$$

Polypropylene

Draw a structural formula for a comparable section of

(a) Poly(vinyl chloride) **(b)** Polytetrafluoroethylene **(c)** Poly(methyl methacrylate)

15.9 Following are structural formulas for sections of three polymers. From what alkene monomer is each derived?

(a) 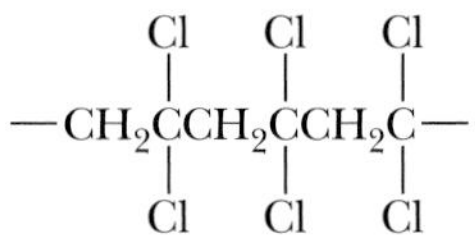 $-CH_2\underset{\underset{\displaystyle Cl}{|}}{\overset{\overset{\displaystyle Cl}{|}}{C}}CH_2\underset{\underset{\displaystyle Cl}{|}}{\overset{\overset{\displaystyle Cl}{|}}{C}}CH_2\underset{\underset{\displaystyle Cl}{|}}{\overset{\overset{\displaystyle Cl}{|}}{C}}-$ **(b)** 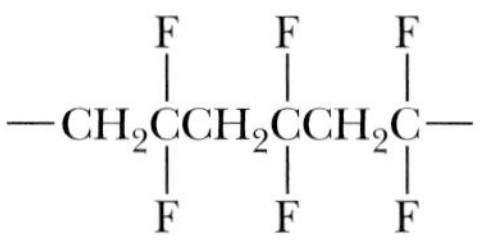 $-CH_2\underset{\underset{\displaystyle F}{|}}{\overset{\overset{\displaystyle F}{|}}{C}}CH_2\underset{\underset{\displaystyle F}{|}}{\overset{\overset{\displaystyle F}{|}}{C}}CH_2\underset{\underset{\displaystyle F}{|}}{\overset{\overset{\displaystyle F}{|}}{C}}-$

15.10 Draw the structure of the alkene monomer used to make each chain-growth polymer.

(a) 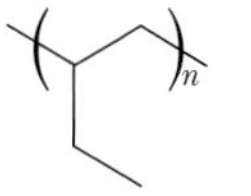 **(b)** 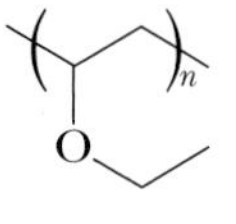 O **(c)** 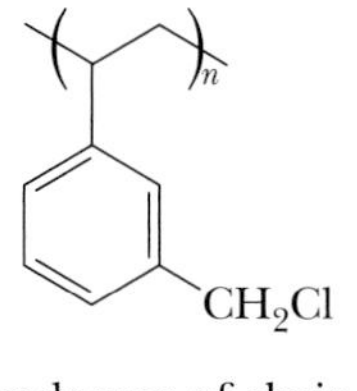 CH_2Cl **(d)** 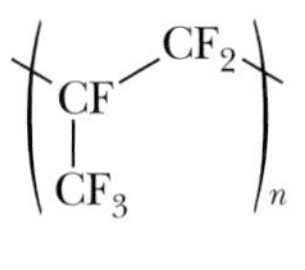 $\left(CF(CF_3)-CF_2\right)_n$

n

15.11 Low-density polyethylene (LDPE) has a higher degree of chain branching than high-density polyethylene (HDPE). Explain the relationship between chain branching and density.

15.12 Compare the densities of low-density polyethylene (LDPE) and high-density polyethylene (HDPE) with the densities of the liquid alkanes listed in Table 2.4. How might you account for the differences between them?

15.13 Polymerization of vinyl acetate gives poly(vinyl acetate). Hydrolysis of this polymer in aqueous sodium hydroxide gives poly(vinyl alcohol). Draw the repeat units of both poly(vinyl acetate) and poly(vinyl alcohol).

$$\text{Vinyl acetate}\quad CH_3-\overset{\overset{\displaystyle O}{\|}}{C}-O-CH{=}CH_2$$

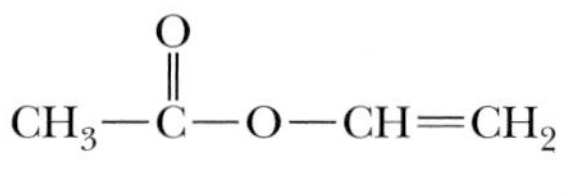

15.14 As seen in the previous problem, poly(vinyl alcohol) is made by polymerization of vinyl acetate followed by hydrolysis in aqueous sodium hydroxide. Why is poly(vinyl alcohol) not made instead by polymerization of vinyl alcohol, $CH_2{=}CHOH$?

A Conversation with . . .

ROALD HOFFMANN

Roald Hoffmann is a remarkable individual. When he was only 44 years old, he shared the 1981 Nobel Prize in Chemistry with Kenichi Fukui of Japan for work in applied theoretical chemistry. In addition, he has received awards from the American Chemical Society in both organic chemistry and inorganic chemistry, the only person to have achieved this honor. And, in 1990, he was awarded the Priestley Medal, the highest award given by the American Chemical Society.

The numerous honors celebrating his achievements in chemistry tell only part of the story of his life. He was born to a Polish Jewish family in Zloczow, Poland, in 1937 and was named Roald after the famous Norwegian explorer Roald Amundsen. Shortly after World War II began in 1939, the Nazis first forced him and his parents into a ghetto and then into a labor camp. However, his father smuggled Hoffmann and his mother out of the camp, and they were hidden for more than a year in the attic of a schoolhouse in the Ukraine. His father was later killed by the Nazis after trying to organize an attempt to break out of the labor camp. After the war, Hoffmann, his mother, and his stepfather made their way west to Czechoslovakia, Austria, and then Germany. They finally emigrated to the United States, arriving in New York in 1949. That Hoffmann and his mother survived these years is our good fortune. Of the 12,000 Jews living in Zloczow in 1941 when the Nazis took over, only 80 people, three of them children, survived the Holocaust. One of those three children was Roald Hoffmann.

On arriving in New York, Hoffmann learned his sixth language, English. He went to public schools in New York City and then to Stuyvesant High School, one of the city's select science schools. From there he went to Columbia University and then on to Harvard University, where he earned his Ph.D. in 1962. Shortly thereafter, he began the work with Professor R. B. Woodward that eventually led to the Nobel Prize. Since 1965 he has been a professor at Cornell University.

In addition to his work in chemistry, Professor Hoffmann also writes popular articles on science for the *American Scientist* and other magazines, and he has published two volumes of his poetry. Finally, he appeared in a series of 26 half-hour television programs for a chemistry course called "The World of Chemistry," which airs on public television and cable channels.

From Medicine to Cement to Theoretical Chemistry

Professor Hoffmann's office at Cornell University is full of mineral samples, molecular models, and Japanese art. When asked what brought him into chemistry he said, "I came rather late to chemistry; I was not interested in it from childhood." However, he clearly feels that one can come late to chemistry, and that it can be a very positive thing. "I am always worried about fields in which people exhibit precocity, like music and mathematics. Precocity is some sort of evidence that you have to have talent. I don't like that. I like the idea that human beings can do anything they want to. They need to be trained sometimes. They need a teacher to awaken the intelligence within them. But to be a chemist requires no special talent, I'm glad to say. Anyone can do it, with hard work."

When he was only 44 years old, Roald Hoffmann shared the 1981 Nobel Prize in Chemistry with Kenichi Fukui of Japan for work in applied theoretical chemistry.

He took a standard chemistry course in high school. He recalls that it was a fine course, but apparently he found biology more enjoyable because, in his high school yearbook, "under the picture of me with a crew cut, it says 'medical research' under my name." Indeed, he says that "medical research was a compromise between my interest in science and the typical Jewish middle class family pressures to become a medical doctor. The same kind of pressures seem to apply to Asian-Americans today."

When he went to Columbia University, Hoffmann enrolled as a premed student, but says that there were several factors that shifted him away from a career in medicine. One of these was his work at the National Bureau of Standards in Washington, D.C., for two summers and then at Brookhaven National Laboratory for a third summer. He says that these experiences gave him a feeling for the excitement of chemical research. Nonetheless, during his first summer at the Bureau of Standards, he "did some not very exciting work on the thermochemistry of cement." During his second summer there, he went over to the National Institutes of Health to find out what medical research was about. "To my amazement," he says, "most of the people had Ph.D.s and not M.D.s. I just didn't know. Young people do not often know what is required for a given profession. Once I found that out, and found that I did well in chemistry, it made me feel that I didn't really have to do medicine, that I could do some research in chemistry or biology. Later, what influenced me to decide on theoretical chemistry was an excellent instructor.

"At the very same time I was being exposed to the humanities, in part because of Columbia's core curriculum—which I think is a great idea—that had so-called contemporary civilization and humanities courses. I took advantage of the liberal arts education to the hilt, and that has remained with me all my life. The humanities teachers have remained permanently fixed in my mind and have changed my ways of thinking. These were the people who really had the intellectual impact on me and helped to shape my life.

"To trace the path, I was a latecomer to chemistry and was inspired by research. I think *research* is the way in. It just gives you a different perception."

"What's beautiful when you make a molecule is that you can make derivatives in which you can vary substituents, the pieces of a molecule, and we know that those substituents give a molecule function, give it complexity and richness."

A Love for Complexity

Having discussed what brought him into chemistry, I was interested in his view of the qualities that a student should possess to pursue a career in the field. He said, "One thing one needs to be a chemist is a love for complexity and richness. To some extent that is true of biology and natural history, too. I think one of the things that is beautiful about chemistry is that there are 10,000,000 compounds, each with different properties. What's beautiful when you make a molecule is that you can make derivatives in which you can vary substituents, the pieces of a molecule, and we know that those substituents give a molecule function, give it complexity and richness. That's why a protein or nucleic acid with all its variety is essential for life. That's why to me, intellectually, isomerism and stereochemistry in organic chemistry are at the heart of chemistry. I think we should teach that much earlier. It requires no mathematics, only a little model building; you can do this without theory. I think it is no accident that organic chemistry drew to it the intellects of its time."

Experiment and Theory

Professor Hoffmann has spent his career immersed in the theories of chemistry. However, he believes that fundamentally "chemistry is an experimental science, in spite of some of my colleagues saying otherwise. However, the educational process certainly favors theory. It's in the nature of things for teacher and student both to want to understand and then give primacy to the soluble and the understood at the expense of other things. We also have this reductionist philosophy of science, the idea that the social sciences derive from biology, that biology follows from chemistry, chemistry from physics, and so on. This notion gives an inordinate amount of importance to theoretical thinking, the more mathematical the better. Of course this is not true in reality, but it's an ideology; it is a religion of science."

There is of course a role for theory. "You can't report just the facts and nothing but the facts; by themselves they are dull. They have to be woven into a framework so that there is understanding. That's accomplished usually by a theory. It may not be mathematical, but a qualitative network of relationships." Indeed, Hoffmann believes that the incorporation of theory into chemistry "is what made American science better than that in many other countries. The emphasis in chemistry on theory and theoretical understanding is very important, but not nearly as important as the syntheses and reactions of molecules.

"Although I think chemists need to like to do experiments, that doesn't mean there is no role for people like me. It turns out that I am really an experimental chemist hiding as a theoretician. I think that is the key to my success. That is, I think I can empathize with what bothers the experimentalists. In another day I could have become an experimentalist."

Major Issues in Chemistry and Science Today

Professor Hoffmann has worked, and is at present working, at the forefront of several major areas of chemistry. He is at present quite intrigued by surface science. "For instance, there is the Fischer-Tropsch process, a pretty incredible thing in detail. Carbon monoxide and hydrogen gas come onto a metal catalyst, a surface of some sort, and off come long chain hydrocarbons and alcohols. The richness of all these things happening is intriguing, and we are on the verge of understanding. We now have structural information on surfaces that's reliable, and we are just beginning to get kinetic information. Surface science is at a crossing of chemistry, physics, and engineering. The field is in some danger of being spun off on its own, but I would like to keep it in chemistry.

"Bioinorganic chemisty is another such field. In my research group we are doing some work trying to understand the mechanism of oxygen production in photosynthesis, the last steps. What is known is very little. There is an enzyme in photosystem II that involves 3 to 4 manganese atoms, and they are at oxidation state 3 to 4. And they somehow take oxide or hydroxide to peroxide and eventually to molecular oxygen. That's all we know. Experimentally, not theoretically, I think bioinorganic chemistry is a very interesting field."

Finally, Hoffmann remarked, "There are going to be finer and finer ways of controlling the synthesis of molecules, the most essential activity of chemists. If I were to point to a single thing that chemists do, it would be that they make molecules. Chemistry is the science of molecules and their transformations. The transformations are the essential part. I think there are exciting possibilities for chemical intervention into biological systems with an ever finer degree of control. We need not be afraid of nature. We can mimic it, and even surpass its synthetic capabilities. And find a way to cooperate with it."

Scientific Literacy and Democracy

Roald Hoffmann is very concerned not only about science in general and chemistry in particular, but also about our society. One of his concerns is scientific literacy because "some degree of scientific literacy is absolutely necessary today for the population at large as part of a democratic system of government. People have to make intelligent decisions about all kinds of technological issues." He recently offered his comments on this important issue in the *New York Times*. He wrote, "What concerns me about scientific, or humanistic, illiteracy is the barrier it poses to rational democratic governance. Democracy occasionally gives in to *technocracy*—a reliance on experts on matters such as genetic engineering, nuclear waste disposal, or the cost of medical care. That is fine, but the people must be able to vote intelligently on these issues. The less we know as a nation, the more we must rely on experts, and the more likely we are to be misled by demagogues. We must know more."

"That's why to me, intellectually, isomerism and stereochemistry in organic chemistry are at the heart of chemistry."

The Responsibility of Scientists

"Scientists have a great obligation to speak to the public," Hoffmann says. "We have an obligation as educators to train the next generation of people. We should pay as much attention to those people who are *not* going to be chemists, and sometimes need to make compromises about what is to be taught and what is the nature of our courses. I think scientists have an obligation to speak to the public broadly, and here I think they have been negligent. I think society is paying scientists money to do research, and can demand an accounting in plain language. That's why I put in a lot of time on that television show ['The World of Chemistry']."

A Teacher of Chemistry—and Proud of It

In the Nobel Yearbook, Professor Hoffmann wrote that the technical description of his work "does not communicate what I think is my major contribution. I am a teacher, and I am proud of it. At Cornell University I have taught primarily undergraduates. . . . I have also taught chemistry courses to nonscientists and graduate courses in bonding theory and quantum mechanics. To the chemistry community at large, and to my fellow scientists, I have tried to teach 'applied theoretical chemistry': a special blend of computations stimulated by experiment and coupled to the construction of general models—frameworks for understanding." His success in this area is unquestioned.

16

Carbohydrates

16.1 Monosaccharides

16.2 The Cyclic Structure of Monosaccharides

16.3 Physical Properties

16.4 Reactions of Monosaccharides

16.5 Glucose Assay—The Search for Specificity

16.6 L-Ascorbic Acid (Vitamin C)

16.7 Disaccharides and Oligosaccharides

16.8 Polysaccharides

Breads, grains, and pasta are sources of carbohydrates. Inset: A model of glucose. *(Charles D. Winters)*

Carbohydrates are the most abundant organic compounds in the plant world. Among the many vital functions of these compounds are storehouses of chemical energy (glucose, starch, glycogen); components of supportive structures in plants (cellulose) and bacterial cell walls (mucopolysaccharides); and essential components of nucleic acids (D-ribose and 2-deoxy-D-ribose). Furthermore, bound to plasma membranes of animal cells are large numbers of relatively small carbohydrates that mediate interactions between cells. For example, A, B, and O blood types are determined by specific membrane-bound carbohydrates.

Saccharide A simpler member of the carbohydrate family, such as glucose.

The simpler members of the carbohydrate family are often referred to as **saccharides** because of the sweet taste of sugars (Latin: *saccharum,* sugar). The name "carbohydrate," or hydrate of carbon, derives from the formula $C_n(H_2O)_m$ for many members of this class. Two examples of carbohydrates with molecular formulas that can be written alternatively as hydrates of carbon are shown in the table.

Carbohydrate	Molecular Formula	Molecular Formula as a Hydrate of Carbon
glucose (blood sugar)	$C_6H_{12}O_6$	$C_6(H_2O)_6$
sucrose (table sugar)	$C_{12}H_{22}O_{11}$	$C_{12}(H_2O)_{11}$

Not all carbohydrates, however, have this general formula. Some contain too few oxygen atoms to fit this formula, whereas some contain too many. Some also contain nitrogen. But the term "carbohydrate" has become firmly rooted in chemical nomenclature, and although not completely accurate, it persists as the name for this class of compounds.

Carbohydrate A polyhydroxyaldehyde or polyhydroxyketone or a substance that gives these compounds on hydrolysis.

At the molecular level, most **carbohydrates** are polyhydroxyaldehydes, polyhydroxyketones, or compounds that yield them after hydrolysis. Therefore, the chemistry of carbohydrates is essentially the chemistry of hydroxyl groups and carbonyl groups and of acetal bonds formed between these two functional groups.

Carbohydrates are classified as monosaccharides, oligosaccharides, or polysaccharides depending on their structure. Most of this chapter is devoted to the discussion of monosaccharides.

16.1 Monosaccharides

A. Structure

Monosaccharide A carbohydrate that cannot be hydrolyzed to a simpler compound.

Aldose A monosaccharide containing an aldehyde group.

Ketose A monosaccharide containing a ketone group.

Monosaccharides are the monomers from which more complex carbohydrates are constructed. They have the general formula $C_nH_{2n}O_n$; the most common monosaccharides have values of n in the range 3 to 8. The suffix -ose indicates that a molecule is a carbohydrate, and the prefixes tri-, tetr-, pent-, and so forth, indicate the number of carbon atoms in the chain. Monosaccharides containing an aldehyde group are classified as **aldoses;** those containing a ketone group are classified as **ketoses.**

There are only two trioses: glyceraldehyde, which is an aldotriose, and dihydroxyacetone, which is a ketotriose.

$$\begin{array}{cc} \mathrm{CHO} & \mathrm{CH_2OH} \\ | & | \\ \mathrm{CHOH} & \mathrm{C{=}O} \\ | & | \\ \mathrm{CH_2OH} & \mathrm{CH_2OH} \end{array}$$

Glyceraldehyde (an aldotriose) Dihydroxyacetone (a ketotriose)

Often the designations aldo- and keto- are omitted, and these molecules are referred to simply as trioses, tetroses, and the like. Although these designations do not tell the nature of the carbonyl group, at least they indicate that the monosaccharide contains three and four carbon atoms, respectively.

Monosaccharides Classified by Number of Carbon Atoms

Name	Formula
triose	$C_3H_6O_3$
tetrose	$C_4H_8O_4$
pentose	$C_5H_{10}O_5$
hexose	$C_6H_{12}O_6$
heptose	$C_7H_{14}O_7$
octose	$C_8H_{16}O_8$

B. Nomenclature

Glyceraldehyde is a common name; the IUPAC name for this monosaccharide is 2,3-dihydroxypropanal. Similarly, dihydroxyacetone is a common name; its IUPAC name is 1,3-dihydroxypropanone. The common names for these and other monosaccharides, however, are so firmly rooted in the literature of organic chemistry and biochemistry that they are used almost exclusively whenever these compounds are referred to. Therefore, throughout our discussions of the chemistry of carbohydrates, we use the names most common in the literature of chemistry and biochemistry.

C. Stereoisomerism

Glyceraldehyde contains a stereocenter and exists as a pair of enantiomers. The stereoisomer shown on the left has the R configuration and is named (R)-glyceraldehyde; its enantiomer, shown on the right, is named (S)-glyceraldehyde.

$$\begin{array}{cc} \mathrm{CHO} & \mathrm{CHO} \\ \mathrm{H{-}C{-}OH} & \mathrm{HO{-}C{-}H} \\ \mathrm{CH_2OH} & \mathrm{CH_2OH} \end{array}$$

(R)-Glyceraldehyde (S)-Glyceraldehyde

D. Fischer Projection Formulas

Chemists commonly use two-dimensional representations called **Fischer projections** to show the configuration of carbohydrates. To write a Fischer projection, orient the stereocenter of the carbohydrate so that the vertical bonds from the stereocenter are directed away from you and the horizontal bonds from it are directed toward you. Then write the molecule as a two-dimensional figure with the

Fischer projection A two-dimensional representation for showing the configuration of a stereocenter; horizontal lines represent bonds projecting forward from the stereocenter and vertical lines represent bonds projecting to the rear.

stereocenter indicated by the point at which the bonds cross. You now have a Fischer projection. The two horizontal segments of this Fischer projection represent bonds directed toward you, and the two vertical segments represent bonds directed away from you. The only atom in the plane of the paper is the stereocenter.

CHO
H►C◄OH
CH_2OH

convert to a Fischer projection →

CHO
H—+—OH
CH_2OH

(R)-Glyceraldehyde (three-dimensional representation)

(R)-Glyceraldehyde (Fischer projection)

E. D- and L-Monosaccharides

Even though the R,S system is widely accepted today as a standard for designating configuration, the system of carbohydrates is still commonly designated by the D,L system, proposed by Emil Fischer in 1891. At that time, it was known that one enantiomer of glyceraldehyde has a specific rotation of $+13.5°$; the other has a specific rotation of $-13.5°$. Fischer proposed that these enantiomers be designated D and L. The question, then, was which enantiomer has which specific rotation? Because there was no experimental way to answer the question at that time, Fischer did the only possible thing—he made an arbitrary assignment. He assigned the dextrorotatory enantiomer the following configuration and named it D-glyceraldehyde. He named its enantiomer L-glyceraldehyde. Fischer could have been wrong, but by a stroke of good fortune, he wasn't. In 1952, his assignment of configuration to the enantiomers of glyceraldehyde was proven correct by a special application of x-ray crystallography.

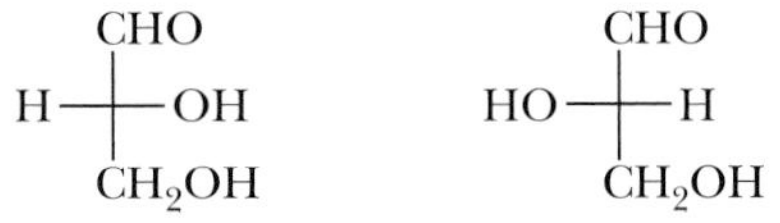

CHO
H—+—OH
CH_2OH
D-Glyceraldehyde
$[\alpha]^{25}_{D} = +13.5°$

CHO
HO—+—H
CH_2OH
L-Glyceraldehyde
$[\alpha]^{25}_{D} = -13.5°$

D-glyceraldehyde and L-glyceraldehyde serve as reference points for the assignment of relative configuration to all other aldoses and ketoses. The reference point is the stereocenter farthest from the carbonyl group. Because this stereocenter is the next to the last carbon on the chain, it is called the **penultimate carbon.** A **D-monosaccharide** is a monosaccharide that has the same configuration at the penultimate carbon as D-glyceraldehyde; an **L-monosaccharide** is a monosaccharide that has the same configuration at the penultimate carbon as L-glyceraldehyde.

Penultimate carbon The stereocenter of a monosaccharide farthest from the carbonyl group, as for example carbon-5 of glucose.

D-Monosaccharide A monosaccharide that, when written as a Fischer projection, has the —OH on its penultimate carbon to the right.

L-Monosaccharide A monosaccharide that, when written as a Fischer projection, has the —OH on its penultimate carbon to the left.

Shown in Tables 16.1 and 16.2 are names and Fischer projection formulas for all D-aldo- and D-2-ketotetroses, pentoses, and hexoses. Each name consists of three parts. The letter D specifies the configuration at the stereocenter farthest from the carbonyl group. Prefixes, such as rib-, arabin-, and gluc-, are given to these compounds without regard to R or S configuration at any stereocenter. These prefixes, however, do specify the relative configuration of one stereocenter to another in a monosaccharide. The suffix -ose shows that the compound is a carbohydrate.

TABLE 16.1 Configurational Relationships Among the Isomeric D-Aldotetroses, D-Aldopentoses, and D-Aldohexoses

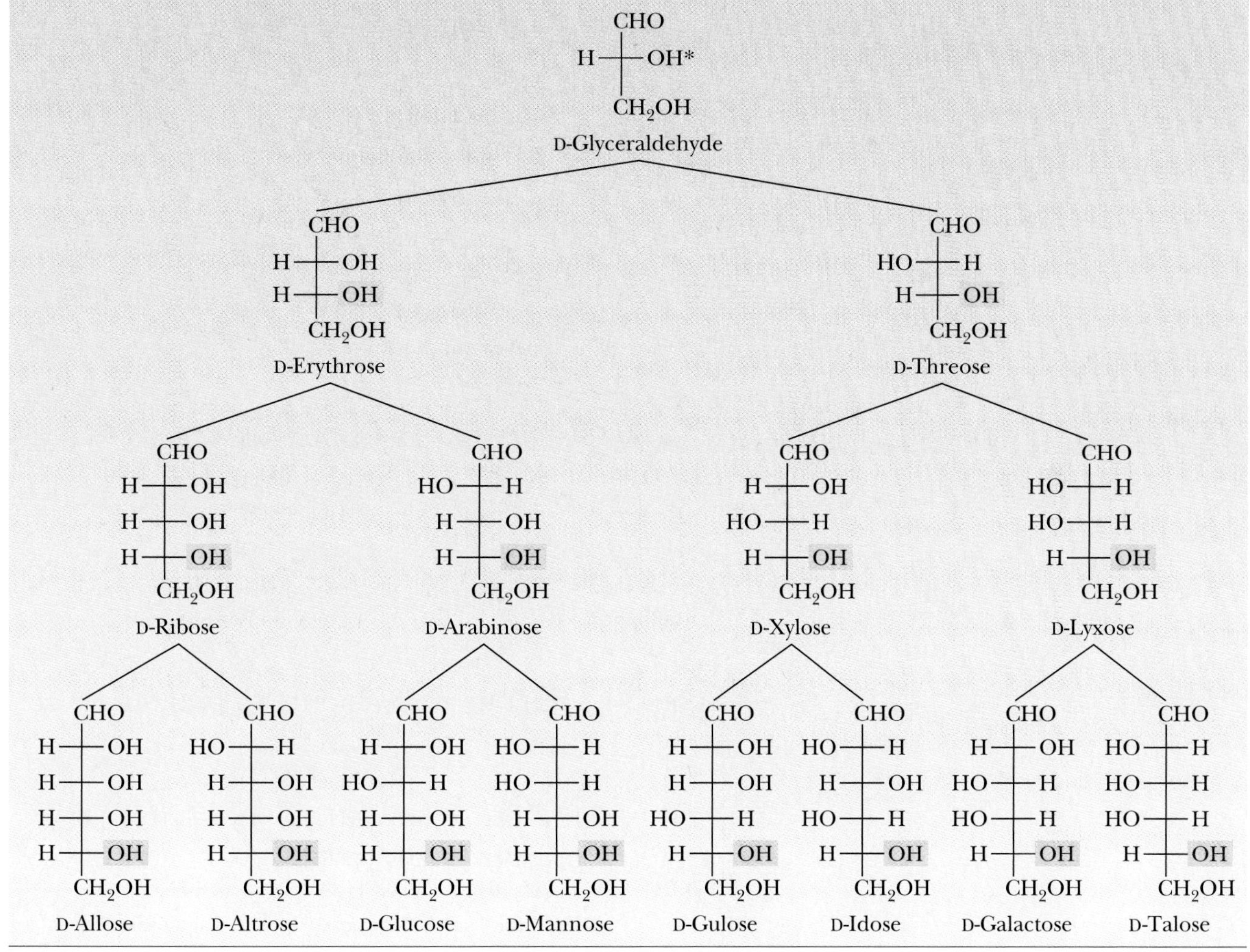

* The configuration of the reference —OH on the penultimate carbon is shown in color.

EXAMPLE 16.1

(a) Draw Fischer projections for the four aldotetroses.
(b) Show which are D-monosaccharides, which are L-monosaccharides, and which are enantiomers.
(c) Refer to Table 16.1 and name each aldotetrose you have drawn.

SOLUTION

Following are Fischer projections for the four aldotetroses. The D- and L- refer to the configuration of the penultimate carbon that, in the case of aldotetroses, is carbon 3. In the Fischer projection of a D-aldotetrose, the —OH on carbon 3 is on the right, and, in an L-aldotetrose, it is on the left.

TABLE 16.2 Configurational Relationships Among the D-2-Ketopentoses and D-2-Ketohexoses

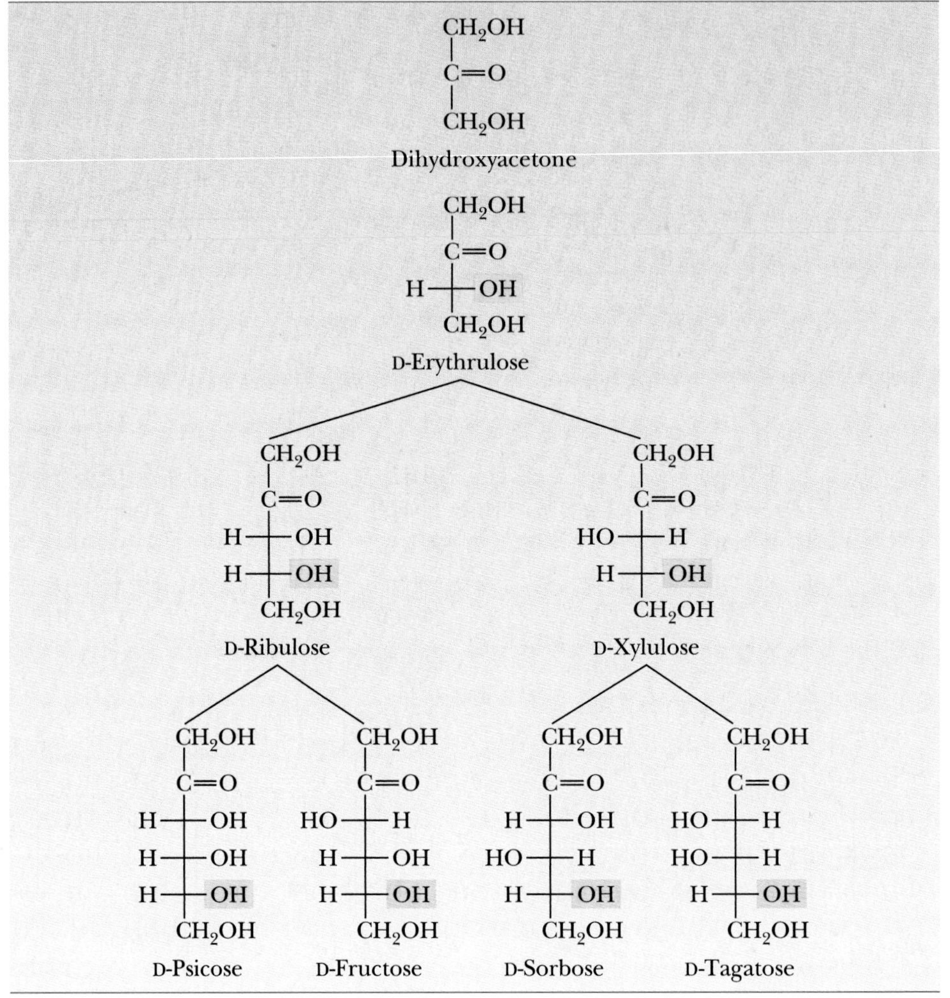

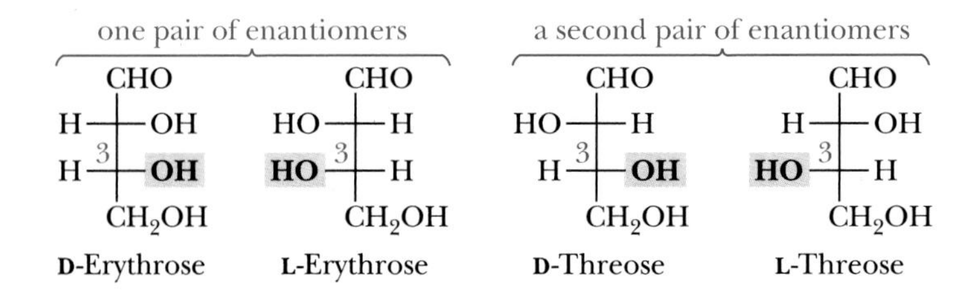

Practice Problem 16.1

(a) Draw Fischer projections for all 2-ketopentoses.
(b) Show which are D-ketopentoses, which are L-ketopentoses, and which are enantiomers.
(c) Refer to Table 16.2 and write names of the ketopentoses you have drawn.

CHEMICAL CONNECTIONS

The Genius of Emil Fischer

Emil Fischer was born in a German village near Cologne in 1852. He was educated in German public schools and continued on to the University of Berlin where, against the wishes of his businessman father, he entered a degree program in the sciences. He graduated with a doctorate in chemistry in 1874. Among his first accomplishments were the synthesis of phenylhydrazine, $C_6H_5NHNH_2$, and the discovery that, when treated with this compound, carbohydrates give crystalline derivatives. This discovery made carbohydrates easier to work with. Until that time, they had been very difficult to work with because of their tendency to form syrupy mixtures. (Think of honey, which is a concentrated mixture of glucose and fructose.) By using the phenylhydrazine reaction and reaction sequences by which monosaccharide chains can be extended by one carbon at a time, Fischer was able to assign relative configurations to each of the stereocenters in glucose. He published his results in 1891.

Underlying Fischer's work was the theory proposed in 1874 by the Dutch physical chemist Jacobus H. van't Hoff and the French organic chemist Joseph-Achille Le Bel. They suggested that carbon is tetrahedal and that a compound containing a carbon atom bonded to four different groups can exist in enantiomeric forms. This theory was by no means universally accepted. Fischer's establishing the structure of glucose, based as it was on the concept of a tetrahedral carbon atom, offered proof of the theory of van't Hoff and Le Bel and moved it quickly to general acceptance. Thus, in establishing the relative configurations of the stereocenters of glucose, Fischer helped unravel the complexities of carbohydrate structure and at the same time helped establish the validity of a tetrahedral carbon atom. Each accomplishment was as significant as the other. In 1902, Fischer was awarded the Nobel Prize for Chemistry for his work on the chemistry of carbohydrates and purines.

Fischer once said that his goal was to be the first to synthesize an "artificial ferment" (what we today call an enzyme) and that he would consider his mission in life complete when that was accomplished. He did not reach that goal, nor could he have realized the difficulties and the decades of research it would take before it was accomplished. Fischer died in 1919 of chronic poisoning from his first "chemical love," phenylhydrazine.

F. Amino Sugars

Amino sugars contain an $-NH_2$ group in place of an $-OH$ group. Only three amino sugars are common in nature: D-glucosamine, D-mannosamine, and D-galactosamine.

D-Glucosamine	D-Mannosamine (C-2 stereoisomer of D-glucosamine)	D-Galactosamine (C-4 stereoisomer of D-glucosamine)	*N*-Acetyl-D-glucosamine
CHO	CHO	CHO	CHO
H—NH_2	**H_2N**—H (2)	H—NH_2	H—$NHC(=O)CH_3$
HO—H	HO—H	HO—H	HO—H
H—OH	H—OH	**HO**—H (4)	H—OH
H—OH	H—OH	H—OH	H—OH
CH_2OH	CH_2OH	CH_2OH	CH_2OH

N-Acetyl-D-glucosamine, a derivative of D-glucosamine, is a component of many polysaccharides, including chitin, the hard shell-like exoskeleton of lobsters, crabs, shrimp, and other crustaceans. Several other amino sugars are components of naturally occurring antibiotics. It is the synthesis of these amino sugars that is often the difficult part of antibiotic synthesis in the laboratory.

16.2 The Cyclic Structure of Monosaccharides

Hemiacetal A molecule containing a carbon bonded to an —OH and an —OR group; the product of adding one molecule of alcohol to the carbonyl group of an aldehyde or ketone.

We saw in Section 11.6 that aldehydes and ketones react with alcohols to form **hemiacetals.** We also saw that cyclic hemiacetals form very readily when hydroxyl and carbonyl groups are parts of the same molecule and their interaction can form a five- or six-membered ring. For example, 4-hydroxypentanal forms a five-membered cyclic hemiacetal. Note that 4-hydroxypentanal contains one stereocenter and that a second stereocenter is generated at carbon 1 as a result of hemiacetal formation.

$$\underset{\text{OH}}{CH_3\overset{4}{C}HCH_2CH_2\overset{1}{C}H{=}O}$$

new stereocenter

4-Hydroxypentanal (redrawn to show how the cyclic hemiacetal forms) A cyclic hemiacetal

Monosaccharides have hydroxyl and carbonyl groups in the same molecule, and they exist almost exclusively as five- and six-membered cyclic hemiacetals.

A. Haworth Projections

Haworth projection A way to view furanose and pyranose forms of monosaccharides. The ring is drawn flat and viewed through its edge with the anomeric carbon on the right and the oxygen atom of the ring in the rear to the right.

Anomeric carbon The hemiacetal carbon of the cyclic form of a monosaccharide.

Anomers Monosaccharides that differ in configuration only at their anomeric carbons.

A common way for representing the cyclic structure of monosaccharides is the **Haworth projection,** named after the English chemist Sir Walter N. Haworth (Nobel laureate, 1937). In a Haworth projection, a five- or six-membered cyclic hemiacetal is represented as a planar pentagon or hexagon, as the case may be, lying perpendicular to the plane of the paper. Groups attached to the carbons of the ring then lie either above or below the plane of the ring. The new carbon stereocenter created in forming the cyclic structure is called the **anomeric carbon.** The **diastereomers** thus formed are given the special name **anomers.** Haworth projections are most commonly written with the anomeric carbon to the right and the hemiacetal oxygen to the back right (Figure 16.1). In Haworth projections, the —OH on the anomeric carbon of the cyclic hemiacetal is either cis or trans to the terminal —CH_2OH, depending on the orientation of the aldehyde group when it reacts with the —OH on carbon 5.

In the terminology of carbohydrate chemistry, the designation β means that the —OH on the anomeric carbon of the cyclic hemiacetal is up, that is, on the same side as the terminal —CH_2OH. Conversely, the designation α means that the —OH on the anomeric carbon of the cyclic hemiacetal is down, on the side opposite the terminal —CH_2OH.

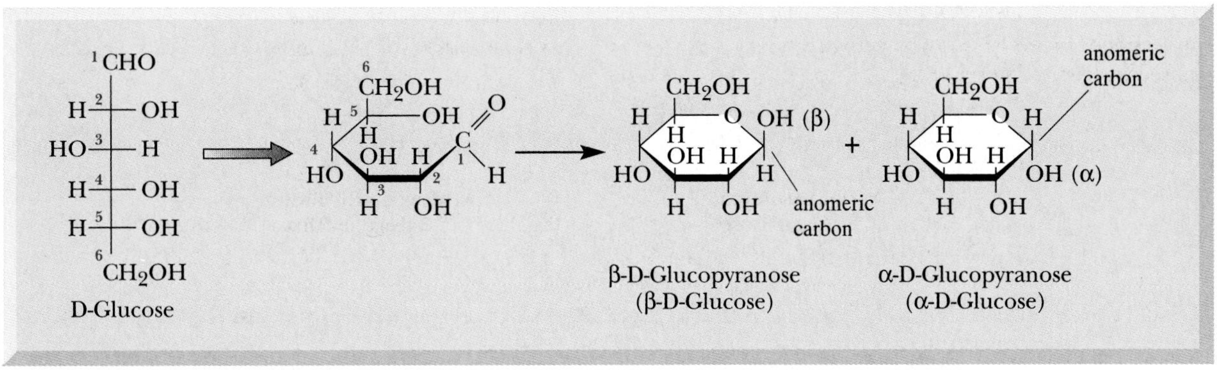

Figure 16.1 Haworth projections for α-D-glucopyranose and β-D-glucopyranose.

A six-membered hemiacetal ring is shown by the infix -pyran-, and a five-membered hemiacetal ring is shown by the infix -furan-. The terms **furanose** and **pyranose** are used because monosaccharide five- and six-membered rings correspond to the heterocyclic compounds furan and pyran.

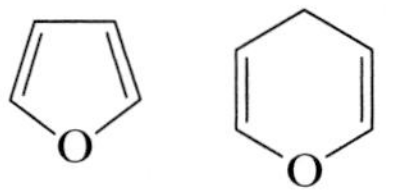

Furanose A five-membered cyclic hemiacetal form of a monosaccharide.

Pyranose A six-membered cyclic hemiacetal form of a monosaccharide.

Because the *α* and *β* forms of glucose are six-membered cyclic hemiacetals, they are named *α*-D-glucopyranose and *β*-D-glucopyranose. However, for convenience, they are often named simply *α*-D-glucose and *β*-D-glucose. You would do well to remember the configuration of groups on the Haworth projection of both *α*-D-glucopyranose and *β*-D-glucopyranose as reference structures. By knowing how the open-chain configuration of any other monosaccharide differs from that of D-glucose, you can then construct its Haworth projection by reference to the Haworth projection of D-glucose.

EXAMPLE 16.2

Draw Haworth projections for the *α* and *β* anomers of D-galactopyranose.

SOLUTION

One way to arrive at the structures for the *α* and *β* anomers of D-galactopyranose is to use the *α* and *β* forms of D-glucopyranose as reference and to remember, or discover by looking at Table 16.1, that D-galactose differs from D-glucose only in the configuration at carbon 4. Following are Haworth projections for *α*-D-galactopyranose and *β*-D-galactopyranose.

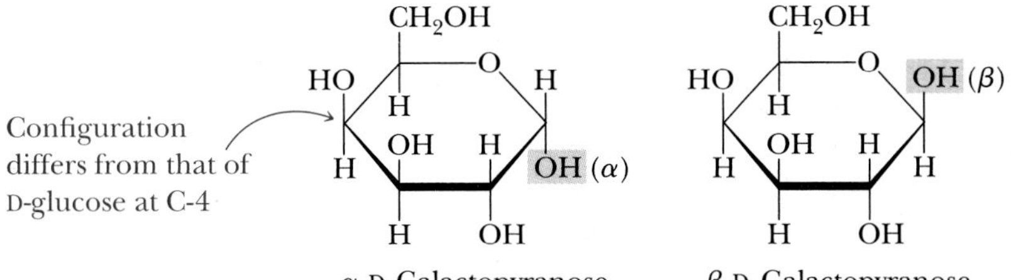

α-D-Galactopyranose
(*α*-D-Galactose)

β-D-Galactopyranose
(*β*-D-Galactose)

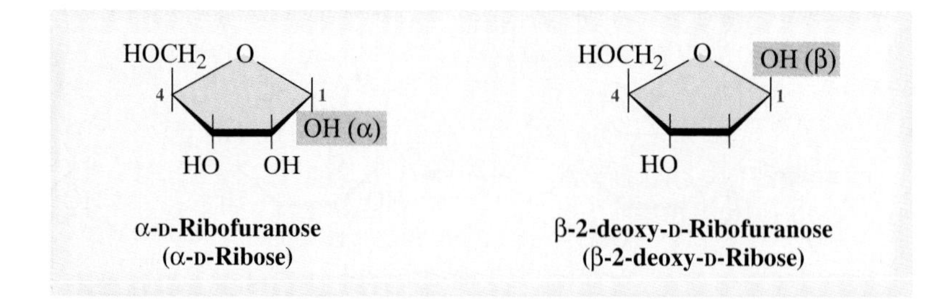

Figure 16.2 Haworth projections for two D-ribofuranoses.

Practice Problem 16.2

Mannose exists in aqueous solution as a mixture of α-D-mannopyranose and β-D-mannopyranose. Draw Haworth projections for these molecules.

The most prevalent forms of D-ribose and other pentoses in the biological world are furanoses. Shown in Figure 16.2 are Haworth projections for α-D-ribofuranose (α-D-ribose) and β-2-deoxy-D-ribofuranose (β-2-deoxy-D-ribose). Units of D-ribose and 2-deoxy-D-ribose in nucleic acids and most other biological molecules are found almost exclusively in the β-configuration. The prefix 2-deoxy indicates the absence of oxygen at carbon 2.

B. Conformation Representations

A five-membered ring is so close to being planar that Haworth projections are adequate to represent furanoses. For pyranoses, however, the six-membered ring is more accurately represented as a **chair conformation** in which strain is a minimum. Structural formulas for α-D-glucopyranose and β-D-glucopyranose are drawn in this manner in Figure 16.3. Also shown is the open-chain or free aldehyde form

Figure 16.3 Chair conformations of α-D-glucopyranose and β-D-glucopyranose.

CH$_2$OH anomeric carbon
HO
HO O OH (β)
OH
β-D-Glucopyranose
$[\alpha]^{25}_{D}$ +18.7°

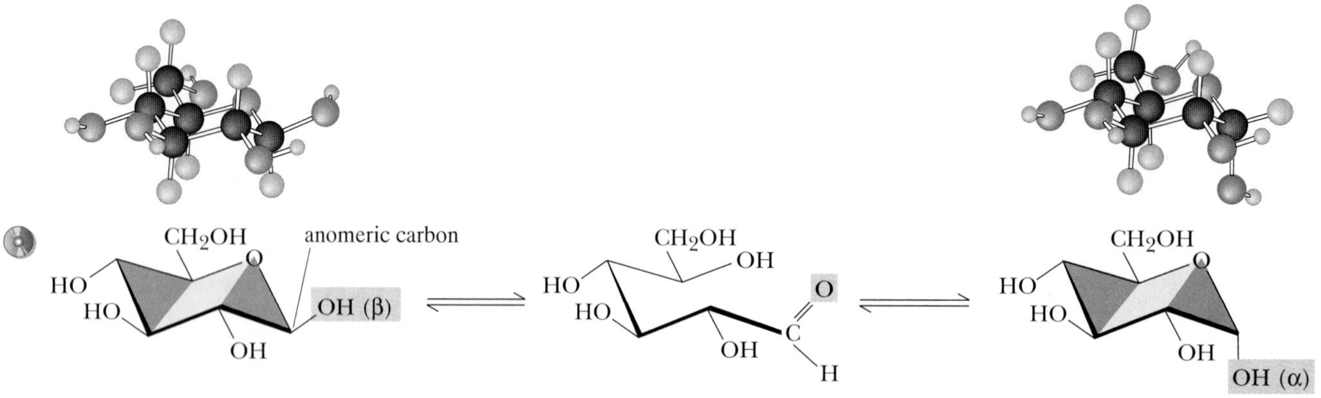

CH$_2$OH
OH
HO
HO O
C
OH H
D-Glucose

CH$_2$OH
HO O
HO
OH
OH (α)
α-D-Glucopyranose
$[\alpha]^{25}_{D}$ +112°

with which the cyclic hemiacetal forms are in equilibrium in aqueous solution. Notice that each group, including the anomeric —OH, on the chair conformation of β-D-glucopyranose is equatorial. Notice also that the —OH group on the anomeric carbon is axial in α-D-glucopyranose. Because of the equatorial orientation of the —OH on its anomeric carbon, β-D-glucopyranose is more stable and predominates in aqueous solution.

At this point, you should compare the relative orientations of groups on the D-glucopyranose ring in the Haworth projection and chair conformation. Notice that the orientations of groups on carbons 1 through 5 in the Haworth projection of β-D-glucopyranose are up, down, up, down, and up, respectively. The same is the case in the chair conformation.

CH_2OH O **OH** (β) H H OH H HO H H OH

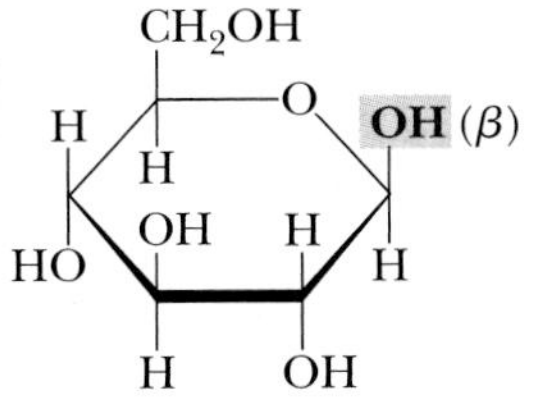

β-D-Glucopyranose
(Haworth projection)

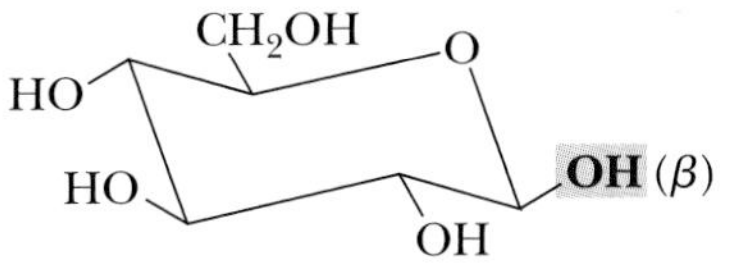

CH_2OH O HO HO **OH** (β) OH

β-D-Glucopyranose
(chair conformation)

Other monosaccharides also form cyclic hemiacetals. As with the examples we have studied, five- and six-membered cyclic hemiacetals are by far the most common. Shown in Figure 16.4 are the five-membered cyclic hemiacetals of fructose. The β-D-fructofuranose form is found in the disaccharide sucrose (Section 16.7C).

EXAMPLE 16.3

Draw chair conformations for α-D-galactopyranose and β-D-galactopyranose. Label the anomeric carbon in each cyclic hemiacetal.

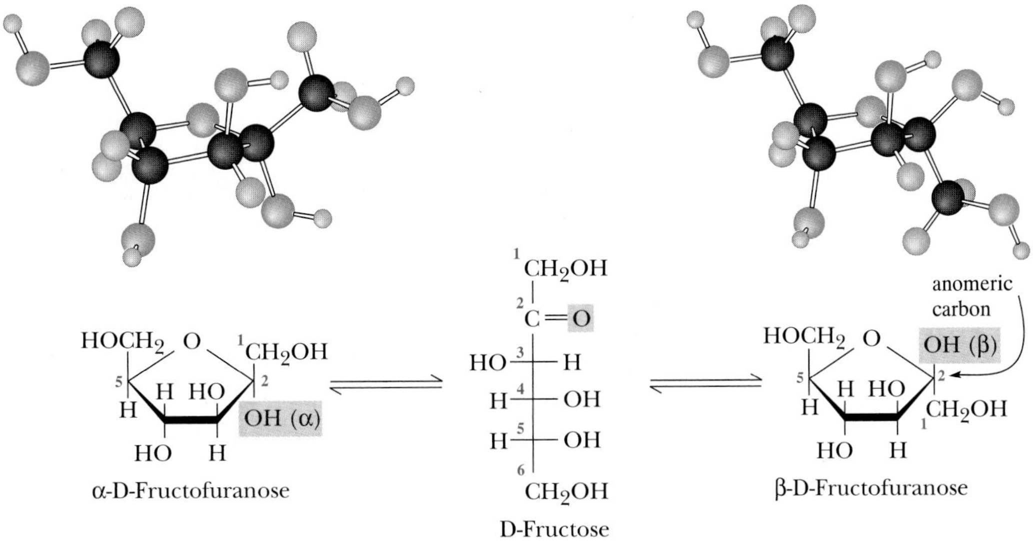

Figure 16.4
Furanose forms of D-fructose at equilibrium in aqueous solution.

$HOCH_2$ O $^1CH_2OH$ 5 H HO 2 H OH (α) HO H

α-D-Fructofuranose

$^1CH_2OH$
$^2C{=}O$
$HO\overset{3}{—}H$
$H\overset{4}{—}OH$
$H\overset{5}{—}OH$
$^6CH_2OH$

D-Fructose

anomeric carbon
$HOCH_2$ O OH (β) 5 H HO 2 H CH_2OH 1 HO H

β-D-Fructofuranose

SOLUTION

D-Galactose differs in configuration from D-glucose only at carbon 4. Therefore, draw the α and β forms of D-glucopyranose and then interchange the positions of the —OH and —H groups on carbon 4. Shown also are the specific rotations of each anomer.

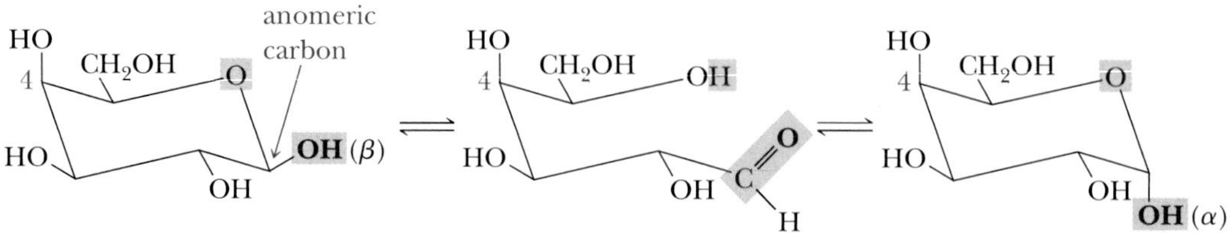

anomeric carbon

β-D-Galactopyranose
(β-D-Galactose)
$[\alpha]_D^{25} = +52.8°$

D-Galactose

α-D-Galactopyranose
(α-D-Galactose)
$[\alpha]_D^{25} = +150.7°$

Practice Problem 16.3

Draw chair conformations for α-D-mannopyranose and β-D-mannopyranose. Label the anomeric carbon atom in each.

C. Mutarotation

Mutarotation The change in optical activity that occurs when an α or β form of a carbohydrate is converted to an equilibrium mixture of the two forms.

Mutarotation is the change in specific rotation that accompanies the interconversion of α- and β-anomers in aqueous solution. As an example, a solution prepared by dissolving crystalline α-D-glucopyranose in water shows an initial rotation of +112° (Table 16.3), which gradually decreases to an equilibrium value of +52.7° as α-D-glucopyranose reaches an equilibrium with β-D-glucopyranose. A solution of β-D-glucopyranose also undergoes mutarotation, during which the specific rotation changes from an initial value of +18.7° to the same equilibrium value of +52.7°. Furthermore, it has been determined with modern techniques, most notably ^{13}C-NMR spectroscopy, that only traces of furanose and open-chain forms are present at equilibrium in aqueous solution. The equilibrium mixture consists of

TABLE 16.3 Specific Rotations for α- and β-Anomers of D-Glucopyranose and D-Galactopyranose Before and After Mutarotation

Monosaccharide	Specific Rotation (degrees)	Specific Rotation after Mutarotation (degrees)	Percent Present at Equilibrium
α-D-glucose	+112.0	+52.7	36
β-D-glucose	+18.7	+52.7	64
α-D-galactose	+150.7	+80.2	28
β-D-galactose	+52.8	+80.2	72

64% β-D-glucopyranose and 36% α-D-glucopyranose. It contains only traces (0.003%) of the open-chain form.

Mutarotation is common to all carbohydrates that exist in hemiacetal forms. Shown also in Table 16.3 are specific rotations for freshly prepared solutions of the α and β forms of D-galactopyranose along with equilibrium values for the specific rotation of each after equilibration. Only traces of the furanose and open-chain forms are present at equilibrium in aqueous solution.

From analyses of the composition of the equilibrium mixtures of monosaccharides in water, we can make the following generalizations. For most monosaccharides in aqueous solution:

1. Little free aldehyde or ketone is present.
2. Pyranose forms predominate over furanose forms. It is important not to confuse this statement about the form present at equilibrium in aqueous solution with a statement about the form that predominates in a biological system. They may be quite different. For example, although D-ribose and 2-deoxy-D-ribose exist in aqueous solution mainly in the pyranose form, each is found in nucleic acids exclusively in the β-furanose form.

16.3 PHYSICAL PROPERTIES

Monosaccharides are colorless, crystalline solids. Because hydrogen bonding is possible between their polar —OH groups and water, all monosaccharides are very soluble in water. They are only slightly soluble in alcohol and are insoluble in nonpolar solvents such as diethyl ether, chloroform, and benzene.

Although all monosaccharides are sweet to the taste, some are sweeter than others (Table 16.4). D-Fructose tastes the sweetest, even sweeter than sucrose (table sugar, Section 16.7C). In the production of sucrose, sugar cane or sugar beet is boiled with water, and the resulting solution is cooled. Sucrose crystals separate and are collected. Subsequent boiling to concentrate the solution followed by cooling yields a dark, thick syrup known as molasses. The sweet taste of honey is due largely to D-fructose and D-glucose, and that of corn syrup is due to D-glucose. Lactose (Section 16.7B) has almost no sweetness and is sometimes added to foods as a filler. Some people cannot tolerate lactose well and should avoid these foods.

TABLE 16.4 Relative Sweetness of Some Carbohydrate Sweetening Agents

Monosaccharide		Disaccharide		Other Carbohydrate Sweetening Agents	
D-fructose	174	sucrose (table sugar)*	100	honey	97
D-glucose	74	lactose (milk sugar)	0.16	molasses	74
D-xylose	0.40			corn syrup	74
D-galactose	0.22				

* Sucrose is taken as a standard for relative sweetness and is assigned a value of 100.

16.4 Reactions of Monosaccharides

In this section, we discuss reactions of monosaccharides with alcohols, reducing agents, and oxidizing agents.

A. Formation of Glycosides (Acetals)

Treatment of a monosaccharide hemiacetal with an alcohol forms an acetal as illustrated by the reaction of β-D-glucopyranose (β-D-glucose) with methanol.

β-D-Glucopyranose (β-D-Glucose) + CH_3OH $\xrightarrow[-H_2O]{H^+}$ Methyl β-D-glucopyranoside (Methyl β-D-glucoside) + Methyl α-D-glucopyranoside (Methyl α-D-glucoside)

anomeric carbon

Glycoside A carbohydrate in which the —OH on its anomeric carbon is replaced by —OR.

Glycoside bond The bond from the anomeric carbon of a glycoside to an —OR group.

A cyclic acetal derived from a monosaccharide is called a **glycoside,** and the bond from the anomeric carbon to the —OR group is called a **glycoside bond.** Mutarotation is no longer possible in a glycoside because, unlike a hemiacetal, an acetal is no longer in equilibrium with the open-chain carbonyl-containing compound in neutral or alkaline solution.

Glycosides are named by listing the alkyl or aryl group attached to oxygen followed by the name of the carbohydrate involved in which the ending -e is replaced by -ide. For example, glycosides derived from β-D-glucopyranose are named β-D-glucopyranosides; those derived from β-D-ribofuranose are named β-D-ribofuranosides.

Example 16.4

Draw a structural formula for each glycoside. On each label the anomeric carbon and the glycoside bond.

(a) Methyl β-D-ribofuranoside (methyl β-D-riboside)
(b) Methyl α-D-galactopyranoside (methyl α-D-galactoside)

Solution

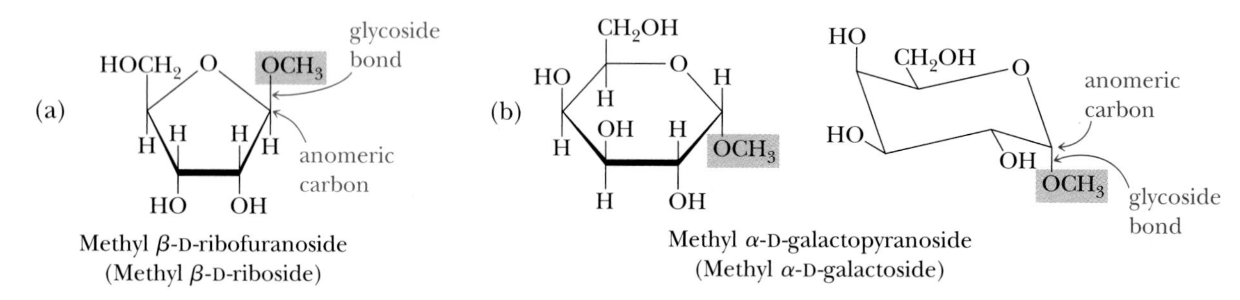

Methyl β-D-ribofuranoside (Methyl β-D-riboside)

Methyl α-D-galactopyranoside (Methyl α-D-galactoside)

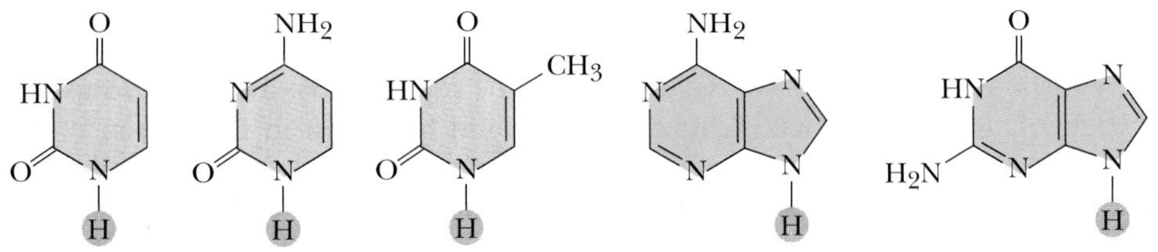

Uracil Cytosine Thymine Adenine Guanine

Figure 16.5 Structural formulas of the five most important purine and pyrimidine bases found in DNA and RNA. The hydrogen atom shown in color is lost in forming an *N*-glycoside.

Practice Problem 16.4

Draw a structural formula for each glycoside. On each label the anomeric carbon and the glycoside bond.

(a) Methyl β-D-fructofuranoside (methyl β-D-fructoside)
(b) Methyl α-D-mannopyranoside (methyl α-D-mannoside)

Glycosides are stable in water and aqueous base, but like other acetals (Section 11.6B), they are hydrolyzed in aqueous acid to an alcohol and a monosaccharide.

Just as the anomeric carbon of a cyclic hemiacetal undergoes reaction with the —OH group of an alcohol to form a glycoside, it also undergoes reaction with the —NH group of an amine to form an *N*-glycoside. Especially important in the biological world are the *N*-glycosides formed between D-ribose and 2-deoxy-D-ribose, each as a furanose, and the heterocyclic aromatic amines uracil, cytosine, thymine, adenine, and guanine (Figure 16.5). *N*-Glycosides of these compounds are structural units of nucleic acids (Chapter 19).

EXAMPLE 16.5

Draw a structural formula for the β-*N*-glycoside formed between D-ribofuranose and cytosine. Label the anomeric carbon and the *N*-glycoside bond.

SOLUTION

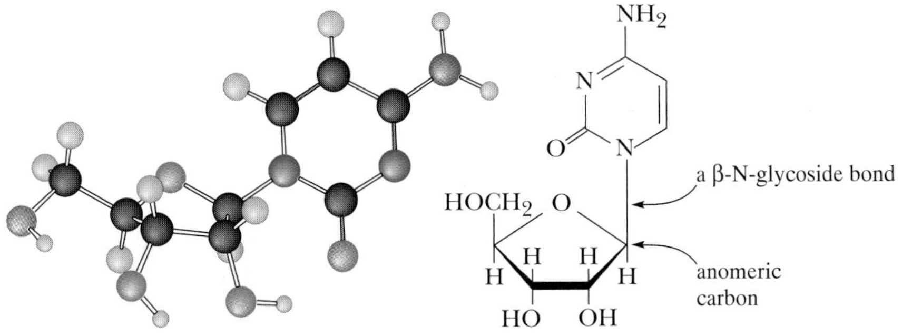

a β-N-glycoside bond

anomeric carbon

Practice Problem 16.5

Draw a structural formula for the β-*N*-glycoside formed between β-D-ribofuranose and adenine.

B. Reduction to Alditols

Alditol The product formed when the C=O group of a monosaccharide is reduced to a CHOH group.

The carbonyl group of a monosaccharide can be reduced to a hydroxyl group by a variety of reducing agents, including $NaBH_4$ and hydrogen in the presence of a transition metal catalyst. The reduction products are known as **alditols.** Reduction of D-glucose gives D-glucitol, more commonly known as D-sorbitol. Note that D-glucose is shown here in the open-chain form. Only a small amount of this form is present in solution, but, as it is reduced, the equilibrium between cyclic hemiacetal forms and the open-chain form shifts to replace it.

$$\begin{array}{rcl} & \text{CHO} & \\ \text{H} & - & \text{OH} \\ \text{HO} & - & \text{H} \\ \text{H} & - & \text{OH} \\ \text{H} & - & \text{OH} \\ & \text{CH}_2\text{OH} & \end{array} + \text{H}_2 \xrightarrow{\text{Ni}} \begin{array}{rcl} & \text{CH}_2\text{OH} & \\ \text{H} & - & \text{OH} \\ \text{HO} & - & \text{H} \\ \text{H} & - & \text{OH} \\ \text{H} & - & \text{OH} \\ & \text{CH}_2\text{OH} & \end{array}$$

D-Glucose D-Glucitol (D-Sorbitol)

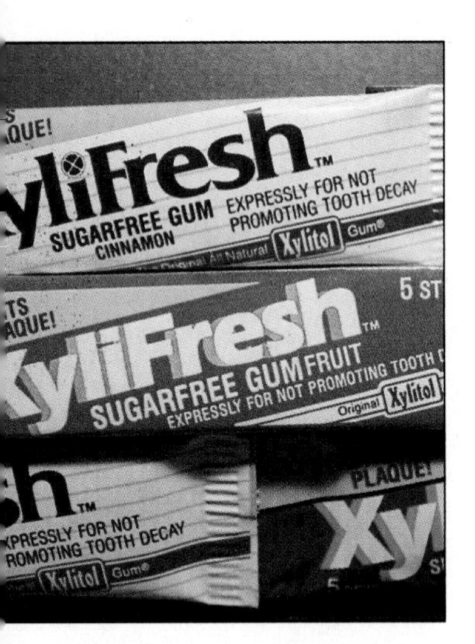

Many "sugar-free" products contain sugar alcohols, such as D-sorbitol and xylitol. *(Gregory Smolin)*

Sorbitol is found in the plant world in many berries (except grapes) and in cherries, plums, pears, apples, seaweed, and algae. It is about 60% as sweet as sucrose (table sugar) and is used in the manufacture of candies and as a sugar substitute for diabetics.

Also common in the biological world are erythritol, D-mannitol, and xylitol.

$$\begin{array}{rcl} & \text{CH}_2\text{OH} & \\ \text{H} & - & \text{OH} \\ \text{H} & - & \text{OH} \\ & \text{CH}_2\text{OH} & \end{array} \qquad \begin{array}{rcl} & \text{CH}_2\text{OH} & \\ \text{HO} & - & \text{H} \\ \text{HO} & - & \text{H} \\ \text{H} & - & \text{OH} \\ \text{H} & - & \text{OH} \\ & \text{CH}_2\text{OH} & \end{array} \qquad \begin{array}{rcl} & \text{CH}_2\text{OH} & \\ \text{H} & - & \text{OH} \\ \text{HO} & - & \text{H} \\ \text{H} & - & \text{OH} \\ & \text{CH}_2\text{OH} & \end{array}$$

Erythritol D-Mannitol Xylitol

Xylitol has been used as a sweetening agent in "sugarless" gum, candy, and sweet cereals.

EXAMPLE 16.6

D-Glucose is reduced by $NaBH_4$ to D-glucitol. Do you expect the alditol formed under these conditions to be optically active or optically inactive? Explain.

SOLUTION

D-Glucitol is a chiral substance. Given the fact that reduction by $NaBH_4$ does not affect any of the stereocenters, nor does the product have a plane of symmetry, predict the product to contain only one stereoisomer and to be optically active.

Practice Problem 16.6

D-Erythrose is reduced by $NaBH_4$ to erythritol. Do you expect the alditol formed under these conditions to be optically active or optically inactive? Explain.

C. Oxidation to Aldonic Acids

Monosaccharides (and carbohydrates in general) are classified as reducing or nonreducing sugars according to their behavior toward Ag(I) (Tollens' reagent, Section 11.9A) or Cu(II) (Benedict's solution and Fehling's solution). Those that reduce copper(II) ion to Cu_2O or silver(I) to metallic silver are classified as **reducing sugars.** Those that do not reduce these reagents are classified as **nonreducing sugars.** A positive Tollens' test is indicated by the formation of a silver mirror.

Reducing sugar A carbohydrate that reduces Ag(I) to Ag or Cu(II) to Cu(I).

$$\underset{}{RCH{=}O} + \underset{\text{Tollens' solution}}{Ag(NH_3)_2^+} \xrightarrow{NH_3,\ H_2O} RC(=O)O^-NH_4^+ + \underset{\text{Precipitates as a silver mirror}}{Ag}$$

Benedict's solution is prepared by dissolving copper(II) sulfate in aqueous sodium citrate. Fehling's solution is prepared by dissolving copper(II) sulfate in aqueous sodium tartrate. The function of the sodium citrate and sodium tartrate is to buffer the solution and to form a complex ion with copper(II). A positive Benedict's or Fehling's test is indicated by formation of copper(I) oxide, which precipitates as a red solid.

$$RCH{=}O + Cu^{2+} \xrightarrow{\text{citrate or tartrate buffer}} RC(=O)O^- + \underset{\text{Precipitates as a red solid}}{Cu_2O}$$

Ketoses also reduce these solutions. Carbon 1 of a ketose is not oxidized directly. Rather the basic conditions of the test catalyze isomerization of a 2-ketose to an aldose by way of an enediol intermediate. The aldose then gives the positive test with Tollens', Fehling's, or Benedict's solutions.

$$\underset{\text{A 2-ketose}}{CH_2OH{-}C(=O){-}(CHOH)_n{-}CH_2OH} \rightleftharpoons \underset{\text{An enediol}}{CHOH{=}C(OH){-}(CHOH)_n{-}CH_2OH} \rightleftharpoons \underset{\text{An aldose}}{CHO{-}CHOH{-}(CHOH)_n{-}CH_2OH}$$

16.5 GLUCOSE ASSAY—THE SEARCH FOR SPECIFICITY

The analytical procedure most often performed in a clinical chemistry laboratory is the determination of glucose in blood, urine, or other biological fluid. The need for a rapid and reliable test for glucose stems from the high incidence of diabetes mellitus. Approximately 2 million known diabetics live in the United States, and it is estimated that another million diabetics are undiagnosed.

Diabetes mellitus is characterized by insufficient blood levels of the polypeptide hormone insulin (Section 18.5C). If the concentration of insulin is insufficient, muscle and liver cells do not absorb glucose, which, in turn, leads to increased levels of blood glucose (hyperglycemia), impaired metabolism of fats and proteins, ketosis, and possible diabetic coma. Thus, a rapid procedure for the determination of blood glucose levels is critical for early diagnosis and effective management of this disease. In addition to being rapid, a test must also be specific for D-glucose; it must give a positive test for glucose but not react with any other substance normally present in biological fluids.

To this end, determination of blood glucose levels is now carried out by an enzyme-based assay procedure using the enzyme **glucose oxidase.** This enzyme catalyzes the oxidation of β-D-glucose to D-gluconic acid.

$$\beta\text{-D-Glucopyranose} + O_2 + H_2O \xrightarrow[\text{oxidase}]{\text{glucose}} \text{D-Gluconic acid} + H_2O_2$$

β-D-Glucopyranose (ring: CH_2OH, O, HO, HO, OH, OH) — D-Gluconic acid (CO_2H; H–OH; HO–H; H–OH; H–OH; CH_2OH) — Hydrogen peroxide

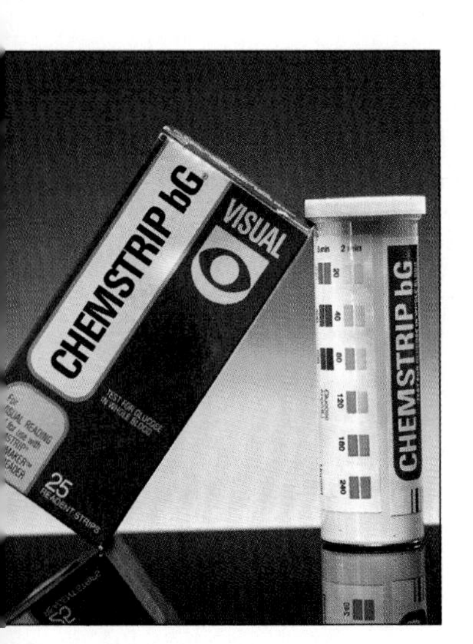

Chemstrip kit for blood glucose. *(Charles D. Winters)*

Glucose oxidase is specific for β-D-glucose. Therefore, complete oxidation of any sample containing both β-D-glucose and α-D-glucose requires conversion of the α form to the β form. Fortunately, this interconversion is rapid and complete in the short time required for the test. Molecular oxygen, O_2, is the oxidizing agent in this reaction and is reduced to hydrogen peroxide, H_2O_2, the concentration of which can be determined spectrophotometrically.

In one procedure, hydrogen peroxide formed in the glucose oxidase-catalyzed reaction is used to oxidize *o*-toluidine to a colored product in a reaction catalyzed by the enzyme, peroxidase. The concentration of the colored oxidation product is determined spectrophotometrically and is proportional to the concentration of glucose in the test solution. Several commercially available test kits use the glucose oxidase reaction for qualitative determination of glucose in urine.

$$o\text{-toluidine} + H_2O_2 \xrightarrow{\text{peroxidase}} \text{colored product} + H_2O$$

16.6 L-ASCORBIC ACID (VITAMIN C)

The structural formula of L-ascorbic acid (vitamin C) resembles that of a monosaccharide. In fact, this vitamin is synthesized both biochemically by plants and some animals and commercially from D-glucose. Humans do not have the enzyme systems required for the synthesis of L-ascorbic acid and, therefore, for us, it is a vitamin. Approximately 66 million kilograms of vitamin C are synthesized every year in the United States.

$$\text{D-Glucose} \xrightarrow{\text{Biochemical and industrial synthesis}} \text{L-Ascorbic acid (Vitamin C)}$$

D-Glucose (CHO; H–OH; HO–H; H–OH; H–OH; CH_2OH) — L-Ascorbic acid (Vitamin C) (CH_2OH; H–OH; O; =O; H; HO; OH)

L-Ascorbic acid is very easily oxidized to L-dehydroascorbic acid, a diketone. Both L-ascorbic acid and L-dehydroascorbic acid are physiologically active and are found together in most body fluids.

CH_2OH / H—OH / O / =O / H / HO / OH → (oxidation ⇌ reduction) → CH_2OH / H—OH / O / =O / H / O / O

L-Ascorbic acid (Vitamin C) L-Dehydroascorbic acid

16.7 DISACCHARIDES AND OLIGOSACCHARIDES

Most carbohydrates in nature contain more than one monosaccharide unit. Those that contain two units are called **disaccharides,** those that contain three units are called **trisaccharides,** and so forth. The more general term, **oligosaccharide,** is often used for carbohydrates that contain from four to ten monosaccharide units. Carbohydrates containing larger numbers of monosaccharide units are called **polysaccharides.**

Disaccharide A carbohydrate containing two monosaccharide units joined by a glycoside bond.

Oligosaccharide A carbohydrate containing from four to ten monosaccharide units, each joined to the next by a glycoside bond.

Polysaccharide A carbohydrate containing a large number of monosaccharide units, each joined to the next by one or more glycoside bonds.

In a disaccharide, two monosaccharide units are joined together by a glycoside bond between the anomeric carbon of one unit and an —OH of the other. Three important disaccharides are maltose, lactose, and sucrose.

A. Maltose

Maltose derives its name from its presence in malt, the juice from sprouted barley and other cereal grains. Maltose consists of two molecules of D-glucopyranose joined by a glycoside bond between carbon 1 (the anomeric carbon) of one glucopyranose unit and carbon 4 of the second unit. Because the oxygen atom on the anomeric carbon of the first glucopyranose unit is alpha, the bond joining the two is called an α-1,4-glycoside bond. Shown in Figure 16.6 is a chair conformation for

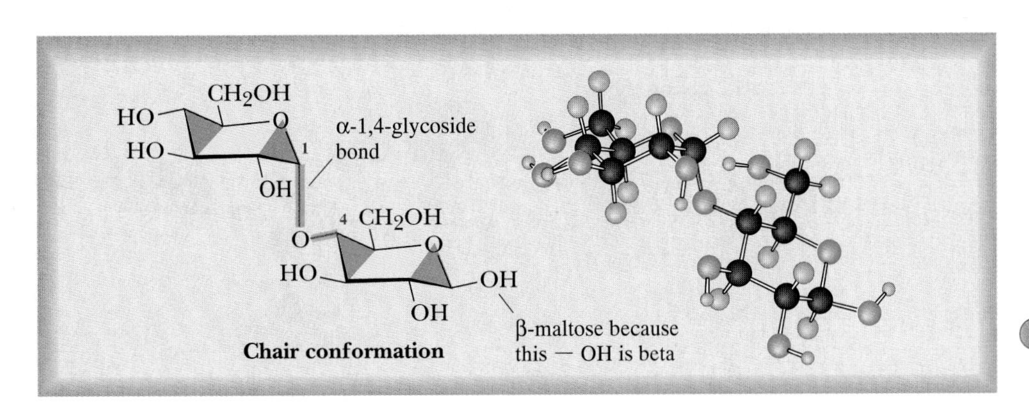

Figure 16.6 β-Maltose (derived from hydrolysis of starch).

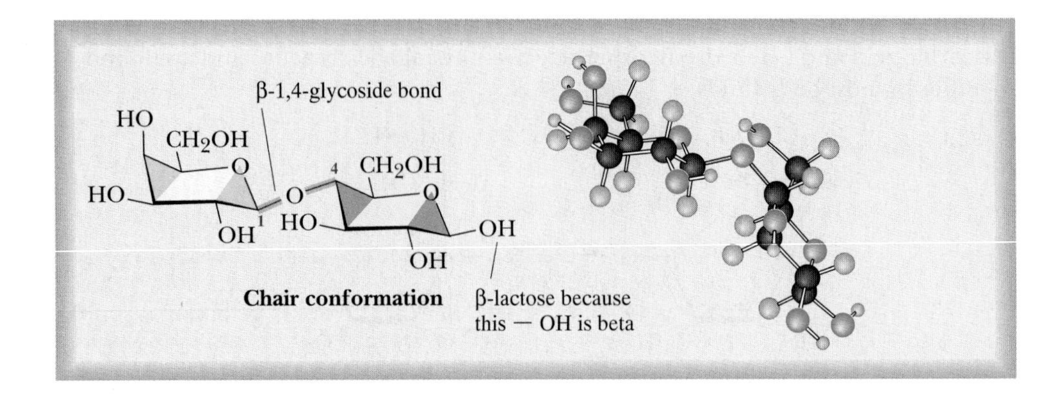

Figure 16.7 Lactose (from the milk of mammals).

β-maltose, so named because the —OH on the anomeric carbon of the glucose unit on the right is beta. Maltose is a reducing sugar because the hemiacetal group on the right unit of D-glucopyranose is in equilibrium with the free aldehyde and can be oxidized to a carboxylic acid.

B. Lactose

Lactose (Figure 16.7) is the principal sugar present in milk. It makes up about 5–8% of human milk and 4–6% of cow's milk. Hydrolysis of lactose yields D-glucose and D-galactose. In lactose, D-galactopyranose is joined by a β-1,4-glycoside bond to carbon 4 of D-glucopyranose. Lactose is a reducing sugar.

Sugar cane, Hawaii *(D. E. Cox/Tony Stone Images)*

C. Sucrose

Sucrose (table sugar) is the most abundant disaccharide in the biological world (Figure 16.8). It is obtained principally from the juice of sugar cane and sugar beets. In sucrose, carbon 1 of D-glucopyranose is joined to carbon 2 of D-fructofuranose by an α-1,2-glycoside bond. Note that glucose is in a six-membered (pyranose) ring form, whereas fructose is in a five-membered (furanose) ring form. Because the anomeric carbons of both the glucopyranose and fructofuranose units are involved in formation of the glycoside bond, sucrose is a nonreducing sugar.

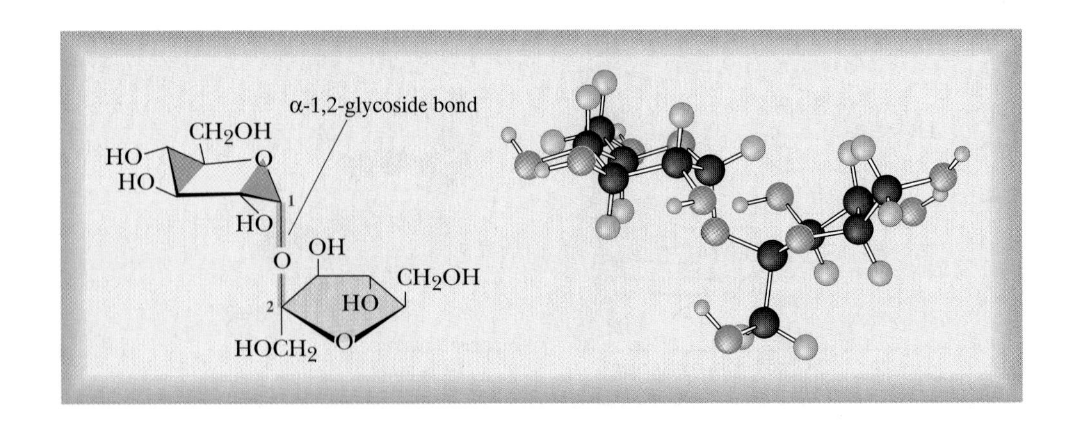

Figure 16.8 Sucrose (from sugar cane and sugar beet).

EXAMPLE 16.7

Draw a chair formula for the β anomer of a disaccharide in which two units of D-glucopyranose are joined by an α-1,6-glycoside bond.

SOLUTION

First draw the structural formula of α-D-glucopyranose. Then connect the anomeric carbon of this monosaccharide to carbon 6 of a second D-glucopyranose unit by an α-glycoside bond. The resulting molecule is either α or β depending on the orientation of the —OH group on the reducing end of the disaccharide. The disaccharide shown here is β.

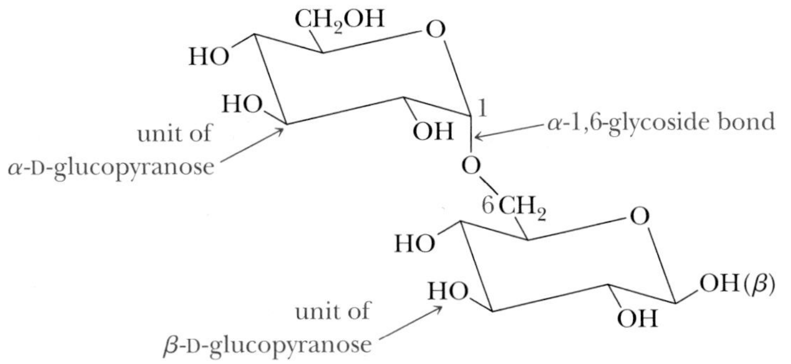

CH$_2$OH, O, HO, HO, 1, OH, O, 6 CH$_2$, O, HO, HO, OH(β), OH

unit of α-D-glucopyranose

α-1,6-glycoside bond

unit of β-D-glucopyranose

Practice Problem 16.7

Draw Haworth and chair formulas for the α form of a disaccharide in which two units of D-glucopyranose are joined by a β-1,3-glycoside bond.

D. Blood Group Substances

Membranes of animal plasma cells have large numbers of relatively small carbohydrates bound to them. In fact, it appears that the outsides of most plasma cell membranes are literally "sugar-coated." These membrane-bound carbohydrates are part of the mechanism by which cell types recognize each other and, in effect, act as biochemical markers (antigenic determinants). Typically, these membrane-bound carbohydrates contain from 4 to 17 monosaccharide units consisting primarily of relatively few monosaccharides, including D-galactose (Gal), D-mannose (Man), L-fucose (Fuc), *N*-acetyl-D-glucosamine (NAGlu), and *N*-acetyl-D-galactosamine (NAGal). L-Fucose is a 6-deoxyaldohexose.

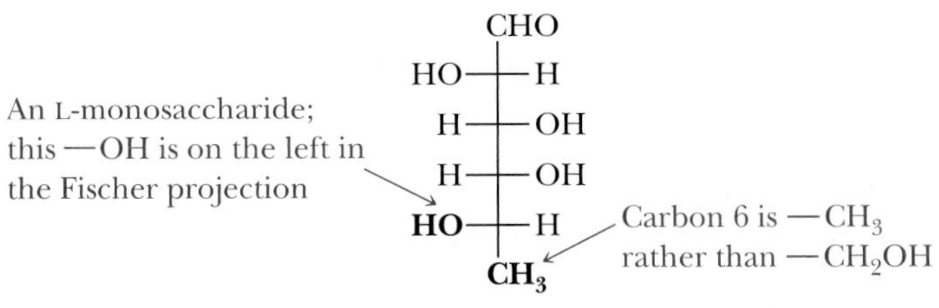

CHO
HO—H
H—OH
H—OH
HO—H
CH$_3$

An L-monosaccharide; this —OH is on the left in the Fischer projection

Carbon 6 is —CH$_3$ rather than —CH$_2$OH

L-Fucose

CHEMICAL CONNECTIONS

Glycosidase Inhibitors—Natural Defensive Weapons of Plants

Plants and their herbivores coevolve. The continuing adjustments of one to the other reflect the biosynthesis of defensive compounds by plants and the development of detoxification mechanisms by herbivores. Examples of defensive compounds are a group of polyhydroxypiperidines and polyhydroxypyrrolidines recently discovered in several members of the pea family. We discuss one of these compounds, which is a powerful inhibitor of glycosidase activity in insects, mammals, and microorganisms. **Glycosidases** are a class of enzymes that catalyze the hydrolysis of glycoside bonds. These enzymes enable plants and animals to convert di-, tri-, and polysaccharides into monosaccharides, which they then use as sources of energy and sources of carbon for the synthesis of other biomolecules.

Deoxynojirimycin, isolated not only from peas but also from species of bacteria and mulberry trees, is an inhibitor of both α- and β-glycosidase activity in susceptible insects (*The Merck Index,* 12th ed., #2952). A possible basis for this inhibition is deoxynojirimycin's similarity in structure to D-glucopyranose, which you can see by drawing deoxynojirimycin in a chair conformation and comparing it with the chair conformation of β-D-glucopyranose. In deoxynojirimycin, —O— of the cyclic hemiacetal of D-glucopyranose is replaced by —NH—, and —OH is missing from carbon 2 of the ring (hence the prefix deoxy-). Thus, deoxynojirimycin is a monosaccharide "look-alike." The glycosidase accepts this look-alike molecule as a monosaccharide, upon which the inhibitor blocks the enzyme from catalyzing glycoside hydrolysis.

Among the first discovered and best understood of these membrane-bound carbohydrates are the **blood group substances.** Although blood group substances are found chiefly on the surface of erythrocytes (red blood cells), they are also found on proteins and lipids in other parts of the body. In the ABO system, first described in 1900, individuals are classified according to four blood types: A, B, AB, and O. Blood from individuals of the same type can be mixed without clumping (agglutination) of erythrocytes. However, if serum of type A blood is mixed with type B blood, or vice versa, the erythrocytes clump. Serum from a type O individual is compatible with types A, B, and AB. At the cellular level, the chemical basis for this classification is a group of relatively small, membrane-bound carbohydrates. Following is the composition of the tetrasaccharide found on the cell membrane of erythrocytes of individuals with type A blood. The configurations of glycoside bonds between the monosaccharide units and of the glycoside bond between the tetrasaccharide and the erythrocyte cell wall are shown in parentheses. This membrane-bound carbohydrate has several distinctive features. First, it contains a monosaccharide of the unnatural, or L series, namely L-fucose. Second, it contains D-galactose to which two other monosaccharides are bonded, one by an α-1,2-glycoside bond and the second by an α-1,4-glycoside bond. This second monosaccharide is what determines the ABO classification. In blood of type A, the chain terminates in *N*-acetyl-D-galactosamine (NAGal); in type B blood, it terminates instead in D-galactose (Gal), and in type O blood, the fourth monosaccha-

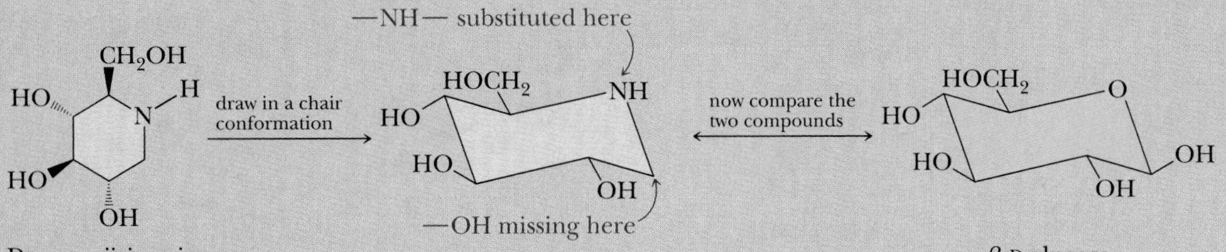

—NH— substituted here

draw in a chair conformation

now compare the two compounds

—OH missing here

Deoxynojirimycin

β-D-glucose

How this glycosidase inhibitor functions as a weapon in the chemical defense systems of higher plants is not fully understood. One suggestion is that insect glycosidases are activated when the insect comes in contact with a suitable plant source of food. These enzymes, thus activated, catalyze the release of free sugars in the plant, which then trigger the feeding response. Because deoxynojirimycin inhibits glycosidase activity, its defensive role may be to inhibit the release of free sugars and, thereby, act as a feeding deterrent. Or, it may be a feeding deterrent because it tastes unpleasant to the bruchid beetles that are serious economic pests to pea crops.

Whatever the biochemical mechanism that deters feeding, many insects have developed a resistance to such inhibitors. Some insects probably have even developed the ability to use them as sources of carbon and nitrogen atoms for their own growth! See *Natural Resistance of Plants to Pests,* M. B. Green and P. A. Hedin editors, ACS Symposium Series No. 296, American Chemical Society, Washington, D.C., 1986.

ride is missing completely. The saccharides of type AB blood contain both A and B tetrasaccharides.

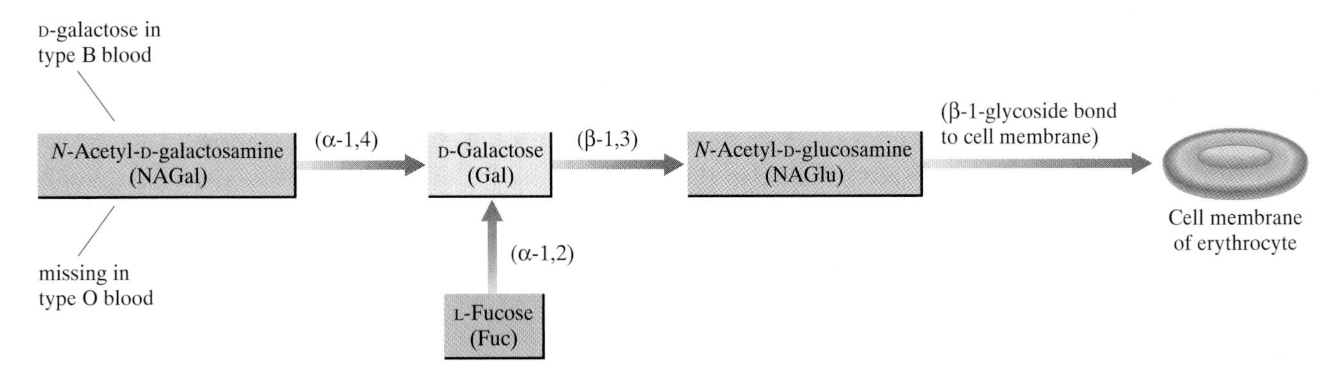

16.8 POLYSACCHARIDES

Polysaccharides consist of a large number of monosaccharide units joined together by glycoside bonds. Three important polysaccharides, all made up of glucose units, are starch, glycogen, and cellulose.

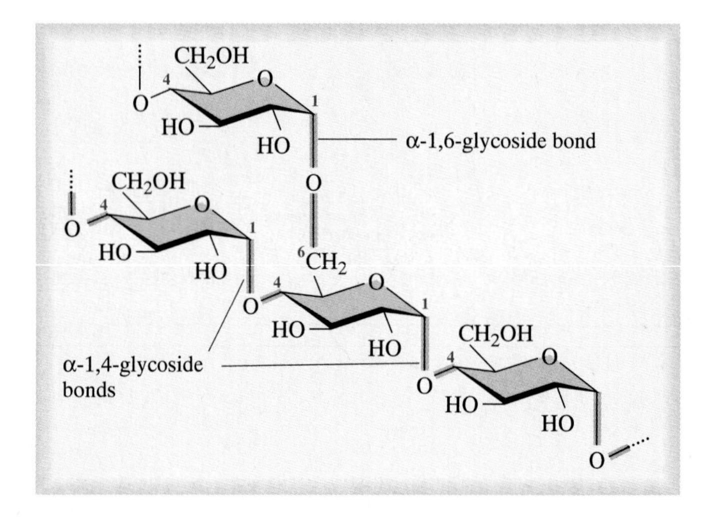

Figure 16.9
Amylopectin is a highly branched polymer of D-glucose. Chains consist of 24–30 units of D-glucose joined by α-1,4-glycoside bonds and branches created by α-1,6-glycoside bonds.

A. Starch—Amylose and Amylopectin

Starch is used for energy storage in plants. This polysaccharide is found in all plant seeds and tubers and is the form in which glucose is stored for later use by plants. Starch can be separated into two principal polysaccharides: amylose and amylopectin. Although the starch from each plant is unique, most starches contain 20–25% amylose and 75–80% amylopectin. Complete hydrolysis of both amylose and amylopectin yields only D-glucose. Amylose is composed of continuous, unbranched chains of up to 4000 D-glucose units joined by α-1,4-glycoside bonds.

Amylopectin has a highly branched structure and contains two types of glycoside bonds. This polymer contains the same type of chains of D-glucose joined by α-1,4-glycoside bonds as amylose does, but polymer chains vary in length from only 24 to 30 units (Figure 16.9). In addition, considerable branching occurs from this linear network. At branch points, new chains are started by α-1,6-glycoside bonds. In fact, amylopectin has such a highly branched structure that it is hardly possible to distinguish between main chains and branch chains.

Why are carbohydrates stored in plants as polysaccharides rather than monosaccharides, a more directly usable source of energy? The answer has to do with **osmotic pressure** which is proportional to the molar concentration, not the molecular weight, of a solute. If 1000 molecules of glucose are assembled into one starch macromolecule, a solution containing 1 g of starch per 10 mL will have only one one-thousandth the osmotic pressure relative to a solution of 1 g of glucose in the same volume. This feat of packaging is a tremendous advantage because it reduces the strain on various membranes enclosing solutions of such macromolecules.

B. Glycogen

Glycogen is the reserve carbohydrate for animals. Like amylopectin, glycogen is a nonlinear polymer of D-glucose joined by α-1,4- and α-1,6-glycoside bonds, but it has a lower molecular weight and an even more highly branched structure. The total amount of glycogen in the body of a well-nourished adult is about 350 g, divided almost equally between liver and muscle.

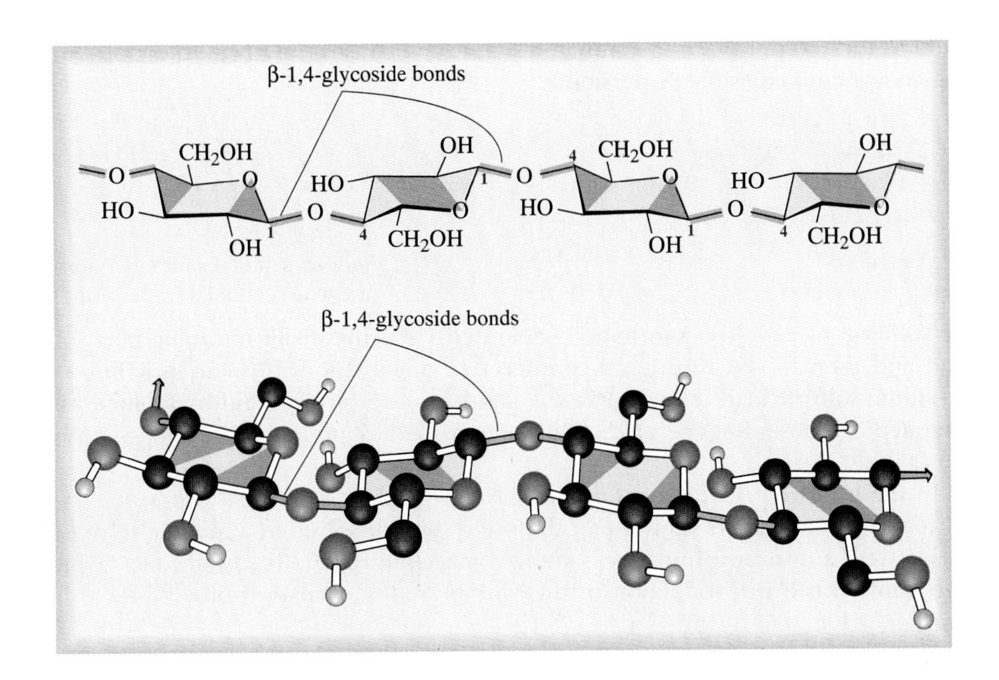

Figure 16.10 Cellulose is a linear polymer of up to 3000 units of D-glucose joined by β-1,4-glycoside bonds.

C. Cellulose

Cellulose, the most widely distributed plant skeletal polysaccharide, constitutes almost half of the cell wall material of wood. Cotton is almost pure cellulose. Cellulose is a linear polymer of D-glucose units joined by β-1,4-glycoside bonds (Figure 16.10). It has an average molecular weight of 400,000, corresponding to approximately 2800 glucose units per molecule. Cellulose fibers consist of bundles of parallel polysaccharide chains held together by hydrogen bonding between hydroxyl groups on adjacent chains. This arrangement of parallel chains in bundles gives cellulose fibers their high mechanical strength.

Humans and other animals cannot use cellulose as food because our digestive systems do not contain β-glucosidases, enzymes that catalyze hydrolysis of β-glucoside bonds. Instead we have only α-glucosidases; hence, the polysaccharides we use as sources of glucose are starch and glycogen. On the other hand, many bacteria and microorganisms do contain β-glucosidases and can digest cellulose. Termites are fortunate (much to our regret) to have such bacteria in their intestines and can use wood as their principal food. Ruminants (cud-chewing animals) and horses can also digest grasses and hay because β-glucosidase-containing microorganisms are present within their alimentary systems.

D. Textile Fibers from Cellulose

Both rayon and acetate rayon are made from chemically modified cellulose and were the first commercially important synthetic textile fibers. In the production of rayon, cellulose fibers are treated with carbon disulfide, CS_2, in aqueous sodium hydroxide. In this reaction, some of the —OH groups on a cellulose fiber are converted to the sodium salt of a xanthate ester, which causes the fibers to dissolve in alkali as a viscous colloidal dispersion.

An —OH group in a cellulose fiber

$$\text{Cellulose—OH} \xrightarrow{\text{NaOH}} \text{Cellulose—O}^-\text{Na}^+ \xrightarrow{\text{S=C=S}} \text{Cellulose—O}\overset{\overset{\text{S}}{\|}}{\text{C}}\text{S}^-\text{Na}^+$$

Cellulose (insoluble in water)

Sodium salt of a xanthate ester (a viscous colloidal suspension)

The solution of cellulose xanthate is separated from the alkali insoluble parts of wood and then forced through a spinneret (a metal disc with many tiny holes) into dilute sulfuric acid to hydrolyze the xanthate ester groups and precipitate regenerated cellulose. Regenerated cellulose extruded as a filament is called viscose rayon thread.

In the industrial synthesis of acetate rayon, cellulose is treated with acetic anhydride (Section 13.4B). Acetylated cellulose is then dissolved in a suitable solvent, precipitated, and drawn into fibers known as acetate rayon. Today, acetate rayon fibers rank fourth in production in the United States, surpassed only by Dacron polyester, nylon, and rayon.

$$\text{(glucose unit)} + 3CH_3\overset{O}{\overset{\|}{C}}O\overset{O}{\overset{\|}{C}}CH_3 \longrightarrow \text{(acetylated glucose unit)} + 3CH_3\overset{O}{\overset{\|}{C}}OH$$

A glucose unit in a cellulose fiber

Acetic anhydride

A fully acetylated glucose unit

SUMMARY

Monosaccharides (Section 16.1A) are polyhydroxyaldehydes or polyhydroxyketones. The most common have the general formula $C_nH_{2n}O_n$ where n varies from 3 to 8. Their names contain the suffix -ose. The prefixes tri-, tetr-, pent-, and so on show the number of carbon atoms in the chain. The prefix aldo- shows an aldehyde and the prefix keto- shows a ketone. In a **Fischer projection** of a carbohydrate, the carbon chain is written vertically with the most highly oxidized carbon toward the top. Horizontal lines show groups projecting above the plane of the page; vertical lines show groups projecting behind the plane of the page.

The **penultimate carbon** of a monosaccharide is the next to last carbon on the carbon chain of a Fischer projection of the monosaccharide. A monosaccharide that has the same configuration at the penultimate carbon as D-glyceraldehyde is called a **D-monosaccharide;** one that has the same configuration at the penultimate carbon as L-glyceraldehyde is called an **L-monosaccharide.**

Monosaccharides exist primarily as cyclic hemiacetals (Section 16.2A). The new stereocenter resulting from hemiacetal formation is referred to as an **anomeric carbon.** The stereoisomers thus formed are called **anomers.** A six-membered cyclic hemiacetal is called a **pyranose;** a five-membered cyclic hemiacetal is called a **furanose.** The symbol ***β*-** indicates that the —OH on the anomeric carbon is on the same side of the ring as the terminal $—CH_2OH$. The symbol α- indicates that —OH on the anomeric carbon is on the opposite from the terminal $—CH_2OH$. Furanoses and pyranoses can be drawn as **Haworth projections.** Pyranoses can also be shown as strain-free **chair conformations** (Section 16.2B).

Mutarotation (Section 16.2C) is the change in specific rotation that accompanies formation of an equilibrium mixture of α- and β-anomers in aqueous solution.

A **glycoside** (Section 16.4A) is an acetal derived from a monosaccharide. The name of the glycoside is composed of the name of the alkyl or aryl group bonded to the acetal oxygen atom followed by the name of the monosaccharide in which the terminal -e has been replaced by -ide.

An **alditol** (Section 16.4B) is a polyhydroxy compound formed by reduction of the carbonyl group of a monosaccharide to a hydroxyl group. An **aldonic acid** (Section 16.4C) is a carboxylic acid formed by oxidation of the aldehyde group of an aldose. **Reducing sugars** (Section 16.4C) reduce Tollens', Fehling's, and Benedict's solutions. When acting as a reducing agent, the sugar is oxidized to an aldonic acid.

L-Ascorbic acid (Section 16.6) is synthesized in nature from D-glucose by a series of enzyme-catalyzed steps.

A **disaccharide** (Section 16.7) contains two monosaccharide units joined by a glycoside bond. Terms applied to carbohydrates containing larger numbers of monosaccharides are **trisaccharide, tetrasaccharide, oligosaccharide,** and **polysaccharide. Maltose** is a disaccharide of two molecules of D-glucose joined by an α-1,4-glycoside bond. **Lactose** is a disaccharide consisting of D-galactose joined to D-glucose by a β-1,4-glycoside bond. **Sucrose** is a disaccharide consisting of D-glucose joined to D-fructose by an α-1,2-glycoside bond.

At the cellular level, the structural basis for the A, B, AB, and O blood type classification is a group of relatively small, membrane-bound carbohydrates called **blood groups substances.**

Starch (Section 16.8A) can be separated into two fractions given the names amylose and amylopectin. **Amylose** is a linear polymer of up to 4000 units of D-glucopyranose joined by α-1,4-glycoside bonds. **Amylopectin** is a highly branched polymer of D-glucose joined by α-1,4-glycoside bonds and, at branch points, by α-1,6-glycoside bonds. **Glycogen** (Section 16.8B), the reserve carbohydrate of animals, is a highly branched polymer of D-glucopyranose joined by α-1,4-glycoside bonds and, at branch points, by α-1,6-glycoside bonds. **Cellulose** (Section 16.8C), the skeletal polysaccharide of plants, is a linear polymer of D-glucopyranose joined by β-1,4-glycoside bonds. **Rayon** (Section 16.8D) is made from chemically modified and regenerated cellulose. **Acetate rayon** is made by acetylation of cellulose.

KEY REACTIONS

1. Formation of Cyclic Hemiacetals (Section 16.2)

A monosaccharide existing as a five-membered ring is a furanose; one existing as a six-membered ring is a pyranose. A pyranose is most commonly drawn as a Haworth projection or as a chair conformation.

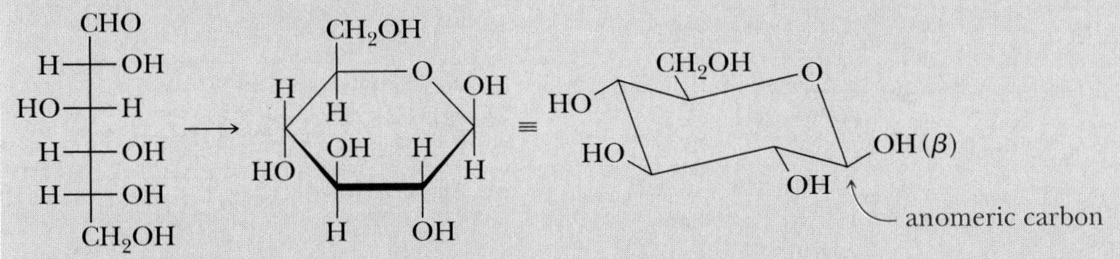

D-Glucose → β-D-Glucopyranose (β-D-Glucose)

OH (β) — anomeric carbon

2. Mutarotation (Section 16.2C)

Anomeric forms of a monosaccharide are in equilibrium in aqueous solution. Mutarotation is the change in specific rotation that accompanies this equilibration.

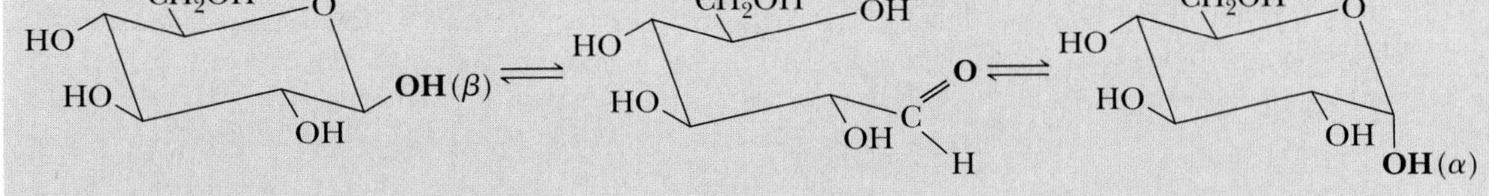

β-D-Glucopyranose $[\alpha]_D^{25}$ + 18.7° ⇌ Open-chain form ⇌ α-D-Glucopyranose $[\alpha]_D^{25}$ + 112°

3. Formation of Glycosides (Section 16.4A)

Treatment of a monosaccharide with an alcohol in the presence of an acid catalyst forms a cyclic acetal called a glycoside. The bond to the new —OR group is called a glycoside bond.

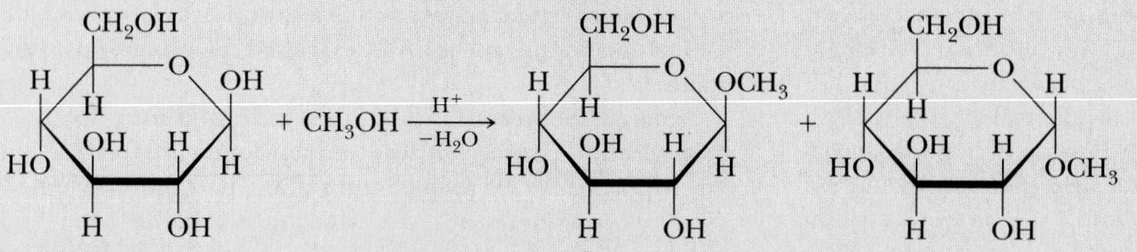

$$\text{(glucopyranose)} + CH_3OH \xrightarrow[-H_2O]{H^+} \text{(methyl glycoside, } \beta\text{)} + \text{(methyl glycoside, } \alpha\text{)}$$

4. Formation of N-Glycosides (Section 16.4A)

The *N*-glycosides formed between a monosaccharide and a heterocyclic aromatic amine are especially important in the biological world.

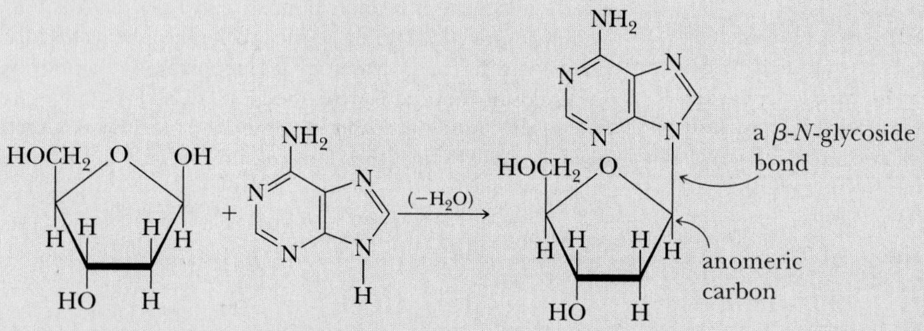

a β-*N*-glycoside bond

anomeric carbon

5. Reduction (Section 16.4B)

Reduction of the carbonyl group of an aldose or ketose to a hydroxyl group yields a polyhydroxy compound called an alditol.

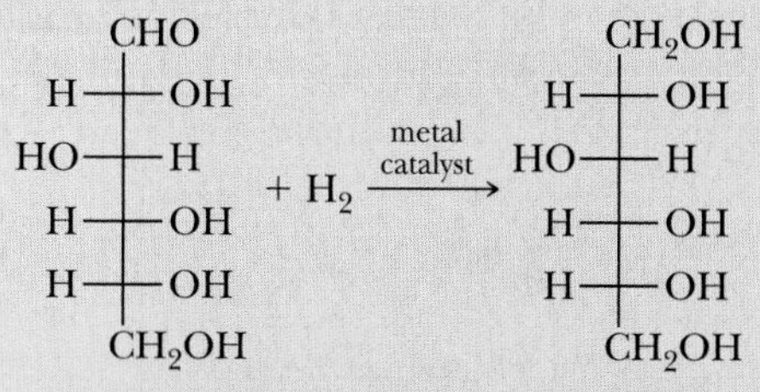

$$\text{D-Glucose} + H_2 \xrightarrow{\text{metal catalyst}} \text{D-Glucitol (D-Sorbitol)}$$

D-Glucose

D-Glucitol (D-Sorbitol)

6. Oxidation to an Aldonic Acid (Section 16.4C)

Oxidation of the aldehyde group of an aldose to a carboxyl group by a mild oxidizing agent gives a polyhydroxycarboxylic acid called an aldonic acid.

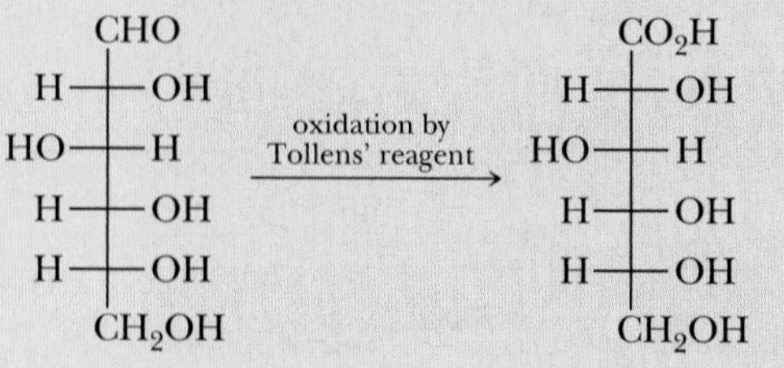

$$\text{D-Glucose} \xrightarrow{\text{oxidation by Tollens' reagent}} \text{D-Gluconic acid}$$

D-Glucose

D-Gluconic acid

ADDITIONAL PROBLEMS

Monosaccharides

16.8 Explain the meaning of the designations D and L as used to specify the configuration of carbohydrates.

16.9 Which compounds are D-monosaccharides and which are L-monosaccharides?

(a)
CHO
H—OH
HO—H
H—OH
H—OH
CH_2OH

(b)
CHO
HO—H
H—OH
HO—H
CH_2OH

(c)
CH_2OH
C=O
H—OH
H—OH
CH_2OH

16.10 Write Fischer projections for L-ribose and L-arabinose.

16.11 2,6-Dideoxy-D-altrose, known alternatively as D-digitoxose (*The Merck Index,* 12th ed., #3207), is a monosaccharide obtained on hydrolysis of digitoxin, a natural product extracted from purple foxglove *(Digitalis purpurea).* Digitoxin has found wide use in cardiology because it reduces pulse rate, regularizes heart rhythm, and strengthens heart beat (see *The Merck Index,* 12th ed., #3206). Draw the structural formula of 2,6-dideoxy-D-altrose.

The foxglove plant produces the important cardiac medication digitalis. *(© Lois Moulton, Tony Stone Images)*

The Cyclic Structure of Monosaccharides

16.12 Define the term "anomeric carbon."

16.13 Explain the conventions for using α and β to designate the configuration of cyclic forms of monosaccharides.

16.14 Draw α-D-glucopyranose (α-D-glucose) as a Haworth projection. Now, using only the information given here, draw Haworth projections for these monosaccharides.

(a) α-D-mannopyranose (α-D-mannose). The configuration of D-mannose differs from the configuration of D-glucose only at carbon 2.

(b) α-D-gulopyranose (α-D-gulose). The configuration of D-gulose differs from the configuration of D-glucose at carbons 3 and 4.

16.15 Convert each Haworth projection to an open-chain form and then to a Fischer projection. Name the monosaccharide you have drawn.

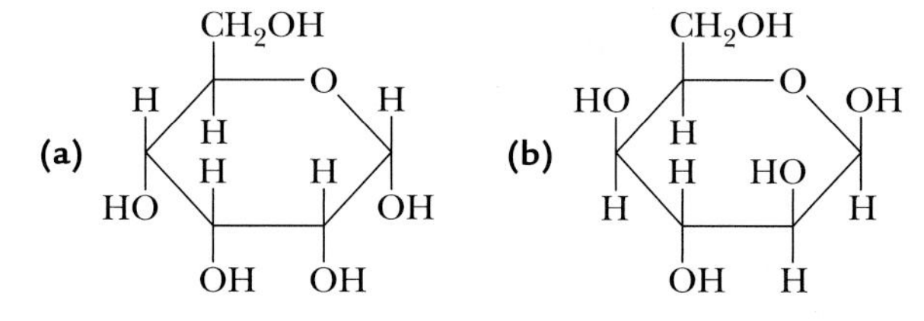

16.16 Convert each chair conformation to an open-chain form and then to a Fischer projection. Name the monosaccharide you have drawn.

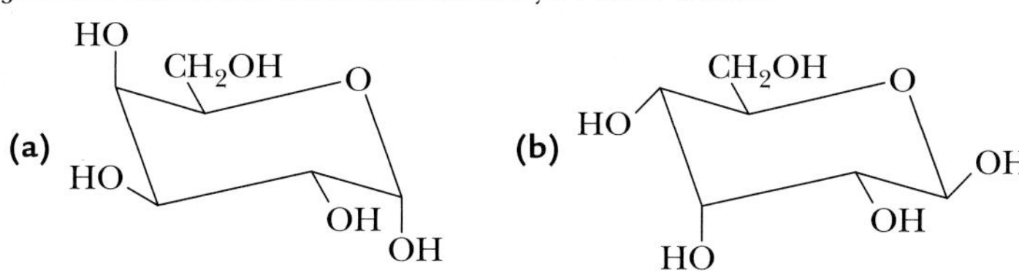

16.17 The configuration of D-arabinose differs from the configuration of D-ribose only at carbon 2. Using this information, draw a Haworth projection for α-D-arabinofuranose (α-D-arabinose).

16.18 Explain the phenomenon of mutarotation with reference to carbohydrates. By what means is it detected?

Reactions of Monosaccharides

16.19 Draw Fischer projections for the product(s) formed by reaction of D-galactose with the following. In addition, state whether each product is optically active or optically inactive.

(a) $NaBH_4$ in H_2O **(b)** H_2/Pt **(c)** $AgNO_3$ in NH_3, H_2O

16.20 Repeat Problem 16.19 using D-ribose.

16.21 There are four D-aldopentoses (Table 16.1). If each is reduced with $NaBH_4$, which yield optically active alditols? Which yield optically inactive alditols?

16.22 Account for the observation that reduction of D-glucose with $NaBH_4$ gives an optically active alditol, whereas reduction of D-galactose with $NaBH_4$ gives an optically inactive alditol.

16.23 Which two D-aldohexoses give optically inactive (meso) alditols on reduction with $NaBH_4$?

16.24 Name the two alditols formed by $NaBH_4$ reduction of D-fructose.

16.25 One pathway for the metabolism of glucose-6-phosphate is its enzyme-catalyzed conversion to fructose-6-phosphate. Show that this transformation can be regarded as two enzyme-catalyzed keto-enol tautomerizations.

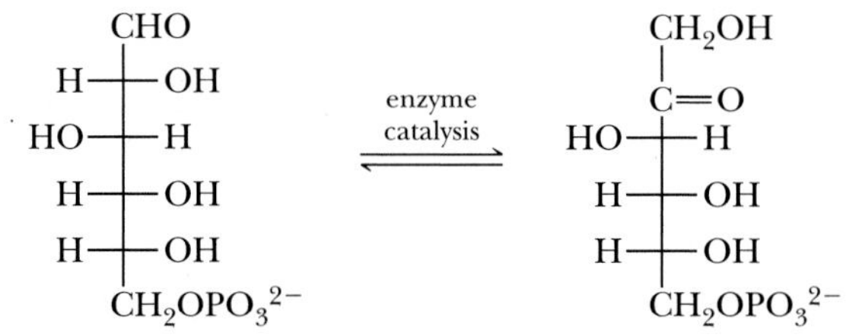

CHO; H—OH; HO—H; H—OH; H—OH; $CH_2OPO_3^{2-}$
D-Glucose-6-phosphate

enzyme catalysis ⇌

CH_2OH; C=O; HO—H; H—OH; H—OH; $CH_2OPO_3^{2-}$
D-Fructose-6-phosphate

16.26 L-Fucose, one of several monosaccharides commonly found in the surface polysaccharides of animal cells (Section 16.7D), is synthesized biochemically from D-mannose in the following eight steps.

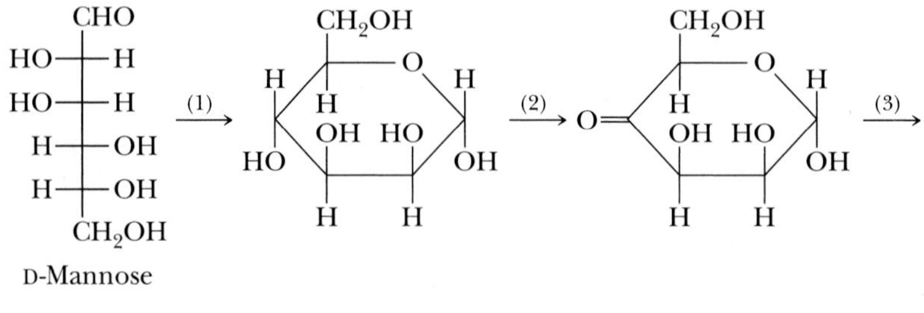

CHO; HO—H; HO—H; H—OH; H—OH; CH_2OH
D-Mannose

(1) →

CH_2OH, O, H, H, H, OH HO, HO, OH, H H

(2) →

CH_2OH, O, O=, H, H, OH HO, OH, H H

(3) →

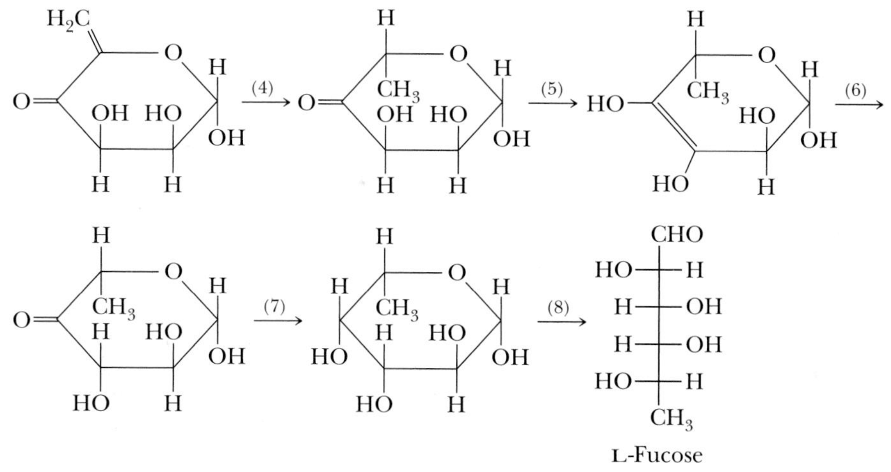

L-Fucose

(a) Describe the type of reaction (oxidation, reduction, hydration, dehydration, and the like) involved in each step.

(b) Explain why this monosaccharide, which is derived from D-mannose, now belongs to the L series.

16.27 Draw structural formulas for the products formed by hydrolysis at pH 7.4 (the pH of blood plasma) of all ester, thioester, amide, anhydride, and glycoside bonds in acetyl coenzyme A (acetyl-CoA). Name as many of the hydrolysis products as you can.

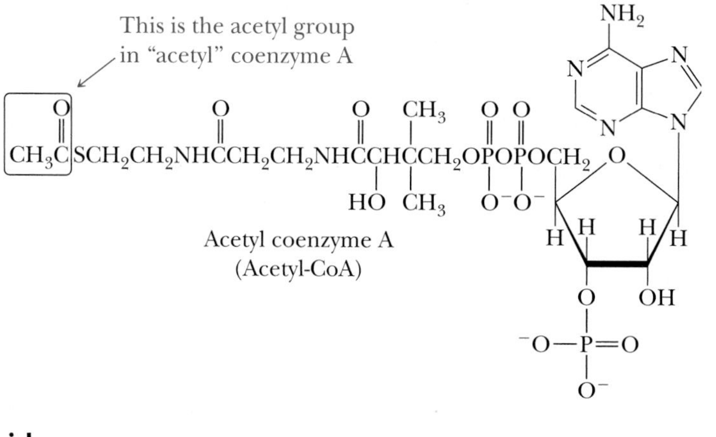

This is the acetyl group in "acetyl" coenzyme A

Acetyl coenzyme A
(Acetyl-CoA)

Ascorbic Acid

16.28 Write a balanced half-reaction to show that conversion of L-ascorbic acid to L-dehydroascorbic acid is an oxidation. How many electrons are involved in this oxidation? Is ascorbic acid a biological oxidizing agent or a biological reducing agent?

16.29 Ascorbic acid is a diprotic acid with the following acid ionization constants.

$$pK_{a1} = 4.10 \qquad pK_{a2} = 11.79$$

The two acidic hydrogens are those connected with the enediol part of the molecule. Which hydrogen has which ionization constant? (*Hint:* Draw separately the anion derived by loss of one of these hydrogens and that formed by loss of the other hydrogen. Which anion has the greater degree of resonance stabilization?)

Disaccharides and Oligosaccharides

16.30 Define the term "glycoside bond."

16.31 What is the difference in meaning between the terms "glycoside bond" and "glucoside bond"?

16.32 In making candy or sugar syrups, sucrose is boiled in water with a little acid, such as lemon juice. Why does the product mixture taste sweeter than the starting sucrose solution?

16.33 Which disaccharides are reduced by $NaBH_4$?

(a) Maltose **(b)** Lactose **(c)** Sucrose

16.34 Trehalose is found in young mushrooms and is the chief carbohydrate in the blood of certain insects. Trehalose is a disaccharide consisting of two D-monosaccharide units, each joined to the other by an α-1,1-glycoside bond (*The Merck Index,* 12th ed., #9713).

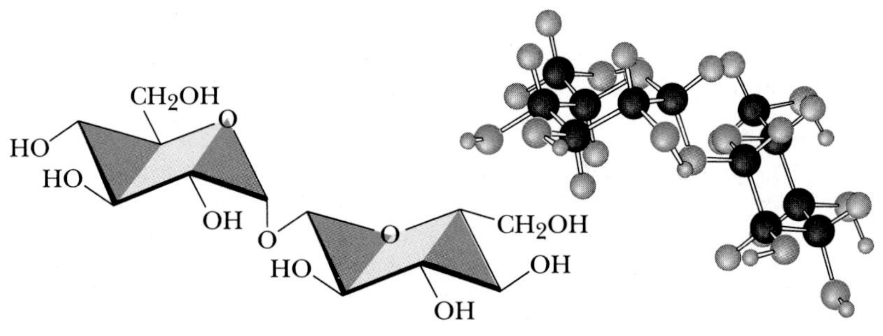

Trehalose

(a) Is trehalose a reducing sugar?

(b) Does trehalose undergo mutarotation?

(c) Name the two monosaccharide units of which trehalose is composed.

16.35 Hot water extracts of ground willow bark are an effective pain reliever. Unfortunately, the liquid is so bitter that most persons refuse it. The pain reliever in these infusions is salicin (*The Merck Index,* 12th ed., #8476). Name the monosaccharide unit in salicin.

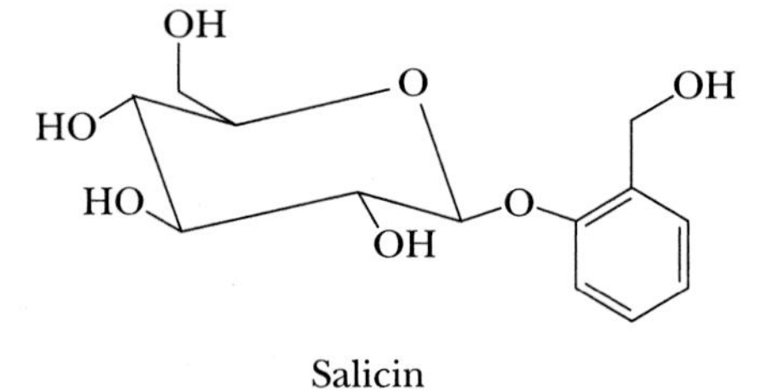

Salicin

Polysaccharides

16.36 A Fischer projection of *N*-acetyl-D-glucosamine is given in Section 16.1F.

(a) Draw Haworth and chair structures for the α- and β-pyranose forms of this monosaccharide.

(b) Draw Haworth and chair structures for the disaccharide formed by joining two units of the pyranose form of *N*-acetyl-D-glucosamine by a β-1,4-glycoside bond.

If you drew this correctly, you have the structural formula for the repeating dimer of chitin, the structural polysaccharide component of the shell of lobster and other crustaceans.

The polysaccharide Chitin is the structural polysaccharide of the shell of lobsters and other crustaceans. *(David Hall/Tony Stone Images)*

16.37 Propose structural formulas for the repeating disaccharide unit in these polysaccharides.

(a) Alginic acid (*The Merck Index,* 12th ed., #241), isolated from seaweed, is used as a thickening agent in ice cream and other foods. Alginic acid is a polymer of D-mannuronic acid in the pyranose form joined by β-1,4-glycoside bonds.

(b) Pectic acid is the main component of pectin (*The Merck Index,* 12th ed., #7194), which is responsible for the formation of jellies from fruits and berries. Pectic acid is a polymer of D-galacturonic acid in the pyranose form joined by α-1,4-glycoside bonds.

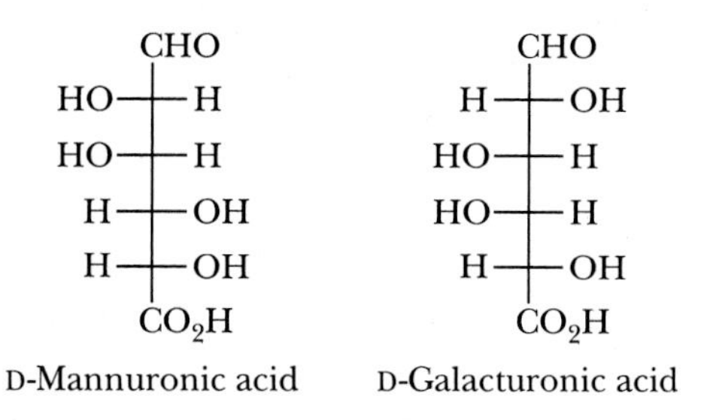

17

Lipids

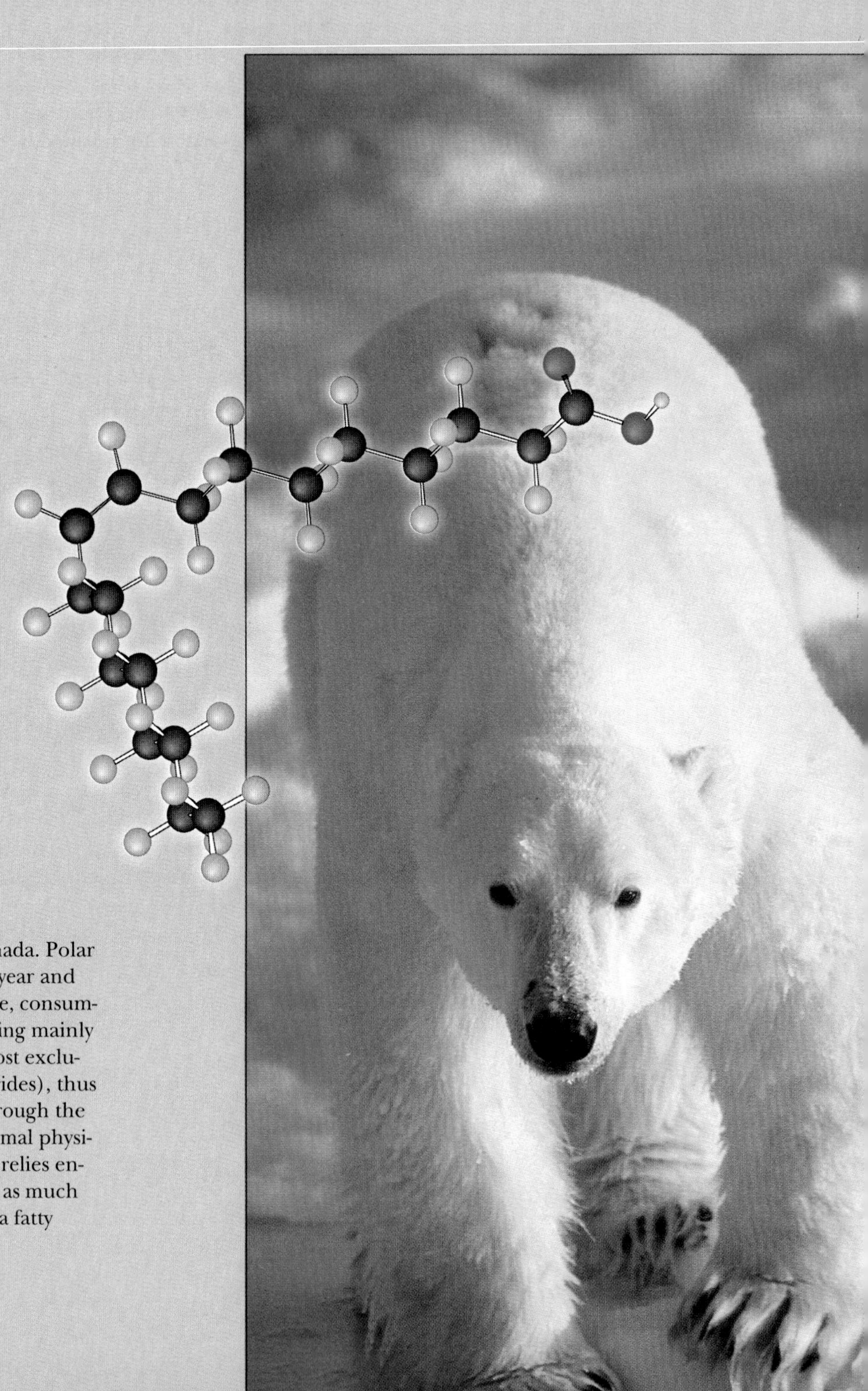

A polar bear in snow-covered landscape, Canada. Polar bears eat only during a few weeks out of the year and then fast for periods of eight months or more, consuming no food or water during that period. Eating mainly in the winter, the adult polar bear feeds almost exclusively on seal blubber (composed of triglycerides), thus building up its own triglyceride reserves. Through the Arctic summer, the polar bear maintains normal physical activity, roaming over long distances, but relies entirely on its body fat for sustenance, burning as much as 1 – 1.5 kg of fat per day. Inset: A model of a fatty acid. *(Daniel J. Cox/Tony Stone Images)*

Lipids are a heterogeneous group of naturally occurring organic compounds, classified together on the basis of their common solubility properties. Lipids are insoluble in water but soluble in nonpolar aprotic organic solvents, including diethyl ether, methylene chloride, and acetone.

Lipid A class of biomolecules isolated from plant or animal sources by extraction with nonpolar organic solvents, such as diethyl ether and acetone.

Lipids are divided into two main groups. First are those lipids that contain both a relatively large nonpolar hydrophobic region, most commonly aliphatic in nature, and a polar hydrophilic region. Found among this group are fatty acids, triglycerides, phospholipids, prostaglandins, and the fat-soluble vitamins. Second are those lipids that contain the tetracyclic ring system called the steroid nucleus, including cholesterol, steroid hormones, and bile acids. In this chapter, we describe the structures and biological functions of each group of lipids.

17.1 TRIGLYCERIDES

Animal fats and vegetable oils, the most abundant naturally occurring lipids, are triesters of glycerol and long-chain carboxylic acids. Fats and oils are also referred to as **triglycerides** or **triacylglycerols.** Hydrolysis of a triglyceride in aqueous base followed by acidification gives glycerol and three fatty acids.

Triglyceride (triacylglycerol) An ester of glycerol with three fatty acids.

$$\begin{array}{c} CH_2OCR \\ | \\ R'COCH \\ | \\ CH_2OCR'' \end{array} \xrightarrow[2.\ HCl,\ H_2O]{1.\ NaOH,\ H_2O} \begin{array}{c} CH_2OH \\ | \\ HOCH \\ | \\ CH_2OH \end{array} + \begin{array}{c} RCO_2H \\ R'CO_2H \\ R''CO_2H \end{array}$$

A triglyceride — 1,2,3-Propanetriol (Glycerol, glycerine) — Fatty acids

A. Fatty Acids

More than 500 different **fatty acids** have been isolated from various cells and tissues. Given in Table 17.1 are common names and structural formulas for the most abundant of these. The number of carbons in a fatty acid and the number of carbon-carbon double bonds in its hydrocarbon chain are shown by two numbers separated by a colon. In this notation, linoleic acid, for example, is designated as an 18:2 fatty acid; its 18-carbon chain contains two carbon-carbon double bonds. Following are several characteristics of the most abundant fatty acids in higher plants and animals.

Fatty acid A long, unbranched-chain carboxylic acid, most commonly of 12 to 20 carbons, derived from the hydrolysis of animal fats, vegetable oils, or the phospholipids of biological membranes.

1. Nearly all fatty acids have an even number of carbon atoms, most between 12 and 20, in an unbranched chain.
2. The three most abundant fatty acids in nature are palmitic acid (16:0), stearic acid (18:0), and oleic acid (18:1).
3. In most unsaturated fatty acids, the cis isomer predominates; the trans isomer is rare.
4. Unsaturated fatty acids have lower melting points than their saturated counterparts. The greater the degree of unsaturation, the lower the melting point. Compare, for example, the melting points of linoleic acid, a **polyunsaturated fatty acid,** and stearic acid, a saturated fatty acid.

Polyunsaturated fatty acid A fatty acid with two or more carbon-carbon double bonds in its hydrocarbon chain.

Among the components of beeswax is triacontyl palmitate, $CH_3(CH_2)_{14}CO_2(CH_2)_{29}CH_3$, an ester of palmitic acid. *(Charles D. Winters)*

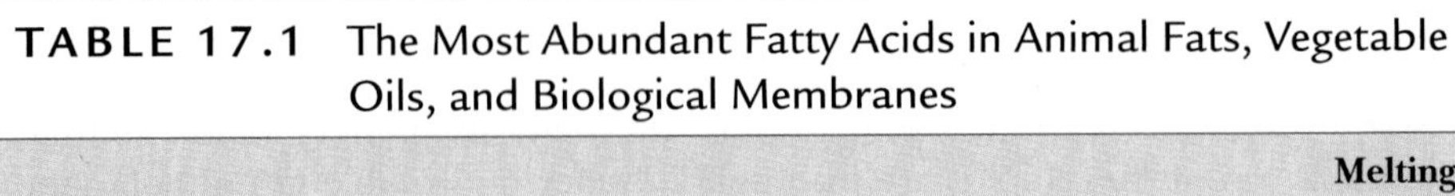

TABLE 17.1 The Most Abundant Fatty Acids in Animal Fats, Vegetable Oils, and Biological Membranes

Carbon Atoms/ Double Bonds*	Structure	Common Name	Melting Point (°C)
Saturated Fatty Acids			
12:0	$CH_3(CH_2)_{10}CO_2H$	lauric acid	44
14:0	$CH_3(CH_2)_{12}CO_2H$	myristic acid	58
16:0	$CH_3(CH_2)_{14}CO_2H$	palmitic acid	63
18:0	$CH_3(CH_2)_{16}CO_2H$	stearic acid	70
20:0	$CH_3(CH_2)_{18}CO_2H$	arachidic acid	77
Unsaturated Fatty Acids			
16:1	$CH_3(CH_2)_5CH{=}CH(CH_2)_7CO_2H$	palmitoleic acid	1
18:1	$CH_3(CH_2)_7CH{=}CH(CH_2)_7CO_2H$	oleic acid	16
18:2	$CH_3(CH_2)_4(CH{=}CHCH_2)_2(CH_2)_6CO_2H$	linoleic acid	−5
18:3	$CH_3(CH_2)_4(CH{=}CHCH_2)_3(CH_2)_6CO_2H$	linolenic acid	−11
20:4	$CH_3(CH_2)_4(CH{=}CHCH_2)_4(CH_2)_2CO_2H$	arachidonic acid	−49

* The first number is the number of carbons in the fatty acid; the second is the number of carbon-carbon double bonds in its hydrocarbon chain.

EXAMPLE 17.1

Draw the structural formula of a triglyceride derived from one molecule each of palmitic acid, oleic acid, and stearic acid, the three most abundant fatty acids in the biological world.

SOLUTION

In this structure, palmitic acid is esterified at carbon 1 of glycerol, oleic acid at carbon 2, and stearic acid at carbon 3.

$$\begin{array}{l} \qquad\qquad\qquad\qquad\qquad\quad CH_2OC(=O)(CH_2)_{14}CH_3 \quad \leftarrow \text{palmitate (16:0)} \\ \text{oleate (18:1)} \rightarrow CH_3(CH_2)_7CH{=}CH(CH_2)_7C(=O)OCH \\ \qquad\qquad\qquad\qquad\qquad\quad CH_2OC(=O)(CH_2)_{16}CH_3 \quad \leftarrow \text{stearate (18:0)} \end{array}$$

A triglyceride

Some vegetable oils. *(Charles D. Winters)*

Practice Problem 17.1

(a) How many constitutional isomers are possible for a triglyceride containing one molecule each of palmitic acid, oleic acid, and stearic acid?

(b) Which of these constitutional isomers are chiral?

TABLE 17.2 Grams of Fatty Acid per 100 g of Triglyceride of Several Fats and Oils*

	Saturated Fatty Acids			Unsaturated Fatty Acids	
Fat or Oil	**Lauric (12:0)**	**Palmitic (16:0)**	**Stearic (18:0)**	**Oleic (18:1)**	**Linoleic (18:2)**
human fat	—	24.0	8.4	46.9	10.2
beef fat	—	27.4	14.1	49.6	2.5
butter fat	2.5	29.0	9.2	26.7	3.6
coconut oil	45.4	10.5	2.3	7.5	trace
corn oil	—	10.2	3.0	49.6	34.3
olive oil	—	6.9	2.3	84.4	4.6
palm oil	—	40.1	5.5	42.7	10.3
peanut oil	—	8.3	3.1	56.0	26.0
soybean oil	0.2	9.8	2.4	28.9	50.7

* Only the most abundant fatty acids are given; other fatty acids are present in lesser amounts.

Figure 17.1 Tripalmitin, a saturated triglyceride. *(Brent Iverson, University of Texas)*

B. Physical Properties

The physical properties of a triglyceride depend on its fatty acid components. In general, the melting point of a triglyceride increases as the number of carbons in its hydrocarbon chains increases and as the number of carbon-carbon double bonds decreases. Triglycerides rich in oleic acid, linoleic acid, and other unsaturated fatty acids are generally liquids at room temperature and are called **oils,** as for example corn oil and olive oil. Triglycerides rich in palmitic, stearic, and other saturated fatty acids are generally semisolids or solids at room temperature and are called **fats,** as for example human fat and butter fat. Fats of land animals typically contain approximately 40–50% saturated fatty acids by weight (Table 17.2). Most plant oils, on the other hand, contain 20% or less saturated fatty acids and 80% or more unsaturated fatty acids. The notable exception to this generalization about plant oils are the **tropical oils** (as for example coconut and palm oils), which are considerably richer in low-molecular-weight saturated fatty acids.

Oil A triglyceride that is liquid at room temperature.

Fat A triglyceride that is semisolid or solid at room temperature.

The lower melting points of triglycerides rich in unsaturated fatty acids are related to differences in three-dimensional shape between the hydrocarbon chains of their unsaturated and saturated fatty acid components. Shown in Figure 17.1 is a space-filling model of tripalmitin, a saturated triglyceride. In this model, the hydrocarbon chains lie parallel to each other, giving the molecule an ordered, compact shape. Because of this compact three-dimensional shape and the resulting strength of the dispersion forces between hydrocarbon chains of adjacent molecules, triglycerides rich in saturated fatty acids have melting points above room temperature.

The three-dimensional shape of an unsaturated fatty acid is quite different from that of a saturated fatty acid. Recall from Section 17.1A that unsaturated fatty acids of higher organisms are predominantly of the cis configuration; trans configurations are rare. Figure 17.2 shows a space-filling model of a **polyunsaturated triglyceride** derived from one molecule each of stearic acid, oleic acid, and linoleic acid. Each double bond in this polyunsaturated triglyceride has the cis configuration.

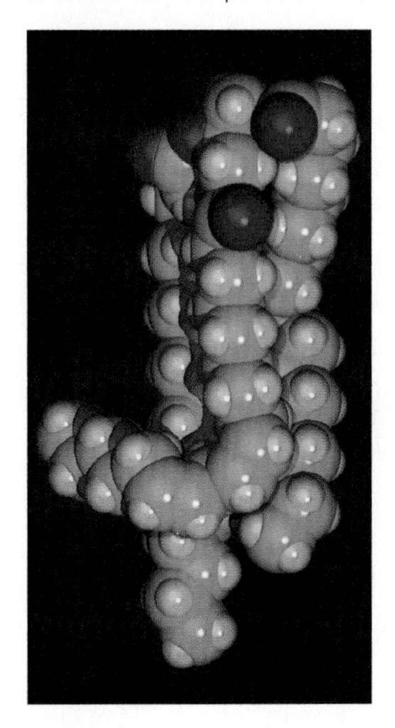

Figure 17.2 A polyunsaturated triglyceride. *(Brent Iverson, University of Texas)*

Polyunsaturated triglyceride A triglyceride having several carbon-carbon double bonds in the hydrocarbon chains of its three fatty acids.

Polyunsaturated triglycerides have a less ordered structure and do not pack together so closely or so compactly as saturated triglycerides. Intramolecular and intermolecular dispersion forces are weaker, with the result that polyunsaturated triglycerides have lower melting points than their saturated counterparts.

C. Reduction of Fatty Acid Chains

The hydrogenation of an oil to a fat is called hardening, because the liquid oil is converted to a semi-solid or solid fat. *(Charles D. Winters)*

For a variety of reasons, in part convenience and in part dietary preference, conversion of oils to fats has become a major industry. The process is called **hardening** of oils and involves catalytic reduction (Section 6.5A) of some or all the carbon-carbon double bonds. In practice, the degree of hardening is carefully controlled to produce fats of a desired consistency. The resulting fats are sold for kitchen use (Crisco, Spry, and others). Margarine and other butter substitutes are produced by partial hydrogenation of polyunsaturated oils derived from corn, cottonseed, peanut, and soybean oils. To the hardened oils are added β-carotene to give the final product a yellow color and make it look like butter, salt, and about 15% milk by volume to form the final emulsion. Also often added are vitamins A and D. Because the product to this stage is tasteless, acetoin (*The Merck Index,* 12th ed., #61) and diacetyl (*The Merck Index,* 12th ed., #3010) are often added. These two compounds mimic the characteristic flavor of butter.

$$CH_3-CH(OH)-C(=O)-CH_3 \qquad CH_3-C(=O)-C(=O)-CH_3$$

3-Hydroxy-2-butanone (Acetoin) 2,3-Butanedione (Diacetyl)

17.2 Soaps and Detergents

A. Structure and Preparation of Soaps

Soap A sodium or potassium salt of a fatty acid.

Natural soaps are prepared most commonly from a blend of tallow and coconut oils. In the preparation of tallow, the solid fats of cattle are melted with steam, and the tallow layer formed on the top is removed. The preparation of soaps begins by boiling these triglycerides with sodium hydroxide. The reaction that takes place is called **saponification** (Latin: *saponem,* soap). At the molecular level, saponification corresponds to base-promoted hydrolysis of the ester groups in triglycerides (Section 13.3C). The resulting soaps contain mainly the sodium salts of palmitic, stearic, and oleic acids from tallow and the sodium salts of lauric and myristic acids from coconut oil.

$$\begin{array}{l} CH_2OC(=O)R \\ RC(=O)OCH \\ CH_2OC(=O)R \end{array} + 3NaOH \xrightarrow{\text{saponification}} \begin{array}{l} CH_2OH \\ CHOH \\ CH_2OH \end{array} + 3RC(=O)O^-Na^+$$

A triglyceride 1,2,3-Propanetriol (Glycerol; glycerin) Sodium soaps

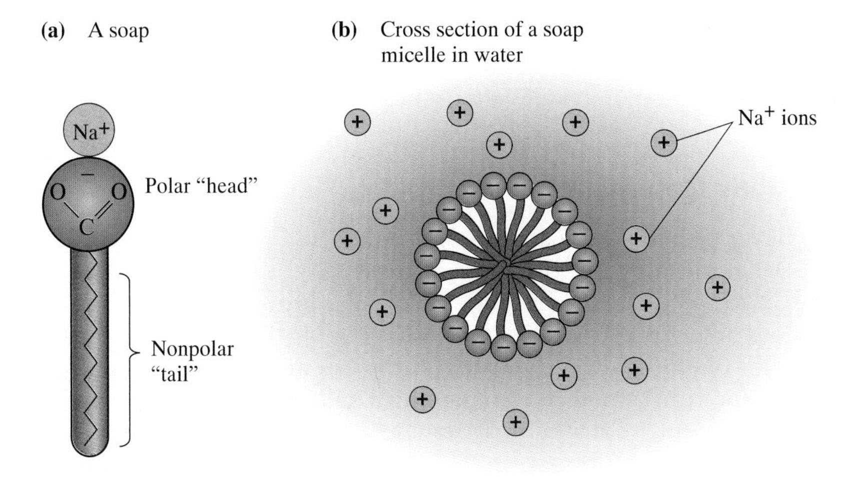

Figure 17.3
Soap micelles. Nonpolar (hydrophobic) hydrocarbon chains are clustered in the interior of the micelle and polar (hydrophilic) carboxylate groups are on the surface of the micelle. Soap micelles repel each other because of their negative surface charges.

After hydrolysis is complete, sodium chloride is added to precipitate the soap as thick curds. The water layer is then drawn off, and glycerol is recovered by vacuum distillation. The crude soap contains sodium chloride, sodium hydroxide, and other impurities. These are removed by boiling the curd in water and reprecipitating with more sodium chloride. After several purifications, the soap can be used without further processing as an inexpensive industrial soap. Other treatments transform the crude soap into pH-controlled cosmetic soaps, medicated soaps, and the like.

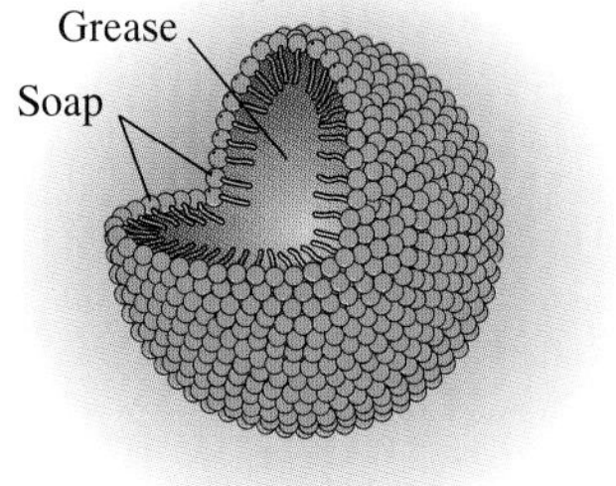

Soap micelle with "dissolved" grease

Grease

Soap

Figure 17.4
A soap micelle with a "dissolved" oil or grease droplet.

B. How Soaps Clean

Soap owes its remarkable cleansing properties to its ability to act as an emulsifying agent. Because the long hydrocarbon chains of natural soaps are insoluble in water, they tend to cluster in such a way as to minimize their contact with surrounding water molecules. The polar carboxylate groups, on the other hand, tend to remain in contact with the surrounding water molecules. Thus, in water, soap molecules spontaneously cluster into **micelles** (Figure 17.3).

Micelle A spherical arrangement of organic molecules in water solution clustered so that their hydrophobic parts are buried inside the sphere and their hydrophilic parts are on the surface of the sphere and in contact with water.

Most of the things we commonly think of as dirt (such as grease, oil, and fat stains) are nonpolar and insoluble in water. When soap and this type of dirt are mixed together, as in a washing machine, the nonpolar hydrocarbon inner parts of the soap micelles "dissolve" the nonpolar dirt molecules. In effect, new soap micelles are formed, this time with nonpolar dirt molecules in their centers (Figure 17.4). In this way, nonpolar organic grease, oil, and fat are "dissolved" and washed away in the polar wash water.

Soaps are not without their disadvantages. Foremost among these, they form insoluble salts when used in water containing Ca(II), Mg(II), or Fe(III) ions (hard water).

$$2CH_3(CH_2)_{14}CO_2^-Na^+ + Ca^{2+} \longrightarrow [CH_3(CH_2)_{14}CO_2^-]_2Ca^{2+} + 2Na^+$$

A sodium soap (soluble in water as micelles) — Calcium salt of a fatty acid (insoluble in water)

These calcium, magnesium, and iron salts of fatty acids create problems, including rings around the bathtub, films that spoil the luster of hair, and grayness and roughness that build up on textiles after repeated washings.

C. Synthetic Detergents

After the cleansing action of soaps was understood, a synthetic detergent could be designed. Molecules of a good detergent must have a long hydrocarbon chain, preferably 12 to 20 carbon atoms long, and a polar group at one end of the molecule that does not form insoluble salts with Ca(II), Mg(II), or Fe(III) ions present in hard water. Chemists recognized that these essential characteristics of a soap could be produced in a molecule containing a sulfate or sulfonate group instead of a carboxylate group. Calcium, magnesium, and iron salts of monoalkylsulfuric and sulfonic acids are much more soluble in water than comparable salts of fatty acids.

The most widely used synthetic detergents are the linear alkylbenzenesulfonates (LAS). One of the most common of these is sodium 4-dodecylbenzenesulfonate. To prepare this type of detergent, a linear alkylbenzene is treated with sulfuric acid (Section 9.7B) to form an alkylbenzenesulfonic acid. The sulfonic acid is then neutralized with NaOH, the product is mixed with builders, and spray-dried to give a smooth flowing powder. The most common builder is sodium silicate.

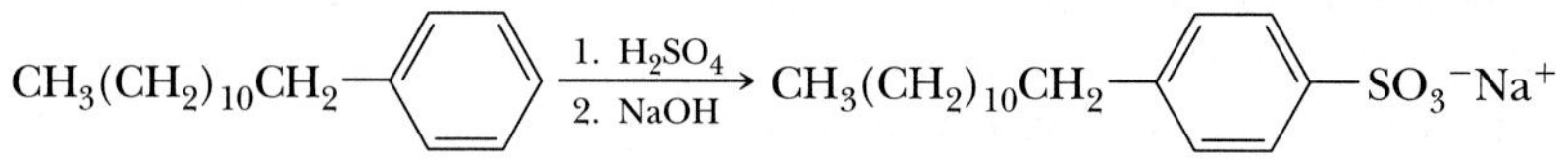

$$CH_3(CH_2)_{10}CH_2-C_6H_5 \xrightarrow[2.\ NaOH]{1.\ H_2SO_4} CH_3(CH_2)_{10}CH_2-C_6H_4-SO_3^-Na^+$$

Dodecylbenzene

Sodium 4-dodecylbenzenesulfonate (SDS)
(an anionic detergent)

Alkylbenzenesulfonate detergents were introduced in the late 1950s, and today they command close to 90% of the market once held by natural soaps.

Among the most common additives to detergent preparations are foam stabilizers, bleaches, and optical brighteners. A common foam stabilizer added to liquid soaps but not laundry detergents (for obvious reasons: think of a top-loading washing machine with foam spewing out the lid!) is the amide prepared from dodecanoic acid (lauric acid) and 2-aminoethanol (ethanolamine). The most common bleach is sodium perborate tetrahydrate (*The Merck Index,* 12th ed., #8797), which decomposes at temperatures above 50°C to give hydrogen peroxide, the actual bleaching agent.

$$CH_3(CH_2)_{10}\overset{O}{\overset{\|}{C}}NHCH_2CH_2OH$$

N-(2-Hydroxyethyl)dodecanamide
(a foam stabilizer)

$$O{=}B{-}O{-}O^-Na^+ \cdot 4H_2O$$

Sodium perborate tetrahydrate
(a bleach)

Also added to laundry detergents are optical brighteners, known also as optical bleaches, that are absorbed into fabrics and, after absorbing ambient light, fluoresce with a blue color, offsetting the yellow color caused by fabric aging. Quite literally, these optical brighteners produce a "whiter-than-white" appearance. You most certainly have observed the effects of optical brighteners if you have seen the glow of "white" T-shirts or blouses when exposed to black light (UV radiation).

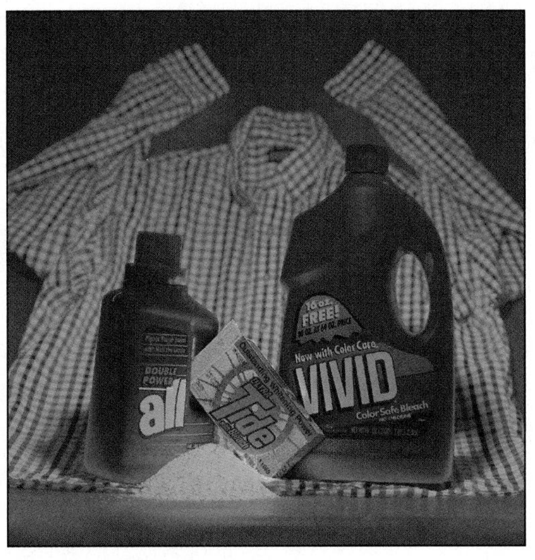

Effects of optical bleaches: *(left)* ordinary light; *(right)* black light. *(Charles D. Winters)*

17.3 PROSTAGLANDINS

The **prostaglandins** are a family of compounds all having the 20-carbon skeleton of prostanoic acid.

Prostaglandin A member of the family of compounds having the 20-carbon skeleton of prostanoic acid.

Prostanoic acid

The story of the discovery and structure determination of these remarkable compounds began in 1930 when gynecologists Raphael Kurzrok and Charles Lieb reported that human seminal fluid stimulates contraction of isolated uterine muscle. A few years later in Sweden, Ulf von Euler confirmed this report and noted that human seminal fluid also produces contraction of intestinal smooth muscle and lowers blood pressure when injected into the bloodstream. Von Euler proposed the name *prostaglandin* for the mysterious substance(s) responsible for these diverse effects because it was believed at the time that they were synthesized in the prostate gland. Although we now know that prostaglandin production is by no means limited to the prostate gland, the name nevertheless has stuck.

Prostaglandins are not stored as such in target tissues. Rather, they are synthesized in response to specific physiological triggers. Starting materials for the biosynthesis of prostaglandins are polyunsaturated fatty acids of 20 carbon atoms, stored until needed as membrane phospholipid esters. In response to a physiological trigger, the ester is hydrolyzed, the fatty acid is released, and the synthesis of prostaglandins is initiated. Figure 17.5 outlines the steps in the synthesis of several prostaglandins from arachidonic acid. A key step in this biosynthesis is the enzyme-

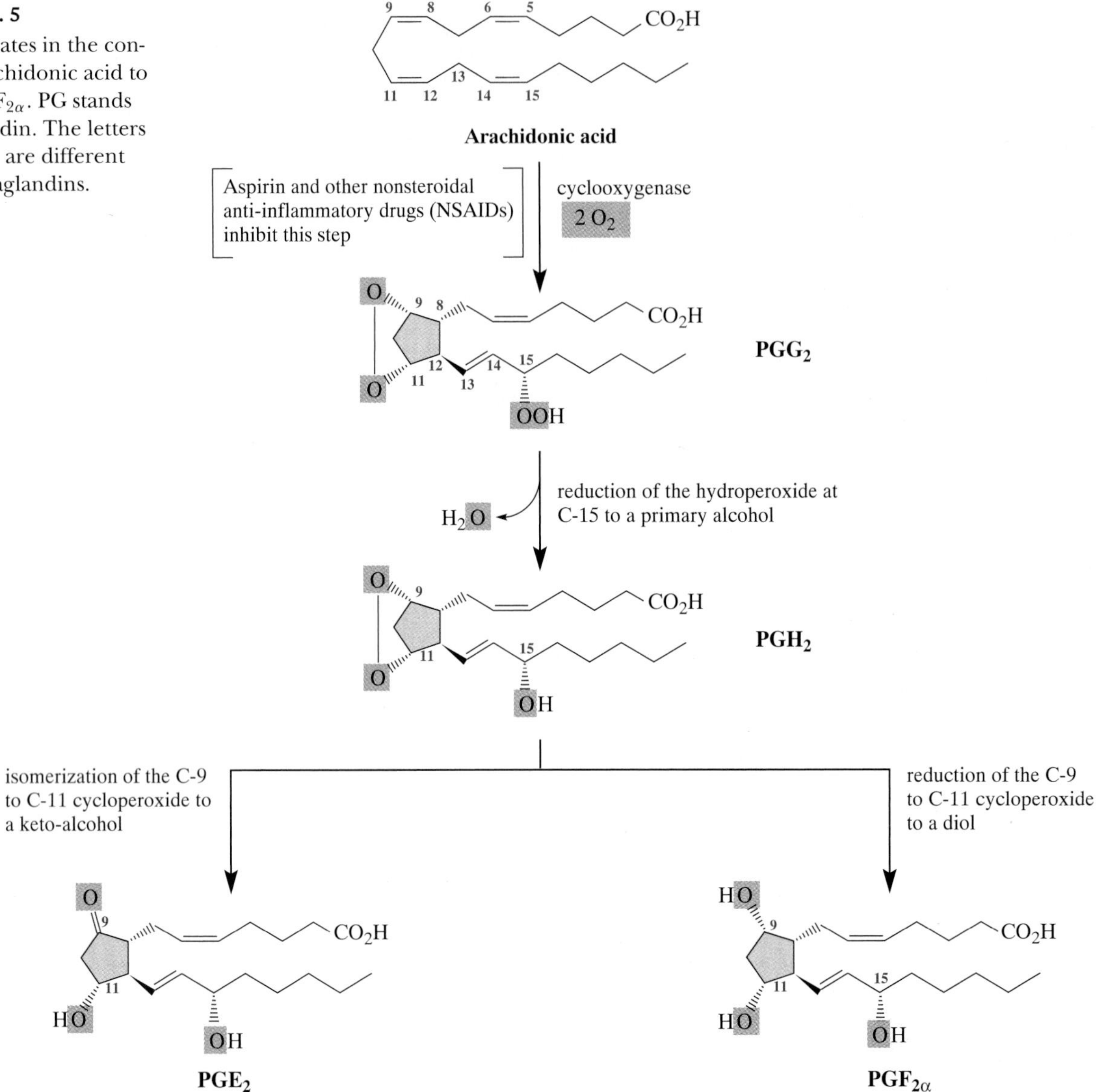

Figure 17.5
Key intermediates in the conversion of arachidonic acid to PGE$_2$ and PGF$_{2\alpha}$. PG stands for prostaglandin. The letters E, F, G, and H are different types of prostaglandins.

catalyzed reaction of arachidonic acid with two molecules of O_2 to form prostaglandin G_2(PGG$_2$). The anti-inflammatory effect of aspirin and other nonsteroidal anti-inflammatory drugs (NSAIDs) results from their ability to inhibit the enzyme that catalyzes this step.

Research on the involvement of prostaglandins in reproductive physiology and the inflammatory process has produced several clinically useful prostaglandin derivatives. The observations that PGF$_{2\alpha}$ stimulates contractions of uterine smooth muscle led to a synthetic derivative that is used as a therapeutic abortifacient. A problem with the use of the natural prostaglandins for this purpose is that they are very rapidly degraded within the body. In the search for less rapidly degraded prostaglandins, a number of analogs have been prepared, one of the most effective of which is carboprost (*The Merck Index,* 12th ed., #1871). This synthetic prostaglandin is 10 to 20 times more potent than the natural $PGF_{2\alpha}$ and is only slowly degraded in the body. The comparison of these two prostaglandins illustrates how a simple change in structure of a drug can make a significant change in its effectiveness.

extra methyl group at carbon-15

$PGF_{2\alpha}$

Carboprost
(15S)-15-Methyl-$PGF_{2\alpha}$

The PGEs along with several other PGs suppress gastric ulceration and appear to heal gastric ulcers. The PGE_1 analog, misoprostol (see *The Merck Index,* 12th ed., #6297), is currently used primarily for prevention of ulceration associated with aspirinlike NSAIDs.

PGE_1

Misoprostol

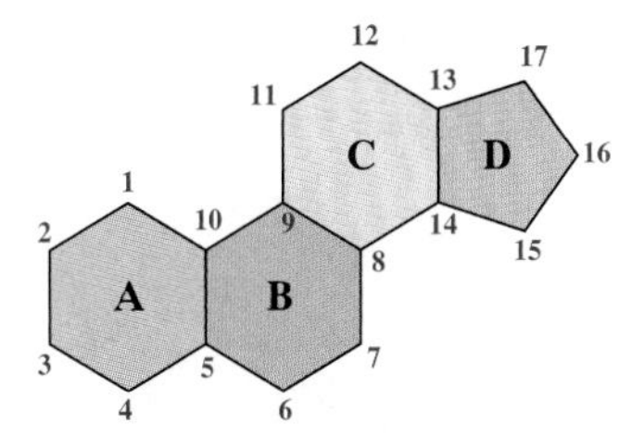

Figure 17.6
The tetracyclic ring system characteristic of steroids.

17.4 STEROIDS

Steroids are a group of plant and animal lipids that have the tetracyclic ring system shown in Figure 17.6. The features common to the tetracyclic ring system of most naturally occurring steroids are illustrated in Figure 17.7.

Steroid A plant or animal lipid having the characteristic tetracyclic ring structure of the steroid nucleus, namely three six-membered rings and one five-membered ring.

1. The fusion of rings is trans, and each atom or group at a ring junction is axial. Compare, for example, the orientations of —H at C-5 and $—CH_3$ at C-10.

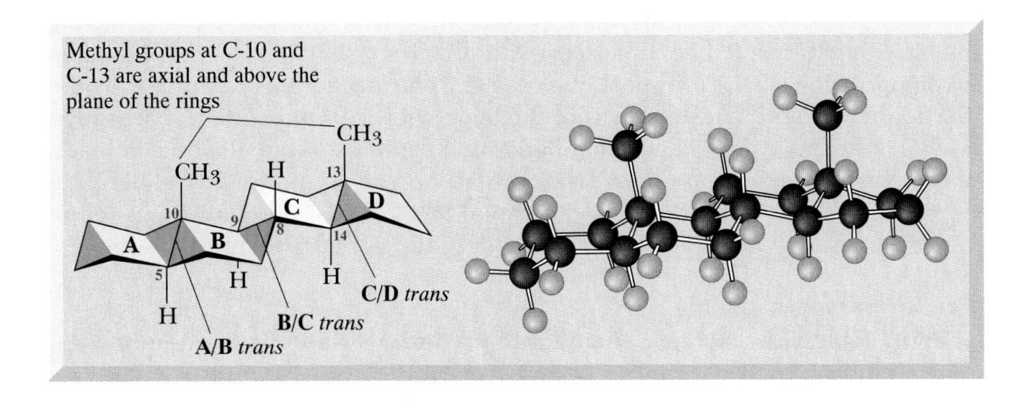

Figure 17.7
Features common to the tetracyclic ring system of many steroids.

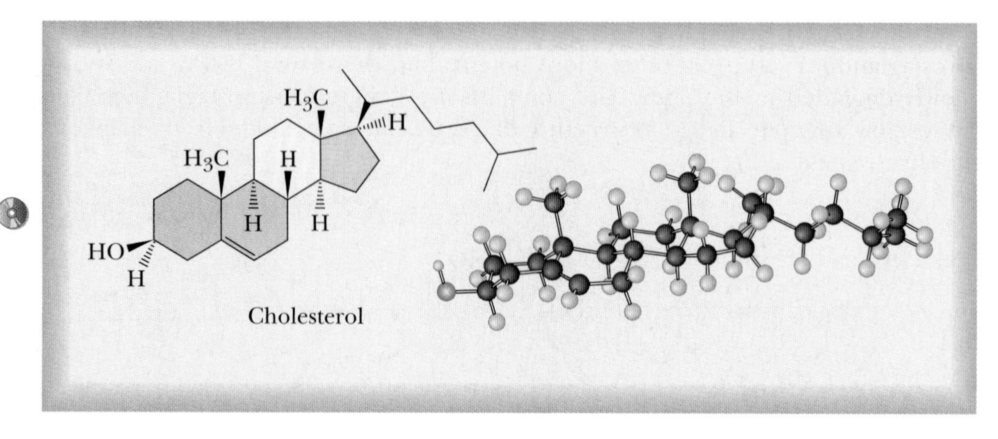

Figure 17.8
Cholesterol is found in blood plasma and in all animal tissues.

2. The pattern of atoms or groups along the points of ring fusion (carbons 5 to 10 to 9 to 8 to 14 to 13) is nearly always trans-anti-trans-anti-trans.
3. Because of the trans-anti-trans-anti-trans arrangement of atoms or groups along the points of ring fusion, the tetracyclic steroid ring system is nearly flat and quite rigid.
4. Many steroids have axial methyl groups at C-10 and C-13 of the tetracyclic ring system.

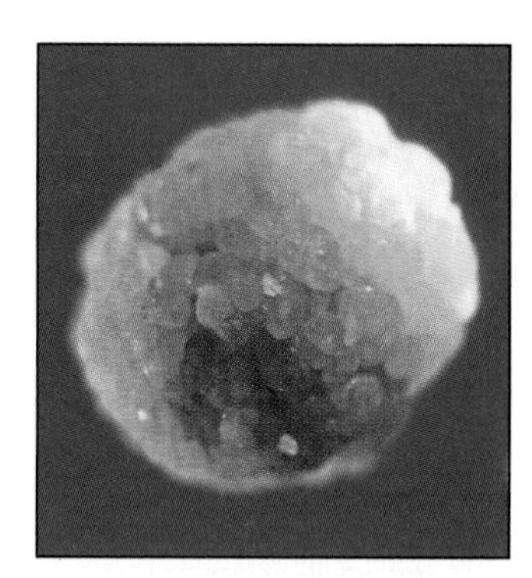

Human gallstones are almost pure cholesterol; this gallstone is about 0.5 cm in diameter. *(© Carolina Biological Supply Company, Phototake, NYC)*

A. Structure of the Major Classes of Steroids

Cholesterol

Cholesterol is a white, water-insoluble, waxy solid found in blood plasma and in all animal tissues. This substance is an integral part of human metabolism in two ways: (1) It is an essential component of biological membranes. The body of a healthy adult contains approximately 140 g of cholesterol, about 120 g of which are present in membranes. Membranes of the central and peripheral nervous systems, for example, contain about 10% cholesterol by weight. (2) It is the compound from which sex hormones, adrenocorticoid hormones, bile acids, and vitamin D are synthesized. Thus, cholesterol is, in a sense, the parent steroid.

Cholesterol has eight stereocenters, and a molecule with this structural feature can exist as 2^8, or 256, stereoisomers (128 pairs of enantiomers). Only one of these stereoisomers is known to exist in nature: the stereoisomer with the configuration shown in Figure 17.8.

Cholesterol is insoluble in blood plasma but can be transported as a plasma-soluble complex formed by cholesterol with proteins called lipoproteins. **Low-density lipoproteins (LDL)** transport cholesterol from the site of its synthesis in the liver to the various tissues and cells of the body where it is to be used. It is primarily cholesterol attached to LDLs that builds up in atherosclerotic deposits in blood vessels. **High-density lipoproteins (HDL)** transport excess and unused cholesterol from cells back to the liver for its degradation to bile acids and eventual excretion in the feces. It is thought that HDLs retard or reduce atherosclerotic deposits.

Low-density lipoprotein (LDL) Plasma particles, density 1.02–1.06 g/mL, consisting of approximately 25% proteins, 50% cholesterol, 21% phospholipids, and 4% triglycerides.

High-density lipoprotein (HDL) Plasma particles, density 1.06–1.21 g/mL, consisting of approximately 33% proteins, 30% cholesterol, 29% phospholipids, and 8% triglycerides.

Steroid Hormones

Given in Table 17.3 are representations of each major class of steroid hormones, along with the principal functions of each.

TABLE 17.3 Selected Steroid Hormones

Structure	Source and Major Effects
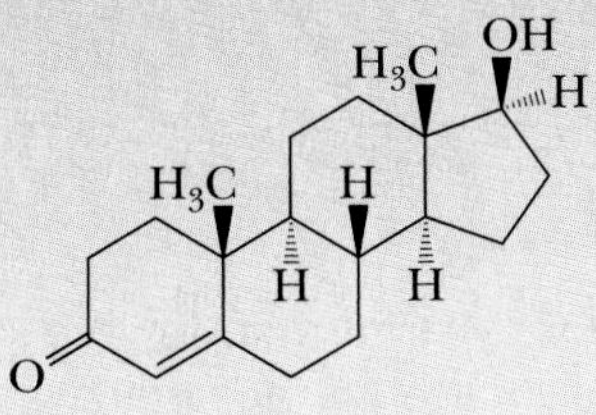 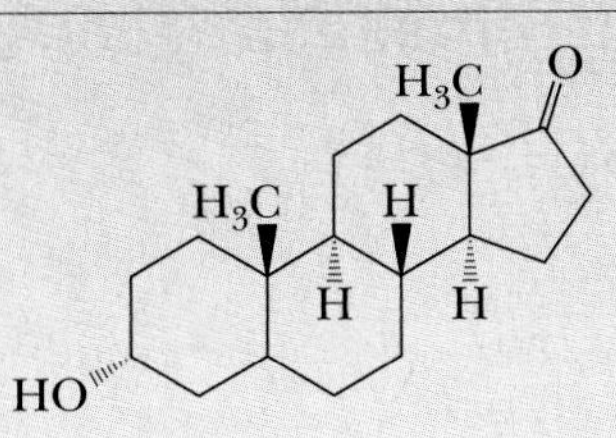 Testosterone; Androsterone	androgens (male sex hormones)—synthesized in the testes; responsible for development of male secondary sex characteristics
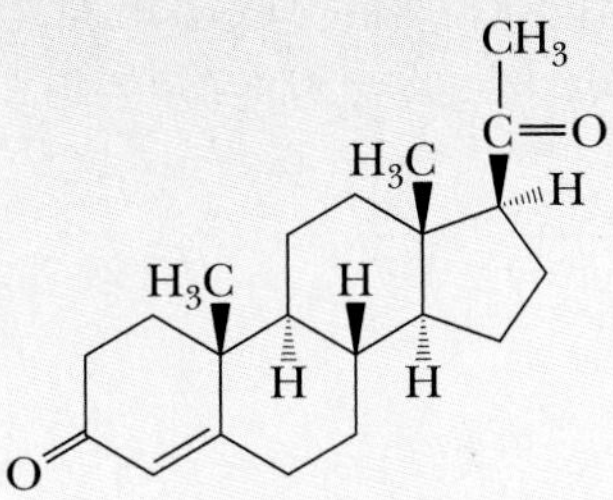 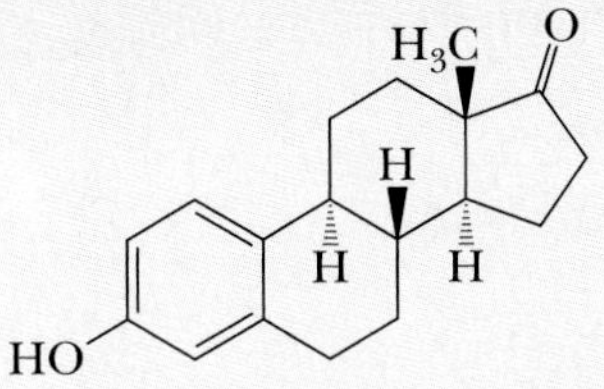 Progesterone; Estrone	estrogens (female sex hormones)—synthesized in the ovaries; responsible for development of female secondary sex characteristics and control of the menstrual cycle
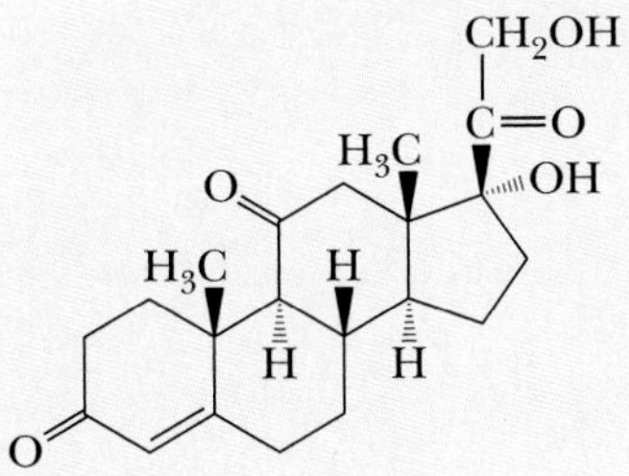 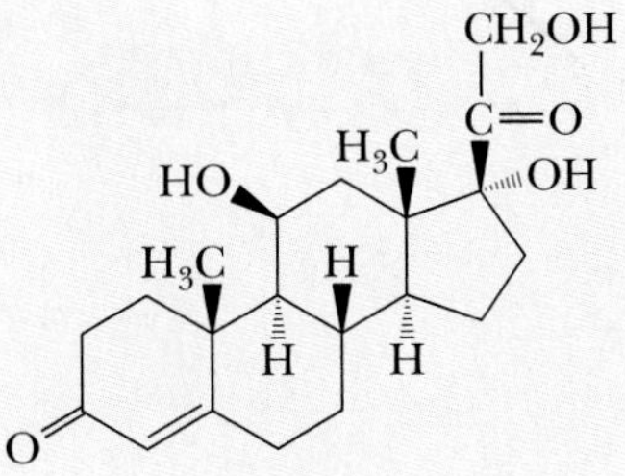 Cortisone; Cortisol	glucocorticoid hormones—synthesized in the adrenal cortex; regulate metabolism of carbohydrates, decrease inflammation, and involved in the reaction to stress
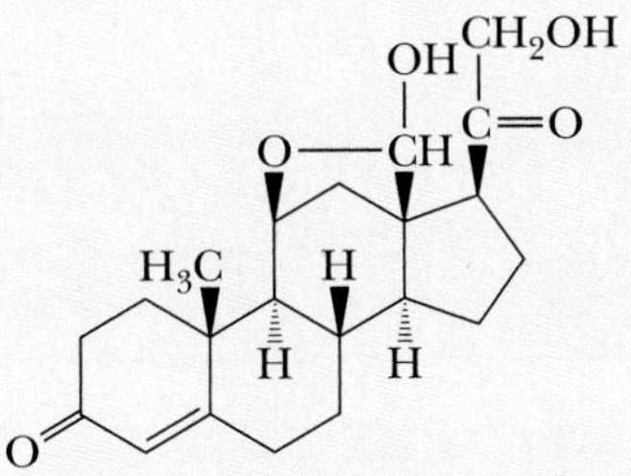 Aldosterone	a mineralocorticoid hormone—synthesized in the adrenal cortex; regulates blood pressure and volume by stimulating the kidneys to absorb Na^+, Cl^-, and HCO_3^-

After the role of **progesterone** in inhibiting ovulation was understood, its potential as a possible contraceptive was realized. Progesterone itself is relatively ineffective when taken orally. As a result of a massive research program in both industrial and academic laboratories, many synthetic progesterone-mimicking steroids became available in the 1960s. (See "A Conversation with Carl Djerassi.") When taken regularly, these drugs prevent ovulation yet allow women to maintain a normal menstrual cycle. Some of the most effective of these preparations contain a

Estrogen A steroid hormone, such as estradiol, which mediates the development and sexual characteristics in females.

CHEMICAL CONNECTIONS

Nonsteroidal Estrogen Antagonists

Estrogens are female sex hormones, the most important of which are estrone, estradiol, and estriol. Of these three, β-estradiol is the most potent. (*Note:* As per convention in steroid nomenclature, the designation beta means toward the reader, on the top side as the rings are viewed here; alpha means away from the reader, on the bottom side.) Notice from the cylindrical bond model of β-estriol that, for the most part, these molecules are flat, with the angular methyl groups projecting vertical to the plane of the molecule. Also notice that ring C is puckered into a chair conformation, and ring D is puckered into an envelope conformation.

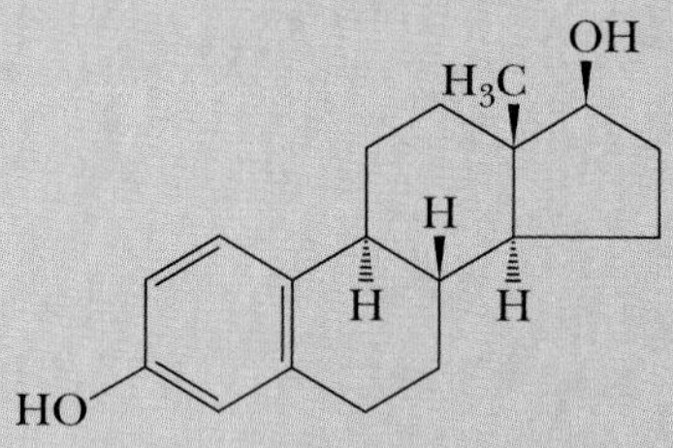

OH H$_3$C H H H HO

β-Estradiol
(*The Merk Index,* 12th ed., #3746)

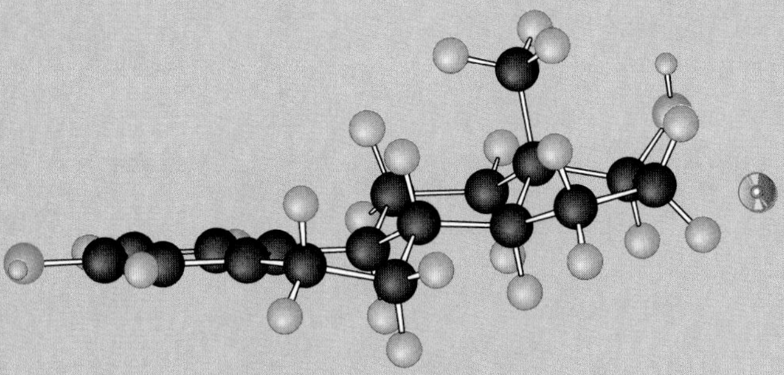

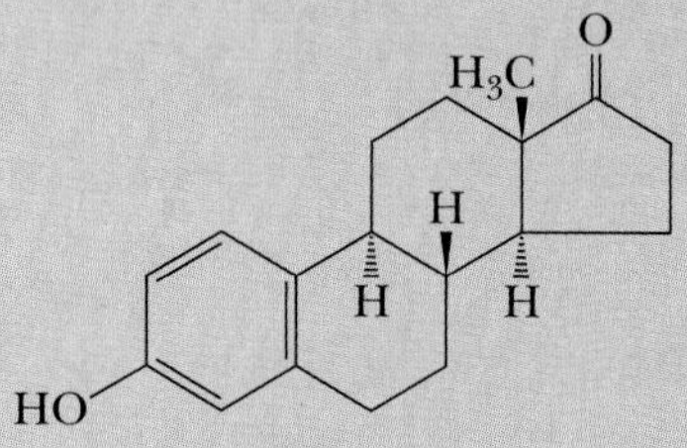

O H$_3$C H H H HO

Estrone
(*The Merk Index,* 12th ed., #3751)

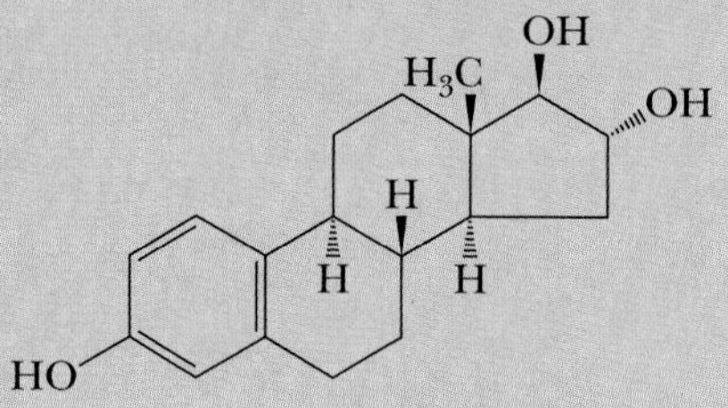

OH H$_3$C OH H H H HO

Estriol
(*The Merk Index,* 12th ed., #3750)

progesterone analog, such as norethindrone (*The Merck Index,* 12th ed., #6790), combined with a smaller amount of an estrogenlike material to help prevent irregular menstrual flow during prolonged use of contraceptive pills.

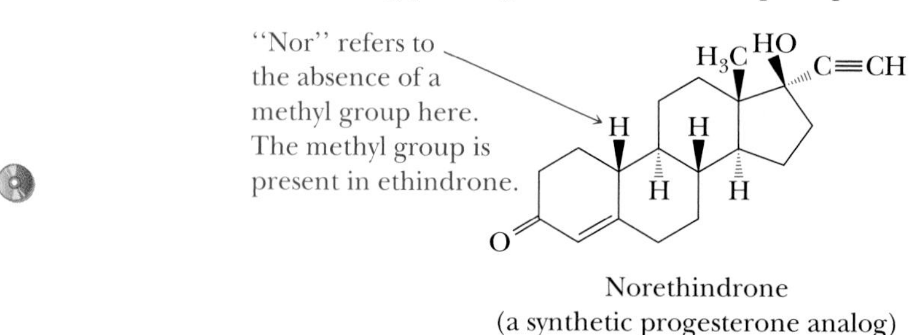
"Nor" refers to the absence of a methyl group here. The methyl group is present in ethindrone.
H$_3$C
HO
C≡CH
H
H
H
H
O

Norethindrone
(a synthetic progesterone analog)

As soon as these compounds were isolated in the early 1930s and their pharmacology studied, it became clear that they are extremely potent. In recent years, there have been intense efforts to design and synthesize molecules that would bind to the estrogen receptor. One target of this research has been nonsteroidal estrogen antagonists, compounds that would interact with the estrogen receptor as antagonists, that is, compounds that would block the effect of endogenous or exogenous estrogens. A feature common to many of those developed is the presence of a 1,2-diphenylethylene with one of the benzene rings bearing a dialkylaminoethoxyl substituent. The first nonsteroidal estrogen antagonist of this type to achieve clinical importance was tamoxifen (*The Merck Index,* 12th ed., #9216), now an important drug in the prevention and treatment of breast cancer.

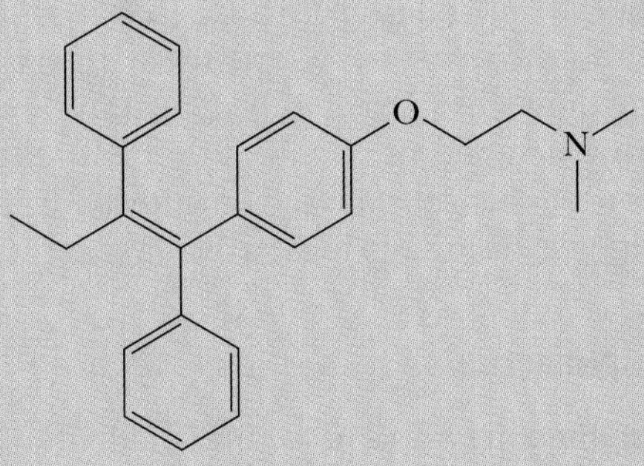

O N

Tamoxifen

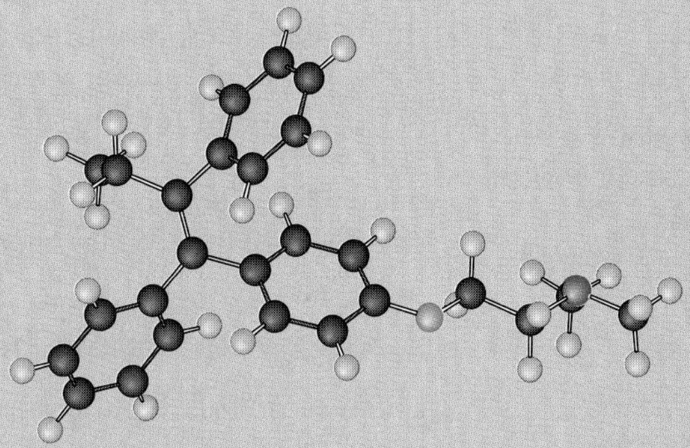

The chief function of testosterone and other **androgens** is to promote normal growth of male reproductive organs (primary sex characteristics) and development of the characteristic deep voice, pattern of body and facial hair, and musculature (secondary sex characteristics). Although testosterone produces these effects, it is not active when taken orally because it is metabolized in the liver to an inactive steroid. A number of oral **anabolic steroids** have been developed for use in rehabilitation medicine, particularly when muscle atrophy occurs during recovery from an injury. Among the synthetic anabolic steroids most widely prescribed for this purpose are methandrostenolone and stanozolol (*The Merck Index,* 12th ed., #6018 and #8951). The structural formula of methandrostenolone differs from that of testosterone by introduction of (1) a methyl group at C-17, and (2) an ad-

Androgen A steroid hormone, such as testosterone, which mediates the development and sexual characteristics of males.

Anabolic steroid A steroid hormone, such as testosterone, which promotes tissue and muscle growth and development.

ditional carbon-carbon double bond between C-1 and C-2. In stanozolol, ring A is modified by attachment of a pyrazole ring.

Methandrostenolone

Stanozolol

Among certain athletes, the misuse of anabolic steroids to build muscle mass and strength, particularly for sports that require explosive action, is common. The risks associated with abuse of anabolic steroids for this purpose are enormous: heightened aggressiveness, sterility, impotence, and risk of premature death from complications of diabetes, coronary artery disease, and liver cancer.

Bile Acids

Shown in Figure 17.9 is a structural formula for cholic acid, a constituent of human bile. The molecule is shown as an anion, as it is ionized in bile and intestinal fluids. **Bile acids,** or more properly bile salts, are synthesized in the liver, stored in the gallbladder, and secreted into the intestine, where their function is to emulsify dietary fats and thereby aid in their absorption and digestion. Furthermore, bile salts are the end products of the metabolism of cholesterol and, thus, are a principal pathway for the elimination of this substance from the body. A characteristic structural feature of bile salts is a cis fusion of rings A/B.

Bile acid A cholesterol-derived detergent molecule, such as cholic acid, which is secreted by the gallbladder into the intestine to assist in the absorption of dietary lipids.

B. Biosynthesis of Cholesterol

The biosynthesis of cholesterol illustrates a point we first made in our introduction to the structure of terpenes (Section 5.4): in building large molecules, one of the common patterns in the biological world is to begin with one or more smaller subunits, join them by an iterative process, and then chemically modify the completed carbon skeleton by oxidation, reduction, cross-linking, addition, elimination, or related processes to give a biomolecule with a unique identity.

The building block from which all carbon atoms of steroids are derived is the two-carbon acetyl group of acetyl-CoA (Additional Problem 13.27). The American

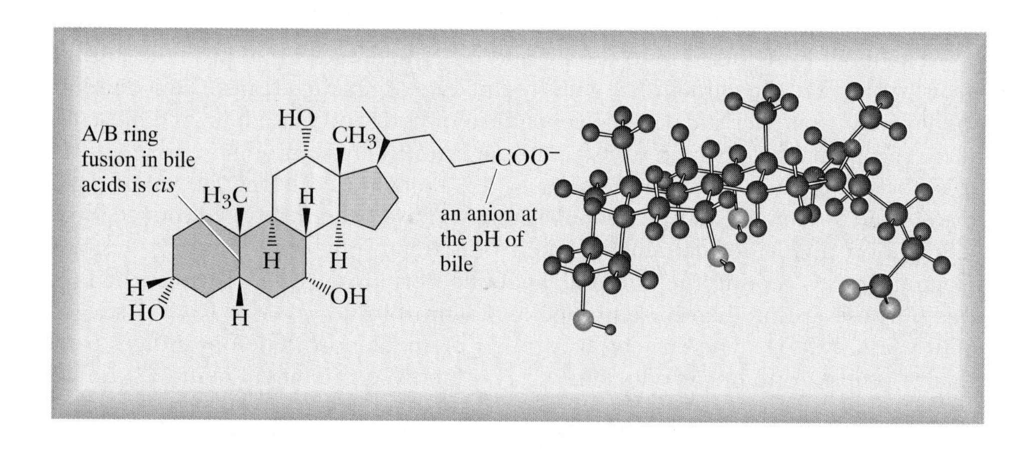

Figure 17.9 Cholic acid, an important constituent of human bile. Each six-membered ring is in a chair conformation.

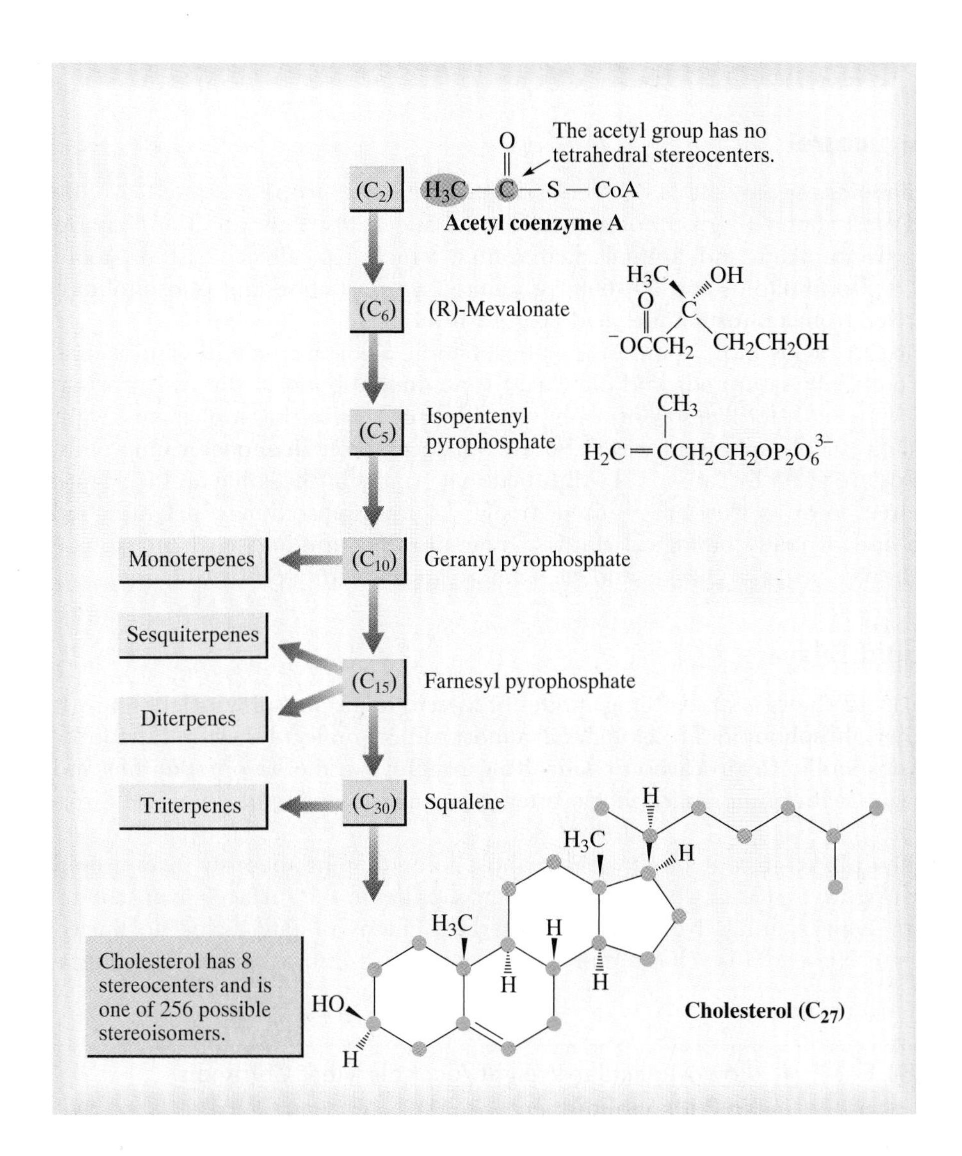

Figure 17.10 Several key intermediates in the synthesis of cholesterol from acetyl groups of acetyl CoA. Eighteen moles of acetyl CoA are required for the synthesis of 1 mole of cholesterol.

biochemist, Konrad Bloch, who shared the 1964 Nobel Prize for Medicine and Physiology with German biochemist Feodor Lynen for their discoveries concerning the biosynthesis of cholesterol and fatty acids, showed that 15 of the 27 carbon atoms of cholesterol are derived from the methyl group of acetyl-CoA; the remaining 12 carbon atoms are derived from the carbonyl group of acetyl-CoA (Figure 17.10).

A remarkable feature of this synthetic pathway is that the biosynthesis of cholesterol from acetyl-CoA is completely stereoselective; it is synthesized as only one of 256 possible stereoisomers. We cannot duplicate this exquisite degree of stereoselectivity in the laboratory. Cholesterol is, in turn, the key intermediate in the synthesis of all other steroids.

cholesterol →
- bile acids (e.g., cholic acid)
- sex hormones (e.g., testosterone and estrone)
- mineralocorticoid hormones (e.g., aldosterone)
- glucocorticoid hormones (e.g., cortisone)

17.5 PHOSPHOLIPIDS

A. Structure

Phospholipid A lipid containing glycerol esterified with two molecules of fatty acid and one molecule of phosphoric acid.

Phospholipids, or phosphoacylglycerols as they are more properly named, are the second most abundant group of naturally occurring lipids. They are found almost exclusively in plant and animal membranes, which typically consist of about 40–50% phospholipids and 50–60% proteins. The most abundant phospholipids are derived from a phosphatidic acid (Figure 17.11).

The fatty acids most common in phosphatidic acids are palmitic and stearic acids (both fully saturated) and oleic acid (one double bond in the hydrocarbon chain). Further esterification of a phosphatidic acid with a low-molecular-weight alcohol gives a phospholipid. Several of the most common alcohols forming phospholipids are given in Table 17.4. All functional groups in this table and in Figure 17.11 are shown as they are ionized at pH 7.4, the approximate pH of blood plasma and of many biological fluids. Under these conditions, each phosphate group bears a negative charge, and each amino group bears a positive charge.

B. Lipid Bilayers

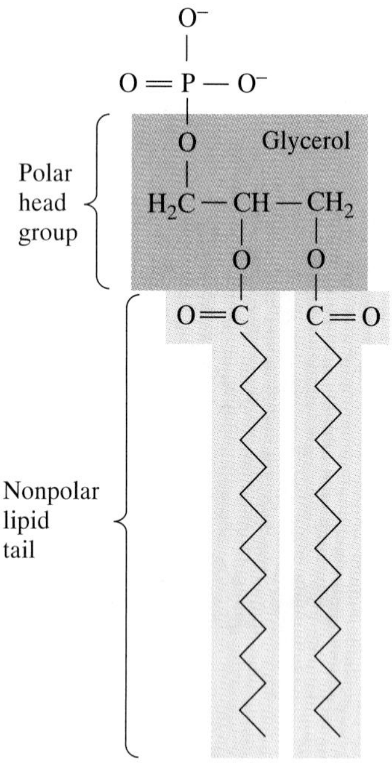

Figure 17.12 shows a space-filling model of a lecithin (a phosphatidylcholine). It and other phospholipids are elongated, almost rodlike molecules, with the nonpolar (hydrophobic) hydrocarbon chains lying roughly parallel to one another and the polar (hydrophilic) phosphoric ester group pointing in the opposite direction.

When placed in aqueous solution, phospholipids spontaneously form a **lipid bilayer** (Figure 17.13) in which polar head groups lie on the surface, giving the bilayer an ionic coating. Nonpolar hydrocarbon chains of fatty acids lie buried within the bilayer. This self-assembly of phospholipids into a bilayer is a sponta-

O^-
$O=P-O^-$
O Glycerol
$H_2C-CH-CH_2$
O O
$O=C$ $C=O$

Polar head group

Nonpolar lipid tail

Figure 17.11
In a phosphatidic acid, glycerol is esterified with two molecules of fatty acid and one molecule of phosphoric acid. Further esterification of the phosphoric acid group with a low-molecular-weight alcohol gives a phospholipid.

TABLE 17.4 Low-Molecular-Weight Alcohols Most Common to Phospholipids

Alcohols Found in Phospholipids		
Structural Formula	**Name**	**Name of Phospholipid**
$\mathbf{HO}CH_2CH_2NH_2$	ethanolamine	phosphatidylethanolamine (cephalin)
$\mathbf{HO}CH_2CH_2\overset{+}{N}(CH_3)_3$	choline	phosphatidylcholine (lecithin)
$\mathbf{HO}CH_2CH(NH_3^+)CO_2^-$	serine	phosphatidylserine
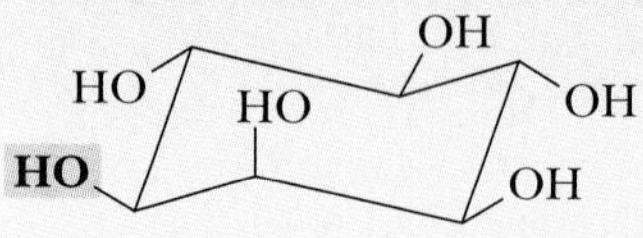 (cyclohexane ring with HO, HO, OH, OH, OH; **HO**)	inositol	phosphatidylinositol

neous process, driven by two types of noncovalent forces: (1) hydrophobic effects, which result when nonpolar hydrocarbon chains cluster together and exclude water molecules, and (2) electrostatic interactions, which result when polar head groups interact with water and other polar molecules in the aqueous environment.

Recall from Section 17.2B that formation of soap micelles is driven by these same noncovalent forces; the polar (hydrophilic) carboxylate groups of soap molecules lie on the surface of the micelle and associate with water molecules, and the nonpolar (hydrophobic) hydrocarbon chains cluster within the micelle and thus are removed from contact with water.

The arrangement of hydrocarbon chains in the interior of a phospholipid bilayer varies from rigid to fluid, depending on the degree of unsaturation of the hydrocarbon chains themselves. Saturated hydrocarbon chains tend to lie parallel and closely packed, leading to a rigidity of the bilayer. Unsaturated hydrocarbon chains, on the other hand, have one or more cis double bonds, which cause "kinks" in the chains, and, as a result, they do not pack as closely and with as great an order as saturated chains. The disordered packing of unsaturated hydrocarbon chains leads to fluidity of the bilayer.

Biological membranes are made of lipid bilayers. The most satisfactory current model for the arrangement of phospholipids, proteins, and cholesterol in plant and animal membranes is the **fluid-mosaic** model proposed in 1972 by S. J. Singer and G. Nicolson (Figure 17.13). The term "mosaic" signifies that the various components in the membrane coexist side by side, as discrete units, rather than combining to form new molecules or ions. "Fluid" signifies that the same sort of fluidity exists in membranes that we have already seen for lipid bilayers. Furthermore, the protein components of membranes "float" in the bilayer and can move laterally along the plane of the membrane.

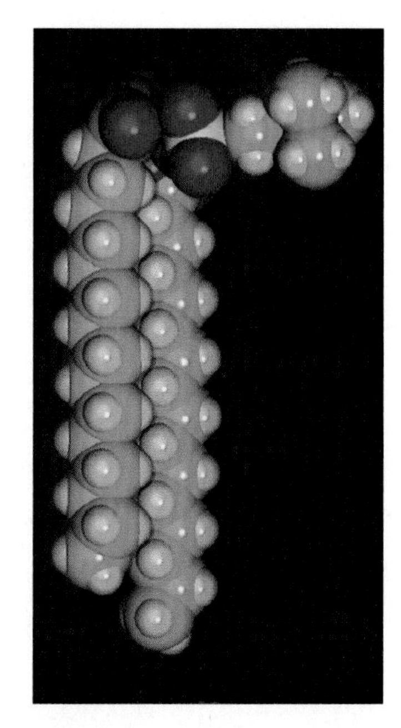

Figure 17.12
Space-filling model of a lecithin. *(Brent Iverson, University of Texas)*

Lipid bilayer A back-to-back arrangement of phospholipid monolayers.

Fluid-mosaic model A biological membrane consists of a phospholipid bilayer with proteins, carbohydrates, and other lipids embedded in and on the surface of the bilayer.

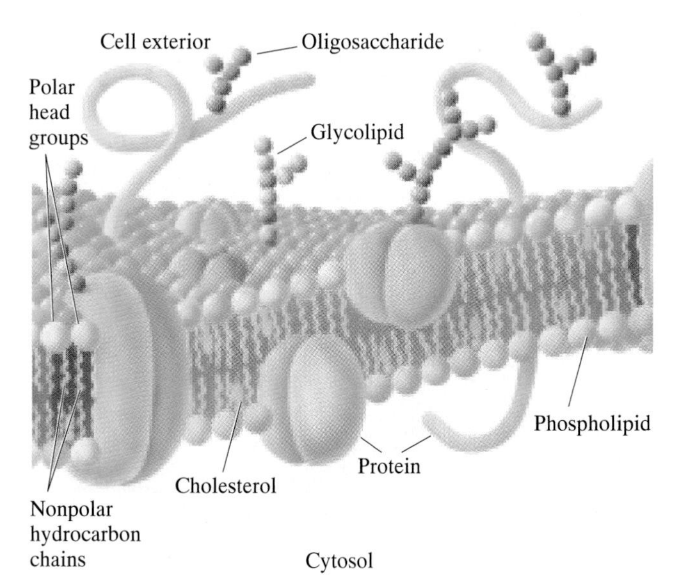

Figure 17.13
Fluid-mosaic model of a biological membrane, showing the lipid bilayer and membrane proteins oriented on the inner and outer surfaces of the membrane and penetrating the entire thickness of the membrane.

All of these products contain lecithin. *(Charles D. Winters)*

CHEMICAL CONNECTIONS

Snake Venom Phospholipases

The venoms of certain snakes contain enzymes called phospholipases. These enzymes catalyze the hydrolysis of carboxylic ester bonds of phospholipids. The venom of the eastern diamondback rattlesnake *(Crotalus adamanteus)* and the Indian cobra *(Naja naja)* both contain phospholipase PLA_2, which catalyzes the hydrolysis of esters at carbon-2 of phospholipids. The breakdown product of this hydrolysis, a lysolecithin, acts as a detergent and dissolves the membranes of red blood cells causing them to rupture. Indian cobras kill several thousand people each year.

$$\underset{\text{A phospholipid}}{\begin{array}{c} CH_2CH_2NH_3^+ \\ | \\ O \\ | \\ O{=}P{-}O^- \\ | \\ O \\ | \\ CH_2{-}CH{-}CH_2 \\ | \quad\;\; | \\ O \quad\; O \\ | \quad\;\; | \\ O{=}C \;\; O{=}C \\ | \quad\;\; | \\ R_1 \quad\; R_2 \end{array}} + H_2O \xrightarrow{PLA_2} \underset{\text{A lysolecithin}}{\begin{array}{c} CH_2CH_2NH_3^+ \\ | \\ O \\ | \\ O{=}P{-}O^- \\ | \\ O \\ | \\ CH_2{-}CH{-}CH_2 \\ | \quad\;\; | \\ O \quad\; OH \\ | \\ O{=}C \\ | \\ R_1 \end{array}} + R_2{-}\overset{\overset{\displaystyle O}{\|}}{C}{-}O^- + H^+$$

PLA_2 catalyzes hydrolysis of this ester bond

Milking an Indian cobra for its venom. *(Dan McCoy/Rainbow)*

The eastern diamondback rattlesnake. *(Leonard Lee Rule III/Photo Researchers, Inc.)*

17.6 FAT-SOLUBLE VITAMINS

Vitamins are divided into two broad classes on the basis of solubility: those that are fat-soluble (and hence classed as lipids) and those that are water-soluble. The fat-soluble vitamins include A, D, E, and K.

A. Vitamin A

Vitamin A, or retinol, occurs only in the animal world, where the best sources are cod-liver oil and other fish-liver oils, animal liver, and dairy products. Vitamin A in the form of a precursor, or provitamin, is found in the plant world in a group of tetraterpene (C_{40}) pigments called carotenes. The most common of these is β-carotene, abundant in carrots but also found in some other vegetables, particularly yellow and green ones. β-Carotene has no vitamin A activity; however, after ingestion, it is cleaved at the central carbon-carbon double bond to give retinol (vitamin A).

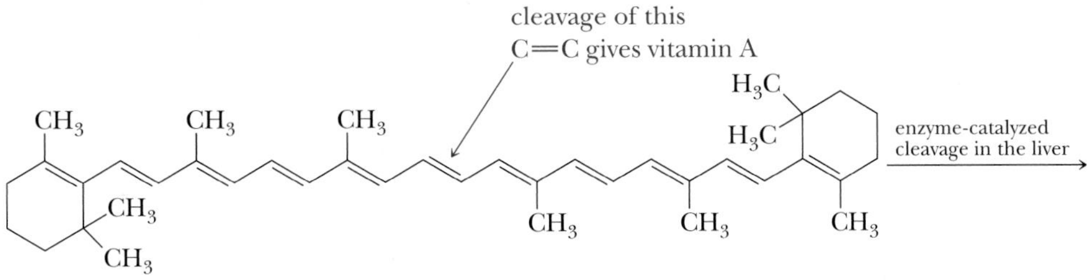

cleavage of this C=C gives vitamin A

enzyme-catalyzed cleavage in the liver

β-Carotene

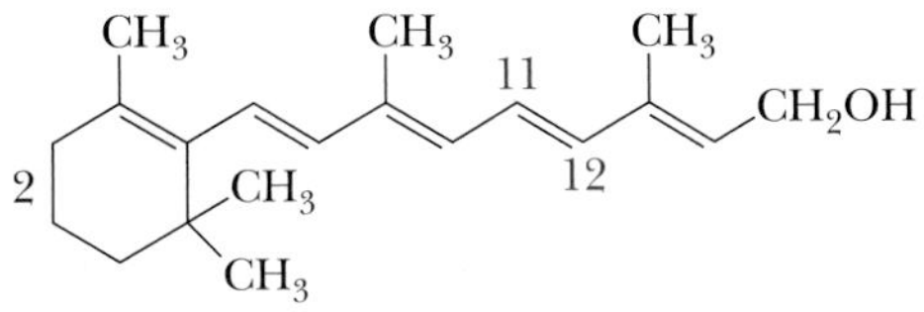

Retinol
(Vitamin A)

Probably the best understood role of vitamin A is its participation in the visual cycle in rod cells. In a series of enzyme-catalyzed reactions (Figure 17.14), retinol undergoes a two-electron oxidation to all-*trans*-retinal, isomerization about the C-11 to C-12 double bond to give 11-*cis*-retinal, and formation of an imine (Section 11.7) with the $—NH_2$ from a lysine unit of the protein, opsin. The product of these reactions is rhodopsin, a highly conjugated pigment that shows intense absorption in the blue-green region of the visual spectrum.

The primary event in vision is absorption of light by rhodopsin in rod cells of the retina of the eye to produce an electronically excited molecule. Within several picoseconds (1 picosec = 10^{-12} sec), the excess electronic energy is converted to vibrational and rotational energy, and the 11-*cis* double bond is isomerized to the more stable 11-*trans* double bond. This isomerization triggers a conformational change in the protein, opsin, that causes firing of neurons in the optic nerve and produces a visual image. Coupled with this light-induced change is hydrolysis of rhodopsin to give 11-*trans*-retinal and free opsin. At this point, the visual pigment is bleached and in a refractory period. Rhodopsin is regenerated by a series of en-

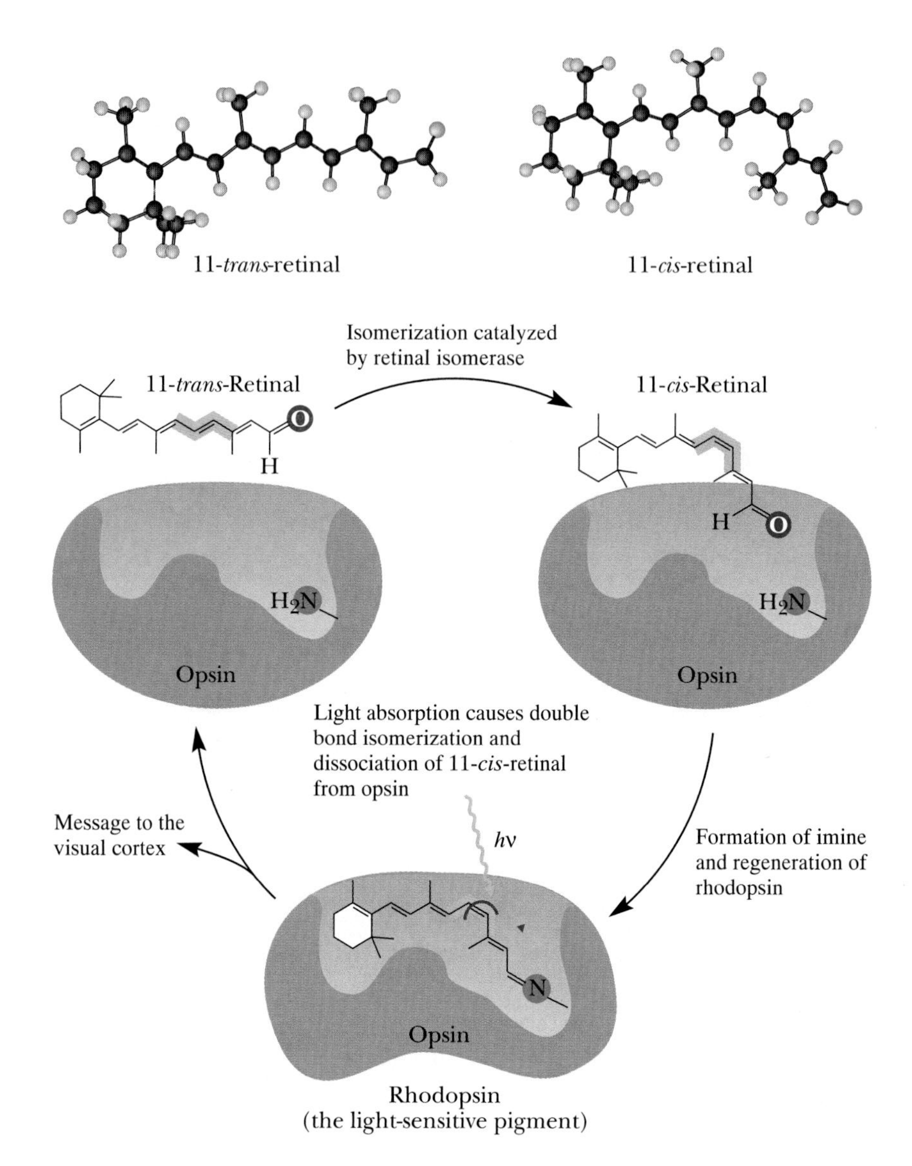

Figure 17.14 The primary chemical reaction of vision in rod cells is absorption of light by rhodopsin followed by isomerization of a carbon-carbon double bond from a cis configuration to a trans configuration.

zyme-catalyzed reactions that converts 11-*trans*-retinal to 11-*cis*-retinal and then to rhodopsin. The **visual cycle** is shown in abbreviated form in Figure 17.14.

B. Vitamin D

Vitamin D is the name for a group of structurally related compounds that plays a major role in the regulation of calcium and phosphorus metabolism. A deficiency of vitamin D in childhood is associated with rickets, a mineral-metabolism disease that leads to bowlegs, knock-knees, and enlarged joints. Vitamin D_3, the most abundant form of the vitamin in the circulatory system, is produced in the skin of mammals by the action of ultraviolet radiation on 7-dehydrocholesterol (cholesterol with a double bond between carbons 7 and 8). In the liver, vitamin D_3 undergoes an enzyme-catalyzed, two-electron oxidation at carbon 25 of the side chain to form 25-hydroxyvitamin D_3; the oxidizing agent is molecular oxygen, O_2. 25-Hydroxyvitamin D_3 undergoes further oxidation in the kidneys, also by O_2, to form 1,25-dihydroxyvitamin D_3, the hormonally active form of the vitamin.

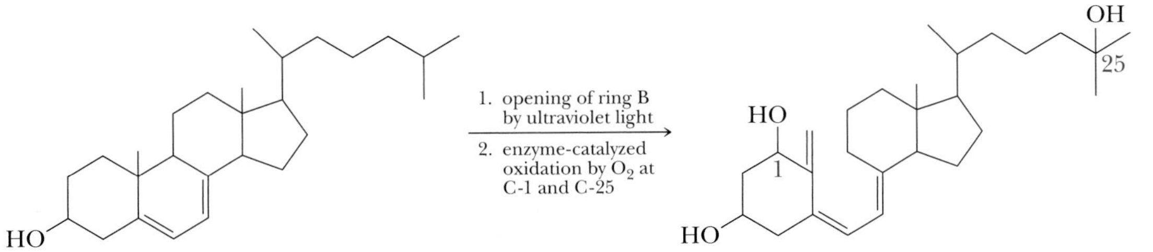

HO

7-Dehydrocholesterol

1. opening of ring B by ultraviolet light
2. enzyme-catalyzed oxidation by O_2 at C-1 and C-25

OH 25 HO 1 HO

1,25-Dihydroxyvitamin D_3

C. Vitamin E

Vitamin E was first recognized in 1922 as a dietary factor essential for normal reproduction in rats, hence its name tocopherol from the Greek: *tocos,* birth, and *pherein,* to bring about. Vitamin E is a group of compounds of similar structure, the most active of which is α-tocopherol. This vitamin occurs in fish oil, in other oils such as cottonseed and peanut, and in leafy green vegetables. The richest source of vitamin E is wheat germ oil.

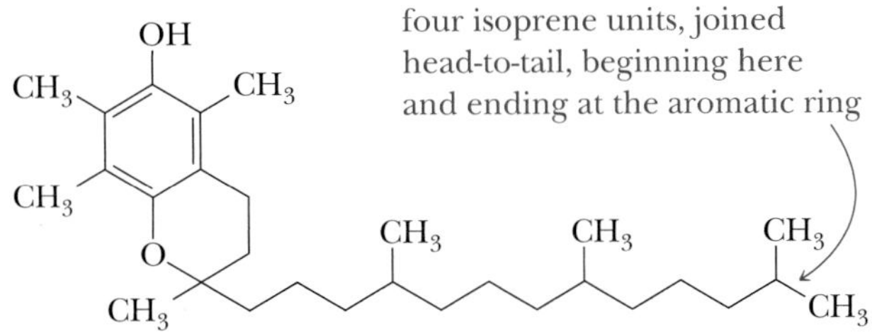

OH CH_3 CH_3 CH_3 O CH_3 CH_3 CH_3 CH_3 CH_3

four isoprene units, joined head-to-tail, beginning here and ending at the aromatic ring

Vitamin E
(α-Tocopherol)

In the body, vitamin E functions as an antioxidant; it traps peroxy radicals of the type HOO· and ROO· formed as a result of enzyme-catalyzed oxidation by molecular oxygen of the unsaturated hydrocarbon chains in membrane phospholipids. There is speculation that peroxy radicals play a role in the aging process and that vitamin E and other antioxidants may retard that process. Vitamin E is also necessary for the proper development and function of the membranes of red blood cells.

D. Vitamin K

The name of this vitamin comes from the German word *koagulation,* signifying its important role in the blood-clotting process. A deficiency of vitamin K results in slowed blood clotting.

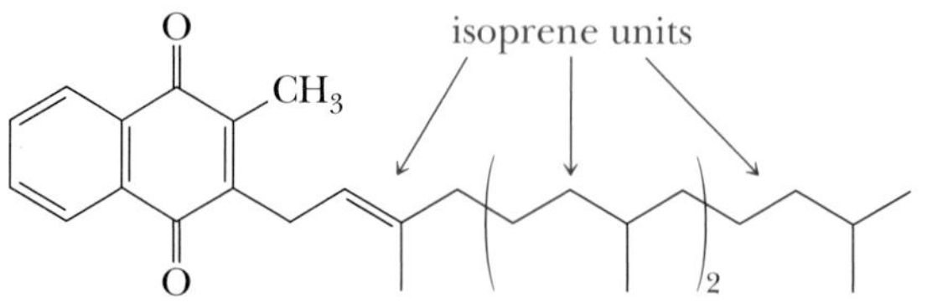

isoprene units

Vitamin K_1

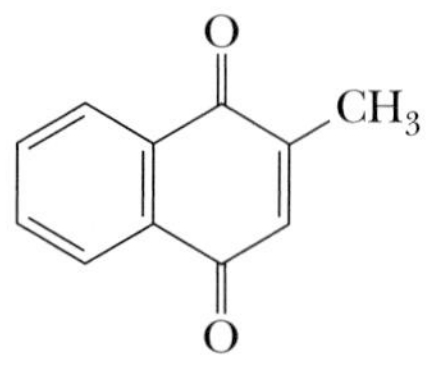

Menadione
(a synthetic vitamin K analog)

Natural vitamins of the K family have, for the most part, been replaced in vitamin supplements by synthetic preparations. Menadione, one such synthetic material with vitamin-K activity, has only hydrogen in the place of the alkyl chain.

SUMMARY

Lipids are a heterogeneous class of compounds grouped together on the basis of their solubility properties; they are insoluble in water and soluble in diethyl ether, acetone, and methylene chloride. Carbohydrates, amino acids, and proteins are largely insoluble in these organic solvents.

Triglycerides (triacylglycerols), the most abundant lipids, are triesters of glycerol and fatty acids (Section 17.1). **Fatty acids** (Section 17.1A) are long-chain carboxylic acids derived from the hydrolysis of fats, oils, and the phospholipids of biological membranes. The melting point of a triglyceride increases as (1) the length of hydrocarbon chains increases and (2) the degree of saturation increases. Triglycerides rich in saturated fatty acids are generally solids at room temperature; those rich in unsaturated fatty acids are generally oils at room temperature (Section 17.1B).

Soaps are sodium or potassium salts of fatty acids (Section 17.2A). In water, soaps form **micelles,** which "dissolve" nonpolar organic grease and oil. Natural soaps precipitate as water-insoluble salts with Mg^{2+}, Ca^{2+}, and Fe^{3+} ions in hard water. The most common and most widely used **synthetic detergents** (Section 17.2C) are linear alkylbenzenesulfonates.

Prostaglandins are a group of compounds having the 20-carbon skeleton of prostanoic acid (Section 17.3). They are synthesized, in response to physiological triggers, from phospholipid-bound arachidonic acid (20:4) and other 20-carbon fatty acids.

Steroids are a group of plant and animal lipids that have a characteristic tetracyclic structure of three six-membered rings and one five-membered ring (Section 17.4). **Cholesterol** is an integral part of animal membranes, and it is the compound from which human sex hormones, adrenocorticoid hormones, bile acids, and vitamin D are synthesized. **Low-density lipoproteins (LDLs)** transport cholesterol from the site of its synthesis in the liver to tissues and cells where it is to be used. **High-density lipoproteins (HDLs)** transport cholesterol from cells back to the liver for its degradation to bile acids and eventual excretion in the feces.

Oral contraceptive pills contain a synthetic progestin, for example norethindrone, which prevents ovulation, yet allows women to maintain an otherwise normal menstrual cycle. A variety of synthetic **anabolic steroids** are available for use in rehabilitation medicine where muscle tissue has weakened or deteriorated due to injury. **Bile acids** differ from most other steroids in that they have a cis configuration at the junction of rings A and B.

The carbon skeleton of cholesterol and those of all biomolecules derived from it originate with the acetyl group (a C_2 unit) of **acetyl-CoA** (Section 17.4B).

Phospholipids (Section 17.5A), the second most abundant group of naturally occurring lipids, are derived from phosphatidic acids, compounds containing glycerol esterified with two molecules of fatty acid and a molecule of phosphoric acid. Further esterification of the phosphoric acid part with a low-molecular-weight alcohol, most commonly ethanolamine, choline, serine, or inositol, gives a phospholipid. When placed in aqueous solution, phospholipids spontaneously form **lipid bilayers** (Section 17.5B).

According to the **fluid-mosaic model** (Section 17.5B), membrane phospholipids form lipid bilayers with membrane proteins associated with the bilayer as both peripheral and integral proteins.

Vitamin A (Section 17.6A) occurs only in the animal world. The carotenes of the plant world are tetraterpenes (C_{40}) and are cleaved, after ingestion, into vitamin A. The best-understood role of vitamin A is its participation in the **visual cycle.**

Vitamin D (Section 17.6B) is synthesized in the skin of mammals by the action of ultraviolet radiation on 7-dehydrocholesterol. This vitamin plays a major role in the regulation of calcium and phosphorus metabolism. **Vitamin E** (Section 17.6C) is a group of compounds of similar structure, the most active of which is α-tocopherol. In the body, vitamin E functions as an antioxidant. **Vitamin K** (Section 17.6D) is required for the clotting of blood.

ADDITIONAL PROBLEMS

Fatty Acids and Triglycerides

17.2 Define the term "hydrophobic."

17.3 Identify the hydrophobic and hydrophilic region(s) of a triglyceride.

17.4 Explain why the melting points of unsaturated fatty acids are lower than those of saturated fatty acids.

17.5 Which would you expect to have the higher melting point, glyceryl trioleate or glyceryl trilinoleate?

17.6 Draw a structural formula for methyl linoleate. Be certain to show the correct configuration of groups about each carbon-carbon double bond.

17.7 Explain why coconut oil is a liquid triglyceride, even though most of its fatty acid components are saturated.

17.8 It is common now to see "contains no tropical oils" on cooking oil labels, meaning that the oil contains no palm or coconut oil. What is the difference between the composition of tropical oils and that of vegetable oils, such as corn oil, soybean oil, and peanut oil?

17.9 What is meant by the term "hardening" as applied to oils?

17.10 How many moles of H_2 are used in the catalytic hydrogenation of 1 mole of a triglyceride derived from glycerol, stearic acid, linoleic acid, and arachidonic acid?

17.11 Characterize the structural features necessary to make a good synthetic detergent.

17.12 Following are structural formulas for a cationic detergent and a neutral detergent. Account for the detergent properties of each.

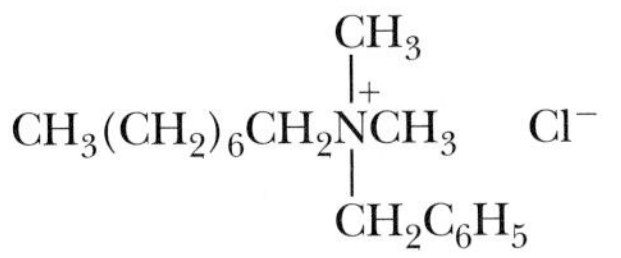

$$CH_3(CH_2)_6CH_2\overset{+}{N}(CH_3)(CH_3)(CH_2C_6H_5)\ Cl^-$$

Benzyldimethyloctylammonium chloride
(a cationic detergent)

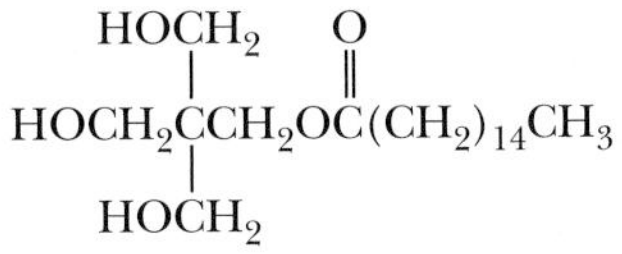

$$(HOCH_2)_3CCH_2OC(=O)(CH_2)_{14}CH_3$$

Pentaerythrityl palmitate
(a neutral detergent)

17.13 Identify some of the detergents used in shampoos and dish washing liquids. Are they primarily anionic, neutral, or cationic detergents?

17.14 Show how to convert palmitic acid (hexadecanoic acid) into the following.

(a) Ethyl palmitate **(b)** Palmitoyl chloride **(c)** 1-Hexadecanol (cetyl alcohol)

(d) 1-Hexadecanamine **(e)** *N,N*-Dimethylhexadecanamide

17.15 Palmitic acid (hexadecanoic acid) is the source of the hexadecyl (cetyl) group in the following compounds. Each is a mild surface-acting germicide and fungicide and is used as a topical antiseptic and disinfectant (*The Merck Index,* 12th ed., #2074 and #2059).

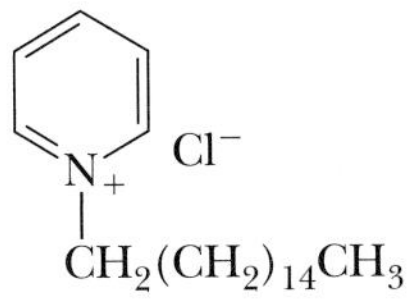

$$C_5H_5\overset{+}{N}{-}CH_2(CH_2)_{14}CH_3\ Cl^-$$

Cetylpyridinium chloride

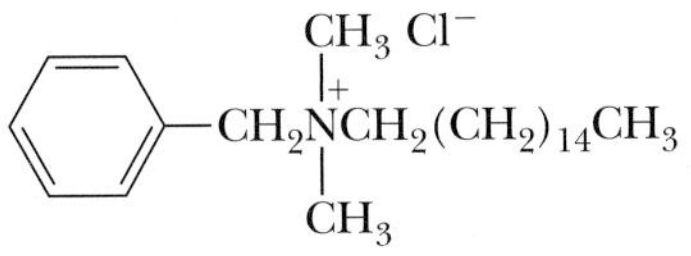

$$C_6H_5{-}CH_2\overset{+}{N}(CH_3)_2CH_2(CH_2)_{14}CH_3\ Cl^-$$

Benzylcetyldimethylammonium chloride

(a) Cetylpyridinium chloride is prepared by treating pyridine with 1-chlorohexadecane (cetyl chloride). Show how to convert palmitic acid to cetyl chloride.

(b) Benzylcetyldimethylammonium chloride is prepared by treating benzyl chloride with *N,N*-dimethyl-1-hexadecanamine. Show how this tertiary amine can be prepared from palmitic acid.

Prostaglandins

17.16 Examine the structure of $PGF_{2\alpha}$ and

(a) Identify all stereocenters.

(b) Identify all double bonds about which cis-trans isomerism is possible.

(c) State the number of stereoisomers possible for a molecule of this structure.

17.17 Following is the structure of unoprostone (*The Merck Index,* 12th ed., #9984), a compound patterned after the natural prostaglandins (Section 17.3). Rescula, the isopropyl ester of unoprostone, is an antiglaucoma drug used to treat ocular hypertension. Compare the structural formula of this synthetic prostaglandin with that of $PGF_{2\alpha}$.

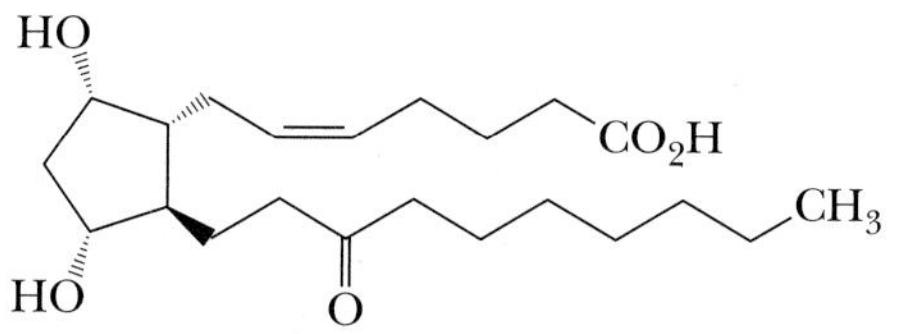

HO CO_2H CH_3 HO O

Unoprostone

Steroids

17.18 Draw the structural formula for the product formed by treatment of cholesterol with H_2/Pd; with Br_2.

17.19 List several ways in which cholesterol is necessary for human life. Why do many people find it necessary to restrict their dietary intake of cholesterol?

17.20 Both low-density lipoproteins (LDL) and high-density lipoproteins (HDL) consist of a core of triacylglycerols and cholesterol esters surrounded by a single phospholipid layer. Draw the structural formula of cholesteryl linoleate, one of the cholesterol esters found in this core.

17.21 Examine the structural formulas of testosterone (a male sex hormone) and progesterone (a female sex hormone). What are the similarities in structure between the two? What are the differences?

17.22 Examine the structural formula of cholic acid and account for the ability of this and other bile salts to emulsify fats and oils and thus aid in their digestion.

17.23 Following is a structural formula for cortisol (hydrocortisone). Draw a stereorepresentation of this molecule showing the conformation of the five- and six-membered rings.

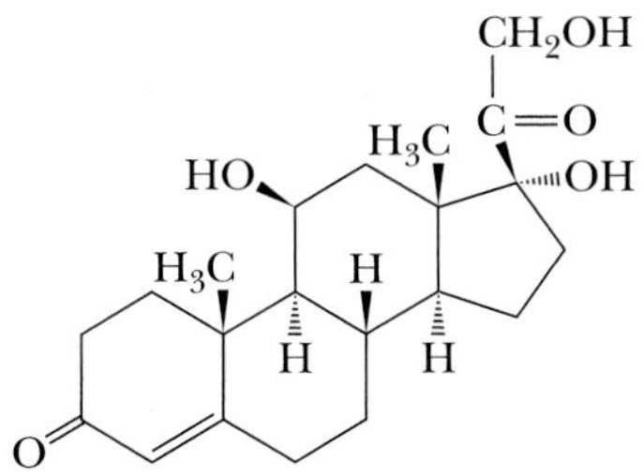

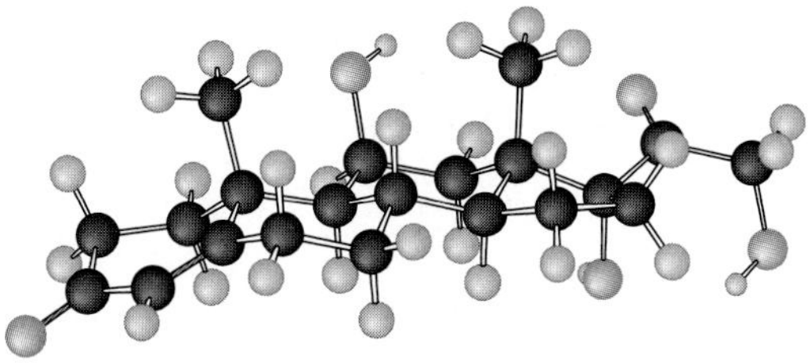

Cortisol
(Hydrocortisone)

17.24 Because some types of tumors need an estrogen to survive, compounds that compete with the estrogen receptor on tumor cells are useful anticancer drugs. The compound tamoxifen is one such drug. To what part of the estrone molecule is the shape of tamoxifen similar?

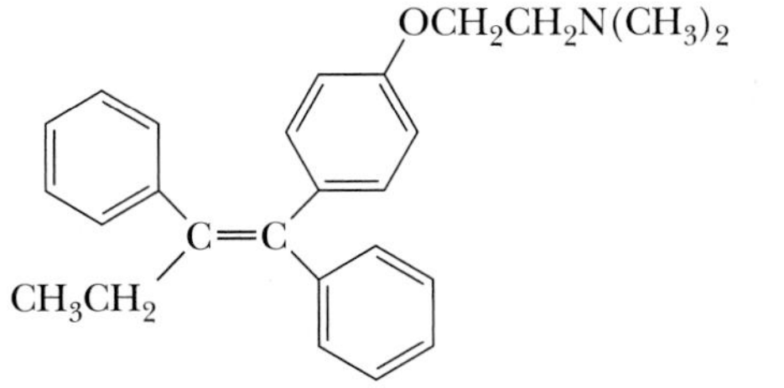

Tamoxifen

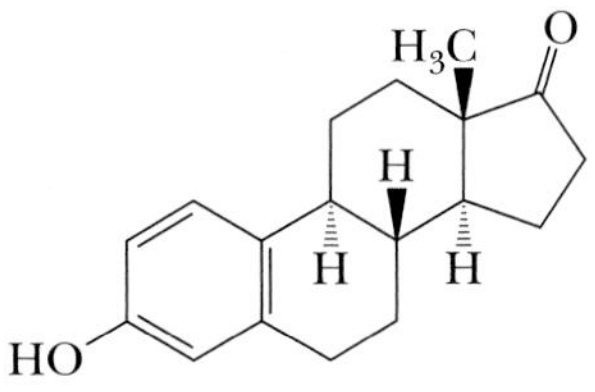

Estrone

Phospholipids

17.25 Draw the structural formula of a lecithin containing one molecule each of palmitic acid and linoleic acid.

17.26 Identify the hydrophobic and hydrophilic region(s) of a phospholipid.

17.27 The hydrophobic effect is one of the most important noncovalent forces directing the self-assembly of biomolecules in aqueous solution. The hydrophobic effect arises from tendencies of biomolecules (1) to arrange polar groups so that they interact with the aqueous environment by hydrogen bonding and (2) to arrange nonpolar groups so that they are shielded from the aqueous environment. Show how the hydrophobic effect is involved in directing

(a) Formation of micelles by soaps and detergents.

(b) Formation of lipid bilayers by phospholipids.

17.28 How does the presence of unsaturated fatty acids contribute to the fluidity of biological membranes?

17.29 Lecithins can act as emulsifying agents. The lecithin of egg yolk, for example, is used to make mayonnaise. Identify the hydrophobic part(s) and the hydrophilic part(s) of a lecithin. Which parts interact with the oils used in making mayonnaise? Which parts interact with the water?

Fat-Soluble Vitamins

17.30 Examine the structural formula of vitamin A, and state the number of cis-trans isomers possible for this molecule.

17.31 The form of vitamin A present in many food supplements is vitamin A palmitate. Draw the structural formula of this molecule.

17.32 Examine the structural formulas of vitamin A, 1,25-dihydroxy-D_3, vitamin E, and vitamin K_1 (Section 17.6). Do you expect them to be more soluble in water or in dichloromethane? Do you expect them to be soluble in blood plasma?

18

Amino Acids and Proteins

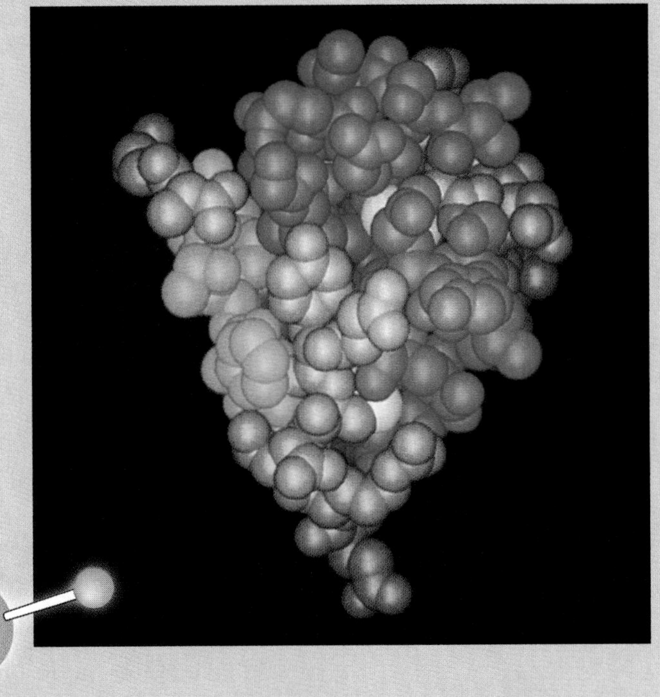

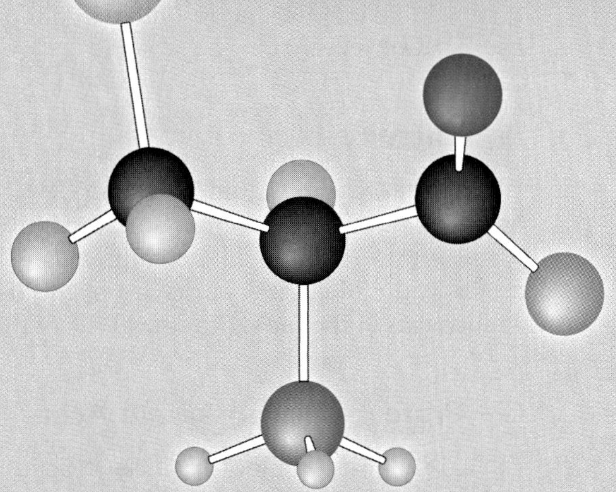

A molecular model of the protein insulin held together by disulfide bonds. Inset: A model cysteine. *(Charles Grisham)*

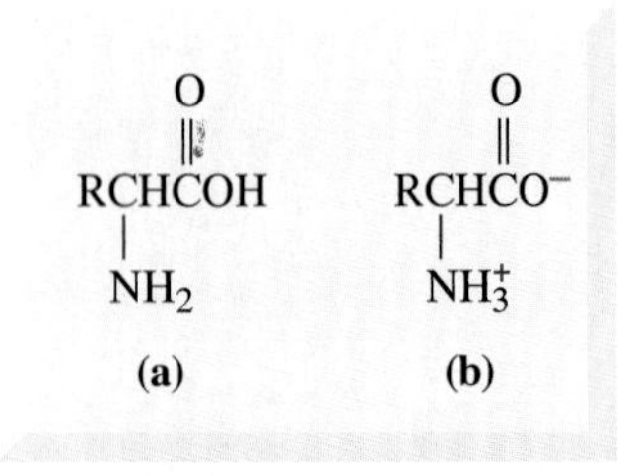

Figure 18.1
An α-amino acid. (a) Un-ionized form and (b) internal salt (zwitterion) form.

We begin this chapter with a study of amino acids, compounds whose chemistry is built on amines (Chapter 10) and carboxylic acids (Chapter 12). We concentrate in particular on the acid-base properties of amino acids because these properties are so important in determining many of the properties of proteins, including the catalytic functions of enzymes. With this understanding of the chemistry of amino acids, we then examine the structure of proteins themselves.

18.1 Amino Acids

A. Structure

Amino acid A compound that contains both an amino group and a carboxyl group.

α-Amino acid An amino acid in which the amino group is on the carbon adjacent to the carboxyl group.

An **amino acid** is a compound that contains both a carboxyl group and an amino group. Although many types of amino acids are known, the **α-amino acids** are the most significant in the biological world because they are the monomers from which proteins are constructed. A general structural formula of an α-amino acid is shown in Figure 18.1.

Although Figure 18.1(a) is a common way of writing structural formulas for amino acids, it is not accurate because it shows an acid ($—CO_2H$) and a base ($—NH_2$) within the same molecule. These acidic and basic groups react with each other to form a dipolar ion or internal salt [Figure 18.1(b)]. The internal salt of an amino acid is given the special name **zwitterion.** Note that a zwitterion has no net charge; it contains one positive charge and one negative charge.

Zwitterion An internal salt of an amino acid.

Because they exist as zwitterions, amino acids have many of the properties associated with salts. They are crystalline solids with high melting points and are fairly soluble in water but insoluble in nonpolar organic solvents such as ether and hydrocarbon solvents.

B. Chirality

With the exception of glycine, $H_2NCH_2CO_2H$, all protein-derived amino acids have at least one stereocenter and, therefore, are chiral. Figure 18.2 shows Fischer projection formulas for the enantiomers of alanine. The vast majority of carbohydrates in the biological world are of the D-series, whereas the vast majority of α-amino acids in the biological world are of the L-series.

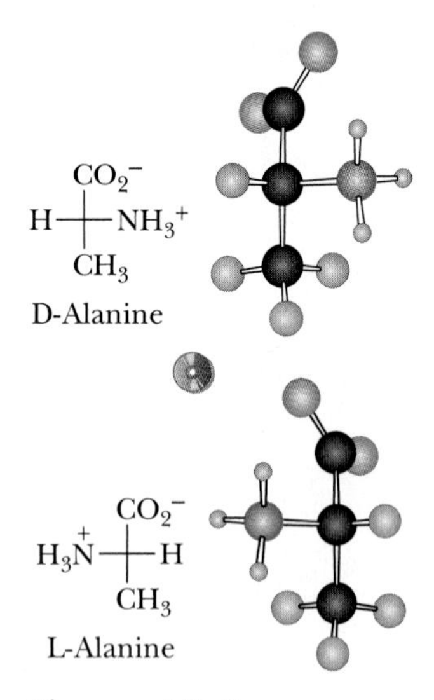

Figure 18.2
The enantiomers of alanine. The vast majority of α-amino acids in the biological world have the L-configuration of the α-carbon.

C. Protein-Derived Amino Acids

Table 18.1 gives common names, structural formulas, and standard three-letter and one-letter abbreviations for the 20 common L-amino acids found in proteins. The amino acids in this table are divided into four categories: those with nonpolar side chains, polar but un-ionized side chains, acidic side chains, and basic side chains. The following structural features of these amino acids should be noted.

1. All 20 of these protein-derived amino acids are α-amino acids, meaning that the amino group is located on the carbon alpha to the carboxyl group.

TABLE 18.1 The 20 Common Amino Acids Found in Proteins

Name	Structure	Name	Structure
Nonpolar Side Chains			
alanine (Ala, A)	$CH_3CHCO_2^-$ NH_3^+	phenylalanine (Phe, F)	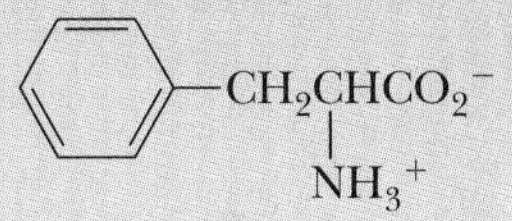 $-CH_2CHCO_2^-$ NH_3^+
glycine (Gly, G)	$HCHCO_2^-$ NH_3^+	proline (Pro, P)	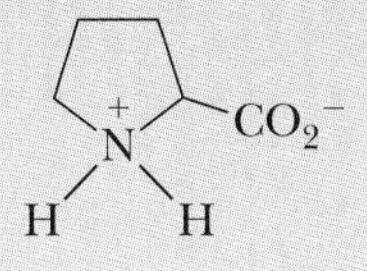 CO_2^- N^+ H H
isoleucine (Ile, I)	CH_3 $CH_3CH_2CHCHCO_2^-$ NH_3^+	tryptophan (Trp, W)	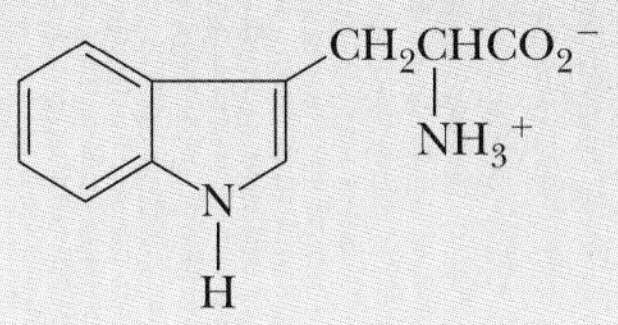 $CH_2CHCO_2^-$ NH_3^+ N H
leucine (Leu, L)	$(CH_3)_2CHCH_2CHCO_2^-$ NH_3^+		
methionine (Met, M)	$CH_3SCH_2CH_2CHCO_2^-$ NH_3^+	valine (Val, V)	$(CH_3)_2CHCHCO_2^-$ NH_3^+
Polar Side Chains			
asparagine (Asn, N)	O $H_2NCCH_2CHCO_2^-$ NH_3^+	serine (Ser, S)	$HOCH_2CHCO_2^-$ NH_3^+
glutamine (Gln, Q)	O $H_2NCCH_2CH_2CHCO_2^-$ NH_3^+	threonine (Thr, T)	OH $CH_3CHCHCO_2^-$ NH_3^+
Acidic Side Chains		**Basic Side Chains**	
aspartic acid (Asp, D)	O $^-OCCH_2CHCO_2^-$ NH_3^+	arginine (Arg, R)	NH_2^+ $H_2NCNHCH_2CH_2CH_2CHCO_2^-$ NH_3^+
glutamic acid (Glu, E)	O $^-OCCH_2CH_2CHCO_2^-$ NH_3^+		
cysteine (Cys, C)	$HSCH_2CHCO_2^-$ NH_3^+	histidine (His, H)	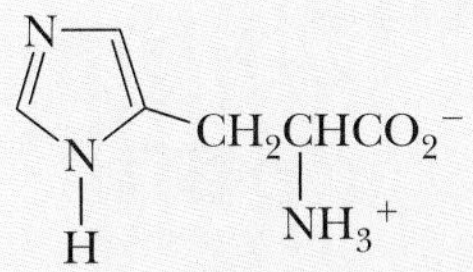 N N $CH_2CHCO_2^-$ H NH_3^+
tyrosine (Tyr, Y)	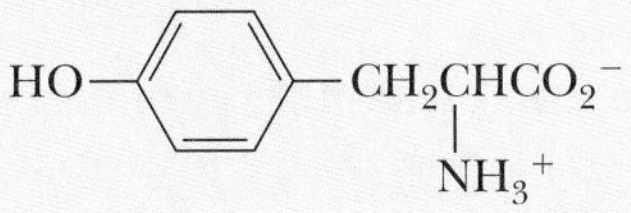 HO— —$CH_2CHCO_2^-$ NH_3^+	lysine (Lys, K)	$H_3\overset{+}{N}CH_2CH_2CH_2CH_2CHCO_2^-$ NH_3^+

* Each ionizable functional group is shown in the form present in highest concentration at pH 7.0.

2. For 19 of the 20 amino acids, the α-amino group is primary. Proline is different; its α-amino group is secondary.
3. With the exception of glycine, the α-carbon of each amino acid is a stereocenter. Although not shown in this table, all 19 chiral amino acids have the same relative configuration at the α-carbon. In the D,L convention, all are L-amino acids.
4. Isoleucine and threonine contain a second stereocenter. Four stereoisomers are possible for each amino acid, but only one is found in proteins.
5. The sulfhydryl group of cysteine, the imidazole group of histidine, and the phenolic hydroxyl of tyrosine are partially ionized at pH 7.0, but the ionic form is not the major form present at this pH.

EXAMPLE 18.1

Of the 20 protein-derived amino acids shown in Table 18.1, how many contain (a) aromatic rings, (b) side-chain hydroxyl groups, (c) phenolic —OH groups, and (d) sulfur?

SOLUTION

(a) Phenylalanine, tryptophan, tyrosine, and histidine contain aromatic rings.
(b) Serine and threonine contain side-chain hydroxyl groups.
(c) Tyrosine contains a phenolic —OH group.
(d) Methionine and cysteine contain sulfur.

Practice Problem 18.1

Of the 20 protein-derived amino acids shown in Table 18.1, which contain (a) no stereocenter, (b) two stereocenters?

D. Some Other Common L-Amino Acids

Although the vast majority of plant and animal proteins are constructed from just these 20 α-amino acids, many other amino acids are also found in nature. Ornithine and citrulline, for example, are found predominantly in the liver and are an integral part of the urea cycle, the metabolic pathway that converts ammonia to urea.

$$H_3\overset{+}{N}CH_2CH_2CH_2\underset{\underset{NH_3^+}{|}}{C}HCO_2^-$$

L-Ornithine

carboxamide derivative of L-ornithine

$$H_2N\overset{\overset{O}{\|}}{C}NHCH_2CH_2CH_2\underset{\underset{NH_3^+}{|}}{C}HCO_2^-$$

L-Citrulline

Thyroxine and triiodothyronine, two of several hormones derived from the amino acid tyrosine, are found in thyroid tissue. Their principal function is to stimulate metabolism in other cells and tissues.

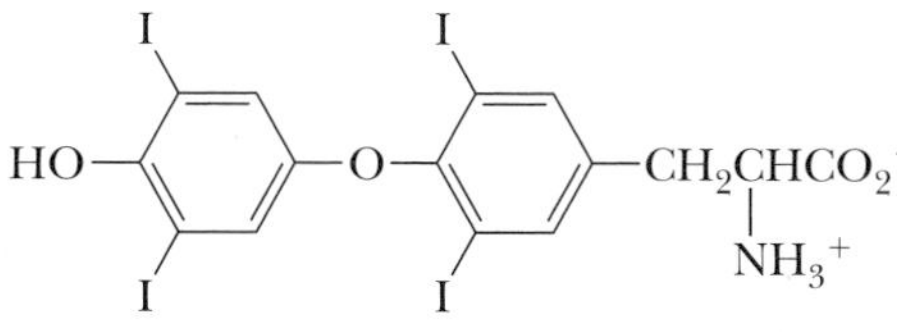

L-Thyroxine, T_4 — HO–(3,5-diiodophenyl)–O–(3,5-diiodophenyl)–$CH_2CH(NH_3^+)CO_2^-$

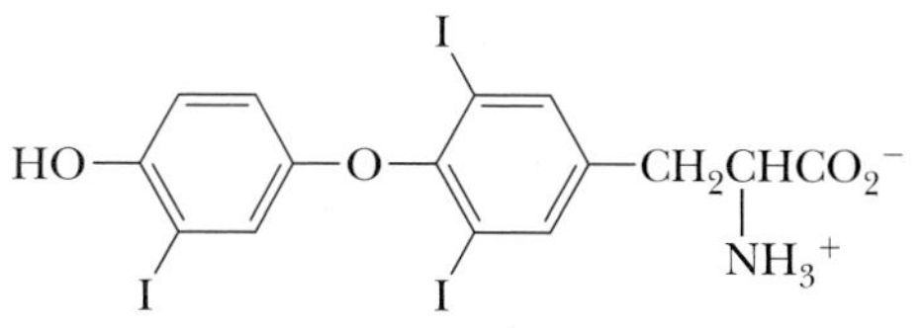

L-Triiodothyronine, T_3 — HO–(3-iodophenyl)–O–(3,5-diiodophenyl)–$CH_2CH(NH_3^+)CO_2^-$

4-Aminobutanoic acid (γ-aminobutyric acid, GABA) is found in high concentration (0.8 mM) in the brain but in no significant amounts in any other mammalian tissue. It is synthesized in neural tissue by decarboxylation of the α-carboxyl group of glutamic acid and is a neurotransmitter in the central nervous system of invertebrates and possibly in humans as well.

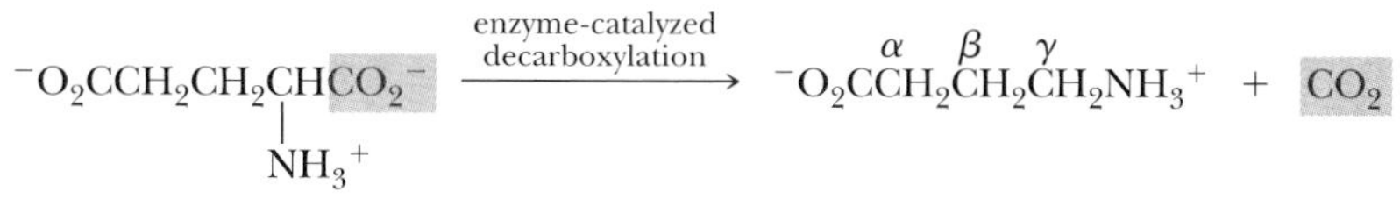

$$\underset{\text{Glutamic acid}}{{}^-O_2CCH_2CH_2CH(NH_3^+)CO_2^-} \xrightarrow[\text{decarboxylation}]{\text{enzyme-catalyzed}} \underset{\substack{\text{4-Aminobutanoic acid}\\(\gamma\text{-Aminobutyric acid, GABA})}}{{}^-O_2CCH_2\overset{\alpha}{C}H_2\overset{\beta}{C}H_2\overset{\gamma}{C}H_2NH_3^+} + CO_2$$

Only L-amino acids are found in proteins, and only rarely are D-amino acids a part of the metabolism of higher organisms. Several D-amino acids, however, along with their L-enantiomers, are found in lower forms of life. D-Alanine and D-glutamic acid, for example, are structural components of the cell walls of certain bacteria. Several D-amino acids are also found in peptide antibiotics.

18.2 ACID-BASE PROPERTIES OF AMINO ACIDS

A. Acidic and Basic Groups of Amino Acids

Among the most important chemical properties of amino acids are their acid-base properties; all are weak polyprotic acids because of their $—CO_2H$ and $—NH_3^+$ groups. Given in Table 18.2 are pK_a values for each ionizable group of the 20 protein-derived amino acids.

Acidity of α-Carboxyl Groups

The average value of pK_a for an α-carboxyl group of a protonated amino acid is 2.19. Thus, the α-carboxyl group is a considerably stronger acid than acetic acid (pK_a 4.76) and other low-molecular-weight aliphatic carboxylic acids. This greater acidity is accounted for by the electron-withdrawing inductive effect of the adjacent $—NH_3^+$ group. Recall that we used similar reasoning in Section 12.4A to account for the relative acidities of acetic acid and its mono-, di-, and trichloroderivatives.

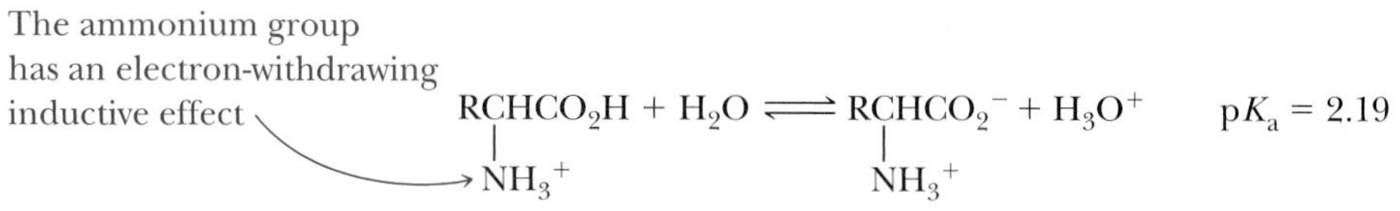

The ammonium group has an electron-withdrawing inductive effect

$$RCH(NH_3^+)CO_2H + H_2O \rightleftharpoons RCH(NH_3^+)CO_2^- + H_3O^+ \qquad pK_a = 2.19$$

TABLE 18.2 pK_a Values for Ionizable Groups of Amino Acids

Amino Acid	pK_a of α-CO_2H	pK_a of α-NH_3^+	pK_a of Side Chain	Isoelectric Point (pI)
alanine	2.35	9.87	—	6.11
arginine	2.01	9.04	12.48	10.76
asparagine	2.02	8.80	—	5.41
aspartic acid	2.10	9.82	3.86	2.98
cysteine	2.05	10.25	8.00	5.02
glutamic acid	2.10	9.47	4.07	3.08
glutamine	2.17	9.13	—	5.65
glycine	2.35	9.78	—	6.06
histidine	1.77	9.18	6.10	7.64
isoleucine	2.32	9.76	—	6.04
leucine	2.33	9.74	—	6.04
lysine	2.18	8.95	10.53	9.74
methionine	2.28	9.21	—	5.74
phenylalanine	2.58	9.24	—	5.91
proline	2.00	10.60	—	6.30
serine	2.21	9.15	—	5.68
threonine	2.09	9.10	—	5.60
tryptophan	2.38	9.39	—	5.88
tyrosine	2.20	9.11	10.07	5.63
valine	2.29	9.72	—	6.00

Acidity of Side-Chain Carboxyl Groups

Owing to the electron-withdrawing inductive effect of the α-NH_3^+ group, the side-chain carboxyl groups of protonated aspartic acid and glutamic acid are also stronger acids than acetic acid (pK_a 4.76). Notice that this acid-strengthening inductive effect decreases with increasing distance of the —CO_2H from the α-NH_3^+. Compare the acidities of the α-CO_2H of alanine (pK_a 2.35), the β-CO_2H of aspartic acid (pK_a 3.86), and the γ-CO_2H of glutamic acid (pK_a 4.07).

Acidity of α-Ammonium Groups

The average value of pK_a for an α-ammonium group, α-NH_3^+ is 9.47, compared with an average value of 10.76 for primary aliphatic ammonium ions (Section 10.4). Thus, the α-ammonium group of an amino acid is a slightly stronger acid than a primary aliphatic ammonium ion. Conversely, an α-amino group is a slightly weaker base than a primary aliphatic amine.

$$RCH(NH_3^+)CO_2^- + H_2O \rightleftharpoons RCH(NH_2)CO_2^- + H_3O^+ \qquad pK_a = 9.47$$

$$CH_3CH(NH_3^+)CH_3 + H_2O \rightleftharpoons CH_3CH(NH_2)CH_3 + H_3O^+ \qquad pK_a = 10.60$$

Basicity of the Guanidine Group of Arginine

The side-chain guanidine group of arginine is a considerably stronger base than an aliphatic amine. As we saw in Section 10.4, guanidine (pK_b 0.4) is the strongest base of any neutral compound. The remarkable basicity of the guanidine group of arginine is attributed to the large resonance stabilization of the protonated form relative to the neutral form.

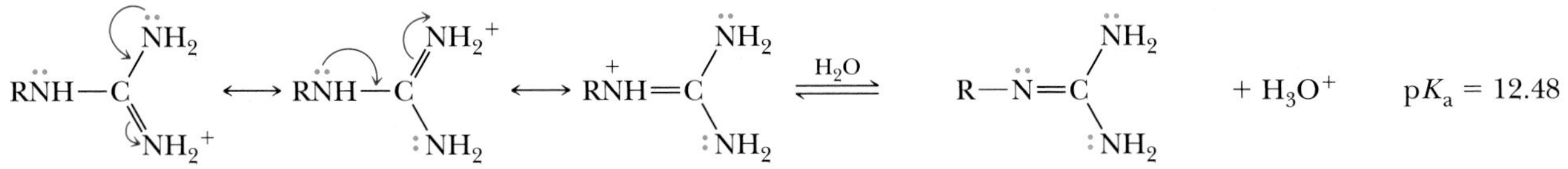

$$RNH{-}C(NH_2)({=}NH_2^+) \longleftrightarrow RNH{-}C({=}NH_2^+)(NH_2) \longleftrightarrow R\overset{+}{N}H{=}C(NH_2)(NH_2) \xrightleftharpoons{H_2O} R{-}N{=}C(NH_2)(NH_2) + H_3O^+ \qquad pK_a = 12.48$$

The guanidinium ion side chain of arginine is a hybrid of three contributing structures

No resonance stabilization without charge separation

Basicity of the Imidazole Group of Histidine

Because the imidazole group on the side chain of histidine contains six π electrons in a planar, fully conjugated ring, imidazole is classified as a heterocyclic aromatic amine (Section 9.2). The unshared pair of electrons on one nitrogen is a part of the aromatic sextet, whereas that on the other nitrogen is not. It is the pair of electrons not part of the aromatic sextet that is responsible for the basic properties of the imidazole ring. Protonation of this nitrogen produces a resonance-stabilized cation.

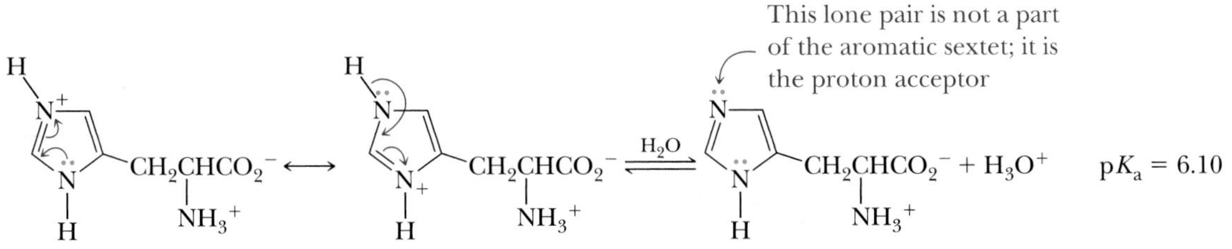

This lone pair is not a part of the aromatic sextet; it is the proton acceptor

$$CH_2CHCO_2^- (NH_3^+) \longleftrightarrow CH_2CHCO_2^- (NH_3^+) \xrightleftharpoons{H_2O} CH_2CHCO_2^- (NH_3^+) + H_3O^+ \qquad pK_a = 6.10$$

Resonance-stabilized imidazolium cation

B. Titration of Amino Acids

Values of pK_a for the ionizable groups of amino acids are most commonly obtained by acid-base titration and measuring the pH of the solution as a function of added base (or added acid, depending on how the titration is done). To illustrate this experimental procedure, consider a solution containing 1.00 mole of glycine to which has been added enough strong acid so that both the amino and carboxyl groups are fully protonated. Next, this solution is titrated with 1.00 *M* NaOH; the volume of base added and the pH of the resulting solution are recorded and then plotted as shown in Figure 18.3. The most acidic group and the one to react first with added sodium hydroxide is the carboxyl group. When exactly 0.50 mole of NaOH has been added, the carboxyl group is half neutralized. At this point, the concentration of the zwitterion equals that of the positively charged ion, and the pH of 2.35 equals the pK_a of the carboxyl group (pK_{a1}).

$$\text{at pH} = pK_{a1} \qquad [H_3\overset{+}{N}CH_2CO_2H] = [H_3\overset{+}{N}CH_2CO_2^-]$$

Positive ion — Zwitterion

Figure 18.3
Titration of glycine with sodium hydroxide.

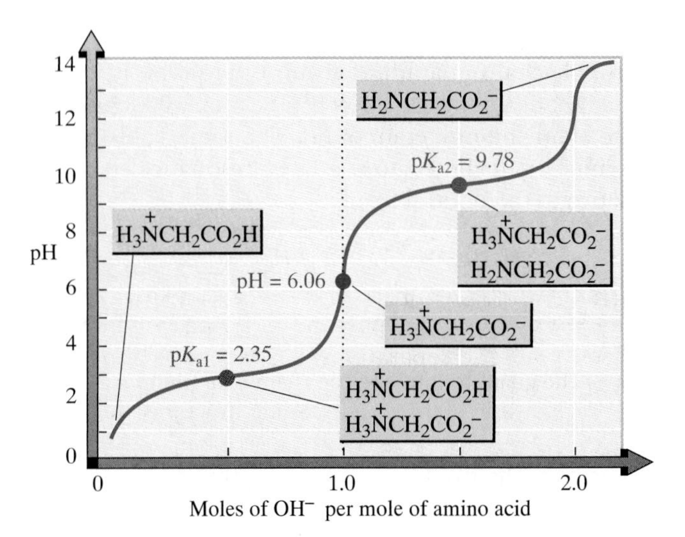

The end point of the first part of the titration is reached when 1.00 mole of NaOH has been added. At this point, the predominant species present is the zwitterion, and the observed pH of the solution is 6.06.

The next section of the curve represents titration of the $—NH_3^+$ group. When another 0.50 mole of NaOH has been added (bringing the total to 1.50 moles), half of the $—NH_3^+$ groups are neutralized and converted to $—NH_2$. At this point, the concentrations of the zwitterion and negatively charged ion are equal, and the observed pH is 9.78, the pK_a of the amino group of glycine (pK_{a2}).

$$\text{at pH} = pK_{a2} \qquad \underset{\text{Zwitterion}}{[H_3\overset{+}{N}CH_2CO_2^-]} = \underset{\text{Negative ion}}{[H_2NCH_2CO_2^-]}$$

The second end point of the titration is reached when a total of 2.00 moles of NaOH have been added and glycine is converted entirely to an anion.

C. Isoelectric Point

Isoelectric point (pI) The pH at which an amino acid, polypeptide, or protein has no net charge.

Titration curves such as that for glycine permit us to determine pK_a values for the ionizable groups of an amino acid. They also permit us to determine another important property: isoelectric point. **Isoelectric point, pI,** for an amino acid is the pH at which the majority of molecules in solution have a net charge of zero (they are zwitterions). By examining the titration curve, you can see that the isoelectric point for glycine falls half way between the pK_a values for the carboxyl and amino groups.

$$\begin{aligned} pI &= \tfrac{1}{2}(pK_a\ \alpha\text{-}CO_2H + pK_a\ \alpha\text{-}NH_3^+) \\ &= \tfrac{1}{2}\ (2.35 + 9.78) = 6.06 \end{aligned}$$

At pH 6.06, the predominant form of glycine molecules is the dipolar ion; furthermore, at this pH, the concentration of positively charged glycine molecules equals the concentration of negatively charged glycine molecules.

Given a value for the isoelectric point of an amino acid, it is possible to estimate the charge on that amino acid at any pH. For example, the charge on tyrosine at pH 5.63, its isoelectric point, is zero. A small fraction of tyrosine molecules are positively charged at pH 5.00 (0.63 unit less than its pI), and virtually all are positively charged at pH 3.63 (2.00 units less than its pI). As another example, the net charge on lysine is zero at pH 9.74. At pH values smaller than 9.74, an increasing fraction of lysine molecules are positively charged.

D. Electrophoresis

Electrophoresis The process of separating compounds on the basis of their electric charge.

Electrophoresis, a process of separating compounds on the basis of their electric charges, is used to separate and identify mixtures of amino acids and proteins. Electrophoretic separations can be carried out using paper, starch, agar, certain plastics, and cellulose acetate as solid supports. In paper electrophoresis, a paper strip saturated with an aqueous buffer of predetermined pH serves as a bridge between two electrode vessels (Figure 18.4). Next, a sample of amino acids is applied as a spot on the paper strip. When an electrical potential is then applied to the electrode vessels, amino acids migrate toward the electrode carrying the charge opposite to their own. Molecules having a high charge density move more rapidly than those with a lower charge density. Any molecule already at its isoelectric point remains at the origin. After separation is complete, the strip is dried and sprayed with a dye to make the separated components visible.

A dye commonly used to detect amino acids is ninhydrin (1,2,3-indanetrione monohydrate). Ninhydrin reacts with α-amino acids to produce an aldehyde, carbon dioxide, and a purple-colored anion. This reaction is used very commonly in both qualitative and quantitative analysis of amino acids.

$$RCH(NH_3^+)CO^- + 2\ \text{Ninhydrin} \longrightarrow \text{Purple-colored anion} + RCH{=}O + CO_2 + H_3O^+$$

An α-amino acid | Ninhydrin | Purple-colored anion

Nineteen of the 20 protein-derived α-amino acids have primary amino groups and give the same purple-colored ninhydrin-derived anion. Proline, a secondary amine, gives a different, orange-colored compound.

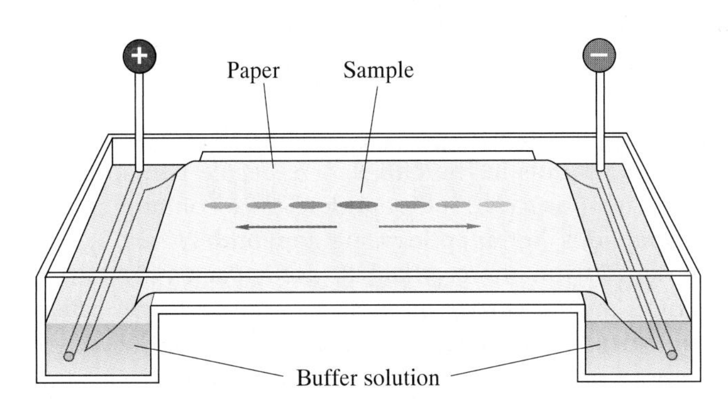

Figure 18.4
An apparatus for electrophoresis of a mixture of amino acids. Those with a negative charge move toward the positive electrode; those with a positive charge move toward the negative electrode; those with no charge remain at the origin.

EXAMPLE 18.2

The isoelectric point of tyrosine is 5.63. Toward which electrode does tyrosine migrate on paper electrophoresis at pH 7.0?

SOLUTION

On paper electrophoresis at pH 7.0 (more basic than its isoelectric point), tyrosine has a net negative charge and migrates toward the positive electrode.

Practice Problem 18.2

The isoelectric point of histidine is 7.64. Toward which electrode does histidine migrate on paper electrophoresis at pH 7.0?

EXAMPLE 18.3

Electrophoresis of a mixture of lysine, histidine, and cysteine is carried out at pH 7.64. Describe the behavior of each amino acid under these conditions.

SOLUTION

The isoelectric point of histidine is 7.64. At this pH, histidine has a net charge of zero and does not move from the origin. The pI of cysteine is 5.02; at pH 7.64 (more basic than its isoelectric point), cysteine has a net negative charge and moves toward the positive electrode. The pI of lysine is 9.74; at pH 7.64 (more acidic than its isoelectric point), lysine has a net positive charge and moves toward the negative electrode.

Practice Problem 18.3

Describe the behavior of a mixture of glutamic acid, arginine, and valine on paper electrophoresis at pH 6.0.

18.3 POLYPEPTIDES AND PROTEINS

In 1902, Emil Fischer proposed that proteins are long chains of amino acids joined together by amide bonds between the α-carboxyl group of one amino acid and the α-amino group of another. For these amide bonds, Fischer proposed the special name **peptide bond.** Figure 18.5 shows the peptide bond formed between serine and alanine in the dipeptide serylalanine.

Peptide is the name given to a short polymer of amino acids. Peptides are classified by the number of amino acid units in the chain. A molecule containing 2 amino acids joined by an amide bond is called a **dipeptide.** Those containing 3 to 10 amino acids are called **tripeptides, tetrapeptides, pentapeptides,** and so on. Molecules containing more than 10 but fewer than 20 amino acids are called **oligopeptides.** Those containing several dozen or more amino acids are called **polypeptides. Proteins** are biological macromolecules of molecular weight 5000 or greater, consisting of one or more polypeptide chains. The distinctions in this terminology are not at all precise.

Peptide bond The special name given to the amide bond formed between the α-amino group of one amino acid and the α-carboxyl group of another amino acid.

Dipeptide A molecule containing two amino acid units joined by a peptide bond.

Tripeptide A molecule containing three amino acid units, each joined to the next by a peptide bond.

Polypeptide A macromolecule containing ten or more amino acid units, each joined to the next by a peptide bond.

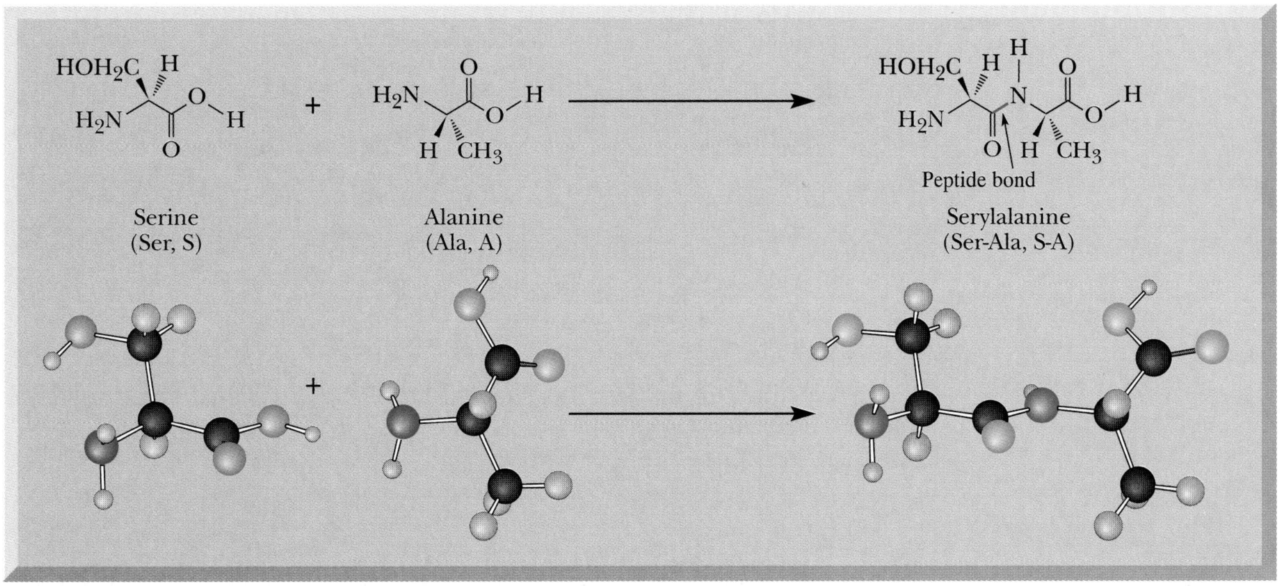

Figure 18.5
The peptide bond in serylalanine.

By convention, polypeptides are written from the left, beginning with the amino acid having the free $—NH_3^+$ group and proceeding to the right toward the amino acid with the free $—CO_2^-$ group. The amino acid with the free $—NH_3^+$ group is called the ***N*-terminal amino acid** and that with the free $—CO_2^-$ group is called the ***C*-terminal amino acid.**

***N*-Terminal amino acid** The amino acid at the end of a polypeptide chain having the free $—NH_2$ group.

***C*-Terminal amino acid** The amino acid at the end of a polypeptide chain having the free $—CO_2H$ group.

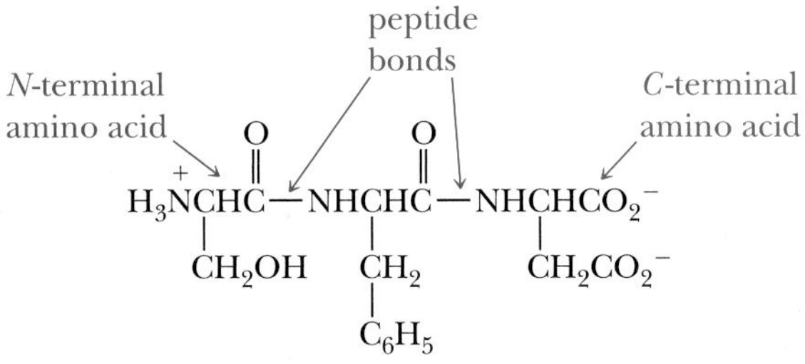

$$\overset{+}{H_3N}CH(CH_2OH)C(=O)—NHCH(CH_2C_6H_5)C(=O)—NHCH(CH_2CO_2^-)CO_2^-$$

N-terminal amino acid; peptide bonds; *C*-terminal amino acid

Ser-Phe-Asp

EXAMPLE 18.4

Draw a structural formula for Cys-Arg-Met-Asn. Label the *N*-terminal amino acid and the *C*-terminal amino acid. What is the net charge on this tetrapeptide at pH 6.0?

SOLUTION

The backbone of this tetrapeptide is a repeating sequence of nitrogen-α-carbon-carbonyl. The net charge on this tetrapeptide at pH 6.0 is + 1.

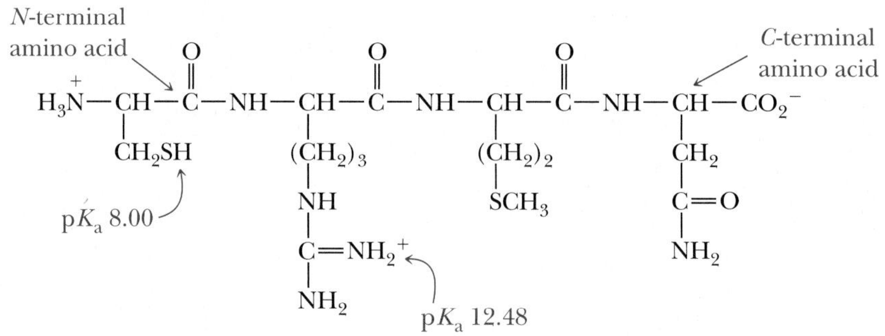

N-terminal amino acid

C-terminal amino acid

$$H_3\overset{+}{N}-CH(CH_2SH)-\overset{O}{\overset{\|}{C}}-NH-CH((CH_2)_3NHC(=NH_2^+)NH_2)-\overset{O}{\overset{\|}{C}}-NH-CH((CH_2)_2SCH_3)-\overset{O}{\overset{\|}{C}}-NH-CH(CH_2C(=O)NH_2)-CO_2^-$$

pK_a 8.00

pK_a 12.48

Practice Problem 18.4

Draw a structural formula for Lys-Phe-Ala. Label the *N*-terminal amino acid and the *C*-terminal amino acid. What is the net charge on this tripeptide at pH 6.0?

18.4 Primary Structure of Polypeptides and Proteins

Primary structure of proteins The sequence of amino acids in the polypeptide chain; read from the *N*-terminal amino acid to the *C*-terminal amino acid.

The **primary (1°) structure** of a polypeptide or protein refers to the sequence of amino acids in its polypeptide chain. In this sense, primary structure is a complete description of all covalent bonding in a polypeptide or protein.

In 1953, Frederick Sanger of Cambridge University, England, reported the primary structure of the two polypeptide chains of the hormone insulin. Not only was this a remarkable achievement in analytical chemistry, but also it clearly established that the molecules of a given protein all have the same amino acid composition and the same amino acid sequence. Today, the amino acid sequences of over 20,000 different proteins are known.

A. Amino Acid Analysis

The first step for determining the primary structure of a polypeptide is hydrolysis and quantitative analysis of its amino acid composition. Recall from Section 13.3D that amide bonds are very resistant to hydrolysis. Typically, samples of protein are hydrolyzed in 6 *M* HCl in sealed glass vials at 110°C for 24 to 72 hours. This hydrolysis can be done in a microwave oven in a shorter time. After the polypeptide is hydrolyzed, the resulting mixture of amino acids is analyzed by ion-exchange chromatography. Amino acids are detected as they emerge from the column by reaction with ninhydrin (Section 18.2D) followed by absorption spectroscopy. Current procedures for hydrolysis of polypeptides and analysis of amino acid mixtures have been refined to the point where it is possible to obtain amino acid composition from as little as 50 nanomoles (50×10^{-9} mole) of polypeptide. Figure 18.6 shows the analysis of a polypeptide hydrolysate by ion-exchange chromatography. Note that during hydrolysis, the side-chain amide groups of asparagine and glutamine are hydrolyzed, and these amino acids are detected as aspartic acid and glutamic acid. For each glutamine or asparagine hydrolyzed, an equivalent amount of ammonium chloride is formed.

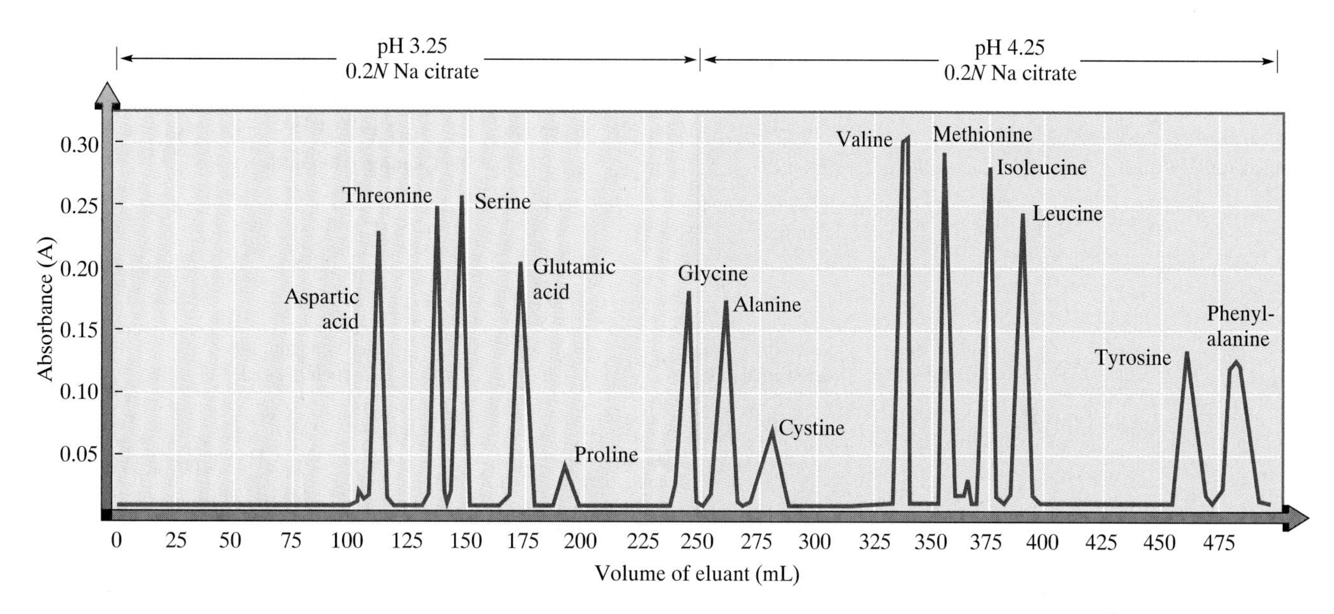

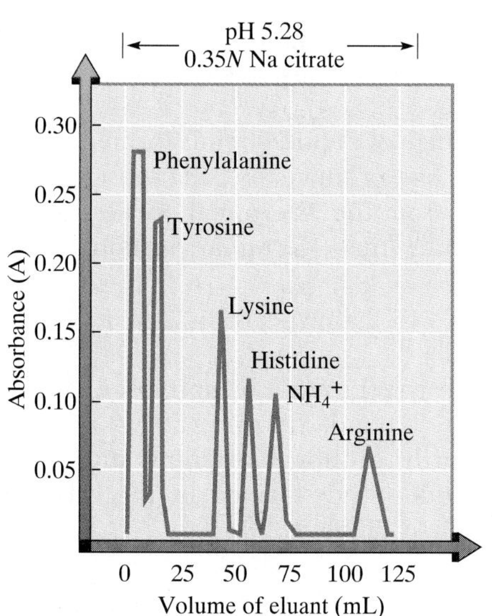

Figure 18.6
Analysis of a mixture of amino acids by ion-exchange chromatography using Amberlite IR-120, a sulfonated polystyrene resin. The resin contains phenyl—$SO_3^-Na^+$ groups. The amino acid mixture is applied to the column at low pH (3.25) under which conditions the acidic amino acids (Asp, Glu) are weakly bound to the resin and the basic amino acids (Lys, His, Arg) are tightly bound. Sodium citrate buffers at two different concentrations and three different values of pH are used to elute the amino acids from the column. Cysteine is determined as cystine, Cys—S—S—Cys, the disulfide of cysteine.

B. Sequence Analysis

Once the amino acid composition of a polypeptide has been determined, the next step is to determine the order in which the amino acids are joined in the polypeptide chain. The most common sequencing strategy is to cleave the polypeptide at specific peptide bonds (using, for example, cyanogen bromide or certain proteolytic enzymes), determine the sequence of each fragment (using, for example, the Edman degradation), and then match overlapping fragments to arrive at the sequence of the polypeptide.

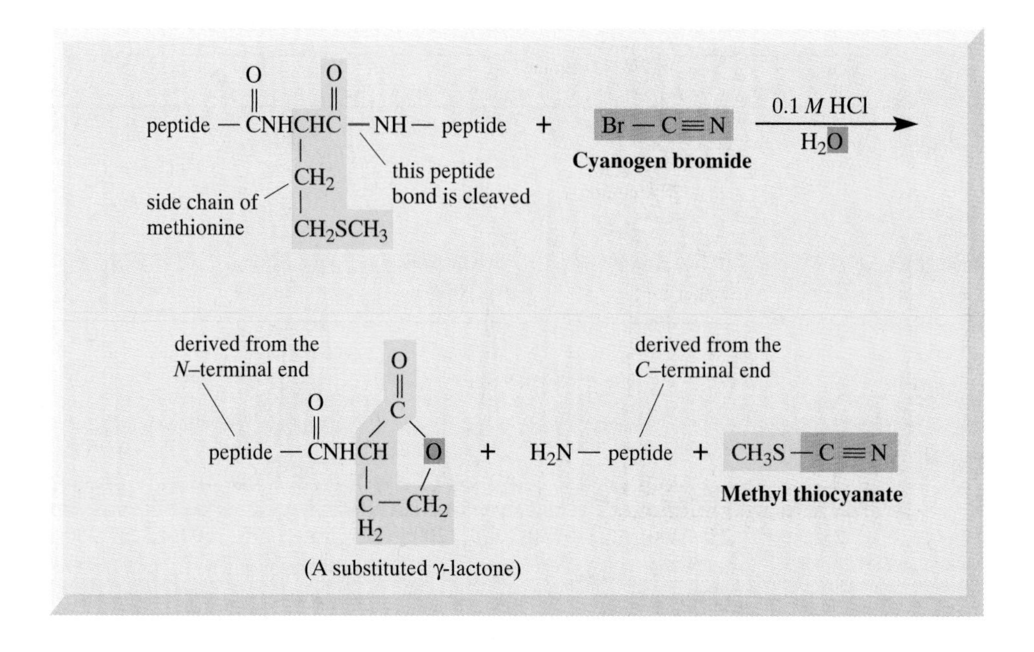

Figure 18.7 Cleavage by cyanogen bromide, BrCN, of a peptide bond formed by the carboxyl group of methionine.

Cyanogen Bromide

Cyanogen bromide (BrCN) is specific for cleavage of peptide bonds formed by the carboxyl group of methionine (Figure 18.7). The products of this cleavage are a substituted γ-lactone (Section 13.1C) derived from the *N*-terminal portion of the polypeptide, and a second fragment containing the *C*-terminal portion of the polypeptide.

Enzyme-Catalyzed Hydrolysis of Peptide Bonds

A group of proteolytic enzymes, among them trypsin and chymotrypsin, can be used to catalyze the hydrolysis of specific peptide bonds. Trypsin catalyzes the hydrolysis of peptide bonds formed by the carboxyl groups of arginine and lysine; chymotrypsin catalyzes the hydrolysis of peptide bonds formed by the carboxyl groups of phenylalanine, tyrosine, and tryptophan (Table 18.3).

Example 18.5

Which of these dipeptides are hydrolyzed by trypsin? By chymotrypsin?

(a) Arg-Glu-Ser (b) Phe-Gly-Lys

SOLUTION

(a) Trypsin catalyzes the hydrolysis of peptide bonds formed by the carboxyl groups of lysine and arginine. Therefore, the peptide bond between arginine and glutamic acid is hydrolyzed in the presence of trypsin.

$$\text{Arg-Glu-Ser} + H_2O \xrightarrow{\text{trypsin}} \text{Arg} + \text{Glu-Ser}$$

Chymotrypsin catalyzes the hydrolysis of peptide bonds formed by the carboxyl groups of phenylalanine, tyrosine, and tryptophan. Because none of

TABLE 18.3 Cleavage of Specific Peptide Bonds Catalyzed by Trypsin and Chymotrypsin

Enzyme	Catalyzes Hydrolysis of Peptide Bond Formed by Carboxyl Group of
trypsin	arginine, lysine
chymotrypsin	phenylalanine, tyrosine, tryptophan

these three aromatic amino acids is present, tripeptide (a) is not affected by chymotrypsin.

(b) Tripeptide (b) is not affected by trypsin. Although lysine is present, its carboxyl group is at the *C*-terminal end and not involved in peptide bond formation. Tripeptide (b) is hydrolyzed in the presence of chymotrypsin.

$$\text{Phe-Gly-Lys} + H_2O \xrightarrow{\text{chymotrypsin}} \text{Phe} + \text{Gly-Lys}$$

Practice Problem 18.5

Which of these tripeptides are hydrolyzed by trypsin? By chymotrypsin?

(a) Tyr-Gln-Val (b) Thr-Phe-Ser (c) Thr-Ser-Phe

Edman Degradation

Of the various chemical methods developed for determining the amino acid sequence of a polypeptide, the one most widely used today is the **Edman degradation,** introduced in 1950 by Pehr Edman of the University of Lund, Sweden. In this procedure, a polypeptide is treated with phenyl isothiocyanate, $C_6H_5N{=}C{=}S$ and then with acid. The effect of Edman degradation is to remove the *N*-terminal amino acid selectively as a substituted phenylthiohydantoin (Figure 18.8), which is then separated and identified.

Edman degradation A method for selectively cleaving and identifying the *N*-terminal amino acid of a polypeptide chain.

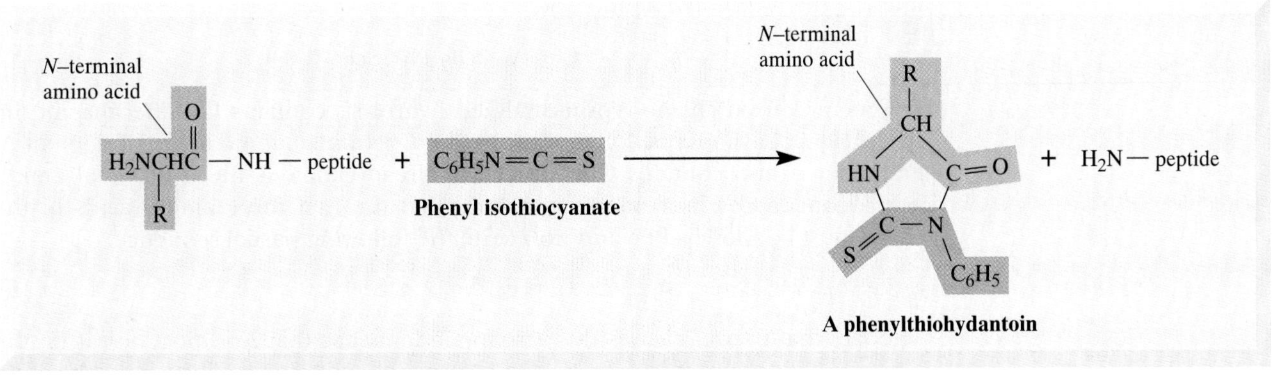

N–terminal amino acid
$H_2NCHC(=O)(R)$ — NH — peptide + $C_6H_5N{=}C{=}S$ ⟶ A phenylthiohydantoin + H_2N— peptide
Phenyl isothiocyanate
N–terminal amino acid
A phenylthiohydantoin

Figure 18.8
Edman degradation. Treatment of a polypeptide with phenyl isothiocyanate followed by acid selectively cleaves the *N*-terminal amino acid as a substituted phenylthiohydantoin.

The special value of the Edman degradation is that it cleaves the *N*-terminal amino acid from a polypeptide without affecting any other bonds in the chain. Furthermore, Edman degradation can be repeated on the shortened polypeptide, causing the next amino acid in the sequence to be cleaved and identified. In practice, it is now possible to sequence as many as the first 20 to 30 amino acids in a polypeptide by this method using as little as a few milligrams of material.

Most polypeptides in nature are longer than 20 to 30 amino acids, the practical limit to the number of amino acids that can be sequenced by repetitive Edman degradation. The special value of cleavage with cyanogen bromide, trypsin, and chymotrypsin is that a long polypeptide chain can be cleaved at specific peptide bonds into smaller polypeptide fragments, and each fragment can then be sequenced separately.

Example 18.6

Deduce the amino acid sequence of a pentapeptide from the following experimental results. Note that under the column Amino Acid Composition, the amino acids are listed in alphabetical order. In no way does this listing give any information about primary structure.

Experimental Procedure	Amino Acid Composition
Pentapeptide	Arg, Glu, His, Phe, Ser
Edman degradation	Glu
Hydrolysis catalyzed by chymotrypsin	
Fragment A	Glu, His, Phe
Fragment B	Arg, Ser
Hydrolysis catalyzed by trypsin	
Fragment C	Arg, Glu, His, Phe
Fragment D	Ser

Solution

Edman degradation cleaves Glu from the pentapeptide; therefore, glutamic acid must be the *N*-terminal amino acid.

Glu-(Arg, His, Phe, Ser)

Fragment A from chymotrypsin-catalyzed hydrolysis contains Phe. Because of the specificity of chymotrypsin, Phe must be the *C*-terminal amino acid of fragment A. Fragment A also contains Glu, which we already know is the *N*-terminal amino acid. From these observations, conclude that the first three amino acids in the chain must be Glu-His-Phe and now write the following partial sequence.

Glu-His-Phe-(Arg, Ser)

The fact that trypsin cleaves the pentapeptide means that Arg must be within the pentapeptide chain; it cannot be the *C*-terminal amino acid. Therefore, the complete sequence must be

Glu-His-Phe-Arg-Ser

Practice Problem 18.6

Deduce the amino acid sequence of an undecapeptide (11 amino acids) from the experimental results shown in the table.

Experimental Procedure	Amino Acid Composition
Undecapeptide	Ala, Arg, Glu, Lys_2, Met, Phe, Ser, Thr, Trp, Val
Edman degradation	Ala
Trypsin-catalyzed hydrolysis	
Fragment E	Ala, Glu, Arg
Fragment F	Thr, Phe, Lys
Fragment G	Lys
Fragment H	Met, Ser, Trp, Val
Chymotrypsin-catalyzed hydrolysis	
Fragment I	Ala, Arg, Glu, Phe, Thr
Fragment J	Lys_2, Met, Ser, Trp, Val
Treatment with cyanogen bromide	
Fragment K	Ala, Arg, Glu, Lys_2, Met, Phe, Thr, Val
Fragment L	Trp, Ser

18.5 THREE-DIMENSIONAL SHAPES OF POLYPEPTIDES AND PROTEINS

A. Geometry of a Peptide Bond

In the late 1930s, Linus Pauling began a series of studies to determine the geometry of a peptide bond. One of his first discoveries was that a peptide bond itself is planar. As shown in Figure 18.9, the four atoms of a peptide bond and the two α-carbons joined to it all lie in the same plane. Had you been asked in Chapter 1 to describe the geometry of a peptide bond, you probably would have predicted bond angles of 120° about the carbonyl carbon and 109.5° about the amide nitro-

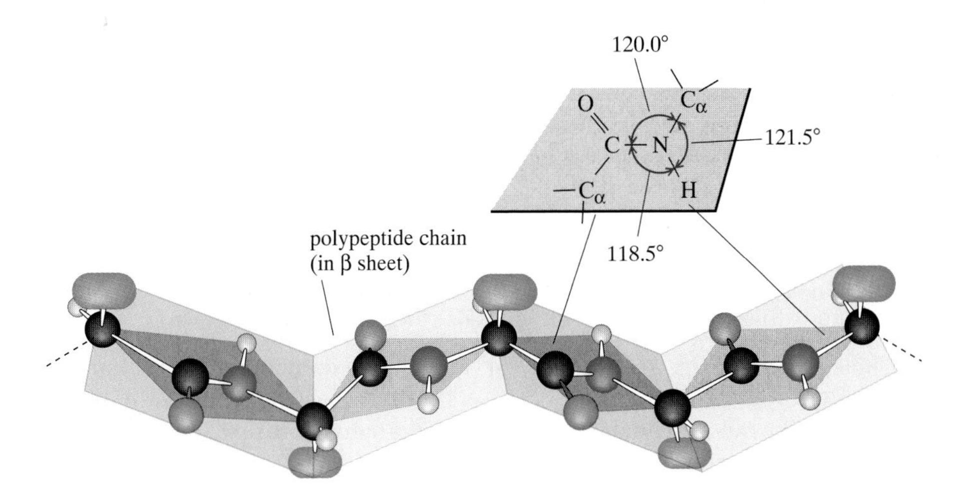

Figure 18.9
Planarity of a peptide bond. Bond angles about the carbonyl carbon and the amide nitrogen are approximately 120°.

gen. This prediction agrees with the observed bond angles of approximately 120° about the carbonyl carbon. It does not agree, however, with bond angles of 120° about the amide nitrogen. To account for this observed geometry, Pauling proposed that a peptide bond is more accurately represented as a resonance hybrid of these two contributing structures.

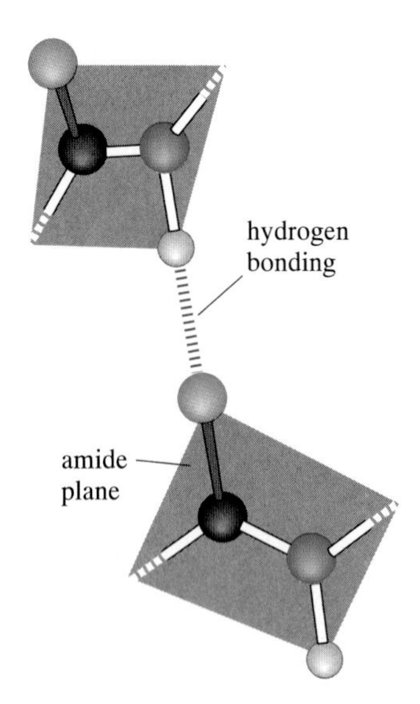

Figure 18.10 Hydrogen bonding between amide groups.

(1) (2)

Contributing structure (1) shows a carbon-oxygen double bond, and structure (2) shows a carbon-nitrogen double bond. The hybrid, of course, is neither of these; in the real structure, the carbon-nitrogen bond has considerable double-bond character. Accordingly, in the hybrid, the six-atom group is planar.

Two configurations are possible for the atoms of a planar peptide bond. In one, the two α-carbons are cis to each other; in the other, they are trans to each other. The trans configuration is more favorable because the α-carbons with the bulky groups attached to them are farther from each other than they are in the cis configuration. Virtually all peptide bonds in naturally occurring proteins studied to date have the trans configuration.

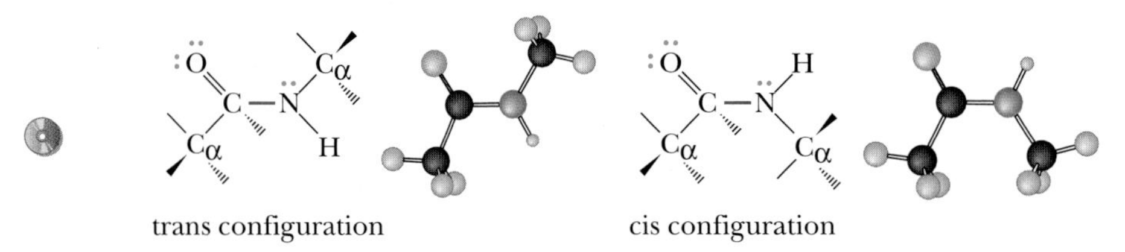

B. Secondary Structure

Secondary structure of proteins The ordered arrangements (conformations) of amino acids in localized regions of a polypeptide or protein.

Secondary (2°) structure refers to ordered arrangements (conformations) of amino acids in localized regions of a polypeptide or protein molecule. The first studies of polypeptide conformations were carried out by Linus Pauling and Robert Corey, beginning in 1939. They assumed that in conformations of greatest stability, all atoms in a peptide bond lie in the same plane, and there is hydrogen bonded between the N—H of one peptide bond and the C=O of another as shown in Figure 18.10.

On the basis of model building, Pauling proposed that two types of secondary structure should be particularly stable: the α-helix and the antiparallel β-pleated sheet.

The α-Helix

α-Helix A type of secondary structure in which a section of polypeptide chain coils into a spiral, most commonly a right-handed spiral.

In an **α-helix** pattern shown in Figure 18.11, a polypeptide chain is coiled in a spiral. As you study this section of α-helix in Figure 18.11, note the following.

1. The helix is coiled in a clockwise, or right-handed, manner. Right-handed means that if you turn the helix clockwise, it twists away from you. In this sense, a right-handed helix is analogous to the right-handed thread of a common wood or machine screw.

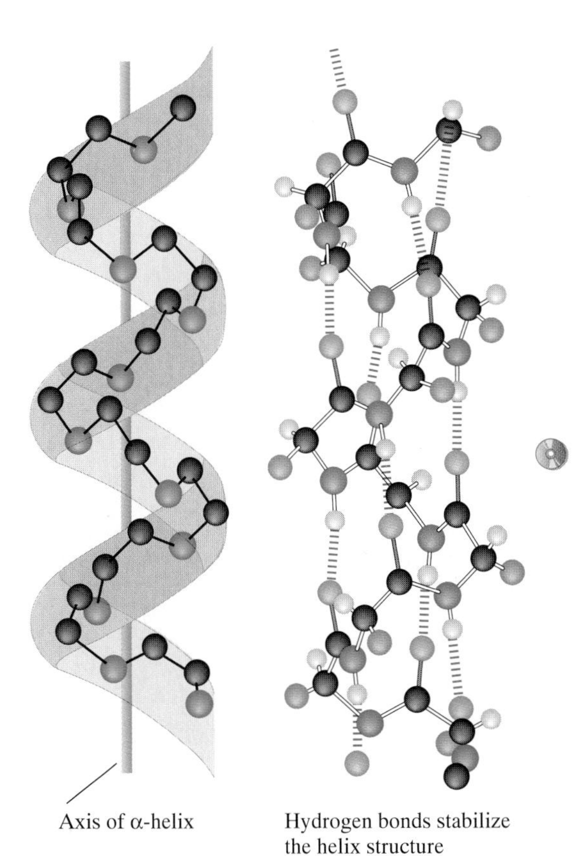

Figure 18.11
An α-helix. *(left)* Ball-and-stick model showing the carbon-nitrogen backbone of the α-helix. *(right)* Ball-and-stick model showing intrachain hydrogen bonding along the carbon-nitrogen backbone.

2. There are 3.6 amino acids per turn of the helix.
3. Each peptide bond is trans and planar.
4. The N—H group of each peptide bond points roughly downward, parallel to the axis of the helix, and the C=O of each peptide bond points roughly upward, also parallel to the axis of the helix.
5. The carbonyl group of each peptide bond is hydrogen-bonded to the N—H group of the peptide bond four amino acid units away from it. Hydrogen bonds are shown as dotted lines.
6. All R- groups point outward from the helix.

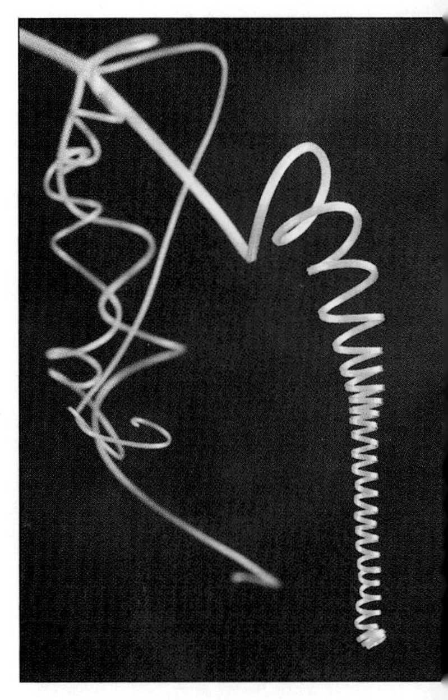

The star cucumber, *Sicyos angulatus,* uses left-handed helical tendrils to attach itself to climbing vines. Its helical pattern is analogous, but in reverse, to the right-handed α-helix of polypeptides. *(Runk/Schoenberger from Grant Heilman)*

Almost immediately after Pauling proposed the α-helix conformation, other researchers proved the presence of α-helix conformations in keratin, the protein of hair and wool. It soon became obvious that the α-helix is one of the fundamental folding patterns of polypeptide chains.

The β-Pleated Sheet

An antiparallel **β-pleated sheet** consists of extended polypeptide chains with neighboring chains running in opposite (antiparallel) directions. In a parallel β-pleated sheet, the polypeptide chains run in the same direction. Unlike the α-helix arrangement, N—H and C=O groups lie in the plane of the sheet and are roughly perpendicular to the long axis of the sheet. The C=O group of each peptide bond is hydrogen-bonded to the N—H group of a peptide bond of a neighboring chain (Figure 18.12).

β-Pleated sheet A type of secondary structure in which two sections of polypeptide chain are aligned parallel or antiparallel to one another.

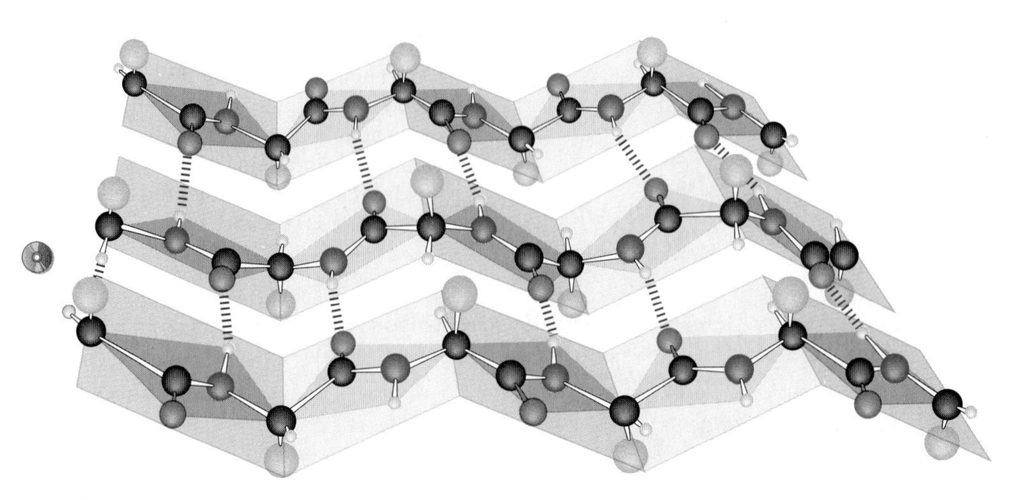

Figure 18.12
β-Pleated sheet conformation with three polypeptide chains running in opposite (antiparallel) directions. Hydrogen bonding between chains is indicated by dashed lines. *(Irving Geis)*

As you study this section of β-pleated sheet, note the following.

1. The three polypeptide chains lie adjacent to each other and run in opposite (antiparallel) directions.
2. Each peptide bond is planar, and the α-carbons are trans to each other.
3. The C=O and N—H groups of peptide bonds from adjacent chains point at each other and are in the same plane so that hydrogen bonding is possible between adjacent polypeptide chains.
4. The R-groups on any one chain alternate, first above, then below the plane of the sheet, and so on.

The β-pleated sheet conformation is stabilized by hydrogen bonding between N—H groups of one chain and C=O groups of an adjacent chain. By comparison, the α-helix is stabilized by hydrogen bonding between N—H and C=O groups within the same polypeptide chain.

C. Tertiary Structure

Tertiary structure of proteins The three-dimensional arrangement in space of all atoms in a single polypeptide chain.

Tertiary (3°) structure refers to the overall folding pattern and arrangement in space of all atoms in a single polypeptide chain. No sharp dividing line exists between secondary and tertiary structures. Secondary structure refers to the spatial arrangement of amino acids close to one another on a polypeptide chain, whereas tertiary structure refers to the three-dimensional arrangement of all atoms of a polypeptide chain. Among the most important factors in maintaining 3° structure are disulfide bonds, hydrophobic interactions, hydrogen bonding, and salt linkages.

Disulfide bond A covalent bond between two sulfur atoms; an —S—S— bond.

Disulfide bonds play an important role in maintaining tertiary structure. Disulfide bonds are formed between side chains of two cysteine units by oxidation of their thiol groups (—SH) to form a disulfide bond (Section 8.8B). Treatment of a disulfide bond with a reducing agent regenerates the thiol groups.

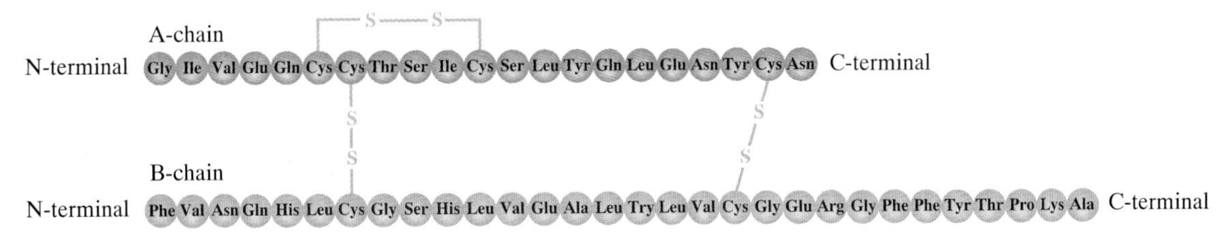

Figure 18.13
Human insulin. The A chain of 21 amino acids and B chain of 30 amino acids are connected by interchain disulfide bonds between A7 and B7 and between A20 and B19. In addition, a single intrachain disulfide bond occurs between A6 and A11.

$$\text{---NH—CH(—C(=O)—NH---)—}CH_2\text{—SH} \quad + \quad \text{HS—}CH_2\text{—CH(—C(=O)—NH---)—NH---} \underset{\text{reduction}}{\overset{\text{oxidation}}{\rightleftharpoons}} \text{---NH—CH(—C(=O)—NH---)—}CH_2\text{—S—S—}CH_2\text{—CH(—C(=O)—NH---)—NH---}$$

side chains of cysteine

a disulfide bond

Figure 18.13 shows the amino acid sequence of human insulin. This protein consists of two polypeptide chains: an A chain of 21 amino acids and a B chain of 30 amino acids. The A chain is bonded to the B chain by two interchain disulfide bonds. An intrachain disulfide bond also connects the cysteine units at positions 6 and 11 of the A chain.

As an example of 2° and 3° structure, let us look at the three-dimensional structure of myoglobin—a protein found in skeletal muscle and particularly abundant in diving mammals, such as seals, whales, and porpoises. Myoglobin and its structural relative, hemoglobin, are the oxygen transport and storage molecules of vertebrates. Hemoglobin binds molecular oxygen in the lungs and transports it to myoglobin in muscles. Myoglobin stores molecular oxygen until it is required for metabolic oxidation.

Dorothy Crowfoot Hodgkin. A recent biography states that "more than any other scientist, she personified the transformation of crystallography from a black art into an indispensable scientific tool. . . . She made not one brilliant breakthrough but a series of them, deciphering the structure of one important structure after another. Her first major achievement was the determination of the structure of penicillin, which she worked on from 1942 to 1949. This was followed by a determination of the structure of a cephalosporin and insulin. It was her determination of the structure of vitamin B_{12} in 1948 that earned her the 1964 Nobel Prize for Chemistry." *(The Bettmann Archive)*

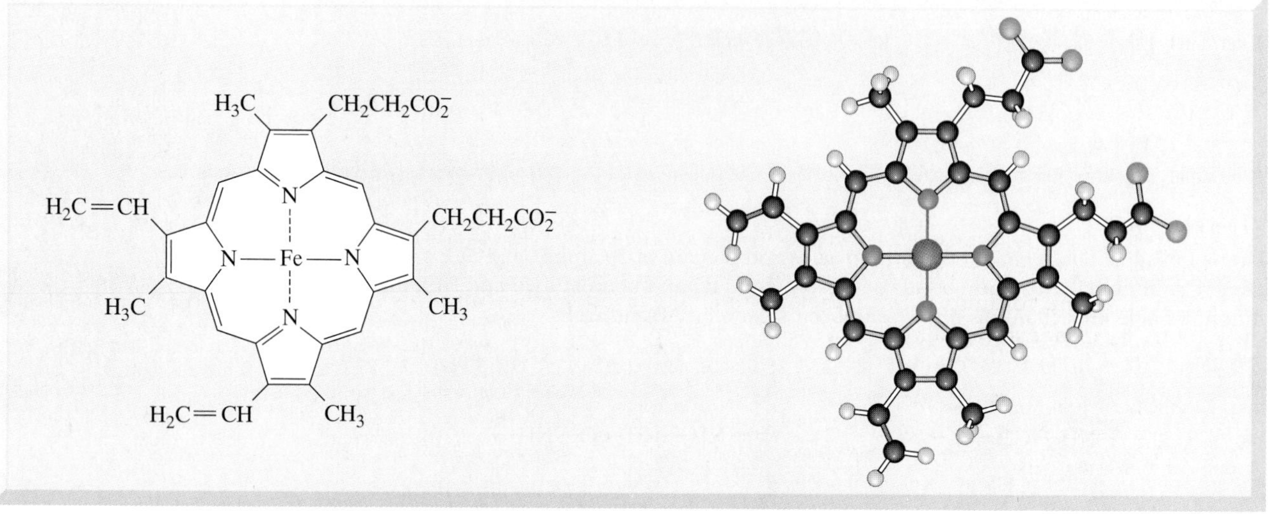

Figure 18.14
The structure of heme, found in myoglobin and hemoglobin.

Myoglobin consists of a single polypeptide chain of 153 amino acids. Myoglobin also contains a single heme unit. Heme consists of one Fe^{2+} ion coordinated in a square planar array with the four nitrogen atoms of a molecule of porphyrin (Figure 18.14).

Determination of the three-dimensional structure of myoglobin represented a milestone in the study of molecular architecture. For their contribution to this research, John C. Kendrew and Max F. Perutz, both of Britain, shared the 1962 Nobel Prize for Chemistry. The secondary and tertiary structures of myoglobin are shown in Figure 18.15. The single polypeptide chain is folded into a complex, al-

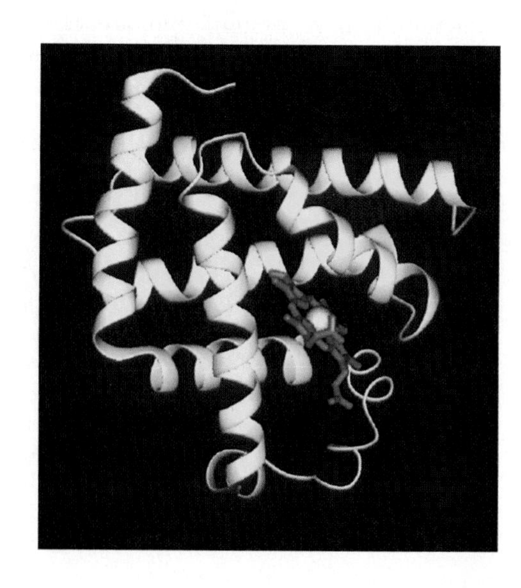

Figure 18.15
Ribbon model of myoglobin. The polypeptide chain is shown in yellow, the heme ligand in red, and the Fe atom as a white sphere. *(Brent Iverson, University of Texas)*

most boxlike shape. Important structural features of the three-dimensional shape of myoglobin are that

1. The backbone consists of eight relatively straight sections of α-helix, each separated by a bend in the polypeptide chain. The longest section of α-helix has 24 amino acids, the shortest has 7. Some 75% of the amino acids are found in these eight regions of α-helix.
2. Hydrophobic side chains of phenylalanine, alanine, valine, leucine, isoleucine, and methionine are clustered in the interior of the molecule where they are shielded from contact with water. **Hydrophobic interactions** are a major factor in directing the folding of the polypeptide chain of myoglobin into this compact, three-dimensional shape.
3. The outer surface of myoglobin is coated with hydrophilic side chains, such as those of lysine, arginine, serine, glutamic acid, histidine, and glutamine, which interact with the aqueous environment by **hydrogen bonding.** The only polar side chains that point to the interior of the myoglobin molecule are those of two histidine units, which point inward toward the heme group.
4. Oppositely charged amino acid side chains close to each other in the three-dimensional structure interact by electrostatic attractions called **salt linkages.** An example of a salt linkage is the attraction of the side chains of lysine ($—NH_3^+$) and glutamic acid ($—CO_2^-$).

The humpback whale relies on myoglobin as a storage form of oxygen. *(Stuart Westmoreland/Tony Stone Images)*

The tertiary structures of hundreds of proteins have also been determined. It is clear that proteins contain α-helix and β-pleated sheet structures, but that wide variations exist in the relative amounts of each. Lysozyme, with 129 amino acids in a single polypeptide chain, has only 25% of its amino acids in α-helix regions. Cytochrome, with 104 amino acids in a single polypeptide chain, has no α-helix structure but does contain several regions of β-pleated sheet. Yet, whatever the proportions of α-helix, β-pleated sheet, or other periodic structure, virtually all nonpolar side chains of water-soluble proteins are directed toward the interior of the molecule, whereas polar side chains are on the surface of the molecule and in contact with the aqueous environment. Note that this arrangement of polar and nonpolar groups in water-soluble proteins very much resembles the arrangement of polar and nonpolar groups of soap molecules in micelles (Figure 17.3). It also resembles the arrangement of phospholipids in lipid bilayers (Figure 17.13).

EXAMPLE 18.7

With which of the following amino acid side chains can the side chain of threonine form hydrogen bonds?

(a) Valine (b) Asparagine (c) Phenylalanine
(d) Histidine (e) Tyrosine (f) Alanine

SOLUTION

The side chain of threonine contains a hydroxyl group that can participate in hydrogen bonding in two ways: its oxygen has a partial negative charge and can function as a hydrogen bond acceptor; its hydrogen has a partial positive charge and can function as a hydrogen bond donor. Therefore, the side chain of threonine can form hydrogen bonds with the side chains of tyrosine, asparagine, and histidine.

Practice Problem 18.7

At pH 7.4, with what amino acid side chains can the side chain of lysine form salt linkages?

D. Quaternary Structure

Quaternary structure of proteins The arrangement of polypeptide monomers into a noncovalently bonded aggregation.

Hydrophobic effect The tendency of nonpolar groups to cluster in such a way as to be shielded from contact with an aqueous environment.

Most proteins of molecular weight greater than 50,000 consist of two or more noncovalently linked polypeptide chains. The arrangement of protein monomers into an aggregation is known as **quaternary (4°) structure.** A good example is hemoglobin, a protein that consists of four separate polypeptide chains: two α-chains of 141 amino acids each and two β-chains of 146 amino acids each. The quaternary structure of hemoglobin is shown in Figure 18.16.

The major factor stabilizing the aggregation of protein subunits is the **hydrophobic effect.** When separate polypeptide chains fold into compact three-dimensional shapes to expose polar side chains to the aqueous environment and shield nonpolar side chains from water, hydrophobic "patches" still appear on the surface, in contact with water. These patches can be shielded from water if two or more monomers assemble so that their hydrophobic patches are in contact. The numbers of subunits of several proteins of known quaternary structure are shown in Table 18.4.

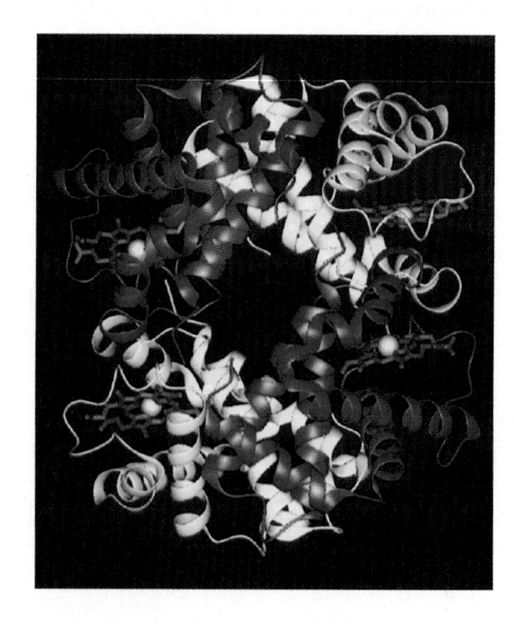

Figure 18.16
Ribbon model of hemoglobin. The α-chains are shown in purple, the β-chains in yellow, the heme ligands in red, and the Fe atoms as white spheres. *(Brent Iverson, University of Texas)*

TABLE 18.4 Quaternary Structure of Selected Proteins

Protein	Number of Subunits
alcohol dehydrogenase	2
aldolase	4
hemoglobin	4
lactate dehydrogenase	4
insulin	6
glutamine synthetase	12
tobacco mosaic virus protein disc	17

SUMMARY

α-Amino acids are compounds that contain an amino group to a carboxyl group (Section 18.1A). A **zwitterion** is an internal salt of an amino acid. With the exception of glycine, all protein-derived amino acids are chiral (Section 18.1B). In the D,L convention, all are L-amino acids. Isoleucine and threonine contain a second stereocenter. The 20 protein-derived amino acids are commonly divided into four categories: 9 with nonpolar side chains, 4 with polar but un-ionized side chains, 4 with acidic side chains, and 3 with basic side chains.

The **isoelectric point, pI,** of an amino acid, polypeptide, or protein is the pH at which it has no net charge (Section 18.2C). **Electrophoresis** is the process of separating compounds on the basis of their electric charge (Section 18.2D). Compounds having a high charge density move more rapidly than those with a lower charge density. Any amino acid or protein in a solution with a pH that equals the pI of the compound remains at the origin.

A **peptide bond** is the special name given to the amide bond formed between α-amino acids (Section 18.3). A **polypeptide** is a biological macromolecule containing 10 or more amino acids joined by peptide bonds. By convention, the sequence of amino acids in a polypeptide is written beginning with the ***N*-terminal amino acid** toward the ***C*-terminal amino acid. Primary (1°) structure** of a polypeptide is the sequence of amino acids in its polypeptide chain.

A **peptide bond** is planar (Section 18.5A), that is, the four atoms of the amide bond and the two α-carbons of a peptide bond lie in the same plane. Bond angles about the amide nitrogen and the amide carbonyl carbon are approximately 120°. **Secondary (2°) structure** (Section 18.5B) refers to the ordered arrangement (conformations) of amino acids in localized regions of a polypeptide or protein. Two types of secondary structure are the α-helix and the β-pleated sheet. **Tertiary (3°) structure** (Section 18.5C) refers to the overall folding pattern and arrangement in space of all atoms in a single polypeptide chain. **Quaternary (4°) structure** (Section 18.5D) is the arrangement of individual polypeptide chains into a noncovalently bonded aggregate.

KEY REACTIONS

1. Acidity of an α-Carboxyl Group (Section 18.2A)

An α-CO_2H (pK_a approximately 2.19) of a protonated amino acid is a considerably stronger acid than acetic acid (pK_a 4.76) or other low-molecular-weight aliphatic carboxylic acid due to the electron-withdrawing inductive effect of the α-NH_3^+ group.

$$\underset{\displaystyle NH_3^+}{\overset{}{RCHCO_2H}} + H_2O \rightleftharpoons \underset{\displaystyle NH_3^+}{RCHCO_2^-} + H_3O^+ \qquad pK_a = 2.19$$

2. Acidity of an α-Ammonium Group (Section 18.2A)

An α-NH_3^+ group (pK_a approximately 9.47) is a slightly stronger acid than a primary aliphatic ammonium ion (pK_a approximately 10.76).

$$\underset{\large NH_3^+}{RCHCO_2^-} + H_2O \rightleftharpoons \underset{\large NH_2}{RCHCO_2^-} + H_3O^+ \qquad pK_a = 9.47$$

3. Reaction of an α-Amino Acid with Ninhydrin (Section 18.2D)

Treatment of an α-amino acid with ninhydrin gives a purple-colored solution. Treatment of proline with ninhydrin gives an orange-colored solution.

$$\underset{\large NH_3^+}{RCH\overset{O}{\overset{\|}{C}}O^-} + 2\ \text{Ninhydrin} \longrightarrow \text{Purple-colored anion} + R\overset{O}{\overset{\|}{C}}H + CO_2 + H_3O^+$$

An α-amino acid Ninhydrin Purple-colored anion

4. Cleavage of a Peptide Bond by Cyanogen Bromide (Section 18.4B)

Cleavage is regioselective for a peptide bond formed by the carboxyl group of methionine.

This peptide bond is cleaved

$$\text{peptide}-\overset{O}{\overset{\|}{C}}NH\underset{\large CH_2CH_2SCH_3}{CH}\overset{O}{\overset{\|}{C}}-NH-\text{peptide} \xrightarrow{Br-CN} \text{peptide}-\overset{O}{\overset{\|}{C}}NHCH\ (\gamma\text{-lactone}) + H_2N-\text{peptide} + CH_3SCN$$

side chain of methionine → $CH_2CH_2SCH_3$

A substituted γ-lactone

This peptide is derived from the *C*-terminal portion

5. Edman Degradation (Section 18.4B)

Treatment with phenyl isothiocyanate followed by acid removes the *N*-terminal amino acid as a substituted phenylthiohydantoin, which is then separated and identified.

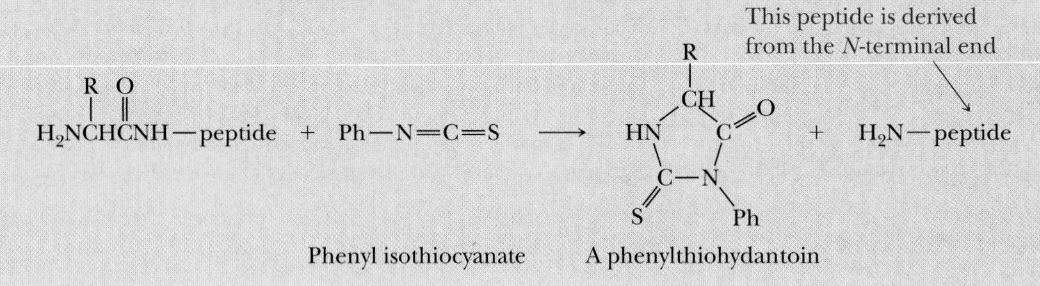

Phenyl isothiocyanate A phenylthiohydantoin

ADDITIONAL PROBLEMS

Amino Acids

18.8 What amino acid does each abbreviation stand for?

(a) Phe **(b)** Ser **(c)** Asp **(d)** Gln **(e)** His **(f)** Gly **(g)** Tyr

18.9 Configuration of the stereocenter in α-amino acids is most commonly specified using the D,L convention. It can also be identified using the R,S convention (Section 3.3). Does the stereocenter in L-serine have the R or the S configuration?

18.10 Assign an R or S configuration to the stereocenter in each amino acid.

(a) L-Phenylalanine **(b)** L-Glutamic acid **(c)** L-Methionine

18.11 The amino acid threonine has two stereocenters. The stereoisomer found in proteins has the configuration 2S,3R about the two stereocenters. Draw a Fischer projection of this stereoisomer and also a three-dimensional representation.

18.12 Define the term "zwitterion."

18.13 Draw zwitterion forms of these amino acids.

(a) Valine **(b)** Phenylalanine **(c)** Glutamine

18.14 Why are Glu and Asp often referred to as acidic amino acids?

18.15 Why is Arg often referred to as a basic amino acid? Which two other amino acids are also basic amino acids?

18.16 What is the meaning of the alpha as it is used in α-amino acid?

18.17 Several β-amino acids exist. There is a unit of β-alanine, for example, contained within the structure of coenzyme A (Problem 13.27). Write the structural formula of β-alanine.

18.18 Although only L-amino acids occur in proteins, D-amino acids are often a part of the metabolism of lower organisms. The antibiotic actinomycin D, for example, contains a unit of D-valine, and the antibiotic bacitracin A contains units of D-asparagine and D-glutamic acid. Draw Fischer projections and three-dimensional representations for these three D-amino acids.

18.19 Histamine (*The Merck Index,* 12th ed., #4756) is synthesized from one of the 20 protein-derived amino acids. Suggest which amino acid is its biochemical precursor and the type of organic reaction(s) involved in its biosynthesis (e.g., oxidation, reduction, decarboxylation, nucleophilic substitution).

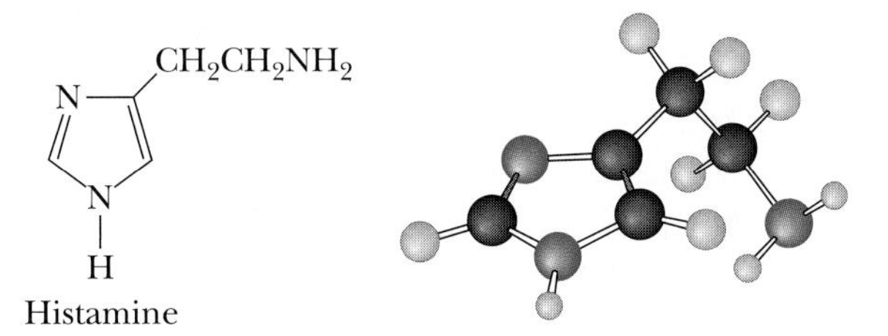

Histamine

18.20 Both norepinephrine and epinephrine (*The Merck Index,* 12th ed., #6788 and #3656) are synthesized from the same protein-derived amino acid. From which amino acid are they synthesized and what types of reactions are involved in their biosynthesis?

(a) HO, HO, H, OH, NH_2 — Norepinephrine

(b) HO, HO, H, OH, H, N, CH_3 — Epinephrine (Adrenaline)

18.21 From which amino acid are serotonin and melatonin (*The Merck Index*, 12th ed., #8607 and #5857) synthesized and what types of reactions are involved in their biosynthesis?

(a) Serotonin (b) Melatonin

Acid-Base Behavior of Amino Acids

18.22 Draw a structural formula for the form of each amino acid most prevalent at pH 1.0.

(a) Threonine **(b)** Arginine **(c)** Methionine **(d)** Tyrosine

18.23 Draw a structural formula for the form of each amino most prevalent at pH 10.0.

(a) Leucine **(b)** Valine **(c)** Proline **(d)** Aspartic acid

18.24 Write the zwitterion form of alanine and show its reaction with

(a) 1 mole NaOH **(b)** 1 mole HCl

18.25 Write the form of lysine most prevalent at pH 1.0 and then show its reaction with the following. Consult Table 18.2 for pK_a values of the ionizable groups in lysine.

(a) 1 mole NaOH **(b)** 2 moles NaOH **(c)** 3 moles NaOH

18.26 Write the form of aspartic acid most prevalent at pH 1.0 and then show its reaction with the following. Consult Table 18.2 for pK_a values of the ionizable groups in aspartic acid.

(a) 1 mole NaOH **(b)** 2 moles NaOH **(c)** 3 moles NaOH

18.27 Given pK_a values for ionizable groups from Table 18.2, sketch curves for the titration of (a) glutamic acid with NaOH and (b) histidine with NaOH.

18.28 Draw a structural formula for the product formed when alanine is treated with the following reagents.

(a) Aqueous NaOH **(b)** Aqueous HCl

(c) CH_3CH_2OH, H_2SO_4 **(d)** $(CH_3CO)_2O$, CH_3CO_2Na

18.29 Account for the fact that the isoelectric point of glutamine (pI 5.65) is higher than the isoelectric point of glutamic acid (pI 3.08).

18.30 Enzyme-catalyzed decarboxylation of glutamic acid gives 4-aminobutanoic acid (Section 18.1D). Estimate the pI of 4-aminobutanoic acid.

18.31 Guanidine and the guanidino group present in arginine are two of the strongest organic bases known. Account for their basicity.

18.32 At pH 7.4, the pH of blood plasma, do the majority of protein-derived amino acids bear a net negative charge or a net positive charge?

18.33 Do the following compounds migrate to the cathode or to the anode on electrophoresis at the specified pH?

(a) Histidine at pH 6.8 **(b)** Lysine at pH 6.8

(c) Glutamic acid at pH 4.0 **(d)** Glutamine at pH 4.0

(e) Glu-Ile-Val at pH 6.0 **(f)** Lys-Gln-Tyr at pH 6.0

18.34 At what pH would you carry out an electrophoresis to separate the amino acids in each mixture?

(a) Ala, His, Lys **(b)** Glu, Gln, Asp **(c)** Lys, Leu, Tyr

18.35 Examine the amino acid sequence of human insulin (Figure 18.14) and list each Asp, Glu, His, Lys, and Arg in this molecule. Do you expect human insulin to have an isoelectric point nearer that of the acidic amino acids (pI 2.0–3.0), the neutral amino acids (pI 5.5–6.5), or the basic amino acids (pI 9.5–11.0)?

Primary Structure of Polypeptides and Proteins

18.36 If a protein contains four different SH groups, how many different disulfide bonds are possible if only a single disulfide bond is formed? How many different disulfides are possible if two disulfide bonds are formed?

18.37 How many different tetrapeptides can be made if

(a) The tetrapeptide contains one unit each of Asp, Glu, Pro, and Phe?

(b) All 20 amino acids can be used, but each only once?

18.38 A decapeptide has the following amino acid composition

$$\text{Ala}_2\text{, Arg, Cys, Glu, Gly, Leu, Lys, Phe, Val}$$

Partial hydrolysis yields the following tripeptides

Cys-Glu-Leu + Gly-Arg-Cys + Leu-Ala-Ala + Lys-Val-Phe + Val-Phe-Gly

One round of Edman degradation yields a lysine phenylthiohydantoin. From this information, deduce the primary structure of this decapeptide.

18.39 Following is the primary structure of glucagon, a polypeptide hormone of 29 amino acids. Glucagon is produced in the *α*-cells of the pancreas and helps maintain blood glucose levels in a normal concentration range.

1 5 10 15
His-Ser-Glu-Gly-Thr-Phe-Thr-Ser-Asp-Tyr-Ser-Lys-Tyr-Leu-Asp-Ser-Arg-Arg-

20 25 29
Ala-Gln-Asp-Phe-Val-Gln-Trp-Leu-Met-Asn-Thr

Glucagon

Which peptide bonds are hydrolyzed when this polypeptide is treated with each reagent?

(a) Phenyl isothiocyanate **(b)** Chymotrypsin **(c)** Trypsin **(d)** Br—CN

18.40 A tetradecapeptide (14 amino acid residues) gives the following peptide fragments on partial hydrolysis. From this information, deduce the primary structure of this polypeptide. Fragments are grouped according to size.

Pentapeptide Fragments	Tetrapeptide Fragments
Phe-Val-Asn-Gln-His	Gln-His-Leu-Cys
His-Leu-Cys-Gly-Ser	His-Leu-Val-Glu
Gly-Ser-His-Leu-Val	Leu-Val-Glu-Ala

18.41 Draw a structural formula of these tripeptides. Mark each peptide bond, the *N*-terminal amino acid, and the *C*-terminal amino acid.

(a) Phe-Val-Asn **(b)** Leu-Val-Gln

18.42 Estimate the pI of each tripeptide on Problem 18.41.

18.43 Glutathione (G—SH, *The Merck Index,* 12th ed., #4483), one of the most common tripeptides in animals, plants, and bacteria, is a scavenger of oxidizing agents. In reacting with oxidizing agents, glutathione is converted to G—S—S—G.

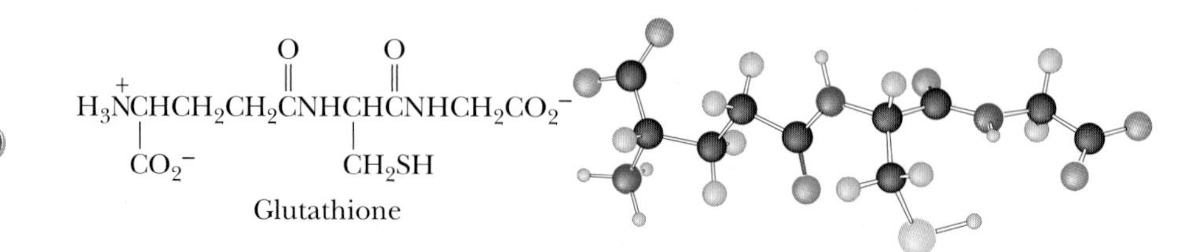

(a) Name the amino acids in this tripeptide.

(b) What is unusual about the peptide bond formed by the *N*-terminal amino acid?

(c) Write a balanced half-reaction for the reaction of two molecules of glutathione to form a disulfide bond. Is glutathione a biological oxidizing agent or a biological reducing agent?

(d) Write a balanced equation for reaction of glutathione with molecular oxygen, O_2, to form G—S—S—G and H_2O. Is molecular oxygen oxidized or reduced in this process?

18.44 Following is a structural formula and ball-and-stick model for the artificial sweetener aspartame (*The Merck Index,* 12th ed., #874). Each amino acid has the L-configuration.

Aspartame is present in many artificially sweetened foods and beverages. *(Charles D. Winters)*

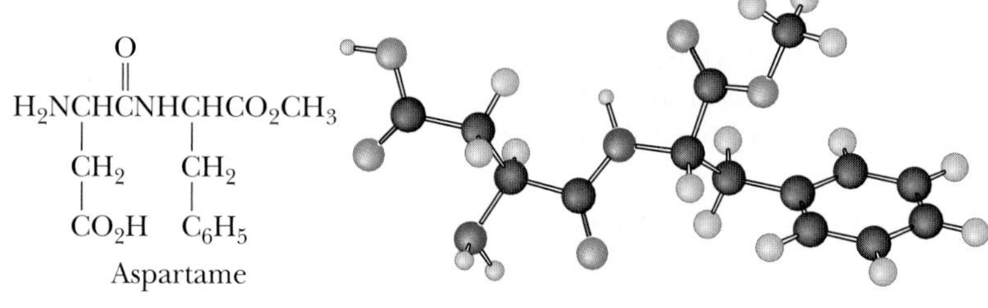

(a) Name the two amino acids in this molecule.

(b) Estimate the isoelectric point of aspartame.

(c) Write structural formulas for the products of hydrolysis of aspartame in 1 *M* HCl.

Three-Dimensional Shapes of Polypeptides and Proteins

18.45 Examine the α-helix conformation. Are amino acid side chains arranged all inside the helix, all outside the helix, or randomly?

18.46 Distinguish between intermolecular and intramolecular hydrogen bonding between the backbone groups on polypeptide chains. In what type of secondary structure do you find intermolecular hydrogen bonds? In what type do you find intramolecular hydrogen bonding?

18.47 Many plasma proteins found in an aqueous environment are globular in shape. Which amino acid side chains would you expect to find on the surface of a globular protein and in contact with the aqueous environment? Which would you expect to find inside, shielded from the aqueous environment? Explain.

(a) Leu **(b)** Arg **(c)** Ser **(d)** Lys **(e)** Phe

19

Nucleic Acids

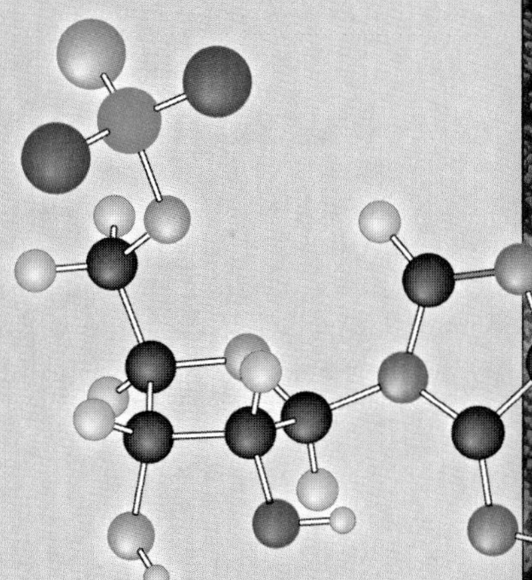

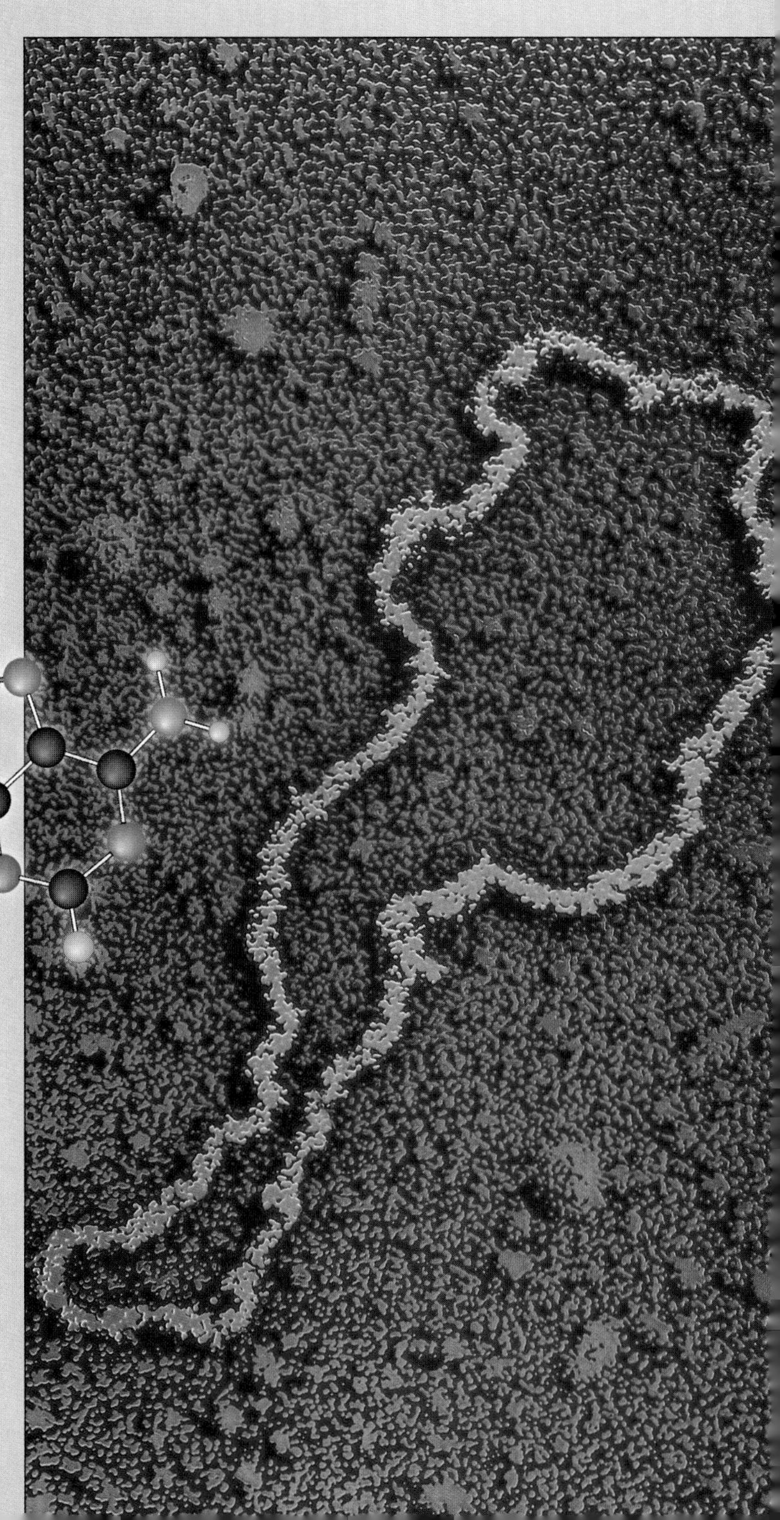

False-colored transmission electron micrograph of plasmid of bacterial DNA. If the cell wall of a bacterium such as *Escherichia coli* is partially digested and the cell then osmotically shocked by dilution with water, its contents are extruded to the exterior. Shown in this electron micrograph is the bacterial chromosome surrounding the cell. *(Professor Stanley Cohen/Science Photo Library/Photo Researchers, Inc.)*

The organization, maintenance, and regulation of cellular function requires a tremendous amount of information, all of which must be processed each time a cell is replicated. With very few exceptions, **genetic information** is stored and transmitted from one generation to the next in the form of **deoxyribonucleic acids (DNA).** Genes, the hereditary units of chromosomes, are long stretches of double-stranded DNA. If the DNA in a human chromosome in a single cell were uncoiled, it would be approximately 1.8 meters long!

Genetic information is expressed in two stages: transcription from DNA to **ribonucleic acids (RNA)** and then translation for the synthesis of proteins.

$$\text{DNA} \xrightarrow{\text{transcription}} \text{RNA} \xrightarrow{\text{translation}} \text{proteins}$$

Thus, DNA is the repository of genetic information in cells, whereas RNA serves in the transcription and translation of this information, which is then expressed through the synthesis of proteins.

In this chapter, we examine the structure of nucleosides and nucleotides and the manner in which these monomers are covalently bonded to form **nucleic acids.** Then, we examine the manner in which genetic information is encoded on molecules of DNA, the function of the three types of ribonucleic acids, and finally how the primary structure of a DNA molecule is determined.

Nucleic acid A biopolymer containing three types of monomer units: heterocyclic aromatic amine bases derived from purine and pyrimidine, the monosaccharides D-ribose or 2-deoxy-D-ribose, and phosphate.

19.1 Nucleosides and Nucleotides

Controlled hydrolysis of nucleic acids yields three components: heterocyclic aromatic amine bases, the monosaccharides D-ribose or 2-deoxy-D-ribose (Section 16.1), and phosphate ions. The five heterocyclic aromatic amine bases most common to nucleic acids are shown in Figure 19.1. Uracil, cytosine, and thymine are referred to as pyrimidine bases after the name of the parent base; adenine and guanine are referred to as purine bases.

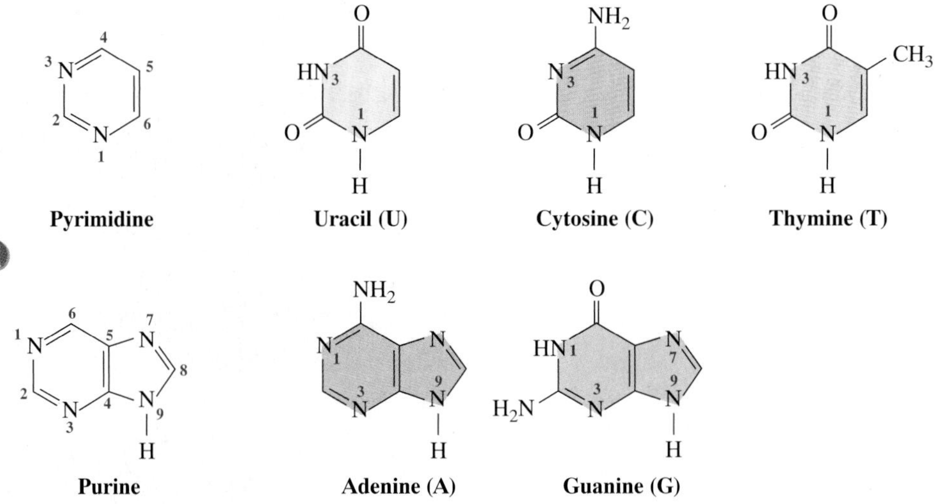

Pyrimidine

Uracil (U)

Cytosine (C)

Thymine (T)

Purine

Adenine (A)

Guanine (G)

Figure 19.1
Names and one-letter abbreviations for the heterocyclic aromatic amine bases most common to DNA and RNA. Bases are numbered according to the patterns of the parent compounds, pyrimidine and purine.

CHEMICAL CONNECTIONS

Retroviruses—From RNA to DNA

An important exception to the DNA-RNA-protein expression of genetic information is the retroviruses, simple organisms that store their genetic information in the form of RNA instead of DNA. Viruses are unable to reproduce themselves and must rely, instead, on the biosynthetic machinery of host cells for reproduction. Because host cells do not recognize RNA as a storage form of genetic information, the information encoded in viral RNA must first be transcribed to DNA by a process called reverse transcription. This viral information is then transcribed into forms of RNA recognized by the protein-synthesizing machinery of the host cell. The **human immunodeficiency virus (HIV)** is a retrovirus and produces the condition known as **acquired immune deficiency syndrome (AIDS).**

A **nucleoside** is a compound containing D-ribose or 2-deoxy-D-ribose bonded to a heterocyclic aromatic amine base by a *β-N*-glycoside bond (Section 16.4A). The monosaccharide component of DNA is 2-deoxy-D-ribose, whereas that of RNA is D-ribose. The glycoside bond is between *C*-1′ (the anomeric carbon) of ribose or 2-deoxyribose and *N*-1 of a pyrimidine base or *N*-9 of a purine base. Figure 19.2 shows the structural formula for a nucleoside derived from ribose and uracil.

Nucleoside A building block of nucleic acids, consisting of D-ribose or 2-deoxy-D-ribose bonded to a heterocyclic aromatic amine base by a *β-N*-glycoside bond.

A **nucleotide** is a nucleoside in which a molecule of phosphoric acid is esterified with a free hydroxyl of the monosaccharide, most commonly either the 3′-hydroxyl or the 5′-hydroxyl. A nucleotide is named by giving the name of the parent nucleoside followed by the word "monophosphate." The position of the phosphoric ester is specified by the number of the carbon to which it is bonded. Figure 19.3 shows a structural formula and space-filling model of 5′-adenosine monophosphate. Monophosphoric esters are diprotic acids with pK_a values of approximately 1 and 6. Therefore, at pH 7, the two hydrogens of a phosphoric monoester are fully ionized giving a nucleotide a charge of −2.

Nucleotide A nucleoside in which a molecule of phosphoric acid is esterified with an —OH of the monosaccharide, most commonly either the 3′-OH or the 5′-OH.

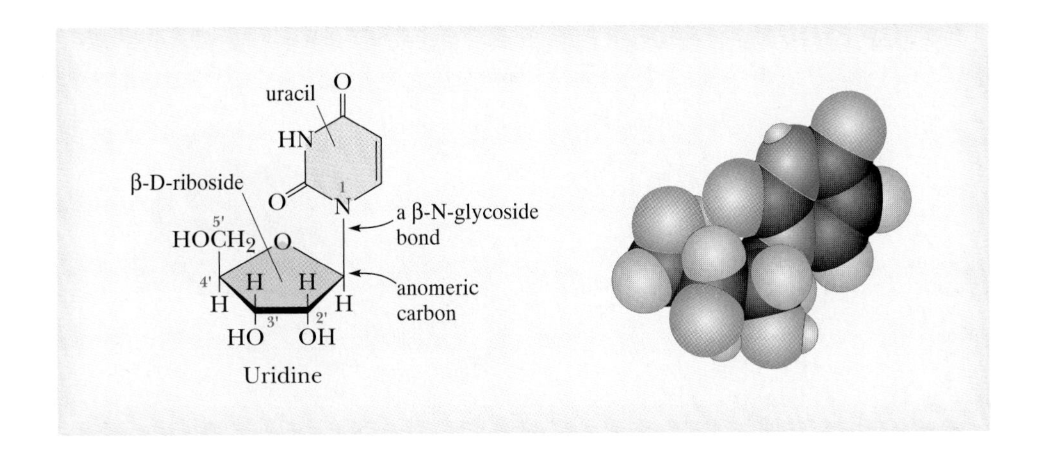

Figure 19.2
Uridine, a nucleoside. Atom numbers on the monosaccharide rings are primed to distinguish them from atom numbers on the heterocyclic aromatic amine bases.

CHEMICAL CONNECTIONS

The Search for Antiviral Drugs

The search for antiviral drugs has been more difficult than the search for antibacterial drugs primarily because viral replication depends on the metabolic processes of the invaded cell. Thus, antiviral drugs are also likely to cause harm to the cells that harbor the virus. The challenge in developing antiviral drugs has been to understand the biochemistry of viruses and to develop drugs that target processes specific to them. Compared with the large number of antibacterial drugs that are available, there are only a handful of antiviral drugs, and they have nowhere near the effectiveness that antibiotics have on bacterial infections.

Acyclovir (*The Merck Index,* 12th ed., #148) is one of the first of a new family of drugs for the treatment of infectious diseases caused by DNA viruses called herpesvirus. Herpes infections in humans are of two kinds: herpes simplex type 1, which gives rise to mouth and eye sores, and herpes simplex type 2, which gives rise to serious genital infections. Acyclovir is highly effective against herpesvirus-caused genital infections. The drug is activated *in vivo* by conversion of the primary —OH (which corresponds to the 5′-OH of a riboside or a deoxyriboside) to a triphosphate. Because of its close resemblance to deoxyguanosine triphosphate, an essential precursor for DNA synthesis, acyclovir triphosphate is taken up by viral DNA polymerase to form an enzyme-substrate complex on which no 3′-OH exists for replication to continue. Thus, the enzyme-substrate complex is no longer active (it is a dead-end complex), viral replication is disrupted, and the virus is destroyed.

Nucleoside monophosphates can be further phosphorylated to form nucleoside diphosphates and nucleoside triphosphates. Shown in Figure 19.4 is a structural formula for adenosine 5′-triphosphate (ATP).

Nucleoside diphosphates and triphosphates are also polyprotic acids and are extensively ionized at pH 7.0. pK_a values of the first three ionization steps for adenosine triphosphate are less than 5.0. The value of pK_{a4} is approximately 7.0. Therefore, at pH 7.0, approximately 50% of adenosine triphosphate is present as ATP^{4-}, and 50% is present as ATP^{3-}.

Figure 19.3 Adenosine 5′-monophosphate, a nucleotide. The phosphate group is fully ionized at pH 7.0 giving this nucleotide a charge of −2.

Perhaps the best known of the new viral antimetabolites is zidovudine (AZT, *The Merck Index*, 12th ed., #10254), an analog of deoxythymidine in which the 3′-OH has been replaced by an azido group, N_3. AZT is effective against HIV-1, a retrovirus that is the causative agent of AIDS. It is converted *in vivo* by cellular enzymes to the 5′-triphosphate, recognized as deoxythymidine 5′-triphosphate by viral RNA-dependent DNA polymerase (reverse transcriptase), and added to a growing DNA chain. There it stops chain elongation because no 3′-OH exists on which to add the next deoxynucleotide. AZT owes its effectiveness to the fact that it binds more strongly to viral reverse transcriptase than it does to human DNA polymerase.

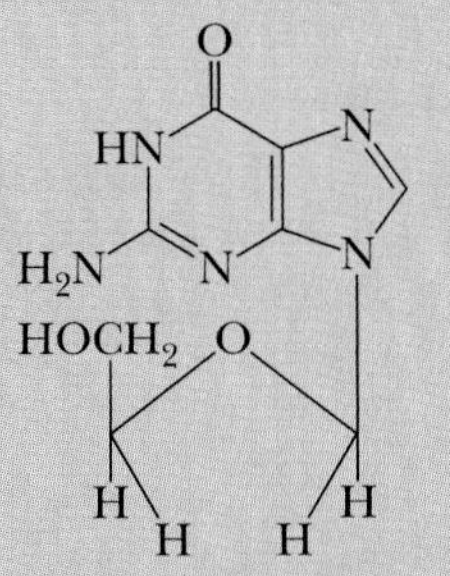

Acyclovir (drawn to show its structural relationship to 2-deoxyguanosine)

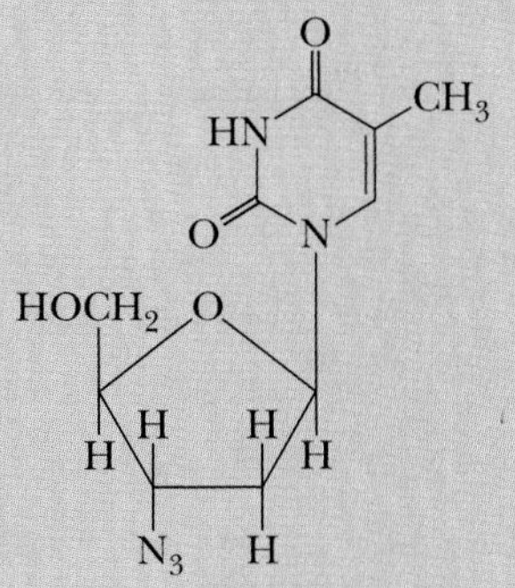

Zidovudine (Azidothymidine; AZT)

EXAMPLE 19.1

Draw a structural formula for each nucleotide.

(a) 2′-Deoxycytidine 5′-diphosphate
(b) 2′-Deoxyguanosine 3′-monophosphate

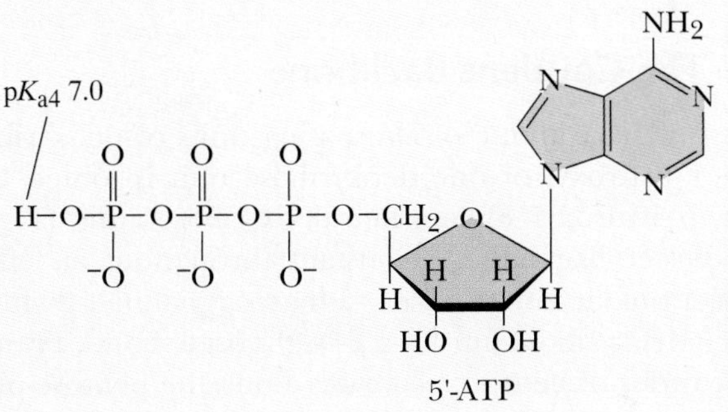

5'-ATP

Figure 19.4
Adenosine triphosphate, ATP.

SOLUTION

(a) Cytosine is joined by a β-*N*-glycoside bond between *N*-1 of cytosine and *C*-1 of the cyclic hemiacetal form of 2-deoxy-D-ribose. The 5′-hydroxyl of the pentose is bonded to a phosphate group by an ester bond, and this phosphate is, in turn, bonded to a second phosphate group by an anhydride bond.

(b) Guanine is joined by a β-*N*-glycoside bond between *N*-9 of guanine and *C*-1 of the cyclic hemiacetal form of 2-deoxy-D-ribose. The 3′-hydroxyl group of the pentose is joined to a phosphate group by an ester bond.

(a) phosphoric ester bond; phosphoric anhydride bond; β-*N*-glycoside bond

(b) β-*N*-glycoside bond; phosphoric ester bond

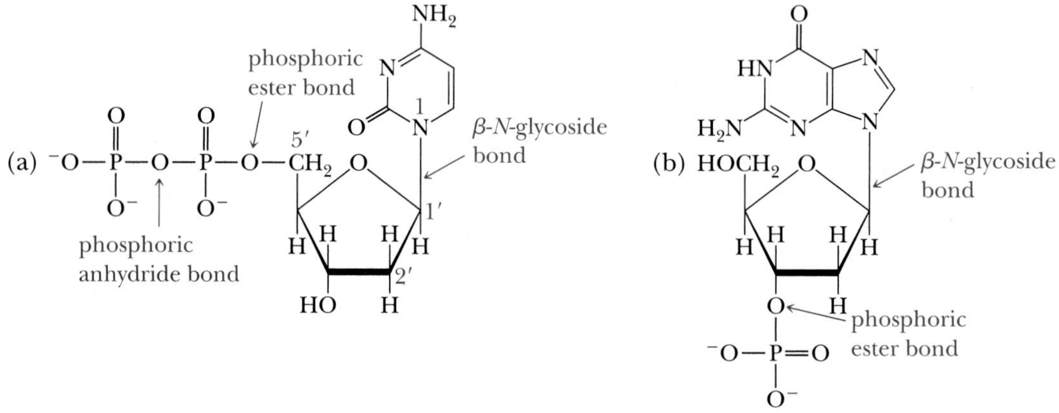

Practice Problem 19.1

Draw a structural formula for each nucleotide.

(a) 2′-Deoxythymidine 5′-monophosphate
(b) 2′-Deoxythymidine 3′-monophosphate

19.2 THE STRUCTURE OF DNA

In Chapter 18 we saw that the four levels of structural complexity in polypeptides and proteins are primary, secondary, tertiary, and quaternary structures. There are three levels of structural complexity in nucleic acids, and although these levels are somewhat comparable to those in polypeptides and proteins, they also differ in significant ways.

A. Primary Structure—The Covalent Backbone

Primary structure of nucleic acids The sequence of bases along the pentose-phosphodiester backbone of a DNA or RNA molecule read from the 5′ end to the 3′ end.

5′ End The end of a polynucleotide in which the 5′-OH of the terminal pentose unit is free.

3′ End The end of a polynucleotide at which the 3′-OH of the terminal pentose unit is free.

Deoxyribonucleic acids consist of a backbone of alternating units of deoxyribose and phosphate in which the 3′-hydroxyl of one deoxyribose unit is joined by a phosphodiester bond to the 5′-hydroxyl of another deoxyribose unit (Figure 19.5). This pentose-phosphodiester backbone is constant throughout an entire DNA molecule. A heterocyclic aromatic amine base—adenine, guanine, thymine, or cytosine—is bonded to each deoxyribose unit by a β-*N*-glycoside bond. **Primary structure** of DNA refers to the order of heterocyclic bases along the pentose-phosphodiester backbone. The sequence of bases is read from the **5′ end** to the **3′ end.**

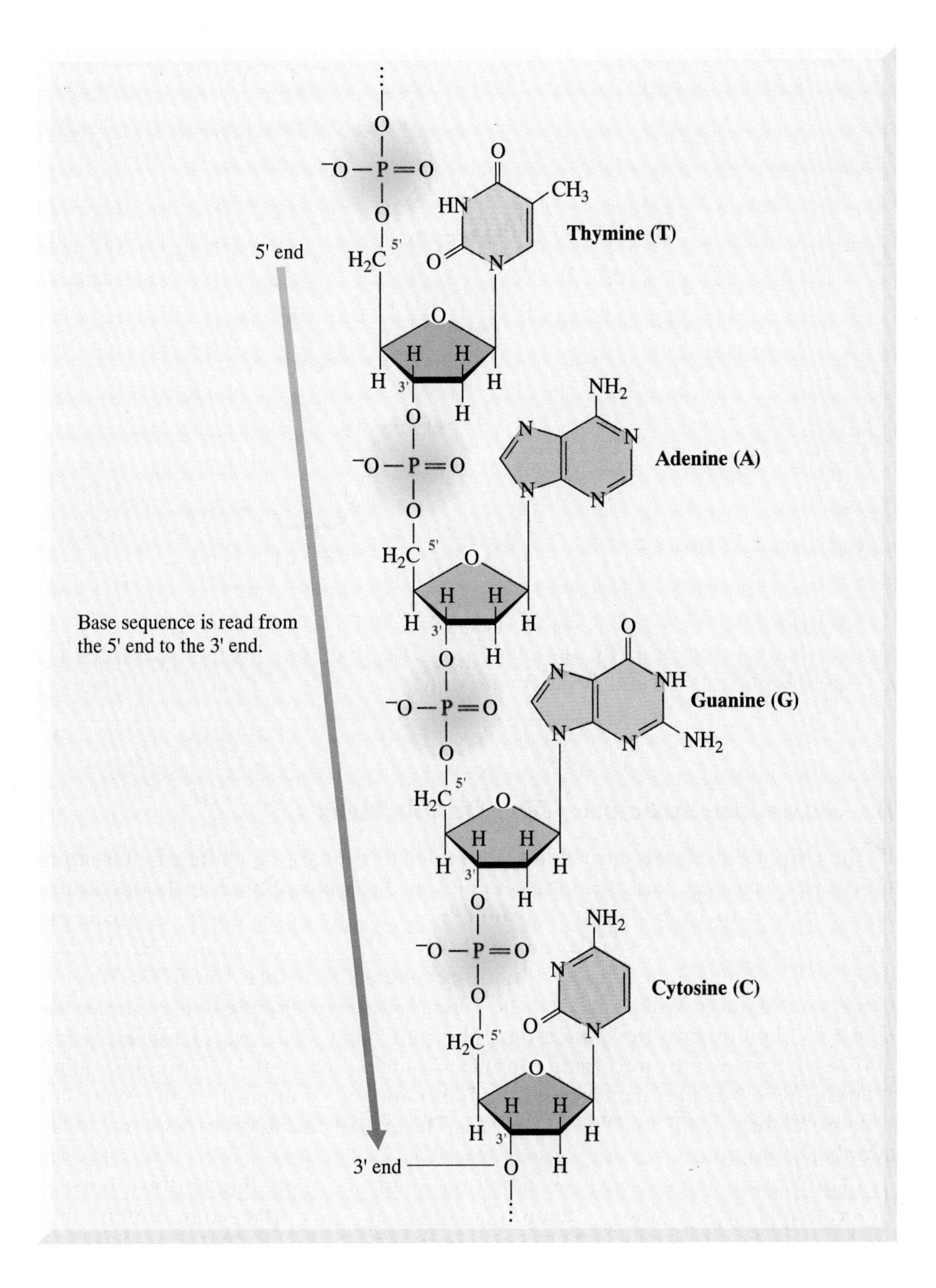

Figure 19.5
A tetranucleotide section of a single-stranded DNA.

EXAMPLE 19.2

Draw a structural formula for the DNA dinucleotide TG that is phosphorylated at the 5′ end only.

SOLUTION

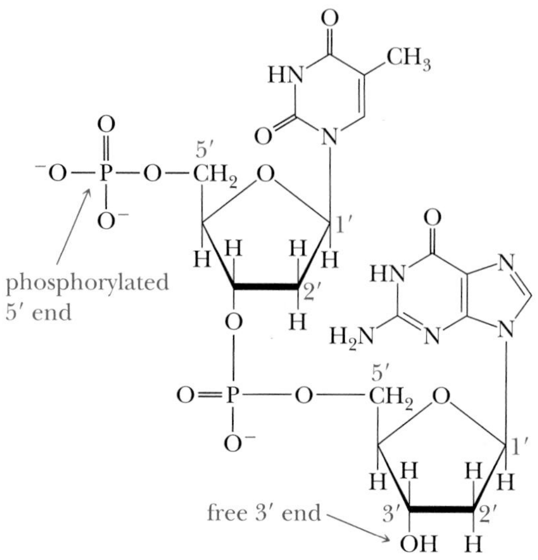

phosphorylated 5′ end

free 3′ end

Practice Problem 19.2

Draw a structural formula for the section of DNA that contains the base sequence CTG and is phosphorylated at the 3′ end only.

B. Secondary Structure—The Double Helix

Secondary structure of nucleic acids The ordered arrangement of nucleic acid strands.

By the early 1950s, it was clear that DNA molecules consist of chains of alternating units of deoxyribose and phosphate joined by 3′,5′-phosphodiester bonds with a base attached to each deoxyribose unit by a β-*N*-glycoside bond. In 1953, the American biologist James D. Watson and the British physicist Francis H. C. Crick proposed a double-helix model for the **secondary structure for DNA.** Watson, Crick, and Maurice Wilkins shared the 1962 Nobel Prize for Physiology and Medicine for "their discoveries concerning the molecular structure of nucleic acids, and its significance for information transfer in living material." Although Rosalind Franklin also took part in this research, her name was omitted from the Nobel list because of her death in 1958 at age 37. The Nobel foundation does not make awards posthumously.

The Watson-Crick model was based on molecular modeling and two lines of experimental observations: chemical analyses of DNA base compositions and mathematical analyses of x-ray diffraction patterns of crystals of DNA.

Base Composition

At one time it was thought that the four principal bases occur in the same ratios and perhaps repeat in a regular pattern along the pentose-phosphodiester backbone of DNA for all species. However, more precise determinations of base composition by Erwin Chargaff revealed that bases do not occur in the same ratios (Table 19.1).

TABLE 19.1 Comparison in Base Composition, in Mole-Percent, of DNA from Several Organisms

	Purines		Pyrimidines				
Organism	**A**	**G**	**C**	**T**	**A/T**	**G/C**	**Purines/ Pyrimidines**
human	30.4	19.9	19.9	30.1	1.01	1.00	1.01
sheep	29.3	21.4	21.0	28.3	1.04	1.02	1.03
yeast	31.7	18.3	17.4	32.6	0.97	1.05	1.00
E. coli	26.0	24.9	25.2	23.9	1.09	0.99	1.04

Rosalind Franklin (1920–1958). In 1951, she joined the Biophysical Laboratory at King's College, London, where she began her studies on the application of x-ray diffraction methods to the study of DNA. She is credited with discoveries that established the density of DNA, its helical conformation, and other significant aspects. Her work was, thus, important to the model of DNA developed by Watson and Crick. She died in 1958 at the age of 37 and, because the Nobel Prize is never awarded posthumously, she did not share in the 1962 Nobel Prize for Physiology and Medicine with Watson, Crick, and Wilkins. Although the relation between Watson, Crick, and Franklin was initially strained, Watson later said that "we later came to appreciate . . . the struggles the intelligent woman faces to be accepted by the scientific world which often regards women as mere diversions from serious thinking." *(Vittorio Luzzati/Centre Nationale de Génétique Moléculaire)*

Researchers drew the following conclusions from this and related data. To within experimental error,

1. The mole-percent base composition in any organism is the same in all cells of the organism and is characteristic of the organism.
2. The mole-percents of adenine (a purine base) and thymine (a pyrimidine base) are equal. The mole-percents of guanine (a purine base) and cytosine (a pyrimidine base) are also equal.
3. The mole-percents of purine bases (A + G) and pyrimidine bases (C + T) are equal.

Watson and Crick with their model of a DNA molecule. *(From M. M. Jones* et al., *Chemistry, Man, and Society, 4th ed. Philadelphia: Saunders College Publishing, 1983, p. 256)*

Analyses of X-Ray Diffraction Patterns

Additional information about the structure of DNA emerged when x-ray diffraction photographs taken by Rosalind Franklin and Maurice Wilkins were analyzed. These diffraction patterns revealed that, even though the base composition of DNA isolated from different organisms varies, DNA molecules themselves are remarkably uniform in thickness. They are long and fairly straight, with an outside diameter of approximately 20 Å, and not more than a dozen atoms thick. Furthermore, the crystallographic pattern repeats every 34 Å. Herein lay one of the chief problems to be solved. How could the molecular dimensions of DNA be so regular even though the relative percentages of the various bases differ so widely? With this accumulated information, the stage was set for the development of a hypothesis about DNA structure.

The Double Helix

Watson-Crick model A double-helix model for the secondary structure of a DNA molecule.

Double helix A type of secondary structure of DNA molecules in which two antiparallel polynucleotide strands are coiled in a right-handed manner about the same axis.

The heart of the **Watson-Crick model** is the postulate that a molecule of DNA is a complementary **double helix.** It consists of two antiparallel polynucleotide strands coiled in a right-handed manner about the same axis to form a double helix. As illustrated in the ribbon models in Figure 19.6, chirality is associated with a double helix; left-handed and right-handed double helices are related by reflection just as enantiomers are related by reflection.

To account for the observed base ratios and uniform thickness of DNA, Watson and Crick postulated that purine and pyrimidine bases project inward toward the axis of the helix and are always paired in a very specific manner. According to scale models, the dimensions of an adenine-thymine base pair are almost identical to the dimensions of a guanine-cytosine base pair, and the length of each pair is consistent with the core thickness of a DNA strand (Figure 19.7). Thus, if the purine base in one strand is adenine, then its complement in the antiparallel strand must be thymine. Similarly, if the purine in one strand is guanine, its complement in the antiparallel strand must be cytosine.

A significant feature of Watson and Crick's model is that no other base-pairing is consistent with the observed thickness of a DNA molecule. A pair of pyrimidine bases is too small to account for the observed thickness, whereas a pair of purine bases is too large. Thus, according to the Watson-Crick model, the repeating units in a double-stranded DNA molecule are not single bases of differing dimensions but rather base pairs of almost identical dimensions.

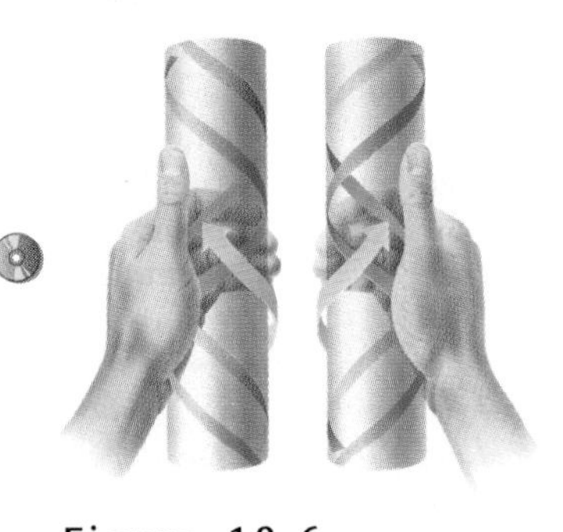

Figure 19.6
A DNA double helix has a chirality associated with the helix. Right-handed and left-handed double helices of otherwise identical DNA chains are nonsuperposable mirror images.

To account for the periodicity observed from x-ray data, Watson and Crick postulated that base pairs are stacked one on top of the other with a distance of 3.4 Å between base pairs and with ten base pairs in one complete turn of the helix. There is one complete turn of the helix every 34 Å. Shown in Figure 19.8 is a ribbon model of double-stranded **B-DNA,** the predominant form of DNA in dilute aqueous solution and thought to be the most common form in nature.

In the double helix, the bases in each base pair are not directly opposite from one another across the diameter of the helix, but rather are slightly displaced. This displacement and the relative orientation of the glycoside bonds linking each base to the sugar-phosphate backbone leads to two differently sized grooves, a major groove and a minor groove (Figure 19.8). Each groove runs along the length of the cylindrical column of the double helix. The major groove is approximately 22 Å wide; the minor groove is approximately 12 Å wide.

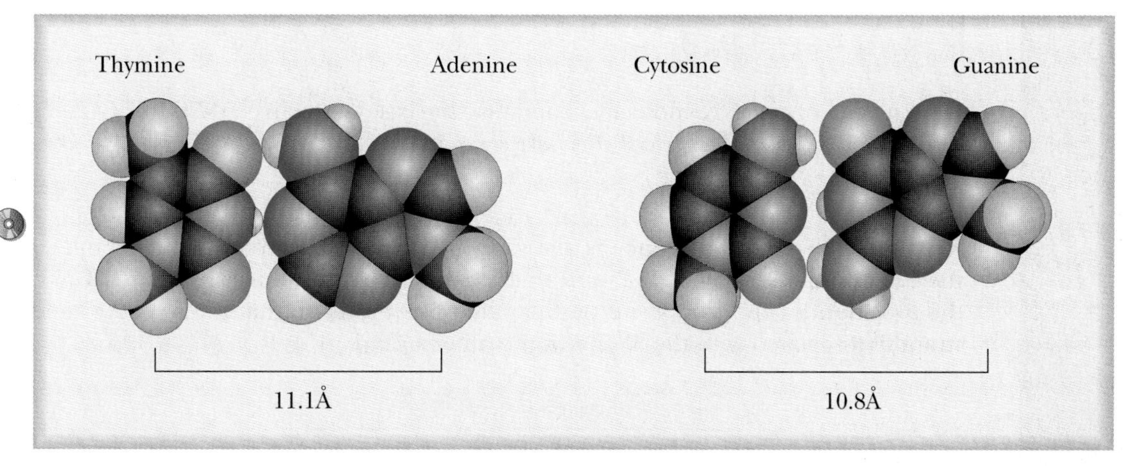

Figure 19.7
Base-pairing between adenine and thymine (A-T) and between guanine and cytosine (G-C). An A-T base pair is held by two hydrogen bonds, whereas a G-C base pair is held by three hydrogen bonds.

Figure 19.9 shows more detail of an idealized B-DNA double helix. The major and minor grooves are clearly recognizable in this model.

Other forms of secondary structure are known that differ in the distance between stacked base pairs and in the number of base pairs per turn of the helix. One of the most common of these, **A-DNA,** also a right-handed helix, is thicker than B-DNA and has a repeat distance of only 29 Å. There are ten base pairs per turn of the helix with a spacing of 2.9 Å between base pairs.

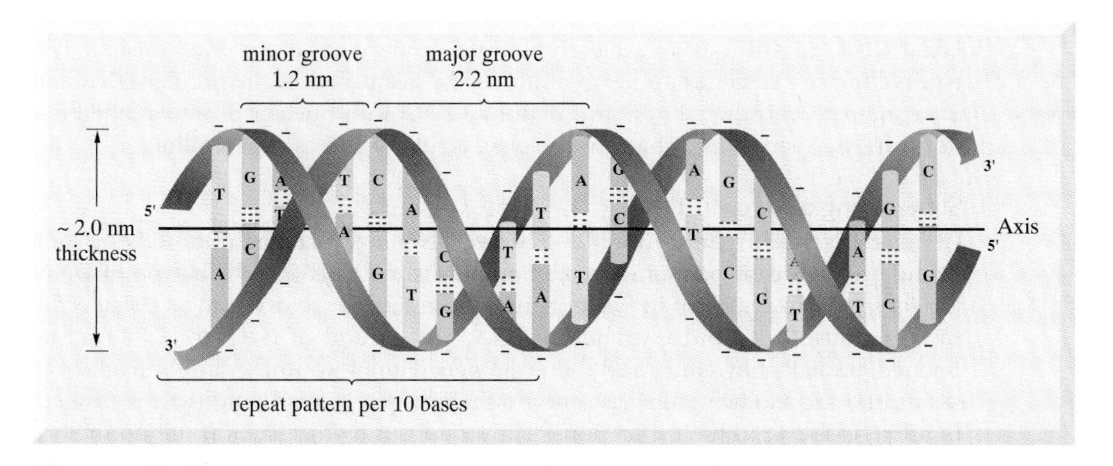

Figure 19.8
Ribbon model of double-stranded B-DNA. Each ribbon shows the pentose-phosphodiester backbone of a single-stranded DNA molecule. The strands are antiparallel; one strand runs to the left from the 5′ end to the 3′ end, the other runs to the right from the 5′ end to the 3′ end. Hydrogen bonds are shown by three dotted lines between each G-C base pair and two dotted lines between each A-T base pair.

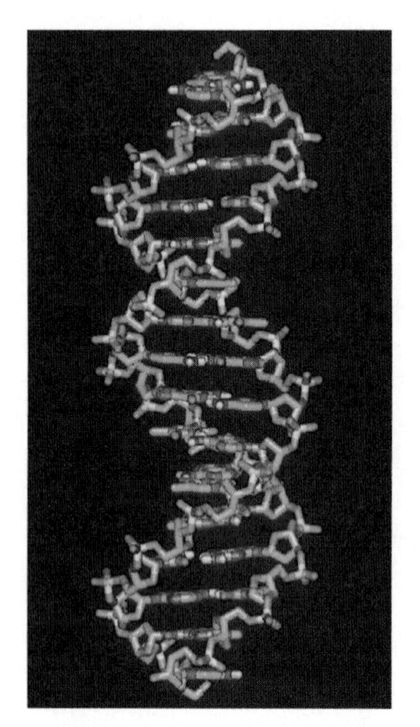

Figure 19.9
B-DNA. An idealized model of B-DNA.

Example 19.3

One strand of a DNA molecule contains the base sequence 5′-ACTTGCCA-3′. Write the DNA complement of this base sequence.

Solution

Remember that base sequence is always written from the 5′ end of the strand to the 3′ end, that A pairs with T, and that G pairs with C. In double-stranded DNA, the two strands run in opposite (antiparallel) directions, so that the 5′ end of one strand is associated with the 3′ end of the other strand.

Direction of strand →

original strand ⟶ 5′—A—C—T—T—G—C—C—A—3′

3′—T—G—A—A—C—G—G—T—5′ ⟵ complementary strand

← Direction of strand

Written from the 5′ end, the complementary strand is 5′-TGGCAAGT-3′.

Practice Problem 19.3

Write the complementary DNA base sequence for 5′-CCGTACGA-3′.

C. Tertiary Structure—Supercoiled DNA

The length of a DNA molecule is considerably greater than its diameter, and the extended molecule is quite flexible. A DNA molecule is said to be relaxed if it has no twists other than those imposed by its secondary structure. Said another way, relaxed DNA does not have a clearly defined tertiary structure. We consider two types of **tertiary structure,** one type induced by perturbations in circular DNA and a second type introduced by coordination of DNA with nuclear proteins called histones. Tertiary structure, whatever the type, is referred to as **supercoiling.**

Tertiary structure of nucleic acids The three-dimensional arrangement of all atoms of a nucleic acid, commonly referred to as supercoiling.

Supercoiling of Circular DNA

Circular DNA is a type of double-stranded DNA in which the two ends of each strand are joined by phosphodiester bonds [Figure 19.10(a)]. This type of DNA, the most prominent form in bacteria and viruses, is also referred to as circular duplex (because it is double-stranded) DNA. One strand of circular DNA may be opened, partially unwound, and then rejoined. The unwound section introduces a strain into the molecule because the nonhelical gap is less stable than hydrogen-bonded, base-paired helical sections. The strain can be localized in the nonhelical gap. Alternatively, it may be spread uniformly over the entire circular DNA by introduction of **superhelical twists,** one twist for each turn of a helix unwound. The circular DNA shown in Figure 19.10(b) has been unwound by four complete turns of the helix. The strain introduced by this unwinding is spread uniformly over the entire molecule by introduction of four superhelical twists [Figure 19.10(c)]. Interconversion of relaxed and supercoiled DNA is catalyzed by groups of enzymes called topoisomerases and gyrases.

Circular DNA A type of double-stranded DNA in which the 5′ and 3′ ends of each strand are joined by phosphodiester groups.

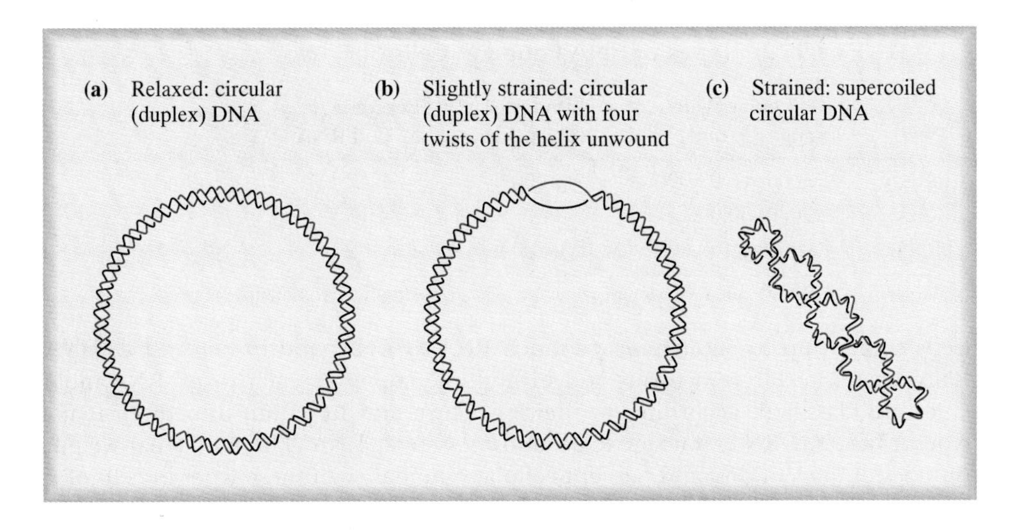

Figure 19.10 Relaxed and supercoiled DNA. (a) Circular DNA is relaxed. (b) One strand is broken, unwound by four turns, and the ends rejoined. The strain of unwinding is localized in the nonhelical gap. (c) Supercoiling by four twists distributes the strain of unwinding uniformly over the entire molecule of circular DNA.

Supercoiling of Linear DNA

Supercoiling of linear DNA in plants and animals takes another form and is driven by interaction between negatively charged DNA molecules and a group of positively charged proteins called **histones.** Histones are particularly rich in lysine and arginine and, at the pH of most body fluids, have an abundance of positively charged sites along their length. The complex between negatively charged DNA and positively charged histones is called **chromatin.** Histones associate to form core particles about which double-stranded DNA then wraps. Further coiling of DNA produces the chromatin found in cell nuclei.

Histone A protein, particularly rich in the basic amino acids lysine and arginine, that is found associated with DNA molecules.

Chromatin A complex formed between negatively charged DNA molecules and positively charged histones.

19.3 Ribonucleic Acids

Ribonucleic acids (RNA) are similar to deoxyribonucleic acids in that they, too, consist of long, unbranched chains of nucleotides joined by phosphodiester groups between the 3′-hydroxyl of one pentose and the 5′-hydroxyl of the next. There are, however, three major differences in structure between RNA and DNA.

1. The pentose unit in RNA is β-D-ribose rather than β-2-deoxy-D-ribose.
2. The pyrimidine bases in RNA are uracil and cytosine rather than thymine and cytosine.
3. RNA is single-stranded rather than double-stranded.

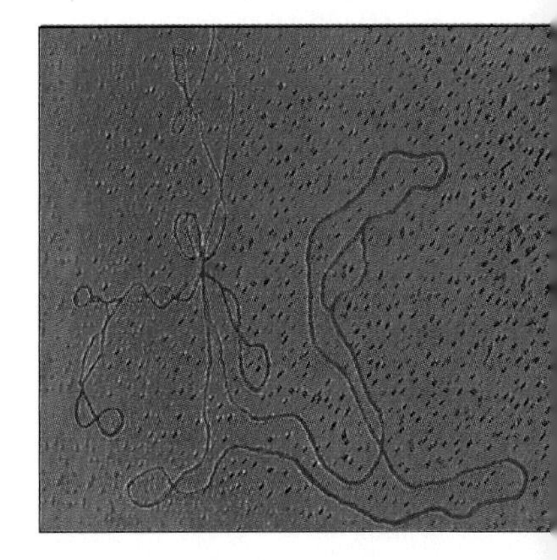

Supercoiled DNA from a mitochondrion. *(Fran Heyl Associates)*

Following are structural formulas for the furanose form of D-ribose and for uracil.

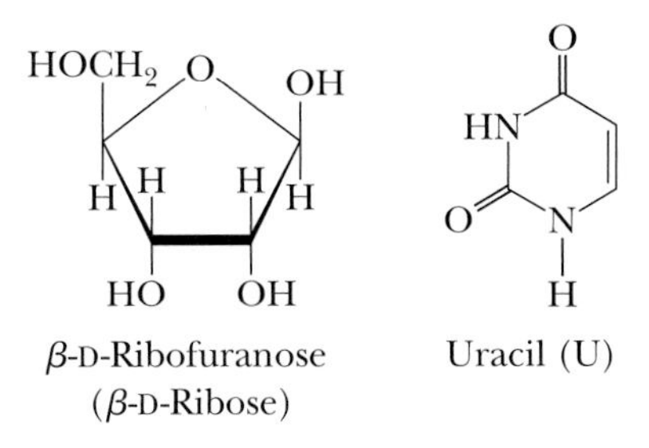

β-D-Ribofuranose (β-D-Ribose)   Uracil (U)

TABLE 19.2 Types of RNA Found in Cells of *E. coli*

Type	Molecular Weight Range (g/mol)	Number of Nucleotides	Percentage of Cell RNA
mRNA	25,000–1,000,000	75–3000	2
tRNA	23,000–30,000	73–94	16
rRNA	35,000–1,100,000	120–2904	82

Cells contain up to eight times as much RNA as DNA, and in contrast to DNA, RNA occurs in different forms and in multiple copies of each form. RNA molecules are classified, according to their structure and function, into three major types: ribosomal RNA, transfer RNA, and messenger RNA. The molecular weight, number of nucleotides, and percent cellular abundance of these types in cells of *E. coli* are summarized in Table 19.2.

A. Ribosomal RNA

Ribosomal RNA (rRNA) A ribonucleic acid found in ribosomes, the sites of protein synthesis.

The bulk of **ribosomal RNA (rRNA)** is found in the cytoplasm in subcellular particles called ribosomes, which contain about 60% RNA and 40% protein. Ribosomes are the sites in cells at which protein synthesis takes place.

B. Transfer RNA

Transfer RNA (tRNA) A ribonucleic acid that carries a specific amino acid to the site of protein synthesis on ribosomes.

Transfer RNA (tRNA) molecules have the lowest molecular weight of all nucleic acids. They consist of 73–94 nucleotides in a single chain. The function of tRNA is to carry amino acids to the sites of protein synthesis on the ribosomes. Each amino acid has at least one tRNA dedicated specifically to this purpose. Several amino acids have more than one. In the transfer process, the amino acid is joined to its specific tRNA by an ester bond between the α-carboxyl group of the amino acid and the 3′ hydroxyl group of the ribose at the 3′ end of the tRNA.

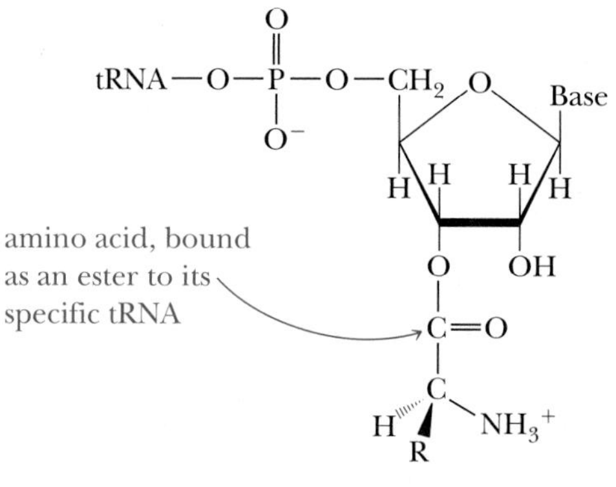

tRNA—O—P(=O)(O⁻)—O—CH_2 (ribose, Base) ... O—C(=O)—C(H)(R)—NH_3^+

amino acid, bound as an ester to its specific tRNA

C. Messenger RNA

Messenger RNA (mRNA) A ribonucleic acid that carries coded genetic information from DNA to the ribosomes for the synthesis of proteins.

Messenger RNAs (mRNAs) are present in cells in relatively small amounts and are very short-lived. They are single-stranded, and their synthesis is directed by information encoded on DNA molecules. Double-stranded DNA is unwound, and a

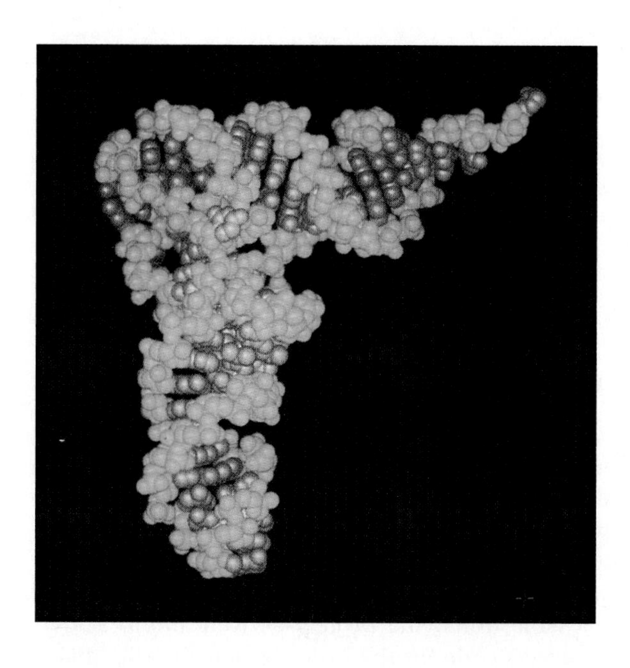

A space-filling model of yeast phenylalanine tRNA. *(After S. H. Kim, in P. Schimmel, D. Söll, and J. N. Abelson, eds., Transfer RNA: Structure, Properties, and Recognition, New York: Cold Spring Harbor Laboratory, 1979)*

complementary strand of mRNA is synthesized along one strand of the DNA template, beginning from the 3′ end. The synthesis of mRNA from a DNA template is called transcription, because genetic information contained in a sequence of bases of DNA is transcribed into a complementary sequence of bases on mRNA. The name "messenger" is derived from the function of this type of RNA, which is to carry coded genetic information from DNA to the ribosomes for the synthesis of new proteins.

Example 19.4

Following is a base sequence from a portion of DNA. Write the sequence of bases of the mRNA synthesized using this section of DNA as a template.

3′-A-G-C-C-A-T-G-T-G-A-C-C-5′

Solution

RNA synthesis begins at the 3′ end of the DNA template and proceeds toward the 5′ end. The complementary mRNA strand is formed using the bases C, G, A, and U. Uracil (U) is the complement of adenine (A) on the DNA template.

Direction of strand

3′—A—G—C—C—A—T—G—T—G—A—C—C—5′ ⟵ DNA template

mRNA ⟶ 5′—U—C—G—G—U—A—C—A—C—U—G—G—3′

Direction of strand

Reading from the 5′ end, the sequence of mRNA is 5′-UCGGUACACUGG-3′

Practice Problem 19.4

Here is a portion of the nucleotide sequence in phenylalanine tRNA.

3′-ACCACCUGCUCAGGCCUU-5′

Write the nucleotide sequence of its DNA complement.

19.4 The Genetic Code

A. Triplet Nature of the Code

It was clear by the early 1950s that the sequence of bases in DNA molecules constitutes the store of genetic information and that the sequence of bases directs the synthesis of messenger RNA, which, in turn, directs the synthesis of proteins. However, the statement that "the sequence of bases in DNA directs the synthesis of proteins" presents the following problem: How can a molecule containing only four variable units (adenine, cytosine, guanine, and thymine) direct the synthesis of molecules containing up to 20 variable units (the protein-derived amino acids)? How can an alphabet of only four letters code for the order of letters in the 20-letter alphabet that occurs in proteins?

An obvious answer is that there is not one base but rather a combination of bases coding for each amino acid. If the code consists of nucleotide pairs, there are $4^2 = 16$ combinations; this is a more extensive code, but it is still not extensive enough to code for 20 amino acids. If the code consists of nucleotides in groups of three, there are $4^3 = 64$ combinations; this is more than enough to code for the primary structure of a protein. This appears to be a very simple solution to a system that must have taken eons of evolutionary trial and error to develop. Yet proof now exists, from comparison of gene (nucleic acid) and protein (amino acid) sequences, that nature does indeed use this simple three-letter or triplet code to store genetic information. A triplet of nucleotides is called a **codon.**

Codon A triplet of nucleotides on mRNA that directs incorporation of a specific amino acid into a polypeptide sequence.

B. Deciphering the Genetic Code

The next question is, Which of the 64 triplets code for which amino acid? In 1961, Marshall Nirenberg provided a simple experimental approach to the problem, based on the observation that synthetic polynucleotides direct polypeptide synthesis in much the same manner as do natural mRNAs. Nirenberg incubated ribosomes, amino acids, tRNAs, and appropriate protein-synthesizing enzymes. With only these components, there was no polypeptide synthesis. However, when he added synthetic polyuridylic acid (poly U), a polypeptide of high molecular weight was synthesized. What was more important, the synthetic polypeptide contained only phenylalanine. With this discovery, the first element of the genetic code was deciphered: the triplet UUU codes for phenylalanine.

Similar experiments were carried out with different synthetic polyribonucleotides. It was found, for example, that polyadenylic acid (poly A) leads to the synthesis of polylysine, and that polycytidylic acid (poly C) leads to the synthesis of polyproline. By 1964, all 64 codons had been deciphered (Table 19.3).

TABLE 19.3 The Genetic Code—mRNA Codons and the Amino Acid Each Codon Directs

First Position (5′ end)	Second Position								Third Position (3′ end)
	U		C		A		G		
U	UUU	Phe	UCU	Ser	UAU	Tyr	UGU	Cys	U
	UUC	Phe	UCC	Ser	UAC	Tyr	UGC	Cys	C
	UUA	Leu	UCA	Ser	UAA	Stop	UGA	Stop	A
	UUG	Leu	UCG	Ser	UAG	Stop	UGG	Trp	G
C	CUU	Leu	CUU	Pro	CAU	His	CGU	Arg	U
	CUC	Leu	CCC	Pro	CAC	His	CGC	Arg	C
	CUA	Leu	CCA	Pro	CAA	Gln	CGA	Arg	A
	CUG	Leu	CCG	Pro	CAG	Gln	CGG	Arg	G
A	AUU	Ile	ACU	Thr	AAU	Asn	AGU	Ser	U
	AUC	Ile	ACC	Thr	AAC	Asn	AGC	Ser	C
	AUA	Ile	ACA	Thr	AAA	Lys	AGA	Arg	A
	AUG*	Met	ACG	Thr	AAG	Lys	AGG	Arg	G
G	GUU	Val	GCU	Ala	GAU	Asp	GGU	Gly	U
	GUC	Val	GCC	Ala	GAC	Asp	GGC	Gy	C
	GUA	Val	GCA	Ala	GAA	Glu	GGA	Gly	A
	GUG	Val	GCG	Ala	GAG	Glu	GGG	Gly	G

* AUG also serves as the principal initiation codon.

C. Properties of the Genetic Code

Several features of the genetic code are evident from a study of Table 19.3.

1. Only 61 triplets code for amino acids. The remaining three (UAA, UAG, and UGA) are signals for chain termination; they signal to the protein-synthesizing machinery of the cell that the primary sequence of the protein is complete. The three chain termination triplets are indicated in Table 19.3 by "Stop."
2. The code is degenerate, which means that several amino acids are coded for by more than one triplet. Only methionine and tryptophan are coded for by just one triplet. Leucine, serine, and arginine are coded for by six triplets, and the remaining amino acids are coded for by two, three, or four triplets.
3. For the 15 amino acids coded for by two, three, or four triplets, it is only the third letter of the code that varies. For example, glycine is coded for by the triplets GGA, GGG, GGC, and GGU.
4. There is no ambiguity in the code, meaning that each triplet codes for only one amino acid.

Finally, we must ask one last question about the genetic code: Is the code universal, that is, is it the same for all organisms? Every bit of experimental evidence available today from the study of viruses, bacteria, and higher animals, including humans, indicates that the code is universal. Furthermore, the fact that it is the same for all these organisms means that it has been the same over millions of years of evolution.

EXAMPLE 19.5

During transcription, a portion of mRNA is synthesized with the following base sequence.

5′-AUG-GUA-CCA-CAU-UUG-UGA-3′

(a) Write the nucleotide sequence of the DNA from which this portion of mRNA was synthesized.
(b) Write the primary structure of the polypeptide coded for by this section of mRNA.

SOLUTION

(a) During transcription, mRNA is synthesized from a DNA strand, beginning from the 3′ end of the DNA template. The DNA strand must be the complement of the newly synthesized mRNA strand.

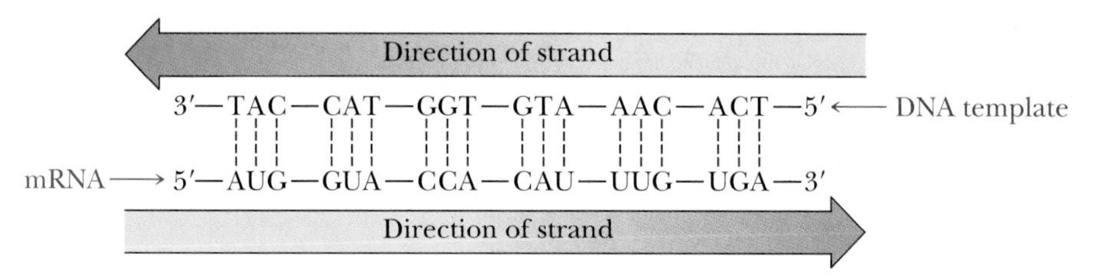

Note that the codon UGA codes for termination of the growing polypeptide chain; therefore, the sequence given in this problem codes for a pentapeptide only.

(b) The sequence of amino acids is shown in the following mRNA strand.

5′-AUG-GUA-CCA-CAU-UUG-UGA-3′

Met—Val—Pro—His—Leu—Stop

Practice Problem 19.5

The following section of DNA codes for oxytocin, a polypeptide hormone.

3′-ACG-ATA-TAA-GTT-TTA-ACG-GGA-GAA-CCA-ACT-5′

(a) Write the base sequence of the mRNA synthesized from this section of DNA.
(b) Given the sequence of bases in part (a), write the primary structure of oxytocin.

19.5 SEQUENCING NUCLEIC ACIDS

As recently as 1975, the task of determining the primary structure of a nucleic acid was thought to be far more difficult than determining the primary structure of a protein. Nucleic acids, it was reasoned, contain only four different units, whereas proteins contain 20 different units. With only four different units, there are fewer specific sites for selective cleavage, distinctive sequences are more difficult to recognize, and there is greater chance of ambiguity in the assignment of sequence. Two breakthroughs reversed this situation. First was the development of a type of electrophoresis called **polyacrylamide gel electrophoresis,** a technique so sensitive that it is possible to separate nucleic acid fragments that differ from one another in only a single nucleotide. The second breakthrough was the discovery of a class of enzymes called **restriction endonucleases,** isolated chiefly from bacteria.

A. Restriction Endonucleases

A **restriction endonuclease** recognizes a set pattern of four to eight nucleotides and cleaves a DNA strand by hydrolysis of the linking phosphodiester bonds at any site that contains that particular sequence. Close to 1000 restriction endonucleases have been isolated and their specificities characterized; each cleaves DNA at a different site and produces a different set of restriction fragments. *E. coli,* for example, has a restriction endonuclease, EcoRI, that recognizes the hexanucleotide sequence, GAATTC and cleaves it between G and A.

Restriction endonuclease An enzyme that catalyzes hydrolysis of a particular phosphodiester bond within a DNA strand.

cleavage here

$$5'\text{----G-A-A-T-T-C----}3' \xrightarrow{\text{EcoRI}} 5'\text{----G} + 5'\text{-A-A-T-T-C----}3'$$

Note that the action of restriction endonucleases is analogous to the action of trypsin (Section 18.4B), which catalyzes hydrolysis of amide bonds formed by the carboxyl groups of Lys and Arg, and of chymotrypsin, which catalyzes cleavage of amide bonds formed by the carboxyl groups of Phe, Tyr, and Trp.

EXAMPLE 19.6

The following is a section of the gene coding for bovine rhodopsin along with several restriction endonucleases, their recognition sequences, and their hydrolysis sites. Which endonucleases will catalyze cleavage of this section of DNA?

5′-GCCGTCTACAACCCGGTCATCTACTATCATGATCAACAAGCAGTTCCGGAACT-3′

Enzyme	Recognition Sequence	Enzyme	Recognition Sequence
A*lu*I	AG ↓ CT	H*pa*II	C ↓ CGG
B*al*I	TGG ↓ CCA	M*bo*I	↓ GATC
F*nu*DII	CG ↓ CG	N*ot*I	GC ↓ GGCCGC
H*ea*III	GG ↓ CC	S*ac*I	GAGCT ↓ C

SOLUTION

Only restriction endonucleases H*pa*II and M*bo*I catalyze cleavage of this polynucleotide: H*pa*II at two sites and M*bo*I at one site.

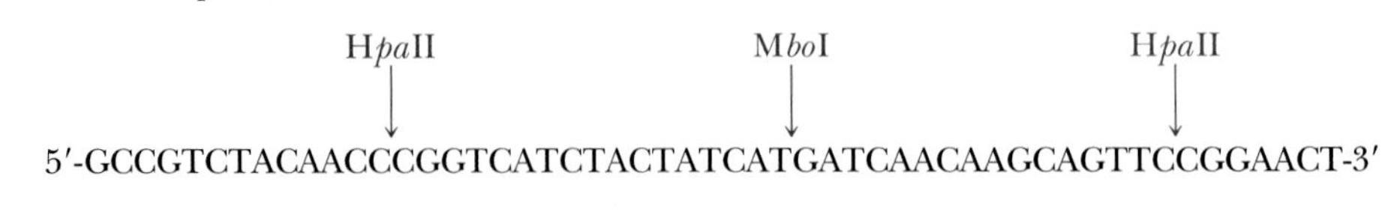

Practice Problem 19.6

The following is another section of the bovine rhodopsin gene. Which of the endonucleases given in Example 19.6 will catalyze cleavage of this section?

5′-ACGTCGGGTCGTCGTCCTCTCGCGGTGGTGAGTCTTCCGGCTCTTCT-3′

B. Methods for Sequencing Nucleic Acids

Any sequencing of DNA begins with site-specific cleavage of double-stranded DNA by one or more restriction endonucleases into smaller fragments called **restriction fragments.** Each restriction fragment is then sequenced separately, overlapping base sequences are identified, and the entire sequence of bases is then deduced.

Two methods for sequencing restriction fragments have been developed. The first of these, developed by Allan Maxam and Walter Gilbert and known as the **Maxam-Gilbert method,** depends on base-specific chemical cleavage. The second method, developed by Frederick Sanger and known as the **chain termination** or **dideoxy method,** depends on interruption of DNA-polymerase catalyzed synthesis. Sanger and Gilbert shared the 1980 Nobel Prize for Biochemistry for their "development of chemical and biochemical analysis of DNA structure." Sanger's dideoxy method is currently more widely used, and it is on this method that we concentrate.

Sanger dideoxy method A method, developed by Frederick Sanger, for sequencing DNA molecules.

C. DNA Replication *in Vitro*

To appreciate the rationale for the dideoxy method, we must first understand certain aspects of the biochemistry of DNA replication. During replication, the sequence of nucleotides in one strand is copied as a complementary strand to form the second strand of a double-stranded DNA molecule. Synthesis of the complementary strand is catalyzed by the enzyme DNA polymerase. DNA polymerase will also carry out this synthesis *in vitro* using single-stranded DNA as a template, provided that both the four deoxynucleotide triphosphate (dNTP) monomers and a primer are present. A **primer** is an oligonucleotide capable of forming a short section of double-stranded DNA (dsDNA) by base-pairing with its complement on a single-stranded DNA (ssDNA). Because a new DNA strand grows from its 5′ to 3′ end, the primer must have a free 3′-OH group to which the first nucleotide of the growing chain is added (Figure 19. 11).

CHEMICAL CONNECTIONS

DNA Fingerprinting

Each human being has a genetic makeup consisting of approximately 3 billion pairs of nucleotides, and, except for identical twins, the base sequence of DNA in one individual is different from that of every other individual. As a result, each person has a unique DNA fingerprint. To determine a DNA fingerprint, a sample of DNA from a trace of blood, skin, or other tissue is treated with a set of restriction endonucleases, and the 5′ end of each restriction fragment is labeled with phosphorus-32. The resulting ^{32}P-labeled restriction fragments are then separated by polyacrylamide gel electrophoresis and visualized by placing a photographic plate over the developed gel.

In the DNA fingerprint patterns shown in the figure, lanes 1, 5, and 9 represent internal standards or control lanes. They contain the DNA fingerprint pattern of a standard virus treated with a standard set of restriction endonucleases. Lanes 2, 3, and 4 were used in a paternity suit. The DNA fingerprint of the mother in lane 4 contains five bands, which match with five of the six bands in the DNA fingerprint of the child in lane 3. The DNA fingerprint of the alleged father in lane 2 contains six bands, three of which match with bands in the DNA fingerprint of the child. Because the child inherits only half of its genes from the father, only half of the child's and father's DNA fingerprints are expected to match. In this instance, the paternity suit was won on the basis of the DNA fingerprint matching.

Lanes 6, 7, and 8 contain DNA fingerprint patterns used as evidence in a rape case. Lanes 6 and 7 are DNA fingerprints of semen obtained from the rape victim. Lane 8 is the DNA fingerprint pattern of the alleged rapist. The DNA fingerprint patterns of the semen do not match that of the alleged rapist and excluded the suspect from the case.

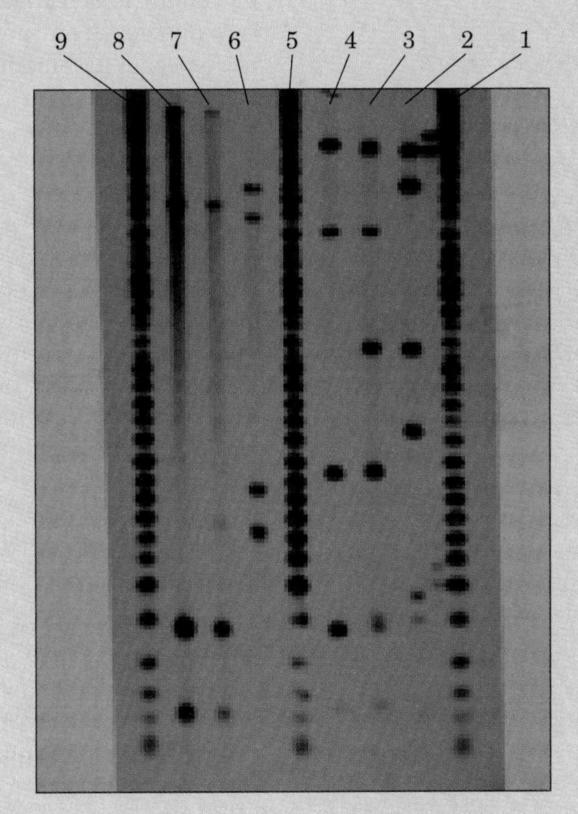

DNA fingerprint. *(Courtesy of Dr. Lawrence Kobilinsky)*

D. The Chain Termination or Dideoxy Method

The key to the chain termination method is the addition to the synthesizing medium of a 2′,3′-dideoxynucleoside triphosphate (ddNTP). Because a ddNTP has no —OH group at the 3′ position, it cannot serve as an acceptor for the next

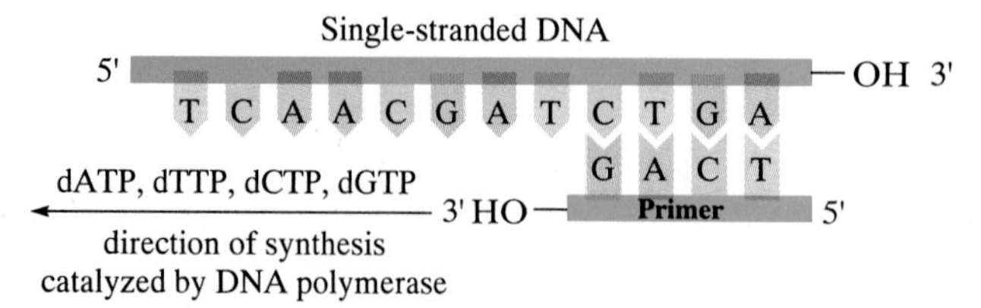

Figure 19.11
DNA polymerase catalyzes the synthesis *in vitro* using single-stranded DNA as a template provided that both the four deoxynucleotide triphosphate (dNTP) monomers and a primer are present. The primer provides a short stretch of double-stranded DNA by base-pairing with its complement on the single-stranded DNA.

nucleotide to be added to the growing polynucleotide chain. Thus, chain synthesis is terminated at any point where a ddNTP becomes incorporated, hence the designation chain termination method.

A 2′,3′-dideoxynucleoside triphosphate
(ddNTP)

In the chain termination method, a single-stranded DNA of unknown sequence is mixed with primer and divided into four separate reaction mixtures. To each reaction mixture are added all four deoxynucleoside triphosphates (dNTPs), one of which is labeled in the 5′ phosphoryl group with ^{32}P so that the newly synthesized fragments can be visualized by autoradiography.

$$^{32}_{15}P \longrightarrow ^{32}_{16}S + \text{Beta particle} + \text{Gamma rays}$$

Also added to each reaction mixture are DNA polymerase and one of the four ddNTPs. The ratio of dNTPs to ddNTP in each reaction mixture is adjusted so that incorporation of a ddNTP takes place infrequently. In each reaction mixture, DNA synthesis takes place but, in a given population of molecules, synthesis is interrupted at every possible site (Figure 19.12).

When gel electrophoresis of each reaction mixture is completed, a piece of x-ray film is placed over the gel, and gamma rays released by radioactive decay of ^{32}P darken the film and create a pattern on it that is an image of the resolved oligonucleotides. The base sequence of the complement to the original single-stranded template is then read directly from bottom to top of the developed film.

A variation on this method is to use a single reaction mixture with each of the four ddNTPs labeled with a different fluorescent indicator. Each label is then detected by its characteristic spectrum. Automated DNA sequencing machines using this variation are capable of sequencing up to 10,000 base pairs per day. One of the ultimate challenges now is to sequence the entire human genome with its estimated 3 billion base pairs.

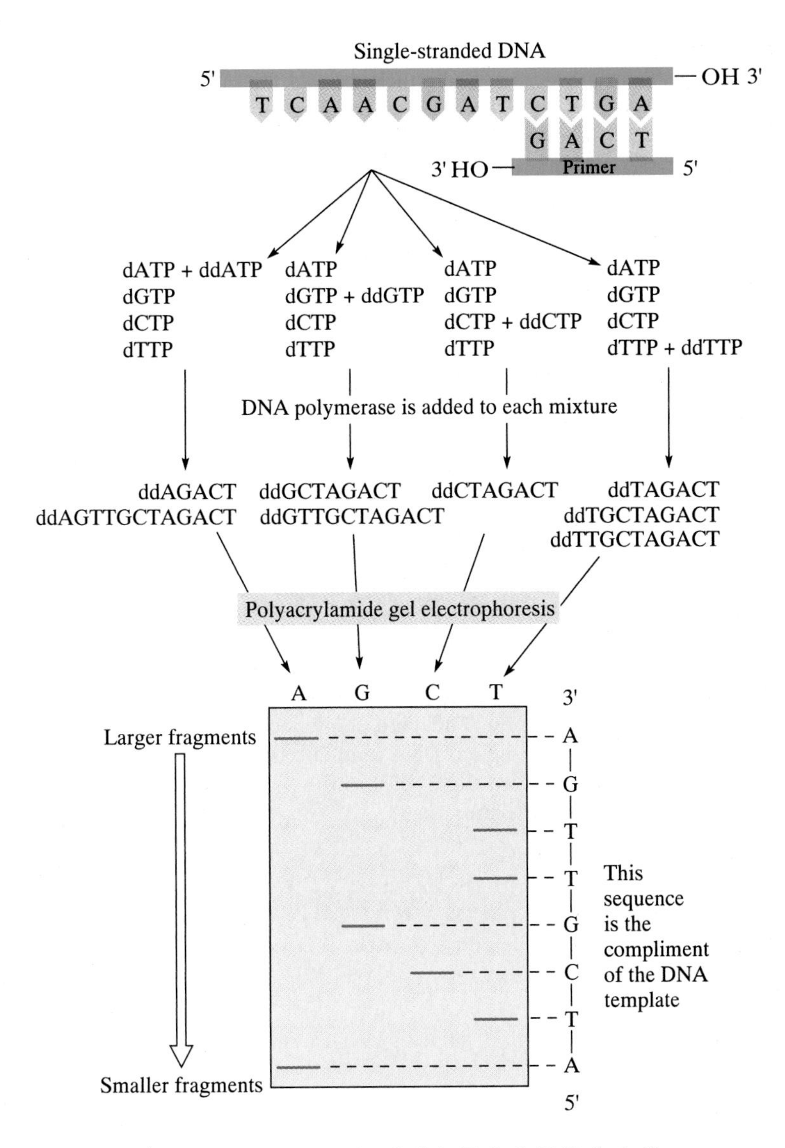

Figure 19.12

The chain termination or dideoxy method of DNA sequencing. The primer-DNA template is divided into four separate reaction mixtures. To each is added the four dNTPs, DNA polymerase, and one of the four ddNTPs. Synthesis is interrupted at every possible site. The mixtures of oligonucleotides are separated by polyacrylamide gel electrophoresis. The base sequence of the DNA complement is read from the bottom to the top (from the 5′ end to the 3′ end) of the developed gel.

SUMMARY

Nucleic acids are composed of three types of monomer units: heterocyclic aromatic amine bases derived from purine and pyrimidine, the monosaccharides D-ribose or 2-deoxy-D-ribose, and phosphate ions (Section 19.1). A **nucleoside** is a compound containing D-ribose or 2-deoxy-D-ribose bonded to a heterocyclic aromatic amine base by a *β-N*-glycoside bond. A **nucleotide** is a nucleoside in which a molecule of phosphoric acid is esterified with an —OH of the monosaccharide, most commonly either the 3′-OH or the 5′-OH. Nucleoside mono-, di- and triphosphates are strong polyprotic acids and are extensively ionized at pH 7.0. At pH 7.0, adenosine triphosphate, for example, is a 50:50 mixture of ATP^{3-} and ATP^{4-}.

The **primary structure** of **deoxyribonucleic acids (DNAs)** consists of units of 2-deoxyribose bonded by 3′,5′-phosphodiester bonds (Section 19.2A). A heterocyclic aromatic amine base is attached to each deoxyribose unit by a β-*N*-glycoside bond. The sequence of bases is read from the 5′ end of the polynucleotide strand to the 3′ end.

The heart of the **Watson-Crick model** is the postulate that a molecule of DNA consists of two antiparallel polynucleotide strands coiled in a right-handed manner about the same axis to form a **double helix** (Section 19.2B). Purine and pyrimidine bases point inward toward the axis of the helix and are always paired G-C and A-T. In **B-DNA,** base pairs are stacked one on top of another with a spacing of 3.4 Å and ten base pairs per 34-Å helical repeat. In **A-DNA,** bases are stacked with a spacing of 2.9 Å between base pairs and ten base pairs per 29-Å helical repeat.

Tertiary structure of DNA is commonly referred to as **supercoiling** (Section 19.2C). **Circular DNA** is a type of double-stranded DNA in which the ends of each strand are joined by phosphodiester groups. Opening of one strand followed by partial unwinding and rejoining the ends introduces strain in the nonhelical gap. This strain can be spread over the entire molecule of circular DNA by introduction of **superhelical twists. Histones** are particularly rich in lysine and arginine and, therefore, have an abundance of positive charges. The association of DNA and histones produces a pigment called **chromatin.**

There are two important differences between the primary structure of **ribonucleic acids (RNAs)** and DNA (Section 19.3). (1) The monosaccharide unit in RNA is D-ribose. (2) Both RNA and DNA contain the purine bases adenine (A) and guanine (G), and the pyrimidine base cytosine (C). As the fourth base, however, RNA contains uracil (U), whereas DNA contain thymine (T).

The **genetic code** (Section 19.4) consists of nucleosides in groups of three; that is, it is a triplet code. Only 61 triplets code for amino acids; the remaining three code for termination of polypeptide synthesis.

Restriction endonucleases recognize a set pattern of four to eight nucleotides and cleave a DNA strand by hydrolysis of the linking phosphodiester bonds at any site that contains that particular sequence (Section 19.5A). In the **chain termination** or **dideoxy method** of DNA sequencing developed by Frederick Sanger (Section 19.5D), a primer-DNA template is divided into four separate reaction mixtures. To each is added the four dNTPs, one of which is labeled with ^{32}P. Also added are DNA polymerase and one of the four ddNTPs. Synthesis is interrupted at every possible site. The mixtures of newly synthesized oligonucleotides are separated by polyacrylamide gel electrophoresis and are visualized by autoradiography. The base sequence of the DNA complement to the original DNA template is read from the bottom to the top (from the 5′ end to the 3′ end) of the developed photographic plate.

Additional Problems

Nucleosides and Nucleotides

19.7 Two drugs used in the treatment of acute leukemia are 6-mercaptopurine (*The Merck Index,* 12th ed., #5919) and 6-thioguanine (*The Merck Index,* 12th ed., #9473). In each drug, the oxygen at carbon 6 of the parent molecule is replaced by divalent sulfur. Draw structural formulas for the enethiol (the sulfur equivalent of an enol) forms of 6-mercaptopurine and 6-thioguanine.

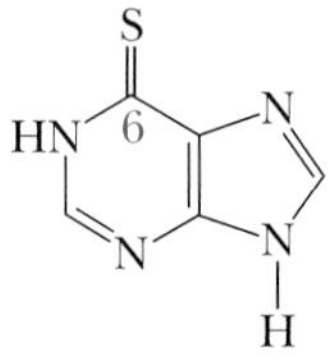

S, HN, 6, N, N, N, H

6-Mercaptopurine

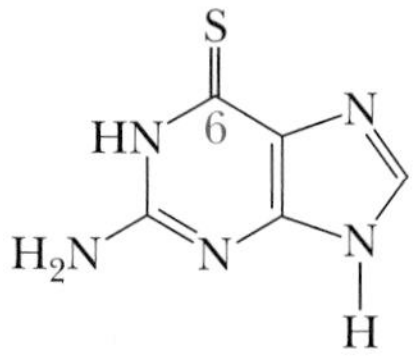

S, HN, 6, N, H_2N, N, N, H

6-Thioguanine

19.8 Following are structural formulas for cytosine and thymine. Draw two additional tautomeric forms for cytosine and three additional tautomeric forms for thymine.

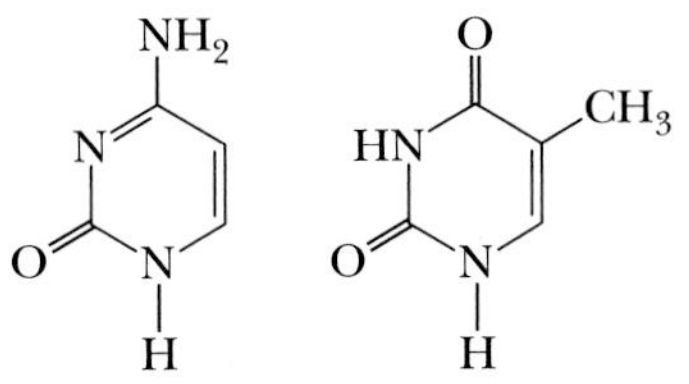

Cytosine (C) Thymine (T)

19.9 Draw a structural formula for a nucleoside composed of

(a) β-D-Ribose and adenine **(b)** β-2-Deoxy-D-ribose and cytosine

19.10 Nucleosides are stable in water and in dilute base. In dilute acid, however, the glycoside bond of a nucleoside undergoes hydrolysis to give a pentose and a heterocyclic aromatic amine base. Propose a mechanism for this acid-catalyzed hydrolysis.

19.11 Explain the difference in structure between a nucleoside and a nucleotide.

19.12 Draw a structural formula for each nucleotide and estimate its net charge at pH 7.4, the pH of blood plasma.

(a) 2′-Deoxyadenosine 5′-triphosphate (dATP)

(b) Guanosine 3′-monophosphate (GMP)

(c) 2′-Deoxyguanosine 5′-diphosphate (dGDP)

19.13 Cyclic-AMP, first isolated in 1959, is involved in many diverse biological processes as a regulator of metabolic and physiological activity. In it, a single phosphate group is esterified with both the 3′ and 5′ hydroxyls of adenosine. Draw a structural formula of cyclic-AMP.

The Structure of DNA

19.14 Why are deoxyribonucleic acids called acids? What are the acidic groups in their structure?

19.15 Human DNA contains approximately 30.4% A. Estimate the percentages of G, C, and T and compare them with the values presented in Table 19.1.

19.16 Draw a structural formula of the DNA tetranucleotide 5′-A-G-C-T-3′. Estimate the net charge on this tetranucleotide at pH 7.0. What is the complementary tetranucleotide to this sequence?

19.17 List the postulates of the Watson-Crick model of DNA secondary structure.

19.18 The Watson-Crick model is based on certain experimental observations of base composition and molecular dimensions. Describe these observations and show how the Watson-Crick model accounts for each.

19.19 Compare the α-helix of proteins and the double helix of DNA in these ways.

(a) The units that repeat in the backbone of the polymer chain.

(b) The projection in space of substituents along the backbone (the R groups in the case of amino acids; purine and pyrimidine bases in the case of double-stranded DNA) relative to the axis of the helix.

19.20 Discuss the role of the hydrophobic interactions in stabilizing

(a) Double-stranded DNA **(b)** Lipid bilayers **(c)** Soap micelles

19.21 Name the type of covalent bond(s) joining monomers in these biopolymers.

(a) Polysaccharides **(b)** Polypeptides **(c)** Nucleic acids

19.22 In terms of hydrogen bonding, which is more stable, an A-T base pair or a G-C base pair?

19.23 At elevated temperatures, nucleic acids become denatured, that is, they unwind into single-stranded DNA. Account for the observation that the higher the G-C content of a nucleic acid, the higher the temperature required for its thermal denaturation.

19.24 Write the DNA complement for 5′-ACCGTTAAT-3′. Be certain to label which is the 5′ end and which is the 3′ end of the complement strand.

19.25 Write the DNA complement for 5′-TCAACGAT-3′.

Ribonucleic Acids

19.26 Compare the degree of hydrogen bonding in the base pair A-T found in DNA with that in the base pair A-U found in RNA.

19.27 Compare DNA and RNA is these ways.

(a) Monosaccharide units **(b)** Principal purine and pyrimidine bases

(c) Primary structure **(d)** Location in the cell

(e) Function in the cell

19.28 What type of RNA has the shortest lifetime in cells?

19.29 Write the mRNA complement for 5′-ACCGTTAAT-3′. Be certain to label which is the 5′ end and which is the 3′ end of the mRNA strand.

19.30 Write the mRNA complement for 5′-TCAACGAT-3′.

The Genetic Code

19.31 What does it mean to say that the genetic code is degenerate?

19.32 Write the mRNA codons for (a) valine, (b) histidine, and (c) glycine.

19.33 Aspartic acid and glutamic acid have carboxyl groups on their side chains and are called acidic amino acids. Compare the codons for these two amino acids.

19.34 Compare the structural formulas of the aromatic amino acids phenylalanine and tyrosine. Compare also the codons for these two amino acids.

19.35 Glycine, alanine, and valine are classified as nonpolar amino acids. Compare their codons. What similarities do you find? What differences do you find?

19.36 Codons in the set CUU, CUC, CUA, and CUG all code for the amino acid leucine. In this set, the first and second bases are identical, and the identity of the third base is irrelevant. For what other sets of codons is the third base also irrelevant, and for what amino acid(s) does each set code?

19.37 Compare the codons with a pyrimidine, either U or C, as the second base. Do the majority of the amino acids specified by these codons have hydrophobic or hydrophilic side chains?

19.38 Compare the codons with a purine, either A or G, as the second base. Do the majority of the amino acids specified by these codons have hydrophilic or hydrophobic side chains?

19.39 What polypeptide is coded for by this mRNA sequence?

5′-GCU-GAA-GUC-GAG-GUG-UGG-3′

19.40 The alpha chain of human hemoglobin has 141 amino acids in a single polypeptide chain. Calculate the minimum number of bases on DNA necessary to code for the alpha chain. Include in your calculation the bases necessary for specifying termination of polypeptide synthesis.

19.41 In HbS, the human hemoglobin found in individuals with sickle-cell anemia, glutamic acid at position 6 in the beta chain is replaced by valine.

(a) List the two codons for glutamic acid and the four codons for valine.

(b) Show that one of the glutamic acid codons can be converted to a valine codon by a single substitution mutation, that is by changing one letter in one codon.

20

The Organic Chemistry of Metabolism

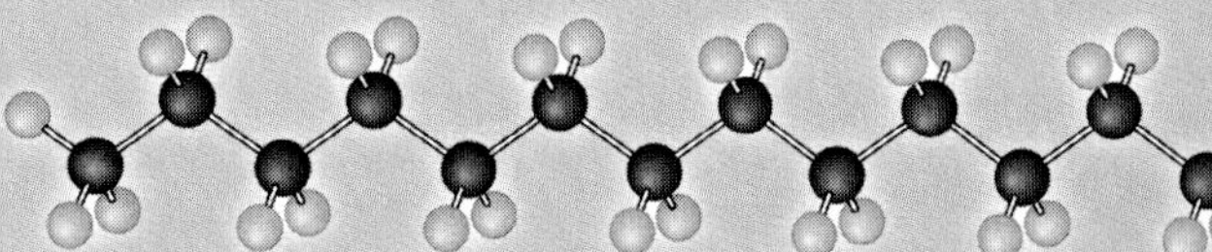

Important stages in the production of metabolic energy are glycolysis and β-oxidation. Insets: Models of stearic acid and glucose. *(Andrea Levskosek/Photo Researchers, Inc.)*

We have now studied the structure and typical reactions of the major types of organic functional groups. Further, we have studied the structure of carbohydrates, lipids, amino acids and proteins, and nucleic acids. Now let us see how this background can be applied to the study of the organic chemistry of metabolism. In this chapter, we study two key metabolic pathways, namely β-oxidation of fatty acids and glycolysis. The first of these is a pathway by which the hydrocarbon chains of fatty acids are degraded, two carbons at a time, to acetyl coenzyme A. The second is a pathway by which glucose is converted to pyruvate and then to acetyl coenzyme A.

Those of you who go on to courses in biochemistry will undoubtedly study these metabolic pathways in considerable detail, including their role in energy production and conservation, their regulation, and the diseases associated with errors of particular metabolic steps. Our concern in this chapter is more limited. It is our purpose to show that reactions of these pathways are biochemical equivalents of organic functional group reactions we have already studied in detail. In these pathways, we find examples of keto-enol tautomerism; oxidation of an aldehyde to a carboxylic acid; oxidation of a secondary alcohol to a ketone; a reverse aldol reaction; a reverse Claisen condensation; and formation and hydrolysis of esters, imines, thioesters, and mixed anhydrides. In this chapter, we use the mechanisms we have studied earlier to give us insights into the mechanisms of these reactions, all of which are enzyme-catalyzed.

20.1 FIVE KEY PARTICIPANTS IN β-OXIDATION AND GLYCOLYSIS

In order to understand what happens in β-oxidation of fatty acids and glycolysis, we first need to introduce five of the principal compounds participating in these and a great many other metabolic pathways. Three of these compounds (ATP, ADP, and AMP) are central to the storage and transfer of phosphate groups. The other two, NAD^+/NADH and FAD/$FADH_2$, are **coenzymes** involved in the oxidation/reduction of metabolic intermediates.

A. ATP, ADP, and AMP—Agents for Storage and Transfer of Phosphate Groups

Following is a structural formula of adenosine triphosphate (Section 19.1), the most important of the compounds involved in the storage and transport of phosphate groups. A building block for ATP is adenosine, which consists of a unit of adenine bonded to a unit of D-ribofuranose by a β-*N*-glycoside bond. Bonded to the terminal —CH_2OH of ribose are three units of phosphate: one joined by a phosphoric ester bond, the remaining two by phosphoric anhydride bonds.

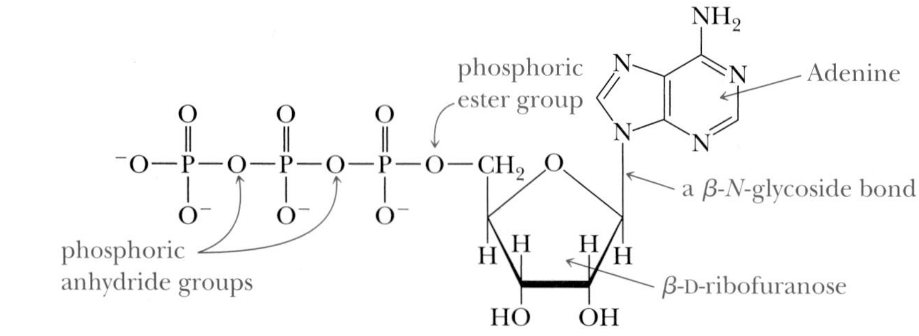

phosphoric ester group

Adenine

a β-N-glycoside bond

phosphoric anhydride groups

β-D-ribofuranose

Adenosine triphosphate (ATP)

Hydrolysis of the terminal phosphate group of ATP gives ADP. In the following abbreviated structural formulas, adenosine and its single phosphoric ester group are represented by the symbol AMP.

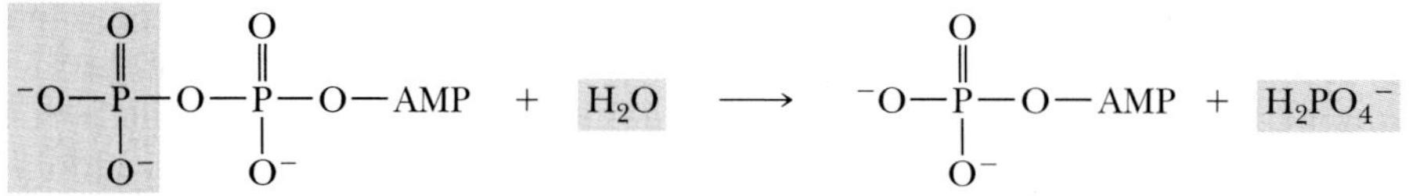

$$^{-}O-P(=O)(O^{-})-O-P(=O)(O^{-})-O-AMP + H_2O \longrightarrow {}^{-}O-P(=O)(O^{-})-O-AMP + H_2PO_4^{-}$$

Adenosine triphosphate (ATP) | Water (a phosphate acceptor) | Adenosine diphosphate (ADP)

The reaction shown is hydrolysis of a phosphoric anhydride; the phosphate acceptor is water. In the first two reactions of glycolysis, the phosphate acceptors are —OH groups of glucose and fructose, respectively, to form phosphoric esters of these molecules. In glycolysis are two reactions in which the phosphate acceptor is ADP, which is converted to ATP.

B. NAD^+/NADH—Agents for Electron Transfer in Biological Oxidation-Reductions

Nicotinamide adenine dinucleotide (NAD^+) A biological oxidizing agent. When acting as an oxidizing agent, NAD^+ is reduced to NADH.

Nicotinamide adenine dinucleotide (NAD^+, *The Merck Index,* 12th ed., #6429) is one of the central agents for the transfer of electrons in metabolic oxidations and reductions. NAD^+ is constructed of a unit of ADP joined by a phosphoric ester bond to the terminal $—CH_2OH$ of β-D-ribofuranose, which is in turn joined to the pyridine ring of nicotinamide by a β-N-glycoside bond.

NAD^+ is a two-electron oxidizing agent, as seen in the following balanced half-reaction, and is reduced to NADH. NADH is, in turn, a two-electron reducing agent and is oxidized to NAD^+. In these abbreviated structural formulas, the adenine dinucleotide part of the molecule is represented by the symbol Ad.

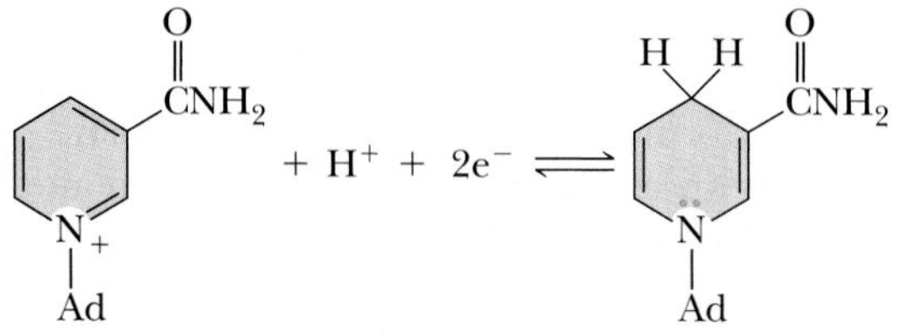

$$NAD^+ + H^+ + 2e^- \rightleftharpoons NADH$$

NAD^+ (oxidized form) | NADH (reduced form)

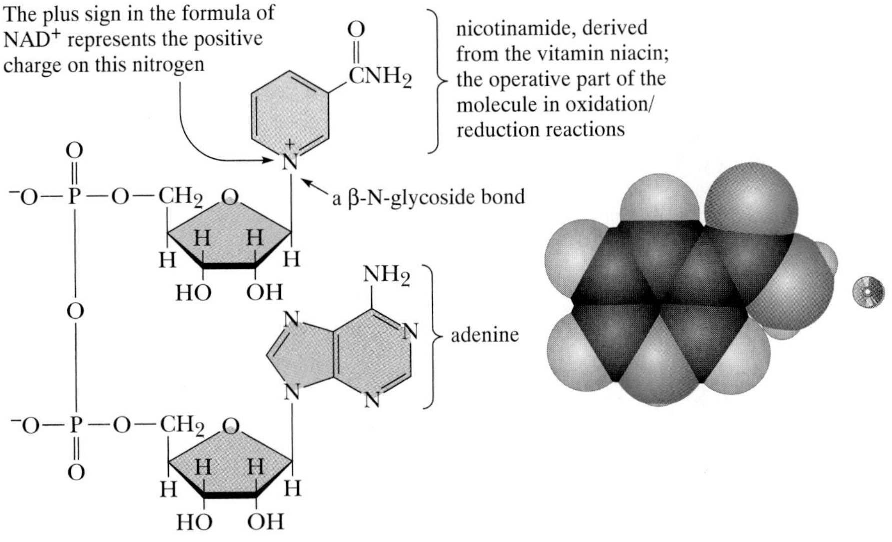
The plus sign in the formula of NAD+ represents the positive charge on this nitrogen
nicotinamide, derived from the vitamin niacin; the operative part of the molecule in oxidation/reduction reactions
a β-N-glycoside bond
adenine
CNH2
NH2

Nicotinamide adenine dinucleotide (NAD^+)

NAD^+ is involved in a variety of enzyme-catalyzed oxidation/reduction reactions. The two types of oxidations we deal with in this chapter are oxidation of a secondary alcohol to a ketone and oxidation of an aldehyde to a carboxylic acid. Each is a two-electron oxidation.

$$\underset{\text{A secondary alcohol}}{-\overset{\overset{\displaystyle OH}{|}}{\underset{\underset{\displaystyle H}{|}}{C}}-} \longrightarrow \underset{\text{A ketone}}{-\overset{\overset{\displaystyle O}{\|}}{C}-} + 2H^+ + 2e^-$$

$$\underset{\text{An aldehyde}}{-\overset{\overset{\displaystyle O}{\|}}{C}-H} + H_2O \longrightarrow \underset{\text{A carboxylic acid}}{-\overset{\overset{\displaystyle O}{\|}}{C}-OH} + 2H^+ + 2e^-$$

Oxidation of each functional group involves transfer of a hydride ion to NAD^+.

MECHANISM Oxidation of an Alcohol by NAD^+

Step 1: A basic group, B^-, on the surface of the enzyme removes H^+ from the —OH group.

Step 2: Electrons of the H—O sigma bond become the pi electrons of the C=O bond.

Step 3: A hydride ion is transferred from carbon to NAD^+ to create a new C—H bond.

Step 4: Electrons within the ring flow to the positively charged nitrogen.

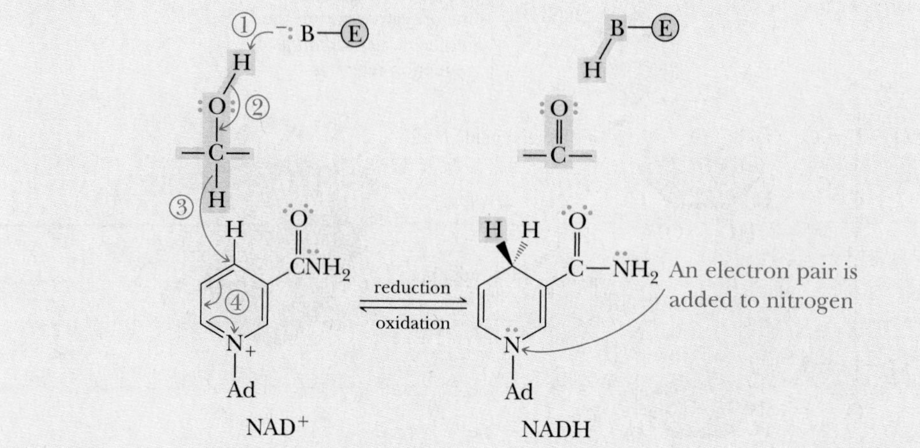

The hydride ion, H:$^-$, which is transferred from the secondary alcohol to NAD$^+$, contains two electrons and, thus, NAD$^+$ and NADH function exclusively in two-electron oxidations and two-electron reductions.

C. FAD/FADH$_2$—Agents for Electron Transfer in Biological Oxidation–Reductions

Flavin adenine dinucleotide (FAD) A biological oxidizing agent. When acting as an oxidizing agent, FAD is reduced to FADH$_2$.

Flavin adenine dinucleotide (FAD, *The Merck Index,* 12th ed., #4131) is also a central component in the transfer of electrons in metabolic oxidations and reductions. In FAD, flavin is bonded to the five-carbon monosaccharide ribitol, which is, in turn, bonded to the terminal phosphate group of ADP.

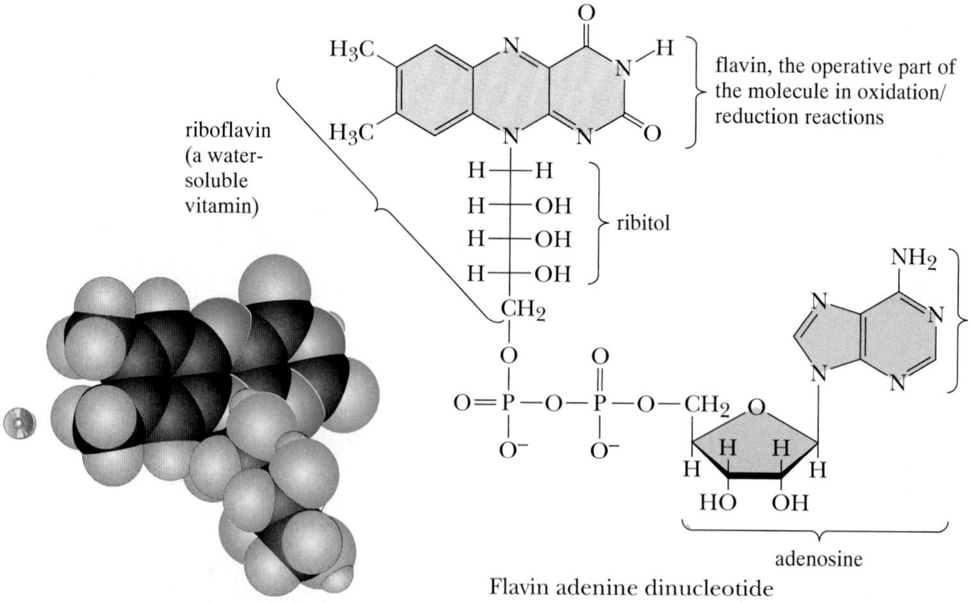

Flavin adenine dinucleotide (FAD)

FAD participates in several types of enzyme-catalyzed oxidation-reduction reactions. Our concern in this chapter is with its role in the oxidation of a carbon-carbon single bond in the hydrocarbon chain of a fatty acid to a carbon-carbon double bond. As seen from balanced half-reactions, the two-electron oxidation of the hydrocarbon chain is coupled with the two-electron reduction of FAD.

Balanced half-reactions:

Oxidation of the hydrocarbon chain: $—CH_2—CH_2— \longrightarrow —CH{=}CH— + 2H^+ + 2e^-$

Reduction of FAD: $FAD + 2H^+ + 2e^- \longrightarrow FADH_2$

The mechanism by which FAD oxidizes $—CH_2—CH_2—$ to $—CH{=}CH—$ involves transfer of a hydride ion from the hydrocarbon chain of the fatty acid to FAD as shown in the following diagram. The individual curved arrows in this figure are numbered 1–6 to help you follow the flow of electrons in this transformation.

MECHANISM Oxidation of a Fatty Acid $—CH_2—CH_2—$ to $—CH{=}CH—$ by FAD

Step 1: A basic group on the surface of the enzyme removes a hydrogen from the carbon alpha to the carboxyl group (shown here as R_2).

Step 2: Electrons from this C—H sigma bond become the pi electrons of the new C—C double bond.

Step 3: A hydride ion is transferred from the carbon beta to the carboxyl group to a nitrogen atom of flavin.

Step 4: The pi electrons within flavin are redistributed.

Step 5: Electrons of the C=N bond remove a hydrogen from the enzyme.

Step 6: A new basic group is created on the enzyme.

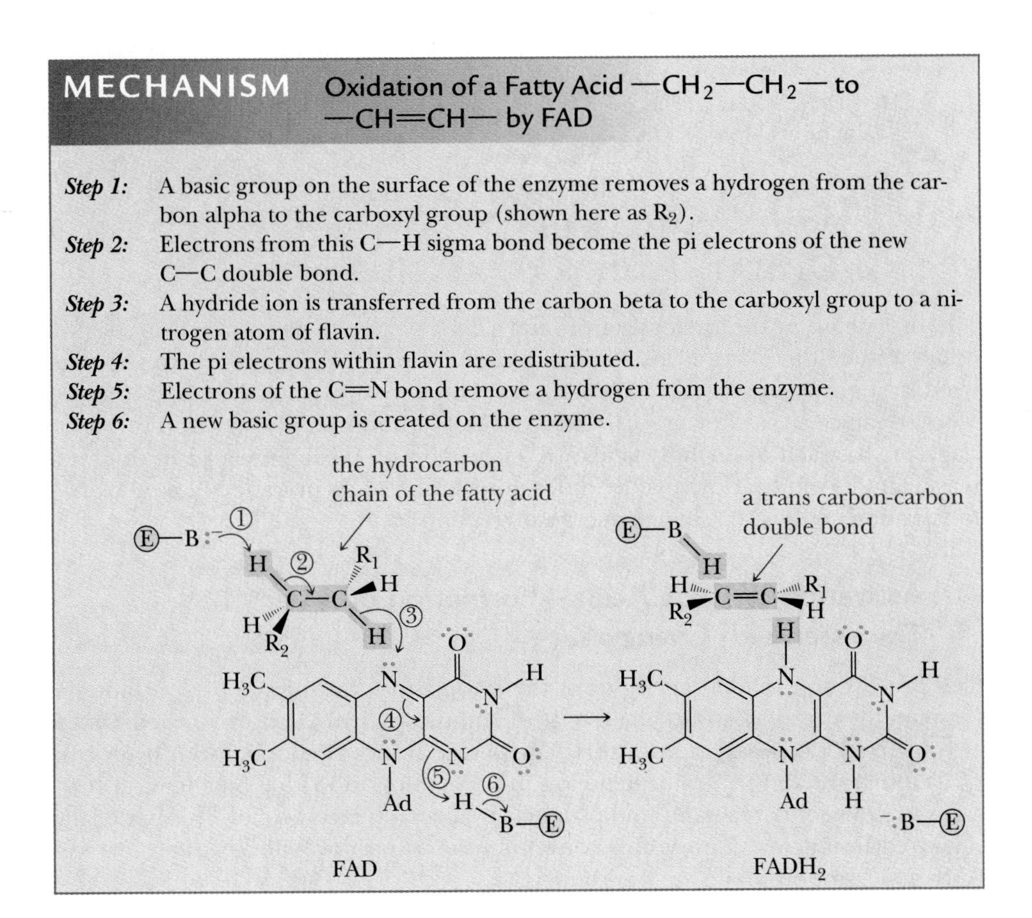

Note that one of the hydrogen atoms added to FAD to produce $FADH_2$ comes from the hydrocarbon chain; the other comes from an acidic group on the surface of the enzyme catalyzing this oxidation. Also note that one group on the enzyme functions as a proton acceptor and that another functions as a proton donor.

20.2 Fatty Acids as a Source of Energy

Fatty acids in the form of triglycerides are the principal storage form of energy for most organisms. The principal advantage of storing energy in this form is that the hydrocarbon chains of fatty acid, consisting mostly of $-CH_2-$ groups, are a more highly reduced form of carbon than the oxygenated chains of carbohydrates; therefore, the energy yield per gram of fatty acid oxidized is greater than that per gram of carbohydrate. Complete oxidation of 1 g of palmitic acid, for example, yields almost 2.5 times the energy obtained from 1 g of glucose.

	Yield of Energy	
	(kcal/mol)	(kcal/g)
$C_6H_{12}O_6 + 6O_2 \longrightarrow 6CO_2 + 6H_2O$ Glucose	−686	−3.8
$CH_3(CH_2)_{14}CO_2H + 23O_2 \longrightarrow 16CO_2 + 16H_2O$ Palmitic acid	−2340	−9.3

20.3 β-Oxidation of Fatty Acids

β-Oxidation A series of four enzyme-catalyzed reactions that cleaves carbon atoms, two at a time, from the carboxyl end of a fatty acid.

The first phase in the catabolism of fatty acids involves their release from triglycerides, either those stored in adipose tissue or from the diet, by hydrolysis catalyzed by a group of enzymes called lipases. The free fatty acids then pass into the bloodstream, where they are carried to sites of utilization. There are two major stages in **β-oxidation of fatty acids:** (A) activation of a free fatty acid in the cytoplasm and its transport across the inner mitochondrial membrane followed by (B) β-oxidation, a repeated sequence of four reactions.

A. Activation of Fatty Acids—Formation of a Thioester with Coenzyme A

Thioester An ester in which one atom of oxygen in the carboxylate group is replaced by an atom of sulfur.

Coenzyme A low-molecular-weight, nonprotein molecule or ion that binds reversibly to an enzyme, functions as a second substrate for the enzyme, and is regenerated by further reaction.

The process of β-oxidation begins in the cytoplasm with formation of a **thioester** between the carboxyl group of a fatty acid and the sulfhydryl group of **coenzyme A** (Problem 13.27). Formation of this acyl-CoA derivative is coupled with hydrolysis of ATP to AMP and pyrophosphate ion. It is common in writing biochemical reactions to show some reactants and products by a curved arrow set over or under the main reaction arrow. We use this convention here to show ATP as a reactant, and AMP and pyrophosphate ion as products.

$$\underset{\substack{\text{Fatty acid}\\\text{(as anion)}}}{R-\overset{\overset{\displaystyle O}{\|}}{C}-O^-} + \underset{\text{Coenzyme A}}{HS-CoA} \xrightarrow{\text{ATP} \quad \text{AMP} + P_2O_7^{4-}} \underset{\substack{\text{An acyl-CoA}\\\text{derivative}}}{R-\overset{\overset{\displaystyle O}{\|}}{C}-S-CoA} + OH^-$$

The mechanism of this reaction involves attack by the fatty acid carboxylate anion on P=O of a phosphoric anhydride group of ATP to form an intermediate analogous to the tetrahedral carbonyl addition intermediate formed in C=O chemistry; in the intermediate formed in the fatty acid-ATP reaction, the phosphorus attacked by the carboxylate anion becomes bonded to five groups. This intermediate then collapses to give an acyl-AMP, which is a highly reactive mixed anhydride of the carboxyl group of the fatty acid and the phosphate group of AMP.

$$\underset{\text{Fatty acid (as anion)}}{R{-}\overset{O}{\overset{\|}{C}}{-}O^-} + \underset{\text{ATP}}{Ad{-}O{-}\overset{O}{\overset{\|}{\underset{O^-}{\underset{|}{P}}}}{-}O{-}\overset{O}{\overset{\|}{\underset{O^-}{\underset{|}{P}}}}{-}O{-}\overset{O}{\overset{\|}{\underset{O^-}{\underset{|}{P}}}}{-}O^-} \rightleftharpoons \left[\begin{array}{c}\text{intermediate}\\ \text{with one}\\ \text{phosphorus}\\ \text{bonded to}\\ \text{five groups}\end{array}\right] \longrightarrow$$

$$\underset{\text{An acyl-AMP (a mixed anhydride)}}{R{-}\overset{O}{\overset{\|}{C}}{-}O{-}\overset{O}{\overset{\|}{\underset{O^-}{\underset{|}{P}}}}{-}O{-}Ad} + \underset{\text{Pyrophosphate ion}}{{}^-O{-}\overset{O}{\overset{\|}{\underset{O^-}{\underset{|}{P}}}}{-}O{-}\overset{O}{\overset{\|}{\underset{O^-}{\underset{|}{P}}}}{-}O^-}$$

This mixed anhydride then undergoes a carbonyl addition reaction with the sulfhydryl group of coenzyme A to form a tetrahedral carbonyl addition intermediate, which collapses to give AMP and an acyl-CoA (a fatty acid thioester of coenzyme A).

$$\underset{\text{Coenzyme A}}{CoA{-}SH} + \underset{\text{An acyl-AMP}}{R{-}\overset{O}{\overset{\|}{C}}{-}O{-}\overset{O}{\overset{\|}{\underset{O^-}{\underset{|}{P}}}}{-}O{-}Ad} \rightleftharpoons \left[\begin{array}{c}\text{tetrahedral}\\ \text{carbonyl}\\ \text{addition}\\ \text{intermediate}\end{array}\right] \longrightarrow$$

$$\underset{\text{An acyl-CoA}}{R{-}\overset{O}{\overset{\|}{C}}{-}S{-}CoA} + \underset{\text{AMP}}{{}^-O{-}\overset{O}{\overset{\|}{\underset{O^-}{\underset{|}{P}}}}{-}O{-}Ad}$$

At this point, the activated fatty acid is transported into the mitochondrion where its carbon chain is degraded by the reactions of β-oxidation.

B. The Four Reactions of β-Oxidation

Reaction 1: Oxidation of a Carbon-Carbon Single Bond to a Carbon-Carbon Double Bond

In the first reaction of β-oxidation, the carbon chain is oxidized, and a double bond is formed between the alpha- and beta-carbons of the fatty acid chain. The oxidizing agent is FAD, which is reduced to $FADH_2$.

$$R-\overset{\beta}{C}H_2-\overset{\alpha}{C}H_2-\overset{O}{\overset{\|}{C}}-SCoA + FAD \longrightarrow \underset{\text{A }trans\text{-enoyl-CoA}}{(R)(H)C=C(H)(C(=O)-SCoA)} + FADH_2$$

An acyl-CoA — A *trans*-enoyl-CoA

Reaction 2: Hydration of the Carbon-Carbon Double Bond

The second reaction of β-oxidation is enzyme-catalyzed hydration of the carbon-carbon double bond to give an (R)-β-hydroxyacyl-CoA. The reaction is regioselective in that —OH is added to carbon 3 of the chain. It is stereoselective in that only the R enantiomer is formed.

$$(R)(H)C=C(H)(C(=O)-SCoA) + H_2O \longrightarrow H\text{-}C(OH)(R)-CH_2-\overset{O}{\overset{\|}{C}}-SCoA$$

A *trans*-enoyl-CoA — (R)-β-Hydroxyacyl-CoA

Reaction 3: Oxidation of the β-Hydroxy Group to a Carbonyl Group

In the second oxidation step of β-oxidation, the secondary alcohol group is oxidized to a ketone group. The oxidizing agent is NAD^+, which is reduced to NADH.

$$H\text{-}C(OH)(R)-CH_2-\overset{O}{\overset{\|}{C}}-SCoA + NAD^+ \longrightarrow R-\overset{O}{\overset{\|}{C}}-CH_2-\overset{O}{\overset{\|}{C}}-SCoA + NADH$$

(R)-β-Hydroxyacyl-CoA — β-Ketoacyl-CoA

Reaction 4: Cleavage of the Carbon Chain by a Reverse Claisen Condensation

The final step of β-oxidation is a reverse Claisen condensation, which results in cleavage between carbons 2 and 3 of the chain to form a molecule of acetyl coenzyme A and a new acyl-CoA, the hydrocarbon chain of which is now shortened by two carbon atoms.

$$R-\overset{O}{\overset{\|}{C}}-CH_2-\overset{O}{\overset{\|}{C}}-SCoA + HS-CoA \longrightarrow R-\overset{O}{\overset{\|}{C}}-SCoA + CH_3\overset{O}{\overset{\|}{C}}-SCoA$$

β-Ketoacyl-CoA — Coenzyme A — An acyl-CoA — Acetyl-CoA

MECHANISM A Reverse Claisen Condensation in β-Oxidation of Fatty Acids

Step 1: A sulfhydryl group of the enzyme thiolase attacks the carbonyl carbon of the ketone to form a tetrahedral carbonyl addition intermediate.

Step 2: The addition intermediate collapses to give an enzyme-bound thioester, which is now shortened by two carbons.

Step 3: The enolate anion reacts with a proton donor to give acetyl-CoA.

Step 4: The enzyme-thioester intermediate undergoes reaction with a molecule of coenzyme A to regenerate sulfhydryl group on the surface of the enzyme and liberate the fatty acyl-CoA, now shortened by two carbon atoms.

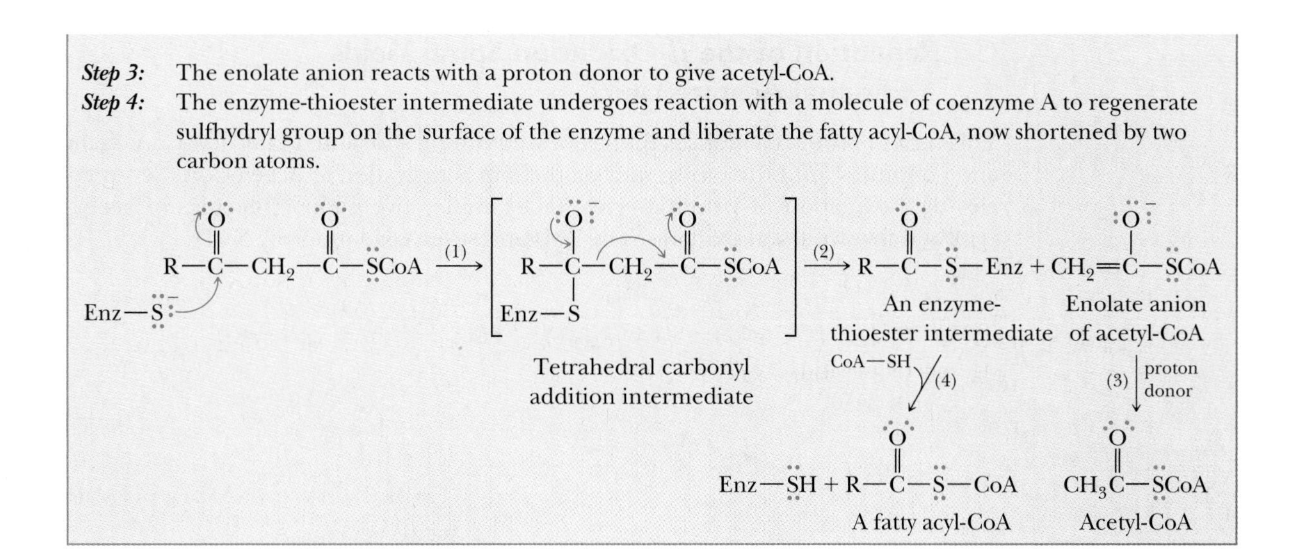

If Steps 1–3 of this mechanism are read in reverse, it is seen as an example of a **Claisen condensation** (Section 14.3A): the attack of the enolate anion of acetyl-CoA on the carbonyl group of a thioester to form a tetrahedral carbonyl addition intermediate, followed by its collapse to give a β-ketothioester.

The four steps in β-oxidation are summarized in Figure 20.1.

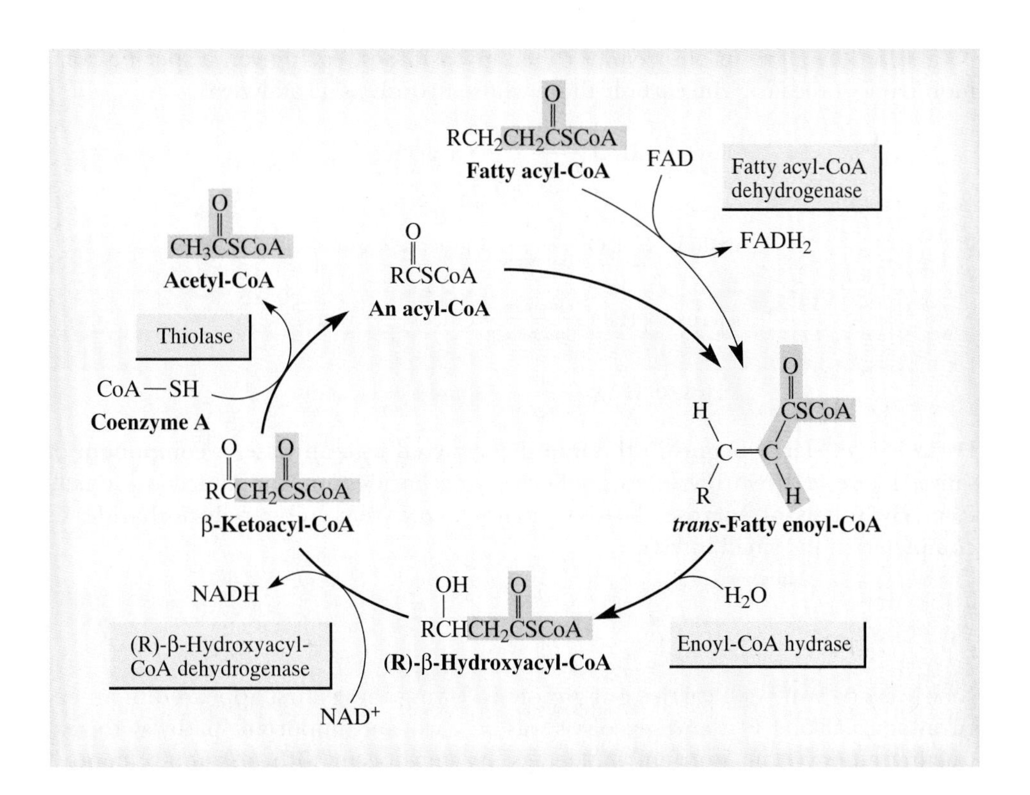

Figure 20.1
The four reactions of β-oxidation. The steps of β-oxidation are called a spiral because, after each series of four reactions, the carbon chain is shortened by two carbon atoms.

C. Repetition of the β-Oxidation Spiral Yields Additional Acetate Units

This series of four reactions is then repeated on the shortened fatty acyl-CoA chain and continues until the entire fatty acid chain is degraded to acetyl-CoA. Seven cycles of β-oxidation of palmitic acid, for example, give eight molecules of acetyl-CoA and involve seven oxidations by FAD and seven oxidations by NAD^+.

$$CH_3(CH_2)_{14}\overset{\displaystyle O}{\overset{\|}{C}}OH + 8\,CoA{-}SH + 7\,NAD^+ + 7\,FAD \xrightarrow{ATP \;\; AMP + P_2O_7^{4-}} 8CH_3\overset{\displaystyle O}{\overset{\|}{C}}SCoA + 7\,NADH + 7\,FADH_2$$

Hexadecanoic acid (Palmitic acid) → Acetyl coenzyme A

20.4 Digestion and Absorption of Carbohydrates

The main function of dietary carbohydrates is as a source of energy. In the typical American diet, carbohydrates provide about 50–60% of daily energy needs. The remainder is supplied by fats and proteins. During digestion of carbohydrates, glycoside (acetal) bonds of disaccharides and polysaccharides are hydrolyzed (Section 11.6) to the hemiacetal and hydroxyl groups of monosaccharides. Hydrolysis is catalyzed by a group of enzymes called **glycosidases,** each given a specific common name indicating the carbohydrate whose hydrolysis it catalyzes.

$$\text{starch} + n\text{H}_2\text{O} \xrightarrow[\text{(mouth and intestine)}]{\alpha\text{-amylase}} n \text{ maltose}$$

$$\text{maltose} + \text{H}_2\text{O} \xrightarrow[\text{(intestine)}]{\text{maltase}} \text{glucose} + \text{glucose}$$

$$\text{sucrose} + \text{H}_2\text{O} \xrightarrow[\text{(intestine)}]{\text{sucrase}} \text{glucose} + \text{fructose}$$

$$\text{lactose} + \text{H}_2\text{O} \xrightarrow[\text{(intestine)}]{\text{lactase}} \text{glucose} + \text{galactose}$$

Hydrolysis of starch begins in the mouth, catalyzed by α-amylase, a component of saliva. There, it is hydrolyzed to smaller polysaccharides and the disaccharide maltose. Hydrolysis of sucrose, lactose, maltose, and remaining polysaccharides is completed in the small intestine.

20.5 Glycolysis

Nearly every living cell carries out glycolysis. Living things first appeared in an environment lacking O_2, and glycolysis was an early and important pathway for extracting energy from nutrient molecules; its steps occur with no requirement of

CHEMICAL CONNECTIONS

Lactose Intolerance

Lactose is the principal carbohydrate in milk; human milk and cow's milk contains about 5% lactose by weight. Human babies are born with the digestive enzymes necessary to hydrolyze lactose to glucose and galactose, but, as they mature, many lose the ability to hydrolyze lactose, a condition known as lactose intolerance. For them, lactose passes through the digestive system to the large intestine. There it increases the osmotic pressure of intestinal fluids, which in turn interferes with reabsorption of water and leads to diarrhea. Further, intestinal bacteria ferment lactose to gases, chiefly carbon dioxide, methane, and hydrogen, which further irritate the intestinal lining and lead to nausea and vomiting. Lactose intolerance typically develops around the age of four in susceptible individuals.

For individuals who cannot tolerate any lactose, milk may be treated with lactase before drinking it. This treatment breaks lactose into glucose and galactose, producing a somewhat sweeter-tasting but lactose-free milk. Lactaid tablets, which can be consumed with or before milk products, also contain lactase.

These products help individuals with lactose intolerance meet their calcium needs. *(Charles D. Winters)*

oxygen. Glycolysis played a central role in anaerobic metabolic processes for the first billion or so years of biological evolution on earth. Modern organisms still employ it to provide precursor molecules for aerobic pathways, such as the tricarboxylic acid cycle, and as a short-term energy source when the supply of oxygen is limited.

Glycolysis is a series of ten enzyme-catalyzed reactions that brings about the oxidation of glucose to two molecules of pyruvate.

Glycolysis From the Greek *glyko,* sweet, and *lysis,* splitting. A series of ten enzyme-catalyzed reactions by which glucose is oxidized to two molecules of pyruvate.

$$\underset{\text{Glucose}}{C_6H_{12}O_6} \xrightarrow[\text{ten enzyme-catalyzed steps}]{\text{glycolysis}} \underset{\text{Pyruvate}}{2CH_3\overset{\overset{\displaystyle O}{\|}}{C}CO_2^-} + 2H^+$$

EXAMPLE 20.1

Show by a balanced half-reaction that the net reaction of glycolysis involves oxidation. Do not worry at this point what the oxidizing agent is; we will come to that later in this section.

SOLUTION

The balanced half-reaction requires four electrons on the right; therefore, the net reaction of glycolysis is a four-electron oxidation. As we will see, this occurs in two separate two-electron oxidation steps, each requiring NAD^+ as the oxidizing agent.

$$\underset{\text{Glucose}}{C_6H_{12}O_6} \xrightarrow{\text{glycolysis}} \underset{\text{Pyruvate}}{2\,CH_3\overset{\overset{\displaystyle O}{\|}}{C}CO_2^-} + 6H^+ + 4e^-$$

Practice Problem 20.1

Under anaerobic (without oxygen) conditions, glucose is converted to lactate by a metabolic pathway called anaerobic glycolysis or, alternatively, lactate fermentation. Is anaerobic glycolysis a net oxidation, a net reduction, or neither?

$$\underset{\text{Glucose}}{C_6H_{12}O_6} \xrightarrow[\text{glycolysis}]{\text{anaerobic}} \underset{\text{Lactate}}{2\,CH_3\overset{\overset{\displaystyle OH}{|}}{C}HCO_2^-} + 2H^+$$

20.6 The Ten Reactions of Glycolysis

Although writing the net reaction of glycolysis is simple, it took several decades of patient, intensive work by scores of scientists to discover the separate reactions by which glucose is converted to pyruvate. Glycolysis is frequently called the Embden-Meyerhof pathway, in honor of the two German biochemists Gustav Embden and Otto Meyerhof, who contributed so greatly to our present knowledge of it. The ten reactions of glycolysis are summarized in Figure 20.2.

Reaction 1: Phosphorylation of α-D-Glucose

Transfer of a phosphate group from ATP to glucose is an example of reaction of an anhydride with an alcohol to form an ester (Section 13.4B), in this case reaction of a phosphoric anhydride with the primary alcohol group of glucose to form a phosphoric ester. In Section 13.6, we saw how a more reactive carboxyl functional group can be transformed into a less reactive carboxyl functional group. The same principle applies to functional derivatives of phosphoric acid. In Reaction 1 of glycolysis, a phosphoric anhydride, a more reactive functional group, is converted to a phosphoric ester, a less reactive functional group.

α-D-Glucose + ATP $\xrightarrow{\text{hexokinase, } Mg^{2+}}$ α-D-Glucose 6-phosphate + ADP

CH₂OH, HO, HO, OH, OH, O; $^-O-P(=O)(O^-)-O-P(=O)(O^-)-O-AMP$; $CH_2OPO_3^{2-}$; $^-O-P(=O)(O^-)-O-AMP$

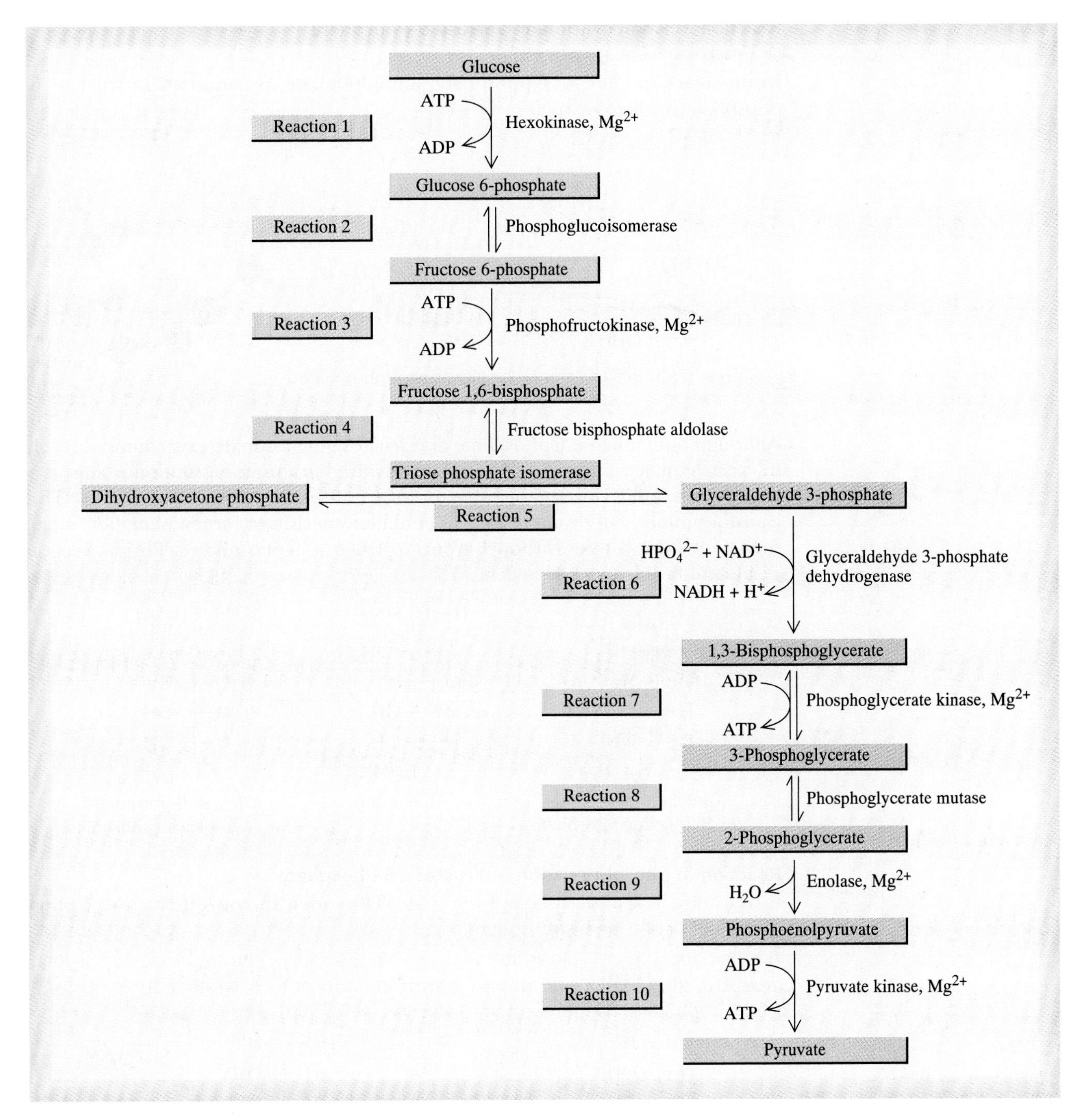

Figure 20.2
The ten reactions of glycolysis.

This enzyme-catalyzed reaction requires divalent magnesium ion, Mg^{2+}, whose function is to coordinate with two negatively charged oxygens of the terminal phosphate group of ATP and to facilitate attack by the —OH group of glucose on the phosphorus atom of the P═O group.

Reaction 2: Isomerization of Glucose 6-Phosphate to Fructose 6-Phosphate

In this reaction, glucose 6-phosphate, an aldohexose, is converted to fructose 6-phosphate, a 2-ketohexose.

phosphogluco-isomerase

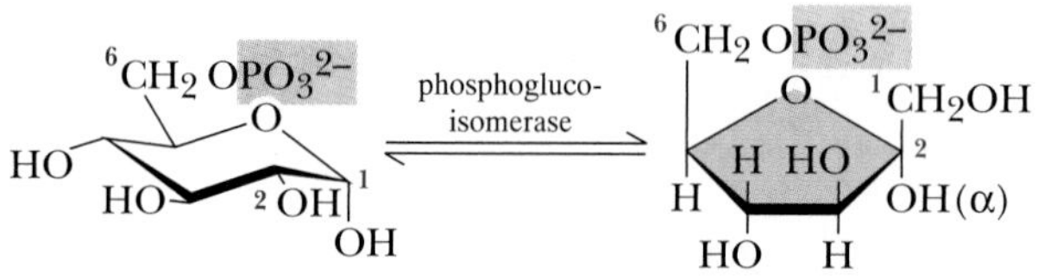

α-D-Glucose 6-phosphate

α-D-Fructose 6-phosphate

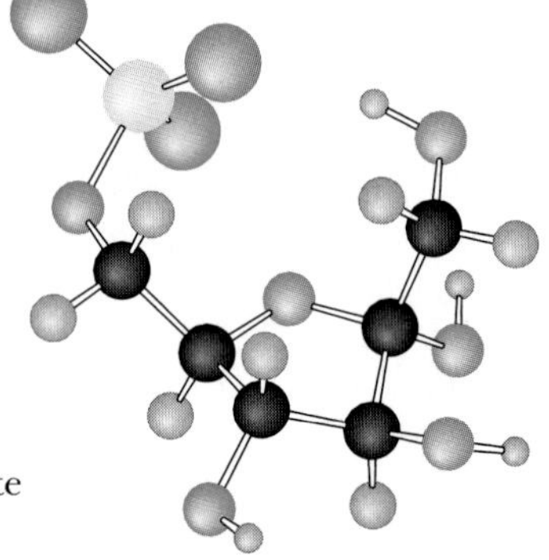

Although both glucose 6-phosphate and fructose 6-phosphate exist almost entirely in cyclic hemiacetal forms, the chemistry involved in this isomerization is most easily seen by considering the open chain forms of these two monosaccharides. This transformation is an example of keto-enol tautomerism to form an enediol, which then forms the ketone carbonyl group in fructose 6-phosphate. (Review Section 11.8B and Problems 11.36 and 11.37.)

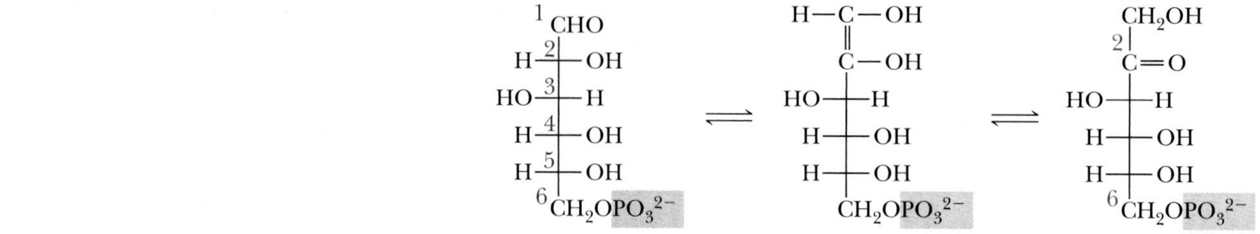

Glucose 6-phosphate (an aldohexose phosphate) ⇌ An enediol ⇌ Fructose 6-phosphate (a 2-ketohexose phosphate)

Reaction 3: Phosphorylation of Fructose 6-Phosphate

In the third reaction, a second mole of ATP is used to convert fructose 6-phosphate to fructose 1,6-bisphosphate.

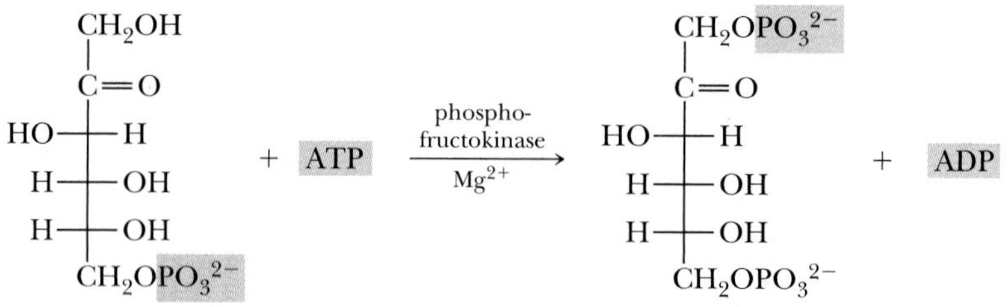

Fructose 6-phosphate + ATP $\xrightarrow[Mg^{2+}]{\text{phosphofructokinase}}$ Fructose 1,6-bisphosphate + ADP

Reaction 4: Cleavage of Fructose 1,6-Bisphosphate to Two Triose Phosphates

In the fourth reaction, fructose 1,6-bisphosphate is cleaved to dihydroxyacetone phosphate and glyceraldehyde 3-phosphate by a reaction that is the reverse of an aldol reaction. Recall that an aldol reaction takes place between the α-carbon of one carbonyl-containing compound and the carbonyl carbon of another and that the functional group of the product of an aldol reaction is a β-hydroxyaldehyde or ketone (Section 14.2).

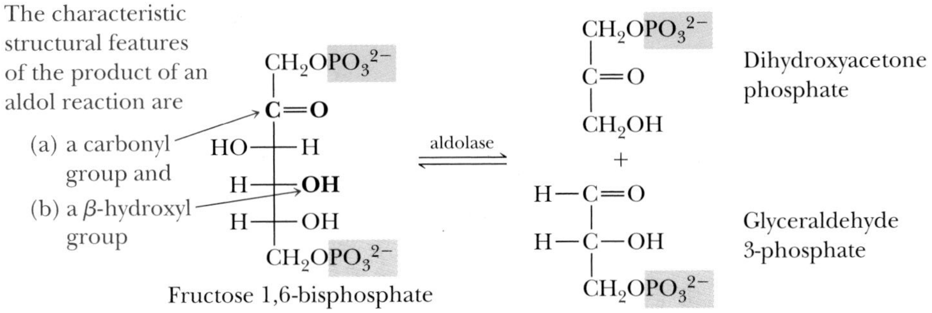

The characteristic structural features of the product of an aldol reaction are (a) a carbonyl group and (b) a β-hydroxyl group

$CH_2OPO_3^{2-}$ | **C=O** | HO—H | H—**OH** | H—OH | $CH_2OPO_3^{2-}$

Fructose 1,6-bisphosphate

$\xrightleftharpoons{\text{aldolase}}$

$CH_2OPO_3^{2-}$ | C=O | CH_2OH — Dihydroxyacetone phosphate

\+

H—C=O | H—C—OH | $CH_2OPO_3^{2-}$ — Glyceraldehyde 3-phosphate

A key intermediate in this enzyme-catalyzed reverse aldol reaction is an imine (Section 11.7) formed between the carbonyl group of fructose 1,6-bisphosphate and the side chain amino group of lysine (Section 18.1A), one of the amino acid building blocks of the enzyme catalyzing this reaction. In the following formulas, the symbol B represents the group on the enzyme involved in proton transfer. As you study this sequence, note the formation and hydrolysis of the imine and the concerted nature of bond breaking and bond forming in the reverse aldol reaction. Also note that the enzyme is regenerated unchanged after its catalytic role is accomplished.

MECHANISM A Reverse Aldol Reaction in Step 4 of Glycolysis

Step 1: Reaction of an ammonium group on the surface of the enzyme with the carbonyl group of fructose 1,6-bisphosphate gives a protonated imine.

Step 2: An enzyme-catalyzed reverse aldol reaction cleaves the six-carbon monosaccharide into two three-carbon units.

Step 3: Proton transfer from the enzyme accompanied by a redistribution of valence electrons gives a new protonated imine.

Step 4: Hydrolysis of the new protonated imine gives dihydroxyacetone phosphate and regenerates the active enzyme, which is now ready to catalyze the cleavage of another molecule of fructose 1,6-bisphosphate.

Fructose 1,6-bisphosphate + $H_3\overset{+}{N}$—Enz (with :B^-) $\xrightleftharpoons[(1)]{(-H_2O)}$ Protonated imine $\xrightleftharpoons[(2)]{\text{reverse aldol reaction}}$ Glyceraldehyde 3-phosphate + $\overset{+}{C}$=NH—Enz intermediate $\xrightleftharpoons{(3)}$

Protonated imine ($CH_2OPO_3^{2-}$ | $\overset{+}{C}$=NH—Enz | CH_2OH, :B^-) $\xrightleftharpoons[(4)]{(+H_2O)}$ Dihydroxyacetone phosphate ($CH_2OPO_3^{2-}$ | C=O | CH_2OH) + $H_3\overset{+}{N}$—Enz (:B^-)

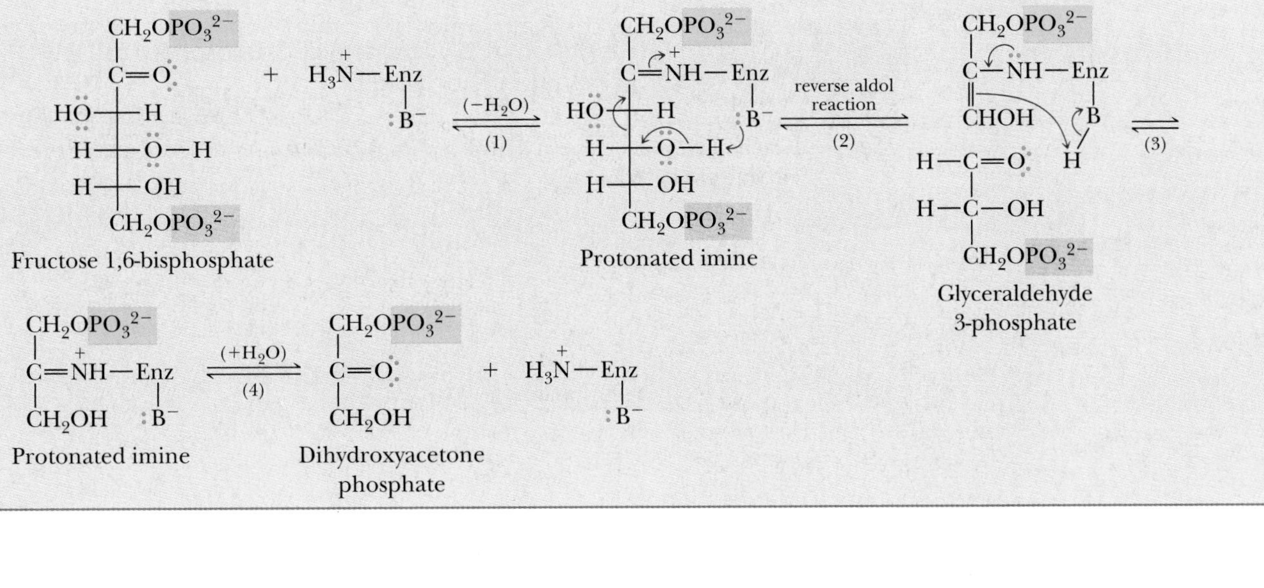

Reaction 5: Isomerization of Dihydroxyacetone Phosphate to Glyceraldehyde 3-Phosphate

This interconversion of triose phosphates occurs by the same type of keto-enol tautomerism and enediol intermediate we have already seen in the isomerization of glucose 6-phosphate to fructose 6-phosphate.

$$\begin{array}{c}CH_2OH\\|\\C{=}O\\|\\CH_2OPO_3^{2-}\end{array} \rightleftharpoons \begin{array}{c}CHOH\\\|\\C{-}OH\\|\\CH_2OPO_3^{2-}\end{array} \rightleftharpoons \begin{array}{c}CHO\\|\\H{-}C{-}OH\\|\\CH_2OPO_3^{2-}\end{array}$$

Dihydroxyacetone phosphate — An enediol intermediate — Glyceraldehyde 3-phosphate

Reaction 6: Oxidation of the Aldehyde Group of Glyceraldehyde 3-Phosphate

To simplify structural formulas in Reaction 6, glyceraldehyde 3-phosphate is abbreviated G—CHO. Two changes occur in this molecule. First the aldehyde group is oxidized to a carboxyl group, which is, in turn, converted to a mixed anhydride. Oxidation of the aldehyde group is a two-electron oxidation. The oxidizing agent is NAD^+, which is reduced to NADH.

$$G{-}\overset{\overset{\displaystyle O}{\|}}{C}{-}H + H_2O \longrightarrow G{-}\overset{\overset{\displaystyle O}{\|}}{C}{-}OH + 2H^+ + 2e^- \qquad \text{A two-electron oxidation}$$

$$NAD^+ + H^+ + 2e^- \longrightarrow NADH \qquad \text{A two-electron reduction}$$

The reaction is considerably more complicated than might appear from combination of the balanced half-reactions. As shown in the Mechanism box, it involves (1) formation of a thiohemiacetal, (2) hydride ion transfer to form a thioester, and (3) conversion of a thioester to a mixed anhydride.

MECHANISM Oxidation of Glyceraldehyde 3-Phosphate to 1,3-Bisphosphoglycerate

Step 1: Reaction between glyceraldehyde 3-phosphate and a sulfhydryl group of the enzyme gives a thiohemiacetal (Section 11.6).

$$G{-}\overset{\overset{\displaystyle \ddot{O}}{\|}}{C}{-}H + H\ddot{S}{-}Enz \rightleftharpoons G{-}\underset{\underset{\displaystyle H}{|}}{\overset{\overset{\displaystyle :\ddot{O}H}{|}}{C}}{-}\ddot{S}{-}Enz$$

Glyceraldehyde 3-phosphate — A thiohemiacetal

Step 2: Oxidation occurs by transfer of a hydride ion from the thiohemiacetal to NAD^+.

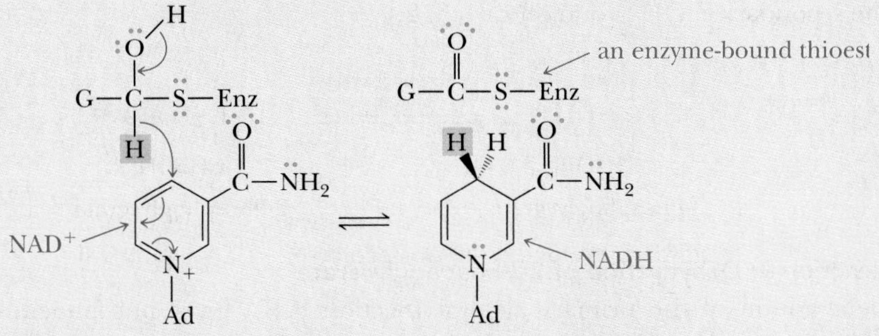

Step 3: Reaction of the thioester with phosphate ion gives a tetrahedral carbonyl addition intermediate, which then collapses to regenerate the enzyme and give a mixed anhydride of phosphoric acid and glyceric acid.

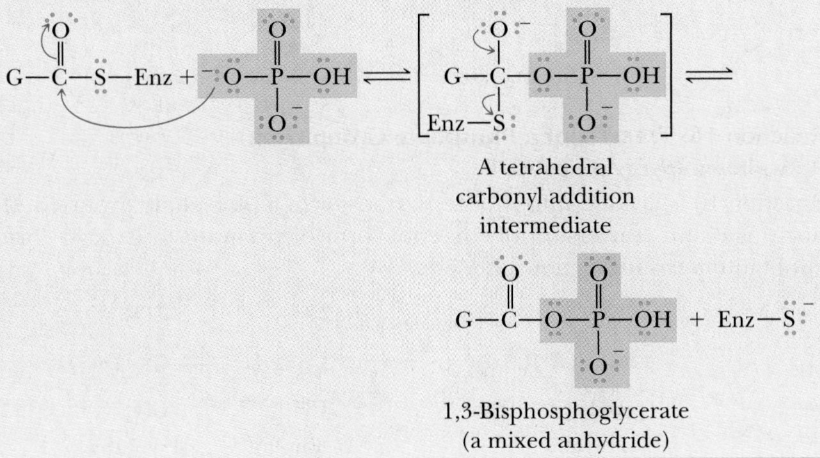

Reaction 7: Transfer of a Phosphate Group from 1,3-Bisphosphoglycerate to ADP

Transfer of a phosphate group in this reaction involves exchange of one anhydride group for another, namely the mixed anhydride of 1,3-bisphosphoglycerate for the new phosphoric anhydride in ATP.

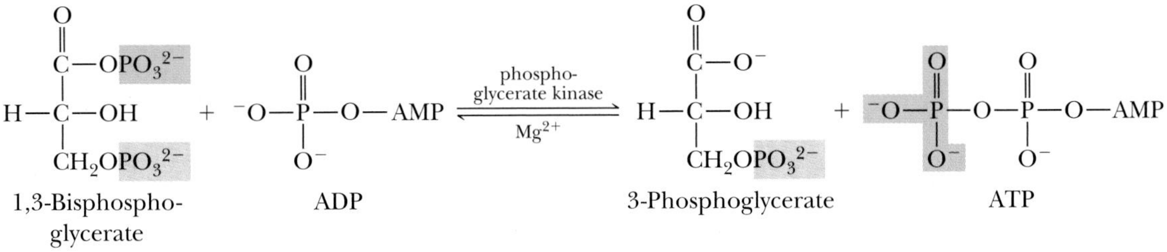

Reaction 8: Isomerization of 3-Phosphoglycerate to 2-Phosphoglycerate

A phosphate group is transferred from the primary —OH group on carbon 3 to the secondary —OH group on carbon 2.

$$\underset{\text{3-Phosphoglycerate}}{\begin{array}{c} CO_2^- \\ | \\ H-C-OH \\ | \\ CH_2OPO_3^{2-} \end{array}} \xrightleftharpoons{\text{phosphoglycerate mutase}} \underset{\text{2-Phosphoglycerate}}{\begin{array}{c} CO_2^- \\ | \\ H-C-OPO_3^{2-} \\ | \\ CH_2OH \end{array}}$$

Reaction 9: Dehydration of 2-Phosphoglycerate

Dehydration of the primary alcohol (Section 8.4E) gives phosphoenolpyruvate, which is the ester of phosphoric acid and the enol form of pyruvic acid.

$$\underset{\text{2-Phosphoglycerate}}{\begin{array}{c} CO_2^- \\ | \\ H-C-OPO_3^{2-} \\ | \\ CH_2OH \end{array}} \xrightarrow[Mg^{2+}]{\text{enolase}} \underset{\text{Phosphoenolpyruvate}}{\begin{array}{c} CO_2^- \\ | \\ C-OPO_3^{2-} \\ \| \\ CH_2 \end{array}} + H_2O$$

Reaction 10: Transfer of a Phosphate Group from Phosphoenolpyruvate to ADP

Reaction 10 is divided into two steps: transfer of a phosphate group to ADP to produce ATP and conversion of the enol form of pyruvate to its keto form by keto-enol tautomerism (Section 11.8A).

$$\underset{\text{Phosphoenol-pyruvate}}{\begin{array}{c} CO_2^- \\ | \\ C-OPO_3^{2-} \\ \| \\ CH_2 \end{array}} \xrightarrow[\text{ADP} \quad \text{ATP}]{\text{pyruvate kinase } Mg^{2+}} \underset{\text{Enol of pyruvate}}{\begin{array}{c} CO_2^- \\ | \\ C-OH \\ \| \\ CH_2 \end{array}} \rightleftharpoons \underset{\text{Pyruvate}}{\begin{array}{c} CO_2^- \\ | \\ C=O \\ | \\ CH_3 \end{array}}$$

Summing these ten reactions gives a balanced equation for the net reaction of glycolysis:

$$\underset{\text{Glucose}}{C_6H_{12}O_6} + 2NAD^+ + 2HPO_4^{2-} + 2ADP \xrightarrow{\text{glycolysis}} \underset{\text{Pyruvate}}{2CH_3\overset{O}{\overset{\|}{C}}CO_2^-} + 2NADH + 2ATP + 2H_3O^+$$

20.7 The Fates of Pyruvate

Pyruvate does not accumulate in cells but rather undergoes one of three possible enzyme-catalyzed reactions, depending on the state of oxygenation and the type of cell in which it is produced. A key to understanding the biochemical logic responsible for two of the possible fates of pyruvate is to recognize that it is produced by the oxidation of glucose through the reactions of glycolysis. NAD^+ is the oxidizing agent and is reduced to NADH. Glycolysis requires a continuing supply of NAD^+; therefore, under anaerobic conditions (where there is no oxygen present for the reoxidation of NADH), two of the metabolic pathways we describe use pyruvate in ways that regenerate NAD^+.

A. Reduction to Lactate—Lactate Fermentation

In vertebrates, the most important pathway for regeneration of NAD^+ under anaerobic conditions is reduction of pyruvate to lactate, catalyzed by the enzyme lactate dehydrogenase.

$$\underset{\text{Pyruvate}}{CH_3\overset{\overset{\large O}{\|}}{C}CO_2^-} + NADH + H_3O^+ \underset{}{\xrightleftharpoons{\text{lactate dehydrogenase}}} \underset{\text{Lactate}}{CH_3\overset{\overset{\large OH}{|}}{C}HCO_2^-} + NAD^+ + H_2O$$

Even though **lactate fermentation** does allow glycolysis to continue in the absence of oxygen, it also brings about an increase in the concentration of lactate, and, perhaps more importantly, it increases the concentration of hydrogen ion, H_3O^+, in muscle tissue and in the bloodstream. This buildup of lactate and H_3O^+ is associated with muscle fatigue. When blood lactate reaches a concentration of about 0.4 mg/100 mL, muscle tissue becomes almost completely exhausted.

Lactate fermentation A metabolic pathway that converts glucose to two molecules of pyruvate.

EXAMPLE 20.2

Show by writing a balanced equation that glycolysis followed by reduction of pyruvate to lactate (lactate fermentation) leads to an increase in the hydrogen ion concentration in the bloodstream.

SOLUTION

Lactate fermentation produces lactic acid, which is completely ionized at pH 7.4, the normal pH of blood plasma. Therefore, the hydrogen ion concentration increases.

$$\underset{\text{Glucose}}{C_6H_{12}O_6} + 2H_2O \xrightarrow{\text{lactate fermentation}} 2\underset{\text{Lactate}}{CH_3\overset{\overset{\large OH}{|}}{C}HCO_2^-} + 2H_3O^+$$

Practice Problem 20.2

Does lactate fermentation result in an increase or decrease in blood pH?

B. Reduction to Ethanol—Alcoholic Fermentation

Yeast and several other organisms have developed an alternative pathway to regenerate NAD^+ under anaerobic conditions. In the first step of this pathway, pyruvate undergoes enzyme-catalyzed decarboxylation to give acetaldehyde.

$$\underset{\text{Pyruvate}}{CH_3\overset{\overset{\large O}{\|}}{C}CO_2^-} + H_3O^+ \xrightarrow{\text{pyruvate decarboxylase}} \underset{\text{Acetaldehyde}}{CH_3\overset{\overset{\large O}{\|}}{C}H} + CO_2 + H_2O$$

The carbon dioxide produced in this reaction is responsible for the foam on beer and the carbonation of naturally fermented wines and champagnes. In a second step, acetaldehyde is reduced by NADH to ethanol.

$$\underset{\text{Acetaldehyde}}{CH_3\overset{\overset{O}{\|}}{C}H} + NADH + H_3O^+ \xrightarrow[\text{dehydrogenase}]{\text{alcohol}} \underset{\text{Ethanol}}{CH_3CH_2OH} + NAD^+ + H_2O$$

Adding the reactions for decarboxylation of pyruvate and reduction of acetaldehyde to the net reaction of glycolysis gives the overall reaction of **alcoholic fermentation.**

Alcoholic fermentation A metabolic pathway that converts glucose to two molecules of ethanol and two molecules of CO_2.

$$\underset{\text{Glucose}}{C_6H_{12}O_6} + 2HPO_4^{2-} + 2ADP + 2H^+ \xrightarrow[\text{fermentation}]{\text{alcoholic}} \underset{\text{Ethanol}}{2CH_3CH_2OH} + 2CO_2 + 2ATP$$

C. Oxidation and Decarboxylation to Acetyl-CoA

Under aerobic conditions, pyruvate undergoes oxidative decarboxylation. The carboxylate group is converted to carbon dioxide, and the remaining two carbons are converted to the acetyl group of acetyl-CoA.

$$\underset{\text{Pyruvate}}{CH_3\overset{\overset{O}{\|}}{C}CO_2^-} + NAD^+ + CoA{-}SH \xrightarrow[\text{decarboxylation}]{\text{oxidative}} \underset{\text{Acetyl-CoA}}{CH_3\overset{\overset{O}{\|}}{C}SCoA} + CO_2 + NADH$$

The oxidative decarboxylation of pyruvate is considerably more complex than is suggested by the preceding equation. In addition to NAD^+ and coenzyme A, this transformation also requires FAD, thiamine pyrophosphate (*The Merck Index,* 12th ed., #9431), which is derived from thiamine (vitamin B_1), and lipoic acid (*The Merck Index,* 12th ed., #9462).

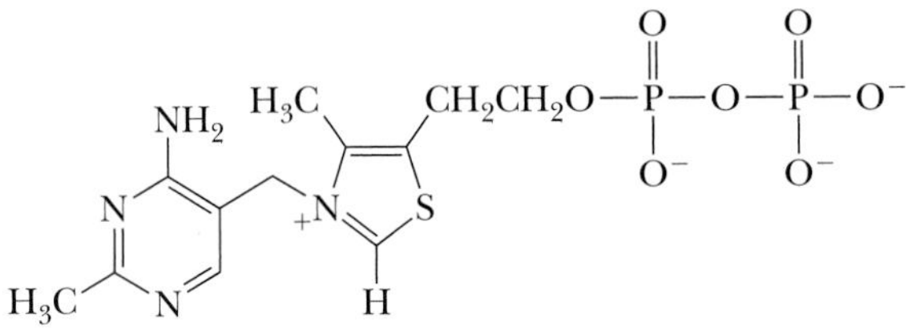

Thiamine pyrophosphate

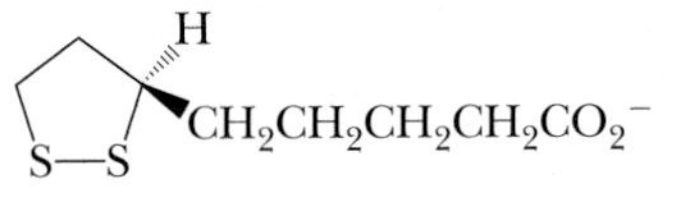

Lipoic acid (shown as the carboxylate anion)

Acetyl coenzyme A then becomes a fuel for the tricarboxylic acid cycle, which results in oxidation of the two-carbon chain of the acetyl group to CO_2 with the production of NADH and $FADH_2$. These reduced coenzymes are, in turn, oxidized to NAD^+ and FAD during respiration with O_2 as the oxidizing agent.

SUMMARY

ATP, ADP, and **AMP** (Section 20.1A) are agents for the storage and transport of phosphate groups. **Nicotinamide adenine dinucleotide (NAD^+)** (Section 20.1B) and **flavin adenine dinucleotide (FAD)** (Section 20.1C) are agents for the storage and transport of electrons in metabolic oxidations and reductions. NAD^+ is a two-electron oxidizing agent and is reduced to NADH. NADH is, in turn, a two-electron reducing agent and is oxidized to NAD^+. In the reactions of FAD involved in β-oxidation of fatty acids, it is a two-electron oxidizing agent and is reduced to $FADH_2$.

Fatty acids in the form of triglycerides are the principal storage form of energy for most organisms (Section 20.2). The hydrocarbon chains of fatty acids are a more highly reduced form of carbon than the oxygenated chains of carbohydrates. The energy yield per gram of fatty acid oxidized is greater than that per gram of carbohydrate.

There are two major stages in the metabolism of fatty acids (Section 20.3): (A) activation of free fatty acids in the cytoplasm by formation of thioesters with **coenzyme A** and transport of the activated fatty acids across the inner mitochondrial membrane followed by (B) β-oxidation. **β-Oxidation of fatty acids** (Section 20.3B) is a series of four enzyme-catalyzed reactions by which a fatty acid is degraded to acetyl-CoA.

During digestion of carbohydrates, hydrolysis of glycoside bonds is catalyzed by a group of enzymes called **glycosidases** (Section 20.4). **Glycolysis** is a series of ten enzyme-catalyzed reactions that brings about the oxidation of glucose to two molecules of pyruvate. The ten reactions of glycolysis (Section 20.6) can be grouped in the following way.

- Transfer of a phosphate group from ATP to an —OH group of a monosaccharide to form a phosphoric ester (Reactions 1 and 3).
- Interconversion of constitutional isomers by keto-enol tautomerism (Reactions 2 and 5).
- Reverse aldol reaction (Reaction 4).
- Oxidation of an aldehyde group to the mixed anhydride of a carboxylic acid and phosphoric acid (Reaction 6).
- Transfer of a phosphate group from a monosaccharide intermediate to ADP to form ATP (Reactions 7 and 10).
- Transfer of a phosphate group from a 1° alcohol to a 2° alcohol (Reaction 8).
- Dehydration of a 1° alcohol to form a carbon-carbon double bond (Reaction 9).

Pyruvate, the product of anaerobic glycolysis, does not accumulate in cells but rather undergoes one of three possible enzyme-catalyzed reactions, depending on the state of oxygenation and the type of cell in which it is produced (Section 20.7). In **lactate fermentation,** pyruvate is reduced to lactate by NADH. In **alcoholic fermentation,** pyruvate is converted to acetaldehyde, which is reduced to ethanol by NADH. Under aerobic conditions, pyruvate is oxidized to acetyl coenzyme A by NAD^+.

KEY REACTIONS

1. β-Oxidation of Fatty Acids (Section 20.3)

A series of four enzyme-catalyzed reactions, after each set of which the carbon chain of a fatty acid is shortened by two carbon atoms.

$$\underset{\text{Hexadecanoic acid (Palmitic acid)}}{CH_3(CH_2)_{14}\overset{\overset{O}{\|}}{C}OH} + 8\,CoA{-}SH + 7\,NAD^+ + 7\,FAD \xrightarrow{ATP \;\rightarrow\; AMP + P_2O_7^{4-}} \underset{\text{Acetyl coenzyme A}}{8\,CH_3\overset{\overset{O}{\|}}{C}SCoA} + 7\,NADH + 7\,FADH_2$$

2. Glycolysis (Section 20.6)

A series of ten enzyme-catalyzed reactions by which glucose is converted to pyruvate.

$$\underset{\text{Glucose}}{C_6H_{12}O_6} + 2NAD^+ + 2HPO_4^{2-} + 2ADP \xrightarrow{\text{glycolysis}} 2\underset{\text{Pyruvate}}{CH_3\overset{\overset{O}{\|}}{C}CO_2^-} + 2NADH + 2ATP + 2H_3O^+$$

3. Reduction of Pyruvate to Lactate—Lactate Fermentation (Section 20.7A)

$$\underset{\text{Pyruvate}}{CH_3\overset{\overset{O}{\|}}{C}CO_2^-} + NADH + H^+ \underset{}{\overset{\text{lactate dehydrogenase}}{\rightleftharpoons}} \underset{\text{Lactate}}{CH_3\overset{\overset{OH}{|}}{C}HCO_2^-} + NAD^+$$

4. Reduction of Pyruvate to Ethanol—Alcohol Fermentation (Section 20.7B)

The carbon dioxide formed in this reaction is responsible for the foam on beer and the carbonation of naturally fermented wines and champagnes.

$$\underset{\text{Pyruvate}}{CH_3\overset{\overset{O}{\|}}{C}CO_2^-} + 2H^+ + NADH \xrightarrow{\text{alcoholic fermentation}} \underset{\text{Ethanol}}{CH_3CH_2OH} + CO_2 + NAD^+$$

5. Oxidative Decarboxylation of Pyruvate to Acetyl-CoA (Section 20.7C)

$$\underset{\text{Pyruvate}}{CH_3\overset{\overset{O}{\|}}{C}CO_2^-} + NAD^+ + CoA{-}SH \xrightarrow{\text{oxidative decarboxylation}} \underset{\text{Acetyl CoA}}{CH_3\overset{\overset{O}{\|}}{C}SCoA} + CO_2 + NADH$$

ADDITIONAL PROBLEMS

β-Oxidation

20.3 Write structural formulas for palmitic, oleic, and stearic acids, the three most abundant fatty acids.

20.4 A fatty acid must be activated before it can be metabolized in cells. Write a balanced equation for the activation of palmitic acid.

20.5 Name three coenzymes necessary for β-oxidation of fatty acids. From what vitamin is each derived?

20.6 We have examined β-oxidation of saturated fatty acids, such as palmitic acid and stearic acid. Oleic acid, an unsaturated fatty acid, is also a common component of dietary fats and oils. This unsaturated fatty acid is degraded by β-oxidation but, at one stage in its degradation, requires an additional enzyme named enoyl-CoA isomerase. Why is this enzyme necessary, and what isomerization does it catalyze? (*Hint:* Consider both the configuration of the carbon-carbon double bond in oleic acid and its position in the carbon chain.)

Glycolysis

20.7 Name one coenzyme required for glycolysis. From what vitamin is it derived?

20.8 Number the carbon atoms of glucose 1 through 6 and show from which carbon atom of glucose the carboxyl group of each molecule of pyruvate is derived.

20.9 How many moles of lactate are produced from 3 moles of glucose?

20.10 Although glucose is the principal source of carbohydrates for glycolysis, fructose and galactose are also metabolized for energy.

(a) What is the main dietary source of fructose? of galactose?

(b) Propose a series of reactions by which fructose might enter glycolysis.

(c) Propose a series of reactions by which galactose might enter glycolysis.

20.11 How many moles of ethanol are produced per mole of sucrose through the reactions of glycolysis and alcoholic fermentation? How many moles of CO_2 are produced?

20.12 Glycerol derived from hydrolysis of triglycerides and phospholipids is also metabolized for energy. Propose a series of reactions by which the carbon skeleton of glycerol might enter glycolysis and be oxidized to pyruvate.

20.13 Ethanol is oxidized in the liver to acetate ion by NAD^+.

(a) Write a balanced equation for this oxidation.

(b) Do you expect the pH of blood plasma to increase, decrease, or remain the same as a result of metabolism of a significant amount of ethanol?

20.14 Write a mechanism to show the role of NADH in the reduction of acetaldehyde to ethanol.

20.15 When pyruvate is reduced to lactate by NADH, two hydrogens are added to pyruvate: one to the carbonyl carbon, the other to the carbonyl oxygen. Which of these hydrogens is derived from NADH?

20.16 Review the oxidation reactions of glycolysis and β-oxidation and compare the types of functional groups oxidized by NAD^+ with those oxidized by FAD.

20.17 Why is glycolysis called an anaerobic pathway?

20.18 Which carbons of glucose end up in CO_2 as a result of alcoholic fermentation?

20.19 Which steps in glycolysis require ATP? Which steps produce ATP?

20.20 The respiratory quotient (RQ) is used in studies of energy metabolism and exercise physiology. It is defined as the ratio of the volume of carbon dioxide produced to the volume of oxygen used.

$$RQ = \frac{\text{Volume } CO_2}{\text{Volume } O_2}$$

(a) Show that RQ for glucose is 1.00. (*Hint:* Look at the balanced equation for complete oxidation of glucose to carbon dioxide and water.)

(b) Calculate RQ for triolein, a triglyceride of molecular formula $C_{57}H_{104}O_6$.

(c) For an individual on a normal diet, RQ is approximately 0.85. Would this value increase or decrease if ethanol were to supply an appreciable portion of caloric needs?

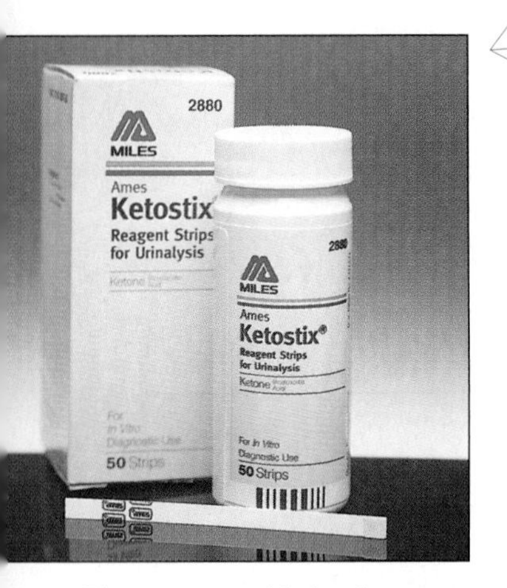

Pharmacy test kit for the presence of ketone bodies in the urine. *(Charles D. Winters)*

20.21 Acetoacetate, β-hydroxybutyrate, and acetone are commonly known within the health sciences as ketone bodies, in spite of the fact that one of them is not a ketone at all. They are products of human metabolism and are always present in blood plasma. Most tissues, with the notable exception of the brain, have the enzyme systems necessary to use them as energy sources. Synthesis of ketone bodies occurs by the following enzyme-catalyzed reactions. Enzyme names are (1) thiolase, (2) β-hydroxy-β-methylglutaryl-CoA synthase, (3) β-hydroxy-β-methylglutaryl-CoA lyase, and (5) β-hydroxybutyrate dehydrogenase. Reaction (4) is spontaneous and uncatalyzed.

$$\underset{\text{Acetyl-CoA}}{2\,CH_3\overset{O}{\overset{\|}{C}}SCoA} \xrightarrow[\text{(1)}]{\;\to CoA-SH\;} \underset{\text{Acetoacetyl-CoA}}{CH_3\overset{O}{\overset{\|}{C}}CH_2\overset{O}{\overset{\|}{C}}SCoA} \xrightarrow[\text{(2)}]{CH_3\overset{O}{\overset{\|}{C}}SCoA \;\;\to\;\; CoA-SH}$$

$$\underset{\text{(S)-3-Hydroxy-3-methylglutaryl-CoA}}{{}^{-}O\overset{O}{\overset{\|}{C}}CH_2-\overset{H_3C\;\;OH}{C}-CH_2\overset{O}{\overset{\|}{C}}SCoA} \xrightarrow[\text{(3)}]{\to CH_3\overset{O}{\overset{\|}{C}}SCoA} \underset{\text{Acetoacetate}}{CH_3\overset{O}{\overset{\|}{C}}CH_2\overset{O}{\overset{\|}{C}}O^{-}}$$

$$\text{Acetoacetate} \xrightarrow[\text{(4)}]{H^+,\;\to CO_2} \underset{\text{Acetone}}{CH_3\overset{O}{\overset{\|}{C}}CH_3}$$

$$\text{Acetoacetate} \xrightarrow[\text{(5)}]{NADH \to NAD^+} \underset{\beta\text{-Hydroxybutyrate}}{CH_3\overset{OH}{\overset{|}{C}}HCH_2\overset{O}{\overset{\|}{C}}O^{-}}$$

Describe the type of reaction involved in each step.

20.22 A connecting point between anaerobic glycolysis and β-oxidation is formation of acetyl-CoA. Which carbon atoms of glucose appear as methyl groups of acetyl-CoA? Which carbon atoms of palmitic acid appear as methyl groups of acetyl-CoA?

A Conversation with . . .

JONATHAN SESSLER

Jonathan Sessler, chemistry professor at the University of Texas, Austin, has spent his life learning different languages. To him, science is international and he has tried to work that element into his education and career. At age six, he began to learn French, and shortly thereafter, Hebrew. When he was about 10, his father, a theoretical physics professor at Lawrence Berkeley National Laboratory in California, went on sabbatical leave to Geneva, Switzerland. Sessler spent a year there in a public school where French was spoken.

After graduating from the University of California, Berkeley, in 1977 with a B.S. degree in chemistry (with highest honors), Sessler attended graduate school at Stanford University where he was in professor James P. Collman's research group. After his first year in graduate school, Sessler held a summer research position at the Max Planck Institute for Medical Research in Heidelberg, Germany.

Sessler received a Ph.D. degree in organic chemistry from Stanford in 1982 for his work with Collman on models for hemoglobin and myoglobin. He then spent one and a half years as a National Science Foundation Fellow with Nobel Laureate Jean-Marie Lehn at Louis Pasteur University in Strasbourg, France, where he helped generate the first artificial systems that could complex both organic and inorganic substrates within the same molecular framework.

For the first six months of 1984, Sessler served as a Japanese Society for the Promotion of Science visiting scientist in professor Iwao Tabushi's group at Kyoto University in Japan. There he worked on quinone-based electron-transport model systems. In September 1984, Sessler began his career as an assistant professor of chemistry at the University of Texas, Austin, and was promoted to full professor in 1992 (http://www.cm.utexas.edu / faculty / profiles / Sessler.html).

Sessler has received a number of honors and fellowships during his academic career. He was awarded a Camille & Henry Dreyfus Foundation Award for Newly Appointed Faculty in 1984 and received an NSF Presidential Young Investigator Award in 1986. He also has received an Arthur C. Cope Scholar Award from the American Chemical Society in 1991.

One of Sessler's main loves is teaching, and he has received several teaching awards from the University of Texas, Austin. These include the College of Natural Science Teaching Excellence Award (1985 and 1990), the President's Associates Teaching Excellence Award (1988), and Outstanding Professor Award in the College of Natural Sciences (1992).

When Sessler joined the faculty at UT-Austin, he was introduced to the culture of the Lone Star State and quickly learned that in Texas everything is supposed to be bigger. It was natural then that one of his first projects involved finding ways to expand porphyrins, making these natural blood pigments bigger. Sessler aptly named the first such compound he made texaphyrin to reflect both its Texas-like size and its Lone Star five nitrogen atom central core. Texaphyrin has since become the unofficial state molecule of Texas.

He saw right away that texaphyrins would have important biomedical applications, especially in the area of cancer. The interest in developing texaphyrin systems as anticancer therapeutic drugs stems from Sessler's own experience with cancer. When he was a senior at UC-Berkeley, Sessler was diagnosed with Hodgkin's disease, a type of lymphoma. After radiation treatment, the disease appeared to have gone into remission, but he unfortunately suffered a relapse after his second year of graduate school at Stanford. One of his physicians at Stanford, Richard A. Miller, took an interest in Sessler's vocation as a chemist and challenged him to use his chemical skills to develop a better treatment for cancer.

In 1989, after Sessler and his coworkers developed the first set of texaphyrins, Sessler and Miller began contacting venture-capital firms to raise the funds needed to start a company to develop, test, and commercialize texaphyrins for use in both the diagnosis and therapy of cancer as well as in the treatment of other diseases. In 1991, they cofounded a new company, Pharmacyclics Inc. (http://www.pcyc.com).

Pharmacyclics now has several texaphyrins and other drugs in different stages of clinical development. Two examples are xcytrin, a gadolinium texaphyrin-based radiation sensitizer designed to enhance the efficacy of radiation therapy; and antrin, a lutetium texaphyrin-based product formulated to permit the light-based removal of atherosclerotic plaque, which can cause coronary or peripheral vascular disease. In work separate from Pharmacyclics, Sessler is beginning to explore whether a class of anion binding agents he terms calixpyrroles may be used to treat cystic fibrosis.

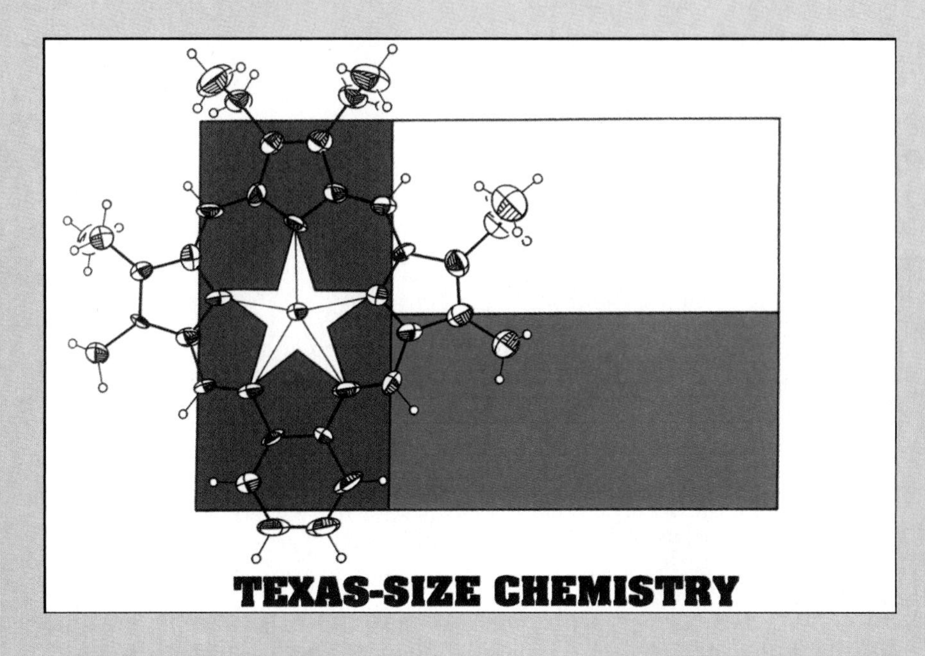

Early Career Choices

"My father is a theoretical physicist and my mother, also schooled in physics, loves the academic life. So it was always assumed from an early age that both my brother and I would be academicians. We clearly weren't going to be physicists—we had to do something different—so my brother chose medicine and I chose chemistry."

Postdoctoral Work

"The postdoctoral stage in one's career is a wonderful time. It provides a great opportunity to work in a foreign country if one so desires. In my case, loving traveling as I do, it was just a question of trying to find an overseas situation that would fit within the goals of developing my career. I was fortunate to have a wonderful research director at Stanford, Jim Collman, and we were blessed in my third year of graduate school to have Jean-Marie Lehn come to give a lecture. I was impressed by his science, his personality, and his obvious intellect. I was lucky enough to get a couple of postdoctoral fellowships, which took me to Strasbourg, France, where I spent a year and a half working hard. I found learning technical French to be a challenge since what I learned as a 10 year old was very different from what I needed every day in the lab. (Once I used the French word for 'condom' to describe the protection of a porphyrin binding site!)

"My decision to go on to Japan after France was again part of a recognition that the postdoctoral stage is perhaps the last chance for travel-related freedom before becoming engaged in what can be an all-encompassing career. I decided to carve out a year or so and went to work with another one of the world's great chemists, the late Iwao Tabushi. Professor Tabushi was one of the most creative people to work in the area called molecular recognition. Unfortunately, he died very young. Many great ideas can be traced back to his really innovative thinking."

Setting Up a Research Program at UT-Austin

"One of our most fun and most successful projects—that related to texaphyrin—came about because I arrived in Texas straight off the plane from Japan and had to take a four-day seminar on how to teach in Texas. One of the themes that came across to me is that things in Texas are bigger. As a graduate student and a postdoc I had been working on a class of molecules called porphyrins that are recognized as being, among other things, important blood pigments. While attending that seminar, I had the idea to make these molecules bigger—to make them Texas-sized."

A Love of Teaching

"Teaching is my main job and one of my main loves—it is why I always wanted to be an academician. Many of my students are certainly more intelligent than I—they may not know as much, but they have enthusiasm and energy and excitement, and being around the best and brightest of the next generation is a real joy and a real honor. I realize that most students will not be chemists, but that doesn't mean that they can't share some of our excitement and, in the process of seeing how we think, become technically informed citizens."

Recognizing Natural Structural Patterns of Compounds Such As Porphyrins and Developing Strategies To Explore Them

"We know that nature uses porphyrins and similar pyrrole-containing macrocycles to coordinate metals such as magnesium in the case of chlorophyll, cobalt in the case of vitamin B_{12}, and iron in the case of hemoglobin. A number of prominent scientists, including my Ph.D. mentor, Jim Collman, are trying to understand the function of these naturally occurring systems and are mimicking them with ever-increasing success. My approach has been to try to understand what happens to the properties of porphyrins when they are made Texas-sized; that is, increasing the number of nitrogen atoms in the central core, raising the number of π electrons in the porphyrin periphery from 18 to 22 or 26, and expanding the whole framework in a way that is not aromatic. Part of the fun is seeing how these changes in structure and electronic character affect the chemistry of the systems."

Synthesis and Characterization of Porphyrins

"The way to do this is to do what we teach in organic chemistry class, namely first carry out a retrosynthesis. We try and break down these rather big molecules into simpler pieces. In the case of porphyrins and expanded porphyrins, the simpler pieces are pyrroles—five-membered, nitrogen-containing ring compounds that are analogs of benzene. Using these as building blocks, we then try to develop ways to connect them synthetically to produce the final desired structure. Sometimes we can just mix the building blocks and some other elements together and get the target porphyrin or expanded porphyrin to come out directly. Other times we have to put sets of building blocks together first, and then react that intermediate product in such a way that we get what we want. We characterize our new systems using standard methods in organic chemistry, such as nuclear magnetic resonance spectroscopy, mass spectrometry, UV-visible spectroscopy, and sometimes x-ray diffraction. We also look carefully at the colors. One of the joys of working with porphyrins and porphyrin-like compounds is that we get to play with systems that are endowed with lovely colors. Furthermore, and this is important from the perspective of analysis, these colors can often teach us a lot in a qualitative sense about molecular structure, size, and shape."

The Unique Properties Gained by Making the Expanded Porphyrins Such As Texaphyrin

"In the case of texaphyrin, we have increased the electronic periphery of the ring from 18 to 22 π electrons. This changes the color from the characteristic red or purple of porphyrins—the name porphyrin comes from the Greek *porphyra,* for purple—to green. Another difference is that we have changed the number of nitrogen atoms in the ring from four to five. We also have increased the size of the ring core by about 20%, enlarging the radius from 2.0 Å to about 2.4 Å. These latter two changes allow the texaphyrins to bind larger metal cations, such as those of the trivalent lanthanide series. This ability to form stable 1:1 complexes with lanthanide(III) cations is not a property of normal porphyrins (those complexes are far less stable). Texaphyrins do, however, share one critical feature with porphyrins: When injected intravenously, they localize to and are retained selectively in neoplastic (abnormal) tissues, including many cancerous tumors."

I realize that most students will not be chemists, but that doesn't mean that they can't share some of our excitement.

The Clinical Development of Texaphyrins

"When we made texaphyrins, the combination of being able to bind lanthanides and to localize in cancerous tissues led me to think they could be very useful in the diagnosis and treatment of cancer and other diseases, including, possibly, AIDS. As is normal in this business, we fired off a proposal to the National Institutes of Health requesting funding to pursue these ideas. It was resoundingly rejected. Shortly after hearing this news from NIH, I went back to the San Francisco Bay area for a checkup with Dr. Miller. I told him about our great cancer-fighting ideas and how poorly they had fared at NIH. He said he would like to read the proposal. This was the day before Thanksgiving in 1989. The next day, Thanksgiving, he called me and said, Jonathan, this stuff is great. We could start a company. We met the next day with a potential investor and started to discuss the possibilities. We discussed how to start gathering biological data and arrange collaborations and illustrate proof of principle. Inspired by these discussions and armed with better data, I later resubmitted the proposal to NIH. Fortunately, it was funded and, in 1991, Pharmacyclics opened its doors as a real company. Since that time, my role in the company has been to act as the chairman of the scientific advisory board and as a consultant and collaborator while keeping my focus firmly on the research side of things."

Human Testing with Texaphyrins

"We started with the idea that putting gadolinium in texaphyrin would allow for magnetic resonance imaging. We quickly confirmed that texaphyrins indeed localize in cancer tissues, and that they do so better than porphyrins. Under the aegis of Pharmacyclics, the gadolinium texaphyrin we are targeting is now in the final stage of human testing. Should these tests come up positive, oncologists and radiologists would have at their disposal a new therapeutic agent that would allow them to enhance the efficacy of the radiation therapy. That means people getting radiation—like me way back when—could receive a lower dose of radiation for the same bang, or rather for the same dose of radiation would enjoy an increased therapeutic effect. Although it is tough to predict the future, we like to think that if we succeed it would be a big, Texas-sized hit. There are about a million cancer patients a year, three-quarters of whom get radiation. Unfortunately, as most of us are probably all too aware from some level of personal experience, most of these latter patients do not enjoy a curative effect. Thus, if we could increase even slightly the percentage that do become cured by radiation therapy, the large number of patients who might benefit would mean that we could end up making a significant contribution to the cancer treatment field."

21

Nuclear Magnetic Resonance Spectroscopy

Anethole is the chief constituent of anise, star anise, and fennel oils, which are used as a flavoring agent in cooking. See its ^{1}H-NMR spectrum in Additional Problem 21.35. Inset: model of anethole. *(Charles D. Winters)*

Determination of molecular structure is a central theme in organic chemistry. For this purpose, chemists now rely almost exclusively on instrumental methods of analysis; we discuss the two most commonly used of these in this text. We begin in this chapter with a discussion of nuclear magnetic resonance (NMR) spectroscopy. In the following chapter, we discuss infrared (IR) spectroscopy. In order to understand the fundamentals of spectroscopy, we must first review some of the fundamentals of electromagnetic radiation.

21.1 ELECTROMAGNETIC RADIATION

Electromagnetic radiation Light and other forms of radiant energy.

Wavelength (λ) The distance between two consecutive peaks on a wave.

Frequency (ν) A number of full cycles of a wave that pass a point in a fixed time period.

Hertz (Hz) The unit in which wave frequency is reported; s^{-1} (read *per second*).

Gamma rays, x-rays, ultraviolet light, visible light, infrared radiation, microwaves, and radio waves are all part of the electromagnetic spectrum. Because **electromagnetic radiation** behaves as a wave traveling at the speed of light, it can be described in terms of its wavelength and its frequency. Wavelengths, frequencies, and energies of some regions of the electromagnetic spectrum are summarized in Table 21.1.

Wavelength is the distance between any two consecutive identical points on a wave. Wavelength is given the symbol λ (Greek lambda) and is usually expressed in the SI base unit of meters. Other derived units commonly used to express wavelength are given in Table 21.2.

Frequency, the number of full cycles of a wave that pass a given point in a second, is given the symbol ν (Greek nu) and is reported in **hertz** (Hz), where 1 MHz = 10^6 Hz. Wavelength and frequency are inversely proportional; one is calculated from the other using the following relationship:

$$\nu\lambda = c$$

where ν is frequency in hertz; c is the velocity of light, 3.00×10^8 m/s; and λ is the wavelength in meters. For example, consider infrared radiation, or heat radiation as it is also called, of wavelength 1.5×10^{-5} m. The frequency of this radiation is 2.0×10^{13} Hz.

$$\nu = \frac{3.0 \times 10^8 \text{ m/s}}{1.5 \times 10^{-5} \text{ m}} = 2.0 \times 10^{13} \text{ Hz}$$

An alternative way to describe electromagnetic radiation is in terms of its properties as a stream of particles. We call these particles **photons.** The energy in a mole of photons is related to the frequency of radiation by the equation

$$E = h\nu = h\frac{c}{\lambda}$$

where E is the energy in kcal/mol and h is Planck's constant, 9.54×10^{-14} $\text{kcal}\cdot\text{s}\cdot\text{mol}^{-1}$ (3.99×10^{-13} $\text{kJ}\cdot\text{s}\cdot\text{mol}^{-1}$). This equation tells us that high-energy radiation corresponds to short wavelengths, and vice versa. Thus, ultraviolet light (higher energy) has a shorter wavelength (approximately 10^{-7} m) than infrared radiation (lower energy), which has a wavelength of approximately 10^{-5} m.

TABLE 21.1 Wavelength, Frequency, and Energy Relationships of Some Regions of the Electromagnetic Spectrum

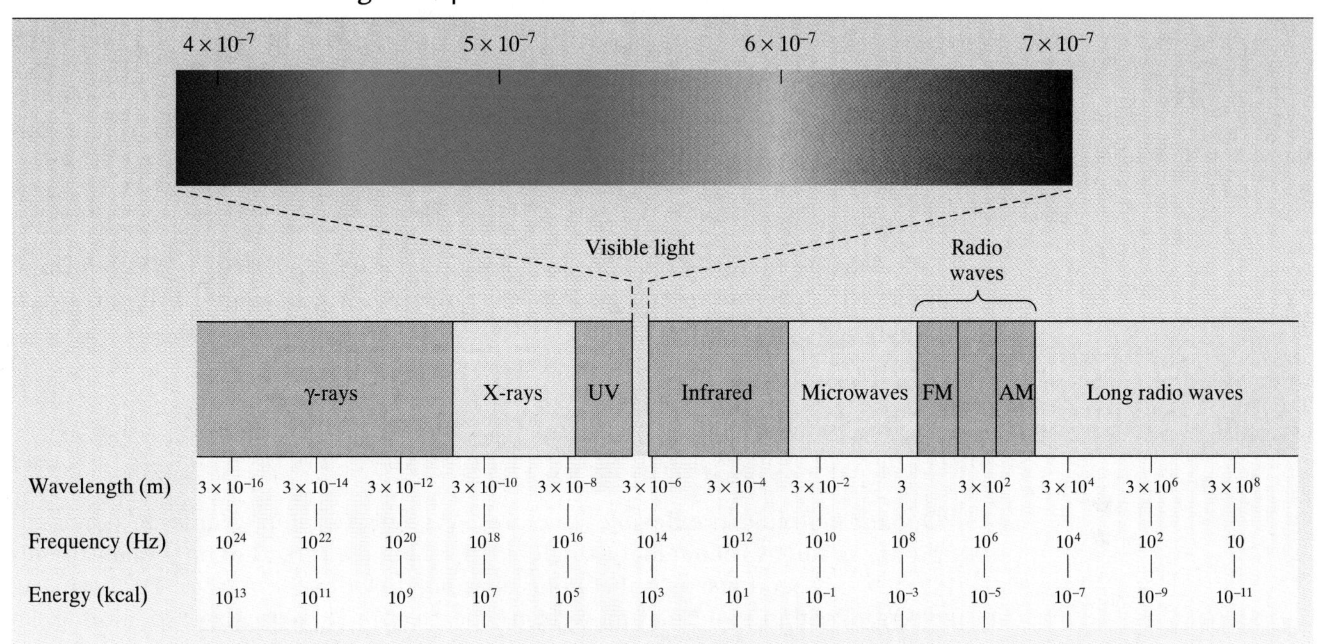

Wavelength (m)	3×10^{-16}	3×10^{-14}	3×10^{-12}	3×10^{-10}	3×10^{-8}	3×10^{-6}	3×10^{-4}	3×10^{-2}	3	3×10^{2}	3×10^{4}	3×10^{6}	3×10^{8}
Frequency (Hz)	10^{24}	10^{22}	10^{20}	10^{18}	10^{16}	10^{14}	10^{12}	10^{10}	10^{8}	10^{6}	10^{4}	10^{2}	10
Energy (kcal)	10^{13}	10^{11}	10^{9}	10^{7}	10^{5}	10^{3}	10^{1}	10^{-1}	10^{-3}	10^{-5}	10^{-7}	10^{-9}	10^{-11}

EXAMPLE 21.1

Calculate the energy in kilocalories per mole of radiation of wavelength 2.50 μm. What type of radiant energy is this? (Refer to Table 21.1.)

SOLUTION

Use the relationship $E = hc/\lambda$. Make certain that dimensions for distance are consistent; if the dimension of wavelength is meters, then express the velocity of light in meters per second. First convert 2.50 μm to meters, using the relationship 1 μm = 10^{-6} m (Table 21.2).

$$2.50\ \mu\text{m} \times \frac{10^{-6}\ \text{m}}{1\ \mu\text{m}} = 2.50 \times 10^{-6}\ \text{m}$$

TABLE 21.2 Common Units Used to Express Wavelength (λ)

Unit	Relation to Meter	Diagram of Wavelength
meter (m)	—	
millimeter (mm)	1 mm = 10^{-3}m	
micrometer (μm)	1 μm = 10^{-6}m	
nanometer (nm)	1 nm = 10^{-9}m	
Angstrom (Å)	1 Å = 10^{-10}m	

Now substitute this value in the equation $E = hc/\lambda$.

$$E = \frac{hc}{\lambda} = 9.54 \times 10^{-14} \frac{\text{kcal} \cdot \text{s}}{\text{mol}} \times 3.00 \times 10^{8} \frac{\text{m}}{\text{s}} \times \frac{1}{2.50 \times 10^{-6}\ \text{m}} = 11.4\ \text{kcal/mol}$$

$$= 47.7\ \text{kJ/mol}$$

Electromagnetic radiation with energy of 11.4 kcal/mol corresponds to radiation in the infrared region.

Practice Problem 21.1

Calculate the energy of red light (680 nm) in kilocalories per mole. Which form of radiation carries more energy, infrared radiation of wavelength 2.50 μm or red light of wavelength 680 nm?

21.2 Molecular Spectroscopy

Organic molecules are flexible structures. As we discussed in Chapter 3, atoms and groups of atoms rotate about single covalent bonds. In addition, covalent bonds stretch and bend just as if the atoms are joined by flexible springs. Furthermore, electrons within molecules move from one electronic energy level to another. Finally, certain nuclei are spinning charged particles that generate magnetic fields and change from one spin energy level to another. We know from experimental observations and from theories of molecular structure that all energy changes within a molecule are quantized: Vibrations of bonds within molecules, for example, can undergo transitions only between allowed vibrational energy levels.

An atom or molecule can be made to undergo a transition from energy state E_1 to a higher energy state E_2 by irradiating it with electromagnetic radiation corresponding to the energy difference between states E_1 and E_2 as illustrated schematically in Figure 21.1. When the atom or molecule returns to the ground state E_1, an equivalent amount of energy is emitted.

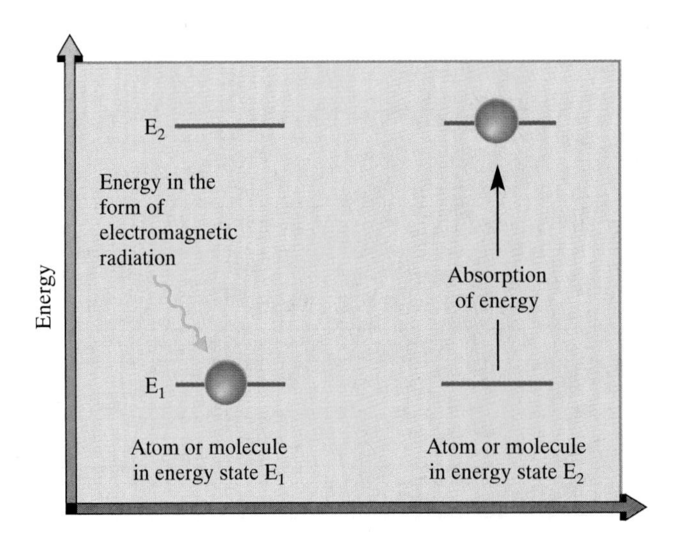

Figure 21.1 Absorption of energy in the form of electromagnetic radiation causes an atom or molecule in energy state E_1 to change to a higher energy state E_2.

Molecular spectroscopy is the experimental process of measuring which frequencies of radiation are absorbed or emitted by particular substances and then correlating these frequencies with specific types of molecular structures.

Molecular spectroscopy The study of which frequencies of electromagnetic radiation are absorbed or emitted by substances and the correlation between these frequencies and specific types of molecular structure.

In this chapter, we concentrate on absorption of radio-frequency radiation, which causes transitions between nuclear spin energy levels; that is, we concentrate on nuclear magnetic resonance spectroscopy.

The phenomenon of nuclear magnetic resonance was first detected in 1946 by Felix Bloch and Edward Purcell, both of the United States, who shared the 1952 Nobel Prize for Physics. The particular value of **nuclear magnetic resonance (NMR) spectroscopy** is that it gives us information about the number and types of atoms in a molecule, for example, about the number and types of hydrogens using **^{1}H-NMR spectroscopy** and about the number and types of carbons using **^{13}C-NMR spectroscopy.**

21.3 THE ORIGIN OF NUCLEAR MAGNETIC RESONANCE

You may already be familiar from general chemistry with the concept that an electron has a spin and that a spinning charge creates an associated magnetic field. In effect, an electron behaves as if it is a tiny bar magnet. Nuclei of ^{1}H and ^{13}C, isotopes of the two elements most common to organic compounds, also have a spin and behave as if they are tiny bar magnets. Note that nuclei of ^{12}C and ^{16}O do not have a spin and do not behave as tiny bar magnets. Thus, in this sense, nuclei of ^{1}H and ^{13}C are quite different from nuclei of ^{12}C and ^{16}O.

Within a collection of ^{1}H and ^{13}C atoms, the spins of their tiny nuclear bar magnets are completely random in orientation. When placed between the poles of a powerful magnet, however, interactions between their nuclear spins and the applied magnetic field are quantized, and only two orientations are allowed (Figure 21.2). By convention, nuclei with spin $+1/2$ are aligned with the applied field and in the lower energy state; nuclei with spin $-1/2$ are aligned against it and in the higher energy state.

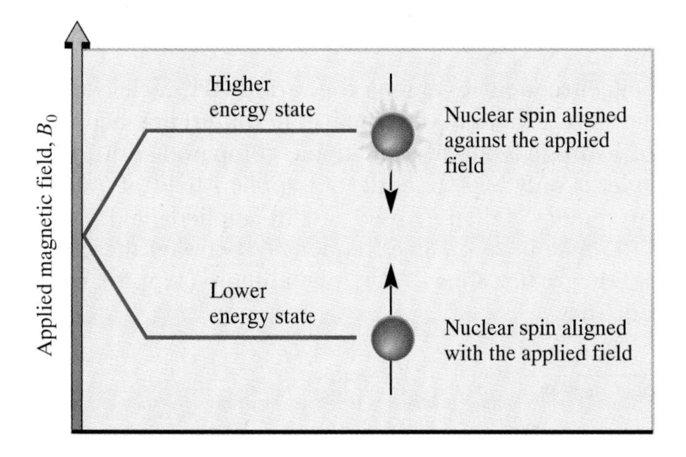

Figure 21.2
^{1}H and ^{13}C nuclei with spin $+1/2$ are aligned with the applied magnetic field and are in the lower spin energy state; those with spin $-1/2$ are aligned against the applied magnetic field and are in the higher spin energy state.

At an applied field strength of 7.05 T, which is readily available with present-day superconducting electromagnets, the difference in energy between nuclear spin states for ^1H is 0.0286 cal/mol (0.120 kJ/mol), which corresponds to electromagnetic radiation of approximately 300 MHz (300,000,000 Hz). At the same magnetic field strength, the difference in energy between nuclear spin states for ^{13}C is 0.0072 cal/mol (0.035 kJ/mol), which corresponds to electromagnetic radiation of 75 MHz. Thus, we can use electromagnetic radiation in the radio-frequency range to detect changes in nuclear spin states for ^1H and ^{13}C. In the next several sections, we describe how these measurements are made for nuclear spin states of these two isotopes and then how this information can be correlated with molecular structure.

21.4 Nuclear Magnetic "Resonance"

When hydrogen nuclei are placed in an applied magnetic field, a small majority of their nuclear spins are aligned with the applied field in the lower energy state. When nuclei in the lower energy spin state are irradiated with a radio frequency of the appropriate energy, they absorb energy, and their nuclear spins flip from the lower energy state to the higher energy state, as illustrated in Figure 21.2. **Resonance** in this context is the absorption of electromagnetic radiation by a spinning nucleus and the resulting flip of its nuclear spin state. The instrument used to detect this absorption and resulting flip of nuclear spin state records it as a resonance **signal.**

Resonance (NMR) The absorption of electromagnetic radiation by a spinning nucleus and the resulting "flip" of its spin from a lower energy state to a higher energy state.

Signal A recording in an NMR spectrum of a nuclear magnetic resonance.

If we were dealing with ^1H nuclei isolated from all other atoms and electrons, any combination of applied field and electromagnetic radiation that produces a resonance signal for one hydrogen nucleus would produce the same resonance signal for all other hydrogen nuclei. In other words, hydrogens would be indistinguishable one from another. Hydrogens in an organic molecule, however, are surrounded by electrons, which themselves have spin and thereby create **local magnetic fields** that oppose the applied field. Although these local magnetic fields created by electrons are orders of magnitude weaker than the applied magnetic fields used in NMR spectroscopy, they are nonetheless significant at the molecular level. The result of these local magnetic fields is to shield hydrogens from the applied field. The greater the **shielding** of a particular hydrogen by local magnetic fields, the greater the strength of the applied field required to bring that hydrogen into resonance.

Shielding In NMR spectroscopy, electrons around a nucleus create their own local magnetic fields and thereby shield the nucleus from the applied magnetic field.

The differences in resonance frequencies among the various ^1H nuclei within a molecule caused by shielding are generally very small. The difference between the resonance frequencies of hydrogens in chloromethane compared with those in fluoromethane, for example, is only 360 Hz under an applied field of 7.05 T. Considering that the radio-frequency radiation used at this applied field is approximately 300 MHz, the difference in resonance frequencies between these two sets of hydrogens is only slightly greater than 1 part per million (1 ppm) compared with the irradiating frequency.

$$\frac{360 \text{ Hz}}{300 \times 10^6 \text{ Hz}} = \frac{1.2}{10^6} = 1.2 \text{ ppm}$$

It is customary to measure the resonance frequencies of individual nuclei relative to the resonance frequency of the same nuclei in a reference compound. The reference compound now universally accepted for ^{1}H-NMR and ^{13}C-NMR spectroscopy is **tetramethylsilane (TMS).**

$$
\begin{array}{c} CH_3 \\ | \\ H_3C-Si-CH_3 \\ | \\ CH_3 \end{array}
$$

Tetramethylsilane (TMS)

When a ^{1}H-NMR spectrum of a compound is determined, the resonance signals of its hydrogens are reported by how far they are shifted from the resonance signal of the 12 equivalent hydrogens in TMS. When a ^{13}C-NMR spectrum is determined, the resonance frequencies of its carbons are reported by how far they are shifted from the resonance signal of the four equivalent carbons in TMS.

To standardize reporting of NMR data, workers have adopted the quantity called **chemical shift (δ)**, expressed in parts per million.

Chemical shift, δ The position of a signal on an NMR spectrum relative to the signal of tetramethylsilane (TMS); expressed in delta (δ) units, where 1 δ equals 1 ppm.

$$
\delta = \frac{\text{Shift in frequency of a signal from TMS (Hz)}}{\text{Operating frequency of the spectrometer (Hz)}}
$$

21.5 AN NMR SPECTROMETER

The essential elements of an NMR spectrometer are a powerful magnet, a radio-frequency generator, a radio-frequency detector, and a sample tube (Figure 21.3).

The sample is dissolved in a solvent having no hydrogens, most commonly carbon tetrachloride (CCl_4), deuterochloroform ($CDCl_3$), or deuterium oxide

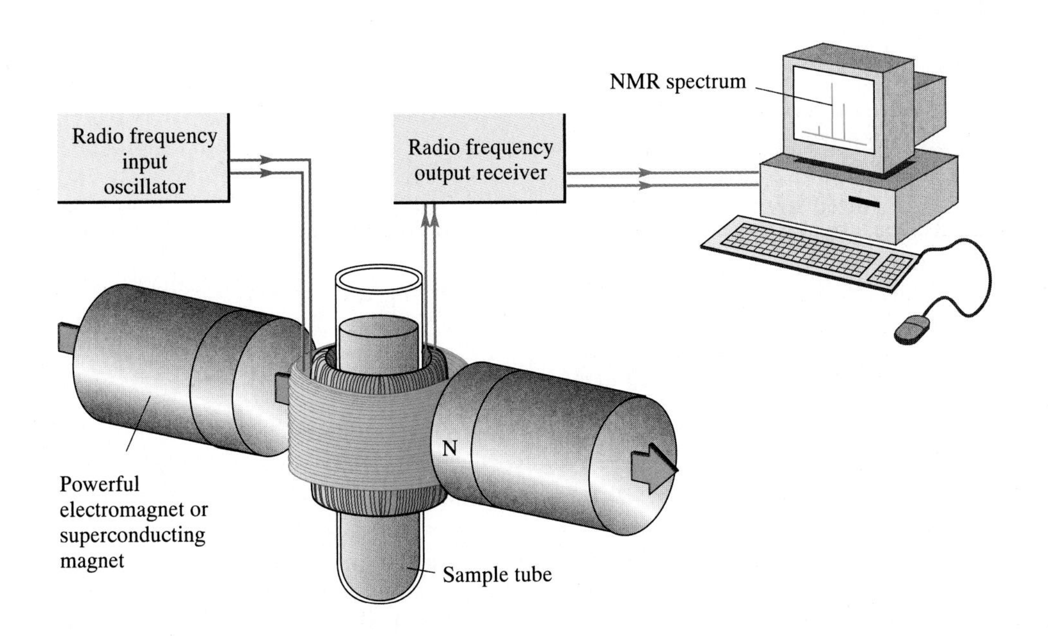

Figure 21.3 Schematic diagram of a nuclear magnetic resonance spectrometer.

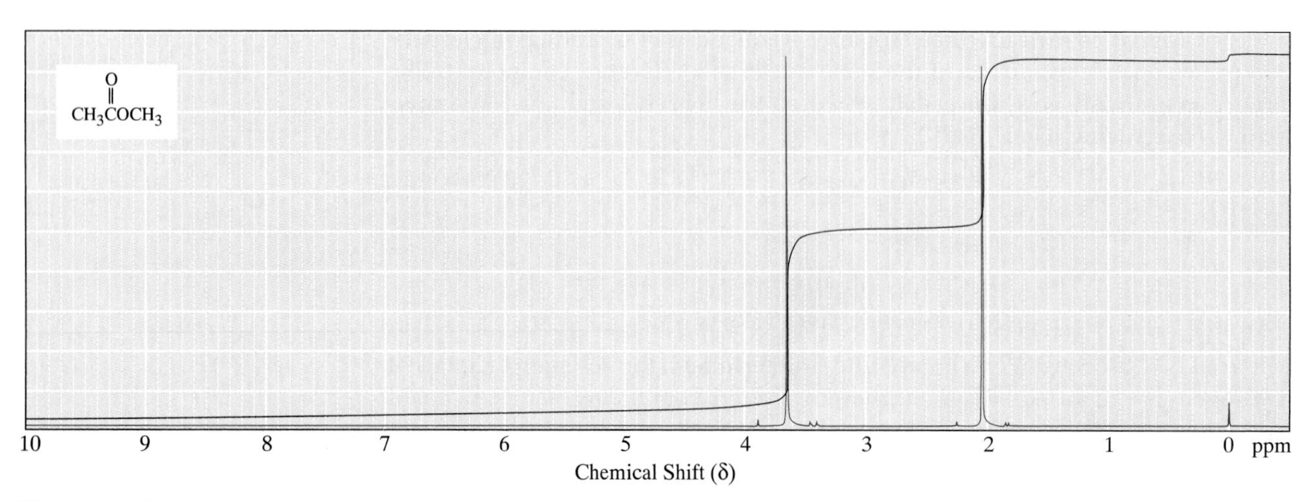

Figure 21.4
^{1}H-NMR spectrum of methyl acetate.

(D_2O). The sample cell is a small glass tube suspended in the gap between the pole pieces of the magnet and set spinning on its long axis to ensure that all parts of the sample experience a homogeneous applied field. In a typical ^{1}H-NMR spectrum, the horizontal axis represents the δ (delta) scale with values from 0 on the right to 10 on the left. The vertical axis represents the intensity of the resonance signal. All ^{1}H-NMR spectra shown in this text were recorded at an applied magnetic field of 7.05 T and a radio frequency of 300 MHz. All ^{13}C-NMR spectra were recorded at 7.05 T and 75 MHz.

Figure 21.4 shows a 300 MHz ^{1}H-NMR spectrum of methyl acetate. The small signal at δ 0 in this spectrum is caused by the hydrogens of the reference compound, TMS. The remainder of the spectrum consists of two signals: one for the three hydrogens on the methyl adjacent to oxygen and one for the three hydrogens on the methyl adjacent to the carbonyl group. It is not our purpose at the moment to determine which hydrogens give rise to which signal but only to recognize the form in which an NMR spectrum is recorded and the origin of the calibration marks.

A note on terminology: If a signal is shifted toward the left on the chart paper, we say that it is shifted **downfield,** meaning that nuclei giving rise to that signal are less shielded and come into resonance at a weaker applied field. Conversely, if a signal is shifted toward the right on the chart paper, we say that it is shifted **upfield,** meaning that nuclei giving rise to that signal are more shielded and come into resonance at a stronger applied field.

21.6 EQUIVALENT HYDROGENS

Equivalent hydrogens Hydrogens that have the same chemical environment.

Given the structural formula of a compound, how do we know how many signals to expect? The answer is that **equivalent hydrogens** give the same ^{1}H-NMR signal; nonequivalent hydrogens give different ^{1}H-NMR signals. A direct way to determine which hydrogens in a molecule are equivalent is to replace each in turn by a test atom, as for example a halogen atom. If replacement of two hydrogens being tested by a test atom gives the same compound, the two hydrogens are equivalent. If replacement gives different compounds, the two hydrogens are nonequivalent.

Using this substitution test, we can show that propane contains two sets of equivalent hydrogens: a set of six equivalent primary hydrogens and a set of two equivalent secondary hydrogens. In a ¹H-NMR spectrum of propane, we would expect to see two signals, one for the six equivalent $—CH_3$ hydrogens and one for the two equivalent $—CH_2—$ hydrogens.

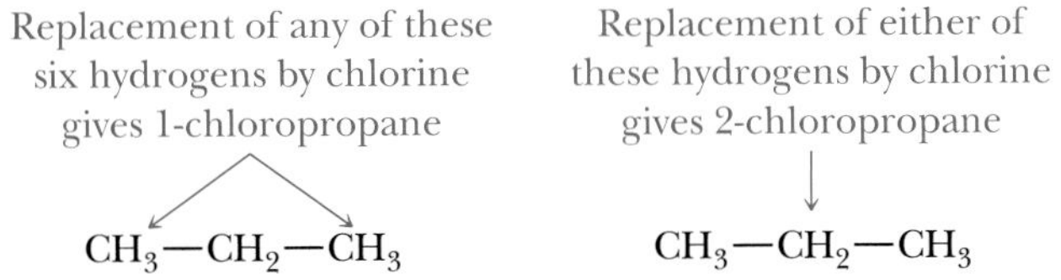

Replacement of any of these six hydrogens by chlorine gives 1-chloropropane

$CH_3—CH_2—CH_3$

Propane

Replacement of either of these hydrogens by chlorine gives 2-chloropropane

$CH_3—CH_2—CH_3$

Propane

EXAMPLE 21.2

State the number of sets of equivalent hydrogens in each compound and the number of hydrogens in each set.

(a) 2-Methylpropane (b) 2-Methylbutane

SOLUTION

(a) 2-Methylpropane contains two sets of equivalent hydrogens: a set of nine equivalent primary hydrogens and one tertiary hydrogen.

(b) 2-Methylbutane contains four sets of equivalent hydrogens. Nine primary hydrogens are in this molecule: three in one set and six in a second set. Replacement by chlorine of any hydrogen in the set of three gives 1-chloro-3-methylbutane. Replacement of any hydrogen in the set of six gives 1-chloro-2-methylbutane.

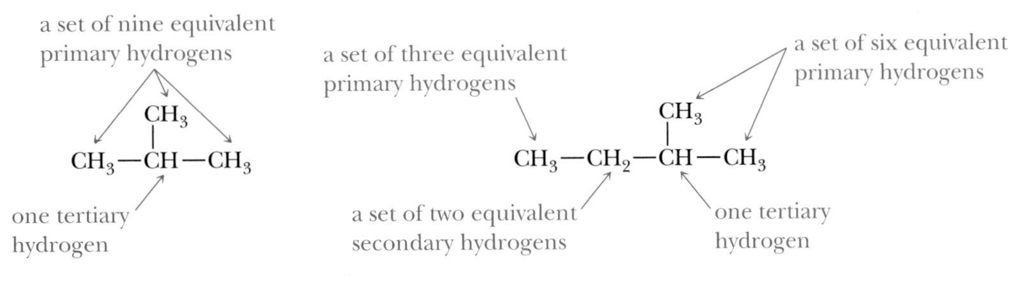

Practice Problem 21.2

State the number of sets of equivalent hydrogens in each compound and the number of hydrogens in each set.

(a) 3-Methylpentane (b) 2,2,4-Trimethylpentane

Here are four organic compounds, each of which has one set of equivalent hydrogens and gives one signal in its ¹H-NMR spectrum.

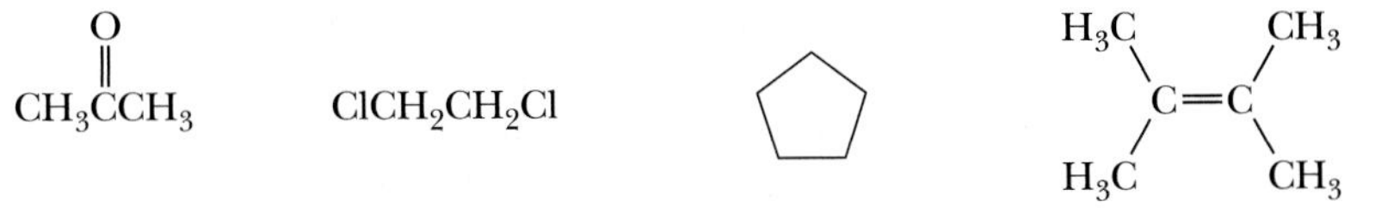

$CH_3C(=O)CH_3$ — Propanone (Acetone)

$ClCH_2CH_2Cl$ — 1,2-Dichloroethane

Cyclopentane

$(H_3C)_2C=C(CH_3)_2$ — 2,3-Dimethyl-2-butene

Molecules with two or more sets of equivalent hydrogens give rise to a different resonance signal for each set. 1,1-Dichloroethane, for example, has one set of three equivalent 1° hydrogens and one 2° hydrogen; there are two resonance signals in its ^{1}H-NMR spectrum.

CH_3CHCl_2 (Cl on CH)

1,1-Dichloroethane (2 signals)

Cyclopentanone (2 signals)

(Z)-1-Chloropropene (3 signals)

Cyclohexene (3 signals)

You should see immediately that valuable information about molecular structure can be obtained simply by counting the number of signals in the ^{1}H-NMR spectrum of a compound. Consider, for example, the two constitutional isomers of molecular formula $C_2H_4Cl_2$. The compound 1,2-dichloroethane has one set of equivalent hydrogens and one signal in its ^{1}H-NMR spectrum. Its constitutional isomer 1,1-dichloroethane has two sets of equivalent hydrogens and two signals in its ^{1}H-NMR spectrum. Thus, simply counting signals allows you to distinguish between these two compounds.

EXAMPLE 21.3

Each compound gives only one signal in its ^{1}H-NMR spectrum. Propose a structural formula for each.

(a) C_2H_6O (b) $C_3H_6Cl_2$ (c) C_6H_{12}

SOLUTION

Following are structural formulas for each part.

(a) CH_3OCH_3 (b) $CH_3CCl_2CH_3$ (c) cyclohexane or $(H_3C)_2C{=}C(CH_3)_2$

Practice Problem 21.3

Each compound gives only one signal in its ^{1}H-NMR spectrum. Propose a structural formula for each.

(a) C_3H_6O (b) C_5H_{10} (c) C_5H_{12} (d) $C_4H_6Cl_4$

21.7 SIGNAL AREAS

We have just seen that the number of signals in a ^{1}H-NMR spectrum gives us information about the number of sets of equivalent hydrogens. The relative areas of the signals provide additional information in the following way. As a spectrum is being run, the instrument's computer numerically adds the area under each signal. In the spectra shown in this text, this information is displayed in the form of a **line of**

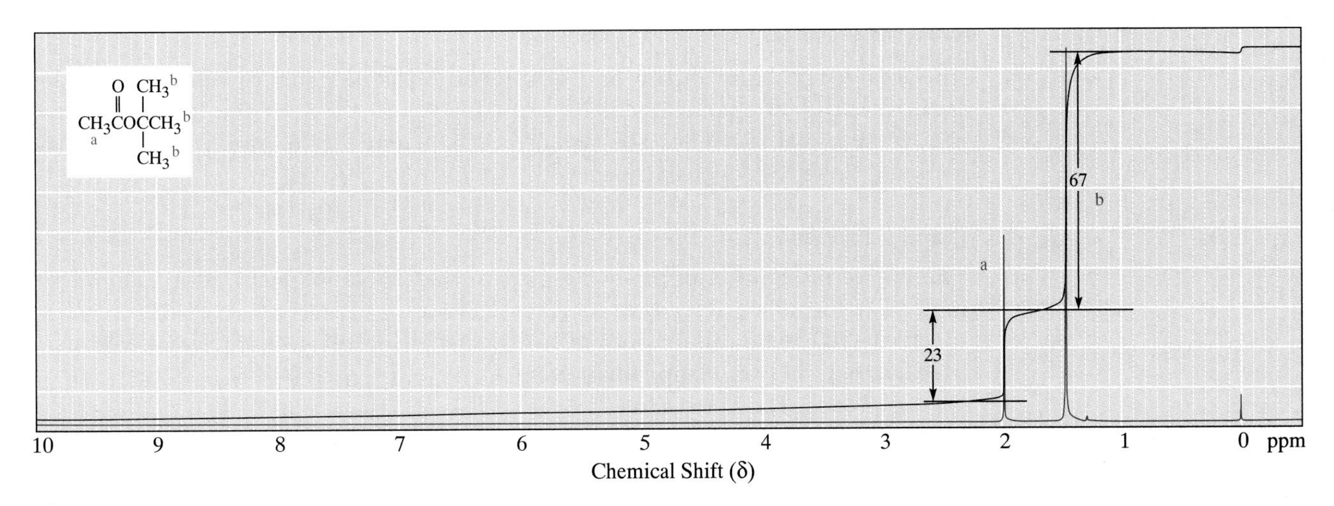

Figure 21.5
^{1}H-NMR spectrum of *tert*-butyl acetate, $C_6H_{12}O_2$, showing a line of integration. The total vertical rise of 90 chart divisions corresponds to 12 hydrogens, 9 in one set and 3 in the other.

integration superposed on the original spectrum. The vertical rise of the line of integration over each signal is proportional to the area under that signal, which, in turn, is proportional to the number of hydrogens giving rise to that signal. Figure 21.5 shows an integrated ^{1}H-NMR spectrum of *tert*-butyl acetate. The spectrum shows signals at δ 1.44 and 1.95. The integrated signal heights are in the ratio 3:1, consistent with the presence of one set of 9 equivalent hydrogens and one set of 3 equivalent hydrogens.

EXAMPLE 21.4

Following is a ^{1}H-NMR spectrum for a compound of molecular formula $C_9H_{10}O_2$. From an analysis of the integration line, calculate the number of hydrogens giving rise to each signal.

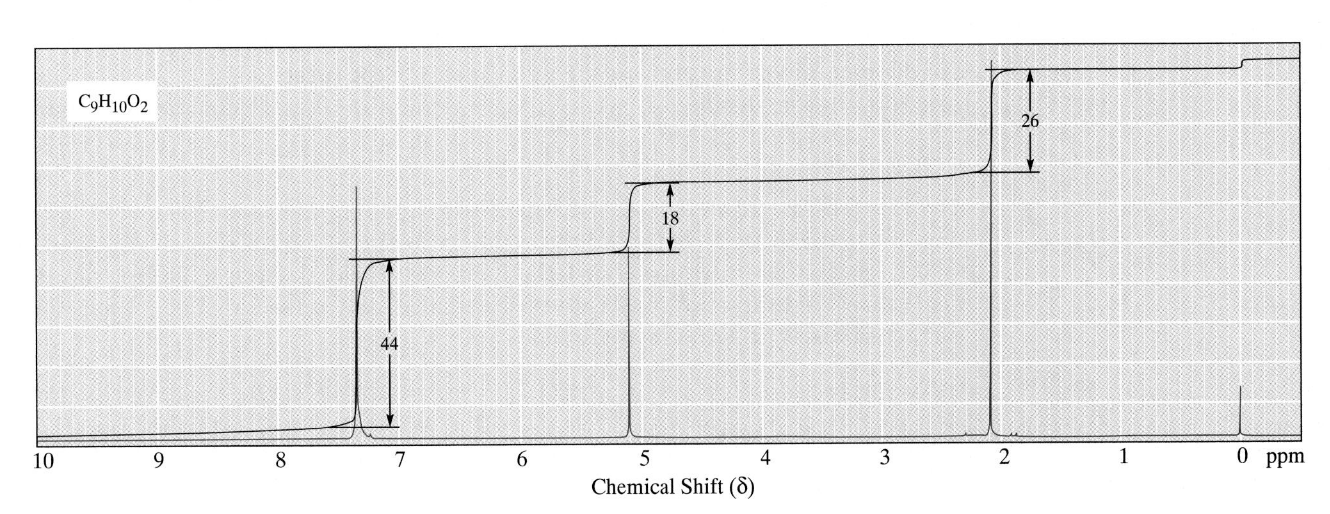

SOLUTION

The total vertical rise in the line of integration is 88 chart divisions and corresponds to 10 hydrogens. Calculate that $44/88 \times 10$, or 5, of the hydrogens give rise to the signal at δ 7.34. The signals at δ 5.08 and δ 2.06 arise from two hydrogens and three hydrogens, respectively.

Practice Problem 21.4

The line of integration of the two signals in the ^{1}H-NMR spectrum of a ketone of molecular formula $C_7H_{14}O$ shows a vertical rise of 62 and 10 chart divisions, respectively. Calculate the number of hydrogens giving rise to each signal, and propose a structural formula for this ketone.

21.8 CHEMICAL SHIFT

The chemical shift for a signal in a ^{1}H-NMR spectrum can give valuable information about the type of hydrogens giving rise to that absorption. Hydrogens on methyl groups bonded to sp^3 hybridized carbons, for example, give signals near δ 0.8–1.0 (compare Figure 21.5). Hydrogens on methyl groups bonded to a carbonyl carbon give signals near δ 2.1–2.3 (compare Figures 21.4 and 21.5), and hydrogens on methyl groups bonded to oxygen give signals near δ 3.7–3.9 (Figure 21.4). Tabulated in Table 21.3 are average chemical shifts for most of the types of hydrogens we deal with in this course. Notice that most of these values fall within a rather narrow range of 0–8 δ units (ppm).

EXAMPLE 21.5

Following are two constitutional isomers of molecular formula $C_6H_{12}O_2$.

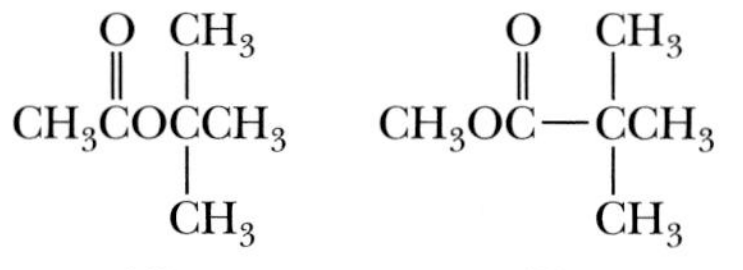

(1) $CH_3C(=O)OC(CH_3)_3$ — drawn as $CH_3\overset{O}{\overset{\|}{C}}O\overset{CH_3}{\overset{|}{\underset{\underset{CH_3}{|}}{C}}}CH_3$

(2) $CH_3OC(=O)-C(CH_3)_3$ — drawn as $CH_3O\overset{O}{\overset{\|}{C}}-\overset{CH_3}{\overset{|}{\underset{\underset{CH_3}{|}}{C}}}CH_3$

(a) Predict the number of signals in the ^{1}H-NMR spectrum of each isomer.
(b) Predict the ratio of areas of the signals in each spectrum.
(c) Show how to distinguish between these isomers on the basis of chemical shift.

SOLUTION

The ^{1}H-NMR spectrum of each consists of two signals in the ratio 9:3, or 3:1. Distinguish between these constitutional isomers by the chemical shift of the single $-CH_3$ group. The hydrogens of CH_3O are less shielded (appear farther downfield) than the hydrogens of $CH_3C{=}O$. See Table 21.3 for approximate values for each chemical shift. Experimental values are

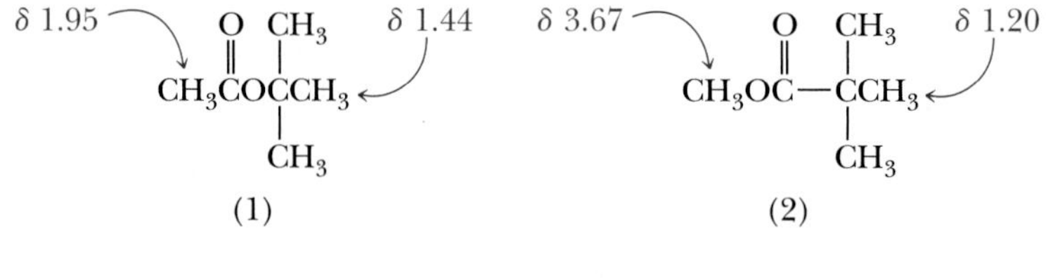

(1) $CH_3C(=O)OC(CH_3)_3$: δ 1.95 ($CH_3C{=}O$), δ 1.44 ($C(CH_3)_3$)

(2) $CH_3OC(=O)-C(CH_3)_3$: δ 3.67 (CH_3O), δ 1.20 ($C(CH_3)_3$)

TABLE 21.3 Average Values of Chemical Shifts of Representative Types of Hydrogens

Type of Hydrogen (R = alkyl, Ar = aryl)	Chemical Shift (δ)*	Type of Hydrogen (R = alkyl, Ar = aryl)	Chemical Shift (δ)*
$(CH_3)_4Si$	0 (by definition)	$RC(=O)OCH_3$	3.7–3.9
RCH_3	0.8–1.0	$RC(=O)OCH_2R$	4.1–4.7
RCH_2R	1.2–1.4	RCH_2I	3.1–3.3
R_3CH	1.4–1.7	RCH_2Br	3.4–3.6
$R_2C{=}CRCHR_2$	1.6–2.6	RCH_2Cl	3.6–3.8
$RC{\equiv}CH$	2.0–3.0	RCH_2F	4.4–4.5
$ArCH_3$	2.2–2.5	ArOH	4.5–4.7
$ArCH_2R$	2.3–2.8	$R_2C{=}CH_2$	4.6–5.0
ROH	0.5–6.0	$R_2C{=}CHR$	5.0–5.7
RCH_2OH	3.4–4.0	ArH	6.5–8.5
RCH_2OR	3.3–4.0	$RC(=O)H$	9.5–10.1
R_2NH	0.5–5.0	$RC(=O)OH$	10–13
$RC(=O)CH_3$	2.1–2.3		
$RC(=O)CH_2R$	2.2–2.6		

* Values are approximate. Other atoms within the molecule may cause the signal to appear outside these ranges.

Practice Problem 21.5

Following are two constitutional isomers of molecular formula $C_4H_8O_2$.

$$\underset{(1)}{CH_3CH_2OC(=O)CH_3} \qquad \underset{(2)}{CH_3CH_2C(=O)OCH_3}$$

(a) Predict the number of signals in the ^{1}H-NMR spectrum of each isomer.
(b) Predict the ratio of areas of the signals in each spectrum.

21.9 SIGNAL SPLITTING AND THE (n + 1) RULE

We have now seen three kinds of information that can be derived from examination of a ^{1}H-NMR spectrum.

1. From the number of signals, we can determine the number of sets of equivalent hydrogens.
2. From integration of signal areas, we can determine the relative numbers of hydrogens giving rise to each signal.
3. From the chemical shift of each signal, we can derive information about the types of hydrogens in each set.

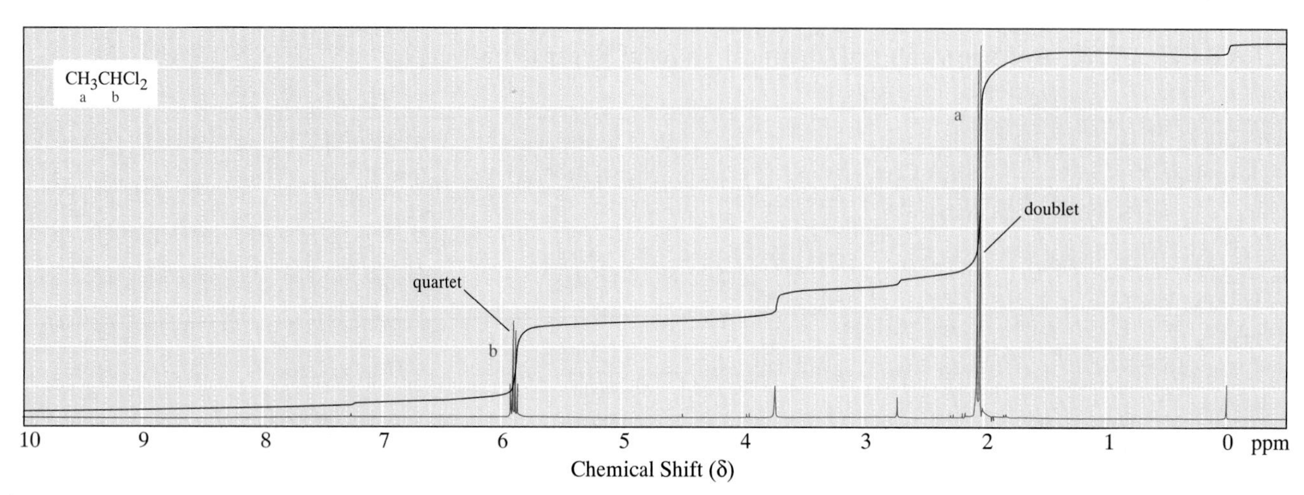

Figure 21.6
^{1}H-NMR spectrum of 1,1-dichloroethane.

Peak (NMR) The units into which an NMR signal is split; two peaks in a doublet, three peaks in a triplet, and so on.

Signal splitting Splitting of an NMR signal into a set of peaks by the influence of neighboring nuclei.

$(n + 1)$ Rule The ^{1}H-NMR signal of a hydrogen or set of equivalent hydrogens is split into $(n + 1)$ peaks by a set of n hydrogens nonequivalent to it but equivalent among themselves on the same or adjacent carbons.

A fourth kind of information can be derived from the splitting pattern of each signal. Consider, for example, the ^{1}H-NMR spectrum of 1,1-dichloroethane (Figure 21.6). This molecule contains two sets of hydrogens and, according to what we have learned so far, we predict two signals with relative areas 3:1 corresponding to the three hydrogens of the —CH_3 group and the one hydrogen of the —$CHCl_2$ group. You see from the spectrum, however, that there are in fact six **peaks.** A peak is a unit into which an NMR signal is split: two peaks in a doublet, three peaks in a triplet, and so on. The grouping of two peaks at δ 2.1 is the signal for the three hydrogens of the —CH_3 group, and the grouping of four peaks at δ 5.9 is the signal for the single hydrogen of the —$CHCl_2$ group. We say that the CH_3 signal at δ 2.1 is split into a doublet and that the CH signal at δ 5.9 is split into a quartet. This phenomenon is called signal splitting. In **signal splitting,** the ^{1}H-NMR signal from one set of hydrogens is split by the influence of neighboring nonequivalent hydrogens.

The degree of signal splitting can be predicted on the basis of the **$(n + 1)$ rule.** According to this rule, if a hydrogen has n hydrogens nonequivalent to it but equivalent among themselves on the same or adjacent atom(s), its ^{1}H-NMR signal is split into $(n + 1)$ peaks.

Let us apply the $(n + 1)$ rule to the analysis of the spectrum of 1,1-dichloroethane. The three hydrogens of the —CH_3 group have one nonequivalent neighboring hydrogen ($n = 1$); their signal is split into a doublet. The single hydrogen of the —$CHCl_2$ group has a set of three nonequivalent neighboring hydrogens ($n = 3$); its signal is split into a quartet.

For these hydrogens, $n = 1$; their signal is split into $(1 + 1)$ or 2 peaks—a **doublet**

For this hydrogen, $n = 3$; its signal is split into $(3 + 1)$ or 4 peaks—a **quartet**

$CH_3—CH(Cl)—Cl$

EXAMPLE 21.6

Predict the number of signals and the splitting pattern of each signal in the ^{1}H-NMR spectrum of each compound.

(a) $CH_3\overset{O}{\overset{\|}{C}}CH_2CH_3$ (b) $CH_3CH_2\overset{O}{\overset{\|}{C}}CH_2CH_3$ (c) $CH_3\overset{O}{\overset{\|}{C}}CH(CH_3)_2$

SOLUTION

The sets of equivalent hydrogens in each molecule are labeled a, b, and c. Molecule (a) shows a singlet, a quartet, and a triplet in the ratio 3:2:3. Molecule (b) shows a triplet and a quartet in the ratio 3:2. Molecule (c) shows a singlet, a septet, and a doublet in the ratio 3:1:6.

(a) $\overset{a}{CH_3}-\overset{O}{\overset{\|}{C}}-\overset{b}{CH_2}-\overset{c}{CH_3}$ (a: singlet, b: quartet, c: triplet)

(b) $\overset{a}{CH_3}-\overset{b}{CH_2}-\overset{O}{\overset{\|}{C}}-\overset{b}{CH_2}-\overset{a}{CH_3}$ (a: triplet, b: quartet)

(c) $\overset{a}{CH_3}-\overset{O}{\overset{\|}{C}}-\overset{b}{CH}(\overset{c}{CH_3})_2$ (a: singlet, b: septet, c: doublet)

Practice Problem 21.6

Following are pairs of constitutional isomers. Predict the number of signals and the splitting pattern of each signal in a ^{1}H-NMR spectrum of each isomer.

(a) $CH_3OCH_2\overset{O}{\overset{\|}{C}}CH_3$ and $CH_3CH_2\overset{O}{\overset{\|}{C}}OCH_3$ (b) $CH_3C(Cl)_2CH_3$ and $ClCH_2CH_2CH_2Cl$

21.10 ^{13}C-NMR SPECTROSCOPY

Nuclei of carbon-12, the most abundant (98.89%) natural isotope of carbon, do not have nuclear spin and are not detected by NMR spectroscopy. Nuclei of carbon-13 (natural abundance 1.11%), however, do have nuclear spin and are detected by NMR spectroscopy in the same manner as hydrogens are detected.

Because both ^{13}C and ^{1}H have spinning nuclei and generate magnetic fields, ^{13}C couples with each ^{1}H attached to it and gives a signal split according to the $(n + 1)$ rule. In the most common mode for recording a ^{13}C spectrum, this coupling is eliminated so as to simplify the spectrum. In a hydrogen-decoupled spectrum, all ^{13}C signals appear as singlets. The hydrogen-decoupled ^{13}C-NMR spectrum of 1-bromobutane (Figure 21.7) consists of four singlets.

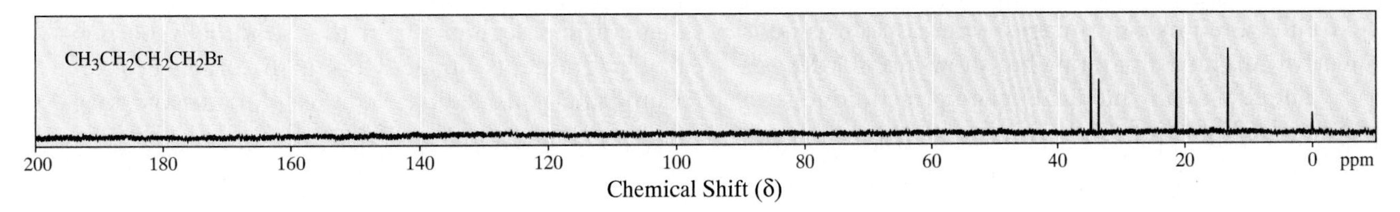

Figure 21.7
Hydrogen-decoupled ^{13}C-NMR spectrum of 1-bromobutane.

CHEMICAL CONNECTIONS

Magnetic Resonance Imaging

The NMR phenomenon was discovered and explained by physicists in the 1950s, and, by the 1960s, it had become an invaluable analytical tool for chemists. It was realized by the early 1970s that imaging of parts of the body using NMR could be a valuable addition to diagnostic medicine. Because the term "nuclear magnetic resonance" sounds to many people as if the technique might involve radioactive material, health care personnel call the technique "magnetic resonance imaging (MRI)."

The body contains several nuclei that, in principle, could be used for MRI. Of these, hydrogens, most of which come from the hydrogens of water, triglycerides, and membrane phospholipids, give the most useful signals. Phosphorus MRI is also used in diagnostic medicine.

Recall that in NMR spectroscopy, energy in the form of radio-frequency radiation is absorbed by nuclei in the sample. The relaxation time is the characteristic time at which excited nuclei give up this energy and relax to their ground state.

In 1971, it was discovered that relaxation of water in certain cancerous tumors takes much longer than the relaxation of water in normal cells. Thus, it was reasoned if a relaxation image of a body could be obtained, it might be possible to identify tumors at an early stage. Subsequent work demonstrated that many tumors can be identified in this way. Another important application of MRI is in the examination of the brain and spinal cord. White and gray matter are easily distinguished by MRI, which is useful in the study of such diseases as multiple sclerosis. Magnetic resonance imaging and x-ray imaging are in many cases complementary. The hard, outer layer of bone is essentially invisible to MRI but shows up extremely well in x-ray images, whereas soft tissue is nearly transparent to x-rays but shows up in MRI.

The key to any medical imaging technique is to know which part of the body gives rise to which signal. In MRI, spatial information is encoded using magnetic field gradients. We know that a linear relationship exists between the frequency at which a nucleus resonates and the intensity of the magnetic field. In ^{1}H-NMR spectroscopy, we use a homogeneous magnetic field, in which all equivalent hydrogens absorb at the same radio frequency and have the same chemical shift. In MRI, a patient is placed in a magnetic field gradient that can be varied from place to place. Nuclei in a weaker magnetic field gradient absorb at a lower frequency. Nuclei elsewhere in a stronger magnetic field absorb at a higher frequency. A gradient along a single axis images a plane. Two mutually perpendicular gradients image a line segment, and three mutually perpendicular gradients image a point. In practice, more complicated procedures are used to obtain magnetic resonance images, but they are all based on the idea of magnetic field gradients.

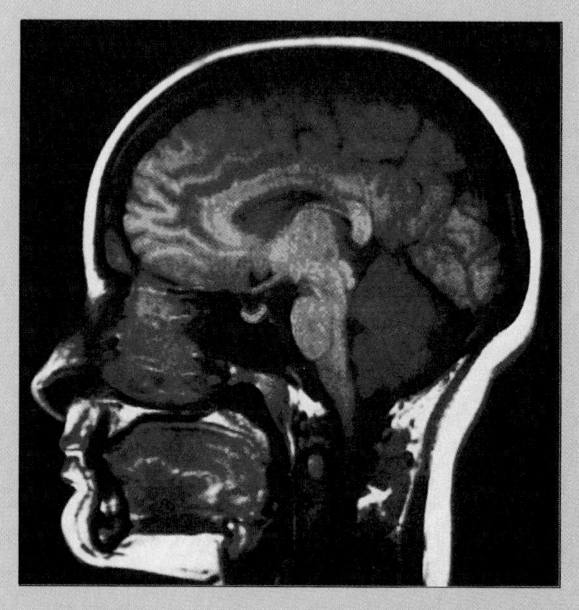

Computer-enhanced MRI scan of a normal human brain with pituitary gland highlighted *(Scott Camazine/Photo Researchers)*

TABLE 21.4 ^{13}C-NMR Chemical Shifts

Type of Carbon	Chemical Shift (δ)	Type of Carbon	Chemical Shift (δ)
RCH_3	0–40	C_6H_5–C–R (aromatic)	110–160
RCH_2R	15–55		
R_3CH	20–60		
RCH_2I	0–40		
RCH_2Br	25–65	RC(=O)OR	160–180
RCH_2Cl	35–80		
R_3COH	40–80	$RC(=O)NR_2$	165–180
R_3COR	40–80		
$RC \equiv CR$	65–85	RC(=O)OH	175–185
$R_2C{=}CR_2$	100–150		
		RC(=O)H, RC(=O)R	180–210

Table 21.4 shows approximate chemical shifts for carbon-13 spectroscopy. Notice how much broader the range of chemical shifts is for ^{13}C-NMR spectroscopy than for ^{1}H-NMR spectroscopy. Whereas most chemical shifts for ^{1}H-NMR spectroscopy fall within a rather narrow range of 0–10 ppm, those for ^{13}C-NMR spectroscopy cover 0–220 ppm. Because of this expanded scale, it is very unusual to find any two nonequivalent carbons in the same molecule with identical chemical shifts. Most commonly, each different type of carbon within a molecule has a distinct signal clearly resolved from all other signals. Notice further that the chemical shift of carbonyl carbons is quite distinct from those of sp^3 hybridized carbons and of other types of sp^2 hybridized carbons. The presence or absence of a carbonyl carbon is quite easy to recognize in a ^{13}C-NMR spectrum.

A great advantage of ^{13}C-NMR spectroscopy is that it is generally possible to count the number of types of carbon atoms in a molecule. There is one caution here, however. Because of the particular manner in which spin-flipped ^{13}C nuclei return to their lower energy states, integration of signal heights is often unreliable, and it is generally not possible to determine the number of carbons of each type based on signal heights.

EXAMPLE 21.7

Predict the number of signals in a proton-decoupled ^{13}C-NMR spectrum of each compound.

(a) $CH_3C(=O)OCH_3$ (b) $CH_3CH_2CH_2C(=O)CH_3$ (c) $CH_3CH_2C(=O)CH_2CH_3$

SOLUTION

Here are the number of signals in each spectrum, along with the chemical shift of each. The chemical shifts of the carbonyl carbons are quite distinctive (Table 21.4), and in these examples occur at δ 171.37, 208.85, and 211.97.

(a) Methyl acetate: three signals δ(171.37, 51.53, and 20.63).
(b) 2-Pentanone: five signals δ(208.85, 45.68, 29.79, 17.35, and 13.68).
(c) 3-Pentanone: three signals δ(211.97, 35.45, and 7.92).

Practice Problem 21.7

Explain how to distinguish between the members of each pair of constitutional isomers based on the number of signals in the ^{13}C-NMR spectrum of each member.

(a) CH_2 (methylenecyclohexane) and CH_3 (1-methylcyclohexene)

(b) $CH_3CH{=}CHCH_2CH_2CH_3$ and $CH_3CH_2CH{=}CHCH_2CH_3$

21.11 INTERPRETING NMR SPECTRA

A. Alkanes

Because all hydrogens in alkanes are in very similar chemical environments, ^{1}H-NMR chemical shifts of their hydrogens fall within a narrow range of δ 0.8–1.7. Chemical shifts for alkane carbons in ^{13}C-NMR spectroscopy fall within the considerably wider range of δ 0–60.

B. Alkenes

The ^{1}H-NMR chemical shifts of vinylic hydrogens (hydrogens on a carbon of a carbon-carbon double bond) are larger than those of alkane hydrogens and typically fall in the range δ 4.6–5.7. Figure 21.8 shows a ^{1}H-NMR spectrum of 1-methylcyclohexene. The signal for the one vinylic hydrogen appears at δ 5.7, split to a triplet by the two hydrogens of the neighboring —CH_2— group of the ring.

The sp^2 hybridized carbons of alkenes come into resonance in ^{13}C-NMR spectroscopy in the range δ 100–150 (Table 21.4), which is considerably downfield from resonances of sp^3 hybridized carbons.

C. Alcohols

The chemical shift of a hydroxyl hydrogen in an ^{1}H-NMR spectrum is variable and depends on the purity of the sample, the solvent, and the temperature. It normally appears in the range δ 3.0–4.5, but, depending on experimental conditions, it may appear as high as δ 0.5. Hydrogens on the carbon bearing the —OH group are deshielded by the electron-withdrawing inductive effect of the oxygen atom, and their absorptions typically appear in the range δ 3.4–4.0. Figure 21.9 shows a

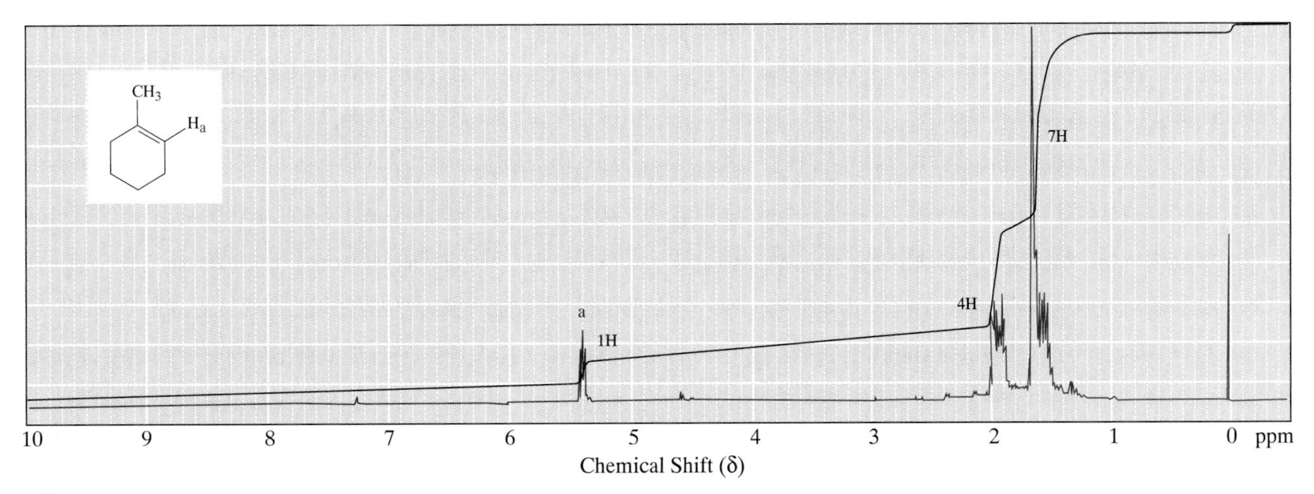

Figure 21.8
^{1}H-NMR spectrum of 1-methylcyclohexene.

^{1}H-NMR spectrum of 1-propanol. This spectrum consists of four signals. The hydroxyl hydrogen appears at δ 3.18 as a narrowly spaced triplet. The signal of the hydrogens on the carbon bearing the hydroxyl group in 1-propanol appears as a narrowly spaced multiplet at δ 3.56.

Signal splitting between the hydrogen on O—H and its neighbors on the adjacent $—CH_2—$ group is seen in a ^{1}H-NMR spectrum of 1-propanol. This splitting, however, is rarely seen. The reason is that most samples of alcohol contain traces of acid, base, or other impurities that catalyze the transfer of the hydroxyl proton from the oxygen of one alcohol molecule to the oxygen of another alcohol molecule. This transfer, which is very fast compared to the time scale required to make

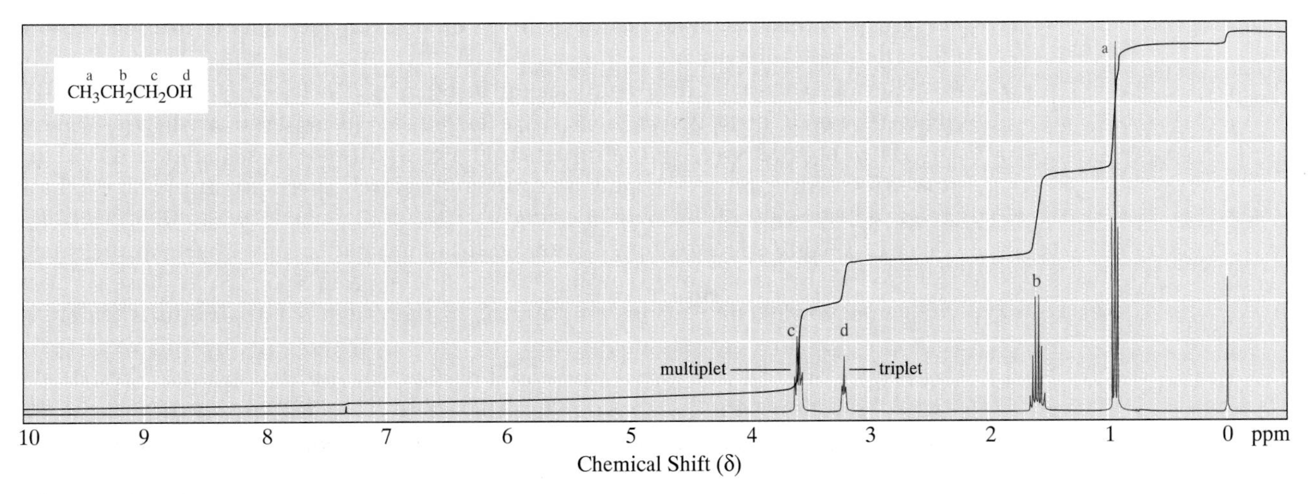

Figure 21.9
^{1}H-NMR spectrum of 1-propanol.

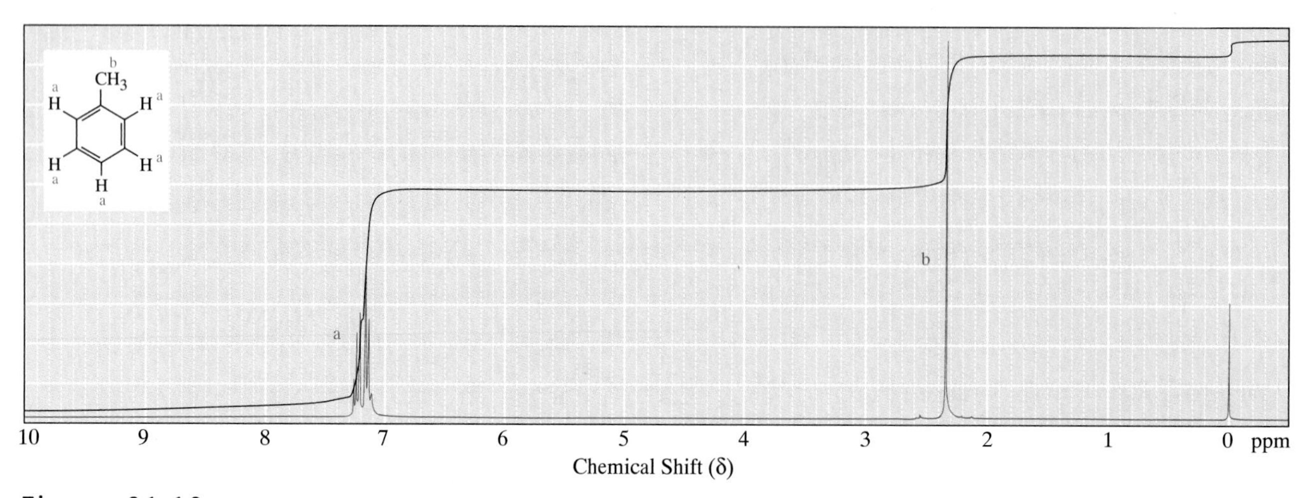

Figure 21.10
^{1}H-NMR spectrum of toluene.

an NMR measurement, decouples the hydroxyl proton from all other protons in the molecule. For this same reason, the hydroxyl proton does not usually split the signal of any α-hydrogens.

hydrogen bonding

These protons have been exchanged

D. Benzene and Its Derivatives

All six hydrogens of benzene are equivalent, and their signal appears in its ^{1}H-NMR spectrum as a sharp singlet at δ 7.27. Hydrogens attached to a substituted benzene ring appear in the region δ 6.5–8.5. Few other hydrogens absorb in this region, and, thus, aromatic hydrogens are quite easily identifiable by their distinctive chemical shifts.

Recall that vinylic hydrogens are in resonance at δ 4.6–5.7 (Section 21.11B). Thus, aromatic hydrogens absorb even farther downfield than vinylic hydrogens.

The ^{1}H-NMR spectrum of toluene (Figure 21.10) shows a singlet at δ 2.32 for the three hydrogens of the methyl group and a closely spaced multiplet at δ 7.3 for the five hydrogens of the aromatic ring.

In ^{13}C-NMR spectroscopy, carbon atoms of aromatic rings appear in the range δ 110–160. Benzene, for example, shows a single signal at δ 128. Because carbon-13 signals for alkene carbons also appear in the range δ 110–160, it is generally not possible to establish the presence of an aromatic ring by ^{13}C-NMR spectroscopy alone. ^{13}C-NMR spectroscopy is particularly useful, however, in establishing substitution patterns of aromatic rings. The ^{13}C-NMR spectrum of 2-chlorotoluene (Figure 21.11) shows six signals in the aromatic region; its more symmetric isomer 4-chlorotoluene (Figure 21.12) shows only four signals in the aromatic region. Thus, all one needs to do is count signals to distinguish between these constitutional isomers.

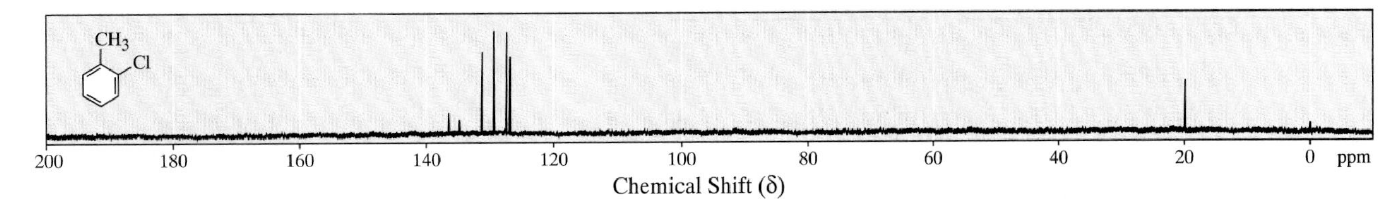

Figure 21.11
^{13}C-NMR spectrum of 2-chlorotoluene.

E. Amines

The chemical shifts of amine hydrogens, like those of alcohol hydrogens, are variable and may be found in the region δ 0.5–5.0, depending on the solvent, the concentration, and the temperature. Furthermore, the rate of intermolecular exchange of amine hydrogens is sufficiently rapid compared with the time scale of an NMR measurement that signal splitting between amine hydrogens and hydrogens on an adjacent α-carbon is prevented. Thus, amine hydrogens generally appear as singlets. The NH_2 hydrogens in benzylamine, $C_6H_5CH_2NH_2$, for example, appear as a singlet at δ 1.34.

F. Aldehydes and Ketones

^{1}H-NMR spectroscopy is an important means for identifying aldehydes and for distinguishing between aldehydes and other carbonyl-containing compounds. Just as a carbon-carbon double bond causes a downfield shift in the signal of a vinylic hydrogen (Section 21.11B), a carbon-oxygen double bond also causes a downfield shift in the signal of an aldehyde hydrogen, typically to δ 9.5–10.1. Signal splitting between this hydrogen and those on the adjacent α-carbon is slight; consequently, the aldehyde hydrogen signal often appears as a singlet rather than a doublet or triplet. The spectrum of butanal, for example, shows a singlet at δ 9.78 for the aldehyde hydrogen (Figure 21.13).

Just as the signal for an aldehyde hydrogen is not split by the adjacent nonequivalent α-hydrogens, the α-hydrogens are not split by the aldehyde hydrogen. Hydrogens on an α-carbon of an aldehyde or ketone typically appear around δ 2.1–2.6. The carbonyl carbons of aldehydes and ketones are readily identifiable in ^{13}C-NMR spectroscopy by the position of their signal between δ 180 and 210.

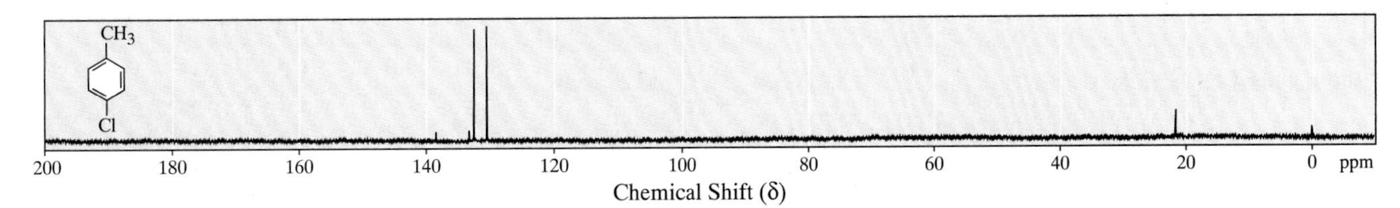

Figure 21.12
^{13}C-NMR spectrum of 4-chlorotoluene.

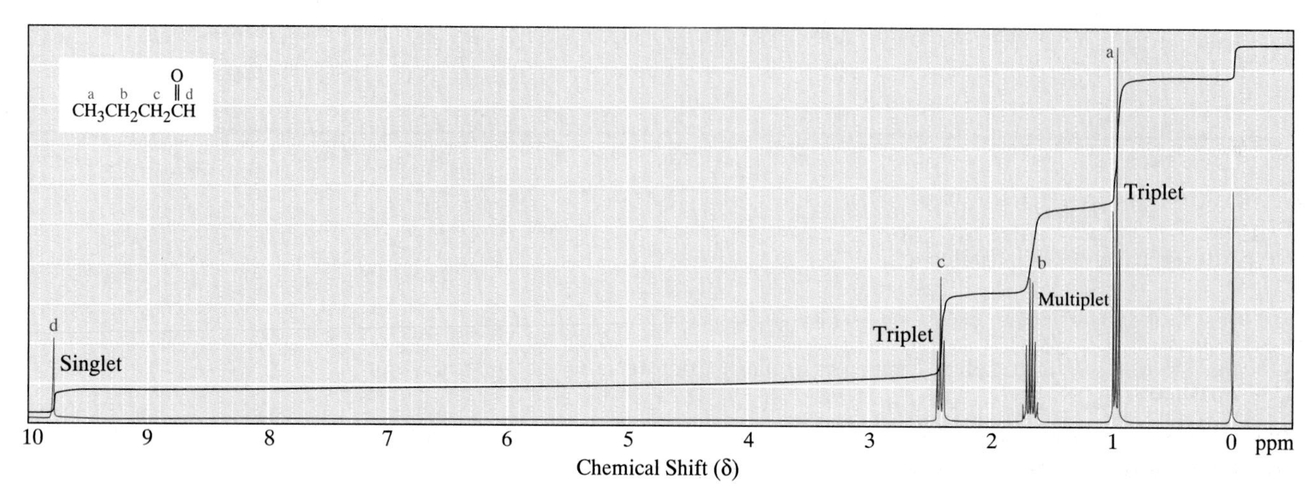

Figure 21.13
^{1}H-NMR spectrum of butanal.

G. Carboxylic Acids

The hydrogen of the carboxyl group gives a signal in the range δ 10–13. The chemical shift of a carboxyl hydrogen is so large, even larger than the chemical shift of an aldehyde hydrogen (δ 9.5–10.1), that it serves to distinguish carboxyl hydrogens from most other types of hydrogens. The signal for the carboxyl hydrogen of 2-methylpropanoic acid is at the left of the ^{1}H-NMR spectrum in Figure 21.14 and has been offset by δ 2.4. The chemical shift of this hydrogen is δ 12.2.

H. Esters

Hydrogens on the α-carbon to the carbonyl group of an ester are slightly deshielded and give signals at δ 2.1–2.6. Hydrogens on the carbon attached to the ester oxygen are more strongly deshielded and give signals at δ 3.7–4.7. It is possible to distinguish between ethyl acetate and its constitutional isomer, methyl propanoate, by the chemical shifts of either the singlet $—CH_3$ absorption (com-

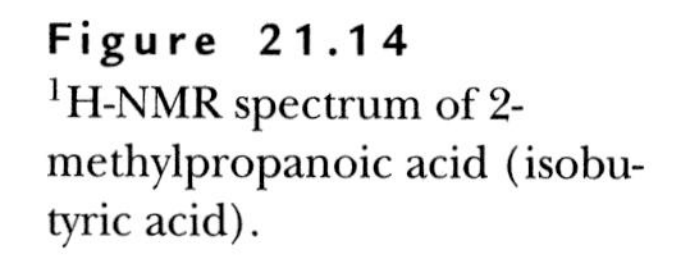

Figure 21.14
^{1}H-NMR spectrum of 2-methylpropanoic acid (isobutyric acid).

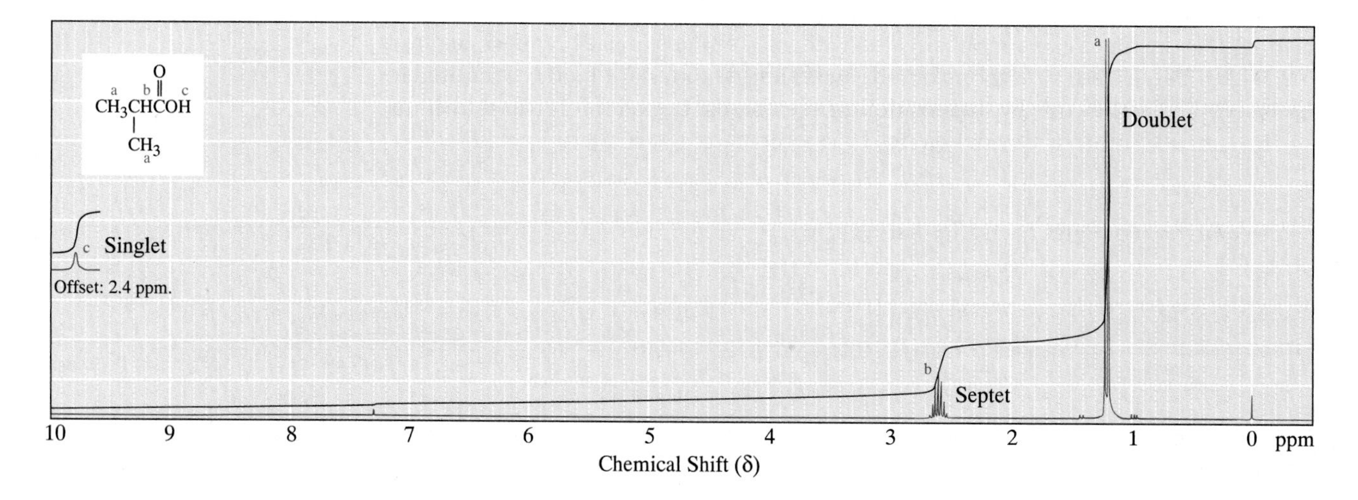

pare δ 2.04 and 3.68) or the quartet —CH_2— absorption (compare δ 4.11 and 2.33).

δ 2.04(s) → $CH_3-\overset{O}{\overset{\|}{C}}-O-CH_2-CH_3$ ← δ 4.11(q)

Ethyl acetate

δ 2.33(q) → $CH_3-CH_2-\overset{O}{\overset{\|}{C}}-O-CH_3$ ← δ 3.6

Methyl propanoate

21.12 Solving NMR Problems

One of the first steps in determining molecular structure is to establish the molecular formula. In the past, this was most commonly done by elemental analysis, combustion to determine percent composition, and so forth. More commonly today, molecular weight and molecular formula are determined by mass spectrometry. In the examples that follow, we assume that the molecular formula of any unknown compound has already been determined, and we proceed from there using spectral analysis to determine a structural formula.

A. Index of Hydrogen Deficiency

Valuable information about the structural formula of an unknown compound can be obtained from inspection of its molecular formula. In addition to learning the number of atoms of carbon, hydrogen, oxygen, nitrogen, and so forth in a molecule of the compound, we can also determine what is called its **index of hydrogen deficiency,** which is the sum of the number of rings and pi bonds in a molecule. This quantity is determined by comparing the number of hydrogens in the molecular formula of a compound of unknown structure with the number of hydrogens in a **reference compound** of the same number of carbon atoms and with no rings or pi bonds. The molecular formula of a reference hydrocarbon is C_nH_{2n+2} (Section 2.1).

$$\text{Index of hydrogen deficiency} = \frac{(H_{\text{reference}} - H_{\text{molecule}})}{2}$$

Example 21.8

Calculate the index of hydrogen deficiency for 1-hexene, molecular formula C_6H_{12}, and account for this deficiency by reference to its structural formula.

Solution

The molecular formula of the reference hydrocarbon of six carbon atoms is C_6H_{14}. The index of hydrogen deficiency of 1-hexene is $(14 - 12)/2 = 1$ and is accounted for by the one pi bond in 1-hexene.

Practice Problem 21.8

Calculate the index of hydrogen deficiency of cyclohexene, and account for this deficiency by reference to its structural formula.

To determine the molecular formula for a reference compound containing elements besides carbon and hydrogen, write the formula of the reference hydrocarbon, add to it other elements contained in the unknown compound, and make the following adjustments to the number of hydrogen atoms.

1. For each atom of a Group VII element (F, Cl, Br, I) added to the reference hydrocarbon, subtract one hydrogen; halogen substitutes for hydrogen and reduces the number of hydrogens by one per halogen. The general formula of an acyclic monochloroalkane, for example, is $C_nH_{2n+1}Cl$.
2. No correction is necessary for the addition of divalent atoms of Group VI elements (O, S, Se) to the reference hydrocarbon. Insertion of a divalent Group VI element into a reference hydrocarbon does not change the number of hydrogens.
3. For each atom of a trivalent Group V element (N, P, As) added to the formula of the reference hydrocarbon, add one hydrogen. Insertion of a trivalent Group V element adds one hydrogen to the molecular formula of the reference compound. The general molecular formula for an acyclic alkylamine, for example, is $C_nH_{2n+3}N$.

EXAMPLE 21.9

Isopentyl acetate, a compound with a banana-like odor, is a component of the alarm pheromone of honey bees. The molecular formula of isopentyl acetate is $C_7H_{14}O_2$. Calculate the index of hydrogen deficiency of this compound.

SOLUTION

The molecular formula of the reference hydrocarbon is C_7H_{16}. Adding oxygens to this formula does not require any correction in the number of hydrogens. The molecular formula of the reference compound is $C_7H_{16}O_2$, and the index of hydrogen deficiency is $(16 - 14)/2 = 1$, indicating either one ring or one pi bond. Following is the structural formula of isopentyl acetate. It contains one pi bond, in this case, in the carbon-oxygen double bond.

$$CH_3\overset{\overset{\displaystyle O}{\|}}{C}OCH_2CH_2CH(CH_3)_2$$

Isopentyl acetate

Practice Problem 21.9

The index of hydrogen deficiency of niacin is 5. Account for this index of hydrogen deficiency by reference to the structural formula of niacin.

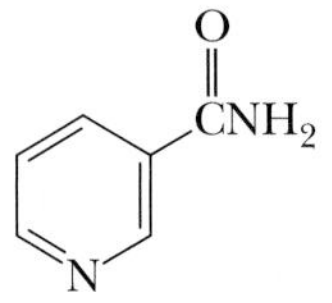

O
‖
CNH$_2$
N

Nicotinamide
(Niacin)

B. From an ^{1}H-NMR Spectrum to a Structural Formula

The following steps may prove helpful as a systematic approach to solving ^{1}H-NMR spectral problems.

Step 1: *Molecular formula and index of hydrogen deficiency.* Examine the molecular formula, calculate the index of hydrogen deficiency, and deduce what information you can about the presence or absence of rings or pi bonds.

Step 2: *Number of signals.* Count the number of signals to determine the minimum number of sets of equivalent hydrogens in the compound.

Step 3: *Integration.* Use the signal integration and the molecular formula to determine the number of hydrogens in each set.

Step 4: *Pattern of chemical shifts.* Examine the NMR spectrum for signals characteristic of the following types of equivalent hydrogens. Keep in mind that these are broad ranges and that hydrogens of each type may be shifted either farther upfield or farther downfield depending on details of molecular structure.

Types of Hydrogens	Descriptive Name	Typically Absorb in the Range (δ)
RCH_3 RCH_2R R_3CH	alkyl hydrogens	0.8–1.7
$R_2C{=}CRCHR_2$	allylic hydrogens	1.6–2.6
RCH_2OH RCH_2OR	hydrogens on a carbon adjacent to an sp^3 hybridized oxygen	3.3–4.0
$R_2C{=}CH_2$ $R_2C{=}CHR$	vinylic hydrogens	4.6–5.7
$RCH{=}O$ (O ‖ RCH)	aldehyde hydrogens	9.5–10.1
$RC(=O)OH$ (O ‖ RCOH)	carboxyl hydrogens	10–13

Step 5: *Splitting patterns.* Examine splitting patterns for information about the number of nonequivalent hydrogen neighbors.

Step 6: *Structural formula.* Write a structural formula consistent with the previous information.

Spectral Problem 1

A colorless liquid of molecular formula $C_5H_{10}O$.

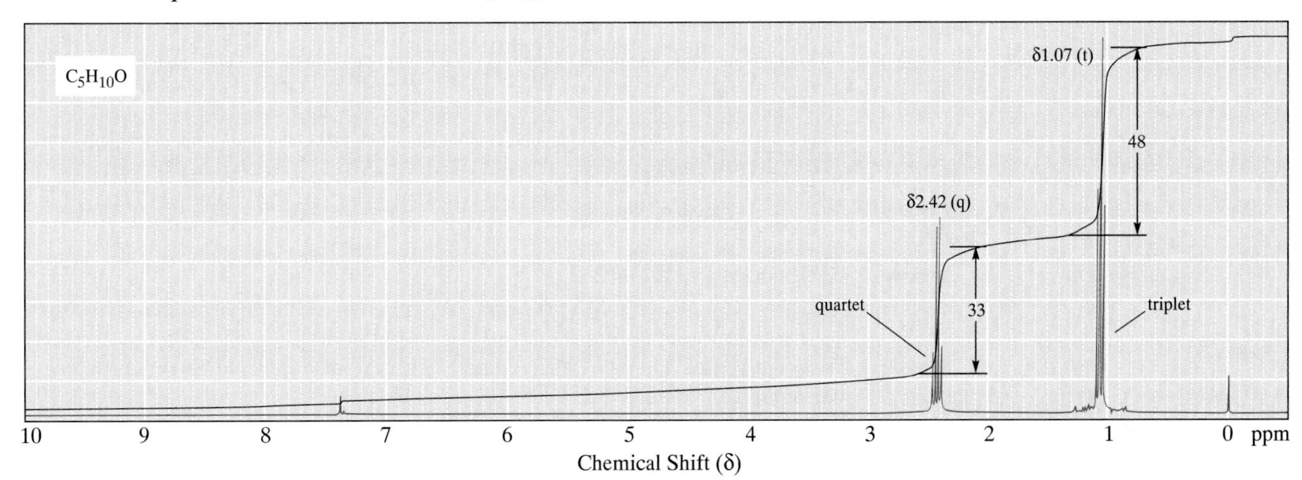

Analysis of Spectral Problem 1

Step 1: *Molecular formula and index of hydrogen deficiency.* The reference compound is $C_5H_{12}O$, and, therefore, the index of hydrogen deficiency is 1; the molecule contains either one ring or one pi bond.

Step 2: *Number of signals.* There are two signals (a triplet and a quartet) and, therefore, two sets of equivalent hydrogens.

Step 3: *Integration.* From the signal integration, calculate that the number of hydrogens giving rise to each signal are in the ratio 3:2. Because there are 10 hydrogens, conclude that 6H give rise to the signal at δ 1.07 and 4H to the signal at δ 2.42.

Step 4: *Pattern of chemical shifts.* The signal at δ 1.07 is in the alkyl region and, based on its chemical shift, most probably represents a methyl group. No signal occurs at δ 4.6–5.7; there are no vinylic hydrogens. If a carbon-carbon double bond is in the molecule, no hydrogens are on it (that is, it is tetrasubstituted).

Step 5: *Splitting pattern.* The methyl signal at δ 1.07 is split into a triplet (t); it must have two neighbors, indicating —CH_2CH_3. The signal at δ 2.42 is split into a quartet (q); it must have three neighbors. An ethyl group accounts for these two signals. No other signals occur in the spectrum, and, therefore, there are no other types of hydrogens in the molecule.

Step 6: *Structural formula.* Put this information together to arrive at the following structural formula. The chemical shift of the methylene group (—CH_2—) at δ 2.42 is consistent with an alkyl group adjacent to a carbonyl group.

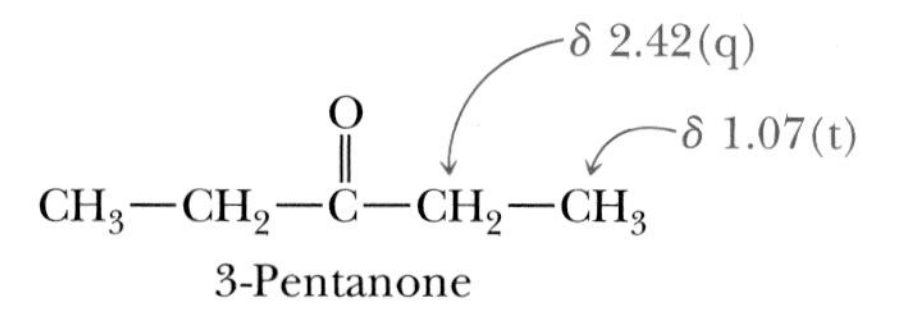

3-Pentanone

Spectral Problem 2

A colorless liquid of molecular formula $C_7H_{14}O$.

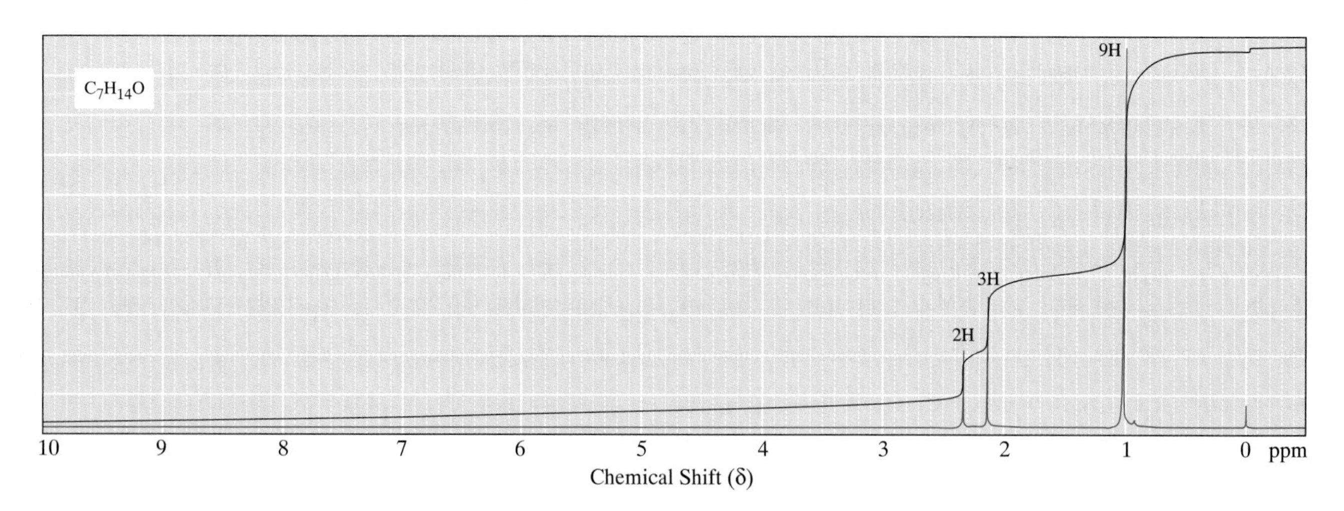

Analysis of Spectral Problem 2

Step 1: *Molecular formula and index of hydrogen deficiency.* The index of hydrogen deficiency is 1; the compound contains one ring or one pi bond.

Step 2: *Number of signals.* There are three signals and, therefore, three sets of equivalent hydrogens.

Step 3: *Integration.* Reading from right to left, the number of hydrogens giving rise to these signals are 9, 3, and 2.

Step 4: *Pattern of chemical shifts.* The singlet at δ 1.01 is characteristic of a methyl group adjacent to an sp^3 hybridized carbon. The singlets at δ 2.11 and 2.32 are characteristic of alkyl groups adjacent to a carbonyl group.

Step 5: *Splitting pattern.* All signals are singlets (s).

Step 6: *Structural formula.* The compound is 4,4-dimethyl-2-pentanone.

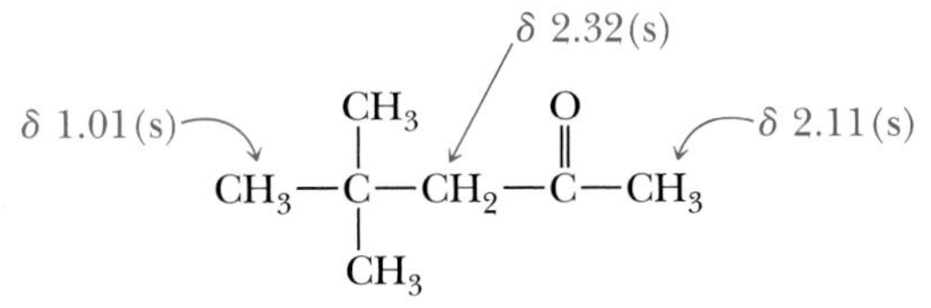

δ 2.32(s)

δ 1.01(s) → CH_3 groups on C4; δ 2.11(s) → terminal CH_3 of C=O

$$CH_3-\underset{CH_3}{\overset{CH_3}{\underset{|}{\overset{|}{C}}}}-CH_2-\overset{O}{\overset{\|}{C}}-CH_3$$

4,4-Dimethyl-2-pentanone

SUMMARY

Electromagnetic radiation (Section 21.1) can be described in terms of its **wavelength (λ)** and its **frequency (ν)**. Frequency is reported in hertz (Hz). An alternative way to describe electromagnetic radiation is in terms of its energy where $E = h\nu$ (Section 21.1).

Molecular spectroscopy (Section 21.2) is the experimental process of measuring which frequencies of radiation are absorbed or emitted by a substance and correlating these patterns with details of molecular structure. Interaction of molecules with **radio-frequency radiation** gives us information about nuclear spin energy levels.

Nuclei of ^1H and ^{13}C, isotopes of the two elements most common to organic compounds, have a spin and behave like tiny bar magnets (Section 21.3). When placed between the poles of a powerful magnet, the nuclear spins of these elements become aligned either with the applied field or against it. Nuclear spins aligned with the applied field are in the lower energy state; those aligned against the applied field are in the higher energy state. **Resonance** is the absorption of electromagnetic radiation by a nucleus and the resulting "flip" of its nuclear spin from a lower energy spin state to a higher energy spin state. An NMR spectrometer (Section 21.5) records such a resonance as a **signal.**

The experimental conditions required to cause nuclei to resonate are affected by the local chemical and magnetic environment. Electrons around a hydrogen also have spin (Section 21.4) and create a local magnetic field that **shields** the hydrogen from the applied field.

A resonance signal in a ^1H-NMR spectrum is reported by how far it is shifted from the resonance signal of the 12 equivalent hydrogens in **tetramethylsilane (TMS).** A resonance signal in a ^{13}C-NMR spectrum is reported by how far it is shifted from the resonance signal of the four equivalent carbons in TMS. **Chemical shift (δ)** (Section 21.4) is the frequency shift from TMS divided by the operating frequency of the spectrometer.

Equivalent hydrogens within a molecule have identical chemical shifts (Section 21.6). The area of a ^1H-NMR signal is proportional to the number of equivalent hydrogens giving rise to that signal (Section 21.7). In **signal splitting,** the ^1H-NMR signal from one hydrogen or set of equivalent hydrogens is split by the influence of nonequivalent hydrogens on the same or adjacent carbon atoms (Section 21.9). According to the **(n + 1) rule,** if a hydrogen has n hydrogens nonequivalent to it but equivalent among themselves on the same or adjacent carbon atom(s), its ^1H-NMR signal is split into (n + 1) peaks. Splitting patterns are commonly referred to as singlets (s), doublets (d), triplets (t), quartets (q), quintets, and multiplets (m).

^{13}C-NMR spectra (Section 21.10) are commonly recorded in a hydrogen-decoupled instrumental mode. In this mode, all ^{13}C signals appear as singlets.

ADDITIONAL PROBLEMS

Index of Hydrogen Deficiency

21.10 Complete the following table.

Class of Compound	Molecular Formula	Index of Hydrogen Deficiency	Reason for Hydrogen Deficiency
alkane	C_nH_{2n+2}	0	(reference hydrocarbon)
alkene	C_nH_{2n}	1	one pi bond
alkyne	______	______	______
alkadiene	______	______	______
cycloalkane	______	______	______
cycloalkene	______	______	______

21.11 Calculate the index of hydrogen deficiency of each compound.

(a) Aspirin, $C_9H_8O_4$ **(b)** Ascorbic acid (vitamin C), $C_6H_8O_6$

(c) Pyridine, C_5H_5N **(d)** Urea, CH_4N_2O

(e) Cholesterol, $C_{27}H_{46}O$ **(f)** Trichloroacetic acid, $C_2HCl_3O_2$

Interpretation of ^{1}H-NMR and ^{13}C-NMR Spectra

21.12 Following are structural formulas for the cis isomers of 1,2-, 1,3-, and 1,4-dimethylcyclohexanes and three sets of ^{13}C-NMR spectral data. Assign each constitutional isomer its correct spectral data.

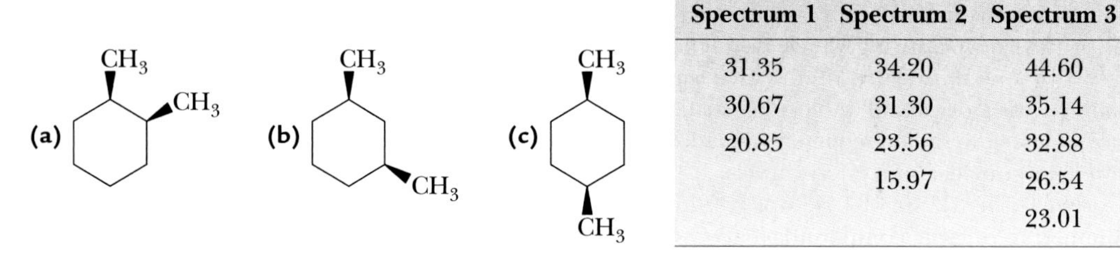

Spectrum 1	Spectrum 2	Spectrum 3
31.35	34.20	44.60
30.67	31.30	35.14
20.85	23.56	32.88
	15.97	26.54
		23.01

21.13 Following is a ^{1}H-NMR spectrum for compound A, molecular formula C_6H_{12}. Compound A decolorizes a solution of bromine in carbon tetrachloride. Propose a structural formula of compound A.

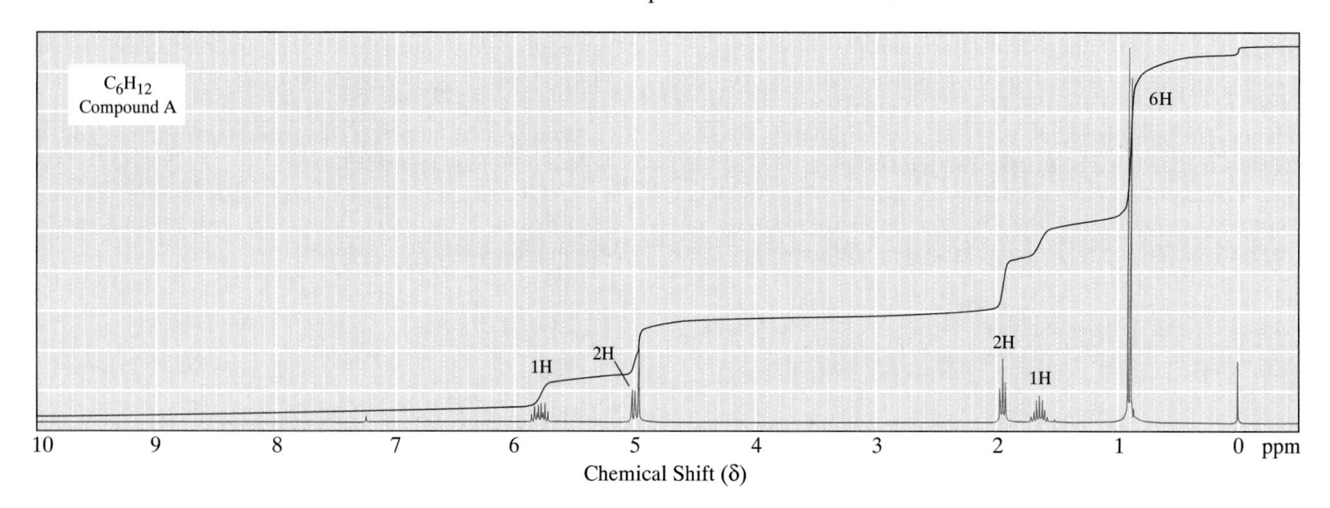

21.14 Following is a ^{1}H-NMR spectrum of compound B, C_7H_{12}. Compound B decolorizes a solution of Br_2 in CCl_4. Propose a structural formula for compound B.

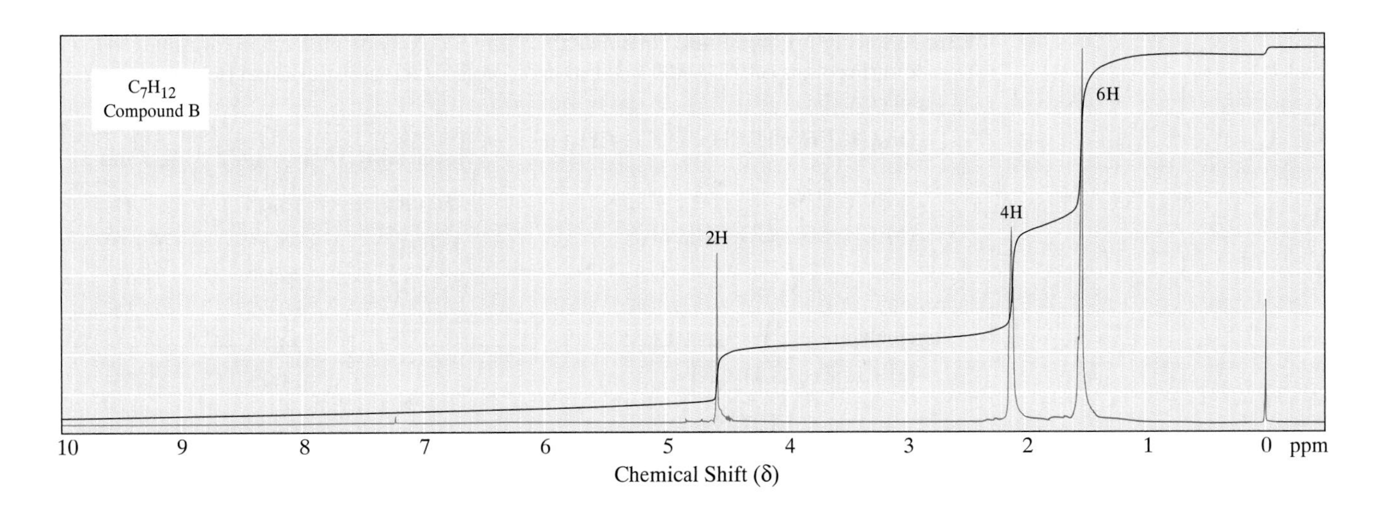

21.15 Following are ^{1}H-NMR spectra for compounds C and D, each of molecular formula $C_5H_{12}O$. Each is a liquid at room temperature, is slightly soluble in water, and reacts with sodium metal with the evolution of a gas. Propose structural formulas for compounds C and D.

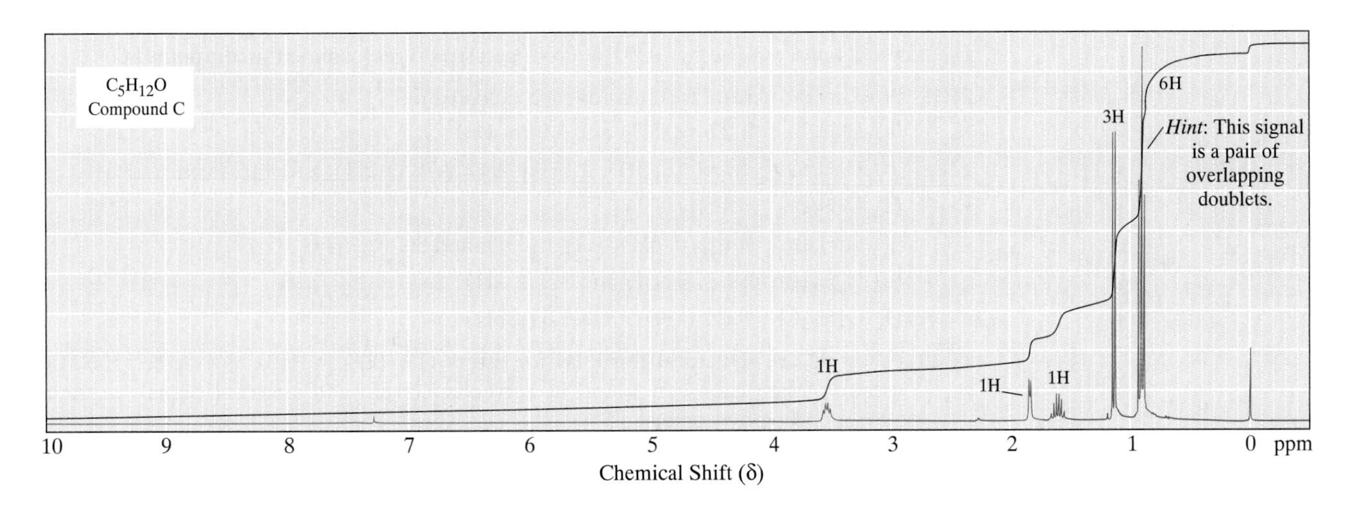

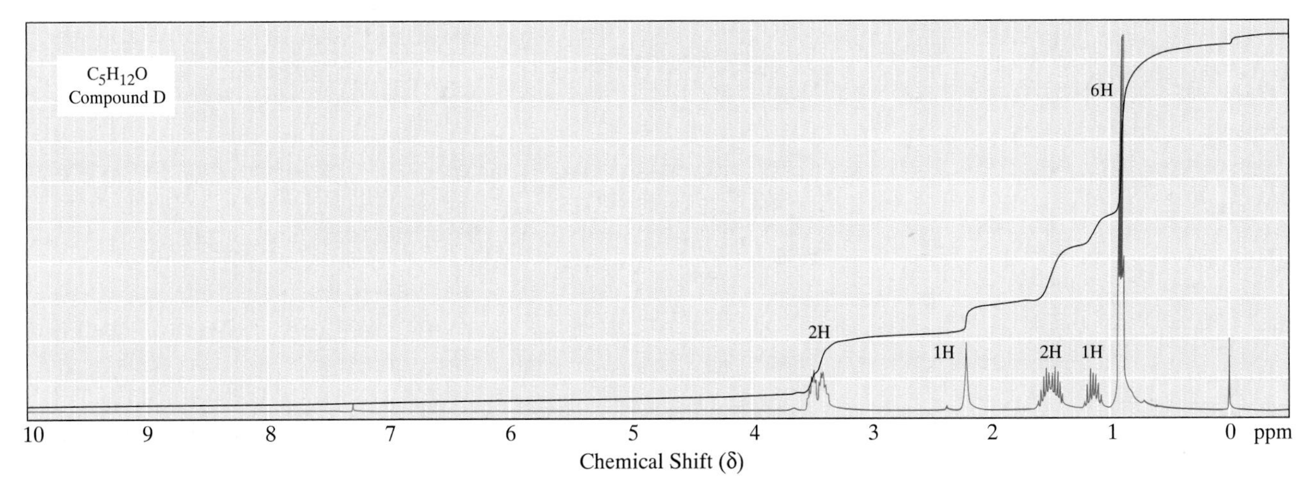

21.16 Following are structural formulas for three alcohols of molecular formula $C_7H_{16}O$ and three sets of ^{13}C-NMR spectral data. Assign each constitutional isomer its correct spectral data.

(a) $CH_3CH_2CH_2CH_2CH_2CH_2CH_2OH$

(b) $CH_3\underset{CH_3}{\overset{OH}{C}}CH_2CH_2CH_2CH_3$

(c) $CH_3CH_2\underset{CH_2CH_3}{\overset{OH}{C}}CH_2CH_3$

Spectrum 1	Spectrum 2	Spectrum 3
74.66	70.97	62.93
30.54	43.74	32.79
7.73	29.21	31.86
	26.60	29.14
	23.27	25.75
	14.09	22.63
		14.08

21.17 Alcohol E, molecular formula $C_6H_{14}O$, undergoes acid-catalyzed dehydration when warmed with phosphoric acid to give compound F, molecular formula C_6H_{12}, as the major product. The ^{1}H-NMR spectrum of compound E shows peaks at δ 0.89 (t, 6H), 1.12 (s, 3H), 1.38 (s, 1H), and 1.48 (q, 4H). The ^{13}C-NMR spectrum of compound E shows peaks at δ 72.98, 33.72, 25.85, and 8.16. Propose structural formulas for compounds E and F.

21.18 Compound G, $C_6H_{14}O$, does not react with sodium metal and does not discharge the color of Br_2 in CCl_4. Its ^{1}H-NMR spectrum consists of only two signals, a 12H doublet at δ 1.1 and a 2H septet at δ 3.6. Propose a structural formula for compound G.

21.19 Propose a structural formula for each haloalkane.

(a) $C_2H_4Br_2$ δ 2.5 (d, 3H) and 5.9 (q, 1H)

(b) $C_4H_8Cl_2$ δ 1.60 (d, 3H), 2.15 (q, 2H), 3.72 (t, 2H), and 4.27 (sextet, 1H)

(c) $C_5H_8Br_4$ δ 3.6 (s, 8H)

(d) C_4H_9Br δ 1.1 (d, 6H), 1.9 (m, 1H), and 3.4 (d, 2H)

(e) $C_5H_{11}Br$ δ 1.1 (s, 9H) and 3.2 (s, 2H)

(f) $C_7H_{15}Cl$ δ 1.1 (s, 9H) and 1.6 (s, 6H)

21.20 Following are structural formulas for esters (1), (2), and (3) and three ^{1}H-NMR spectra. Assign each compound its correct spectrum and assign all signals to their corresponding hydrogens.

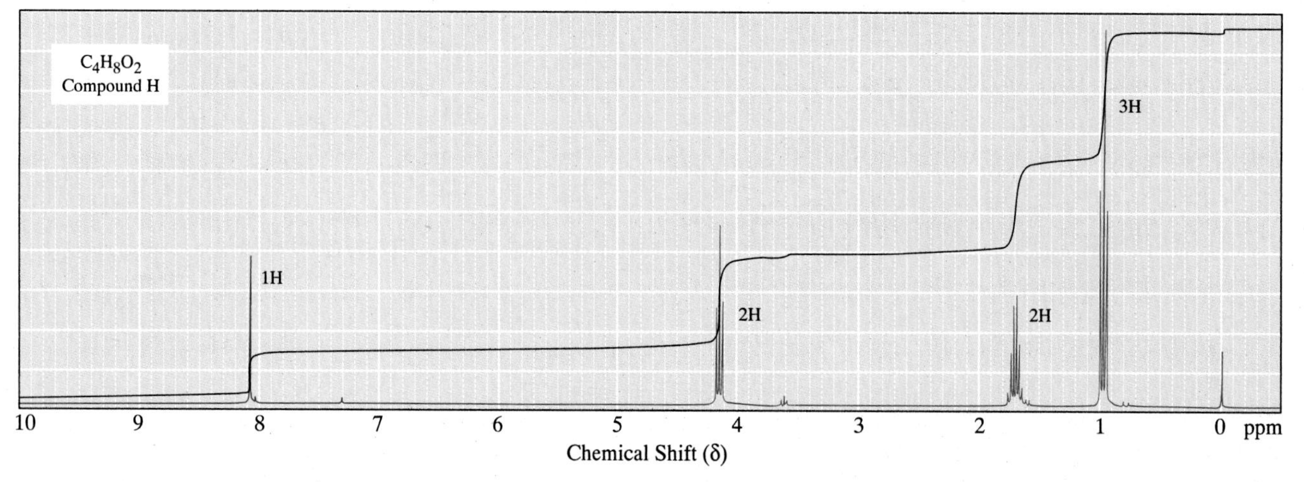

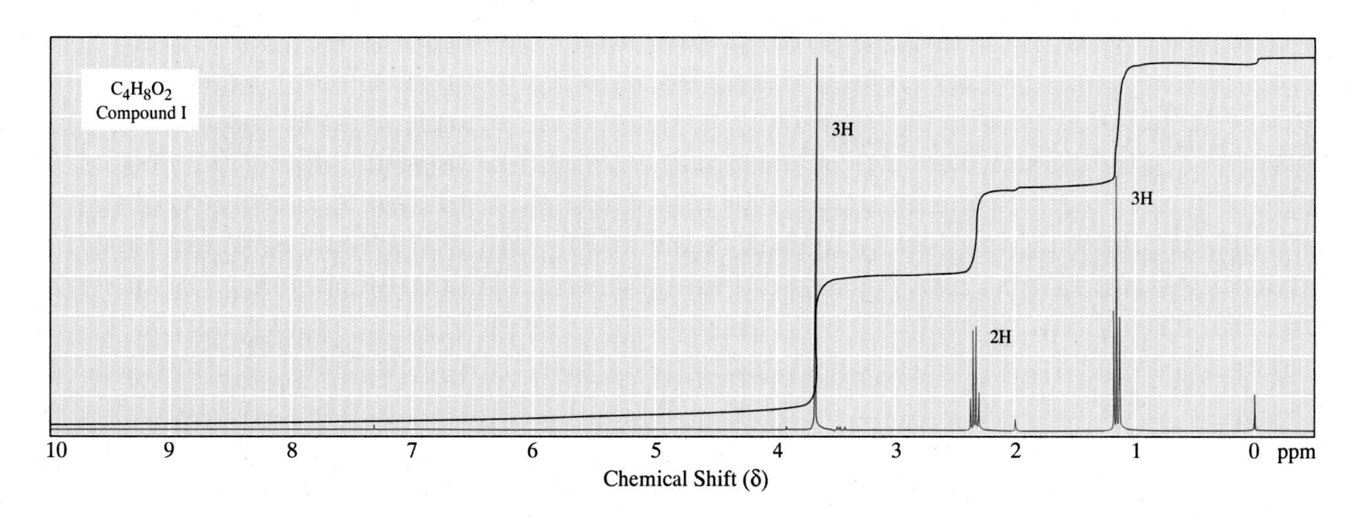

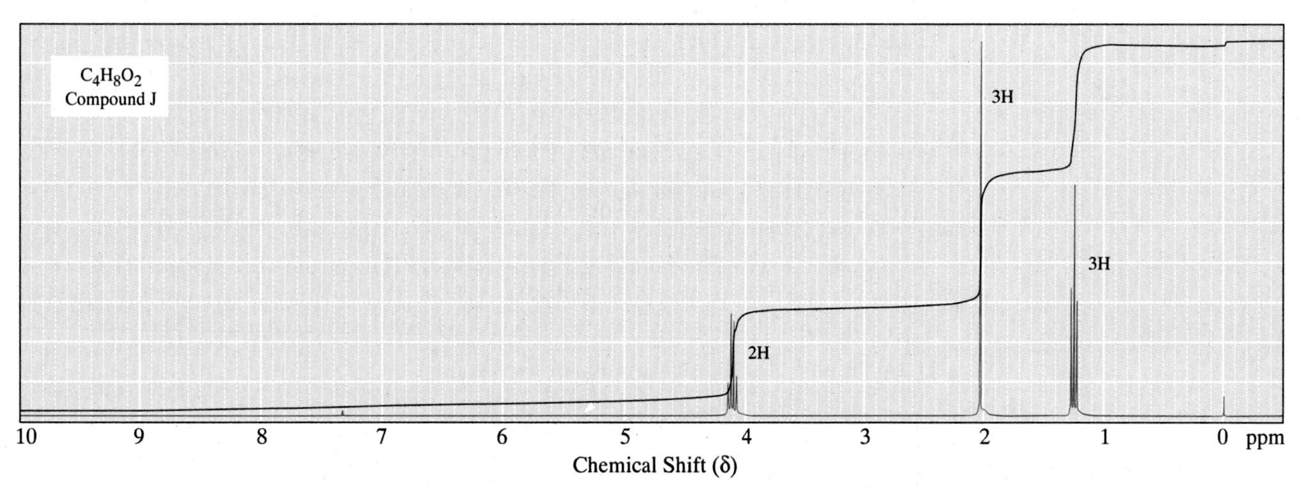

21.21 Compound K, $C_{10}H_{12}O_2$, is insoluble in water, 10% NaOH, and 10% HCl. Compound K is reduced by sodium borohydride to compound L, $C_{10}H_{14}O_2$. Given are a ^{1}H-NMR spectrum and ^{13}C-NMR spectral data for compound K. Propose structural formulas for compounds K and L.

^{13}C-NMR	
206.51	114.17
158.67	55.21
130.33	50.07
126.31	29.03

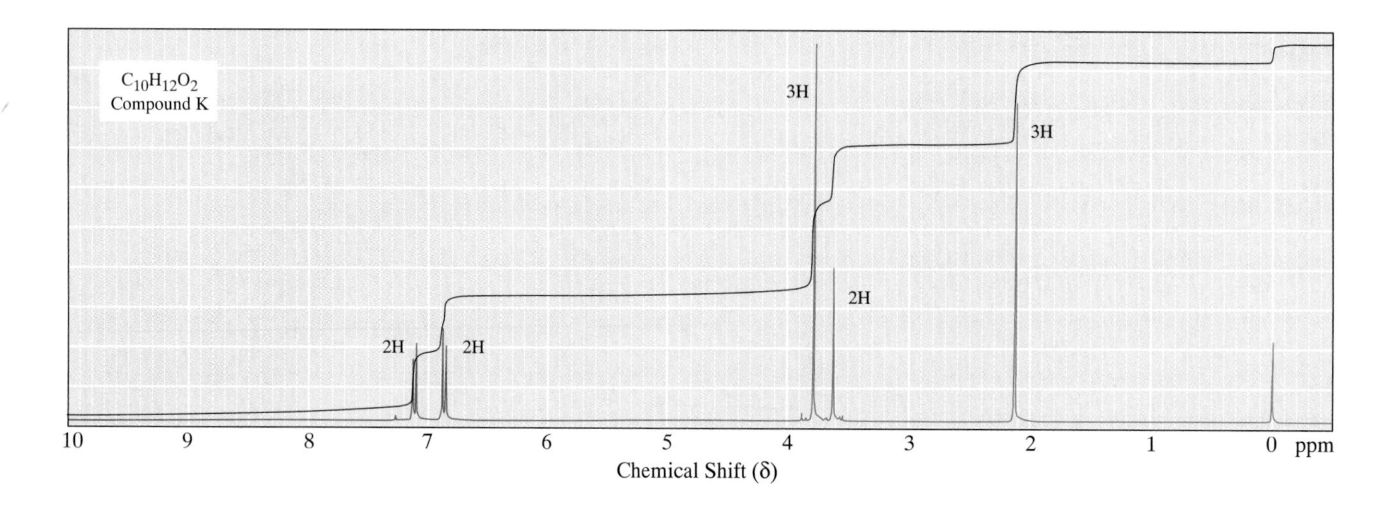

21.22 Propose a structural formula for each compound. Each contains an aromatic ring.

(a) $C_9H_{10}O$ δ 1.2 (t, 3H), 3.0 (q, 2H), and 7.4–8.0 (m, 5H)

(b) $C_{10}H_{12}O_2$ δ 2.2 (s, 3H), 2.9 (t, 2H), 4.3 (t, 2H), and 7.3 (s, 5H)

(c) $C_{10}H_{14}$ δ 1.2 (d, 6H), 2.3 (s, 3H), 2.9 (septet, 1H), and 7.0 (s, 4H)

(d) C_8H_9Br δ 1.8 (d, 3H), 5.0 (q, 1H), and 7.3 (s, 5H)

21.23 Compound M, molecular formula $C_9H_{12}O$, readily undergoes acid-catalyzed dehydration to give compound N, C_9H_{10}. A ^{1}H-NMR spectrum of compound M shows signals at δ 0.91 (t, 3H), 1.78 (m, 2H), 2.26 (d, 1H), 4.55 (m, 1H), and 7.31 (m, 5H). From this information, propose structural formulas for compounds M and N.

21.24 Propose a structural formula for each ketone.

(a) C_4H_8O δ 1.0 (t, 3H), 2.1 (s, 3H), and 2.4 (q, 2H)

(b) $C_7H_{14}O$ δ 0.9 (t, 6H), 1.6 (sextet, 4H), and 2.4 (t, 4H)

21.25 Propose a structural formula for compound O, a ketone of molecular formula $C_{10}H_{12}O$.

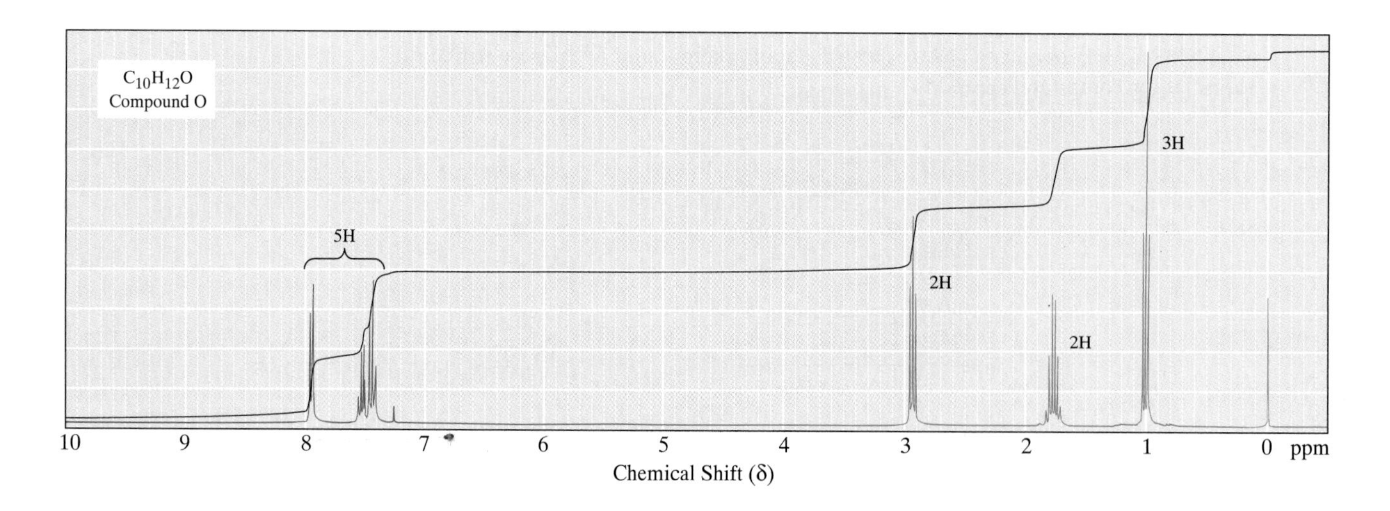

21.26 Following is a ^{1}H-NMR spectrum for compound P, $C_6H_{12}O_2$. Compound P undergoes acid-catalyzed dehydration to give compound Q, $C_6H_{10}O$. Propose structural formulas for compounds P and Q.

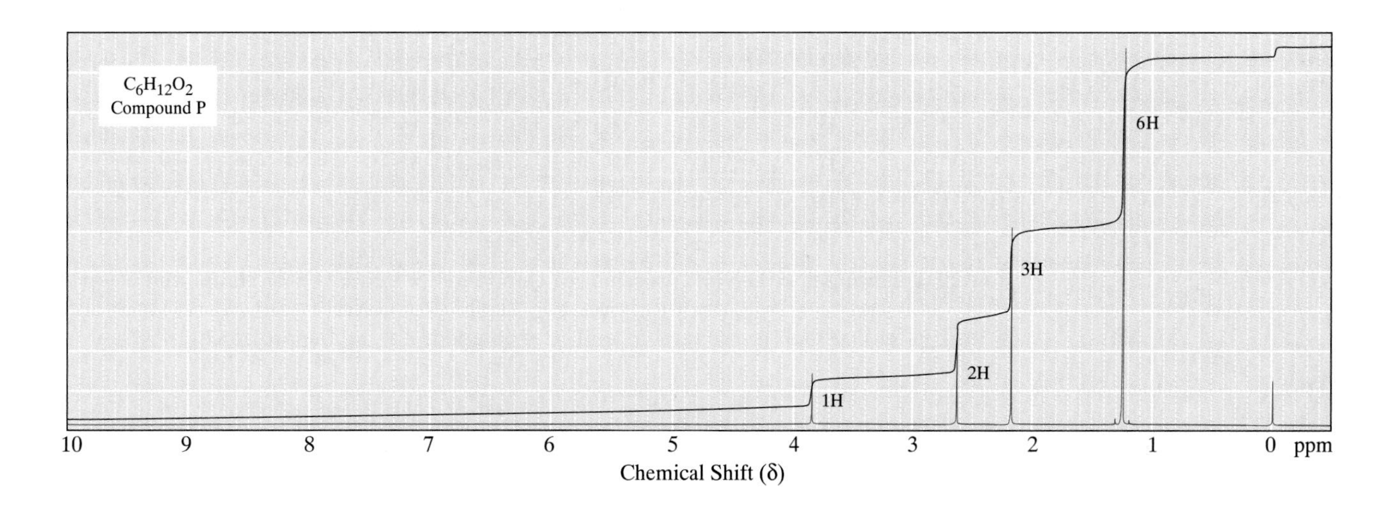

21.27 Propose a structural formula for compound R, $C_{12}H_{16}O$. Following is its ^{1}H-NMR spectrum and the position of signals in its ^{13}C-NMR spectrum.

^{13}C-NMR	
207.82	50.88
134.24	50.57
129.36	24.43
128.60	22.48
126.86	

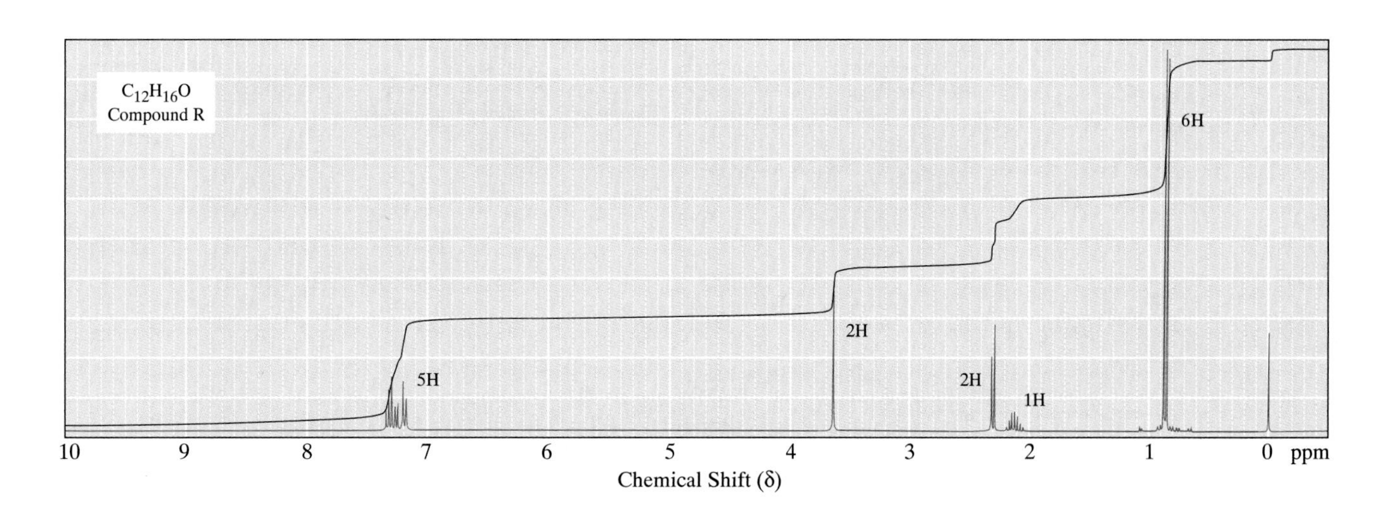

21.28 Propose a structural formula for each carboxylic acid.

(a) $C_5H_{10}O_2$

^{1}H-NMR	^{13}C-NMR
0.94 (t, 3H)	180.7
1.39 (m, 2H)	33.89
1.62 (m, 2H)	26.76
2.35 (t, 2H)	22.21
12.0 (s, 1H)	13.69

(b) $C_6H_{12}O_2$

^{1}H-NMR	^{13}C-NMR
1.08 (s, 9H)	179.29
2.23 (s, 2H)	46.82
12.1 (s, 1H)	30.62
	29.57

(c) $C_5H_8O_4$

^{1}H-NMR	^{13}C-NMR
0.93 (t, 3H)	170.94
1.80 (m, 2H)	53.28
3.10 (t, 1H)	21.90
12.7 (s, 2H)	11.81

21.29 Following is a ^{1}H-NMR spectrum of compound S. Reduction of compound S using lithium aluminum hydride gives two alcohols, molecular formulas $C_4H_{10}O$ and C_3H_8O. Propose structural formulas for compound S and the two alcohols.

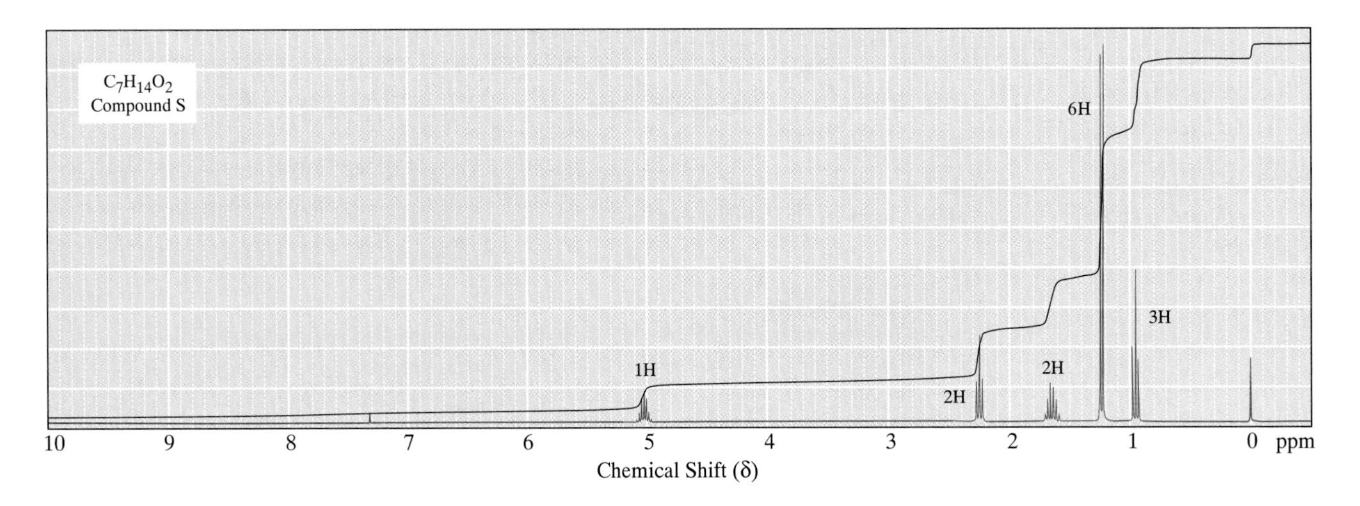

21.30 Propose a structural formula for each ester.

(a) $C_6H_{12}O_2$

^{1}H-NMR	^{13}C-NMR
1.18 (d, 6H)	177.16
1.26 (t, 3H)	60.17
2.51 (m, 1H)	34.04
4.13 (q, 2H)	19.01
	14.25

(b) $C_7H_{12}O_4$

^{1}H-NMR	^{13}C-NMR
1.28 (t, 6H)	166.52
3.36 (s, 2H)	61.43
4.21 (q, 4H)	41.69
	14.07

(c) $C_7H_{14}O_2$

^{1}H-NMR	^{13}C-NMR
0.92 (d, 6H)	171.15
1.52 (m, 2H)	63.12
1.70 (m, 1H)	37.31
2.09 (s, 3H)	25.05
4.10 (t, 2H)	22.45
	21.06

21.31 Following is a ^{1}H-NMR spectrum of compound T, $C_{11}H_{14}O_3$. Propose a structural formula for this compound. (*Hint:* The signal at δ 1.4 is actually two signals, each a closely spaced triplet.)

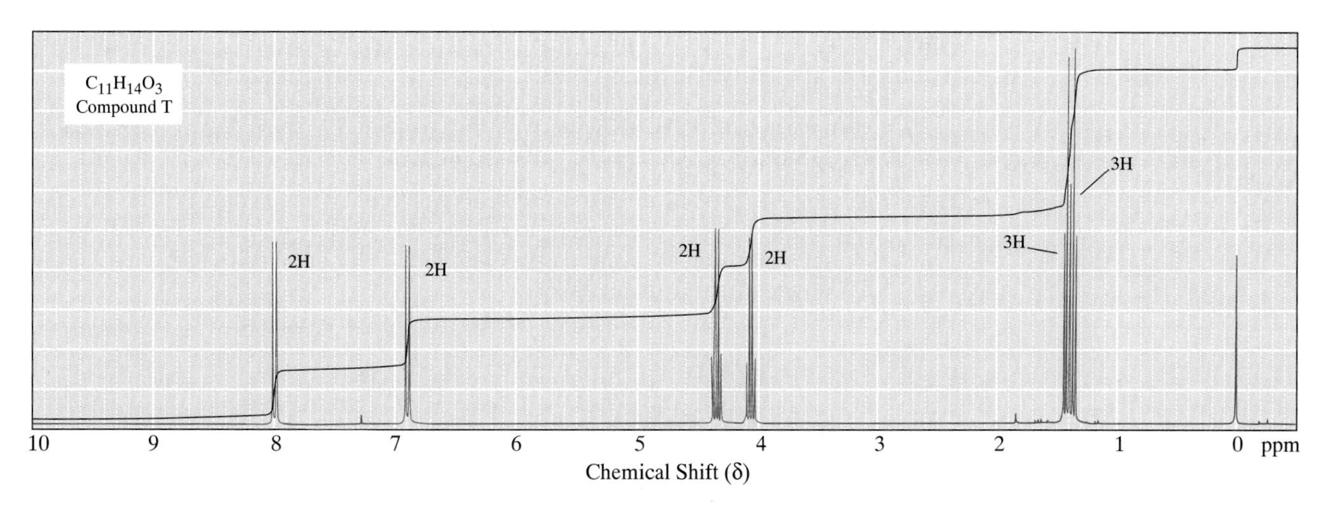

21.32 Propose a structural formula for amide U, molecular formula $C_6H_{13}NO$.

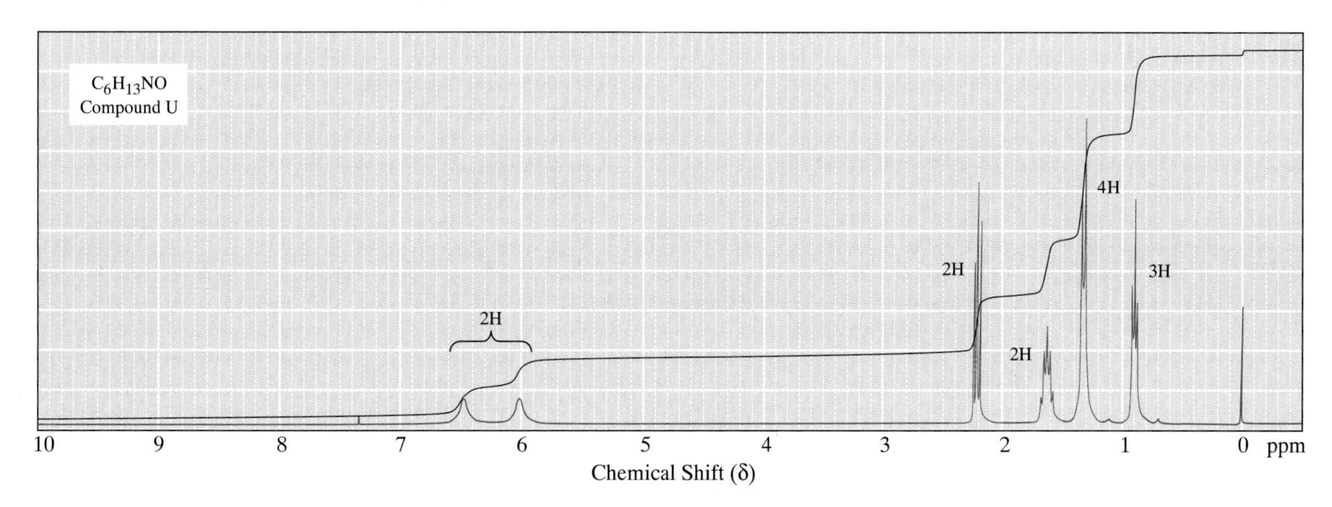

21.33 Propose a structural formula for the analgesic phenacetin, molecular formula $C_{10}H_{13}NO_2$, based on its ^{1}H-NMR spectrum.

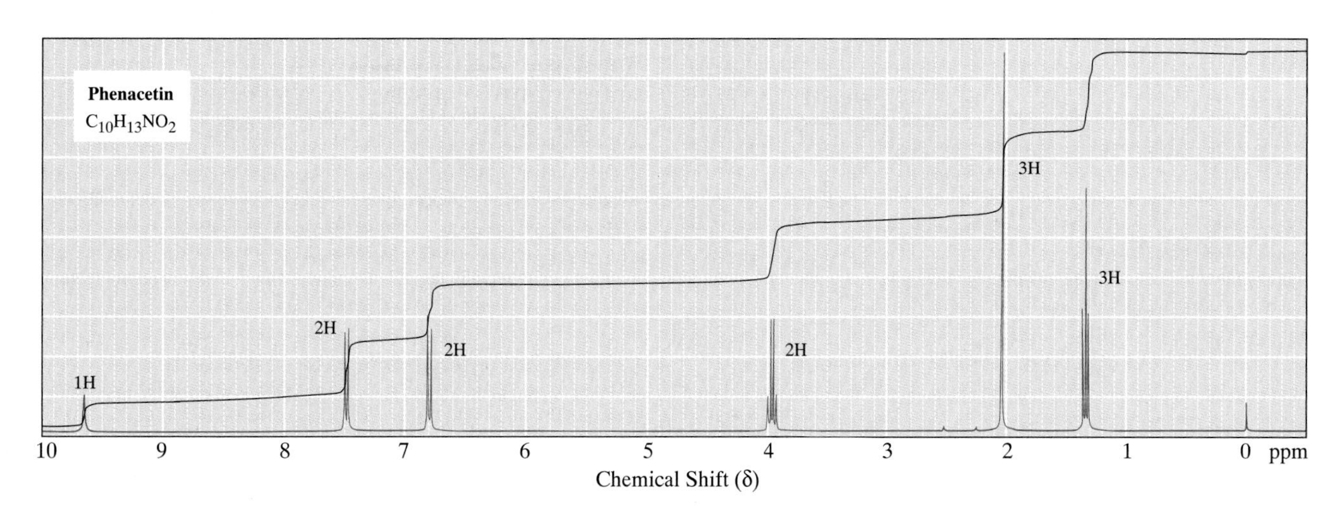

21.34 Propose a structural formula for compound V, an oily liquid of molecular formula $C_8H_9NO_2$. Compound V is insoluble in water and aqueous NaOH but dissolves in 10% HCl. When its solution in HCl is neutralized with NaOH, compound V is recovered unchanged. A ^{1}H-NMR spectrum of compound V shows signals at δ 3.84 (s, 3H), 4.18 (s, 2H), 7.60 (d, 2H), and 8.70 (d, 2H).

21.35 Following is a ^{1}H-NMR spectrum and structural formula for anethole, $C_{10}H_{12}O$, a fragrant natural product obtained from anise. Using the line of integration, determine the number of protons giving rise to each signal. Show that this spectrum is consistent with the structure of anethole.

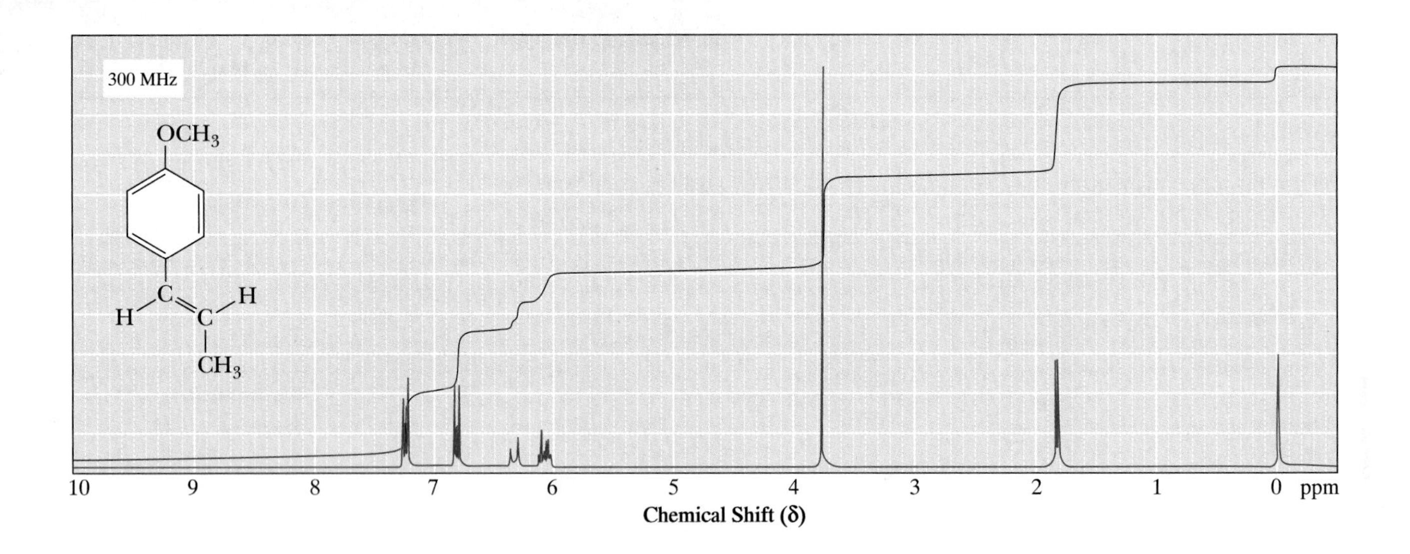

22

Infrared Spectroscopy

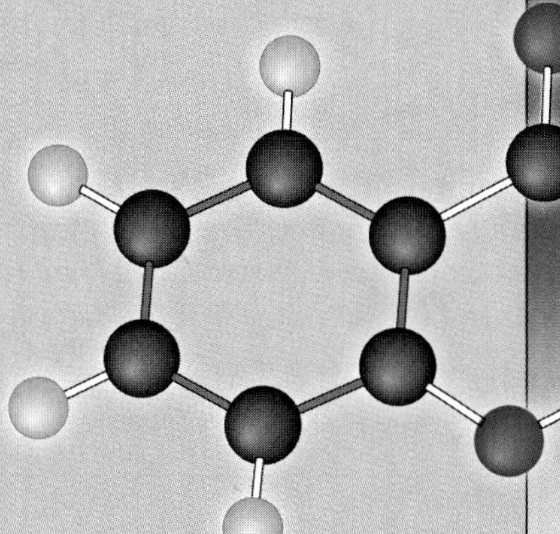

Oil of wintergreen. Inset: A model of methyl salicylate. See its IR spectrum in Additional Problem 22.14. *(Courtesy of Perkin-Elmer Co.)*

In **infrared (IR) spectroscopy,** a compound is irradiated with infrared radiation, absorption of which causes covalent bonds to change from a lower vibrational energy level to a higher one. Thus, in infrared spectroscopy, we detect functional groups by the vibrations of their bonds.

22.1 Infrared Spectroscopy

A. The Vibrational Infrared Spectrum

Vibrational infrared The portion of the infrared region that extends from 4000 to 400 cm^{-1}.

Wavenumber ($\overline{\nu}$) A characteristic of electromagnetic radiation equal to the number of waves per centimeter.

The infrared region of the electromagnetic spectrum (Table 21.1) covers the range 7.8×10^{-7} m (just above the visible region) to 2.0×10^{-3} m (just below the microwave region). Only the middle portion of this range, however, called the vibrational infrared region, is routinely used in organic chemistry. The **vibrational infrared region** extends from 2.5×10^{-6} to 2.5×10^{-5} m. Radiation in this region is most commonly referred to by its **wavenumber ($\overline{\nu}$),** the number of waves per centimeter.

$$\overline{\nu} = \frac{1}{\lambda\ (\text{cm})} = \frac{10^{-2}\ (\text{m} \cdot \text{cm}^{-1})}{\lambda\ (\text{m})}$$

When expressed in wavenumbers, the vibrational region of the infrared spectrum extends from 4000 to 400 cm^{-1}. The unit cm^{-1} is read "reciprocal centimeter."

$$\overline{\nu} = \frac{10^{-2}\,\text{m} \cdot \text{cm}^{-1}}{2.5 \times 10^{-6}\,\text{m}} = 4000\ \text{cm}^{-1} \qquad \overline{\nu} = \frac{10^{-2}\,\text{m} \cdot \text{cm}^{-1}}{2.5 \times 10^{-5}\,\text{m}} = 400\ \text{cm}^{-1}$$

An advantage of using wavenumbers is that they are directly proportional to energy; the higher the wavenumber, the higher the energy of the radiation.

Shown in Figure 22.1 is an infrared spectrum of 3-methyl-2-butanone. The horizontal axis at the bottom of the chart paper is calibrated in wavenumbers

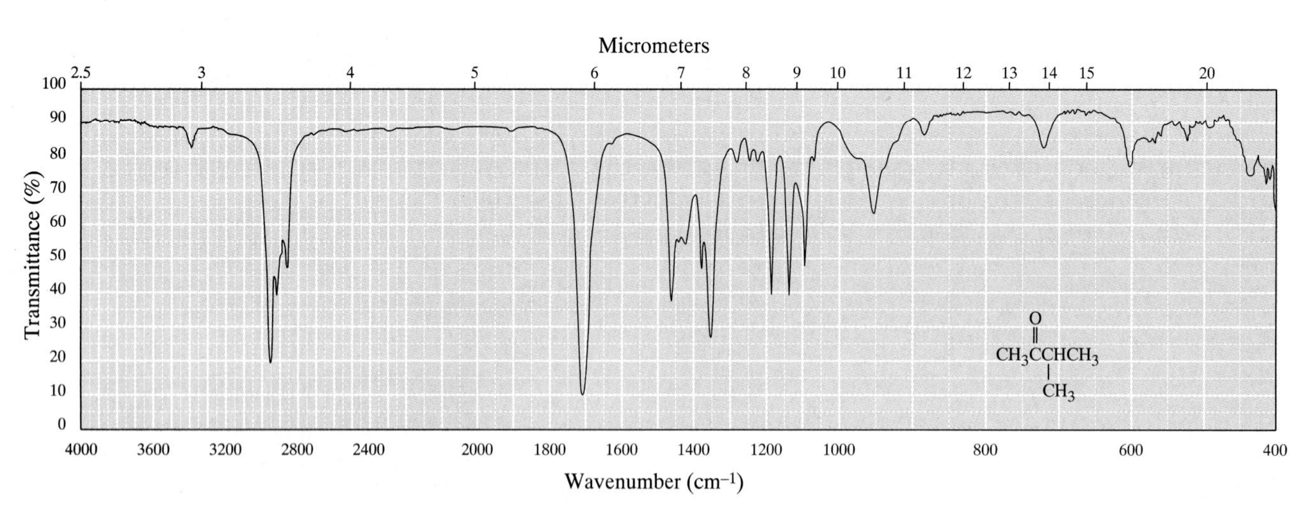

Figure 22.1
Infrared spectrum of 3-methyl-2-butanone.

(cm^{-1}); that at the top is calibrated in wavelength (micrometers, μm). The wavenumber scale is often divided into two or more linear regions. For all spectra reproduced in this text, it is divided into three linear regions: 4000–2200 cm^{-1}, 2200–1000 cm^{-1}, and 1000–400 cm^{-1}. The vertical axis measures transmittance with 100% transmittance at the top and 0% transmittance at the bottom. Thus, the baseline for an infrared spectrum (100% transmittance of radiation through the sample, 0% absorption) is at the top of the chart paper, and absorption of radiation corresponds to a trough or valley. Strange as it may seem, we commonly refer to infrared absorptions as peaks, even though they are upside down for peaks.

The spectrum in Figure 22.1 was recorded on a **neat** sample, which means using the pure liquid. A few drops of the liquid were placed between two sodium chloride discs and spread to give a thin film through which infrared radiation was then passed. Liquid and solid samples may also be dissolved in carbon tetrachloride or another solvent with minimal infrared absorption and analyzed in a liquid sampling cell. Still another way to obtain the infrared spectrum of a solid is to mix it with potassium bromide and compact the mixture under high pressure to a thin wafer, which is then placed in the beam of the spectrophotometer. Both NaCl and KBr are transparent to infrared radiation. Infrared spectra of gas samples are determined using specially constructed gas-handling cells.

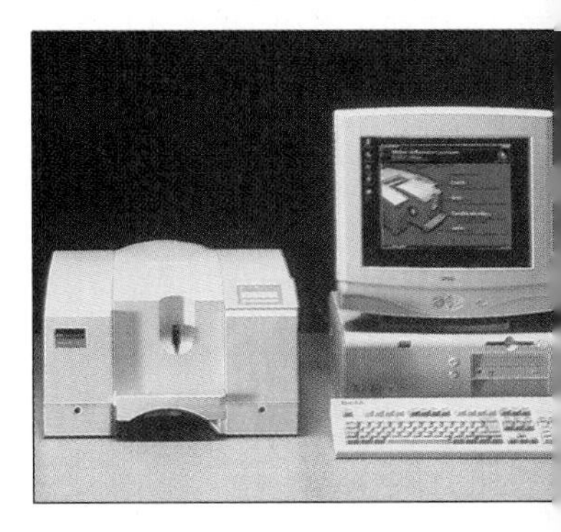

A Perkin-Elmer Paragon 1000 Fourier Transform infrared spectrophotometer. Spectra are shown in the monitor. *(Courtesy of the Perkin-Elmer Corporation)*

B. Molecular Vibrations

Atoms joined by covalent bonds are not permanently fixed in one position but, instead, undergo continual vibrations relative to each other. The energies associated with these vibrations are quantized, which means that, within a molecule, only specific vibrational energy levels are allowed. The energies associated with transitions between vibrational energy levels in most covalent molecules range from 2 to 10 kcal/mol (8.4 to 42 kJ/mol). Such transitions can be induced by absorption of radiation in the infrared region of the electromagnetic spectrum.

For a molecule to absorb infrared radiation, the bond undergoing vibration must be polar; the greater its polarity, the more intense the absorption. Any vibration that meets this criterion is said to be **infrared-active.** Covalent bonds in homonuclear diatomic molecules, such as H_2, Br_2, and some carbon-carbon double bonds in symmetrical alkenes do not absorb infrared radiation because they are not polar bonds.

For a nonlinear molecule containing n atoms, $3n - 6$ allowed fundamental vibrations exist. For a molecule as simple as ethanol, C_2H_6O, there are 21 fundamental vibrations, and for hexanoic acid, $C_6H_{12}O_2$, there are 54. Thus, for even relatively simple molecules, a large number of vibrational energy levels exists, and the patterns of energy absorption for these and larger molecules are very complex.

The simplest vibrational motions in molecules giving rise to absorption of infrared radiation are **stretching** and **bending** motions. Illustrated in Figure 22.2 are the fundamental stretching and bending vibrations for a methylene group.

To one skilled in the interpretation of infrared spectra, the absorption patterns can yield an enormous amount of information about chemical structure. We, however, have neither the time nor the need to develop this level of competence. The value of infrared spectra for us is that they can be used to determine the presence or absence of particular functional groups. A carbonyl group, for example, typically shows strong absorption at approximately 1630–1800 cm^{-1}. The position

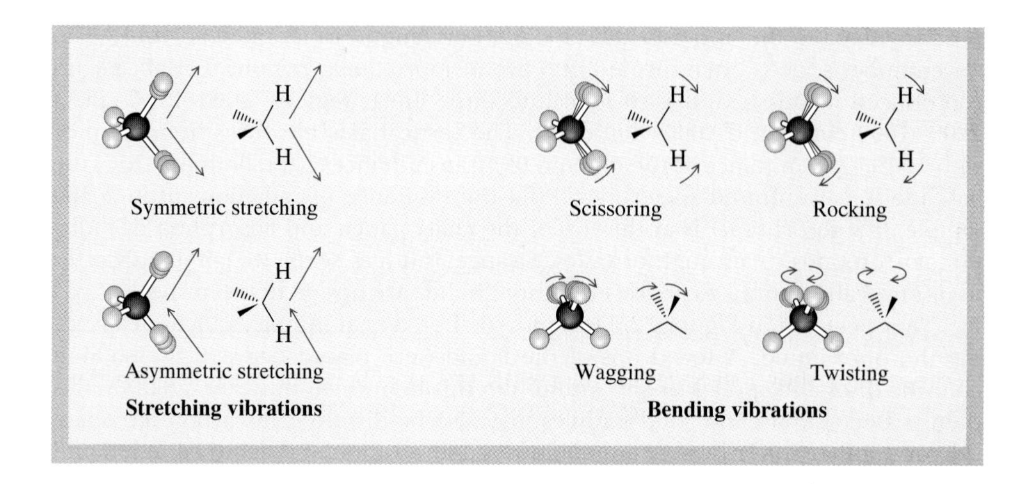

Figure 22.2 Fundamental modes of vibration for a methylene group.

of absorption for a particular carbonyl group depends on whether it is that of an aldehyde, a ketone, a carboxylic acid, an ester, or an amide, and, if the carbonyl carbon is in a ring, the size of the ring.

C. Correlation Tables

Data on absorption patterns of selected functional groups are collected in tables called **correlation tables.** The correlation table given in Table 22.1 is a listing of characteristic infrared absorptions for the types of bonds and functional groups we deal with most often. A cumulative correlation chart can be found in Appendix 5. In these tables, the intensity of a particular absorption is often referred to as **strong (s), medium (m),** or **weak (w).**

In general, we pay most attention to the region from 3650 to 1000 cm^{-1} because the characteristic stretching vibrations for most functional groups are found in this region. Vibrations in the region 1000–400 cm^{-1} are much more complex and far more difficult to analyze. Because even slight variations in molecular structure and absorption patterns are most obvious in this region, it is often called the **fingerprint region.** If two compounds have even slightly different structures, the differences in their infrared spectra are most clearly discernible in this region.

Fingerprint region The portion of the vibrational infrared region that extends from 1000 to 400 cm^{-1}.

TABLE 22.1 Characteristic IR Absorptions of Selected Functional Groups

Bond	Frequency (cm^{-1})	Intensity
O—H	3200–3500	strong and broad
N—H	3100–3500	medium
C—H	2850–3100	medium to strong
C=O	1630–1800	strong
C=C	1600–1680	weak
C—O	1050–1250	strong

EXAMPLE 22.1

What functional group is most likely present if a compound shows IR absorption at

(a) 1705 cm^{-1} (b) 2950 cm^{-1}

SOLUTION

(a) A C═O group (b) An aliphatic C—H group

Practice Problem 22.1

A compound shows strong, very broad IR absorption in the region 3200–3500 cm^{-1} and strong absorption at 1715 cm^{-1}. What functional group accounts for both of these absorptions?

EXAMPLE 22.2

Propanone and 2-propen-1-ol are constitutional isomers. Show how to distinguish between them by IR spectroscopy.

$$CH_3—\overset{\overset{\Large O}{\|}}{C}—CH_3 \qquad CH_2{=}CH—CH_2—OH$$

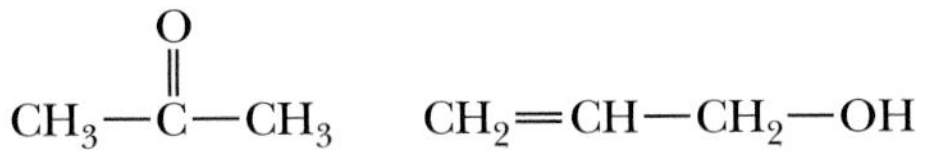

Propanone (Acetone) 2-Propen-1-ol (Allyl alcohol)

SOLUTION

Only propanone shows strong absorption in the C═O stretching region, 1630–1800 cm^{-1}. Alternatively, only 2-propen-1-ol shows strong absorption in the O—H stretching region, 3200–3500 cm^{-1}.

Practice Problem 22.2

Propanoic acid and methyl ethanoate are constitutional isomers. Show how to distinguish between them by IR spectroscopy.

$$CH_3CH_2\overset{\overset{\Large O}{\|}}{C}OH \qquad CH_3\overset{\overset{\Large O}{\|}}{C}OCH_3$$

Propanoic acid Methyl ethanoate (Methyl acetate)

22.2 INTERPRETING INFRARED SPECTRA

A. Alkanes

Infrared spectra of alkanes are usually simple with few peaks, the most common of which is given in Table 22.2.

TABLE 22.2 Characteristic IR Absorptions of Alkanes and Alkenes

Hydrocarbon	Vibration	Frequency (cm^{-1})	Intensity
Alkane			
C—H	stretching	2850–3000	strong
CH_2	bending	1450	medium
CH_3	bending	1375 and 1450	weak to medium
Alkene			
C—H	stretching	3000–3100	weak to medium
C=C	stretching	1600–1680	weak to medium

Figure 22.3 shows an infrared spectrum of decane. The strong peak with multiple splittings between 2850 and 3000 cm^{-1} is characteristic of C—H stretching. The other prominent peaks in this spectrum are methylene bending absorption at 1465 cm^{-1} and methyl bending absorption at 1380 cm^{-1}.

B. Alkenes

An easily recognized alkene absorption is the vinylic C—H stretching slightly to the left (greater wavenumber) of 3000 cm^{-1}. Also characteristic of alkenes is C=C stretching at 1600–1680 cm^{-1}. This vibration, however, is often weak and difficult to observe. Both vinylic C—H stretching and C=C stretching can be seen in the infrared spectrum of cyclohexene (Figure 22.4). Also visible are the aliphatic C—H stretching near 2900 cm^{-1} and the methylene bending near 1440 cm^{-1}.

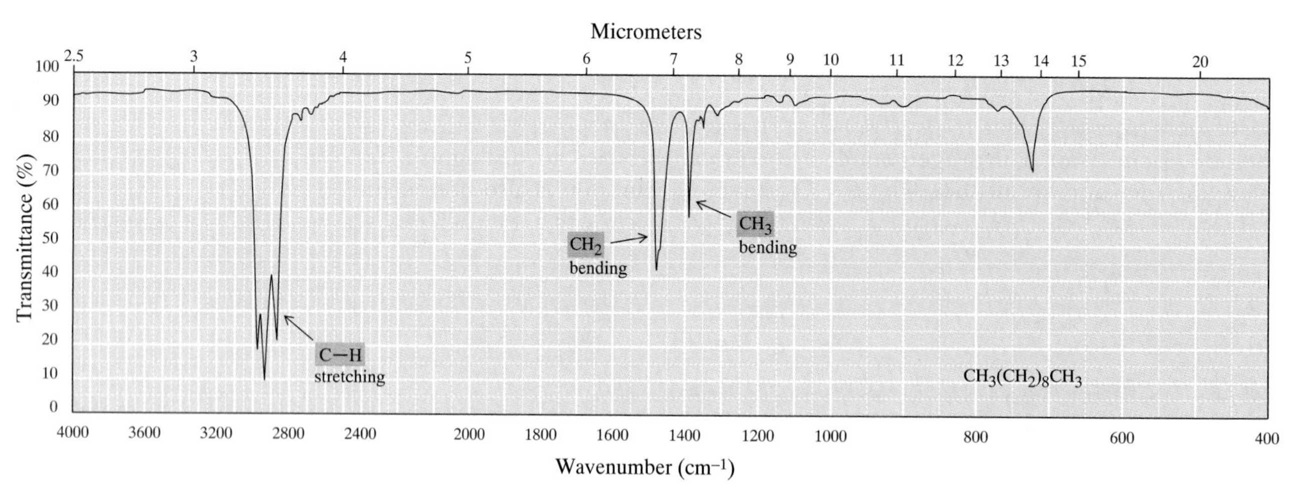

Figure 22.3
Infrared spectrum of decane (neat, salt plates).

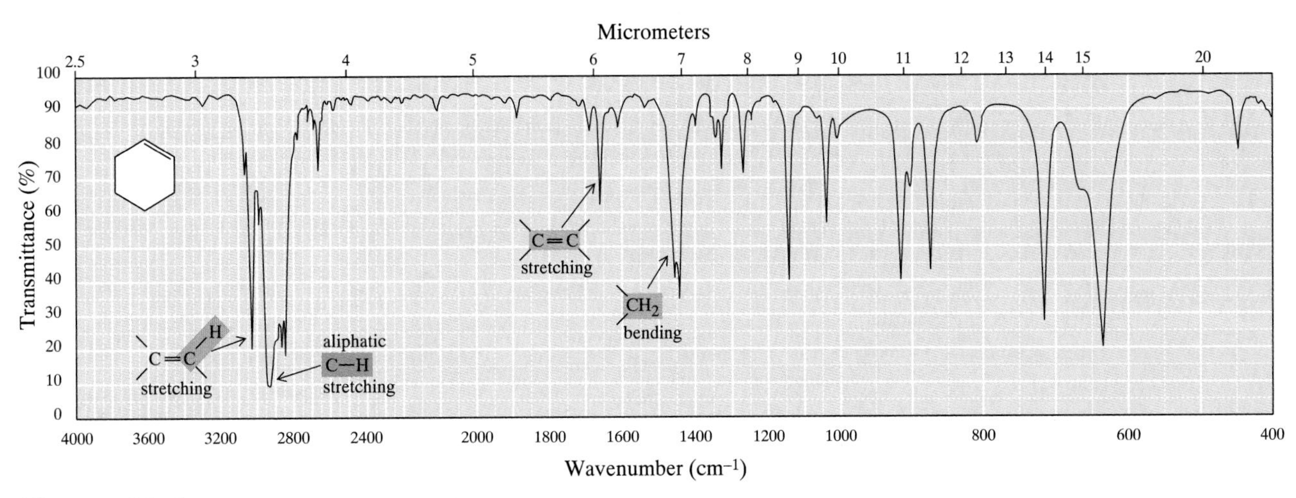

Figure 22.4
Infrared spectrum of cyclohexene (neat, salt plates).

C. Alcohols

Both the position of the O—H stretching absorption and its intensity depend on the extent of hydrogen bonding (Section 8.3A). Under normal conditions, where there is extensive hydrogen bonding between alcohol molecules, O—H stretching occurs as a broad peak at 3200–3500 cm^{-1}. The C—O stretching vibration of alcohols appears in the range 1050–1250 cm^{-1} (Table 22.3).

Figure 22.5 shows an infrared spectrum of 1-hexanol. The hydrogen-bonded O—H stretching appears as a strong, broad peak of medium intensity centered at 3340 cm^{-1}. The C—O stretching appears near 1050 cm^{-1}, a value characteristic of primary alcohols.

D. Ethers

The C—O stretching frequencies of ethers are similar to those observed in alcohols and esters. Dialkyl ethers typically show a single absorption in this region between 1070 and 1150 cm^{-1}. The presence or absence of O—H stretching at 3200–3500 cm^{-1} for a hydrogen-bonded O—H can be used to distinguish between an ether and an alcohol. The C—O stretching vibration is also present in esters. In this case, the presence or absence of C=O stretching can be used to distinguish between an ether and an ester.

TABLE 22.3 Characteristic IR Absorptions of Alcohols

Bond	Frequency (cm^{-1})	Intensity
O—H (hydrogen bonded)	3200–3500	medium, broad
C—O	1050–1250	medium

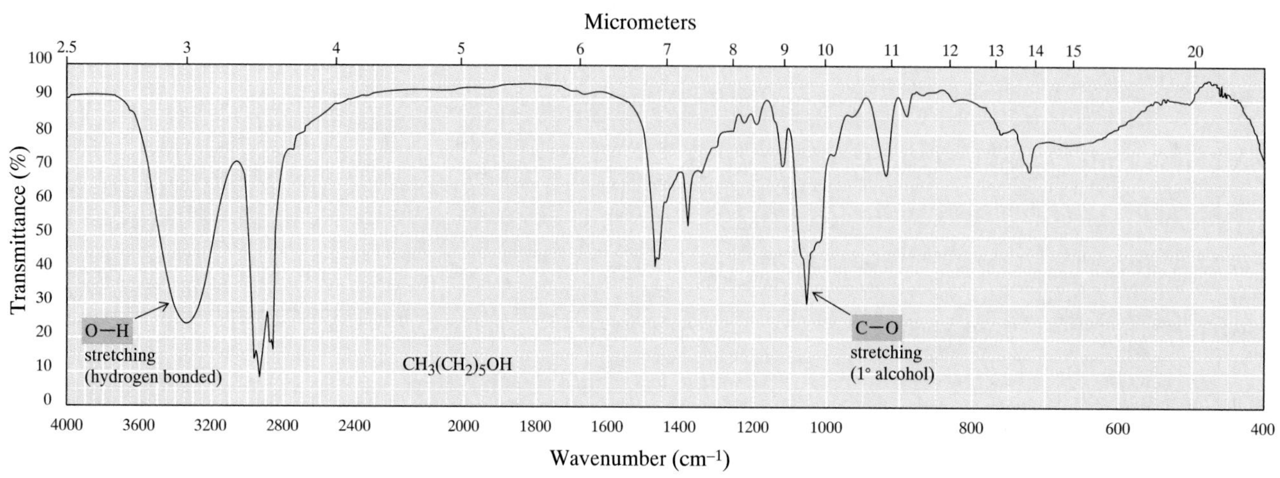

Figure 22.5
Infrared spectrum of 1-hexanol (neat, salt plates).

E. Benzene and Its Derivatives

Aromatic rings show a medium to weak peak in the C—H stretching region at approximately 3030 cm^{-1} characteristic of sp^2 C—H bonds. In addition, aromatic rings show strong absorption in the region 690–900 cm^{-1} due to C—H bending. Finally, aromatic rings show several absorptions due to C═C stretching between 1450 and 1600 cm^{-1} (Table 22.4). The C—H and C═C absorption patterns characteristic of aromatic rings can be seen in the infrared spectrum of toluene (Figure 22.6).

F. Amines

The most important and readily observed infrared absorptions of primary and secondary amines are due to N—H stretching vibrations, and appear in the region 3100–3500 cm^{-1}. Primary amines have two bands in this region, one due to a symmetric stretching vibration and the other due to an asymmetric stretching. The two N—H stretching absorptions characteristic of a primary amine can be seen in

TABLE 22.4 Characteristic IR Absorptions of Aromatic Hydrocarbons

Bond	Vibration	Frequency (cm^{-1})	Intensity
C—H	stretching	3030	medium to weak
C—H	bending	690–900	strong
C═C	stretching	1475 and 1600	strong to medium

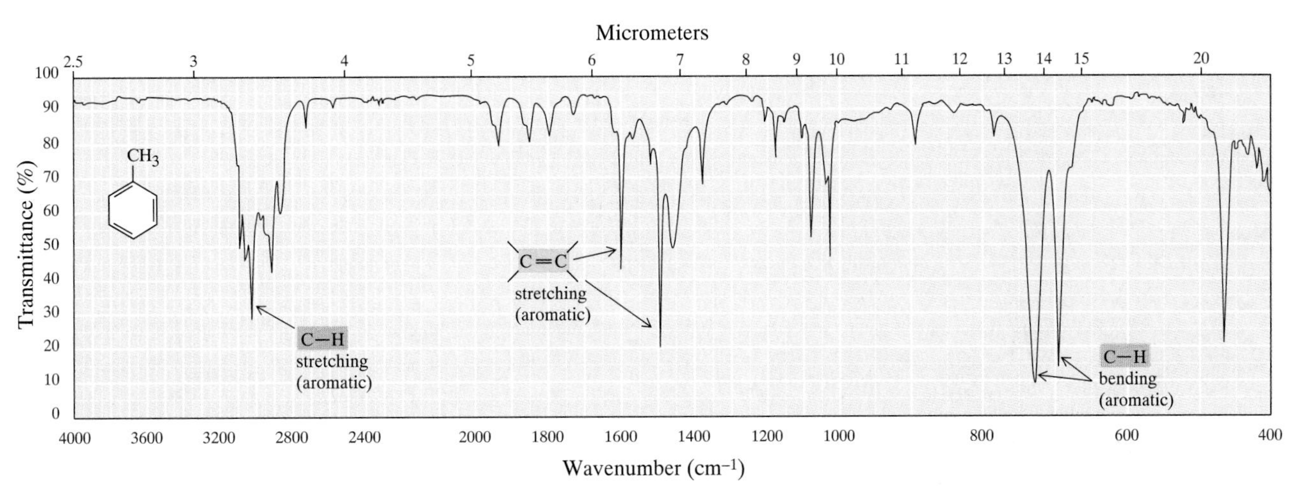

Figure 22.6
Infrared spectrum of toluene.

the IR spectrum of butylamine (Figure 22.7). Secondary amines give only one absorption in this region. Tertiary amines have no N—H and, therefore, are transparent in this region of the infrared spectrum.

G. Aldehydes and Ketones

Aldehydes and ketones show characteristic strong infrared absorption between 1705 and 1780 cm^{-1} associated with the stretching vibration of the carbon-oxygen

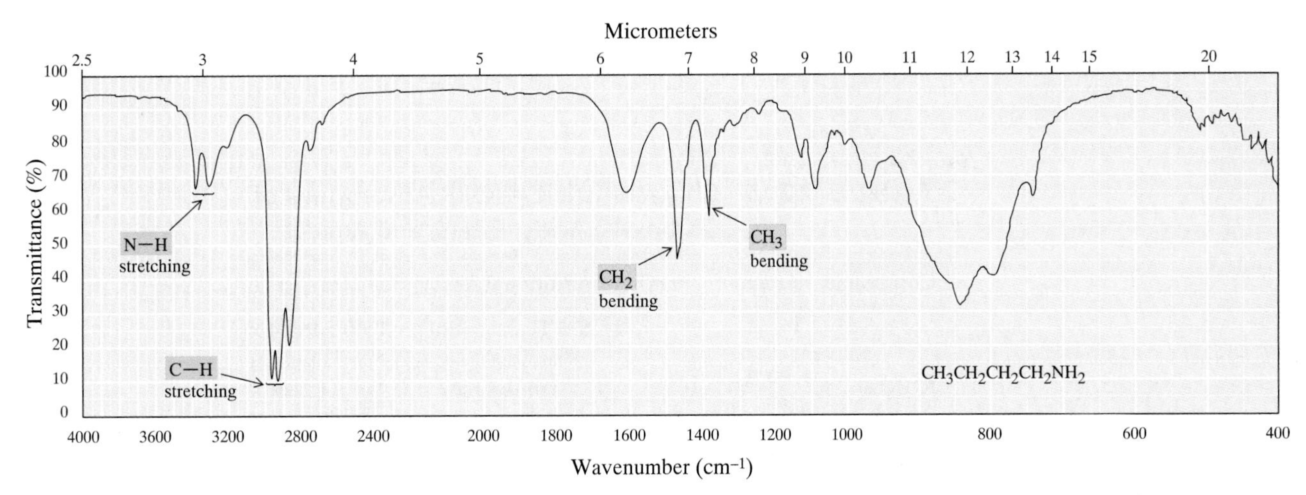

Figure 22.7
Infrared spectrum of butylamine, a primary amine.

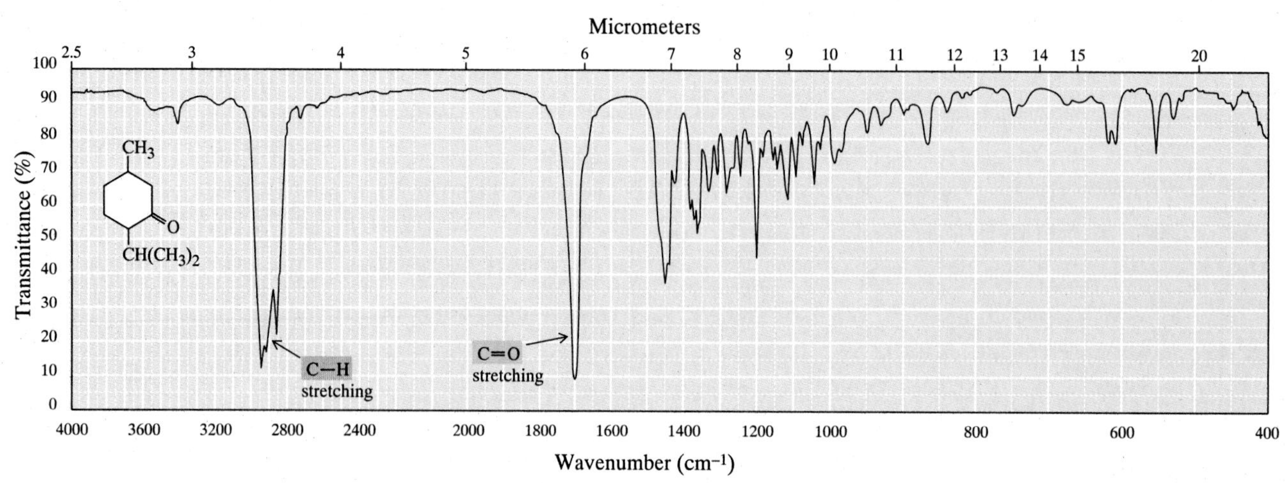

Figure 22.8
Infrared spectrum of menthone.

double bond. The stretching vibration for the carbonyl group of menthone occurs at 1705 cm^{-1} (Figure 22.8).

Because several different functional groups contain a carbonyl group, however, it is often not possible to tell from absorption in this region alone whether the carbonyl-containing compound is an aldehyde, ketone, carboxylic acid, or ester.

H. Carboxylic Acids and Their Derivatives

The most important infrared absorptions of carboxylic acids and their functional derivatives are due to the C=O stretching vibration; they are summarized in Table 22.5.

The carboxyl group of a carboxylic acid gives rise to two characteristic absorptions in the infrared spectrum. One of these occurs in the region 1700–1725 cm^{-1}

TABLE 22.5 Characteristic C=O IR Absorptions of Carboxylic Acids, Esters, and Amides

Compound	Frequency (cm^{-1})	Additional Absorptions (cm^{-1})
RC(=O)NH$_2$	1630–1680	N—H stretching at 3200 and 3400 (1° amides have two N—H peaks) (2° amides have one N—H peak)
RC(=O)OH	1700–1725	O—H stretching at 2400–3400 C—O stretching at 1210–1320
RC(=O)OR	1735–1800	C—O stretching at 1000–1100 and 1200–1250

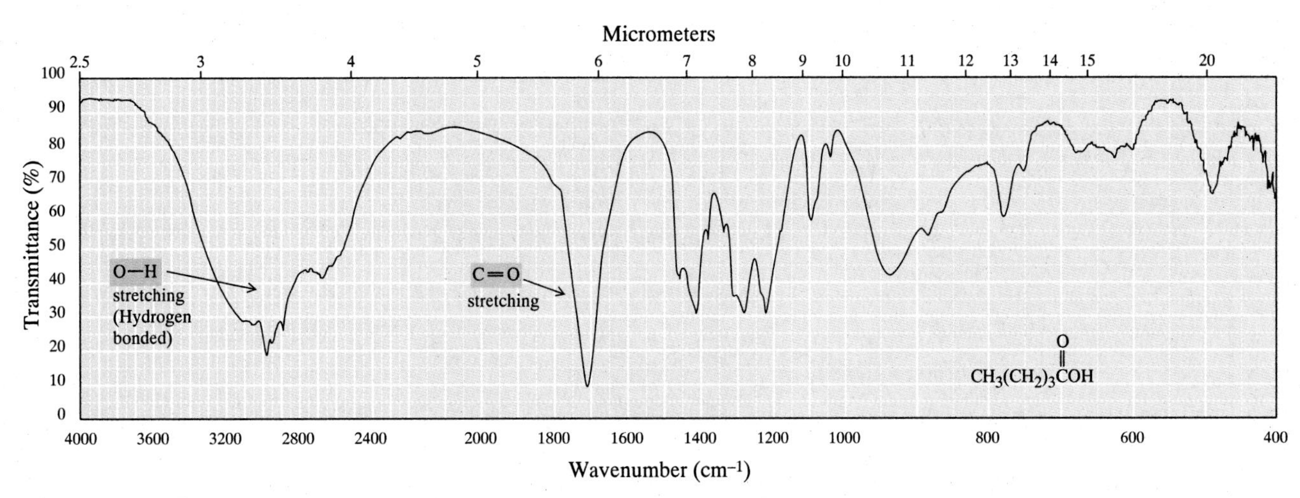

Figure 22.9
Infrared spectrum of pentanoic acid.

and is associated with the stretching vibration of the carbonyl group. This is essentially the same range of absorption as that for the carbonyl group of aldehydes and ketones. The other infrared absorption characteristic of a carboxyl group is a peak between 2400 and 3400 cm^{-1} due to the stretching vibration of the O—H group and often overlaps the C—H stretching absorptions. This absorption is generally very broad due to hydrogen bonding between molecules of the carboxylic acid. Both C=O and O—H stretchings can be seen in the infrared spectrum of pentanoic acid in Figure 22.9.

Esters display strong C=O stretching absorption in the region between 1735 and 1780 cm^{-1}. In addition, they also display strong C—O stretching absorption in the region 1000–1250 cm^{-1} (Figure 22.10). The carbonyl stretching of amides

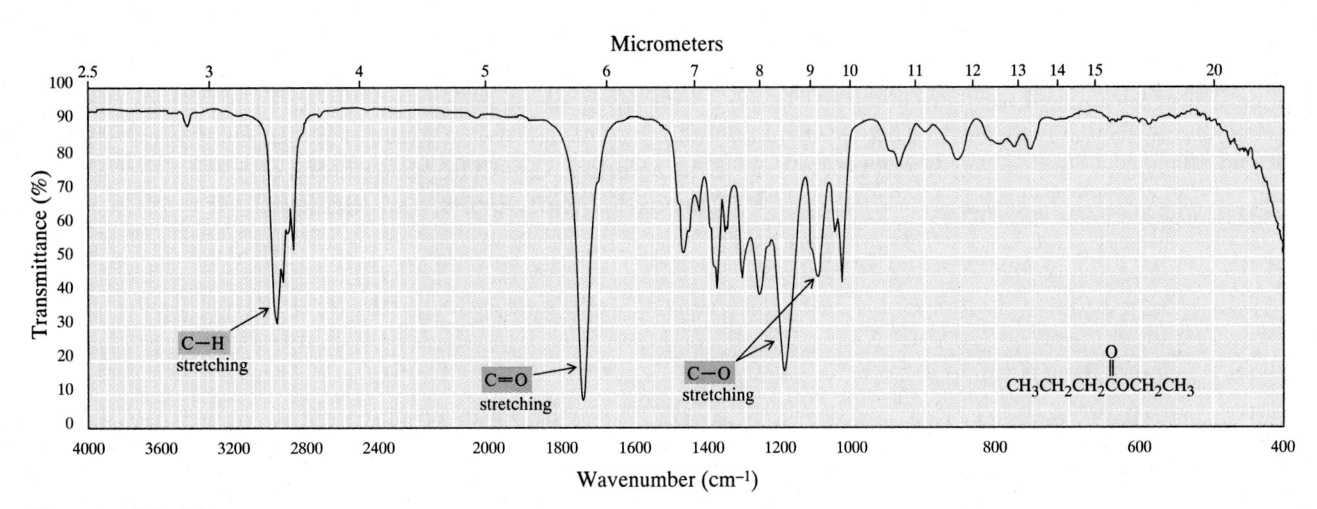

Figure 22.10
Infrared spectrum of ethyl butanoate.

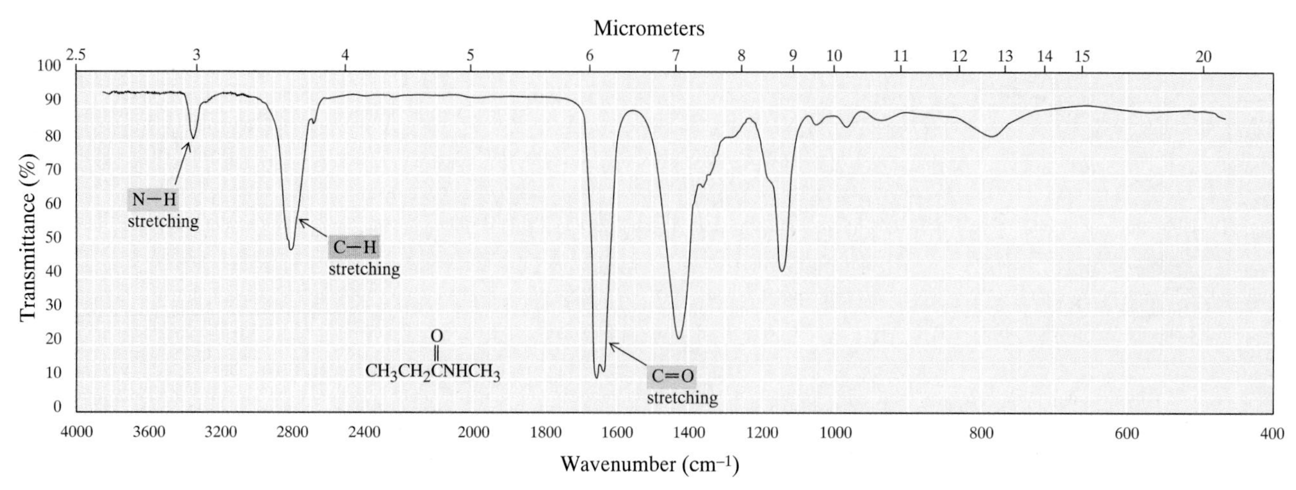

Figure 22.11
Infrared spectrum of *N*-methylpropanamide (a secondary amide).

occurs at 1630–1680 cm^{-1}, a lower wavenumber than for other carbonyl compounds. Primary and secondary amides show N—H stretching in the region 3200–3400 cm^{-1}; primary amides ($RCONH_2$) show two N—H absorptions, whereas secondary amides (RCONHR) show only a single N—H absorption (Figure 22.11).

SUMMARY

The **vibrational infrared** spectrum (Section 22.1A) extends from 4000 to 400 cm^{-1}. Radiation in this region is referred to by its wavenumber $\bar{\nu}$ in reciprocal centimeters (cm^{-1}). To be **infrared-active** (Section 22.1B), a bond must be polar; the more polar it is, the stronger its absorption of IR radiation. There are $3n - 6$ allowed fundamental vibrations for a nonlinear molecule containing n atoms. The simplest vibrations that give rise to absorption of infrared radiation are **stretching** and **bending** vibrations. Stretching may be symmetrical or asymmetrical. Scissoring, rocking, wagging, and twisting are names given to types of bending vibrations.

A **correlation table** is a list of the absorption patterns of functional groups. The intensity of a peak is referred to as strong (s), medium (m), or weak (w). Stretching vibrations for most functional groups appear in the region 3400–1000 cm^{-1}. The region 1000–400 cm^{-1} is called the **fingerprint region** (Section 22.1C).

ADDITIONAL PROBLEMS

22.3 Compound A, molecular formula C_6H_{10}, reacts with H_2/Ni to give compound B, C_6H_{12}. See also the IR spectrum of compound A. From this information about compound A, tell

(a) Its index of hydrogen deficiency.

(b) The number of its rings and/or pi bonds.

(c) What structural feature(s) would account for its index of hydrogen deficiency.

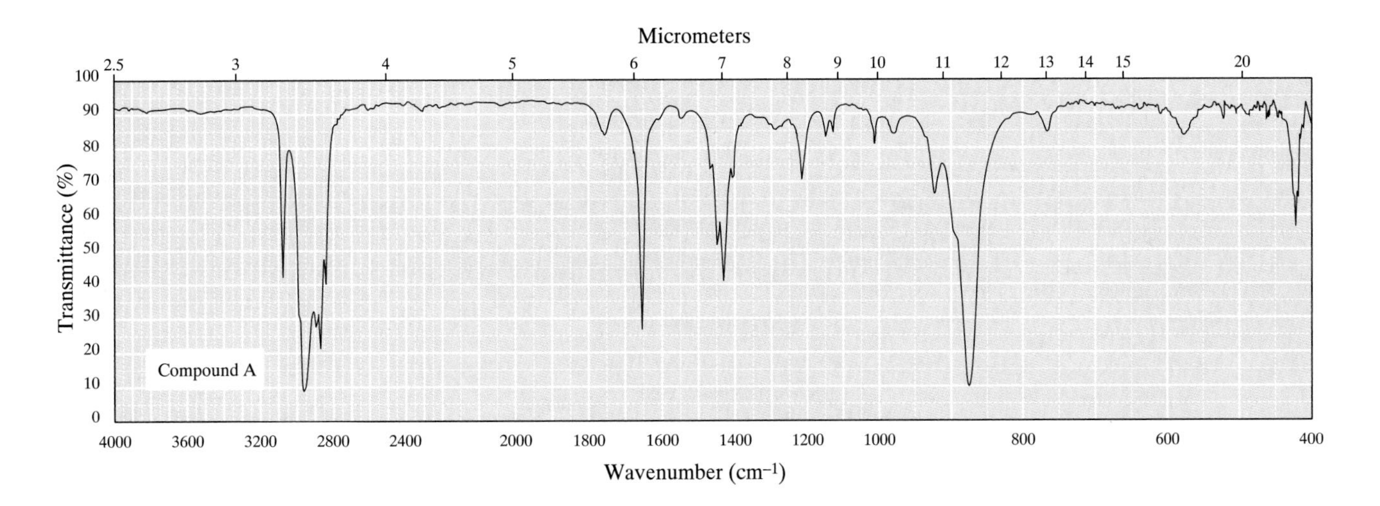

22.4 Compound C, molecular formula C_6H_{12}, reacts with H_2/Ni to give compound D, C_6H_{14}. See also the IR spectrum of compound C. From this information about compound C, tell

(a) Its index of hydrogen deficiency.

(b) The number of its rings and/or pi bonds.

(c) What structural feature(s) would account for its index of hydrogen deficiency.

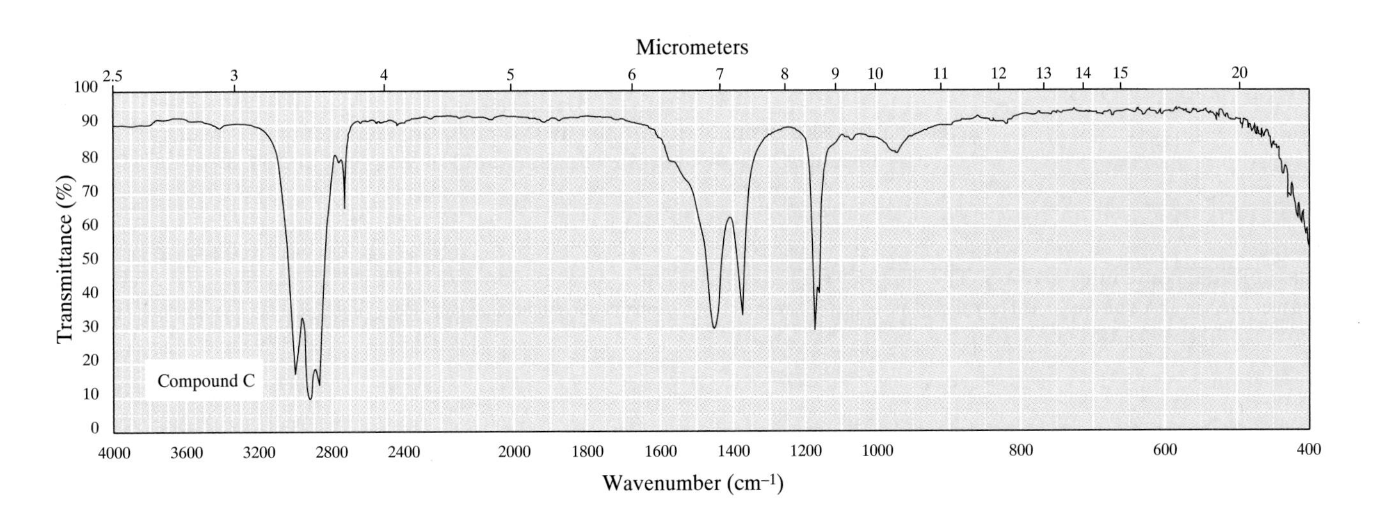

22.5 Following are infrared spectra of compounds E and F. One spectrum is of 1-hexanol and the other of nonane. Assign each compound its correct spectrum.

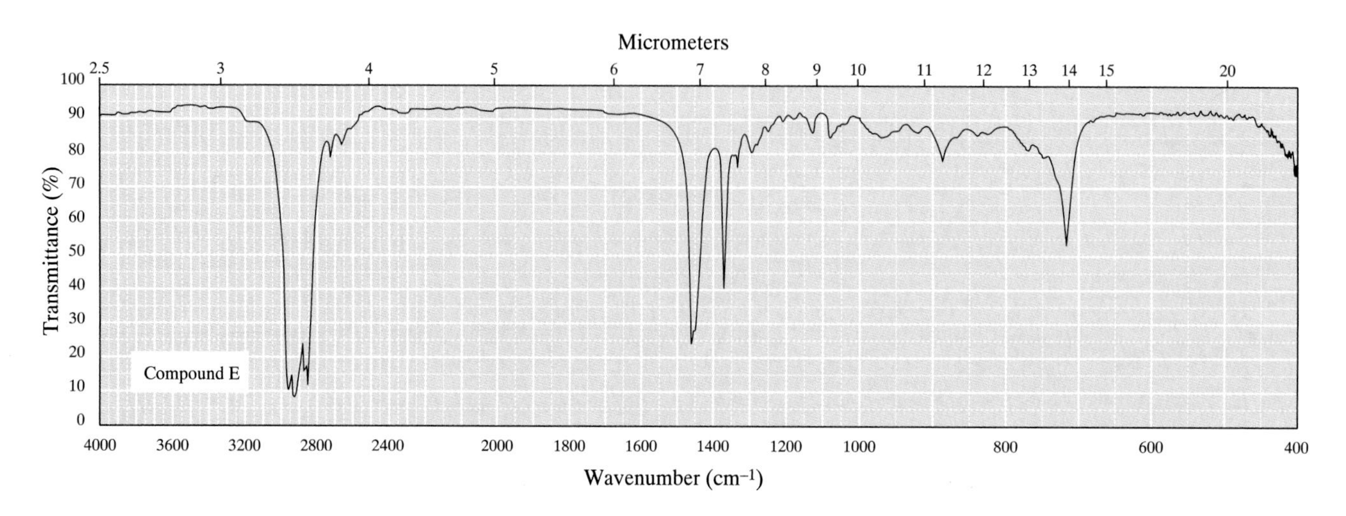

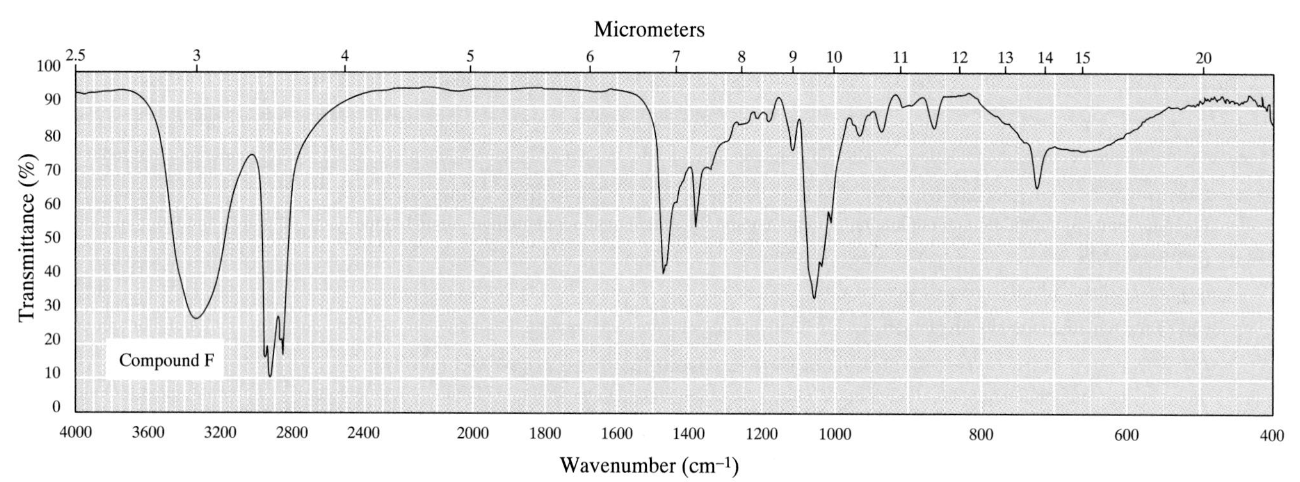

22.6 2-Methyl-1-butanol and *tert*-butyl methyl ether are constitutional isomers of molecular formula $C_5H_{12}O$. Assign each compound its correct infrared spectrum, G or H.

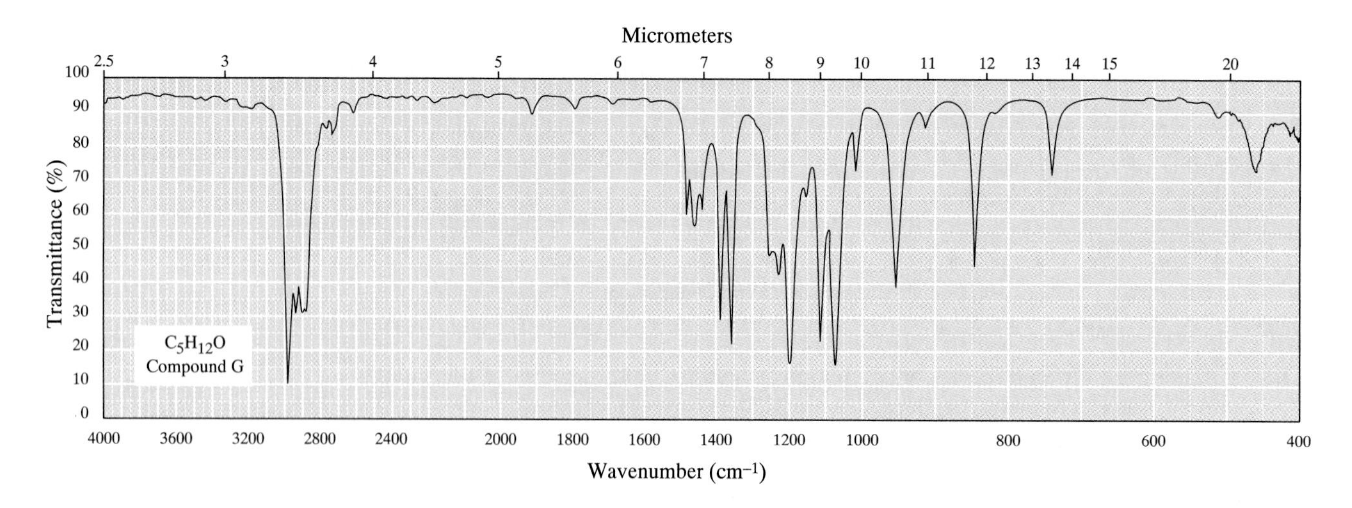

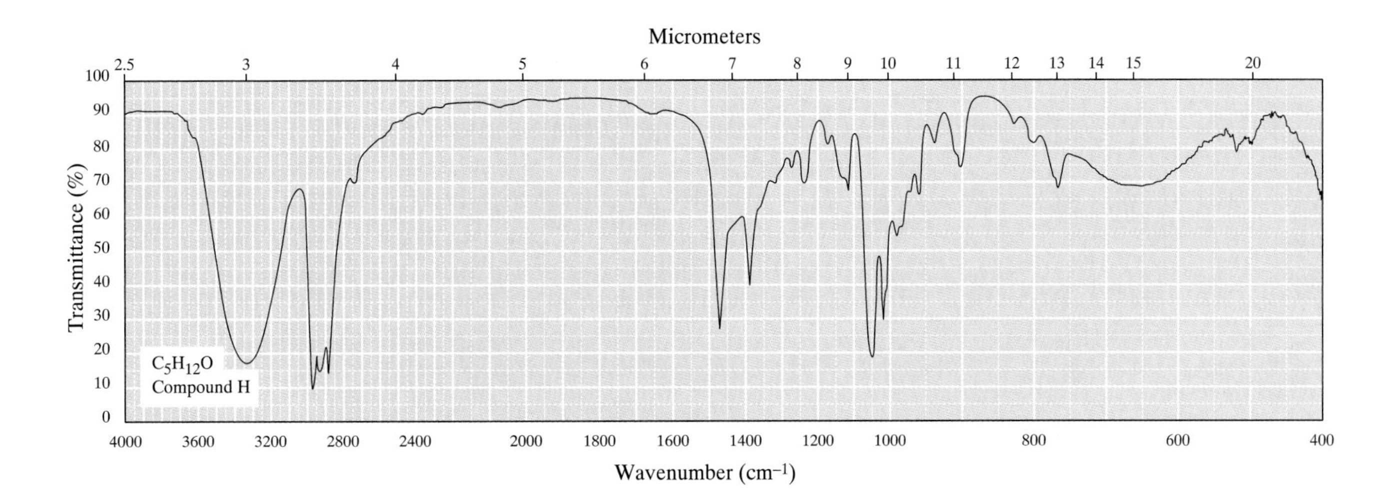

22.7 From examination of the molecular formula and IR spectrum of compound I, $C_9H_{12}O$, tell

(a) Its index of hydrogen deficiency.

(b) The number of its rings and/or pi bonds.

(c) What one structural feature would account for this index of hydrogen deficiency.

(d) What oxygen-containing functional group it contains.

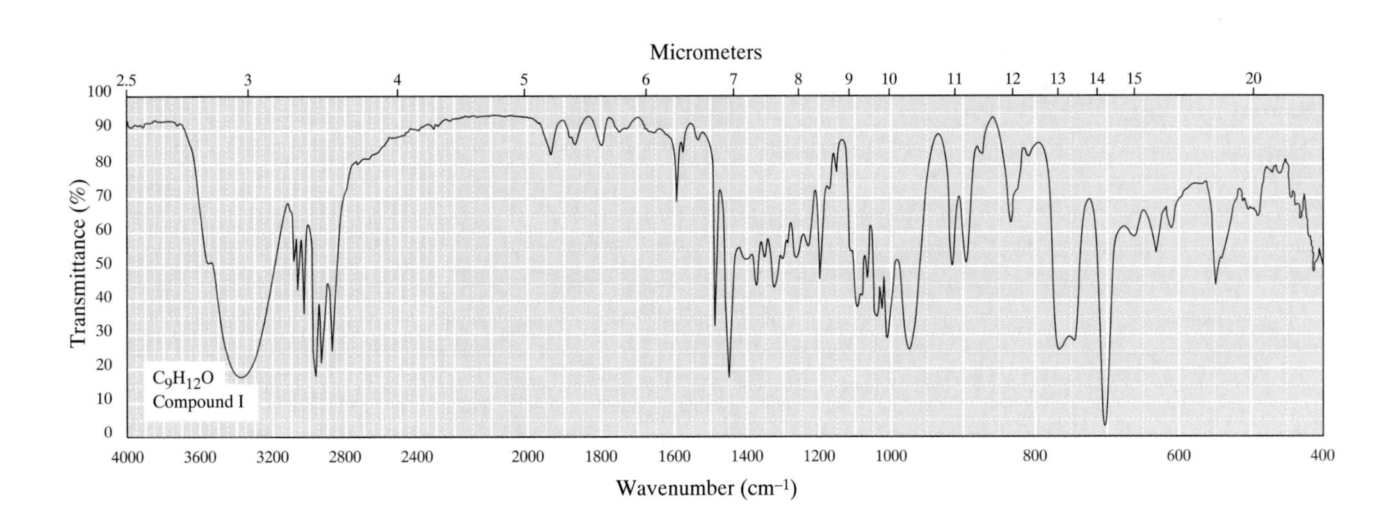

22.8 From examination of the molecular formula and IR spectrum of compound J, $C_5H_{13}N$, tell

(a) Its index of hydrogen deficiency.

(b) The number of its rings and/or pi bonds.

(c) The nitrogen-containing functional group(s) it might contain.

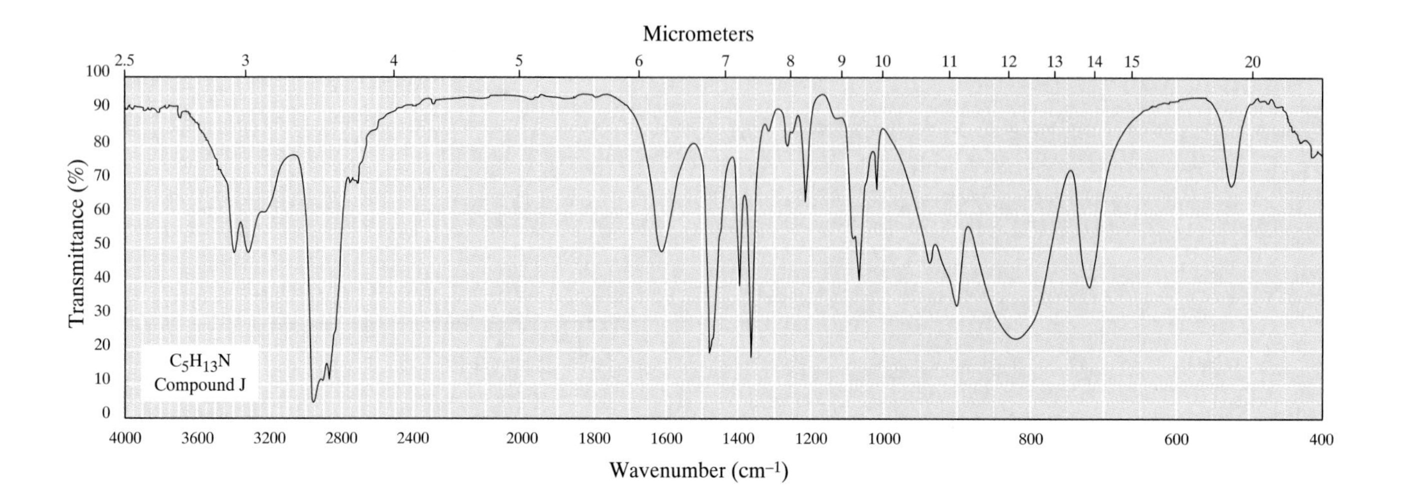

22.9 From examination of the molecular formula and IR spectrum of compound K, $C_{10}H_{12}O$, tell

(a) Its index of hydrogen deficiency.

(b) The number of its rings and/or pi bonds.

(c) What structural features would account for this index of hydrogen deficiency.

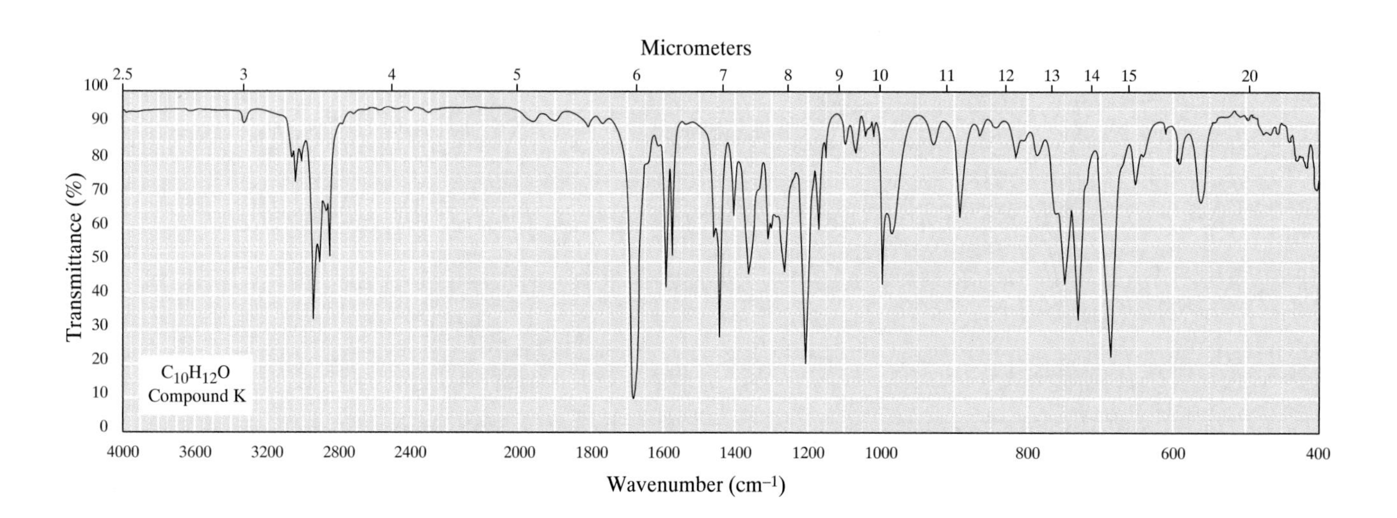

22.10 From examination of the molecular formula and IR spectrum of compound L, $C_7H_{14}O_2$, tell

(a) Its index of hydrogen deficiency.

(b) The number of its rings and/or pi bonds.

(c) The oxygen-containing functional group(s) it might contain.

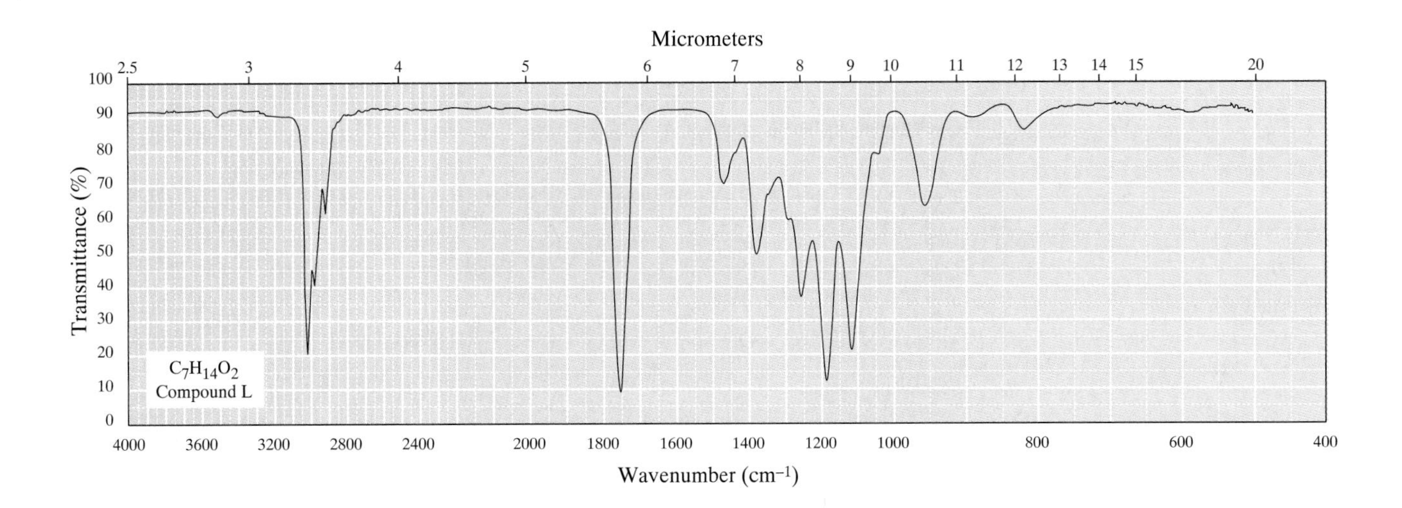

22.11 From examination of the molecular formula and IR spectrum of compound M, $C_6H_{13}NO$, tell

(a) Its index of hydrogen deficiency.

(b) The number of its rings and/or pi bonds.

(c) The oxygen and nitrogen-containing functional group(s).

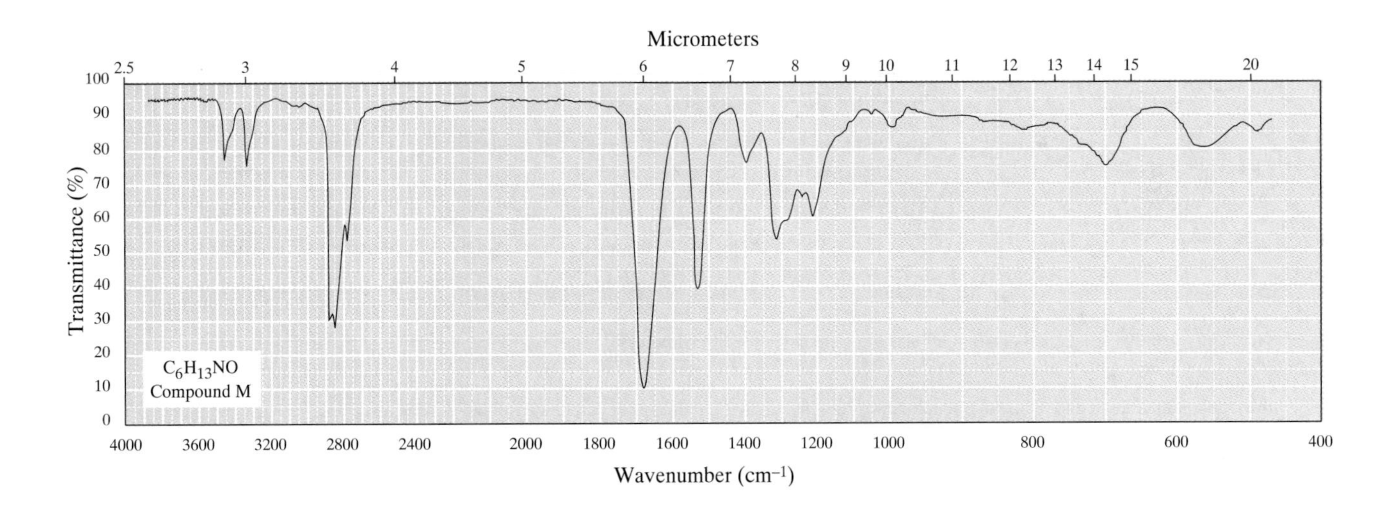

22.12 Show how IR spectroscopy can be used to distinguish between the compounds in each set.

(a) 1-Butanol and diethyl ether

(b) Butanoic acid and 1-butanol

(c) Butanoic acid and 2-butanone

(d) Butanal and 1-butene

(e) 2-Butanone and 2-butanol

(f) Butane and 2-butene

22.13 For each set of compounds, list one major feature that appears in the IR spectrum of one compound but not the other. Your answer should state what type of bond vibration is responsible for the spectral feature you list and its approximate position in the IR spectrum.

(a) $C_6H_5\overset{O}{\overset{\|}{C}}H$ and $C_6H_5\overset{O}{\overset{\|}{C}}OH$

(b) $C_6H_{11}\overset{O}{\overset{\|}{C}}N(CH_3)_2$ and $C_6H_{11}CH_2N(CH_3)_2$

(c) (six-membered lactone) and $HO(CH_2)_4\overset{O}{\overset{\|}{C}}OH$

(d) $C_6H_{11}\overset{O}{\overset{\|}{C}}NH_2$ and $C_6H_{11}\overset{O}{\overset{\|}{C}}N(CH_3)_2$

22.14 Following is an infrared spectrum and a structural formula for methyl salicylate, the fragrant component of oil of wintergreen. The spectrum was recorded using the pure liquid between KBr plates. The dashed line between O of the C=O group and H of the —OH group indicates intramolecular hydrogen bonding. On this spectrum, locate the absorption peak(s) due to

(a) O—H stretching of the hydrogen-bonded —OH group (very broad and of medium intensity).

(b) C—H stretching of the aromatic ring (sharp and of weak intensity).

(c) C=O stretching of the ester group (sharp and of strong intensity).

(d) C=C stretching of the aromatic ring (sharp and of medium intensity).

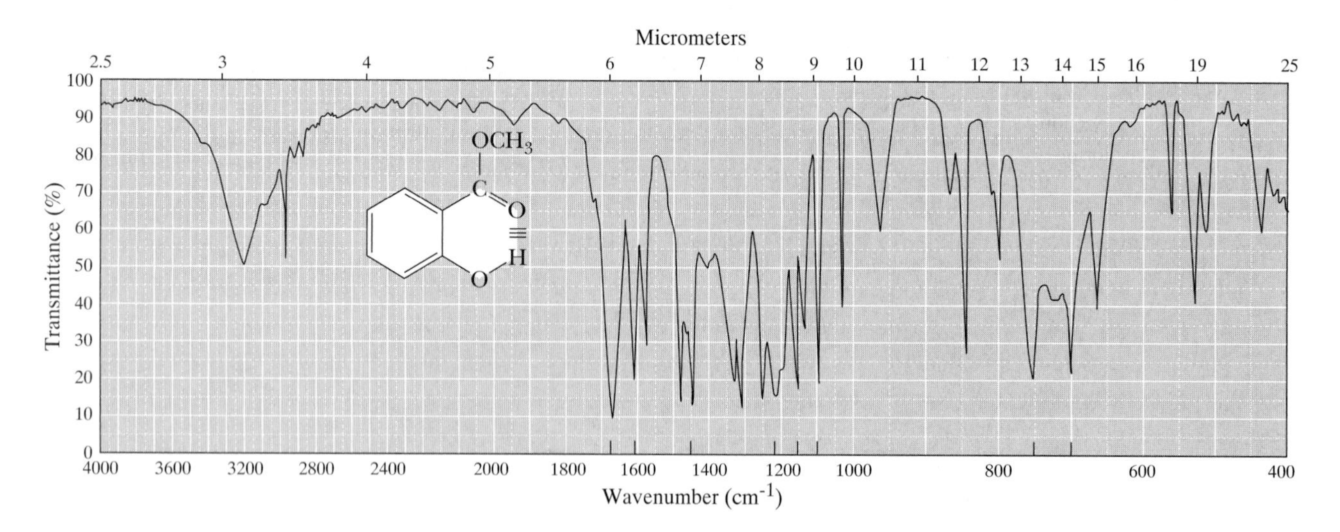

Appendix 1

Reagents and Their Uses

Reagent	Common Names and Uses
$Ag(NH_3)_2^+$	Tollens' reagent With aldehydes; oxidation to carboxylic acids (Section 11.9A) With aldoses and ketones; oxidation to aldonic acids (Section 16.4C)
Ag_2O	Silver oxide With aldehydes; oxidation to carboxylic acids (Section 11.9A)
$AlCl_3$	Aluminum chloride Lewis acid catalyst for Friedel-Crafts alkylations and acylations (Section 9.7)
$ArCO_3H$	A peroxycarboxylic acid; alternatively, a peracid With alkenes; stereoselective oxidation to epoxides (Section 8.6B)
BrCN	Cyanogen bromide With polypeptides; cleavage of peptide bonds at carboxyl group of methionine (Section 18.4B)
Br_2	Bromine With alkenes; stereoselective addition (Section 6.3C) With arenes and a Lewis acid catalyst; bromination (Section 9.7A)
C_6H_5NCS	Phenyl isothiocyanate With polypeptides; Edman degradation of *N*-terminal amino acid (Section 18.4B)
$(C_6H_5CO_2)_2$	Benzoyl peroxide With alkenes; initiation of radical polymerization (Section 15.5A)
Cl_2	Chlorine With alkenes; stereoselective addition (Section 6.3C) With arenes and a Lewis acid catalyst; chlorination (Section 9.7A)
CO_2	Carbon dioxide With Grignard reagents; formation of carboxylic acids (Section 11.5B)
$FeCl_3$	Iron(III) chloride; alternatively, ferric chloride Lewis acid catalyst for chlorination and bromination of arenes (Section 9.7)
HBr	Hydrogen bromide; when in aqueous solution, hydrobromic acid
HBr	With alkenes; regioselective addition (Section 6.3A) With 1°, 2°, and 3° alcohols; formation of alkyl bromides (Section 8.4D)
HCl	Hydrogen chloride; when in aqueous solution, hydrochloric acid With alkenes; regioselective addition (Section 6.3A) With 2° and 3° alkyl halides; formation of alkyl chlorides (Section 8.4D) With amines; formation of water-soluble salts (Section 10.5)
HI	Hydrogen iodide; when in aqueous solution, hydriodic acid With alkenes; regioselective addition (Section 6.3A) With 1°, 2°, and 3° alcohols; formation of alkyl iodides (Section 8.4D)
HNO_3	Nitric acid With arenes; nitration (Section 9.7B)
H_2	Hydrogen; alternatively, hydrogen gas With alkenes; catalytic reduction to alkanes (Section 6.5A) With nitro groups; reduction to 1° amines (Section 9.7B) With aldehydes and ketones; catalytic reduction to alcohols (Section 11.10A) With imines; reduction to amines (Section 11.10C) With aldoses and ketoses; catalytic reduction (Section 16.4B)
H_2CrO_4	Chromic acid With 1° alcohols; oxidation to carboxylic acids (Section 8.4F) With 2° alcohols; oxidation to ketones (Section 8.4F) With aldehydes; oxidation to carboxylic acids (Section 11.9A) With ketones; oxidation to carboxylic acids (Section 11.9A) With alkylarenes; oxidation to arenecarboxylic acids (Section 9.5)

Reagent	Common Names and Uses
H_2O_2	Hydrogen peroxide With aldehydes; oxidation to carboxylic acids (Section 11.9A)
H_2SO_4	Sulfuric acid With alkenes; catalyst for regioselective hydration (Section 6.3B) With alcohols; catalyst for regioselective dehydration (Section 8.4E) With arenes; sulfonation (Section 9.7B)
H_3PO_4	Phosphoric acid With alcohols; catalyst for regioselective dehydration (Section 8.4E) With alkenes and arenes; catalyst for alkylation of arenes (Section 9.7D) With alcohols and arenes; catalyst for alkylation of arenes (Section 9.7D)
K	Potassium; alternatively, potassium metal With alcohols; formation of potassium alkoxides (Section 8.4C)
Li	Lithium; alternatively, lithium metal With alcohols; formation of lithium alkoxides (Section 8.4C)
$LiAlH_4$	Lithium aluminum hydride (LAH) With aldehydes and ketones; reduction to alcohols (Section 11.10B) With carboxylic acids; reduction to primary alcohols (Section 12.5A) With esters; reduction to two alcohols (Section 13.8A) With amides; reduction to amines (Section 13.8B)
Mg	Magnesium; alternatively, magnesium metal With alkyl and aryl halides; formation of Grignard reagents (Section 11.5A)
Na	Sodium; alternatively, sodium metal With alcohols; formation of sodium alkoxides (Section 8.4C)
$NaBH_4$	Sodium borohydride With aldehydes and ketones; reduction to alcohols (Section 11.10B)
NH_3	Ammonia With aldehydes and ketones; formation of imines (Section 11.7) With acid chlorides; formation of amides (Section 13.5A) With acid anhydrides; formation of amides (Section 13.5B) With esters; formation of amides (Section 13.5C)
O_2	Oxygen; alternatively, molecular oxygen With alkanes; combustion (Section 3.9) With thiols; oxidation to disulfides (Section 8.8B) With aldehydes; oxidation to carboxylic acids (Section 11.9A)
OsO_4	Osmium tetroxide With alkenes; syn stereoselective oxidation to a glycol (Section 6.4B)
PCC	Pyridinium chlorochromate With 1° alcohols; oxidation to aldehydes (Section 8.4F) With 2° alcohols; oxidation to ketones (Section 8.4F)
RCO_3H	A peroxycarboxylic acid; alternatively, a peracid With alkenes; stereoselective oxidation to epoxides (Section 8.6B)
RMgX	An organomagnesium reagent; alternatively, a Grignard reagent With aldehydes and ketones; formation of alcohols (Section 11.5B) With carbon dioxide; formation of carboxylic acids (Section 11.5B) With ethylene oxide; formation of a 1° alcohol (Section 11.5B) With esters; formation of alcohols (Section 13.7)
R_2NH	A secondary amine With acid chlorides; formation of amides (Section 13.5A) With acid anhydrides; formation of amides (Section 13.5B) With esters; formation of amides (Section 13.5C)
RNH_2	A primary amine With aldehydes and ketones; formation of imines (Section 11.7) With acid chlorides; formation of amides (Section 13.5A) With acid anhydrides; formation of amides (Section 13.5B) With esters; formation of amides (Section 13.5C)
$SOCl_2$	Thionyl chloride With alcohols; formation of alkyl chlorides (Section 8.4D) With carboxylic acids; formation of acid chlorides (Section 12.7)

APPENDIX 2

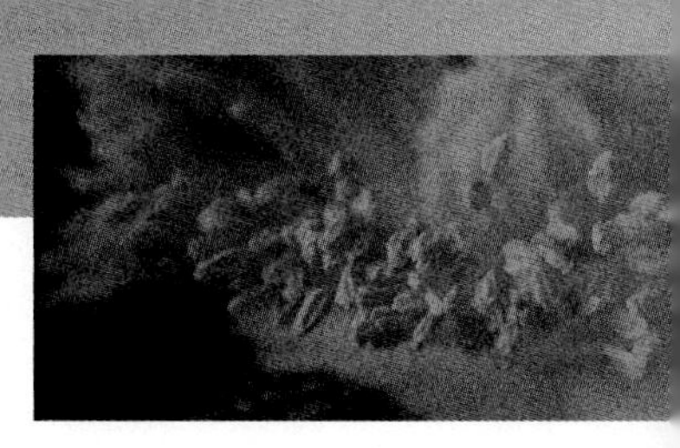

ACID IONIZATION CONSTANTS FOR THE MAJOR CLASSES OF ORGANIC ACIDS

Name and Example	Typical pK_a	Name and Example	Typical pK_a
carboxylic acid $CH_3C(=O)O—\mathbf{H}$	3–5	alkylammonium ion $(CH_3CH_2)_3N^+—\mathbf{H}$	10–12
arylammonium ion $C_6H_5—NH_2^+—\mathbf{H}$	4–5	β-ketoester $CH_3C(=O)C\mathbf{H}C(=O)OCH_2CH_3$	11
thiol $CH_3CH_2S—\mathbf{H}$	8–12	water $HO—\mathbf{H}$	15.7
phenol $C_6H_5—O\mathbf{H}$	9–10	alcohol $CH_3CH_2O—\mathbf{H}$	15–19
ammonium ion $NH_3—\mathbf{H}^+$	9.24	aldehyde, ketone $CH_3C(=O)CH_2—\mathbf{H}$	18–20
β-diketone $CH_3C(=O)C\mathbf{H}C(=O)CH_3$	10	ester $\mathbf{H}—CH_2C(=O)OCH_2CH_3$	23–25
		ammonia $NH_2—\mathbf{H}$	38

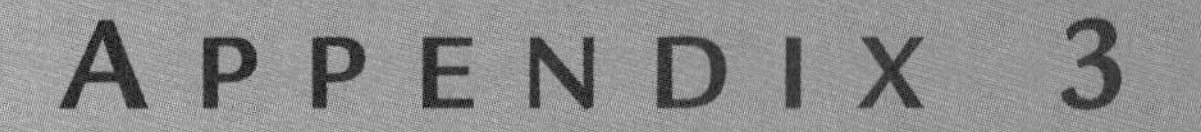

CHARACTERISTIC ^{1}H-NMR CHEMICAL SHIFTS

Type of Hydrogen (R = alkyl, Ar = aryl)	Chemical Shift (δ)*	Type of Hydrogen (R = alkyl, Ar = aryl)	Chemical Shift (δ)*
$(C\mathbf{H}_3)_4Si$	0 (by definition)	$RC(=O)OC\mathbf{H}_3$	3.7–3.9
$RC\mathbf{H}_3$	0.8–1.0	$RC(=O)OC\mathbf{H}_2R$	4.1–4.7
$RC\mathbf{H}_2R$	1.2–1.4	$RC\mathbf{H}_2I$	3.1–3.3
$R_3C\mathbf{H}$	1.4–1.7	$RC\mathbf{H}_2Br$	3.4–3.6
$R_2C{=}CRC\mathbf{H}R_2$	1.6–2.6	$RC\mathbf{H}_2Cl$	3.6–3.8
$RC{\equiv}C\mathbf{H}$	2.0–3.0	$RC\mathbf{H}_2F$	4.4–4.5
$ArC\mathbf{H}_3$	2.2–2.5	$ArO\mathbf{H}$	4.5–4.7
$ArC\mathbf{H}_2R$	2.3–2.8	$R_2C{=}C\mathbf{H}_2$	4.6–5.0
$RO\mathbf{H}$	0.5–6.0	$R_2C{=}C\mathbf{H}R$	5.0–5.7
$RC\mathbf{H}_2OH$	3.4–4.0	$Ar\mathbf{H}$	6.5–8.5
$RC\mathbf{H}_2OR$	3.3–4.0	$RC(=O)\mathbf{H}$	9.5–10.1
$R_2N\mathbf{H}$	0.5–5.0	$RC(=O)O\mathbf{H}$	10–13
$RC(=O)C\mathbf{H}_3$	2.1–2.3		
$RC(=O)C\mathbf{H}_2R$	2.2–2.6		

* Values are approximate. Other atoms within the molecule may cause the signal to appear outside these ranges.

APPENDIX 4

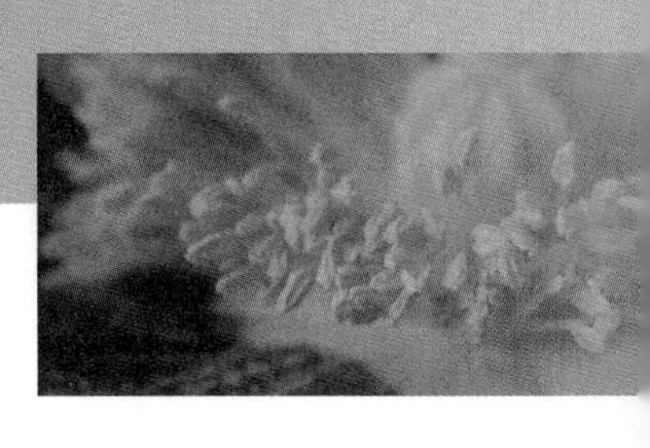

CHARACTERISTIC ^{13}C-NMR CHEMICAL SHIFTS

Type of Carbon	Chemical Shift (δ)	Type of Carbon	Chemical Shift (δ)
RCH_3	0–40	C_6H_5**C**—R (benzene ring carbon)	110–160
RCH_2R	15–55		
R_3CH	20–60		
RCH_2I	0–40		
RCH_2Br	25–65	R**C**(=O)OR	160–180
RCH_2Cl	35–80		
R_3**C**OH	40–80	R**C**(=O)NR_2	165–180
R_3**C**OR	40–80		
R**C**≡CR	65–85		
R_2**C**=CR_2	100–150	R**C**(=O)OH	175–185
		R**C**(=O)H, R**C**(=O)R	180–210

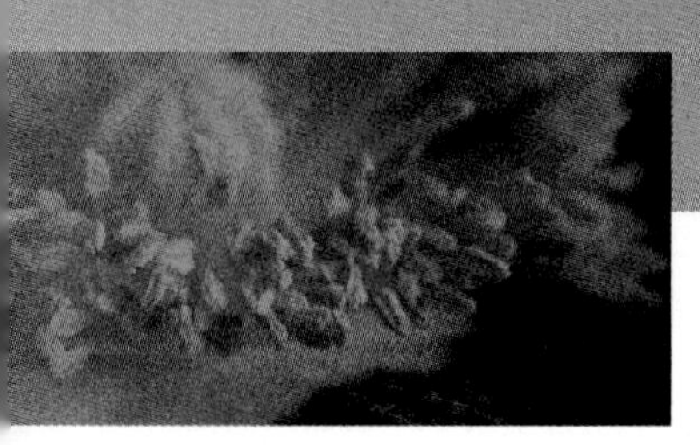
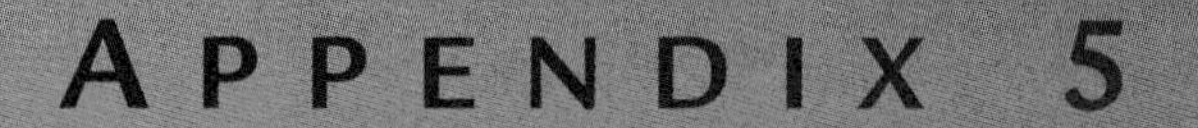

APPENDIX 5

CHARACTERISTIC INFRARED ABSORPTION FREQUENCIES

Bonding		Frequency (cm^{-1})	Intensity*
C—H	alkane	2850–3000	w-m
	alkene	3000–3100	w-m
	alkyne	3300	s
	aromatic	3000–3100	s
C—C	alkane	(not interpretatively useful)	
C=C	alkene	1600–1680	w-m
	aromatic	1450 and 1600	w-m
C—O	alcohol, ether, ester carboxylic acid	1050–1250	s
C=O	amide	1630–1680	s
	carboxylic acid	1700–1750	s
	ketone	1705–1780	s
	aldehyde	1705–1740	s
	ester	1735–1800	s
O—H	alcohol, phenol hydrogen bonded	3200–3500	m
	carboxylic acid	2400–3400	m
N—H	amine and amide	3100–3500	m-s

* m = medium, s = strong, w = weak.

GLOSSARY

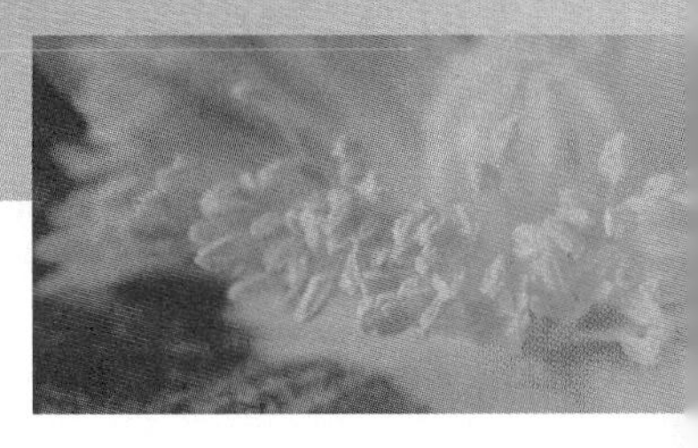

Acetal (Section 11.6) A molecule containing two —OR or —OAr groups bonded to the same carbon.

Aceto group (Section 12.2B) A CH_3CO— group.

Achiral (Section 4.2) An object that lacks chirality; an object that has no handedness.

Activating group (Section 9.8A) Any substituent on a benzene ring that causes the rate of electrophilic aromatic substitution to be greater than that for benzene.

Activation energy (Section 6.2A) The difference in potential energy between reactants and the transition state.

Acid halide (Sections 9.7C, 13.1A) A derivative of a carboxylic acid in which the —OH of the carboxyl group is replaced by a halogen, most commonly chloride.

Alcohol (Section 8.1A) A compound containing an —OH (hydroxyl) group bonded to an sp^3 hybridized carbon.

Alcoholic fermentation (Section 20.7B) A metabolic pathway that converts glucose to two molecules of ethanol and two molecules of CO_2.

Aldehyde (Section 11.1) A compound containing a carbonyl group bonded to hydrogen (a CHO group).

Alditol (Section 16.4B) The product formed when the C=O group of a monosaccharide is reduced to a CHOH group.

Aldol reaction (Section 14.2) A carbonyl condensation reaction between two aldehydes or ketones to give a *β*-hydroxyaldehyde or *β*-hydroxyketone.

Aldose (Section 16.2A) A monosaccharide containing an aldehyde group.

Aliphatic amine (Section 10.1) An amine in which nitrogen is bonded only to alkyl groups.

Aliphatic hydrocarbon (Section 3.1) An alternative word to describe an alkane.

Alkane (Section 3.1) A saturated hydrocarbon whose carbon atoms are arranged in an open chain.

Alkene (Section 5.1) An unsaturated hydrocarbon that contains a carbon-carbon double bond.

Alkoxy group (Section 8.2B) An —OR group, where R is an alkyl group.

Alkyl group (Section 3.3A) A group derived by removing a hydrogen from an alkane; given the symbol R—.

Alkyl halide (Section 7.1) A compound containing a halogen atom covalently bonded to an alkyl group; given the symbol RX.

Alkyne (Section 5.1) An unsaturated hydrocarbon that contains a carbon-carbon triple bond.

Amino acid (Section 18.1A) A compound that contains both an amino group and a carboxyl group.

α-Amino acid (Section 18.1A) An amino acid in which the amino group is on the carbon adjacent to the carboxyl group.

Amino group (Section 1.7D) An sp^3 hybridized nitrogen atom bonded to one, two, or three carbon groups.

Amorphous domains (Section 15.3) Disordered, noncrystalline regions in the solid state of a polymer.

Anabolic steroid (Section 17.4A) A steroid hormone, such as testosterone, which promotes tissue and muscle growth and development.

Androgen (Section 17.4A) A steroid hormone, such as testosterone, which mediates the development and sexual characteristics of males.

Angle strain (Section 3.6A) The strain that arises when a bond angle is either compressed or expanded compared to its normal value.

Anion (Section 1.2B) An atom or group of atoms bearing a negative charge.

Anomeric carbon (Section 16.2A) The hemiacetal carbon of the cyclic form of a monosaccharide.

Anomers (Section 16.2A) Monosaccharides that differ in configuration only at their anomeric carbons.

Anti addition (Section 6.3C) Addition of atoms or groups of atoms from opposite sides or faces of a carbon-carbon double bond.

Aprotic solvent (Section 7.4D) A solvent that cannot serve as a hydrogen bond donor, as for example acetone, diethyl ether, and dichloromethane.

Ar— (Section 9.1) The symbol used for an aryl group, by analogy with R— for an alkyl group.

Aramid (Section 15.4A) A poly**ar**omatic **amid**e; a polymer in which the monomer units are an aromatic diamine and an aromatic dicarboxylic acid.

Arene (Section 5.1) A compound containing one or more benzene rings.

Aromatic amine (Section 10.1) An amine in which nitrogen is bonded to one or more aryl groups.

Aromatic compound (Section 9.1) A term used to classify benzene and its derivatives.

Aryl group (Section 9.1) A group derived from an aromatic compound (an arene) by removal of an H; given the symbol Ar—.

Average degree of polymerization, *n* (Section 15.2) A subscript placed outside the parentheses of the simplest nonredundant unit of a polymer to indicate that the unit repeats *n* times in the polymer.

Axial position (Section 3.6B) A position on a chair conformation of a cyclohexane ring that extends from the ring parallel to the imaginary axis of the ring.

Axial-axial interactions (Section 3.6B) Interactions between groups in parallel axial positions on the same side of a chair conformation of a cyclohexane ring.

Benzyl group (Section 9.3A) $C_6H_5CH_2$—, the alkyl group derived by removing a hydrogen from the methyl group of toluene.

Benzylic carbon (Section 9.5) An sp^3 hybridized carbon bonded to a benzene ring.

Bile acid (Section 17.4A) A cholesterol-derived detergent molecule, such as cholic acid, which is secreted by the gallbladder into the intestines to assist in the absorption of dietary lipids.

Bimolecular reaction (Section 7.3A) A reaction in which two species are involved in the reaction leading to the transition state of the rate-limiting step.

Boat conformation (Section 3.6B) A puckered conformation of a cyclohexane ring in which carbons 1 and 4 of the ring are bent toward each other.

Bond length (Section 1.2C) The distance between atoms in a covalent bond.

Bonding electrons (Section 1.2D) Valence electrons involved in forming a covalent bond, that is, shared electrons.

Brønsted-Lowry acid (Section 2.1) A proton donor.

Brønsted-Lowry base (Section 2.1) A proton acceptor.

Carbanion (Section 11.5A) An anion in which carbon has an unshared pair of electrons and bears a negative charge.

Carbocation (Section 6.3A) A species containing a carbon atom with only three bonds to it and bearing a positive charge.

Carbohydrate (Section 16.1) A polyhydroxyaldehyde or polyhydroxyketone or a substance that gives these compounds on hydrolysis.

α-Carbon (Section 11.8A) A carbon atom adjacent to a carbonyl group.

Carbonyl group (Section 1.7B) A C=O group.

Carboxyl group (Section 1.7C) A $—CO_2H$ group.

Carboxylic acid anhydride (Section 13.1B) A compound in which two acyl groups are bonded to an oxygen.

Cation (Section 1.2B) An atom or group of atoms bearing a positive charge.

Chain-growth polymerization (Section 15.5) A polymerization that involves sequential addition reactions, either to unsaturated monomers or to monomers possessing other reactive functional groups.

Chain-initiation step (Section 15.5A) In radical polymerization, the formation of radicals from molecules containing only paired electrons.

Chain-propagation step (Section 15.5A) In radical polymerization, a reaction of a radical and a molecule to give a new radical.

Chain-termination step (Section 15.5A) In radical polymerization, a reaction in which two radicals combine to form a covalent bond.

Chain-transfer reaction (Section 15.5A) In radical polymerization, the transfer of reactivity of an endgroup from one chain to another during a polymerization.

Chair conformation (Section 3.6B) The most stable puckered conformation of a cyclohexane ring; all bond angles are approximately 109.5°, and bonds to all adjacent carbons are staggered.

Chemical shift, δ (Section 21.4) The position of a signal on an NMR spectrum relative to the signal of tetramethylsilane (TMS); expressed in delta (δ) units, where 1 δ equals 1 ppm.

Chiral (Section 4.2) From the Greek, *cheir,* meaning hand; objects that are not superposable on their mirror images.

Chromatin (Section 19.2C) A complex formed between negatively charged DNA molecules and positively charged histones.

Circular DNA (Section 19.2C) A type of double-stranded DNA in which the 5′ and 3′ ends of each strand are joined by phosphodiester groups.

Cis (Section 3.7) A prefix meaning on the same side.

Cis,trans isomerism (Section 5.1C) Isomers that have the same order of attachment of their atoms but a different arrangement of their atoms in space due to the presence of either a ring (Chapter 3) or a carbon-carbon double bond (Chapter 5).

Claisen condensation (Section 14.3A) A carbonyl condensation reaction between two esters to give a β-ketoester.

Codon (Section 19.4A) A triplet of nucleotides on mRNA that directs incorporation of a specific amino acid into a polypeptide sequence.

Coenzyme (Section 20.1) A low-molecular-weight, nonprotein molecule or ion that binds reversibly to an enzyme, functions as a second substrate for the enzyme, and is regenerated by further reaction.

Condensation polymerization (Section 15.4) A polymerization in which chain growth occurs in a stepwise manner between difunctional monomers. Also called step-growth polymerization.

Conformation (Section 3.6A) Any three-dimensional arrangement of atoms in a molecule that results by rotation about a single bond.

Conjugate acid (Section 2.1) The species formed when a base accepts a proton.
Conjugate base (Section 2.1) The species formed when an acid donates a proton.
Constitutional isomers (Section 3.2) Compounds with the same molecular formula but a different order of attachment of their atoms.
Contributing structure (Section 1.5A) Representations of a molecule or ion that differ only in the distribution of valence electrons.
Covalent bond (Section 1.2B) A chemical bond formed by sharing one or more pairs of electrons.
Crossed aldol reaction (Section 14.2) An aldol reaction between two different aldehydes and/or ketones.
Crossed Claisen condensation (Section 14.3C) A Claisen condensation between two different esters.
Crystalline domains (Section 15.3) Ordered crystalline regions in the solid state of a polymer; also called crystallites.
Curved arrow (Section 1.5B) A symbol used to show the redistribution of valence electrons.
Cyclic ether (Section 8.2B) An ether in which the ether oxygen is one of the atoms of a ring.
Cycloalkane (Section 3.4) A saturated hydrocarbon that contains carbon atoms joined to form a ring.

Deactivating group (Section 9.8A) Any substituent on a benzene ring that causes the rate of electrophilic aromatic substitution to be lower than that for benzene.
Decarboxylation (Section 12.8) Loss of CO_2 from an organic molecule.
Dehydration (Section 8.4E) Elimination of a molecule of water from a compound.
Dehydrohalogenation (Section 7.6) Removal of —H and —X from adjacent carbons; a type of β-elimination.
Dextrorotatory (Section 4.7B) Rotation of the plane of polarized light in a polarimeter to the right.
Diastereomers (Section 4.1) Stereoisomers that are not mirror images of each other; refers to relationships among objects.
Dieckmann condensation (Section 14.3B) An intramolecular Claisen condensation of an ester of a dicarboxylic acid to give a five- or six-membered ring.
Dipeptide (Section 18.3) A molecule containing two amino acid units joined by a peptide bond.
Disaccharide (Section 16.7A) A carbohydrate containing two monosaccharide units joined by a glycoside bond.
Dispersion forces (Section 3.8) Very weak intermolecular forces of attraction.
Disulfide bond (Section 18.5C) A covalent bond between two sulfur atoms; an —S—S— bond.
Double-headed arrow (Section 1.5A) A symbol used to connect contributing structures.
Double helix (Section 19.2B) A type of secondary structure of DNA molecules in which two antiparallel polynucleotide strands are coiled in a right-handed manner about the same axis.

E (Section 5.2C) From the German *entgegen*, meaning opposite; specifies that groups of higher priority on the carbons of a double bond are on opposite sides.
E,Z system (Section 5.2C) A system to specify the configuration of groups about a carbon-carbon double bond.
Eclipsed conformation (Section 3.6A) A conformation about a carbon-carbon single bond where the atoms on one carbon are as close as possible to the atoms on the adjacent carbon.
Edman degradation (Section 18.4B) A method for selectively cleaving and identifying the *N*-terminal amino acid of a polypeptide chain.
Elastomer (Section 15.3) A material that, when stretched or otherwise distorted, returns to its original shape when the distorting force is released.
Electromagnetic radiation (Section 21.1) Light and other forms of radiant energy.
Electronegativity (Section 1.2C) A measure of the force of an atom's attraction for the electrons it shares in a chemical bond with another atom.
Electrophile (Section 6.3A) Any molecule or ion that can accept a pair of electrons to form a new covalent bond; a Lewis acid.
Electrophilic aromatic substitution (Section 9.7) A reaction in which there is substitution of an electrophile for a hydrogen on an aromatic ring.
Electrophoresis (Section 18.2D) The process of separating compounds on the basis of their electric charge.
β-Elimination reaction (Section 7.6) Removal of atoms or groups of atoms from two adjacent carbon atoms, as for example removal of H and X from an alkyl halide or H and OH from an alcohol to form a carbon-carbon double bond.
Enantiomers (Section 4.1) Stereoisomers that are nonsuperposable mirror images; refers to a relationship between pairs of objects.
3′ End (Section 19.2A) The end of a polynucleotide at which the 3′-OH of the terminal pentose unit is free.
5′ End (Section 19.2A) The end of a polynucleotide at which the 5′-OH of the terminal pentose unit is free.
Endothermic reaction (Section 6.2A) A reaction in which the energy of the products is higher than the energy of the reactants; a reaction in which heat is absorbed.
Enol (Section 11.8A) A molecule containing an —OH group bonded to a carbon of a carbon-carbon double bond.
Enolate anion (Section 14.1A) An anion formed by removal of an α-hydrogen from a carbonyl-containing compound.
Epoxide (Section 8.6A) A cyclic ether in which oxygen is one atom of a three-membered ring.

Epoxy resin (Section 15.4E) A material prepared by a polymerization in which one monomer contains at least two epoxy groups.

Equatorial position (Section 3.6B) A position on a chair conformation of a cyclohexane ring that extends from the ring roughly perpendicular to the imaginary axis of the ring.

Equivalent hydrogens (Section 21.6) Hydrogens that have the same chemical environment.

Estrogen (Section 17.4A) A steroid hormone, such as estradiol, that mediates the development and sexual characteristics in females.

Ether (Section 8.1B) A compound containing an oxygen atom bonded to two carbon atoms.

Exothermic reaction (Section 6.2A) A reaction in which the energy of the products is lower than the energy of the reactants; a reaction in which heat is liberated.

Fat (Section 17.1B) A triglyceride that is semisolid or solid at room temperature.

Fatty acid (Section 17.1A) A long, unbranched-chain carboxylic acid, most commonly of 12 to 20 carbons, derived from the hydrolysis of animal fats, vegetable oils, or the phospholipids of biological membranes.

Fingerprint region (Section 22.1C) The portion of the vibrational infrared region that extends from 1600 to 400 cm^{-1}.

Fischer esterification (Section 12.6A) The process of forming an ester by refluxing a carboxylic acid and an alcohol in the presence of an acid catalyst, commonly sulfuric acid.

Fischer projection (Section 16.2D) A two-dimensional representation for showing the configuration of a stereocenter; horizontal lines represent bonds projecting forward from the stereocenter, and vertical lines represent bonds projecting to the rear.

Fishhook arrow (Section 15.5A) A single-barbed curved arrow used to show the change in position of a single electron.

Flavin adenine dinucleotide (FAD) (Section 20.1C) A biological oxidizing agent. When acting as an oxidizing agent, FAD is reduced to $FADH_2$.

Fluid-mosaic model (Section 17.5B) A biological membrane consists of a phospholipid bilayer with proteins, carbohydrates, and other lipids embedded in and on the surface of the bilayer.

Formal charge (Section 1.2E) The charge on an atom in a molecule or polyatomic ion.

Frequency (ν) (Section 21.1) The number of full cycles of a wave that pass a point in a fixed time period.

Functional group (Section 1.7) An atom or group of atoms within a molecule that shows a characteristic set of physical and chemical properties.

Furanose (Section 16.2A) A five-membered cyclic hemiacetal form of a monosaccharide.

Glass transition temperature, T_g (Section 15.3) The temperature at which a polymer undergoes the transition from a hard glass to a rubbery state.

Glycol (Sections 6.4B, 8.2A) A compound with two hydroxyl (—OH) groups on adjacent carbons.

Glycolysis (Section 20.5) From the Greek *glyko,* sweet, and *lysis,* splitting; a series of ten enzyme-catalyzed reactions by which glucose is oxidized to two molecules of pyruvate.

Glycoside (Section 16.4A) A carbohydrate in which the —OH on its anomeric carbon is replaced by —OR.

Glycoside bond (Section 16.4A) The bond from the anomeric carbon of a glycoside to an —OR group.

Grignard reagent (Section 11.5A) An organomagnesium compound of the type RMgX or ArMgX.

Ground-state electron configuration (Section 1.1B) The electron configuration of lowest energy for an atom, molecule, or ion.

Halonium ion (Section 6.3C) An ion in which a halogen atom bears a positive charge.

Haworth projection (Section 16.2A) A way to view furanose and pyranose forms of monosaccharides. The ring is drawn flat and viewed through its edge with the anomeric carbon on the right and the oxygen atom of the ring in the rear to the right.

Heat of reaction (Section 6.2A) The difference in potential energy between reactants and products.

α-Helix (Section 18.5B) A type of secondary structure in which a section of polypeptide chain coils into a spiral, most commonly a right-handed spiral.

Hemiacetal (Section 11.6) A molecule containing an —OH and an —OR or —OAr group bonded to the same carbon.

Hertz (Hz) (Section 21.1) The unit in which wave frequency is reported; s^{-1} (read *per second*).

Heterocyclic amine (Section 10.1) An amine in which nitrogen is one of the atoms of a ring.

Heterocyclic aromatic amine (Section 10.1) An amine in which nitrogen is one of the atoms of an aromatic ring.

Heterocyclic compound (Section 9.2) An organic compound that contains one or more atoms other than carbon in its ring.

High-density lipoprotein (HDL) (Section 17.4A) Plasma particles, density 1.06 to 1.21 g/mL, consisting of approximately 33% proteins, 30% cholesterol, 29% phospholipids, and 8% triglycerides.

Histone (Section 19.2C) A protein, particularly rich in the basic amino acids lysine and arginine, that is found associated with DNA molecules.

Hybridization (Section 1.6B) The combination of two or more atomic orbitals to form a new set of atomic orbitals.

Hydration (Section 6.3B) Addition of water.

Hydride ion (Section 11.10B) A hydrogen atom with two electrons in its valence shell; $H{:}^-$.

Hydrocarbon (Section 3.1) A compound that contains only carbon and hydrogen atoms.

α-Hydrogen (Section 11.8A) A hydrogen on an α-carbon.

Hydrogen bonding (Section 8.3A) The attractive force between a partial positive charge on hydrogen and partial negative charge of a nearby oxygen, nitrogen, or fluorine atom.

Hydrophilic (Section 12.3) From the Greek meaning water-loving.

Hydrophobic (Section 12.3) From the Greek meaning water-hating.

Hydrophobic effect (Section 18.5C) The tendency of nonpolar groups to cluster in such a way as to be shielded from contact with an aqueous environment.

Hydroxyl group (Section 1.7A) An —OH group.

Imide (Section 13.1D) A compound containing two acyl groups, RCO— or ArCO—, bonded to a nitrogen atom.

Imine (Section 11.7) A compound containing a carbon-nitrogen double bond; also called a Schiff base.

Ionic bond (Section 1.2B) A chemical bond resulting from the electrostatic attraction of an anion and a cation.

Isoelectric point (pI) (Section 18.2C) The pH at which an amino acid, polypeptide, or protein has no net charge.

Ketone (Section 11.1) A compound containing a carbonyl group bonded to two carbons.

Ketose (Section 16.2A) A monosaccharide containing a ketone group.

Lactam (Section 13.1D) A cyclic amide.

Lactate fermentation (Section 20.7A) A metabolic pathway that converts glucose to two molecules of pyruvate.

Lactone (Section 13.1C) A cyclic ester.

Levorotatory (Section 4.7B) Rotation of the plane of polarized light in a polarimeter to the left.

Lewis acid (Section 2.5) Any molecule or ion that can form a new covalent bond by accepting a pair of electrons.

Lewis base (Section 2.5) Any molecule or ion that can form a new covalent bond by donating a pair of electrons.

Lewis structure of an atom (Section 1.1C) The symbol of an element surrounded by a number of dots equal to the number of electrons in the valence shell of the atom.

Line-angle drawing (Section 3.4) An abbreviated way to draw structural formulas in which each line ending represents a carbon atom and a line represents a bond.

Lipid (Section 17.1) A class of biomolecules isolated from plant or animal sources by extraction with nonpolar organic solvents, such as diethyl ether and acetone.

Lipid bilayer (Section 17.5B) A back-to-back arrangement of phospholipid monolayers.

Low-density lipoprotein (LDL) (Section 17.4A) Plasma particles, density 1.02 to 1.06 g/mL, consisting of approximately 25% proteins, 50% cholesterol, 21% phospholipids, and 4% triglycerides.

Markovnikov's rule (Section 6.3A) In the addition of HX or H_2O to an alkene, hydrogen adds to the carbon of the double bond having the greater number of hydrogens.

Melt transition temperature, T_m (Section 15.3) The temperature at which crystalline regions of a polymer melt.

Mercaptan (Section 8.2C) A common name for any molecule containing an —SH group.

Meso compound (Section 4.4B) An achiral compound possessing two or more stereocenters.

Messenger RNA (mRNA) (Section 19.3C) A ribonucleic acid that carries coded genetic information from DNA to the ribosomes for the synthesis of proteins.

Meta (m) (Section 9.3B) Refers to groups occupying 1 and 3 positions on a benzene ring.

Meta director (Section 9.8A) Any substituent on a benzene ring that directs electrophilic aromatic substitution preferentially to a meta position.

Micelle (Section 17.2B) A spherical arrangement of organic molecules in water solution clustered so that their hydrophobic parts are buried inside the sphere and their hydrophilic parts are on the surface of the sphere and in contact with water.

Mirror image (Section 4.1) The reflection of an object in a mirror.

Molecular spectroscopy (Section 21.2) The study of which frequencies of electromagnetic radiation are absorbed or emitted by substances and the correlation between these frequencies and specific types of molecular structure.

Monomer (Section 15.1) From the Greek *mono,* single, and *meros,* part; the simplest nonredundant unit from which a polymer is synthesized.

Monosaccharide (Section 16.2A) A carbohydrate that cannot be hydrolyzed to a simpler compound.

D-Monosaccharide (Section 16.2E) A monosaccharide that, when written as a Fischer projection, has the —OH on its penultimate carbon to the right.

L-Monosaccharide (Section 16.2E) A monosaccharide that, when written as a Fischer projection, has the —OH on its penultimate carbon to the left.

Mutarotation (Section 16.2C) The change in optical activity that occurs when an α or β form of a carbohydrate is converted to an equilibrium mixture of the two forms.

(n + 1) Rule (Section 21.9) The ^{1}H-NMR signal of a hydrogen or set of equivalent hydrogens is split into $(n + 1)$ peaks by a nonequivalent set of n neighboring hydrogens.

Newman projection (Section 3.6A) A way to view a molecule by looking along a carbon-carbon bond.

Nicotinamide adenine dinucleotide (NAD^+) (Section 20.1B) A biological oxidizing agent. When acting as an oxidizing agent, NAD^+ is reduced to NADH.
Nonbonded interaction strain (Section 3.6A) The strain that arises when atoms not bonded to each other are forced abnormally close to one another.
Nonbonding electrons (Section 1.2D) Valence electrons not involved in forming covalent bonds, that is, unshared electrons.
Nonpolar covalent bond (Section 1.2C) A covalent bond between atoms whose difference in electronegativity is less than 0.5.
Nucleic acid (Section 19.1) A biopolymer containing three types of monomer units: heterocyclic aromatic amine bases derived from purine and pyrimidine, the monosaccharides D-ribose or 2-deoxy-D-ribose, and phosphate.
Nucleophile (Section 7.2) An atom or group of atoms that can donate a pair of electrons to another atom or group of atoms to form a new covalent bond.
Nucleophilic acyl substitution (Section 13.2) A reaction in which a nucleophile bonded to a carbonyl carbon is replaced by another nucleophile.
Nucleophilic substitution (Section 7.2) A reaction in which one nucleophile is substituted for another.
Nucleoside (Section 19.1) A building block of nucleic acids, consisting of D-ribose or 2-deoxy-D-ribose bonded to a heterocyclic aromatic amine base by a β-*N*-glycoside bond.
Nucleotide (Section 19.1) A nucleoside in which a molecule of phosphoric acid is esterified with an —OH of the monosaccharide, most commonly either the 3′-OH or the 5′-OH.

Observed rotation (Section 4.7B) The number of degrees through which a compound rotates the plane of polarized light.
Octane rating (Section 3.10B) The percentage of isooctane in a mixture of isooctane and heptane that has knock properties equivalent to a test gasoline.
Octet rule (Section 1.2A) The tendency among atoms of Group IA–VIIA elements to react in ways that achieve an outer shell of eight valence electrons.
Oil (Section 17.1B) A triglyceride that is liquid at room temperature.
Oligosaccharide (Section 16.7A) A carbohydrate containing from four to ten monosaccharide units, each joined to the next by a glycoside bond.
Optically active (Section 4.7A) Showing that a compound rotates the plane of polarized light.
Orbital (Section 1.1) A region of space where an electron or pair of electrons spends 90–95% of its time.
Order of precedence of functional groups (Section 11.2B) A system for ranking functional groups in order of priority for the purposes of IUPAC nomenclature.
Organometallic compound (Section 11.5A) A compound containing a carbon-metal bond.
Ortho (o) (Section 9.3B) Refers to groups occupying 1 and 2 positions on a benzene ring.
Ortho-para director (Section 9.8A) Any substituent on a benzene ring that directs electrophilic aromatic substitution preferentially to ortho and para positions.
Oxidation (Section 6.4A) The loss of electrons.
β-Oxidation (Section 20.3A) A series of four enzyme-catalyzed reactions that cleaves carbon atoms, two at a time, from the carboxyl end of a fatty acid.
Oxonium ion (Section 6.3B) An ion in which oxygen is bonded to three other atoms and bears a positive charge.

Para (p) (Section 9.3B) Refers to groups occupying 1 and 4 positions on a benzene ring.
Peak (NMR) (Section 21.9) The units into which an NMR signal is split; two peaks in a doublet, three peaks in a triplet, and so on.
Penultimate carbon (Section 16.2E) The stereocenter of a monosaccharide farthest from the carbonyl group, as for example carbon-5 of glucose.
Peptide bond (Section 18.3) The special name given to the amide bond formed between the α-amino group of one amino acid and the α-carboxyl group of another amino acid.
Phenol (Section 9.4A) A compound that contains an —OH bonded to a benzene ring.
Phenyl group (Section 9.3A) C_6H_5—, the aryl group derived by removing a hydrogen from benzene.
Phospholipid (Section 17.5A) A lipid containing glycerol esterified with two molecules of fatty acid and one molecule of phosphoric acid.
Pi (π) bond (Section 1.6D) A covalent bond formed by the overlap of parallel *p* orbitals.
Plane of symmetry (Section 4.2) An imaginary plane passing through an object dividing it such that one half is the mirror image of the other half.
Plane polarized light (Section 4.7A) Light vibrating in only parallel planes.
Plastic (Section 15.1) A polymer that can be molded when hot and retain its shape when cooled.
β-Pleated sheet (Section 18.5B) A type of secondary structure in which two sections of polypeptide chain are aligned parallel or antiparallel to one another.
Polar covalent bond (Section 1.2C) A covalent bond between atoms whose difference in electronegativity is between 0.5 and 1.9.
Polarimeter (Section 4.7B) A device for measuring the ability of a compound to rotate the plane of polarized light.
Polyamide (Section 15.4A) A polymer in which each monomer unit is joined to the next by an amide bond, as for example nylon 66.

Polycarbonate (Section 15.4C) A polyester in which the carboxyl groups are derived from carbonic acid.

Polyester (Section 15.4B) A polymer in which each monomer unit is joined to the next by an ester bond, as for example poly(ethylene terephthalate).

Polymer (Section 15.1) From the Greek *poly,* many, and *meros,* parts; any long-chain molecule synthesized by linking together many single parts called monomers.

Polynuclear aromatic hydrocarbon (Section 9.3C) A hydrocarbon containing two or more fused aromatic rings.

Polypeptide (Section 18.3) A macromolecule containing ten or more amino acid units, each joined to the next by a peptide bond.

Polysaccharide (Section 16.7A) A carbohydrate containing a large number of monosaccharide units, each joined to the next by one or more glycoside bonds.

Polyunsaturated fatty acid (Section 17.1A) A fatty acid with two or more carbon-carbon double bonds in its hydrocarbon chain.

Polyunsaturated triglyceride (Section 17.1B) A triglyceride having several carbon-carbon double bonds in the hydrocarbon chains of its three fatty acids.

Polyurethane (Section 15.4D) A polymer containing the —$NHCO_2$— group as a repeating unit.

Potential energy (PE) diagram (Section 6.2A) A graph showing the changes in energy that occur during a chemical reaction; energy is plotted on the *y* axis, and reaction progress is plotted on the *x* axis.

Primary (1°) amine (Section 1.7D) An amine in which one hydrogen of ammonia has been replaced by an alkyl or aryl group.

Primary (1°) carbon (Section 3.3C) A carbon bonded to one other carbon atom.

Primary structure of nucleic acids (Section 19.2A) The sequence of bases along the pentose-phosphodiester backbone of a DNA or RNA molecule read from the 5′ end to the 3′ end.

Primary structure of proteins (Section 18.4A) The sequence of amino acids in the polypeptide chain, read from the *N*-terminal amino acid to the *C*-terminal amino acid.

Prostaglandin (Section 17.3) A member of the family of compounds having the 20-carbon skeleton of prostanoic acid.

Protic solvent (Section 7.4D) A hydrogen bond donor solvent, as for example water, ethanol, and acetic acid.

Pyranose (Section 16.2A) A six-membered cyclic hemiacetal form of a monosaccharide.

Quaternary (4°) ammonium ion (Section 10.2B) An ion in which nitrogen is bonded to four alkyl or aryl groups and bears a positive charge.

Quaternary (4°) carbon (Section 3.3C) A carbon bonded to four other carbon atoms.

Quaternary structure (Section 18.5C) The arrangement of polypeptide monomers into a noncovalently bonded aggregation.

R (Section 4.3) From the Latin *rectus,* meaning right; used in the R,S convention to show that the order of priority of groups on a stereocenter is clockwise.

R— (Section 3.3A) A symbol used to represent an alkyl group.

R,S convention (Section 4.3) A set of rules for specifying configuration about a stereocenter; also called the Cahn-Ingold-Prelog convention.

Racemic mixture (Section 4.7C) A mixture of equal amounts of two enantiomers.

Racemization (Section 11.8B) The conversion of a pure enantiomer into a racemic mixture.

Radical (Section 15.5A) Any molecule that contains one or more unpaired electrons.

Rate-limiting step (Section 6.2A) The step in a reaction sequence that crosses the highest potential energy barrier; the slowest step in a multistep reaction.

Reaction coordinate (Section 6.2A) A measure of the progress of a reaction, plotted on the *x* axis in a potential energy diagram.

Reaction intermediate (Section 6.2A) An unstable species, formed during a two-step reaction, that lies in a potential energy minimum between two transition states.

Reaction mechanism (Section 6.2) A step-by-step description of how a chemical reaction occurs.

Reducing sugar (Section 16.4B) A carbohydrate that reduces Ag(I) to Ag or Cu(II) to Cu(I).

Reduction (Section 6.4A) The gain of electrons.

Reductive amination (Section 11.10C) The formation of an imine from an aldehyde or ketone followed by its reduction to an amine.

Regioselective reaction (Section 6.3A) A reaction in which one direction of bond forming or bond breaking occurs in preference to all other directions.

Relative nucleophilicity (Section 7.4A) The relative rate at which a nucleophile reacts in a reference nucleophilic substitution reaction.

Resolution (Section 4.8A) Separation of a racemic mixture into its enantiomers.

Resonance (NMR) (Section 21.4) The absorption of electromagnetic radiation by a spinning nucleus and the resulting "flip" of its spin from a lower energy state to a higher energy state.

Resonance energy (Section 9.1D) The difference in energy between a resonance hybrid and the most stable of its hypothetical contributing structures.

Resonance hybrid (Section 1.5A) A molecule or ion that is best described as a composite of a number of contributing structures.

Restriction endonuclease (Section 19.4B) An enzyme that catalyzes hydrolysis of a particular phosphodiester bond within a DNA strand.

Ribosomal RNA (rRNA) (Section 19.3A) A ribonucleic acid found in ribosomes, the sites of protein synthesis.

S (Section 4.3) From the Latin *sinister,* meaning left; used in the R,S convention to show that the order of priority of groups on a stereocenter is counterclockwise.

Saccharide (Section 16.1) A simpler member of the carbohydrate family, such as glucose.

Sanger dideoxy method (Section 19.4D) A method, developed by Frederick Sanger, for sequencing DNA molecules.

Saponification (Section 13.3C) Hydrolysis of an ester in aqueous NaOH or KOH to an alcohol and the sodium or potassium salt of a carboxylic acid.

Saturated hydrocarbon (Section 3.1) A hydrocarbon containing only carbon-carbon single bonds.

Secondary (2°) amine (Section 1.7D) An amine in which two hydrogens of ammonia have been replaced by alkyl or aryl groups.

Secondary (2°) carbon (Section 3.3C) A carbon bonded to two other carbon atoms.

Secondary structure of nucleic acids (Section 19.2B) The ordered arrangement of nucleic acid strands.

Secondary structure of proteins (Section 18.5B) The ordered arrangements (conformations) of amino acids in localized regions of a polypeptide or protein.

Shell (Section 1.1) A region of space around a nucleus where electrons are found.

Shielding in NMR spectroscopy (Section 21.4) Electrons around a nucleus create their own local magnetic fields and thereby shield the nucleus from the applied magnetic field.

Sigma (σ) bond (Section 1.6A) A covalent bond in which the overlap of atomic orbitals is concentrated along the bond axis.

Signal (Section 21.4) A recording in an NMR spectrum of a nuclear magnetic resonance.

Signal splitting (Section 21.9) Splitting of an NMR signal into a set of peaks by the influence of neighboring nuclei.

Soap (Section 17.2A) A sodium or potassium salt of a fatty acid.

***sp* Hybrid orbital** (Section 1.6E) A hybrid atomic orbital produced by the combination of one *s* atomic orbital and one *p* atomic orbital.

sp^2 Hybrid orbital (Section 1.6D) A hybrid atomic orbital produced by the combination of one *s* atomic orbital and two *p* atomic orbitals.

sp^3 Hybrid orbital (Section 1.6C) A hybrid atomic orbital produced by the combination of one *s* atomic orbital and three *p* atomic orbitals.

Specific rotation (Section 4.7B) Observed rotation of the plane of polarized light when a sample is placed in a tube 1.0 dm in length and at a concentration of 1 g/mL.

Staggered conformation (Section 3.6A) A conformation about a carbon-carbon single bond, where the atoms on one carbon are as far apart as possible from atoms on the adjacent carbon.

Step-growth polymerization (Section 15.4) A polymerization in which chain growth occurs in a stepwise manner between difunctional monomers, as for example between adipic acid and hexamethylenediamine to form nylon 66.

Stereocenter (Section 4.2) An atom that has four different groups attached to it.

Stereoisomers (Section 4.1) Isomers that have the same molecular formula and the same connectivity but different orientations of their atoms in space.

Stereoselective reaction (Section 6.3C) A reaction in which one stereoisomer is formed or destroyed in preference to all others that may be formed or destroyed.

Steric hindrance (Section 7.4B) The ability of an atom or group of atoms, because of their size, to hinder access to a reaction site.

Steroid (Section 17.4A) A plant or animal lipid having the characteristic tetracyclic ring structure of the steroid nucleus, namely three six-membered rings and one five-membered ring.

Strong acid (Section 2.2) An acid that is completely ionized in aqueous solution.

Strong base (Section 2.2) A base that is completely ionized in aqueous solution.

Syn addition (Section 6.4A) Addition of atoms or groups of atoms from the same side or face of a carbon-carbon double bond.

Tautomers (Section 11.8A) Constitutional isomers that differ in the location of hydrogen and a double bond relative to O, N, or S.

C-Terminal amino acid (Section 18.3) The amino acid at the end of a polypeptide chain having the free —CO_2H group.

***N*-Terminal amino acid** (Section 18.3) The amino acid at the end of a polypeptide chain having the free —NH_2 group.

Terpene (Section 5.4) A compound whose carbon skeleton can be divided into two or more units identical with the carbon skeleton of isoprene.

Tertiary (3°) amine (Section 1.7D) An amine in which three hydrogens of ammonia have been replaced by alkyl or aryl groups.

Tertiary (3°) carbon (Section 3.3C) A carbon bonded to three other carbon atoms.

Tertiary structure of nucleic acids (Section 19.2C) The three-dimensional arrangement of all atoms of a nucleic acid; commonly referred to as supercoiling.

Tertiary structure of proteins (Section 18.5C) The three-dimensional arrangement in space of all atoms in a single polypeptide chain.
Thermoplastic (Section 15.1) A polymer that can be melted and molded into a shape that is retained when it is cooled.
Thermosetting plastic (Section 15.1) A polymer that can be molded when it is first prepared, but once cooled, hardens irreversibly and cannot be remelted.
Thioester (Section 20.3A) An ester in which one atom of oxygen in the carboxylate group is replaced by an atom of sulfur.
Thiol (Section 8.1C) A compound containing an —SH (sulfhydryl) group.
Trans- (Section 3.7) A prefix meaning across from.
Transfer RNA (tRNA) (Section 19.3B) A ribonucleic acid that carries a specific amino acid to the site of protein synthesis on ribosomes.
Transition state (Section 6.2A) An unstable species of maximum energy formed during the course of a reaction; an energy maximum on a potential energy diagram.
Triglyceride (triacylglycerol) (Section 17.1) An ester of glycerol with three fatty acids.
Tripeptide (Section 18.3) A molecule containing three amino acid units, each joined to the next by a peptide bond.

Unimolecular reaction (Section 7.3B) A reaction in which only one species is involved in the reaction leading to the transition state of the rate-limiting step.

Valence electrons (Section 1.1C) Electrons in the valence (outermost) shell of an atom.
Valence shell (Section 1.1C) The outermost electron shell of an atom.
Vibrational infrared (Section 22.1A) The portion of the infrared region that extends from 4000 to 400 cm^{-1}.

Watson-Crick model (Section 19.2B) A double-helix model for the secondary structure of a DNA molecule.
Wavenumber ($\overline{\nu}$) (Section 22.1A) A characteristic of electromagnetic radiation equal to the number of waves per centimeter.
Wavelength (λ) (Section 21.1) The distance between two consecutive peaks on a wave.

Z (Section 5.2C) From the German *zusammen,* meaning together. Specifies that groups of higher priority on the carbons of a double bond are on the same side.
Zaitsev's rule (Section 7.6) The major product from a β-elimination reaction contains the more highly substituted carbon-carbon double bond.
Zwitterion (Section 18.1A) An internal salt of an amino acid.

INDEX

D

I

K

Q

R

S

T